LINEAR

McGraw-Hill Series in Electrical Engineering

Consulting Editor
Stephen W. Director, Carnegie-Mellon University

Circuits and Systems
Communications and Signal Processing
Control Theory
Electronics and Electronic Circuits
Power and Energy
Electromagnetics
Computer Engineering
Introductory
Radar and Antennas
VLSI

Circuits and Systems

Consulting Editor
Stephen W. Director, Carnegie-Mellon University

Antoniou: *Digital Filters: Analysis and Design*
Belove and Drossman: *Systems and Circuits for Electrical Engineering Technology*
Belove, Schachter, and Schilling: *Digital and Analog Systems, Circuits, and Devices*
Bracewell: *The Fourier Transform and Its Applications*
Cannon: *Dynamics of Physical Systems*
Chirlian: *Basic Network Theory*
Chua, Desoer, and Kuh: *Linear and Nonlinear Circuits*
Desoer and Kuh: *Basic Circuit Theory*
Glisson: *Introduction to Systems Analysis*
Hammond and Gehmlich: *Electrical Engineering*
Hayt and Kemmerly: *Engineering Circuit Analysis*
Hilburn and Johnson: *Manual of Active Filter Design*
Jong: *Methods of Discrete Signal and System Analysis*
Liu and Liu: *Linear Systems Analysis*
Papoulis: *The Fourier Integral and Its Application*
Reid: *Linear System Fundamentals: Continuous and Discrete, Classic and Modern*
Sage: *Methodology for Large Scale Systems*
Schwartz and Friedland: *Linear Systems*
Temes and LaPatra: *Introduction to Circuit Synthesis*
Truxal: *Introductory System Engineering*

To Diana, Jackie, and Bettine

LINEAR AND NONLINEAR CIRCUITS

Leon O. Chua

Charles A. Desoer

Ernest S. Kuh

University of California—Berkeley

McGraw-Hill Book Company

New York St. Louis San Francisco Auckland Bogotá Hamburg
London Madrid Mexico Milan Montreal New Delhi
Panama Paris São Paulo Singapore Sydney Tokyo Toronto

LINEAR AND NONLINEAR CIRCUITS
INTERNATIONAL EDITION

2 3 4 5 6 7 8 9 0 CMO PMP 9 5 4 3 2 1 0

This book was set in Times Roman
by Intercontinental Photocomposition Limited.
The editors were Sanjeev Rao and Alar E. Elken;
the cover was designed by Scott Chelius;
the production supervisor was Diana Renda.
Project supervision was done by The Total Book.

Library of Congress Cataloging-in-Publication Data

Chua, Leon O. (date)
Linear and nonlinear circuits.

(McGraw-Hill series in electrical engineering.
Circuits and systems)
Includes bibliographical references and index.
1. Electric circuits, Linear. 2. Electric circuits,
Nonlinear. I. Desoer, Charles A. II. Kuh, Ernest S.
III. Title. IV. Series.
TA454.C56 1987 621.319'21 86-12498
ISBN 0-07-010898-6
ISBN 0-07-010899-4 (solutions manual)

When ordering this title use ISBN 0-07-100685-0

Printed in Singapore

ABOUT THE AUTHORS

Leon O. Chua received the S.M. degree from Massachusetts Institute of Technology in 1961 and his Ph.D. from the University of Illinois, Urbana in 1964. He was awarded *Doctor Honoris Causa* from the Ecole Polytechnique Fédérale de Lausanne (Switzerland) in 1983 and an Honorary Doctorate from the University of Tokushima (Japan) in 1984. In 1971 he joined the faculty of the University of California, Berkeley, where he is currently Professor of Electrical Engineering and Computer Sciences. Dr. Chua is the author of *Introduction to Nonlinear Network Theory* and coauthor of *Computer-Aided Analysis of Electronic Circuits: Algorithms and Computational Techniques*. His research interests are in the areas of general nonlinear network and system theory. He has published numerous research papers and has been a consultant to various electronic industries.

Dr. Chua is a Fellow of the IEEE (1973), the Editor of the *IEEE Transactions on Circuits and Systems* (1973–75), and the President of the IEEE Circuits and Systems Society (1976). He has been awarded four patents and received several honors and awards, including the IEEE Browder J. Thompson Prize (1967), the IEEE W.R.G. Baker Prize (1973), the Frederick Emmons Terman Award (1974), the Miller Professor (1976), the Senior Visiting Fellow at Cambridge University (1982), the Alexander von Humboldt Senior U.S. Scientist award (1983) at the Technische Universität (München), the Japan Society for Promotion of Science Senior Visiting Fellowship (1983) at Waseda University (Tokyo), the IEEE Centennial Medal (1984), the Guillemin-Cauer Prize (1985), and the Professeur Invité award (Université de Paris-SUD) from the French National Ministry of Education (1986).

Charles A. Desoer received his Ph.D. from the Massachusetts Institute of Technology in 1953. He then worked for several years at Bell Telephone Laboratories on circuit and communication system problems. In 1958 he joined the Department of Electrical Engineering and Computer Sciences at the University of California, Berkeley, where he has taught and researched in the areas of circuits, systems, and control.

Dr. Desoer has received a number of awards for his research, including the Medal of the University of Liege (1970), the Guggenheim Fellowship (1970–1971), the Prix Montefiore (1975), and the 1986 IEEE Control Systems Science and Engineering Award. He received the 1971 Distinguished Teaching Award at the University of California, Berkeley and the 1975 IEEE Education Medal. Dr. Desoer's interests lie in system theory with emphasis on control systems and circuits. He has coauthored several books, including *Linear System Theory* with L. A. Zadeh, *Basic Circuit Theory* with E. S. Kuh, and *Multivariable Feedback Systems* with F. M. Callier. He is author of *Notes for a Second Course on Linear Systems*.

Ernest S. Kuh is author of over 80 technical papers in circuits, systems, electronics, and computer-aided design, and is a coauthor of three books, including *Basic Circuit Theory*, which he wrote with one of his present coauthors, Charles A. Desoer. He received the B.S. degree from the University of Michigan, Ann Arbor in 1949, the S.M. degree from the Massachusetts Institute of Technology in 1950, and the Ph.D from Stanford University in 1952.

Dr. Kuh was a member of the Technical Staff of Bell Telephone Laboratories from 1952 until 1956. He joined the faculty of the University of California, Berkeley in 1956, and there, he served as Chairman of the Department of Electrical Engineering and Computer Sciences from 1968 to 1972. From 1973 to 1980 he was Dean of the College of Engineering. Dr. Kuh is a Fellow of IEEE and AAAS and a member of the National Academy of Engineering and of the Academia Sinica. He has been the recipient of a number of awards and honors, including the Lamme Medal of the American Society of Engineering Education, the IEEE Education Medal, and the IEEE Centennial Medal. He has served on several committees of the National Academy of Engineering and the National Academy of Sciences.

CONTENTS

PREFACE

Electrical engineering is a discipline driven by inventions and technological breakthroughs. To mention one: 20 years ago, engineers barely knew how to produce an IC chip: now some chips have one million devices; it is expected that with foreseeable developments in silicon technology during the next decade chips will have 10^8 devices. Also the ubiquitous presence of the computer terminal reminds us of the enormous impact of computers on engineering design. Clearly such tremendous changes would have considerable influence on engineering education.

In teaching a course on introductory electrical circuits, the traditional approach has been to teach exclusively linear time-invariant passive RLC circuits. Admittedly they constitute a good vehicle to learn the dynamics of such simple circuits. Clearly such an approach is obsolete.

It is clear that circuit theory is one of the basic disciplines of electrical engineering; a well designed circuit theory course should cover the basic concepts and the basic results used in circuit design. It should serve as a foundation course to be followed by courses in various fields of electrical engineering, e.g., communication and signal processing, electronic devices and circuits, control and power systems, microwaves and optoelectronics, etc. . . . The concept of device modeling and its applications to currently used devices are crucial in a course on linear and nonlinear circuits: many examples of device modeling are given in the text. Furthermore, the course should be designed so that the graduate from such a curriculum knows how to approach the devices and circuits yet to be invented but that he or she will encounter, say, 10 to 15 years from now. With these goals in mind, the present book presents material with sufficient breadth, depth, and rigor to give a solid foundation to the student's future professional life.

At the University of California, Berkeley, as in most American engineering schools, there is a sophomore 45-lecture-hour course called Introductory Electrical Engineering. Its purpose is to give a broad introduction to most of the aspects of electrical engineering.

This book is intended as a textbook for the junior course that follows.

Since it is a junior course, it takes advantage of the greater competence and maturity of the students: in particular, physics, linear algebra, and differential equations. This course is the electrical analog of the typical junior physics course in say, mechanics, electromagnetism, and so on.

This book differs from many other texts on circuit theory by the following features:

1. Due to the ubiquitous op amp and similar devices, it views a circuit as an interconnection of *multiterminal* elements rather than of *two-terminal* elements.
2. *Active* and *passive* circuits are given equal emphasis.
3. *Linear* and *nonlinear* elements are treated together. (Note that computers simulate nonlinear circuits almost as easily as linear ones.)
4. The concept of *operating point* and the topic of *small-signal analysis* are covered thoroughly.
5. Switching, triggering, and memory circuits as well as oscillators are illustrated with first-order and second-order examples.
6. Tableau analysis is used to greatly simplify the proof of many network theorems in linear and nonlinear circuits.
7. Modified node analysis is introduced in view of its complete generality and importance in the design of computer circuit simulators, such as SPICE.
8. Some numerical methods are introduced and implemented via equivalent circuits: in particular, solution of nonlinear algebraic equations (Newton-Raphson) and integration of the circuit differential equations (forward and backward Euler method).
9. Stability issues are met head on; in particular oscillators are analyzed and an elementary version of the Nyquist criterion (useful in the design of op amps circuits) is introduced.

CONTENTS OF THE BOOK

Chapter One treats Kirchhoff's Laws and Tellegen's theorem. The next four chapters introduce two-terminal and multiterminal *resistive* elements and *resistive* circuits; linear, nonlinear, passive, and active circuits; op amp circuits with linear and nonlinear models; operating points and small signal analysis; and network theorems and the Newton-Raphson procedure for solving nonlinear dc resistive circuits.

Chapters Six and Seven cover first- and second-order linear and nonlinear dynamic circuits: our goal is to exhibit their properties and illustrate them by numerous examples, including flip flops and oscillators. General dynamic circuits, analyzed by Tableau or modified node analysis, are covered in Chapter Eight.

The next three chapters build up the fundamentals of linear time-invariant circuits: sinusoidal steady-state analysis; very brief treatment of Laplace transforms; properties of natural frequencies and network functions such as poles, zeros, stability, and convolution; Nyquist criterion; and stability of terminated one-ports.

A brief Chapter Twelve broadens the background on network topology and treats the usual general circuit analysis methods.

Chapter Thirteen covers two-ports, *n*-ports, and their properties—reciprocity in particular.

The last chapter brings out design issues such as the approximation problem, design of active Butterworth filters, and sensitivity analysis.

We believe that the topics covered in this text constitute an excellent background for further education in electronic circuits, computer-aided design, communications, control, and power.

ACKNOWLEDGMENT

Even though this book is a systematic introduction to circuit theory, it uses many concepts and techniques which were developed by people doing research in circuits, communications and control. In fact, without our own deep involvement in research, this book could not have been written. It is a pleasure to publicly acknowledge the research support of the University of California, the National Science Foundation, the Department of Defense, the State of California, and the support of industry.

We are also indebted to many people who have taught the course while the preliminary notes were written and revised; they have given us valuable suggestions. It is a pleasure to mention in particular F. Ayrom, G. Bernstein, C. C. Chang, A. C. Deng, A. Dervisoglu, N. A. Gruñdes, H. Haneda, K. Inan, P. Kennedy, R. W. Liu, H. Narayanan, M. Odyniec, T. S. Parker, A. Sangiovanni-Vincentelli, E. W. Szeto-Lee, L. Yang, and Quing-jian Yu.

We also benefited greatly from the judicious comments of our teaching assistants and the penetrating questions of our students.

We wish to record our gratitude for the high professional skills and care that B. Fuller, I. Stanczyk-Ng, and T. Sticpewich gave to the preparation of the manuscript.

Leon O. Chua
Charles A. Desoer
Ernest S. Kuh

CHAPTER

ONE

KIRCHHOFF'S LAWS

As an electrical engineer, one needs to analyze and design circuits. Electric circuits are present almost everywhere, in home computers, television and hi-fi sets, electric power networks, transcontinental telecommunication systems, etc. Circuits in these applications vary a great deal in nature and in the ways they are analyzed and designed. The purpose of this book is to give an introductory treatment of circuit theory which covers considerable breadth and depth. This differs from a traditional introductory course on circuits, which is restricted to "linear" circuits and covers mainly circuits containing the classical *RLC* elements.

The first chapter deals with the fundamental postulates of lumped-circuit theory, namely, Kirchhoff's laws. Naturally, we need to explain the word "lumped" first. It is also important to understand the concept of "modeling." For example, in circuit theory we first model a "physical circuit" made of electric devices by a "circuit" which is an interconnection of circuit elements. Since Kirchhoff's laws hold for any lumped circuit, the discussion can be dissociated with the electrical properties of circuit elements, which will be treated in the succeeding chapters.

A key concept introduced in this chapter is the representation of a circuit by a graph. This allows us to deal with multiterminal devices in the same way as we would with a conventional two-terminal device. In addition, it enables us to give a formal treatment of Kirchhoff's laws and a related fundamental theorem, Tellegen's theorem.

1 THE DISCIPLINE OF CIRCUIT THEORY

Circuit theory is the fundamental engineering discipline that pervades all electrical engineering. For the present, by *physical circuit* we mean any *interconnection* of (physical) *electric devices*. Familiar examples of electric devices are resistors, coils, condensers, diodes, transistors, operational amplifiers (op amps), batteries, transformers, electric motors, electric generators, etc.

The goal of circuit theory is to *predict* the electrical *behavior* of physical circuits. The purpose of these predictions is to improve their design: in particular, to decrease their cost and improve their performance under all conditions of operation (e.g., temperature effects, aging effects, possible fault conditions, etc.).

Circuit theory is an *engineering discipline* whose domain of application is extremely broad. For example, the *size* of the circuits varies enormously: from large-scale integrated circuits which include hundreds of thousands of components and which fit on a fingernail to circuits found in radios, TV sets, electronic instruments, small and large computers, and finally, to telecommunications circuits and power networks that span continents. The *voltages* encountered in the study of circuits vary from the microvolt (μV) [e.g., in noise studies of precision instruments—to megavolts (MV) of power networks]. The *currents* vary from femtoamperes ($1 \text{ fA} = 10^{-15}$ A) [e.g., in electrometers—to megaamperes (MA)] encountered in studies of power networks under fault conditions. The *frequencies* encountered in circuit theory vary from zero frequency [direct current (dc) conditions] to tens of gigahertz ($1 \text{ GHz} = 10^9$ Hz) encountered in microwave circuits. The *power levels* vary greatly from 10^{-14} watts (W) for the incoming signal to a sensitive receiver (e.g., faint radio signals from distant galaxies) to electric generators producing $10^9 \text{ W} = 1000$ megawatts (MW).

Circuit theory focuses on the *electrical* behavior of circuits. For example, it does not concern itself with thermal, mechanical, or chemical effects. Its aim is to predict and explain the (terminal) *voltages* and (terminal) *currents* measured at the device *terminals*. It does not concern itself with the physical phenomena occurring inside the device (e.g., in a transistor or in a motor). These considerations are covered in device physics courses and in electrical machinery courses.

The goal of circuit theory is to make quantitative and qualitative predictions on the electrical behavior of circuits; consequently the tools of circuit theory will be mathematical, and the *concepts* and *results* pertaining to circuits will be expressed in terms of circuit equations and *circuit variables*, each with an obvious operational interpretation.

2 LUMPED-CIRCUIT APPROXIMATION

Throughout this book we shall consider only *lumped circuits*. For a physical circuit to be considered *lumped*, its physical dimension must be small enough so that, for the problem at hand, electromagnetic waves propagate across the circuit virtually instantaneously. Consider the following two examples:

Example 1 Consider a small computer circuit on a chip whose extent is, say, 1 millimeter (mm); let the shortest signal time of interest be 0.1 nanosecond [$\frac{1}{10}$ of a nanosecond (ns) $= 10^{-10}$ of a second (s)]. Electromagnetic waves travel at the velocity of light, i.e., 3×10^8 meters per second (m/s); to travel 1 mm, the time elapsed is 10^{-3} m/(3×10^8 m/s) $= 3.3 \times 10^{-12}$ s $= 0.0033$ ns. Therefore the propagation time in comparison with the shortest signal time of interest is negligible. More generally, let d be the largest dimension of the circuit, Δt the shortest time of interest, and c the velocity of light. If $d \ll c \cdot \Delta t$, then the circuit may be considered to be lumped.

Example 2 Consider an audio circuit: The highest frequency of interest is, say, $f = 25$ kHz. For electromagnetic waves, this corresponds to a wavelength of $\lambda = c/f = (3 \times 10^8 \text{ m/s})/(2.5 \times 10^4 \text{ s}^{-1}) = 1.2 \times 10^4 \text{ m} =$ 12 km $\cong$ 7.5 miles. So even if the circuit is spread across a football stadium, the size of the circuit is very small compared to the shortest wavelength of interest λ. More generally, if $d \ll \lambda$, the circuit may be considered to be lumped.

When these conditions are satisfied, electromagnetic theory proves[1] and experiments show that the lumped-circuit approximation holds; namely, throughout the physical circuit the current $i(t)$ through any device terminal and the voltage difference $v(t)$ across any pair of terminals, at any time t, are well-defined. A circuit that satisfies these conditions is called a *lumped circuit*.

From an electromagnetic theory point of view, a lumped circuit reduces to a point since it is based on the approximation that electromagnetic waves propagate through the circuit instantaneously. For this reason, in lumped-circuit theory, the respective locations of the elements of the circuit will not affect the behavior of the circuit. The approximation of a *physical* circuit by a *lumped* circuit is analogous to the modeling of a rigid body as a particle: In doing so, all the data relating to the extent (shape, size, orientation, etc.) of the body are ignored by the theory.

Thus, lumped-circuit theory is related to the more general electromagnetic theory by an approximation (propagation effects are neglected). This is analogous to the relation of classical mechanics to the more exact relativistic mechanics: Classical mechanics delivers excellent predictions provided the velocities are much smaller than the velocity of light. Similarly, when the above conditions hold, lumped-circuit theory delivers excellent predictions of physical circuit behavior.

In situations where lumped approximation is not valid, the physical dimensions of the circuit must be considered. To distinguish such circuits from

[1] R. M. Fano, L. J. Chu, and R. M. Adler, *Electromagnetic Fields, Energy and Forces*, John Wiley and Sons, New York, 1960.

lumped circuits we call them *distributed circuits*. Typical examples of distributed circuits are circuits made of waveguides and transmission lines. In distributed circuits the current and voltage variables would depend not only on time, but also on space variables such as length and width. We need electromagnetic theory for predictions of the behavior of distributed circuits and for analysis and design. In this book we restrict our treatment to lumped circuits.

3 ELECTRIC CIRCUITS, MODELS, AND CIRCUIT ELEMENTS

By *electric device* we mean the physical object in the laboratory or in the factory, for example, the coil, the capacitor, the battery, the diode, the transistor, the motor, etc. *Physical circuits* are obtained by connecting electric devices by wires. Most of the time, these wires will be assumed to be perfectly conducting. We think of these electric devices in terms of *idealized models* like the resistor ($v = Ri$), the inductor ($v = L\, di/dt$), the capacitor ($i = C\, dv/dt$), etc., that you have studied in physics.

Note that these *idealized models* are precisely defined; to distinguish them from electric devices we call them *circuit elements*. It is important to distinguish between a coil made of a fine wire wrapped around a ferrite torus—an *electric device*—and its model as an inductor, or as a resistor in series with an inductor—a *circuit element*, or a combination of circuit elements.

Every model is an approximation. Depending on the application or the problem under consideration, the same physical device may be approximated by several different models. Each of these models is an interconnection of (idealized) circuit elements. For example, we will encounter several different models for the operational amplifier (op amp).[2]

Any *interconnection* of *circuit elements* is called a *circuit*. Thus a circuit is an interconnection of (idealized) models of the corresponding physical devices. The relation between physical circuits and circuits is illustrated in Fig. 3.1. If the (theoretical) predictions based on *analysis* of the circuit do not agree with the measurements, the cause of the disagreement may lie at any step of the process (e.g., erroneous measurement, faulty analysis, etc.). One frequent cause is a poor choice of model, e.g., using a low-frequency model outside of its frequency range of validity, or a linear model outside its amplitude range of validity.

Our subject is circuit theory, consequently we consider the models of the electric devices constituting the physical circuit as given at the outset; our goal is to develop methods to predict the behavior of the circuit. Note that we say "circuit," not "physical circuit": Past experience, however, does give us the

[2] Analogously, in classical mechanics a communications satellite circling the earth may be modeled as a particle, or a rigid body, or an elastic body depending on the problem being studied.

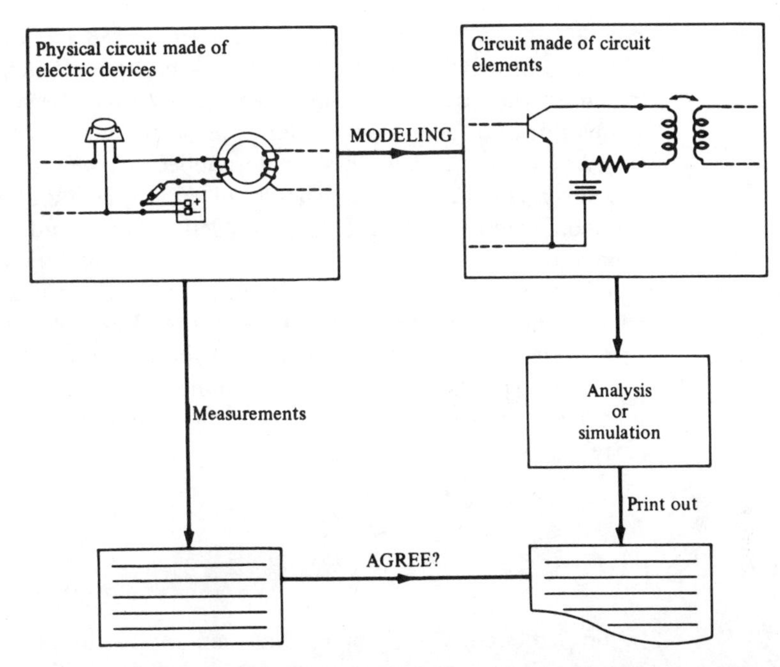

Figure 3.1 Illustration of the relation between physical circuits and circuits, between physical devices and circuit elements, and between laboratory measurements and circuit analysis.

confidence that given any physical circuit we can model it by a circuit which will adequately predict its behavior.

In Fig. 3.2*a* we show a *physical circuit* made up of *electric devices*: a generator, resistor, transistor, battery, transformer, and load. To analyze the physical circuit, we first model it with the circuit shown in Fig. 3.2*b*, which is an interconnection of *circuit elements*: voltage sources, resistors, a capacitor,

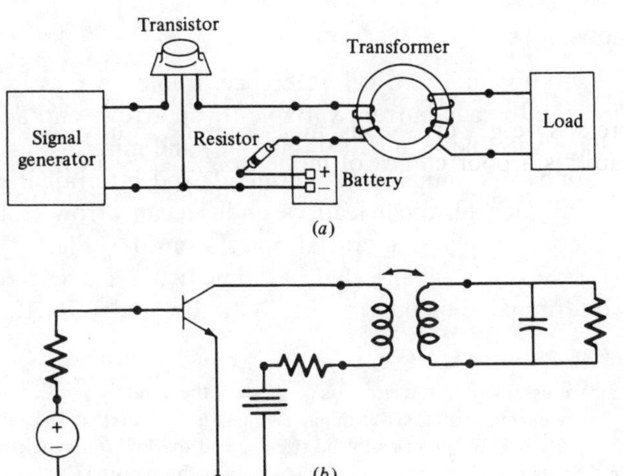

Figure 3.2 (*a*) Physical circuit made of electric devices and (*b*) its circuit model made of circuit elements.

coupled inductors, and a transistor represented by their usual symbols. The electrical properties of some of the two-terminal elements (voltage sources and resistors) will be discussed in Chap. 2, and that of the multiterminal elements (transistor and ideal transformer) will be treated in Chap. 3.

When electric devices are interconnected, we use conducting wires to tie the terminals together as shown in Fig. 3.2*a*. When circuit elements are interconnected, we delete the conducting wires and merge the terminals to obtain the circuit in Fig. 3.2*b*. A *node* is any junction in a circuit where terminals are joined together or any isolated terminal of a circuit element, which is not connected. The circuit in Fig. 3.2*b* has eight nodes (marked with heavy dots). With the introduction of the concept of a node, we are ready to formally treat the subject of interconnection and state the two fundamental postulates of circuit theory, namely, Kirchhoff's voltage law and Kirchhoff's current law.

4 KIRCHHOFF'S LAWS

In lumped circuits, the voltage between any two nodes and the current flowing into any element through a node are well-defined.[3] Since the *actual* direction of current flow and the *actual* polarity of voltage difference in a circuit can *vary* from one instant to another, it is generally impossible to specify in advance the *actual* current direction and voltage polarity in a given circuit. Just as in classical mechanics where it is essential to set up a "frame of reference" from which the actual instantaneous positions of a system of particles can be uniquely specified, so too must we set up an "electrical frame of reference" in a circuit in order that currents and voltages may be *unambiguously* measured.

4.1 Reference Directions

To set up an electrical reference frame, we assign *arbitrarily* a *reference direction* to each current variable by an arrow, and a *reference polarity* to each voltage variable by a pair of plus (+) and minus (−) signs, as illustrated in Fig. 4.1 for two-terminal, three-terminal, and *n*-terminal elements.[4]

On each terminal lead we indicate an arrow called the *current reference direction*. It plays a crucial role. Consider Fig. 4.1*a*. If at some time t_0, $i_2(t_0) = 2\,\mathrm{A}$, it means that, at time t_0, a current of 2 A flows *out* of the two-terminal element of Fig. 4.1*a* by node ②. If, at some later time t_1,

[3] We assume that the circuit is connected; the definition of "connectedness" will be given later.

[4] An example of a six-terminal element is the filter at the output of an audio amplifier: It directs the high frequencies to the tweeter and the low frequencies to the woofer. (Later we shall see that such a filter may also be viewed as a three-port.)

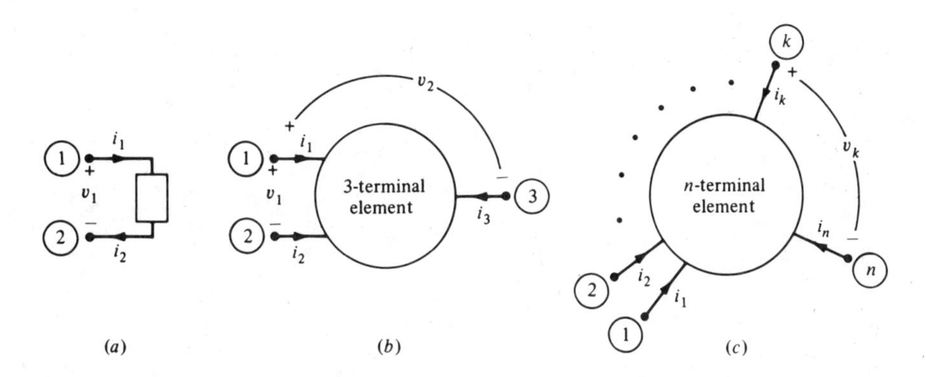

Figure 4.1 Illustration of reference directions using two-, three-, and n-terminal elements.

$i_2(t_1) = -25$ mA, it means that, at time t_1, a current of 25 mA flows *into* the two-terminal element by node ②.

The point is that the current reference direction together with the sign of $i(t)$ determines the actual direction of the flow of electric charges.

On Fig. 4.1 we assign + and − signs to pairs of terminals, e.g., in Fig. 4.1*b* the pair ①, ② and the pair ①, ③. These signs indicate the *voltage reference direction*. Consider Fig. 4.1*a*. If, at some time t_0, $v_1(t_0) = 3$ millivolts (mV), it means that, at time t_0, the electric potential of terminal ① is 3 mV *larger* than the electric potential of terminal ②. Similarly, considering Fig. 4.1*c*, if at time t_1, $v_k(t_1) = -320$ V, it means that the electric potential of terminal ⓚ is, at time t_1, 320 V *smaller* than the electric potential of terminal ⓝ.

Exercise Write down the physical meaning of the following statements in Fig. 4.1*c*: $i_k(t_1) = -2$ mA, $i_2(t_1) = 4$ A, $-v_k(t_1) = 5$ V.

4.2 Kirchhoff's Voltage Law (KVL)

Given any *connected* lumped circuit having n nodes, we may choose (arbitrarily) one of these nodes as a *datum node*, i.e., as a reference for measuring electric potentials. By *connected*, we mean that any node can be reached from any other node in the circuit by traversing a path through the circuit elements. Note that the circuit in Fig. 3.2*b* is not connected. With respect to the chosen datum node, we define $n-1$ *node-to-datum* voltages as shown in Fig. 4.2. Since the circuit is a connected lumped circuit, these $n-1$ node-to-datum voltages are well-defined and, in principle, physically measurable quantities. Henceforth, we shall label them $e_1, e_2, \ldots, e_{n-1}$, and dispense with the + and − signs indicating the voltage reference direction. Note that $e_n = 0$ since node ⓝ is the chosen datum node.

Let v_{k-j} denote the voltage difference between node ⓚ and node ⓙ (see Fig. 4.2). Kirchhoff's voltage law states:

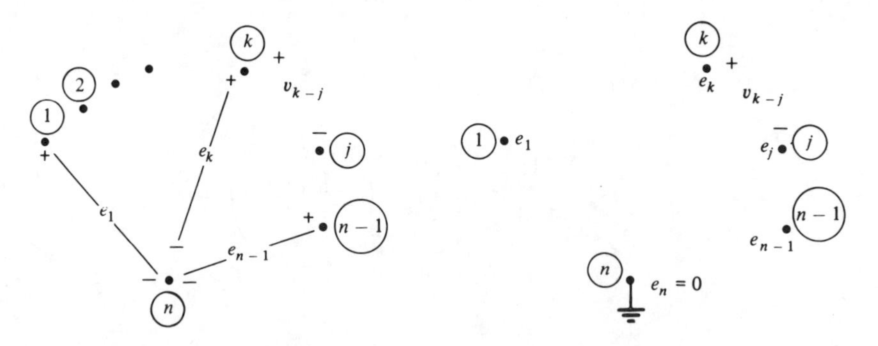

Figure 4.2 Labeling node-to-datum voltages for a circuit with n nodes.

> *KVL* For all lumped connected circuits, for all choices of datum node, for all times t, for all pairs of nodes Ⓚ and Ⓙ,
> $$v_{k-j}(t) = e_k(t) - e_j(t)$$

REMARK Clearly,

$$v_{j-k}(t) = e_j(t) - e_k(t) = -v_{k-j}(t) \tag{4.1}$$

Example The connected circuit in Fig. 4.3 is made of five 2-terminal elements and one 3-terminal element labeled T. There are five nodes, labeled ① through ⑤. Choosing (arbitrarily) node ⑤ as datum, we define the four node-to-datum voltages, e_1, e_2, e_3, and e_4. Therefore by KVL, we may write the following seven equations[5] (for convenience, we drop the dependence on t):

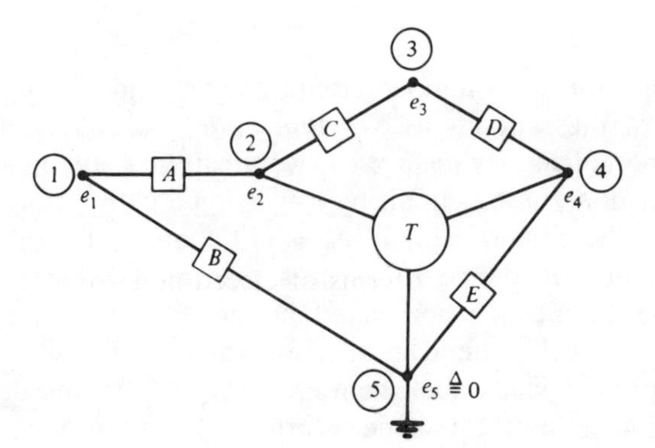

Figure 4.3 A connected circuit with five nodes.

[5] In view of Eq. (4.1) there are altogether two out of five, i.e., $C_2^5 = 10$ nontrivial equations which can be written.

$$
\begin{aligned}
v_{1-5} &= e_1 - e_5 = e_1 \\
v_{1-2} &= e_1 - e_2 \\
v_{2-3} &= e_2 - e_3 \\
v_{3-4} &= e_3 - e_4 \\
v_{4-5} &= e_4 - e_5 = e_4 \\
v_{2-4} &= e_2 - e_4 \\
v_{5-2} &= e_5 - e_2 = -e_2
\end{aligned}
\tag{4.2}
$$

Note that $v_{1\text{-}5}$ and v_{1-2} are the voltages across the two-terminal elements B and A, respectively; v_{2-4}, v_{4-5}, and v_{5-2} are the voltages across the node pairs ②, ④; ④, ⑤; and ⑤, ② of the three-terminal element T, respectively.

If we add the last three equations in (4.2), we find that

$$v_{4-5} + v_{2-4} + v_{5-2} = 0$$

Let us consider the *closed node sequence* ②–④–⑤–②. It is *closed* because the sequence starts and ends at the same node ②. Thus for this particular closed node sequence, the sum of the voltages is equal to zero.

Let us consider a different closed node sequence ①–②–③–④–⑤–①. From the first five equations of (4.2) and using Eq. (4.1), we find that

$$v_{1-2} + v_{2-3} + v_{3-4} + v_{4-5} + v_{5-1} = 0$$

The closed node sequence ①–②–③–④–⑤–① is identified as a *loop* in the circuit, i.e., it is a closed path starting from any node, traversing through *two-terminal elements*, and ending at the same node. The closed node sequence ②–④–⑤–② is not a loop, neither is the closed node sequence ②–③–⑤–②.

Exercise Show that for the closed node sequence ②–③–⑤–② the sum of the voltages, v_{2-3}, v_{3-5}, and v_{5-2} is equal to zero.

We can state KVL in terms of closed node sequences:

> *KVL* (*closed node sequences*) For all lumped connected circuits, for all closed node sequences, for all times t, the algebraic sum of all node-to-node voltages around the chosen closed node sequence is equal to zero.

Theorem KVL in terms of node voltages is equivalent to KVL in terms of closed node sequences.

PROOF

1. We assume that KVL in terms of node voltages holds. Consider any closed node sequence, say ⓐ–ⓑ–ⓒ–ⓓ–ⓐ, and write the algebraic sum of all voltages around that sequence.

$$v_{a-b} + v_{b-c} + v_{c-d} + v_{d-a}$$

By KVL in terms of node voltages this sum can be expressed as

$$(e_a - e_b) + (e_b - e_c) + (e_c - e_d) + (e_d - e_a) = 0$$

so the first statement implies the second.

2. Now assume that KVL in terms of closed node sequences is true. Consider any closed node sequence, say ⓟ–ⓠ–ⓡ–ⓟ then

$$v_{p-q} + v_{q-r} + v_{r-p} = 0 \tag{4.3}$$

Choosing (arbitrarily) ⓡ as the datum node, we have $v_{q-r} = e_q$ and $v_{r-p} = -e_p$ by definition of the node-to-datum voltages. Therefore from Eq. (4.3), we obtain

$$v_{p-q} = e_p - e_q$$

So KVL in terms of closed node sequences implies KVL in terms of node voltages. ■

REMARK For any given connected circuit with n nodes, let us choose (arbitrarily) node ⓝ as the datum node; then the $n-1$ node-to-datum voltages $e_1, e_2, \ldots, e_{n-1}$ specify uniquely and unambiguously the voltage v_{j-k} from any node ⓙ to any other node ⓚ in the circuit. This fact is of crucial importance in circuit theory and is the key concept in node analysis of Chap. 5.

4.3 Kirchhoff's Current Law (KCL)

A fundamental law of physics asserts that electric charge is conserved: There is no known experiment in which a net electric charge is either created or destroyed. Kirchhoff's current law (KCL) expresses this fundamental law in the context of lumped circuits.

To express KCL we shall use gaussian surfaces. A *gaussian surface* is by definition a two-sided "balloon-like" closed surface. Since it is two-sided, it has an "inside" and an "outside." To express the fact that the sum of the charges inside the gaussian surface $\mathcal{S}$ is constant, we shall require that at all times, the algebraic sum of all the currents leaving the surface $\mathcal{S}$ is equal to zero. Let us choose $\mathcal{S}$ so that it cuts only the connecting wires which connect the circuit elements as shown in Fig. 4.4. In the circuit, we have shown a four-terminal element: an operational amplifier, which is connected to the rest of the circuit

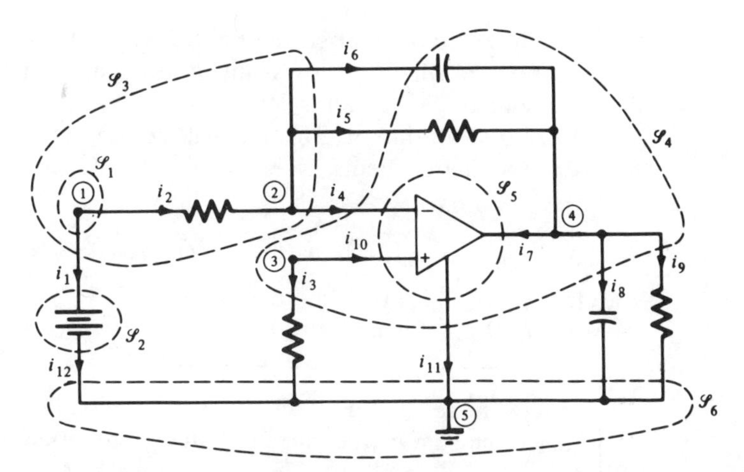

Figure 4.4 An op-amp circuit illustrating gaussian surfaces and KCL.

at nodes ②, ③, ④, and ⑤. The properties of the op amp will be treated in Chap. 4. In the figure we draw six gaussian surfaces: $\mathscr{S}_1, \mathscr{S}_2, \ldots, \mathscr{S}_6$. We will use these surfaces to illustrate Kirchhoff's current law:

> *KCL* For all lumped circuits, for all gaussian surfaces $\mathscr{S}$, for all times t, the algebraic sum of all the currents *leaving* the gaussian surface $\mathscr{S}$ at time t is equal to zero.

For $\mathscr{S}_1$, KCL states:

$$i_1(t) + i_2(t) = 0 \qquad \text{for all } t$$

Note that $\mathscr{S}_1$ contains only node ① in its "inside"; thus a node may be considered as a special case of a gaussian surface, i.e., the surface is shrunk to a point.

For $\mathscr{S}_2$, KCL states:

$$-i_1(t) + i_{12}(t) = 0 \qquad \text{or} \qquad i_1(t) = i_{12}(t)$$

Note that $\mathscr{S}_2$ encloses the two-terminal element, namely, the battery. Thus we make the conclusion that for a *two-terminal element*, the current entering the element from one node at any time t is equal to the current leaving the element from the other node at t.

For $\mathscr{S}_3$, KCL states:

$$i_1(t) + i_4(t) + i_5(t) + i_6(t) = 0$$

For $\mathscr{S}_4$, KCL states:

$$i_3(t) + i_{11}(t) + i_8(t) + i_9(t) - i_6(t) - i_5(t) - i_4(t) = 0$$

For $\mathscr{S}_5$, KCL states:

$$i_{11}(t) - i_{10}(t) - i_4(t) - i_7(t) = 0$$

Note that these are the four currents pertaining to the op amp. Thus choosing a gaussian surface which encloses any n-terminal element, we state that the algebraic sum of the currents leaving or entering the n-terminal element is equal to zero at all times t. This fact will be used in the next section when we discuss n-terminal elements.

For $\mathscr{S}_6$, we have

$$-i_{12}(t) - i_3(t) - i_{11}(t) - i_8(t) - i_9(t) = 0$$

Note that $\mathscr{S}_6$ contains only the datum node ⑤.

We state KCL for nodes:

> *KCL* (*node law*) For all lumped circuits, for all times t, the algebraic sum of the currents *leaving* any node is equal to zero.

REMARK Although a node is a special case of a gaussian surface, KCL for nodes is far more useful than the general statement in terms of gaussian surfaces. Equations written for nodes from the node law are subsets of the equations written for gaussian surfaces of a given circuit. Yet as we shall see in Sec. 6, KCL equations for nodes lead easily to simple analytic formulation of KCL and are the key idea in the node analysis of Chap. 5.

4.4 Three Important Remarks

1. KVL and KCL are the two fundamental postulates of lumped-circuit theory.
2. KVL and KCL hold irrespective of the *nature* of the elements constituting the circuit. Hence, we may say that Kirchhoff's laws reflect the *interconnection* properties of the circuit.
3. KVL and KCL always lead to *homogeneous linear algebraic* equations with *constant real coefficients*, 0, 1, and −1, if written in the fashion given in this section.

5 FROM CIRCUITS TO GRAPHS

The interconnection properties of a circuit can best be exhibited by way of a graph, called a *circuit graph*. In this section, we will demonstrate how a graph can be obtained from a circuit. The graph retains all the interconnection properties of the circuit but suppresses the information on the circuit elements. Therefore, as far as KVL and KCL are concerned, the circuit graph is all that we need.

A *graph* $\mathscr{G}$ is specified by a set of nodes {①, ②, . . . , ⓝ} together with a set of branches $\{\beta_1, \beta_2, \ldots, \beta_b\}$. If each branch is given an orientation, indicated by an arrow on the branch, we call the graph directed, or, simply, a *digraph*. In Fig. 5.1, we show a connected digraph with five nodes and seven

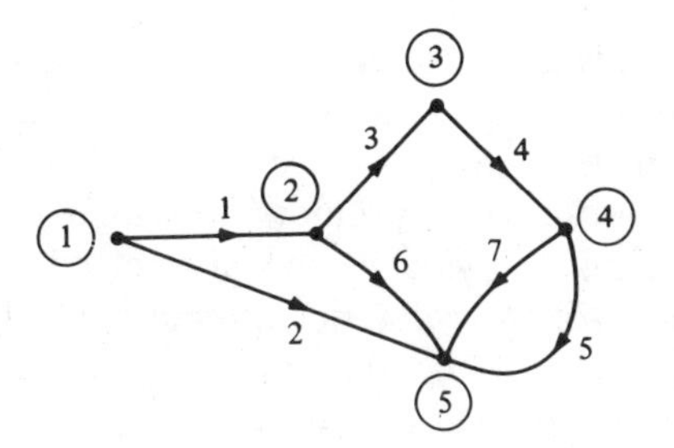

Figure 5.1 A digraph with five nodes and seven branches.

branches, i.e., $n=5$ and $b=7$. The *arrows* on the branches are used to denote the reference directions of the currents.

5.1 The Element Graph: Branch Currents, Branch Voltages, and the Associated Reference Directions

A two-terminal element, shown in Fig. 5.2*a*, can be represented by a graph with two nodes and one branch. This graph is called the *element graph* of the two-terminal element. By KCL, the current i flowing from node ① into the element is equal to the current leaving the element by node ②. We therefore represent a two-terminal element by a digraph with the arrow on the branch indicating the reference direction of the current a shown in Fig. 5.2*b*. By doing so we have suppressed the circuit element; and, as such, the current i is called the *branch current* of the two-terminal element.

The voltage across the element is the voltage v between the node-pair ①, ② shown in Fig. 5.2*a*. The voltage v is called the *branch voltage* of the two-terminal element. The reference direction is specified by the + and − signs associated with node-pair ①, ②. Thus the branch voltage $v(t)>0$ if and only if, at time t, the potential of node ① is larger than that of node ②. Similarly, the branch current $i(t)>0$ if and only if, at time t, the current enters the element by node ① and leaves it by node ②. When, for the two-terminal elements shown, the current and voltage reference directions are chosen as in Fig. 5.2*a*, we say that we have chosen associated reference directions for that two-terminal element.

More precisely, the *associated reference directions* are defined as follows: Suppose that the voltage reference direction is chosen; then the current

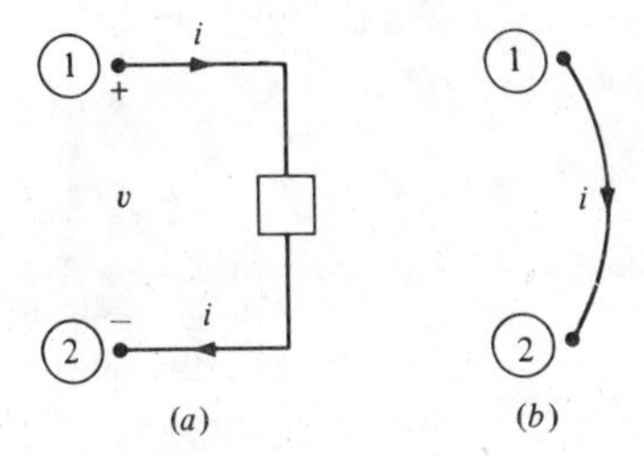

Figure 5.2 (*a*) A two-terminal element and (*b*) its digraph representation.

reference direction is always selected so that the arrow is directed from the + sign toward the − sign through the element. Or, if the reference direction for the current is chosen, the voltage reference direction is specified with the + sign at the node where the current enters the element. This is the convention we will follow throughout, giving us the distinct advantage of not having to mark the signs for the voltage reference direction any more. Therefore in Fig. 5.2*b*, we show only the arrow on the digraph.

Associated reference directions have a very useful property, namely, they make the accounting of power flow quite easy. For the two-terminal element of Fig. 5.2:

$$p(t) \triangleq v(t)i(t) \tag{5.1}$$

= *power* delivered at time *t to* the two-terminal element *by* the remainder of the circuit to which it is connected

If the voltage $v(t)$ is expressed in *volts* and the current in *amperes*, then the power is expressed in *watts*.

Three-terminal elements The digraph representation of two-terminal elements discussed above can be extended to three-terminal elements. For a three-terminal element as shown in Fig. 5.3, there are three node currents i_1, i_2, and i_3, and three voltages v_{1-3}, v_{3-2}, and v_{2-1}. However, from KVL we know that $v_{1-3} + v_{3-2} + v_{2-1} = 0$; and therefore only two voltages can be specified independently. So let us choose arbitrarily node ③ as the datum node and use the node-to-datum voltages for nodes ① and ② as the two independent voltages. Similarly, from KCL, we know that $i_1 + i_2 + i_3 = 0$. Therefore, for the datum node chosen at ③, we use i_1 and i_2 as the two independent currents.

The digraph representation of a three-terminal element with node ③ as datum is shown in Fig. 5.4. Note that it contains *two* branches and three nodes. The arrows indicate the current reference directions for i_1 and i_2. The two currents i_1 and i_2 are called the *branch currents* of the three-terminal element. Using the associated reference directions for the voltages, we redraw the three-terminal element as shown in Fig. 5.5 and define $v_1 = v_{1-3}$ and $v_2 = v_{2-3}$

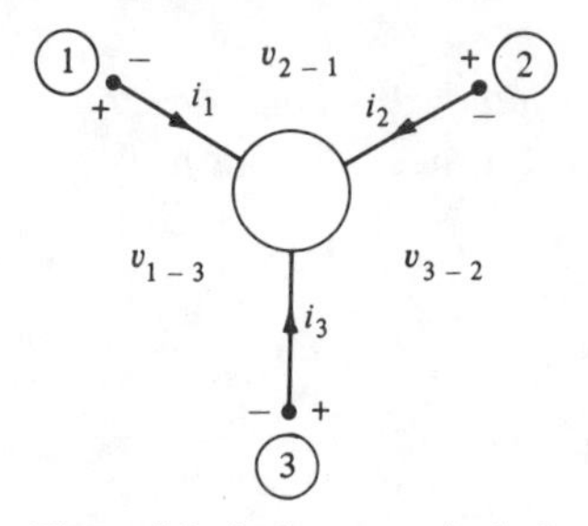

Figure 5.3 A three-terminal element.

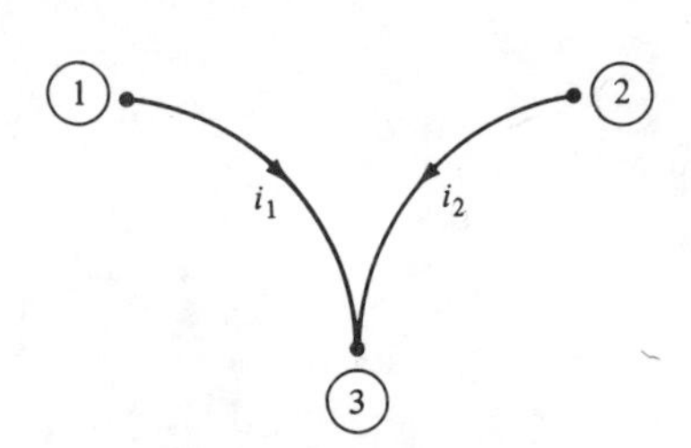

Figure 5.4 The digraph representation of a three-terminal element with node ③ chosen as datum.

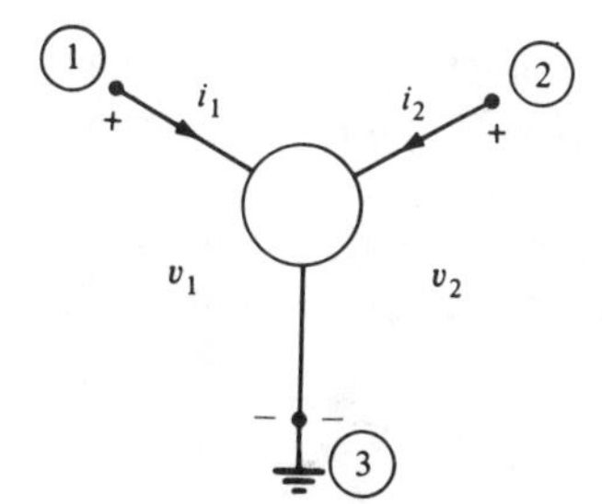

Figure 5.5 A three-terminal element with branch currents i_1, i_2 and branch voltages, v_1, v_2 using associated reference directions.

as the two *branch voltages* of the three-terminal element. Thus by using the digraph representation, we have extended the *circuit variables*: branch voltages and branch currents from two-terminal elements to three-terminal elements.

Obviously, for a three-terminal element, there exist altogether three possible digraph representations depending on which node is chosen as the datum node. In addition to the digraph in Fig. 5.4 we have two other digraphs as shown in Fig. 5.6.

***n*-Terminal elements** We can easily generalize the above to *n*-terminal elements as shown in Fig. 5.7. Thus for an *n*-terminal element, we have an

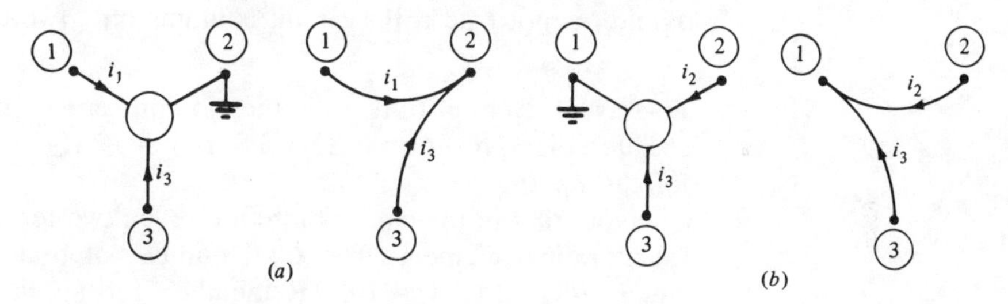

Figure 5.6 Other digraph representations of a three-terminal element: (*a*) Datum node, ②; (*b*) datum node, ①.

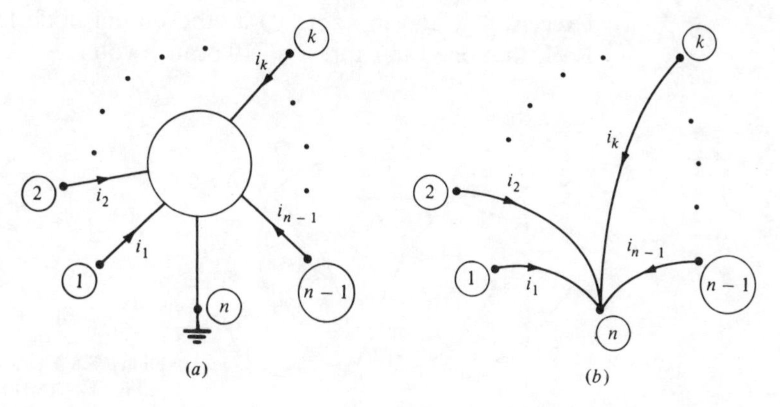

Figure 5.7 An *n*-terminal element and its element graph with node ⓝ as datum node.

element graph with $n-1$ branches and n nodes. There are $n-1$ branch currents and $n-1$ branch voltages; and we always use the associated reference directions and choose the current reference directions as shown, i.e., with arrows entering the element at the nodes. The *power* delivered to the element from the outside to the element at time t is therefore

$$p(t) = \sum_{k=1}^{n-1} v_k(t) i_k(t) \tag{5.2}$$

5.2 The Circuit Graph: Digraph

For a given circuit, if we replace each element by its element graph, the result is a directed *circuit graph*, or simply a *digraph*.

For example, a digraph associated with the circuit in Fig. 4.3 is the one shown in Fig. 5.1. We may now use the digraph instead of the circuit to write equations of KVL and KCL. It is interesting to note that since the circuit contains a three-terminal element, the digraph bears little resemblance to the circuit. In fact, given the digraph, without specifying which nodes belong to the three-terminal element, it is not possible to reconstruct the circuit. This observation is not true if the circuit contains only two-terminal elements.

Exercise 1 Demonstrate that the op-amp circuit in Fig. 4.4 has its associated digraph shown in Fig. 5.8 if node ⑤ is chosen as the datum node for the op amp.

Note that in the circuit there are seven two-terminal elements and one four-terminal element. The total number of branches in the digraph is equal to $7 + (4-1) = 10$. (Remember for an n-terminal element, the element graph has $n-1$ branches.)

Exercise 2 Choosing note ⑤ as the datum node for the circuit, show by KVL that one can express all 10 branch voltages $v_1, v_2, \ldots, v_{10}$ in terms of

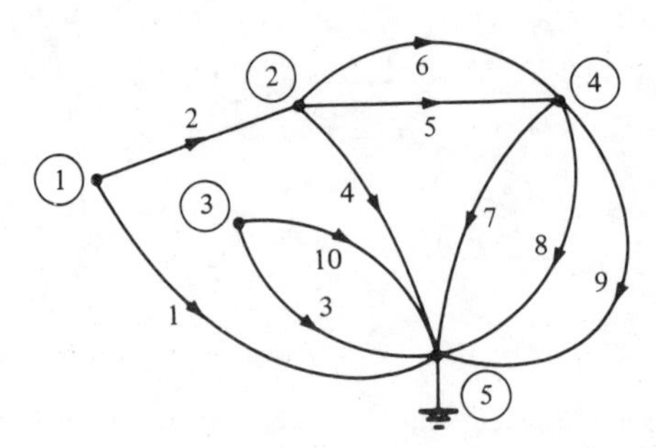

Figure 5.8 Digraph associated with the circuit in Fig. 4.4. The branches are numbered according to the corresponding currents in Fig. 4.4.

the four node-to-datum voltages e_1, e_2, e_3, and e_4 as follows:

$$
\begin{aligned}
v_1 &= e_1 \\
v_2 &= e_1 - e_2 \\
v_3 &= e_3 \\
v_4 &= e_2 \\
v_5 &= e_2 - e_4 \\
v_6 &= e_2 - e_4 \\
v_7 &= e_4 \\
v_8 &= e_4 \\
v_9 &= e_4 \\
v_{10} &= e_3
\end{aligned}
\tag{5.3}
$$

Exercise 3 Show that KCL equations written for the four nodes ① to ④ are

$$
\begin{aligned}
i_1 + i_2 &= 0 \\
-i_2 + i_4 + i_5 + i_6 &= 0 \\
i_3 + i_{10} &= 0 \\
-i_5 - i_6 + i_7 + i_8 + i_9 &= 0
\end{aligned}
\tag{5.4}
$$

Exercise 4 Express Eqs. (5.3) and (5.4) in matrix form using the vectors **v**, **e**, and **i**, e.g., $\mathbf{v} = [\mathbf{v}_1, \mathbf{v}_2, \ldots, \mathbf{v}_{10}]^T$, where the superscript T denotes matrix transposition.

Remark The fundamental concept of using a circuit graph instead of the circuit itself in writing KVL and KCL equations is the following:

1. We convert circuit elements whether two-terminal, three-terminal, or n-terminal into *branches*, thus we were able to define *branch voltages* and *branch currents* for any element in a circuit.
2. With a circuit graph we can define precisely the interconnection properties of a circuit using the branch-node incidence relation of a graph to be discussed in Sec. 6.

Exercise 5 Show that if branch 3 in Fig. 4.4 is replaced by a short circuit thereby coalescing nodes ③ and ⑤ into one node, then the digraph in Fig. 5.8 will contain a *self-loop*, i.e., a loop made of one branch and one node.

5.3 Two-Ports, Multiports, and Hinged Graphs

Up to now we have assumed that the circuit is connected. In Fig. 3.2*b* the circuit, because of the presence of a two-winding transformer, is not connected. It turns out that we can easily take care of the situation; but before we do so, we need to introduce a special class of four-terminal elements called two-ports. A *two-port* is a circuit element or a circuit with two *pairs* of *accessible* terminals. Thus a two-port may contain many circuit elements.

Two-ports In many engineering situations the terminals of a multiterminal device are naturally associated in pairs: For example, in a hi-fi chain the input pair is connected, say, to a microphone and the output pair to a loudspeaker system. These pairs of associated terminals are called *ports*. Another example is a two-winding transformer: The two input terminals constitute a natural *input port* and the two output terminals constitute a natural *output port*. In either case, the typical connections to the four-terminal element have the form shown in Fig. 5.9. Note the labeling of the nodes and the currents: the input pair is ①, ①′ and the output pair is ②, ②′.

When we view the four-terminal element of Fig. 5.9 as a *two-port*, we consider *only* the voltages v_1 and v_2 and the four terminal currents i_1, i_1', i_2, i_2'. Naturally, v_k is called the *port voltage* at port ⓚ, $k = 1, 2$. Now the gaussian surfaces $\mathcal{S}_1$ and $\mathcal{S}_2$ shown in Fig. 5.9 and KCL impose the two current constraints:

$$i_1 = i_1' \qquad \text{and} \qquad i_2 = i_2'$$

The point is that these two port constraints reduce the number of current variables from four to two: i_1 and i_2. The current i_k is called the *port current* at port ⓚ.

Note that at each port the port voltage v_k and the port current i_k have associated reference directions: Hence $v_k(t)i_k(t)$ is the power entering port k at time t. For example, the power delivered at time t, *by* the remainder of the circuit *to* the two-port of Fig. 5.9 is given by

$$v_1(t)i_1(t) + v_2(t)i_2(t)$$

Naturally, a two-terminal element may be viewed as a one-port. Thus, in generalizing the digraph representation from a one-port to a two-port, we use two branches and four nodes for its element graph as shown in Fig. 5.10.

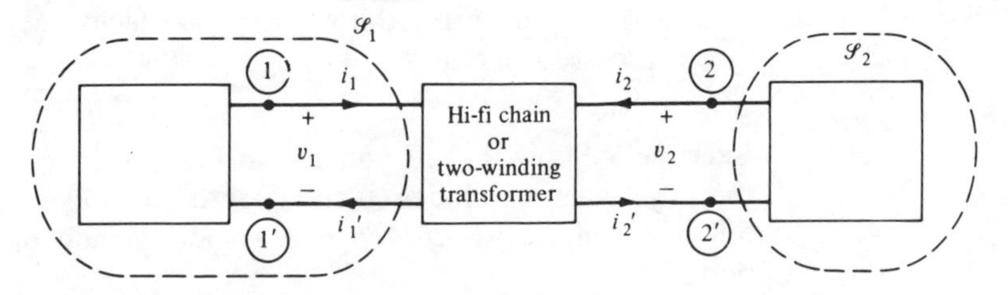

Figure 5.9 Example of a two-port.

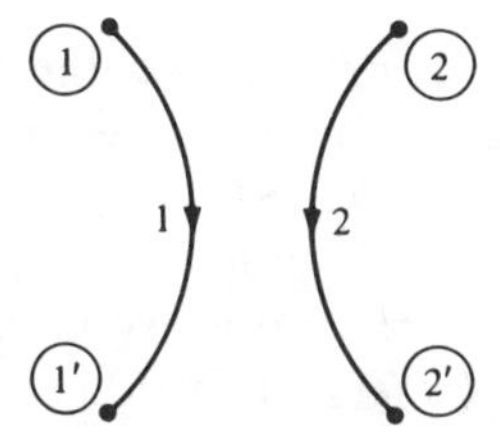

Figure 5.10 The element graph of a two-port.

Therefore the port voltages v_1 and v_2 are also referred to as the branch voltages of the two-port. Similarly, we can also call the port currents i_1 and i_2 the branch currents of the two-port. This is in contrast to a four-terminal element where there are three branches in its element graph, thus three branch voltages and three branch currents.

Multiports We can generalize the concept of two-ports to multiports. For example, a three-winding transformer is a three-port as shown in Fig. 5.11. Its element graph has three branches and six nodes as shown in Fig. 5.11*c*. The three branch voltages and three branch currents are the port voltages and port currents, respectively, for the three-port.

Hinged graphs The element graph of a two-port consists of two branches which are not connected. It signifies that the port voltages or port currents at different ports are not related because of connections but rather are *coupled* because of physical phenomena within the element. For example, the trans-

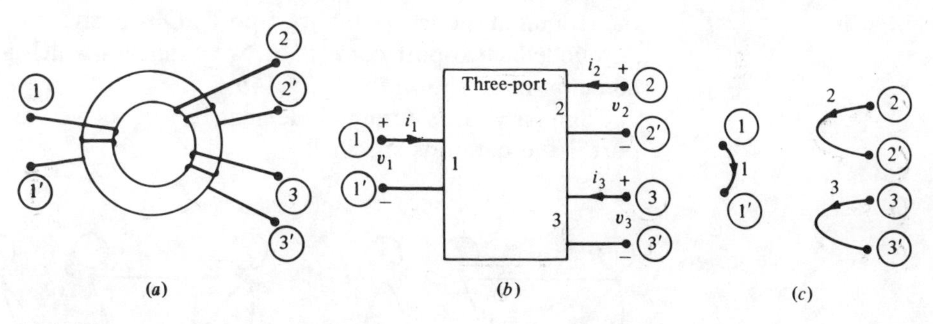

Figure 5.11 (*a*) A three-winding transformer. (*b*) the corresponding three-port, and (*c*) its element graph.

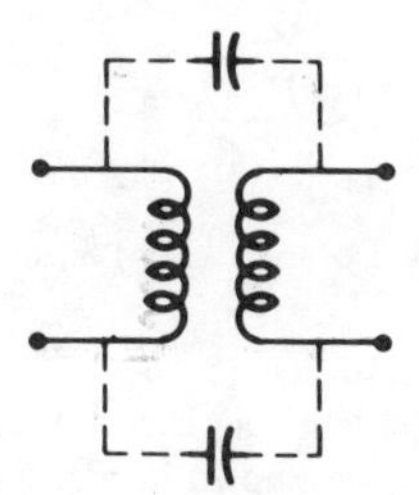

Figure 5.12 A model of a physical transformer which includes two parasitic capacitors.

former port voltages are coupled magnetically via the flux linkages among the various windings. Therefore circuits containing two-ports or multiports have circuit graphs which are often unconnected.[6]

To avoid an "unconnected" circuit graph, we can tie together the two separate ports of a circuit graph at two arbitrary nodes by a branch. This is illustrated in Fig. 5.13*a*, where nodes ③ and ⑤ are tied together by a branch k. This connection does not change any branch voltage or current in the original circuit. This is easily seen because, by using KCL with a gaussian surface which encloses one of the separate parts of the graph and which cuts branch k, the current i_k is zero. If $i_k = 0$, it amounts to an open circuit or no connection; thus we have not changed the behavior of the circuit. Next, since voltages are measured between nodes, we choose a datum node for each separate part. If we choose nodes ③ and ⑤ as the datum nodes for the separate parts, we may "solder" together node ③ and node ⑤ as shown in Fig. 5.13*b* to make them the common datum. The graph so obtained is called a *hinged graph*. With the introduction of the concept of a hinged graph, we have generalized our treatment so far to include two-ports and multiports, that is, we can always assume without loss of generality that any lumped circuit and its circuit graph are *connected*.

"Grounded" two-ports If a common connection exists between nodes ①' and ②' of a two-port as shown by the low-pass filter in Fig. 5.14*a*, we call it, by tradition, a *"grounded" two-port*. The word "grounded" does not necessarily mean that the node is always set to zero potential. Rather, a "grounded" two-port is essentially a three-terminal element with its datum node specified as the common node of the two-port. Obviously, the element graph for a "grounded" two-port consists of two branches which are tied together at the common node shown in Fig. 5.14*b*.

Similarly, an n-terminal element can be viewed as a "grounded" $(n-1)$-port if the datum is specified.

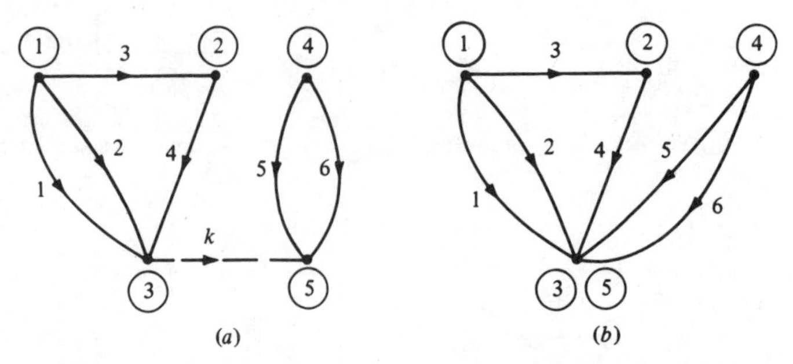

Figure 5.13 (*a*) Connecting nodes ③ and ⑤ by a branch k. (*b*) Soldering together nodes ③ and ⑤ to obtain a hinged graph.

[6] An exception to this is, for example, in modeling a physical transformer; we may need to use additional elements to tie the windings together as shown in Fig. 5.12.

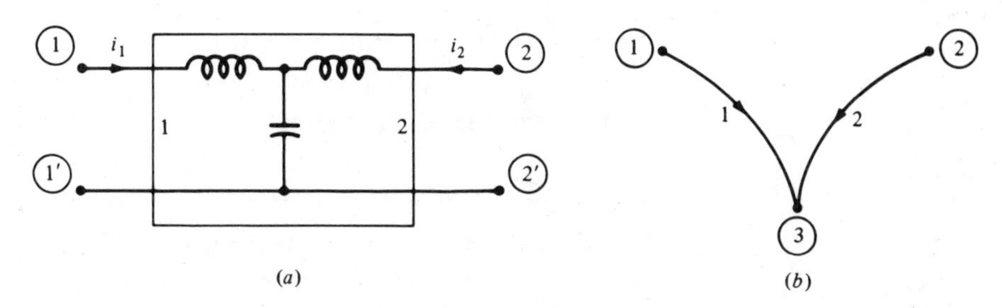

Figure 5.14 (*a*) A "grounded" two-port and (*b*) its element graph.

5.4 Cut Sets and KCL

A very useful graph-theoretic concept is the cut set. Given a *connected* digraph $\mathscr{G}$, a set of branches $\mathscr{C}$ of $\mathscr{G}$ is called a *cut set* iff[7] (*a*) the removal of all the branches of the cut set results in an *unconnected digraph*, which means that the resulting digraph is no longer connected, and (*b*) the removal of all but any one branch of $\mathscr{C}$ leaves the digraph *connected*. Stated in another way, (*b*) implies that if any branch in the set is left intact, the digraph remains connected.

For the digraph of Fig. 5.15, $\mathscr{C}_1 = \{\beta_1, \beta_3\}$, $\mathscr{C}_2 = \{\beta_4, \beta_5, \beta_6\}$, and $\mathscr{C}_3 = \{\beta_4, \beta_5, \beta_7\}$ form cut sets. Here, β_k denotes "branch k."

Exercise Refer to Fig. 5.15.
(*a*) Is $\{\beta_1, \beta_3, \beta_4, \beta_5, \beta_6\}$ a cut set?
(*b*) List all cut sets of the digraph shown in Fig. 5.15.

REMARKS
1. Any cut set creates a partition of the set of nodes in the graph into two subsets.
2. To any cut set corresponds a gaussian surface which cuts precisely the same branches.
3. Similarly, to any gaussian surface corresponds *either* one cut set *or* a union of cut sets (see $\mathscr{S}_1$ in Fig. 5.15).
4. To each cut set we can define arbitrarily a *reference direction*, as shown by the arrows attached to the cut sets in Fig. 5.15.

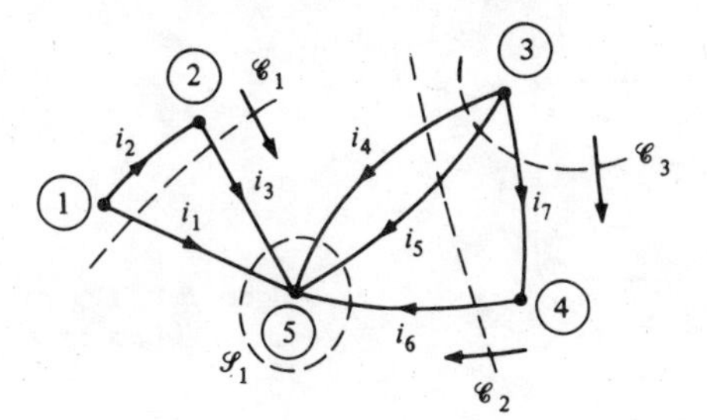

Figure 5.15 Digraph illustrating cut sets.

[7] "iff" means "if and only if."

> *KCL (cut-set law)* For all lumped circuits, for all time t, the algebraic sum of the currents associated with any cut set is equal to zero.

Example For the digraph shown in Fig. 5.16, the cut set $\mathscr{C} = \{\beta_1, \beta_2, \beta_3\}$ is indicated by the dashed line cutting through these branches. Let us assign a reference direction to $\mathscr{C}$ as shown by the arrow; then the KCL applied to $\mathscr{C}$ gives

$$i_1(t) + i_2(t) - i_3(t) = 0$$

The $-i_3$ comes about because the reference direction of i_3 disagrees with the reference direction of the cut set $\mathscr{C}$.

By now we have learned three forms of KCL, namely, in terms of (1) gaussian surfaces, (2) nodes, and (3) cut sets.

KCL theorem The three forms of the KCL are equivalent. Symbolically,[8]

$$\begin{pmatrix}\text{KCL}\\ \text{gaussian surface}\end{pmatrix} \Leftrightarrow \begin{pmatrix}\text{KCL}\\ \text{node law}\end{pmatrix} \Leftrightarrow \begin{pmatrix}\text{KCL}\\ \text{cut sets}\end{pmatrix}$$

PROOF

(1)$\Rightarrow$(2) Simply use a gaussian surface that surrounds only the node in question. For example, consider node ⑤ in Fig. 5.15: For gaussian surface $\mathscr{S}_1$, KCL applied to $\mathscr{S}_1$ is identical with KCL applied to node ⑤, namely,

$$-i_1 - i_3 - i_4 - i_5 - i_6 = 0$$

(2)$\Rightarrow$(3) Any cut set partitions the set of nodes into two subsets. Writing the KCL equation for each node in such a subset and adding the results, we obtain the cut-set equation, except for maybe a -1 factor. For example, consider the cut set $\mathscr{C}_2$ in Fig. 5.15: If we

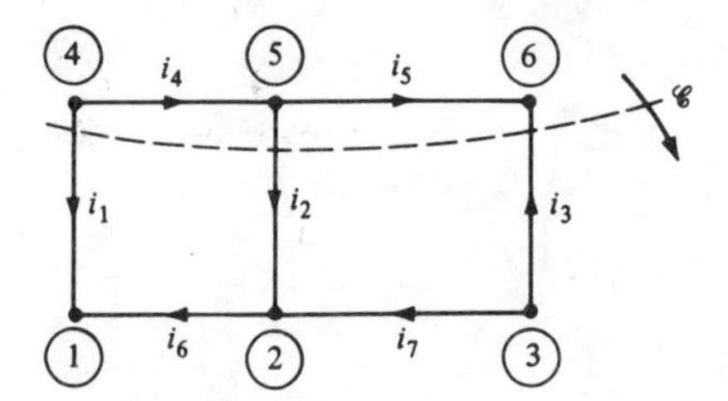

Figure 5.16 Digraph illustrating the reference direction of a cut set.

[8] $\Rightarrow$ means "implies"; $\Leftarrow$ means "is implied by"; $\Leftrightarrow$ means "is equivalent to."

add the KCL equations applied to nodes ③ and ④, we obtain

$$i_4 + i_5 + i_6 = 0$$

(note that i_7 cancels out in the addition!), which is the cut-set equation for $\mathscr{C}_2$.

(3)⇒(1) It is easy to demonstrate that the set of branches cut by a gaussian surface is either a cut set or a disjoint union of cut sets. So given any gaussian surface, let us write the KCL equation for each of these cut sets; then adding or subtracting these equations, we obtain the KCL equation for the gaussian surface. For example, consider gaussian surface $\mathscr{S}_1$ of Fig. 5.15. It is the union of cut set $\{\beta_1, \beta_3\}$ and cut set $\{\beta_4, \beta_5, \beta_6\}$ whose equations are, respectively,

$$-i_1 - i_3 = 0$$

$$+i_4 + i_5 + i_6 = 0$$

Subtracting the second equation from the first gives

$$-i_1 - i_3 - i_4 - i_5 - i_6 = 0$$

which is the KCL equation for gaussian surface $\mathscr{S}_1$. ■

6 MATRIX FORMULATION OF KIRCHHOFF'S LAWS

6.1 Linear Independence

Consider a set of m linear algebraic equations in n unknowns: For $j = 1, 2, \ldots, m$

$$f_j(x_1, x_2, \ldots, x_n) = \alpha_{j1}x_1 + \alpha_{j2}x_2 + \cdots + \alpha_{jn}x_n = 0 \tag{6.1}$$

where the α_{jk}'s are real or complex numbers. It is important to decide whether or not each equation brings new information not contained in the others; equivalently, it is important to decide whether the equations are linearly independent. These m equations are said to be *linearly dependent* iff there are constants $k_1, k_2, \ldots, k_m$ and *not all zero* such that

$$\sum_{j=1}^{m} k_j f_j(x_1, x_2, \ldots, x_n) = 0 \qquad \text{for all } x_1, x_2, \ldots, x_n \tag{6.2}$$

Clearly if these m equations are linearly dependent, then at least one equation may be written as a linear combination of the others; in other words, that equation repeats the information contained in the others!

It is crucial to note that the left-hand side of Eq. (6.2) must be zero *for all* values of $x_1, x_2, \ldots, x_n$.

Example Consider an example where $m = 3$ and $n = 4$:

$$\begin{aligned} x_1 - x_2 + x_3 + 3x_4 &= 0 \\ 2x_1 + 3x_2 - x_3 - 4x_4 &= 0 \\ -4x_1 - 11x_2 + 5x_3 + 18x_4 &= 0 \end{aligned}$$

Direct calculation shows that with $k_1 = 2$, $k_2 = -3$, and $k_3 = -1$ the condition for Eq. (6.2) holds; in other words, these three equations are linearly dependent.

The set of m linear algebraic equations (6.1) is said to be *linearly independent* iff it is not linearly dependent.

In practice, we use gaussian elimination to decide whether or not a given set of linear equations is linearly dependent.

6.2 Independent KCL Equations

For a given circuit, we can write many KCL equations by the node law, the cut-set law, or using gaussian surfaces. How many of them are linearly independent and how to write a complete set that contains all the necessary information as far as KCL is concerned are the subjects of this subsection. We will give a systematic treatment by means of the digraph of the circuit under consideration: in particular, a list of nodes, a list of branches, and for each branch the specification of the node it leaves and of the node it enters. This is done by the *incidence matrix* $\mathbf{A}_a$ of the digraph.

Let digraph $\mathcal{G}$ have n nodes and b branches, then $\mathbf{A}_a$ has n rows—one row to each node—and b columns—one column to each branch. To see how the matrix is built up consider the four-node six-branch digraph shown in Fig. 6.1. Let us write the KCL equations for each node:

$$\begin{array}{rrrrrrl} i_1 & + i_2 & & & & - i_6 & = 0 \\ -i_1 & & - i_3 & + i_4 & & & = 0 \\ & - i_2 & + i_3 & & + i_5 & & = 0 \\ & & & - i_4 & - i_5 & + i_6 & = 0 \end{array} \tag{6.3}$$

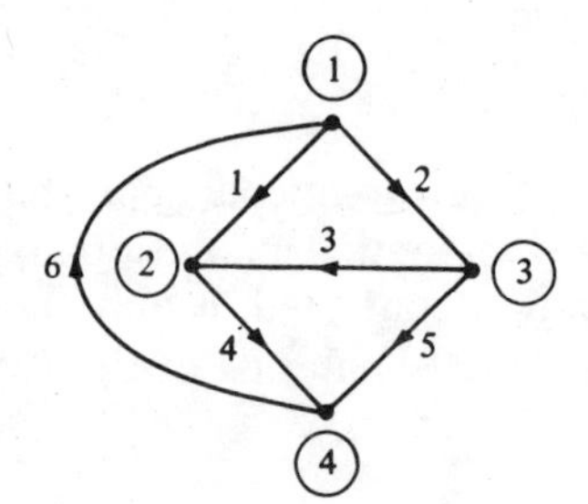

Figure 6.1 A digraph with four nodes and six branches.

In matrix form it reads

$$\begin{array}{c} \text{①}\rightarrow \\ \text{②}\rightarrow \\ \text{③}\rightarrow \\ \text{④}\rightarrow \end{array} \begin{bmatrix} 1 & 1 & 0 & 0 & 0 & -1 \\ -1 & 0 & -1 & 1 & 0 & 0 \\ 0 & -1 & 1 & 0 & 1 & 0 \\ 0 & 0 & 0 & -1 & -1 & 1 \end{bmatrix} \begin{bmatrix} i_1 \\ i_2 \\ i_3 \\ i_4 \\ i_5 \\ i_6 \end{bmatrix} = \begin{bmatrix} 0 \\ 0 \\ 0 \\ 0 \end{bmatrix} \tag{6.4}$$

(arrows below the matrix: branch 1 under the first column, branch 6 under the last column)

The 4×6 matrix just obtained is called the *incidence matrix* $\mathbf{A}_a$ of $\mathcal{G}$.

Exercise

(*a*) Demonstrate that the four equations in (6.3) are linearly dependent.
(*b*) Demonstrate that any three of the four equations in (6.3) are linearly independent.

In general, for an n-node b-branch *connected* digraph $\mathcal{G}$ which does *not* contain *self-loops*[9] the matrix $\mathbf{A}_a$ is specified as follows: For $i = 1, 2, \ldots, n$ and $k = 1, 2, \ldots, b$

$$a_{ik} = \begin{cases} +1 \text{ if branch } k \textit{ leaves} \text{ node } (i) \\ -1 \text{ if branch } k \textit{ enters} \text{ node } (i) \\ \ \ 0 \text{ if branch } k \text{ does not touch node } (i) \end{cases} \tag{6.5}$$

and the n node equations of $\mathcal{G}$ read

$$\mathbf{A}_a \mathbf{i} = \mathbf{0} \tag{6.6}$$

where $\mathbf{i} = (i_1, i_2, \ldots, i_b)^T$ is called the *branch current vector*.

REMARK Each column of $\mathbf{A}_a$ has precisely a single $+1$ and a single -1; consequently, if we add together the n equations in (6.6), all the variables $i_1, i_2, \ldots, i_b$ cancel out; equivalently the n KCL equations are linearly dependent.

Suppose that for the *connected* digraph $\mathcal{G}$ we choose a datum node and we throw away the corresponding KCL equation, then the remaining $n-1$ equations are linearly independent. Since this is important we state it formally:

Independence property of KCL equations For any *connected* digraph $\mathcal{G}$ with n nodes, the KCL equations for *any* $n-1$ of these nodes form a set of $n-1$ *linearly independent equations*.

[9] The digraph of circuits containing multiterminal elements will contain *self-loops* whenever one or more terminals are connected to the datum, as in the last exercise of Sec. 5.2.

PROOF We prove it by contradiction. Suppose that the first k of these $n-1$ equations are *linearly dependent*. More precisely, there are k real constants $\gamma_1, \gamma_2, \ldots, \gamma_k$, *not all zero*, such that

$$\sum_{j=1}^{k} \gamma_j f_j(i_1, i_2, \ldots, i_n) = 0 \qquad \text{for all } i_1, i_2, \ldots, i_n \tag{6.7}$$

Without loss of generality, we may assume that $\gamma_j \neq 0$ for $j = 1, 2, \ldots, k$, i.e., there are exactly k equations in the sum of Eq. (6.7).

Consider the two sets of nodes in $\mathcal{G}$, namely, the set which corresponds to the k equations and that of the remaining nodes. Since the digraph is connected, there is at least one branch which connects a node in the first set to a node in the second set. Clearly the current in that branch appears only *once* in the first k node equations, hence that current cannot cancel out in the sum of Eq. (6.7). This contradiction shows that for any $k \leq n-1$ it is not the case that a subset of k of the KCL equations is linearly dependent. That is, these $n-1$ equations are linearly independent. ■

If in $\mathbf{A}_a$, the incidence matrix of the connected digraph $\mathcal{G}$, we delete the row corresponding to the datum node, we obtain the *reduced incidence matrix* $\mathbf{A}$ which is of dimension $(n-1) \times b$. The corresponding KCL equations read

$$\mathbf{Ai} = \mathbf{0} \tag{6.8}$$

As a consequence of the independence property just proved, we may state that the $(n-1) \times b$ matrix $\mathbf{A}$ is full rank, i.e., its $n-1$ rows are linearly independent vectors in the b-dimensional space. Stated in another way, (6.8) consists of $n-1$ linearly independent KCL equations.

6.3 Independent KVL Equations

Similarly, to write a set of complete linearly independent KVL equations in a systematic way is of crucial importance. Let us write KVL for the four-node six-branch digraph of Fig. 6.1. Using *associated reference directions* and choosing node ④ as the datum node, we obtain

$$\begin{aligned}
v_1 &= e_1 - e_2 \\
v_2 &= e_1 \qquad - e_3 \\
v_3 &= \qquad -e_2 + e_3 \\
v_4 &= \qquad\ \ e_2 \\
v_5 &= \qquad\qquad\ \ e_3 \\
v_6 &= -e_1
\end{aligned} \tag{6.9}$$

or in matrix form

$$\mathbf{v} = \mathbf{M}\mathbf{e} \tag{6.10}$$

where $\mathbf{v} = (v_1, v_2, \ldots, v_b)^T$ is the *branch voltage vector*, $\mathbf{e} = (e_1, e_2, \ldots, e_{n-1})^T$ is the *node-to-datum voltage vector*, and $\mathbf{M}$ is a $b \times (n-1)$ matrix. Thinking in terms of KVL, we see that for $k = 1, 2, \ldots, b$ and $i = 1, 2, \ldots, n-1$

$$m_{ki} = \begin{cases} +1 & \text{if branch } k \textit{ leaves} \text{ node } \textcircled{i} \\ -1 & \text{if branch } k \textit{ enters} \text{ node } \textcircled{i} \\ 0 & \text{if branch } k \text{ does not touch node } \textcircled{i} \end{cases} \tag{6.11}$$

Comparing Eq. (6.11) with (6.5), we conclude that

$$\mathbf{M} = \mathbf{A}^T$$

and more usefully, KVL is expressed by the equation

$$\mathbf{v} = \mathbf{A}^T\mathbf{e} \tag{6.12}$$

With a *connected* digraph $\mathcal{G}$ $\mathbf{A}$ has $n-1$ linearly independent rows, and consequently $\mathbf{A}^T$ has $n-1$ linearly independent columns.

REMARKS

1. Note that, in the digraph, (*a*) we choose current reference directions, (*b*) we choose a datum node and define the reduced incidence matrix $\mathbf{A}$, (*c*) we write KCL as $\mathbf{Ai} = \mathbf{0}$, (*d*) then we use associated reference directions to find that KVL reads $\mathbf{v} = \mathbf{A}^T\mathbf{e}$. Thus whenever we invoke this last equation, we automatically use associated reference directions for the branch voltages. We also assume the same datum node is used in writing KCL and KVL.
2. When we deal with digraphs which are not connected, we could either use the concept of the hinged graph to make the digraph connected or treat each separate part independently. In the latter, each separate part will have its own incidence matrix and datum node.

7 TELLEGEN'S THEOREM

Tellegen's theorem is a very general and very useful theorem. We'll use it repeatedly in this text. Tellegen's theorem is a direct consequence of Kirchhoff's laws.

7.1 Theorem, Proof, and Remarks

Example Consider the digraph shown in Fig. 7.1. Choose *arbitrarily* the

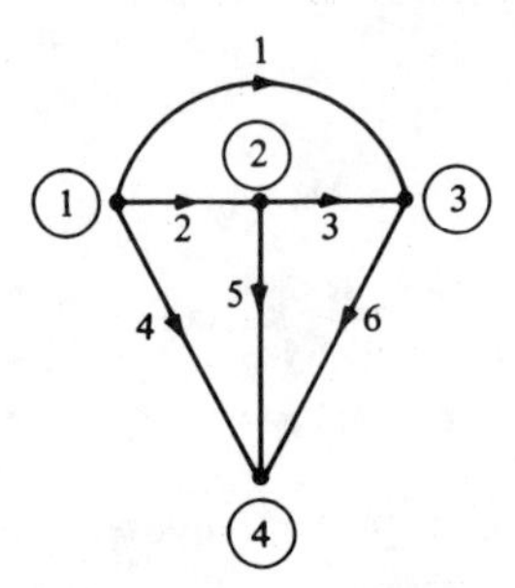

Figure 7.1 A digraph with four nodes and six branches.

values of the currents i_1, i_2, i_3 and calculate i_4, i_5, i_6 so that KCL is satisfied: Let

$$i_1 = 1 \qquad i_2 = 2 \qquad i_3 = 3$$

hence
$$i_4 = -3 \qquad i_5 = -1 \qquad i_6 = 4$$

Now choose *arbitrarily* v_4, v_5, and v_6 and calculate v_1, v_2, v_3 so that KVL is satisfied (note that we use associated reference directions). Let

$$v_4 = 4 \qquad v_5 = 5 \qquad v_6 = 6$$

hence
$$v_1 = -2 \qquad v_2 = -1 \qquad v_3 = -1$$

Note that $i_1, i_2, \ldots, i_6$ obey KCL and $v_1, v_2, \ldots, v_6$ obey KVL for the circuit under consideration. Now it is easy to verify that

$$\sum_{k=1}^{6} v_k i_k = 0$$

This result is surprising since the i_k's and the v_k's seem to bear so little relation to each other.

Tellegen's theorem Consider an arbitrary circuit. Let the digraph $\mathcal{G}$ have b branches. Let us use associated reference directions. Let $\mathbf{i} = (i_1, i_2, \ldots, i_b)^T$ be *any* set of branch currents satisfying KCL for $\mathcal{G}$ and let $\mathbf{v} = (v_1, v_2, \ldots, v_b)^T$ be *any* set of branch voltages satisfying KVL for $\mathcal{G}$, then

$$\sum_{k=1}^{b} v_k i_k = 0 \qquad \text{or equivalently} \qquad \mathbf{v}^T\mathbf{i} = 0 \tag{7.1}$$

Proof For the connected digraph $\mathcal{G}$,[10] choose a datum node; hence its reduced matrix $\mathbf{A}$ is defined unambiguously. Since $\mathbf{i}$ satisfies KCL, we have

$$\mathbf{Ai} = \mathbf{0} \tag{7.2}$$

[10] We again use a hinged graph to take care of graphs which are not connected.

Since $\mathbf{v}$ satisfies KVL and since we use associated reference directions, for some node-to-datum voltage vector $\mathbf{e}$, we have

$$\mathbf{v} = \mathbf{A}^T \mathbf{e} \tag{7.3}$$

Using these two equations we obtain successively,

$$\mathbf{v}^T \mathbf{i} = (\mathbf{A}^T \mathbf{e})^T \mathbf{i} = \mathbf{e}^T (\mathbf{A}^T)^T \mathbf{i} = \mathbf{e}^T (\mathbf{A}\mathbf{i}) = 0 \tag{7.4}$$

where in the last step we used Eq. (7.2). ■

REMARKS

1. The $\mathbf{v}$ and the $\mathbf{i}$ in the theorem need not bear any relation to each other: $\mathbf{v}$ must *only* satisfy KVL and $\mathbf{i}$ must *only* satisfy KCL, and we must use associated reference directions.
2. Suppose that for the given connected digraph $\mathcal{G}$, let $\mathbf{v}'$ and $\mathbf{v}''$ satisfy KVL, and let $\mathbf{i}'$ and $\mathbf{i}''$ satisfy KCL. Then Tellegen's theorem asserts that

$$\mathbf{v}'^T \mathbf{i}' = 0 \qquad \mathbf{v}'^T \mathbf{i}'' = 0 \qquad \mathbf{v}''^T \mathbf{i}' = 0 \qquad \mathbf{v}''^T \mathbf{i}'' = 0 \tag{7.5}$$

Equation (7.5) is of particular interest. Note that $\mathbf{v}'$, $\mathbf{v}''$, $\mathbf{i}'$, and $\mathbf{i}''$ are not related other than by the fact that they pertain to the same digraph and that they each independently satisfy Kirchhoff's laws. Clearly, Tellegen's theorem depicts only the interconnection properties of the circuit or the *topology* of the digraph. We will demonstrate later that this general form of Tellegen's theorem can be used to prove some general results in circuit theory.

7.2 Tellegen's Theorem and Conservation of Energy

Consider a lumped connected circuit and let us measure, at some time t, all its branch voltages $v_k(t)$ and all its branch currents $i_k(t)$, $k = 1, 2, \ldots, b$. Obviously $\mathbf{v}(t)$ and $\mathbf{i}(t)$ satisfy KVL and KCL, hence, by Tellegen's theorem

$$\mathbf{v}(t)^T \mathbf{i}(t) = \sum_{k=1}^{b} v_k(t)\, i_k(t) = 0 \tag{7.6}$$

Now, since we use associated reference directions, $v_k(t) i_k(t)$ is the power delivered, at time t, *to* branch k *by* the remainder of the circuit; equivalently, $v_k(t) i_k(t)$ is the *rate* at which energy is delivered, at time t, *to* branch k *by* the remainder of the circuit. Hence Eq. (7.6) asserts that the energy is conserved. Thus, for *lumped circuits*, conservation of energy is a consequence of Kirchhoff's laws.

To appreciate the fact that Tellegen's theorem is far more general than conservation of energy, work out the following exercise:

Exercise Consider an arbitrary circuit with digraph $\mathscr{G}$. Suppose that, for all $t \geq 0$, $\mathbf{v}(t)$ satisfies KVL for $\mathscr{G}$ and $\mathbf{i}(t)$ satisfies KCL for $\mathscr{G}$. Show that for all $t_1, t_2 \geq 0$

$$\sum_{k=1}^{b} v_k(t_1) i_k(t_2) = 0 \qquad \sum_{k=1}^{b} v_k(t_2) i_k(t_1) = 0 \tag{7.7a}$$

$$\sum_{k=1}^{b} v_k(t_1) \dot{i}_k(t_2) = 0 \qquad \sum_{k=1}^{b} \dot{v}_k(t_1) i_k(t_2) = 0 \tag{7.7b}$$

$$\sum_{k=1}^{b} v_k(t_2) \dot{i}_k(t_1) = 0 \qquad \sum_{k=1}^{b} \dot{v}_k(t_2) i_k(t_1) = 0 \tag{7.7c}$$

where $\dot{v}_k(t)$ denotes $dv_k/dt(t)$ and $\dot{i}_k(t)$ denotes $di_k/dt(t)$.

7.3 The Relation between Kirchhoff's Laws and Tellegen's Theorem

In circuit theory there are two fundamental postulates: KCL and KVL. We have proved that KCL and KVL imply Tellegen's theorem. It is interesting to note that any one of Kirchhoff's laws together with Tellegen's theorem implies the other. More precisely we have the following properties:

Properties

1. If, for all $\mathbf{v}$ satisfying KVL, $\mathbf{v}^T\mathbf{i} = 0$ then $\mathbf{i}$ satisfies KCL.
2. If, for all $\mathbf{i}$ satisfying KCL, $\mathbf{v}^T\mathbf{i} = 0$, then $\mathbf{v}$ satisfies KVL.

PROOF

1. For all $\mathbf{e}$ let $\mathbf{v} = \mathbf{A}^T\mathbf{e}$, and thus $\mathbf{v}$ satisfies KVL. By assumption,

$$0 = \mathbf{v}^T\mathbf{i} = \mathbf{e}^T\mathbf{A}\mathbf{i}$$

Now since $\mathbf{e}$ is an arbitrary node-to-datum voltage vector, the last equality implies $\mathbf{Ai} = 0$, i.e., $\mathbf{i}$ satisfies KCL.

2. Let ℓ be an arbitrary loop in the graph $\mathscr{G}$. Consider the $\mathbf{i}$ obtained by assigning zero current to all branches of $\mathscr{G}$ except for those of loop ℓ; depending on whether the reference direction of branch j in loop ℓ agrees with that of loop ℓ, we assign i_j to be 1 A or -1 A. The resulting $\mathbf{i}$ satisfies KCL at all nodes of $\mathscr{G}$. Tellegen's theorem gives

$$\sum_{j=1}^{b} v_j i_j = \sum_{\substack{\text{over} \\ \text{branches} \\ \text{in loop } \ell}} \pm v_j = 0$$

thus the algebraic sum of the branch voltages around loop ℓ is zero, i.e., KVL holds for loop ℓ. Since ℓ is arbitrary, we have shown that KVL holds for all loops of $\mathscr{G}$. ∎

7.4 Geometric Interpretation[11]

In this section we shall use linear vector space to interpret the significance of Kirchhoff's laws and Tellegen's theorem. We will use the standard notations. For example, "$\mathbb{R}^b$" means "a b-dimensional vector space," "$\in$" means "is a member of," etc.

Tellegen's theorem requires that $\mathbf{v}$ satisfy KVL and $\mathbf{i}$ satisfy KCL for the given digraph $\mathcal{G}$. Let $\mathcal{G}$ be connected and have b branches and n nodes. From Sec. 6.3, we have

KCL: $$\mathbf{Ai} = \mathbf{0} \tag{7.8}$$

KVL: $$\mathbf{v} = \mathbf{A}^T\mathbf{e} \tag{7.9}$$

We state the following properties based on the discussion of linear independence of equations.

KCL properties

1. The $(n-1) \times b$ matrix $\mathbf{A}$ is full rank, i.e., its $n-1$ rows are linearly independent vectors in the b-dimensional space $\mathbb{R}^b$. (7.10)
2. $\mathbf{Ai}(t) = \mathbf{0} \Leftrightarrow$ the b-dimensional current vector $\mathbf{i}(t)$ satisfies KCL. (7.11)
3. The set of all branch current vectors $\mathbf{i}$ that satisfy KCL form a subspace, called the *KCL solution space*, and we label it K_i. (7.12)
4. Since K_i is obtained by imposing $n-1$ linearly independent constraints on the b-dimensional current vector $\mathbf{i}$, the *dimension* of K_i is $b-n+1$. (7.13)

The above implies

$$\begin{pmatrix} \mathbf{i} \in \mathbb{R}^b \\ \text{satisfies} \\ \text{KCL} \end{pmatrix} \Leftrightarrow (\mathbf{Ai} = \mathbf{0}) \Leftrightarrow (\mathbf{i} \in K_i) \tag{7.14}$$

KVL properties

1. $\mathbf{A}^T$ has full column rank, i.e., its $n-1$ columns are linearly independent vectors in the b-dimensional space $\mathbb{R}^b$. (7.15)
2. For some $(n-1)$-dimensional vector $\mathbf{e}(t)$, $\mathbf{v}(t) = \mathbf{A}^T\mathbf{e}(t) \Leftrightarrow$ the b-dimensional vector $\mathbf{v}(t)$ satisfies KVL. (7.16)
3. The set of all $\mathbf{v}$'s satisfying KVL form a $(n-1)$-dimensional subspace which we call the *KVL solution space* K_v. (7.17)
4. Since the subspace K_v is spanned by $n-1$ linearly independent vectors, the dimension of K_v is $n-1$. (7.18)

[11] Advanced topic, may be omitted without loss of continuity.

The above implies

$$\begin{pmatrix} \mathbf{v} \in \mathbb{R}^b \\ \text{satisfies} \\ \text{KVL} \end{pmatrix} \Leftrightarrow \begin{pmatrix} \mathbf{v} = \mathbf{A}^T\mathbf{e} \\ \text{for some } \mathbf{e} \\ \text{in } \mathbb{R}^{n-1} \end{pmatrix} \Leftrightarrow (\mathbf{v} \in K_v) \tag{7.19}$$

Now Tellegen's theorem says that for *any* such $\mathbf{v} \in \mathbb{R}^b$ and *any* such $\mathbf{i} \in \mathbb{R}^b$, $\mathbf{v}^T\mathbf{i} = 0$, i.e., the vectors $\mathbf{v}$ and $\mathbf{i}$ are *orthogonal.*

So viewing the subspaces K_v and K_i as subspaces of the *same* vector space $\mathbb{R}^b$, Tellegen's theorem asserts that *every vector in K_v is orthogonal to every vector of K_i*. This is denoted by

$$K_v \perp K_i \tag{7.20}$$

i.e., the subspaces K_v and K_i are orthogonal. The orthogonality of K_v and K_i is illustrated in Fig. 7.2.

Recalling that the dimension of K_i is $b - n + 1$ and that of K_v is $n - 1$, the sum of their dimensions is b. Consequently the subspaces K_i and K_v are not only orthogonal, but also have their direct sum equal to $\mathbb{R}^b$. In other words, any vector in $\mathbb{R}^b$ can be written uniquely as the sum of a vector in K_i and a vector in K_v.

To illustrate the equivalences in Eqs. (7.14) and (7.19) we consider two simple examples.

Example 1 $\mathscr{G}$ is the digraph of a two-node three-branch circuit shown in Fig. 7.3; we see that $\mathbf{A}$ is a 1×3 matrix, namely,

$$\mathbf{A} = [1 \;\; 1 \;\; 1]$$

So

$$\mathbf{Ai} = \mathbf{0} \Leftrightarrow i_1 + i_2 + i_3 = 0 \tag{7.21}$$

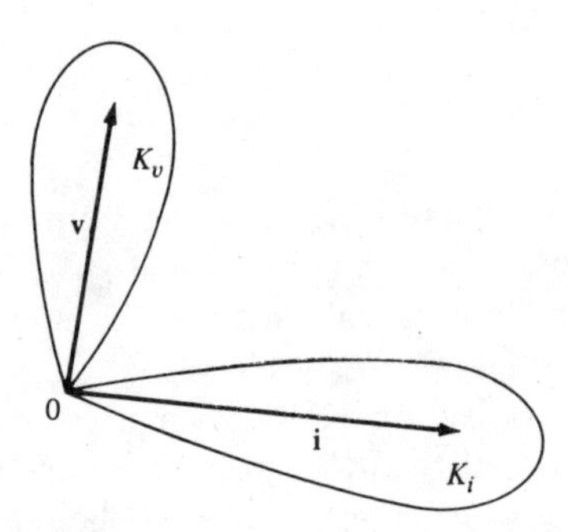

Figure 7.2 Figure illustrating the orthogonality of the subspaces K_i and K_v, where K_i is the set of all **i**'s satisfying KCL and K_v is the set of all **v**'s satisfying KVL.

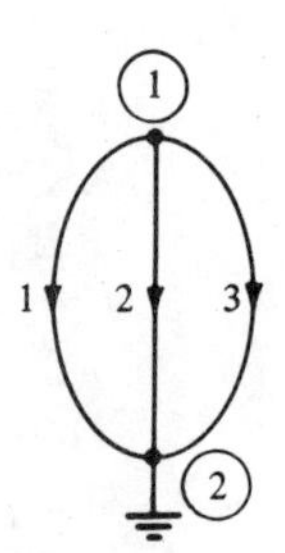

Figure 7.3 A digraph with two nodes and three branches.

$$\mathbf{v} = \mathbf{A}^T\mathbf{e} \Leftrightarrow \begin{bmatrix} v_1 \\ v_2 \\ v_3 \end{bmatrix} = \begin{bmatrix} 1 \\ 1 \\ 1 \end{bmatrix} e_1 \tag{7.22}$$

K_i is a two-dimensional subspace; i_1, i_2, i_3 are constrained by one equation, the KCL at node ①, Eq. (7.21). K_i is shown in Fig. 7.4.

K_v is a one-dimensional subspace: There is only one degree of freedom, namely, the node voltage e_1. [See Eq. (7.22).] K_v is shown in Fig. 7.5. Note that the vector $(1, 1, 1)^T$ which spans K_v is orthogonal to K_i, as required by Tellegen's theorem.

Example 2 $\mathscr{G}$ is the digraph of a three-node four-branch circuit shown in Fig. 7.6. Now **A** is a 2×4 matrix, namely

$$\mathbf{A} = \begin{bmatrix} 1 & 0 & 1 & 1 \\ 0 & 1 & -1 & -1 \end{bmatrix} \tag{7.23}$$

KCL, namely $\mathbf{Ai} = \mathbf{0}$, reads

$$\begin{aligned} i_1 + i_3 + i_4 &= 0 \\ i_2 - i_3 - i_4 &= 0 \end{aligned} \tag{7.24}$$

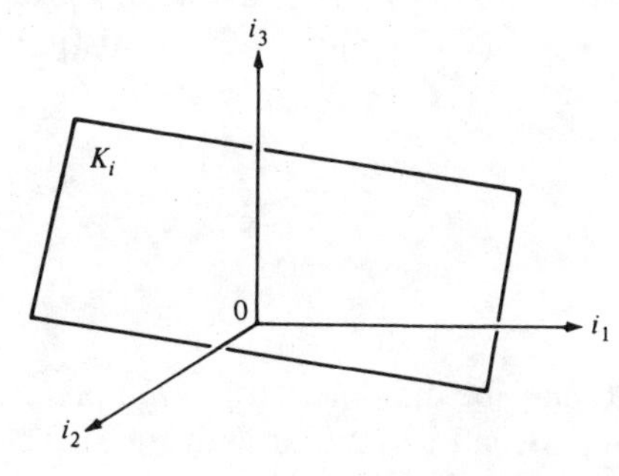

Figure 7.4 The two-dimensional KCL solution space of Example 1.

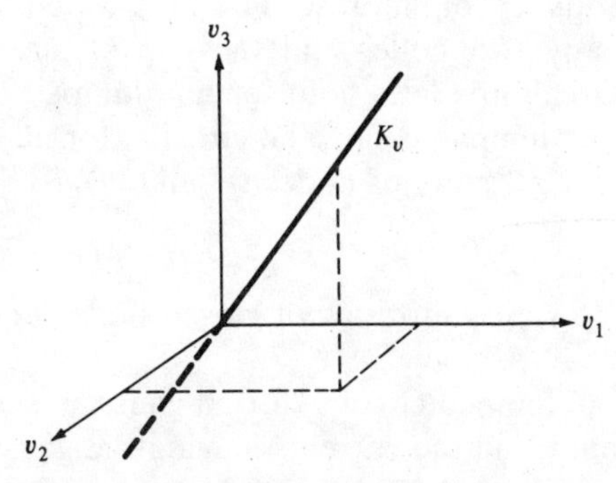

Figure 7.5 The one-dimensional KVL solution space of Example 1.

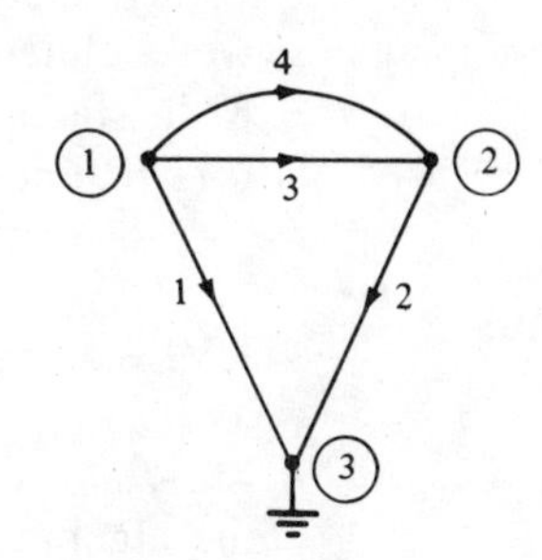

Figure 7.6 A digraph with three modes and four branches considered in Example 2.

The two scalar equations (7.24) are linearly independent; furthermore, their coefficient matrix is in row echelon form. Thus if we consider i_3 and i_4 as parameters, Eq. (7.24) gives

$$\mathbf{i} = (i_1, i_2, i_3, i_4) \in K_i \Leftrightarrow \begin{cases} i_1 = -(i_3 + i_4) \\ i_2 = i_3 + i_4 \end{cases} \quad \left. \begin{matrix} i_3 \text{ and } i_4 \text{ are} \\ \text{arbitrary} \end{matrix} \right\} \tag{7.25}$$

The right-hand side of Eq. (7.25) is the general solution of Eq. (7.24). Physically, it asserts: Any $\mathbf{i} \in K_i$ is the superposition of an arbitrary loop current i_3 going through the loop $\{\beta_3, \beta_2, \beta_1\}$ and of an arbitrary loop current i_4 going through the loop $\{\beta_4, \beta_2, \beta_1\}$. K_i is clearly a two-dimensional subspace in $\mathbb{R}^4$.

KVL, namely $\mathbf{v} = \mathbf{A}^T\mathbf{e}$, reads in the present case

$$\mathbf{v} = \begin{bmatrix} 1 \\ 0 \\ 1 \\ 1 \end{bmatrix} e_1 + \begin{bmatrix} 0 \\ 1 \\ -1 \\ -1 \end{bmatrix} e_2 \tag{7.26}$$

Thus K_v is the two-dimensional subspace formed by all linear combinations of the vectors $(1, 0, 1, 1)^T$ and $(0, 1, -1, -1)^T$.

Exercise Verify that for all i_3, i_4 and for all e_1, e_2, the $\mathbf{i}$ and the $\mathbf{v}$ given respectively by Eqs. (7.25) and (7.26) satisfy $\mathbf{v}^T\mathbf{i} = 0$, as expected by Tellegen's theorem.

SUMMARY

- A physical circuit may be modeled by a *lumped* circuit provided that its physical dimension is small compared with the wavelength corresponding to the highest frequency of interest. For any *connected* lumped circuit, the voltage across any two nodes and the current anywhere in the circuit is well-defined. Kirchhoff's laws hold for any lumped circuit.
- KVL states for all lumped connected circuits, for all choices of datum node, for all times t, for all pairs of nodes ⓚ and ⓙ,

$$v_{k-j}(t) = e_k(t) - e_j(t)$$

  KVL can also be stated in terms of closed node sequences and in terms of loops.
- KCL states for all lumped circuits, for all gaussian surfaces $\mathcal{S}$, for all times t, the algebraic sum of all the currents leaving the gaussian surface $\mathcal{S}$ at time t is equal to zero. KCL can also be stated in terms of nodes and in terms of cut sets.

- Any circuit element can be represented by an element graph. A two-terminal element is represented by a digraph which contains two nodes and one branch. An n-terminal element is represented by a digraph with n nodes and $n-1$ branches. Therefore, there are $n-1$ branch voltages and $n-1$ branch currents for an n-terminal element.
- A two-port is represented by a digraph with four nodes and two unconnected branches. There are two independent branch voltages and two independent branch currents.
- Any lumped circuit can be represented by a circuit graph, called a digraph, which depicts the interconnection properties of the circuit and the chosen reference directions. For a connected digraph the node-branch incidence relation is given by a reduced incidence matrix $\mathbf{A}$ of $n-1$ rows and b columns, where n is the number of nodes and b is the number of branches.
- KCL: The b-dimensional branch current vector $\mathbf{i}$ is constrained by $\mathbf{Ai}=\mathbf{0}$.
- KVL: The branch voltage vector $\mathbf{v}$ and the node-to-datum voltage vector $\mathbf{e}$ are related by $\mathbf{v}=\mathbf{A}^T\mathbf{e}$.
- Tellegen's theorem: For any digraph, for all $\mathbf{v}$'s satisfying KVL, and for all $\mathbf{i}$'s satisfying KCL, $\mathbf{v}^T\mathbf{i}=0$, provided that the associated reference directions are used for each branch.
- Kirchhoff's laws and Tellegen's theorem are valid for any lumped circuit regardless of the nature of the circuit elements. They reflect the interconnection properties of the circuit.

PROBLEMS

Digraph

1 Consider the circuit shown in Fig. P1.1. Choose the following datum for each device except D_2, which is a three-port device.

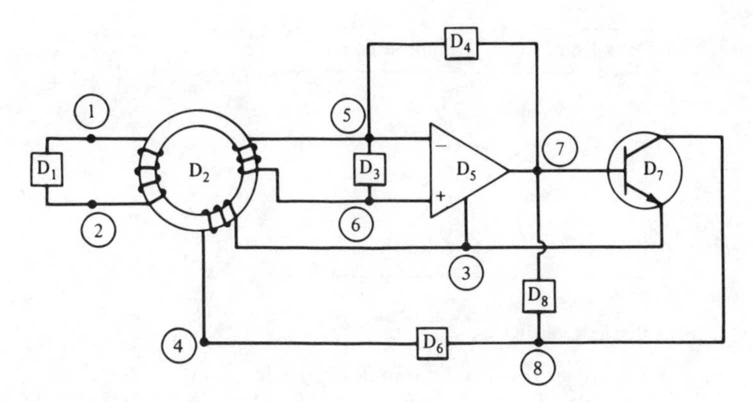

Figure P1.1

Device No.	Datum terminal no.
1	①
3	⑥
4	⑤
5	③
6	⑧
7	⑦
8	⑦

(*a*) Using the datum terminal specified, draw the element graph for each device in Fig. P1.1.

(*b*) Using the element graphs from (*a*), draw the digraph of the circuit in Fig. P1.1.

(*c*) Repeat (*b*) but with the datum terminal for the op amp changed to ⑦ and that of the transistor changed to ⑧.

2 Consider the circuit shown in Fig. P1.2. Choose the following datum for each device, except D_8, which is a three-port device.

(*a*) Draw the element graph with the datum terminal specified above for each device in Fig. P1.2.

(*b*) Using the element graphs from (*a*), draw the digraph for the circuit in Fig. P1.2.

(*c*) Repeat (*b*) but with the datum terminal for the transistor changed to ③, and that of the op amp changed to ⑧.

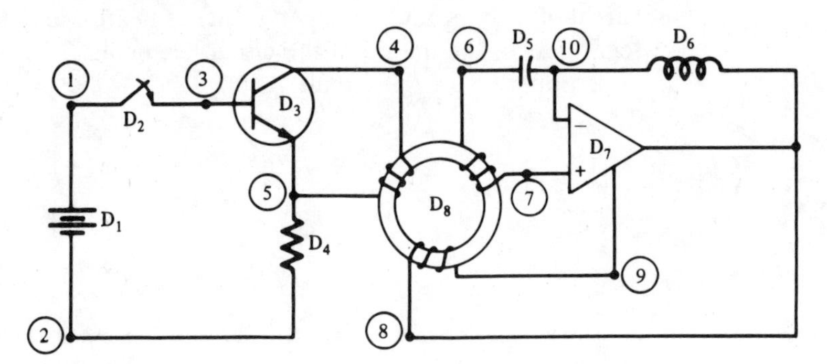

Figure P1.2

Device no.	Datum terminal no.
1	②
2	①
3	⑤
4	⑤
5	⑩
6	⑩
7	⑨

Digraph and KVL

3 Consider the circuit shown in Fig. P1.3.

(*a*) Given $v_{2-3} = 10$ V, $v_{6-3} = 6$ V, and $v_{4-1} = 2$ V, find v_{6-1}, v_{4-6}, and v_{4-2}.

(*b*) Draw the digraph with terminal ③ chosen as the datum of the op amp and terminal ④ chosen as the datum of the transistor. Repeat (*a*) using this digraph.

(*c*) Repeat (*b*) but with terminal ⑤ chosen as the datum terminal for both the op amp and the transistor.

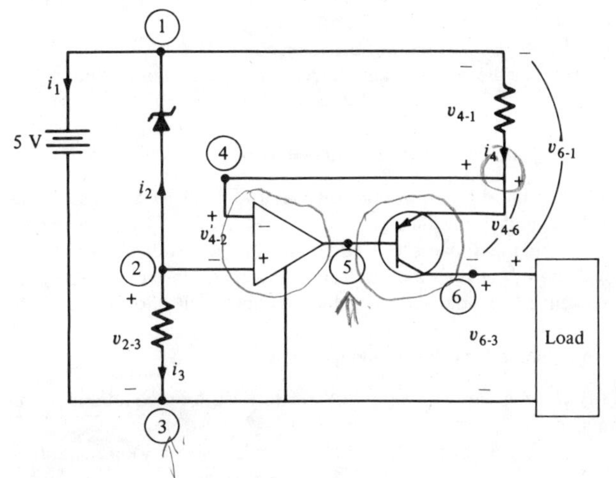

Figure P1.3

Digraph and closed node sequence

4 For the circuit shown in Fig. P1.4:

(*a*) Draw the digraph. Choose node ⑤ as the datum terminal of the op amp and use the associated reference convention.

(*b*) Write 10 KVL equations for each closed node sequence which contains four nodes.

(*c*) Are the equations from (*b*) linearly independent?

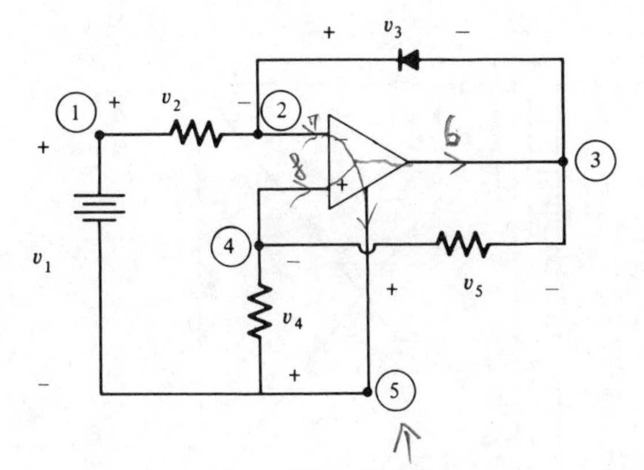

Figure P1.4

Incidence matrix, KVL, and KCL

5 (*a*) For the circuit shown in Fig. 4.4 in the text, draw the digraph by choosing node ④ as the datum node of the op amp.

(*b*) Determine the incidence matrix $\mathbf{A}_a$. What is the rank of $\mathbf{A}_a$?

(*c*) Choosing node ⑤ as the datum node, write down the KCL and KVL equations based on the incidence matrix $\mathbf{A}_a$ from (*b*).

Degree of freedom, KVL, and KCL

6 (*a*) For the circuit in Fig. 4.4 in the text, but using the digraph in Fig. 5.8, how many branch voltages can you specify independently? Specify such a set arbitrarily and demonstrate that the remaining voltages can be determined.

(*b*) Repeat (*a*) with branch currents.

(*c*) For the voltages and currents determined in (*a*) and (*b*), verify that Tellegen's theorem is valid.

Cut-set equations and linear independence

7 (*a*) For the digraph in Fig. 5.8, determine all KCL cut-set equations which are not included in the KCL node equations.

(*b*) Is the above set of equations linearly independent? If it is, prove it. If it is not, delete a minimum subset such that the remaining equations are linearly independent. Does the resulting set represent a maximal set, i.e., does it contain all information of the digraph?

Loop equations and linear independence

8 (*a*) For the digraph in Fig. 5.8, write KVL loop equations for all loops containing four or more branches.

(*b*) Repeat part (*b*) of Prob. 7 for the above loop equations.

Gaussian surface and cut sets

9 (*a*) Draw the digraph for the "full-wave rectifier circuit" shown in Fig. P1.9.

(*b*) Consider the following subsets of branches from this digraph: $\{1,2\}$, $\{1,2,3,4\}$, $\{4,5,6,7\}$, $\{3,6\}$, $\{3,4,5,6\}$, $\{1,2,4,5\}$, and $\{1,2,3,5,7\}$. Identify those subsets which qualify as branches cutting a gaussian surface and write the associated KCL equation. Explain why the remaining subsets do not qualify.

(*c*) Identify those subsets which qualify as cut sets and write the associated KCL equations. Explain why the remaining subsets do not qualify.

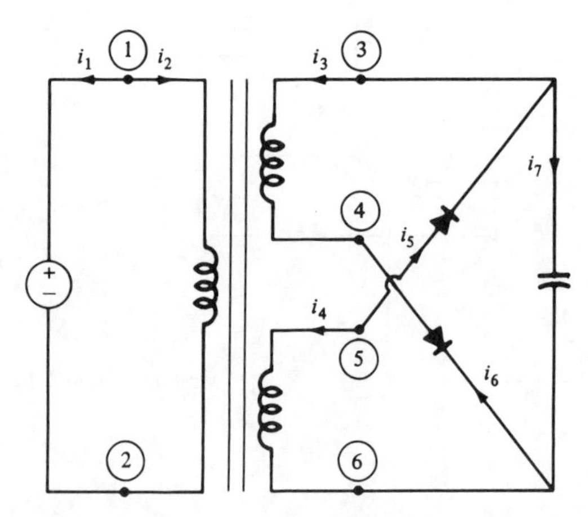

Figure P1.9

Incidence matrix and hinged graphs

10 (*a*) Write the incidence matrix $\mathbf{A}_a$ associated with the digraph from Prob. 9(*a*) and verify that the rows are not linearly independent. Why?

(*b*) Delete any row from $\mathbf{A}_a$ and verify that the remaining rows are still not linearly independent. Why?

(*c*) Hinge nodes ② and ⑥ and write the associated reduced incidence matrix **A**. Verify that the rows of **A** are linearly independent.

(*d*) Write a system of linearly independent KCL and KVL equations using **A**.

Incidence matrix and cut sets

11 For the digraph shown in Fig. P1.11:

(*a*) Write the incidence matrix $\mathbf{A}_a$.

(*b*) With node ③ as the datum node, write the KVL equations

$$\mathbf{v} = \mathbf{A}^T\mathbf{e}$$

(*c*) Give all the cut sets not already represented by $\mathbf{A}_a$.

(*d*) Select a maximum subset of the above which leads to linearly independent KCL equations.

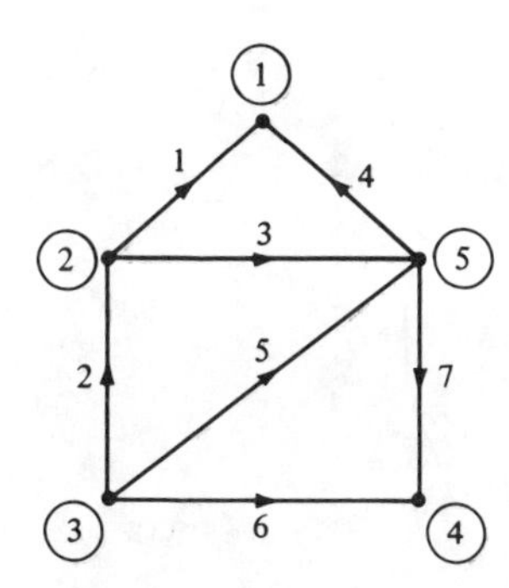

Figure P1.11

Degree of freedom, KVL, and KCL

12 In the circuit shown in Fig. P1.12, the branch current reference directions are indicated.

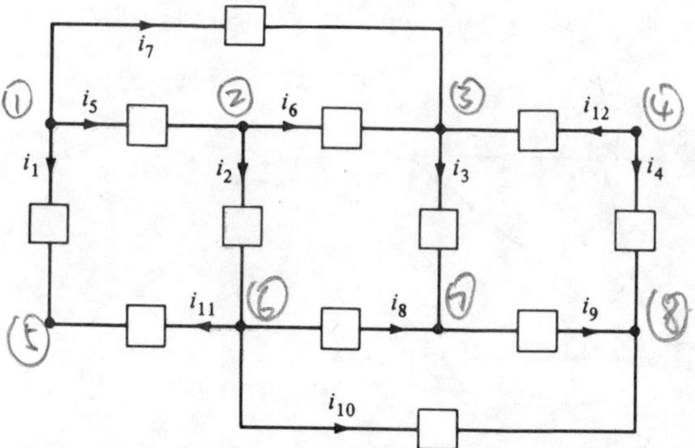

Figure P1.12

(*a*) How many branch currents can be independently specified? Why?

(*b*) If the following currents are given in amperes:

$$i_7 = -5 \qquad i_4 = 5 \qquad i_{10} = -3 \qquad i_3 = 1 \qquad i_1 = 24$$

is it possible to determine the remaining currents? Determine as many as you can.

(*c*) Let the branch voltages be measured in the associated reference directions. How many of the branch voltages can be independently specified? Why?

(*d*) If the following voltages are given in volts:

$$v_1 = 10 \qquad v_2 = 5 \qquad v_4 = -3 \qquad v_6 = 2 \qquad v_7 = -3 \qquad v_{12} = 8$$

determine as many branch voltages as possible.

Cut-set equations and linear independence

13 (*a*) Enumerate all cut sets for the digraph in Fig. P1.13.

(*b*) Write a KCL equation corresponding to each cut set from (*a*).

(*c*) Show that the equations from (*b*) are linearly dependent.

(*d*) Extract a subset of KCL equations from (*b*) containing a maximum number, ρ, of linearly independent equations. Verify that $\rho = n - p$, where n = number of nodes and p = number of connected components (or separate parts) in the digraph.

(*e*) Hinge nodes ⑤ and ⑦ and verify that the same KCL equations from (*b*) also hold for the resulting "connected" digraph.

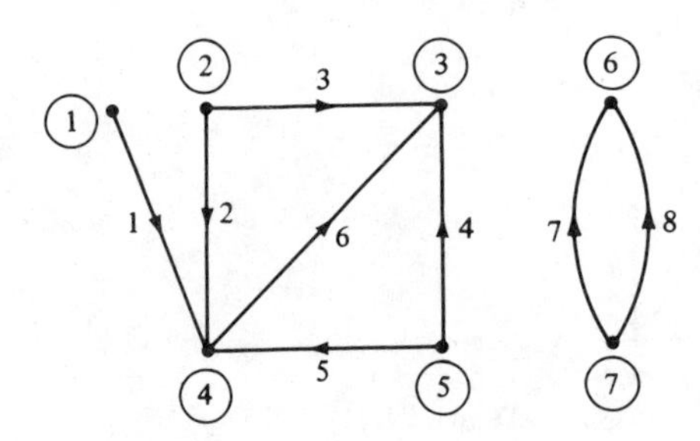

Figure P1.13

Incidence matrix, KVL, and KCL

14 (*a*) Write the incidence matrix $\mathbf{A}_a$ for the digraph in Fig. P1.14.

(*b*) Write the reduced incidence matrix $\mathbf{A}$ with node ⑤ as the datum node.

(*c*) Using $\mathbf{A}$, write a system of linearly independent KVL and KCL equations.

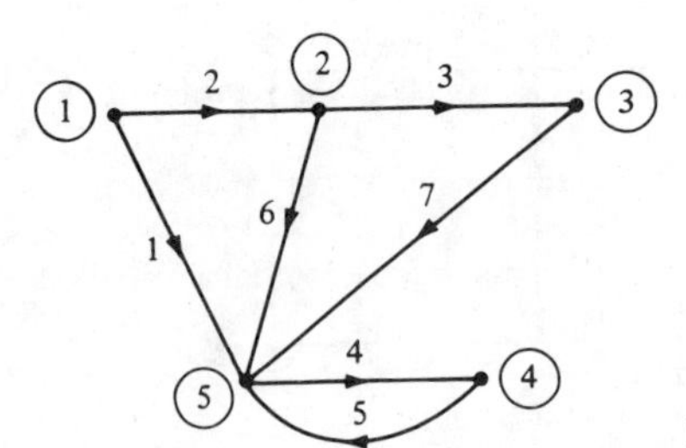

Figure P1.14

Degree of freedom, Tellegen's theorem

15 (*a*) Specify the degree of freedom (i.e., dimensions) of the KCL and KVL solution sets associated with the digraph from Fig. P1.14.

(*b*) Find two (among infinitely many) branch voltage distributions $\{v_1', v_2', \ldots, v_7'\}$ and $\{v_1'', v_2'', \ldots, v_7''\}$ which satisfy KVL for the digraph in Fig. P1.14 by assigning, arbitrarily, two distinct sets of node-to-datum voltages $\{e_1', e_2', e_3', e_4'\}$ and $\{e_1'', e_2'', e_3'', e_4''\}$, respectively.

(*c*) Find two (among infinitely many) branch current distributions $\{i_1', i_2', \ldots, i_7'\}$ and $\{i_1'', i_2'', \ldots, i_7''\}$ which satisfy KCL for the digraph in Fig. P1.14.

(*d*) Use your two sets of voltage and current distributions from (*b*) and (*c*) to verify Tellegen's theorem.

Incidence matrix and cut sets

16 Given the reduced incidence matrix of a digraph:

$$\mathbf{A} = \begin{bmatrix} 1 & 0 & 0 & 1 & 0 \\ 0 & -1 & 0 & 0 & -1 \\ 0 & 0 & 1 & -1 & 1 \end{bmatrix} \begin{matrix} \text{Node} \\ \textcircled{1} \\ \textcircled{2} \\ \textcircled{3} \end{matrix}$$
$$\text{Branch} \quad 1 \quad 2 \quad 3 \quad 4 \quad 5$$

(*a*) Draw the associated digraph and mark all nodes and branches.
(*b*) List all cut sets which are not already included in **A**.

Incidence matrix, digraphs, cut-set and node equations

17 (*a*) Draw a digraph whose incidence matrix $\mathbf{A}_a$ is given by

$$\mathbf{A}_a = \begin{bmatrix} -1 & 1 & 0 & 0 & 0 & 0 \\ 0 & -1 & 1 & 0 & -1 & 1 \\ 1 & 0 & 0 & -1 & 1 & -1 \\ 0 & 0 & -1 & 1 & 0 & 0 \end{bmatrix} \begin{matrix} \text{Node} \\ \textcircled{1} \\ \textcircled{2} \\ \textcircled{3} \\ \textcircled{4} \end{matrix}$$
$$\text{Branch} \quad 1 \quad 2 \quad 3 \quad 4 \quad 5 \quad 6$$

(*b*) Given the following subgraphs of the digraph obtained in (*a*), identify the ones which form cut sets and the ones which are associated with gaussian surfaces.

$$\{1, 2, 3, 4\}, \ \{5, 6\}, \ \{2, 4, 5, 6\}, \ \{1, 3, 5, 6\}$$

(*c*) For the cut sets from (*b*), write down the corresponding KCL equations and also express these equations as the sum of appropriate node equations.

Cut sets, rank, and linear independence

18 A circuit has a digraph, as shown in Fig. P1.18.

(*a*) Choose node ④ as datum; write down the reduced incidence matrix **A**.
(*b*) Establish the rank of **A**.
(*c*) Show that the cut-set equation for the cut set {2, 3, 4, 5} is linearly dependent on the node equations.
(*d*) Given $i_1 = 1\,\text{A}$, $i_3 = 3\,\text{A}$, and $i_5 = 5\,\text{A}$, determine, if you can, all other currents.

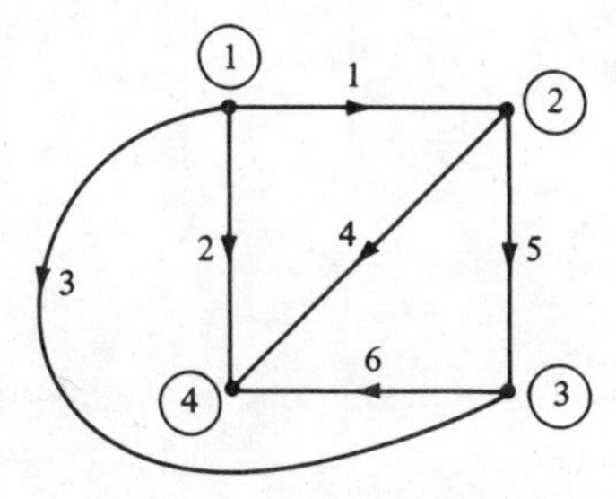

Figure P1.18

Loop equations and linear independence

19 A given circuit leads to the digraph of Fig. P1.19. For the digraph shown:

(*a*) List all possible loops and, for each one, write a KVL equation in terms of branch voltages.

(*b*) Are the equations from (*a*) linearly independent? If not, show that they are not.

(*c*) How many linearly independent loop equations are there in the list above? Justify your answer.

(*d*) Is it true that $i_7 + i_9 = 0$? Prove or disprove it.

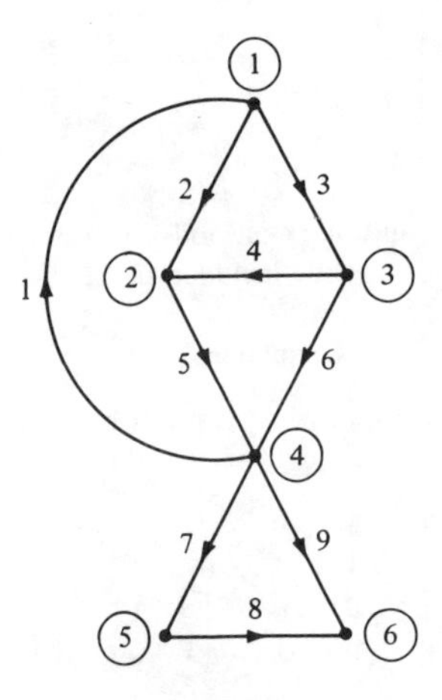

Figure P1.19

Tellegen's theorem

20 Let N be a two-port made of arbitrary interconnections of linear two-terminal linear resistors (i.e., each resistor satisfies Ohm's law: $v_j = R_j i_j$). Consider the following two experiments:

Experiment 1. Drive N with two voltage sources v_1' and v_2' and measure the resulting port currents i_1' and i_2', respectively. (See Fig. P1.20.)

Experiment 2. Drive N with two voltage sources v_1'' and v_2'' and measure the resulting port currents i_1'' and i_2'', respectively.

(*a*) Prove that the two sets of measurements are related as follows:

$$v_1'i_1'' + v_2'i_2'' = v_1''i_1' + v_2''i_2'$$

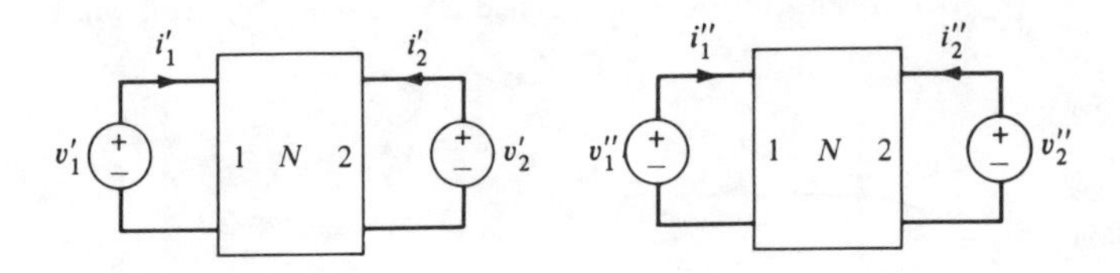

Figure P1.20

21 Let N be the two-port described in Prob. 20. Consider the following experiments

Experiment 1. Drive port 1 with a voltage source with voltage E and measure the current i_2' in the short circuit across port 2. (See Fig. P1.21.)

Experiment 2. Drive port 2 with a voltage source with an identical voltage E and measure the current i_1'' in the short circuit across port 1.

Prove that $i_2' = i_1''$, and state this remarkable property in words (reciprocity property).

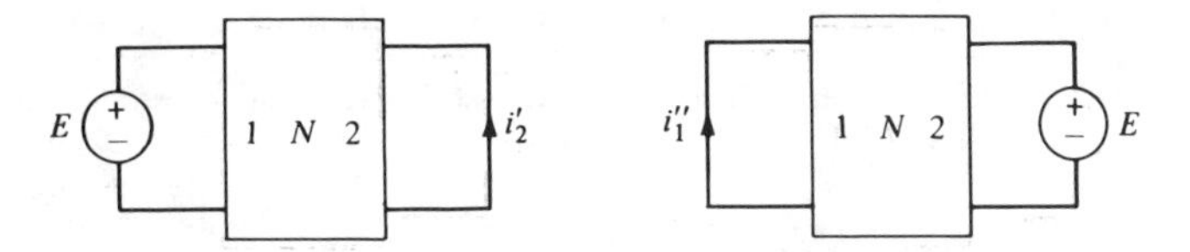

Figure P1.21

22 Without drawing the digraph, verify that Tellegen's theorem holds for any v_0 in the circuit shown in Fig. P1.22.

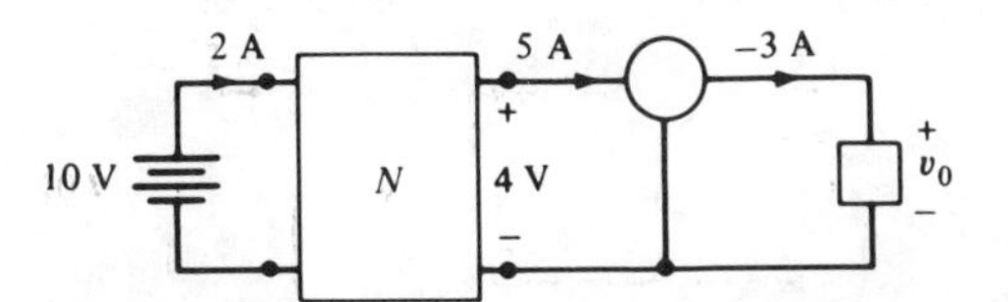

Figure P1.22

23 Let N denote a one-port made of n-terminal and n-port elements. Prove that the instantaneous power entering N at time t is equal to the sum of the instantaneous power entering each element inside N at time t.

24 In the circuits shown in Fig. P1.24, let v_k and i_k be the branch voltage and current in circuit A and $\hat{v}_k$ and $\hat{i}_k$ be the branch voltage and current in circuit B. The following measurements have been obtained:

$$i_s = 1\text{ A} \qquad \hat{v}_s = 3\text{ V} \qquad v_L = 2\text{ V}$$

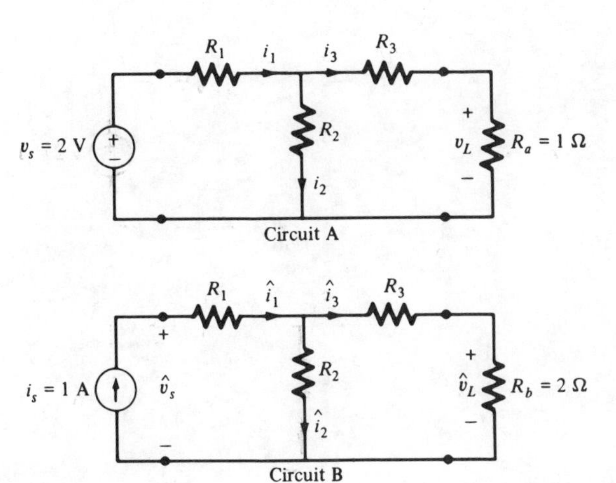

Figure P1.24

Use a particular form of Tellegen's theorem to determine $\hat{v}_L$, where R_1, R_2, and R_3 are unknown resistors satisfying Ohm's law.

25 Consider the circuit shown in Fig. P1.25. Two sets of measurements on this circut give the following results:

(i) When $R_L = 2\ \Omega$: $\quad v_1 = 8\text{ V} \quad i_1 = -2\text{ A} \quad v_L = 2\text{ V}$

(ii) When $R_L = 4\ \Omega$: $\quad \hat{v}_1 = 12\text{ V} \quad \hat{i}_1 = -2.4\text{ A} \quad \hat{v}_L = ?$

Determine $\hat{v}_L$, given that R_1, R_2, R_3, and R_4 are linear resistors satisfying Ohm's law.

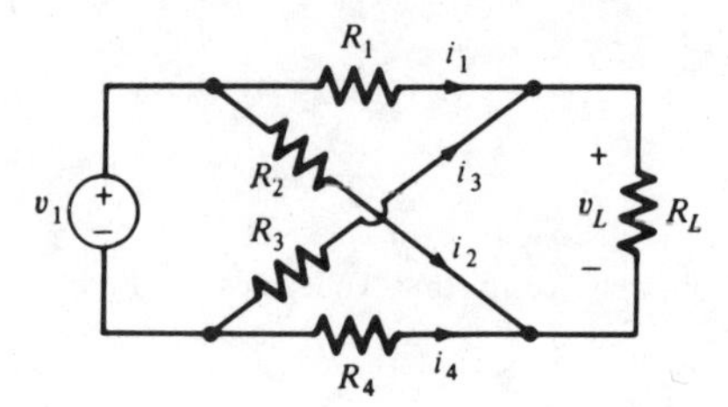

Figure P1.25

CHAPTER

TWO

TWO-TERMINAL RESISTORS

Two-terminal elements play a major role in electric circuits. As a matter of fact, many introductory texts on electric circuits consider circuits consisting only of two-terminal elements exclusively. In this chapter we give a comprehensive treatment of two-terminal resistors. However, unlike the usual terminology, a resistor may be linear, nonlinear, time-invariant, or time-varying. It is characterized by a relation between the branch voltage and the branch current. We speak of the *v-i* characteristic of a resistor, and we discuss the characteristics of various types of resistors such as a linear resistor which satisfies Ohm's law, an ideal diode, a dc current source, a *pn*-junction diode, and a periodically operating switch. All of these are resistors.

By interconnecting two-terminal resistors, we form a resistive circuit. The simplest forms of interconnection, i.e., series, parallel, and series-parallel interconnections, will be treated and illustrated with examples. These require the use of Kirchhoff's laws together with branch equations which characterize the elements. A one-port formed by the interconnection of resistors is characterized by its driving-point characteristics relating its port voltage and its port current. We introduce the concepts of equivalence and duality of one-ports by simple examples. These will be generalized in later chapters.

An important problem in nonlinear circuits is the determination of the dc operating points, i.e., the solutions with dc inputs. Various methods and techniques are introduced and illustrated.

Another important problem in nonlinear circuits is the small-signal analysis. Its relation to dc operating points and the derivation of the small-signal equivalent circuit are treated by way of a simple example. This subject will be discussed in a more general fashion in later chapters.

Finally, we discuss the transfer characteristic of resistive circuits and demonstrate the usefulness of the graphic method in analyzing nonlinear resistive circuits.

1 v-i CHARACTERISTIC OF TWO-TERMINAL RESISTORS

1.1 From Linear Resistor to Resistor

The most familiar circuit element that one encounters in physics or in an elementary electrical engineering course is a two-terminal resistor which satisfies Ohm's law; i.e., the voltage across such an element is proportional to the current flowing through it. We call such an element a *linear resistor*. We represent it by the symbol shown in Fig. 1.1, where the current i through the resistor and the voltage v across it are measured using the *associated reference directions*. *Ohm's law* states that, at all times

$$v(t) = Ri(t) \qquad \text{or} \qquad i(t) = G\,v(t) \tag{1.1}$$

where the constant R is the *resistance* of the linear resistor measured in the unit of ohms (Ω), and G is the *conductance* measured in the unit of siemens (S). The voltage $v(t)$ and the current $i(t)$ in Eq. (1.1) are expressed in volts (V) and amperes (A), respectively. Equation (1.1) can be plotted on the i-v plane or the v-i plane[1] as shown in Fig. 1.2*a* and *b*, where the slope in each is the resistance and the conductance, respectively.

While the linear resistor is perhaps the most prevalent circuit element in electrical engineering, nonlinear devices which can be modeled with nonlinear resistors have become increasingly important. Thus it is necessary to define the concept of nonlinear resistor in a most general way.

Consider a two-terminal element as shown in Fig. 1.3. The voltage v across the element and the current i which enters the element through one terminal and leaves from the other are shown using the associated reference directions. A two-terminal element will be called a *resistor* if its voltage v and current i

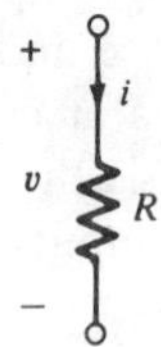

Figure 1.1 Symbol for a linear resistor with resistance R.

[1] When we say x-y plane, we denote specifically x as the horizontal axis and y as the vertical axis of the plane. This is consistent with the conventional usage where the first variable denotes the abscissa and the second variable denotes the ordinate.

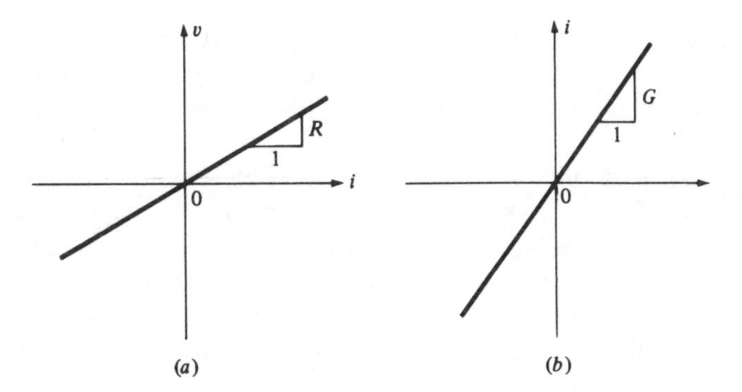

Figure 1.2 Linear resistor characteristic plotted (*a*) on the *i*-*v* plane and (*b*) on the *v*-*i* plane.

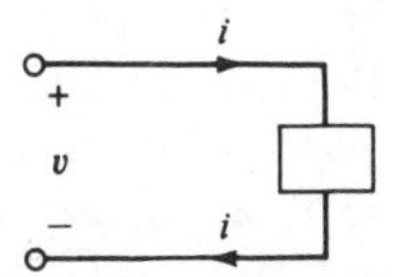

Figure 1.3 A two-terminal element with v and i in the associated reference directions.

satisfy the following relation:

$$\mathcal{R}_R = \{(v, i): f(v, i) = 0\} \tag{1.2}$$

This relation is called the *v-i characteristic* of the resistor and can be plotted graphically in the v-i plane (or i-v plane). The equation $f(v, i) = 0$ represents a curve in the v-i plane (or i-v plane) and specifies completely the two-terminal resistor. The key idea of a resistor is that in Eq. (1.2) the relation is between $v(t)$, the instantaneous value of the voltage $v(\cdot)$ and $i(t)$ the instantaneous value of the current $i(\cdot)$ at time t.

The dc voltage versus current characteristics of devices can be measured using a curve tracer.[2]

The linear resistor is a special case of a resistor in which

$$f(v, i) = v - Ri = 0 \qquad \text{or} \qquad f(v, i) = i - Gv = 0 \tag{1.3}$$

A resistor which is not linear is called *nonlinear*.

Before considering nonlinear resistors, we should first understand linear resistors. Equations (1.1) and (1.3) state that, for a linear resistor, the relation between the voltage v and current i is expressed by *linear* functions. The first equation in (1.1) expresses v as a linear function of i, and the second equation in (1.1) expresses i as a linear function of v. Figure 1.2 shows that the v-i

[2] See for example:
J. Mulvey, *Semiconductor Device Measurements*, Tektronix Inc., Beaverton, Oregon, 1968.
L. O. Chua and G. Q. Zhong, "Negative Resistance Curve Tracer," *IEEE Transactions on Circuits and Systems*, vol. CAS-32, pp. 569–582, June 1985.

characteristic of a linear resistor is a *straight line passing through the origin at all times*. The single number R (or G), i.e., the slope of the characteristic in the i-v plane (or v-i plane), specifies completely the *linear* two-terminal resistor.

Open circuits and short circuits There are two special cases of linear resistors which deserve special mention, namely, the open circuit and the short circuit. A two-terminal resistor is called an *open circuit* iff its current i is identically zero irrespective of the voltage v; i.e., $f(v, i) = i = 0$. The characteristic of an open circuit is the v axis in the v-i plane, or in the i-v plane, as shown in Fig. 1.4. In the i-v plane, it has an infinite slope, i.e., $R = \infty$; and in the v-i plane, it has a zero slope, i.e., $G = 0$.

Similarly, a two-terminal resistor is called a *short circuit* iff its voltage is identically zero irrespective of its current i; i.e., $f(v, i) = v = 0$. The characteristic of a short circuit is the i axis in the i-v plane, or in the v-i plane, as shown in Fig. 1.5. In the i-v plane, the characteristic has a zero slope, i.e., $R = 0$; and, equivalently, $G = \infty$. Comparing Figs. 1.4 and 1.5, we see that the curve of the open circuit in one plane is identical to the curve of the short circuit in the other plane. For this reason, the *open circuit* is said to be the *dual* of the short circuit; and, conversely, the short circuit is the dual of the open circuit. Generalizing to the nonlinear case, we say that the *dual* of a given resistor is another resistor whose v-i characteristic in the v-i plane is the same

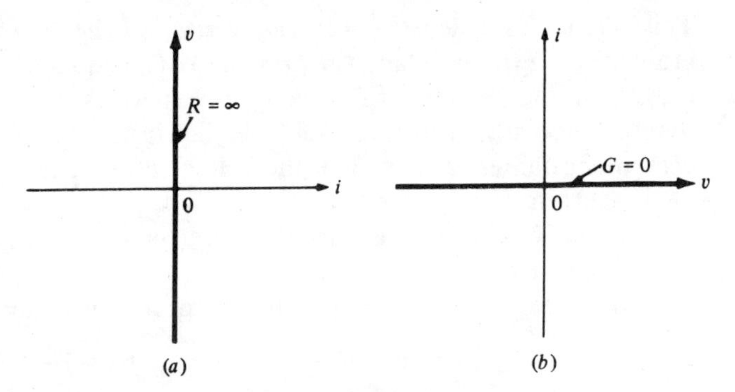

Figure 1.4 Characteristic of an open circuit.

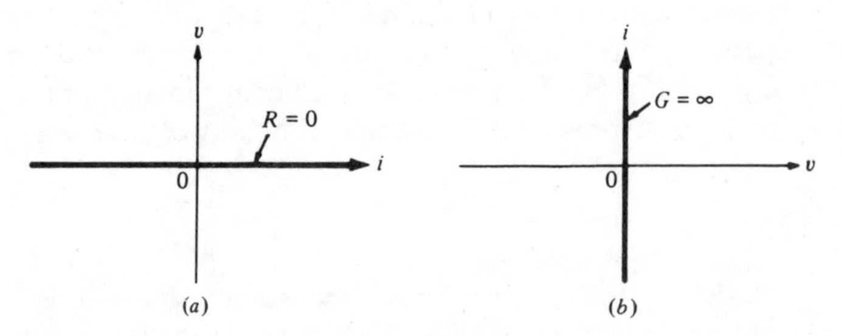

Figure 1.5 Characteristic of a short circuit.

curve as that of the given resistor in the i-v plane. The concept of duality is of utmost importance in circuit theory. It helps us in understanding and analyzing circuits of great generality. We will encounter duality throughout this book.

Exercises

1. A linear resistor of 100 Ω is given; what is its dual?
2. If $\mathcal{R}_R = \{(v, i): f(v, i) = v - i^3 = 0\}$ specifies a resistor, write down the relation of the dual resistor.
3. Given the v-i characteristic Γ of a resistor $\mathcal{R}$ on the v-i plane, show that the dual characteristic is obtained by reflecting Γ about the 45° line through the origin.

Power, passive resistors, active resistors, and modeling The symbol for a two-terminal nonlinear resistor $\mathcal{R}$ is shown in Fig. 1.6. The *power* delivered *to* the *resistor* at time t *by* the remainder of the circuit to which it is connected is, from Chap. 1,

$$p(t) = v(t)i(t) \tag{1.4a}$$

If the resistor is linear having resistance R

$$p(t) = Ri^2(t) = Gv^2(t) \tag{1.4b}$$

Thus, the power delivered to a linear resistor is always nonnegative if $R \geq 0$. We say that a *linear* resistor is *passive* iff its resistance is nonnegative. Thus a passive resistor always absorbs energy from the remainder of the circuit.

Also from Eq. (1.4b), the power delivered to a linear resistor is negative if $R < 0$; i.e., as current flows through it, the resistor delivers energy to the remainder of the circuit. Therefore, we call such a *linear* resistor with negative resistance an *active* resistor. The characteristic of a linear active resistor is shown in Fig. 1.7; note that the slope is negative. While linear passive resistors are familiar to everyone, linear active resistors are perhaps new to some readers. They are one of the basic circuit elements in the design of negative-

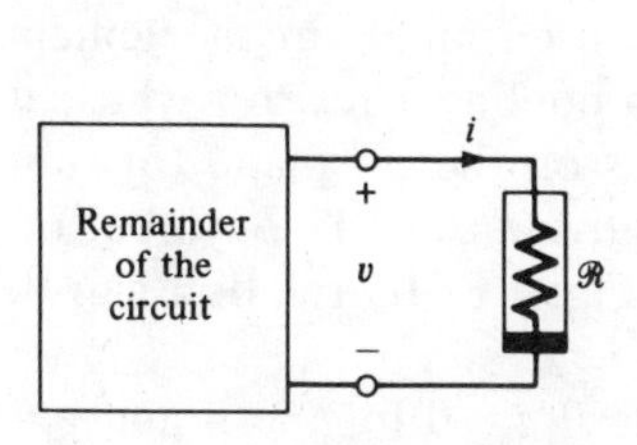

Figure 1.6 Illustrating power delivered to a nonlinear resistor from the remainder of the circuit.

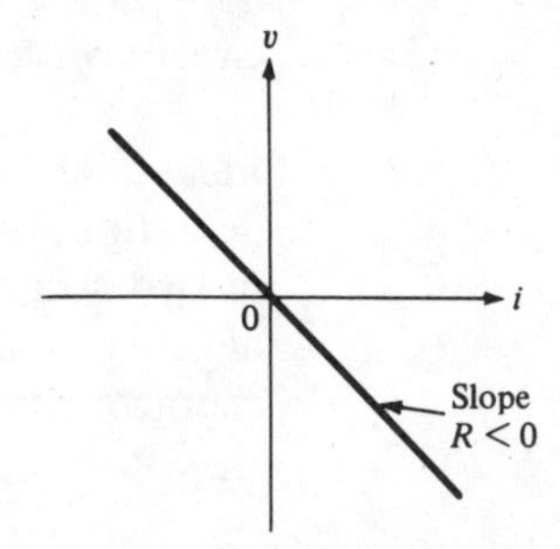

Figure 1.7 Characteristic of a linear active resistor with resistance $R < 0$.

resistance amplifiers and oscillators. We will illustrate their applications later. For the present we only wish to mention that the linear active resistor is useful in modeling nonlinear devices and circuits over certain ranges of voltages, currents, and frequencies.

We can easily generalize the above concept to nonlinear resistors. Obviously, from Eq. (1.4*a*), $p(t) \geq 0$ if and only if $v(t)$ and $i(t)$ have the same sign for all t. Thus we call a two-terminal resistor *passive* iff its v-i characteristic lies in the closed first and third quadrants of the v-i plane or the i-v plane. A resistor is said to be *active* if it is not passive.

At this juncture we wish to recall the concept of modeling introduced in Chap. 1. Let us use the term "physical resistor" to refer to the electric device in the laboratory or in a piece of equipment. This is not to be confused with the resistor we defined as a circuit element in Eq. (1.2). What is remarkable is that for most physical resistors made of metallic material, we can use the circuit element, a linear passive resistor, to model them almost precisely, i.e., they satisfy Ohm's law. The model is good over a large operating range. Only for excessive voltages or currents, or at very high frequencies, is a better model necessary. Often in such a situation the physical resistor fails to be of ordinary use; for example, a physical resistor will burn out if the current exceeds the specified normal operating range. Historically, and in most engineering usage, the term "resistor" is often loosely used to mean the physical resistor, a device which satisfies Ohm's law. In circuit theory we depart from the traditional practice and define a resistor as a *circuit element* which is specified by a voltage-current relation called the v-i characteristic. This way of defining a resistor has a special significance in modeling electric and electronic devices which, at low frequencies, behave like nonlinear resistors. We next turn our attention to nonlinear resistors.

1.2 The Nonlinear Resistor

Recall that a resistor that is not linear is said to be nonlinear. In this section we will first introduce some typical examples of nonlinear resistors and illustrate their properties. We will then point out some essential differences between circuits with linear resistors and those with nonlinear resistors.

Ideal diode A very useful two-terminal circuit element is the ideal diode. By definition, an *ideal diode* is a nonlinear resistor whose v-i characteristic consists of two straight line segments on the v-i plane (or the i-v plane), namely, the negative v axis and the positive i axis. The symbol of the ideal diode and its characteristic are shown in Fig. 1.8. Its relation can be expressed by

$$\mathcal{R}_{\text{ID}} = \{(v, i) \colon vi = 0,\ i = 0 \text{ for } v < 0 \text{ and } v = 0 \text{ for } i > 0\} \tag{1.5}$$

Thus, if the diode is *reversed biased* ($v < 0$), the current is zero, i.e., the diode acts as an open circuit; if the diode is *conducting* ($i > 0$), the voltage is zero, i.e., the diode acts like a short circuit. Clearly, the power delivered to an ideal diode is identically zero at all times. In circuit theory there exist several ideal circuit elements having such a property, and we call them *non-energic* circuit elements.

The ideal diode is a very useful circuit element which plays a crucial role in device modeling and helps us in understanding how various electronic devices and circuits function. It is also essential in the method of piecewise-linear analysis to be introduced later.

There are many two-terminal electronic devices, called diodes, whose characteristics resemble that of an ideal diode to some extent. We will consider a few of them and discuss their characteristics. We shall consider the low-frequency properties of these devices because, at low frequencies, these devices can be modeled fairly accurately with nonlinear resistors alone.

***pn*-Junction diode** The *pn*-junction diode and its *v-i* characteristic are shown in Fig. 1.9. For most applications the device is operated to the right of point A, where A is near the "knee" of the curve. In the normal operating range, i.e., to the right of A, the current obeys the "diode junction law,"

$$i = I_s\left[\exp\left(\frac{v}{V_T}\right) - 1\right] \tag{1.6}$$

where I_s is a constant of the order of microamperes, and it represents the *reverse saturation current*, i.e., the current in the diode when it is reversed biased with a large voltage. The parameter $V_T = kT/q$ is called the *thermal voltage*, where q is the charge of an electron, k is Boltzmann's constant, and T is temperature in Kelvins. At room temperature V_T is approximately 0.026 V. In Eq. (1.6) we have a nonlinear resistor whose current i is expressed as a function of its voltage v. This means that, for any given voltage v, the current i is *uniquely* specified. Any nonlinear resistor having this property is called a *voltage-controlled nonlinear resistor*. While Eq. (1.6) represents a good model

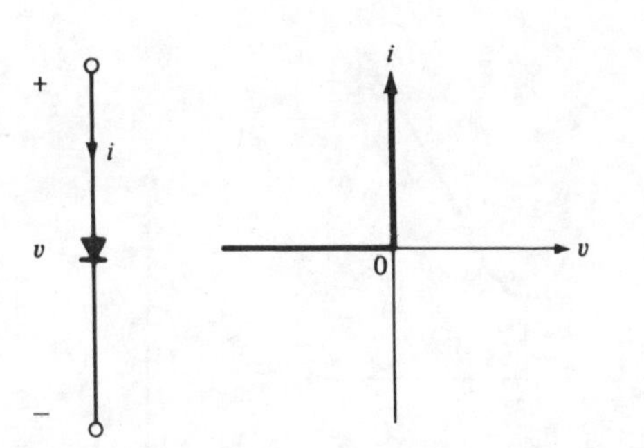

Figure 1.8 Symbol for an ideal diode and its characteristic.

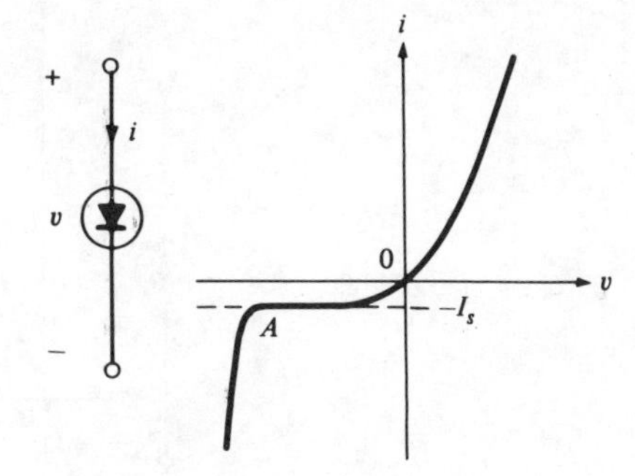

Figure 1.9 Symbol for a *pn*-junction diode and its characteristic.

for the *pn*-junction diode at low frequencies, we need to use additional circuit elements, capacitors, inductors, and linear resistors to model it at higher frequencies.

Tunnel diode The symbol of the tunnel diode and its *v-i* characteristic are shown in Fig. 1.10. The current i can be expressed as a function of the voltage v, hence we may write

$$i = \hat{i}(v) \tag{1.7a}$$

or, in terms of Eq. (1.2),

$$f(v, i) = i - \hat{i}(v) = 0 \tag{1.7b}$$

Note that the function $\hat{i}(\cdot)$ is *single-valued*, hence the nonlinear resistor given by Eq. (1.7) is voltage-controlled. Note, for $V_1 < v < V_2$, the slope of the curve is negative just like that of a linear active resistor having negative resistance introduced earlier. We shall see later that this negative slope makes the device useful in such applications as amplifiers and oscillators. Note also that, for $I_2 < i < I_1$, a given current i corresponds to three different values of v on the characteristic. This multivalued property makes the device useful in memory and switching circuits.

Glow tube The symbol of the glow tube and its characteristic are shown in Fig. 1.11. We can express the characteristic of such a nonlinear resistor by a single-valued function

$$v = \hat{v}(i) \tag{1.8a}$$

or, in terms of Eq. (1.2),

$$f(v, i) = v - \hat{v}(i) = 0 \tag{1.8b}$$

A nonlinear resistor whose voltage is a "single-valued" function of the current

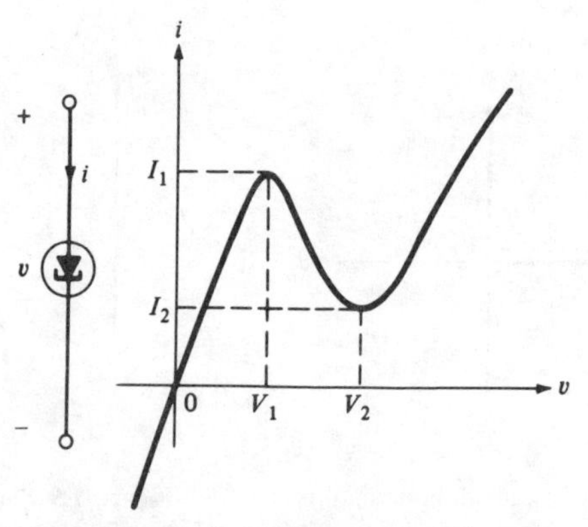

Figure 1.10 Symbol for a tunnel diode and its characteristic.

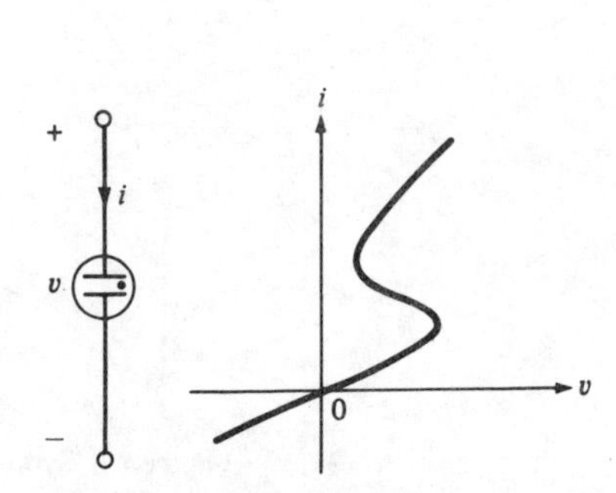

Figure 1.11 Symbol for a glow tube and its characteristic.

is called a *current-controlled resistor*. Thus, the glow tube is a current-controlled nonlinear resistor. Since the current is not a (single-valued) function of the voltage, it is not voltage-controlled.

Note that the ideal diode is neither voltage-controlled nor current-controlled.

Bilateral property In contrast to linear resistors, a nonlinear resistor in general has a v-i characteristic which is not symmetric with respect to the origin of the v-i plane. (See Figs. 1.9 and 1.10.) Consider the tunnel diode in Fig. 1.10. Let us change the reference direction of the current and of the voltage. We redraw the circuit as shown in Fig. 1.12, where $i_1 = -i$ and $v_1 = -v$.

The characteristic in the v_1-i_1 plane, which corresponds to that of the original two-terminal resistor with the two terminals interchanged, is shown in the figure. For this reason, it is important that the symbol for a nonlinear resistor indicate its orientation. Note that the general symbol for a nonlinear resistor in Fig. 1.13 is dissymmetric with respect to its two terminals; consequently, it is possible to specify the correct connection of the two terminals of a nonlinear resistor to a circuit.

The characteristic of a *linear* resistor is always symmetric with respect to the origin. A circuit element with this kind of symmetry is called *bilateral*. A *bilateral resistor* satisfies the property $f(v, i) = f(-v, -i)$ for all (v, i) on its characteristic. A nonlinear resistor may be bilateral, e.g., have the characteristic shown in Fig. 1.14.

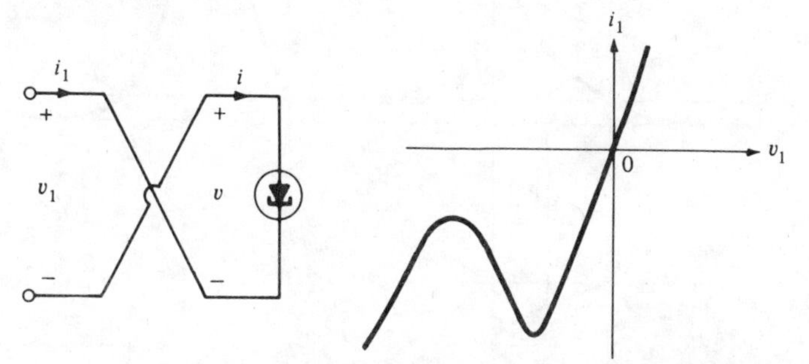

Figure 1.12 Characteristic of a tunnel diode with its terminals turned around.

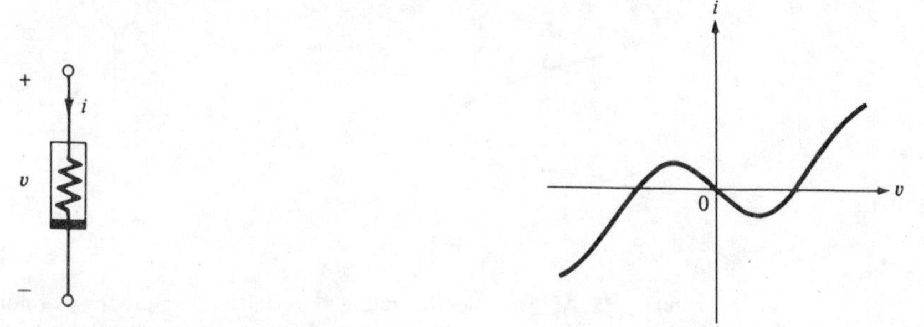

Figure 1.13 Symbol for a nonlinear resistor.

Figure 1.14 v-i Characteristic of a bilateral nonlinear resistor.

Simple circuits Circuits containing nonlinear resistors have properties totally different from those which have only linear resistors. The following examples illustrate some of the differences.

Example 1 (nonlinear resistors can produce harmonics) Consider a sinusoidal voltage waveform,

$$v(t) = 2 \sin \omega t \text{ (in volts)} \qquad t \geq 0$$

where the constant ω is the angular frequency in radians per second, i.e., $\omega = 2\pi f$ where f is frequency in hertz. If the waveform is applied to a linear resistor of $10\,\Omega$, the current is $i(t) = 0.2 \sin \omega t$ (in amperes), $t \geq 0$.

Let us apply the same voltage waveform to a nonlinear resistor which has the v-i characteristic shown in Fig. 1.15a, where

$$i = \hat{i}(v)$$

We wish to determine the current waveform $i(t)$ for $t \geq 0$. In simple

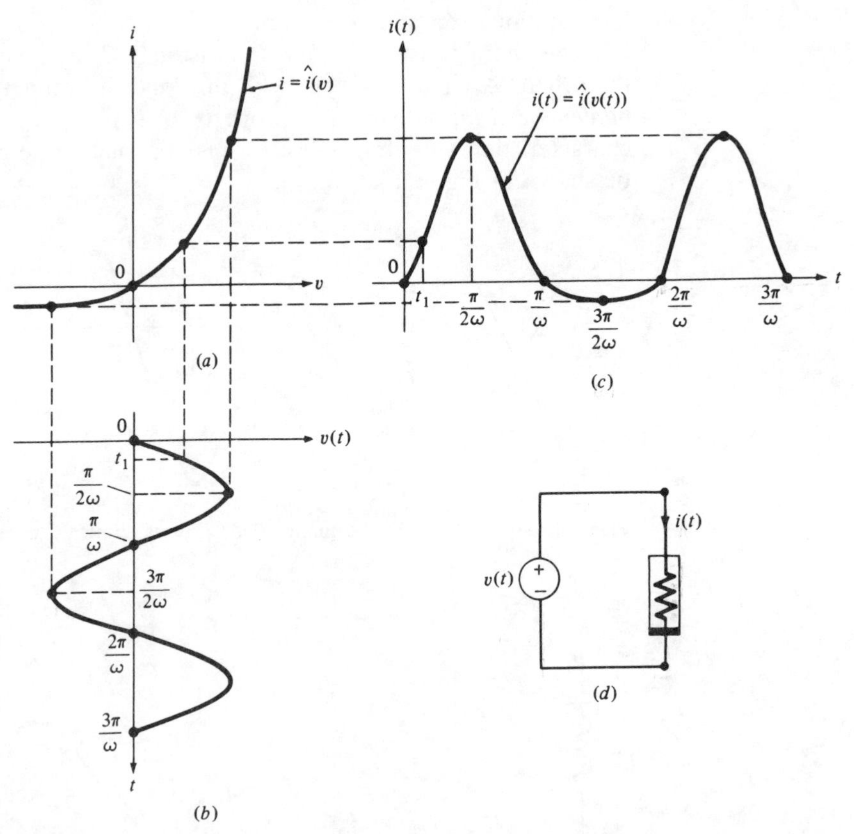

Figure 1.15 An example illustrating a special clipping property of nonlinear resistors; the negative half of the waveform has been clipped.

nonlinear circuit analysis a graphic method is often useful. Let us plot the voltage waveform $v(\cdot)$ as shown in Fig. 1.15*b* with the t axis lined up with the i axis of Fig. 1.15*a*. Then it is easy to obtain the current $i(t)$ shown in Fig. 1.15*c* by transcribing points on the voltage waveform $v(t)$ at $t=0$, t_1, $\pi/2\omega$, π/ω, $3\pi/2\omega$, etc., through the v-i curve of the nonlinear resistor. The resulting current waveform $i(\cdot)$ is again a periodic function with period $2\pi/\omega$; hence it can be expressed in terms of a Fourier series

$$i(t) = I_0 + \sum_k I_k \sin k\omega t \tag{1.9}$$

It contains, besides the fundamental sinusoid, higher harmonics, i.e., the sinusoidal terms in Eq. (1.9) with $k>1$. What is especially important to note is that, unlike the voltage waveform $v(\cdot)$, the current waveform $i(\cdot)$ has a *direct current* (*dc*) component I_0. The term I_0 in Eq. (1.9) is equal to the average value of the current waveform $i(\cdot)$ over a complete period. This dc component can be filtered out with a simple low-pass filter. This, of course, is the basis of all rectifier circuits which convert alternating current to direct current.

Example 2 (piecewise-linear approximation) Another basic method of nonlinear circuit analysis is the use of piecewise-linear approximation. The v-i curve in Fig. 1.15*a* can be roughly approximated by two linear segments as shown in Fig. 1.16. With this simple approximation, the current waveform can be obtained immediately. For v negative, i is identically zero. For v positive, the nonlinear resistor behaves like a linear resistor with conductance G. Thus the waveform is given by

$$i(t) = \begin{cases} 2G \sin \omega t & \text{for} \quad \dfrac{2n\pi}{\omega} \le t \le \dfrac{(2n+1)\pi}{\omega} \\[2ex] 0 & \text{for} \quad \dfrac{(2n+1)\pi}{\omega} \le t \le \dfrac{(2n+2)\pi}{\omega} \end{cases}$$

where n runs through nonnegative integers. The result is of course different from that obtained by the graphic method. However, had we used

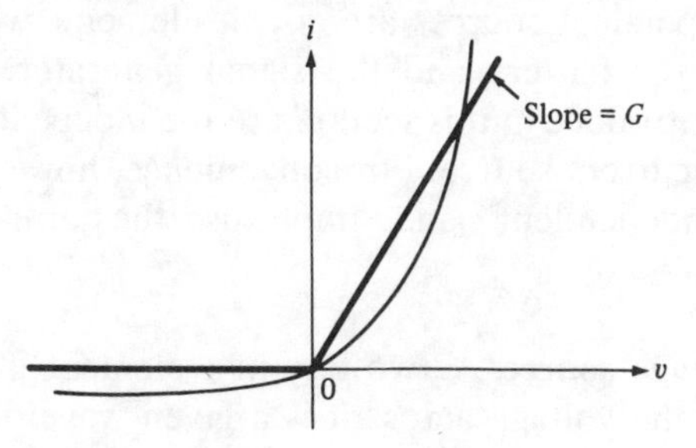

Figure 1.16 Piecewise-linear approximation of a nonlinear characteristic.

more linear segments to approximate the v-i curve, we could have obtained a solution closer to the current waveform shown in Fig. 1.15. The subject of piecewise-linear approximation will be treated in greater detail later in the chapter.

Example 3 (homogeneity and additivity) In this example we will compare the currents in a linear resistor whose characteristic is

$$i_\ell = \hat{i}_\ell(v) = 0.1v \tag{1.10a}$$

and in a nonlinear resistor whose characteristic is

$$i_n = \hat{i}_n(v) = 0.1v + 0.02v^3 \tag{1.10b}$$

The subscript ℓ denotes linear and the subscript n denotes nonlinear. Let us consider three different constant input voltages: $v_1 = 1$ V, $v_2 = k$ V, and $v_3 = v_1 + v_2 = 1 + k$ V. For the linear resistor, from Eq. (1.10a), we have $i_{\ell 1} = 0.1$ A, $i_{\ell 2} = 0.1 \times k$ A, and $i_{\ell 3} = 0.1 + 0.1 \times k$ A. Clearly

$$i_{\ell 2} = \hat{i}_\ell(kv_1) = k\hat{i}_\ell(v_1) \tag{1.11a}$$

and

$$i_{\ell 3} = \hat{i}_\ell(v_1 + v_2) = \hat{i}_\ell(v_1) + \hat{i}_\ell(v_2) \tag{1.11b}$$

Equation (1.11a) states the *homogeneity property* of a linear function, while Eq. (1.11b) states the *additivity property*.

Next, let us consider the nonlinear resistor defined by Eq. (1.10b). Setting $v_1 = 1$ V, $v_2 = k$ V, and $v_3 = v_1 + v_2$ V, we obtain successively $i_{n1} = 0.12$ A, $i_{n2} = 0.1 \times k + 0.02 \times k^3$ A, and $i_{n3} = 0.1(1 + k) + 0.02(1 + k)^3$ A. Clearly

$$i_{n2} = \hat{i}_n(kv_1) \neq k\hat{i}_n(v_1)$$

and

$$i_{n3} = \hat{i}_n(v_1 + v_2) \neq \hat{i}_n(v_1) + \hat{i}_n(v_2)$$

Thus in general for nonlinear resistors, neither the homogeneity property nor the additivity property holds.

1.3 Independent Sources

In circuit theory, independent sources play the same role as external forces in mechanics. Independent sources are circuit elements which are used to model such devices as the battery and the signal generator. The two independent sources we will introduce in this section are the independent voltage source and the independent current source. For convenience, however, we shall often omit the adjective "independent" and simply use the terms "voltage source" and "current source."

Independent voltage source A two-terminal element is called an *independent voltage source* if the voltage across it is a given waveform $v_s(\cdot)$ irrespective of the current flowing through it. Note that the waveform $v_s(\cdot)$ is part of the specification of the voltage source. Commonly used waveforms include the *dc voltage source*, i.e., $v_s(t)$ is a constant E for all t, the sinusoid, the square wave, etc. The symbol of an independent voltage source with waveform $v_s(\cdot)$ is shown in Fig. 1.17*a* where the signs + and − specify the polarity. The symbol for a dc voltage source is shown in Fig. 1.17*b* with $E > 0$.

At any time t, the independent voltage source can be expressed by the relation

$$\mathcal{R}_{vs} = \{(v, i): v = v_s(t) \text{ for } -\infty < i < \infty\} \tag{1.12}$$

Consequently, an independent voltage source is a two-terminal resistor. Its characteristic in the v-i plane is a straight line parallel to the i axis. This is shown in Fig. 1.18*a*, where $v_s(t)$ denotes the voltage of the independent voltage source at time t.

Consider a sinusoidal voltage source $v_s(t) = V_m \sin \omega t$, where V_m, the amplitude of the sinusoid, is a constant and ω is the angular frequency. Then, for each t, the v-i characteristic is a vertical line as shown in Fig. 1.18*b*, depicting the value of the voltage waveform at time t.

Since the v-i characteristic is a straight line parallel to the i axis, the independent voltage source is a current-controlled nonlinear resistor. It is nonlinear because the straight line does not go through the origin unless $v_s = 0$; in that case, it becomes identical to the characteristic of a short circuit. The

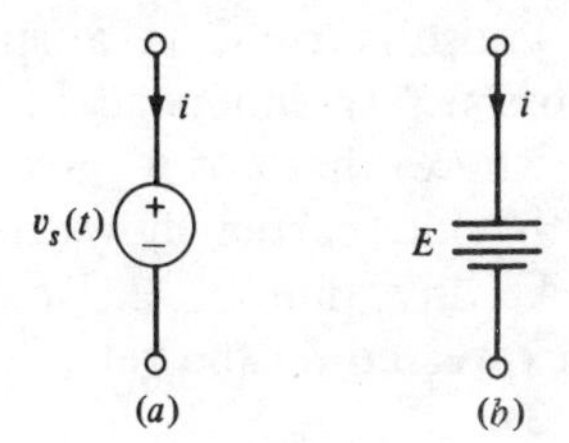

Figure 1.17 Symbols for an independent voltage source and a dc ,voltage source.

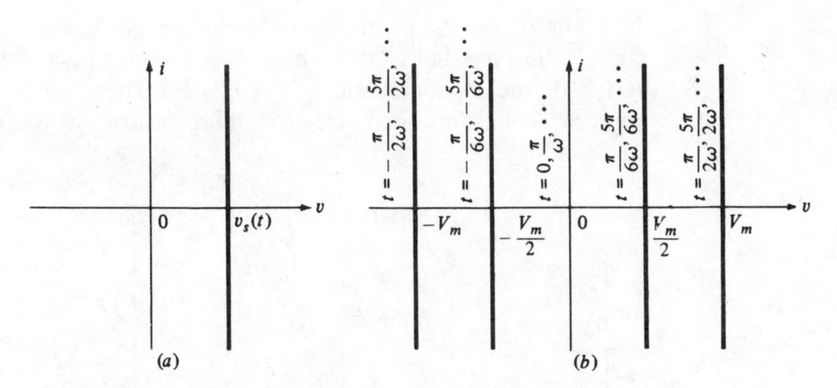

Figure 1.18 (*a*) Characteristic of an independent voltage source at time t, $v_s(t)$. (*b*) Characteristic at different values of t for $v_s(t) = V_m \sin \omega t$.

property that an independent voltage source v_s becomes a short circuit (zero resistance) if $v_s = 0$ is very important in circuit analysis. We shall use this property from time to time throughout the book.

In Fig. 1.19 we show an independent voltage source connected to an arbitrary external circuit. The physical significance of the definition of the independent voltage source is that the voltage across the source is maintained to the prescribed waveform $v_s(\cdot)$ no matter what the external circuit may be. The nature of the external circuit only affects the current i flowing through the source. This is because an independent voltage source has "zero internal resistance" in contrast to a real battery which has a finite nonzero resistance. This will be illustrated later in the chapter after we introduce the series connection of resistors.

So far, we have used the associated reference directions for all circuit elements for consistency. With the associated reference directions, the power delivered *to* the voltage source is $v_s \times i$; therefore, the power delivered *from* the voltage source to the external circuit is

$$p = v_s \times (-i) = v_s i'$$

where i' is the current entering the external circuit. Note that as far as the external circuit is concerned v_s and i' are measured using the associated reference directions.

Independent current source An *independent current source* is defined as a two-terminal circuit element whose current is a specified waveform $i_s(\cdot)$ irrespective of the voltage across it. An independent current source has the symbol shown in Fig. 1.20, where the arrow gives the positive current direction, i.e., $i_s(t) > 0$ means that the current flows through the source from terminal ① to terminal ②. At any time t, the v-i characteristic for an independent current source is expressed by the relation

$$\mathcal{R}_{is} = \{(v, i): i = i_s(t) \text{ for } -\infty < v < \infty\} \tag{1.13}$$

In terms of the v-i plane, an independent current source is represented by a straight line parallel to the v axis. It is a voltage-controlled nonlinear resistor. If $i_s = 0$, the characteristic is the v axis. Therefore an independent current source becomes an open circuit (infinite resistance) when the source current is zero.

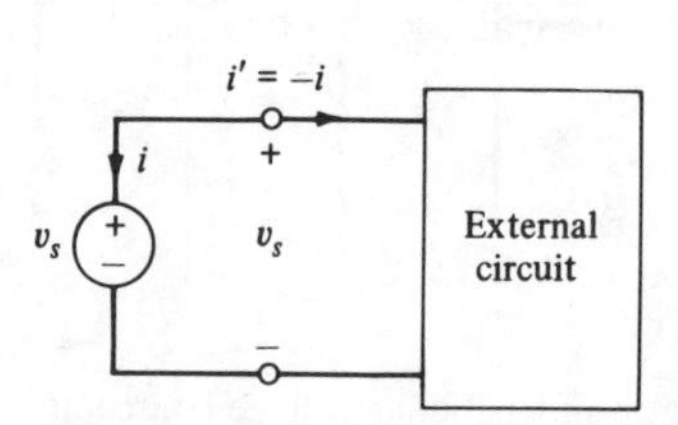

Figure 1.19 An independent voltage source connected to an external circuit.

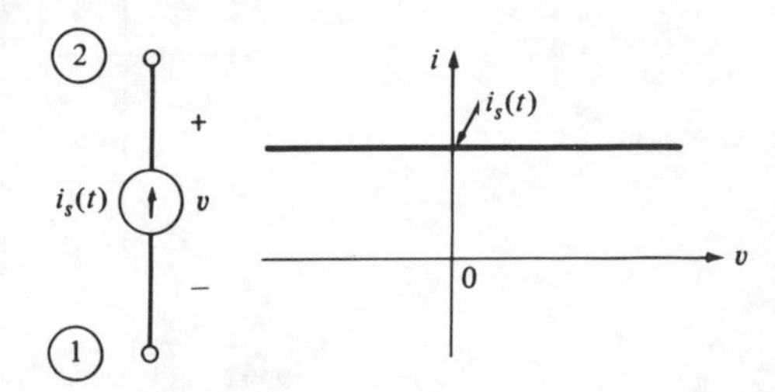

Figure 1.20 Symbol for an independent current source and its characteristic.

In Fig. 1.21 we show an independent current source connected to an arbitrary external circuit. The significance of the definition of the independent current source is that the current of the current source maintains its prescribed waveform $i_s(\cdot)$ and the voltage across it is determined by the external circuit. This is because an independent current source has "infinite internal resistance." The power delivered *from* the source *to* the external circuit is $p(t) = i_s(t)v'(t)$, for all t. Note that as far as the external circuit is concerned, v' and i_s are in the associated reference directions.

Exercise Under what condition is an independent current source i_s the dual of an independent voltage source v_s?

1.4 Time-Invariant and Time-Varying Resistors

We have discussed the independent voltage source, whose voltage is a sinusoidal waveform. It is a nonlinear resistor whose v-i characteristic is a function of time as shown in Fig. 1.18b. We shall introduce a formal definition of a time-varying resistor: A resistor is said to be *time-varying* iff its v-i characteristic varies with time; otherwise, it is said to be *time-invariant*. *A linear time-varying resistor* is characterized by Ohm's law

$$v(t) = R(t)i(t) \qquad \text{or} \qquad i(t) = G(t)v(t) \tag{1.14}$$

where $G(t) = 1/R(t)$. The function $R(\cdot)$ or $G(\cdot)$ is part of the specification of the resistor. Thus $R(t)$ gives the value of the time-varying resistance at time t. From Eq. (1.14) it is clear that an independent voltage source or current source having a nonconstant waveform is a nonlinear time-varying resistor. In this section we will first introduce examples of useful time-varying resistors and then bring out a unique property of time-varying resistors, which is important in communication engineering.

Periodically operating switches A periodically operating ideal switch is a linear time-varying resistor. Its symbol, property, and v-i characteristic are shown in Fig. 1.22. For $0 \le t < t_1$, the switch is open, $S = 1$, the current is zero, and the v-i characteristic is the v axis. For $t_1 \le t < T$, the switch is closed, $S = 0$, the voltage is zero, and the v-i characteristic is the i axis. After a period T, the switch repeats its operation; i.e., $S = 1$ and $S = 0$ alternately for $nT \le t < nT + t_1$ and $nT + t_1 \le t < (n+1)T$, respectively, where n is any integer >1. The periodically operating switch is a key circuit element in digital circuits and in both digital and analog communication systems.

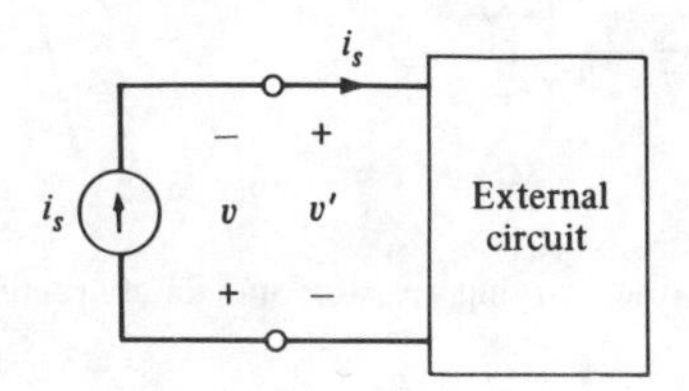

Figure 1.21 An independent current source connected to an external circuit.

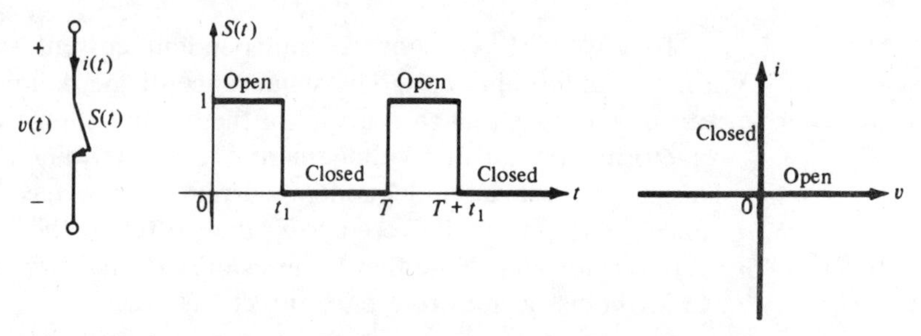

Figure 1.22 Symbol for a periodically operating switch and its characteristic. Here, $S(t)$ denotes the "state" (i.e., position) of the switch.

A practical switch behaves somewhat differently, instead of being an open circuit or a short circuit, the switch has finite and nonzero resistances: very large when it is in the open position and near zero when it is in the closed position. The symbol of a linear time-varying resistor, its property (for a practical periodically operating switch), and its v-i characteristic are shown in Fig. 1.23.

A linear time-varying resistor is characterized for each t by a linear function, and therefore satisfies the homogeneity and additivity properties. Yet it has other properties rather different from that of linear time-invariant resistors. We illustrate one such property with the following example.

Example Consider a linear time-varying resistor with v-i characteristic

$$i(t) = G(t)v(t) \qquad \text{and} \qquad G(t) = G_a + G_b \cos \omega t$$

where the constants G_a and G_b satisfy the condition $G_a > G_b$; consequently $G(t) > 0$ for all t. The angular frequency ω is also a constant. Let the voltage waveform be a sinusoid with angular frequency ω_1. Then

$$v(t) = V_m \cos \omega_1 t$$

$$i(t) = G_a V_m \cos \omega_1 t + G_b V_m \cos \omega_1 t \cos \omega t$$

$$= G_a V_m \cos \omega_1 t + \frac{G_b V_m}{2} \cos(\omega + \omega_1)t + \frac{G_b V_m}{2} \cos(\omega - \omega_1)t$$

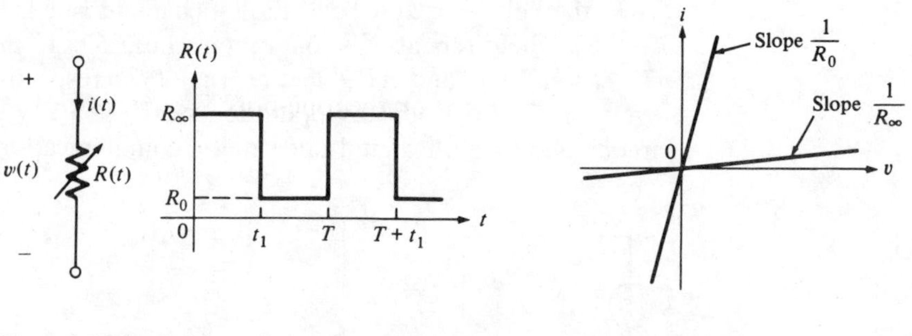

Figure 1.23 Symbol for a linear time-varying resistor and its characteristic.

Thus a sinusoidal voltage with angular frequency ω_1 applied to a linear time-varying resistor generates, in addition to a sinusoidal current with the same angular frequency ω_1, two sinusoids at angular frequencies $\omega + \omega_1$ and $\omega - \omega_1$. This property is the basis of several modulation schemes in communication systems. With linear *time-invariant* resistors, a sinusoidal input can only generate a sinusoidal response at the *same* frequency.

2 SERIES AND PARALLEL CONNECTIONS

In Chap. 1 we considered general circuits with arbitrary circuit elements. The primary objective was to learn Kirchhoff's current law (KCL) and Kirchhoff's voltage law (KVL). KCL and KVL do not depend on the nature of the circuit elements. They lead to two sets of *linear* algebraic equations in terms of two sets of pertinent circuit variables: the *branch currents* and the *branch voltages*. These equations depend on the topology of the circuit, i.e., how the circuit elements are connected to one another. The branch currents and branch voltages are in turn related according to the characteristics of the circuit elements. As seen in the previous section, these characteristics for two-terminal resistors may be linear or nonlinear, time-invariant or time-varying. The equations describing the v-i characteristics are called *element equations* or *branch equations*. Together with the equations from KCL and KVL, they give a complete specification of the circuit. The purpose of circuit analysis is to write down the complete specification of any circuit and to obtain pertinent solutions of interest.

In this section we will consider a special but very important class of circuits: circuits formed by series and parallel connections of two-terminal resistors. First, we wish to generalize the concept of the v-i characteristic of a resistor to that of a two-terminal circuit made of two-terminal resistors, or more succinctly a *resistive one-port*. We will demonstrate that the series and parallel connections of two-terminal resistors will yield a one-port whose v-i characteristic is again that of a resistor. We say that two resistive one-ports are *equivalent* iff their v-i characteristics are the same.

When we talk about resistive one-ports, we naturally use *port voltage* and *port current* as the pertinent variables. The v-i characteristic of a one-port in terms of its port voltage and port current is often referred to as the *driving-point characteristic* of the one-port. The reason we call it the driving-point characteristic is that we may consider the one-port as being driven by an independent voltage source v_s or an independent current source i_s as shown in Fig. 2.1. In the former, the input is $v_s = v$, the port voltage; and the response is the port current i. In the latter, the input is $i_s = i$, the port current; and the response is the port voltage v. In the following subsections we will discuss the driving-point characteristics of one-ports made of two-terminal resistors connected in series, connected in parallel, and connected in series-parallel.

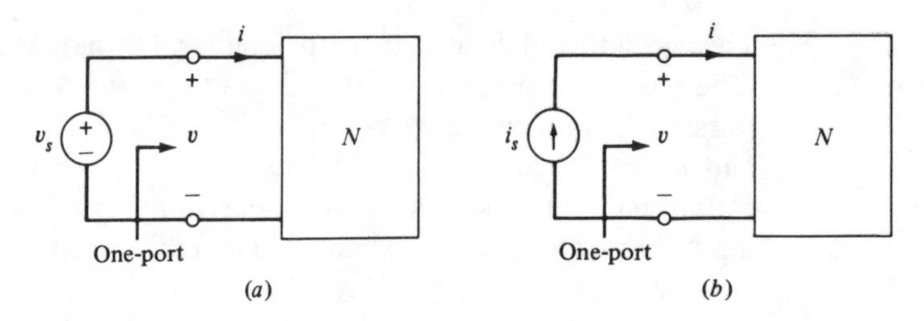

Figure 2.1 A one-port N driven (*a*) by an independent voltage source and (*b*) by an independent current source.

2.1 Series Connection of Resistors

From physics we know that the series connection of linear resistors yields a linear resistor whose resistance is the sum of the resistances of each linear resistor. Let us extend this simple result to the series connections of resistors in general.

Consider the circuit shown in Fig. 2.2 where two *nonlinear* resistors $\mathcal{R}_1$ and $\mathcal{R}_2$ are connected at node ②. Nodes ① and ③ are connected to the rest of the circuit, which is designated by $\mathcal{N}$. Looking toward the right from nodes ① and ③, we have a one-port which is formed by the *series connection of two resistors* $\mathcal{R}_1$ and $\mathcal{R}_2$. For our present purposes the nature of $\mathcal{N}$ is irrelevant. We are interested in obtaining the driving-point characteristic of the one-port with port voltage v and port current i.

Let us assume that both resistors are current-controlled, i.e.,

$$v_1 = \hat{v}_1(i_1) \qquad \text{and} \qquad v_2 = \hat{v}_2(i_2) \tag{2.1}$$

These are the two branch equations. Next, we consider the circuit topology and write the equations using KCL and KVL. KCL applied to nodes ① and ② gives

$$i = i_1 = i_2 \tag{2.2}$$

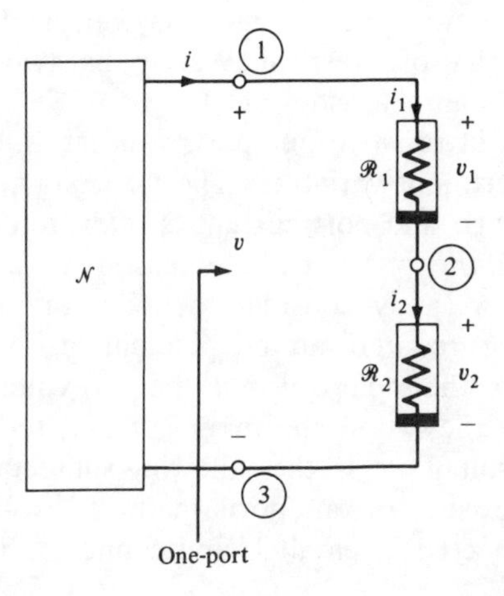

Figure 2.2 Two nonlinear resistors connected in series together with the rest of the circuit $\mathcal{N}$.

The KVL equation for the node sequence ①-②-③-① leads to

$$v = v_1 + v_2 \tag{2.3}$$

Combining Eqs. (2.1), (2.2), and (2.3), we obtain

$$v = \hat{v}_1(i) + \hat{v}_2(i) \tag{2.4}$$

which is the v-i characteristic of the one-port. It states that the driving-point characteristic of the one-port is again a current-controlled resistor

$$v = \hat{v}(i) \tag{2.5a}$$

where

$$\hat{v}(i) \stackrel{\Delta}{=} \hat{v}_1(i) + \hat{v}_2(i) \qquad \text{for all } i \tag{2.5b}$$

Exercise If the two terminals of the nonlinear resistor $\mathcal{R}_1$ in Fig. 2.2 are turned around as shown in Fig. 2.3, show that the series connection gives a one-port which has a driving-point characteristic:

$$\hat{v}(i) = -\hat{v}_1(-i) + \hat{v}_2(i)$$

Example 1 (a battery model) A battery is a physical device which can be modeled by the series connection of a linear resistor and a dc voltage source, as shown in Fig. 2.4. Since both the independent voltage source and the linear resistor are current-controlled resistors, this is a special case of the circuit in Fig. 2.2. The branch equations are

$$v_1 = Ri_1 \qquad \text{and} \qquad v_2 = E \tag{2.6}$$

Adding v_1 and v_2 and setting $i_1 = i$, we obtain

$$v = Ri + E \tag{2.7}$$

Or, we can add the characteristics graphically to obtain the driving-point characteristic of the one-port shown in the i-v plane in the figure.[3] The

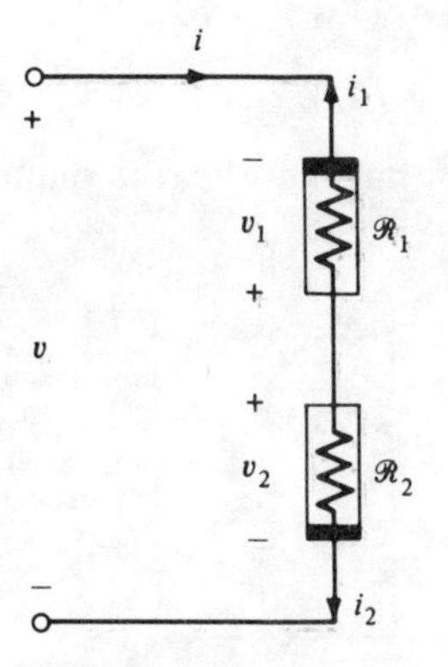

Figure 2.3 Series connection of $\mathcal{R}_1$ and $\mathcal{R}_2$ with the terminals of $\mathcal{R}_1$ turned around.

[3] We use the i-v plane to facilitate the addition of voltages.

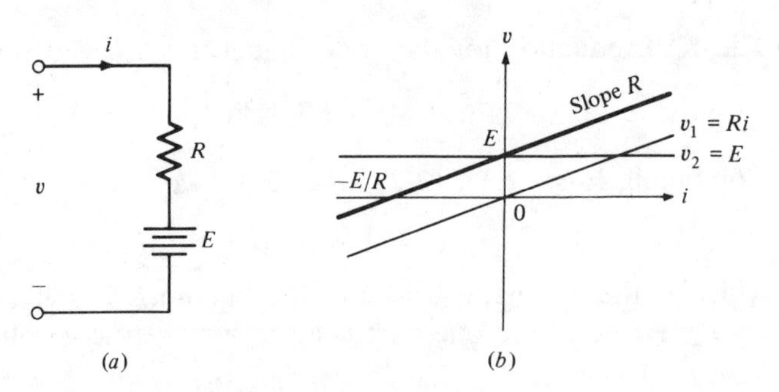

Figure 2.4 (*a*) Series connection of a linear resistor and a dc voltage source, and (*b*) its driving-point characteristic.

heavy line in Fig. 2.4*b* gives the characteristic of a battery with an internal resistance R. Usually R is small, thus the characteristic in the i-v plane is reasonably flat. However, it should be clear that a real battery does not behave like an independent voltage source because the port voltage v depends on the current i. If we connect the real battery to an external load, e.g., a linear resistor with resistance R, the actual voltage across the load will be $E/2$.

Since a battery is used to deliver power to an external circuit, we usually prefer to use the opposite of the associated reference direction when the battery is connected to an external circuit. The characteristic plotted on the i'-v plane where $i' = -i$ is shown in Fig. 2.5 together with the external circuit.

Example 2 (ideal diode circuit) Consider the series connection of a real battery and an ideal diode as shown in Fig. 2.6. Since the ideal diode is not a current-controlled resistor, we cannot use Eq. (2.5) to add the voltages directly. Instead, we consider each segment of the ideal diode characteristic independently. Recall the definition of the ideal diode, for $i_2 > 0$, $v_2 = 0$. Since by KCL, $i_2 = i_1 = i$, we have

$$v = v_1 = Ri + E \qquad \text{for } i > 0 \tag{2.8a}$$

In other words, in the right half of the i-v plane ($i > 0$), the one-port

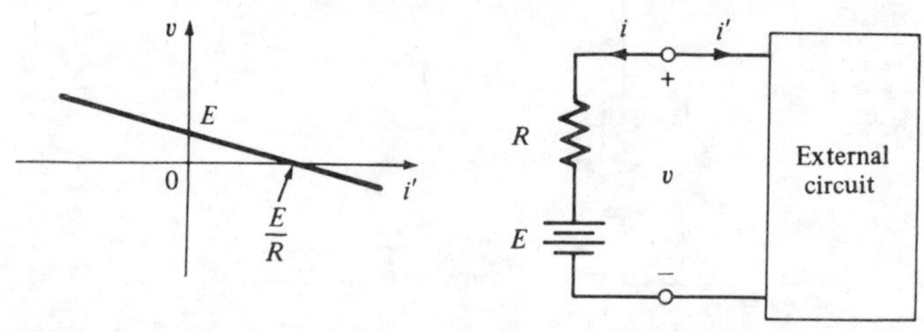

Figure 2.5 Characteristic of a real battery plotted on the i'-v plane and the external circuit connection.

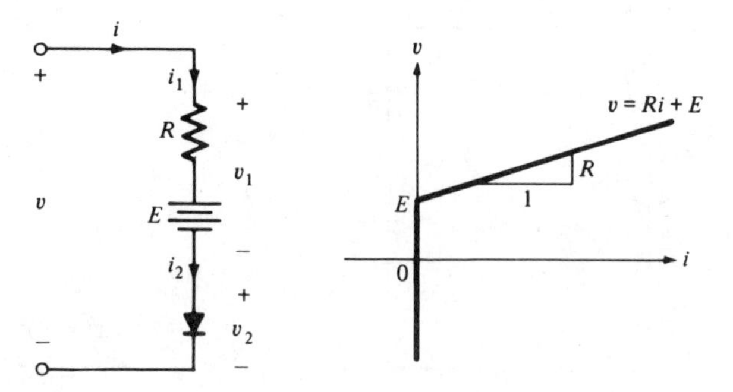

Figure 2.6 Series connection of a battery with an ideal diode and the driving-point characteristic of the resulting one-port.

characteristic is identical with that of Fig. 2.4, i.e., a straight line starting at the point $(0, E)$ with a slope R. Next, for $v_2 < 0$, $i_2 = 0$, i.e., the ideal diode is reversed biased, and its characteristic requires that $i = i_2 = 0$. Therefore there is no current flow. Hence $v_1 = E$, and thus

$$v = E + v_2 \qquad \text{and} \qquad i = 0 \qquad \text{for } v_2 \leq 0 \tag{2.8b}$$

In other words, when the diode is reversed biased, the one-port characteristic consists of the vertical half line lying on the v axis below the point $(0, E)$. (See Fig. 2.6.)

Exercise Determine the v-i characteristic of the series connection of the same three elements as in Fig. 2.6 except that the diode is turned around.

Example 3 (voltage sources in series) Consider m independent voltage sources in series as shown in Fig. 2.7. Since voltage sources are current-controlled, we only need to add the voltages of each independent source to obtain an equivalent one-port, which is an independent voltage source whose voltage is given by the sum of the m voltages.

Example 4 (current sources in series) Next, consider independent current sources in series as shown in Fig. 2.8. Applying KCL at nodes ①, ②, etc.,

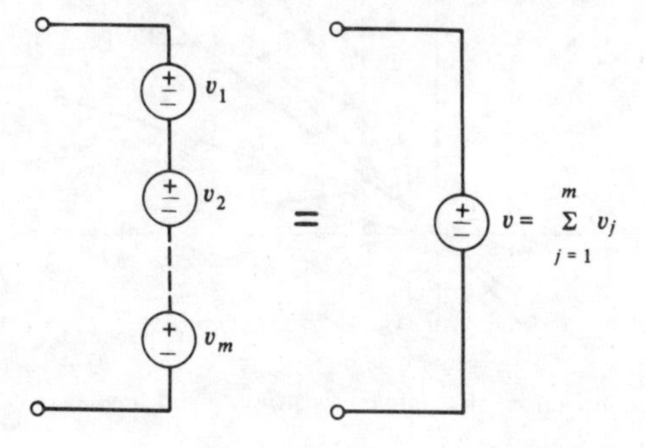

Figure 2.7 m Independent voltage sources in series.

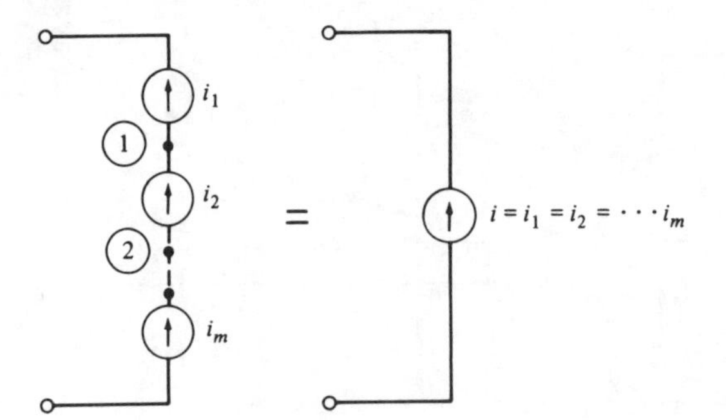

Figure 2.8 *m* Independent current sources in series.

we note that the m independent current sources must have the same current, i.e., $i_1 = i_2 = \cdots = i_m$. Otherwise, we reach a contradiction because (a) by definition, an independent current source has a current i_s irrespective of the external connection and (b) KCL cannot be violated at any node of a circuit. Therefore we conclude that (a) the only possibility for the connection to make sense is that all currents are the same and (b) the resulting one-port is a current source with the same current.

Example 5 (graphic method) Graphic methods are extremely useful in analyzing simple nonlinear circuits. Consider the series connection of a linear resistor and a *voltage-controlled* nonlinear resistor as shown in Fig. 2.9. The branch equations for the two resistors are

$$v_1 = R_1 i_1 \qquad \text{and} \qquad i_2 = \hat{i}_2(v_2) \tag{2.9}$$

First, we wish to find out whether the nonlinear resistor $\mathcal{R}_2$ is also current-controlled. Since, for $I_1 \leq i \leq I_2$, the voltage is multivalued, we know $\mathcal{R}_2$ is not current-controlled. Therefore we cannot solve the problem analytically as before [see Eqs. (2.1) and (2.5)]. However, it is possible to add the voltages v_1 and v_2 graphically point by point on the two curves.

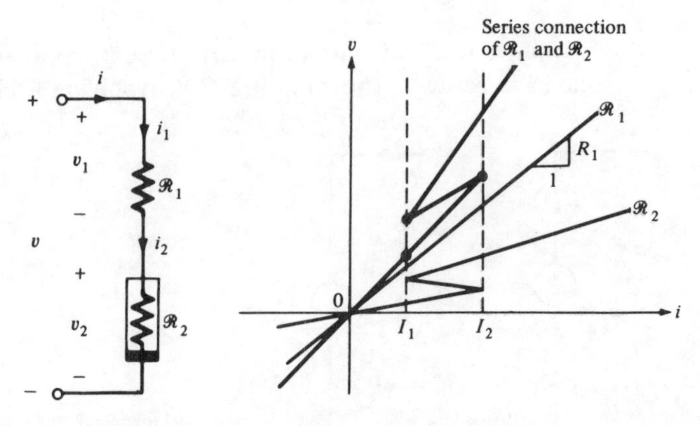

Figure 2.9 Series connection of a linear resistor $\mathcal{R}_1$ and a voltage-controlled resistor $\mathcal{R}_2$.

Thus for $I_1 \le i \le I_2$, the sum of v_1 and v_2 is multivalued. What is of interest is to note that the resulting characteristic is neither voltage-controlled nor current-controlled.

REMARK Figure 2.9 shows that $\mathscr{R}_2$ is voltage-controlled, i.e., $i_2 = \hat{i}_2(v_2)$ and $\mathscr{R}_1$ is both voltage-controlled and current-controlled. The characteristic of the series connection is neither voltage-controlled nor current-controlled but may be described by the *parametric representation*:

$$v = R_1\hat{i}_2(v_2) + v_2$$
$$i = \hat{i}_2(v_2)$$

where v_2 plays the role of a parameter.

Summary of series connection The key concepts used in obtaining the driving-point characteristic of a one-port formed by the series connection of two-terminal resistors are

1. KCL forces all branch currents to be equal to the port current.
2. KVL requires the port voltage v to be equal to the sum of the branch voltages of the resistors.
3. If each resistor is current-controlled, the resulting driving-point characteristic of the one-port is also a current-controlled resistor.

2.2 Parallel Connection of Resistors

Consider the circuit shown in Fig. 2.10 where two resistors $\mathscr{R}_1$ and $\mathscr{R}_2$ are connected in parallel at nodes ① and ② to the rest of the circuit designated $\mathscr{N}$. The nature of $\mathscr{N}$ is immaterial for the present discussion. We wish to determine the driving-point characteristic of the one-port defined by the two nodes ① and ② looking to the right, i.e., the parallel connection of $\mathscr{R}_1$ and $\mathscr{R}_2$. We assume that the resistors are both voltage-controlled, i.e.,

$$i_1 = \hat{i}_1(v_1) \qquad \text{and} \qquad i_2 = \hat{i}_2(v_2) \tag{2.10}$$

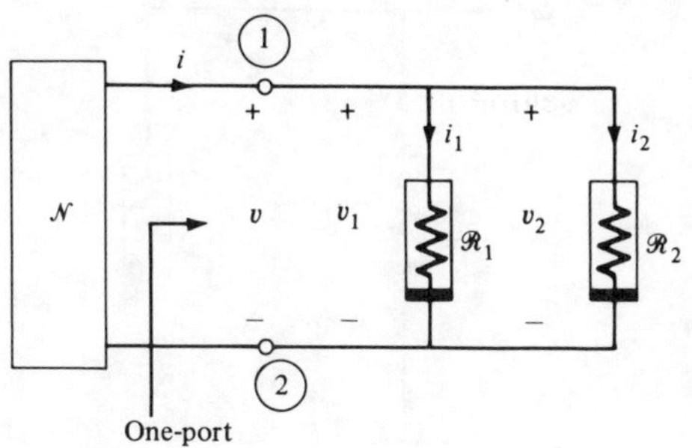

Figure 2.10 Two nonlinear resistors in parallel together with the rest of the circuit $\mathscr{N}$.

Picking node ② as the datum node, we have, from KVL

$$v = v_1 = v_2 \tag{2.11}$$

Applying KCL at node ①, we have

$$i = i_1 + i_2 \tag{2.12}$$

Combining the above equations, we obtain

$$i = \hat{i}_1(v) + \hat{i}_2(v) \tag{2.13}$$

Equation (2.13) states that the driving-point characteristic of the one-port is a voltage-controlled resistor defined by

$$i = \hat{i}(v) \tag{2.14a}$$

where $$\hat{i}(v) \stackrel{\Delta}{=} \hat{i}_1(v) + \hat{i}_2(v) \qquad \text{for all } v \tag{2.14b}$$

Duality It is interesting to compare the two sets of equations: (2.1) to (2.5) for the series connection and (2.10) to (2.14) for the parallel connection. If we make the substitutions for all the v's with i's and for all the i's with v's in one set of equations, we obtain precisely the other set. For this reason, we can extend and generalize the concept of duality introduced earlier for resistors to *circuits*.

Let us redraw the two circuits of the series connection and of the parallel connection of two nonlinear resistors as shown in Fig. 2.11. Let us denote the series-connection circuit by $\mathcal{N}$ and the parallel-connection circuit by $\mathcal{N}^*$. Comparing Eqs. (2.1) to (2.5) with Eqs. (2.10) to (2.14) one by one together with the two circuits in Fig. 2.11, we can learn how to generalize the concept of duality. First, for the branch equations (2.1) and (2.10), note that substituting all the v's with i's and all the i's with v's in one equation, we obtain the other. This, however, requires that the function $\hat{v}_1(\cdot)$ be the same as $\hat{i}_1(\cdot)$ and the function $\hat{v}_2(\cdot)$ be the same as $\hat{i}_2(\cdot)$. In other words, the nonlinear resistor $\mathcal{R}_1^*$ in $\mathcal{N}^*$ must be the dual of the nonlinear resistor $\mathcal{R}_1$ in $\mathcal{N}$, and similarly $\mathcal{R}_2^*$ in $\mathcal{N}^*$ must be the dual of $\mathcal{R}_2$ in $\mathcal{N}$. Next, compare Eqs. (2.2) and (2.3) with Eqs.

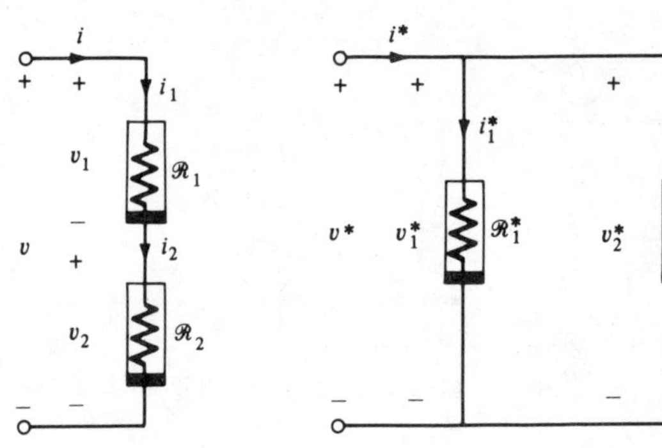

Figure 2.11 A circuit $\mathcal{N}$ and its dual $\mathcal{N}^*$.

(2.11) and (2.12). While Eq. (2.2) states KCL imposed by the series connection of $\mathscr{R}_1$ and $\mathscr{R}_2$ in $\mathcal{N}$, Eq. (2.11) states KVL imposed by the parallel connection of $\mathscr{R}_1^*$ and $\mathscr{R}_2^*$ in $\mathcal{N}^*$. Similarly, Eq. (2.3) is KVL which sums the two voltages v_1 and v_2 in $\mathcal{N}$, while Eq. (2.12) is KCL which sums the two currents i_1^* and i_2^* in $\mathcal{N}^*$. Finally, Eqs. (2.5*a*) and (2.5*b*) specify the driving-point characteristic of the one-port obtained by the series connection of two current-controlled resistors $\mathscr{R}_1$ and $\mathscr{R}_2$ as a current-controlled resistor. Equations (2.14*a*) and (2.14*b*) specify the driving-point characteristic of the one-port obtained by the parallel connection of two voltage-controlled resistors $\mathscr{R}_1^*$ and $\mathscr{R}_2^*$ as a voltage-controlled resistor.

In Table 2.1 we list two sets of terms S and S^* which we have encountered and which are said to be dual to one another. With these terms, we can generalize the concept of duality to define a dual circuit. Two circuits $\mathcal{N}$ and $\mathcal{N}^*$ are said to be *dual* to one another if the equations describing circuit $\mathcal{N}$ are identical to those describing circuit $\mathcal{N}^*$ after substituting every term in S for $\mathcal{N}$ by the corresponding dual term in S^*.

In this section we can take advantage of the duality concept in discussing the parallel connections of resistors. In later chapters, we shall enlarge the set of dual terms as we learn more about circuit theory.

Exercise Show that the parallel connection of m linear resistors gives a linear resistor whose conductance G is equal to the sum of the conductances of the m linear resistors.

Example 1 (dual one-port and equivalent one-port) Consider the parallel connection of a linear resistor with conductance G and an independent current source i_s as shown in Fig. 2.12. The branch equations are

$$i_1 = Gv_1 \qquad \text{and} \qquad i_2 = i_s \tag{2.15}$$

Adding i_1 and i_2, and setting $v_1 = v$, we obtain

$$i = Gv + i_s \tag{2.16}$$

Table 2.1 Dual terms

S	S^*
Branch voltage	Branch current
Current-controlled resistor	Voltage-controlled resistor
Resistance	Conductance
Open circuit	Short circuit
Independent voltage source	Independent current source
Series connection	Parallel connection
KVL	KCL
Port voltage	Port current

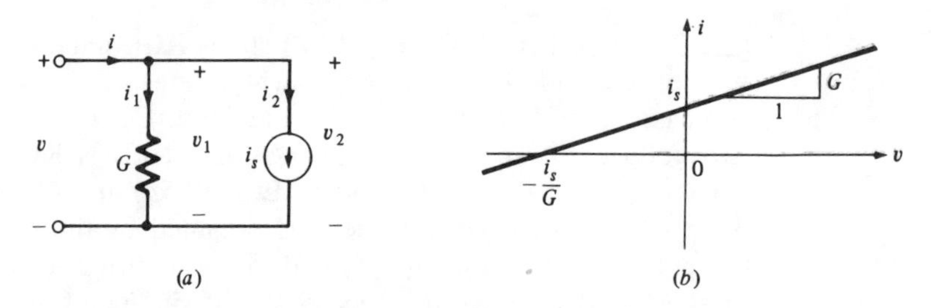

Figure 2.12 (*a*) Parallel connection of a linear resistor and an independent current source. (*b*) The driving-point characteristic of the resulting one-port.

The characteristic is shown in Fig. 2.12. Comparing Eq. (2.7) with Eq. (2.16) and Fig. 2.4 with Fig. 2.12, we know that the two circuits are dual to one another provided $i_s = E$ and $G = R$.

We also wish to use this example to illustrate the concept of *equivalent one-ports*. Recall two resistive one-ports are said to be *equivalent* if they have the same driving-point characteristics. Let $R' = 1/G$. Multiplying both sides of Eq. (2.16) by R' we obtain

$$R'i = v + R'i_s \tag{2.17}$$

Let $v_s' = R'i_s$; then the above equation can be written as

$$v = R'i - v_s' \tag{2.18}$$

This equation can be represented by a series connection of a linear resistor with resistance R' and an independent voltage source v_s' as shown in Fig. 2.13*a*. The driving-point characteristic is plotted in the *i-v* plane, which is shown in Fig. 2.13*b*. In order to compare it with the circuit in Fig. 2.12, we wish to plot the characteristic also in the *v-i* plane. This can be done by first drawing a straight line (dashed) passing through the origin with an angle 45° from the axis as shown in Fig. 2.13*c*. Next, taking the mirror image of the characteristic in Fig. 2.13*b* with respect to the 45° line, we obtain the characteristic in the *v-i* plane, which is exactly the same driving-point characteristic as that of Fig. 2.12. Therefore the two one-ports in Figs. 2.12 and 2.13 are equivalent. This particular equivalence turns out to be extremely important in circuit analysis. It allows us the flexibility of changing an independent voltage source in a circuit to an independent current source yet preserving the property of the circuit. We shall see later that the one-port in Fig. 2.13 is the Thévenin equivalent circuit and the one-port in Fig. 2.12 is the Norton equivalent circuit.

Example 2 (more on the ideal diode) Consider the parallel connection of a linear resistor, an independent current source, and an ideal diode as shown in Fig. 2.14. We wish to determine the driving-point characteristic of the one-port. Figure 2.15*a*, *b*, and *c* shows the branch characteristics. Note

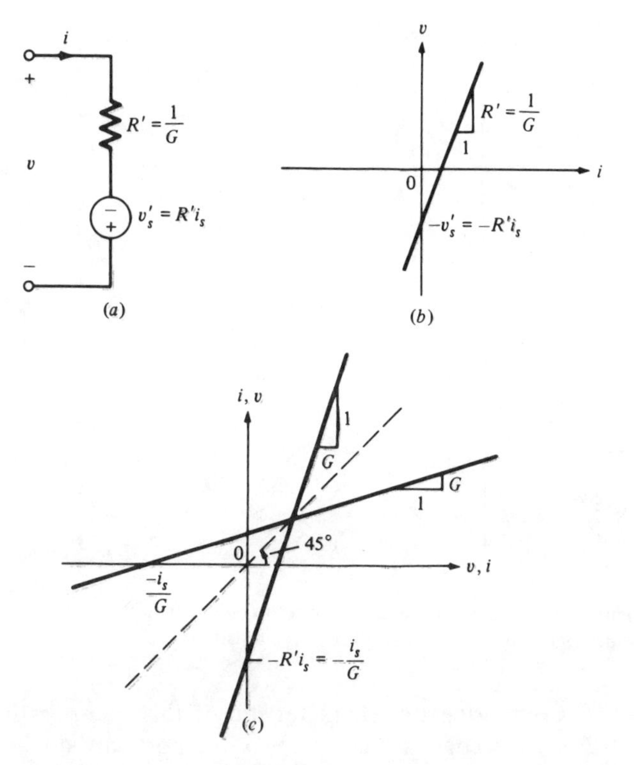

Figure 2.13 The one-port in (*a*) is equivalent to that of Fig. 2.12*a* since they have the same driving-point characteristics.

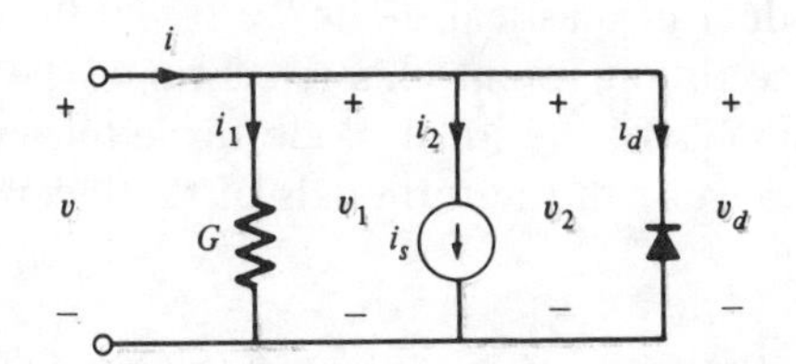

Figure 2.14 Parallel connection of a linear resistor, a current source, and an ideal diode.

that the characteristic of the ideal diode in Fig. 2.15*c* with the diode turned around is the mirror image with respect to the origin of that given in Fig. 1.8. In order to add the three branch currents we again consider the two individual segments of the ideal diode separately. Note that for $v > 0$, the summation of the current yields a half line with slope G, starting at the point $(0, i_s)$ as shown in Fig. 2.15*d*. For $v < 0$, we use a limiting process. First, consider that the ideal diode has a finite but very large slope G_d for the purpose of adding the three currents and observe that i_d dominates the other two currents. Then, we let $G_d \to \infty$, and the resulting characteristic becomes a half line on the i axis, which meets the other portion of the characteristic at $(0, i_s)$, as shown in Fig. 2.15*d*.

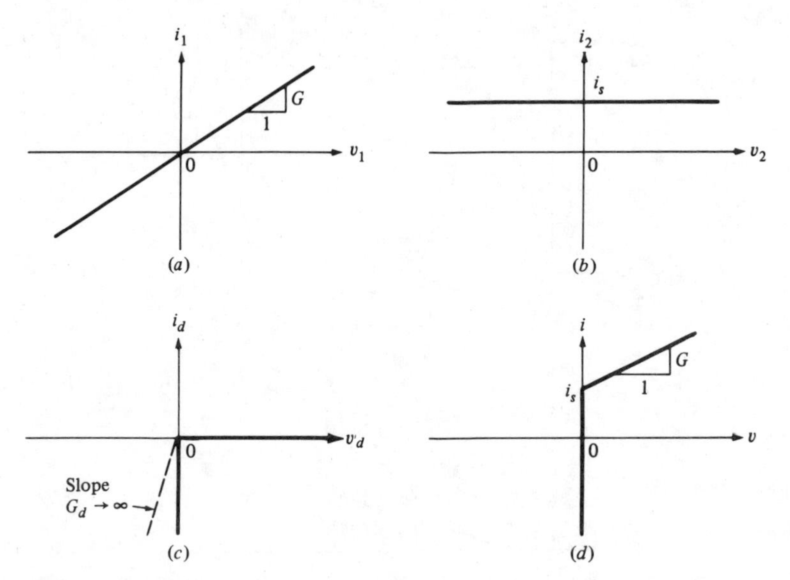

Figure 2.15 Branch v-i characteristics of (a) a linear resistor, (b) a current source, and (c) an ideal diode. (d) The resulting one-port characteristic.

Comparing the characteristic of this one-port with that shown in Fig. 2.6, we recognize that the two one-ports are dual to one another if $E = i_s$ and $R = G$.

Exercises

1. What is the dual of an ideal diode?
2. Determine the driving-point characteristic of the one-port in Fig. 2.14 with the terminals of the ideal diode turned around.
3. Repeat the above with the terminals of the independent current source turned around.

Example 3 (parallel connection of current sources) In Fig. 2.16 there are m independent current sources connected in parallel. This is the dual of m independent voltage sources connected in series. By KCL, it is clear that the parallel connection gives an equivalent one-port which is an independent current source whose current is $i_s = \Sigma_{j=1}^{m} i_j$.

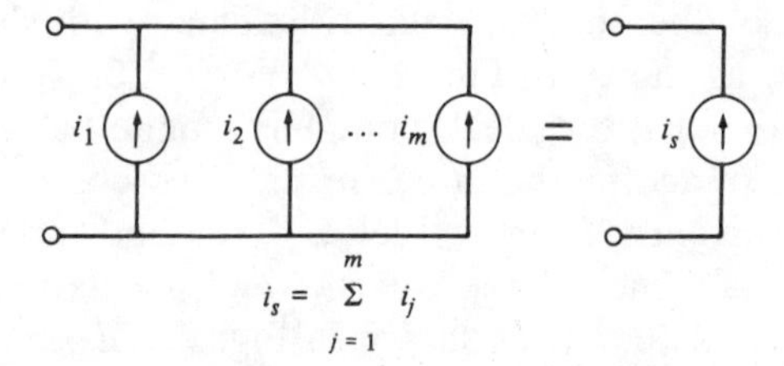

Figure 2.16 m Independent current sources in parallel.

Example 4 (parallel connection of voltage sources) The parallel connection of independent voltage sources violates KVL with the exception of the trivial case where all voltage sources are equal.

Summary of parallel connection The key concepts used in obtaining the driving-point characteristic of a one-port formed by the parallel connection of two-terminal resistors are

1. KVL forces all branch voltages to be equal.
2. KCL requires the port current i to be equal to the sum of the branch currents of the resistors.
3. If each resistor is voltage-controlled, the resulting driving-point characteristic of the one-port is also a voltage-controlled resistor.

2.3 Series-Parallel Connection of Resistors

We now extend the ideas introduced in the last two sections to series-parallel connections of two-terminal resistors. Let us give three examples to illustrate the method of analysis.

Example 1 (series-parallel connection of linear resistors) Consider a *ladder circuit*, i.e., a circuit formed by alternate series and parallel connection of two-terminal circuit elements, such as the one shown in Fig. 2.17. This ladder is made of four linear resistors with resistances, R_1, R_2, R_3, and R_4, respectively. The branch currents and branch voltages are indicated on the figure, and we wish to determine the driving-point characteristic of the one-port defined at terminals ① and ①'. We shall proceed to solve this problem from the back end of the ladder. Since the series connection forces $i_3 = i_4$, we can express the voltage v_2 by adding the voltages:

$$v_2 = v_3 + v_4 = (R_3 + R_4)i_3 \tag{2.19}$$

This equation specifies the characteristic of an equivalent linear resistor with resistance $R_3 + R_4$. We next consider the parallel connection of the resistor with resistance R_2 and the equivalent resistor just obtained. For a parallel connection, we add the currents; using Eq. (2.19), we obtain

$$i_1 = i_2 + i_3 = G_2 v_2 + \frac{1}{R_3 + R_4} v_2 = \left(G_2 + \frac{1}{R_3 + R_4}\right) v_2 \tag{2.20}$$

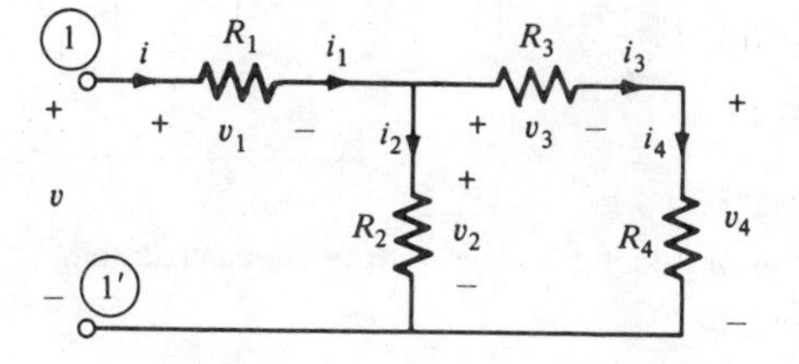

Figure 2.17 A ladder circuit with linear resistors.

where $G_2 = 1/R_2$. We next consider the series connection of the resistor with resistance R_1 and the equivalent resistor specified by Eq. (2.20). Adding the voltages v_1 and v_2, and using Eq. (2.20), we obtain

$$v = v_1 + v_2 = R_1 i_1 + \frac{1}{G_2 + 1/(R_3 + R_4)} i_1 \tag{2.21}$$

Since $i_1 = i$, we conclude that the driving-point characteristic of the one-port at ① and ①' is given by

$$v = Ri$$

where

$$R \triangleq R_1 + \frac{1}{G_2 + 1/(R_3 + R_4)} \tag{2.22}$$

is the resistance of the equivalent linear resistor.

Exercise Determine the resistance R of the one-port shown in Fig. 2.18.

Example 2 (series-parallel connection of nonlinear resistors) Consider the circuit in Fig. 2.19, where $\mathcal{R}_1$ is connected in series with the parallel connection of $\mathcal{R}_2$ and $\mathcal{R}_3$. All three resistors are nonlinear, and the problem is to determine the equivalent one-port resistor, $\mathcal{R}$. We will use the method of successive reduction from the back of the ladder as illustrated in the figure. Thus $\mathcal{R}^*$ is equivalent to the parallel connection of $\mathcal{R}_2$ and $\mathcal{R}_3$, and $\mathcal{R}$ is equivalent to the series connection of $\mathcal{R}_1$ and $\mathcal{R}^*$.

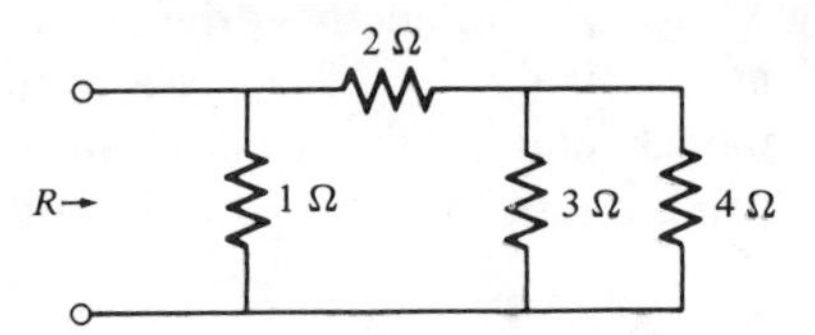

Figure 2.18 A ladder with four linear resistors.

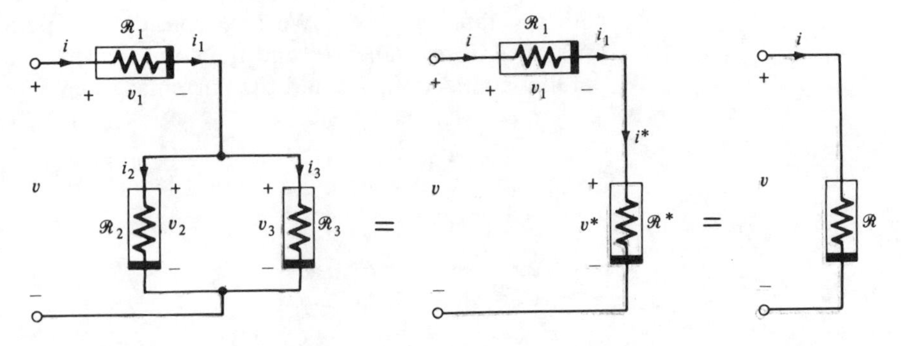

Figure 2.19 Reduction of a ladder circuit with nonlinear resistors.

We assume that the characteristics of $\mathscr{R}_2$ and $\mathscr{R}_3$ are voltage-controlled and specified by

$$i_2 = \hat{i}_2(v_2) \qquad \text{and} \qquad i_3 = \hat{i}_3(v_3)$$

The parallel connection has an equivalent resistor $\mathscr{R}^*$ which is voltage-controlled and specified by

$$i^* = g(v^*) \tag{2.23a}$$

where i^* and v^* are the branch current and branch voltage, respectively, of the resistor $\mathscr{R}^*$. The parallel connection requires $v^* = v_2 = v_3$ and $i^* = i_2 + i_3$. Therefore we have the characteristic of $\mathscr{R}^*$ which is related to that of $\mathscr{R}_2$ and $\mathscr{R}_3$ by

$$g(v^*) = \hat{i}_2(v^*) + \hat{i}_3(v^*) \qquad \text{for all } v^* \tag{2.23b}$$

The next step is to obtain the series connection of $\mathscr{R}_1$ and $\mathscr{R}^*$. Let us assume that the characteristic of $\mathscr{R}_1$ is current-controlled and specified by

$$v_1 = \hat{v}_1(i_1)$$

In order to proceed with the series connection of $\mathscr{R}_1$ and $\mathscr{R}^*$ we must also express $\mathscr{R}^*$ as a current-controlled resistor. This calls for finding just the inverse of the function $g(\cdot)$ in Eqs. (2.23*a*) and (2.23*b*), which is given by

$$v^* = g^{-1}(i^*) \qquad \text{for all } i^*$$

The series connection of $\mathscr{R}^*$ and $\mathscr{R}_1$ requires $i = i_1 = i^*$ and $v = v_1 + v^*$. Thus we obtain the characteristic of $\mathscr{R}$:

$$v = \hat{v}(i) \tag{2.24a}$$

where
$$\hat{v}(i) = \hat{v}_1(i) + g^{-1}(i) \qquad \text{for all } i \tag{2.24b}$$

In this problem, one key step is the determination of the inverse function $g^{-1}(\cdot)$. The question is therefore whether the inverse exists. If it does not, the characteristic of $\mathscr{R}$ cannot be written as in Eq. (2.24) because it is not current-controlled. One simple criterion which guarantees the existence of the inverse is that the v-i characteristic is *strictly monotonically increasing*, i.e., the slope, $g'(v^*)$ is positive for all v^*

REMARK The characteristic of the one-port shown in Fig. 2.19 can always be represented parametrically. Indeed, we have

$$i = i_1 = i^* \qquad \text{and} \qquad v = v_1 + v^*$$

Hence, using v^* as a parameter, we obtain

$$i = g(v^*)$$
$$v = \hat{v}_1[g(v^*)] + v^*$$

Example 3 (zener diode circuit) Consider the circuit shown in Fig. 2.20 where two zener diodes are connected back to back. The symbol of the zener diode and its characteristic are shown in Fig. 2.21, where E_z is called the *breakdown voltage*. The breakdown voltage depends on the doping impurity. For a heavily doped diode, E_z is in the neighborhood of 5 to 6 V.

For this circuit, we will use the approximate characteristic shown in Fig. 2.22*a*. The characteristic of the second diode with the terminals turned around is shown in Fig. 2.22*b*. We add the voltages to get the characteristic of the back-to-back connection of the diodes as shown in Fig. 2.22*c*. The next step is to consider the parallel connection of the linear resistor with negative conductance G, whose characteristic is shown in Fig. 2.22*d*. The resulting curve shown in Fig. 2.22*e* is the characteristic of the one-port. This characteristic is obtained by adding the currents i_a and i_b in Fig. 2.22*c* and *d*, respectively. The vertical portions of the characteristic need some explanation. When we consider the vertical portions of the characteristic in Fig. 2.22*c*, we may use the limiting process by first assuming that they each have a positive slope G_1 and let $G_1 \to \infty$. Thus adding the current $i_b = -Gv$, we note that $G_1 - G$ also approaches ∞. Therefore we have the two vertical portions of the characteristic in Fig. 2.22*e*.

Exercise Consider the one-port shown in Fig. 2.23*a* which consists of two tunnel diodes and two dc voltage sources. The v-i characteristic of the tunnel diode is given in Fig. 2.23*b*. Determine the driving-point characteristic of the one-port in Fig. 2.23*a*.

3 PIECEWISE-LINEAR TECHNIQUES

The method of piecewise-linear analysis is extremely useful in studying circuits with nonlinear resistors. We have seen in the last section that nonlinear v-i characteristics can be approximated by linear segments. This allows us to make simple calculations. In this section we will present some useful techniques which will be used later in piecewise-linear analysis. But, first we will introduce two ideal piecewise-linear models which are used as building blocks.

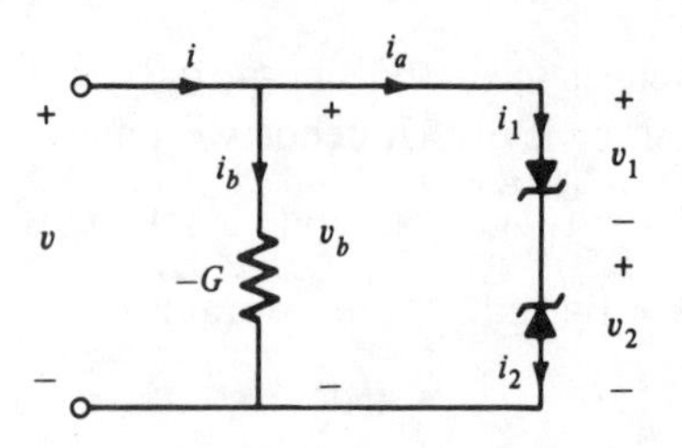

Figure 2.20 A series-parallel circuit with two back-to-back zener diodes.

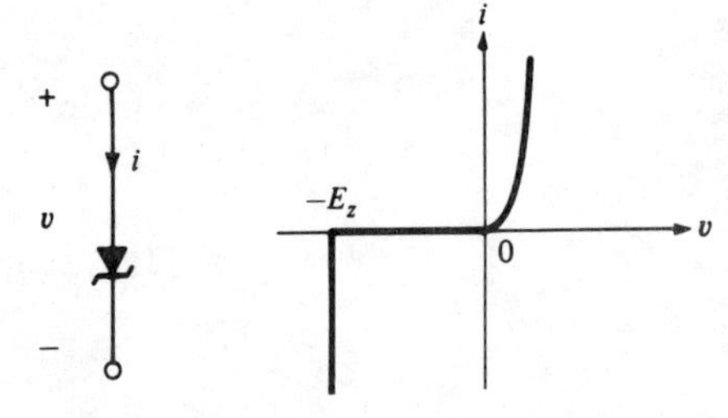

Figure 2.21 Symbol for a zener diode and its characteristic.

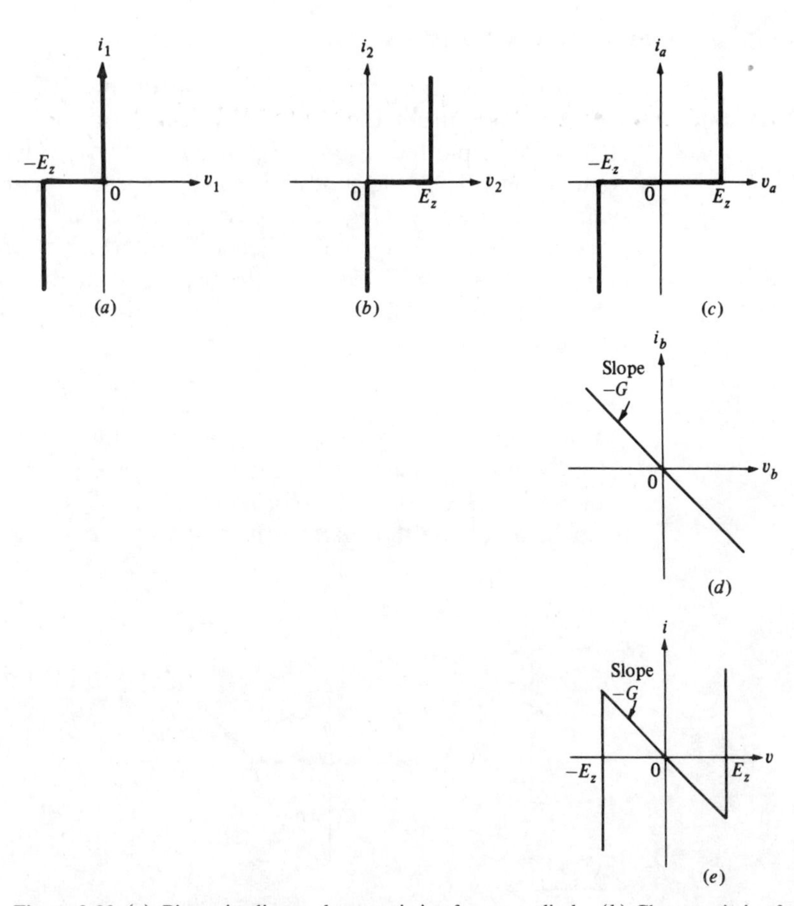

Figure 2.22 (*a*) Piecewise-linear characteristic of a zener diode. (*b*) Characteristic of a zener diode with its two terminals turned around, (*c*) characteristic of the series connection of the two, (*d*) characteristic of the linear resistor, and (*e*) the resulting driving-point characteristic of the one-port.

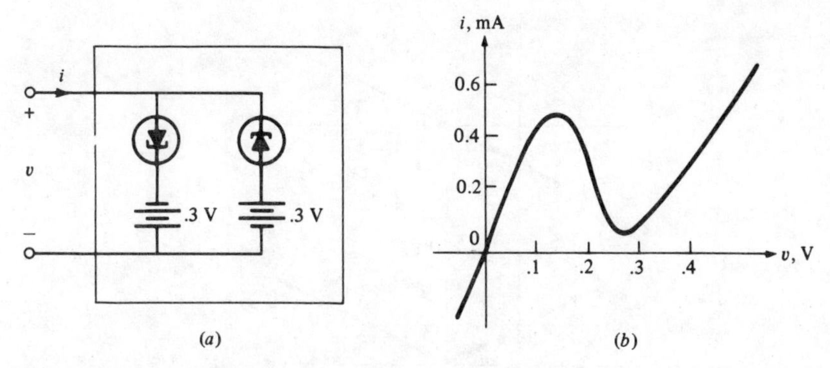

Figure 2.23 (*a*) A one-port composed of two tunnel diodes and two dc voltage sources. (*b*) *v-i* Characteristic of the tunnel diode.

3.1 The Concave and Convex Resistors

The usual convention for plotting characteristics of nonlinear resistors is to use the v-i plane instead of the i-v plane. For the circuit in Fig. 3.1*a*, the characteristic plotted in the v-i plane is shown in Fig. 3.1*b*. (See Fig. 2.6 for comparison.) The shape of the curve suggests a name: concave resistor. Its symbol is shown in Fig. 3.1*c*. Thus, the *concave resistor* in Fig. 3.1 is a piecewise-linear voltage-controlled resistor which is uniquely specified by two parameters: G, the slope of the linear segment, and E, the breakpoint voltage. In terms of a function representation, a concave resistor can be specified by

$$i = \tfrac{1}{2}G[|v - E| + (v - E)] \tag{3.1}$$

Let us demonstrate that Eq. (3.1) indeed leads to the v-i characteristic of Fig. 3.1*b*. We use two approaches. First, consider Fig. 3.2*a* where $i = (G/2)(v - E)$ is plotted. In Fig. 3.2*b*, we plot its absolute value, i.e., $(G/2)|v - E|$. Adding the two, we obtain the characteristic of the concave resistor shown in Fig. 3.1*b*.

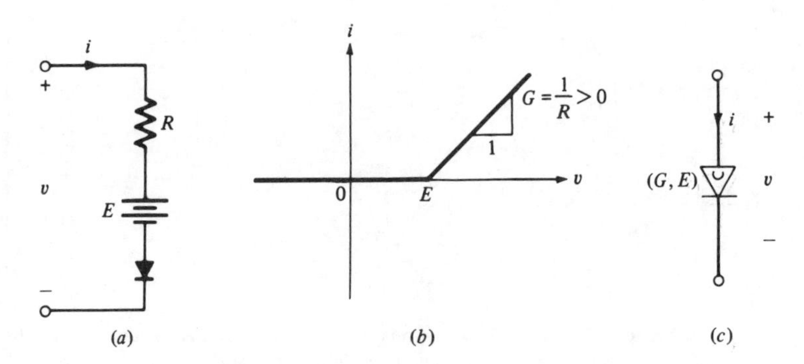

Figure 3.1 (*a*) Equivalent circuit, (*b*) characteristic, and (*c*) symbol for a concave resistor.

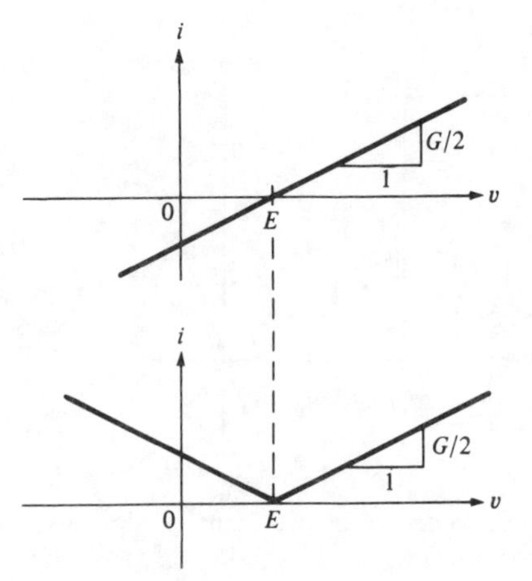

Figure 3.2 Graphic interpretation of Eq. (3.1): (*a*) $i = (G/2)(v - E)$ and (*b*) $i = (G/2)|v - E|$.

Second, we consider directly Eq. (3.1) which states

$$i = G(v - E) \qquad \text{for } v - E \geq 0 \tag{3.2a}$$

This is precisely the characteristic of the one-port in Fig. 3.1*a* when the diode is conducting. Now Eq. (3.1) also states

$$i = 0 \qquad \text{for } v - E < 0 \tag{3.2b}$$

This is the characteristic of the same one-port when the diode is reversed biased.

The circuit shown in Fig. 3.3*a* has the *v-i* characteristic shown in Fig. 3.3*b*. This characteristic defines a convex resistor whose symbol is shown in Fig. 3.3*c*. The *convex resistor* in Fig. 3.3 is a piecewise-linear *current-controlled resistor* which is uniquely specified by two parameters: $G = 1/R$, the slope of the linear segment, and I, the breakpoint current. Being current-controlled, it can be represented by

$$v = \tfrac{1}{2}R[|i - I| + (i - I)] \tag{3.3}$$

Clearly,

$$v = R(i - I) \qquad \text{for } i \geq I \tag{3.4a}$$

and

$$v = 0 \qquad \text{for } i < I \tag{3.4b}$$

Note that concave and convex resistors are dual elements provided $G = R$ and $E = I$.

Exercise Demonstrate that if the circuit in Fig. 3.3*a* has $G < 0$, the *v-i* characteristic of the one-port is given by that of Fig. 3.4*b* and not by that of Fig. 3.4*a*. Note that the characteristic in Fig. 3.4*b* is not current-controlled.

REMARK Concave and convex resistors realized by the circuits in Figs. 3.1*a* and 3.3*a* require that $G > 0$. However, as will be demonstrated in the next section, it is convenient to use concave and convex resistors in device modeling even with G negative. For example, if $G < 0$, the convex resistor

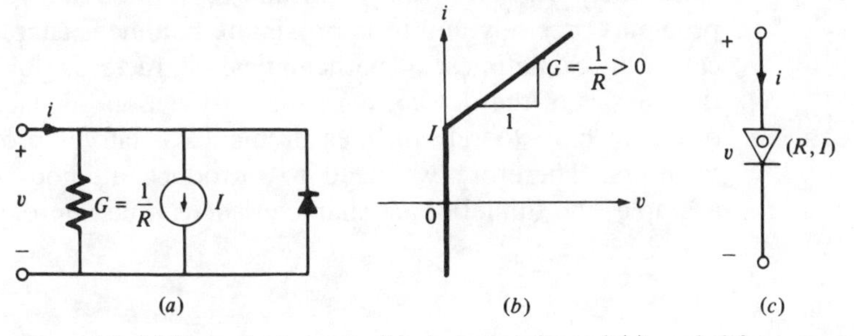

Figure 3.3 (*a*) Equivalent circuit, (*b*) characteristic, and (*c*) symbol for a convex resistor.

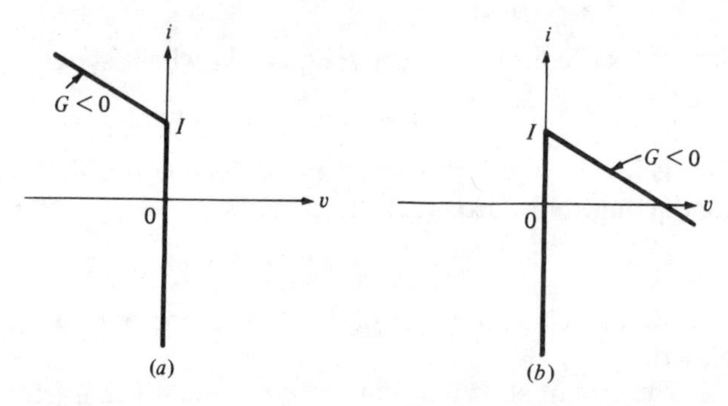

Figure 3.4 (*a*) The characteristic of a convex resistor with G negative. (*b*) The v-i characteristic of the one-port in Fig. 3.3*a* when G is negative.

is defined by the current-controlled characteristic shown in Fig. 3.4*a*. In this case, Eq. (3.3) is still valid but the circuit shown in Fig. 3.3*a* no longer gives the correct v-i characteristic. However, other circuits (using the operational amplifier) exist for realizing a concave or convex resistor having a negative G. For device-modeling purposes, concave and convex resistors are simple two-terminal resistors defined by Eqs. (3.1) and (3.3), respectively, where G may assume either positive or negative values. In such applications, since we do not intend to build the circuit model—models are used mainly for analysis purposes—the question of a physical circuit realization is irrelevant.

3.2 Approximation and Synthesis

We have used graphic methods to study the properties of nonlinear resistors. Graphic methods are useful not only for analyzing simple circuits but also for providing insight and understanding of nonlinear behavior of circuits. We will have further opportunities to illustrate various graphic techniques in this book.

When we deal with complex circuits, graphic methods are either too cumbersome or not applicable. In such cases, we need to rely on computers for calculation. Therefore it is important to develop analytic methods to formulate problems precisely and to approximate nonlinear characteristics in mathematical form. Sometimes, a mathematical characterization can be obtained from the physics of the device, e.g., the *pn*-junction law [see Eq. (1.6)]. However, often we have to rely on measurements or curves provided by device manufacturers. Therefore we need to introduce methods of approximation. For example, the tunnel-diode characteristic can be approximated by a polynomial

$$i \approx \sum_{k=0}^{n} a_k v^k$$

The subject of polynomial approximation or interpolation is well-developed, but any discussion of it would lead us astray. On the other hand, piecewise-linear approximation is useful in dealing with both simple and general circuits made up of nonlinear resistors. Furthermore, like many graphic methods, the piecewise-linear method helps us in understanding qualitatively the nonlinear behavior of circuits. In addition, it is straightforward and effective. We will devote this subsection to piecewise-linear approximation and also introduce the concept of circuit synthesis.

Example The piecewise-linear approximation of a tunnel-diode characteristic is shown in Fig. 3.5. The three linear segments have slopes

Region 1: $\quad G = G_a \quad$ for $v \le E_1 \quad$ (3.5*a*)

Region 2: $\quad G = G_b \quad$ for $E_1 < v \le E_2 \quad$ (3.5*b*)

Region 3: $\quad G = G_c \quad$ for $E_2 < v \quad$ (3.5*c*)

These define the three regions as shown in Fig. 3.5. Beginning from left to right, we can decompose the piecewise-linear characteristic into the sum of three components as shown in Fig. 3.6*a*: a straight line passing through the origin with slope G_0, a concave resistor characteristic which starts at E_1 with a negative slope G_1, and a concave resistor characteristic which starts at E_2 with a positive slope G_2. The corresponding circuit is shown in Fig. 3.6*b*. Adding the three branch currents, we have the driving-point characteristic of the one-port:

$$i = i_0 + i_1 + i_2 = \hat{i}(v) \tag{3.6}$$

To make it identical with the piecewise-linear characteristic of the tunnel

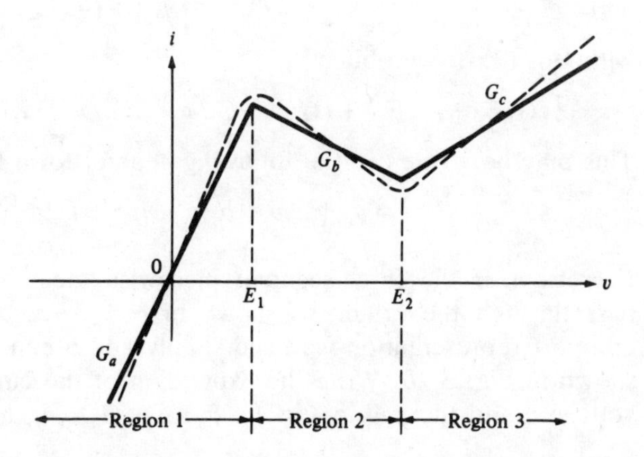

Figure 3.5 Piecewise-linear approximation of the tunnel-diode characteristic. The three-segment approximation defines the three regions indicated.

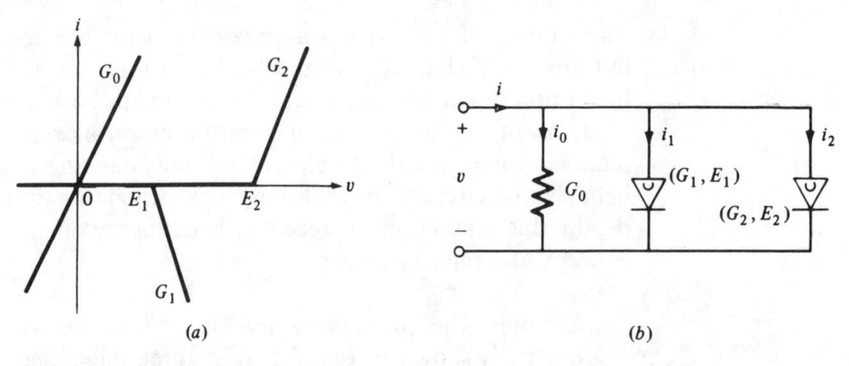

Figure 3.6 (*a*) Decomposition of the piecewise-linear characteristic into three components, and (*b*) the corresponding circuit which has these characteristics.

diode, the three parameters G_0, G_1, and G_2 in Fig. 3.6 must satisfy the following:

Region 1: $$G_0 = G_a \tag{3.7a}$$

Region 2: $$G_0 + G_1 = G_b \tag{3.7b}$$

Region 3: $$G_0 + G_1 + G_2 = G_c \tag{3.7c}$$

Thus $\quad G_0 = G_a \quad G_1 = -G_a + G_b \quad$ and $\quad G_2 = -G_b + G_c$

We may generalize the above by using the function representation of the concave resistor of Eq. (3.1).

Combining

$$i_0 = G_0 v$$

$$i_1 = \tfrac{1}{2}G_1[|v - E_1| + (v - E_1)]$$

and $$i_2 = \tfrac{1}{2}G_2[|v - E_2| + (v - E_2)]$$

with Eq. (3.6), we obtain

$$i = -\tfrac{1}{2}(G_1E_1 + G_2E_2) + (G_0 + \tfrac{1}{2}G_1 + \tfrac{1}{2}G_2)v + \tfrac{1}{2}G_1|v - E_1| + \tfrac{1}{2}G_2|v - E_2|$$

This may be written in the following general form for future use:

$$i = a_0 + a_1 v + b_1|v - E_1| + b_2|v - E_2| \tag{3.8}$$

Exercise 1 If the three-segment piecewise-linear characteristic does not pass through the origin, as shown in Fig. 3.7*a*, then in the equivalent one-port representation we need simply add a constant current source as shown in Fig. 3.7*b*. Write the expression of the current i in terms of the voltage v and the parameters I_0, E_1, E_2, G_a, G_b, and G_c.

Exercise 2 Extending the idea introduced in the preceding example, show that every voltage-controlled $(n + 1)$-segment piecewise-linear characteris-

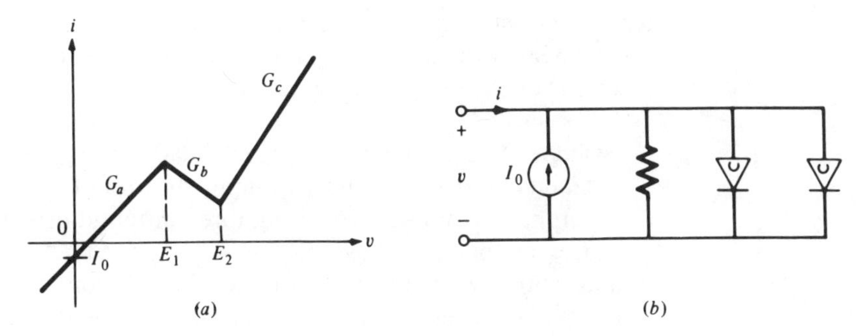

Figure 3.7 A three-segment piecewise-linear characteristic and its corresponding circuit realization.

tic can be represented by

$$i = \hat{i}(v) = a_0 + a_1 v + \sum_{j=1}^{n} b_j |v - E_j| \tag{3.9}$$

where $E_1 < E_2 < E_3 \cdots < E_n$ are the breakpoint voltages, and

$$a_1 = \tfrac{1}{2}(m_0 + m_n)$$
$$b_j = \tfrac{1}{2}(m_j - m_{j-1})$$
$$a_0 = \hat{i}(0) - \sum_{j=1}^{n} b_j |E_j|$$

where m_0 is the slope of the first linear segment from the left and m_j is the slope of the $(j+1)$th linear segment. Here, the segments are labeled consecutively from left to right, starting from zero.

REMARKS
1. The above development can be duplicated for current-controlled resistors by applying the concept of duality.
2. From the above we may state that any v-controlled piecewise-linear characteristic can be synthesized by a current source, a linear resistor, and concave resistors in parallel, and any i-controlled piecewise-linear characteristic can be synthesized by a voltage source, a linear resistor, and convex resistors in series.

4 DC OPERATING POINTS

Given any problem, one is interested in determining a solution. For some problems there exists a *unique* solution. This is the case of a circuit containing two-terminal linear passive resistors and an independent current source connected to any two nodes of the circuit serving as input. For other problems there may exist a unique solution, multiple solutions, or even no solution at all. This happens with circuits containing nonlinear resistors. Let us illustrate these possibilities with the following example:

Example 1 Consider a one-port N made of two-terminal resistors and an independent current source as input shown in Fig. 4.1*a*. Let us study three different situations representing three different driving-point characteristics of the one-port, as shown in Fig. 4.1*b*, *c*, and *d*. Let the current source be a dc source with current $i_s = I$. Let the solution be the port voltage v. The first case corresponds to a linear resistor with a positive resistance R.

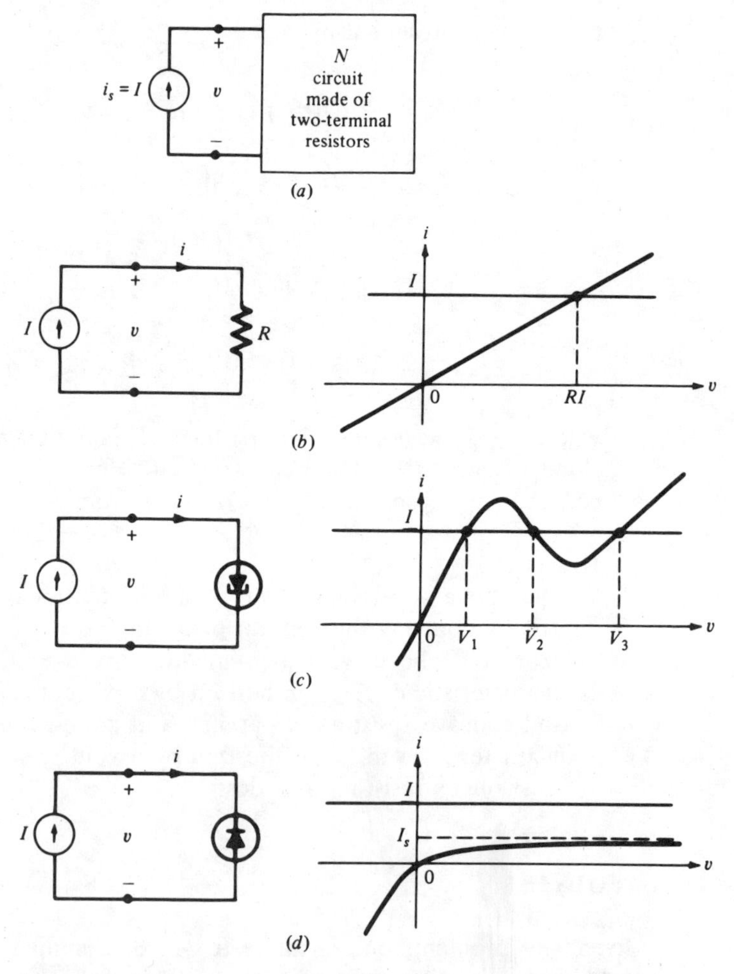

Figure 4.1 (*a*) A one-port N driven by a dc source. (*b*) With a linear resistor in N, there exists a unique solution. (*c*) With a tunnel-diode characteristic, the circuit has three solutions. (*d*) With a *pn*-junction-diode characteristic, the circuit has no solution if $I > I_s$.

Obviously, the solution is $v = RI$ as illustrated by the intersection of the characteristic of the linear resistor and that of the current source plotted in the v-i plane, shown in Fig. 4.1*b*. In the second case, the one-port has the characteristic of a tunnel diode. In Fig. 4.1*c*, we see that there exist three solutions at $v = V_1$, V_2, and V_3. In the third case, the one-port is a *pn*-junction diode connected backwards. If we ignore the breakdown phenomenon and model the diode by Eq. (1.6), the current saturates as v approaches ∞. Since I_s, the saturation current, is less than I, the current of the dc current source, there exists no solution.

Linear circuits and nonlinear circuits We will repeatedly see in this book that circuits containing linear resistors (or linear circuit elements[4] in general) behave quite differently in many respects from those containing nonlinear resistors (or nonlinear circuit elements in general). We shall call a circuit containing linear circuit elements and independent sources a *linear circuit*. The role played by independent sources is important to understand. They are considered as *inputs* to the circuit. Thus the solution is considered as a *response* to these inputs. A circuit is called *nonlinear* iff it is *not linear*.

DC analysis The solutions to a circuit with dc input are called *operating points*. The term *dc analysis* refers to the determination of operating points. It will be shown later that dc analysis of a general *RLC* circuit is equivalent to finding solutions of a *resistive circuit* which can be simply derived from the given circuit. The subject is of major importance in circuit theory and electronics. There exist a number of effective computer programs for general dc analysis. In this section we will consider dc analysis for simple circuits. We will use both the graphic method and piecewise-linear analysis to determine the operating points. An important concept to be introduced is the use of a load line in analyzing or designing a biasing circuit.

The basic concepts of dc analysis can be illustrated with the simple circuit configuration shown in Fig. 4.2, i.e., the back-to-back connection of two one-ports at nodes ① and ①'. What is interesting to note is that this simple configuration, because it includes two unspecified one-ports, covers circuits with great generality. We assume that each one-port is specified by the following driving-point characteristics in terms of its port voltage and port current, v_a, i_a and v_b, i_b, respectively,

$$f_a(v_a, i_a) = 0 \qquad \text{and} \qquad f_b(v_b, i_b) = 0 \tag{4.1}$$

These are generalizations of the branch characteristics since each one-port is formed by an interconnection of resistors. We are not concerned with what is inside of the one-ports N_a and N_b. Therefore we only need to write two

[4] We have so far only defined two-terminal linear resistors. Other linear circuit elements such as linear capacitors, inductors, and multiterminal resistors will be introduced later.

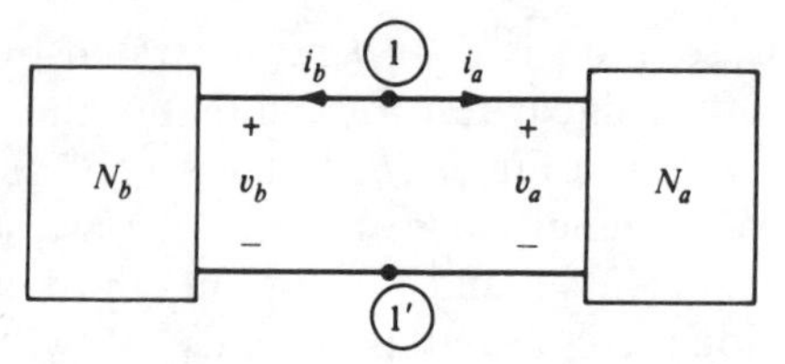

Figure 4.2 Two resistive one-ports connected in parallel.

equations from Kirchhoff's laws to describe the port interconnection at the two nodes ① and ①'. KCL states

$$i_a = -i_b \tag{4.2}$$

and KVL states

$$v_a = v_b \tag{4.3}$$

Therefore we can eliminate one set of voltage and current by combining Eqs. (4.1), (4.2), and (4.3). Let us denote

$$i_a = -i_b \stackrel{\Delta}{=} i \qquad \text{and} \qquad v_a = v_b \stackrel{\Delta}{=} v \tag{4.4}$$

The two resulting equations in terms of v and i are

$$f_a(v, i) = 0 \qquad \text{and} \qquad f_b(v, -i) = 0 \tag{4.5}$$

The solutions of the two equations are the operating points we are looking for. We will give a number of examples to illustrate the analytic, graphic, numerical, and piecewise-linear methods.

Example 2 (analytic approach) Let N_a be a nonlinear voltage-controlled resistor which is specified by

$$f_a(v_a, i_a) = i_a - 4v_a^2 = 0 \tag{4.6}$$

and N_b be a series connection of a dc voltage source and a linear resistor, which can be used to model a real battery connected in series with a resistive load. The specification is

$$f_b(v_b, i_b) = v_b - E_b - R_b i_b = 0 \tag{4.7}$$

The circuit and the v-i characteristics of the two 1-ports are shown in Fig. 4.3. As before we let $i_a = -i_b \stackrel{\Delta}{=} i$ and $v_a = v_b \stackrel{\Delta}{=} v$. Equations (4.6) and (4.7) become

$$i = 4v^2 \tag{4.8a}$$

and

$$v = E_b - R_b i \tag{4.8b}$$

These two equations lead to a quadratic equation in terms of v:

$$4R_b v^2 + v - E_b = 0 \tag{4.9}$$

For $E_b = 2$ V and $R_b = \frac{1}{4}\ \Omega$, the two solutions are $v = 1$ and -2. Thus from

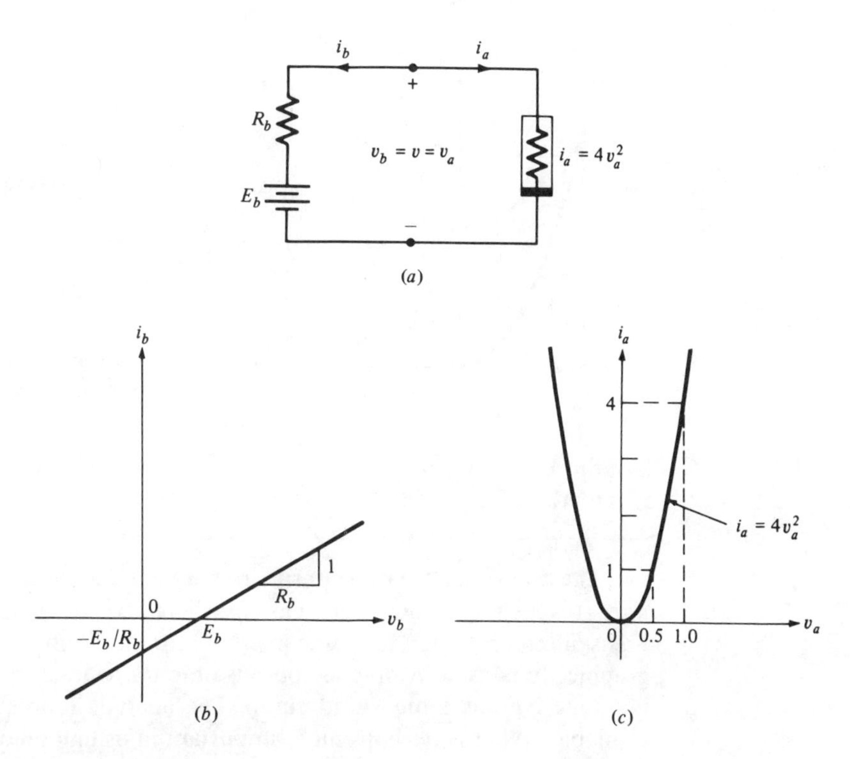

Figure 4.3 Circuit of Fig. 4.2 with given characteristics for N_a and N_b.

Eq. (4.8*a*), $i = 4$ and 16, respectively. Therefore the two operating points are

$$v_1 = 1 \qquad i_1 = 4 \tag{4.10a}$$

and

$$v_2 = -2 \qquad i_2 = 16 \tag{4.10b}$$

In practice one rarely encounters problems in nonlinear circuits which can be solved analytically. Let us use the same example to go through some key ideas of the other methods.

Example 3 (graphic method: load line) The circuit in Fig. 4.3 represents a typical *biasing circuit* in dc design, i.e., a simple nonlinear circuit which consists of a battery, a resistor, and an electronic device modeled by a nonlinear resistor with a specified v-i characteristic. The way to find the solution is to transcribe the characteristic of the battery and the resistor in the v_b-i_b plane to the v_a-i_a plane where the characteristic of the device is plotted. Since $v_b = v_a$ and $i_b = -i_a$, the transcribed curve is the mirror image with respect to the v axis of the curve in the v_b-i_b plane. This is superimposed with the characteristic of the nonlinear one-port N_a as shown in Fig. 4.4. There are two intersections of the two curves, and these give

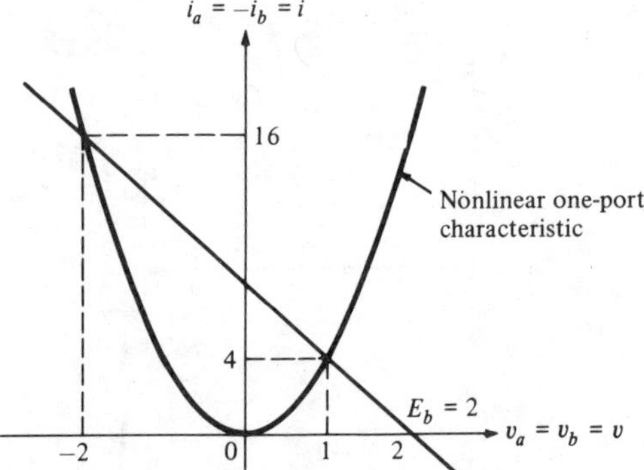

Figure 4.4 The two one-port characteristics are superimposed on the v-i plane.

the operating points: $(1, 4)$ and $(-2, 16)$, which check with the analytic method in Eq. (4.10).

The transcribed battery-resistor characteristic in Fig. 4.4 is called the *load line*. It is a straight line which has E_b as its v-axis intercept and has E_b/R_b as its i-axis intercept. The load-line method for determining the operating points graphically is used in practice because the v-i characteristic of the one-port N_a is often given as a measured curve. The analytic representation is usually not available. What is perhaps more important in using the load-line method is that we can visualize the effects of the dc source E_b and the load resistance R_b on the operating points, thereby helping us in the design of a biasing circuit. The subject of load lines will be encountered again in the next chapter, but in a more complex environment.

Exercise If the linear resistor in Fig. 4.3a is replaced by a pn-junction diode, draw the *nonlinear* load line and use the graphic method to find the operating points.

The graphic method illustrated above can only yield approximate solutions. An analytic approach can rarely be used in real circuits. In order to obtain accurate solutions and especially for general resistive circuits, we need to use numerical methods. The Newton-Raphson method is the most commonly used numerical method for finding dc operating points. The details will be given in Chap. 5. In the following we will give the essence of the method and illustrate it with a simple example.

The Newton-Raphson method is an iterative method for solving nonlinear equations of the form

$$f(x) = 0 \tag{4.11}$$

Let the solution be x^*, i.e., $f(x^*) = 0$. The Newton-Raphson iteration formula states

$$x_n = x_{n-1} - \frac{f(x_{n-1})}{f'(x_{n-1})} \tag{4.12}$$

Thus, starting with an initial guess x_0, one computes x_1 from Eq. (4.12). We use Eq. (4.12) successively until x_n converges to the solution x^*. The idea can best be illustrated with the following example.

Example 4 (Newton-Raphson method) Let us consider the same circuit as in Examples 2 and 3. The pertinent equation to be solved is given by Eq. (4.9) with $R_b = \frac{1}{4}\ \Omega$ and $E_b = 2$ V, thus

$$f(v) = v^2 + v - 2 \tag{4.13}$$

The equation has two solutions at $v = 1$ and $v = -2$. In order to use Eq. (4.12), we need first to obtain the derivative:

$$f'(v) = 2v + 1 \tag{4.14}$$

We next choose arbitrarily an initial guess, say, $v_0 = 0$. From Eqs. (4.13) and (4.14), we have $f(v_0) = -2$ and $f'(v_0) = 1$. Therefore, from Eq. (4.12), we obtain

$$v_1 = v_0 - \frac{f(v_0)}{f'(v_0)} = 2$$

We continue this procedure by evaluating $f(v_1) = 4$ and $f'(v_1) = 5$ to compute v_2:

$$v_2 = v_1 - \frac{f(v_1)}{f'(v_1)} = 1.2$$

The next two steps yield

$$v_3 = 1.012 \qquad \text{and} \qquad v_4 = 1$$

which is one of the solutions.

Exercise Let $f(v) = av + b$. Demonstrate that the Newton-Raphson method leads to the exact solution in *one* step regardless of the initial guess, v_0.

The Newton-Raphson method has a simple geometric interpretation. This is shown in Fig. 4.5.

The iteration formula in Eq. (4.12) comes from the Taylor series expansion:

$$f(x_n) = f(x_{n-1}) + f'(x_{n-1})(x_n - x_{n-1}) + \text{h.o.t.} \tag{4.15}$$

where h.o.t. designates higher-order terms. Thus if we neglect the higher-order

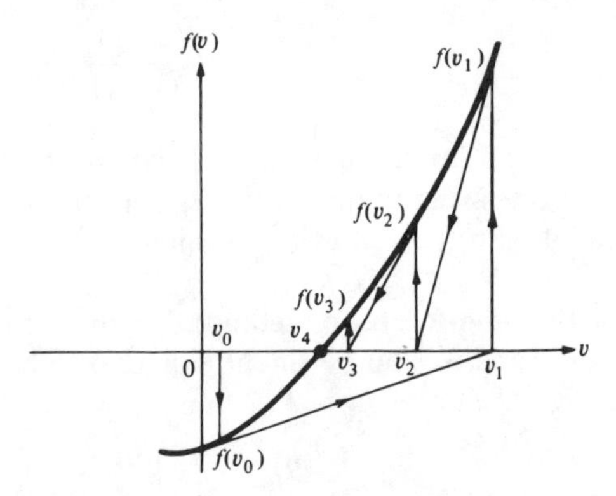

Figure 4.5 Illustrating the Newton-Raphson method of finding the solution of $f(v) = 0$.

terms and if $x_n \approx x^*$, then $f(x_n) \approx 0$; we obtain Eq. (4.12). We should caution the readers that the method works only when the initial guess x_0 is sufficiently near an actual solution and the function is well-behaved. Take the situation shown in Fig. 4.6 where the solution is at x^*. If we choose an initial guess x_0 as shown, using the Newton-Raphson method we find that the process will not converge.

The other numerical method which is very useful in solving nonlinear resistive circuits is the piecewise-linear method. This method also gives considerable physical insight. We will give two examples to illustrate the essence of the method.

Example 5 (piecewise-linear method) Consider the same simple circuit configuration shown in Fig. 4.2, where the one-port N_b is the same battery-resistor combination but the one-port N_a has the characteristic of a tunnel diode as shown in Fig. 4.7. We assume that it is given by a three-segment piecewise-linear specification of Eq. (3.8):

$$i = a_0 + a_1 v + b_1|v - E_1| + b_2|v - E_2| \tag{4.16}$$

Let us further give the parameters as follows: $a_0 = -\frac{1}{2}$, $a_1 = 2$, $b_1 = -\frac{5}{2}$,

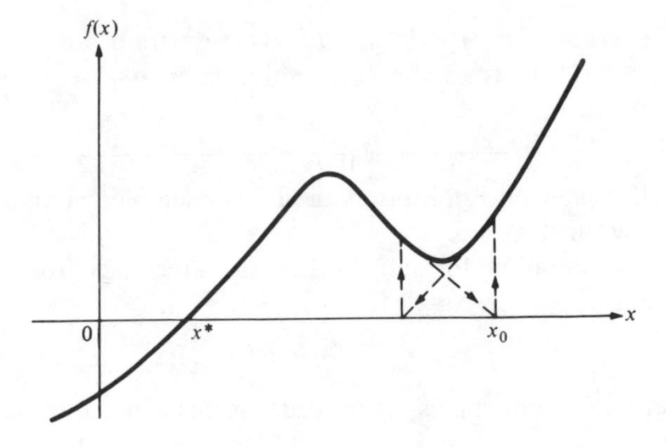

Figure 4.6 The Newton-Raphson method does not converge with x_0 shown as the initial guess.

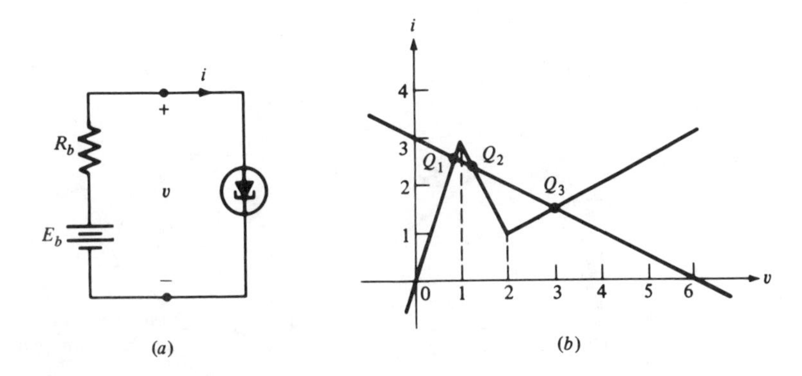

Figure 4.7 Operating points of a tunnel-diode circuit determined by the piecewise-linear method.

$b_2 = \frac{3}{2}$, $E_1 = 1$, and $E_2 = 2$. Let the battery-resistor characteristic be given by $E_b = 6$ and $R_b = 2$. The superimposed curves in the v-i plane are shown in Fig. 4.7*b*. Thus we know that the three operating points are at the three intersections Q_1, Q_2, and Q_3. However, for the present, we wish to determine them analytically by using Eq. (4.16).

As we have shown in Fig. 3.5 and Eq. (3.5), the v axis can be divided into three regions:

Region 1: $$v \le E_1 = 1 \tag{4.17a}$$

Region 2: $$1 < v \le E_2 = 2 \tag{4.17b}$$

Region 3: $$2 < v \tag{4.17c}$$

In the three regions, Eq. (4.16) can be replaced by equations without absolute value signs as follows:

Region 1: $$i = a_0 + a_1 v - b_1(v - E_1) - b_2(v - E_2) \tag{4.18a}$$

Region 2: $$i = a_0 + a_1 v + b_1(v - E_1) - b_2(v - E_2) \tag{4.18b}$$

Region 3: $$i = a_0 + a_1 v + b_1(v - E_1) + b_2(v - E_2) \tag{4.18c}$$

For the battery-resistor combination, the equation is

$$v = E_b - R_b i = 6 - 2i \tag{4.19}$$

First, solving Eqs. (4.18*a*) and (4.19) for the solution in region 1, we obtain $V_{Q1} = \frac{6}{7}$. Similarly, solving Eqs. (4.18*b*) and (4.19) for the solution in region 2, we obtain $V_{Q2} = \frac{4}{3}$. Finally, solving Eqs. (4.18*c*) and (4.19) for the solution in region 3, we obtain $V_{Q3} = \frac{8}{3}$.

It is crucial to remember that we must check these calculated solutions to see whether they fall in the assumed regions. If they indeed do, they are the *valid* solutions, otherwise they are called *virtual solutions*: They do not correspond to reality; they are artifacts of the method. Of course, only the

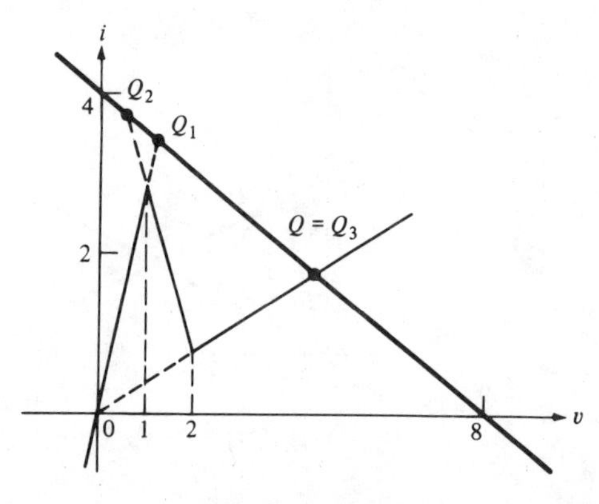

Figure 4.8 The load line has been shifted upward from that of Fig. 4.7 ($E_b = 8$, $R_b = 2$). There exists only one operating point Q.

valid solutions are the true solutions. In the present case, we see that $V_{Q1} = \frac{6}{7}$ falls in region 1 ($v_1 \le 1$), $V_{Q2} = \frac{4}{3}$ falls in region 2 ($1 < v \le 2$), and $V_{Q3} = \frac{8}{3}$ falls in region 3 ($2 < v$). Therefore, they are valid solutions. The corresponding currents for these voltages can be obtained from Eq. (4.19). They are $I_{Q1} = \frac{18}{7}$, $I_{Q2} = \frac{7}{3}$, and $I_{Q3} = \frac{5}{3}$. Therefore the three operating points Q_1, Q_2, and Q_3 in terms of the coordinates of the v-i plane are $(\frac{6}{7}, \frac{18}{7})$, $(\frac{4}{3}, \frac{7}{3})$, and $(\frac{8}{3}, \frac{5}{3})$, respectively.

Example 6 (piecewise-linear method) We next consider the same circuit as in Example 5 with the exception that the load line has been shifted upward with $E_b = 8$ and $R_b = 2$. This is shown in Fig. 4.8, where, as we see, there is only one operating point Q.

We now go through the calculation as before and determine the solution analytically. Equation (4.19) becomes

$$v = E_b - R_b i = 8 - 2i \tag{4.20}$$

Combining Eq. (4.20) with Eqs. (4.18*a*, *b*, and *c*), respectively, we obtain $V_{Q1} = \frac{8}{7}$, $V_{Q2} = \frac{2}{3}$, and $V_{Q3} = \frac{10}{3}$. Next we check the regions and note that $V_{Q1} = \frac{8}{7}$ which falls outside of region 1 ($v_1 \le 1$), $V_{Q2} = \frac{2}{3}$ which falls outside of region 2 ($1 < v_1 \le 2$), and $V_{Q3} = \frac{10}{3}$ which falls inside region 3 ($2 < v_1$). Therefore V_{Q1} and V_{Q2} are virtual solutions, and only V_{Q3} is a valid solution as shown in Fig. 4.8. The current corresponding to V_{Q3} is $I_{Q3} = \frac{7}{3}$. Hence, the only operating point is Q at $(\frac{10}{3}, \frac{7}{3})$ in the v-i plane.

The treatment of dc analysis for general circuits will be given in later chapters.

5 SMALL-SIGNAL ANALYSIS

There is a good reason to call the solutions to dc analysis "operating points." When a circuit is used, some input signal (e.g., a sinusoidal waveform) is applied to it so that we get a useful output. An operating point specifies a region in the v-i plane in the neighborhood of which the actual voltage and current in the circuit vary as a function of time. If the applied signal has a sufficiently small voltage or current (in magnitude), the circuit can be analyzed to a good approximation by using *small-signal analysis*. This is the subject of this section.

Consider the tunnel-diode circuit shown in Fig. 5.1 where, in addition to the circuit elements treated earlier, there is a sinusoidal voltage source

$$v_s(t) = V_m \cos \omega t$$

First we assume that the biasing circuit, i.e., the circuit without the signal source $v_s(t)$, has been designed properly so that there is only one operating

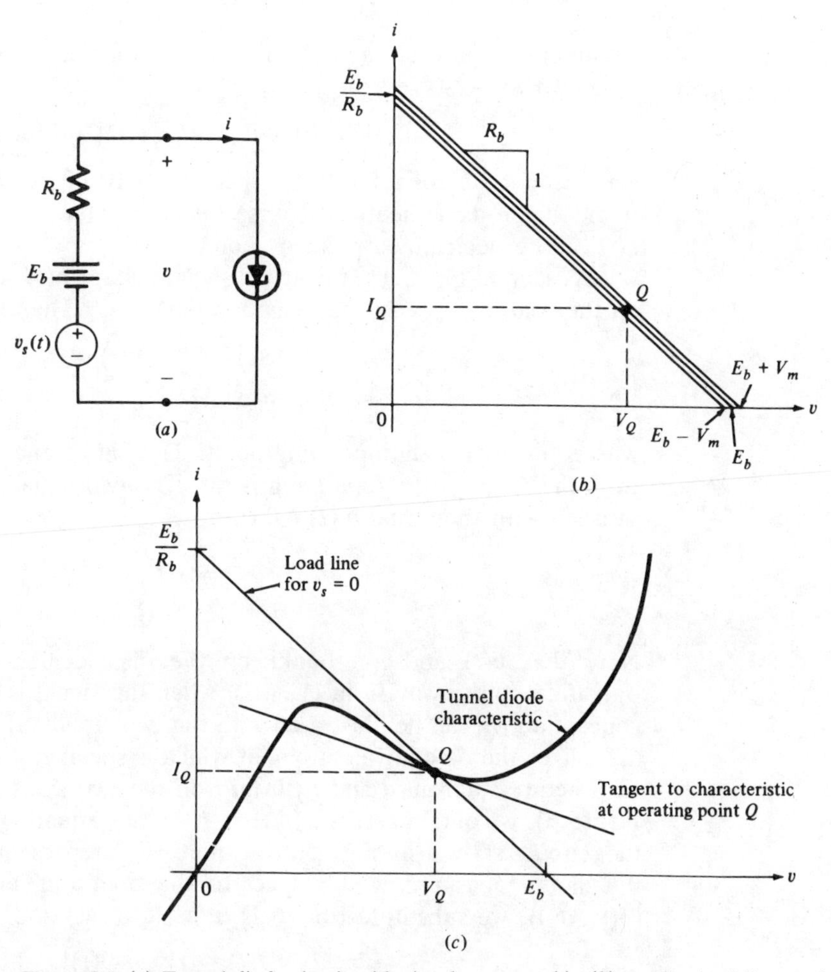

Figure 5.1 (*a*) Tunnel-diode circuit with signal source $v_s(t)$, (*b*) moving load line, and (*c*) linear approximation to the diode characteristic at the operating point Q.

point Q as shown. To be specific, assume that it lies where the slope is negative. As $v_s(t)$ varies with time, we may imagine that the load line is being moved parallel to the biasing load line as shown in the figure. Thus the solution of the circuit driven by the input signal $v_s(t)$ can be determined graphically point by point as the intersection point of the characteristic of the tunnel diode and the moving load line. This gives us a mental picture of the influence of the signal source $v_s(t)$ as t changes.

Let the v-i characteristic of the tunnel diode be specified by

$$i = \hat{i}(v) \tag{5.1}$$

KCL states that all branch currents in the circuit are the same. KVL for the single loop in the circuit yields the following equation:

$$v(t) = v_s(t) + E_b - R_b i(t) \qquad \text{for all } t \tag{5.2}$$

Combining Eqs. (5.1) and (5.2), we obtain a single equation with $v(\cdot)$ as the unknown to be solved for:

$$v(t) = v_s(t) + E_b - R_b \hat{i}[v(t)] \qquad \text{for all } t \tag{5.3}$$

This cannot be solved readily since we only know the curve given by the tunnel-diode data sheet. Of course for each value of t, we can find $v(t)$, thus $v(\cdot)$ can be determined point by point.

As seen in Fig. 5.1b, the actual signal voltage $v(t)$ and signal current $i(t)$ lie on the characteristic in the neighborhood of Q. Therefore, let us denote

$$v(t) \overset{\Delta}{=} V_Q + \tilde{v}(t) \tag{5.4a}$$

$$i(t) \overset{\Delta}{=} I_Q + \tilde{i}(t) \tag{5.4b}$$

where (V_Q, I_Q) is the operating point. This, in essence, shifts the coordinates from the origin to the operating point. The two equations (5.1) and (5.2) are satisfied with the signal $v_s(t) = 0$; i.e.,

$$I_Q = \hat{i}(V_Q) \tag{5.5}$$

$$V_Q = E_b - R_b I_Q \tag{5.6}$$

Note that $\tilde{v}(t)$ and $\tilde{i}(t)$ bookkeep the displacement of the instantaneous operating point away from (V_Q, I_Q) when the signal is applied. The pertinent concept above can be illustrated with the two circuits shown in Fig. 5.2. Figure 5.2a gives the dc equivalent circuit which is specified by Eqs. (5.5) and (5.6). Subtracting V_Q from v and I_Q from i on the two sides of Eq. (5.2) and using Eq. (5.6), we obtain $\tilde{v}(t) = v_s(t) - R_b \tilde{i}(t)$. This equation can be represented by the circuit shown in Fig. 5.2b, where D_{ac} represents the ac behavior of the diode measured with respect to the operating point Q. To determine $(\tilde{v}(t), \tilde{i}(t))$, we substitute Eq. (5.4) into Eq. (5.1) and obtain

$$I_Q + \tilde{i}(t) = \hat{i}[V_Q + \tilde{v}(t)] \tag{5.7}$$

Up to now our analysis is general. At this juncture, let us assume that the

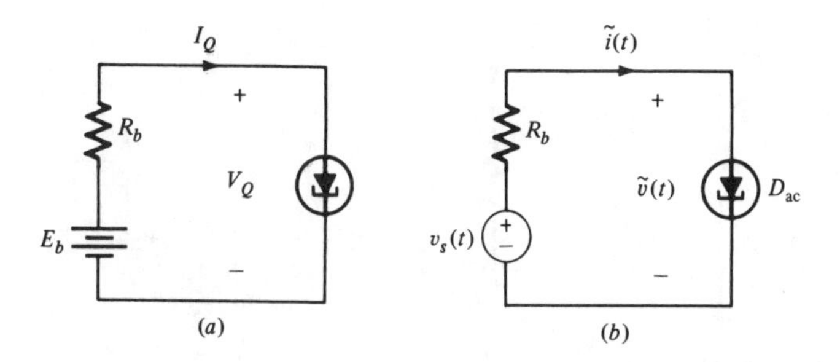

Figure 5.2 The circuit shown in Fig. 5.1 can be viewed in terms of (*a*) its dc equivalent circuit and (*b*) its ac equivalent circuit, where the diode characteristic has its origin at (V_Q, I_Q). D_{ac} denotes the diode with its origin shifted.

amplitude of the sinusoidal voltage $v_s(t)$ is *small*, i.e., $V_m \ll E$. Thus the voltage $\tilde{v}(t)$ is "small" in comparison with V_Q. What follows is the "small-signal analysis." Taking the first two terms of the Taylor series expansion of $\hat{i}[V_Q + \tilde{v}(t)]$ about the points V_Q, we obtain,

$$i(t) = I_Q + \tilde{i}(t) = \hat{i}[V_Q + \tilde{v}(t)] \approx \hat{i}[V_Q] + \left.\frac{d\hat{i}}{dv}\right|_{V_Q} \tilde{v}(t) \qquad \text{for all } t \tag{5.8}$$

Geometrically (see Fig. 5.1*c*), the approximation carried out in Eq. (5.8) amounts to replacing the nonlinear diode characteristic by its linear approximation about the operating point Q. Comparing Eq. (5.5) with Eq. (5.8), we obtain

$$\tilde{i}(t) \approx \left.\frac{d\hat{i}}{dv}\right|_{V_Q} \tilde{v}(t) \qquad \text{for all } t \tag{5.9a}$$

The term $d\hat{i}/dv|_{V_Q}$ is the slope of the diode characteristic at the operating point Q; note that in the present case it is negative. Let us denote

$$\left.\frac{d\hat{i}}{dv}\right|_{V_Q} \stackrel{\Delta}{=} G = \frac{1}{R} \tag{5.9b}$$

where G is negative. The quantity $d\hat{i}/dv|_{V_Q}$ is called the "*small-signal*" *conductance* of the diode at the operating point Q. Then, we have

$$\tilde{i}(t) = G\tilde{v}(t) = \frac{1}{R}\,\tilde{v}(t) \tag{5.10}$$

Next, substituting Eq. (5.4) into Eq. (5.2), we obtain

$$v_Q + \tilde{v}(t) = v_s(t) + E_b - R_b I_Q - R_b \tilde{i}(t) \qquad \text{for all } t \tag{5.11}$$

Comparing Eqs. (5.6) and (5.11), we obtain

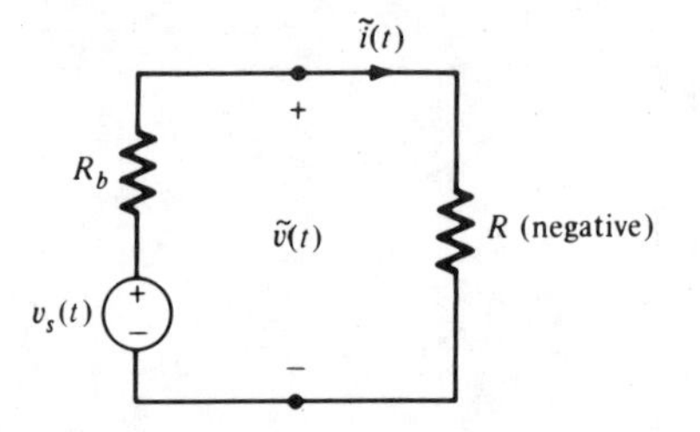

Figure 5.3 Small-signal equivalent circuit about the operating point Q. Note $R = 1/G$, and $G = (di/dv)|_{V_Q} < 0$.

$$\tilde{v}(t) = v_s(t) - R_b \tilde{i}(t) \qquad \text{for all } t \tag{5.12}$$

Equation (5.12) together with Eq. (5.10) can be interpreted as representing a circuit describing the relation between the small-signal voltages and currents; this circuit is shown in Fig. 5.3. It is called the *small-signal equivalent circuit* about the operating point Q; it is a *linear* circuit because the two resistors are linear. Note that the resistance R is negative, thus we have a linear *active* resistor in the circuit. The solution can be obtained immediately from the small-signal equivalent circuit or directly from Eqs. (5.10) and (5.12):

$$\tilde{i}(t) = \frac{v_s(t)}{R_b + R} = \frac{V_m}{R_b + R} \cos \omega t \tag{5.13}$$

and

$$\tilde{v}(t) = R\tilde{i}(t) = \frac{RV_m}{R_b + R} \cos \omega t \tag{5.14}$$

Since R is negative the factor $|R/(R_b + R)|$ can be made very large. From Eqs. (5.13) and (5.14), we can define the *small-signal power gain*[5]

$$\mathcal{G} = \left| \frac{\tilde{v}\tilde{i}}{\tilde{v}_s \tilde{i}} \right| = \left| \frac{R}{R_b + R} \right| \tag{5.15}$$

The expression $|R/(R_b + R)|$ depends on $G = 1/R$, which is the slope of the tunnel-diode characteristic at the operating point Q, and R_b, the resistance of the battery-resistor circuit.

6 TRANSFER CHARACTERISTICS

In Sec. 2 we introduced the concept of a driving-point characteristic of a one-port. The essence of the term "driving-point" is that both the source acting as input and the response variable appear at the same port. If the one-port is driven by a current source $i_s = i$, the port current, the response is the port voltage v; and if the one-port is driven by a voltage source $v_s = v$, the port voltage, the response is the port current i. This was illustrated in Fig. 2.1. Thus

[5] Note that $\tilde{v}(t)\tilde{i}(t)$ is a function of time, and so is $v_s(t)\tilde{i}(t)$; nevertheless, their ratio is constant.

we may also use the term "v-i characteristic of a one-port" to mean the driving-point characteristic of a one-port made of resistors.

The term "transfer characteristic" also describes an input-response relation; however, the word "transfer" means that the response variable does *not* appear at the same port as the source serving as input. As an example, if a voltage source is applied to the series connection of two nonlinear resistors $\mathcal{R}_1$ and $\mathcal{R}_2$, as shown in Fig. 6.1, and the response of interest is the voltage v_2, we can describe a transfer characteristic between the input voltage v_s and the response voltage v_2. Often in discussing transfer characteristics, we prefer to use the term "output" instead of "response." The term output also carries the notion that the variable of interest occurs at a different location from the port where the input source is applied. Thus we describe the "input-output" relation by a transfer characteristic. The transfer characteristic for a resistive circuit is no longer restricted to a voltage-current relation as demonstrated by the example here, where both the input and output are voltages.

We will use the nonlinear voltage divider circuit shown in Fig. 6.1 to illustrate the determination of transfer characteristics.

Example 1 (analytic method) The circuit in Fig. 6.1 with input $v_i = v_s$ and output $v_o = v_2$ is called a *voltage divider*. Again, we assume both nonlinear resistors to be current-controlled, namely:

$$v_1 = \hat{v}_1(i_1) \qquad \text{and} \qquad v_2 = \hat{v}_2(i_2) \tag{6.1}$$

Since by KCL, $i_1 = i_2 = i$, and by KVL, $v_i = v_1 + v_2$, we write

$$v_i = \hat{v}_1(i) + \hat{v}_2(i) \overset{\Delta}{=} f(i) \qquad \text{for all } i \tag{6.2}$$

and

$$v_o = \hat{v}_2(i) \overset{\Delta}{=} h(i) \qquad \text{for all } i \tag{6.3}$$

If we wish to express the transfer characteristic as specifying v_o in terms of v_i, it is necessary to eliminate the variable i from these two equations. Assume that the *inverse* of the function $f(\cdot)$ exists, that is, for all v_i we can solve Eq. (6.2) for i; then

$$i = f^{-1}(v_i) \qquad \text{for all } v_i \tag{6.4}$$

that is, i is expressed as a function of v_i. Substituting Eq. (6.4) into Eq. (6.3), we obtain the transfer characteristic

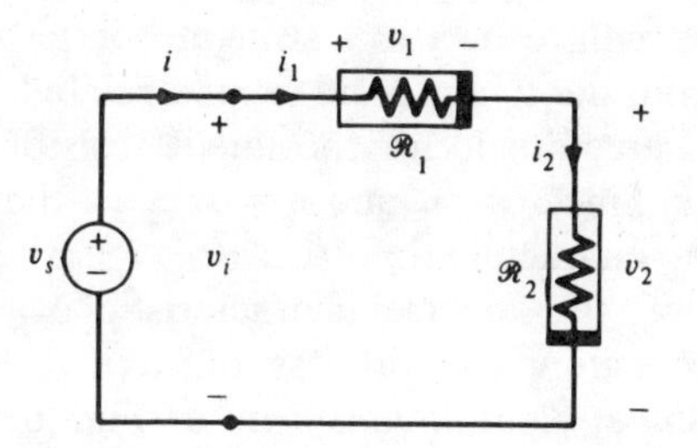

Figure 6.1 Nonlinear circuit with v_s as input and v_2 as response or output.

$$v_o = h[f^{-1}(v_i)] \tag{6.5}$$

Note that Eq. (6.5) involves finding the function of another function, namely, the *composition* of two functions.

Let us consider $\mathscr{R}_1$ to be a *pn*-junction diode characterized by

$$i_1 = I_s(e^{Kv_1} - 1) \tag{6.6}$$

and $\mathscr{R}_2$ be a linear resistor with resistance R_2. The diode equation can be written as

$$v_1 = \frac{1}{K} \ln\left(1 + \frac{i_1}{I_s}\right) \tag{6.7}$$

therefore
$$v_i = \frac{1}{K} \ln\left(1 + \frac{i}{I_s}\right) + R_2 i \stackrel{\Delta}{=} f(i) \tag{6.8}$$

and
$$v_o = R_2 i \stackrel{\Delta}{=} h(i) \tag{6.9}$$

Unfortunately, the function $f(\cdot)$ defined in Eq. (6.8) is such that its inverse function, $f^{-1}(\cdot)$, cannot be expressed in a *closed* form. Thus the analytic approach terminates here, and we therefore resort to a graphic method.

REMARK Equations (6.8) and (6.9) representing the v_o vs. v_i characteristic, namely,

$$v_i = f(i) \qquad \text{and} \qquad v_o = h(i)$$

may be viewed as the parametric equations representing a curve in the v_i-v_o plane with the variable i playing the role of parameter. By running through a list of values of i of interest in the two parametric equations, we can operate an automatic plotter. This is a standard way of representing curves in a graphics terminal.

Example 2 (graphic method) The graphic method of finding the inverse function is straightforward. Continuing with this example, Fig. 6.2*a* gives the *pn*-junction-diode characteristic of Eq. (6.6). Since the diode is connected in series with a linear resistor, the current is the same and we need to add the voltages. Therefore, it is more convenient to plot the characteristics on the *i*-*v* plane. The characteristics plotted on the i_1-v_1 plane for the diode are obtained by first taking the mirror image of the curve in Fig. 6.2*a* with respect to a straight line passing through the origin at an angle 45° from the v_1 axis, and second relabeling the axis as shown in Fig. 6.2*b*. The characteristic for the linear resistor is plotted in the i_2-v_2 plane in Fig. 6.2*c*. Since $i_1 = i_2$ and $v = v_1 + v_2$, the voltages v_1 and v_2 are added to form the characteristic of Eq. (6.8) as shown in Fig. 6.2*d* in the i-v_1 plane. Finally, the inverse characteristic of Eq. (6.8) is obtained graphically by repeating the process of taking the mirror image with respect to the 45° straight line shown in Fig. 6.2*e*. The v_i-v_o transfer

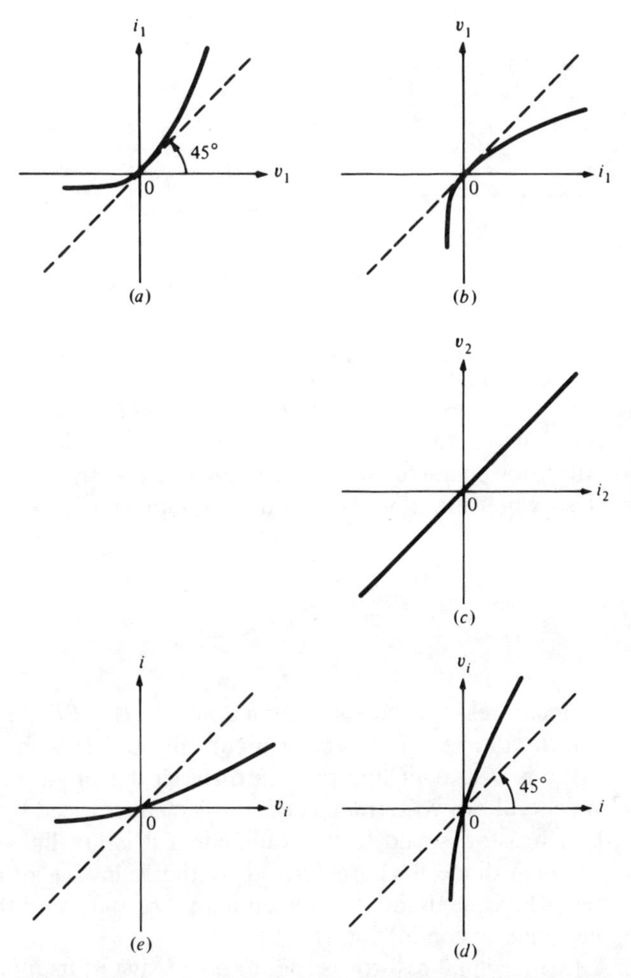

Figure 6.2 (*a*) *pn*-junction-diode characteristic and (*b*) its inverse characteristic obtained by taking the mirror image of the characteristic in (*a*) with respect to a 45° angle line. (*c*) Characteristic of the linear resistor. (*d*) Characteristic of the series connection and (*e*) the inverse characteristic.

characteristic is proportional to the curve in Fig. 6.2*e* since $v_o = Ri$. It is shown in Fig. 6.3 together with the circuit. This circuit is typically used as a half-wave rectifier because any *sinusoidal* input $v_i(t)$ yields an output $v_o(t)$ which is nearly zero for half of the cycle.

Exercises

1. Consider the voltage divider circuit shown in Fig. 6.1. Let $\mathcal{R}_1$ be a linear resistor of 2 Ω and $\mathcal{R}_2$ have the piecewise-linear tunnel-diode characteristic given by Fig. 4.7*b*. Determine the transfer characteristic v_2 vs. v_s.
2. Repeat the above with $\mathcal{R}_1$ and $\mathcal{R}_2$ interchanged.

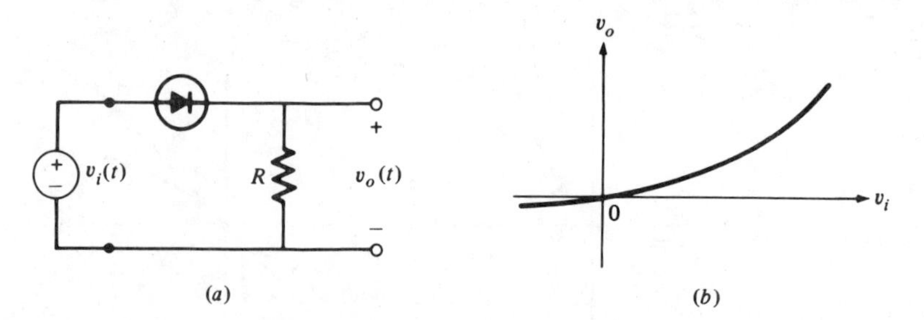

Figure 6.3 (*a*) A rectifier and (*b*) its transfer characteristic.

Like dc analysis there exist other methods for determining transfer characteristics. The most useful one for real circuits is by numerical techniques, such as the Newton-Raphson method. In essence, the determination of driving-point and transfer characteristics of a nonlinear resistive circuit calls for repeated dc analysis when the input is varied continuously over the entire range of voltage or current.

SUMMARY

- A linear resistor satisfies Ohm's law: $v(t) = Ri(t)$ for all t where v is the branch voltage, i is the branch current, and R is the resistance. Its characteristic is a straight line passing through the origin of the v-i plane.
- In general, a two-terminal resistor is characterized by the equation: $f(v, i) = 0$. A resistor is said to be nonlinear if it is not linear.
- An ideal diode is characterized by the following properties: The current is zero when the diode is reversed biased ($v < 0$), and the voltage is zero when the diode is conducting ($i > 0$).

- A two-terminal resistor is said to be passive iff its v-i characteristic lies in the closed first and third quadrants of the v-i plane. A two-terminal resistor is said to be active iff it is not passive.
- A two-terminal resistor is said to be bilateral if its characteristic is symmetric with respect to the origin of the v-i plane.
- A resistor is said to be voltage-controlled if its characteristic can be written $i = \hat{i}(v)$ for all v; similarly, it is said to be current-controlled if its characteristic can be written as $v = \hat{v}(i)$ for all i.
- An independent voltage source has a vertical line parallel to the i axis in the v-i plane as its characteristic; similarly, an independent current source has a horizontal line parallel to the v axis on the v-i plane as its characteristic.
- Two one-ports are said to be equivalent iff they have identical driving-point characteristics.
- Two resistors are said to be dual to each other if the characteristic of one in the v-i plane is identical to that of the other in the i-v plane.

- A circuit is said to be linear if the circuit contains linear circuit elements and independent sources. A circuit is said to be nonlinear if it is not linear.
- In a series connection of resistors, the current in all resistors is the same. The voltage across the series connection is the sum of the voltages across each individual resistor.
- Dually in a parallel connection of resistors, the voltage across all resistors is the same. The current of the parallel connection is the sum of the currents through each individual resistor.
- The solutions of a nonlinear resistive circuit with dc inputs are called operating points. The determination of operating points of a circuit is called dc analysis.
- The small-signal analysis is applicable when the signal voltage or current is sufficiently small in magnitude.
- The small-signal equivalent circuit at operating point Q holds only in some neighborhood of the operating point; it is obtained by setting the dc source to zero and replacing each nonlinear resistor by a linear resistor whose conductance is equal to the slope of its nonlinear characteristic in the v-i plane at the operating point Q.

PROBLEMS

Two-terminal resistors

1 For the two-terminal resistors given by the v-i characteristics shown in Fig. P2.1, state whether each one is (i) linear or nonlinear, (ii) voltage-controlled and/or current-controlled, (iii) passive or active, (iv) bilateral or nonbilateral.

2 The equations below specify the characteristics of some resistors. Indicate whether they are (i) linear or nonlinear, (ii) time-varying or time-invariant, (iii) voltage-controlled or current-controlled, (iv) passive or active, (v) bilateral or nonbilateral.

(*a*) $v + 10i = 0$ (*e*) $i = \tanh v$

(*b*) $v = (\cos 2t)i + 3$ (*f*) $i + 3v = 10$

(*c*) $i = \exp(-v)$ (*g*) $i = 2 + \cos \omega t$

(*d*) $v = i^2$ (*h*) $i = v + (\cos 2t)v/|v|$

Nonlinear resistors

3 Suppose that the nonlinear resistor $\mathcal{R}$ has a characteristic specified by the equation

$$v = 20i + i^2 + \tfrac{1}{2}i^3$$

(*a*) Express v as a sum of sinusoids, given

$$i(t) = \cos \omega_1 t + 2 \cos \omega_2 t$$

(*b*) If $\omega_2 = 2\omega_1$, what frequencies are present in v?

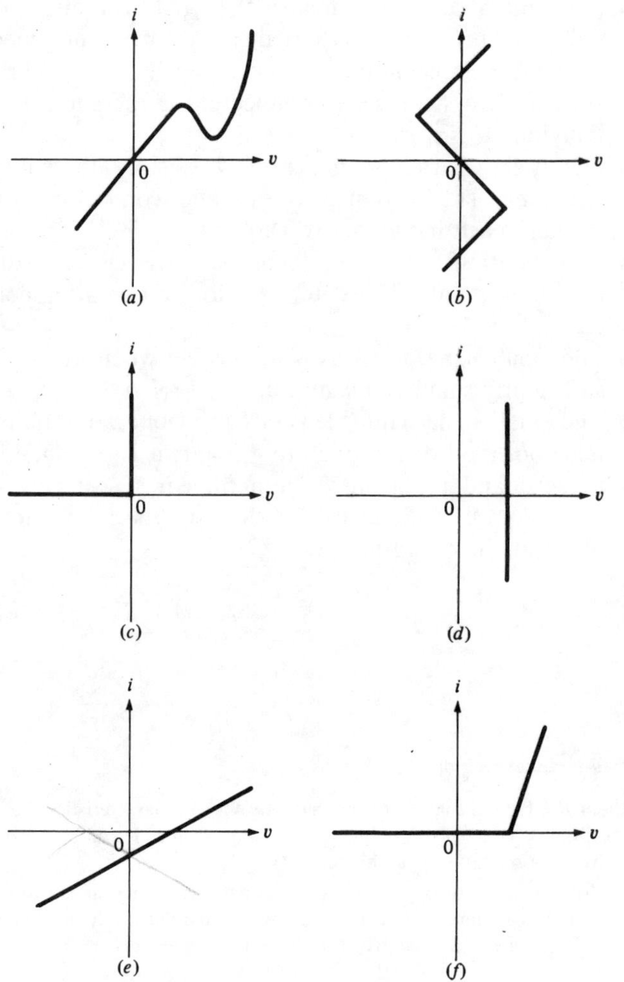

Figure P2.1

Series or parallel connection

4 Use graphic series and parallel addition to find and plot the driving-point characteristics of the one-port shown in Fig. P2.4.

5 (*a*) Draw the driving-point characteristics for the two circuits shown in Fig. P2.5*a* and *b*. The two nonlinear resistors are described by the v-i characteristics shown in Fig. P2.5*c* and *d*, respectively.

(*b*) Will the driving-point characteristic obtained in (*a*) change if the terminals of $\mathscr{R}_1$ are turned around? Explain.

6 Repeat Prob. 5(*a*) with the terminals of $\mathscr{R}_2$ turned around.

Parallel connection

7 Use graphic addition to find the driving-point characteristic of the circuit shown in Fig. P.2.7.

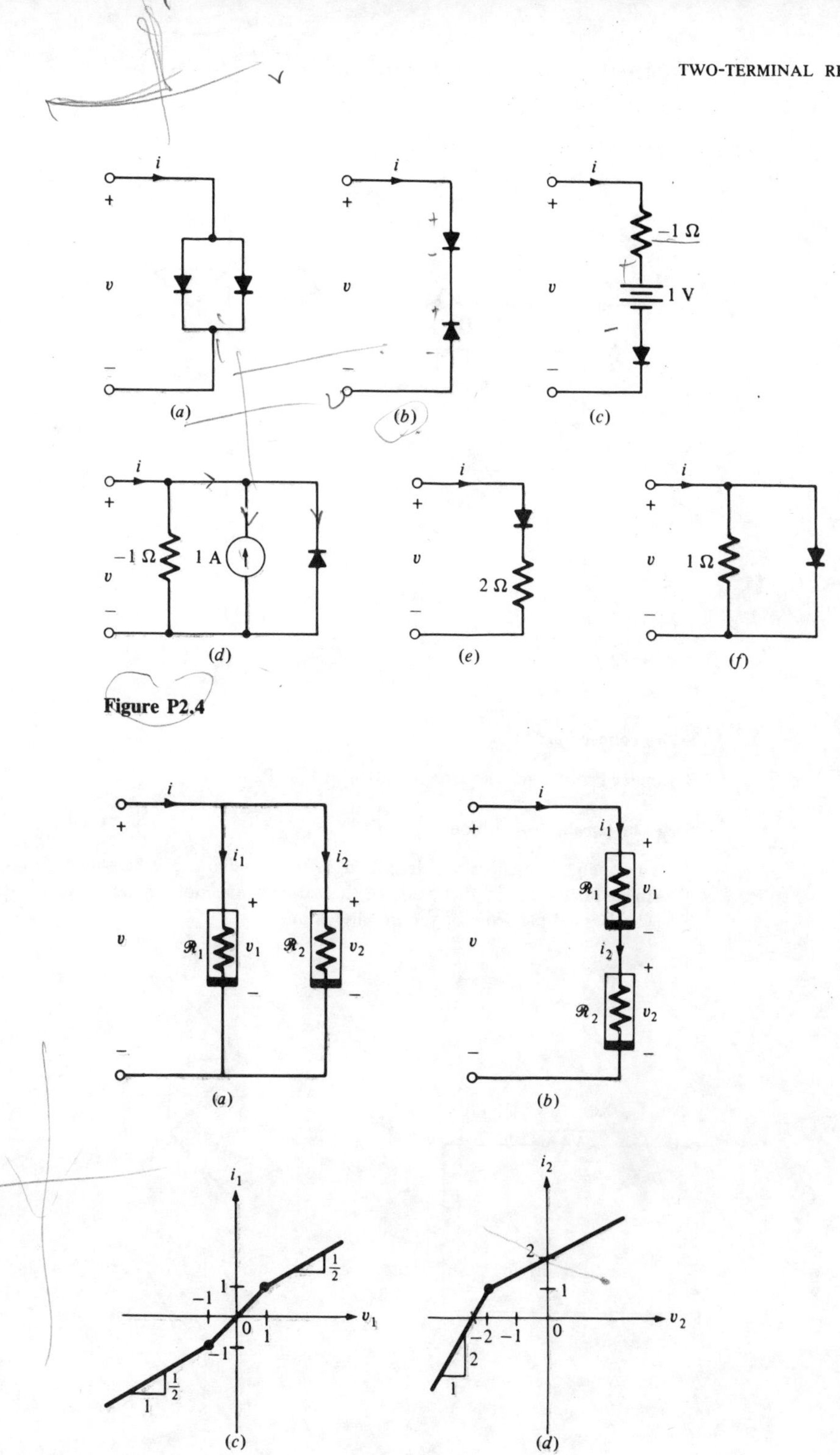

Figure P2.4

Figure P2.5

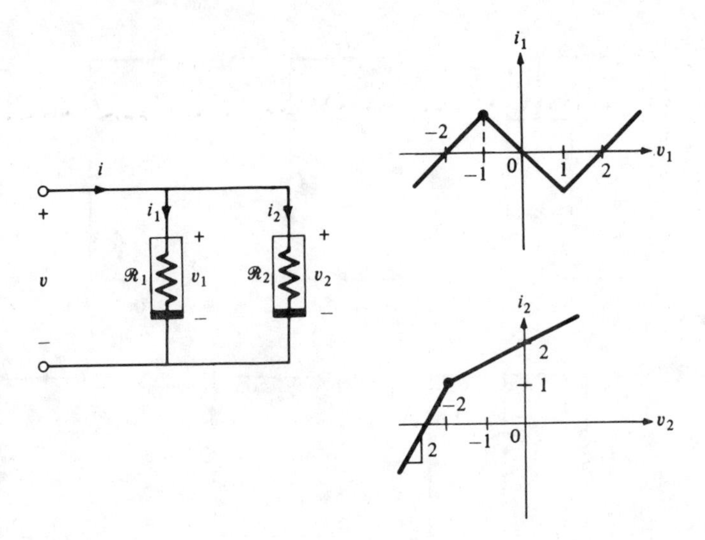

Figure P2.7

Series connection

8 Repeat Prob. 7 for the circuit shown in Fig. P2.8.

Series or parallel connection

9 Two nonlinear resistors are described by the v-i characteristics shown in Fig. P2.9.

(*a*) Find the v-i characteristic of the series connection of these two resistors.

(*b*) Repeat part (*a*) for the parallel connection.

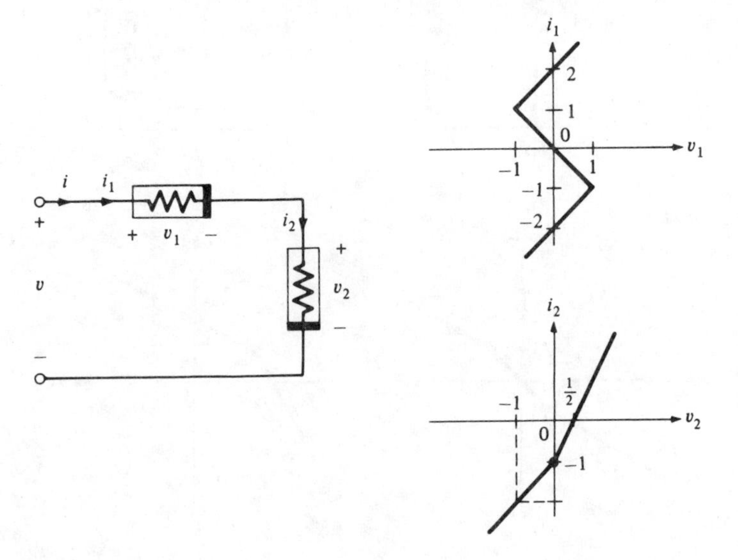

Figure P2.8

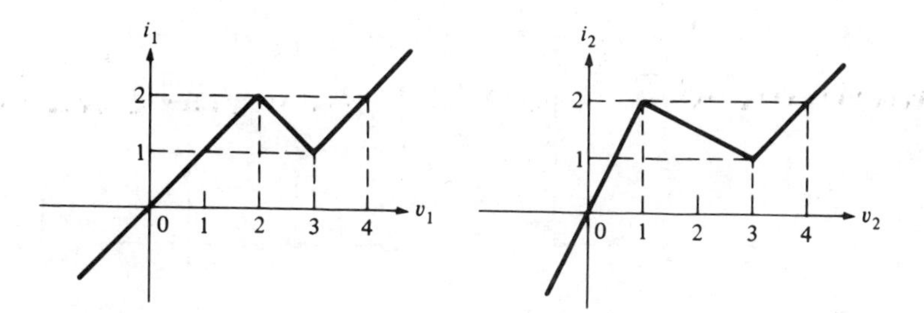

Figure P2.9

10 Repeat Prob. 9 for the nonlinear resistors shown in Fig. P2.10.

11 Suppose we are given two resistive elements, the v-i characteristics for which are shown in Fig. P2.11.

(*a*) Find the v-i characteristic of the series connection of these two elements.

(*b*) Find the v-i characteristic of the parallel connection of these two elements.

12 Use graphic series and parallel addition to derive the driving-point characteristics of the one-ports shown in Fig. P2.12.

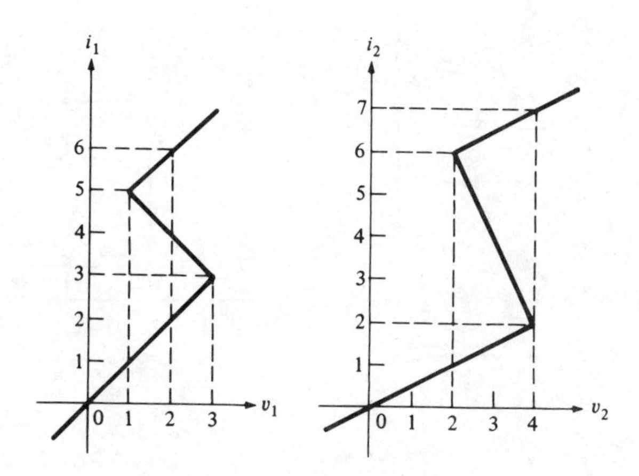

Figure P2.10

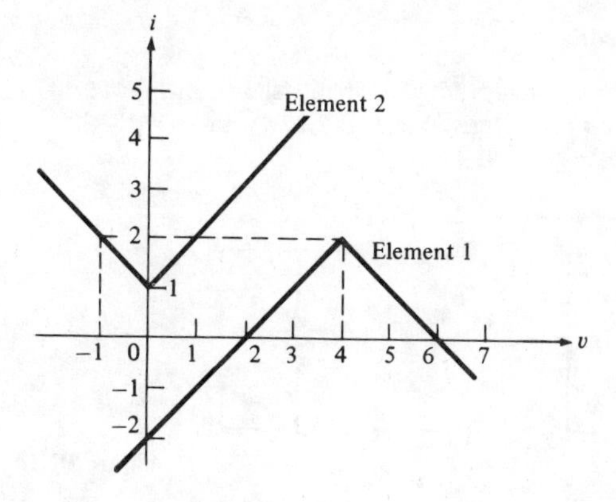

Figure P2.11

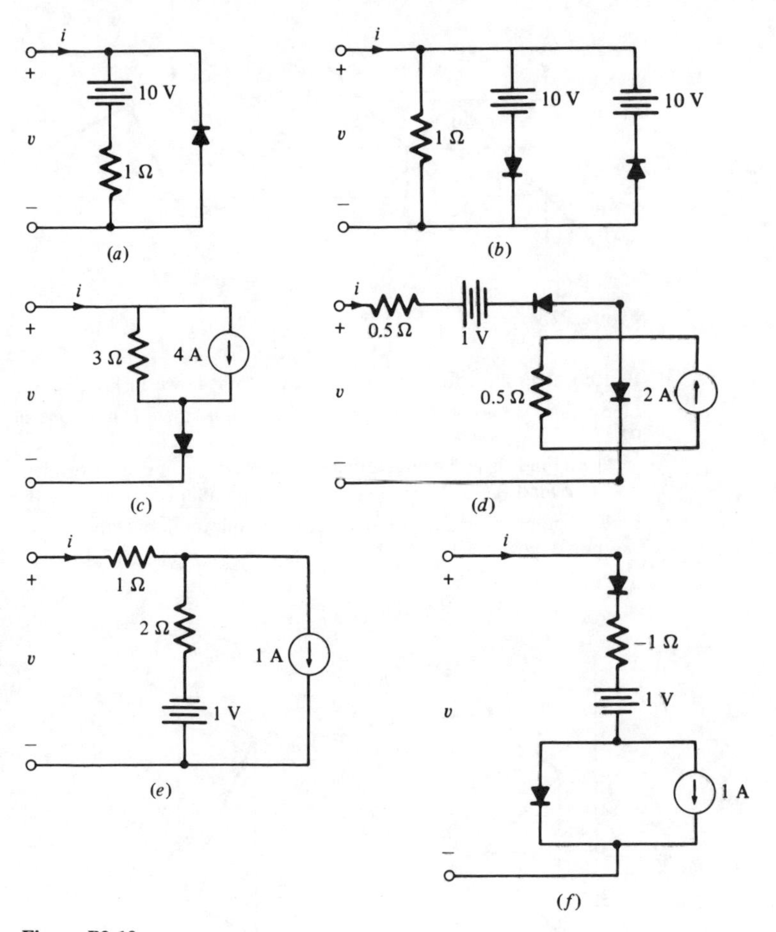

Figure P2.12

Dual circuits

13 Find the dual circuits for the circuits shown in Fig. P2.12.

***v*-*i* Plot, zener diodes**

14 Plot the v-i characteristics for the circuits shown in Fig. P2.14. Assume the zener diodes are described by the v-i characteristic in Fig. 2.22*a* with $E_z = 10$ V.

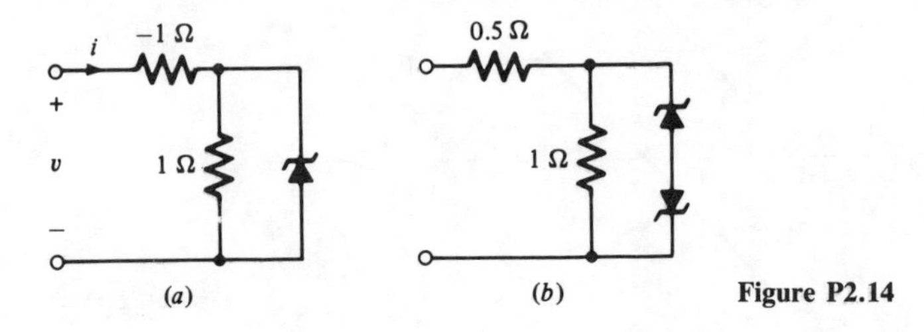

Figure P2.14

v-i Plot, equivalence, and duality

15 For the circuit shown in Fig. P2.15:

(*a*) Determine the driving-point characteristic of the one-port ①, ①′, that is, the equation describing the one-port in terms of the port voltage and the port current.

(*b*) Plot the characteristic in the v-i plane.

(*c*) Determine an equivalent one-port which consists of one independent current source and one linear resistor.

(*d*) Determine the dual one-port of the above.

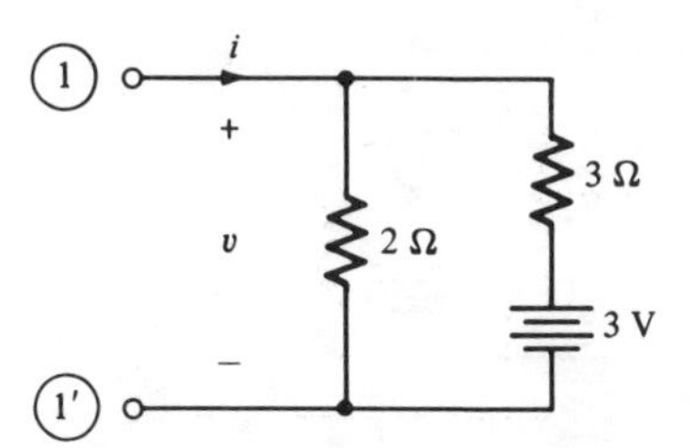

Figure P2.15

16 Repeat Prob. 15 for the circuit shown in Fig. P2.16.

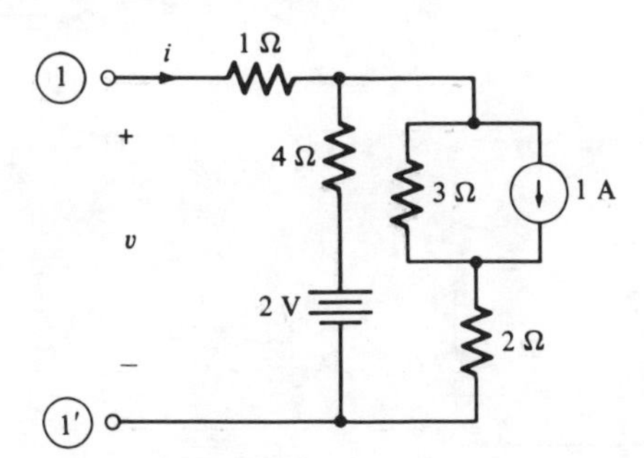

Figure P2.16

v-i Plot and dc solution

17 (*a*) Plot the driving-point characteristic of the one-port shown in Fig. P2.17*a*.

(*b*) A current source i_s is connected to the above one-port as shown in Fig. P2.17*b*. Determine the port voltage v for

(i) $i_s = 2$ A

(ii) $i_s = 1$ A

(iii) $i_s = \frac{1}{2}$ A

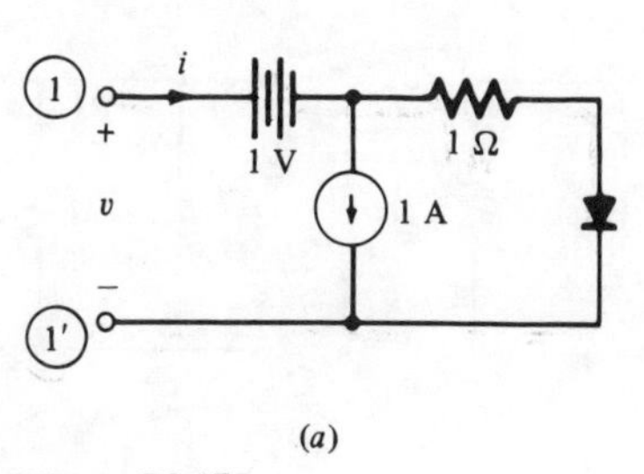

(*a*)

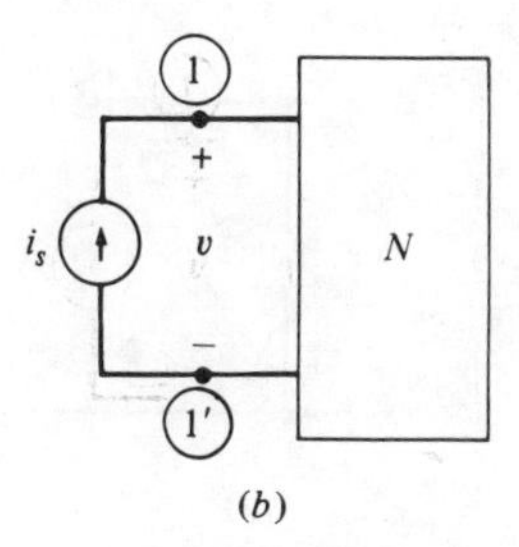

(*b*)

Figure P2.17

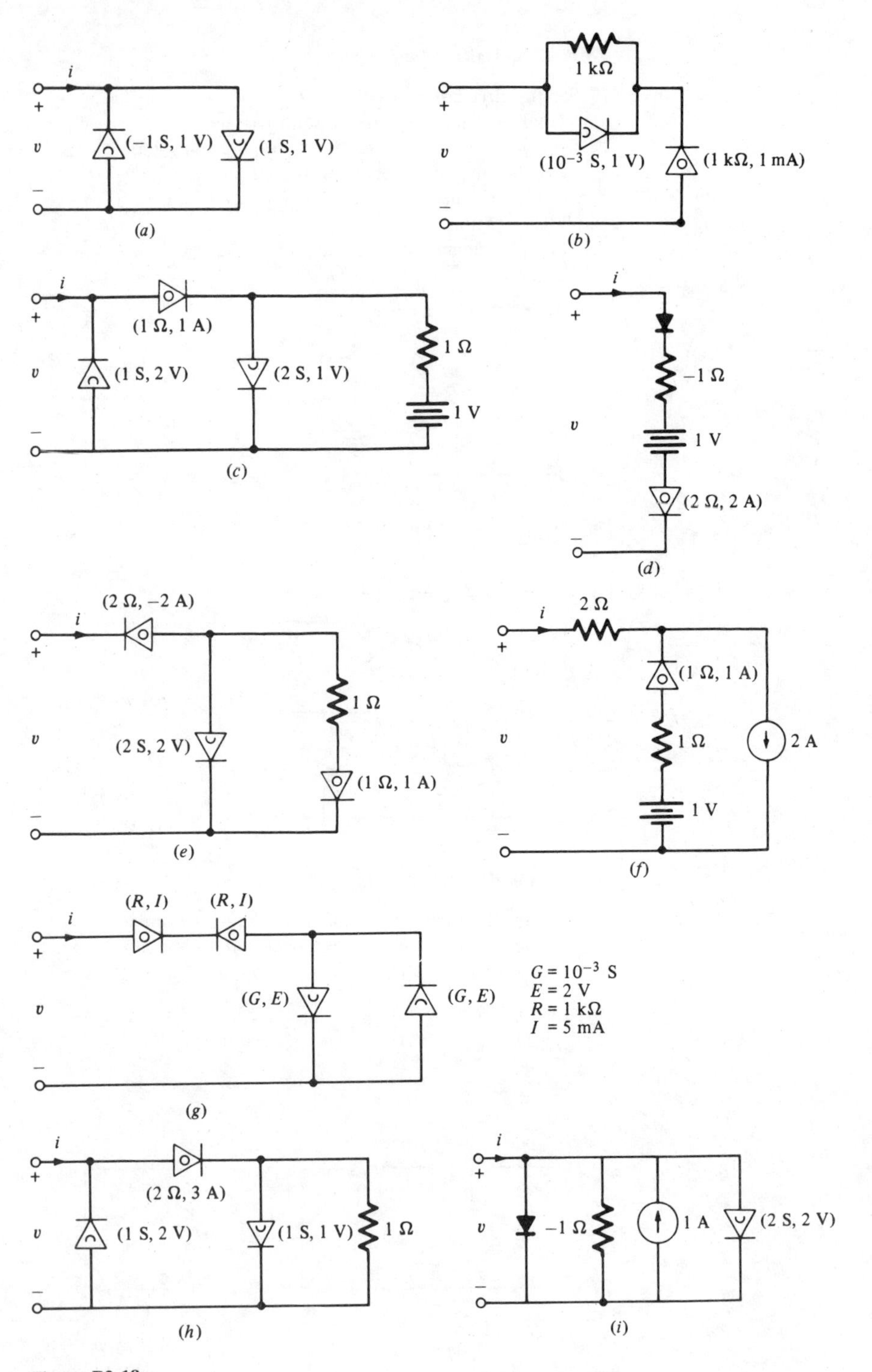

i
+
v
−
(−1 S, 1 V)
(1 S, 1 V)
(a)
1 kΩ
(10⁻³ S, 1 V)
(1 kΩ, 1 mA)
(b)
(1 Ω, 1 A)
(1 S, 2 V)
(2 S, 1 V)
1 Ω
1 V
(c)
−1 Ω
1 V
(2 Ω, 2 A)
(d)
(2 Ω, −2 A)
1 Ω
(2 S, 2 V)
(1 Ω, 1 A)
(e)
2 Ω
(1 Ω, 1 A)
1 Ω
2 A
1 V
(f)
(R, I)
(R, I)
(G, E)
(G, E)
G = 10⁻³ S
E = 2 V
R = 1 kΩ
I = 5 mA
(g)
(2 Ω, 3 A)
(1 S, 2 V)
(1 S, 1 V)
1 Ω
(h)
−1 Ω
1 A
(2 S, 2 V)
(i)

Figure P2.18

Driving-point characteristic, concave and convex resistors, synthesis

18 Use graphic series and parallel addition to find the driving-point characteristics of the circuits shown in Fig. P2.18.

19 Using only concave resistors, convex resistors, and at most one linear resistor and one dc source, synthesize the driving-point characteristics shown in Fig. P2.19.

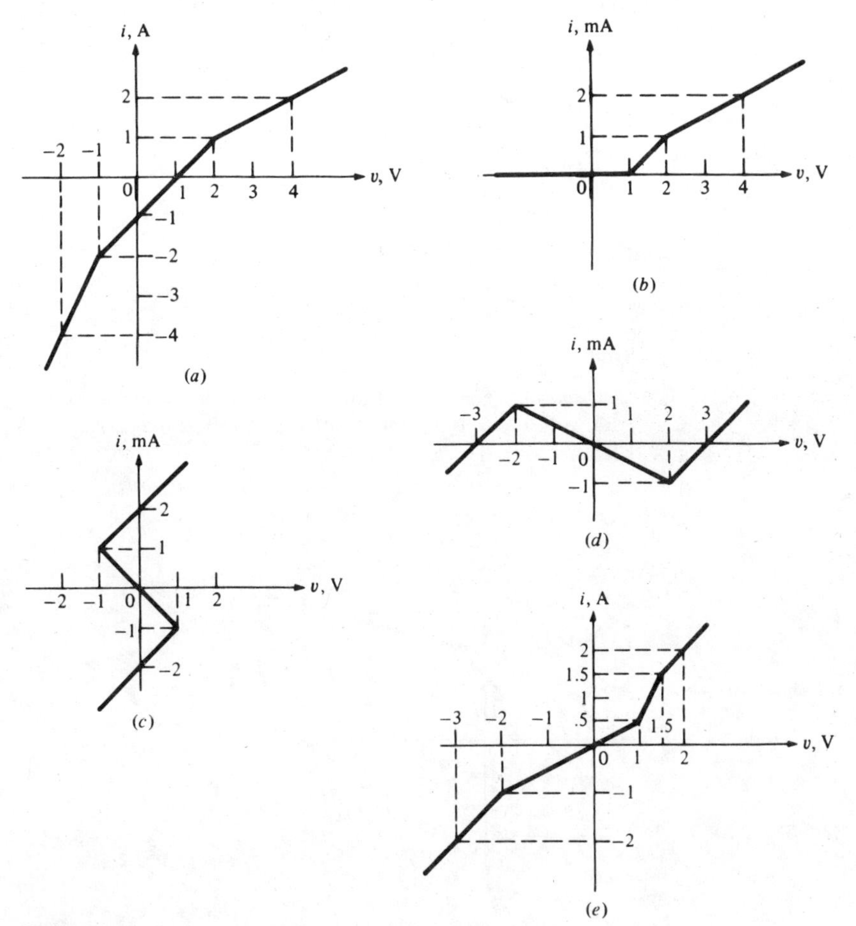

Figure P2.19

Piecewise-linear approximation

20 Use Eq. (3.9) to write the piecewise-linear equation describing each characteristic in Fig. P2.19.

21 (*a*) The piecewise-linear *v*-*i* characteristic in Fig. P2.21*a* can be described by

$$v = a_0 + a_1 i + b_1|i - I_1| + b_2|i - I_2|$$

Specify the coefficients a_0, a_1, b_1, b_2, I_1, and I_2.

(*b*) Specify the parameters R_0, R_a, R_b, I_a, and I_b in Fig. P2.21*b* so that the driving-point characteristic of the one-port *N* is given by Fig. P2.21*a*.

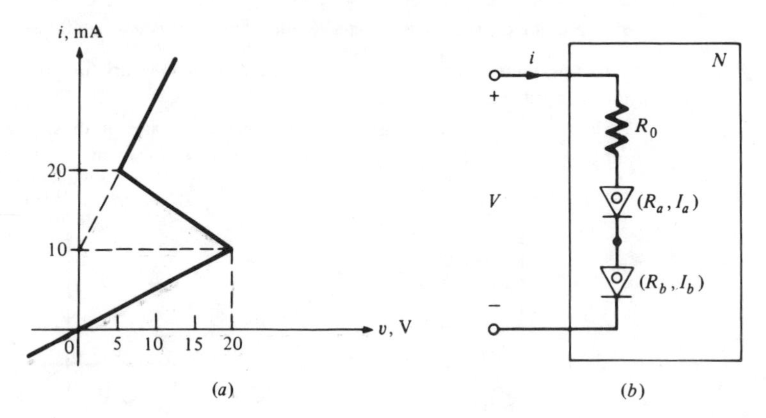

Figure P2.21

Synthesis

22 Using only concave resistors (with $G > 0$), convex resistors (with $R > 0$), and at most one positive linear resistor and a battery, synthesize the driving-point characteristic shown in Fig. P2.22.

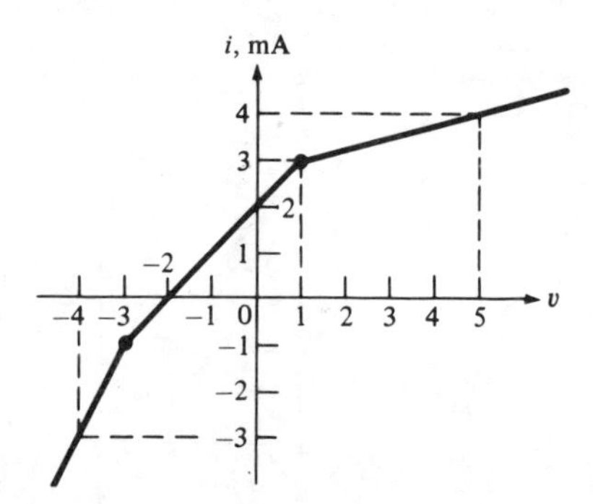

Figure P2.22

Driving-point characteristic, load line

23 The zener-diode circuit shown in Fig. P2.23 functions as an inexpensive voltage regulator which maintains a constant output voltage when the load resistance R_L and/or power supply voltage E change within a prescribed range.

(*a*) Assuming an ideal zener-diode v-i characteristic with $E_z = 5$ V (see Fig. 2.22), find (by the graphic method) the driving-point characteristic of the one-port N.

(*b*) Using the driving-point characteristic from (*a*), find the output voltage v_o when $R_L = 2\ \text{k}\Omega$ and $R_L = 500\ \Omega$, respectively.

(*c*) If E can vary by ± 5 percent, specify the allowable range of R_L in order to maintain a constant 5 V output voltage.

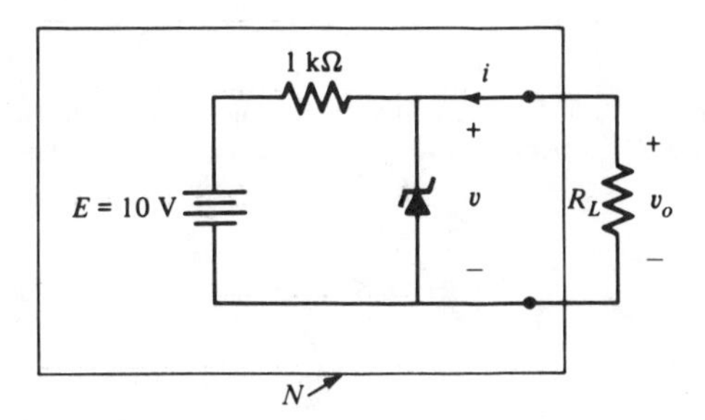

Figure P2.23

Operating point, analytic method

24 (*a*) For the one-port N shown in Fig. P2.24, find the driving-point characteristic analytically.
(*b*) Find the operating point(s) of the circuit shown in Fig. P2.24 using the analytic approach.

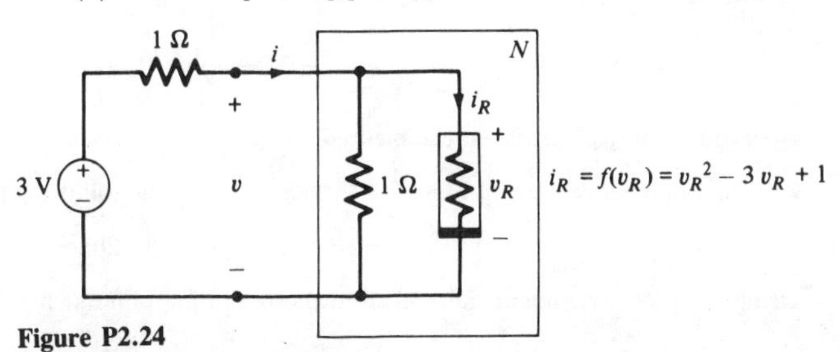

Figure P2.24

Operating point, load line

25 Consider the circuit shown in Fig. P2.25*a*, where the glow tube and the *pn*-junction diode are modeled by the piecewise-linear characteristics in Fig. P2.25*b* and *c*, respectively. Use the load-line method to find the voltage v_1 and the current i_2 at each operating point of the circuit.

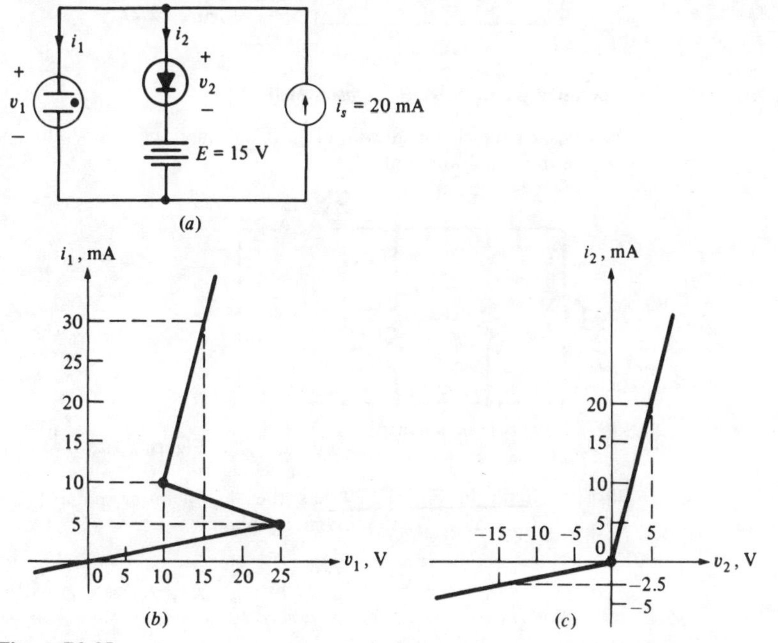

Figure P2.25

26 Replace the glow tube in Fig. P2.25*a* with a tunnel diode characterized by the piecewise-linear curve shown in Fig. P2.26. Repeat Prob. 25 for this circuit with $i_s = 12.5$ mA and $E = 10$ V.

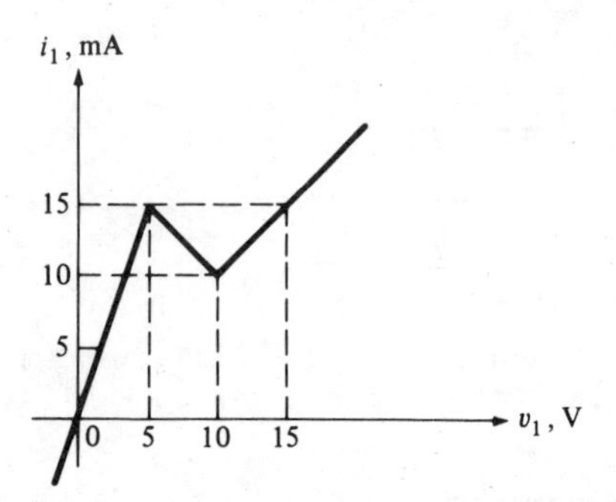

Figure P2.26

Operating point, piecewise-linear method

27 The nonlinear resistor in Fig. P2.27 is described by the following piecewise-linear characteristic:

$$i_1 = -2 + 5v_1 - 2|v_1 + 1| + 2|v_1 - 2|$$

Using the piecewise-linear method, find the operating point(s) for this circuit.

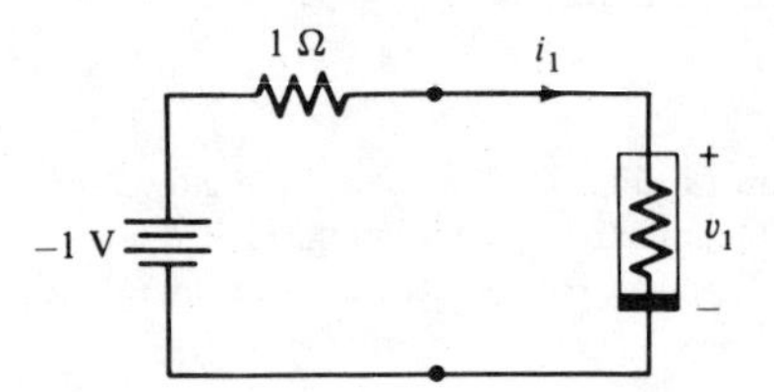

Figure P2.27

Operating point, Newton-Raphson method

28 Consider the circuit shown in Fig. P2.28. Find the dc operating point(s) for this circuit by using the Newton-Raphson method.

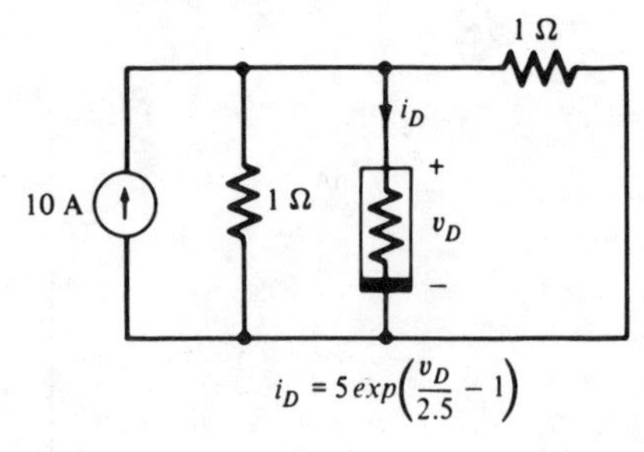

Figure P2.28

29 The circuit in Fig. P2.29 has the dc operating point $(V_{1Q} + \Delta V_1, \ldots, V_{6Q} + \Delta V_6;\ I_{1Q} + \Delta I_1, \ldots, I_{6Q} + \Delta I_6)$. What can you say about

$$\sum_{j=1}^{6} (V_{jQ} \cdot \Delta I_j) \qquad \sum_{j=1}^{6} (\Delta V_j \cdot I_{jQ}) \qquad \sum_{j=1}^{6} (\Delta V_j \cdot \Delta I_j)$$

Justify your answer.

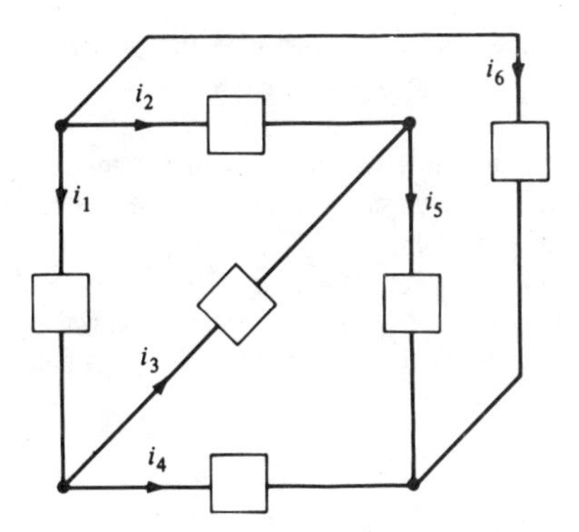

Figure P2.29

Operating point, small-signal equivalent circuits

30 (*a*) Find all possible dc operating points of the circuit shown in Fig. P2.30.
(*b*) For each operating point draw the small-signal equivalent circuit.

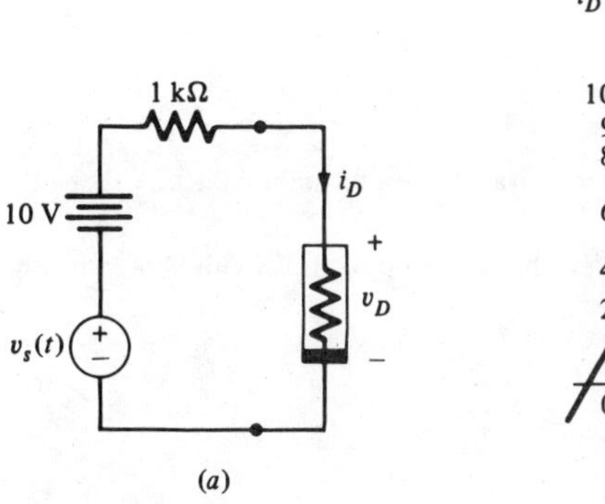

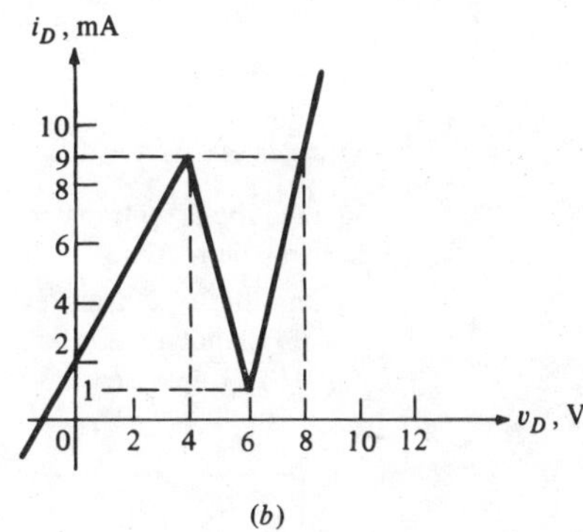

Figure P2.30

31 Consider the nonlinear voltage divider shown in Fig. P2.31*a*. The glow tube is modeled by a three-segment piecewise-linear v-i characteristic as shown in Fig. P2.31*b*.

(*a*) Assuming $v_i = 40$ V, find the operating point(s) using the piecewise-linear method.

(*b*) Assuming $v_i = 40 + (10^{-3}) \sin \omega t$, draw the small-signal equivalent circuit about the operating point on the negative slope segment and calculate the small-signal output voltage $\tilde{v}_o(t)$.

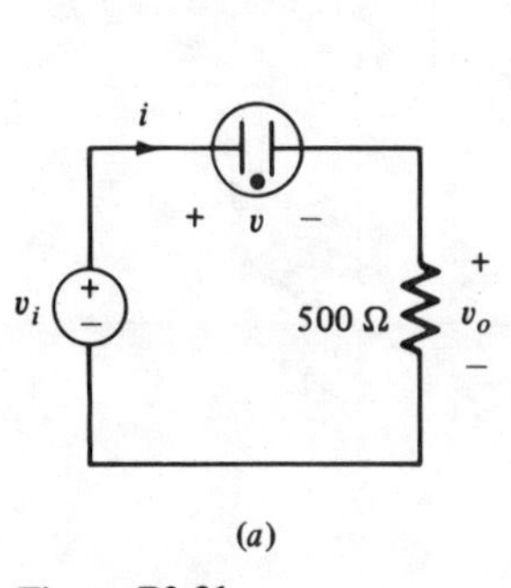

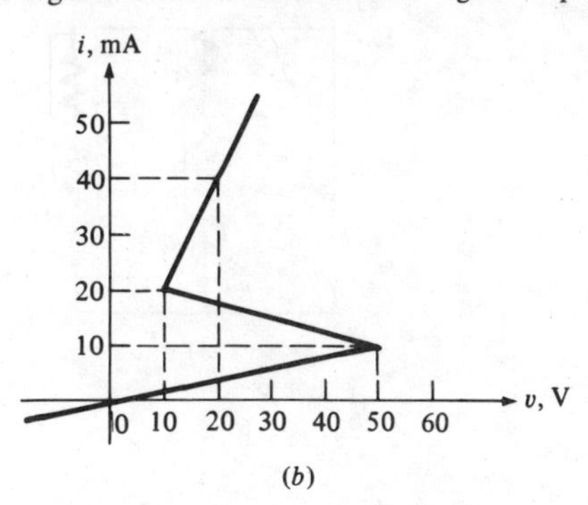

Figure P2.31

32 Insert a small-signal voltage source in series with the battery in the circuit from Fig. P2.25*a*. Use your answer from Prob. 25 to draw a small-signal equivalent circuit about each operating point of this circuit.

33 Insert a small-signal current source in parallel with the independent dc current source in the circuit from Fig. P2.28. Use your answer from Prob. 28 to draw a small-signal equivalent circuit about the operating point of this circuit.

34 Consider the circuit shown in Fig. P2.34 where the *pn*-junction diode is described by Eq. (1.6).

(*a*) Find the small-signal conductance of the diode in terms of v, I_s, and V_T.

(*b*) If V_{Q1} is an operating point for this circuit, draw the small-signal equivalent circuit about this operating point.

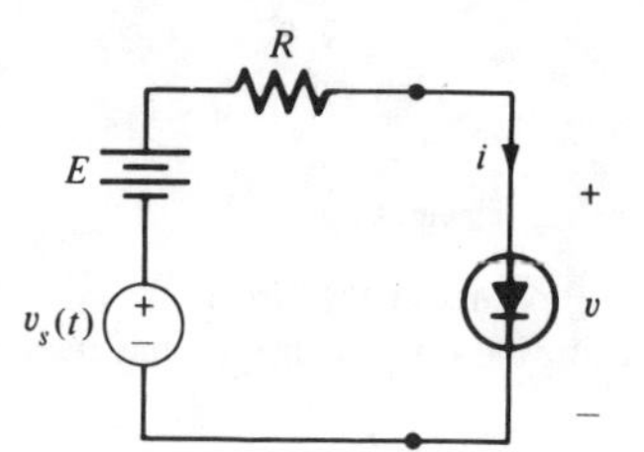

Figure P2.34

Transfer characteristic

35 Find the v_o-v_i transfer characteristics of the nonlinear voltage divider shown in Fig. P2.31 using the graphic method.

36 Find the v_o-v_i transfer characteristic of the circuit shown in Fig. P2.36 by the

(*a*) Analytic method.

(*b*) Graphic method.

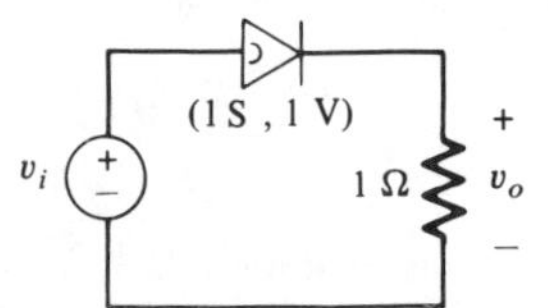

Figure P2.36

37 Plot the i_2-i_s transfer characteristic for the circuit shown in Fig. P2.37.

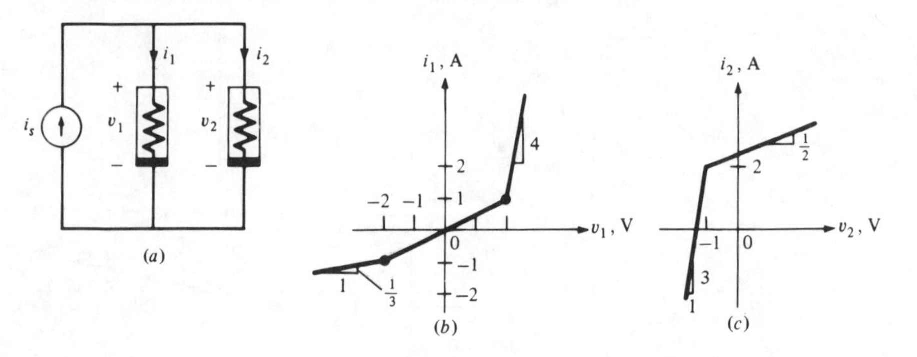

Figure P2.37

38 (*a*) Plot the i_o-v_s transfer characteristic for the circuit shown in Fig. P2.38.
(*b*) Let $v_s(t) = 4 \cos t$, sketch $i_o(t)$.

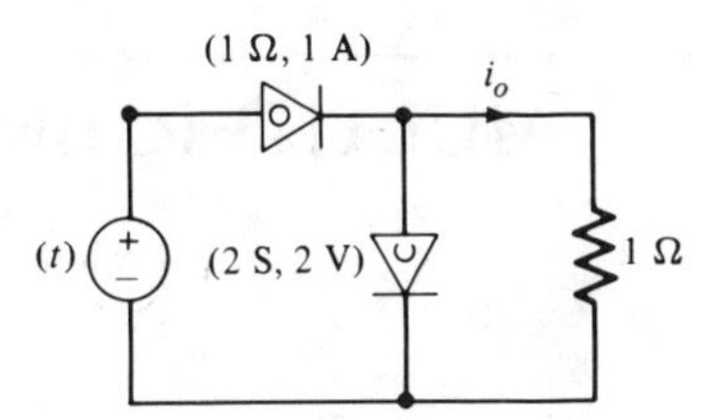

Figure P2.38

39 For the circuit shown in Fig. P2.39 plot the transfer characteristic v_2 as a function of i_s.

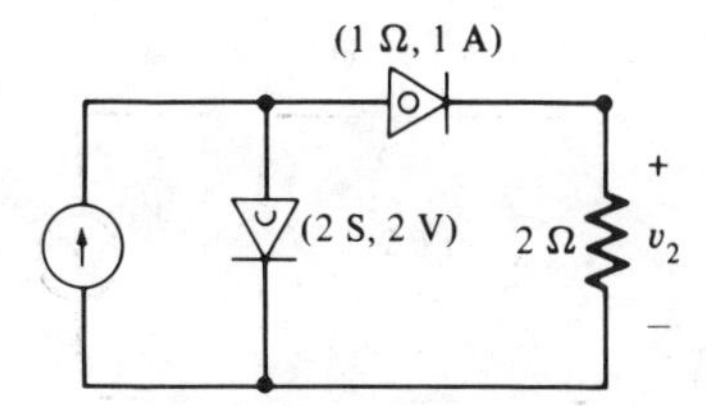

Figure P2.39

40 Find the v_o-v_i transfer characteristic of the circuit shown in Fig. P2.40 using the
(*a*) Graphic method.
(*b*) Analytic method.

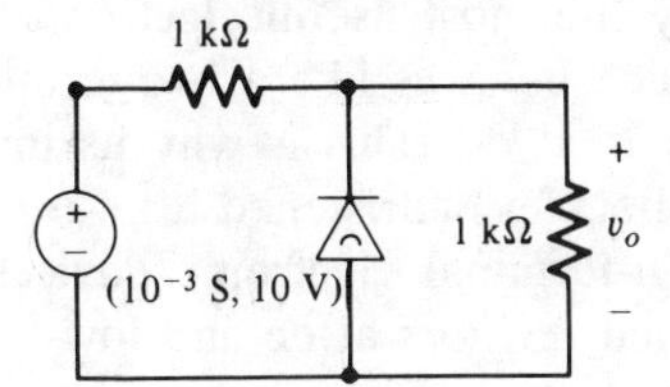

Figure P2.40

CHAPTER

THREE

MULTITERMINAL RESISTORS

While the conventional resistor is probably the most familiar circuit element, the transistor is certainly the most useful electronic device. A transistor is a three-terminal device which behaves like a two-terminal *resistor* when viewed from any pair of terminals at dc. This is why its inventors (Nobel laureates Bardeen, Bratten, and Shockley) christened it a "*transfer resistor*," or transistor in brief. Numerous n-terminal electronic devices available commercially also behave like n-terminal *resistors* at dc and low-frequency operations. The purpose of this chapter is to generalize our study from two-terminal resistors to three-terminal and multiterminal resistors.

A resistive two-port is a two-port which is made of resistive elements only. We will start this chapter with the analysis of a simple linear resistive two-port. Recall that a three-terminal element can be viewed as a grounded two-port. Thus by studying resistive two-ports we automatically bring three-terminal resistors into consideration. Both are specified in terms of a set of two voltages and a set of two currents. Therefore, we need to generalize the v-i characteristic of a two-terminal resistor (or a resistive one-port) to a vector relation among the two voltages and two currents of a three-terminal resistor (or a resistive two-port).

We will introduce some ideal two-port elements which are especially useful in modeling. We will present the dc characteristics of the bipolar and MOS transistors and illustrate them with examples. We conclude the chapter with brief discussions of two useful resistive three-ports and a four-terminal element. A more general discussion of two-ports and n-ports will be given in Chap. 13.

1 RESISTIVE TWO-PORTS

Figure 1.1 shows a three-terminal element with node ③ chosen as the datum node. There are two (terminal) voltages v_1 and v_2 and two (terminal) currents i_1 and i_2, and these are the circuit variables describing the three-terminal element. Figure 1.2 shows a two-port with port voltages v_1 and v_2 and port currents i_1 and i_2 as its circuit variables. Whereas three-terminal resistors often pertain to an intrinsic device in practice, two-port resistors are usually made up of an interconnection of *resistive* elements (e.g., two-terminal resistors, three-terminal resistors, etc.). Therefore, for convenience, we often speak of resistive two-ports in this and succeeding sections instead of three-terminal resistors.

The generalization from a two-terminal resistor to a resistive two-port amounts to extending from scalar voltage and current variables to two-dimensional vector voltage and current variables. In other words, the v-i characteristic of a resistive one-port is generalized to a relation between two vectors: the *port voltage vector* $\mathbf{v}$ and the *port current vector* $\mathbf{i}$, where

$$\mathbf{v} = \begin{bmatrix} v_1 \\ v_2 \end{bmatrix} \quad \text{and} \quad \mathbf{i} = \begin{bmatrix} i_1 \\ i_2 \end{bmatrix}$$

and we need *two* equations in general to express the relation.

A *three-terminal* element or a *two-port* will be called a (time-invariant) resistor if its port voltages and port currents satisfy the following relation:

$$\mathscr{R}_R = \{(v_1, v_2, i_1, i_2): f_1(v_1, v_2, i_1, i_2) = 0 \text{ and } f_2(v_1, v_2, i_1, i_2) = 0\} \tag{1.1}$$

This relation,[1] similar to the two-terminal resistor given by Eq. (1.2) in Chap. 2, will be called the *v-i characteristic of a three-terminal resistor or a resistive two-port*. The difference is that we need *two* scalar functions $f_1(\cdot)$ and $f_2(\cdot)$ to characterize a two-port and there are four scalar variables v_1, v_2, i_1, and i_2; the characteristic is in general a two-dimensional surface in a four-dimensional space.

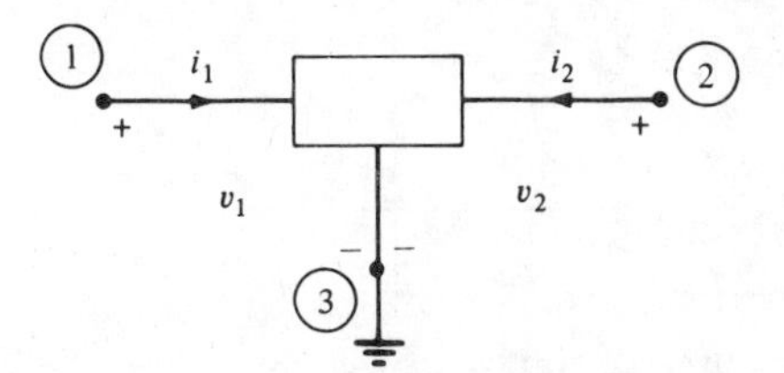

Figure 1.1 Three-terminal element with node ③ chosen as the datum node.

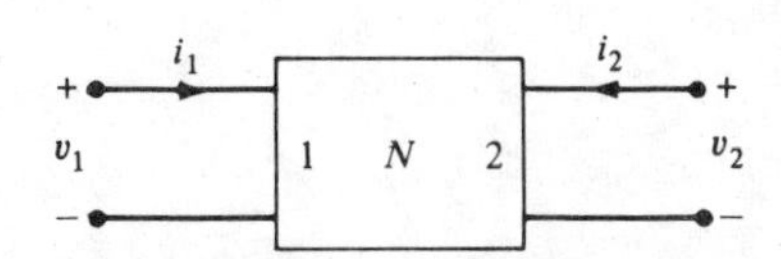

Figure 1.2 A two-port with its port voltage v_1, v_2 and port currents i_1, i_2.

[1] If, in addition, the functions f_1 and f_2 in Eq. (1.1) depend explicitly on time t, $\mathscr{R}_R$ is called *time-varying*.

When we deal with two-ports, we often need to distinguish the ports, so one of them is marked port 1 and the other is marked port 2 as shown in Fig. 1.2. As a tradition port 1 is often referred to as the *input port* and port 2 is often referred to as the *output port*.

In the following we will first consider linear resistors and use them to bring out pertinent concepts in the generalization from a one-port to a two-port.

1.1 A Linear Resistive Two-Port Example

Consider a resistive two-port made up of three linear resistors as shown in Fig. 1.3. Let us apply two independent current sources to the two-port as shown. KCL applied to nodes ①, ②, and ③ yields

$$i_{s1} = i_1 \qquad i_{s2} = i_2 \qquad i_3 = i_1 + i_2$$

Using Ohm's law and KVL for node sequences ①–③–④–① and ②–③–④–②, we obtain the two equations characterizing the resistive two-port:

$$v_1 = R_1 i_1 + R_3(i_1 + i_2) = (R_1 + R_3)i_1 + R_3 i_2 \tag{1.2a}$$

$$v_2 = R_2 i_2 + R_3(i_1 + i_2) = R_3 i_1 + (R_2 + R_3)i_2 \tag{1.2b}$$

In terms of the port voltage vector and the port current vector, we can rewrite the above equations in matrix form as

$$\mathbf{v} = \begin{bmatrix} \mathbf{v}_1 \\ \mathbf{v}_2 \end{bmatrix} = \mathbf{Ri} = \begin{bmatrix} R_1 + R_3 & R_3 \\ R_3 & R_2 + R_3 \end{bmatrix} \begin{bmatrix} i_1 \\ i_2 \end{bmatrix} \tag{1.3}$$

where

$$\mathbf{R} \triangleq \begin{bmatrix} R_1 + R_3 & R_3 \\ R_3 & R_2 + R_3 \end{bmatrix} \tag{1.4}$$

is called the *resistance matrix* of the *linear* resistive two-port. It is linear because $\mathbf{v} = \mathbf{Ri}$ expresses $\mathbf{v}$ as a linear function of $\mathbf{i}$.[2] Equation (1.3) gives the

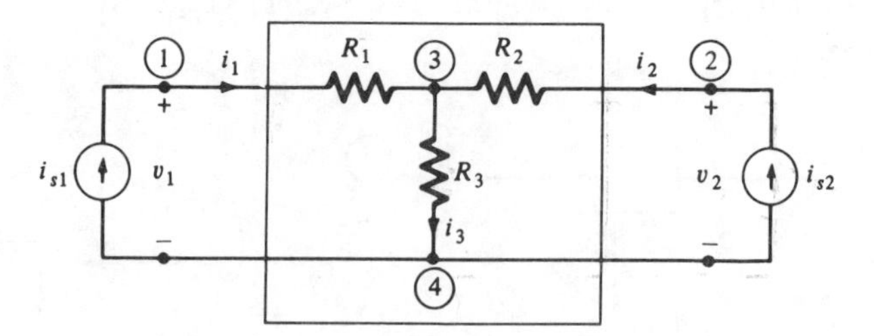

Figure 1.3 A linear resistive two-port.

[2] As will be seen later, a two-port containing independent sources and linear circuit elements is defined as a linear two-port. This is similar to the definition of a linear circuit. If a linear two-port contains independent sources, it will have an *affine* representation. This is discussed in Chap. 5, Sec. 4.

current-controlled representation of the linear resistive two-port because the voltages are expressed as functions of currents. The vector equation $\mathbf{v} = \mathbf{Ri}$ represents two linear constraints imposed by the two-port on the four variables v_1, v_2, i_1, and i_2. [See Eq. (1.1).]

In the circuit in Fig. 1.3, the two currents $i_1 = i_{s1}$ and $i_2 = i_{s2}$ are the *sources* and the two voltages v_1 and v_2 are the *responses*. Thus i_1 and i_2 in Eqs. (1.2*a* and *b*) are the independent variables and v_1 and v_2 are the dependent variables. Of course, we can also solve for i_1 and i_2 in terms of v_1 and v_2 from the two equations (1.2*a* and *b*), or directly from the vector equation (1.3) to obtain

$$\mathbf{i} = \mathbf{Gv} \tag{1.5}$$

where
$$\mathbf{G} \overset{\Delta}{=} \mathbf{R}^{-1} = \frac{1}{R_1R_2 + R_2R_3 + R_3R_1} \begin{bmatrix} R_2 + R_3 & -R_3 \\ -R_3 & R_1 + R_3 \end{bmatrix} \tag{1.6}$$

is called the *conductance matrix* of the *linear* resistive two-port. In scalar form, Eqs. (1.5) and (1.6) can be written as

$$i_1 = \frac{R_2 + R_3}{R_1R_2 + R_2R_3 + R_3R_1} v_1 - \frac{R_3}{R_1R_2 + R_2R_3 + R_3R_1} v_2 \tag{1.7}$$

$$i_2 = \frac{-R_3}{R_1R_2 + R_2R_3 + R_3R_1} v_1 + \frac{R_1 + R_3}{R_1R_2 + R_2R_3 + R_3R_1} v_2 \tag{1.8}$$

The two equations above give an alternative representation of the same two-port. It is called the *voltage-controlled representation*. If we view the resistance matrix $\mathbf{R}$ of a two-port as the generalization of the resistance R of a linear two-terminal resistor, then the conductance matrix $\mathbf{G} = \mathbf{R}^{-1}$ of the same two-port is the generalization of the conductance $G = 1/R$ of the same two-terminal resistor.

1.2 Six Representations

With four scalar variables v_1, v_2, i_1, and i_2 and two equations to characterize a resistive two-port, there exist other representations besides the two just introduced. Since in most instances we may choose any two of the four variables as independent variables, the remaining two are then the dependent variables. Thus, altogether there are $C_2^4 = (4 \times 3)/2 = 6$ possibilities. Table 3.1 gives the classification of the six representations according to independent and dependent variables. Table 3.2 gives the equations of the six possible representations of a linear resistive two-port. As pointed out in Sec. 1.1, $\mathbf{G}$ is the inverse matrix of $\mathbf{R}$. Similarly, from Table 3.2, we also have $\mathbf{H}' = \mathbf{H}^{-1}$ and $\mathbf{T}' = \mathbf{T}^{-1}$. We call $\mathbf{H}$ and $\mathbf{H}'$ the *hybrid matrices* because both the dependent and independent variables are mixtures of a voltage and a current. We call $\mathbf{T}$ and $\mathbf{T}'$ the *transmission matrices* because they relate the variables pertaining to one port to that pertaining to the other and the two-port serves as a

Table 3.1 Six representations of a two-port

Representations	Independent variables	Dependent variables
Current-controlled	i_1, i_2	v_1, v_2
Voltage-controlled	v_1, v_2	i_1, i_2
Hybrid 1	i_1, v_2	v_1, i_2
Hybrid 2	v_1, i_2	i_1, v_2
Transmission 1	v_2, i_2	v_1, i_1
Transmission 2	v_1, i_1	v_2, i_2

Table 3.2 Equations for the six representations of a linear resistive two-port

Representations	Scalar equations	Vector equations
Current-controlled	$v_1 = r_{11}i_1 + r_{12}i_2$ $v_2 = r_{21}i_1 + r_{22}i_2$	$\mathbf{v} = \mathbf{Ri}$
Voltage-controlled	$i_1 = g_{11}v_1 + g_{12}v_2$ $i_2 = g_{21}v_1 + g_{22}v_2$	$\mathbf{i} = \mathbf{Gv}$
Hybrid 1	$v_1 = h_{11}i_1 + h_{12}v_2$ $i_2 = h_{21}i_1 + h_{22}v_2$	$\begin{bmatrix} v_1 \\ i_2 \end{bmatrix} = \mathbf{H}\begin{bmatrix} i_1 \\ v_2 \end{bmatrix}$
Hybrid 2	$i_1 = h'_{11}v_1 + h'_{12}i_2$ $v_2 = h'_{21}v_1 + h'_{22}i_2$	$\begin{bmatrix} i_1 \\ v_2 \end{bmatrix} = \mathbf{H}'\begin{bmatrix} v_1 \\ i_2 \end{bmatrix}$
Transmission 1†	$v_1 = t_{11}v_2 - t_{12}i_2$ $i_1 = t_{21}v_2 - t_{22}i_2$	$\begin{bmatrix} v_1 \\ i_1 \end{bmatrix} = \mathbf{T}\begin{bmatrix} v_2 \\ -i_2 \end{bmatrix}$
Transmission 2†	$v_2 = t'_{11}v_1 + t'_{12}i_1$ $-i_2 = t'_{21}v_1 + t'_{22}i_1$	$\begin{bmatrix} v_2 \\ -i_2 \end{bmatrix} = \mathbf{T}'\begin{bmatrix} v_1 \\ i_1 \end{bmatrix}$

† For historical reasons, a minus sign is used in conjunction with i_2. Because of the reference direction chosen for i_2, $-i_2$ gives the current leaving the output port.

transmission media. Transmission matrices are important in the study of communication networks and will be treated in Chap. 13.

Example Consider the two-port in Fig. 1.3. Let $R_1 = 1\ \Omega$, $R_2 = 2\ \Omega$, and $R_3 = 3\ \Omega$. Equations (1.2*a* and *b*) give the following current-controlled representation:

$$v_1 = 4i_1 + 3i_2 \tag{1.9}$$

$$v_2 = 3i_1 + 5i_2 \tag{1.10}$$

The voltage-controlled representation is given by Eqs. (1.7) and (1.8):

$$i_1 = \tfrac{5}{11}v_1 - \tfrac{3}{11}v_2 \tag{1.11}$$

$$i_2 = -\tfrac{3}{11}v_1 + \tfrac{4}{11}v_2 \tag{1.12}$$

It is straightforward to derive the other four representations from the equations above. The general treatment will be given in Chap. 13; however, it is easy to obtain, for example, the hybrid representations: Using Eqs. (1.10) and (1.11) and solving for i_1 in terms of v_1 and i_2, we have

$$i_1 = \tfrac{1}{4}v_1 - \tfrac{3}{4}i_2 \tag{1.13}$$

which is the first equation of the hybrid 2 representation. Similarly, we obtain the second equation

$$v_2 = \tfrac{3}{4}v_1 + \tfrac{11}{4}i_2 \tag{1.14}$$

Equations (1.13) and (1.14) define the hybrid 2 matrix

$$\mathbf{H}' = \begin{bmatrix} \frac{1}{4} & -\frac{3}{4} \\ \frac{3}{4} & \frac{11}{4} \end{bmatrix}$$

The hybrid 1 matrix can be obtained by simply taking the inverse of $\mathbf{H}'$, thus

$$\mathbf{H} = \mathbf{H}'^{-1} = \begin{bmatrix} \frac{1}{4} & -\frac{3}{4} \\ \frac{3}{4} & \frac{11}{4} \end{bmatrix}^{-1} = \begin{bmatrix} \frac{11}{5} & \frac{3}{5} \\ -\frac{3}{5} & \frac{1}{5} \end{bmatrix}$$

Exercise Determine the transmission 1 and the transmission 2 reptions of the resistive two-port shown in Fig. 1.3.

1.3 Physical Interpretations

In the example in Sec. 1.1, we derived the current-controlled representation by using two current sources at the two ports and determining the two port voltages. This is shown in Fig. 1.4*a*. We can similarly interpret the voltage-controlled and the two hybrid representations by using appropriate sources as shown in Fig. 1.4*b*, *c*, and *d*. It is seen that in the two hybrid representations, we use a current source and a voltage source as inputs, thus the responses are a voltage and a current.

Current-controlled representation In Chap. 2 we defined a linear two-terminal resistor as one having a straight line characteristic passing through the origin in the *v-i* plane. For two-ports we have four variables and two equations, e.g., the current-controlled representation is

$$\begin{aligned} v_1 &= r_{11}i_1 + r_{12}i_2 \\ v_2 &= r_{21}i_1 + r_{22}i_2 \end{aligned} \tag{1.15}$$

These two equations impose two linear constraints on the port voltages and the port currents and hence the point representing the four variables; namely,

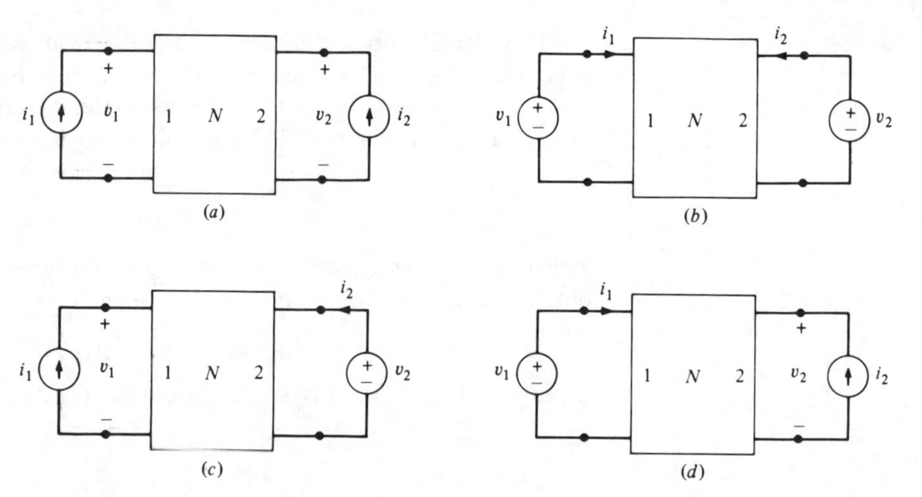

Figure 1.4 Sources and responses to two-ports for (*a*) current-controlled representation, (*b*) voltage-controlled representation, (*c*) hybrid 1 representation, and (*d*) hybrid 2 representation.

(v_1, v_2, i_1, i_2), is constrained to a two-dimensional subspace in the four-dimensional space spanned by v_1, v_2, i_1, i_2. Of course this is difficult to visualize. However, if we take one equation at a time, we can represent it by a family of curves in the appropriate *i-v* planes. Consider plotting, in the i_1-v_1 *plane*, the straight lines

$$v_1 = r_{11}i_1 + r_{12}i_2$$

where i_2 is considered as a *fixed parameter* which is given successively several values. The result is a family of parallel straight lines with a slope equal to r_{11}. These straight lines are as shown in Fig. 1.5*a*. Similarly, in Fig. 1.5*b* we plot, in the i_2-v_2 plane, the straight lines

$$v_2 = r_{21}i_1 + r_{22}i_2$$

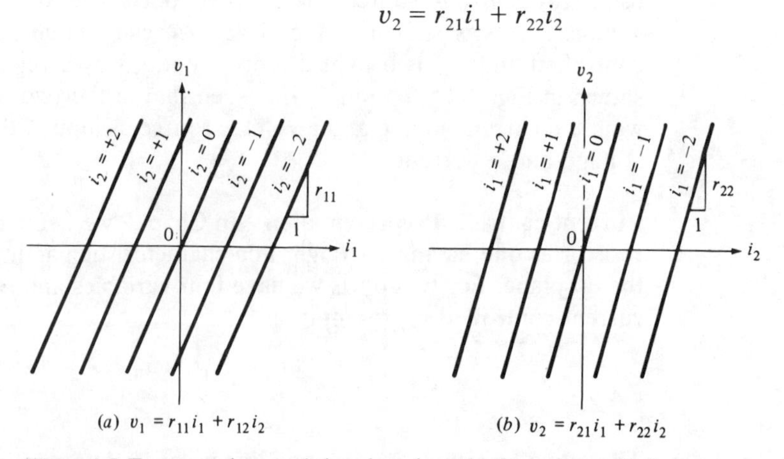

Figure 1.5 Two-port characteristics plotted on the i_1-v_1 plane with i_2 as a parameter and on the i_2-v_2 plane with i_1 as a parameter.

and we use i_1 as a parameter. These two families of parallel straight lines in the i_1-v_1 plane and i_2-v_2 plane depict the current-controlled representation of the linear resistive two-port described by Eq. (1.15).

From the first equation in (1.15) we can give the following interpretations to r_{11} and r_{12}:

$$r_{11} = \left.\frac{v_1}{i_1}\right|_{i_2=0} \tag{1.16}$$

Thus r_{11} is called the *driving-point resistance at port* 1 when $i_2 = 0$, i.e., port 2 is kept open-circuited. Also, from the first equation of (1.15), if i_1 is set to zero, we obtain $v_1 = r_{12}i_2$. Thus the v_1-axis intercept of the $i_2 = 1$ characteristic in the i_1-v_1 plane is equal to r_{12}. Like Eq. (1.16), r_{12} can be interpreted by

$$r_{12} = \left.\frac{v_1}{i_2}\right|_{i_1=0} \tag{1.17}$$

which is called the *transfer resistance* when $i_1 = 0$, i.e., port 1 is kept open-circuited.

Similarly, from the second equation in (1.15), we can give the following interpretations:

$$r_{21} = \left.\frac{v_2}{i_1}\right|_{i_2=0} \tag{1.18}$$

$$r_{22} = \left.\frac{v_2}{i_2}\right|_{i_1=0} \tag{1.19}$$

where r_{22} is the *driving-point resistance at port* 1 when $i_1 = 0$, i.e., port 1 is kept open-circuited; and r_{21} is the *transfer resistance* when $i_2 = 0$, i.e., port 2 is kept open-circuited.

In Fig. 1.6 we give the physical interpretations of r_{11}, r_{12}, r_{21}, and r_{22} according to Eqs. (1.16) to (1.19). Note that in each case the input is a current source and the response is a voltage across a port which is open-circuited. Recalling, from Chap. 2, that an independent current source has infinite internal resistance, we call the resistance matrix $\mathbf{R}$ the *open-circuit resistance matrix* and the four parameters r_{11}, r_{12}, r_{21} and r_{22} *open-circuit resistance parameters* of the linear resistive two-port. More specifically, r_{11} and r_{22} are the *open-circuit driving-point resistances* at port 1 and port 2, respectively; r_{21} and r_{12} are the *open-circuit forward* and *reverse transfer resistances*, respectively.

It is easy to go through a dual treatment to give the corresponding interpretations for the voltage-controlled representation. We shall leave that as an exercise.

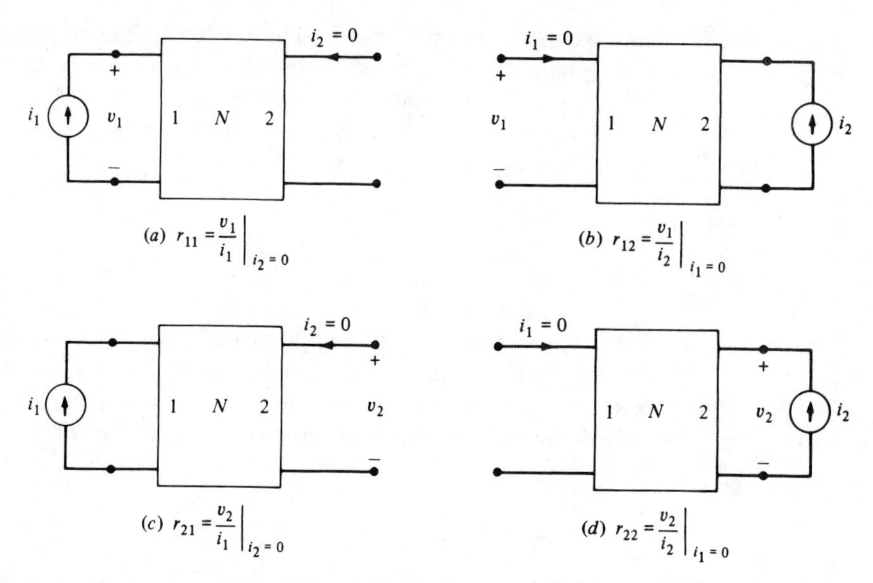

Figure 1.6 Interpretations of (a) r_{11}, (b) r_{12}, (c) r_{21}, and (d) r_{22}.

Exercise Give the physical interpretation of the voltage-controlled representation of a linear resistive two-port. Use Fig. 1.7 to interpret the *short-circuit conductances* g_{11}, g_{12}, g_{21}, and g_{22} in terms of port voltages and port currents.

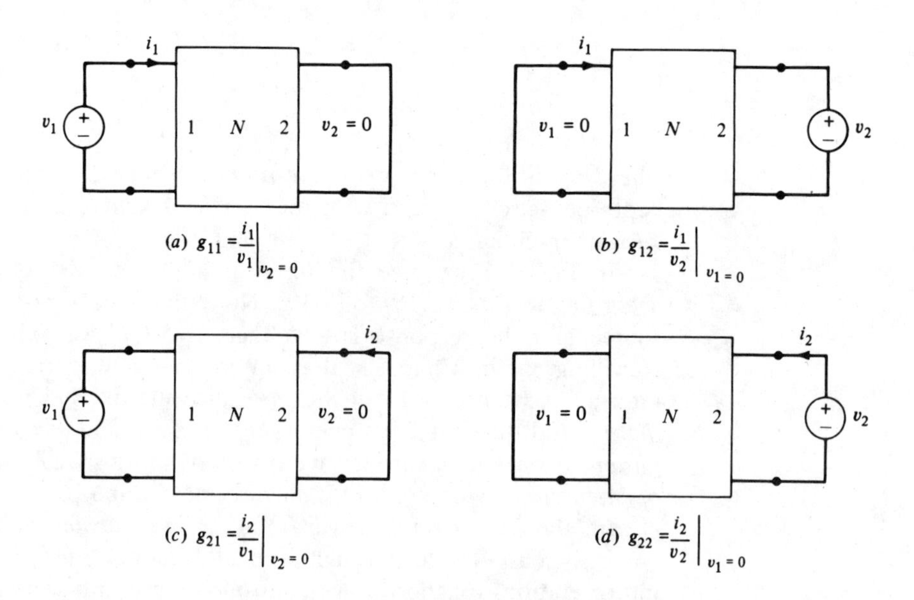

Figure 1.7 Interpretation of (a) g_{11}, (b) g_{12}, (c) g_{21}, and (d) g_{22}.

Hybrid representation The two equations for the hybrid 1 representation read:

$$v_1 = h_{11} i_1 + h_{12} v_2 \tag{1.20}$$

$$i_2 = h_{21} i_1 + h_{22} v_2 \tag{1.21}$$

Following the same treatment as the current-controlled representation, we write

$$h_{11} = \left.\frac{v_1}{i_1}\right|_{v_2=0} \tag{1.22}$$

$$h_{12} = \left.\frac{v_1}{v_2}\right|_{i_1=0} \tag{1.23}$$

$$h_{21} = \left.\frac{i_2}{i_1}\right|_{v_2=0} \tag{1.24}$$

$$h_{22} = \left.\frac{i_2}{v_2}\right|_{i_1=0} \tag{1.25}$$

The physical interpretation in terms of sources, responses, and external connections for the four hybrid parameters is given in Fig. 1.8.

Note that the four hybrid parameters h_{11}, h_{12}, h_{21}, and h_{22} represent a driving-point resistance, a reverse *voltage transfer ratio*, a forward *current transfer ratio*, and a driving-point conductance, respectively. Furthermore, in each case, the external connection for port 1 is either a current source or an open circuit, and the external connection for port 2 is either a voltage source

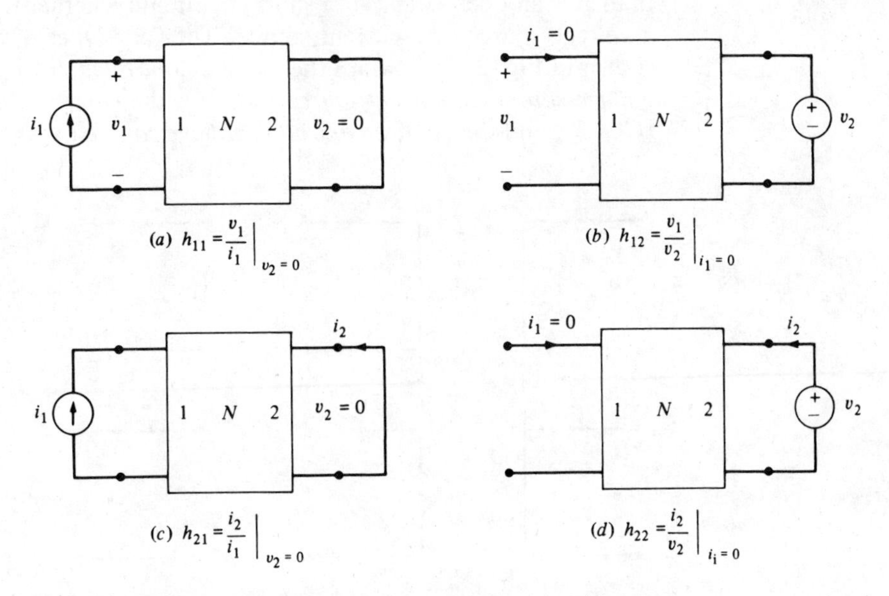

Figure 1.8 Interpretations of (*a*) h_{11}, (*b*) h_{12}, (*c*) h_{21}, and (*d*) h_{22}.

or a short circuit. As will be seen later in the chapter, hybrid representation is commonly used in dealing with the common-emitter configuration of the bipolar transistor.

2 USEFUL RESISTIVE TWO-PORTS

There are a number of (ideal) two-port circuit elements which are extremely useful in modeling and in exhibiting specific properties of devices. We will describe the most important ones in this section, namely, the linear controlled sources, the ideal transformer, and the gyrator. All of them are *linear* circuit elements and are characterized in terms of the four port variables v_1, v_2, i_1, and i_2. Thus they are resistive *two-ports* according to our definition.

2.1 Linear Controlled Sources

Up to this point we have encountered independent voltage sources and current sources. Independent sources are used as inputs to a circuit. In this section we will introduce another type of sources, called *controlled sources* or *dependent sources*. A *controlled source* is a resistive two-port element consisting of two branches: a primary branch which is either an open circuit or a short circuit and a secondary branch which is either a voltage source or a current source. The voltage or current waveform in the secondary branch is *controlled* by (or dependent upon) the voltage or current of the primary branch. Therefore, there exist four types of controlled sources depending on whether the primary branch is an open circuit or a short circuit and whether the secondary branch is a voltage source or a current source. The four types of controlled sources are shown in Fig. 2.1. They are the *current-controlled voltage source* (CCVS), the *voltage-controlled current source* (VCCS), the *current-controlled current source* (CCCS), and the *voltage-controlled voltage source* (VCVS). Note that we use a

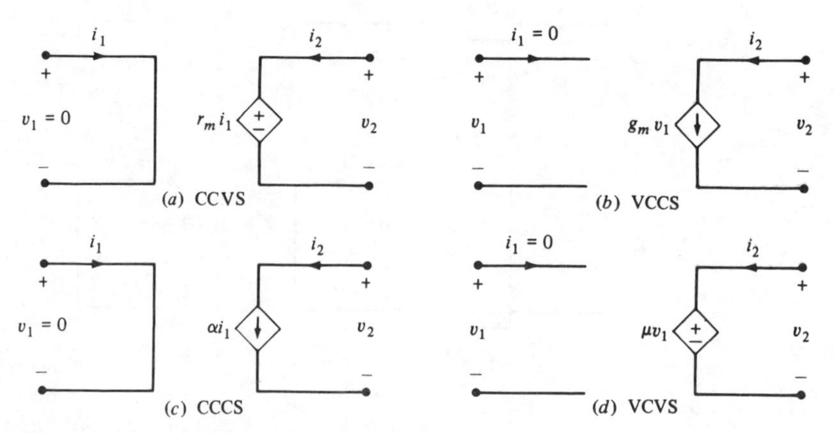

Figure 2.1 Four types of linear controlled sources.

diamond-shaped symbol to denote controlled sources. This is to differentiate them from the independent sources which, as we see repeatedly in the sequel, have very different properties. Each linear controlled source is characterized by two linear equations.

$$\text{CCVS:} \qquad v_1 = 0 \qquad v_2 = r_m i_1 \tag{2.1}$$

$$\text{VCCS:} \qquad i_1 = 0 \qquad i_2 = g_m v_1 \tag{2.2}$$

$$\text{CCCS:} \qquad v_1 = 0 \qquad i_2 = \alpha i_1 \tag{2.3}$$

$$\text{VCVS:} \qquad i_1 = 0 \qquad v_2 = \mu v_1 \tag{2.4}$$

where r_m is called the *transresistance*, g_m is called the *transconductance*, α is called the *current transfer ratio*, and μ is called the *voltage transfer ratio*. They are all constants, thus the four controlled sources are linear time-invariant two-port resistors. More generally, if a CCVS is characterized by the two equations: $v_1 = 0$ and $v_2 = f(i_1)$, where $f(\cdot)$ is a given *nonlinear* function, then that CCVS is a *nonlinear controlled source*. Similarly, if a CCCS is characterized by the two equations: $v_1 = 0$, $i_2 = \alpha(t) i_1$, where $\alpha(\cdot)$ is a given function of time, then this CCCS is a *linear time-varying controlled source*.

In the previous section we demonstrated with an example that a linear resistive two-port has six representations. In the case of linear controlled sources, Eqs. (2.1) to (2.4) can be put in matrix form each corresponding to one representation:

$$\text{CCVS:} \qquad \begin{bmatrix} v_1 \\ v_2 \end{bmatrix} = \begin{bmatrix} 0 & 0 \\ r_m & 0 \end{bmatrix} \begin{bmatrix} i_1 \\ i_2 \end{bmatrix} \tag{2.5}$$

$$\text{VCCS:} \qquad \begin{bmatrix} i_1 \\ i_2 \end{bmatrix} = \begin{bmatrix} 0 & 0 \\ g_m & 0 \end{bmatrix} \begin{bmatrix} v_1 \\ v_2 \end{bmatrix} \tag{2.6}$$

$$\text{CCCS:} \qquad \begin{bmatrix} v_1 \\ i_2 \end{bmatrix} = \begin{bmatrix} 0 & 0 \\ \alpha & 0 \end{bmatrix} \begin{bmatrix} i_1 \\ v_2 \end{bmatrix} \tag{2.7}$$

$$\text{VCVS:} \qquad \begin{bmatrix} i_1 \\ v_2 \end{bmatrix} = \begin{bmatrix} 0 & 0 \\ \mu & 0 \end{bmatrix} \begin{bmatrix} v_1 \\ i_2 \end{bmatrix} \tag{2.8}$$

In Eq. (2.5) we have the current-controlled representation for the CCVS. Since the resistance matrix is singular, its inverse does not exist; therefore, there is no voltage-controlled representation for a CCVS. As a matter of fact, it is easy to see that neither of the hybrid representations exist as well. We can make similar statements for the other three controlled sources, i.e., only one of the representations in the first four rows of Table 3.2 exists.

Linear controlled sources are ideal *coupling* circuit elements, yet they are extremely useful in modeling electronic devices and circuits. This will be illustrated later in Sec. 4 of the chapter. In Chap. 4 we will see that all four controlled sources can be realized physically (to a good approximation) by using operational amplifiers.

Equivalent circuits of linear resistive two-ports Controlled sources are useful in modeling resistive two-ports. Consider the current-controlled representation of a linear resistive two-port:

$$v_1 = r_{11}i_1 + r_{12}i_2 \tag{2.9}$$

$$v_2 = r_{21}i_1 + r_{22}i_2 \tag{2.10}$$

Equation (2.9) can be interpreted as a series connection of a linear resistor with resistance r_{11} and a CCVS whose voltage is dependent on the current i_2 at port 2. Similarly, Eq. (2.10) can be interpreted as a series connection of a linear resistor with resistance r_{22} and a CCVS whose voltage is dependent on the current i_1 at port 1. Therefore, we may use the equivalent circuit shown in Fig. 2.2 to represent the two-port. This equivalent circuit puts in evidence the meanings of the four parameters r_{11}, r_{12}, r_{21}, and r_{22} introduced earlier. The reader should review Eqs. (1.16) to (1.19) and use Fig. 2.2 to give appropriate interpretations of the four resistance parameters.

Exercises

1. Show that the circuit shown in Fig. 2.3 is the equivalent circuit of a voltage-controlled linear resistive two-port.
2. Show that the circuit shown in Fig. 2.4 is the equivalent circuit corresponding to the hybrid 1 representation of a linear resistive two-port.

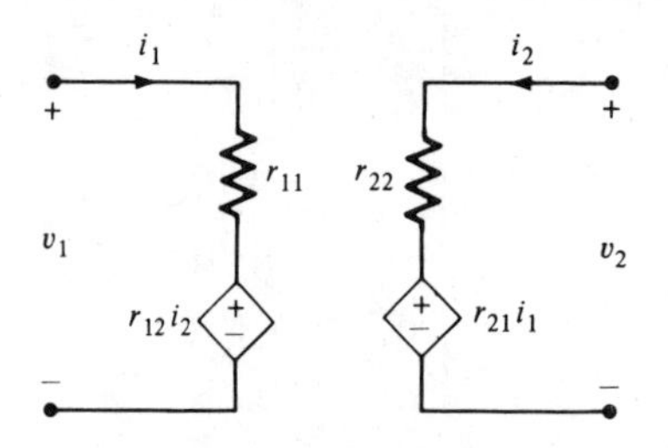

Figure 2.2 Equivalent circuit of a current-controlled linear resistive two-port.

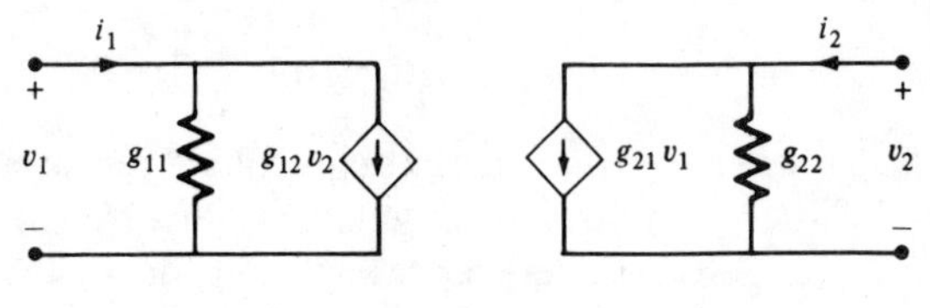

Figure 2.3 Equivalent circuit of a voltage-controlled linear resistive two-port.

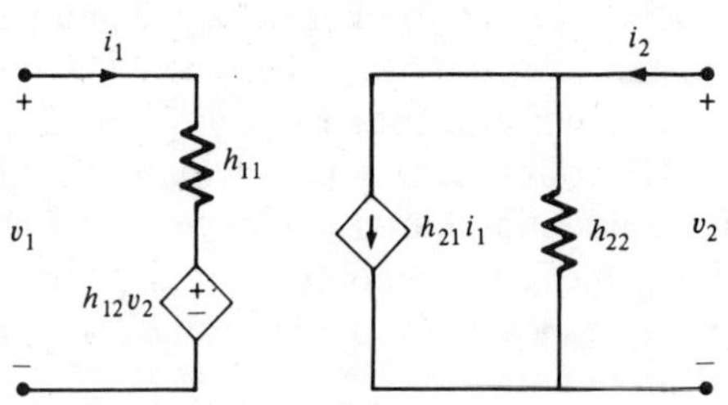

Figure 2.4 Equivalent circuit based on the hybrid 1 representation of a linear resistive two-port: h_{11} is a resistance (in ohms) and h_{22} is a conductance (in siemens).

In the following we will give three examples to illustrate additional properties of controlled sources.

Example 1 Consider the circuit shown in Fig. 2.5 where a battery with internal resistance R_b is connected to the primary branch of a VCVS. Since $i_1 = 0$ and $v_2 = \mu v_1 = \mu E_b$, the output behaves like an independent dc voltage source with zero internal resistance irrespective of the battery resistance R_b.

Example 2 In Fig. 2.6 the primary branch and the secondary branch of a CCVS are connected in such a fashion that the resulting one-port behaves like a linear resistor. Here $i_2 = i_1$ and $v_2 = r_m i_1 = r_m i_2$. Thus looking backwards from port 2, we have a linear resistor with resistance equal to r_m.

Exercise Modify the connection of the circuit in Fig. 2.6 so that the resistance seen at port 2 is $-r_m$.

Example 3 Consider the circuit in Fig. 2.7 where the CCCS has its two branches connected at a common node ① to a linear resistor with resistance R_1. An independent voltage source v_i serves as the input, and

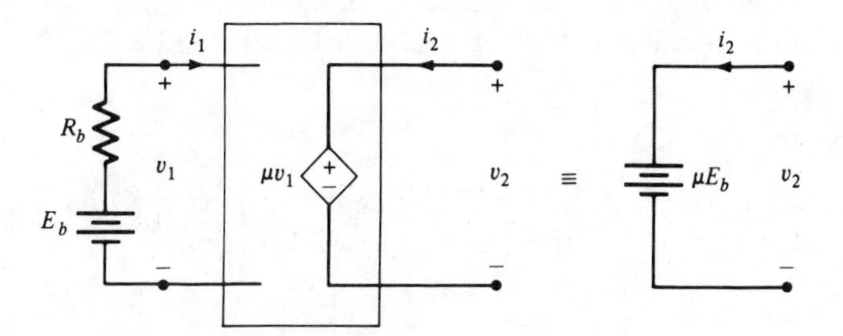

Figure 2.5 A VCVS connected at the primary side by a battery with internal resistance R_b functions as an independent dc voltage source.

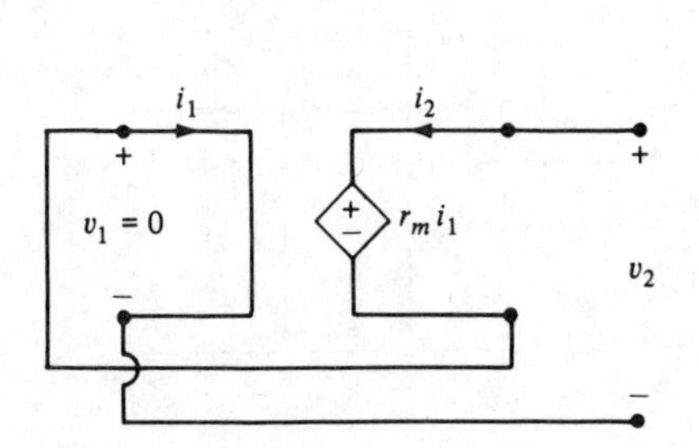

Figure 2.6 The CCVS functions as a linear resistor.

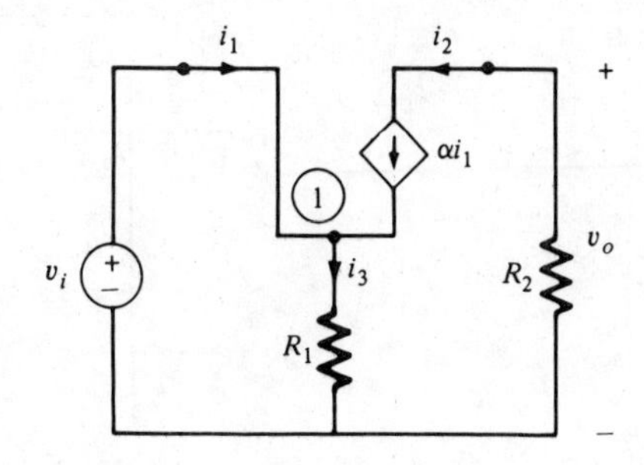

Figure 2.7 A power amplifier using a CCCS.

we wish to determine the output voltage v_o across the second linear resistor with resistance R_2 acting as a load. Using KCL at node ①, we have $i_3 = i_1 + i_2 = (1 + \alpha)i_1$. Thus

$$v_i = R_1(1 + \alpha)i_1$$

$$v_o = R_2(-i_2) = -R_2 \alpha i_1$$

The power delivered by the source v_i to the circuit is

$$p_i = v_i i_1 = R_1(1 + \alpha)i_1^2$$

The power delivered by the circuit to the load resistor R_2 is

$$p_o = v_o(-i_2) = R_2 \alpha^2 i_1^2$$

Therefore the power gain is

$$\frac{p_o}{p_i} = \frac{\alpha^2}{1 + \alpha} \frac{R_2}{R_1}$$

Clearly, by choosing R_1 and R_2 we can obtain an arbitrarily large gain for any given value of α. Thus controlled sources can be used in the design of a power amplifier.

Exercise Demonstrate that a nonlinear controlled source in Fig. 2.8*a* can be realized with the circuit shown in Fig. 2.8*b*, which contains a two-terminal nonlinear resistor, a CCCS, and a CCVS.

2.2 Ideal Transformer

The *ideal transformer* is an ideal two-port resistive circuit element which is characterized by the following two equations:

$$v_1 = nv_2 \tag{2.11}$$

and

$$i_2 = -ni_1 \tag{2.12}$$

where n is a real number called the *turns ratio*. The symbol for the ideal

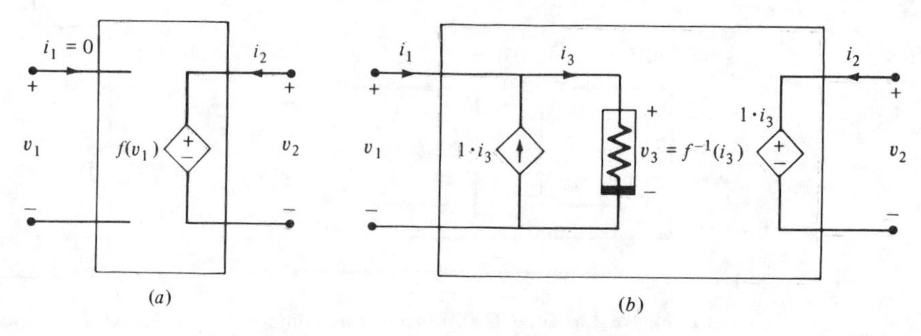

Figure 2.8 (*a*) A nonlinear VCVS and (*b*) its equivalent realization using linear controlled sources.

transformer is shown in Fig. 2.9. The ideal transformer is a *linear* resistive two-port since its equations impose *linear* constraints on its port voltages and port currents. Note that neither the current-controlled representation nor the voltage-controlled representation exists for the ideal transformer. Equations (2.11) and (2.12) can be written in matrix form in terms of the hybrid matrix representation:

$$\begin{bmatrix} v_1 \\ i_2 \end{bmatrix} = \mathbf{H} \begin{bmatrix} i_1 \\ v_2 \end{bmatrix} = \begin{bmatrix} 0 & n \\ -n & 0 \end{bmatrix} \begin{bmatrix} i_1 \\ v_2 \end{bmatrix} \tag{2.13}$$

The ideal transformer is an idealization of a physical transformer that is used in many applications. The properties of the physical transformer will be discussed in Chap. 8.

We wish to stress that because the ideal transformer is an ideal element defined by Eqs. (2.11) and (2.12), the relation between port voltages and between port currents holds for *all* waveforms and for *all* frequencies, including dc.

Two fundamental properties of the ideal transformer

1. The ideal transformer neither dissipates nor stores energy. Indeed, the power entering the two-port at time t from Eq. (2.13) is

$$p(t) = v_1(t)i_1(t) + v_2(t)i_2(t) = 0 \tag{2.14}$$

   Thus, like the ideal diode, the ideal transformer is a *non-energic element.*
2. When an ideal transformer is terminated at the output port with an R-Ω linear resistor, the input port behaves as a linear resistor with resistance $n^2R\,\Omega$. This situation is shown in Fig. 2.10 where

$$v_2 = -Ri_2$$

   Therefore
$$\frac{v_1}{i_1} = \frac{nv_2}{-i_2/n} = n^2R \tag{2.15}$$

   is the resistance of the equivalent linear resistor at the input port.

Mechanical analog The ideal transformer is the electrical analog of an ideal pair of mechanical gears shown in Fig. 2.11. Since the velocity at A, the point

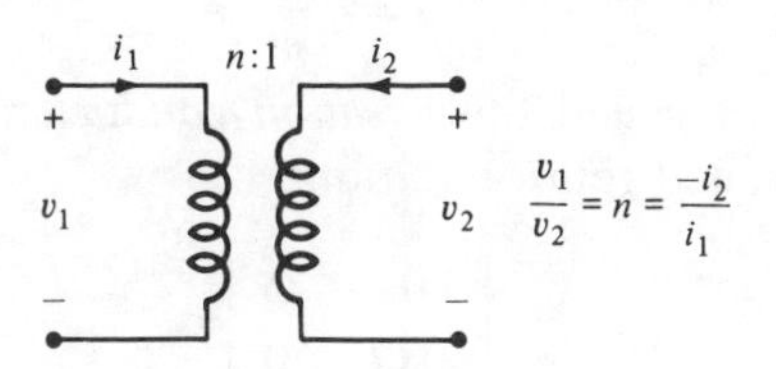

Figure 2.9 An ideal transformer defined by the single parameter n, the turns ratio.

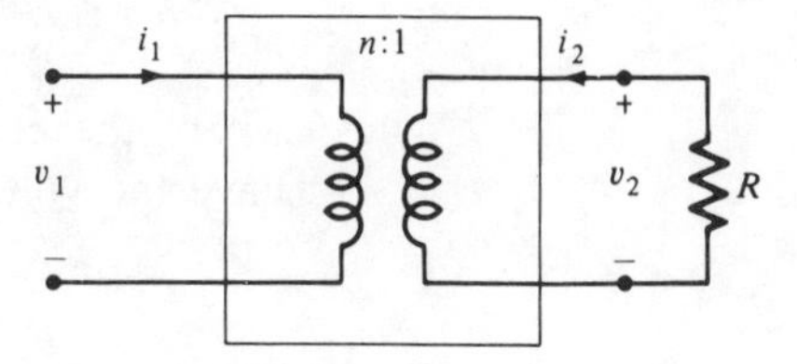

Figure 2.10 An ideal transformer terminated at the output port with a resistor.

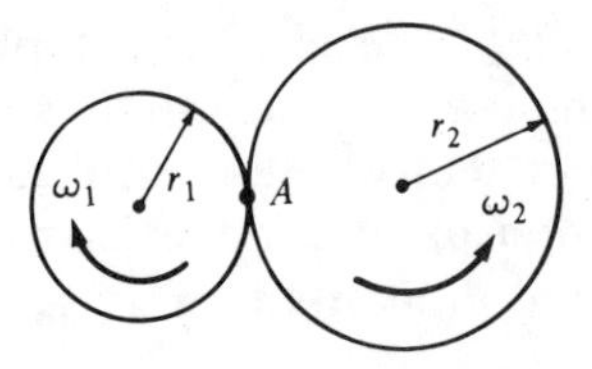

Figure 2.11 A pair of gears.

of contact of the gears, must be the same for both gears, we have

$$\omega_1 r_1 = \omega_2 r_2 = \text{velocity of } A$$

where ω_1 and ω_2 are the angular velocities of the two gears [radians per second (rad/s)], and r_1 and r_2 are the radii. On the other hand the forces applied by one gear to the other at point A must be equal in magnitude and opposite in direction, thus

$$\frac{\tau_1}{r_1} = -\frac{\tau_2}{r_2} = \text{force}$$

where τ_1 and τ_2 are the torques applied to the two gears.

Comparing the two defining equations for the ideal transformer with the two equations above for the pair of gears, we note the following analogs:

Electric circuits	Mechanical systems
Voltage v	Angular velocity ω
Current i	Torque τ
Turns ratio n	Radius ratio r_2/r_1

Mechanical analogs are helpful in understanding the property of electric circuits. In Chap. 7 we will encounter other analogous mechanical systems when we introduce other circuit elements: capacitors and inductors.

2.3 Gyrator

A *gyrator* is an ideal two-port element defined by the following equations:

$$i_1 = Gv_2 \tag{2.16}$$

and

$$i_2 = -Gv_1 \tag{2.17}$$

where the constant G is called the *gyration conductance*. In vector form we have the voltage-controlled representation:

$$\mathbf{i} = \begin{bmatrix} 0 & G \\ -G & 0 \end{bmatrix} \mathbf{v} \tag{2.18}$$

The symbol for a gyrator is shown in Fig. 2.12. It is easy to check that the ideal

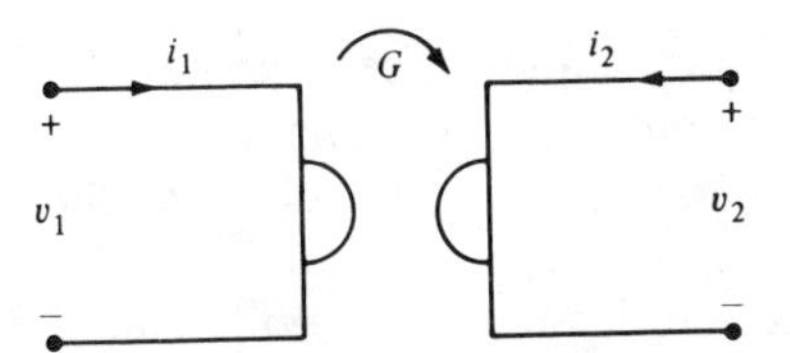

Figure 2.12 An ideal gyrator.

gyrator is also a non-energic element, i.e., at all times the power delivered to the two-port is identically zero.

The fundamental property of an ideal gyrator is given by the following equation:

$$\frac{v_1}{i_1} = \frac{-i_2/G}{Gv_2} = \frac{1}{G^2}\left(\frac{-i_2}{v_2}\right) \tag{2.19}$$

That is, when a gyrator is terminated at the output port with an R_L-Ω linear resistor as shown in Fig. 2.13, the input port behaves as a linear resistor with resistance $G_L/G^2\ \Omega$, where $G_L \stackrel{\Delta}{=} 1/R_L$. As will be clear in Chap. 6, if the output port of an ideal gyrator is terminated with a capacitor, the input port behaves like an inductor. Thus a gyrator is a useful element in the design of inductorless filters.

Another interesting observation is the following: If the output port of a gyrator is connected to a current-controlled two-terminal resistor, i.e., $v_2 = f(-i_2)$, then the input port becomes a voltage-controlled resistor. For example, setting $G = 1$ in Eqs. (2.16) and (2.17), we easily obtain

$$i_1 = v_2 = f(-i_2) = f(v_1)$$

The resulting current-controlled resistor is then the dual of the original voltage-controlled resistor.

Physical gyrators which approximate the property of an ideal gyrator over low operating frequencies (below 10 kHz) are available commercially in the form of integrated circuit modules.

3 NONLINEAR RESISTIVE TWO-PORTS

In the previous two sections we discussed linear resistive two-ports and their various characterizations and properties. In the real world we need to deal with

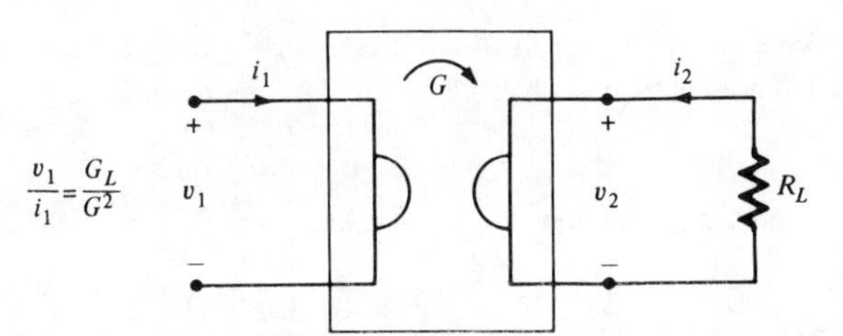

Figure 2.13 A gyrator terminated at the output port with a resistor.

nonlinear resistive two-ports and three-terminal devices such as transistors. Much of the material given in the previous two sections can be extended and generalized to the *nonlinear* case. This section illustrates the basic idea of nonlinear two-port representation with an example. Sections 4 and 5 will treat the modeling and characterizations of bipolar and MOS transistors together with dc analysis and the small-signal analysis.

Recall that the definition of a resistive two-port is expressed by Eq. (1.1) in terms of two functions and four variables. Thus, for example, the following two equations characterize a nonlinear two-port resistor:

$$\begin{aligned} f_1(v_1, v_2, i_1, i_2) &= i_1 + 2i_2 - (v_1 + v_2)^3 - 2(v_2 - v_1)^{1/3} = 0 \\ f_2(v_1, v_2, i_1, i_2) &= 2i_1 - i_2 - 2(v_1 + v_2)^3 + (v_2 - v_1)^{1/3} = 0 \end{aligned} \tag{3.1}$$

Similar to linear resistive two-ports, there are six possible explicit representations for nonlinear resistive two-ports, which express two variables in terms of the two others. This is a contrived example which has the remarkable property that we can find the analytic forms of all six representations. They are given as follows:

$$\begin{array}{ll} 1.\ v_1 = \frac{1}{2}(i_1^{1/3} - i_2^3) & 2.\ i_1 = (v_1 + v_2)^3 \\ \quad v_2 = \frac{1}{2}(i_2^3 + i_1^{1/3}) & \quad i_2 = (v_2 - v_1)^{1/3} \\ 3.\ v_1 = (i_1^{1/3} - v_2) & 4.\ i_1 = (2v_1 + i_2^3)^3 \\ \quad i_2 = (2v_2 - i_1^{1/3})^{1/3} & \quad v_2 = (i_2^3 + v_1) \\ 5.\ v_1 = (v_2 - i_2^3) & 6.\ v_2 = -v_1 + i_1^{1/3} \\ \quad i_1 = (-i_2^3 + 2v_2)^3 & \quad i_2 = (i_1^{1/3} - 2v_1)^{1/3} \end{array} \tag{3.2}$$

These are the six representations of the same nonlinear resistive two-port.

As in the case of the linear two-port (shown in Table 3.2), we summarize the six representations in Table 3.3.

Each of the six representations can be plotted as two families of curves, each parameterized by a third variable as shown in Fig. 3.1*a* through *f*, respectively.

Table 3.3 Equation for the six representations of a nonlinear resistive two-port

Current-controlled representation	Voltage-controlled representation
$v_1 = \hat{v}_1(i_1, i_2)$	$i_1 = \hat{i}_1(v_1, v_2)$
$v_2 = \hat{v}_2(i_1, i_2)$	$i_2 = \hat{i}_2(v_1, v_2)$
Hybrid 1 representation	Hybrid 2 representation
$v_1 = \hat{v}_1(i_1, v_2)$	$i_1 = \hat{i}_1(v_1, i_2)$
$i_2 = \hat{i}_2(i_1, v_2)$	$v_2 = \hat{v}_2(v_1, i_2)$
Transmission 1 representation	Transmission 2 representation
$v_1 = \hat{v}_1(v_2, -i_2)$	$v_2 = \hat{v}_2(v_1, i_1)$
$i_1 = \hat{i}_1(v_2, -i_2)$	$-i_2 = \hat{i}_2(v_1, i_1)$

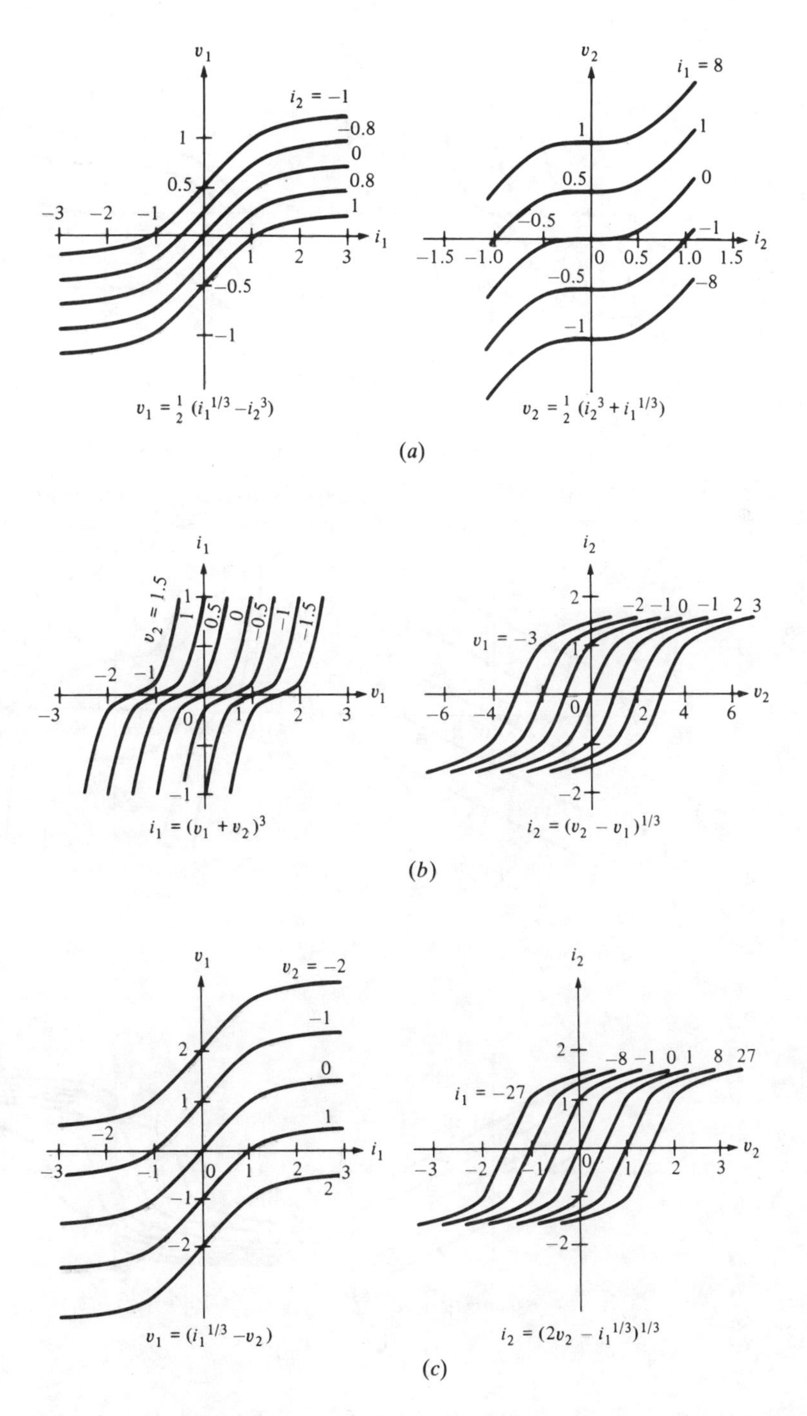

Figure 3.1 Two families of v-i characteristics: (a) current-controlled representation, (b) voltage-controlled representation, (c) hybrid 1 representation.

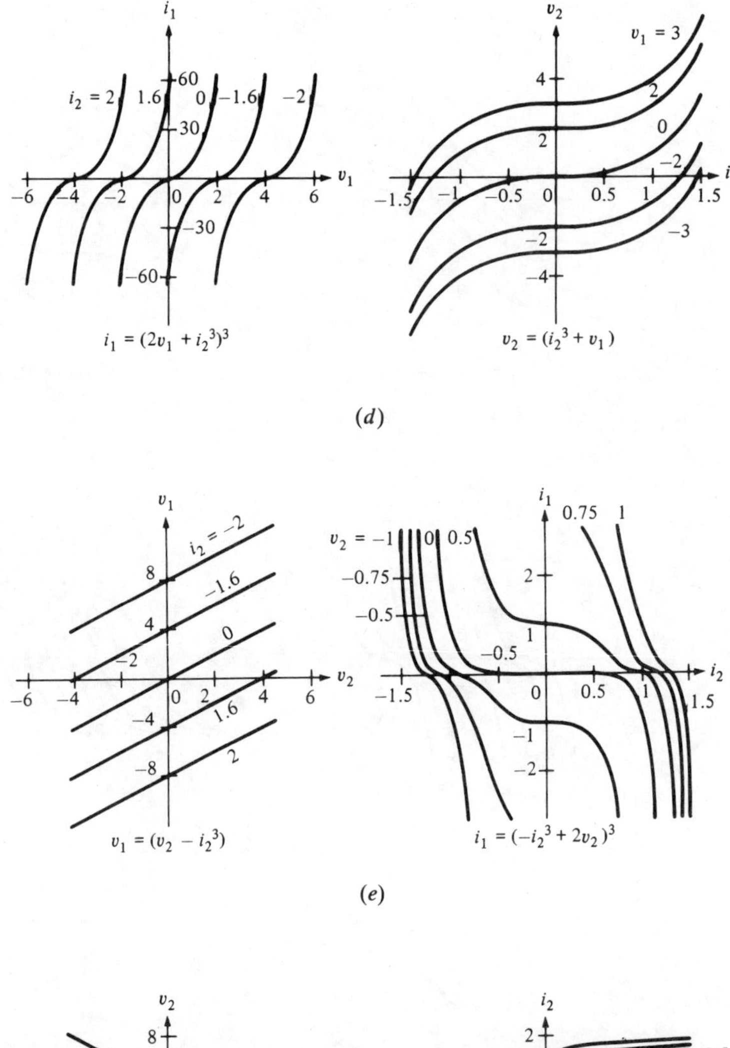

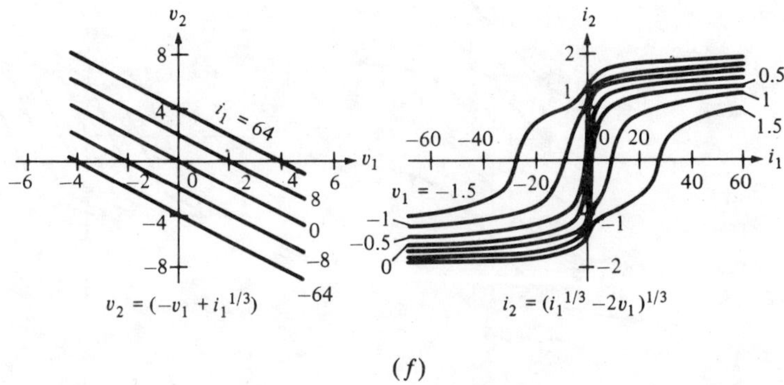

Figure 3.1 (*Continued*) Two families of v-i characteristics: (d) hydrid 2 representation, (e) transmission 1 representation, and (f) transmission 2 representation.

Observe that although these six families of curves are in principle equivalent to each other, some are "smoother" than others. For example, the left family of curves in Fig. 3.1*e* and *f* are straight lines, whereas the associated family of curves on the right are quite "nonlinear" and contain regions having almost abrupt changes. If these curves were to be measured, as would be the case in modeling a three-terminal electric device, serious measurement errors could occur in the region of rapid changes. We conclude: From a measurement accuracy point of view, it is desirable to choose a representation which involves only smoothly varying curves.

4 THE *npn* BIPOLAR TRANSISTOR

The most commonly used three-terminal device is the transistor. In this section we shall first discuss the low-frequency characteristics of the *npn* bipolar transistor together with some aspects of modeling. Then we shall analyze the biasing circuit, determine the operating point, and perform the small-signal analysis.

4.1. *v-i* Characteristics and Modeling

Consider the common-base *npn* transistor shown in Fig. 4.1: The nodes are labeled ⓔ, ⓑ, and ⓒ corresponding to the emitter, base, and collector, respectively. A good low-frequency characterization is given by the one-dimensional diffusion model which yields the *Ebers-Moll equations*:

$$i_e = -I_{\mathrm{ES}}\left(\exp \frac{-v_{eb}}{V_T} - 1\right) + \alpha_R I_{\mathrm{CS}}\left(\exp \frac{-v_{cb}}{V_T} - 1\right) \tag{4.1}$$

$$i_c = \alpha_F I_{\mathrm{ES}}\left(\exp \frac{-v_{eb}}{V_T} - 1\right) - I_{\mathrm{CS}}\left(\exp \frac{-v_{cb}}{V_T} - 1\right) \tag{4.2}$$

Here, I_{ES}, I_{CS}, α_R, and α_F are device parameters, and V_T is the thermal voltage defined earlier in the *pn*-junction diode in Chap. 2. Typically, $\alpha_R = 0.5$ to 0.8, $\alpha_F = 0.99$; I_{ES}, I_{CS} are 10^{-12} to 10^{-10} A at 25°C, and $V_T \approx 26$ mV at 25°C. Note

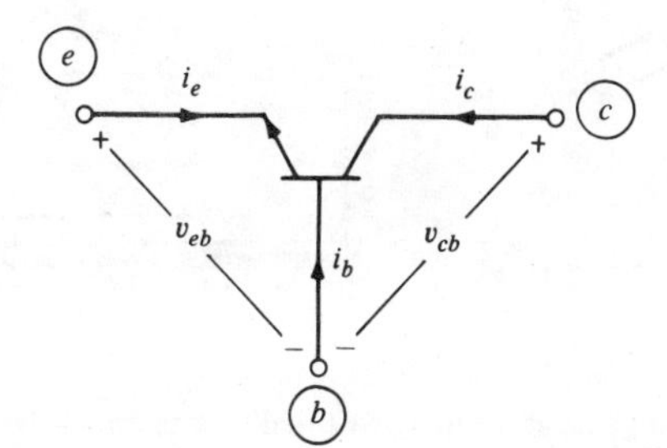

Figure 4.1 The common-base *npn* transistor.

that an *npn* bipolar transistor is, in essence, two *interacting pn*-junction diodes connected back to back to form a three-terminal device. Thus, with the base terminal as the datum node, the currrents i_e and i_c entering the device at the emitter and the collector, respectively, are functions of two node-to-datum voltages v_{eb} and v_{cb}. From Eqs. (4.1) and (4.2), we see that the transistor is a *three-terminal voltage-controlled nonlinear resistor*. It can be represented by the equivalent circuit shown in Fig. 4.2, where the two *pn*-junction diodes are connected at the base node ⓑ to model the terms $-I_{ES}[\exp(-v_{eb}/V_T)-1]$ and $-I_{CS}[\exp(-v_{cb}/V_T)-1]$, and the two CCCS are used to model the terms $\alpha_R I_{CS}[\exp(-v_{cb}/V_T)-1]$ and $\alpha_F I_{ES}[\exp(-v_{eb}/V_T)-1]$ which represent the interaction between the two diodes.

The characteristics of Eqs. (4.1) and (4.2) are shown in Fig. 4.3 in the v_{eb}-i_e plane and the v_{cb}-i_c plane, respectively. Note that v_{cb} serves as a parameter in the family of curves in the v_{eb}-i_e plane; and, similarly, v_{eb} serves as a parameter in the family of curves in the v_{cb}-i_c plane.

Common-emitter configuration In most amplifier circuits, the common-emitter configuration shown in Fig. 4.4 is used. It is possible to derive the equations for the common-emitter configuration directly from those of the common-base

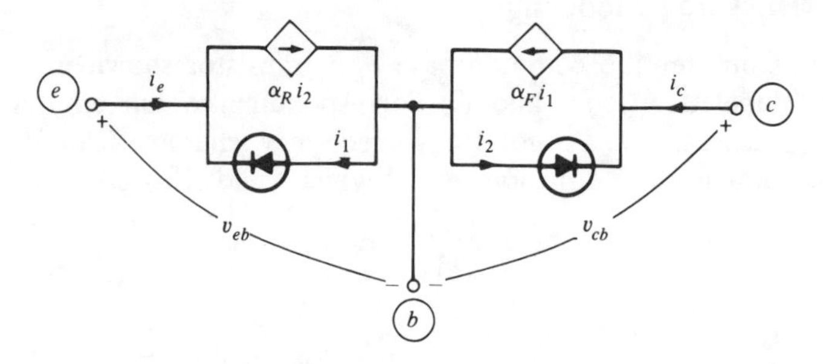

Figure 4.2 Ebers-Moll circuit model of *npn* transistor.

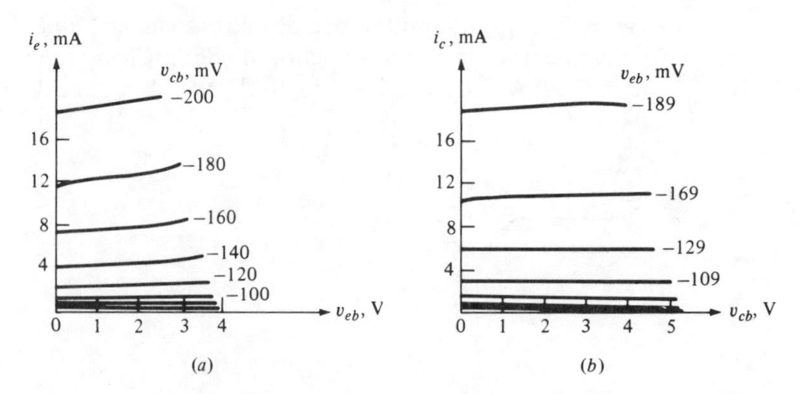

Figure 4.3 *v-i* Characteristics of an *npn* transistor in the common-base configuration.

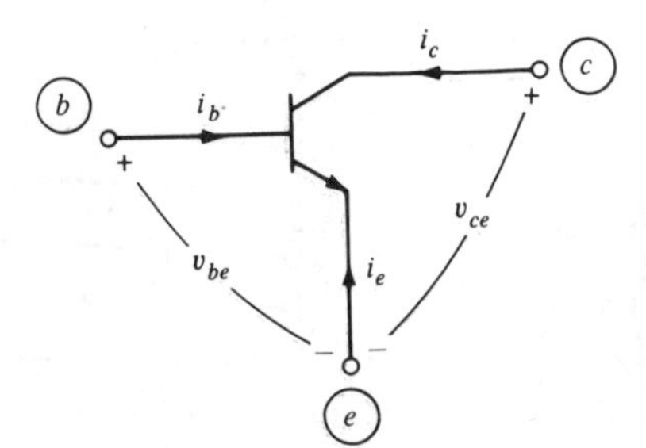

Figure 4.4 The common-emitter configuration of an *npn* transistor.

configuration of Eqs. (4.1) and (4.2). For the common-emitter configuration the two port voltages are v_{be} and v_{ce}, and the two port currents are i_b and i_c. These are related to the variables of the common-base configuration by the following simple equations from Kirchhoff's laws:

$$v_{be} = -v_{eb} \tag{4.3}$$

$$v_{ce} = v_{cb} - v_{eb} \tag{4.4}$$

$$i_b = -(i_e + i_c) \tag{4.5}$$

$$i_c = i_c \tag{4.6}$$

Substituting the above equations in Eqs. (4.1) and (4.2), we can express the port currents i_b and i_c for the common-emitter configuration in terms of the port voltages v_{be} and v_{ce}. They are

$$i_b = (1 - \alpha_F)I_{\text{ES}}\left(\exp\frac{v_{be}}{V_T} - 1\right) + (1 - \alpha_R)I_{\text{CS}}\left(\exp\frac{v_{be} - v_{ce}}{V_T} - 1\right) \tag{4.7}$$

$$i_c = \alpha_F I_{\text{ES}}\left(\exp\frac{v_{be}}{V_T} - 1\right) - I_{\text{CS}}\left(\exp\frac{v_{be} - v_{ce}}{V_T} - 1\right) \tag{4.8}$$

Thus, again, we have a voltage-controlled representation for the common-emitter configuration. This set of equations is not particularly useful, because, in practice, the measured data are usually expressed in terms of the hybrid 1 representation, i.e.,

$$v_{be} = \hat{v}_{be}(i_b, v_{ce}) \tag{4.9}$$

$$i_c = \hat{i}_c(i_b, v_{ce}) \tag{4.10}$$

Furthermore, as a tradition, we usually plot i_b vs. v_{be} with v_{ce} as a parameter, and i_c vs. v_{ce} with i_b as a parameter, as shown in Fig. 4.5.

In Fig. 4.5 we plot the characteristic in the customary fashion, where Eq. (4.9) is plotted in the v_{be}-i_b plane with v_{ce} as a parameter, and Eq. (4.10) is plotted in the v_{ce}-i_c plane with i_b as a parameter. This particular representation (with v_{ce} serving as a parameter in one plane and i_b in the other) is preferred because it gives a smoothly varying family of collector-to-emitter v-i curves.

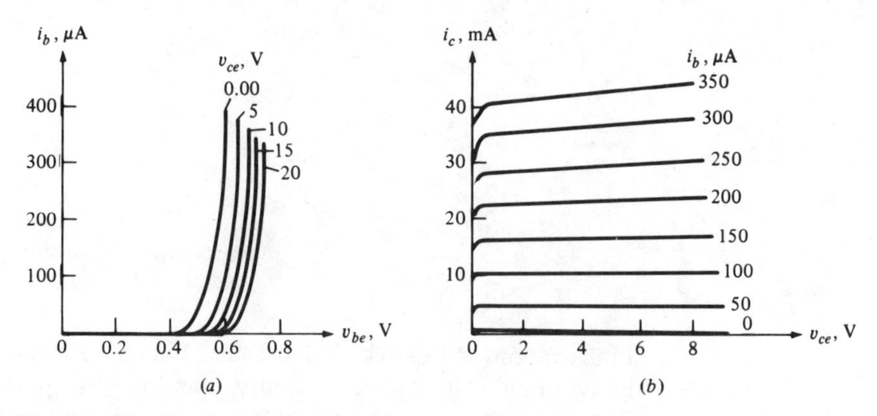

Figure 4.5 Characteristic of common-emitter configuration of an *npn* transistor.

Piecewise-linear approximation Let us obtain the piecewise-linear approximation for the characteristics in Fig. 4.5. This is shown in Fig. 4.6. We may then express i_b and i_c in terms of the v-i characteristics of the concave and convex resistors and appropriate controlled sources. The equivalent circuit for this representation is shown in Fig. 4.7. Note that with $v_{ce} = 0$, the characteris-

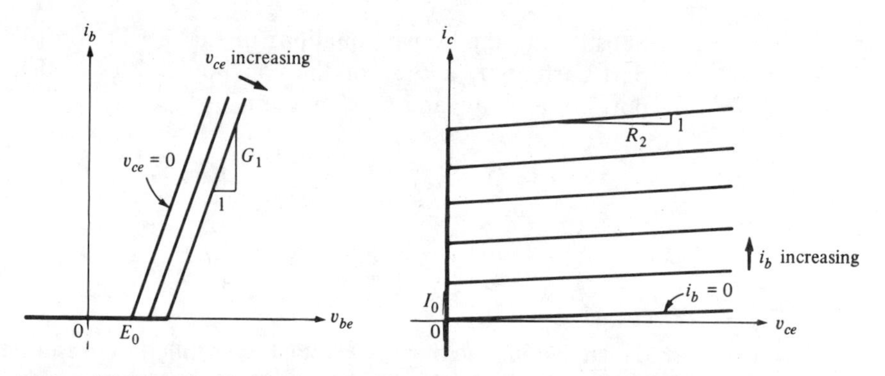

Figure 4.6 Piecewise-linear approximation of the common-emitter characteristics.

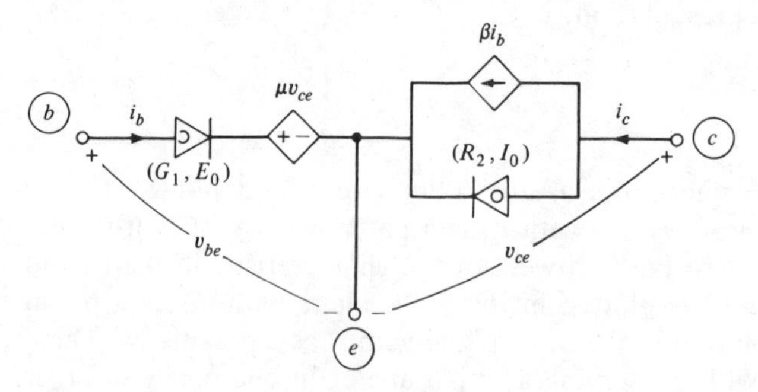

Figure 4.7 Piecewise-linear model of common-emitter transistor configuration.

tic in Fig. 4.6*a* is precisely that of a concave resistor specified by E_0 and slope G_1. As v_{ce} increases, the voltage v_{be} increases, as indicated by the parallel translations toward the right. The proportional shift is modeled with a linear VCVS with transfer voltage ratio μ (see Fig. 4.7). Similarly, in Fig. 4.6*b*, with $i_b = 0$, the characteristic is that of a convex resistor with the i_c-axis intercept equal to I_0 and the slope equal to $1/R_2$. As i_b increases, the current i_c increases, as indicated by the upward parallel translations. The proportional shift of i_c as a function of i_b is modeled by the linear CCCS with transfer current ratio β (see Fig. 4.7).

For many large-signal applications, e.g., digital circuits, these models are unnecessarily complicated and further simplifications are possible. In some approximate dc analysis, it suffices to represent the base emitter by a constant voltage source E_0 in series with an ideal diode and to represent the collector emitter by a CCCS together with an ideal diode as shown in Fig. 4.8; the corresponding characteristics are shown in Fig. 4.9. For silicon transistors at room temperature, the value of E_0 is usually between 0.6 and 0.7 V.

It is important to bear in mind that models are developed with specific applications in mind. Obviously, the simpler the model, the easier the circuit analysis. Thus, for some applications, if only an approximate solution is called for, we should use the simplest but valid model to get an idea of how the circuit functions. In other situations, e.g., in determining the precise operating points using a computer, we need to use a more precise model for the transistor than that of the Ebers-Moll model. Some complicated models have been developed using a combination of physical principles and experimental measurements

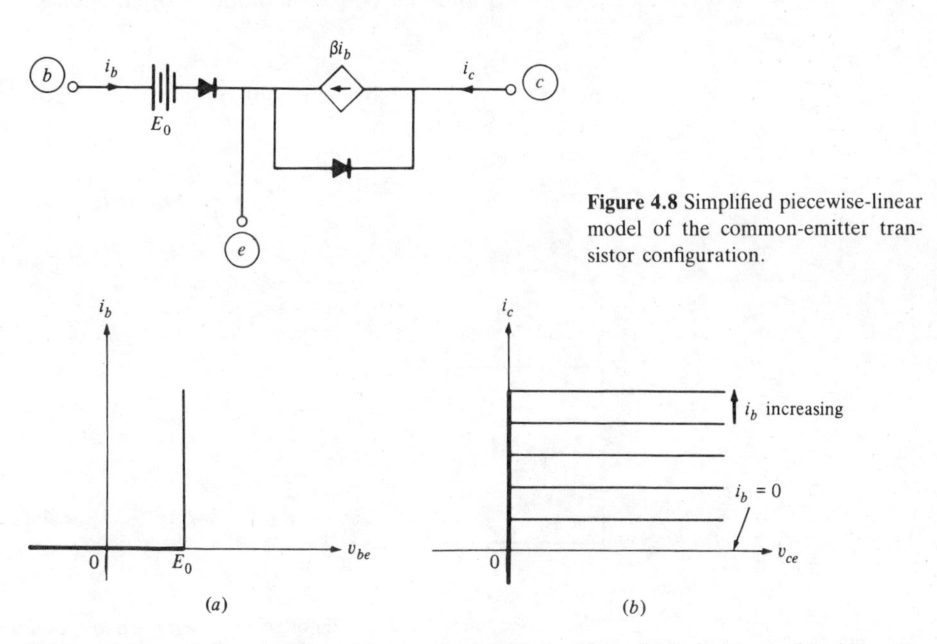

Figure 4.8 Simplified piecewise-linear model of the common-emitter transistor configuration.

Figure 4.9 The *v-i* characteristics of the common-emitter configuration using the model of Fig. 4.8.

(e.g., the Gummel-Poon model[3]). In small-signal analysis, we use a linear model which holds only approximately in the neighborhood of the operating point. Some of these will be discussed in the remaining portions of this section.

4.2 DC Operating Points and Double Load Lines

We shall use the common-emitter configuration to illustrate the determination of the operating point of a three-terminal nonlinear resistor. Consider the circuit in Fig. 4.10. We wish to determine the operating point Q under dc inputs E_1 and E_2.

Analytic method Analytically, we can express the branch equations as follows: For the two battery-resistor branches, we have

$$v_{be} = E_1 - R_1 i_b \tag{4.11}$$

$$v_{ce} = E_2 - R_2 i_c \tag{4.12}$$

For the common-emitter transistor, we have Eqs. (4.7) and (4.8) expressing i_b and i_c in terms of v_{be} and v_{ce}. Therefore, we obtain four equations in four unknowns i_b, i_c, v_{be}, v_{ce}.

Example 1 (analytic method) First, let us consider the simplest model for the transistor shown in Fig. 4.8. Since $E_1 > 0$ and $E_2 > 0$, the emitter diode is shorted and the collector diode is open; hence

$$v_{be} = E_0 \tag{4.13}$$

and i_c is specified by the controlled source. Combining Eqs. (4.11) and (4.13), we obtain

$$i_b = \frac{1}{R_1}(E_1 - E_0) \tag{4.14}$$

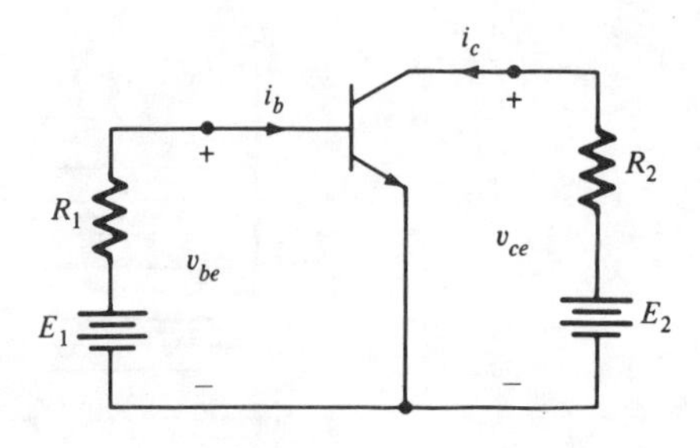

Figure 4.10 Biasing circuit for common-emitter configuration.

[3] H. K. Gummel and H. C. Poon, "An Integral Charge Control Model of Bipolar Transistors," *BSTJ*, vol. 49, p. 827–851, 1970.

For the collector side

$$i_c = \beta i_b = \frac{\beta}{R}(E_1 - E_0) \tag{4.15}$$

Therefore, from Eq. (4.12), we obtain

$$v_{ce} = E_2 - \beta \frac{R_2}{R_1}(E_1 - E_0) \tag{4.16}$$

The four linear equations above, (4.13) to (4.16), define a unique operating point Q; we denote it by (V_{beQ}, I_{bQ}), (V_{ceQ}, I_{cQ}).

Graphic method In Chap. 2 we used the load line to determine the operating point for a two-terminal resistor. We now extend this method to three-terminal resistors. The two load lines corresponding to Eqs. (4.11) and (4.12) are superimposed with the transistor characteristics on the two planes in Fig. 4.11. Equations (4.11) and (4.12) require that the operating point lie on both load lines. We can determine the operating point graphically by first finding the intersection of load line 1 on the v_{be}-i_b plane with the characteristic curve of the transistor, $v_{be} = E_0$. This determines V_{beQ} and I_{bQ}. From the value of I_{bQ}, we locate the intersection of load line 2 on the v_{ce}-i_c plane with the $i_b = I_{bQ}$ line. This determines V_{ceQ} and I_{cQ}.

In the following we will give an example of extending the above graphic method to a more complicated transistor model.

Example 2 (graphic method) In this example we consider the same circuit of Fig. 4.10 with $E_1 = 0.5$ V, $R_1 = 0.5$ kΩ, $E_2 = 10$ V, and $R_2 = 200$ kΩ. We will use the piecewise-linear model of Fig. 4.12 to represent the characteristics of the common-emitter transistor. The load lincs are drawn as shown. Unlike the previous example, load line 1 has many intersections

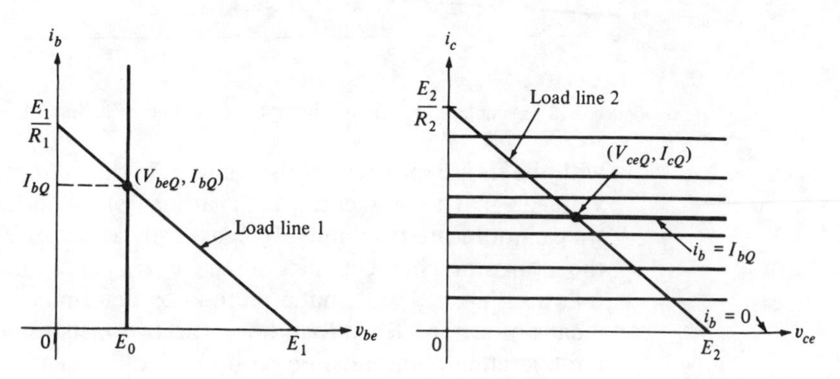

Figure 4.11 Using load lines to determine the operating point for the simplified transistor model of Fig. 4.8.

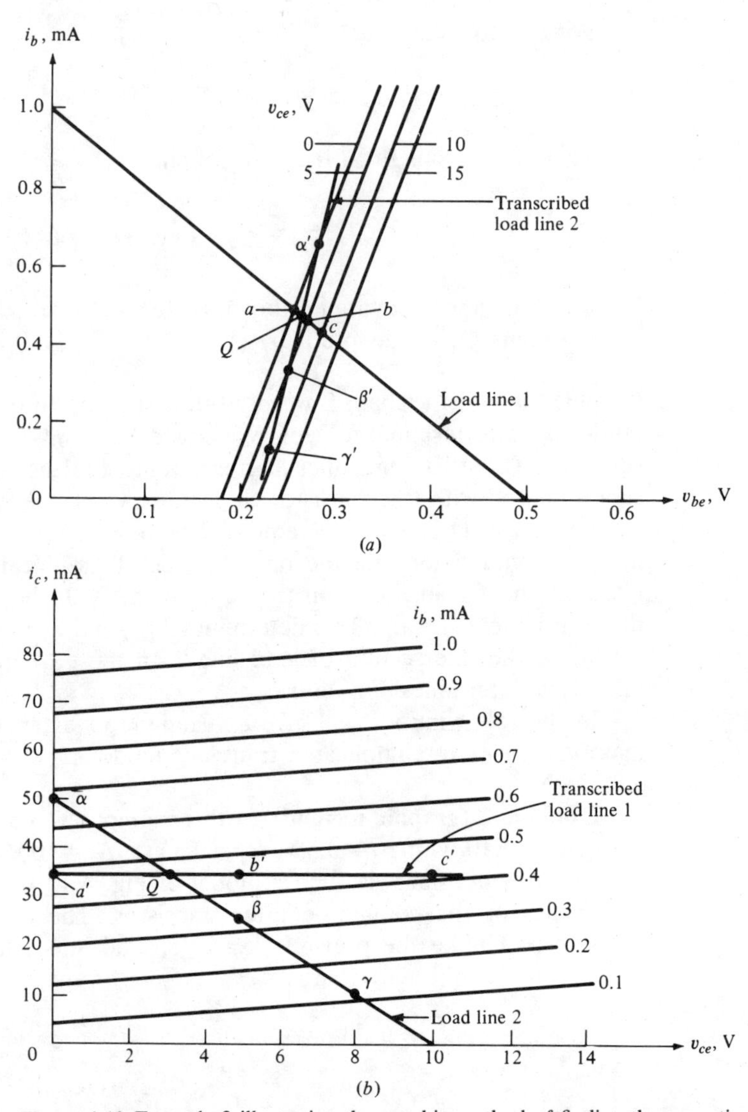

Figure 4.12 Example 2 illustrating the graphic method of finding the operating point.

with the transistor characteristic curves in the v_{be}-i_b plane depending on the voltage v_{ce}. The base current i_b is a function of both v_{be} and v_{ce}. Therefore we cannot locate the value of I_{bQ} as easily as before. We must also consider the transistor characteristics in the v_{ce}-i_c plane and on load line 2. The following gives a systematic method to determine the operating point by means of a transcribed load line. The two facts to be kept in mind are (*a*) the operating point must lie on both load lines and (*b*) the two variables i_b and v_{ce} are common to both planes, and, hence, are the variables which relate the information of the two families of transistor characteristics in the two planes.

Table 3.4 Summary of Example 2 using load line 2

Load line 2	i_b, mA	v_{ce}, V	Transcribed load line 2
α	0.67	0	α'
β	0.32	5	β'
γ	0.12	8	γ'

First, let us consider the v_{be}-i_b plane. The intersections of load line 1 with the curves corresponding to $v_{ce} = 0$, 5, and 10 V are designated as points a, b, and c, respectively. The values of i_b at these points are read off the graph and found to be 0.49, 0.45, and 0.42 mA, respectively. With this information about i_b we can locate the points a', b', and c' on the v_{ce}-i_c plane: For example, from the v_{be}-i_b plane, point a corresponds to $v_{ce} = 0$ V, $i_b = 0.49$ mA, and hence the corresponding point a' in the v_{ce}-i_c plane is located by the same information $i_b = 0.49$ mA and $v_{ce} = 0$ V. We determine b' and c' similarly. A curve is next fitted to pass through a', b', and c' in the v_{ce}-i_c plane, which intersects load line 2 at point Q, the operating point where $I_{bQ} \approx 0.47$ mA and $V_{ceQ} \approx 3$ V. Thus $I_{cQ} \approx 35$ mA. Similarly, using load line 1 on the v_{be}-i_b plane, we locate $V_{beQ} \approx 0.265$ V. The curve which passes through a', b', and c' is the transcribed load line of load line 1. A *transcribed load line* gives the characteristic of a load line plotted in a different plane.

Alternatively, we could transcribe load line 2 to the v_{be}-i_b plane. The coordinates α, β, and γ on load line 2 are transcribed to α', β', and γ' in the v_{be}-i_b plane shown. From these points we fit a curve which is the transcribed load line of load line 2. In this way we determine first the operating point (V_{beQ}, I_{bQ}) in the v_{be}-i_b plane. The result is summarized in Table 3.4.

It is obvious that this graphic method gives an approximate solution. However, an approximate solution is useful for design purposes and can also be used as the starting point to obtain a very accurate solution by means of numerical algorithms.

4.3 Small-Signal Analysis

Figure 4.13 is a simple amplifier circuit where a small signal, $v_s(t) = V_m \cos \omega t$, is applied as input to the biased common-emitter circuit. The output voltage v_{ce} then contains an amplified signal waveform at the same angular frequency ω. We can extend the method of small-signal analysis for two-terminal nonlinear resistors (see Chap. 2) to three-terminal nonlinear resistors.

In this section we will derive the small-signal method using the hybrid representation of the common-emitter transistor described by Eqs. (4.9) and (4.10), which we repeat below:

$$v_{be} = \hat{v}_{be}(i_b, v_{ce}) \tag{4.9}$$

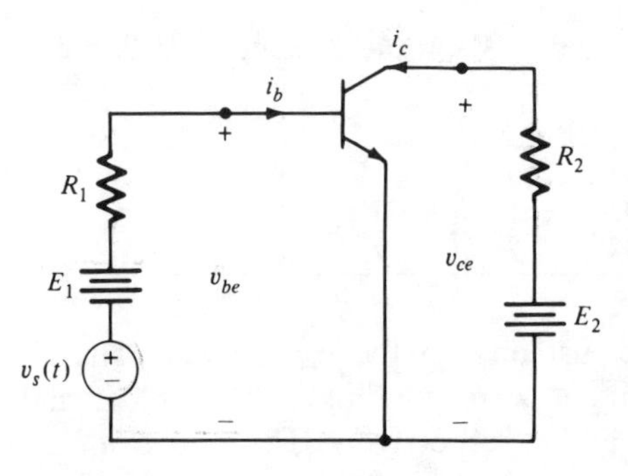

Figure 4.13 Common-emitter transistor amplifier with input $v_S(t)$.

$$i_c = \hat{i}_c(i_b, v_{ce}) \tag{4.10}$$

Obviously, without the small-signal source v_s, the operating point (V_{beQ}, I_{bQ}), (V_{ceQ}, I_{cQ}) satisfies not only the above two equations but also Eqs. (4.11) and (4.12) characterizing the two battery-resistor branches. They are written as follows:

$$V_{beQ} = \hat{v}_{be}(I_{bQ}, V_{ceQ}) \tag{4.17}$$

$$I_{cQ} = \hat{i}_c(I_{bQ}, V_{ceQ}) \tag{4.18}$$

$$V_{beQ} = E_1 - R_1 I_{bQ} \tag{4.19}$$

$$V_{ceQ} = E_2 - R_2 I_{cQ} \tag{4.20}$$

When the signal source $v_s(t)$ is present in the circuit, we may express the four signal variables $v_{be}(t)$, $i_b(t)$, $v_{ce}(t)$, $i_c(t)$ for all t as

$$v_{be}(t) = V_{beQ} + \tilde{v}_1(t) \tag{4.21}$$

$$i_b(t) = I_{bQ} + \tilde{i}_1(t) \tag{4.22}$$

$$v_{ce}(t) = V_{ceQ} + \tilde{v}_2(t) \tag{4.23}$$

$$i_c(t) = I_{cQ} + \tilde{i}_2(t) \tag{4.24}$$

where $\tilde{v}_1(t)$, $\tilde{i}_1(t)$, $\tilde{v}_2(t)$, and $\tilde{i}_2(t)$ represent the *small* displacements of the voltages and currents from the fixed operating point Q. At this juncture it remains only to determine the small-signal voltages and currents $\tilde{v}_1(t)$, $\tilde{i}_1(t)$, $\tilde{v}_2(t)$, and $\tilde{i}_2(t)$.

First substituting Eqs. (4.21) through (4.24) into Eqs. (4.9) and (4.10), we obtain

$$v_{be}(t) = V_{beQ} + \tilde{v}_1(t) = \hat{v}_{be}[I_{bQ} + \tilde{i}_1(t), V_{ceQ} + \tilde{v}_2(t)] \tag{4.25}$$

$$i_c(t) = I_{cQ} + \tilde{i}_2(t) = \hat{i}_c[I_{bQ} + \tilde{i}_1(t), V_{ceQ} + \tilde{v}_2(t)] \tag{4.26}$$

No approximation has been introduced up to this step. Next, we assume that the signal $v_s(t)$ is "small" and take the first two terms of the Taylor series expansions of $\hat{v}_{be}(.\,,.)$ and $\hat{i}_c(.\,,.)$ about the operating point Q and obtain the following approximation:

$$v_{be}(t) \approx \hat{v}_{be}(I_{bQ}, V_{ceQ}) + \left.\frac{\partial \hat{v}_{be}}{\partial i_b}\right|_Q \tilde{i}_1(t) + \left.\frac{\partial \hat{v}_{be}}{\partial v_{ce}}\right|_Q \tilde{v}_2(t) \tag{4.27}$$

$$i_c(t) \approx \hat{i}_c(I_{bQ}, V_{ceQ}) + \left.\frac{\partial \hat{i}_c}{\partial i_b}\right|_Q \tilde{i}_1(t) + \left.\frac{\partial \hat{i}_c}{\partial v_{ce}}\right|_Q \tilde{v}_2(t) \tag{4.28}$$

Comparing Eqs. (4.25) and (4.26) with Eqs. (4.27) and (4.28) and using Eqs. (4.17) and (4.18), we obtain the following approximations:

$$\tilde{v}_1(t) = \left.\frac{\partial \hat{v}_{be}}{\partial i_b}\right|_Q \tilde{i}_1(t) + \left.\frac{\partial \hat{v}_{be}}{\partial v_{ce}}\right|_Q \tilde{v}_2(t)$$

$$\tilde{i}_2(t) = \left.\frac{\partial \hat{i}_c}{\partial i_b}\right|_Q \tilde{i}_1(t) + \left.\frac{\partial \hat{i}_c}{\partial v_{ce}}\right|_Q \tilde{v}_2(t)$$

These two equations give a useful approximation provided that $\tilde{i}_1(\cdot)$ and $\tilde{v}_2(\cdot)$ are sufficiently small.

These two equations can be viewed as hybrid equations relating the small signals $\tilde{i}_1$ and $\tilde{v}_2$ to $\tilde{v}_1$ and $\tilde{i}_2$. Thus we write

$$\begin{bmatrix} \tilde{v}_1 \\ \tilde{i}_2 \end{bmatrix} = \mathbf{H} \begin{bmatrix} \tilde{i}_1 \\ \tilde{v}_2 \end{bmatrix} \tag{4.29}$$

where $\mathbf{H}$ is the jacobian matrix of the functions $\hat{v}_{be}(\cdot\,,\cdot)$ and $\hat{i}_c(\cdot\,,\cdot)$ evaluated at Q. More precisely, we have

$$\mathbf{H} = \begin{bmatrix} h_{11} & h_{12} \\ h_{21} & h_{22} \end{bmatrix} \triangleq \begin{bmatrix} \dfrac{\partial \hat{v}_{be}}{\partial i_b} & \dfrac{\partial \hat{v}_{be}}{\partial v_{ce}} \\ \dfrac{\partial \hat{i}_c}{\partial i_b} & \dfrac{\partial \hat{i}_c}{\partial v_{ce}} \end{bmatrix}_Q \tag{4.30}$$

where $\mathbf{H}$ is called the *small-signal hybrid matrix*[4] about the operating point Q. Equation (4.29) is a linear equation, and the small-signal hybrid matrix $\mathbf{H}$ is a constant matrix. $\mathbf{H}$ depends on the transistor charactereistics and the operating point. Thus, as far as the small signal is concerned, we have a *linear resistive two-port* specified by its hybrid matrix $\mathbf{H}$. Refer to Fig. 4.14: Eqs. (4.29) and (4.30) can be interpreted geometrically as representing two tangent planes to the two surfaces defined by Eqs. (4.9) and (4.10) at the operating point Q. The approximation obtained by dropping the higher-order terms from Eqs. (4.25) and (4.26) amounts to replacing, in the neighborhood of the operating point Q, the curved characteristic surfaces by their tangent planes, as shown in Fig. 4.14.

[4] Also called the "incremental" hybrid matrix.

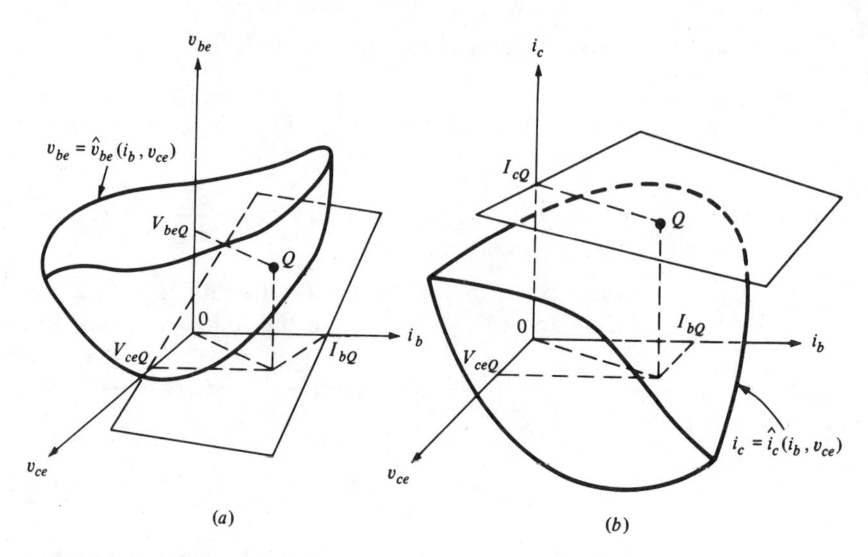

Figure 4.14 Geometric interpretation of the transistor *v-i* characteristics and operating point using the hybrid representation.

The elements of the hybrid matrix, namely,

$$h_{11} = \left.\frac{\partial \hat{v}_{be}}{\partial i_b}\right|_Q \quad \text{and} \quad h_{22} = \left.\frac{\partial \hat{i}_c}{\partial v_{ce}}\right|_Q \tag{4.31}$$

are called the *small-signal input resistance and the output conductance*, respectively, of the transistor about the operating point Q, and

$$h_{12} = \left.\frac{\partial \hat{v}_{be}}{\partial v_{ce}}\right|_Q \quad \text{and} \quad h_{21} = \left.\frac{\partial \hat{i}_c}{\partial i_b}\right|_Q \tag{4.32}$$

are called the *small-signal reverse voltage ratio and forward current ratio*, respectively, of the transistor about the operating point Q. The small-signal equivalent circuit of the transistor together with its source and external resistors are shown in Fig. 4.15.

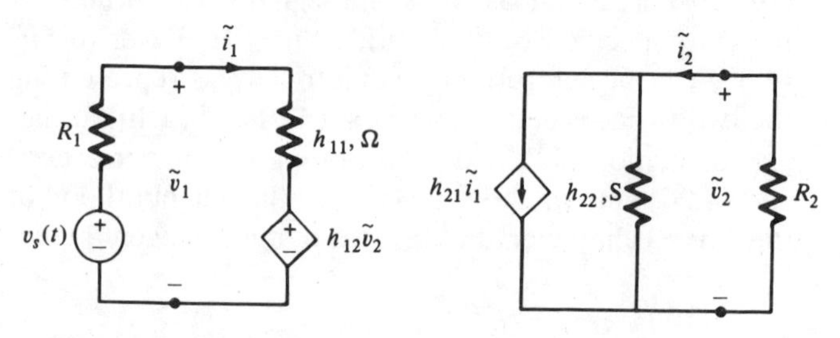

Figure 4.15 Small-signal equivalent circuit of the amplifier.

Note that the small-signal equivalent circuit bookkeeps only the relation between the small-signal voltages and currents: All the biasing data, E_1, E_2, V_{beQ}, V_{ceQ}, etc., have been eliminated. As we shall see later, the small-signal equivalent circuit is obtained by (*a*) setting to zero all the constant biasing sources, (*b*) leaving all *linear* elements unchanged, and (*c*) replacing all *nonlinear* elements by their small-signal equivalent circuit about the operating point.

Exercises

1. Prove from the given circuit in Fig. 4.13 together with Eqs. (4.19) to (4.24) and Eq. (4.29) that the circuit in Fig. 4.15 is indeed the small-signal equivalent circuit.
2. Show that the voltage gain of the amplifier is given by

$$\frac{\tilde{v}_2}{v_s} = \frac{-h_{21}}{(h_{11} + R_1)(h_{22} + 1/R_2) - h_{12}h_{21}}$$

Interpretation of the small-signal hybrid parameters Since in practice the common-emitter transistor characteristic is not given in the function form of Eqs. (4.9) and (4.10) but by two families of curves in the v_{be}-i_b plane and v_{ce}-i_c plane, we need to determine the small-signal hybrid parameters directly from these data. Let us use Fig. 4.12 to indicate the meaning of the four hybrid parameters. From Eq. (4.31) it is clear that h_{11} and h_{22} are simply related to the slopes of the characteristics at the operating point Q in the v_{be}-i_b plane and v_{ce}-i_c plane, respectively. Thus, using Fig. 4.12, we obtain

$$h_{11} = \left.\frac{\partial \hat{v}_{be}}{\partial i_b}\right|_Q = \frac{1}{\text{slope at } Q} \approx 160\ \Omega$$

$$h_{22} = \left.\frac{\partial \hat{i}_c}{\partial v_{ce}}\right|_Q = \text{slope at } Q \approx 0.56 \times 10^{-6}\ \text{S}$$

To determine h_{12} and h_{21} from the two families of characteristics specified by Eqs. (4.9) and (4.10), we need to calculate the small-signal transfer voltage ratio $h_{12} = (\partial \hat{v}_{be}/\partial v_{ce})|_Q$ and the small-signal transfer current ratio $h_{21} = (\partial \hat{i}_c/\partial i_b)|_Q$ at the operating point Q. This can only be done approximately unless the relation is strictly linear as in Fig. 4.12. When we compute the partial derivatives from

$$v_{be} = \hat{v}_{be}(i_b, v_{ce}) \tag{4.9}$$

$$i_c = \hat{i}_c(i_b, v_{ce}) \tag{4.10}$$

at the operating point Q, we use the approximations

$$h_{12} = \left.\frac{\partial \hat{v}_{be}}{\partial v_{ce}}\right|_Q \approx \left.\frac{\Delta v_{be}}{\Delta v_{ce}}\right|_{i_b = \text{constant}} \tag{4.33}$$

$$h_{21} = \left.\frac{\partial \hat{i}_c}{\partial i_b}\right|_Q \approx \left.\frac{\Delta i_c}{\Delta i_b}\right|_{v_{ce}=\text{constant}} \tag{4.34}$$

Therefore, by drawing a horizontal line passing through the operating point Q on the v_{be}-i_b plane, we can determine the change of v_{be} with respect to v_{ce} for a constant $i_b = I_{bQ}$. In the example, using Fig. 4.12, we obtain $h_{12} \approx 0.004$. Similarly, by drawing a vertical line passing through the operating point Q on the v_{ce}-i_c plane, we can determine the change of i_c with respect to i_b for a constant $v_{ce} = V_{ceQ}$. In the example, again using Fig. 4.12, we obtain $h_{21} \approx 0.1$.

5 THE MOS TRANSISTOR AND THE BASIC INVERTER

With the advent of VLSI technology the metal-oxide-semiconductor (MOS) transistor has become a dominant device in microelectronics. The MOS transistor symbol is shown in Fig. 5.1 with the three terminals indicated: Ⓖ (gate), Ⓓ (drain), and Ⓢ (source). Usually, Ⓢ is chosen as the datum node. Therefore, the two port voltages are v_{gs}, the gate-to-source voltage, and v_{ds}, the drain-to-source voltage, and the two port currents are i_g, the gate current, and i_d, the drain current.

For the *n*-channel MOS transistor, the voltage v_{ds} is normally positive. The *v-i* characteristics for the two basic types of MOS transistors, the *enhancement type* and the *depletion type*, are shown in Fig. 5.2. Note that i_g is approximately zero for all v_{gs}, thus the input port behaves like an open circuit. The only difference between the enhancement and depletion types lies in their output-port characteristic. In the enhancement type, i_d is approximately zero for $v_{gs} = 0$; in the depletion type, i_d is large for $v_{gs} = 0$. Let us consider the enhancement type. The output-port characteristic can be separately considered for two regions of the first quadrant. The boundary is shown by the dotted line defined by

$$v_{ds} = v_{gs} - V_{\text{th}} \tag{5.1}$$

where V_{th} is the threshold voltage which is typically equal to $0.2V_{\text{DD}}$ where V_{DD} is the positive supply voltage for the particular technology used. To the right of the dotted line the current i_d is somewhat independent of the drain-to-source

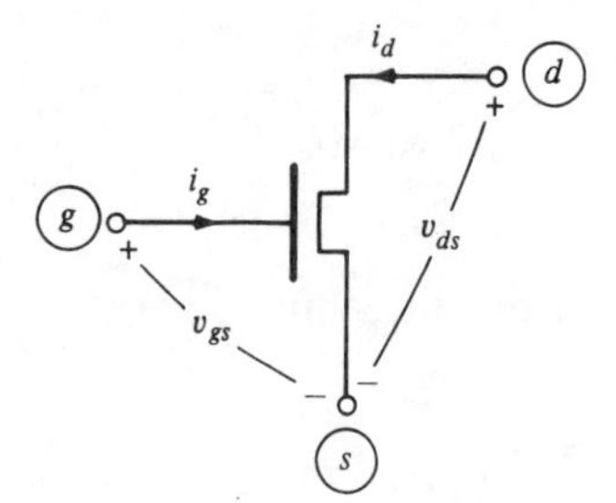

Figure 5.1 Symbol for the enhancement-type MOS transistor.

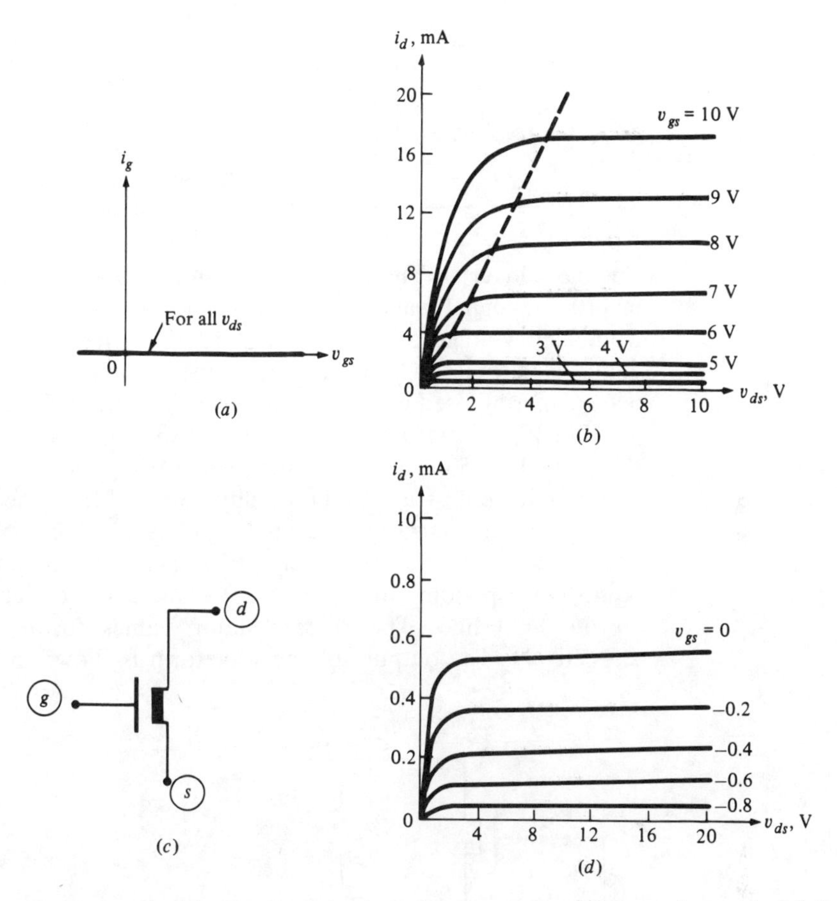

Figure 5.2 (*a*), (*b*) Characteristics for enhancement-type MOS transistor. (*c*), (*d*) Symbol for depletion-type MOS transistor and its output-port characteristic.

voltage v_{ds} but is a nonlinear function of the gate-to-source v_{gs}. Approximately.

$$i_d = \tfrac{1}{2}\beta(v_{gs} - V_{\text{th}})^2 \qquad \text{for } v_{ds} \geq v_{gs} - V_{\text{th}} \tag{5.2}$$

where β is a parameter which depends on the dimension of the device in addition to other physical constants. We can model the MOS transistor in the nonlinear region by a nonlinear CCVS as shown in Fig. 5.3.

To the left of the dotted line, called the linear[5] region,

$$i_d = f(v_{gs}, v_{ds}) \approx \beta[(v_{gs} - V_{\text{th}})v_{ds} - \tfrac{1}{2}v_{ds}^2] \qquad \text{for } v_{ds} < v_{gs} - V_{\text{th}} \tag{5.3}$$

Thus, the MOS transistor at low frequencies is a three-terminal voltage-controlled nonlinear resistor characterized by $i_g = 0$ and Eqs. (5.2) and (5.3).

[5] Approximately linear with respect to v_{gs}.

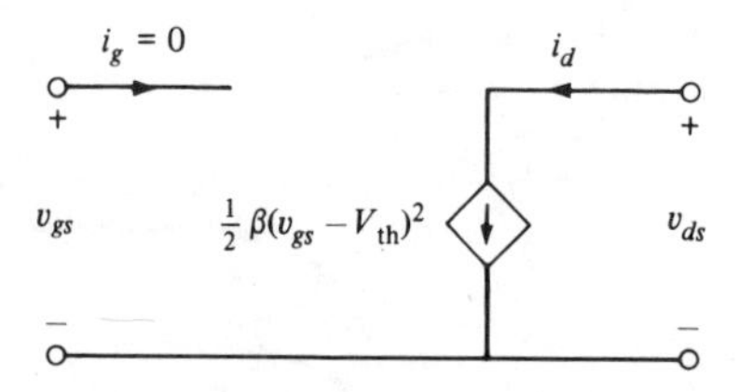

Figure 5.3 A nonlinear CCVS model for a MOS transistor in the nonlinear region.

The basic inverter The circuit shown in Fig. 5.4 is called an *inverter* because it inverts the signal waveform from low voltage to high voltage and conversely from high voltage to low voltage. The inverter is a basic building block in digital electronics. The load line for the battery-resistor is drawn in the v_{ds}-i_d plane superimposed with the characteristic of the MOS transistor shown in Fig. 5.5. Let $V_{DD} = 10$ V and $R = 200\ \Omega$. Let the signal voltage v_i be a square wave shown in Fig. 5.6*a*, where T represents the period of the square wave. The output voltage $v_o = v_{ds}$ can be readily obtained from the load line drawn in the v_{ds}-i_d plane. In contrast to the small-signal analysis before, we now use the load line to determine the output voltage for a *large-signal* input. The two operating points corresponding to $v_i = v_{gs} = 10$ V and 2 V are marked as points A and B on the load line. The corresponding values for $v_o = v_{ds}$ are 1 and 8 V, respectively. The output voltage waveform is shown in Fig. 5.6*b*.

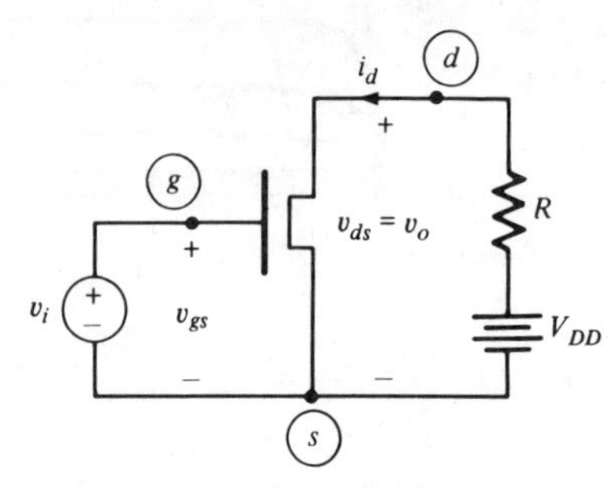

Figure 5.4 An inverter circuit using an enhancement-type MOS transistor.

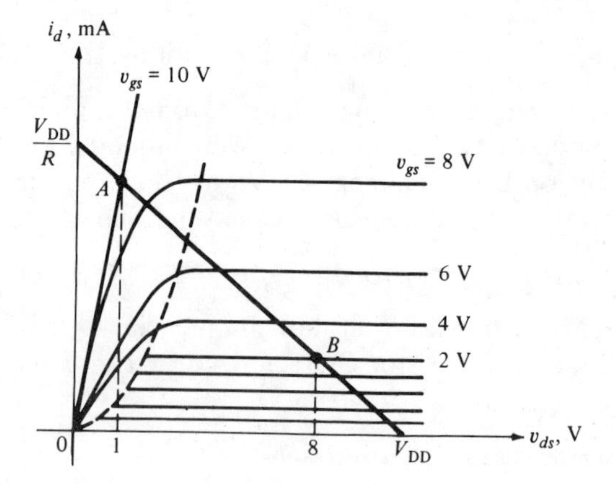

Figure 5.5 Load line for the inverter circuit.

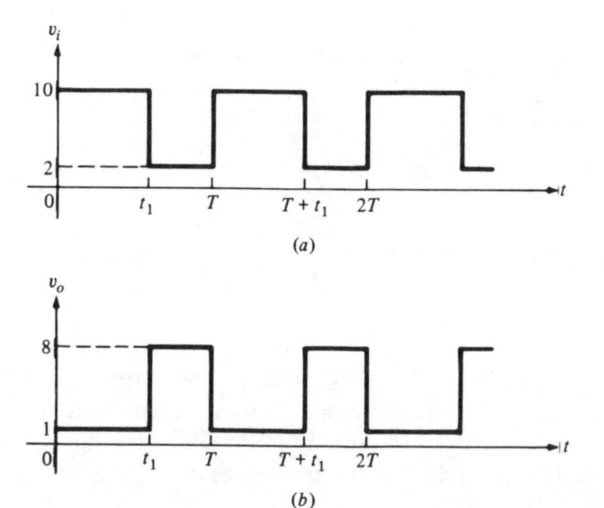

Figure 5.6 (a) Input voltage waveform. (b) Output voltage waveform.

The depletion-type transistor as a nonlinear load On an integrated circuit chip, an n-channel MOS transistor is formed wherever a polysilicon path crosses a diffusion path. Thus it is much simpler and economical to produce a transistor than a resistor! In the inverter circuit in Fig. 5.4, the two-terminal linear resistor can be replaced with a nonlinear device formed by a depletion-type MOS transistor.

Consider the circuit shown in Fig. 5.7a where the gate and the source terminals (g') and (s') of a depletion-type MOS transistor are tied together. Thus $v'_{gs} = 0$. The v-i characteristic of the two-terminal resistor seen at (d') and (s') is given by the $v'_{gs} = 0$ curve in Fig. 5.2d and is shown in Fig. 5.7b. It is a voltage-controlled nonlinear resistor.

Let us now replace the linear resistor in the circuit in Fig. 5.4 by this two-terminal nonlinear resistor as shown in Fig. 5.8. Consider the one-port to

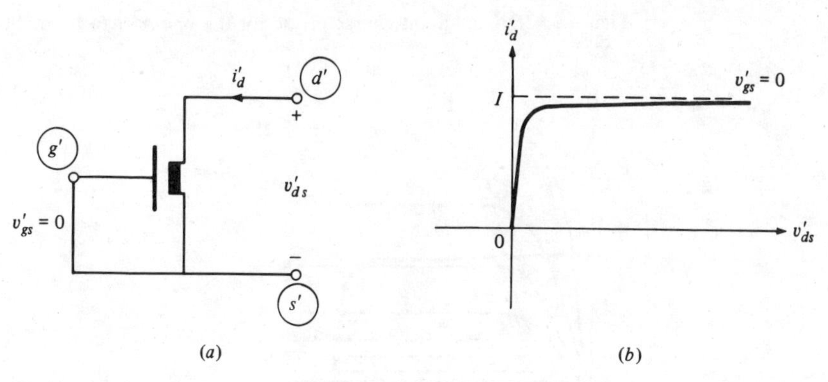

Figure 5.7 The depletion-type MOS transistor as a two-terminal resistor together with its v-i characteristic.

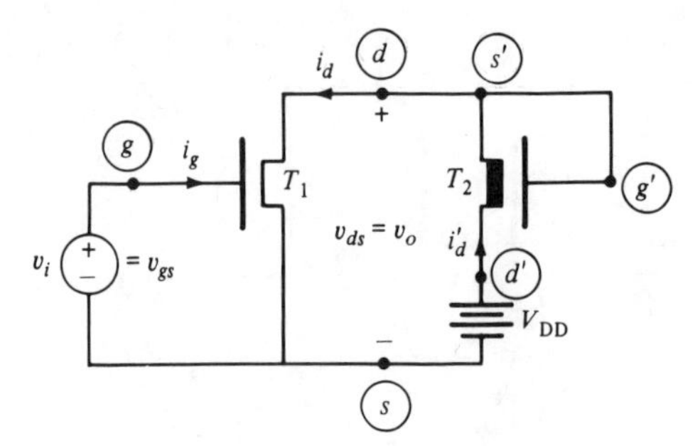

Figure 5.8 A basic inverter circuit with two MOS transistors.

the right of nodes ⓓ and ⓢ, which is the series connection of the two-terminal nonlinear resistor just created using the depletion-type transistor T_2 and a dc voltage source V_{DD}. Its driving-point characteristic is easily obtained from Fig. 5.7, and is shown in Fig. 5.9 plotted in the v_o-i_d plane. Note that $v_o = V_{DD} - v'_{ds}$ and $i_d = i'_d$.

We can now draw the *nonlinear load line* on the v_{ds}-i_d plane superimposed on the characteristic of the transistor T_1. This is shown in Fig. 5.10. The intersections of the load line and the characteristic curves for different values of v_{gs} specify the transfer characteristic from the input voltage v_i to the output voltage v_o. These points are marked on Fig. 5.10. From them we obtain the transfer characteristic of the inverter shown in Fig. 5.11.

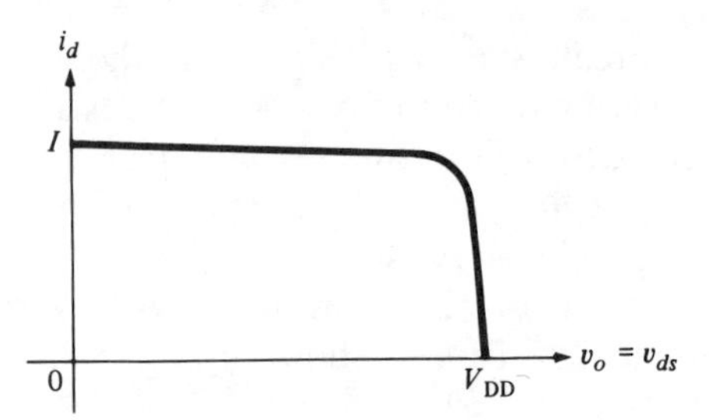

Figure 5.9 Driving-point characteristic for the one-port to the right of nodes ⓓ and ⓢ in Fig. 5.8.

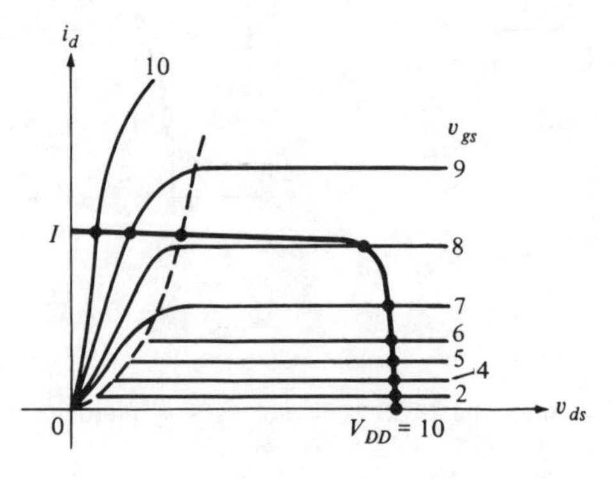

Figure 5.10 Nonlinear load line.

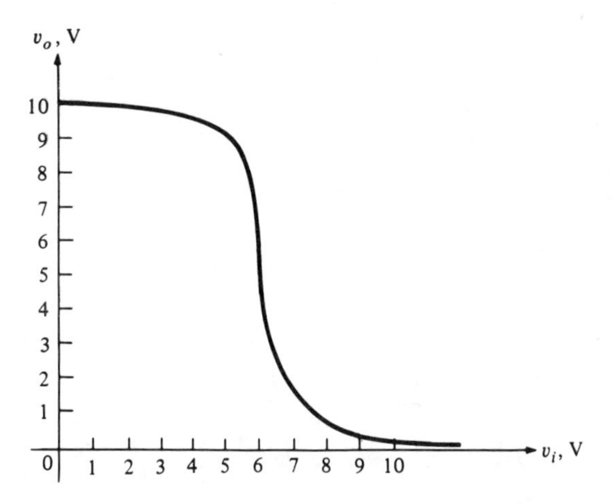

Figure 5.11 Transfer characteristic of the basic inverter circuit.

6 MULTIPORT AND MULTITERMINAL RESISTORS

We conclude this chapter with a brief discussion of two linear three-port resistors, the three-port ideal transformer and ideal circulator, and a nonlinear four-terminal resistor, the analog multiplier. In Chap. 4 we shall study thoroughly a most important four-terminal element, namely, the operational amplifier.

Since an $(n+1)$-terminal resistor may be considered as a special case of an n-port resistor having a common terminal among its n ports, i.e., a "grounded" n-port, the treatment of a $(n+1)$-terminal resistor and a resistive n-port is the same. Both involve n port voltages and n port currents, and in both cases we need n scalar equations to specify the v-i characteristics. In the most general form the characteristics are given by n scalar equations:

$$\begin{aligned} f_1(v_1, v_2, \ldots, v_n, i_1, i_2, \ldots, i_n) &= 0 \\ f_2(v_1, v_2, \ldots, v_n, i_1, i_2, \ldots, i_n) &= 0 \\ &\vdots \\ f_n(v_1, v_2, \ldots, v_n, i_1, i_2, \ldots, i_n) &= 0 \end{aligned} \tag{6.1}$$

6.1 The Three-Port Ideal Transformer

Many communication systems and measurement systems make use of a device called the *hybrid coil*. A hybrid coil is a multiwinding transformer. We consider only the three-winding transformer shown in Fig. 6.1. Note that it is a three-port which has six port variables $(v_1, v_2, v_3, i_1, i_2, i_3)$, and thus we need

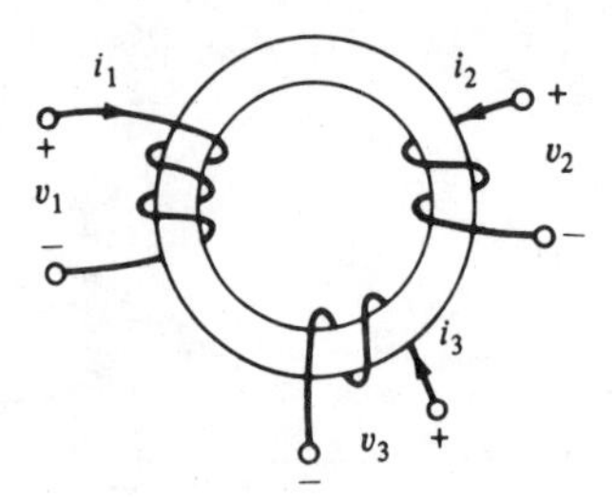

Figure 6.1 A three-winding hybrid coil.

three equations to define its characteristics. A three-port ideal transformer is a non-energic element characterized by:

$$\begin{aligned} f_1(v_1, v_2, v_3, i_1, i_2, i_3) &\stackrel{\Delta}{=} \frac{v_1}{n_1} - \frac{v_3}{n_3} = 0 \\ f_2(v_1, v_2, v_3, i_1, i_2, i_3) &\stackrel{\Delta}{=} \frac{v_2}{n_2} - \frac{v_3}{n_3} = 0 \\ f_3(v_1, v_2, v_3, i_1, i_2, i_3) &\stackrel{\Delta}{=} n_1 i_1 + n_2 i_2 + n_3 i_3 = 0 \end{aligned} \tag{6.2}$$

Like the two-port ideal transformer, neither the resistance matrix nor the conductance matrix exists. However, the hybrid matrices do exist.

Exercise 1 Derive from Eq. (6.2) the hybrid representation below:

$$\begin{bmatrix} v_1 \\ v_2 \\ i_3 \end{bmatrix} = \begin{bmatrix} 0 & 0 & \dfrac{n_1}{n_3} \\ 0 & 0 & \dfrac{n_2}{n_3} \\ -\dfrac{n_1}{n_3} & -\dfrac{n_2}{n_3} & 0 \end{bmatrix} \begin{bmatrix} i_1 \\ i_2 \\ v_3 \end{bmatrix} \tag{6.3}$$

Exercise 2 Show, that at all times, the power entering a three-port ideal transformer is identically zero. It is non-energic.

6.2 The Three-Port Circulator

Circulators are very useful microwave devices used in communication systems and measurements. An *ideal three-port circulator* is a linear circuit element specified by the following three equations:

$$\begin{aligned} f_1(v_1, v_2, v_3, i_1, i_2, i_3) &\stackrel{\Delta}{=} v_1 - Ri_2 + Ri_3 = 0 \\ f_2(v_1, v_2, v_3, i_1, i_2, i_3) &\stackrel{\Delta}{=} v_2 + Ri_1 - Ri_3 = 0 \\ f_3(v_1, v_2, v_3, i_1, i_2, i_3) &\stackrel{\Delta}{=} v_3 - Ri_1 + Ri_2 = 0 \end{aligned} \tag{6.4}$$

where R is a real constant called the *reference resistance*. We can recast (6.4) in terms of the open-circuit resistance matrix as follows:

$$\begin{bmatrix} v_1 \\ v_2 \\ v_3 \end{bmatrix} = \begin{bmatrix} 0 & R & -R \\ -R & 0 & R \\ R & -R & 0 \end{bmatrix} \begin{bmatrix} i_1 \\ i_2 \\ i_3 \end{bmatrix} \tag{6.5}$$

Observe that a three-port circulator is *non-energic* because the instantaneous power *entering* the three-port is identically zero:

$$\begin{aligned} p &= v_1 i_1 + v_2 i_2 + v_3 i_3 \\ &= (Ri_2 - Ri_3)i_1 + (-Ri_1 + Ri_3)i_2 + (Ri_1 - Ri_2)i_3 \\ &\equiv 0 \end{aligned} \tag{6.6}$$

Hence energy is neither stored nor dissipated in the circulator. To demonstrate how energy is being *redistributed*, suppose we connect three identical resistors whose values are chosen equal to R in the setup shown in Fig. 6.2. Since $v_2 = -Ri_2$ and $v_3 = -Ri_3$, it follows from Eq. (6.5) that

$$\begin{aligned} v_1 &= Ri_2 - Ri_3 \\ -Ri_2 &= -Ri_1 + Ri_3 \\ -Ri_3 &= Ri_1 - Ri_2 \end{aligned}$$

Solving these equations, we obtain

$$v_1 = Ri_1 \tag{6.7}$$

$$i_1 = i_2 \quad \text{and} \quad i_3 = 0 \tag{6.8}$$

Applying Tellegen's theorem to this circuit, we obtain[6]

$$v_s(-i_1) + v_R i_1 + \underbrace{(v_1 i_1 + v_2 i_2 + v_3 i_3)}_{0} + v_2(-i_2) + \underbrace{v_3(-i_3)}_{0} = 0 \tag{6.9}$$

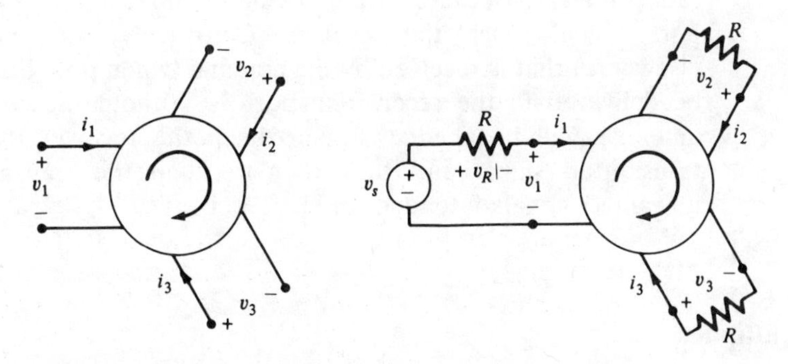

Figure 6.2 Three-port circulator.

[6] The negative sign is needed at three locations in Eq. (6.9) because the *associated reference directions* must be used when applying Tellegen's theorem.

If we let $p_s = v_s i_1$, the power supplied by the voltage source, Eq. (6.9) reduces to[7]

$$p_s = v_R i_1 + v_2(-i_2) = Ri_1^2 + Ri_2^2 = 2(Ri_1^2) \tag{6.10}$$

We conclude that half of the power supplied by the voltage source is dissipated in its associated series resistor, while the other half is dissipated in the resistor across port 2. In other words, all power entering port 1 is *redirected* to port 2 (to be dissipated in the terminating resistor) with nothing left for port 3. We can verify this conclusion using Eqs. (6.5), (6.7), and (6.8) directly.

The power entering port 1 is given by

$$p_1 = v_1 i_1 = (Ri_2 - Ri_3)i_1 = Ri_1^2 \tag{6.11}$$

The power dissipated in the resistors across port 2 and port 3 are

$$p_2 = v_2(-i_2) = (-Ri_1 + Ri_3)(-i_2) = Ri_1^2 \tag{6.12}$$

$$p_3 = v_3(-i_3) = (Ri_1 - Ri_2)(-i_3) = 0 \tag{6.13}$$

If we repeat the preceding analysis but with the voltage source inserted in port 2, instead of port 1, we will find that all power entering port 2 gets delivered to port 3 with nothing left for port 1. Similarly, inserting the voltage source in port 3, we find that all the power entering port 3 gets delivered to port 1 with nothing left for port 2. Hence the circulator functions by "circulating" the energy entering one port into the next port *whenever all ports are terminated by resistors equal to the reference resistance R.*

This property is widely exploited in many communication systems for diverting power into various desired channels. For example, the setup in Fig. 6.2 can be used to model the following situation: Let the voltage-source resistor combination model a portable radio *transmitter*. Let the resistor $\mathcal{R}$ across port 2 model an *antenna*, and let the resistor $\mathcal{R}$ across port 3 model a radio *receiver*. Because of the circulator, no outgoing signal transmitted from port 1 will reach the receiver. Conversely, any incoming signal (from elsewhere) that is received by the antenna (when port 1 is not transmitting) will be delivered to the receiver in port 3. Without the circulator, two separate antennas will be needed, one to keep the receiver from receiving its own transmitted signal and the other to keep the transmitter from receiving unwanted signals intended for the receiver.

6.3 Analog Multiplier

The symbol of an *analog multiplier* is shown in Fig. 6.3*a*. For low operating frequencies, this device can be modeled by the equations:

[7] Observe that in applying Tellegen's theorem, the terms corresponding to the ports of a *non-energic n*-port resistor always sum to zero and *need* not be included in the calculation.

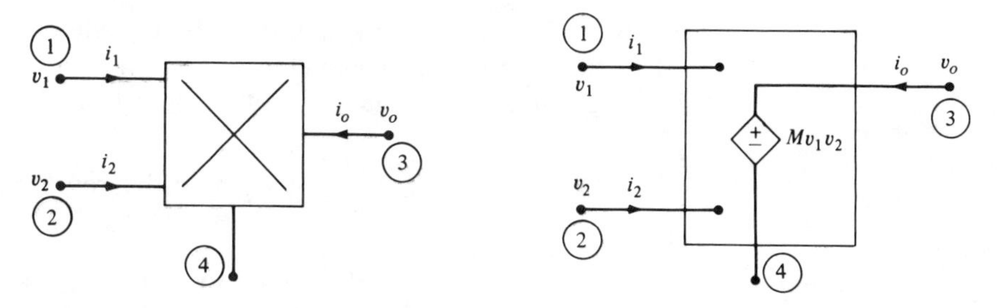

Figure 6.3 (*a*) Symbol for an analog multiplier. (*b*) A two-input controlled source model of an analog multiplier.

$$i_1 = 0$$
$$i_2 = 0 \qquad (6.14)$$
$$v_o = Mv_1v_2$$

where v_1, v_2, and v_o are node voltages with respect to the datum node ④ and M is a *scaling* factor having a dimension of V^{-1}. The range of v_1 and v_2 for typical IC analog multipliers is ± 10 V, and the associated scale factor M is $\frac{1}{10}$ so that the output voltage is also restricted to ± 10 V.

The analog multiplier is commercially available in the form of low-cost IC modules. It is widely used in the industry for numerous low-frequency signal processing applications.

We can model an analog multiplier as a *two-input nonlinear controlled voltage source* as shown in Fig. 6.3*b*. This model is extremely simple and therefore cannot account for various second-order effects. Nor does it impose a maximum voltage operating range. However, for the majority of low-frequency applications where the input and output voltages are restricted, by design, to ± 10 V, this model is more than adequate.

SUMMARY[8]

- A three-terminal resistor or a resistive two-port is characterized by two equations in terms of two voltages v_1 and v_2 and two currents i_1 and i_2:

$$f_1(v_1, v_2, i_1, i_2) = 0$$
$$f_2(v_1, v_2, i_1, i_2) = 0$$

A linear resistive two-port which is current-controlled has a representation $\mathbf{v} = \mathbf{R}\mathbf{i}$, or in scalar form:

$$v_1 = r_{11}i_1 + r_{12}i_2$$
$$v_2 = r_{21}i_1 + r_{22}i_2$$

The constant matrix $\mathbf{R}$ is called the open-circuit resistance matrix of the

[8] Since we consider exclusively time-invariant elements in the summary, we omit the qualifier "time-invariant."

two-port. The dual representation is $\mathbf{i} = \mathbf{G}\mathbf{v}$ where the constant matrix $\mathbf{G}$ is called the short-circuit conductance matrix.

- The two hybrid representations of a linear resistive two-port are

$$\begin{bmatrix} v_1 \\ i_2 \end{bmatrix} = \mathbf{H} \begin{bmatrix} i_1 \\ v_2 \end{bmatrix} \qquad \text{hybrid 1 representation}$$

and

$$\begin{bmatrix} i_1 \\ v_2 \end{bmatrix} = \mathbf{H}' \begin{bmatrix} v_1 \\ i_2 \end{bmatrix} \qquad \text{hybrid 2 representation}$$

where $\mathbf{H}' = \mathbf{H}^{-1}$.

- The four linear controlled sources can be characterized by the four two-port equations:

CCVS:	$v_1 = 0$	$v_2 = r_m i_1$
VCCS:	$i_1 = 0$	$i_2 = g_m v_1$
CCCS:	$v_1 = 0$	$i_2 = \alpha i_1$
VCVS:	$i_1 = 0$	$v_2 = \mu v_1$

where r_m, g_m, α, and μ are constants.

- A two-port ideal transformer is a non-energic linear element characterized by

$$\begin{bmatrix} v_1 \\ i_2 \end{bmatrix} = \begin{bmatrix} 0 & n \\ -n & 0 \end{bmatrix} \begin{bmatrix} i_1 \\ v_2 \end{bmatrix}$$

where n is a constant.

- A gyrator is a linear non-energic two-port characterized by

$$\mathbf{i} = \begin{bmatrix} 0 & G \\ -G & 0 \end{bmatrix} \mathbf{v}$$

where G is a constant.

- At dc, a bipolar transistor can be modeled as a three-terminal resistor. Its characteristics can be expressed graphically using two sets of curves plotted on appropriate v-i planes. (See Figs. 4.3 and 4.5.)
- Like the study of two-terminal nonlinear resistors, the two important problems encountered in three-terminal nonlinear resistors are the dc analysis (i.e., finding the operating point) and the small-signal analysis about a given operating point.
- The small-signal parameters of a three-terminal nonlinear resistor are the elements of the jacobian matrix evaluated at the operating point. The four parameters are obtained by taking the partial derivatives of pertinent nonlinear functions with respect to the appropriate variable and evaluating them at the operating point.

- The characterizations of resistive two-ports and three-terminal resistors can be extended to resistive n-ports and $(n+1)$-terminal resistors: it requires n scalar functions involving n voltages and n currents.

PROBLEMS

Two-port representations

1 For the linear two-ports specified by the following equations, find as many representations as you can

(*a*) $-i_1 + 2i_2 + v_2 = 0$
$v_1 + v_2 = 0$

(*b*) $v_1 + i_2 + v_2 = 0$
$i_1 = 0$

(*c*) $v_1 = 0$
$i_2 - i_1 = 0$

(*d*) $v_1 + v_2 = 0$
$i_1 + i_2 = 0$

2 Find the six representations (whenever they exist) of the two-ports shown in Fig. P3.2.

Conductance matrix

3 (*a*) For the linear two-port shown in Fig. P3.3*a* determine the conductance matrix.

(*b*) The two-port is driven by a source and terminated by a load as shown in Fig. P3.3*b*. Determine the output voltage v_2 in terms of v_s, R_s, R_L, and the conductances g_{11}, g_{12}, g_{21}, and g_{22}.

Transmission matrix, cascade connection

4 Let N_1 and N_2 be two linear two-ports described by transmission 1 matrices $\mathbf{T}_1$ and $\mathbf{T}_2$, respectively. Show that the two-port N obtained from the cascade connection of N_1 and N_2 (Fig. P3.4) is specified by the transmission 1 matrix $\mathbf{T} = \mathbf{T}_1 \cdot \mathbf{T}_2$.

Equivalent circuit of linear two-ports

5 Using only linear resistors and linear controlled sources, draw an equivalent circuit for the two-ports described in Prob. 1.

6 Using only linear resistors and linear controlled sources, find an equivalent circuit for the two-ports described by the following matrices:

(*a*) $\mathbf{R} = \begin{bmatrix} 3 & 1 \\ 2 & 5 \end{bmatrix}$ (*d*) $\mathbf{H}' = \begin{bmatrix} 1 & -1 \\ 2 & 3 \end{bmatrix}$

(*b*) $\mathbf{G} = \begin{bmatrix} 3 & -2 \\ -1 & 2 \end{bmatrix}$ (*e*) $\mathbf{T} = \begin{bmatrix} 0 & 0 \\ 1 & -1 \end{bmatrix}$

(*c*) $\mathbf{H} = \begin{bmatrix} 2 & 5 \\ -2 & \frac{1}{2} \end{bmatrix}$ (*f*) $\mathbf{T}' = \begin{bmatrix} 1 & -1 \\ -2 & 1 \end{bmatrix}$

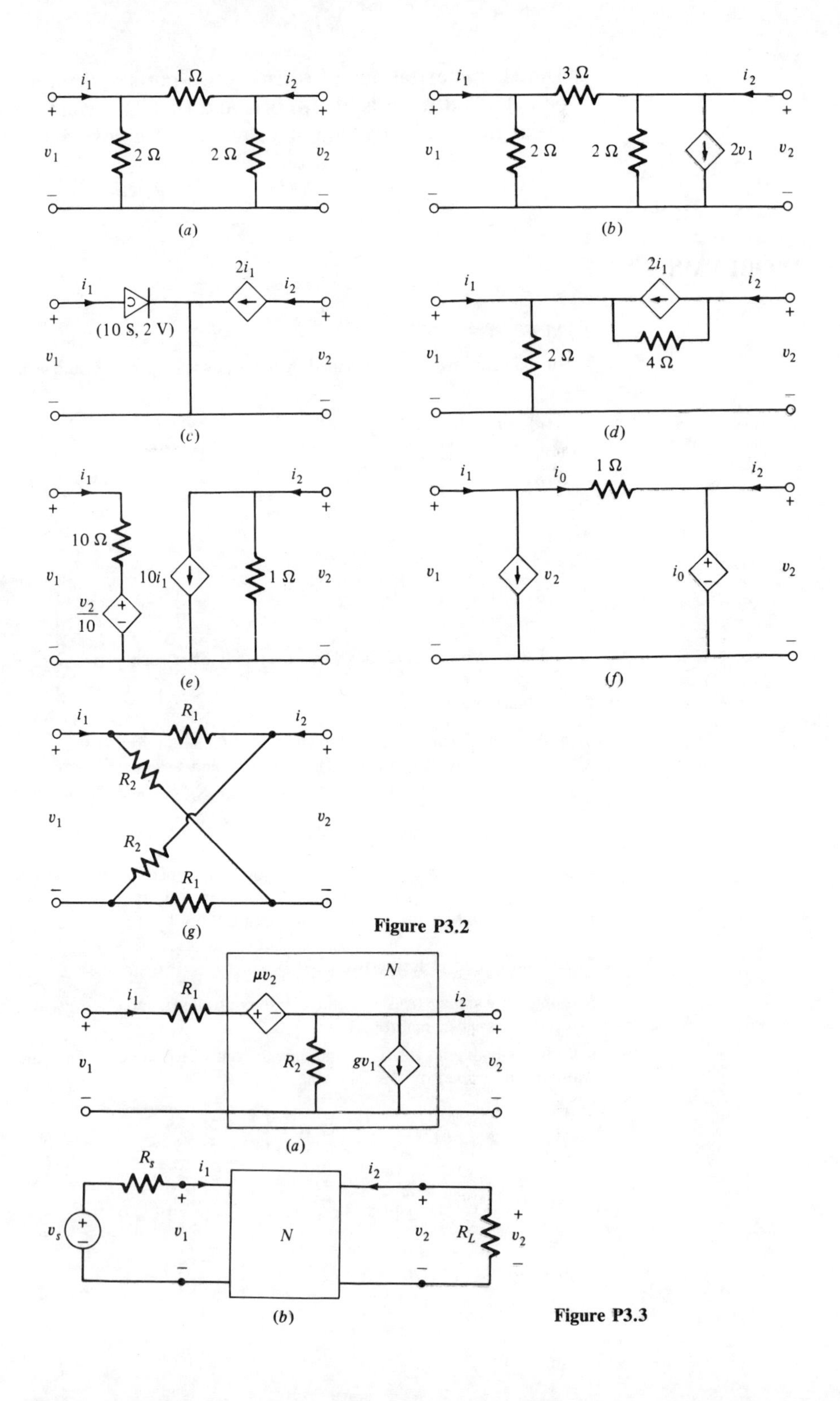

Figure P3.2

Figure P3.3

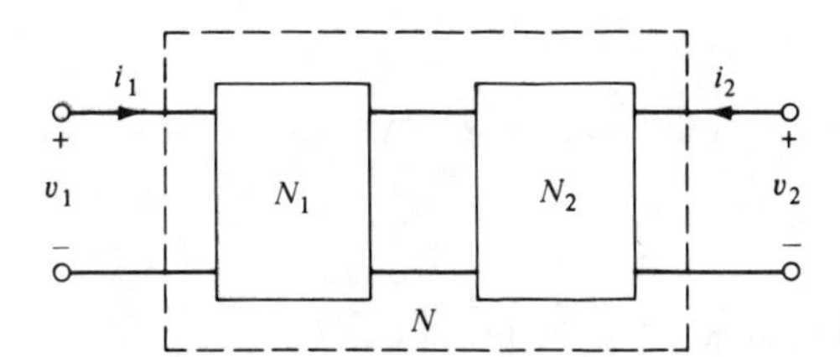

Figure P3.4

Two-port representations, gyrators, and ideal transformers

7 Write down as many two-port representation matrices as possible for the two-ports shown in Fig. P3.7.

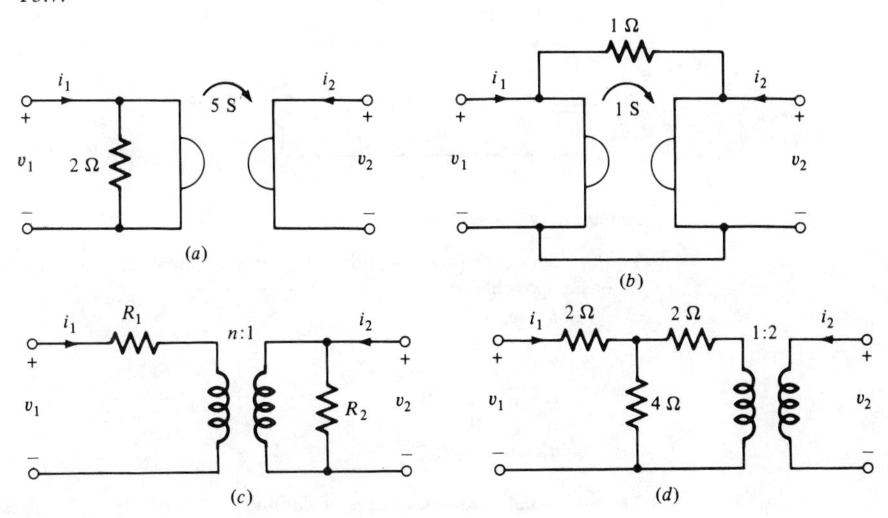

Figure P3.7

Gyrator

8 Consider the gyrator-tunnel-diode circuit shown in Fig. P3.8a, where the tunnel-diode v-i curve is approximated as shown in Fig. P3.8b.

(a) Write an analytic expression describing this v-i curve.

(b) Derive the analytic relationship between v_1 and i_1.

(c) Plot the v_1-i_1 relationship from (b).

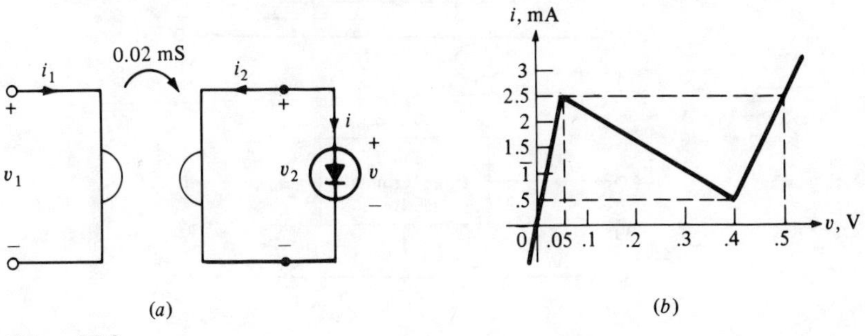

Figure P3.8

Synthesis

9 Show that every linear CCCS or linear VCVS can be synthesized using only a linear CCVS and a linear VCCS.

Power

10 For the circuit shown in Fig. P3.10 calculate

(*a*) i_1 and v_2

(*b*) The power delivered by the battery

(*c*) The power dissipated in R_1 and R_2

(*d*) Explain the balance of energy flow in the circuit.

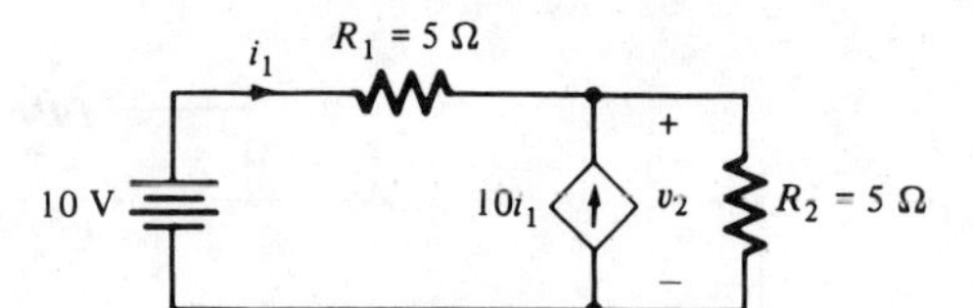

Figure P3.10

Nonlinear two-ports

11 For each of the following nonlinear three-terminal resistors, obtain as many two-port representations as possible.

(*a*) $i_1 + v_2 = 0$
$v_1 + i_2 - i_1^3 = 0$

(*b*) $i_1 - v_1 v_2 - v_2 = 0$
$i_2 - v_1^2 - v_1 = 0$

(*c*) $i_1 + i_2 = 0$
$v_1 + v_2 - i_1^2 = 0$

Three-terminal resistors, change of datum

12 (*a*) The linear three-terminal resistor shown in Fig. P3.12*a* is connected as a grounded two-port. Determine the conductance matrix **G**.

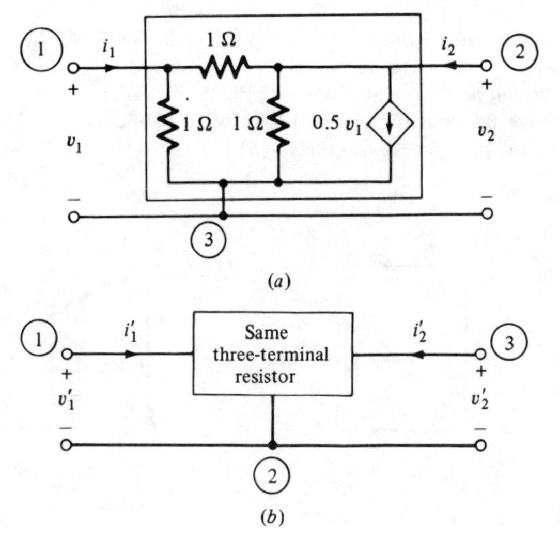

Figure P3.12

(*b*) If terminal ② of the same three-terminal resistor is used as the datum node to form a grounded two-port, as shown in Fig. P3.12b, determine the conductance matrix **G** where

$$\begin{bmatrix} i_1' \\ i_2' \end{bmatrix} = \mathbf{G}' \begin{bmatrix} v_1' \\ v_2' \end{bmatrix}$$

Two-port representation, parallel connection

13 The two-port N in Fig. P3.13 is obtained by connecting two two-ports N_a and N_b in parallel. N is described by the following representation:

$$i_1 = v_1 + v_2$$

$$i_2 = v_1 + 2v_2$$

If N_a has the following representation:

$$v_{a1} = v_{a2} + i_{a2}$$

$$i_{a1} = v_{a2} - 3i_{a2}$$

find the voltage-controlled representation of N_b.

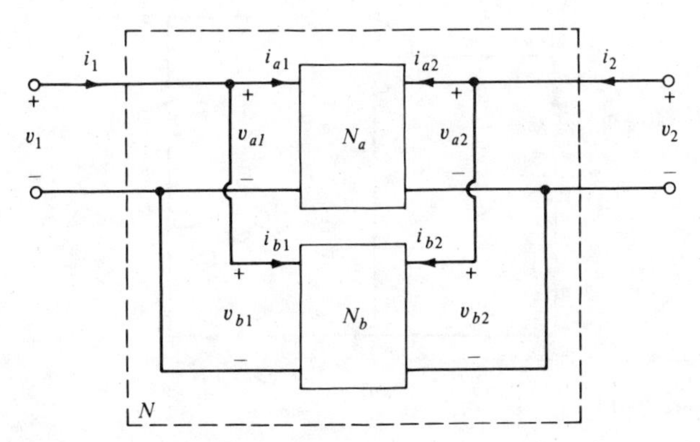

Figure P3.13

DC operating point, small-signal analysis

14 The common-emitter characteristics of a typical *npn* transistor are shown in Fig. P3.14*a*. For the circuit shown in Fig. P3.14*b*:

(*a*) Find the dc operating point using the double load-line method.
(*b*) Determine the small-signal hybrid parameters of the transistor at the operating point.
(*c*) For $v_s(t) = 0.01 \cos t$ determine the output voltage $v_o(t)$.

npn **Transistor, piecewise-linear model**

15 Using the transistor model in Fig. 4.7 with

$$G_1 = 0.1\ \text{S} \qquad \mu = 0.1$$
$$\beta = 100 \qquad R_2 = 3\ \text{M}\Omega$$
$$E_0 = 0.25\ \text{V} \qquad I_0 = 0$$

compute, analytically, the operating point for the circuit shown in Fig. 4.10 with

$$E_1 = 5\ \text{V} \qquad R_1 = 50\ \text{K}\Omega$$
$$E_2 = 9\ \text{V} \qquad R_2 = 500\ \Omega$$

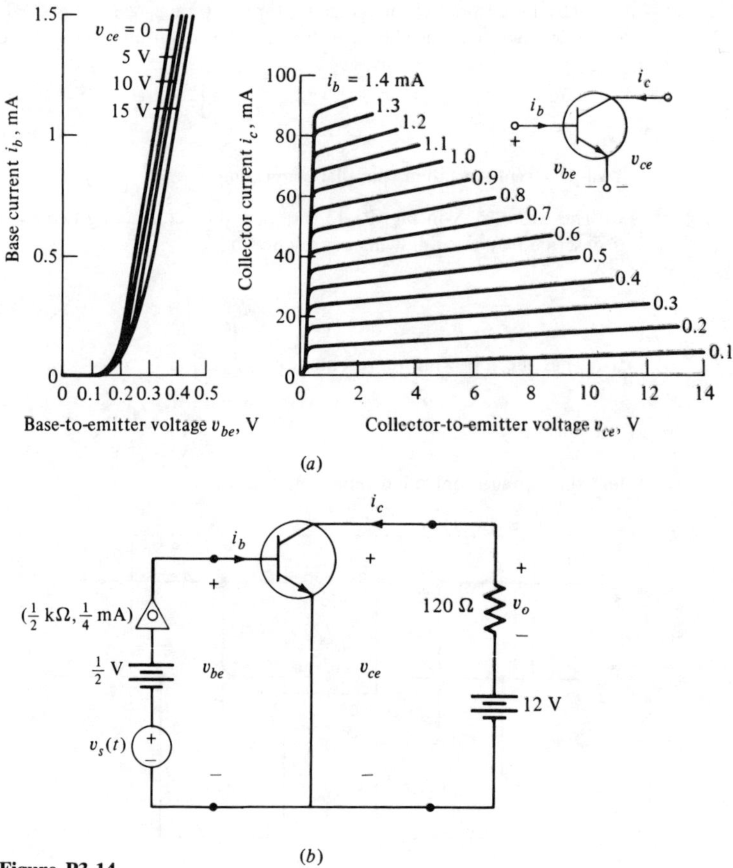

Figure P3.14

DC operating point

16 Using the v-i characteristic of the solar cell shown in Fig. P3.16a and the transistor characteristics in Fig. P3.14a, find the operating point of the circuit shown in Fig. P3.16b.

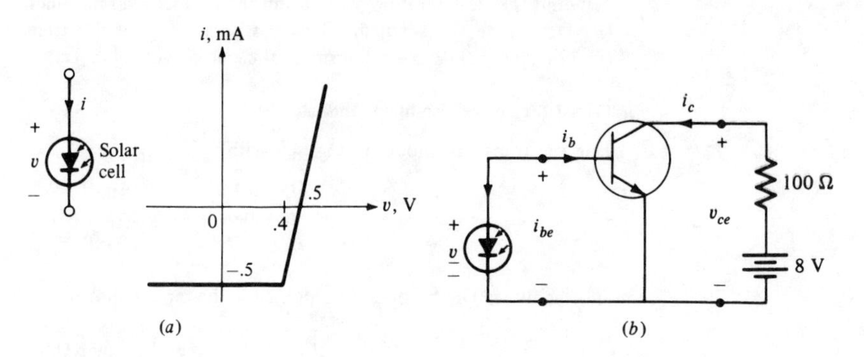

Figure P3.16

DC operating point, small-signal analysis

17 Consider the circuit shown in Fig. P3.17*a*. The two-port N is specified by

$$v_1 = i_1 + i_1^2$$

and the curves in Fig. P3.17*b*.

(*a*) Determine the dc operating point(s).

(*b*) Determine the small-signal equivalent circuit at each operating point.

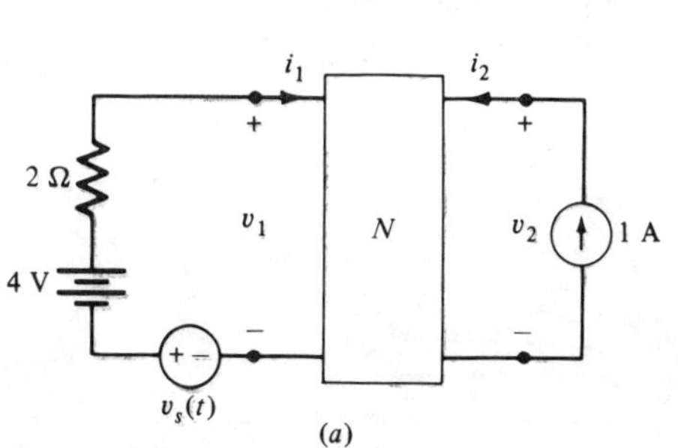

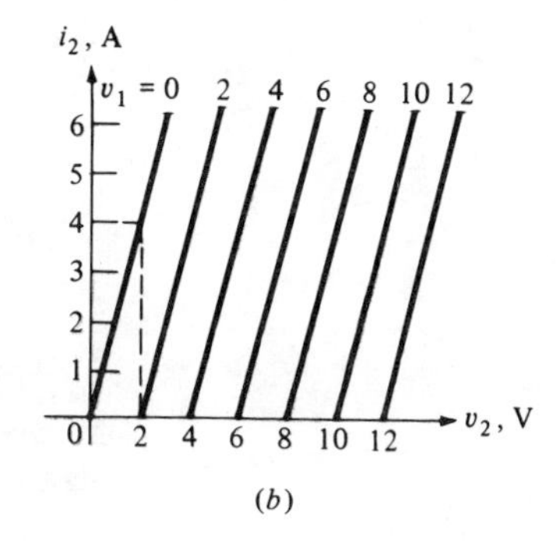

Figure P3.17

18 The three-terminal device in Fig. P3.18 is modeled by

$$i_1 = v_1 + v_1^2$$

$$i_2 = -4v_i + \tfrac{1}{2}(v_2 - 1) + \tfrac{1}{2}|v_2 - 1|$$

(*a*) Draw the equivalent circuit of the three-terminal device using two-terminal elements.

(*b*) Determine the operating point of the circuit with $i_s = 2$ A and $v_1 > 0$.

(*c*) Draw the small-signal equivalent circuit at this operating point.

(*d*) Let $i_s = 2 + 0.1 \cos t$; determine $v_2(t)$.

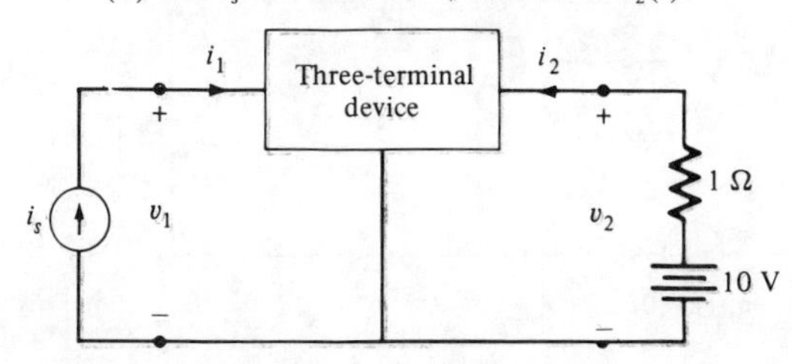

Figure P3.18

19 Two-port N in Fig. P3.19 is described by the following current-controlled representation:

$$v_1 = e^{i_1} + i_2 \qquad v_2 = i_1 i_2 + i_2$$

(*a*) Determine the dc operating point.

(*b*) Draw a small-signal equivalent circuit about this operating point.

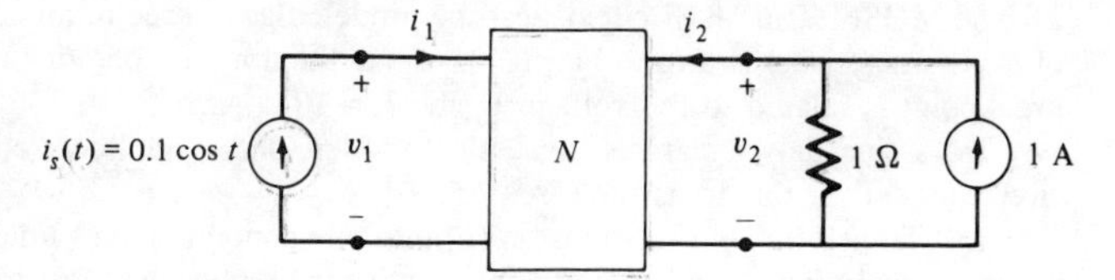

Figure P3.19

MOS transistor

20 The characteristic of a typical MOS transistor is shown in Fig. P3.20*a*. For the circuit shown in Fig. P3.20*b*:

(*a*) Find the operating point *Q*.

(*b*) Synthesize a small-signal circuit model for the transistor about the operating point *Q* using only one linear resistor and one linear controlled source.

$$\beta = 0.64\ \text{mA/V}^2 \qquad V_{\text{th}} = -5\ \text{V}$$

(*c*) Draw the small-signal equivalent circuit.

(*d*) Calculate the small-signal voltage $\tilde{v}_2(t)$.

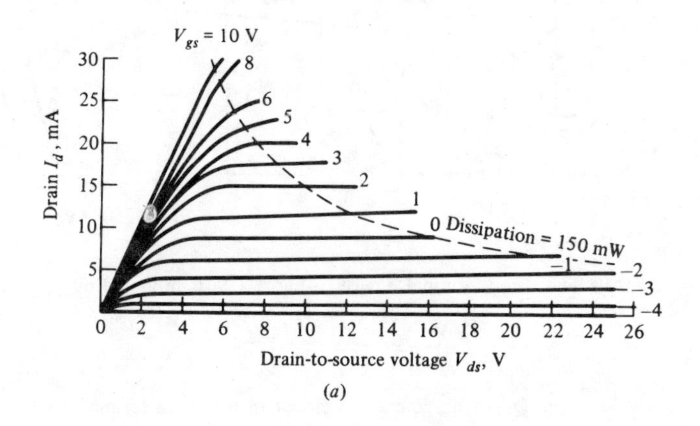

(*a*)

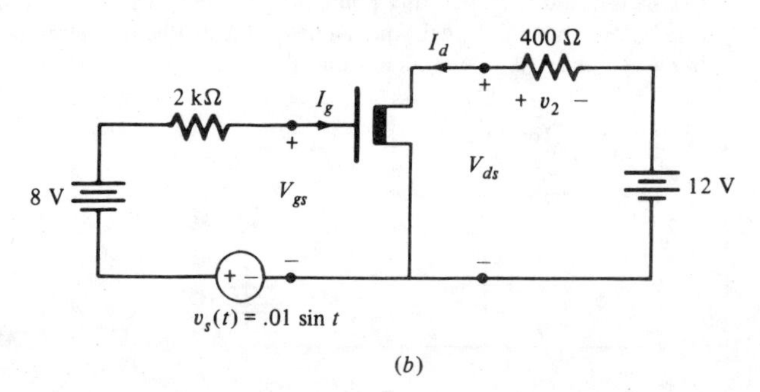

(*b*)

Figure P3.20

MOS transistor, modeling

21 The MOSFET in VLSI circuits can be modeled as a three-terminal resistor by the two families of *v*-*i* characteristics shown in Fig. P3.21. Over the normal operating region, the current I_d at each breakpoint is related to the voltage V_{gs} by $I_d = 10\,V_{gs}^2$ mA.

(*a*) Using only linear resistors, ideal diodes, and a nonlinear controlled source synthesize a circuit model for this transistor.

(*b*) Transform your model from (*a*) into one containing only linear resistors, linear controlled sources, ideal diodes, and at most one nonlinear two-terminal resistor.

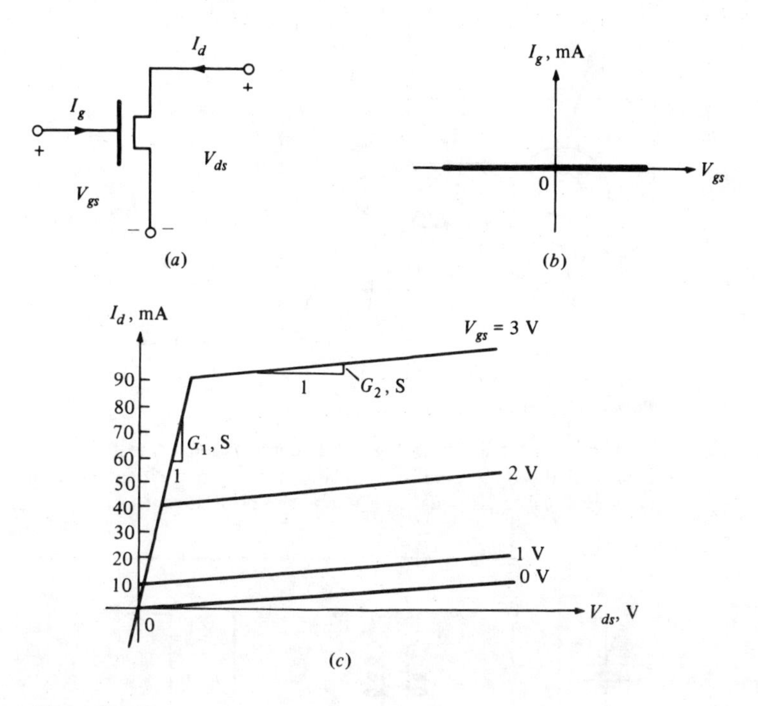

Figure P3.21

Circulator

22 Show that the one-port shown in Fig. P3.22*a* is equivalent to the one-port in Fig. P3.22*b*, where R is the characteristic resistance of the circulator.

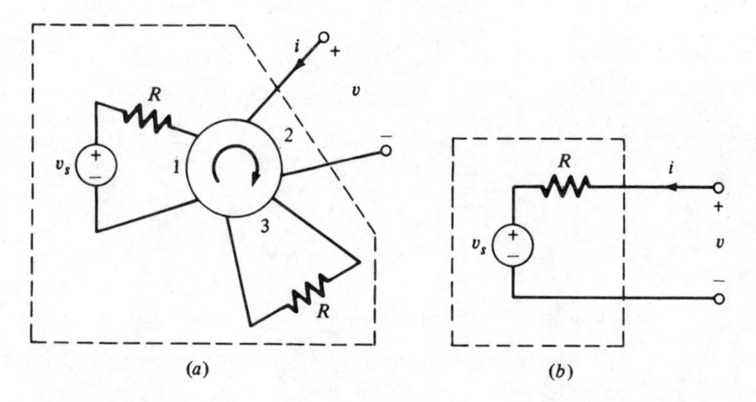

Figure P3.22

23 A four-port circulator can be realized by the circuit shown in Fig. P3.23. Describe the property of this circulator.

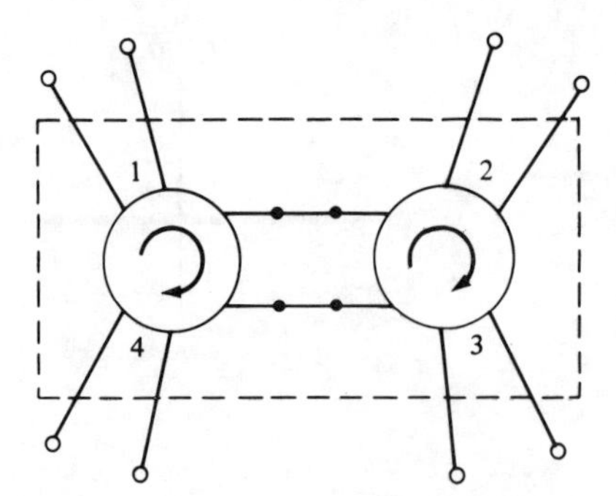

Figure P3.23

Power, circulator

24 The two-port shown in Fig. P3.24 contains ideal transformers, gyrators, circulators, and ideal diodes. If the battery delivers 10 W to the external circuit, and the output voltage v_o across the nonlinear resistor is equal to 5 V, find the port current i_o.

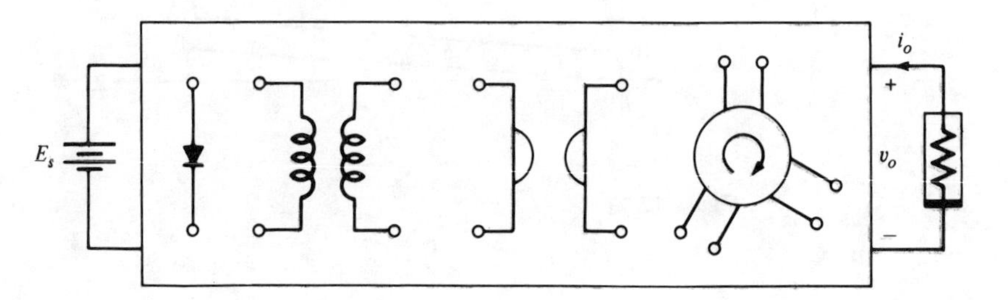

Figure P3.24

CHAPTER

FOUR

OPERATIONAL-AMPLIFIER CIRCUITS

The *operational amplifier* (op amp) is an extremely versatile and inexpensive semiconductor device. It is the workhorse of the communication, control, and instrumentation industry.

For *low-frequency* applications, the op amp behaves like a *four-terminal nonlinear resistor* which can often be represented by an *ideal op-amp model*. This model greatly simplifies the analysis and design of op-amp circuits. In fact, one of the reasons why op amps are so popular is that, at low frequencies, they behave almost like the *ideal* model! Consequently, except in the last section, our methods of analysis in this chapter will be based on the *ideal* model. This choice is justified in Sec. 4 by analyzing a typical op-amp circuit (operating at low frequency) using a more complicated (*finite-gain*) op-amp model and then comparing the results with those predicted by the ideal op-amp model.

Depending on the *dynamic range* of the input signals, an op amp may operate in the *linear* or *nonlinear* region. Section 2 is devoted to those circuits where the op amp is operating only in the linear region. This restriction allows us to simplify the (nonlinear) ideal op-amp model into a *linear* model, called the *virtual short-circuit model*. This model is used exclusively in Sec. 2 for analyzing both simple circuits *by inspection*, as well as complicated circuits via a *systematic method*.

The organization in Sec. 2 is followed in Sec. 3 for op amps operating in the nonlinear region. Here, it is necessary to use the (nonlinear) ideal op-amp model.

1 DEVICE DESCRIPTION, CHARACTERISTICS, AND MODEL

Operational amplifiers (op amps) are multiterminal devices sold in several standard packages, two of which are shown in Figs. 1.1 and 1.2. Because they

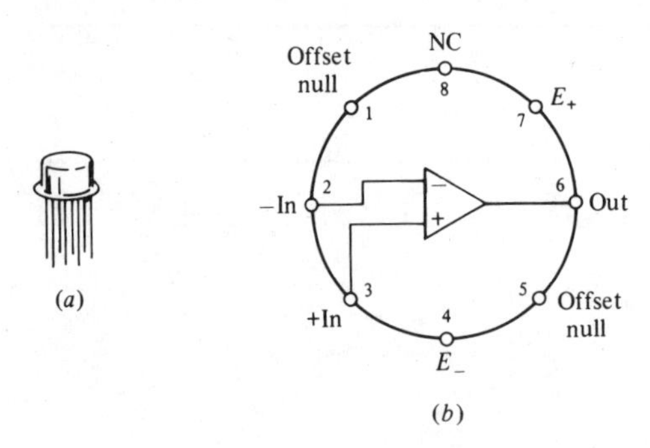

Figure 1.1 Eight-lead metal can. (*a*) Side view. (*b*) Top view.

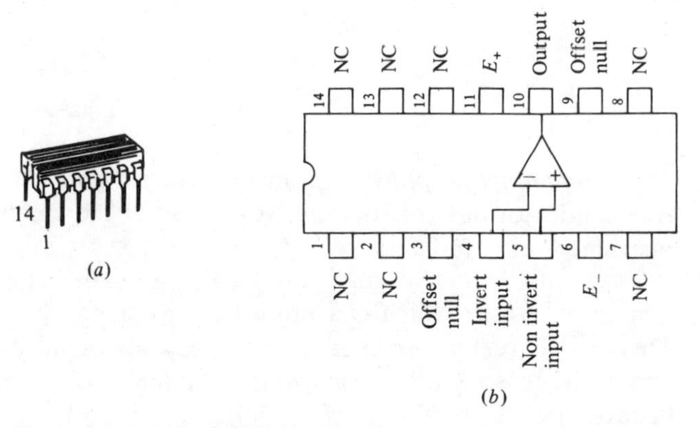

Figure 1.2 A 14-lead DIP (dual in-line package). (*a*) Side view (*b*) Top view.

are inexpensive (some cost less than 25 cents a piece), reliable, and extremely versatile, op amps have become the workhorse of the electronics industry.

Over 2000 types of integrated circuits (*IC*) op amps are currently available, each containing nearly two dozen transistors. Figure 1.3 gives the schematic of the popular μA741, a second-generation op amp introduced by Fairchild Semiconductor in 1968. The seven terminals brought out through the package leads are labeled *inverting input*, *noninverting input*, *output*, E_+, E_-, and *offset null* (two of them). The remaining terminals of the package in Figs. 1.1*b* and 1.2*b* not connected to the *IC* are labeled *NC* (for no connection).

Some op amps have more than seven terminals; others have less. For most applications, however, only the five terminals indicated in the standard op-amp symbol in Fig. 1.4*a* are essential. The additional terminals are usually connected to some external nulling or compensation circuit for improving the performance of the op amp. In order for the op amp to function properly its internal transistors must be biased at appropriate operating points. Terminals E_+ and E_- are provided for this purpose. In general, they are connected to a split power supply as shown in Fig. 1.4*b*, where E_+ and E_- denote the voltage with respect to the *external* ground. Typically, $E_+ = +15$ V and $E_- = -15$ V.

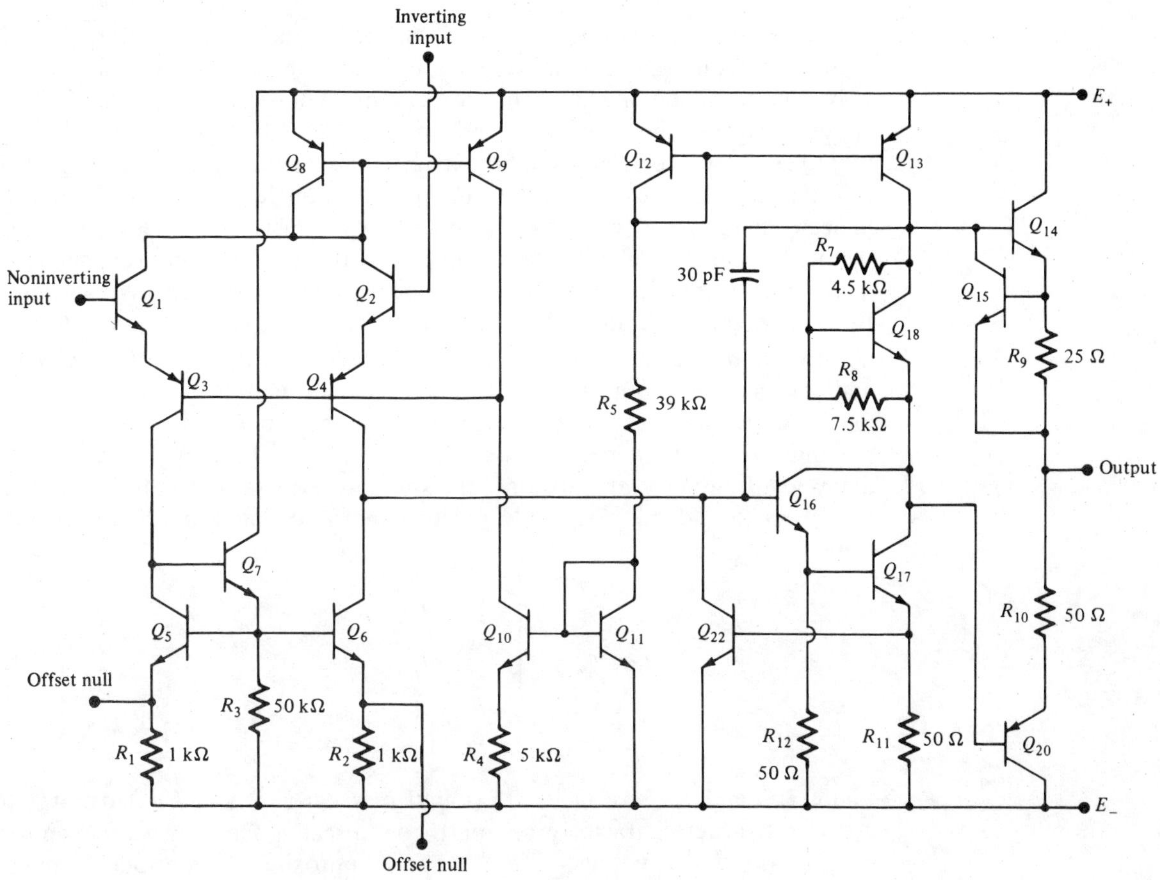

Figure 1.3 Schematic of the μA741 op amp.

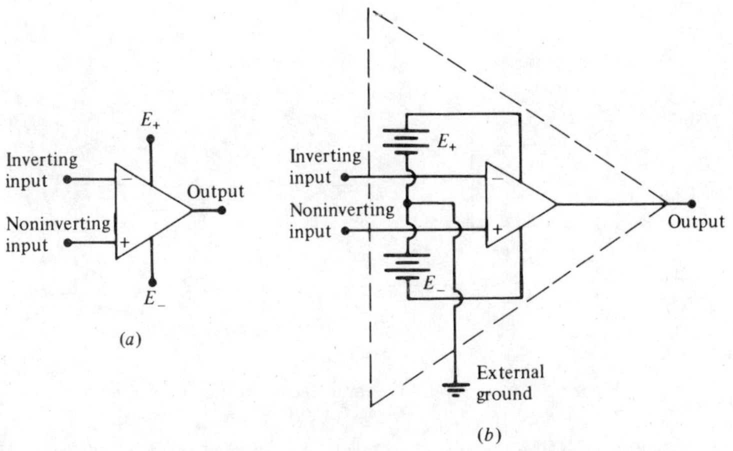

Figure 1.4 Standard op-amp symbol and a typical biasing scheme. (*a*) The − and + signs inside the triangle denote the inverting and noninverting input terminals, respectively. (*b*) A "biased" op amp (enclosed within the triangle) can be considered as a 4-terminal device for circuit analysis and design purposes.

After the power supply has been connected and after an external nulling and/or compensation circuit has been connected to any additional terminals, only four terminals are available for external connections. Hence, from the circuit designer's point of view, an op amp is really a *four-terminal device*, regardless of the original number of terminals in the op-amp package. This four-terminal device lies inside the dotted triangle in Fig. 1.4*b* and will henceforth be denoted by the symbol shown in Fig. 1.5*a*.[1] Here, i_- and i_+ denote the current entering the op-amp "inverting" and "noninverting" terminals, respectively. Similarly, v_-, v_+, and v_o denote respectively the voltage from the inverting terminal ⊖, noninverting terminal ⊕, and output terminal ⓞ to *ground*. The variable $v_d \triangleq v_+ - v_-$ is called the *differential input voltage* and will play an important role in op-amp circuit analysis.

To derive an exact characterization of an op amp would require analyzing the entire integrated circuit, such as the one shown in Fig. 1.3. Fortunately, for many low-frequency applications, the op-amp terminal currents and voltages have been found *experimentally* to obey the following *approximate* relationships:

$$\boxed{\begin{aligned} i_- &= I_{B-} \\ i_+ &= I_{B+} \\ v_o &= f(v_d) \end{aligned}} \tag{1.1}$$

where I_{B-} and I_{B+} are called the *input bias currents* and $f(v_d)$ denotes the v_o-vs.-v_d transfer characteristic. Apart from a scaling factor which depends on the power supply voltage, $f(v_d)$ follows approximately an odd-symmetric

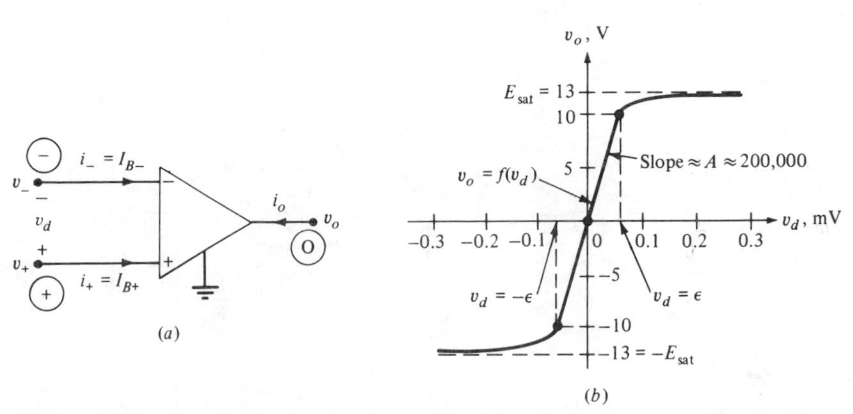

Figure 1.5 Experimental characterization of a typical op amp.

[1] The op-amp symbol given in most electronics literature shows only three terminals with the ground terminal omitted. This is because the ground terminal in Fig. 1.4*b* does not exist physically as a pin in most modern op-amp packages, but is rather created externally through the dual power supply. We added this terminal because without it, KCL would give the erroneous relationship $i_- + i_+ + i_o = 0$.

function as shown in Fig. 1.5*b* (drawn for a ±15-V supply voltage). Moreover, this function has been found to be rather insensitive to changes in the output current i_o.

The transfer characteristic in Fig. 1.5*b* displays three remarkable properties:

1. v_o and v_d have different scales: one in volts, the other in millivolts.
2. In a small interval $-\epsilon < v_d < \epsilon$ of the origin, $f(v_d) \approx Av_d$ is nearly *linear* with a very steep slope A—called the *open-loop voltage gain*.
3. $f(v_d)$ *saturates* at $v_o = \pm E_{sat}$, where E_{sat} is typically 2 V less than the power supply voltage ($E_{sat} = 13$ V in Fig. 1.5*b*).

In most op amps using bipolar input transistors, such as in Fig. 1.3, I_{B-} and I_{B+} represent the dc *base* currents used to bias the transistors (typically, less than 0.2 mA). For op amps using FET input transistors, the input bias currents are much smaller. For example, the *average input bias current* $I_B \triangleq \frac{1}{2}(|I_{B+}| + |I_{B-}|)$ is equal to 0.1 mA for the μA741 but only 0.1 nA for the μA740 (which uses a pair of FET input transistors).

The open-loop voltage gain A is typically equal to at least 100,000 (200,000 for the μA741). On the other hand, the voltage ϵ at the end of the *linear* region in Fig. 1.5*b* is typically less than 0.1 mV.

An ideal op-amp model In view of the typical magnitudes of I_{B-}, I_{B+}, A, and ϵ, little accuracy is lost by assuming $I_{B-} = I_{B+} = \epsilon = 0$ and $A = \infty$. This simplifying assumption leads to the *ideal op-amp model* shown in Fig. 1.6*a* and *b*. Note that the transfer characteristic $f(v_d)$ in this *ideal* model has been approximated by a three-segment piecewise-linear characteristic. For future reference, the three distinct operating regions are labeled *Linear*, + *Saturation*, and − *Saturation*, respectively, in Fig. 1.6.

To emphasize that $A = \infty$ in the linear region, we add ∞ inside the triangle to distinguish the *ideal op-amp symbol* in Fig. 1.6*a* from other models. Unless otherwise stated, this ideal op-amp model will be used throughout this book.

The ideal op-amp model can be described analytically as follows:

Equations describing the ideal op-amp model

$$i_- = 0 \tag{1.2a}$$

$$i_+ = 0 \tag{1.2b}$$

$$v_o = E_{sat} \frac{|v_d|}{v_d}, \qquad v_d \neq 0 \tag{1.2c}$$

$$v_d = 0, \qquad -E_{sat} < v_o < E_{sat} \tag{1.2d}$$

Because these equations are rather cumbersome and difficult to manipulate analytically, it is much more practical to represent each region by a simple *equivalent circuit*, as shown in Fig. 1.6*c*, *d*, and *e*, respectively.

Note that these three equivalent circuits contain exactly the same information as Eq. (1.2). In particular, when the op amp is operating in the *linear*

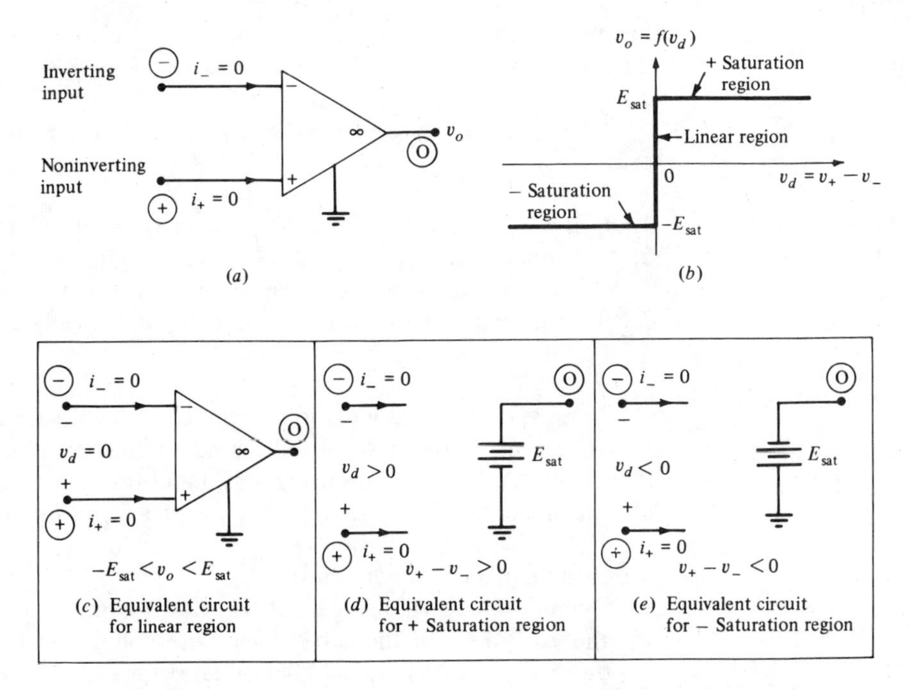

Figure 1.6 Ideal op-amp model.

region, the ideal op-amp model reduces to that shown in Fig. 1.6c. Note that here, v_d is constrained to be *zero* at all times while $|v_o|$ is constrained to be *less* than the saturation voltage E_{sat}. Hence, this circuit is described by Eqs. (1.2a), (1.2b), and (1.2d).

The circuit in Fig. 1.6d is described by Eqs. (1.2a), (1.2b), and (1.2c) with $v_d > 0$. Likewise, the circuit in Fig. 1.6e is described by Eqs. (1.2a), (1.2b), and (1.2c) with $v_d < 0$.

The ideal op-amp model is therefore described by three equivalent circuits, one for each operating region. If an op amp is known, a priori, to be operating in only one of these three regions in a given circuit, then we sometimes abuse our language by referring to the corresponding equivalent circuit in Fig. 1.6 as the ideal op-amp model.

For most low-frequency applications, the ideal op-amp model has been found to be quite realistic. For some specialized low-frequency applications (such as precision instrumentation) or high-frequency applications (such as filters), various op-amp imperfections may become important. In that case, the ideal model can be refined by introducing additional circuit elements.

Most op-amp circuits are designed so that the op amps operate *only in the linear region*. These circuits may contain both linear and nonlinear elements, and are studied in Sec. 2 using the equivalent circuit in Fig. 1.6c. Other op-amp circuits are designed to take advantage of the abrupt nonlinearities and are studied in Sec. 3 using all three equivalent circuits in Fig. 1.6.

Exercises

1. The op-amp manufacturers' data sheets usually specify the typical value of the average *input bias current* $I_B \triangleq \frac{1}{2}(|I_{B+}| + |I_{B-}|)$ and the *offset current* $I_{os} \triangleq |I_{B+}| - |I_{B-}|$. Express $|I_{B+}|$ and $|I_{B-}|$ in terms of I_B and I_{os}.
2. Calculate I_{B+} and I_{B-} for the following op amps:

	Op amp				
	μA709	LM101	μA741	LM301A	LM101A
Typical input bias current at 25°C	200 nA	120 nA	80 nA	70 nA	30 nA
Typical offset current at 25°C	50 nA	40 nA	20 nA	3.0 nA	1.5 nA

3. The data sheet for the μA741 shows a typical open-loop voltage gain of 200,000. Calculate the value of ϵ for the following power supply voltages (assume E_{sat} = magnitude of power supply voltage − 2 V): (*a*) ±15 V and (*b*) ±20 V.

2 OP-AMP CIRCUITS OPERATING IN THE LINEAR REGION

The methods to be developed in this section are valid *only* if the op-amp output voltage satisfies

$$-E_{sat} < v_o(t) < E_{sat} \tag{2.1}$$

for all times t (see Fig. 1.6*b*). We will henceforth refer to the expression (2.1) as the *validating inequality* for the *linear* region. If this inequality is violated over any time interval $[t_1, t_2]$, the solution in this interval is incorrect and must be recalculated using the method in Sec. 3.

2.1 Virtual Short Circuit Model

Recall from Chap. 3 that a three-port or four-terminal resistor is characterized by three relationships among the associated voltage and current variables. In the *linear* region, the ideal op-amp model in Fig. 1.6*a* and *b* can be described analytically by three equations:[2]

Virtual short circuit model

$$\boxed{\begin{aligned} i_- &= 0 \\ i_+ &= 0 \\ v_+ - v_- &= 0 \end{aligned}} \tag{2.2}$$

Consequently, we can think of the ideal op-amp model in Fig. 1.6*c* as a three-port or four-terminal resistor. For purposes of analysis, Eq. (2.2) is

[2] These correspond to Eqs. (1.2*a*), (1.2*b*), and (1.2*d*).

equivalent to (*a*) connecting a *short circuit* across the op-amp input terminals, and (*b*) stipulating that *the current through it is zero at all times*. To emphasize the special nature of this short circuit, we will henceforth refer to Eq. (2.2) as the *virtual short circuit* model. Using this equivalent circuit, many op-amp circuits can be analyzed by inspection.

2.2 Inspection Method

This method usually requires no more than three calculations and is often implemented by invoking KCL and Eq. (2.2) mentally with perhaps an occasional scribble on the "back of the envelope." It is best illustrated via some useful op-amp circuits as examples.

A. Voltage follower (buffer) The simplest op-amp circuit operating in the linear region is the voltage follower shown in Fig. 2.1*a*. To illustrate the *inspection method*, we first apply KCL at node ② and obtain

$$i_{\text{in}} = i_+ = 0 \tag{2.3}$$

Applying next KVL around the closed node sequence ④–③–②–①–④. we obtain $v_o - v_{\text{in}} + v_d = 0$. Since $v_d = 0$, we have

$$v_o = v_{\text{in}} \tag{2.4}$$

To complete the analysis, we apply the *validating inequality* (2.1) and obtain

$$-E_{\text{sat}} < v_{\text{in}} < E_{\text{sat}} \tag{2.5}$$

This gives the dynamic range of input voltages beyond which the op amp no longer operates in the linear region.

Note that Eqs. (2.3) and (2.4) define a unity-gain VCVS (Fig. 2.1*b*). This circuit has an *infinite* input resistance because $i_{\text{in}} = 0$ and its output "duplicates" the input voltage, regardless of the external load. Consequently, it is

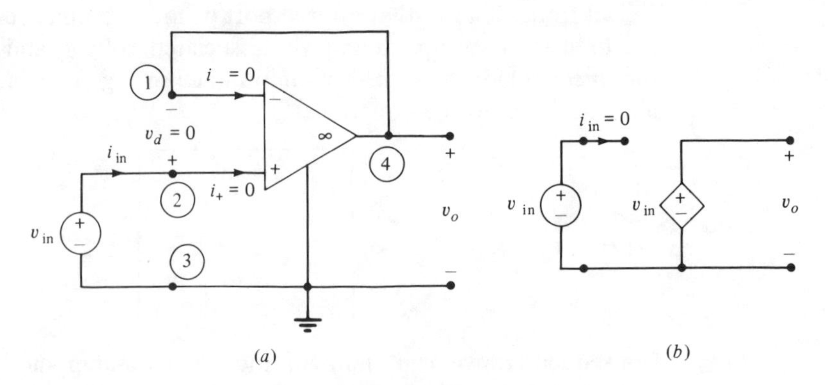

Figure 2.1 The voltage follower circuit in (*a*) is equivalent to the unity-gain VCVS in (*b*).

usually called a *voltage follower*, a *buffer*, or an *isolation amplifier*. It is widely used between 2 two-ports as shown in Fig. 2.2 to prevent N_2 from "loading down" N_1. This isolation technique is one of the most useful tools in the designer's bag of tricks.

Exercises

1. (*a*) Let N_1 and N_2 denote two identical *linear* voltage dividers made of resistances R_1 and R_2. Find the transfer characteristic $v_o = f(v_{in})$. (*b*) Repeat (*a*) without the buffer.
2. (*a*) Let N_1 and N_2 denote the "half-wave rectifier circuit" (Fig. 6.3*a*) analyzed earlier in Chap. 2. Find the v_o vs. v_{in} transfer characteristic. (*b*) Does your answer from (*a*) remain valid if N_1 is connected directly to N_2?

B. Inverting amplifier To illustrate the *inspection method* for op-amp circuits containing linear resistors, consider the circuit shown in Fig. 2.3. Since $v_d = 0$, we have $v_1 = v_{in}$, and hence $i_1 = v_{in}/R_1$. Since $i_- = 0$, we have $i_2 = i_1$, and hence $v_2 = R_f i_1 = R_f(v_{in}/R_1)$. Applying KVL around the closed node sequence ④–②–①–④, we obtain

$$v_o = -\left(\frac{R_f}{R_1}\right) v_{in} \tag{2.6}$$

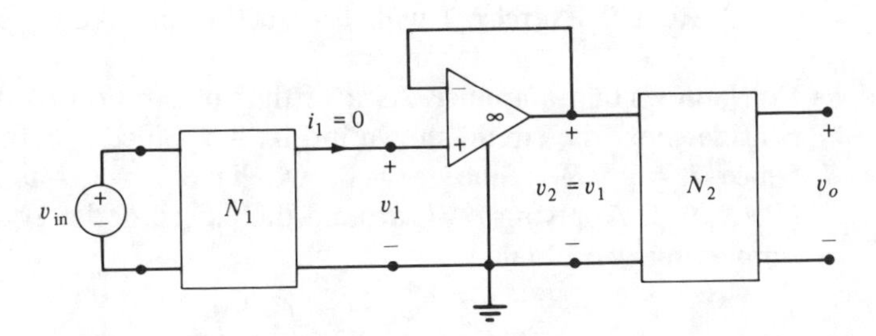

Figure 2.2 The above buffer greatly simplifies analysis and allows N_1 and N_2 to be designed separately.

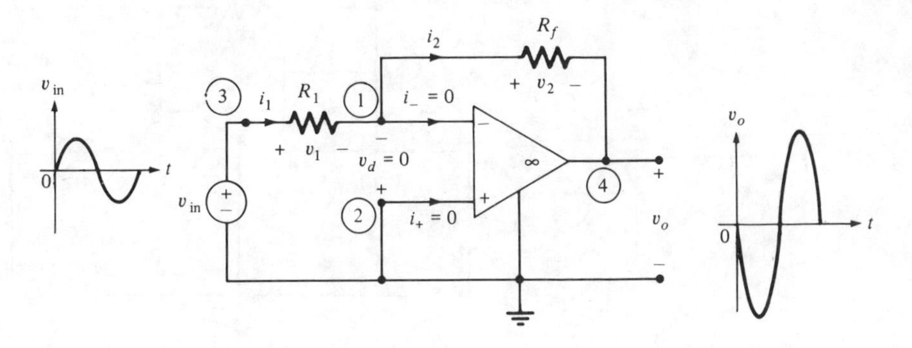

Figure 2.3 An inverting amplifier.

Substituting Eq. (2.6) into the validating inequality (2.1) and solving for v_{in}, we obtain the dynamic range

$$-\left(\frac{R_1}{R_f}\right)E_{\text{sat}} < v_{\text{in}} < \left(\frac{R_1}{R_f}\right)E_{\text{sat}} \tag{2.7}$$

for which Eq. (2.6) is valid.

Hence, so long as the input signal satisfies Eq. (2.7), this circuit functions as a voltage amplifier with a *voltage gain* equal to $-R_f/R_1$ (assuming $R_f > R_1$). Note that the negative sign means that for a sinusoidal input, the output is shifted in phase by 180°. Consequently, this circuit is called an *inverting amplifier*. In the special case where $R_1 = R_f$, it is called a *phase inverter*. (Why?)

Note that whereas $i_- = 0$ and $i_+ = 0$ are imposed by the op-amp v-i characteristics, the "virtual short circuit" $v_d = 0$ is achieved externally by "feeding back" the output voltage v_o to the op-amp inverting terminal through the feedback resistor R_f. The physical mechanism which automatically adjusts v_d to a nearly zero voltage is discussed in Sec. 3.2B.

Exercises

1. Using a buffer and the circuit in Fig. 2.3 (assume $R_1 = 10\text{ K}$), design a VCVS ($v_o = \mu v_{\text{in}}$) with a controlling coefficient $\mu = -1000$.
2. Repeat Exercise 1 with $\mu = 1000$. Hint: Add a phase inverter.

C. Noninverting amplifier As a further illustration of the inspection method, consider next the circuit shown in Fig. 2.4. Since $v_d = 0$, we have $v_1 = v_{\text{in}}$, and hence $i_1 = v_{\text{in}}/R_1$. Since $i_- = 0$, we have $i_2 = i_1 = v_{\text{in}}/R_1$, and hence $v_2 = (R_f/R_1)v_{\text{in}}$. Applying KVL around the closed node sequence ④–③–①–④ and simplifying, we obtain

$$v_o = \left(1 + \frac{R_f}{R_1}\right)v_{\text{in}} \tag{2.8}$$

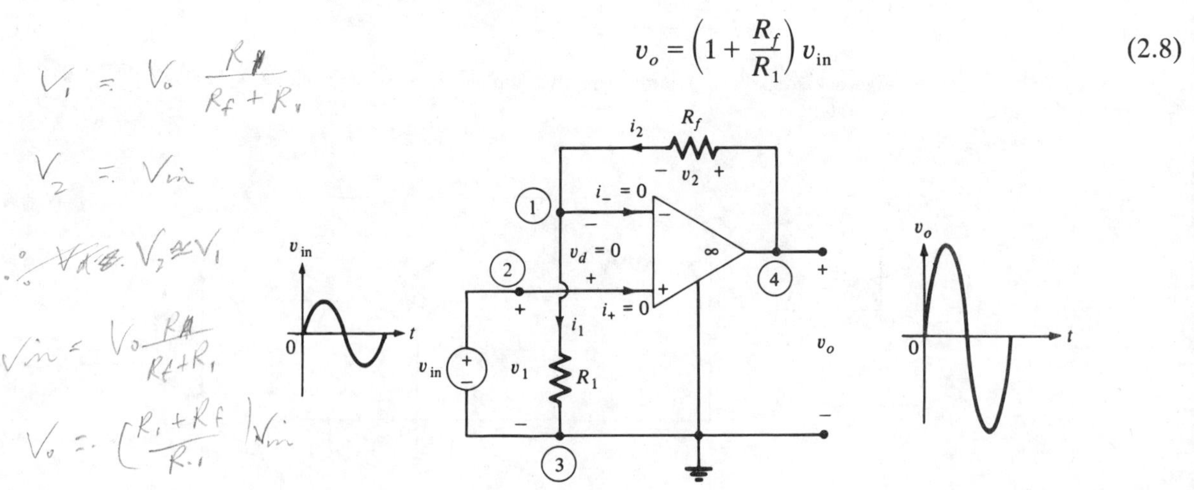

Figure 2.4 A noninverting amplifier.

Substituting Eq. (2.8) in the validating inequality (2.1) and solving for v_{in}, we obtain the dynamic range

$$-\left(\frac{R_1}{R_1 + R_f}\right) E_{\text{sat}} < v_{\text{in}} < \left(\frac{R_1}{R_1 + R_f}\right) E_{\text{sat}} \tag{2.9}$$

for which Eq. (2.8) is valid.

Hence, so long as the input signal satisfies Eq. (2.9), this circuit functions as a voltage amplifier with a *positive* voltage gain $(R_1 + R_f)/R_1$. It is usually called a *noninverting amplifier*. Note that a voltage follower is simply a unity-gain noninverting amplifier obtained by choosing $R_1 = \infty$ and $R_f = 0$.

Exercises

1. The circuit in Fig. 2.5 is called an *algebraic summer* because $v_o = k_1 v_1 + k_2 v_2$. Find k_1 and k_2 and identify the region in the v_1-v_2 plane for which this relationship is valid.
2. Using exactly two op amps and $n + 3$ resistors, design an n-input summer giving $v_o = v_1 + v_2 + \cdots + v_n$.
3. (*a*) Explain why the resistor R_f in Figs. 2.3 and 2.4 can be replaced by *any* one-port (except an open circuit) without affecting the value of i_2. (*b*) Using a 3-V battery and either circuit in Figs. 2.3 and 2.4, design a *dc current source* having a terminal current of 30 mA. Hint: Use the property from (*a*). (*c*) Repeat (*b*) for a terminal current of −30 mA.
4. Using only one op amp and one resistor, design a VCCS described by $i_2 = kv_{\text{in}}$, where $k > 0$. Specify the maximum range of permissible "load" voltage across the current source.

D. Resistance measurement without surgery To show that the "virtual short circuit" is not just a powerful tool for simplifying *analysis*, Fig. 2.6 gives a circuit which exploits this remarkable property in a practical *design*. The linear resistive circuit enclosed within the circle represents the portion of a circuit where the value of each resistance is to be measured *without cutting any wires*. This problem usually arises when a circuit breaks down and a faulty resistor is to be identified by comparing its resistance with the nominal value.

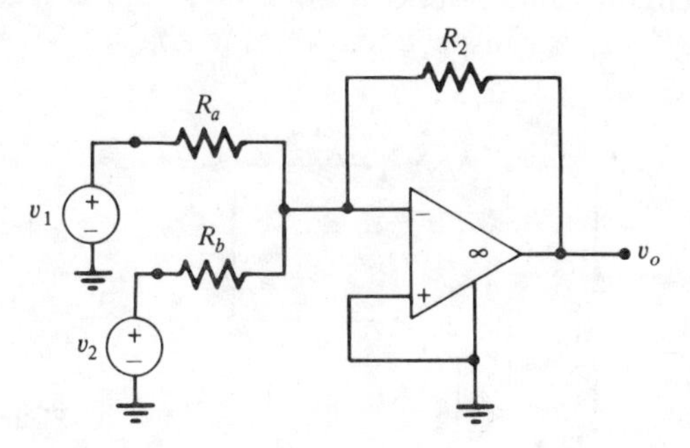

Figure 2.5 An algebraic summer.

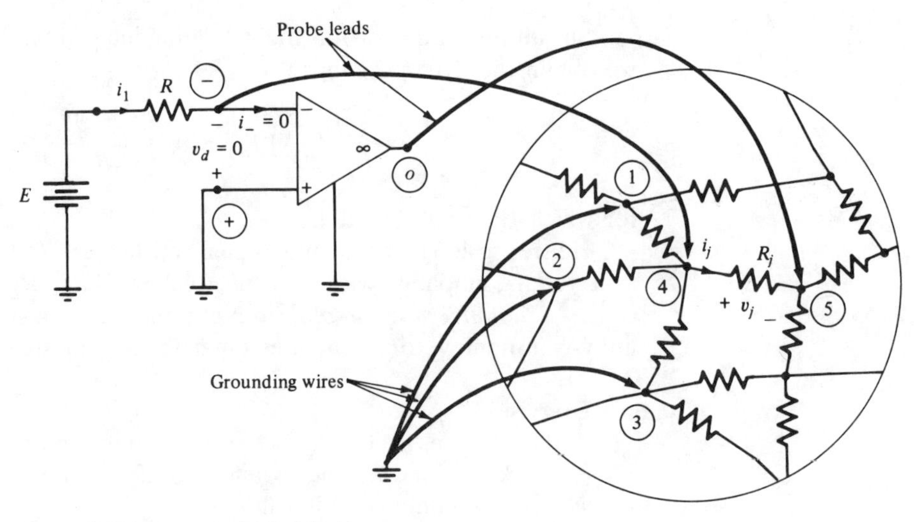

Figure 2.6 An op-amp fault detector.

To show how this circuit works, suppose resistor R_j is to be measured. (*a*) Connect the op-amp inverting terminal ⊖ to one terminal of R_j (node ④ in Fig. 2.6) and ground the second terminal of all other resistors connected to node ④ (nodes ①, ②, and ③ in Fig. 2.6). (*b*) Connect the op-amp output terminal ⓞ to the second terminal of R_j (node ⑤ in Fig. 2.6). It follows from the virtual short-circuit property that except for R_j, the current through all resistors connected to node ④ is zero. Moreover, since $i_1 = E/R$ and $i_- = 0$, we have $i_j = E/R$ and $v_j = (E/R)R_j$. Hence, by measuring the voltage v_j, we can calculate

$$R_j = \frac{R}{E} v_j \tag{2.10}$$

Note that without the virtual short circuit, R_j would have to be cut before its value can be measured.

E. Nonlinear feedback To illustrate that the inspection method holds even if the op-amp circuit contains one or more *nonlinear* resistors, consider the circuit shown in Fig. 2.7. By inspection, we note that $i_2 = i_1 = v_{\text{in}}/R_1$ and $v_o = -v_2$.

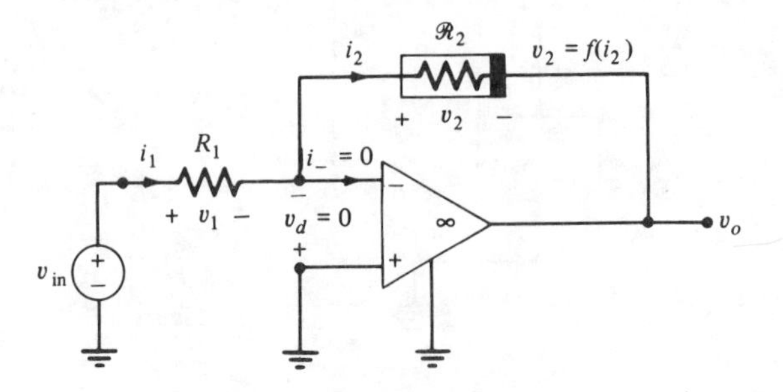

Figure 2.7 An op-amp circuit containing a nonlinear resistor.

Consequently

$$v_o = -f\left(\frac{v_{\text{in}}}{R_1}\right) \tag{2.11}$$

To determine the dynamic range of v_{in} for which Eq. (2.11) holds, we apply the validating inequality (2.1) and obtain

$$-E_{\text{sat}} < -f\left(\frac{v_{\text{in}}}{R_1}\right) < E_{\text{sat}} \tag{2.12}$$

Equations (2.11) and (2.12) give the *nonlinear* transfer characteristic with the op amp operating in the *linear* region. Since this configuration is widely used in nonlinear applications, we will consider an example.

Example Let $R_1 = 1\ \text{k}\Omega$ in Fig. 2.7. Let the nonlinear resistor represent the one-port shown in Fig. 2.8*a*. Using the graphic method from Chap. 2, we obtain the driving-point characteristic in Fig. 2.8*b*, where we have chosen v_2 as the vertical axis so that the curve represents $v_2 = f(i_2)$. It follows from Eq. (2.11) that the transfer characteristic is obtained by flipping this curve about the horizontal axis and then relabeling v_3 and i_2 with v_o and v_{in}, respectively. The result is shown in Fig. 2.8*c*.

Assuming a 15-V supply voltage for the op amp so that $E_{\text{sat}} = 13$ V, we note that Eq. (2.12) is satisfied for *all* values of v_{in} because $|v_o| = |-f(v_{\text{in}}/R_1)| < 10$ V in Fig. 2.8*c*. Hence, we have demonstrated that the op amp can operate in the linear region for all values of input voltages, even though the circuit contains two nonlinear devices (zener diodes in this example).

An examination of Fig. 2.8*c* shows that all input signal amplitudes exceeding 5 V will give a constant output of ±10 V. Consequently, the circuit in this example is called a *limiter* or *clipper*, and is widely used for overvoltage protection and other applications in communication circuits.

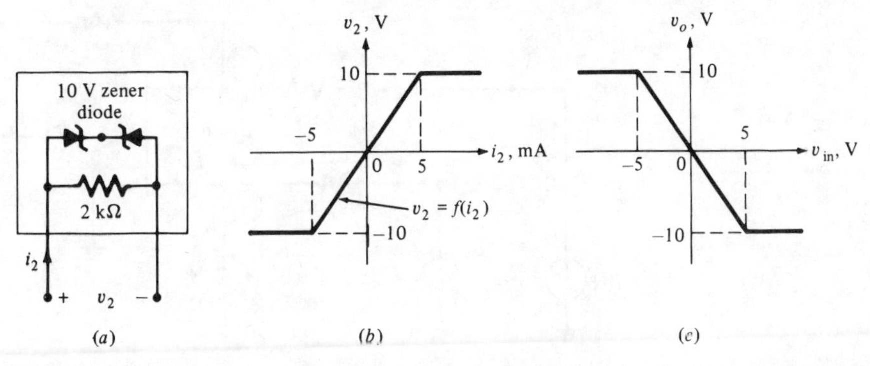

Figure 2.8 (*a*) Circuit for realizing the nonlinear resistor $\mathcal{R}_2$ in Fig. 2.7. (*b*) Driving-point characteristic of the circuit in *a*. (*c*) Transfer characteristic of the circuit in Fig. 2.7 with $\mathcal{R}_2$ replaced by the circuit in *a*, and assuming $R_1 = 1\ \text{k}\Omega$.

2.3 Systematic Method

The *inspection method* often fails whenever it is necessary to solve two or more simultaneous equations. In such cases, it is desirable to develop a *systematic method* for writing a system of linearly independent equations involving *as few variables* as possible. The following example illustrates the basic steps involved.

Example Consider the op-amp circuit shown in Fig. 2.9, where the op amp is modeled by a virtual short circuit (Fig. 1.6*c*). Although this circuit could be solved by inspection, we will solve it by the systematic method, and let the reader verify its answer by the inspection method.

Step 1. Label the nodes consecutively and let e_j denote as usual the voltage from node ⓙ to datum, $j = 1, 2, \ldots, 5$. Express *all* resistor voltages and the differential op-amp voltage v_d in terms of node-to-datum voltages via KVL:

$$v_1 = e_1 - e_3 \tag{2.13a}$$

$$v_2 = e_3 - e_5 \tag{2.13b}$$

$$v_3 = e_2 - e_4 \tag{2.13c}$$

$$v_4 = e_4 \tag{2.13d}$$

$$v_d = e_4 - e_3 \tag{2.13e}$$

Step 2. Express the branch current in each linear resistor in terms of node-to-datum voltages via Ohm's law:

$$i_1 = \frac{e_1 - e_3}{R_1} \tag{2.14a}$$

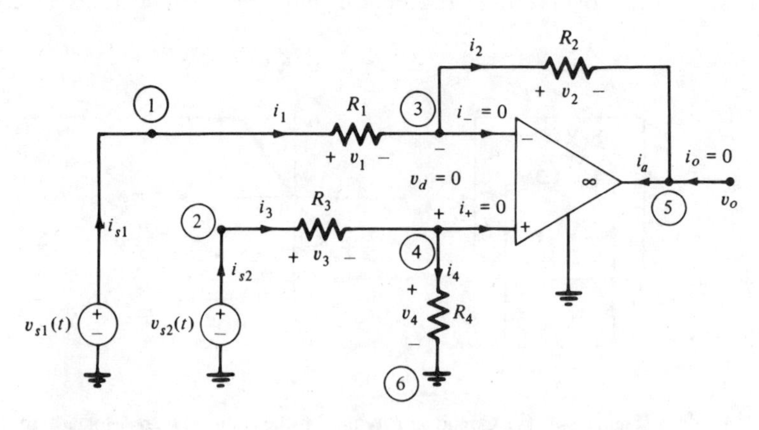

Figure 2.9 An op-amp circuit for illustrating the systematic method.

$$i_2 = \frac{e_3 - e_5}{R_2} \tag{2.14b}$$

$$i_3 = \frac{e_2 - e_4}{R_3} \tag{2.14c}$$

$$i_4 = \frac{e_4}{R_4} \tag{2.14d}$$

Step 3. Identify all other branch *current* variables which *cannot* be expressed in terms of node-to-datum voltages, namely, the currents i_{s1} and i_{s2} of the voltage sources and the current i_a of the op-amp *output* terminal. Note that the op-amp input currents i_- and i_+ are *not* variables (assuming an ideal op-amp model) because they are equal to zero. Our objective is to write a system of linearly independent equations in terms of the node-to-datum voltages $\{e_1, e_2, \ldots, e_5\}$ and the *identified* current variables $\{i_{s1}, i_{s2}, i_a\}$.

Step 4. Write KCL at each node except the datum node in terms of $\{e_1, e_2, e_3, e_4, e_5, i_{s1}, i_{s2}, i_a\}$:

Node ①: $$\frac{e_1 - e_3}{R_1} - i_{s1} = 0 \tag{2.15a}$$

Node ②: $$\frac{e_2 - e_4}{R_3} - i_{s2} = 0 \tag{2.15b}$$

Node ③: $$\frac{e_3 - e_5}{R_2} - \frac{e_1 - e_3}{R_1} = 0 \tag{2.15c}$$

Node ④: $$\frac{e_4}{R_4} - \frac{e_2 - e_4}{R_3} = 0 \tag{2.15d}$$

Node ⑤: $$i_a - \frac{e_3 - e_5}{R_2} = 0 \tag{2.15e}$$

Step 5. Equation (2.15) consists of five equations with eight variables. Hence, we need to write three more independent equations. Since we have already made use of KVL (Step 1), KCL (Step 4), and the resistor characteristics (Step 2), these three equations must come from the characteristics of the voltage sources and the op amp:

Voltage sources:
$$e_1 = v_{s1} = v_{s1}(t) \tag{2.16a}$$
$$e_2 = v_{s2} = v_{s2}(t) \tag{2.16b}$$

Op amp:[3]
$$e_4 - e_3 = 0 \tag{2.16c}$$

[3] Note that $v_d = e_4 - e_3$.

Step 6. Together, Eqs. (2.15) and (2.16) constitute a system of eight linearly independent equations in terms of eight variables. Solving these equations for the desired op-amp output voltage e_5 by elimination and substitution of variables, or by any other method, we obtain

$$v_o = e_5 = \left[\frac{R_4(1 + R_2/R_1)}{R_3 + R_4}\right] v_{s2}(t) - \left(\frac{R_2}{R_1}\right) v_{s1}(t) \tag{2.17}$$

Note that only Eqs. (2.15*c* and *d*) and (2.16) are used to solve for v_0. The remaining equations (2.15*a*, *b*, and *e*) are needed, however, to solve for the remaining variables i_{s1}, i_{s2}, and i_a, respectively.

Step 7. Determine the dynamic range of the input voltages where Eq. (2.17) holds, i.e., where the op amp is operating in the *linear* region:

$$-E_{\text{sat}} < \left[\frac{R_4(1 + R_2/R_1)}{R_3 + R_4}\right] v_{s2}(t) - \left(\frac{R_2}{R_1}\right) v_{s1}(t) < E_{\text{sat}} \tag{2.18}$$

Hence, Eq. (2.17) holds at all times when the expression (2.18) is satisfied.

Exercise Derive Eqs. (2.17) and (2.18) by the inspection method.

Special case (differential amplifier) Suppose $R_1/R_2 = R_3/R_4$ in Fig. 2.9. Then Eqs. (2.17) and (2.18) reduce to the following:

$$v_o = \frac{R_2}{R_1}\left[v_{s2}(t) - v_{s1}(t)\right],$$

$$-\frac{R_1}{R_2} E_{\text{sat}} < v_{s2}(t) - v_{s1}(t) < \frac{R_1}{R_2} E_{\text{sat}} \tag{2.19}$$

Equation (2.19) defines a *differential dc amplifier*, a circuit widely used in instrumentation applications.

The preceding systematic method is applicable to any op-amp circuit containing linear resistors, independent voltage and current sources, and op amps modeled by virtual short circuits. This method will be generalized in Chap. 8 [called the *modified node analysis* (MNA) method] for arbitrary resistive circuits.

Exercises

1. Generalize the steps in the preceding systematic method for a connected n-node circuit containing *linear* resistors, k voltage sources, ℓ current sources, and m op amps.
2. (*a*) Show that in the linear region, the ideal op-amp model is equivalent to a *linear* two-port resistor described by a *transmission matrix* **T** which specifies the port variables v_d and i_- (associated with port 1) in terms of the port variables v_o and i_o (associated with port 2). (*b*) Use the "linearity" property from (*a*) to show any circuit made of linear resistors, independent sources, and ideal op amps operating in the linear region can be analyzed by solving only linear equations.

3 OP-AMP CIRCUITS OPERATING IN THE NONLINEAR REGION

There are many applications where the op amp operates in all three regions of the ideal op-amp model in Fig. 1.6. This occurs whenever the amplitudes of one or more input signals are such that the validating inequality in each region is violated over some time intervals. In this case it is necessary to revert to the nonlinear model in Fig. 1.6 and we say the op amp is operating in the *nonlinear* region. Fortunately, since the characteristic in Fig. 1.6*b* is *piecewise linear*, the circuit in each region can be easily analyzed as a *linear* circuit.

3.1 + Saturation and − Saturation Equivalent Circuits

In the + *Saturation* region, the ideal op-amp model in Fig. 1.6 can be described analytically by three equations:

+ Saturation characteristics

$$\boxed{\begin{aligned} i_- &= 0 \\ i_+ &= 0 \\ v_o &= E_{\text{sat}} \end{aligned}} \tag{3.1}$$

These equations are applicable provided the following *validating inequality* holds:

$$\boxed{v_d = v_+ - v_- > 0} \tag{3.2}$$

Note that the *crucial* difference between the "+ Saturation characteristics" and the previous "linear characteristics" is that here, $v_d \triangleq v_+ - v_- \neq 0$ *and* v_o is now "clamped" at a fixed positive voltage equal to E_{sat}. In this region, we can replace the op amp by the equivalent circuit shown in Fig. 1.6*d*, which is redrawn in Fig. 3.1 for convenience.

In the − *Saturation* region, the ideal op-amp model in Fig. 1.6 can be described analytically as follows:

− Saturation characteristics

$$\boxed{\begin{aligned} i_- &= 0 \\ i_+ &= 0 \\ v_o &= -E_{\text{sat}} \end{aligned}} \tag{3.3}$$

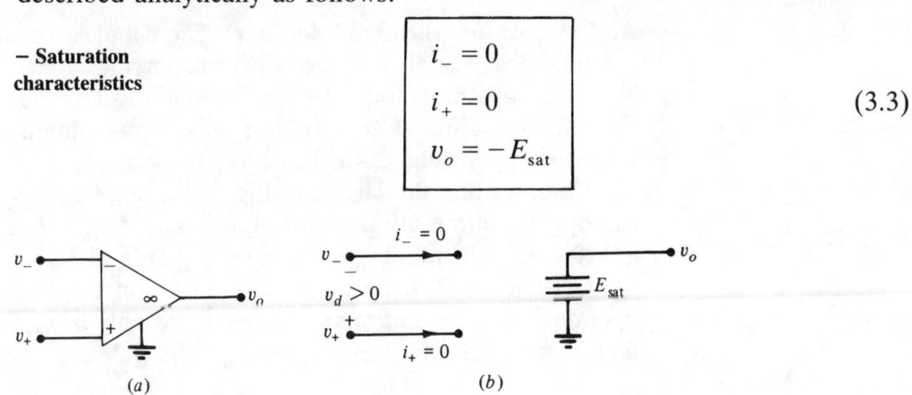

Figure 3.1 The + Saturation model.

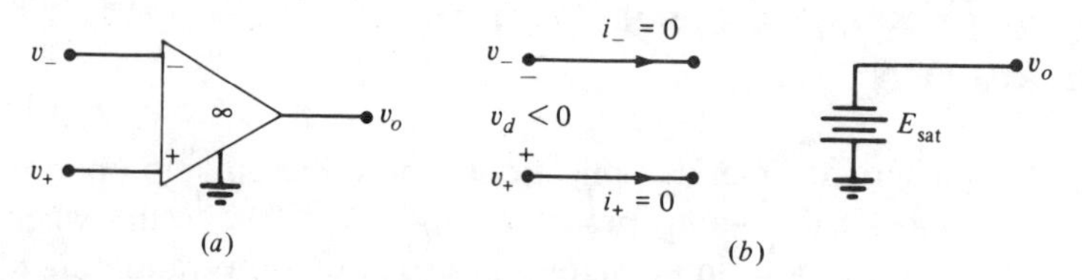

Figure 3.2 The − Saturation model.

These equations are applicable provided the following *validating inequality* holds:

$$\boxed{v_d = v_+ - v_- < 0} \tag{3.4}$$

Again, in sharp contrast to the "linear characteristics," here $v_d \triangleq v_+ - v_- \neq 0$ *and* v_o is "clamped" at a fixed negative voltage equal to $-E_{sat}$. In this region, we can replace the op amp by the equivalent circuit shown in Fig. 1.6*e*, which is redrawn in Fig. 3.2 for convenience.

Analogous to Eq. (2.1), we will henceforth call Eqs. (3.2) and (3.4) the *validating inequality* for the + *Saturation* and − *Saturation* regions, respectively.

Corresponding to the three regions in the ideal op-amp model of Fig. 1.6, we have three simplified equivalent circuits defined by Eqs. (2.1) and (2.2), (3.1) and (3.2), and (3.3) and (3.4), respectively. *The correct equivalent circuit to use in a given situation depends on, and only on, which of the three validating inequalities (2.1), (3.2), or (3.4) holds.*

3.2 Inspection Method

Most op-amp circuits which operate in the *nonlinear* region have a single input and a single output of interest. For this class of circuits, the basic problem is to derive the *driving-point characteristic* or the *transfer characteristic*. Once these characteristics are found, the output waveform due to *any* input waveform can be easily obtained either graphically or by direct substitution. The method for deriving these characteristics is best illustrated via examples.

A. Comparator (threshold detector) The simplest op-amp circuit operating in the nonlinear region is the comparator circuit shown in Fig. 3.3*a*. Replacing the ideal op-amp model by the virtual short circuit, + Saturation, and − Saturation equivalent circuits, respectively, we obtain the corresponding *linear* circuit shown in Fig. 3.4*a*, *b*, and *c*, respectively.

Consider first the circuit in Fig. 3.4*a*. Since $v_d = v_{in} - E_T = 0$, the op amp can operate in the *linear region* if and only if $v_{in} = E_T$. In such a case, we find $i_{in} = 0$ (Fig. 3.3*b*) and $-E_{sat} < v_o < E_{sat}$ (Fig. 3.3*c*).

Consider next the circuit in Fig. 3.4*b*. Since $v_d = v_{in} - E_T > 0$, the op amp operates in the + *Saturation region* if and only if $v_{in} > E_T$. In such a case, we find $i_{in} = 0$ (Fig. 3.3*b*) and $v_o = E_{sat}$ (Fig. 3.3*c*).

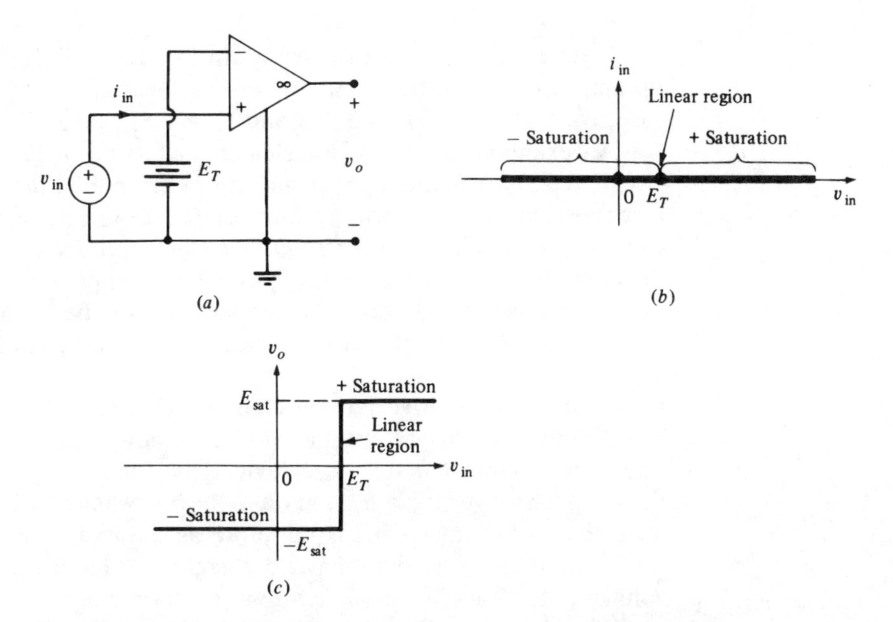

Figure 3.3 (*a*) Comparator. (*b*) Driving-point characteristic. (*c*) Transfer characteristic.

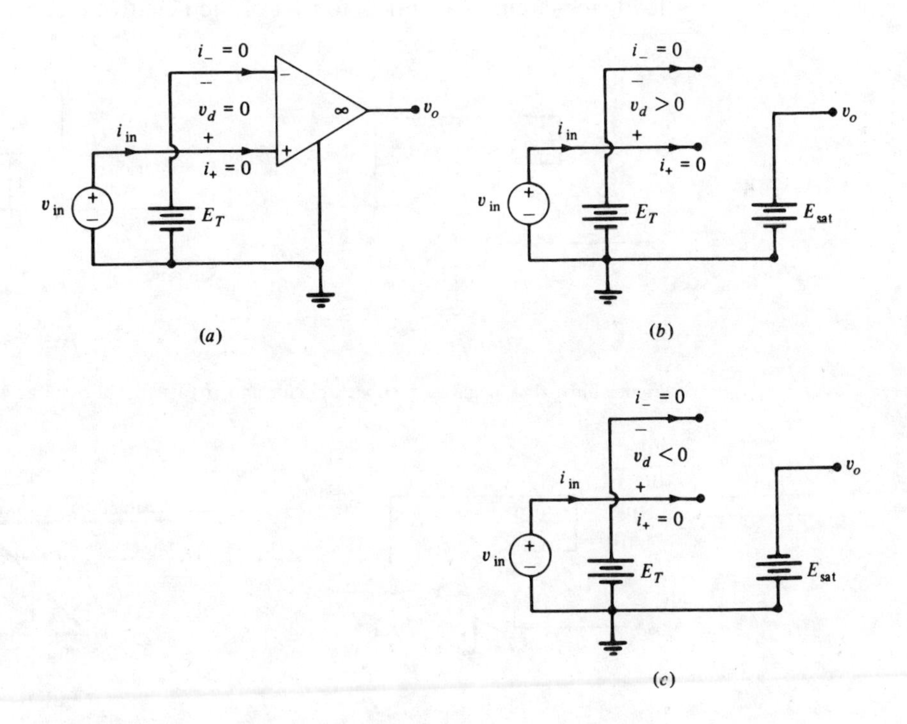

Figure 3.4 Linear circuit for each region.

It remains to consider the circuit in Fig. 3.4*c*. Since $v_d = v_{in} - E_T < 0$, the op amp operates in the − *Saturation region* if and only if $v_{in} < E_T$. In which case, we find $i_{in} = 0$ (Fig. 3.3*b*) and $v_o = -E_{sat}$ (Fig. 3.4*c*).

An examination of the transfer characteristic in Fig. 3.3*c* shows that the circuit "compares" the input signal with a prescribed threshold voltage E_T and responds by jumping abruptly from one level to another. Consequently, it is called a *comparator* or a *threshold detector*. In the special case where $E_T = 0$, the circuit becomes a *zero-crossing detector*. Comparators are so widely used in digital circuits that they are mass produced (with "bells and whistles" added for improved performance) and sold under the name "comparator."

B. Negative vs. positive feedback circuit Consider the circuit shown in Fig. 3.5*a*. Note that this is just the voltage follower in Fig. 2.1 studied earlier. There, we found that $v_o = v_{in}$ provided $|v_{in}| < E_{sat}$. By inspection, we found $v_o = E_{sat}$ whenever $v_{in} > E_{sat}$, and $v_o = -E_{sat}$ whenever $v_{in} < -E_{sat}$. The complete transfer characteristic is therefore as shown in Fig. 3.5*b*.

This circuit is said to have a "negative" feedback because the output voltage is fed back to the *inverting* input terminal.

What happens if we interchange the inverting and noninverting terminals as shown in Fig. 3.6*a*? By inspection, we found $v_o = v_{in}$ provided $|v_{in}| < E_{sat}$. Hence, in the linear region, the transfer characteristic for this "positive" feedback circuit is identical to that of the negative feedback circuit in Fig. 3.5*a*.

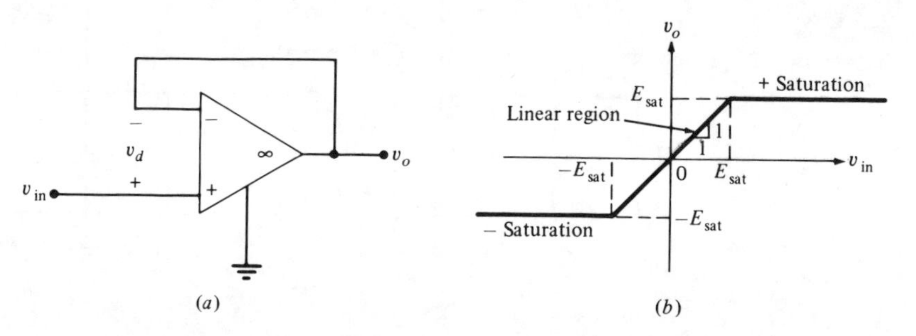

Figure 3.5 (*a*) A negative feedback circuit and (*b*) its transfer characteristic.

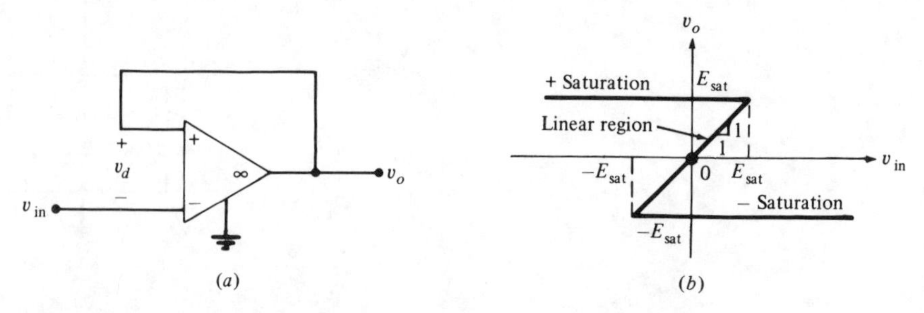

Figure 3.6 (*a*) A positive feedback circuit and (*b*) its transfer characteristic.

In practice, however, they do not behave in the same way: One functions as a voltage follower, the other does *not*. To uncover the reason, let us derive the transfer characteristics in the remaining regions.

When the op amp is in the + Saturation region, we can replace it by the equivalent circuit shown in Fig. 3.7*a*. The validating inequality (3.2) requires that $v_d = E_{sat} - v_{in} > 0$ or $v_{in} < E_{sat}$. Hence, the transfer characteristic in this region is given by $v_o = E_{sat}$ whenever $v_{in} < E_{sat}$, as shown in Fig. 3.6*b*.

Conversely, when the op amp is in the − Saturation region, the equivalent circuit shown in Fig. 3.7*b* holds and hence we obtain $v_o = -E_{sat}$ whenever $v_{in} > -E_{sat}$, as shown in Fig. 3.6*b*.

Note that the complete transfer characteristics in Figs. 3.5 and 3.6 are quite different. Even if the op amp is operating in the linear region ($|v_{in}| < E_{sat}$), there are three distinct output voltages for each value of v_{in} for the positive feedback circuit. Using a more realistic op-amp circuit model augmented by a capacitor, and the method to be developed in Chap. 6, we will show that all operating points on the middle segment (linear region) in Fig. 3.6*b* are *unstable*. The important concept of *stability* and *instability* will be discussed in detail in Chap. 6. In the present context, having *unstable operating points* in the middle region means that even if the initial voltage $v_{in}(0)$ lies on this segment, it will quickly move into the + Saturation region if $v_{in}(0) > 0$, or into the − Saturation region if $v_{in}(0) < 0$.

We can also give an *intuitive* explanation of this *unstable* behavior by referring back to the *nonideal* op-amp characteristics shown in Fig. 1.5*b*, where

$$v_o = A(v_+ - v_-) \tag{3.5}$$

in the linear region. Equation (3.5) shows that the output voltage v_o *decreases* (respectively, *increases*) whenever the potential v_- at the *inverting* (respectively noninverting) node increases, and vice versa.

Since physical signals can only propagate at a *finite* velocity, changes in the input voltage are not felt instantaneously at the output terminal, but at some moments [say 1 picosecond (ps)] later. Now suppose $v_d = v_{in} - v_o = 1$ nV at $t = 0$, whereupon v_{in} is increased slightly in both circuits in Figs. 3.5 and 3.6.

For the negative feedback circuit, v_o will increase initially in accordance with Eq. (3.5). However, since this signal is fed back to the inverting terminal

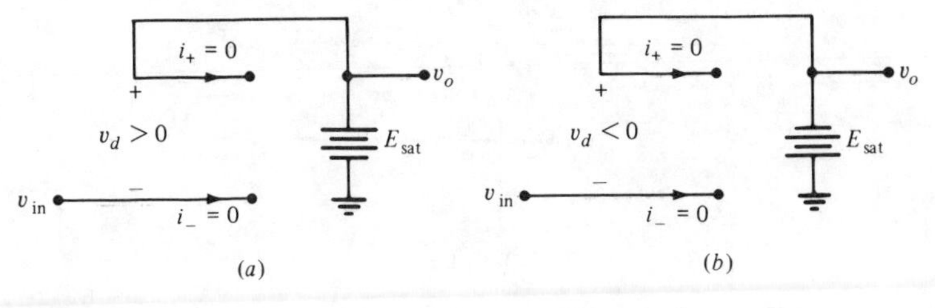

Figure 3.7 (*a*) Equivalent circuit in + Saturation region. (*b*) Equivalent circuit in − Saturation region.

in Fig. 3.5*a*, $v_d = v_+ - v_-$ will decrease a short moment later. Hence, the operating point in the middle segment of Fig. 3.5*b* will tend to return to its original position. The negative feedback circuit is therefore said to be "stable" because it tends to restore the original equilibrium position in the presence of small disturbances.

Exactly the opposite happens in the *positive* feedback circuit in Fig. 3.6*a*. Here, the slightest disturbance is amplified strongly—in view of the high gain *A*—as the signal goes around the feedback loop in finite time. This increase in v_d causes a further increase in v_d the next time around the loop. This "unstable" phenomenon is repeated in rapid order until the output is driven into saturation; thereafter, the model must be replaced by either Fig. 3.1 or 3.2.

C. Negative-resistance converter The circuit shown in Fig. 3.8*a* incorporates both a *negative* feedback path (via R_f) and a *positive* feedback path (via R_1).

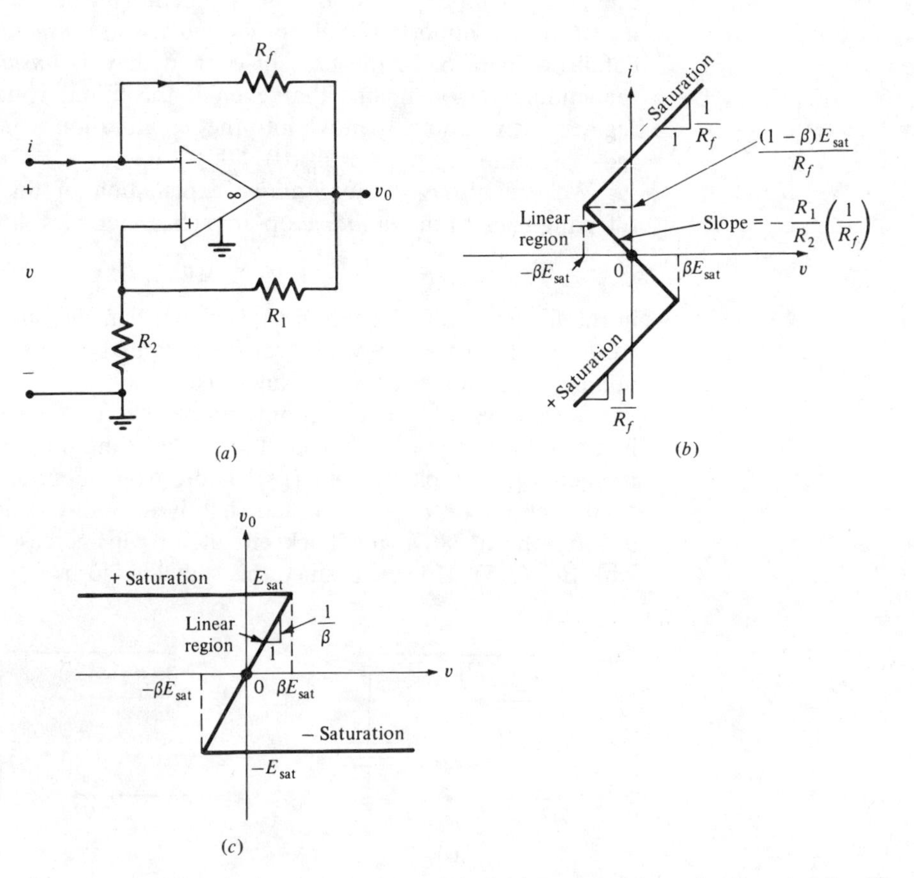

Figure 3.8 A negative-resistance converter and its driving-point and transfer characteristics. Here, $\beta \overset{\Delta}{=} R_2/(R_1 + R_2)$.

Our problem is to derive its driving-point and transfer characteristics. As before, we replace the op amp in Fig. 3.8*a* by its three ideal models as shown in Fig. 3.9.

Linear region By inspection of the equivalent circuit in Fig. 3.9*a*, we note that R_1 and R_2 form a voltage divider so that

$$v_2 = \frac{R_2}{R_1 + R_2} v_o = \beta v_o \tag{3.6}$$

where $\beta \triangleq R_2/(R_1 + R_2)$. Substituting $v_2 = v$ into Eq. (3.6), we obtain

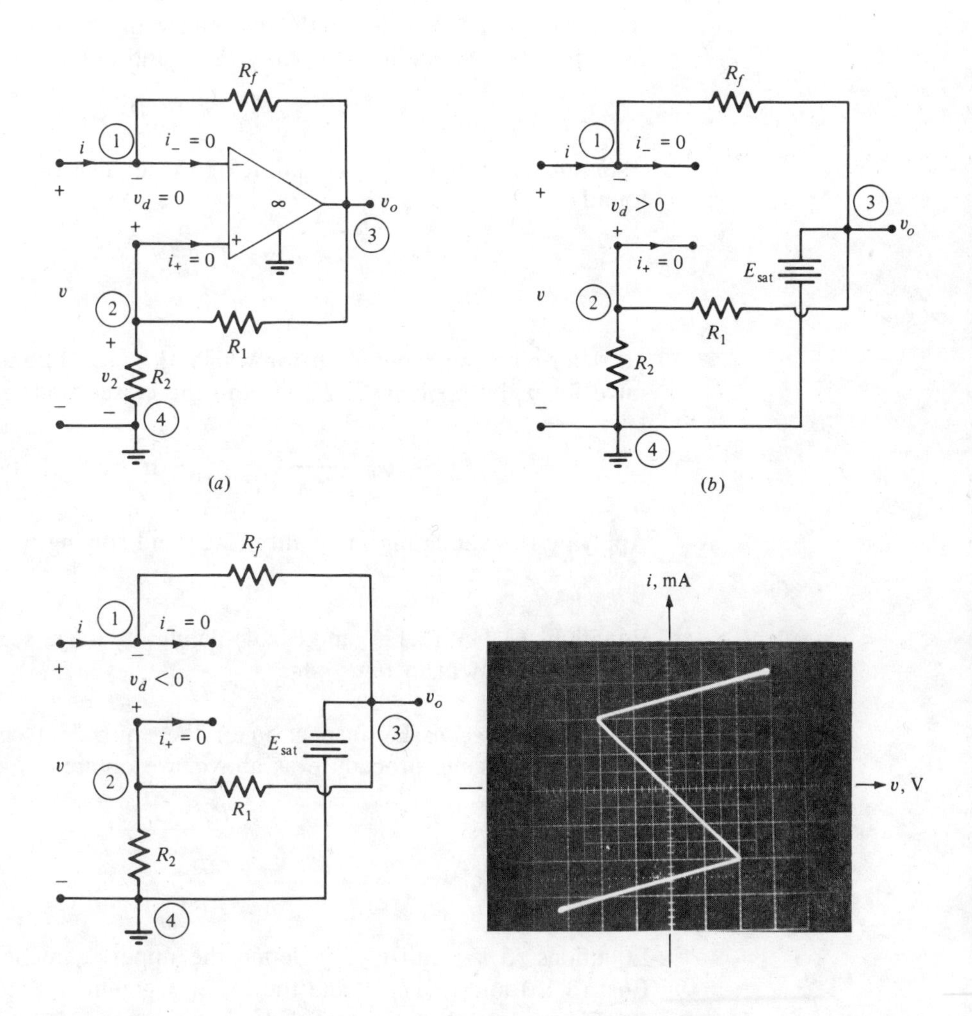

Figure 3.9 Equivalent circuit in (*a*) linear region, (*b*) + Saturation region, (*c*) − Saturation region, and (*d*) typical negative-resistance characteristic measured from an actual op-amp circuit.

$$v_o = \frac{1}{\beta} v \tag{3.7}$$

Applying KVL around the closed node sequence ④–①–③–④, we obtain

$$v = R_f i + v_o \tag{3.8}$$

Substituting Eq. (3.7) into Eq. (3.8) and solving for i, we obtain

$$i = -\left(\frac{R_1}{R_2}\right)\left(\frac{1}{R_f}\right) v \tag{3.9}$$

Equations (3.9) and (3.7) are drawn as the middle segment in Fig. 3.8*b* and *c*, respectively. To determine the boundary of these segments, substitute Eq. (3.7) into the validating inequality (2.1) and obtain

$$-\beta E_{\text{sat}} < v < \beta E_{\text{sat}} \tag{3.10}$$

+ Saturation region By inspection of the equivalent circuit in Fig 3.9*b*, we found

$$v = R_f i + E_{\text{sat}} \tag{3.11}$$

$$v_o = E_{\text{sat}} \tag{3.12}$$

To determine the range of v for which Eqs. (3.11) and (3.12) are valid, we solve for v_d by applying KVL around the closed node sequence ④–①–②–④:

$$v_d = \frac{R_2}{R_1 + R_2} E_{\text{sat}} - v = \beta E_{\text{sat}} - v \tag{3.13}$$

Applying the validating inequality (3.2) and solving for v, we obtain

$$v < \beta E_{\text{sat}} \tag{3.14}$$

Equations (3.11), (3.12), and (3.14) define the lower segment in Fig. 3.8*b* and the upper segment in Fig. 3.8*c*.

– Saturation region By inspection of the equivalent circuit in Fig. 3.9*c* and following the same procedure as above, we obtain

$$v = R_f i - E_{\text{sat}} \tag{3.15}$$

$$v_o = -E_{\text{sat}} \tag{3.16}$$

$$v > -\beta E_{\text{sat}} \tag{3.17}$$

Equations (3.15) and (3.17) define the upper segment in Fig. 3.8*b* whereas Eqs. (3.16) and (3.17) define the lower segment in Fig. 3.8*c*.

Figure 3.9*d* shows a typical driving-point characteristic measured from the op-amp circuit in Fig. 3.8*a*. The slopes and breakpoints of this nearly *piecewise-linear* characteristic have been found to agree remarkably well with those prediced by Eqs. (3.9), (3.11), and (3.15).[4]

The circuit in Fig. 3.8*a* is called a *negative-resistance converter* because it converts *positive* resistances R_1, R_2, and R_f into a *negative resistance* equal to $-(R_2R_f/R_1)\,\Omega$ in the linear region. We will show in Chap. 6 how this circuit can be easily transformed into an oscillator or a flip-flop.

D. Concave and convex resistors The circuit in Fig. 3.10 contains a *pn*-junction diode described by[5]

$$i_D = \hat{i}(v_D) = \begin{cases} I_s\left(\exp\dfrac{v_D}{v_T} - 1\right) & v_D \geq 0 \\ 0 & v_D < 0 \end{cases} \tag{3.18}$$

in its feedback path. Our objective here is to show that when the op amp is operating in the linear and + Saturation regions, the resulting driving-point

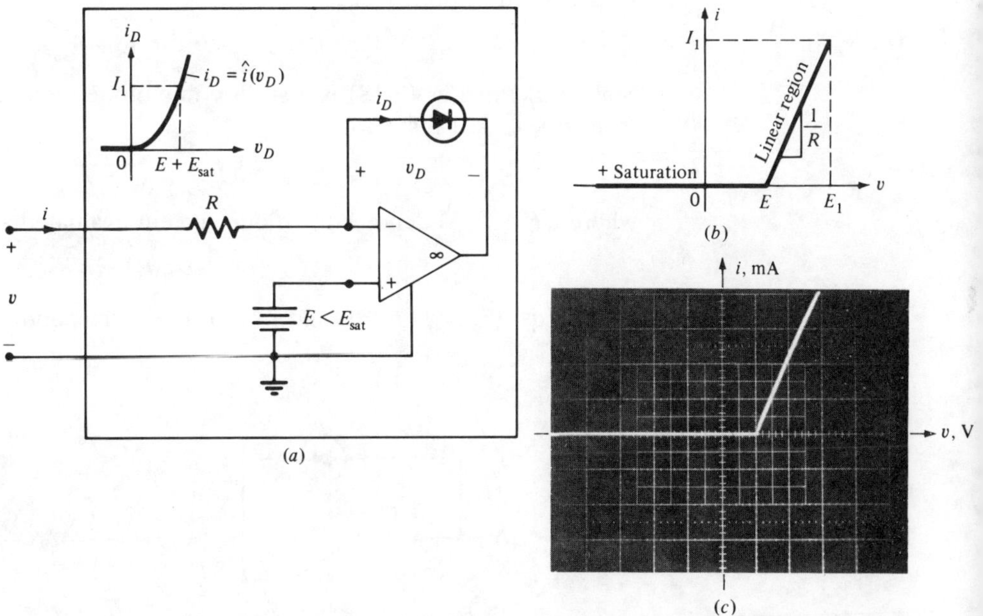

Figure 3.10 (*a*) Practical realization of a nearly ideal concave resistor characteristic. (*b*) Predicted characteristic for $v < E_1$, where $E_1 \triangleq E + R\hat{i}(E + E_{sat})$, provided $E < E_{sat}$. (*c*) Measured characteristic.

[4] For more examples of practical negative-resistance op-amp circuits, see L. O. Chua and F. Ayrom, "Designing Nonlinear Single Op-Amp Circuits: A Cookbook Approach," *Int. J. Circuit Theory Appl.*, pp. 309–326, October 1985.

[5] We assume $i_D = 0$ for $v_D < 0$ to simplify our analysis.

characteristic is identical to that defining a concave resistor for all $v < E_1$, where

$$E_1 \triangleq E + \underbrace{RI_s\left\{\exp\left[\frac{1}{V_T}(E + E_{\text{sat}})\right] - 1\right\}}_{I_1} \tag{3.19}$$

provided $E < E_{\text{sat}}$. The two equivalent circuits corresponding to these regions are shown in Fig. 3.11*a* and *b*, respectively.

Linear region From Fig. 3.11*a*, we note that $v_d = 0$ implies $e_2 = E$, and hence

$$v = Ri + E \tag{3.20}$$

To determine the range of i for which Eq. (3.20) is valid, we note first that $i = i_D \geq 0$ in view of Eq. (3.18). To determine the upper boundary, note that

$$-E_{\text{sat}} < v_o = -v_D + E < E_{\text{sat}} \tag{3.21}$$

in the linear region. Hence,

$$E - E_{\text{sat}} < v_D < E + E_{\text{sat}} \tag{3.22}$$

Since $\hat{i}(v_D)$ in Eq. (3.18) is a strictly monotone increasing function, it follows from Eq. (3.22) that

$$i_D < \hat{i}(E + E_{\text{sat}}) \triangleq I_1 \tag{3.23}$$

where $\hat{i}(E + E_{\text{sat}})$ denotes the diode current evaluated at $v_D = E + E_{\text{sat}}$. Hence

$$0 \leq i < I_1 \tag{3.24}$$

Using Eqs. (3.20), (3.23), and (3.24), the corresponding boundary in terms of

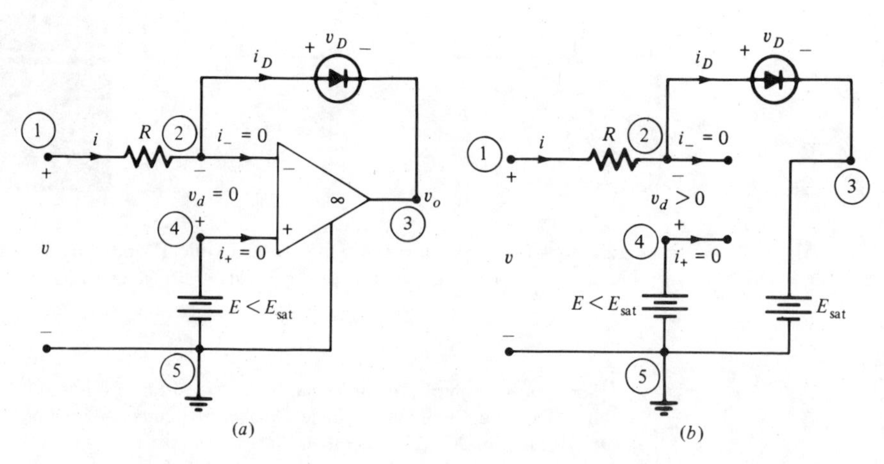

Figure 3.11 Equivalent circuit in (*a*) linear region and (*b*) + Saturation region.

v is seen to be

$$E \leq v < E_1 \tag{3.25}$$

where

$$E_1 \triangleq E + RI_1 = E + R\hat{i}(E + E_{\text{sat}}) \tag{3.26}$$

Equations (3.20) and (3.24) or Eqs. (3.25) and (3.26) define the right segment in Fig. 3.10*b*.

+ Saturation region Consider the equivalent circuit of Fig. 3.11*b* where $v_d = E - v + Ri$. Since $v_d > 0$ in the + Saturation region, we have

$$v < E + Ri \tag{3.27}$$

Applying KVL around the closed node sequence ②–①–⑤–③–② and making use of Eq. (3.27) we obtain

$$\begin{aligned} v_D &= v - Ri - E_{\text{sat}} < (E + Ri) - Ri - E_{\text{sat}} \\ &= E - E_{\text{sat}} < 0 \end{aligned} \tag{3.28}$$

because $E < E_{\text{sat}}$ by *assumption*. Hence, the diode is reversed biased when the op amp is in the + *Saturation* region. It follows from Eqs. (3.18) and (3.27) that

$$i = 0 \tag{3.29}$$

$$v < E \tag{3.30}$$

Equations (3.29) and (3.30) define the left segment in Fig. 3.10*b*.

A typical *v-i* characteristic measured from the op-amp circuit in Fig. 3.10*a* is shown in Fig. 3.10*c*. Note that the "corner" at the breakpoint is remarkably sharp.

Special case Observe that in the limiting case where $R \to 0$ and $E \to 0$, the driving-point characteristic in Fig. 3.10*b* reduces to that of an *ideal diode*, as shown in Fig. 3.12*b*. Laboratory measurements show that even though $A < \infty$ in a real op amp, the resulting driving-point characteristic still nearly approaches that of an ideal diode. Figure 3.12*c* shows a typical *pn*-junction diode characteristic, and Fig. 3.12*d* shows the nearly "ideal" diode characteristic measured from the op-amp circuit in Fig. 3.12*a* with this *pn*-junction diode connected in the negative feedback path.

Convex resistor realization Using the above observation, we can design a nearly ideal convex resistor by first transposing the *pn*-junction diode in Fig. 3.12*a* to obtain the "dual" ideal diode shown in Fig. 3.13.

Substituting this circuit in place of the "transposed" ideal diode in Fig. 2.14 of Chap. 2, we obtain the op-amp circuit in Fig. 3.14*a* which realizes the "ideal" convex resistor characteristic in Fig. 3.14*b*. The driving-point

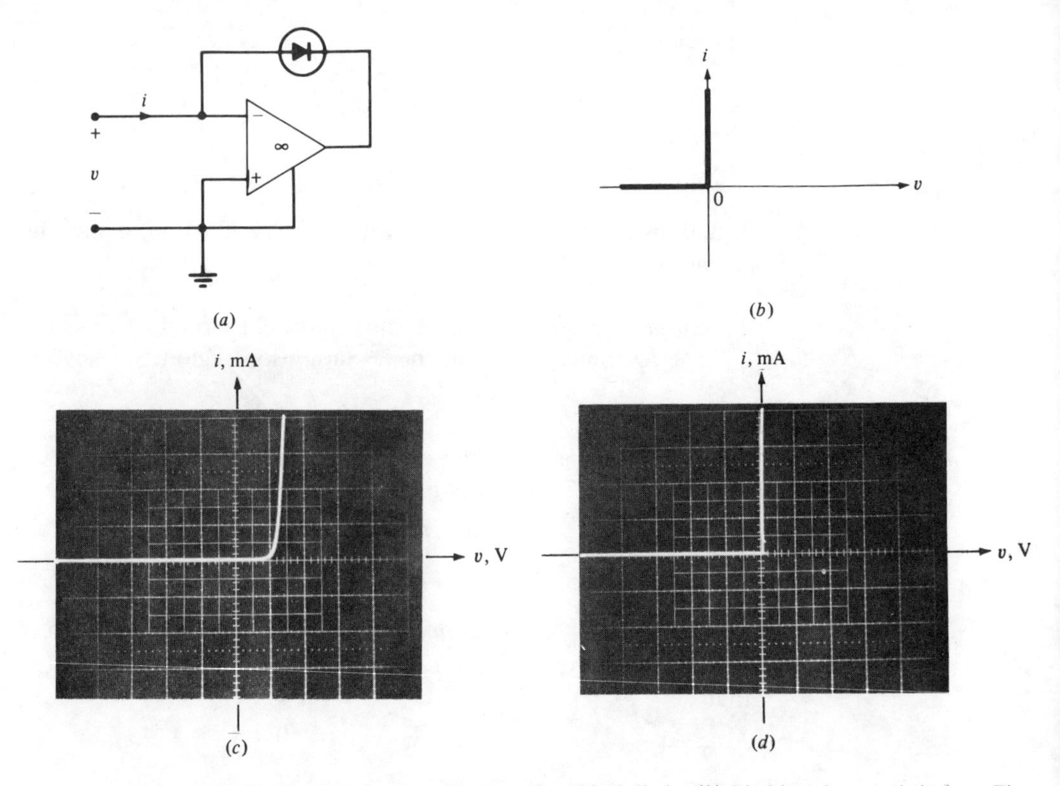

Figure 3.12 (*a*) Op-amp circuit realization of an ideal diode. (*b*) Limiting characteristic from Fig. 3.10*b* when $R \to 0$ and $E \to 0$. (*c*) Characteristic of the *pn*-junction diode in the op-amp circuit. (*d*) Measured driving-point characteristic.

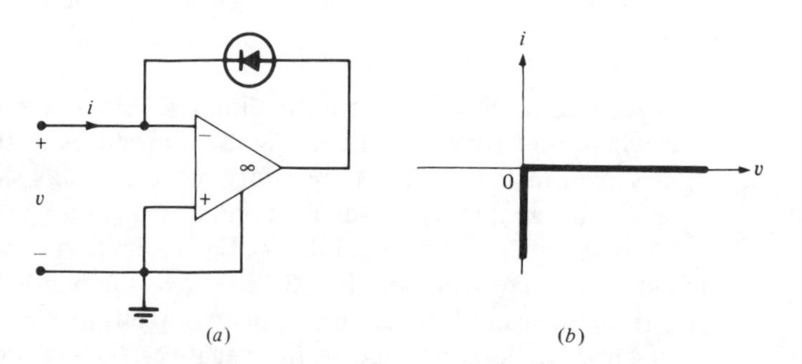

Figure 3.13 Op-amp realization of a transposed ideal diode.

characteristic measured from the circuit in Fig. 3.14*a* is shown in Fig. 3.14*c*. Again, note the sharp corner at the breakpoints.

Using the above concave and convex resistor realizations, any monotone increasing piecewise-linear driving-point characteristic can be designed with high precision.

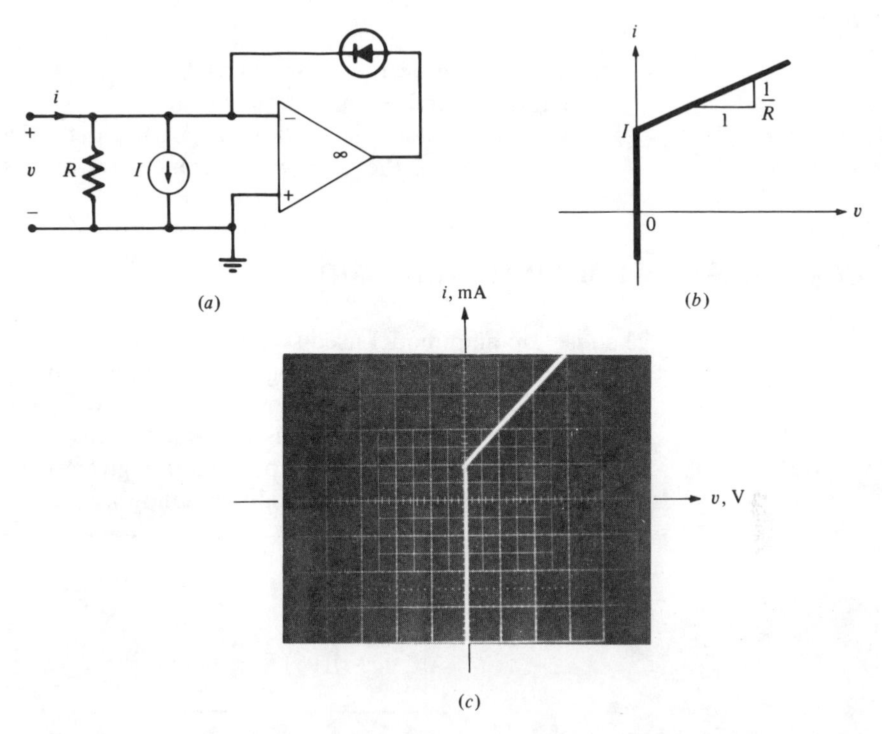

Figure 3.14 (*a*) Op-amp circuit realization of a convex resistor. (*b*) Ideal convex resistor characteristic. (*c*) Measured characteristic.

Exercises

1. Derive the remaining portion of the driving-point characteristic in Fig. 3.10*b* for $v > E_1$.
2. Derive the complete driving-point characteristic of the convex resistor circuit shown in Fig. 3.14*a*.

3.3 Systematic Method

For more complicated circuits which cannot be analyzed by the above method (e.g., feedback circuits containing several op amps), the systematic method presented in Sec. 2.3 can of course be used to derive the segment of the driving-point or transfer characteristic when the op amp is operating in the linear region. This systematic method is a special case of the modified node analysis method to be presented in Chap. 8.

For the + Saturation or − Saturation region, the same procedure can be easily modified for the corresponding equivalent circuits. In fact, the analysis in these regions is easier because the op amp is modeled by a battery.

Exercises

1. Give the procedure for carrying out the systematic method for the + Saturation and − Saturation regions.
2. Use the systematic method to derive the driving-point and transfer characteristics of the negative-resistance converter shown in Fig. 3.8*a*.

4 COMPARISON WITH FINITE-GAIN MODEL

The *ideal* op-amp model used so far in our analysis assumes that the open-loop voltage gain A in the linear region is infinite. When $A < \infty$, the model and equivalent circuits in Fig. 1.6 should be modified as shown in Fig. 4.1. We will henceforth refer to this model as the *finite-gain op-amp model*.

Using the *piecewise-linear* representation given by Eq. (3.9) in Chap. 2, we can describe this model analytically as follows:

Finite-gain op-amp model

$$i_- = 0 \tag{4.1a}$$

$$i_+ = 0 \tag{4.1b}$$

$$v_o = f(v_d) \triangleq \frac{A}{2}|v_d + \epsilon| - \frac{A}{2}|v_d - \epsilon| \tag{4.1c}$$

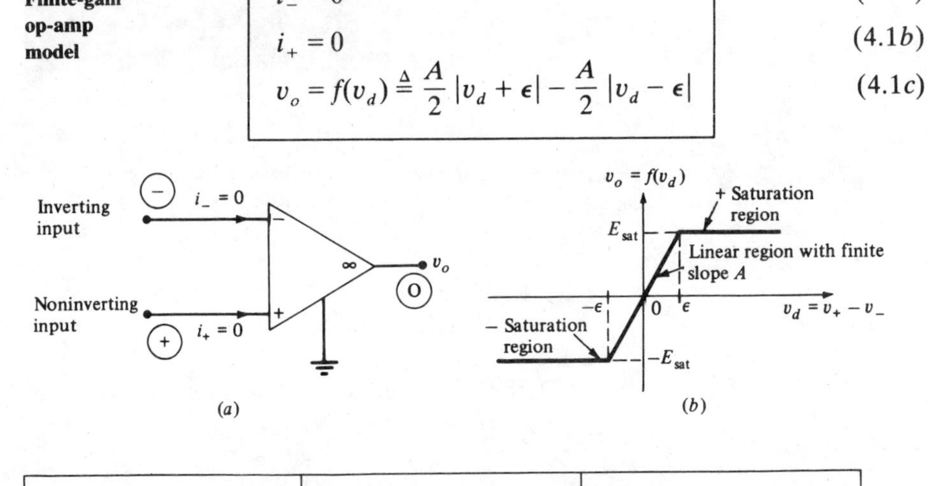

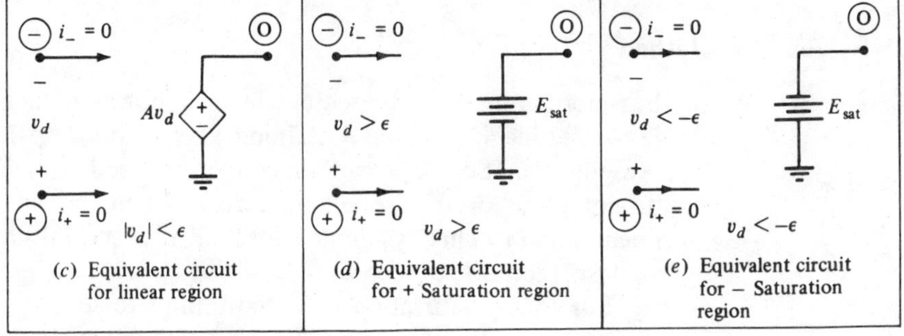

Figure 4.1 Finite-gain op-amp model.

Note that unlike Eq. (1.2*c*) of the ideal model, Eq. (4.1*c*) is a well-defined function for all values of v_d, *including* $v_d = 0$.

Since $i_- = 0$ and $i_+ = 0$ in both Figs. 1.6 and 4.1, we can write $i_- = -i_+$ and interpret both the *ideal* and the *finite-gain* op-amp model as a *two-port*. For example, the model in Fig. 4.1 can be redrawn as shown in Fig. 4.2, where $f(v_d)$ is given by Eq. (4.1*c*). In the linear region, the finite-gain model reduces to a linear *voltage-controlled voltage source* as shown in Fig. 4.3.

In order to compare the answers obtained from using the *ideal* and the finite-gain models, let us consider an example.

Example Let us analyze the *inverting amplifier* circuit in Fig. 2.3 using the finite-gain model. Since the op amp in this circuit is known to be operating in the linear region with a dynamic range given by Eq. (2.7), let us replace the op amp by the linear two-port model shown in Fig. 4.3*b*. The resulting circuit is shown in Fig. 4.4. We can calculate v_o by the *inspection* method as follows:

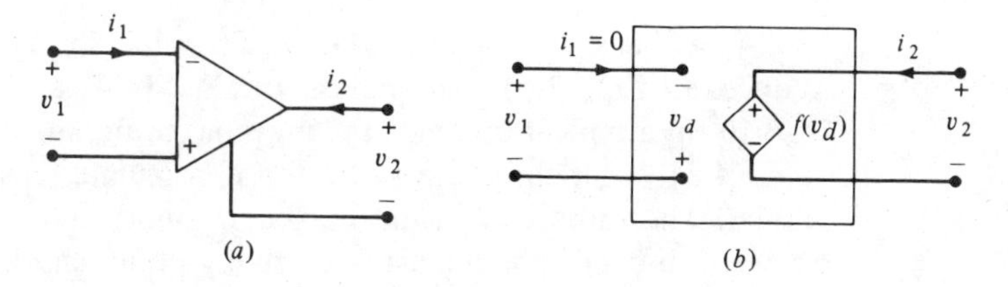

Figure 4.2 Equivalent nonlinear two-port model.

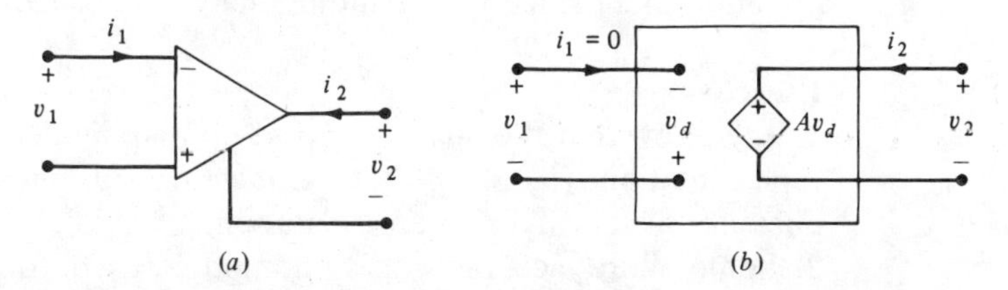

Figure 4.3 Equivalent linear two-port model (valid only if the op amp is operating in the linear region).

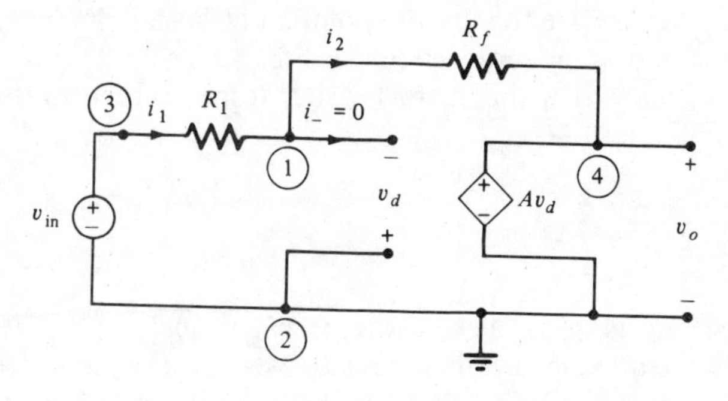

Figure 4.4 Inverting amplifier circuit with op amp modeled by Fig. 4.3*b*.

KCL at node ①:
$$i_2 = \frac{v_{\text{in}} + v_d}{R_1} \tag{4.2}$$

KVL at the closed node sequence ①–④–②–①:

$$R_f \left[\frac{v_{\text{in}} + v_d}{R_1} \right] + Av_d + v_d = 0 \tag{4.3}$$

Solving for v_d in Eq. (4.3), we obtain

$$v_d = -\left[\frac{1}{(R_1/R_f)A + (1 + R_1/R_f)} \right] v_{\text{in}} \tag{4.4}$$

Substituting Eq. (4.4) into $v_o = Av_d$, we obtain

$$v_o = -\left[\frac{1}{(R_1/R_f) + (1/A)(1 + R_1/R_f)} \right] v_{\text{in}} \tag{4.5}$$

As a check, note that as $A \to \infty$, Eq. (4.4) implies $v_d \to 0$ and Eq. (4.5) reduces to Eq. (2.6). An analysis of Eqs. (4.4) and (4.5) shows that since $A > 10^5$ in a typical op amp, the more accurate answers given by Eqs. (4.4) and (4.5) are nearly equal to those calculated using the *ideal* op-amp model. The same conclusion has been found to hold for the other circuits as well. Indeed, the measured driving-point characteristics in Figs. 3.9, 3.10, 3.12, and 3.14 all agree remarkably well with those predicted by the ideal op-amp model. This observation justifies our choice of the *ideal* op-amp model since the resulting analysis is usually much simpler.

Exercises

1. (*a*) Show that the linear two-port op-amp model in Fig. 4.3 has a *hybrid* representation. (*b*) Show that the *ideal* op-amp model in the linear region does *not* have a hybrid representation.
2. Some more accurate op-amp models used for high-precision circuit analysis has $i_- \neq -i_+$ in order to account for the small but nonzero currents entering the inverting and noninverting op-amp terminals. In this case, can you redraw the model as a two-port?
3. (*a*) Derive the driving-point and transfer characteristics of the negative-resistance converter circuit in Fig. 3.8 using the finite-gain op-amp model. (*b*) Show that the characteristics from (*a*) tend to those given in Fig. 3.8*b* and *c* as $A \to \infty$.

SUMMARY

- The op amp is a versatile four-terminal device which behaves like a nonlinear four-terminal resistor at dc. For low-frequency circuit applications, it can be modeled realistically by the *ideal op-amp model*.

- The *ideal op-amp model* is described by the following equations involving only its terminal voltages and currents (hence it is a four-terminal resistor by definition):

Ideal op-amp equations

$$i_- = 0$$

$$i_+ = 0$$

$$v_o = E_{sat} \frac{|v_d|}{v_d}, \qquad v_d \neq 0$$

$$v_d = 0, \quad -E_{sat} < v_o < E_{sat}$$

where $v_d \overset{\Delta}{=} v_+ - v_-$.

- The ideal op-amp model has three distinct *operating regions*:

Linear region: $-E_{sat} < v_o < E_{sat}$
+ Saturation region: $v_d > 0$
− Saturation region: $v_d < 0$

The *ideal op-amp model* can also be uniquely represented by three *equivalent circuits* (see Fig. 1.6), each one corresponding to one operating region.

- In the *linear region*, the ideal op-amp model is described by

Ideal op-amp equations in the linear region

$$i_- = 0$$

$$i_+ = 0$$

$$v_d = 0, \qquad -E_{sat} < v_o < E_{sat}$$

This model is equivalent to a *linear two-port* resistor where the input port behaves like an *open circuit* (zero input current) *and* a *short circuit* (zero input voltage) simultaneously. Hence, it is called a *virtual short-circuit model*. The output port behaves like a VCVS with an *infinite* gain.

- The *output* voltage v_o of the ideal op-amp model is *not* defined at $v_d = 0$ because it can assume *any* value between $-E_{sat}$ and E_{sat}. The actual output voltage is determined only by the *external* circuit constraints. Consequently, we say the *output* voltage v_o of the ideal op-amp model is a *multivalued* function of the input voltage v_d.
- In some situations where it is awkward to deal with *multivalued* functions, it may be more convenient to use the *finite-gain op-amp model* defined by

Finite-gain op-amp model

$$i_- = 0$$

$$i_+ = 0$$

$$v_o = f(v_d) \overset{\Delta}{=} \frac{A}{2}|v_d + \epsilon| - \frac{A}{2}|v_d - \epsilon|$$

- For typical op amps where $A > 10^5$, the answers predicted by using the *ideal* op-amp model are virtually indistinguishable from those predicted by using the finite-gain model.
- Depending on the *dynamic range* of the input signal(s), as well as the interconnecting circuit, an op amp may operate either in the *linear* region, or in the *nonlinear* region consisting of at least two piecewise-linear segments.
- The op amps in many practical circuits do operate only in the linear region. Such circuits can be easily analyzed using *only* the *virtual short-circuit model*.
- Op-amp circuits containing only *one* op amp and a few resistors can often be analyzed *by inspection* using the ad hoc techniques described in Sec. 2.2 (for op amps operating in the linear region) and Sec. 3.2 (for op amps operating in the nonlinear region).

PROBLEMS

Linear op-amp model

1 Assuming the op amp is operating in the linear region, find the voltage gain $A = v_o/v_i$ for the circuit shown in Fig. P4.1. What is the dynamic range of the input?

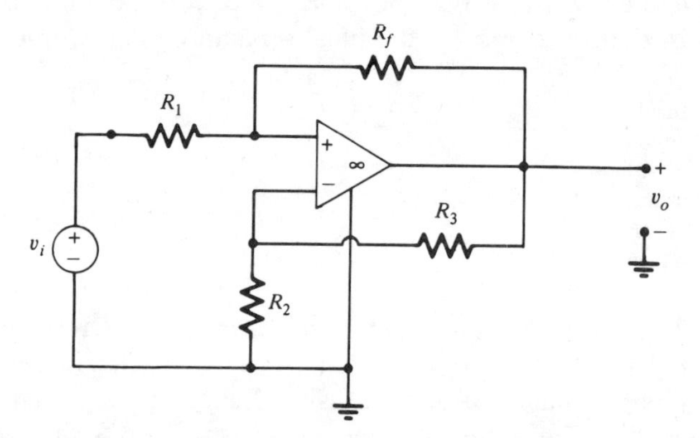

Figure P4.1

2 (*a*) Show that the two-port in Fig. P4.2 realizes a CCCS. Assume the op amp is operating in the linear region.

(*b*) Find the dynamic range of v_2 and i_1 required for linear operation of the op amp.

3 Consider the solar-cell op-amp circuit shown in Fig. P4.3*a* where the *v-i* characteristics of the solar cell are shown in Fig. P4.3*b* as a function of ambient light intensity in lumens (lm). Assuming the op amp is operating in the linear region, find the output voltage when the solar cell is subjected to a light intensity of 0.02 and 0.05 lm, respectively.

4 Consider the two-op-amp instrumentation amplifier shown in Fig. P4.4.

(*a*) Assuming identical op amps operating in their linear region, obtain v_o in terms of v_{s1} and v_{s2}.

(*b*) Specify the dynamic range for v_{s1} and v_{s2} to guarantee linear operation.

5 Show that the two-port in Fig. P4.5 realizes a gyrator with $G = -1/R$ S. Assume the op amp is operating in the linear region.

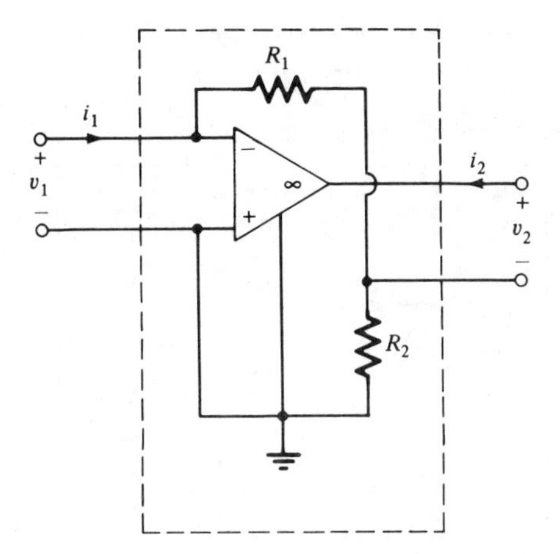

Figure P4.2

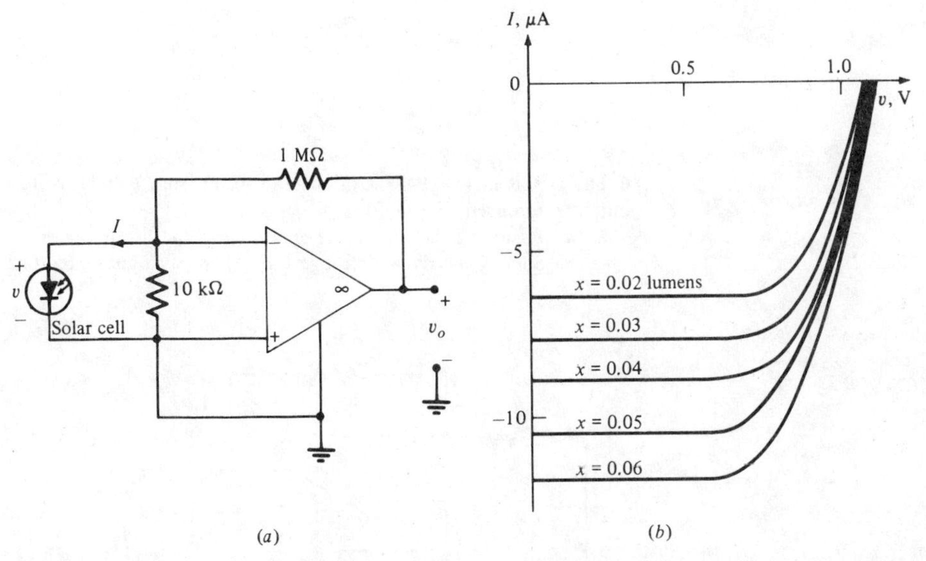

(a)

(b)

Figure P4.3

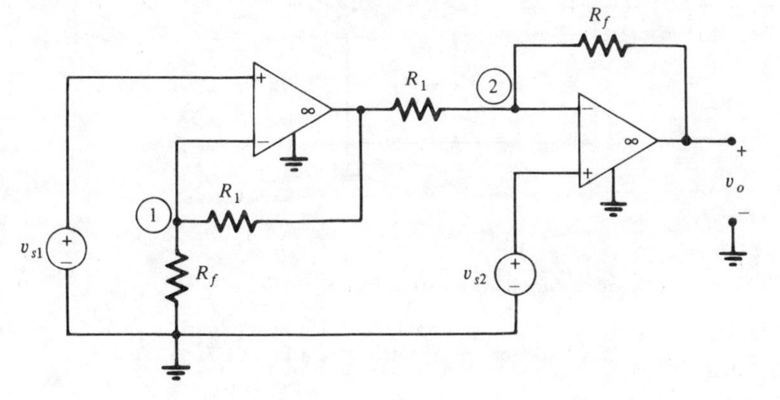

Figure P4.4

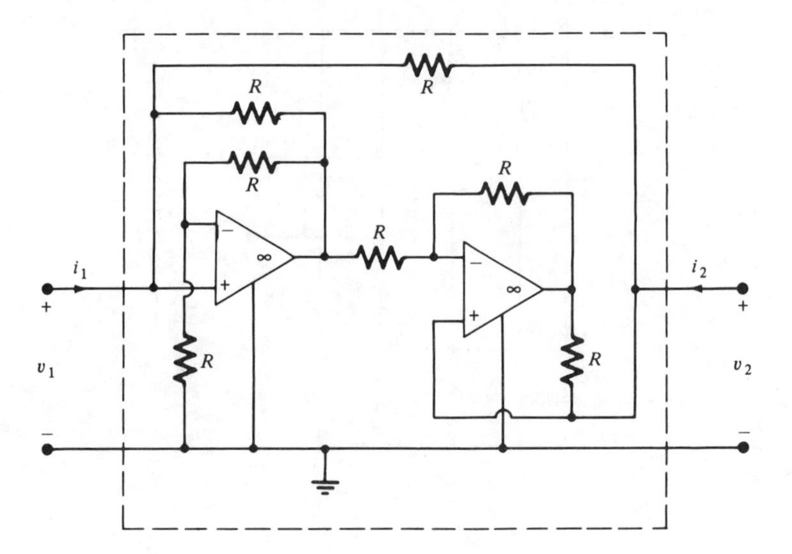

Figure P4.5

6 The circuit in Fig. P4.6 is often used to measure v_2 relative to v_1 and produce a proportional output referenced to ground.

(*a*) Assuming the op amp is operating in the linear region, find v_o in terms of v_1 and v_2.

(*b*) Specify the dynamic range for v_1 and v_2 to guarantee linear operation.

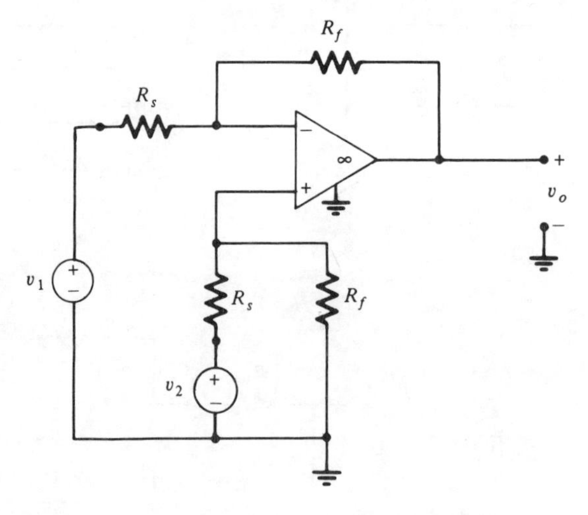

Figure P4.6

Nonlinear feedback

7 Assuming the op amp is operating in the linear region, find and sketch the v_o vs. v_i transfer characteristic for the circuit shown in Fig. P4.7.

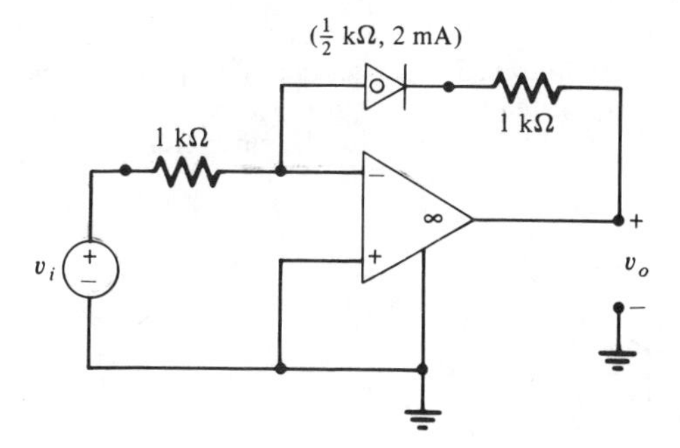

Figure P4.7

8 Consider the circuit shown in Fig. P4.8. The nonlinear resistors $\mathscr{R}_1$ and $\mathscr{R}_2$ are described by the following characteristics:

$$i_1 = g(v_1)$$

$$v_2 = f(i_2)$$

Assuming the op amp is operating in the linear region, find v_o as a function of v_i.

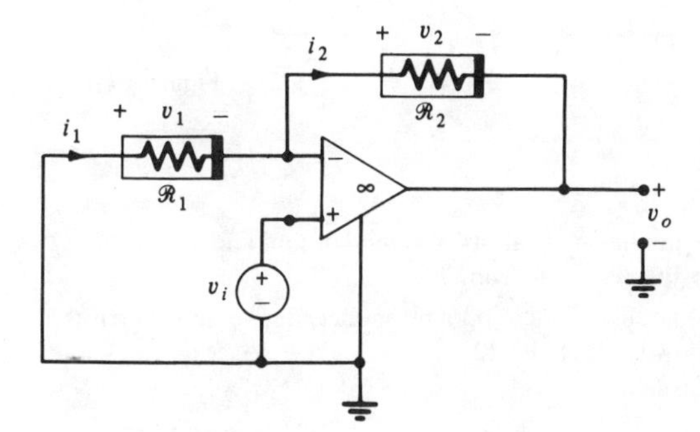

Figure P4.8

9 Consider the circuit shown in Fig. P4.9*a*. The nonlinear resistor $\mathscr{R}_1$ is described by the v-i characteristic shown in Fig. P4.9*b*, and $\mathscr{R}_2$ is characterized by $v_2 = \exp(i_2)$. Assuming the op amp is operating in the linear region find the v_o vs. v_i transfer characteristic.

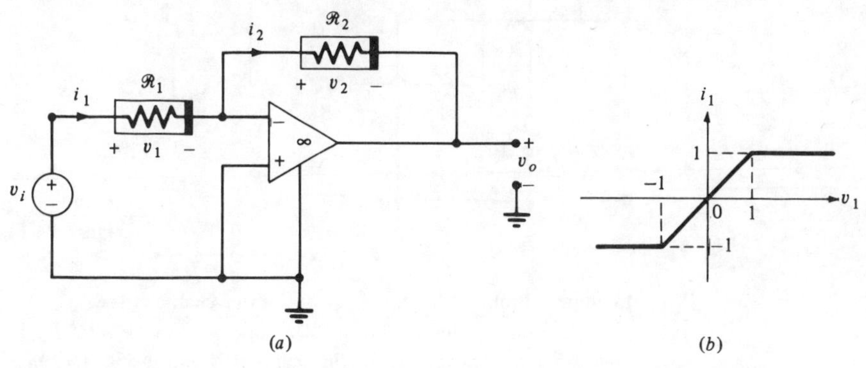

Figure P4.9

10 The circuit shown in Fig. P4.10 contains two "ideal" zener diodes D_1 and D_2 with zener voltages of 5 V and 3 V, respectively. Assuming the op amp is operating in its linear region:

(*a*) Obtain a plot of the v_o vs. v_i transfer characteristic.

(*b*) Obtain an analytic expression for the curve from (*a*).

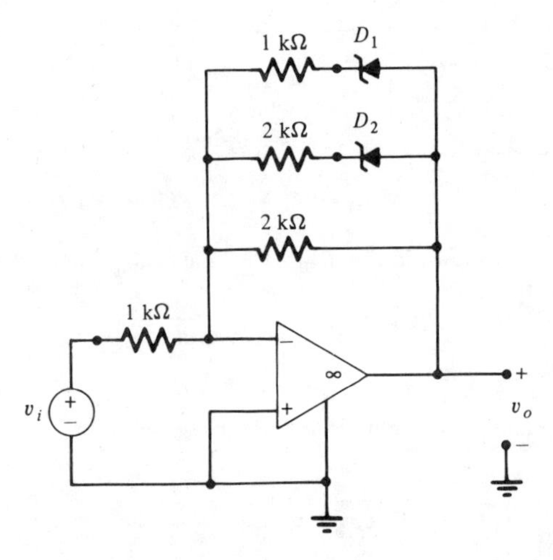

Figure P4.10

Nonlinear op-amp model

11 Using the nonlinear ideal op-amp model, find and sketch the v_o vs. v_i transfer characteristic for the circuit in Fig. P4.1 of Prob. 1.

12 Using the nonlinear ideal op-amp model, derive and sketch the driving-point characteristic for the circuit shown in Fig. P4.12.

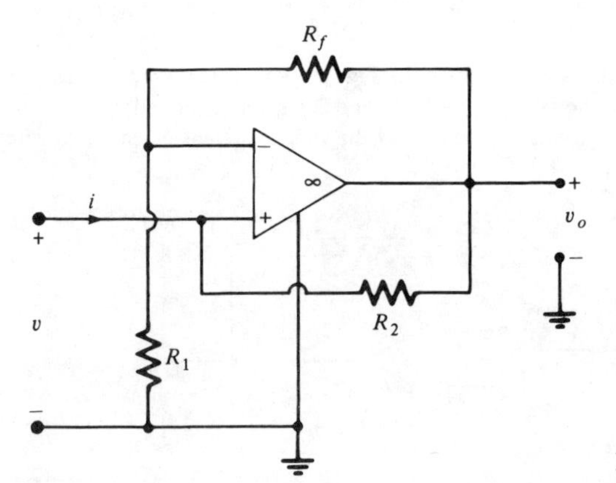

Figure P4.12

13 Repeat Prob. 12 for the v_o vs. v transfer characteristic.

14 (*a*) Assuming the nonlinear ideal op-amp model with $E_{sat} = 15$ V, derive and sketch the driving-point characteristic for the one-port shown in Fig. P4.14.

(*b*) Connect a linear resistor R_1 across the one-port, and find the maximum value of the resistance for which the one-port functions as an independent current source.

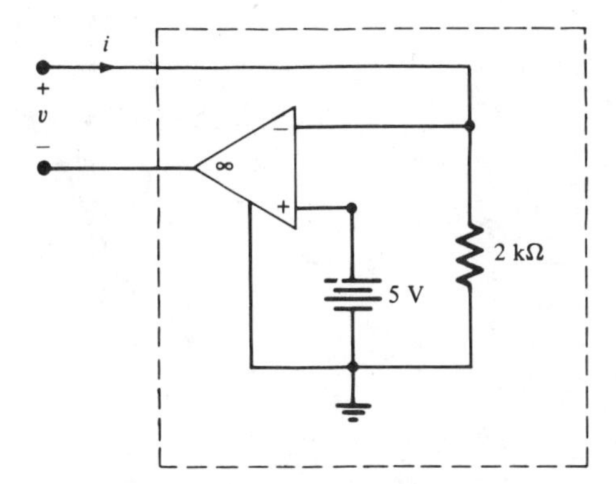

Figure P4.14

15 (*a*) Assuming the nonlinear ideal op-amp model with $E_{sat} = 15$ V, derive and sketch the driving-point characteristic for the one-port shown in Fig. P4.15.

(*b*) If a linear resistor R_1 is connected across the one-port, what is the range of R for which the "composite" one-port operates as an independent voltage source?

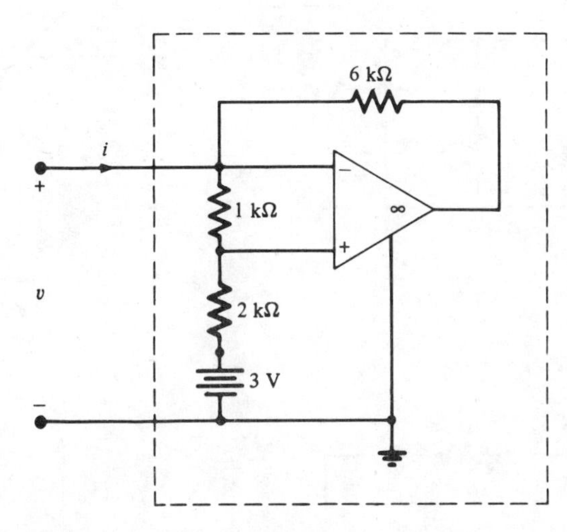

Figure P4.15

Driving-point characteristic, design

16 The circuit shown in Fig. P4.16*a* can be used to synthesize any odd symmetric *v-i* characteristic of the form shown in Fig. P4.16*b*.

(*a*) Use the nonlinear ideal op-amp model to derive the driving-point characteristic of this circuit. Express all pertinent slopes and breakpoints in terms of the resistances in the circuit and the op-amp saturation voltage E_{sat}.

(*b*) Using the results from part (*a*) propose a design procedure, i.e., give an algorithm for finding the resistance values given m_o, m_1, E_B, and E_{sat}. State any conditions which must be satisfied to guarantee positive resistance values.

17 Repeat Prob. 16 for the circuit and characteristic shown in Fig. P4.17*a* and *b*, respectively.

18 Repeat Prob. 16 for the circuit and characteristic shown in Fig. P4.18*a* and *b*, respectively.

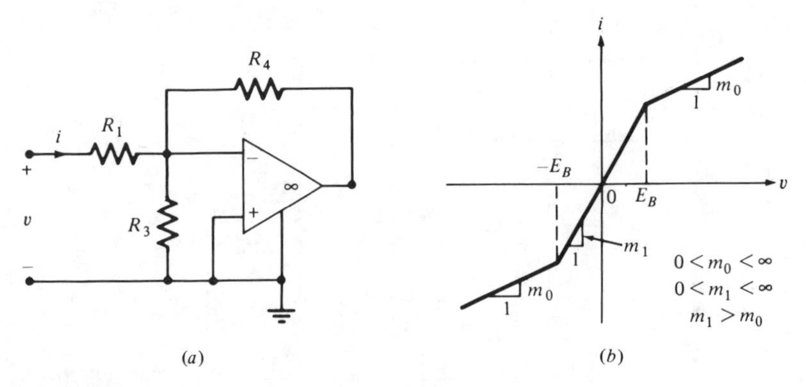

Figure P4.16

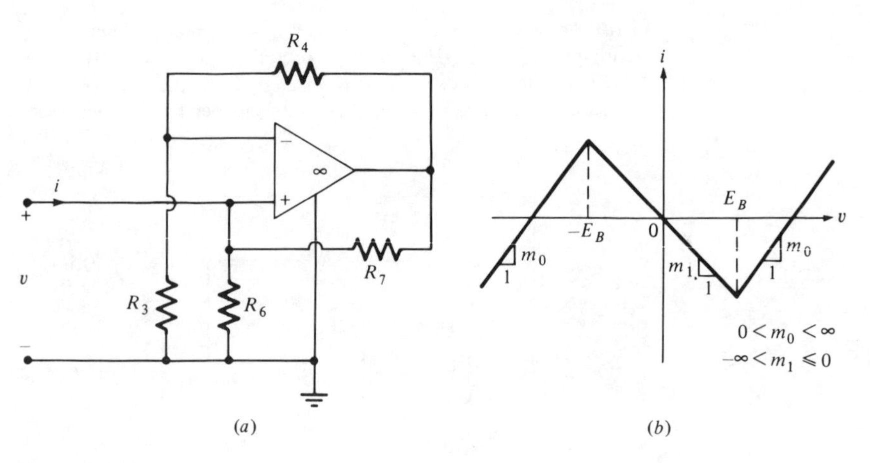

Figure P4.17

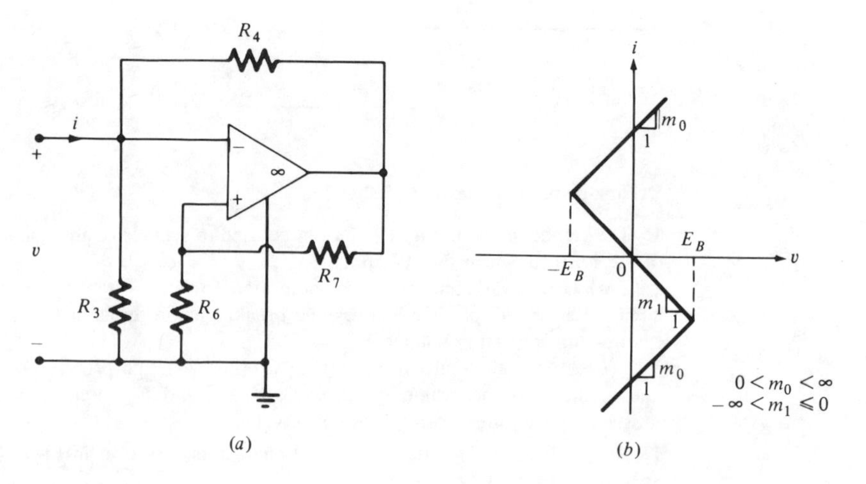

Figure P4.18

19 Repeat Prob. 16 for the circuit and characteristic shown in Fig. P4.19*a* and *b*, respectively.

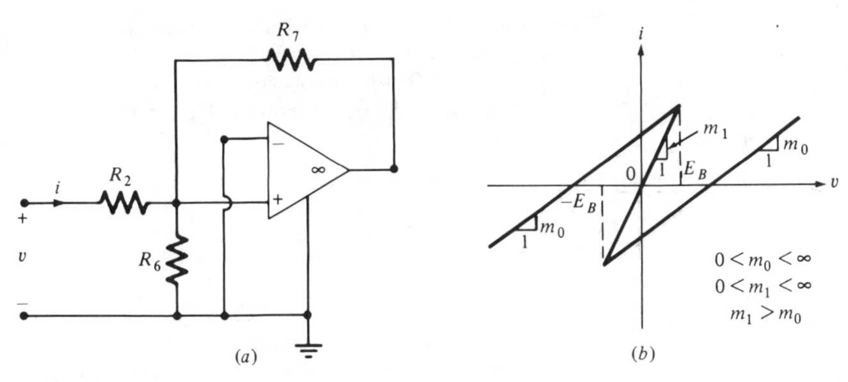

Figure P4.19

20 The circuit shown in Fig. P4.20*a* can be used to synthesize any odd-symmetric *v*-*i* characteristic of the form shown in Fig. P4.20*b*.

(*a*) Use the nonlinear ideal op-amp model to derive the driving-point characteristic of this circuit. Express all pertinent slopes and breakpoints in terms of the resistances in the circuit and the op-amp saturation voltage E_{sat}.

(*b*) Using the results from (*a*) propose a design procedure, i.e., give an algorithm, for finding the resistance values given m_o, m_1, I_B, and E_{sat}. State any conditions which must be satisfied to guarantee positive resistance values.

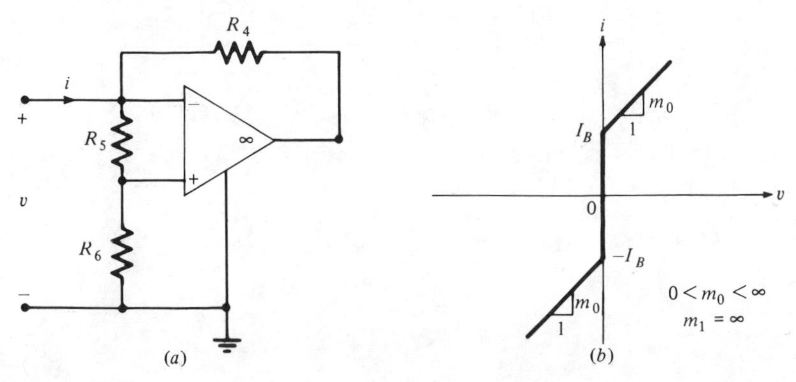

Figure P4.20

21 Repeat Prob. 20 for the circuit and characteristic shown in Fig. P4.21*a* and *b*, respectively.

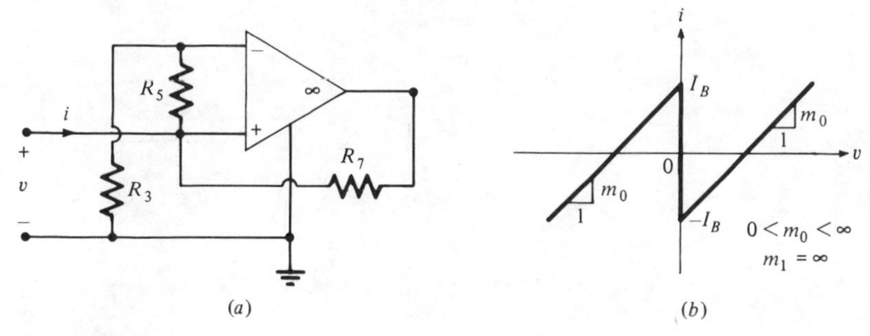

Figure P4.21

Finite-gain op-amp model

22 For the circuit shown in Fig. P4.1 use the finite-gain op-amp model to find
  (*a*) The driving-point characteristic
  (*b*) The v_o vs. v_i transfer characteristic

23 Repeat Prob. 22 for the circuit shown in Fig. P4.12.

CHAPTER

FIVE

GENERAL RESISTIVE CIRCUITS

In the preceding chapters we have studied Kirchhoff's laws and their consequences, two-terminal resistors, and multiterminal resistors. We have also analyzed and studied the properties of *simple* resistive circuits containing both linear and nonlinear, two-terminal and multiterminal resistors, and independent sources.

In this chapter, we will embark on a more general and definitive study of resistive circuits. Our aim is to develop *general* methods of analysis, both for linear and nonlinear resistive circuits, and to derive general properties of resistive circuits. The term *resistive circuit* applies to all circuits containing *two-terminal resistors*, *multiterminal resistors*, multiport resistors, and independent voltage and current sources. Common circuit elements such as ideal transformers, gyrators, controlled sources, transistors and op amps modeled by resistive circuits, etc., *are all included*. To avoid clutter, all of these garden-variety circuit elements will be lumped under the umbrella "multiterminal and multiport resistors." However, independent sources will always be singled out separately because, as will be clear shortly, they play a fundamentally different role.

Both time-invariant and time-varying resistors are of course included. A resistive circuit is said to be *time-invariant* iff when all independent sources are set to zero, it contains only *time-invariant resistors*. A resistive circuit is said to be *time-varying* iff it is not time-invariant.

Almost all results, methods, and properties to be presented in this chapter are valid for both time-invariant and time-varying resistive circuits. For simplicity, we will often state them for time-invariant resistive circuits only.

The importance of resistive circuits has increased considerably in recent years: First, they are used in modeling many engineering problems, second, the analysis of many general nonresistive circuits reduces to the analysis of an associated resistive circuit, and third, many computer algorithms for simulating dynamic circuits require at each step the analysis of a resistive circuit.

Recall that a *physical circuit* is an interconnection of *real electric devices*. For purposes of analysis and design, each electric device is replaced by a *device model* made of ideal circuit elements[1] (e.g., ideal diodes, batteries, linear and nonlinear resistors, controlled sources, etc.). The interconnection of these models gives the (electric) circuit. In other words, a circuit consists of an interconnection of common circuit elements.

Since the important study of *device modeling* is beyond the scope of this book, our point of departure in this book is a *circuit*. Whether it arises from the model of some physical circuit or from the figment of one's imagination is irrelevant. In fact, it is often through the introduction of hypothetical, and sometimes pathological, circuits that one gains an in-depth understanding of this subject.

This chapter contains five sections. For pedagogical reasons, we will always start with specific examples for motivation. We go from linear to nonlinear and from special to general results. Several important theorems and properties are carefully stated, illustrated, and proved. Usually the proof is given last—after the reader has seen several examples which illustrate the result and is all fired-up by its profound applications and implications.

Finally, a few words concerning some *general* technical terms to be used throughout this book.

A resistive circuit is said to be *linear* iff, after setting all independent sources to zero, it contains only *linear* two-terminal, multiterminal, and/or multiport resistors (i.e., those described by a *linear equation* involving the terminal or port voltage $\mathbf{v}$ and the terminal or port current $\mathbf{i}$, at any time t).

A resistive circuit is said to be *nonlinear* iff it contains at least one *nonlinear* resistor besides independent sources.

A resistive circuit is said to be *uniquely solvable* iff Kirchhoff's laws and the branch equations are simultaneously satisfied by a *unique* set of branch voltages $v_1(t), v_2(t), \ldots, v_b(t)$, and a *unique* set of branch currents $i_1(t), i_2(t), \ldots, i_b(t)$, for all t.

1 NODE ANALYSIS FOR RESISTIVE CIRCUITS

The simplest method for analyzing a resistive circuit is to solve for its *node-to-datum* voltages. Once these node voltages have been calculated, we

[1] Of course, *all circuit elements are ideal*. We will, nevertheless, occasionally throw in the word "ideal" to remind the reader that "nonphysical" answers (e.g., a circuit that has no solution) and "paradoxes" (e.g., a resistive circuit having several solutions) are quite possible and even expected. When they do occur, the culprit is not the theory, but the model. Such unrealistic situations can only be rectified by returning to the drawing board to come up with a more realistic circuit model.

can solve for the *branch voltages* trivially via KVL ($\mathbf{v} = \mathbf{A}^T\mathbf{e}$). They in turn can be used to calculate the *branch currents*, provided all elements in the circuit other than current sources are *voltage-controlled*. In this section, we will consider only the subclass of resistive circuits which are amenable to this common analysis method, henceforth called *node analysis*.

1.1 Node Equation Formulation: Linear Resistive Circuits

For simple circuits, the node equation can be formulated almost by inspection, as illustrated in the following examples.

Example 1 (two-terminal resistors) The circuit shown in Fig. 1.1 contains only linear time-invariant resistors and independent current sources. Choosing (arbitrarily) node ④ as the datum node, each branch current can be expressed in terms of at most two node voltages (recall we use v_j to denote the branch voltage for the jth branch and e_k to denote the node voltage at node ⓚ with respect to the datum):

$$\begin{aligned} i_1 &= G_1 v_1 = G_1 e_1 & i_4 &= G_4 v_4 = G_4(e_1 - e_2) \\ i_2 &= G_2 v_2 = G_2(e_2 - e_1) & i_5 &= G_5 v_5 = G_5(e_3 - e_2) \\ i_3 &= G_3 v_3 = G_3(-e_2) & i_6 &= G_6 v_6 = G_6 e_3 \end{aligned} \tag{1.1}$$

It follows from Sec. 6.2 of Chap. 1 that we can write three *linearly independent KCL equations* in terms of e_1, e_2, and e_3, namely

Node ①: $$G_1 e_1 - G_2(e_2 - e_1) + G_4(e_1 - e_2) = i_{s1}(t)$$

Node ②: $$G_2(e_2 - e_1) - G_3(-e_2) - G_4(e_1 - e_2) - G_5(e_3 - e_2) = -i_{s3}(t) \tag{1.2}$$

Node ③: $$G_5(e_3 - e_2) + G_6 e_3 = i_{s3}(t) - i_{s2}(t)$$

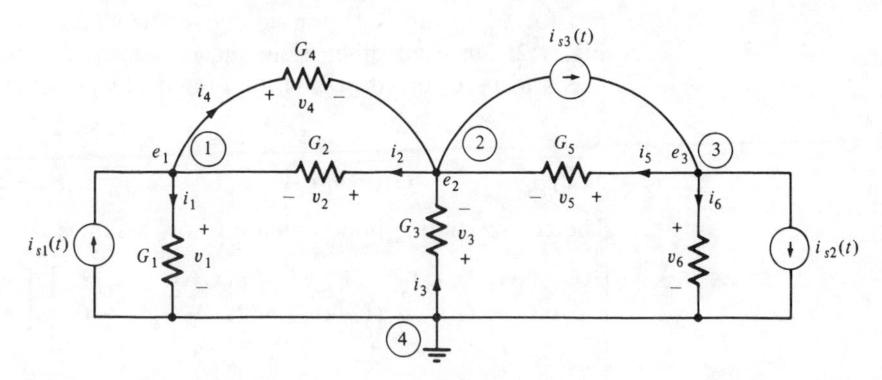

Figure 1.1 Circuit for Example 1. Here, G_j denotes the conductance in siemens for the jth resistor.

Since these equations are linear, we can recast them into the following matrix form:

$$\underbrace{\begin{bmatrix} (G_1+G_2+G_4) & -(G_2+G_4) & 0 \\ -(G_2+G_4) & (G_2+G_3+G_4+G_5) & -G_5 \\ 0 & -G_5 & (G_5+G_6) \end{bmatrix}}_{\mathbf{Y}_n} \underbrace{\begin{bmatrix} e_1 \\ e_2 \\ e_3 \end{bmatrix}}_{\mathbf{e}} = \underbrace{\begin{bmatrix} i_{s1}(t) \\ -i_{s3}(t) \\ i_{s3}(t)-i_{s2}(t) \end{bmatrix}}_{\mathbf{i}_s(t)} \tag{1.3}$$

We will show shortly that a large class of linear resistive circuits is described by an equation of the form

$$\mathbf{Y}_n\mathbf{e} = \mathbf{i}_s(t) \tag{1.4}$$

henceforth called the *node equation*, where $\mathbf{Y}_n$ is a square matrix called the *node-admittance matrix* and $\mathbf{i}_s(t)$ is a vector called the *equivalent source vector*.

An inspection of Fig. 1.1 and Eq. (1.3) reveals the following properties, which we will prove in Sec. 1.2 to be true for any linear circuit containing *only* linear two-terminal resistors and independent current sources:

Table 1.1 Properties of the node equation in Sec. 1.1

For any circuit made of linear two-terminal resistors and independent sources

(*a*) The kth diagonal element of $\mathbf{Y}_n$ is equal to the sum of all conductances attached to node ⓚ.
(*b*) The jkth off-diagonal element of $\mathbf{Y}_n$ is equal to the *negative* of the sum of all conductances between node ⓙ and node ⓚ.
(*c*) The matrix $\mathbf{Y}_n$ is *symmetric*, i.e., $\mathbf{Y}_n = \mathbf{Y}_n^T$.
(*d*) The kth element of $\mathbf{i}_s(t)$ is equal to the algebraic sum of currents of all independent current sources entering node ⓚ

Example 2 (two-terminal resistors and VCCS) Suppose we replace the resistor G_6 in Fig. 1.1 by a voltage-controlled current source as shown in Fig. 1.2. Since all other elements remain unchanged, only one equation needs to be changed in Eqs. (1.1) and (1.2), namely,

$$i_6 = -g_m v_2 = -g_m(e_2 - e_1) \tag{1.5}$$

Node ③: $$G_5(e_3 - e_2) - g_m(e_2 - e_1) = i_{s3}(t) - i_{s2}(t) \tag{1.6}$$

The corresponding node equation now assumes the form:

$$\underbrace{\begin{bmatrix} (G_1+G_2+G_4) & -(G_2+G_4) & 0 \\ -(G_2+G_4) & (G_2+G_3+G_4+G_5) & -G_5 \\ g_m & -(G_5+g_m) & G_5 \end{bmatrix}}_{\mathbf{Y}_n} \underbrace{\begin{bmatrix} e_1 \\ e_2 \\ e_3 \end{bmatrix}}_{\mathbf{e}} = \underbrace{\begin{bmatrix} i_{s1}(t) \\ -i_{s3}(t) \\ i_{s3}(t)-i_{s2}(t) \end{bmatrix}}_{\mathbf{i}_s(t)} \tag{1.7}$$

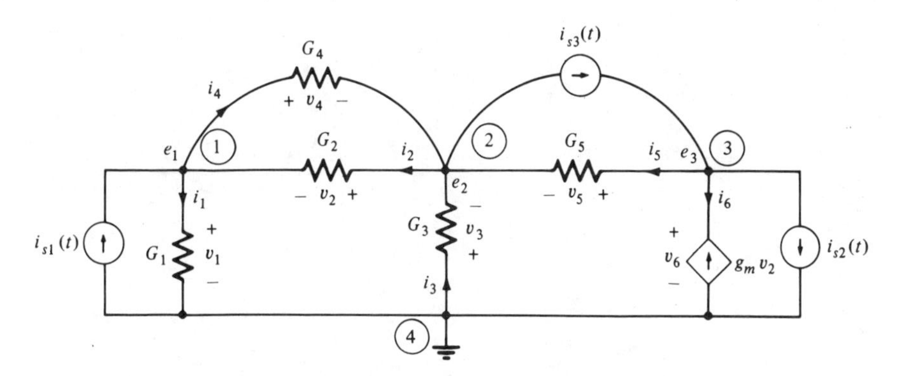

Figure 1.2 Circuit for Example 2. Note that the controlled source depends on node voltages e_1 and e_2 but not e_3.

Note that properties (b) and (c) of Table 1.1 no longer hold for this node equation. If we replace the resistor G_3 (instead of G_6) by the same controlled source we will find that property (a) of Table 1.1 is violated in the resulting node equation. Hence we conclude that properties (a), (b), and (c) in Table 1.1 are generally false if the circuit contains even only one controlled source, in addition to resistors and current sources.

Formulation in terms of the reduced incidence matrix A Let $\mathcal{N}$ denote any *connected* linear resistive circuit containing only time-invariant[2] two-terminal, multiterminal, and/or multiport *linear voltage-controlled* resistors, and independent current sources.

For example, $\mathcal{N}$ may contain *gyrators* because they are defined by a *voltage*-controlled linear equation [see Eq. (2.18) in Chap. 3].

$$\begin{bmatrix} i_1 \\ i_2 \end{bmatrix} = \begin{bmatrix} 0 & G \\ -G & 0 \end{bmatrix} \begin{bmatrix} v_1 \\ v_2 \end{bmatrix} \quad \text{or} \quad \mathbf{i} = \mathbf{Y}\mathbf{v} \tag{1.8}$$

On the other hand, $\mathcal{N}$ may *not* contain ideal transformers because it is *not* voltage-controlled, i.e., it is impossible to solve for i_1 and i_2 from the defining equation (2.13) of Chap. 3 in terms of only v_1 and v_2. Controlled sources other than VCCS are also disallowed for the same reason.

Note that independent voltage sources are not allowed in our present formulation. However, they can be included later through equivalent circuit transformations.

If the terminals and/or ports of all circuit elements which are not independent current sources are labeled consecutively, and if $\mathbf{v} = [v_1, v_2, \ldots, v_b]^T$ and $\mathbf{i} = [i_1, i_2, \ldots, i_b]^T$ denote the respective branch voltage and branch current vectors, then $\mathcal{N}$ is precisely the class where $\mathbf{i}$ can be described as a *linear* function of $\mathbf{v}$; namely,

[2] Exactly the same formulation holds for the time-varying case. We made the time-invariant assumption in order to simplify our notation.

$$\underbrace{\begin{bmatrix} i_1 \\ i_2 \\ \vdots \\ i_b \end{bmatrix}}_{\mathbf{i}} = \underbrace{\begin{bmatrix} y_{11} & y_{12} & \cdots & y_{1b} \\ y_{21} & y_{22} & \cdots & y_{2b} \\ \vdots & \vdots & \vdots & \vdots \\ y_{b1} & y_{b2} & \cdots & y_{bb} \end{bmatrix}}_{\mathbf{Y}_b} \underbrace{\begin{bmatrix} v_1 \\ v_2 \\ \vdots \\ v_b \end{bmatrix}}_{\mathbf{v}} \tag{1.9}$$

or simply

Branch equation

$$\mathbf{i} = \mathbf{Y}_b \mathbf{v} \tag{1.10}$$

where $\mathbf{Y}_b$ is called the *branch admittance matrix*. In general, $\mathbf{Y}_b$ is a $b \times b$ nonsymmetric and nondiagonal real matrix, where b is the number of branches, excluding the independent current sources, in the associated digraph.

For example, $\mathbf{Y}_b$ for the circuits in Figs. 1.1 and 1.2 are given respectively by Eqs. (1.11) and (1.12):

$$\mathbf{Y}_b = \begin{bmatrix} G_1 & & & & & \\ & G_2 & & & \mathbf{0} & \\ & & G_3 & & & \\ & & & G_4 & & \\ & \mathbf{0} & & & G_5 & \\ & & & & & G_6 \end{bmatrix} \tag{1.11}$$

$$\mathbf{Y}_b = \begin{bmatrix} G_1 & 0 & 0 & 0 & 0 & 0 \\ 0 & G_2 & 0 & 0 & 0 & 0 \\ 0 & 0 & G_3 & 0 & 0 & 0 \\ 0 & 0 & 0 & G_4 & 0 & 0 \\ 0 & 0 & 0 & 0 & G_5 & 0 \\ 0 & -g_m & 0 & 0 & 0 & 0 \end{bmatrix} \tag{1.12}$$

Note that $\mathbf{Y}_b$ is a *diagonal* matrix if $\mathcal{N}$ contains *only* two-terminal linear resistors and independent current sources. In particular, if all resistors are strictly *passive* (i.e., $G_j > 0$), then $\mathbf{Y}_b$ is a positive-definite diagonal matrix.

We have deliberately left out the independent current sources because they can be easily accounted for separately. In particular, the contribution of the current sources can be represented by a single vector

$$i_s(t) = [i_{s1}(t), i_{s2}(t), \ldots, i_{s(n-1)}(t)]^T \tag{1.13}$$

where $i_{sk}(t)$ denotes the algebraic sum of currents of *all* independent current sources *entering* node ⓚ, $k = 1, 2, \ldots, n-1$, and n denotes the number of nodes in the *connected* circuit $\mathcal{N}$.

To avoid violating KCL, it is necessary to assume that no cut sets are *made exclusively of independent current sources*.

Equivalently, the *digraph* associated with the *reduced circuit* obtained by open-circuiting all independent current soures is *connected*. Let **A** denote the *reduced incidence matrix* of this connected digraph as defined in Sec. 6.2 of Chap. 1.

It follows from the above assumptions and definition that

$$\mathbf{A}\mathbf{i} = \mathbf{i}_s(t) \tag{1.14}$$

constitutes a system of $n-1$ linearly independent KCL equations. It is important to note that the KCL equation (1.14) differs from the usual form ($\mathbf{A}\mathbf{i} = \mathbf{0}$) because here, the reduced incidence matrix $\mathbf{A}$ pertains to the *reduced* digraph obtained by open-circuiting all independent current sources from the digraph associated with the circuit.

Substituting the branch equation (1.10) in place of $\mathbf{i}$ in Eq. (1.14), we obtain

$$\mathbf{A}\mathbf{Y}_b\mathbf{v} = \mathbf{i}_s(t) \tag{1.15}$$

Now, recall that the *branch voltage vector* $\mathbf{v}$ is related to the *node voltage vector* $\mathbf{e}$ by KVL via the equation [Eq. (7.3) from Chap. 1]:[3]

$$\mathbf{v} = \mathbf{A}^T\mathbf{e} \tag{1.16}$$

Substituting Eq. (1.16) for $\mathbf{v}$ in Eq. (1.15), we obtain the *node equation*

$$(\mathbf{A}\mathbf{Y}_b\mathbf{A}^T)\mathbf{e} = \mathbf{i}_s(t) \tag{1.17}$$

Comparing Eq. (1.4) and Eq. (1.17), we identify the following

Node-admittance matrix

$$\mathbf{Y}_n = \mathbf{A}\mathbf{Y}_b\mathbf{A}^T \tag{1.18}$$

Exercises

1. Derive the node equation (1.3) using Eqs. (1.17) and (1.18).
2. Derive the node equation (1.7) using Eqs. (1.17) and (1.18).
3. Prove that whenever a circuit contains one *cut set* of independent current sources, the digraph associated with the *reduced* circuit obtained by open-circuiting all current sources is *not* connected. In this case, show that Eq. (1.17) is *ill*-defined.

If the circuit contains at least one time-varying resistor, then at least one element y_{jk} of $\mathbf{Y}_b$ in Eq. (1.9) will be a function of time. For example, if branch k is a time-varying resistor described by $i_k(t) = G_k(t)v_k(t)$, then $y_{kk} = G_k(t)$. Hence, the branch equation (1.10) for time-varying circuits should be written as follows:

$$\mathbf{i}(t) = \mathbf{Y}_b(t)\mathbf{v}(t) \tag{1.19}$$

[3] Unlike Eq. (1.14) the form of the KVL equation (1.16) is identical to that written for the original digraph because the *reduced* digraph is *connected* and contains exactly the same nodes as in the original digraph (recall our *standing assumption* that there are no cut sets made of independent current sources).

Since all other steps in the preceding formulation remain unchanged, we have derived the following general result:

Node equation for linear resistive circuits

For any *connected* circuit containing two-terminal, multiterminal, and/or multiport *linear voltage-controlled* (possibly time-varying) resistors, and independent current sources which do not form cut sets, the node equation is given explicitly by

$$\mathbf{Y}_n(t)\mathbf{e}(t) = \mathbf{i}_s(t) \tag{1.20}$$

where

$$\mathbf{Y}_n(t) \overset{\Delta}{=} \mathbf{A}\mathbf{Y}_b(t)\mathbf{A}^T \tag{1.21}$$

denotes the *node-admittance matrix*, $\mathbf{i}_s(t)$ denotes the *equivalent source vector* whose kth element $i_{sk}(t)$ is equal to the algebraic sum of the current of all independent current sources *entering* node $\textcircled{k}$, and $\mathbf{A}$ denotes the reduced incidence matrix of the digraph associated with the *reduced* circuit obtained by deleting all *independent* current sources.

Note that the dimension of $\mathbf{A}$ and $\mathbf{Y}_b(t)$ are $(n-1) \times b$ and $b \times b$ respectively, where n is the number of nodes in the circuit, and b is the number of branches in the digraph associated with the *reduced* circuit. Consequently, the dimension of the node-admittance matrix $\mathbf{Y}_n(t)$ is $(n-1) \times (n-1)$.

In other words, *the node equation (1.20) always contains $n-1$ linear equations in terms of $n-1$ node voltages $e_1, e_2, \ldots, e_{n-1}$.*

To find the solution of the circuit, we simply solve the node equation (1.20) at *each instant of time t* by any convenient method, say Cramer's rule if $n \leq 3$ and gaussian elimination if $n > 3$. If n is very large, say $n > 100$, and if the matrix $\mathbf{Y}_n(t)$ contains only a small percentage of nonzero entries as is typical in practice,[4] there exist specially efficient computer algorithms for solving Eq. (1.19). If the circuit is time-invariant and contains only dc current sources, then $\mathbf{Y}_n(t)$ is a constant matrix and $\mathbf{i}_s(t)$ is a constant vector. In this case, Eq. (1.20) need be solved only once.

Unlike several other methods of analysis (e.g., *tableau analysis*, *modified node analysis*, and *loop analysis*) to be studied later, the number of equations to be solved in *node analysis* does *not* depend on the number of circuit elements. Hence, for a 100-element circuit containing 10 nodes, we only need to solve 9 equations!

Once $\mathbf{e}(t)$ has been found, the branch voltages can be calculated by *substitution* into Eq. (1.16)—$\mathbf{v}(t) = \mathbf{A}^T\mathbf{e}(t)$—and the branch currents can be calculated by *substitution* into the branch equation (1.19)—$\mathbf{i}(t) = \mathbf{Y}_b(t)\mathbf{v}(t)$.

[4] Such a matrix is said to be a *sparse matrix*, and solution methods which exploit the sparsity for increased efficiency are called sparse-matrix techniques.

1.2 Linear Circuits Containing Two-Terminal Resistors and Independent Sources

We will see later that nonlinear resistive or *RLC* circuits can be solved by an iterative numerical method where each iteration involves the analysis of a *linear* time-invariant resistive circuit. If the circuit contains only two-terminal (nonlinear or *RLC*) elements, the associated circuit to be analyzed will contain only linear time-invariant two-terminal resistors and independent current sources. It is important, therefore, that we investigate the properties of the node equations associated with this subclass of resistive circuits, namely, those characterized by a *diagonal* branch-admittance matrix. We will now show that these properties are precisely those listed in Table 1.1.

Let us begin by expanding Eq. (1.18) for a circuit with three nodes and three resistors:

$$\mathbf{Y}_n = \mathbf{A}\mathbf{Y}_b\mathbf{A}^T = \begin{bmatrix} a_{11} & a_{12} & a_{13} \\ a_{21} & a_{22} & a_{23} \end{bmatrix} \begin{bmatrix} G_1 & 0 & 0 \\ 0 & G_2 & 0 \\ 0 & 0 & G_3 \end{bmatrix} \begin{bmatrix} a_{11} & a_{21} \\ a_{12} & a_{22} \\ a_{13} & a_{23} \end{bmatrix}$$

$$= \begin{bmatrix} (a_{11}a_{11}G_1 + a_{12}a_{12}G_2 + a_{13}a_{13}G_3) & (a_{11}a_{21}G_1 + a_{12}a_{22}G_2 + a_{13}a_{23}G_3) \\ (a_{21}a_{11}G_1 + a_{22}a_{12}G_2 + a_{23}a_{13}G_3) & (a_{21}a_{21}G_1 + a_{22}a_{22}G_2 + a_{23}a_{23}G_3) \end{bmatrix} \tag{1.22}$$

If we denote the jkth element of $\mathbf{Y}_n$ by $(\mathbf{Y}_n)_{jk}$ and the kth diagonal element of $\mathbf{Y}_b$ by G_k, then in general, we have

$$(\mathbf{Y}_n)_{jk} = \sum_{\ell=1}^{b} a_{j\ell}a_{k\ell}G_\ell \tag{1.23}$$

provided that $\mathbf{Y}_b$ is a *diagonal* matrix, i.e., provided the circuit contains only *two-terminal* resistors and independent current sources.

If we let $j = k$ in Eq. (1.23), we find the kth diagonal element of $\mathbf{Y}_n$ is given by:

$$(\mathbf{Y}_n)_{kk} = \sum_{\ell=1}^{b} a_{k\ell}^2 G_\ell = \sum_{\substack{\text{over all branches} \\ \text{connected to} \\ \text{node } (k)}} G_\ell \tag{1.24}$$

The last term follows from the observation that $a_{k\ell} = 1, -1$, or 0, and $a_{k\ell} \neq 0$ if and only if branch G_ℓ is connected to node ⓚ. Hence, we have proved that property (*a*) of Table 1.1 holds for any circuit described by a diagonal branch-admittance matrix $\mathbf{Y}_b$.

Observe next that if $a_{j\ell} \neq 0$, i.e., G_ℓ is connected to node ⓙ, then

$$a_{k\ell} = -a_{j\ell}$$

if G_ℓ is connected between nodes ⓙ and ⓚ, and (1.25)

$$a_{k\ell} = 0$$

if G_ℓ is connected between node ⓙ and the datum node. It follows from Eqs. (1.23) and (1.25) that each *off-diagonal* element ($j \neq k$) of $\mathbf{Y}_n$ can be simplified as follows:

$$(\mathbf{Y}_n)_{jk} = -\sum_{\substack{\text{over all branches}\\ \text{connected between}\\ \text{nodes ⓙ and ⓚ}}} G_\ell \tag{1.26}$$

Hence, we have proved property (*b*) of Table 1.1.

Moreover, Eq. (1.26) implies that

$$(\mathbf{Y}_n)_{jk} = (\mathbf{Y}_n)_{kj} \qquad \text{or} \qquad \mathbf{Y}_n = \mathbf{Y}_n^T \tag{1.27}$$

This proves property (*c*) of Table 1.1. Note that this symmetry property has nothing to do with whether the circuit is symmetrical or not. It is actually a consequence of an important circuit-theoretic property called *reciprocity* to be studied in Chap. 13.

The last property (*d*) of Table 1.1 follows by definition and is therefore true regardless of whether $\mathbf{Y}_b$ is diagonal or not.

1.3 Existence and Uniqueness of Solution

When we talked about various methods for solving the *linear* node equation (1.20) in Sec. 1.1, we implicitly assumed that Eq. (1.20) had a *unique* solution for any $\mathbf{i}_s(t)$ at any time t. To show that this assumption is not always satisfied even by simple circuits, consider the circuit shown in Fig. 1.3*a*.

Using the properties from Table 1.1, we obtain the following node equation by inspection (note the resistances are given in ohms):

$$\begin{bmatrix} 1 & -2 \\ -2 & 4 \end{bmatrix} \begin{bmatrix} e_1 \\ e_2 \end{bmatrix} = \begin{bmatrix} i_{s1}(t) \\ i_{s2}(t) \end{bmatrix} \tag{1.28}$$

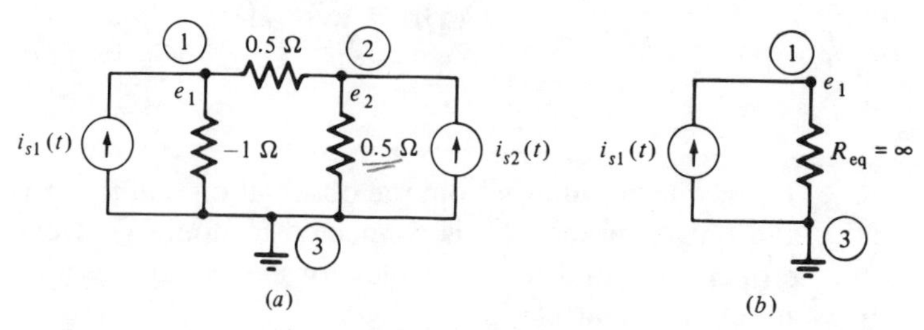

Figure 1.3 A circuit containing a negative resistance.

Since $\det(\mathbf{Y}_n)=0$, Eq. (1.28) either has no solution or has infinitely many solutions. [The latter occurs if and only if, for all t, $i_{s1}(t)=-\frac{1}{2}i_{s2}(t)$.]

To give a circuit interpretation of the above conclusion, let us assume for simplicity that $i_{s2}(t)=0$ for all t so that the current source on the right-hand side can be deleted without affecting the circuit's solution. The resulting circuit can be further simplified to that shown in Fig. 1.3*b*, where the three resistors in Fig. 1.3*a* are replaced by an equivalent resistor with $R_{\text{eq}}=\infty$ in Fig. 1.3*b*. Since the current source $i_{s1}(t)$ flows into an open circuit, it follows that the circuit does *not* have a solution if $i_{s1}(t)\neq 0$. On the other hand, if $i_{s1}(t)=0$ for all t, then the circuit is satisfied by *any* node voltage e_1, and hence it admits an infinite number of solutions.

The following result gives a *sufficient* (but not necessary) condition for a circuit to have a unique solution.

Existence and uniqueness condition Any resistive circuit containing only two-terminal linear *positive* conductances[5] and independent current sources which do not form cut sets has a *unique* solution.

PROOF The above hypotheses guarantee that the node equation given by Eq. (1.20) is well-defined. Moreover, $\mathbf{Y}_b$ is a *positive-definite* diagonal matrix (since for all j, $G_j>0$); i.e., $\mathbf{v}^T\mathbf{Y}_b\mathbf{v}>0$ for all $\mathbf{v}\neq\mathbf{0}$.

Now, for *any* node voltage vector $\mathbf{e}\neq\mathbf{0}$, $\mathbf{e}^T\mathbf{Y}_n\mathbf{e}=\mathbf{e}^T(\mathbf{A}\mathbf{Y}_b\mathbf{A}^T)\mathbf{e}=(\mathbf{A}^T\mathbf{e})^T\mathbf{Y}_b(\mathbf{A}^T\mathbf{e})=\mathbf{v}^T\mathbf{Y}_b\mathbf{v}>0$ for all $\mathbf{v}\neq\mathbf{0}$. Since $\mathbf{v}=\mathbf{A}^T\mathbf{e}=\mathbf{0}$ if and only if $\mathbf{e}=\mathbf{0}$ (recall the rows of $\mathbf{A}$ are linearly independent), it follows that $\mathbf{e}^T\mathbf{Y}_n\mathbf{e}>0$ for all $\mathbf{e}\neq\mathbf{0}$, and hence $\mathbf{Y}_n$ is a positive-definite matrix. Consequently, $\det(\mathbf{Y}_n)>0$, and by Cramer's rule, the node equation (1.20) has a *unique* solution given by

$$\mathbf{e}=\mathbf{Y}_n^{-1}\mathbf{i}_s(t) \qquad \blacksquare$$

1.4 Node Equation Formulation: Nonlinear Resistive Circuits

When the circuit contains one or more *nonlinear* resistors, the procedure for writing the node equation in Sec. 1.1 in terms of the node voltage vector $\mathbf{e}$ still holds provided all nonlinear resistors are *voltage-controlled*. For example, suppose the two linear resistors G_2 and G_5 in Fig. 1.1 are replaced by a *pn*-junction diode described by $i_2=I_s[\exp(v_2/V_T)-1]$ and a nonlinear resistor described by $i_5=v_5^3$, as shown in Fig. 1.4.

Our first step as usual is to express the branch currents of the resistors in terms of the node voltages e_1, e_2, and e_3:

[5] Such resistors are examples of strictly passive resistors to be defined in Sec. 5.3.

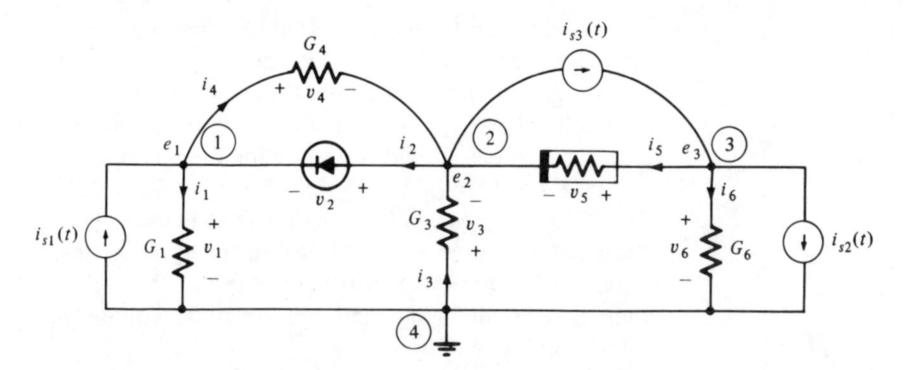

Figure 1.4 A nonlinear circuit.

$$i_1 = G_1 v_1 = G_1 e_1 \qquad i_4 = G_4 v_4 = G_4(e_1 - e_2)$$

$$i_2 = I_s\left(\exp\frac{v_2}{V_T} - 1\right) = I_s\left(\exp\frac{e_2 - e_1}{V_T} - 1\right) \qquad i_5 = v_5^3 = (e_3 - e_2)^3$$

$$i_3 = G_3 v_3 = G_3(-e_2) \qquad i_6 = G_6 v_6 = G_6(e_3) \tag{1.29}$$

Note that this step is always possible so long as the nonlinear resistors are voltage-controlled, i.e., the branch currents are functions of branch voltages.

Our next step is to apply KCL at each node, except the datum node:

Node ①: $$G_1 e_1 - I_s\left(\exp\frac{e_2 - e_1}{V_T} - 1\right) + G_4(e_1 - e_2) = i_{s1}(t)$$

Node ②: $$I_s\left(\exp\frac{e_2 - e_1}{V_T} - 1\right) - G_3(-e_2) - G_4(e_1 - e_2) - (e_3 - e_2)^3 = -i_{s3}(t)$$

Node ③: $$(e_3 - e_2)^3 + G_6(e_3) = i_{s3}(t) - i_{s2}(t) \tag{1.30}$$

We call Eq. (1.30) the *node equation* of the circuit in Fig. 1.4 because the only variables are node voltages as in Sec. 1.1. However, since these equations are nonlinear, they cannot be described by a node-admittance matrix.

Consider now the general case where the circuit may contain two-terminal, multiterminal, and/or multiport *nonlinear voltage-controlled* resistors, in addition to independent current sources. In this case, instead of Eq. (1.9), the branch equations now assume the following form:[6]

$$\begin{aligned} i_1 &= g_1(v_1, v_2, \ldots, v_b) \\ i_2 &= g_2(v_1, v_2, \ldots, v_b) \\ &\vdots \\ i_b &= g_b(v_1, v_2, \ldots, v_b) \end{aligned} \tag{1.31}$$

[6] Again, to simplify notation, we assume all resistors are time-invariant.

In vector notation, Eq. (1.31) becomes simply

Nonlinear branch equation

$$\mathbf{i} = \mathbf{g}(\mathbf{v}) \tag{1.32}$$

Since the independent current sources do not form cut sets (by assumption), Eq. (1.14) remains valid. Substituting Eq. (1.32) for **i** in Eq. (1.14), we obtain

$$\mathbf{A}\mathbf{g}(\mathbf{v}) = \mathbf{i}_s(t) \tag{1.33}$$

Substituting next Eq. (1.16) for **v** in Eq. (1.33), we obtain:

Nonlinear node equation

$$\mathbf{A}\mathbf{g}(\mathbf{A}^T\mathbf{e}) = \mathbf{i}_s(t) \tag{1.34}$$

For each solution **e** of Eq. (1.34), we can calculate the corresponding *branch voltage vector* **v** by direct substitution into Eq. (1.16), namely, $\mathbf{v} = \mathbf{A}^T\mathbf{e}$. This in turn can be used to calculate the *branch current vector* **i** by direct substitution into Eq. (1.32), namely, $\mathbf{i} = \mathbf{g}(\mathbf{v})$. Hence, the basic problem is to solve the nonlinear node equation (1.34). The rest is trivial.

In general, *nonlinear* equations do not have closed form solutions. Consequently, they must be solved by numerical techniques. In Sec. 3, we will study the most widely used method called the *Newton-Raphson algorithm*. Before we do this, however, let us first study how to formulate circuit equations in the most general case.

2 TABLEAU ANALYSIS FOR RESISTIVE CIRCUITS

The only, albeit major, shortcoming of *node analysis* is that it disallows many standard circuit elements from the class of allowable circuits, e.g., the voltage source, ideal transformer, ideal op amp, CCCS, CCVS, VCVS, current-controlled nonlinear resistor (such as neon bulb), etc. In this section, we overcome this objection by presenting a *completely general* analysis method—one that works for *all* resistive circuits. Conceptually, this method is simpler than node analysis: It consists of writing out the complete list of linearly independent KCL equations, linearly independent KVL equations, and the branch equations. For obvious reasons, this list of equations is called *tableau equations*.

Since no variables are eliminated (recall both **v** and **i** must be eliminated in node analysis, leaving **e** as the only variable) in listing the tableau equations, all three vectors **e**, **v**, and **i** are present as variables. Since we must have as many tableau equations as there are variables, it is clear that the price we pay for the increased generality is that the *tableau analysis* involves many more equations than *node analysis* does. In our era of computer-aided circuit analysis, however, this objection turns out to be a blessing in disguise because the matrix associated with tableau analysis is often extremely sparse, thereby allowing highly efficient numerical algorithms to be brought to bear.

The significance of tableau analysis actually transcends the above more mundane numerical considerations. As we will see over and over again in this book, tableau analysis is a powerful *analytic tool* which allows us to derive many profound results with almost no pain at all—at least compared to other approaches.

2.1 Tableau Equation Formulation: Linear Resistive Circuits

To write the tableau equation for any *linear* resistive circuit, we simply use the following algorithm:[7]

Step 1. Draw the digraph of the circuit and hinge it if necessary so that the resulting digraph is connected. Pick an arbitrary datum node and formulate the reduced-incidence matrix **A**.

Step 2. Write a complete set of linearly independent KCL equations:[8]

$$\mathbf{A}\mathbf{i}(t) = \mathbf{0} \tag{2.1}$$

Step 3. Write a complete set of linearly independent KVL equations:

$$\mathbf{v}(t) - \mathbf{A}^T\mathbf{e}(t) = \mathbf{0} \tag{2.2}$$

Step 4. Write the branch equations. Since the circuit is *linear*, these equations can always be recast into the form

$$\mathbf{M}(t)\mathbf{v}(t) + \mathbf{N}(t)\mathbf{i}(t) = \mathbf{u}_s(t) \tag{2.3}$$

Together, Eqs. (2.1), (2.2), and (2.3) constitute the tableau equations. If the digraph has n nodes and b branches, Eqs. (2.1), (2.2), and (2.3) will contain $n-1$, b, and b equations, respectively. Since the vectors **e**, **v**, and **i** also contain $n-1$, b, and b variables, respectively, the tableau equation for a linear resistive circuit always consists of $(n-1)+2b$ linear equations in $(n-1)+2b$ variables.

Example Consider the linear resistive circuit shown in Fig. 2.1*a*. It contains only three elements: a voltage source, an ideal transformer described by $v_1 = (n_1/n_2)v_2$ and $i_2 = -(n_1/n_2)i_1$, and a time-varying linear resistor described by $v_3(t) = R(t)i_3(t)$. Note that the first two elements are not allowed in node analysis because they are not voltage-controlled. The third

[7] The reader may wish to scan the following illustrative example after *each* step in order to get familiarized first with the notations used in writing the tableau equation.

[8] Note that unlike Eq. (1.14), tableau analysis deals with the *original* digraph where each independent current source is represented by a branch.

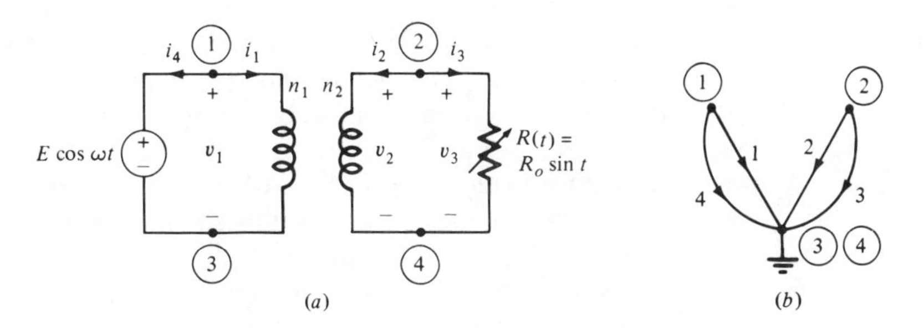

Figure 2.1 All three elements in this circuit are disallowed in node analysis.

element, which would normally be acceptable, is also disallowed here because its conductance $G(t) = 1/(R_o \sin t) = \infty$, at $t = 0, 2\pi, 4\pi, \ldots$, and is therefore not defined for all t.

Applying the preceding recipe, we hinge nodes ③ and ④ and draw the *connected* digraph shown in Fig. 2.1*b*. Choosing the hinged node as datum, the tableau equation is formulated as follows:

$$\text{KCL:} \qquad \mathbf{AI} = \mathbf{0} \Leftrightarrow \underbrace{\begin{bmatrix} 1 & 0 & 0 & 1 \\ 0 & 1 & 1 & 0 \end{bmatrix}}_{\mathbf{A}} \underbrace{\begin{bmatrix} i_1 \\ i_2 \\ i_3 \\ i_4 \end{bmatrix}}_{\mathbf{i}} = \underbrace{\begin{bmatrix} 0 \\ 0 \end{bmatrix}}_{\mathbf{0}} \tag{2.1$'$}$$

$$\text{KVL:} \qquad \mathbf{v} - \mathbf{A}^T\mathbf{e} = \mathbf{0} \Leftrightarrow \underbrace{\begin{bmatrix} v_1 \\ v_2 \\ v_3 \\ v_4 \end{bmatrix}}_{\mathbf{v}} - \underbrace{\begin{bmatrix} 1 & 0 \\ 0 & 1 \\ 0 & 1 \\ 1 & 0 \end{bmatrix}}_{\mathbf{A}^T} \underbrace{\begin{bmatrix} e_1 \\ e_2 \end{bmatrix}}_{\mathbf{e}} = \underbrace{\begin{bmatrix} 0 \\ 0 \\ 0 \\ 0 \end{bmatrix}}_{\mathbf{0}} \tag{2.2$'$}$$

Branch equations:

$$\left.\begin{aligned} n_2 v_1 - n_1 v_2 &= 0 \\ n_1 i_1 + n_2 i_2 &= 0 \\ v_3 - R(t) i_3 &= 0 \\ v_4 &= E \cos \omega t \end{aligned}\right\} \Leftrightarrow \underbrace{\begin{bmatrix} n_2 & -n_1 & 0 & 0 \\ 0 & 0 & 0 & 0 \\ 0 & 0 & 1 & 0 \\ 0 & 0 & 0 & 1 \end{bmatrix}}_{\mathbf{M}(t)} \underbrace{\begin{bmatrix} v_1 \\ v_2 \\ v_3 \\ v_4 \end{bmatrix}}_{\mathbf{v}}$$

$$+ \underbrace{\begin{bmatrix} 0 & 0 & 0 & 0 \\ n_1 & n_2 & 0 & 0 \\ 0 & 0 & -R(t) & 0 \\ 0 & 0 & 0 & 0 \end{bmatrix}}_{\mathbf{N}(t)} \underbrace{\begin{bmatrix} i_1 \\ i_2 \\ i_3 \\ i_4 \end{bmatrix}}_{\mathbf{i}} = \underbrace{\begin{bmatrix} 0 \\ 0 \\ 0 \\ E \cos \omega t \end{bmatrix}}_{\mathbf{u}_s(t)} \tag{2.3$'$}$$

Note that $n=3$ and $b=4$ for the digraph in Fig. 2.1b. Consequently, we expect the tableau equation to contain $(n-1)+2b=10$ equations involving 10 variables, namely, $e_1, e_2, v_1, v_2, v_3, v_4, i_1, i_2, i_3$, and i_4. An inspection of Eqs. (2.1)′, (2.2)′, and (2.3)′ shows that indeed we have 10 equations involving only these 10 variables. Note that had it been possible to write the node equation for this circuit, we would have to solve only two equations in two variables.

Note also that the branch equation (2.3)′ is of the form stipulated in Eq. (2.3), where $\mathbf{M}(t)=\mathbf{M}$ is a *constant* matrix. The second matrix $\mathbf{N}(t)$ is a function of t in view of the entry $-R(t)$. Clearly, if the circuit is time-invariant, then both $\mathbf{M}(t)$ and $\mathbf{N}(t)$ in Eq. (2.3) will be constant matrices.

The vector $\mathbf{u}_s(t)$ on the right-hand side of Eq. (2.3) does *not* depend on any *variable* e_j, v_j, or i_j, and is therefore due only to *independent* voltage and current sources in the circuit. Consequently, element k of $\mathbf{u}_s(t)$ will be *zero* whenever branch k is *not* an *independent* source. Note that *controlled* source coefficients always appear in the matrices $\mathbf{M}(t)$ and/or $\mathbf{N}(t)$, never in $\mathbf{u}_s(t)$.

An inspection of Eq. (2.3)′ reveals that each *row* k of $\mathbf{M}(t)$ and $\mathbf{N}(t)$ contains coefficients or time functions which define uniquely the *linear relation between* v_k and i_k of *branch* k in the digraph, assuming branch k corresponds to a resistor. If branch k happens to be an *independent* source, then the kth diagonal element is equal to *one* in $\mathbf{M}(t)$ (for a *voltage* source) or $\mathbf{N}(t)$ (for a current source), while *all other* elements in row k are zeros. In this case, the kth element of $\mathbf{u}_s(t)$ will contain either a constant (for a dc source) or a time function which specifies uniquely this independent source. On the other hand, if branch k is *not* an *independent* source, then the kth element of $\mathbf{u}_s(t)$ is always zero. It follows from the above interpretation that both $\mathbf{M}(t)$ and $\mathbf{N}(t)$ are $b\times b$ matrices and $\mathbf{u}_s(t)$ is a $b\times 1$ vector, where b is the number of branches in the digraph.

Finally, note that we can state that a *resistive circuit* is *linear* iff its branch equations can be written in the form stipulated in Eq. (2.3), and that it is *time-invariant* iff both $\mathbf{M}(t)$ and $\mathbf{N}(t)$ are constant real matrices.

The tableau matrix Since Eqs. (2.1)′, (2.2)′, and (2.3)′ which constitute the tableau equation consist of a system of *linear* equations, it is convenient and more illuminating to rewrite them as a single matrix equation; Eq. (2.4)′, page 229.

In the general case, Eqs. (2.1), (2.2), and (2.3) can be recast into the following compact matrix form, where **0** and **1** denote a *zero* and a *unit* matrix of appropriate dimension, respectively as shown in Eq. (2.4), page 229.

It is natural to call $\mathbf{T}(t)$ the *tableau matrix* associated with the linear resistive circuit. If the circuit is time-invariant, the tableau matrix $\mathbf{T}(t)=\mathbf{T}$ is a *constant* real matrix.

Every linear resistive circuit is associated with a *unique* $[(n-1)+2b]\times$

$$\underbrace{\begin{bmatrix} 0 & 0 & 0 & 0 & 0 & 0 & 1 & 0 & 0 & 1 \\ 0 & 0 & 0 & 0 & 0 & 0 & 0 & 1 & 1 & 0 \\ \hline -1 & 0 & 1 & 0 & 0 & 0 & 0 & 0 & 0 & 0 \\ 0 & -1 & 0 & 1 & 0 & 0 & 0 & 0 & 0 & 0 \\ 0 & -1 & 0 & 0 & 1 & 0 & 0 & 0 & 0 & 0 \\ -1 & 0 & 0 & 0 & 0 & 1 & 0 & 0 & 0 & 0 \\ \hline 0 & 0 & n_2 & -n_1 & 0 & 0 & 0 & 0 & 0 & 0 \\ 0 & 0 & 0 & 0 & 0 & 0 & n_1 & n_2 & 0 & 0 \\ 0 & 0 & 0 & 0 & 1 & 0 & 0 & 0 & -R(t) & 0 \\ 0 & 0 & 0 & 0 & 0 & 1 & 0 & 0 & 0 & 0 \end{bmatrix}}_{\mathbf{T}(t)} \underbrace{\begin{bmatrix} e_1 \\ e_2 \\ \hline v_1 \\ v_2 \\ v_3 \\ v_4 \\ \hline i_1 \\ i_2 \\ i_3 \\ i_4 \end{bmatrix}}_{\mathbf{w}(t)} = \underbrace{\begin{bmatrix} 0 \\ 0 \\ \hline 0 \\ 0 \\ 0 \\ 0 \\ \hline 0 \\ 0 \\ 0 \\ E\cos\omega t \end{bmatrix}}_{\mathbf{u}(t)} \begin{matrix} \\ \\ \\ \\ \\ \\ \left.\begin{matrix} \\ \\ \\ \\ \end{matrix}\right\} \mathbf{u}_s(t) \end{matrix} \tag{2.4$'$}$$

Linear tableau equation

$$\underbrace{\begin{bmatrix} \mathbf{0} & \mathbf{0} & \mathbf{A} \\ -\mathbf{A}^T & \mathbf{1} & \mathbf{0} \\ \mathbf{0} & \mathbf{M}(t) & \mathbf{N}(t) \end{bmatrix}}_{\mathbf{T}(t)} \underbrace{\begin{bmatrix} \mathbf{e}(t) \\ \mathbf{v}(t) \\ \mathbf{i}(t) \end{bmatrix}}_{\mathbf{w}(t)} = \underbrace{\begin{bmatrix} \mathbf{0} \\ \mathbf{0} \\ \mathbf{u}_s(t) \end{bmatrix}}_{\mathbf{u}(t)} \tag{2.4}$$

$[(n-1)+2b]$ square tableau matrix $\mathbf{T}(t)$, and a *unique* $[(n-1)+2b]\times 1$ vector $\mathbf{u}(t)$.[9] The following result testifies to the significance of the tableau matrix:

Existence and uniqueness theorem: Linear resistive circuits A linear resistive circuit has a *unique* solution at any time t_0 if and only if

$$\det[\mathbf{T}(t_0)] \neq 0 \tag{2.5}$$

PROOF The *inverse* matrix $\mathbf{T}^{-1}(t_0)$ exists at time t_0 if and only if $\det[\mathbf{T}(t_0)] \neq 0$. Hence, the solution at $t = t_0$ is given *uniquely* by

$$\mathbf{w}(t_0) = \mathbf{T}^{-1}(t_0)\mathbf{u}(t_0) \tag{2.6}$$

■

REMARK The linear tableau equation (2.4) also holds for circuits containing $(n+1)$-terminal or n-port resistors described by (see *Exercise 6*):

$$\mathbf{Av} + \mathbf{Bi} + \mathbf{c} = \mathbf{0}$$

where $\mathbf{A}$ and $\mathbf{B}$ are $n \times n$ matrices and $\mathbf{c}$ is an n-vector. Such an element is called an *affine resistor*.[10]

Note that when $\mathbf{c} = \mathbf{0}$, an *affine* resistor reduces to a *linear* resistor. Hence, a two-terminal affine (but *not* linear) resistor is characterized by a *straight line* in the v-i plane which does *not* include the origin.

Affine resistors arise naturally in the analysis of *nonlinear* resistive circuits in Sec. 3.

[9] The "uniqueness" is of course relative to a particular choice of element and node numbers.

[10] In mathematics, $\mathbf{f}(\mathbf{v}, \mathbf{i}) \triangleq \mathbf{Av} + \mathbf{Bi} + \mathbf{c}$ is called an *affine function* if $\mathbf{c} \neq \mathbf{0}$, and a *linear function* if $\mathbf{c} = \mathbf{0}$.

Exercise 1 Write the tableau equation for the circuit in Fig. 1.2, and identify the tableau matrix $\mathbf{T}(t)$.

Exercise 2 Specify the dimension of each submatrix in the tableau matrix $\mathbf{T}(t)$ in Eq. (2.4) for a connected digraph with n nodes and b branches.

Exercise 3 Write the tableau equation for the subclass of linear resistive circuits studied in Sec. 1.1; i.e., specify $\mathbf{M}(t)$, $\mathbf{N}(t)$, and $\mathbf{u}_s(t)$ in terms of $\mathbf{Y}_b(t)$ and $\mathbf{i}_s(t)$ in Eq. (1.17).

Exercise 4 (*a*) Show that the circuit in Fig. 2.1*a* does *not* have a unique solution at $t_0 = 0, 2\pi, 4\pi, \ldots, m2\pi$ for any integer m, by showing $\det[\mathbf{T}(t_0)] = 0$ for the tableau matrix in Eq. (2.4)′.

(*b*) Give a circuit interpretation which explains why the circuit in Fig. 2.1*a* does *not* have a unique solution at the above time instants. Do this for the two cases $E \neq 0$ and $E = 0$, respectively.

(*c*) Show that the circuit has a unique solution if $R(t) \neq 0$ *by inspection*. *Hint*: Start with the determinant expansion on the last column.

Exercise 5 (*a*) Replace the ideal transformer in Fig. 2.1*a* by a two-port resistor described by

$$\begin{bmatrix} i_1 \\ v_2 \end{bmatrix} = \begin{bmatrix} 2 & -3 \\ -5 & 1 \end{bmatrix} \begin{bmatrix} v_1 \\ i_2 \end{bmatrix} + \begin{bmatrix} 4 \\ 6 \end{bmatrix}$$

Write the tableau equation in the form of Eq. (2.4). Explain why the first two elements of $\mathbf{u}_s(t)$ are no longer zero in this case.

(*b*) Show that the above two-port resistor can be realized by a *linear* two-port resistor with port 1 in parallel with a 4-A current source and port 2 in series with a 6-V voltage source.

Exercise 6 Show that the tableau equation (2.4) remains unchanged if the circuit considered in this section also includes *affine* resistors. Explain why an entry k of the vector $\mathbf{u}_s(t)$ need *not* be zero in this case even if the corresponding branch k is not an independent source.

2.2 Tableau Equation Formulation: Nonlinear Resistive Circuits

Exactly the same principle is used to formulate the tableau equation for *nonlinear* resistive circuits: Simply list the linearly independent KCL and KVL equations, and the branch equations, which are now nonlinear. Hence, the first three steps of the algorithm at the beginning of Sec. 2.1 remain unchanged. Only Step 4 needs to be modified because Eq. (2.3) is valid only for *linear* resistive circuits. The following example will illustrate and suggest the modified form of Eq. (2.3).

Example Consider the circuit shown in Fig. 2.2, where the *npn* transistor is modeled by the following nonlinear *Ebers-Moll equation* [see Eqs. (4.1) and (4.2) of Chap. 3]:

$$i_1 = -I_{\text{ES}}\left(\exp\frac{-v_1}{V_T} - 1\right) + \alpha_R I_{\text{CS}}\left(\exp\frac{-v_2}{V_T} - 1\right) \tag{2.7a}$$

$$i_2 = \alpha_F I_{\text{ES}}\left(\exp\frac{-v_1}{V_T} - 1\right) - I_{\text{CS}}\left(\exp\frac{-v_2}{V_T} - 1\right) \tag{2.7b}$$

Since the digraph for this circuit is identical to that shown in Fig. 2.1*b*, the same KCL equation (2.1)′ and KVL equation (2.2)′ also apply for this circuit. However, instead of Eq. (2.3)′, we have the following:

Branch equations:

$$h_1(v_1, v_2, i_1) \triangleq i_1 + I_{\text{ES}}\left(\exp\frac{-v_1}{V_T} - 1\right) - \alpha_R I_{\text{CS}}\left(\exp\frac{-v_2}{V_T} - 1\right) = 0 \tag{2.8a)'}$$

$$h_2(v_1, v_2, i_2) \triangleq i_2 - \alpha_F I_{\text{ES}}\left(\exp\frac{-v_1}{V_T} - 1\right) + I_{\text{CS}}\left(\exp\frac{-v_2}{V_T} - 1\right) = 0 \tag{2.8b)'}$$

$$h_3(v_3, i_3, t) \triangleq v_3 - R(t)i_3 = 0$$

$$h_4(v_4, t) \triangleq v_4 - E\cos\omega t = 0 \tag{2.8d)'}$$

Note that $h_1(\cdot, \cdot, \cdot)$ and $h_2(\cdot, \cdot, \cdot)$ are *nonlinear* functions of (v_1, v_2, i_1) and (v_1, v_2, i_2) respectively; $h_3(\cdot, \cdot, \cdot)$ is a linear function of v_3 and i_3 but a nonlinear function of t; and $h_4(\cdot, \cdot)$ is a function only of v_4 and t. Even for this simple circuit, we see that there is really no simple standard form analogous to that of Eq. (2.3). To avoid keeping track of which variables are present in each function, we will simply denote Eq. (2.8)′ as follows:

$$\begin{aligned} h_1(v_1, v_2, i_1, i_2, t) &= 0 \\ h_2(v_1, v_2, i_1, i_2, t) &= 0 \\ h_3(v_1, v_2, i_1, i_2, t) &= 0 \\ h_4(v_1, v_2, i_1, i_2, t) &= 0 \end{aligned} \tag{2.8}$$

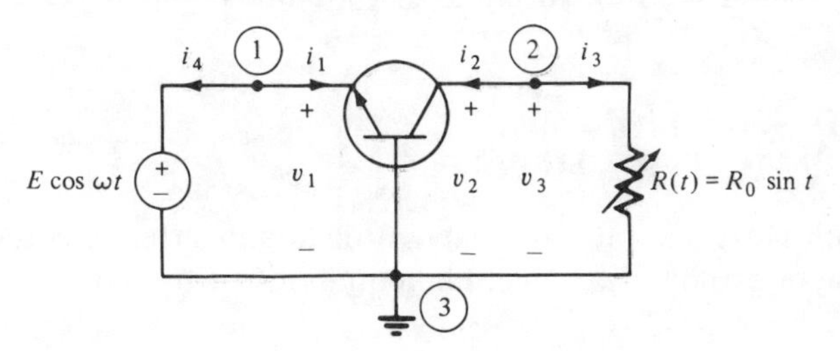

Figure 2.2 A nonlinear resistive circuit.

or, in vector form, we simply write

$$\mathbf{h}(\mathbf{v}, \mathbf{i}, t) = \mathbf{0} \tag{2.9}$$

For a circuit with b branches, Eq. (2.9) can be written as

$$h_j(v_1, v_2, \ldots, v_b, i_1, i_2, \ldots, i_b, t) = 0 \qquad j = 1, 2, \ldots, b \tag{2.9$'$}$$

It is understood that some variables may not be present in each component equation.

It follows from the above discussion that *every* nonlinear resistive circuit is described by a unique system of $(n-1)+2b$ nonlinear algebraic equations in $(n-1)+2b$ variables, namely,

Nonlinear tableau equation

$$\begin{aligned} \mathbf{A}\mathbf{i}(t) &= \mathbf{0} \\ \mathbf{v}(t) - \mathbf{A}^T\mathbf{e}(t) &= \mathbf{0} \\ \mathbf{h}(\mathbf{v}(t), \mathbf{i}(t), t) &= \mathbf{0} \end{aligned} \tag{2.10}$$

Observe that the branch equation (2.3) in the *linear* tableau equation is a special case of $\mathbf{h}(\mathbf{v}(t), \mathbf{i}(t), t) = \mathbf{0}$ in the nonlinear tableau equation (2.10), where

$$\mathbf{h}(\mathbf{v}(t), \mathbf{i}(t), t) \stackrel{\Delta}{=} \mathbf{M}(t)\mathbf{v}(t) + \mathbf{N}(t)\mathbf{i}(t) - \mathbf{u}_s(t) \tag{2.11}$$

Unlike Eq. (2.3), where the source vector $\mathbf{u}_s(t)$ is sorted out and displayed prominently, on the right-hand side of the equation, there is no advantage to doing this in the nonlinear case. Consequently, we simply lump the terms due to the independent sources with the rest of the terms in $\mathbf{h}(\cdot)$ so that the right-hand side of the three component equations in Eq. (2.10) is always zero.

In general, *nonlinear* algebraic equations do not admit of a closed form solution. We will present an effective and general numerical method—the Newton-Raphson algorithm—in the next section for solving nonlinear tableau equations. From the numerical solution point of view, it is unnecessary, and sometimes confusing, to distinguish the three different component equations in Eq. (2.10). Consequently, we will often write the nonlinear tableau equation (2.10) in the *standard form*

$$\mathbf{f}(\mathbf{v}(t), \mathbf{i}(t), \mathbf{e}(t)) = \mathbf{0} \tag{2.12}$$

normally called for in numerical methods.

3 COMPUTER-AIDED SOLUTION OF NONLINEAR ALGEBRAIC EQUATIONS

Both node and tableau analysis for nonlinear circuits require the solution of a system of nonlinear algebraic equations of the form

$$\mathbf{f}(\mathbf{x}(t), t) = \mathbf{0} \tag{3.1}$$

Since the solution of Eq. (3.1) at different time instants are independent of each other, we will henceforth suppress the time variable t and focus our attention on the following system of p nonlinear equations:

$$\left.\begin{aligned} f_1(x_1, x_2, \ldots, x_p) &= 0 \\ f_2(x_1, x_2, \ldots, x_p) &= 0 \\ &\vdots \\ f_p(x_1, x_2, \ldots, x_p) &= 0 \end{aligned}\right\} \quad \text{or} \quad \mathbf{f}(\mathbf{x}) = \mathbf{0} \tag{3.2}$$

In general, Eq. (3.2) can have no solutions, one solution, finitely many solutions, or infinitely many solutions. Nonlinear resistive circuits encountered in practice usually have either a *unique* solution or have *finitely* many "multiple" solutions. Those that have either no solution or infinitely many solutions occur as a result of poor modeling and are therefore pathological.

For series-parallel circuits, or those requiring the solution of no more than two nonlinear equations, the graphic methods presented in Chap. 2 are ideal and often more effective in the case where the circuit has multiple solutions.

For the great majority of the remaining nonlinear circuits, however, the most general approach presently available is to find the solution(s) by *numerical iteration* using a *digital computer*.[11] Many numerical solution algorithms are presently available and most computer centers have at least one subroutine in their program library for solving Eq. (3.2). Most of these subroutines are based on the Newton-Raphson algorithm or its various improved modified versions.

Since the purpose of this book is to teach circuit analysis, rather than numerical analysis, we will present only the basic Newton-Raphson algorithm and its important circuit interpretations. No attempt will be made to study the convergence properties of this algorithm.

3.1 The Newton-Raphson Algorithm

The Newton-Raphson algorithm for solving Eq. (3.2) for $p = 1$ (i.e., one equation in one variable) has already been introduced in Sec. 2.4 of Chap. 2. Here we will give an *intuitive* derivation of the Newton-Raphson formula for higher dimensions, namely, for any $p \geq 2$. Let us assume that the function $\mathbf{f}(\mathbf{x})$ in Eq. (3.2) has a *Taylor expansion* about any point $\mathbf{x}^{(j)} \triangleq [x_1^{(j)}, \ldots, x_p^{(j)}]^T$; namely,

[11] If all nonlinear resistors are modeled by *piecewise-linear* functions, a much more efficient and general method is given in L. O. Chua and A.-C. Deng, "Canonical Piecewise-Linear Analysis," *IEEE Trans. on Circuits and Systems*, vol. CAS-32, pp. 417–444, May 1985.

$$\underbrace{\begin{bmatrix} f_1(\mathbf{x}) \\ f_2(\mathbf{x}) \\ \vdots \\ f_p(\mathbf{x}) \end{bmatrix}}_{\mathbf{f}(\mathbf{x})} = \underbrace{\begin{bmatrix} f_1(\mathbf{x}^{(j)}) \\ f_2(\mathbf{x}^{(j)}) \\ \vdots \\ f_p(\mathbf{x}^{(j)}) \end{bmatrix}}_{\mathbf{f}(\mathbf{x}^{(j)})} + \underbrace{\begin{bmatrix} \frac{\partial f_1(\mathbf{x})}{\partial x_1} & \frac{\partial f_1(\mathbf{x})}{\partial x_2} & \cdots & \frac{\partial f_1(\mathbf{x})}{\partial x_p} \\ \frac{\partial f_2(\mathbf{x})}{\partial x_1} & \frac{\partial f_2(\mathbf{x})}{\partial x_2} & \cdots & \frac{\partial f_2(\mathbf{x})}{\partial x_p} \\ \vdots & \vdots & \vdots & \vdots \\ \frac{\partial f_p(\mathbf{x})}{\partial x_1} & \frac{\partial f_p(\mathbf{x})}{\partial x_2} & \cdots & \frac{\partial f_p(\mathbf{x})}{\partial x_p} \end{bmatrix}_{\mathbf{x}=\mathbf{x}^{(j)}}}_{\mathbf{J}(\mathbf{x}^{(j)})} \underbrace{\begin{bmatrix} x_1 - x_1^{(j)} \\ x_2 - x_2^{(j)} \\ \vdots \\ x_p - x_p^{(j)} \end{bmatrix}}_{\mathbf{x}-\mathbf{x}^{(j)}} + \text{h.o.t.} \tag{3.3}$$

or in matrix form,

$$\mathbf{f}(\mathbf{x}) = \mathbf{f}(\mathbf{x}^{(j)}) + \mathbf{J}(\mathbf{x}^{(j)})(\mathbf{x} - \mathbf{x}^{(j)}) + \text{h.o.t.} \tag{3.4}$$

where h.o.t. denotes higher-order terms involving $(x_1 - x_1^{(j)})^k$, $(x_2 - x_2^{(j)})^k, \ldots, (x_p - x_p^{(j)})^k$, $k = 2, 3, \ldots$, etc. The $p \times p$ square matrix $\mathbf{J}(\mathbf{x}^{(j)})$ of first partial derivatives is called the *jacobian matrix* of $\mathbf{f}(\mathbf{x})$ at $\mathbf{x}^{(j)}$. Note that this matrix is evaluated at $\mathbf{x} = \mathbf{x}^{(j)}$ so that $\mathbf{J}(\mathbf{x}^{(j)})$ is a *constant* real matrix.

Now let $\mathbf{x} = \mathbf{x}^{(j+1)}$ be the value calculated by the Newton-Raphson formula (yet to be derived) at the next iteration (i.e., iteration $j+1$ assuming the preceding iteration is j) and *assume* that $\mathbf{x}^{(j+1)}$ is sufficiently close to $\mathbf{x}^{(j)}$ (calculated in the preceding iteration) so that the h.o.t. in Eq. (3.4) can be neglected. Hence,

$$\mathbf{f}(\mathbf{x}^{(j+1)}) \approx \mathbf{f}(\mathbf{x}^{(j)}) + \mathbf{J}(\mathbf{x}^{(j)})(\mathbf{x}^{(j+1)} - \mathbf{x}^{(j)}) \tag{3.5}$$

Moreover, assume that $\mathbf{x}^{(j+1)}$ is also sufficiently close to *a* solution of Eq. (3.2), say $\mathbf{x} = \mathbf{x}^*$, so that $\mathbf{f}(\mathbf{x}^{j+1}) \approx \mathbf{f}(\mathbf{x}^*) = \mathbf{0}$. Hence, Eq. (3.4) becomes

$$\mathbf{f}(\mathbf{x}^{(j)}) + \mathbf{J}(\mathbf{x}^{(j)})(\mathbf{x}^{(j+1)} - \mathbf{x}^{(j)}) \approx \mathbf{0} \tag{3.6}$$

Now, the Newton-Raphson formula can be derived by replacing the approximation sign in Eq. (3.6) by an equality sign and solving for $\mathbf{x}^{(j+1)}$ in terms of $\mathbf{x}^{(j)}$, namely,[12]

Newton-Raphson formula

$$\mathbf{x}^{(j+1)} = \mathbf{x}^{(j)} - [\mathbf{J}(\mathbf{x}^{(j)})]^{-1}\mathbf{f}(\mathbf{x}^{(j)}) \tag{3.7}$$

This formula is the basis of the following iterative scheme for solving a system of nonlinear algebraic equations:

[12] Rather than inverting the jacobian matrix in Eq. (3.7), it is more efficient numerically to solve the following equivalent system of linear equations for $\mathbf{x}^{(j+1)}$:

$$\mathbf{J}(\mathbf{x}^{(j)})\mathbf{x}^{(j+1)} = \mathbf{J}(\mathbf{x}^{(j)})\mathbf{x}^{(j)} - \mathbf{f}(\mathbf{x}^{(j)}) \tag{3.7$'$}$$

Newton-Raphson Algorithm Choose a small positive number ε so that the algorithm is *terminated* after $N+1$ iterations whenever

$$\|\mathbf{x}^{(N+1)} - \mathbf{x}^{(N)}\| \triangleq \sqrt{\sum_{i=1}^{p} (x_i^{(N+1)} - x_i^{(N)})^2} < \varepsilon \tag{3.8}$$

In this case, $\mathbf{x} = \mathbf{x}^{(N+1)}$ is taken to be a solution of Eq. (3.2). Clearly, ε depends on the desired accuracy as well as on the computer word length.

Step 0. Choose a maximum number of steps m; choose tolerance $\varepsilon > 0$. Pick initial guess $\mathbf{x}^{(0)}$; set $j = 0$.
Step 1. Solve the linear equation (3.7′) in the footnote for $\mathbf{x}^{(j+1)}$.
Step 2. If $\|\mathbf{x}^{(j+1)} - \mathbf{x}^{(j)}\| < \varepsilon$, stop;
else, replace j by $j+1$;
If $j + 1 \geq m$, stop;
else, go to *step 1.*

It can be shown that the above algorithm will terminate in *one* step if $\mathbf{f}(\mathbf{x})$ is *affine* regardless of the initial guess $\mathbf{x}^{(0)}$, and in a *finite* number of steps in general, provided the initial guess $\mathbf{x}^{(0)}$ is sufficiently close to a solution $\mathbf{x}^*$ of $\mathbf{f}(\mathbf{x}) = \mathbf{0}$. If the initial guess $\mathbf{x}^{(0)}$ is far off from $\mathbf{x}^*$, or if the function $\mathbf{f}(\mathbf{x})$ is not smooth, the algorithm may *not* converge in the sense that Eq. (3.8) is never satisfied.

To be practical, the Newton-Raphson algorithm should converge in a few steps, say $N < 10$. Hence, most computer programs for implementing the Newton-Raphson algorithm will terminate after a preset number of iterations and repeat the algorithm with a different initial guess.

Picking an appropriate initial guess is therefore crucial to the success of the Newton-Raphson algorithm. For an *arbitrary* system $\mathbf{f}(\mathbf{x}) = \mathbf{0}$, we can only make *arbitrary* guesses and hope that some would converge. Fortunately, practical problems have realistic bounds on their solutions so that $\mathbf{x}^{(0)}$ can at least be chosen from within some reasonable ball park. For example, the dc voltage solution of most electronic circuits seldom exceeds the sum of the power supply voltages. Hence, for transistor circuits powered by two 15-V dc supplies, we would pick $\mathbf{x}^{(0)}$ such that $\|\mathbf{x}^{(0)}\| < 30$ when solving the associated *node equation* (here, $\mathbf{x}$ denotes the *node* voltage vector). In Sec. 5 of this chapter, we will actually be able to derive an *upper bound* for the node voltages for a certain class of nonlinear resistive circuits. Such results are clearly of great value towards an intelligent choice of $\mathbf{x}^{(0)}$.

Aside from the convergence problem, the Newton-Raphson algorithm can find only one solution—usually the one closest to the initial guess $\mathbf{x}^{(0)}$.

Different initial guesses could of course give additional solutions for nonlinear equations having *multiple* solutions.

3.2 Newton-Raphson Discrete Equivalent Circuit

An examination of the Newton-Raphson formula in Eq. (3.7) shows that it is necessary to calculate p^2 *partial derivatives* of $\mathbf{f}(\mathbf{x})$ with respect to the p variables $x_1, x_2, \ldots, x_p$ *at each iteration*. This calculation is often the most time-consuming task in the computer implementation of the Newton-Raphson algorithm. Our goal in this section is to show that, when applied to a *nonlinear* circuit, this objection can be overcome by defining an associated linear time-invariant resistive circuit at *each* iteration so that its solution is *identical* to that obtained by the Newton-Raphson formula (3.7). Although the *same* circuit applies to all iterations, the circuit parameters must be *updated* from one iteration to another. Since these parameters do not vary smoothly—they jump *discretely* from one iteration to another—we call the associated circuit the *Newton-Raphson discrete equivalent circuit*.

Consider first the simple circuit shown in Fig. 3.1*a*, where the nonlinear resistor is characterized by the current-controlled *i-v* curve shown in Fig. 3.1*b*. Note that the nonlinear resistor shown in Fig. 3.1*a* is short-circuited ($v = 0$). This circuit is described by the single equation

$$f(i) = 0 \tag{3.9}$$

Solving this equation by the Newton-Raphson algorithm, we pick an initial guess $i^{(0)}$ and iterate the Newton-Raphson formula

$$i^{(j+1)} = i^{(j)} - (R^{(j)})^{-1} f(i^{(j)}) \tag{3.10}$$

where

$$R^{(j)} \triangleq \left. \frac{df(i)}{di} \right|_{i=i^{(j)}} \tag{3.11}$$

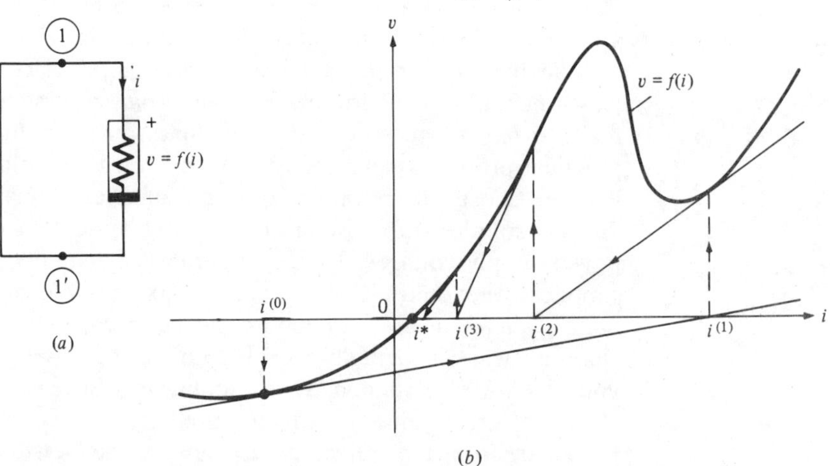

Figure 3.1 Graphic interpretation of the Newton-Raphson algorithm for a simple circuit.

The graphic interpretation of this iterative process for the one-variable case has been given earlier in Fig. 4.5 of Chap. 2. Applying this interpretation to Eq. (3.9), we obtain the graphic constructions shown in Fig. 3.1*b*, beginning from the initial guess $i^{(0)}$ to the left of the origin. The algorithm in this case is clearly seen to converge to the unique solution $i = i^*$ after a few iterations.

Now suppose that, at the jth *iteration*, we replace the *nonlinear function* $v = f(i)$ by the *affine function*

$$v = R^{(j)}i + E^{(j)} \stackrel{\Delta}{=} \hat{f}(i) \tag{3.12}$$

where $R^{(j)}$ and $E^{(j)}$ denote the *slope* and *voltage intercept*, respectively, of the *tangent line* drawn from the point $(i^{(j)}, f(i^{(j)}))$, where $i^{(j)}$ is the latest current calculated using Eq. (3.10), as shown in Fig. 3.2*a* for an arbitrary chosen $i^{(j)}$.

Applying the Newton-Raphson formula *at the* jth *iteration* to the corresponding equation

$$\hat{f}(i) = R^{(j)}i + E^{(j)} = 0 \tag{3.13}$$

we obtain

$$\begin{aligned} i^{(j+1)} &= i^{(j)} - (R^{(j)})^{-1}\hat{f}(i^{(j)}) \\ &= i^{(j)} - (R^{(j)})^{-1}f(i^{(j)}) \end{aligned} \tag{3.14}$$

where the second equation follows because, even though $\hat{f}(\cdot) \neq f(\cdot)$, at $i = i^{(j)}$, we have $\hat{f}(i^{(j)}) = f(i^{(j)})$.

Note that Eqs. (3.10) and (3.14) are *identical*. In other words, applying the Newton-Raphson formula (3.7) at the jth iteration to the original *nonlinear* equation (3.9) and the associated *linear* equation (3.13) gives us the same answer for $i^{(j+1)}$!

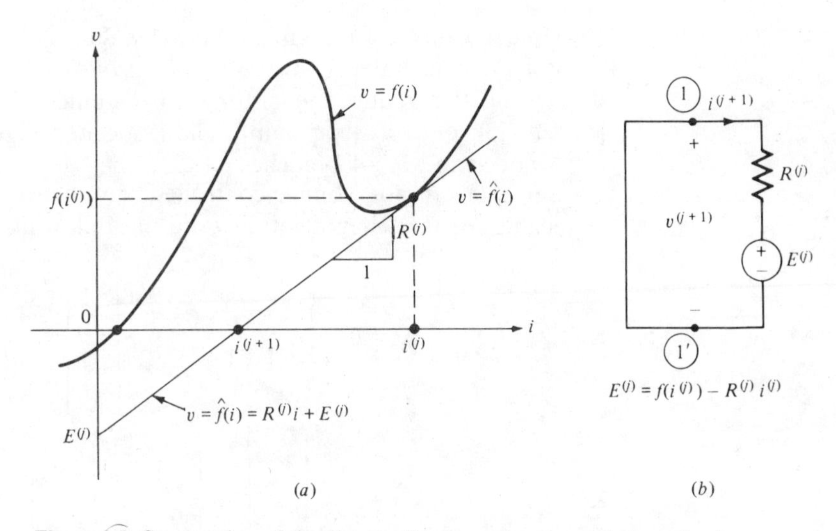

Figure 3.2 Construction of the Newton-Raphson discrete equivalent circuit.

It follows from the above important observation that if we *update* the parameters $R^{(j)}$ and $E^{(j)}$ in Eq. (3.13) *after each iteration* so that $R^{(j)}$ and $E^{(j)}$ represent the *slope* and *voltage intercept*, respectively, *evaluated at the latest calculated value* of $i^{(j)}$, then we would obtain exactly the same iterated values as those calculated from Eqs. (3.10) and (3.11).

Now observe that the affine function $\hat{f}(i)$ in Eq. (3.12) can be interpreted as the equation describing an equivalent one-port resistor consisting of a *linear time-invariant* resistor (with a resistance equal to $R^{(j)}\ \Omega$) and an $E^{(j)}$-V dc voltage source. Hence, solving the *nonlinear* resistive circuit in Fig. 3.1*a* by the Newton-Raphson algorithm is equivalent to solving the *linear* resistive circuit shown in Fig. 3.2*b*, henceforth called the associated Newton-Raphson discrete equivalent circuit at the jth *iteration*.

We cannot overemphasize that the circuit in Fig. 3.2*b* is equivalent *only at the* jth *iteration*. Note that the port current and port voltage in Fig. 3.2*b* are labeled $i^{(j+1)}$ and $v^{(j+1)}$, respectively, because the solution of this circuit will give precisely the value $i^{(j+1)}$ as would be calculated using Eqs. (3.10) and (3.11). To solve $i^{(j+2)}$ for the next iteration, $R^{(j)}$ and $E^{(j)}$ must first be updated to $R^{(j+1)}$ and $E^{(j+1)}$. This is a relatively painless process which a computer can be instructed to do.

Note that we have managed to transform a *nonlinear* resistive circuit into a sequence of *linear* resistive circuits having the same structure and whose parameters can be easily updated after each iteration.

The main significance of the above almost trivial example is that, using *tableau analysis*, it is easy to prove that all of the properties and observations so far presented remain valid for *any* nonlinear resistive circuit! Since this result is the basis of many general purpose computer nonlinear circuit simulation programs, we will illustrate its application by two examples.

Example 1 Consider the circuit shown in Fig. 3.3*a*, where the nonlinear resistor $\mathcal{R}_1$ is assumed to be *voltage-controlled* [$i_1 = \hat{i}_1(v_1)$, e.g., a tunnel diode] and the nonlinear resistor $\mathcal{R}_2$ is assumed to be current-controlled [$v_2 = \hat{v}_2(i_2)$, e.g., a neon bulb]. The associated digraph in Fig. 3.3*b* has $n = 3$ nodes and $b = 4$ branches.

Since $\mathcal{R}_2$ is not voltage-controlled and hence *node analysis* is not applicable for this circuit, let us write the following *tableau equation*:

$$\text{KCL:} \qquad \mathbf{Ai} = \mathbf{0} \Leftrightarrow \begin{bmatrix} 1 & 0 & 0 & 1 \\ -1 & 1 & 1 & 0 \end{bmatrix} \begin{bmatrix} i_1 \\ i_2 \\ i_3 \\ i_4 \end{bmatrix} = \begin{bmatrix} 0 \\ 0 \end{bmatrix} \tag{3.15a}$$

$$\text{KVL:} \qquad \mathbf{v} - \mathbf{A}^T\mathbf{e} = \mathbf{0} \Leftrightarrow \begin{bmatrix} v_1 \\ v_2 \\ v_3 \\ v_4 \end{bmatrix} - \begin{bmatrix} 1 & -1 \\ 0 & 1 \\ 0 & 1 \\ 1 & 0 \end{bmatrix} \begin{bmatrix} e_1 \\ e_2 \end{bmatrix} = \begin{bmatrix} 0 \\ 0 \\ 0 \\ 0 \end{bmatrix} \tag{3.15b}$$

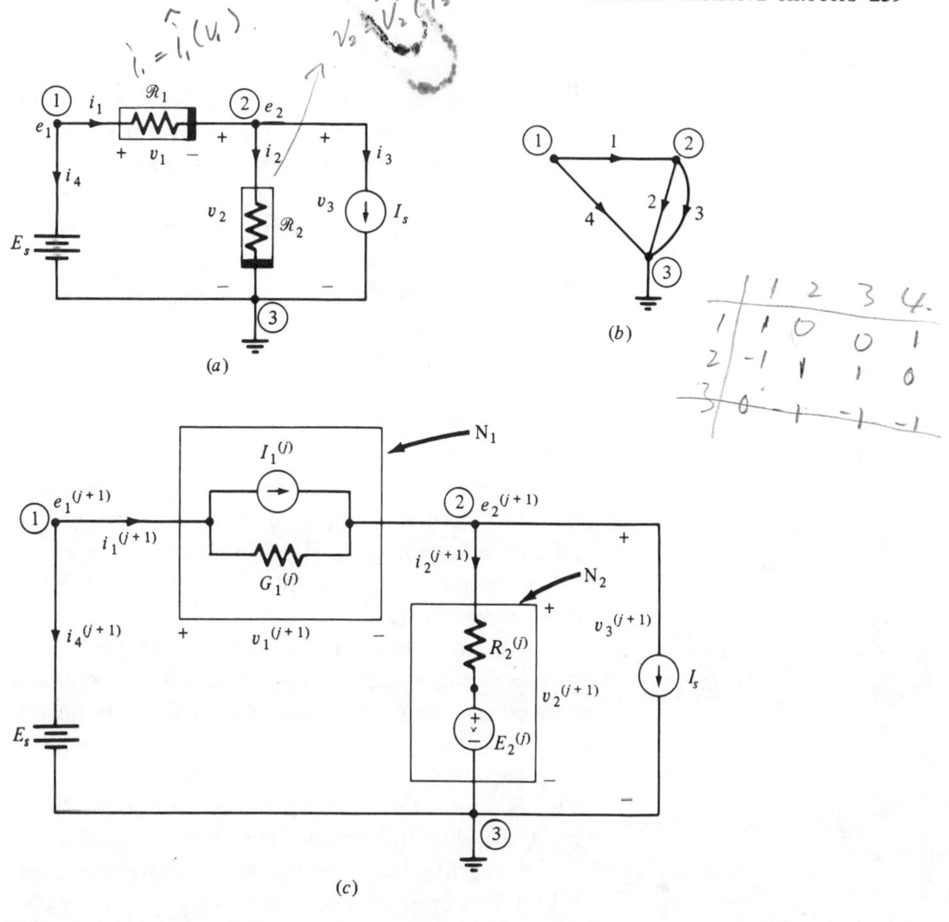

Figure 3.3 A nonlinear circuit and its associated Newton-Raphson discrete equivalent circuit at iteration j.

Branch equations:

$$\mathbf{h}(\mathbf{v}, \mathbf{i}) = \mathbf{0} \Leftrightarrow \begin{cases} i_1 - \hat{i}_1(v_1) = 0 \\ v_2 - \hat{v}_2(i_2) = 0 \\ i_3 - I_s = 0 \\ v_4 - E_s = 0 \end{cases} \tag{3.15c}$$

As expected, the above *nonlinear* tableau equation consists of 10 equations in 10 variables. Solving Eq. (3.15) by the *Newton-Raphson algorithm* directly using a computer would require the numerical evaluation of $10 \times 10 = 100$ partial derivatives *at each iteration* in order to obtain the associated 10×10 jacobian matrix $\mathbf{J}(\mathbf{x}^{(j)})$ in Eq. (3.7), where $\mathbf{x} \triangleq (\mathbf{e}, \mathbf{v}, \mathbf{i})$.[13]

[13] Note that even though the *exact* jacobian matrix associated with Eq. (3.15) is quite sparse—many entries are zero—the computers do not know this and will systematically calculate *each* partial derivative by numerical differentiation.

Now suppose we implement the Newton-Raphson algorithm using the discrete equivalent circuit approach. Since the circuit in Fig. 3.3*a* contains only two *nonlinear* elements, let us replace the voltage-controlled nonlinear resistor $\mathcal{R}_1$ by the "resistor-current source" one-port N_1 shown in Fig. 3.3*c*. Similarly, let us replace the *current-controlled* nonlinear resistor $\mathcal{R}_2$ by the "resistor-voltage source" one-port N_2 shown in Fig. 3.3*c*. Leaving all other elements intact, we obtain the associated Newton-Raphson discrete equivalent circuit shown in Fig. 3.3*c*. Note that

$$G_1^{(j)} \triangleq \hat{i}_1'(v_1^{(j)}) \qquad \text{and} \qquad R_2^{(j)} \triangleq \hat{v}_2'(i_2^{(j)}) \tag{3.16}$$

are *constant conductance* (in siemens) and constant *resistance* (in ohms) representing, respectively, the *slope* of the functions $\hat{i}_1(v_1)$ and $\hat{v}_2(i_2)$, evaluated at $v_1 = v_1^{(j)}$ and $i_2 = i_2^{(j)}$, respectively. Likewise, $I_1^{(j)}$ and $E_2^{(j)}$ denote the *current intercept* of $\mathcal{R}_1$ *and voltage intercept* of $\mathcal{R}_2$, respectively, and are also constants. The only variables are $e_1, e_2, v_1, v_2, v_3, v_4, i_1, i_2, i_3$, and i_4, where a superscript $(j+1)$ is attached to each of these variables to emphasize the resulting solution is valid only for the jth iteration.

It is important to observe that the two parallel branches representing $\mathcal{R}_1$, as well as the two series branches representing $\mathcal{R}_2$, are each being treated as a *one-port resistor* described by an *affine function*, namely,[14]

$$i_1 = G_1^{(j)} v_1 + I_1^{(j)} \qquad v_2 = R_2^{(j)} i_2 + E_2^{(j)} \tag{3.17}$$

This is why we did not introduce additional "internal" variables representing the individual components inside N_1 and N_2.

There are two reasons for treating the discrete equivalent circuits in Fig. 3.3 as one-ports and not as two separate elements: (*a*) Treating them as two branches each would significantly increase the number of variables in the tableau equation, thereby increasing the computation time (since the dimension of the jacobian matrix would increase accordingly); (*b*) by representing each discrete equivalent circuit as a one-port, only one branch is needed in drawing the associated digraph. Consequently, both *the nonlinear circuit and its associated discrete equivalent circuit have an identical digraph*.

It follows from the preceding conclusion that the values of $\mathbf{e}^{(j+1)}$, $\mathbf{v}^{(j+1)}$, and $\mathbf{i}^{(j+1)}$ obtained at the jth iteration of the Newton-Raphson formula (3.7) [when applied to the nonlinear tableau equation (3.15) directly] are identical to those obtained by solving the *linear* resistive circuit in Fig. 3.3*c*. Note that only 2 numerical differentiations—compared to 100 when Eq. (3.7) is applied directly to Eq. (3.15)—are needed in the present discrete equivalent circuit approach as called for in Eq. (3.16), assuming that the respective derivatives are not given in analytic form.

Since the circuit in Fig. 3.3 is *linear* it is described by the *linear tableau equation* (2.4) with a constant 10×10 tableau matrix $\mathbf{T}$. Since the digraphs for the circuits in Fig. 3.3*a* and *c* are identical, Eqs. (3.15*a*) and (3.15*b*)

[14] They are examples of affine resistors defined in Sec. 2.1.

are applicable to both circuits. Only the branch equations are different.

The branch equations for the discrete equivalent circuit in Fig. 3.3*c* are given by

$$\underbrace{\begin{bmatrix} -G_1^{(j)} & 0 & 0 & 0 \\ 0 & 1 & 0 & 0 \\ 0 & 0 & 0 & 0 \\ 0 & 0 & 0 & 1 \end{bmatrix}}_{\mathbf{M}^{(j)}} \underbrace{\begin{bmatrix} v_1^{(j+1)} \\ v_2^{(j+1)} \\ v_3^{(j+1)} \\ v_4^{(j+1)} \end{bmatrix}}_{\mathbf{v}^{(j+1)}} + \underbrace{\begin{bmatrix} 1 & 0 & 0 & 0 \\ 0 & -R_2^{(j)} & 0 & 0 \\ 0 & 0 & 1 & 0 \\ 0 & 0 & 0 & 0 \end{bmatrix}}_{\mathbf{N}^{(j)}} \underbrace{\begin{bmatrix} i_1^{(j+1)} \\ i_2^{(j+1)} \\ i_3^{(j+1)} \\ i_4^{(j+1)} \end{bmatrix}}_{\mathbf{i}^{(j+1)}} = \underbrace{\begin{bmatrix} I_1^{(j)} \\ E_2^{(j)} \\ I_s \\ E_s \end{bmatrix}}_{\mathbf{u}_s^{(j+1)}} \tag{3.18}$$

or

$$\mathbf{M}^{(j)}\mathbf{v}^{(j+1)} + \mathbf{N}^{(j)}\mathbf{i}^{(j+1)} = \mathbf{u}_s^{(j+1)} \tag{3.19}$$

Note that the first two elements in $\mathbf{u}_s^{(j+1)}$ are *not* zeros, as would be the case if all resistive branches are linear. Here, branches 1 and 2 are each represented in Eq. (3.17) by an *affine function*, which is different from *a linear function*. Instead, they are filled with the appropriate sources imbedded within the one-port, namely, $I_1^{(j)}$ and $E_2^{(j)}$ in this case. It is easy to see that this simple modification will account for the presence of any independent sources which arise as part of an affine equation representing a multiterminal or multiport resistor in the general case (see Exercises 5 and 6 in Sec. 2.1).

Exercises

1. Prove that the solution ($\mathbf{e}^{(j+1)}$, $\mathbf{v}^{(j+1)}$, and $\mathbf{i}^{(j+1)}$) of the *linear* circuit in Fig. 3.3*c* is identical to that obtained by solving (3.15) by the Newton-Raphson algorithm *at the* jth *iteration*.
2. Draw the digraph and write thc branch equations for the linear resistive circuit in Fig. 3.3 when each element is treated as a branch. In particular, label the current source $I_1^{(j)}$ and voltage source $E_2^{(j)}$ as branches 5 and 6, respectively. Compare your answers with Fig. 3.3*b* and Eq. (3.19).

Example 2 Consider the *npn*-transistor circuit at time $t = t_0$, as shown earlier in Fig. 2.2. Recall that the nonlinear tableau equation for this circuit is given by Eqs. (2.1)′, (2.2)′, and (2.8) where the transistor was treated as a nonlinear three-terminal resistor described by the Ebers-Moll equation (2.7).

Using the *Ebers-Moll transistor circuit model*[15] given in Fig. 4.2 of Chap. 3, we can replace the circuit in Fig. 3.4*a* by the model shown in Fig. 3.4*b*. The two *pn*-junction diodes described by

$$i_k = I_s\left(\exp\frac{v_k}{V_T} - 1\right) \qquad k = 1, 2 \tag{3.20}$$

[15] For convenience, we have reversed the polarity of v_1 and v_2, so that our present voltages v_1 and v_2 are the *negative* of those defined in Fig. 2.2.

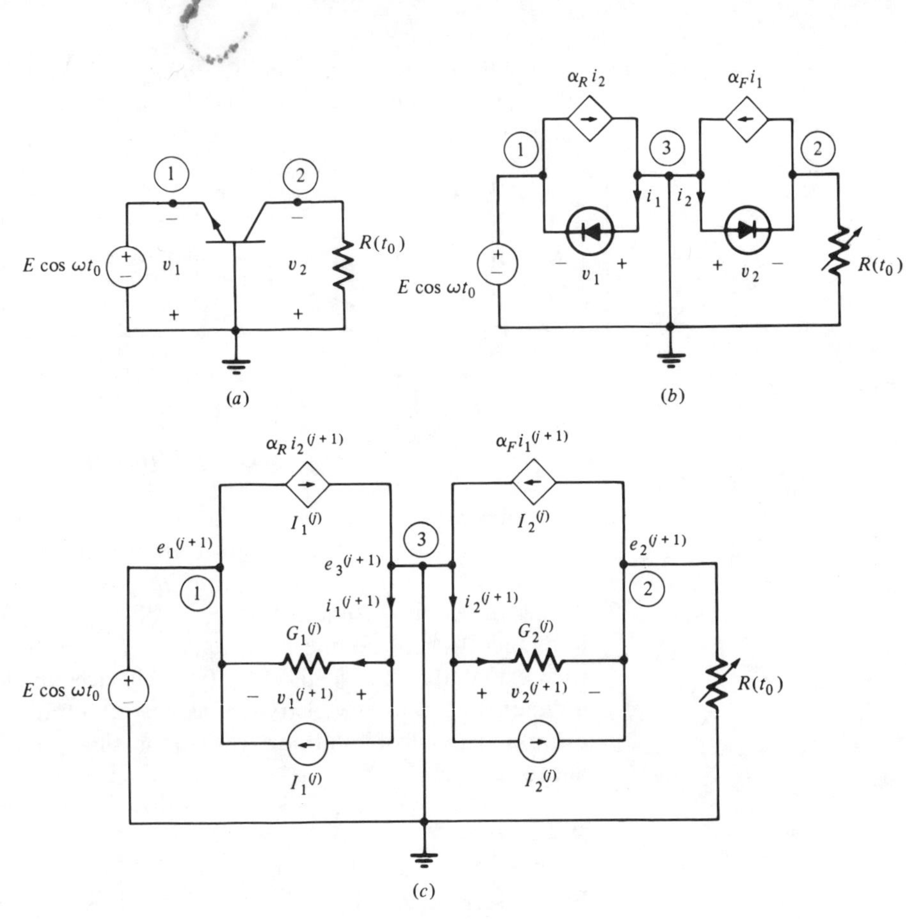

Figure 3.4 (*a*) A transistor circuit. (*b*) Circuit obtained by replacing the transistor from (*a*) by its Ebers-Moll circuit model. (*c*) Associated Newton-Raphson discrete equivalent circuit at $t = t_0$.

The associated discrete equivalent circuit is shown in Fig. 3.4*c*, where

$$G_k^{(j)} \triangleq \frac{I_s}{V_T}\left(\exp \frac{v_k^{(j)}}{V_T}\right) \qquad k = 1, 2 \tag{3.21}$$

Hence, instead of applying the Newton-Raphson formula (3.7) directly to Eqs. (2.1)′, (2.2)′, and (2.8), we only need to solve the *linear* resistive circuit in Fig. 3.4*c*.

Examples 1 and 2 show that the Newton-Raphson discrete equivalent circuit gives us not only an interesting circuit-theoretic interpretation of an otherwise mathematical formula, it actually allows us to analyze nonlinear resistive circuits with a greatly reduced computation time, and hence cost, as compared to blindly solving the nonlinear equations directly. Even more remarkable is that the discrete equivalent circuit is always a *linear* resistive *circuit*. Thus, we only need to develop an efficient computer program for solving linear resitive circuits, and then modify it to implement the discrete equivalent circuits associated with *nonlinear* resistive circuits.

Exercises

1. Prove that the solution of the *linear* circuit in Fig. 3.4*c* is identical to that obtained by solving the nonlinear equation describing the circuit in Fig. 3.4*b* by the Newton-Raphson algorithm *at the* jth *iteration*.
2. Rewrite Eq. (3.20) into the current-controlled form $v_k = \hat{v}_k(i_k)$, and draw the discrete equivalent circuit associated with the transistor circuit in Fig. 3.4*b*.

   Using the nonlinear op-amp circuit model shown in Fig. 3.5*a*, draw the Newton-Raphson discrete equivalent circuit associated with the op-amp circuit shown in Fig. 3.5*b*. Assume $f(\cdot)$ is a differentiable function.

4 GENERAL PROPERTIES OF LINEAR RESISTIVE CIRCUITS

In this section, we state and prove several general theorems for linear time-invariant resistive circuits,[16] namely, the *superposition theorem*, the

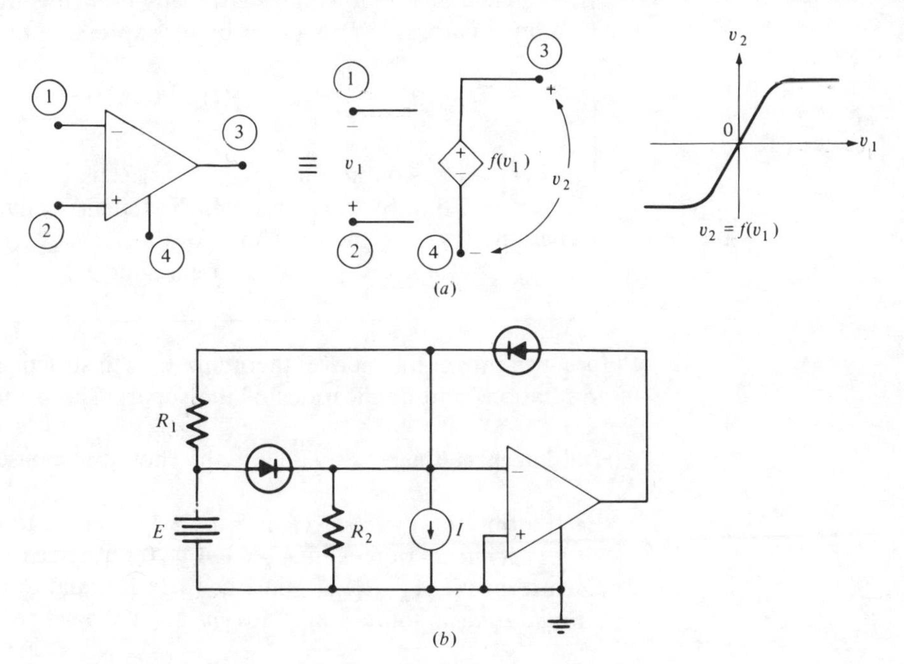

Figure 3.5 Op-amp circuit for Exercise 2.

[16] Recall that a *linear* resistive circuit may contain, in addition to two-terminal linear resistors and independent voltage and current sources, any multiterminal or multiport linear resistors (e.g., ideal transformers, gyrators, and all four types of linear-controlled sources).

Thévenin-Norton theorem, and the *two-port representation theorem*. Intelligent use of these theorems often results in a dramatic simplification of an otherwise much more difficult problem. They also serve as invaluable tools for deriving several results in the subsequent chapters. In short, they are useful tools for *linear* resistive circuit analysis.

All of these theorems are valid if and only if the associated circuit is *uniquely solvable*,[17] equivalently, if and only if the associated *tableau matrix* **T** is *nonsingular*.

Although stated only for *time-invariant* circuits for simplicity, all of these theorems are valid also for *time-varying* circuits by simply allowing all parameters and coefficients to vary with time.

4.1 Superposition Theorem

> Let $\mathcal{N}$ be any *linear time-invariant uniquely solvable* resistive circuit driven by α *independent* voltage sources $v_{s1}(t)$, $v_{s2}(t), \ldots, v_{s\alpha}(t)$, and β *independent* current sources $i_{s1}(t)$, $i_{s2}(t), \ldots, i_{s\beta}(t)$.
>
> Then any node voltage $e_j(t)$, any branch voltage $v_j(t)$, or any branch current $i_j(t)$ is given by an expression of the form
>
> $$H_1 v_{s1}(t) + \cdots + H_\alpha v_{s\alpha}(t) + K_1 i_{s1}(t) + \cdots + K_\beta i_{s\beta}(t) \tag{4.1}$$
>
> where the coefficients H_k, $k = 1, 2, \ldots, \alpha$, and K_k, $k = 1, 2, \ldots, \beta$, are *constants* which depend only on the circuit parameters of $\mathcal{N}$ and the choice of the output variable (i.e., e_j, v_j, or i_j), but *not* on the independent sources.

Before we prove the above theorem, it is instructive to give some circuit interpretations and applications of the superposition theorem.

Circuit interpretations Let y denote the chosen response (i.e., $y = e_j$, v_j, or i_j).

1. Each term $y(v_{sk}) \triangleq H_k v_{sk}(t)$ in Eq. (4.1) is equal to the response of y when all *independent* sources in $\mathcal{N}$ *except* $v_{sk}(t)$ are set to zero.
2. Each term $y(i_{sk}) \triangleq K_k i_{sk}(t)$ in Eq. (4.1) is equal to the response of y when all *independent* sources in $\mathcal{N}$ *except* $i_{sk}(t)$ are set to zero.

[17] For a large class of linear and nonlinear resistive circuits, including those containing all four types of linear-controlled sources, a simple test is available to determine, often by inspection, whether a circuit is uniquely solvable. See, for example: T. Nishi and L. O. Chua, "Topological Criteria for Nonlinear Resistive Circuits Containing Controlled Sources to Have a Unique Solution," *IEEE Trans. on Circuits and Systems*, vol. CAS-31, pp. 722–741, August 1984.

3. Equation (4.1) shows that the response due to several independent voltage and current sources is equal to the *sum* of the responses due to *each* independent source *acting alone*, i.e., with all other *independent* voltage sources replaced by short circuits, and all other *independent* current sources replaced by open circuits.
4. The response at *any time* $t = t_0$ depends *only* on the value of the independent sources at *the same time* $t = t_0$. In other words, linear resistive circuits have no *memory*.

Application 1 Let us use the *superposition theorem* to calculate the node voltage e_1 and the resistor current i_2 in Fig. 4.1*a*.

The contributions to e_1 and i_2 due to $v_{s1}(t)$ acting alone [with $i_{s1}(t) = 0$] can be obtained by inspection of the *voltage-divider circuit* in Fig. 4.1*b* obtained by *open-circuiting* the current source in Fig. 4.1*a*:

$$e_1(v_{s1}) = \frac{R_2}{R_1 + R_2} v_{s1}(t) \tag{4.2}$$

$$i_2(v_{s1}) = \frac{1}{R_1 + R_2} v_{s1}(t) \tag{4.3}$$

Here, the "input" v_{s1} is shown as the "argument" of $e_1(\bullet)$ and $i_2(\bullet)$ to remind the reader that the node voltage e_1 given by Eq. (4.2), and the branch current i_2, given by Eq. (4.3), are due to v_{s1} *acting alone*, and are therefore *functions* of v_{s1} only.

The contributions to e_1 and i_2 due to $i_{s1}(t)$ acting alone [with $v_{s1}(t) = 0$] can be obtained by inspection of the *current-divider circuit* in Fig. 4.1*c* obtained by *short-circuiting* the voltage source in Fig. 4.1*a*:

$$e_1(i_{s1}) = \frac{R_1 R_2}{R_1 + R_2} i_{s1}(t) \tag{4.4}$$

$$i_2(i_{s1}) = \frac{R_1}{R_1 + R_2} i_{s1}(t) \tag{4.5}$$

Adding the respective contributions, we obtain:

$$e_1 = e_1(v_{s1}) + e_1(i_{s1}) = H_1 v_{s1}(t) + K_1 i_{s1}(t) \tag{4.6a}$$

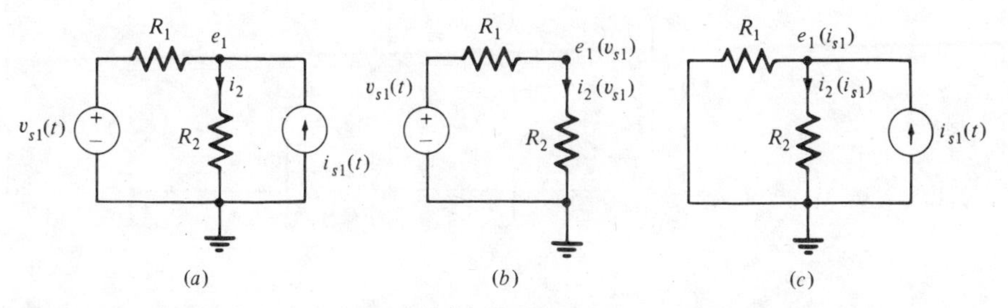

Figure 4.1 (*a*) Circuit for application 1. (*b*) Voltage divider. (*c*) Current divider.

where $$H_1 \triangleq \frac{R_2}{R_1 + R_2} \qquad K_1 \triangleq \frac{R_1 R_2}{R_1 + R_2} \tag{4.6b}$$

$$i_2 = i_2(v_{s1}) + i_2(i_{s1}) = H_1 v_{s1}(t) + K_1 i_{s1}(t) \tag{4.7a}$$

where $$H_1 \triangleq \frac{1}{R_1 + R_2} \qquad K_1 \triangleq \frac{R_1}{R_1 + R_2} \tag{4.7b}$$

As expected, both e_1 and i_2 are of the form specified by Eq. (4.1), where H_1 and K_1 are constants depending *only* on the circuit parameters R_1 and R_2 and the chosen output variable. They do *not* depend on $v_{s1}(t)$ or $i_{s1}(t)$. Of course, for different choices of output variables, we get different H_1's and K_1's, as seen in Eqs. (4.6*b*) and (4.7*b*).

Note also that if the two resistors are *time-varying*, i.e., $R_1 = R_1(t)$ and $R_2 = R_2(t)$, exactly the same solutions held except that H_1 and K_1 are now functions of time.

Application 2 Consider next the circuit shown in Fig. 4.2*a*, which is obtained by replacing resistor R_2 in Fig. 4.1*a* by two *current-controlled voltage sources*. Let us calculate e_1 and i_2 using the superposition theorem.

To solve for the contribution due to $v_{s1}(t)$ acting alone, we set $i_{s1}(t) = 0$ by *open-circuiting* the current source in Fig. 4.2*a* to obtain the circuit in Fig. 4.2*b*. Writing KVL around the loop and setting $i_3 = i_{s1}(t) = 0$, we obtain

$$R_1 i_1 + r_{m2} i_1 = v_{s1}(t) \tag{4.8}$$

Solving for i_1 and noting that $e_1(v_{s1}) = r_{m2} i_1$ and $i_2(v_{s1}) = i_1$, we obtain

$$e_1(v_{s1}) = \frac{r_{m2}}{R_1 + r_{m2}} v_{s1}(t) \tag{4.9}$$

$$i_2(v_{s1}) = \frac{1}{R_1 + r_{m2}} v_{s1}(t) \tag{4.10}$$

To solve for the contribution due to $i_{s1}(t)$ acting alone, we set $v_{s1}(t) = 0$ by *short-circuiting* the *independent* voltage source in Fig. 4.2*a* to obtain the circuit

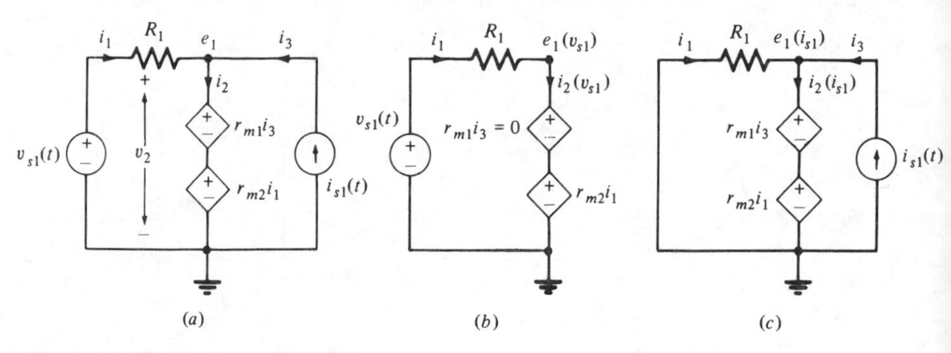

Figure 4.2 Circuits for application 2.

in Fig. 4.2*c*. Applying KVL around the loop on the left and noting that $i_3 = i_{s1}(t)$, we obtain

$$R_1 i_1 + r_{m1} i_{s1}(t) + r_{m2} i_1 = 0 \tag{4.11}$$

Solving for i_1 from Eq. (4.11) and noting that $e_1(i_{s1}) = -R_1 i_1$ and $i_2(i_{s1}) = i_1 + i_{s1}(t)$, we obtain

$$e_1(i_{s1}) = \frac{R_1 r_{m1}}{R_1 + r_{m2}} i_{s1}(t) \tag{4.12}$$

$$i_2(i_{s1}) = \frac{R_1}{R_1 + r_{m2}} i_{s1}(t) \tag{4.13}$$

Adding the respective contributions, we obtain

$$e_1 = \frac{r_{m2}}{R_1 + r_{m2}} v_{s1}(t) + \frac{R_1 r_{m1}}{R_1 + r_{m2}} i_{s1}(t) \tag{4.14}$$

$$i_2 = \frac{1}{R_1 + r_{m2}} v_{s1}(t) + \frac{R_1}{R_1 + r_{m2}} i_{s1}(t) \tag{4.15}$$

As an independent check of the above answer, note that if we pick $r_{m1} = r_{m2} = R_2$, then Eqs. (4.14) and (4.15) become identical to Eqs. (4.6) and (4.7), respectively. Indeed, under this condition, the two controlled sources in Fig. 4.2*a* can be replaced by a single CCVS defined by

$$v_2 = r_{m1} i_3 + r_{m2} i_1 = R_2(i_1 + i_3) = R_2 i_2 \tag{4.16}$$

which in turn is equivalent to an R_2-Ω resistor. Hence, when $r_{m1} = r_{m2} = R_2$, the circuit in Fig. 4.2*a* is equivalent to that Fig. 4.1*a* and hence their answers must be identical.

REMARK In applying the superposition theorem, controlled sources are left intact.

Application 3 The circuit in Figs. 4.1*a* and 4.2*a* could have been solved almost as easily without applying the superposition theorem. For a more impressive application, let us solve for the output voltage v_o of the op-amp circuit in Fig. 2.9 of Chap. 4. Recall that the answer given earlier by Eq. (2.17) of Chap. 4 was obtained by solving a system of eight linear equations, which is a nontrivial task.

Applying the superposition theorem, we draw the simplified circuits shown in Fig. 4.3*a* and *b* by short-circuiting the voltage sources $v_{s1}(t)$ and $v_{s2}(t)$ in Fig. 2.9 of Chap. 4 to zero, one at a time.

To calculate v_{o1}, note that $v_1 = v_{s1}(t)$ and $i_2 = v_{s1}(t)/R_1$ (recall the virtual short-circuit property and observe that $v_4 = 0$). Hence, we obtain *by inspection*:

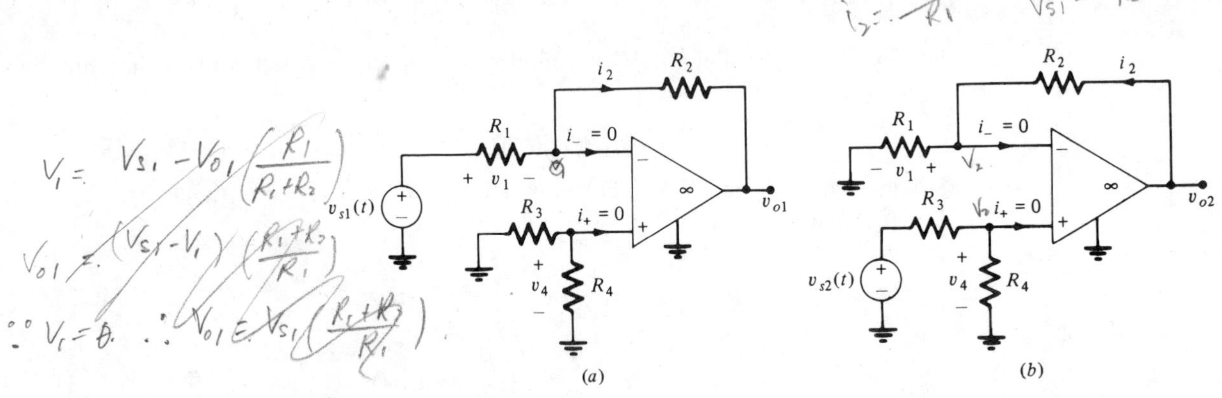

Figure 4.3 (*a*) Circuit due to $v_{s1}(t)$ acting alone. (*b*) Circuit due to $v_{s2}(t)$ acting alone.

$$v_{o1} = -\frac{R_2}{R_1} v_{s1}(t) \tag{4.17}$$

Note that R_3 and R_4 are irrelevant here since $i_+ = 0$.

To calculate v_{o2}, note the voltage-divider relation

$$v_1 = v_4 = \frac{R_4}{R_3 + R_4} v_{s2}(t) \tag{4.18}$$

Applying KVL, we obtain

$$v_{o2} = R_2 i_2 + v_1 = v_1 + R_2 \frac{v_1}{R_1} = \left(1 + \frac{R_2}{R_1}\right) v_1 \tag{4.19}$$

Substituting Eq. (4.18) for v_1 in Eq. (4.19), we obtain

$$v_{o2} = \left(1 + \frac{R_2}{R_1}\right) \frac{R_4}{R_3 + R_4} v_{s2}(t) \tag{4.20}$$

Adding Eqs. (4.17) and (4.20), we obtain

$$v_o(t) = -\left(\frac{R_2}{R_1}\right) v_{s1}(t) + \left[\frac{R_4(1 + R_2/R_1)}{R_3 + R_4}\right] v_{s2}(t) \tag{4.21}$$

which is identical to Eq. (2.17) of Chap. 4, as it should be. Note that the small amount of work involved here compared to that of solving eight simultaneous linear equations is indeed quite significant.

Application 4 Our final example shows how *superposition* and *symmetry* help solve a seemingly impossible problem.

Let $\mathcal{N}$ be a linear resistive circuit consisting of an infinite square grid of 1-Ω resistors. Suppose we apply a 1-A current source across nodes ① and ② as shown in Fig. 4.4 and *assume* that at infinity all nodes are grounded. The problem is to calculate the voltage v across nodes ① and ②.

To solve this seemingly intractable problem let us first replace the current source by a 1-A current source (with reference direction entering node ①) connected between *ground* and node ①, and another 1-A current source (with reference direction leaving node ②) connected between *ground* and node ②. Since KCL remains unchanged at nodes ① and ②, this modified circuit has the same solution as the original circuit.

Now it can be shown that with the assumed boundary condition (all nodes at infinity are grounded), this circuit has a *unique* solution. It follows by *symmetry* that *each* 1-A current source *acting alone* would cause a current of $\frac{1}{4}$ A to flow through the resistor from node ① to node ②. Hence by the *superposition* theorem, $\frac{1}{2}$ A flows through the resistor when both sources are present. It follows that the voltage $v = \frac{1}{2}$ V.

Note that this problem would have been extremely difficult to solve without using the superposition theorem.

PROOF OF THE SUPERPOSITION THEOREM Since $\mathcal{N}$ is linear and time-invariant, it is described by the linear tableau equation (2.4):

$$\mathbf{Tw}(t) = \mathbf{u}(t) \tag{4.22}$$

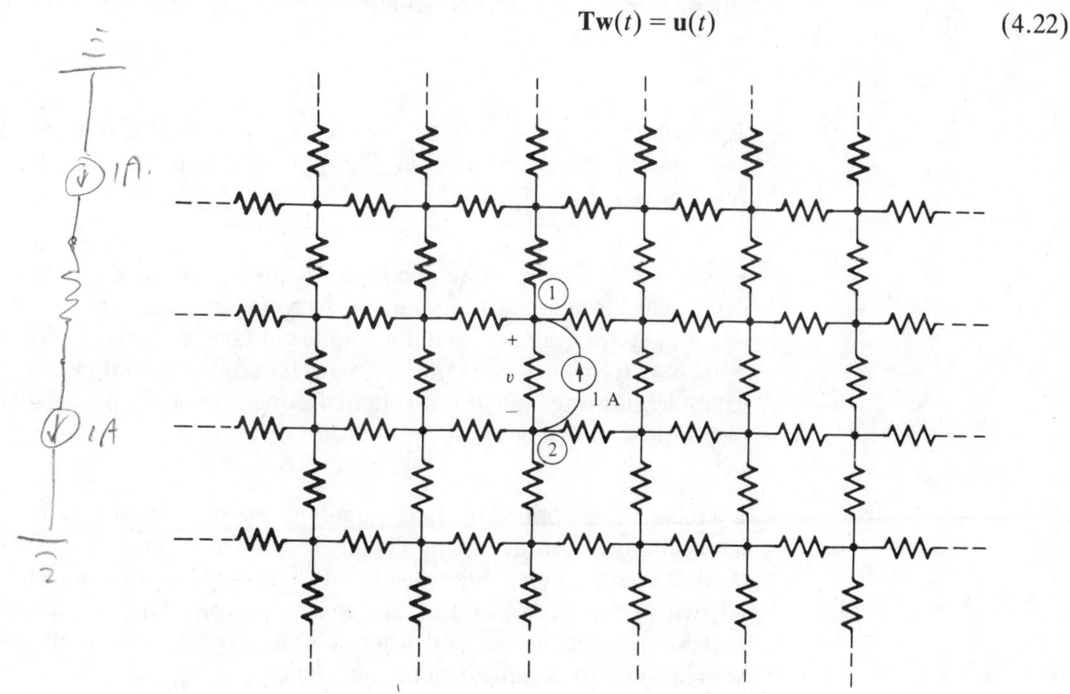

Figure 4.4 A grid of 1-Ω resistors which extends into infinity in all directions. A 1-A current source is connected across nodes ① and ②.

where $\mathbf{T}$ is an $[(n-1)+2b] \times [(n-1)+2b]$ constant real tableau matrix. However, since $\mathcal{N}$ is *uniquely solvable*, $\mathbf{T}^{-1}$ exists and the unique solution is given by

$$\mathbf{w}(t) = \mathbf{T}^{-1}\mathbf{u}(t) \tag{4.23}$$

where

$$\mathbf{u}(t) \triangleq [\underbrace{\mathbf{0}^T}_{\substack{n-1\\ \text{zeros}}} \quad \underbrace{\mathbf{0}^T}_{\substack{b\\ \text{zeros}}} \quad \underbrace{\underbrace{0\,0\cdots 0}_{\text{resistors}} \; \underbrace{v_{s1}(t)\cdots v_{s\alpha}(t)}_{\text{voltage sources}} \; \underbrace{i_{s1}(t)\cdots i_{s\beta}(t)}_{\text{current sources}}}_{\mathbf{u}_s(t)}]^T \tag{4.24}$$

Here, we have assumed, without loss of generality that all independent sources are labeled last in the order depicted above.

Since each component of $\mathbf{w}(t)$ (i.e., e_j, v_j, or i_j) is obtained by multiplying the corresponding row of $\mathbf{T}^{-1}$ with $\mathbf{u}(t)$, it follows that each response e_j, v_j, or i_j is given by an expression in the form of Eq. (4.1). Moreover, since $\mathbf{T}^{-1}$ is a *constant* matrix which does not involve any independent source terms, so are the constant coefficients H_k and K_k.

∎

Exercise 1 Insert a voltage source $v_{s3}(t)$ between R_4 and ground in the op-amp circuit shown in Fig. 2.9 of Chap. 4. Calculate $v_o(t)$ *by inspection*.

Exercise 2 If a linear time-invariant resistive circuit is driven by m distinct sinusoidal frequency component waveforms with input frequencies $\omega_1, \omega_2, \ldots, \omega_m$, prove that any node voltage e_j, branch voltage v_j, or branch current i_j can contain no *new* frequency components. Give an example showing that *not all* input frequencies need be present in some response waveforms.

Exercise 3 To appreciate the benefits of superposition and its domain of applicability, consider a nonlinear resistor $v = \hat{v}(i) = i^3$ driven by two current sources $i_{s1}(t) = I_1 \cos \omega_1 t$ and $i_{s2}(t) = I_2 \cos \omega_2 t$ connected in parallel, where I_1, I_2, ω_1, and ω_2 are constants. Calculate the voltage v when each source acts alone, and when they act together. In each case, reduce your answer to a sum of pure sine waves.

(*a*) Does superposition hold for this circuit?

(*b*) What are the frequency components of the output waveform?

4.2 The Thévenin-Norton Theorem

Any *well-defined*[18] *linear time-invariant resistive* one-port N which satisfies the following *unique solvability* condition can be replaced by the following *equivalent one-ports* N_{eq} without affecting the solution of any *external* circuit (not necessarily linear or resistive) connected across N.

1. *Thévenin equivalent one-port* N_{eq}
   unique solvability condition: The circuit $\mathcal{N}$ obtained by connecting a *current* source i across N has a *unique* solution for all i.

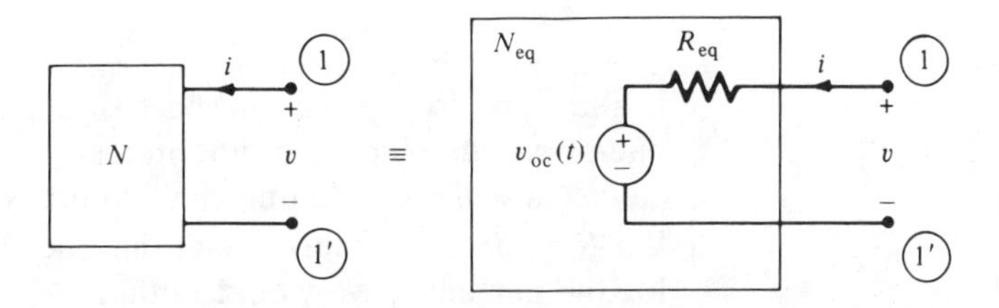

$$R_{eq} \triangleq \textit{Thévenin equivalent resistance} \text{ in ohms}$$
$\triangleq$ driving-point or input resistance across N after *all independent* sources inside N are set to zero

$$v_{oc}(t) \triangleq \textit{open-circuit voltage}$$
$\triangleq$ voltage v across terminals ① and ①′ when the port ①, ①′ is left open-circuited

2. *Norton equivalent one-port* N_{eq}
   unique solvability condition: The circuit $\mathcal{N}$ obtained by connecting a *voltage* source v across N has a *unique* solution for all v.

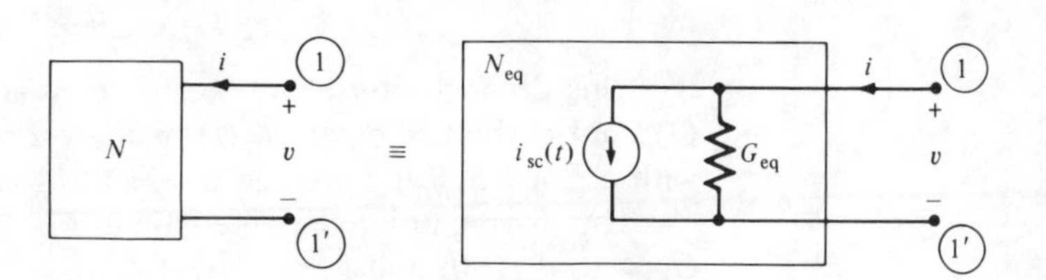

[18] A one-port N is said to be *well-defined* iff it does not contain any circuit element which is *coupled*, electrically or nonelectrically, to some physical variable *outside* of N: e.g., controlled sources depending on a variable external to N, transformer windings coupled magnetically to an external winding, a photoresistor coupled to an external light source, etc.

$G_{eq} \triangleq$ *Norton equivalent conductance* in siemens
$\triangleq$ driving-point or input conductance across N after *all independent* sources inside N are set to zero
$i_{sc}(t) \triangleq$ *short-circuit current*
$\triangleq$ current i entering terminal ① when terminals ① and ①′ are connected by an external short circuit

Before we prove the above theorem, let us consider first some circuit interpretations and applications.

1. The main value of *Thévenin's theorem*, as well as *Norton's theorem*, is that it allows us to replace *any* part of a circuit which forms a *linear* resistive one-port (but which is of no interest in a given situation) by only two circuit elements *without* affecting the solution of the remainder of the circuit.
2. Let $R_{eq} \neq 0$. If we short-circuit the Thévenin equivalent circuit N_{eq} and solve for the current i, we would obtain

$$i_{sc} = -\frac{v_{oc}}{R_{eq}} \tag{4.25}$$

If $i_{sc} \neq 0$, we can calculate the Thévenin equivalent resistance by

$$R_{eq} = -\frac{v_{oc}}{i_{sc}} \tag{4.26}$$

3. When $R_{eq} \neq 0$ and $G_{eq} \neq 0$, the one-port N is *equivalent* to both its Thévenin and its Norton equivalent one-ports: Its *driving-point characteristic* at time t is defined by

$$v = R_{eq}i + v_{oc}(t) \tag{4.27}$$

or

$$i = G_{eq}v + i_{sc}(t) \tag{4.28}$$

This driving-point characteristic consists of a *straight line* with a slope G_{eq} and current intercept $i_{sc}(t)$ in the v-i plane, as shown in Fig. 4.5*a*, or with a slope R_{eq} and voltage intercept $v_{oc}(t)$ in the i-v plane.

The driving-point characteristic in Fig. 4.5*a* is drawn for the case when $G_{eq} > 0$, $v_{oc} > 0$, and $i_{sc} < 0$.
4. The limiting case $R_{eq} = 0$ is shown in Fig. 4.5*b*. The Thévenin equivalent one-port in this case consists of just a battery of v_{oc} volts. The corresponding Norton equivalent one-port does *not* exist because $G_{eq} = \infty$. Indeed, the unique solvability condition fails in this case—KVL is violated when a voltage source $v \neq v_{oc}$ is applied.

The "dual" limiting case $G_{eq} = 0$ is shown in Fig. 4.5*c*. The Thévenin

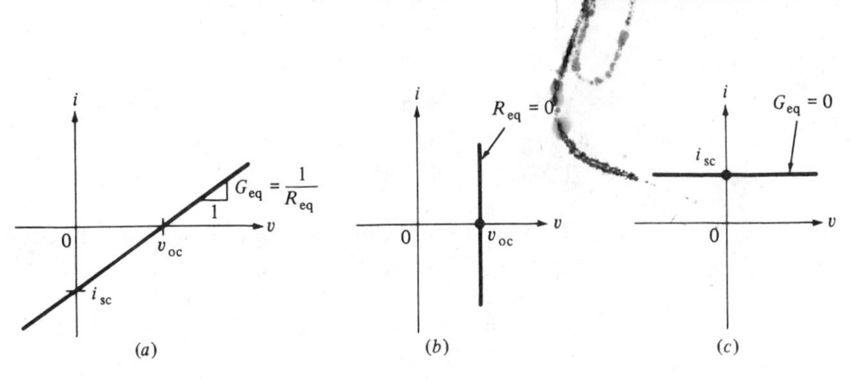

Figure 4.5 (*a*) Driving-point characteristic of N with $v_{oc} > 0$ and $G_{eq} > 0$. (*b*) Driving-point characteristic with $v_{oc} > 0$ and $R_{eq} = 0$. (*c*) Driving-point characteristic with $i_{sc} > 0$ and $G_{eq} = 0$.

equivalent one-port does not exist in this case, whereas the Norton equivalent one-port degenerates into a current source of i_{sc} amperes.

5. A one-port which has neither a Thévenin nor Norton equivalent is shown in Fig. 4.6*a*: Its driving-point characteristic is defined by

$$v = 0 \qquad i = 0 \tag{4.29}$$

and consists therefore of only one point, namely, the origin. Note that the "virtual short circuit" characterizing the input port of an *ideal* op amp operating in the linear region has precisely this property. Such a one-port is called a *nullator*.

Since the driving-point characteristic for both Thévenin and Norton equivalent one-ports consists of a straight line, it is clear that the nullator does not have a Thévenin or Norton equivalent one-port. Indeed, both "unique solvability conditions" are violated by this one-port. Note that we can only drive N with a 0-V voltage source or a 0-A current source.

6. It follows from the above observations that if N is not current-controlled, then it does *not* possess a Thévenin equivalent. Dually, if N is not voltage-controlled, then it does *not* possess a Norton equivalent. Hence, in applying Thévenin's or Norton's theorem, we can ignore checking for the "unique solvability condition" since this generally entails the difficult task of checking the determinant of the associated tableau matrix **T**. Instead, we

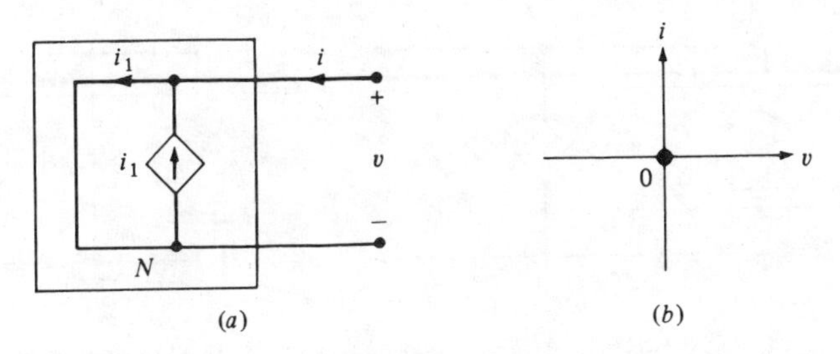

Figure 4.6 A one-port characterized by only one point.

simply proceed to calculate R_{eq} or G_{eq}. Failure to obtain a unique finite value for R_{eq} (respectively, G_{eq}) would then imply that N does not have a Thévenin (respectively, Norton) equivalent.

Example 1 Find the Thévenin and Norton equivalent one-ports of the one-port shown in Fig. 4.7*a*.

Let us calculate R_{eq} and G_{eq} first using the simplified circuit shown in Fig. 4.7*b*. For any applied voltage v, we find $i_1 = v/R$ so that $i = -4i_1 = -(4/R)v$. Hence,

$$R_{eq} = \frac{1}{G_{eq}} = -\frac{R}{4} \tag{4.30}$$

Since both R_{eq} and G_{eq} are finite numbers, we know N has a Thévenin and a Norton equivalent one-port.

We proceed therefore to calculate v_{oc} using the circuit shown in Fig. 4.7*c* (since $i = 0$). Applying KCL we obtain $i_1 - 5i_1 + I_s = 0$ or $i_1 = I_s/4$. Hence,

$$v_{oc} = E + \frac{R}{4} I_s \tag{4.31}$$

To calculate i_{sc}, we could use Eqs. (4.26) and (4.31) to obtain

$$i_{sc} = \frac{-v_{oc}}{R_{eq}} = \frac{4E}{R} + I_s \tag{4.32}$$

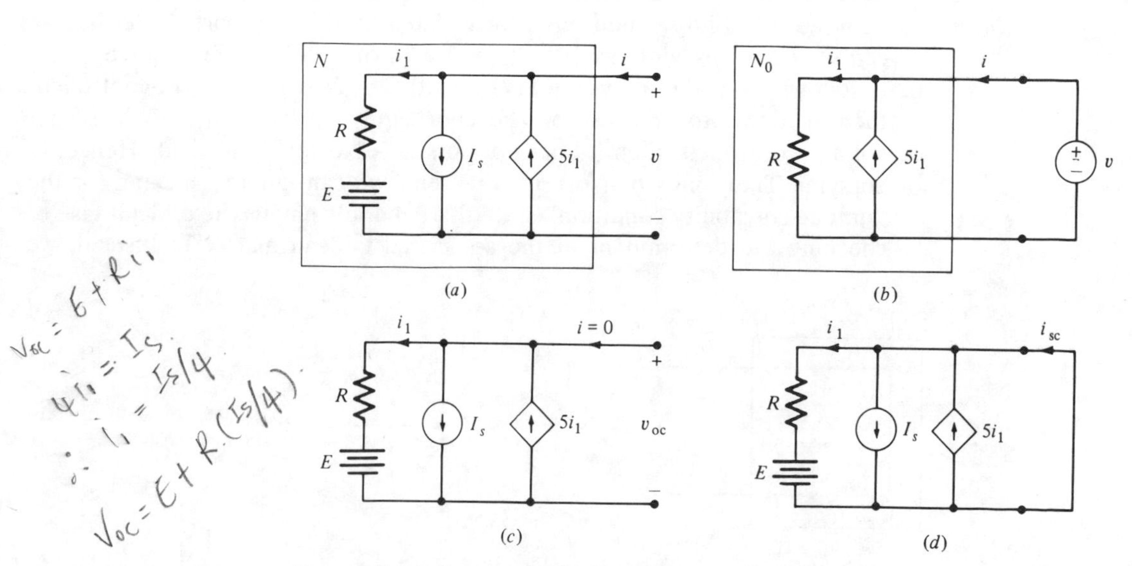

Figure 4.7 (*a*) One-port N. (*b*) Simplified one-port N_0 obtained by setting all *independent* sources inside N to zero. (*c*) Circuit used for calculating v_{oc}. (*d*) Circuit used for calculating i_{sc}.

As an independent check, let us derive i_{sc} using the circuit shown in Fig. 4.7*d*. Since $i_1 = -E/R$ in this case, KCL implies

$$i_{sc} = i_1 - 5i_1 + I_s = \frac{4E}{R} + I_s \tag{4.33}$$

which agrees with Eq. (4.32), as it should.

Example 2 Find the Thévenin and Norton equivalent of the one-port shown in Fig. 4.8*a*.

To calculate R_{eq} and G_{eq}, we use the simplified circuit shown in Fig. 4.8*b*. As a variation from Example 1, let us drive N_0 by a current source i, instead of a voltage source. It follows from the ideal transformer relation [Eq. (2.13) of Chap. 3] that $i_1 = -ni_2 = -ni$. This causes a voltage of $v_1 = R_1 ni$ across R_1. Using the transformer voltage relation $v_2 = nv_1 = R_1 n^2 i$, we find $v = v_2 = n^2 R_1 i$. Hence, we identify

$$R_{eq} = n^2 R_1 \qquad \text{and} \qquad G_{eq} = \frac{1}{n^2 R_1} \tag{4.34}$$

To find $v_{oc}(t)$, we let $i = 0$ in Fig. 4.8*a* to obtain $i_1 = -ni_2 = 0$ and $v_1 = v_s(t)$. Consequently,

$$v_{oc} = nv_s(t) \tag{4.35}$$

It follows from Eqs. (4.34), (4.35), and (4.25) that

$$i_{sc} = \frac{-v_{oc}}{R_{eq}} = \frac{-v_s(t)}{nR_1} \tag{4.36}$$

Application 1 In order to maximize the output power delivered by the power amplifier in a concert hall to the loudspeaker with a voice coil resistance equal to R_L, a transformer with an appropriate turns ratio n is usually sandwiched between the amplifier and the loudspeaker, as shown in Fig. 4.9*a*. The problem

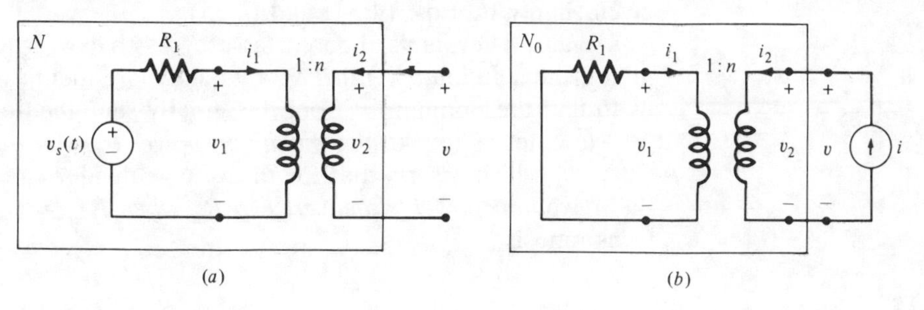

Figure 4.8 (*a*) Ideal transformer one-port N. (*b*) Simplified one-port N_0 obtained by short-circuiting the independent voltage source $v_s(t)$.

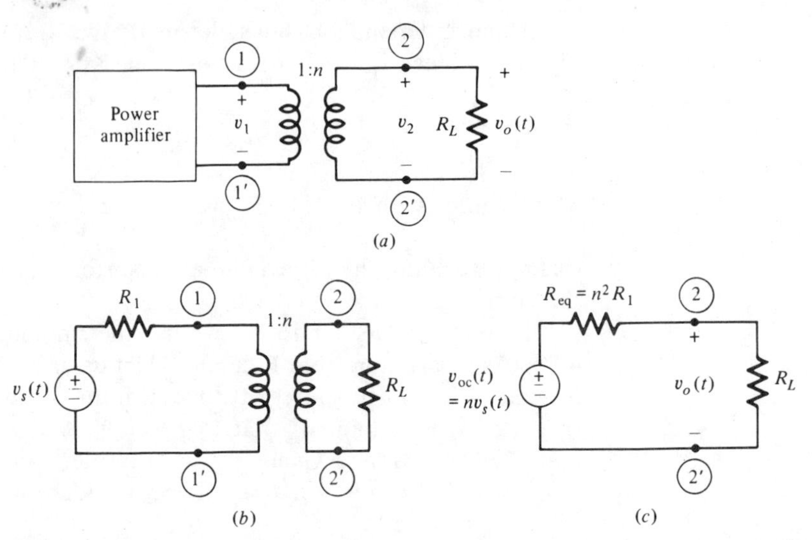

Figure 4.9 (*a*) A transformer sandwiched between a power amplifier and a loudspeaker modeled by R_L. (*b*) Replacing the one-port to the left of terminals ①, ①' by its Thévenin equivalent. (*c*) Replacing the one-port to the left of terminals ②, ②' by its Thévenin equivalent.

is to determine the *optimum* turn ratio n so that *maximum power* is delivered to the load resistor R_L, namely, the loudspeaker.

To solve this problem, our first step is to apply Thévenin's theorem to the power amplifier (which we modeled as a linear time-invariant resistive circuit) to obtain the equivalent circuit shown in Fig. 4.9*b*. Here, R_1 and $v_s(t)$ denote the Thévenin equivalent resistance and the open-circuit voltage of the output stage of the power amplifier. Assuming that R_1 and $v_s(t)$ have been correctly calculated, Thévenin's theorem assures us that the solution to the right of terminals ① and ①' in Fig. 4.9*a* and *b* are identical.

Since our goal is the solution in R_L, let us apply Thévenin's theorem one more time and replace the one-port to the left of terminals ② and ②' in Fig. 4.9*b* by its Thévenin equivalent as shown in Fig. 4.9*c*. Since this one-port is identical to that of Fig. 4.8*a*, in Example 2, the values of R_{eq} and $v_{oc}(t)$ are given simply by Eqs. (4.34) and (4.35).

Again, Thévenin's theorem assures us that with $R_{eq} = n^2 R_1$ and $v_{oc}(t) = nv_s(t)$, the solution $v_o(t)$ in Fig. 4.9*a* and *c* are identical. Hence, it suffices for us to find the optimum n from this greatly simplified equivalent circuit.

Now, let us invoke the *maximum power transfer theorem* to be derived in Chap. 9, which asserts that for the voltage-divider circuit shown in Fig. 4.9*c*, the *maximum power is delivered to R_L when $R_{eq} = R_L$*. Hence, the optimum turns ratio is

$$n = \sqrt{\frac{R_L}{R_1}} \tag{4.37}$$

Note that without Thévenin's theorem, this problem would have been much harder to solve.

Application 2 Consider the linear resistive circuit shown in Fig. 4.10. Since it contains a voltage source, which is not a voltage-controlled circuit element, the *node analysis* method developed in Sec. 1.1 cannot be applied.

However, if we replace the voltage source resistor series one-port by its *Norton equivalent one-port*, we would obtain the circuit shown earlier in Fig. 1.2, where

$$R_1 = \frac{1}{G_1} \qquad i_{s1}(t) = -\frac{v_{s1}(t)}{R_1} \tag{4.38}$$

The *node equation* for this circuit is then given by Eq. (1.7).

Hence, with the help of *Norton's theorem*, we can enlarge the class of linear resistive circuits amenable to node analysis by allowing independent voltage sources, provided each voltage source is in series with a linear resistor.

PROOF OF THÉVENIN AND NORTON'S THEOREM We will prove only Norton's theorem, as the dual proof then applies to Thévenin's theorem.

Let N denote the one-port in question, and let the remaining part of the circuit $\mathcal{N}$ be denoted by N_L, as shown in Fig. 4.11*a*. By hypotheses, N contains only *linear time-invariant resistors* and independent sources, whereas N_L need not be linear or resistive.

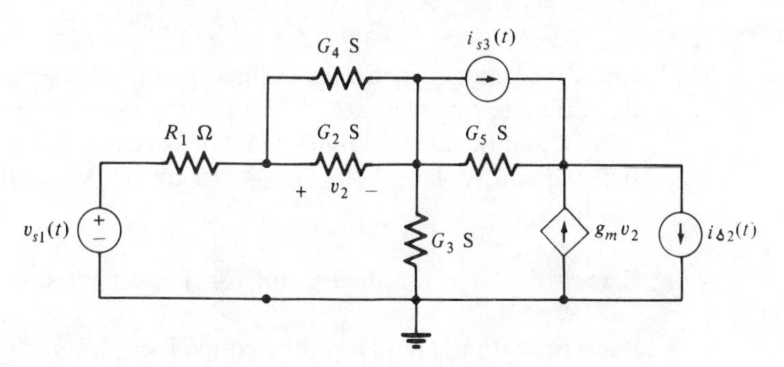

Figure 4.10 Node analysis is not directly applicable to this circuit.

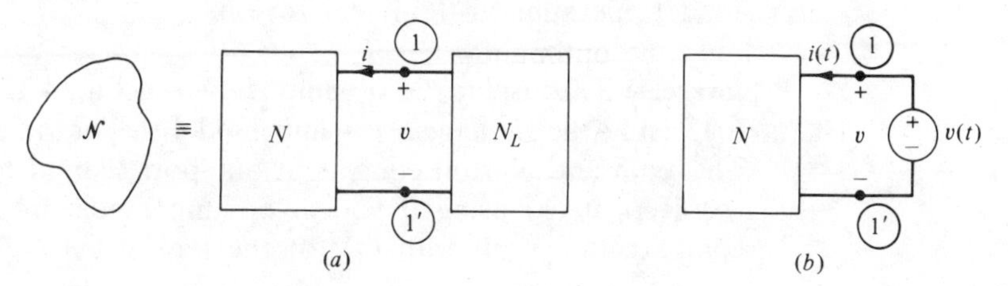

Figure 4.11 (*a*) Partitioning arbitrary circuit $\mathcal{N}$ into a linear resistive one-port N and a not necessarily linear or resistive one-port N_L. (*b*) Driving N with a voltage source $v(t)$.

Since N is purely resistive, it is *completely specified* by its *driving-point characteristic* at each instant of time. Hence, insofar as N_L is concerned, its solution depends only on this driving-point characteristic: The elements inside N which give rise to this driving-point characteristic are completely irrelevant.[19] It suffices therefore for us to prove that both N and its Norton equivalent one-port have an *identical* driving-point characteristic.

Let us drive N with an independent voltage source $v(t)$, as shown in Fig. 4.11*b*. Let us label this voltage source, together with the *independent voltage sources* inside N, by $v_{s0}(t), v_{s1}(t), \ldots, v_{s\alpha}(t)$, respectively, where $v_{s0}(t) \triangleq v(t)$. Similarly, let us label the *independent current sources* inside N by $i_{s1}(t), \ldots, i_{s\beta}(t)$, respectively.

It follows from the "unique solvability condition" that the linear time-invariant resistive circuit in Fig. 4.11*b* has a *unique* solution for all values of the independent sources, at all times. Hence, we can apply the *superposition theorem* and conclude that the port current $i(t)$ in Fig. 4.11*b* must assume the form

$$i(t) = H_0 v(t) + \sum_{k=1}^{\alpha} H_k v_{sk}(t) + \sum_{k=1}^{\beta} K_k i_{sk}(t) \tag{4.39}$$

Now if $v(t) = 0$ for all t, $i(t)$ is by definition $i_{sc}(t)$: Hence the last two sums add up to $i_{sc}(t)$.

If we set to zero all *independent* sources inside N, we are left with $i(t) = H_0 v(t)$, i.e., $H_0 = G_{eq}$. Hence, Eq. (4.39) can be written in the form

$$i(t) = G_{eq} v(t) + i_{sc}(t) \tag{4.40}$$

where G_{eq} and $i_{sc}(t)$ are as defined in the theorem. Equation (4.40) gives the driving-point characteristic of the given one-port N. Since this is the same equation which defines the Norton equivalent one-port N_{eq}, it follows that we can replace N in Fig. 4.11*a* by N_{eq} without affecting the solution inside N_L. ■

Exercise 1 Give the dual proof for Thévenin's theorem.

Exercise 2 Find the Thévenin equivalent circuit of the *negative-resistance converter* circuit given in Fig. 3.8*a* of Chap. 4. Assume an *ideal* op-amp model operating in the linear region.

Exercise 3 Consider the op-amp shown in Fig. 4.12.
(*a*) Using the *finite-gain* op-amp model in Fig. 4.3*b* of Chap. 4, find the Thévenin and Norton equivalent one-port seen at terminals ① and ①'.
(*b*) Repeat (*a*) using the ideal op-amp model in the linear region, and compare the result with (*a*) for the typical value $R_0 = 50\,\Omega$ and $A = 10^5$.

[19] For example, we couldn't care less whether N consists of a 2-Ω resistor, two 1-Ω resistors connected in series, or any interconnection of resistors so long as they all have a 2-Ω equivalent driving-point, or input, resistance.

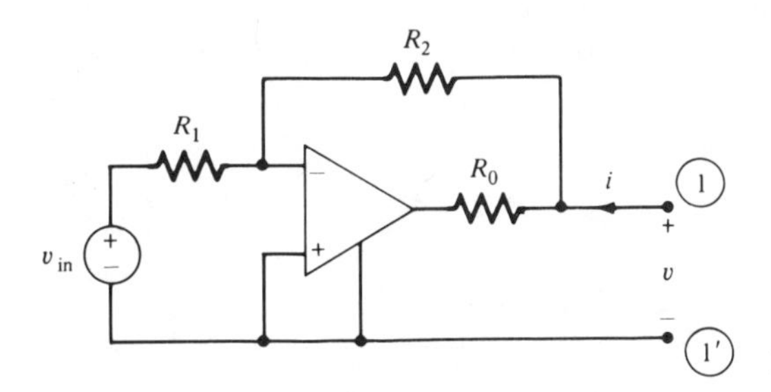

Figure 4.12 Circuit for Exercise 3.

Exercise 4 (*a*) Prove that the Thévenin equivalent resistance R_{eq} is equal numerically to the port voltage v when N is driven by a 1-A current source, provided all *independent* sources inside N are set to zero.
(*b*) Prove the dual statement of (*a*).

4.3 Two-Port Representation Theorem[20]

Let $\mathcal{N}$ denote a well-defined linear time-invariant resistive two-port N terminated in two not necessarily linear or resistive one-ports N_1 and N_2.

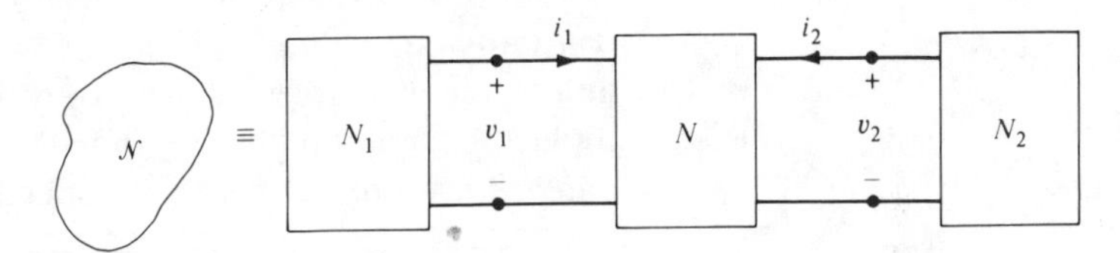

1. *Voltage-controlled representation*:

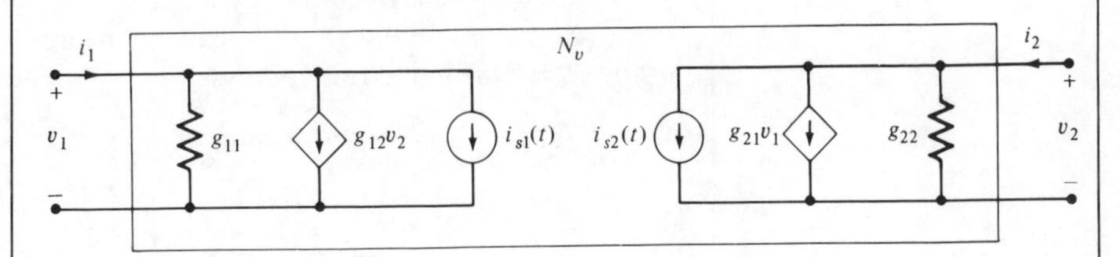

N is *equivalent* to N_v where the parameters g_{jk} are constants depending only on the resistors in N and not on the independent sources, if and only if, the circuit obtained by replacing N_1 and N_2 by arbitrary *independent voltage sources* is *uniquely solvable*.

[20] There is a representation theorem for each two-port representation. We have chosen the three representations here because they will be referred to in Chap. 7.

2. *Current-controlled representation*:

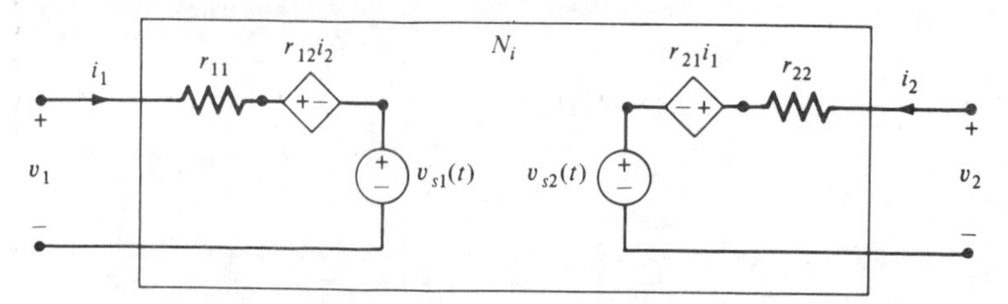

N is *equivalent* to N_i where the parameters r_{jk} are constants depending only on the resistors in N and not on the independent sources, if and only if, the circuit obtained by replacing N_1 and N_2 by arbitrary *independent current sources* is *uniquely solvable*.

3. *Hybrid 2 representation*:

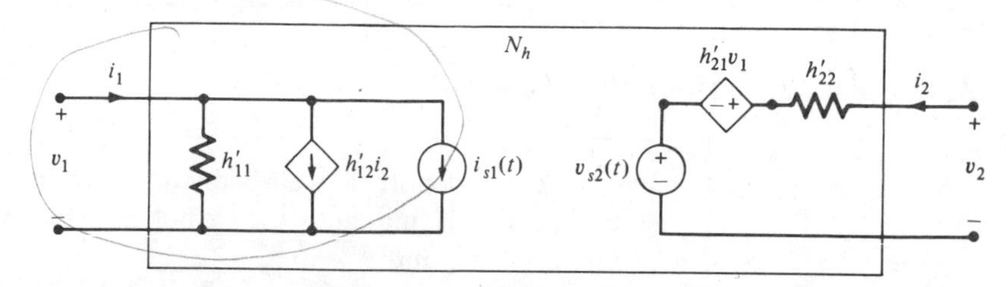

N is *equivalent* to N_h where the parameters h'_{jk} are constants depending only on the resistors in N and not on the independent sources, if and only if, the circuit obtained by replacing N_1 and N_2 by any *independent voltage source* and any *independent current source*, respectively, is *uniquely solvable*.

Moreover, for each of the equivalent two-ports N_v, N_i, or N_h which exists, the associated parameters and *independent-source waveforms* are unique, and N can be represented by a *unique affine function* defined as follows:

1. *Voltage-controlled representation*:

$$\begin{bmatrix} i_1 \\ i_2 \end{bmatrix} = \begin{bmatrix} g_{11} & g_{12} \\ g_{21} & g_{22} \end{bmatrix} \begin{bmatrix} v_1 \\ v_2 \end{bmatrix} + \begin{bmatrix} i_{s1}(t) \\ i_{s2}(t) \end{bmatrix} \tag{4.41}$$

2. *Current-controlled representation*:

$$\begin{bmatrix} v_1 \\ v_2 \end{bmatrix} = \begin{bmatrix} r_{11} & r_{12} \\ r_{21} & r_{22} \end{bmatrix} \begin{bmatrix} i_1 \\ i_2 \end{bmatrix} + \begin{bmatrix} v_{s1}(t) \\ v_{s2}(t) \end{bmatrix} \tag{4.42}$$

3. *Hybrid 2 representation*:

$$\begin{bmatrix} i_1 \\ v_2 \end{bmatrix} = \begin{bmatrix} h'_{11} & h'_{12} \\ h'_{21} & h'_{22} \end{bmatrix} \begin{bmatrix} v_1 \\ i_2 \end{bmatrix} + \begin{bmatrix} i_{s1}(t) \\ v_{s2}(t) \end{bmatrix} \tag{4.43}$$

As usual we give some examples and applications first before proving the theorem.

Example 1 Consider the op-amp two-port N shown in Fig. 4.13. To obtain its equivalent *voltage-controlled* representation, we *must* connect a voltage source v_1 across port 1 and a voltage source v_2 across port 2, and then solve for the associated port currents i_1 and i_2. If the resulting "terminated" circuit is uniquely solvable, then N has a voltage-controlled representation.

Assuming the op amp is ideal and operating in its linear region, and applying KCL at nodes ① and ②, respectively, we obtain

Node ①: $$i_1 = \frac{v_1}{R_3} - \frac{v_2 - v_1}{R_2} - \frac{v_{\text{in}} - v_1}{R_1} \tag{4.44a}$$

Node ②: $$i_2 = \frac{v_1}{R_3} \tag{4.44b}$$

Rewriting Eq. (4.44) in matrix form, we obtain

$$\begin{bmatrix} i_1 \\ i_2 \end{bmatrix} = \begin{bmatrix} \frac{1}{R_1} + \frac{1}{R_2} + \frac{1}{R_3} & -\frac{1}{R_2} \\ \frac{1}{R_3} & 0 \end{bmatrix} \begin{bmatrix} v_1 \\ v_2 \end{bmatrix} + \begin{bmatrix} -\frac{1}{R_1} v_{\text{in}}(t) \\ 0 \end{bmatrix} \tag{4.45}$$

Comparing Eqs. (4.41) and (4.45) and invoking the uniqueness of the parameters and independent source waveforms for the equivalent two-port N_v, we obtain

$$g_{11} = \frac{1}{R_1} + \frac{1}{R_2} + \frac{1}{R_3} \qquad g_{12} = -\frac{1}{R_2} \qquad g_{21} = \frac{1}{R_3} \qquad g_{22} = 0$$

$$i_{s1}(t) = -\frac{1}{R} v_{\text{in}}(t) \qquad \text{and} \qquad i_{s2}(t) = 0$$

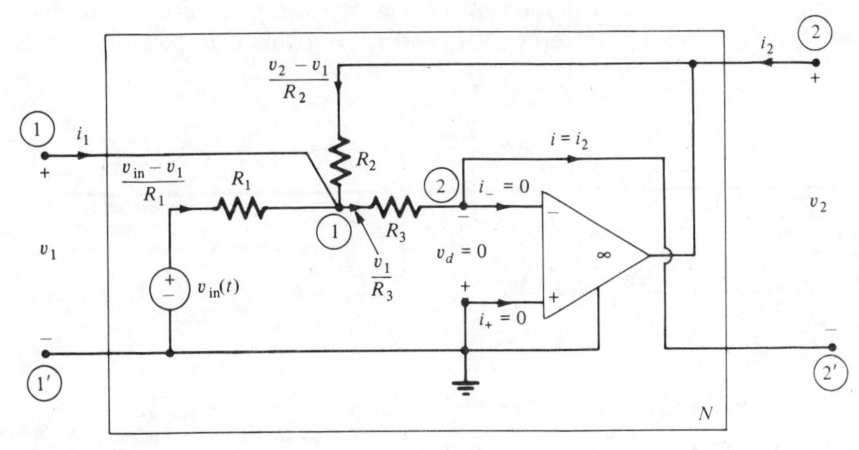

Figure 4.13 A two-port linear op-amp circuit for Example 1.

Note that if R_1, R_2, or R_3 is replaced by a short circuit, some g_{jk} would become *infinite*. In this case, N does *not* have a voltage-controlled representation.

Example 2 Consider the two-port N shown in Fig. 4.14. To obtain its equivalent *hybrid 2 representation*, we *must* connect a voltage source v_1 across port 1 and a current source i_2 across port 2. Assuming the resulting terminated circuit is uniquely solvable, then N has a hybrid 2 representation. Applying KCL at node ① and KVL around port 1, port 2, $v_s(t)$, and the resistor, and noting that $i_3 = i_2$, we obtain

KCL: $$i_1 = -i_2 - 2[v_s(t) - 5i_2] = 9i_2 - 2v_s(t) \tag{4.46a}$$

KVL: $$v_2 = v_1 + 5i_2 - v_s(t) \tag{4.46b}$$

Rewriting Eq. (4.46) in the matrix form

$$\begin{bmatrix} i_1 \\ v_2 \end{bmatrix} = \begin{bmatrix} 0 & 9 \\ 1 & 5 \end{bmatrix} \begin{bmatrix} v_1 \\ i_2 \end{bmatrix} + \begin{bmatrix} -2v_s(t) \\ -v_s(t) \end{bmatrix} \tag{4.47}$$

and comparing with Eq. (4.43), we obtain:

$$h'_{11} = 0 \qquad h'_{12} = 9 \qquad h'_{21} = 1 \qquad h'_{22} = 5$$

$$i_{s1}(t) = -2v_s(t) \qquad \text{and} \qquad v_{s2}(t) = -v_s(t)$$

Note that if we solve for i_2 in Eq. (4.46*b*) in terms of v_1 and v_2 and substitute the result for i_2 in Eq. (4.46*a*), we would obtain the following *equivalent* voltage-controlled representation:

$$\begin{bmatrix} i_1 \\ i_2 \end{bmatrix} = \begin{bmatrix} -\frac{9}{5} & \frac{9}{5} \\ -\frac{1}{5} & \frac{1}{5} \end{bmatrix} \begin{bmatrix} v_1 \\ v_2 \end{bmatrix} + \begin{bmatrix} -\frac{1}{5} & v_s(t) \\ \frac{1}{5} & v_s(t) \end{bmatrix} \tag{4.48}$$

On the other hand, to obtain an equivalent current-controlled representation, it is necessary to replace Eq. (4.46*a*) by an equation involving v_1 in terms of i_1 and i_2. But this is impossible in this case because v_1 is not

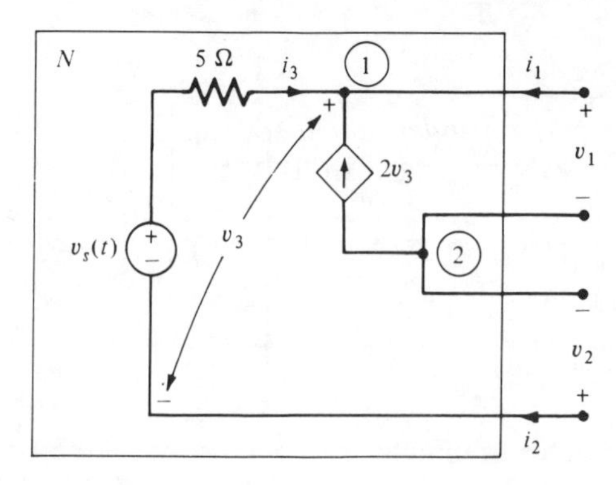

Figure 4.14 Circuit for Example 2.

present in Eq. (4.46*a*) and no variable (other than i_1 and i_2) is available to relate to v_1. We conclude therefore that the two-port in Fig. 4.14 does *not* have a current-controlled representation.

Let us give a circuit interpretation of what happens when a particular representation fails to exist, using the circuit in Fig. 4.14 as a vehicle. Note that in order for N to have a *current-controlled representation*, it is *necessary* to drive both ports of N with two *independent* current sources i_1 and i_2, respectively, say $i_1 = I_1$ and $i_2 = I_2$. Since $i_3 = I_2$, KCL at node ① *constrains* the current flowing *down* the wire connected to the controlled source to be equal to $I_1 + I_2$ at all times. Since the voltage $v_3 = v_s(t) - 5I_2$ is fixed in this case, the current flowing down the VCCS must also be fixed at $-2v_3 = 10I_2 - 2v_s(t)$ at all times. Hence KCL *cannot* be satisfied at node ① for arbitrary values of I_1 and I_2, and the current-source terminated circuit is not solvable.

REMARKS

1. A *linear* resistive two-port N may admit simultaneously several *equivalent* representations. A particular representation will fail to exist if and only if the tableau equation of the associated circuit (terminated by appropriate independent sources) does not have a solution or does not have a unique solution for arbitrary values of independent sources.
2. There exist other equivalent representations (e.g., see Table 3.3 of Chap. 3). The three representations are singled out in the preceding theorem because they will be needed in Chap. 7.
3. The two-ports N_v, N_i, and N_h (assuming they exist) defined in the two-port representation theorem are all *equivalent* in the sense that N can be replaced by *any one* without affecting the solution in the terminated one-ports N_1 and N_2. Note that N_1 and N_2 *are arbitrary* and are not restricted to the independent sources stipulated *in connection with unique solvability*. In fact, N_1 and N_2 can be nonlinear and *not* resistive: In Chaps. 6 and 7, N_1 and N_2 will be capacitors or inductors.
4. We have deliberately refrained from drawing the terminating independent sources in Figs. 4.13 and 4.14 to avoid creating the erroneous impression that the resulting equivalent two-port holds only under such terminations. Once a particular equivalent two-port exists, it is *indistinguishable* externally from its parent two-port N.
5. If N contains no *independent* sources, the vector on the right of Eqs. (4.41) to (4.43) vanishes. In this case, N is represented by a *linear* function.
6. If N contains *time-varying* resistors, the above theorem still applies. In this case, the parameters g_{jk}, r_{jk}, and h'_{jk} become functions of time.

Application 1 (hybrid analysis) Consider a 500-node circuit containing 1000 *linear* two-terminal resistors, 1000 dc voltage sources, 1000 dc current sources, a voltage-controlled nonlinear resistor $\mathcal{R}_a$ described by $i_a = \hat{i}_a(v_a)$, and a

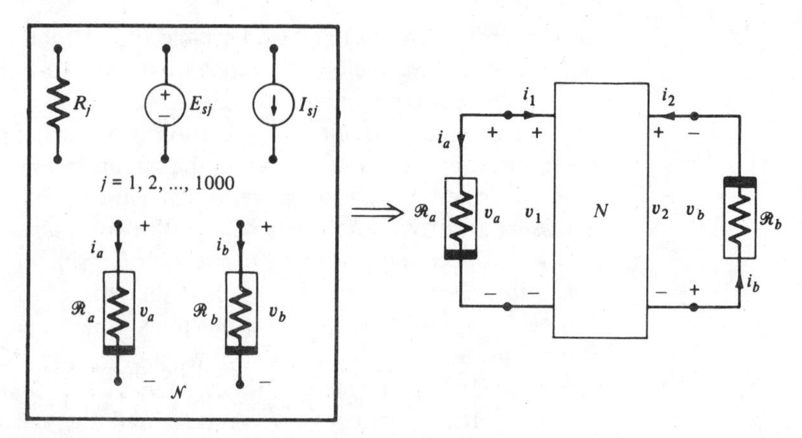

Figure 4.15 A large resistive circuit containing only two nonlinear resistors and its equivalent circuit.

current-controlled nonlinear resistor $\mathcal{R}_b$ described by $v_b = \hat{v}_b(i_b)$, as depicted in Fig. 4.15. The problem is to solve for v_a and i_b.

Since the associated *nonlinear* tableau equation consists of $(n-1)+2b = 6503$ equations, solving them via the Newton-Raphson algorithm would be prohibitively expensive.

A much more efficient method is to extract $\mathcal{R}_a$ and $\mathcal{R}_b$ from $\mathcal{N}$ and to consider the remaining circuit as a two-port N.[21] Assuming N has a *hybrid 2 representation* in the form of Eq. (4.43), we can solve for the six parameters h'_{11}, h'_{12}, h'_{21}, h'_{22}, i_{s1}, and v_{s2} as follows:

1. Since

$$h'_{11} = \left.\frac{i_1}{v_1}\right|_{i_2=0,\ i_{s1}=0} \tag{4.49}$$

   we apply a 1-V battery across port 1 (so that $v_1 = 1$), open-circuit port 2 (so that $i_2 = 0$), set all independent sources in N to zero (so that $i_{s1} = 0$), and solve for i_1. Then h'_{11} is equal numerically to the resulting current i_1 entering port 1. The other hybrid parameters are similarly calculated.

2. Since

$$i_{s1} = i_1\big|_{v_1=0,\ i_2=0} \qquad \text{and} \qquad v_{s2} = v_2\big|_{v_1=0,\ i_2=0} \tag{4.50}$$

   we short-circuit port 1, open-circuit port 2, and solve for $i_1 = i_{s1}$ and $v_2 = v_{s2}$.

Hence, all six parameters can be found by solving three linear resistive circuits. This need be done *only once*.

[21] It will be clear shortly that it is more convenient to define the reference directions of N, $\mathcal{R}_a$, and $\mathcal{R}_b$ such that $v_a = v_1$ and $i_b = i_2$, as shown in Fig. 4.15.

Substituting next $i_1 = -\hat{i}_a(v_1)$ and $v_2 = -\hat{v}_b(i_2)$ into Eq. (4.43), we obtain[22]

$$\begin{aligned} \hat{i}_a(v_1) + h'_{11}v_1 + h'_{12}i_2 + i_{s1} &= 0 \\ \hat{v}_b(i_2) + h'_{21}v_1 + h'_{22}i_2 + v_{s2} &= 0 \end{aligned} \tag{4.51}$$

We can now solve for $v_1 = v_a$ and $i_2 = i_b$ by solving Eq. (4.51) via the Newton-Raphson algorithm. Note that the time-consuming iteration task has been reduced dramatically because we only have to iterate on 2 equations, versus 6503 equations if we were to solve the tableau equation blindly. The preliminary cost involved in carrying out three separate "linear circuit" analyses is insignificant since this need be done only once, at the beginning.

A similar approach will be used in Chap. 7 where $\mathscr{R}_a$ and $\mathscr{R}_b$ become a capacitor and/or an inductor.

Application 2 We can now enlarge the class of allowable *nonlinear* resistors considered earlier in Sec. 3.2 to include three-terminal or two-port *nonlinear* resistors with the help of the equivalent two-port N_v, N_i, or N_h.

For example, consider a three-terminal nonlinear resistor described by a hybrid 2 representation:

$$\begin{aligned} i_1 &= g_1(v_1, i_2) \\ v_2 &= g_2(v_1, i_2) \end{aligned} \tag{4.52}$$

It follows from the derivations in Sec. 3.2 and the two-port representation theorem that the *Newton-Raphson discrete equivalent circuit* for this element is as shown in Fig. 4.16. We leave it as an exercise to define the parameters $h'^{(j)}_{jk}$, $i^{(j)}_{s1}$, and $v^{(j)}_{s2}$ in terms of the coefficients of the Taylor expansion of Eq. (4.52).

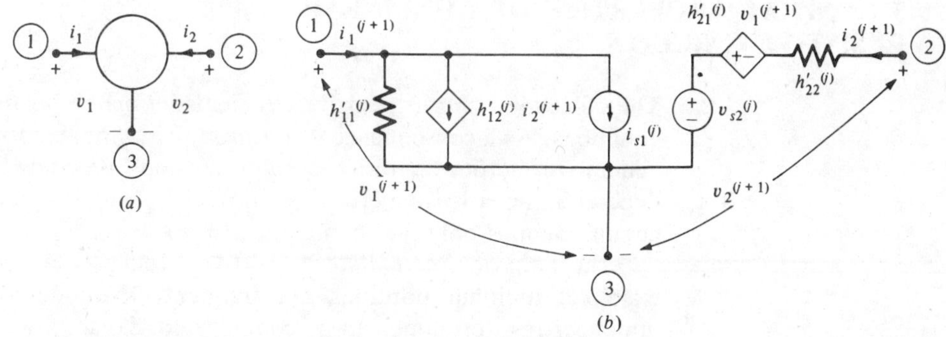

Figure 4.16 (*a*) Three-terminal nonlinear resistor described by Eq. (4.52). (*b*) Newton-Raphson discrete equivalent circuit.

[22] Note that we have drawn the nonlinear resistor-terminated two-port in Fig. 4.15 so that $i_1 = -i_a$, $v_2 = -v_b$, but $v_1 = v_a$ and $i_2 = i_b$. This is done to avoid introducing a negative sign within the parenthesis. Clearly, this entails no loss of generality.

PROOF OF THE TWO-PORT REPRESENTATION THEOREM We will prove only the voltage-controlled case, as the other cases follow similarly. Let us label the independent voltage sources (both external and inside N) by $v_1, v_2, v_{s1}(t), \ldots, v_{s\alpha}(t)$ and the independent current sources by $i_{s1}(t), \ldots, i_{s\beta}(t)$. Since, by hypotheses, the linear resistive circuit obtained by replacing N_1 and N_2 by independent voltage sources v_1 and v_2 has a *unique* solution, it follows from the *superposition theorem* that i_1 and i_2 must assume the form

$$i_1 = H_{1a}v_1 + H_{1b}v_2 + \underbrace{H_{11}v_{s1}(t) + \cdots + H_{1\alpha}v_{s\alpha}(t) + K_{11}i_{s1}(t) + \cdots + K_{1\beta}i_{s\beta}(t)}_{i_{s1}(t)} \tag{4.53a}$$

$$i_2 = H_{2a}v_1 + H_{2b}v_2 + \underbrace{H_{21}v_{s1}(t) + \cdots + H_{2\alpha}v_{s\alpha}(t) + K_{21}i_{s1}(t) + \cdots + K_{2\beta}i_{s\beta}(t)}_{i_{s2}(t)} \tag{4.53b}$$

Observe that Eq. (4.53) has exactly the form as Eq. (4.41). This equation in turn describes the two-port N_v, for all possible values of v_1 and v_2.

Now since the two-port N is *completely characterized* by Eq. (4.53), it is indistinguishable externally from N_v. Consequently N and N_v are equivalent. ■

Exercises

1. Prove the two-port representation theorem for the *hybrid 2 representation*.
2. Derive the parameters in the discrete equivalent circuit in Fig. 4.16*b* in terms of $g_1(v_1, v_2)$ and $g_2(v_1, v_2)$.
3. Derive the equivalent two-port for the *hybrid 1 representation*.

5 GENERAL PROPERTIES OF NONLINEAR RESISTIVE CIRCUITS[23]

The behavior of *linear* resistive circuits is intimately related to linear algebraic equations. As a consequence of *linearity*, we were able to derive several rather general properties in the preceding section. Precisely because their proofs depend on linearity in a crucial way, none of these properties holds even if the circuit contains only one nonlinear resistor.

The behavior of *nonlinear* resistive circuits is far more complicated; for example, multiple solutions are frequent. Even describing a two-terminal nonlinear resistor alone can be complicated: To specify it analytically we need to use a *function* which may require many parameters (e.g., the *pn*-junction diode, the concave resistor, and the convex resistor in Chap. 2 require two parameters; others would require even more).

[23] Except for Sec. 5.1, the rest of this section may be skipped without loss of continuity. For more general properties of nonlinear circuits not covered in this book, see L.O. Chua, "Nonlinear Circuits," *IEEE Trans. on Circuits and Systems*, vol. CAS-31, pp. 69–87, January 1984.

In spite of its greatly increased complexity, many useful properties can be proved for various subclasses of *nonlinear* resistive circuits. Our objective in this section is to state only those properties which we are in a position to prove—some in a remarkably elegant manner. These general properties form only a small albeit important subset of our "nonlinear tool kits." We hope this final section will whet the reader's appetite into a more advanced study of this subject.

5.1 Substitution Theorem

Figure 5.1*a* shows a circuit $\mathcal{N}$ made of a *well-defined resistive* (possibly nonlinear and time-varying) one-port N_R terminated in a one-port N_L which need *not* be resistive or linear.

1. If $\mathcal{N}$ has a *unique* solution[24] $v = \hat{v}(t)$ for all t, then N_L may be substituted by a voltage source $\hat{v}(t)$ without affecting the branch voltage and branch current solution inside N_R, provided the substituted circuit $\mathcal{N}_v$ in Fig. 5.1*b* has a *unique* solution for all t.
2. If $\mathcal{N}$ has a *unique* solution[24] $i = \hat{i}(t)$ for all t, then N_L may be substituted by a current source $\hat{i}(t)$ without affecting the branch voltage and branch current solution inside N_R, provided the substituted circuit $\mathcal{N}_i$ in Fig. 5.1*c* has a *unique* solution for all t.

Circuit interpretation Note that if $\mathcal{N}$ is *not* uniquely solvable, then the conclusion of the *substitution theorem* becomes ambiguous because the *unique* solution of $\mathcal{N}_v$, or $\mathcal{N}_i$, can at best be identical to only one of the multiple solutions of $\mathcal{N}$. To show that the substitution theorem is false if $\mathcal{N}_v$ or $\mathcal{N}_i$ is *not*

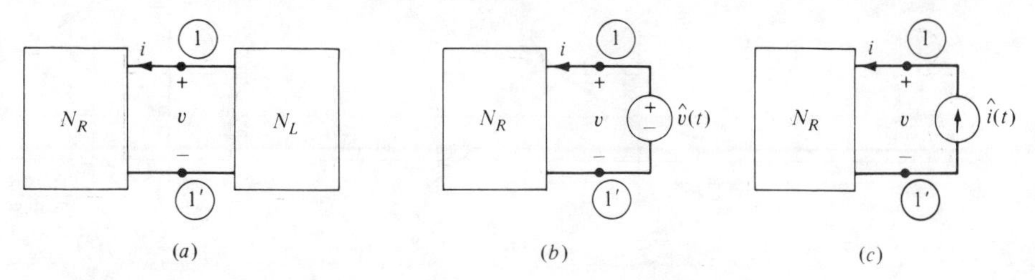

Figure 5.1 (*a*) Circuit $\mathcal{N}$. (*b*) Substituted circuit $\mathcal{N}_v$. (*c*) Substituted circuit $\mathcal{N}_i$.

[24] If N_L contains *dynamic* elements, such as a capacitor, then different initial conditions would of course give different solution waveforms. In this case, the solution $\hat{v}(t)$ is assumed to be *unique for each initial condition*. Note also that we have departed from our usual "^" notation: Throughout this section, $\hat{v}(t)$ and $\hat{i}(t)$ denote a voltage and a current time waveform, respectively.

uniquely solvable, let N_R be a *current-controlled* resistor and let N_L denote a resistor-battery combination whose *load line* intersects the v-i curve at one point (E, I) only, as shown in Fig. 5.2*a*. The circuit $\mathcal{N}_v$ in this case would give rise to three distinct current solutions, as shown in Fig. 5.2*b*, where the original circuit $\mathcal{N}$ has only one. Consequently, any substituted circuit $\mathcal{N}_v$, or $\mathcal{N}_i$, must be assumed to be uniquely solvable in order to exclude these extraneous solutions.

PROOF OF THE SUBSTITUTION THEOREM The digraphs for $\mathcal{N}$, $\mathcal{N}_v$, and $\mathcal{N}_i$ are identical assuming that N_L is treated as a single branch. Let us label all branches in N_R consecutively from 1 to $b-1$. The *last* branch b therefore corresponds to N_L in $\mathcal{N}$, the voltage source in $\mathcal{N}_v$, and the current source in $\mathcal{N}_i$, respectively. The tableau equation for $\mathcal{N}$ is given by:[25]

KCL:
$$\mathbf{A}\begin{bmatrix}\mathbf{i}_R \\ i_b\end{bmatrix} = \mathbf{0} \tag{5.1}$$

KVL:
$$\begin{bmatrix}\mathbf{v}_R \\ v_b\end{bmatrix} - \mathbf{A}^T\mathbf{e} = \mathbf{0} \tag{5.2}$$

Branch equations:

$$\mathbf{h}(\mathbf{v}_R, \mathbf{i}_R, t) = \mathbf{0} \tag{5.3}$$

$$h_b(v_b, i_b, t) = 0 \tag{5.4}$$

The last equation denotes the driving-point characteristic of N_L and *need not* be an algebraic relation.[26]

Now consider the circuit $\mathcal{N}_v$. Its tableau equation is given by Eqs. (5.1) to (5.3), *and*

$$v_b - \hat{v}(t) = 0 \tag{5.4$'$}$$

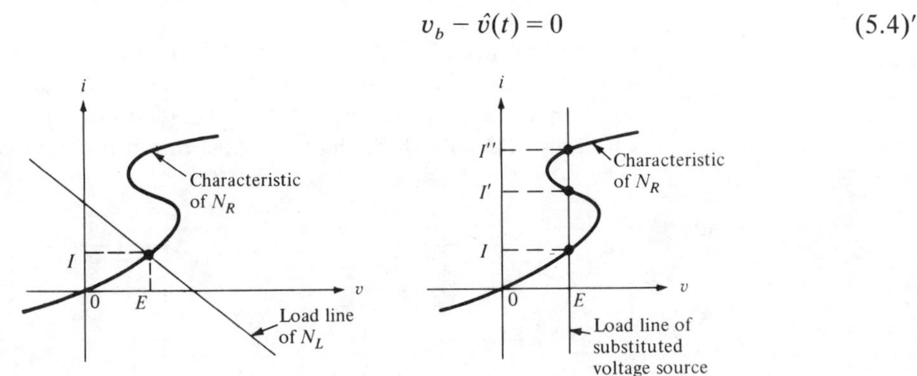

Figure 5.2 (*a*) $\mathcal{N}$ has a unique operating point. (*b*) Substituted circuit $\mathcal{N}_v$ has three operating points.

[25] Here, $\mathbf{v}_R$ and $\mathbf{i}_R$ denote the branch voltages and branch currents of all branches inside N_R.

[26] For example, if branch b is a C-farad capacitor, then this equation becomes $h_b(v_b, i_b, t) \triangleq i_b - C\,dv_b/dt = 0$.

$$\text{Now let} \qquad \hat{\mathbf{e}}(t)\,, \qquad \begin{bmatrix} \hat{\mathbf{v}}_R(t) \\ \hat{v}(t) \end{bmatrix} \qquad \text{and} \qquad \begin{bmatrix} \hat{\mathbf{i}}_R(t) \\ \hat{i}(t) \end{bmatrix} \tag{5.5}$$

denote the unique solution of Eqs. (5.1) to (5.4). Since the first $(n-1)+2b-1$ equations of both $\mathcal{N}$ and $\mathcal{N}_v$ are identical, expression (5.5) is clearly also a solution of Eqs. (5.1) to (5.3), and (5.4)′ because *any* $i_b'(t)$ will satisfy Eq. (5.4)′—it is a voltage source.

But, by hypotheses, $\mathcal{N}_v$ has a *unique* solution, consequently, expression (5.5) is the *only* solution of the substituted circuit $\mathcal{N}_v$.

A similar proof applies to the substituted circuit $\mathcal{N}_i$. ∎

5.2 Loop-Cut-Set Exclusion Property

Our first property will be stated in terms of a *digraph* and is therefore valid for *any* circuit, including nonlinear and nonresistive circuits. To simplify the statement of this property, it is convenient to introduce the following terminologies.

A *loop* in a digraph is said to be *similarly directed* iff all branches in the loop are oriented in the same direction. For example, in the digraph shown in Fig. 5.3, loop {1, 2, 3, 4, 5} is similarly directed, but loop {1, 2, 11, 10, 7, 5} is not.

A *cut set* in a digraph is said to be *similarly directed* iff all branches in the cut set are oriented in the same direction, i.e., either all entering, or all leaving, the associated gaussian surface. For example, in the digraph shown in Fig. 5.3, cut set {7, 8, 9, 11} is similarly directed, but cut set {4, 6, 2, 11} is not.

> *Loop-cut-set exclusion property*[27] Let β^* denote any branch in a connected digraph $\mathcal{G}$. Then there are two *mutually exclusive* possibilities: β^* belongs to either a *similarly directed loop* or a *similarly directed cut set*, but not to both.

Example 1 Pick branch 4 in the digraph of Fig. 5.3 to be β^*. Since branch 4 belongs to the *similarly directed loop* {1, 2, 3, 4, 5}, the above property guarantees that branch 4 does *not* belong to any *similarly directed cut set*, as can be easily verified.

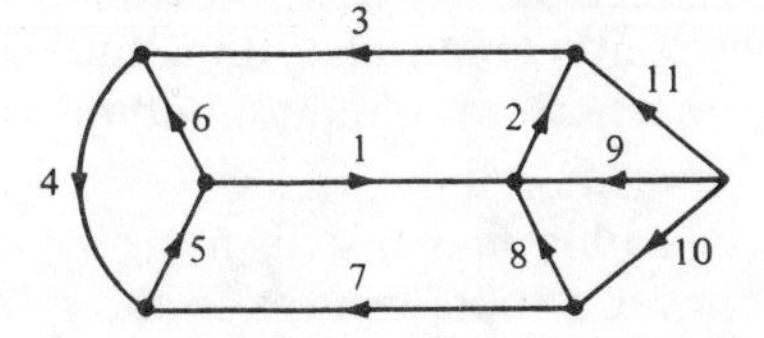

Figure 5.3 Illustration of a similarly directed loop and a similarly directed cut set.

[27] This property is crucial in the proof of many nonlinear circuit results. This will be illustrated in Sec. 5.3. For more applications, see J. Vandewalle and L. O. Chua, "The Colored-Branch Theorem and its Applications in Circuit Theory," *IEEE Trans. on Circuits and Systems*, vol. CAS-27, pp. 816–825, September 1980.

Example 2 Pick branch 7 in the digraph of Fig. 5.3 to be β^*. It is easily verified that branch 7 does *not* belong to any *similarly directed loop*. It follows from the above property that branch 7 *must* belong to at least one *similarly directed cut set*, e.g., cut set $\{7, 8, 9, 11\}$.

PROOF OF THE LOOP-CUT-SET EXCLUSION PROPERTY Suppose the contrary; i.e. suppose branch β^* belongs *simultaneously* to both a *similarly directed loop* formed by branches $\{\beta_1, \beta_2, \ldots, \beta_p\}$, and a *similarly directed cut set* formed by branches $\{\beta'_1, \beta'_2, \ldots, \beta'_q\}$. Let us assign the following current and voltage distributions to all branches in $\mathcal{G}$:

$$(a)\quad i_j = \begin{cases} 1 & \text{if it belongs to a branch in the loop } \{\beta_1, \beta_2, \ldots, \beta_p\} \\ 0 & \text{otherwise} \end{cases}$$

Clearly, this particular choice of branch currents satisfies KCL.

$$(b)\quad v_j = \begin{cases} 1 & \text{if it belongs to a branch in the cut set } \{\beta'_1, \beta'_2, \ldots, \beta'_q\} \\ 0 & \text{otherwise} \end{cases}$$

Clearly, this particular choice of branch voltages satisfies KVL.

It follows from Tellegen's theorem that $\Sigma_{j=1}^{b} v_j i_j = 0$. But this is impossible for the above choice of current and voltage distributions because $v_j \geq 0$ and $i_j \geq 0$ and at least one branch, namely, branch β^*, has $v_j i_j > 0$. Hence, branch β^* can*not* belong to both a similarly directed loop *and* a similarly directed cut set.

To prove that given any branch β^*, there always exists at least one similarly directed loop, or one similarly directed cut set, containing branch β^*, think of each branch in the digraph as a *one-way* street. Label the *beginning* and the *end* node of branch β^* by ① and ②, respectively, i.e., β^* is directed from node ① to node ②. Our algorithm consists of moving *away* from node ② along *all possible* one-way streets until we eventually return to node ①. If we succeed, then the branches along the closed route form a similarly directed loop *containing* branch β^*.

For example, choose branch 4 in Fig. 5.3 to be our desired branch β^*. Starting from the end node of branch 4, we move *away* along branch 5. From the *end* node of branch 5, we move away along branch 6 or branch 1. If we take the former, we would get the *similarly directed loop* $\{4, 5, 6\}$. If we pick the latter, we continue our algorithm and move away along branch 2, and finally to branch 3 to obtain another similarly directed loop $\{4, 5, 1, 2, 3\}$.

Of course, our algorithm may *not* always get us back to node ①. This situation occurs whenever a route returns to the *end* node ② of branch β^* or whenever all possible routes are eventually blocked at some intermediate end nodes where all confluent branches are directed *toward* (and not away from) these nodes.

For example, choose branch 7 in Fig. 5.3 to be our branch β^*. Starting from the *end* node of branch 7, we can only move along branches $\{5, 6, 4\}$ or $\{5, 1, 2, 3, 4\}$. In either case, we return to the end node of branch β^*, and hence β^* is *not* contained in any similarly directed loop. To illustrate the second possibility, let us *reverse* the direction of branch 4 in Fig. 5.3. In this case, our algorithm would terminate at the *end* node of branch 6 and branch 3.

Now let us *coalesce* all nodes that we have traversed, including all blocked end nodes, with node ② of branch β^* into a single "supernode" ②'. This is equivalent to short-circuiting all branches that we have traversed, as well as those branches that formed self-loops along the coalescing process. Note that all remaining branches connected to the coalesced supernode ②' must be directed *toward* node ②. These branches, together with branch β^*, form either a *similarly directed cut set* or a union of two or more similarly directed cut sets.

For example, choosing branch 7 in Fig. 5.3 as our branch β^* as before, the supernode is obtained by short-circuiting branches $\{5, 6, 4, 1, 2, 3\}$. The remaining branches $\{7, 8, 9, 11\}$ connected to this supernode clearly form a *similarly directed cut set*. ∎

Exercise 1 Suppose all streets in a small isolated town are one-way streets, e.g., Fig. 5.3. Suppose there are 15 streets in this road system which form a similarly directed cut set and they are all traversed in the same direction. Explain why people living along these 15 streets do not park their cars at home.

Exercise 2 (*a*) Can branch 9 in Fig. 5.3 belong to some similarly directed loop?

(*b*) Can branch 6 in Fig. 5.3 belong to a similarly directed cut set?

5.3 Consequences of Strict Passivity

A two-terminal resistor is said to be *strictly passive* iff $vi > 0$ for all points (v, i) on its characteristic except the origin $(0, 0)$.

Geometrically, this means that the v-i curve of a *strictly passive* resistor must lie only in the first and third quadrants and stay clear of the v and i axis, except the origin.

Most of the nonlinear resistors encountered in Chap. 2 are strictly passive. However, the ideal diode, zener diode, concave resistor, and convex resistor are passive but *not* strictly passive.

In this section we will state and prove four general properties for circuits containing *only* strictly passive resistors and independent sources.

Strict passivity property A one-port made of strictly passive two-terminal resistors is itself strictly passive.

PROOF Consider the one-port N shown in Fig. 5.4, which is driven by a voltage source. Let N contain m strictly passive resistors. Applying Tellegen's theorem and noting that the current *entering* the positive terminal of the voltage source is equal to $-i$ (associated reference direction convention), we obtain

$$vi = \sum_{\alpha=1}^{m} v_\alpha i_\alpha$$

Since the m resistors are strictly passive for all α, $v_\alpha i_\alpha \geq 0$, hence $vi \geq 0$.

Suppose $v > 0$, then by KVL some of the v_α's must be nonzero; thus by strict passivity the corresponding i_α's are also nonzero and of the same sign. Hence, whenever $v > 0$, at least one term, say $v_k i_k$, is positive. So we have $v > 0$ implies $i > 0$. A similar proof shows that $v < 0$ implies $i < 0$. Hence, $vi > 0$ for all points on the driving-point characteristic except the origin. Therefore N is strictly passive. ∎

Maximum node-voltage property Let $\mathcal{N}$ be a connected circuit made of *strictly passive two-terminal resistors* and driven by a single dc voltage source of E volts, where $E > 0$. Then, with the negative voltage-source terminal chosen as datum, no node-to-datum voltage can exceed E volts.

PROOF Since $\mathcal{N}$ is connected, all node-to-datum voltages $e_1, e_2, \ldots, e_{n-1}$ are well-defined. Suppose there exists a node ⓜ with the *highest* potential $e_m > E$. Then all elements connected to node ⓜ are two-terminal *resistors*, say $\mathcal{R}_a, \mathcal{R}_b, \ldots, \mathcal{R}_k$, as shown in Fig. 5.5.

Since $e_a, e_b, \ldots, e_k \leq e_m$ (recall node ⓜ has the highest potential), we have $v_a, v_b, \ldots, v_k \geq 0$. Since all resistors are *strictly passive*, this implies

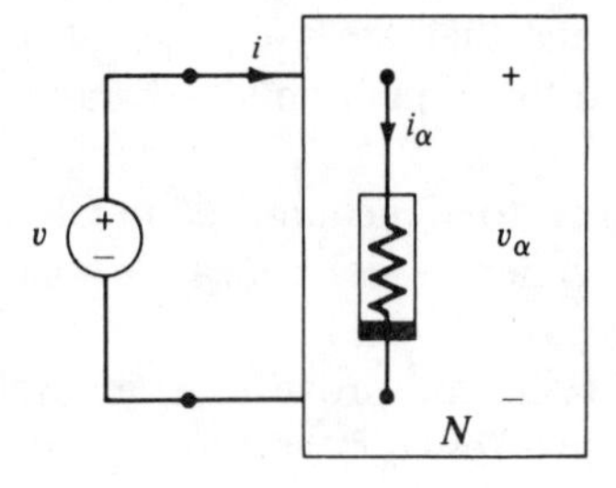

Figure 5.4 One-port N.

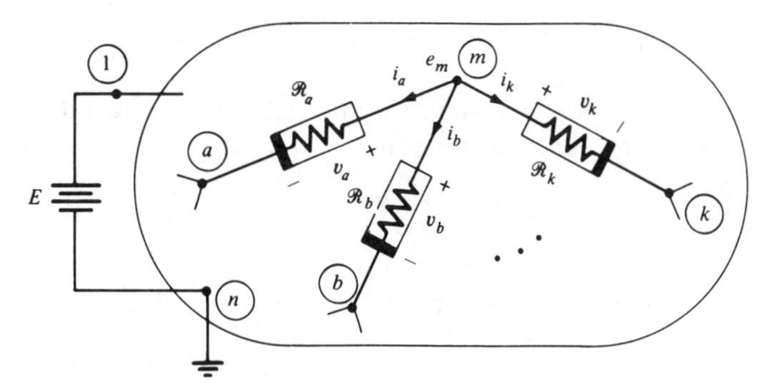

Figure 5.5 KCL at node ⓜ implies $i_a + i_b + \cdots + i_k = 0$.

that $i_a, i_b, \ldots, i_k \geq 0$. Now KCL would be violated at node ⓜ unless $i_a = i_b = \cdots = i_k = 0$. By strict passivity this implies that $v_a = v_b = \cdots = v_k = 0$, and hence $e_a = e_b = \cdots = e_k = e_m > E$.

Hence, we can move on to nodes ⓐ, ⓑ, . . . , ⓚ and repeat the above reasoning to conclude that all resistors tied to these nodes must also have zero current. Repeating this algorithm by leaping from one node to another, we must eventually reach node ① of the voltage source, where our reasoning would still imply that $e_{①} = e_m > E$, which is absurd. We conclude therefore that $e_m \leq E$. ∎

> *Transfer characteristic bounding region* The v_o vs. v_{in} transfer characteristic of any connected circuit made of *strictly passive* two-terminal resistors must lie within the wedge-shaped region
>
> $$|v_o| \leq |v_{\text{in}}| \tag{5.6}$$
>
> as shown in Fig. 5.6.

Proof Consider in Fig. 5.6*b* the right-half plane $v_{\text{in}} > 0$. Suppose the output voltage v_o is measured between node ⓚ and node ⓛ so that

$$v_o = e_k - e_\ell \tag{5.7}$$

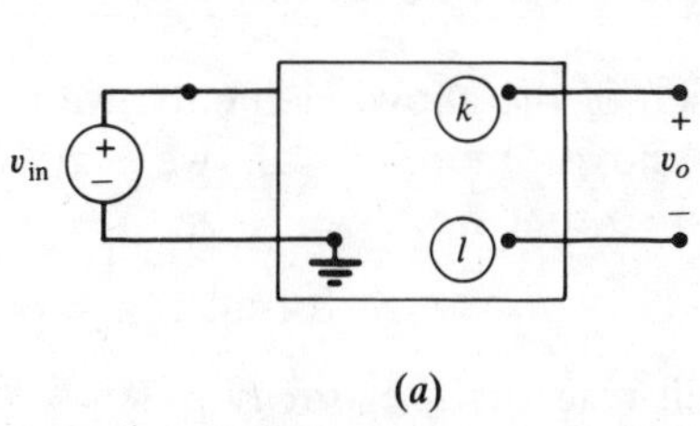

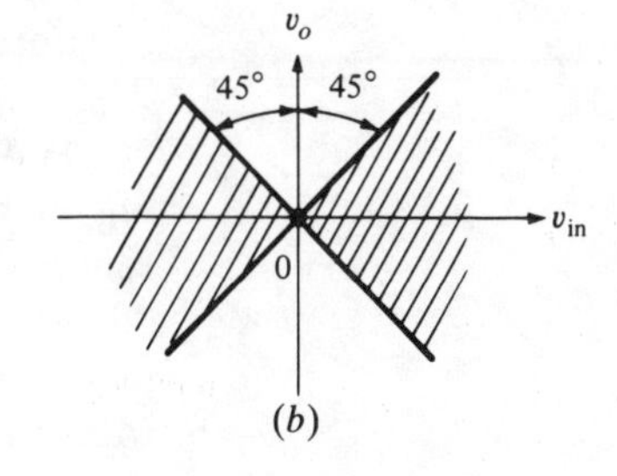

Figure 5.6 Output voltage bounding region.

where e_k and e_ℓ are measured with respect to the datum node shown in Fig. 5.6*a*.

Since $\mathcal{N}$ contains only *strictly passive* two-terminal resistors, it follows from the "maximum node-voltage property" that

$$0 \le e_k \le v_{\text{in}} \tag{5.8}$$

and

$$0 \le e_\ell \le v_{\text{in}} \tag{5.9}$$

Inequality (5.9) can be rewritten as follows:

$$-v_{\text{in}} \le -e_\ell \le 0 \tag{5.10}$$

Adding both sides of Eqs. (5.8) and (5.10), we obtain

$$-v_{\text{in}} \le e_k - e_\ell \le v_{\text{in}} \tag{5.11}$$

It follows from Eqs. (5.7) and (5.11) that

$$|v_o| \le v_{\text{in}} \qquad \text{if} \qquad v_{\text{in}} > 0 \tag{5.12}$$

If $v_{\text{in}} < 0$, a similar proof shows that

$$|v_o| \le -v_{\text{in}} \tag{5.13}$$

From Eqs. (5.12) and (5.13), we obtain Eq. (5.6). ■

> *No current-gain property* Let $\mathcal{N}$ contain only *strictly passive* two-terminal resistors and dc current sources. Then the *magnitude* of the current in any resistor *cannot* exceed the *sum* of the magnitudes of the currents in *all* current sources.

PROOF Without loss of generality, let us first dispose of the following pathological situations:

1. If $v_\ell = 0$ for some resistor $\mathcal{R}_\ell$, then $i_\ell = 0$ (since $\mathcal{R}_\ell$ is strictly passive) and we can remove that resistor from $\mathcal{N}$.
2. If $v_{sk} = 0$ for some current source, we can replace that current source by a short circuit. (If there were a resistor in parallel with that current source it would already have been removed in step 1.)

Having disposed of the above elements, we have $v_j \ne 0$ for *all* remaining elements. If one or more $v_j < 0$, we simply redefine their voltage reference direction so that we have

$$v_j > 0 \qquad \text{for all resistors} \tag{5.14}$$

Moreover, since all resistors are *strictly passive*, we have

$$i_j > 0 \qquad \text{for all resistors} \tag{5.15}$$

For all current sources, we also have $v_{sk} > 0$ but i_{sk} may be of either sign.

Consider now an arbitrary resistor $\mathscr{R}_j$. We claim that there does not exist a *similarly directed* loop containing $\mathscr{R}_j$ in the associated digraph. Indeed, for such a loop, $\Sigma\, v_k = 0$ by KVL, but we have, by construction, $v_k > 0$ for all *elements*: This is a contradiction. Hence no such loop exists.

It follows from the "loop-cut-set exclusion" property that the digraph *must* contain a *similarly directed cut set* containing $\mathscr{R}_j$. Applying KCL to this cut set gives

$$i_j + \underbrace{\sum_{\ell}}_{\substack{\text{over all other resistors}\\ \text{in the cut set}}} i_\ell + \underbrace{\sum_{\alpha}}_{\substack{\text{over all current sources}\\ \text{in the cut set}}} i_{s\alpha} = 0 \tag{5.16}$$

It follows from Eqs. (5.15) and (5.16) that

$$|i_j| + \sum_{\ell} |i_\ell| + \sum_{\alpha} i_{s\alpha} = 0 \tag{5.17}$$

Hence,

$$|i_j| = -\sum_{\alpha} i_{s\alpha} - \sum_{\ell} |i_\ell| \leq -\sum_{\alpha} i_{s\alpha} \leq \sum_{\alpha} |i_{s\alpha}| \tag{5.18}$$

Observe that the last inequality on the right-hand side of Eq. (5.18) extends over only all current sources in the above similarly directed cut set. This implies that

$$|i_j| \leq \underbrace{\sum}_{\substack{\text{over all current}\\ \text{sources in } \mathcal{N}}} |i_{s\alpha}| \tag{5.19}$$

■

Exercise 1 Show that if $\mathcal{N}$ also contains dc voltage sources, then the "no current-gain" property still holds with Eq. (5.19) replaced by

$$|i_j| \leq \underbrace{\sum}_{\substack{\text{over all current}\\ \text{sources in } \mathcal{N}}} |i_{s\alpha}| + \underbrace{\sum}_{\substack{\text{over all voltage}\\ \text{sources in } \mathcal{N}}} |i_{s\beta}| \tag{5.20}$$

Exercise 2 Show that both the "maximum-node voltage" property and the "no current-gain" property do not hold if the circuit contains also ideal diodes which are *passive*, but *not* strictly passive. *Hint*: Connect two ideal diodes *back to back* in series (respectively, parallel) with a voltage (respectively, current) source.

Exercise 3 A two-port resistor is said to be *strictly passive* iff $v_1 i_1 + v_2 i_2 > 0$ for all (v_1, v_2, i_1, i_2) satisfying the resistor characteristics, except the origin $(0, 0, 0, 0)$.
(a) Insert a 1-Ω resistor in series with *each* port of an ideal transformer. Prove that the resulting two-port is *strictly passive*.
(b) Show that in general both the "maximum-node voltage" property and the "no current-gain" property do not hold if the circuit contains also a *strictly passive* two-port, such as the one defined in (a).

5.4 Consequences of Strict Monotonicity

Strict passivity does not impose any constraint on the slope of the resistor characteristic: It only requires that the product vi be positive except at the origin. For example, the *tunnel diode* and the *glow tube* described in Figs. 1.10 and 1.11 of Chap. 2 are both strictly passive. Yet the slope of their v-i characterisics can assume both positive or negative values, depending on the operating point. Such characteristics are said to be *nonmonotonic*. It is clear that resistive circuits made of nonmonotonic resistors would in general also give rise to a nonmonotonic driving-point and transfer characteristic. Hence, in order to derive properties involving constraints on the slope of the driving-point and transfer characteristics, it is necessary to impose stronger conditions on the resistor characteristics. The strictly monotone-increasing, or *strictly increasing* for brevity, property is one such condition which we require in this final section.

Strictly increasing means roughly that the slope of the characteristic is positive everywhere. More precisely, a two-terminal resistor is said to be *strictly increasing* iff, for all pairs of (distinct) points on its characteristic, say (v', i') and (v'', i''), we have

Strictly increasing resistor

$$(v' - v'')(i' - i'') > 0 \tag{5.21}$$

A strictly increasing resistor characteristic is illustrated in Fig. 5.7. It is

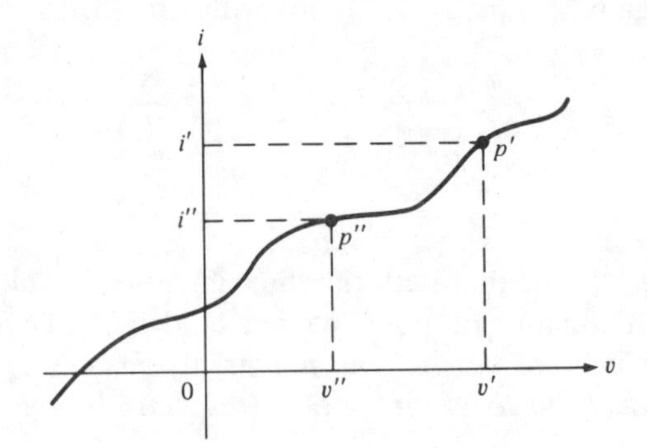

Figure 5.7 Example of a strictly increasing characteristic.

clear that the condition (5.21) is equivalent to requiring the *secant* $p'p''$ to have a positive slope.

Note that a strictly increasing characteristic is not restricted to lie in the first and third quadrants only. Hence, a strictly increasing resistor need not be strictly passive, and a strictly passive resistor need not be strictly increasing.

Uniqueness property Any circuit made of strictly increasing two-terminal resistors and independent sources has *at most one solution*.

PROOF Suppose there are two distinct operating points Q and Q', at some time t, corresponding to $(v_1, v_2, \ldots, v_b;\ i_1, i_2, \ldots, i_b)$ and $(v'_1, v'_2, \ldots, v'_b; i'_1, i'_2, \ldots, i'_b)$, respectively. Here, we assume "associated reference directions" for all elements, including the independent sources. Since each of these two solutions satisfies Tellegen's theorem, so does their difference; i.e.,

$$\sum_{k=1}^{b} (v_k - v'_k)(i_k - i'_k) = 0 \tag{5.22}$$

Observe that each term in Eq. (5.22) which corresponds to either a voltage source $(v_k = v'_k)$ or a current source $(i_k = i'_k)$ vanishes. However, since these are two *distinct* solutions and since all resistors are strictly increasing, there must exist at least one branch such that $(v_k - v'_k)(i_k - i'_k) > 0$ for this branch. This contradicts Eq. (5.22). Hence, there *cannot* be two distinct operating points. ■

REMARK The above hypothesis is *not* sufficient to guarantee that the circuit has a solution. For example, the *pn*-junction diode in the circuit in Fig. 4.1*d* of Chap. 2 is *strictly increasing*, but this circuit has *no* solution if $I > I_s$.

It can be shown that the following additional conditions will guarantee that the circuit always has a unique solution:

(*a*) The characteristic tends to $\pm\infty$ as $v \to \pm\infty$, respectively.

(*b*) There is no loop made up of voltage sources and no cut set made up of current sources.

Strictly increasing closure property A one-port made of strictly increasing two-terminal resistors is itself *strictly increasing*.

PROOF Suppose the one-port N in Fig. 5.4 contains only strictly increasing resistors. Then for any two *distinct* driving-point voltages v and v', let

(v_k, i_k) and (v_k', i_k'), $k = 1, 2, \ldots, b$, denote the corresponding *unique* branch voltage and current solutions, for all b resistors inside N. It follows from Tellegen's theorem that

$$(v - v')(i - i') = \sum_{k=1}^{b} (v_k - v_k')(i_k - i_k') \tag{5.23}$$

where the input term appears on the left of the equation because the input current i in Fig. 5.4 is defined leaving the positive terminal of the voltage source.

Since $v \neq v'$, KVL requires that at least one of the $(v_k - v_k')$ differs from 0; hence, at least one term on the right-hand side of Eq. (5.23) is positive, while all others are ≥ 0 (since all resistors are strictly increasing) where the equality sign holds whenever $v_k = v_k'$ or $i_k = i_k'$. Consequently,

$$(v - v')(i - i') > 0 \tag{5.24}$$

whenever $v \neq v'$; i.e., the driving-point characteristic of N is *strictly increasing*. ■

> *Slope-bounding property* The *magnitude* of the *slope* of the v_o vs. v_{in} transfer characteristic of any circuit made of *strictly increasing* two-terminal resistors with well-defined *positive* slopes can*not* exceed unity.

PROOF Let $(\bar{v}_{\text{in}}, \bar{v}_o)$ be the coordinates of any point Q on the v_o vs. v_{in} transfer characteristic Γ. Since the circuit $\mathcal{N}$ contains only strictly increasing resistors, each resistor $\mathcal{R}_k$ has a corresponding *unique* operating point Q_k with coordinates $(\bar{v}_k, \bar{i}_k)$, $k = 1, 2, \ldots, b$, and a *positive slope* R_k at Q_k.

Now in an arbitrarily small neighborhood of Q_k, each resistor can be modeled by a *Thévenin equivalent one-port* with $R_{\text{eq}} \triangleq R_k$ and v_{oc} equal to the voltage intercept of the tangent drawn through Q_k (see Fig. 4.5*a* of this chapter). Clearly, if we replace each resistor by this model, the v_o vs. v_{in} transfer characteristic of the resulting *linear resistive circuit* $\mathcal{N}_Q$ will consist of a *straight line* tangent to the nonlinear transfer characteristic Γ of $\mathcal{N}$ at the point Q. Indeed, by the *superposition theorem*, this straight line is described by an equation of the form

$$v_o = H_o v_{\text{in}} + E_o \tag{5.25}$$

where E_o accounts for the contribution of all independent sources in $\mathcal{N}_Q$ (they come from the Thévenin one-ports). Hence, the slope of Γ at Q is equal to H_o.

Now if we set all independent sources in $\mathcal{N}_Q$ to zero, Eq. (5.25) reduces to $v_o = H_o v_{\text{in}}$. Since this sourceless circuit now contains only *linear positive* resistors, it follows from the "transfer characteristic bounding region property" that $|H_o| \leq 1$. ■

REMARKS

1. The additional hypothesis that the resistors have "*well-defined positive* slopes" is not superfluous. Indeed, the slope of any strictly increasing piecewise-linear characteristic is undefined at the breakpoints. Observe also that $R_k = 0$ at $i_k = 0$ for the *strictly increasing* resistor $v_k = i_k^3$.
2. Both the "transfer characteristic bounding region" property and the "slope bounding" property remain valid if the nonlinear resistors are *increasing*, but not *strictly*; i.e., $vi \geq 0$ for all points on the resistor characteristic. However, the proof becomes much more involved.

Exercises

1. Prove that a two-terminal resistor characterized by a *continuous* and *strictly increasing v-i* curve is *strictly passive* if and only if the *v-i* curve goes through the origin $(0, 0)$.
2. An $(n+1)$-terminal or n-port resistor is said to be *strictly increasing* iff for all pairs of (distinct) points on its characteristic, say $(\mathbf{v}', \mathbf{i}')$ and $(\mathbf{v}'', \mathbf{i}'')$, we have

$$(\mathbf{v}' - \mathbf{v}'')^T(\mathbf{i}' - \mathbf{i}'') > 0 \tag{5.26}$$

(*a*) Prove that the "uniqueness" property and the "slope-bounding" property remain valid if the circuit also contains strictly increasing multiterminal or multiport resistors. (*b*) Show via an example that the "strictly increasing closure" property is *false* for an n-port $(n > 1)$ made of strictly increasing two-terminal resistors.

SUMMARY

- For resistive circuits, *node analysis* is applicable if the circuit contains only *voltage-controlled* (possibly nonlinear and multiterminal) *resistors* and *independent current sources* (which do not form cut sets among themselves).
- The *node equation* for a linear resistive circuit is given by

$$\mathbf{Y}_n \mathbf{e}(t) = \mathbf{i}_s(t)$$

where $\mathbf{Y}_n \triangleq \mathbf{A}\mathbf{Y}_b\mathbf{A}^T$ is called the *node-admittance matrix*; $\mathbf{A}$ is the *reduced incidence matrix* of the reduced digraph obtained by open-circuiting all branches corresponding to *independent* current sources from the original digraph; $\mathbf{Y}_b$ is the *branch-admittance matrix*; $\mathbf{i}_s(t)$ is the source vector whose kth entry is equal to the algebraic sum of *all* independent current sources entering node ⓚ.

For a *reduced* digraph with n nodes and b branches, $\mathbf{Y}_b$ is a $b \times b$ matrix, $\mathbf{Y}_n$ is an $(n-1) \times (n-1)$ matrix, $\mathbf{A}$ is an $(n-1) \times b$ matrix; both $\mathbf{e}$ and $\mathbf{i}_s(t)$ are $n-1$ vectors.

- For circuits containing only *two-terminal* resistors and independent current sources, the *node-admittance matrix* $\mathbf{Y}_n$ satisfies the following properties:
  1. The kth diagonal element of $\mathbf{Y}_n$ is equal to the sum of all conductances attached to node Ⓚ.
  2. The jkth off-diagonal element of $\mathbf{Y}_n$ is equal to the *negative* of the sum of all conductances between node Ⓙ and node Ⓚ.
  3. $\mathbf{Y}_n$ is *symmetric*.
- Consider a *nonlinear* resistive circuit driven only by independent current sources; its node equation is given by

$$\mathbf{A}\mathbf{g}(\mathbf{A}^T\mathbf{e}) = \mathbf{i}_s(t)$$

where $\mathbf{i} = \mathbf{g}(\mathbf{v})$ denotes the characteristics of all (voltage-controlled) resistors.

- Both linear and nonlinear node equations consist of $n-1$ equations in terms of the *node voltage vector* $\mathbf{e}$, where n is the number of *nodes* in the circuit. Hence, the number of equations in node analysis does *not* depend on the number of branches in the circuit.
- Every *linear time-invariant* resistive circuit has a *tableau equation* of the form:

Linear tableau equation

$$\underbrace{\begin{bmatrix} \mathbf{0} & \mathbf{0} & \mathbf{A} \\ -\mathbf{A}^T & \mathbf{1} & \mathbf{0} \\ \mathbf{0} & \mathbf{M} & \mathbf{N} \end{bmatrix}}_{\mathbf{T}} \underbrace{\begin{bmatrix} \mathbf{e} \\ \mathbf{v} \\ \mathbf{i} \end{bmatrix}}_{\mathbf{w}} = \underbrace{\begin{bmatrix} \mathbf{0} \\ \mathbf{0} \\ \mathbf{u}_s(t) \end{bmatrix}}_{\mathbf{u}(t)}$$

The entries of $\mathbf{M}$ and $\mathbf{N}$ contain constant coefficients defining the resistors (possibly multiterminal); the entries of $\mathbf{u}_s(t)$ contain constant or time functions defining the *independent* sources.

- A linear time-invariant resistive circuit has a *unique* solution if and only if the *tableau matrix* $\mathbf{T}$ is *nonsingular*.
- Every *nonlinear* resistive circuit has a *tableau equation* of the form

Nonlinear tableau equation

$$\begin{aligned} \mathbf{A}\mathbf{i}(t) &= \mathbf{0} \\ \mathbf{v}(t) - \mathbf{A}^T\mathbf{e}(t) &= \mathbf{0} \\ \mathbf{h}(\mathbf{v}(t), \mathbf{i}(t), t) &= \mathbf{0} \end{aligned}$$

- A multiterminal or multiport resistor described by

$$\mathbf{A}\mathbf{v} + \mathbf{B}\mathbf{i} + \mathbf{c} = \mathbf{0}$$

is called an *affine resistor*. When $\mathbf{c} = \mathbf{0}$, the affine resistor reduces to a *linear resistor*.

- The system of nonlinear algebraic equations

$$\mathbf{f}(\mathbf{x}) = \mathbf{0}$$

can be solved by numerical iteration using the *Newton-Raphson formula*

$$\mathbf{x}^{(j+1)} = \mathbf{x}^{(j)} - [\mathbf{J}(\mathbf{x}^{(j)})]^{-1}\mathbf{f}(\mathbf{x}^{(j)})$$

where $\mathbf{J}(\mathbf{x}^{(j)})$ denotes the jacobian matrix of $\mathbf{f}(\mathbf{x})$ evaluated at $\mathbf{x} = \mathbf{x}^{(j)}$.
- It is computationally more efficient to calculate the $(j+1)$th iterate $\mathbf{x}^{(j+1)}$ in the Newton-Raphson formula by solving the following equivalent system of linear equations for $\mathbf{x}^{(j+1)}$:

$$\mathbf{J}(\mathbf{x}^{(j)})\,\Delta\mathbf{x}^{(j)} = -\mathbf{f}(\mathbf{x}^{(j)})$$

$$\mathbf{x}^{(j+1)} = \mathbf{x}^{(j)} + \Delta\mathbf{x}^{(j)}$$

- Each iteration in the Newton-Raphson formula can be implemented more efficiently by solving an associated Newton-Raphson *discrete* equivalent circuit.
- The *structure* of the Newton-Raphson equivalent circuit remains fixed for all iterations. Only the element parameters need to be *updated* in each iteration.
- A resistive circuit is said to be *uniquely solvable* iff Kirchhoff's laws and the branch equations are simultaneously satisfied by a *unique* set of branch voltages and a *unique* set of branch currents for all t.
- The *superposition theorem* is applicable for any *linear uniquely solvable* resistive circuit. It allows us to find the solution by calculating first the solutions due to each independent source acting alone (all other *independent* voltage sources are short-circuited and all other independent current sources are open-circuited) and then adding them.
- The superposition theorem does *not* hold for *nonlinear* circuits.
- A one-port N is said to be *well-defined* iff it does not contain any circuit element which is *coupled*, electrically or nonelectrically, to some physical variable *outside* of N.
- The *Thévenin theorem* allows us to replace any well-defined *linear current-controlled* resistive one-port by an equivalent one-port consisting of an *equivalent* Thévenin *resistance* R_{eq} in series with an *open-circuit voltage source* $v_{oc}(t)$, *without affecting* the solution of the remaining part of the circuit connected to the one-port.
- The *Norton theorem* allows us to replace any well-defined *linear voltage-controlled* resistive one-port by an equivalent one-port consisting of an *equivalent* Norton *conductance* G_{eq} in parallel with a *short-circuit current source* $i_{sc}(t)$, without *affecting* the solution of the remaining part of the circuit connected to the one-port.
- In applying the superposition theorem, Thévenin theorem, and Norton theorem, all *controlled* sources must be left *intact* while the *independent* sources are set to zero.

- If a linear time-invariant resistive two-port terminated by two voltage sources is uniquely solvable, then it has a *voltage-controlled representation* given by the *affine* equation (4.41).
- If a linear time-invariant resistive two-port terminated by two current sources is uniquely solvable, then it has a *current-controlled representation* given by the *affine equation* (4.42).
- If a linear time-invariant resistive two-port terminated by a current source at port 1 and a voltage source at port 2 is uniquely solvable, then it has a *hybrid 2 representation* given by the *affine equation* (4.43).
- The *substitution theorem* allows us to substitute a one-port (not necessarily linear or resistive) by a voltage source or a current source, without affecting the solution of any *resistive* circuit connected across the one-port, provided the circuit has a *unique* solution before and after the substitution.
- *Loop-cut-set exclusion property*: Any branch in a connected digraph belongs to either a *similarly directed loop* or a *similarly directed cut set*, but not to both.
- A two-terminal resistor is *strictly passive* iff

$$vi > 0$$

for all points on its characteristic except the origin.
- A two-terminal resistor is *strictly increasing* iff,

$$(v' - v'')(i' - i'') > 0$$

for all pairs of distinct points (v', i') and (v'', i'') on its characteristic.
- *Strict passivity property*: A one-port made of *strictly passive* resistors is itself *strictly passive*.
- *Maximum node-voltage property*: For a circuit made of strictly passive resistors and a battery of E volts, the voltage between any node and datum cannot exceed E volts.
- The v_o vs. v_{in} transfer characteristic of any connected circuit made of *strictly passive* resistors must lie within the wedge-shaped region $|v_o| \le |v_{\text{in}}|$.
- *No current-gain property*: For a circuit made of strictly passive two-terminal resistors and dc current sources, the *magnitude* of the current in any resistor can*not* exceed the *sum* of the magnitudes of the currents in *all* current sources.
- *Uniqueness property*: Any circuit made of *strictly increasing* two-terminal resistors and independent sources has *at most one solution*.
- *Strictly increasing closure property*: A one-port made of strictly increasing two-terminal resistors is itself *strictly increasing*.
- *Slope-bounding property*: The *magnitude* of the slope of the v_o vs. v_{in} transfer characteristic of any circuit made of *strictly increasing* resistors with well-defined *positive* slopes cannot exceed unity.

PROBLEMS

Node equations: linear circuits

1 For the circuits shown in Fig. P5.1 write node equations in matrix form.

2 Consider the circuit shown in Fig. P5.2.

(*a*) Can you write node equations for this circuit? If not, modify the circuit and then write node equations in matrix form.

(*b*) Determine $\mathbf{Y}_b$ and $\mathbf{A}$ for your modified circuit.

(*c*) Verify that $\mathbf{Y}_n = \mathbf{A}\mathbf{Y}_b\mathbf{A}^T$.

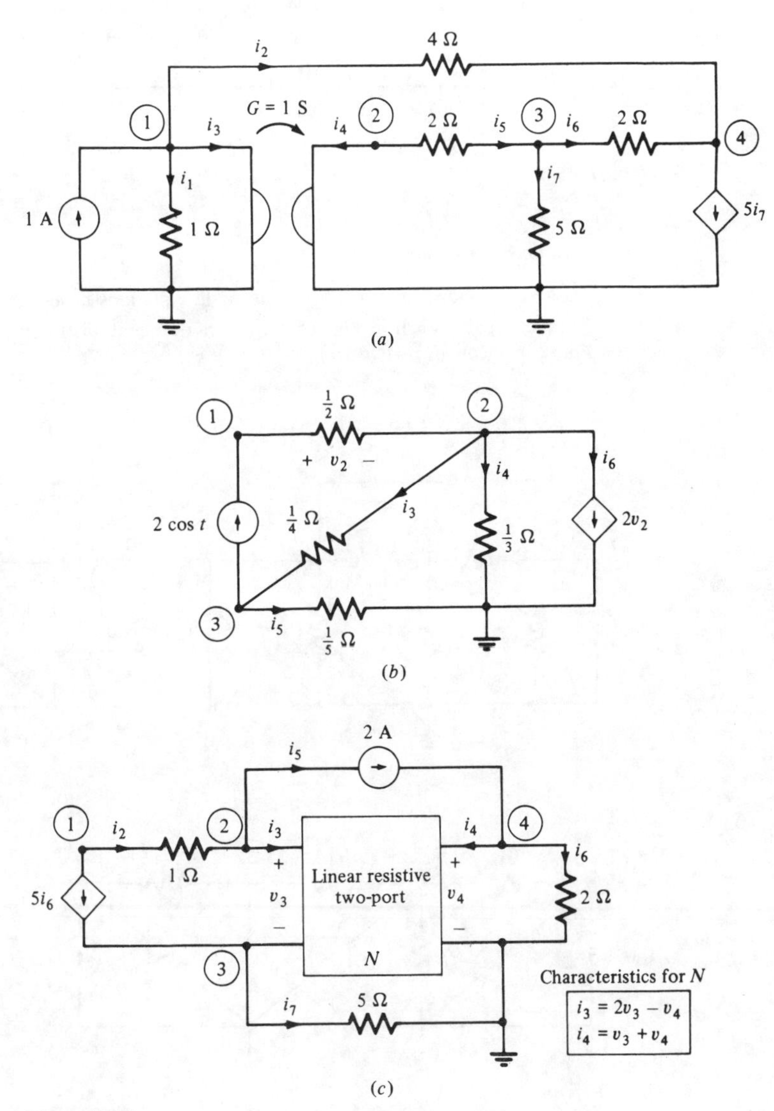

Figure P5.1

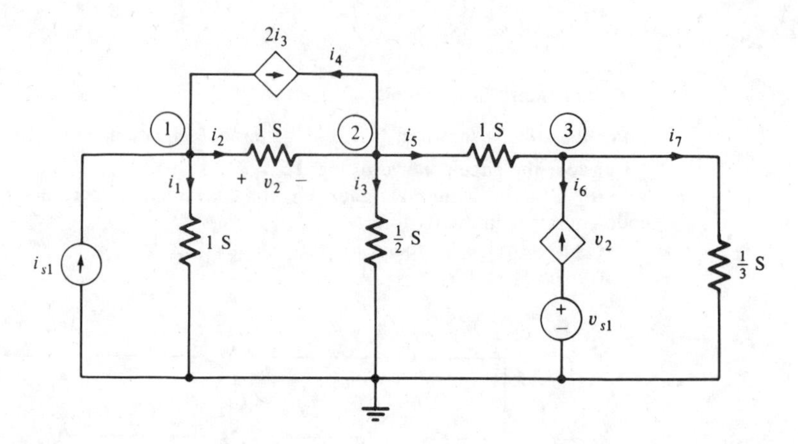

Figure P5.2

Node equations: inspection method

3 For the circuits shown in Fig. P5.3 write node equations in matrix form by inspection.

4 (*a*) For the circuit shown in Fig. P5.4 write node equations by inspection.
(*b*) Find the Jacobian matrix $\mathbf{J}(\mathbf{e})$.

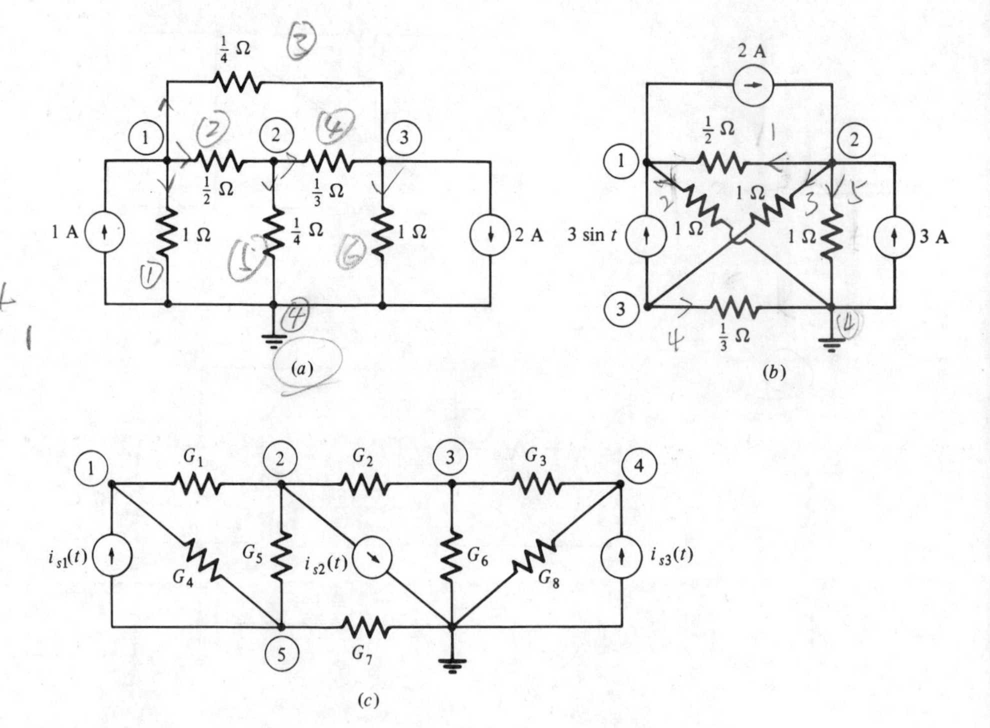

Figure P5.3

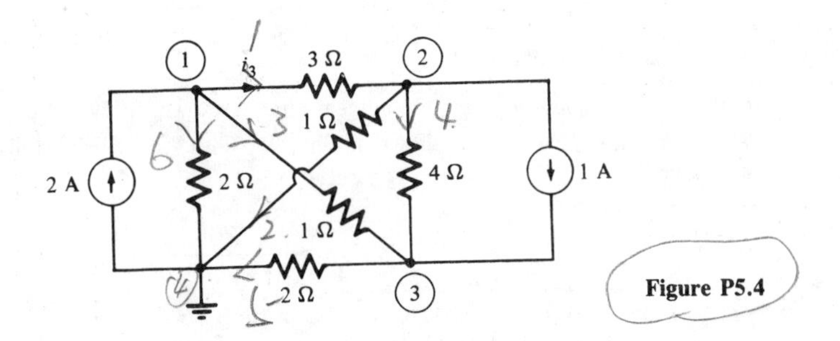

Figure P5.4

Node equations: existence and uniqueness of solution

5 Consider the circuit shown in Fig. P5.5.

(*a*) Write node equations.

(*b*) Determine the conductance G such that the circuit does not have a unique solution.

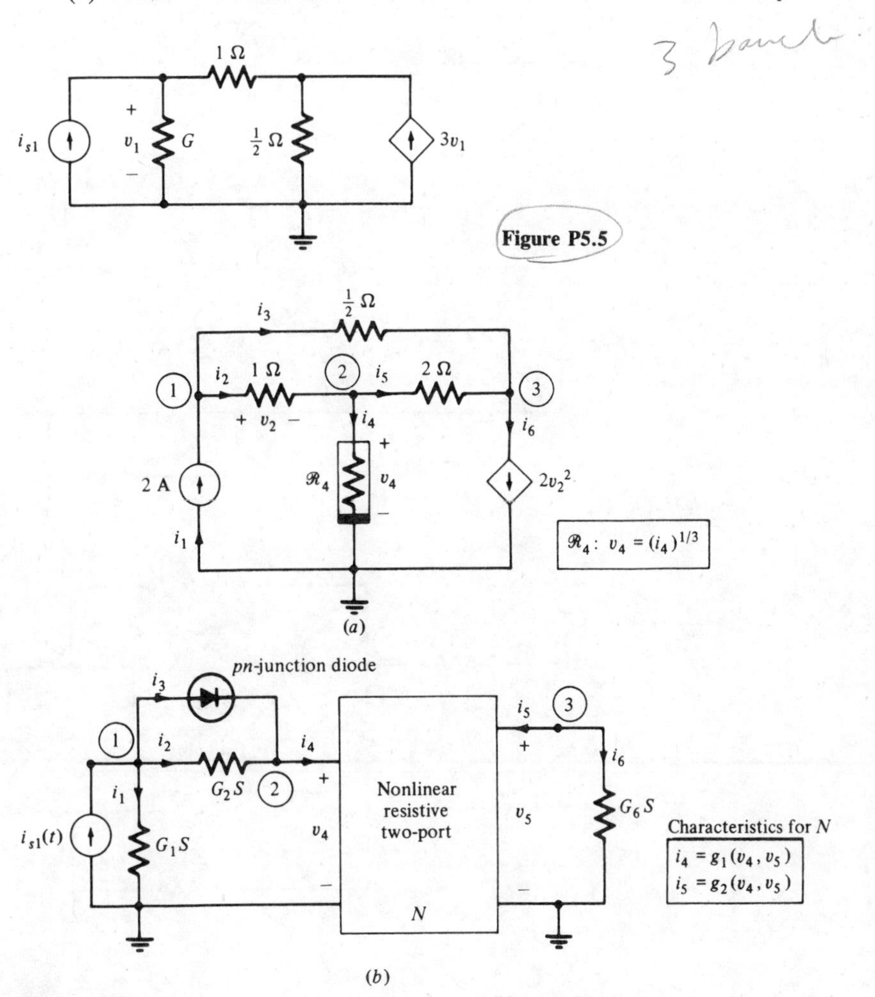

Figure P5.6

Node equations: nonlinear circuits

6 Write node equations for the nonlinear circuits shown in Fig. P5.6.

7 For the nonlinear circuit shown in Fig. P5.7 use the Ebers-Moll equations to write node equations.

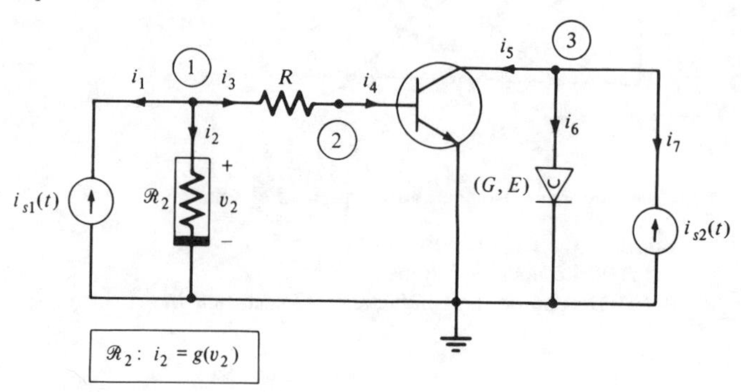

Figure P5.7

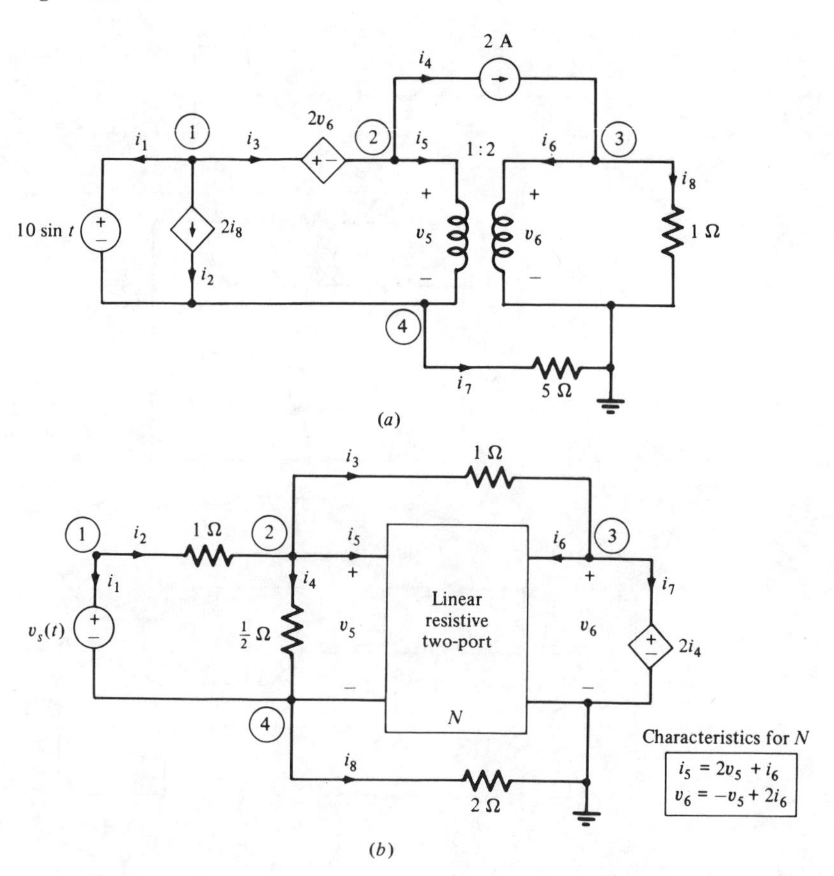

Figure P5.9

Tableau equations: linear circuits

8 Write the tableau equations for the circuits shown in Fig. P5.1.

9 (*a*) Can you write node equations for the circuits shown in Fig. P5.9?

(*b*) Write the tableau equations for the circuits shown in Fig. P5.9. Identify the matrices **M**, **N**, and the vector $\mathbf{u}_s$.

Tableau equations: nonlinear circuits

10 Repeat Prob. 7 for the tableau equations.

11 Write the tableau equations for the circuit shown in Fig. P5.11. The op amp is modeled by Fig. 4.2 of Chap. 4 and the *pn*-junction diode is modeled by Eq. (1.6) of Chap. 2.

Newton-Raphson algorithm

12 For the circuits shown in Fig. P5.12 draw the associated Newton-Raphson discrete equivalent circuit at the jth iteration.

13 Using the nonlinear op-amp circuit model shown in Fig. 3.5*a* of Chap. 4, draw the Newton-Raphson discrete equivalent circuit associated with the circuit shown in Fig. P5.13.

14 Consider the circuit shown in Fig. P5.14.

(*a*) Write the node equations in the form $\mathbf{f}(\mathbf{e}) = \mathbf{0}$.

(*b*) Find the jacobian matrix $\mathbf{J}(\mathbf{e})$.

(*c*) Let $\mathbf{e}^{(0)} = [0\ 1]^T$; determine $\mathbf{e}^{(1)}$ using the Newton-Raphson algorithm.

15 (*a*) Draw the Newton-Raphson discrete equivalent circuit associated with the circuit shown in Fig. P5.14.

(*b*) Let $\mathbf{e}^{(0)} = [0\ 1]^T$; determine $\mathbf{e}^{(1)}$ using the equivalent circuit obtained in (*a*).

Superposition theorem

16 Using the superposition theorem find the current i_1 for the circuit shown in Fig. P5.16.

17 Find the voltage v_o of the circuit shown in Fig. P5.17. Use the superposition theorem.

18 Use the superposition theorem to find the current i_3 for the circuit shown in Fig. P5.4.

19 Assuming the ideal op-amp model in the linear region, find $v_o(t)$ in terms of $v_1(t)$, $v_2(t)$, and $v_3(t)$ for the circuit shown in Fig. P5.19.

20 Find the output v_o for the circuit shown in Fig. P5.20. Assume the ideal op-amp model in the linear region.

21 Use the superposition theorem to find the voltage v_o for the circuit shown in Fig. P5.21.

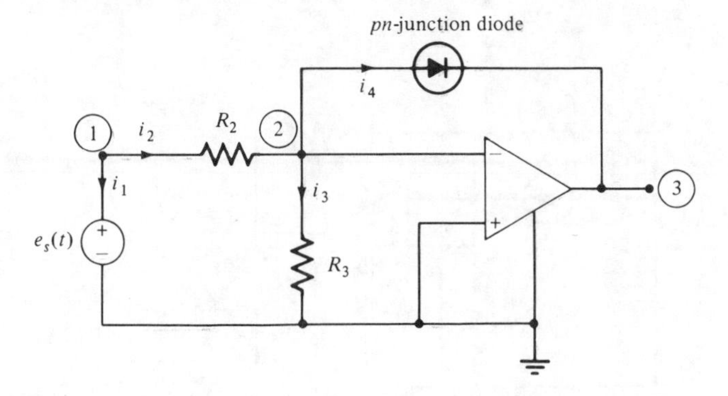

Figure P5.11

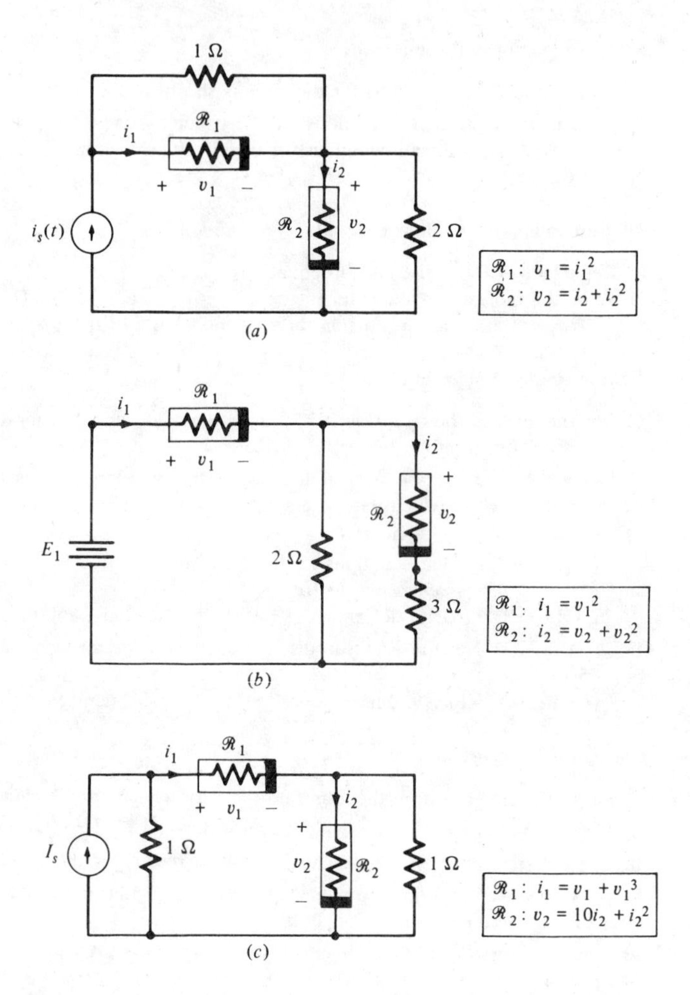

Figure P5.12

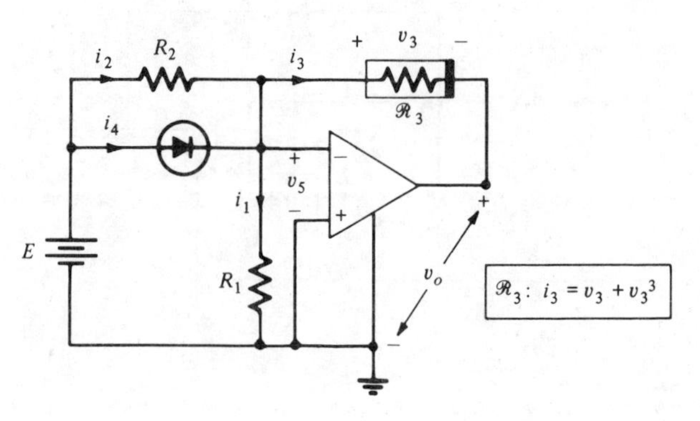

Figure P5.13

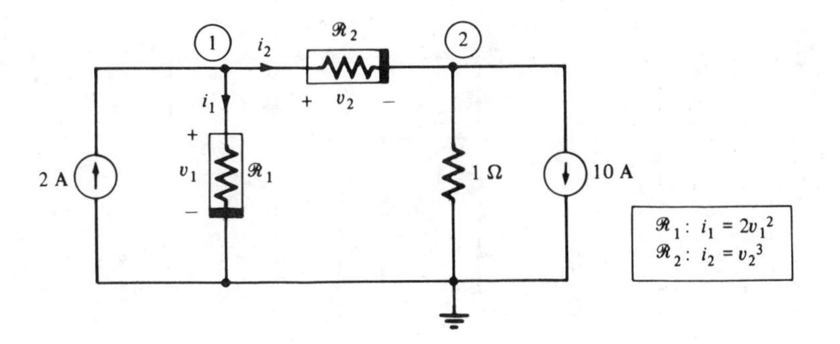

Figure P5.14

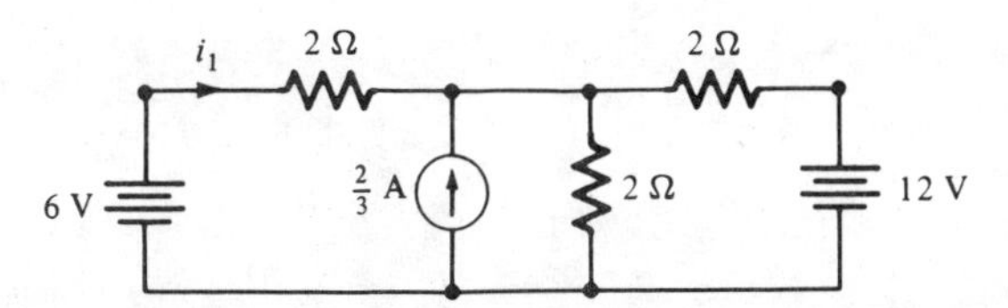

Figure P5.16

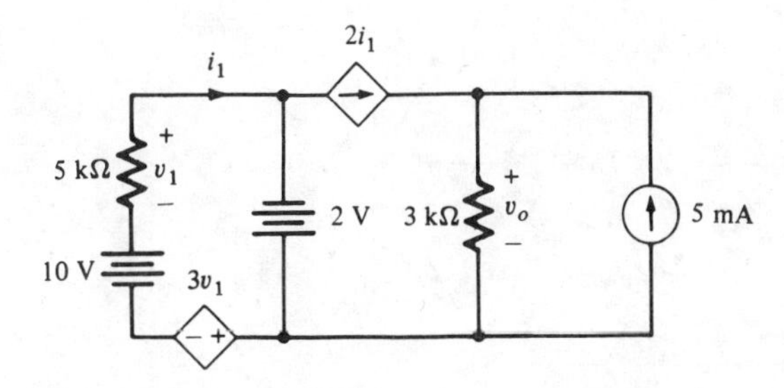

Figure P5.17

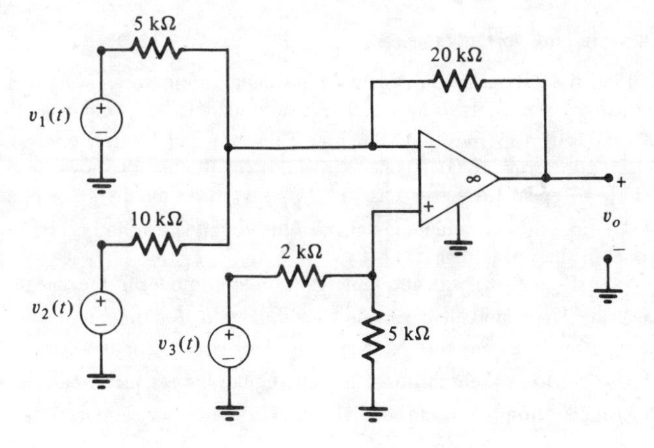

Figure P5.19

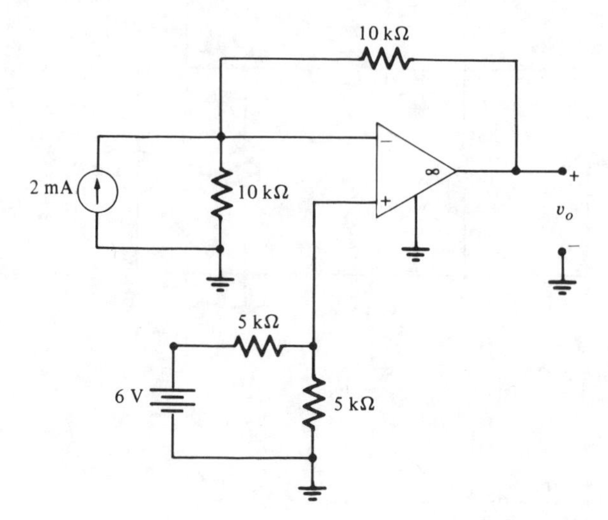

Figure P5.20

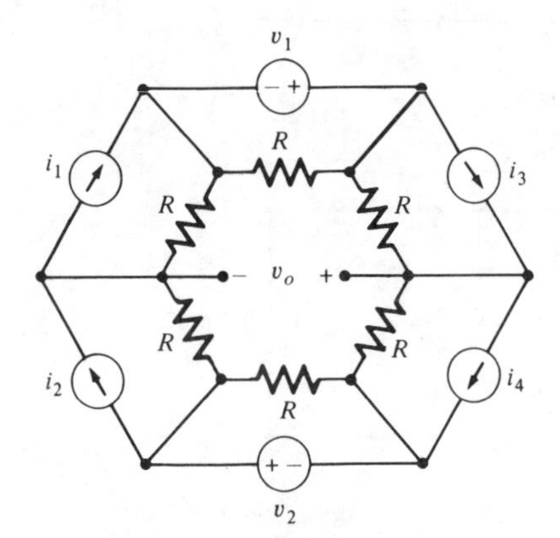

Figure P5.21

Thévenin and Norton's theorem

22 Find the Thévenin and Norton equivalent circuits for the one-ports shown in Fig. P5.22. If a particular circuit fails to have a Thévenin and/or Norton equivalent, explain.

23 (*a*) Using superposition, find the Thévenin and Norton equivalent circuits across terminals Ⓐ–Ⓐ' (with terminals Ⓑ–Ⓑ' open-circuited) for the circuit shown in Fig. P5.23.

(*b*) Repeat (*a*) across terminals Ⓑ–Ⓑ' (with terminals Ⓐ–Ⓐ' open-circuited).

24 (*a*) Find the Thévenin equivalent of the circuit shown in Fig. P5.24*a*. Assume the ideal op-amp model in the linear region.

(*b*) Repeat (*a*) with the finite-gain op-amp model in the linear region.

25 Using Thévenin's theorem, find the voltage v_o for the circuit shown in Fig. P5.25.

26 Find the Thévenin and Norton equivalent circuits for the circuits shown in Fig. P5.26.

27 Use Norton's theorem to write node equations for the circuits shown in Fig. P5.27.

28 Using Norton's theorem, find all possible values of i_R and v_R for the circuit shown in Fig. P5.28.

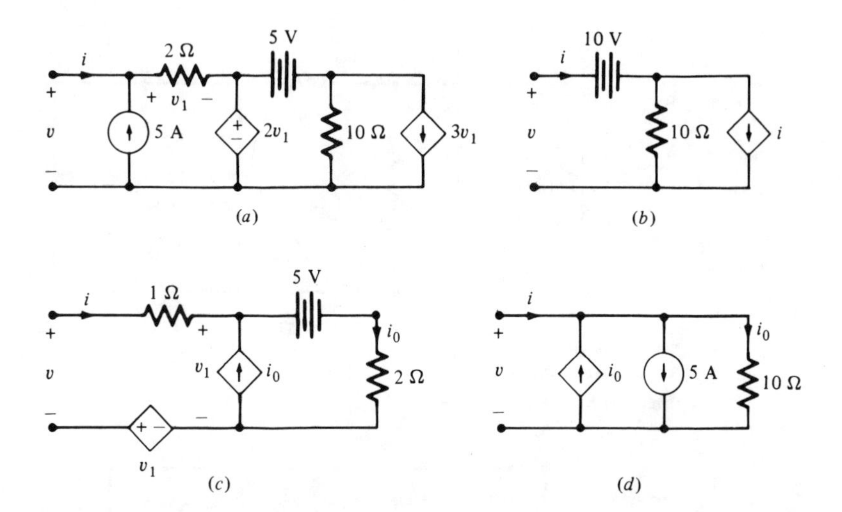

Figure P5.22

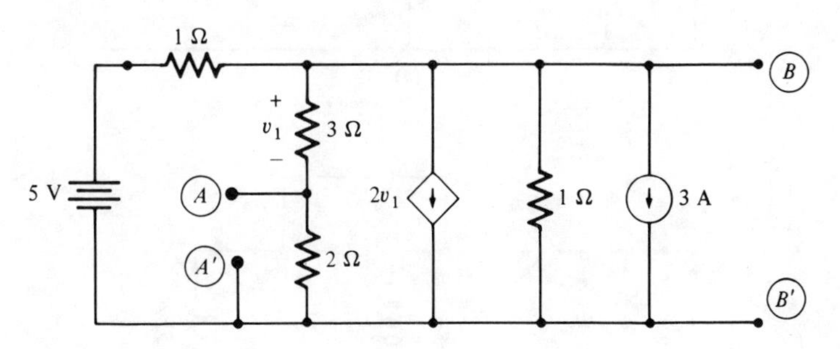

Figure P5.23

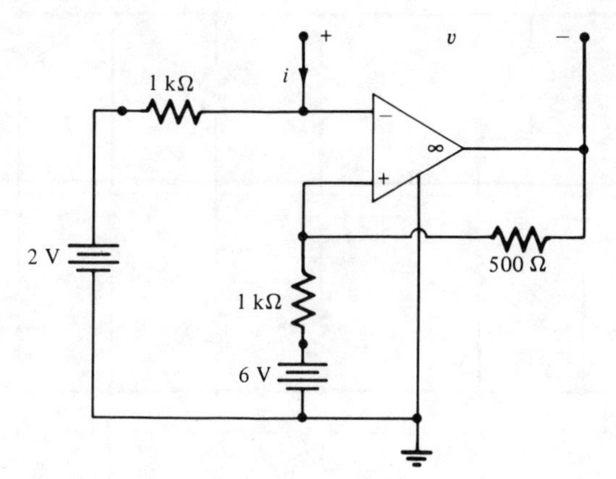

Figure P5.24

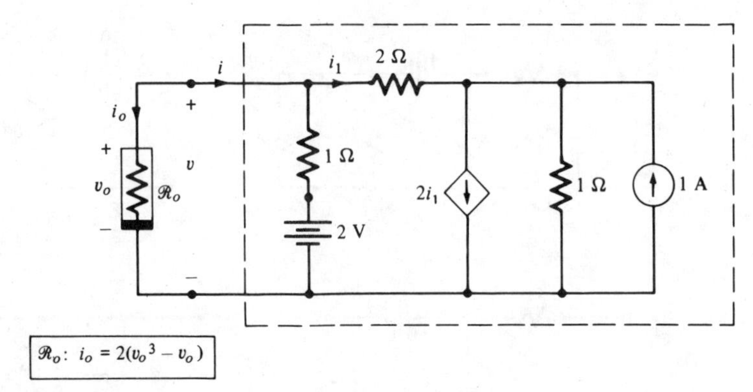

Figure P5.25

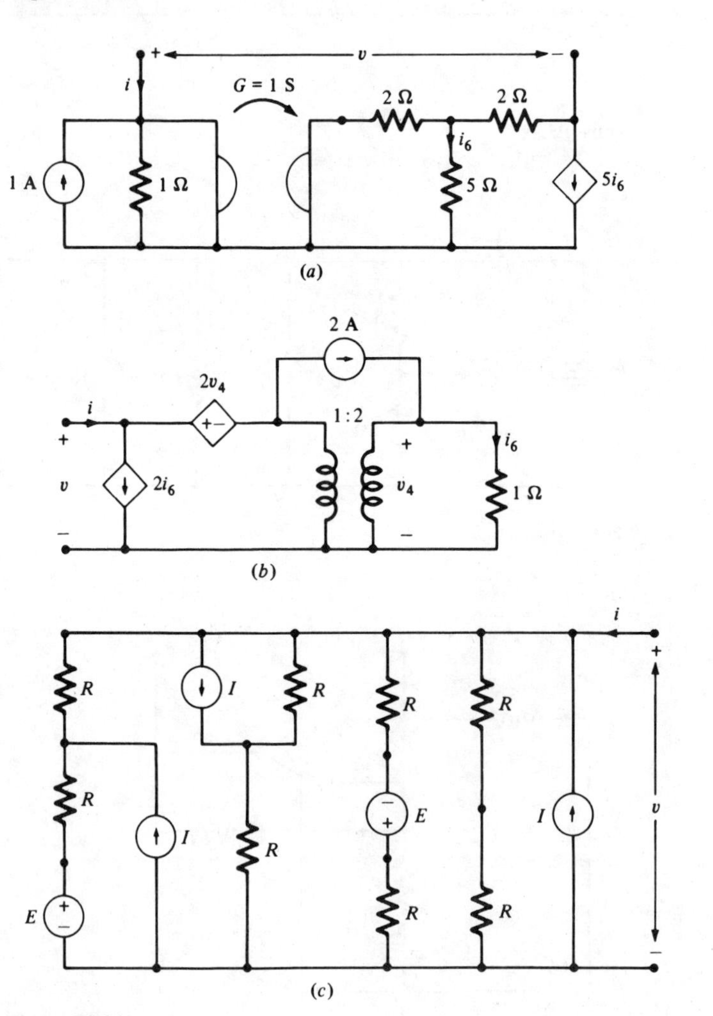

Figure P5.26

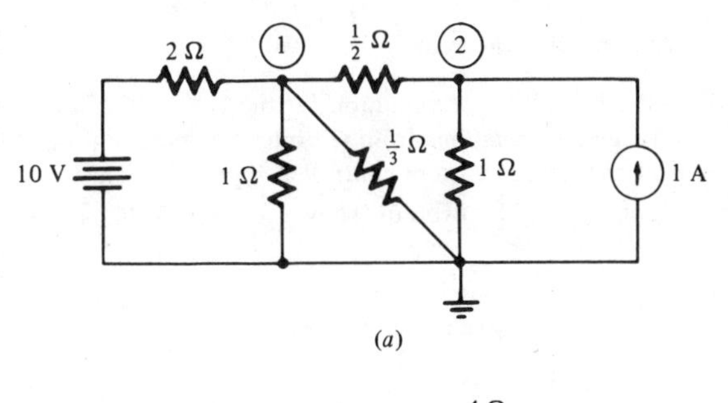

(a)

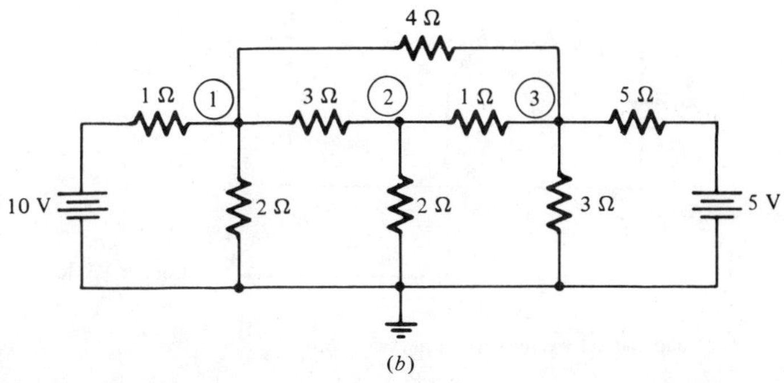

(b)

Figure P5.27

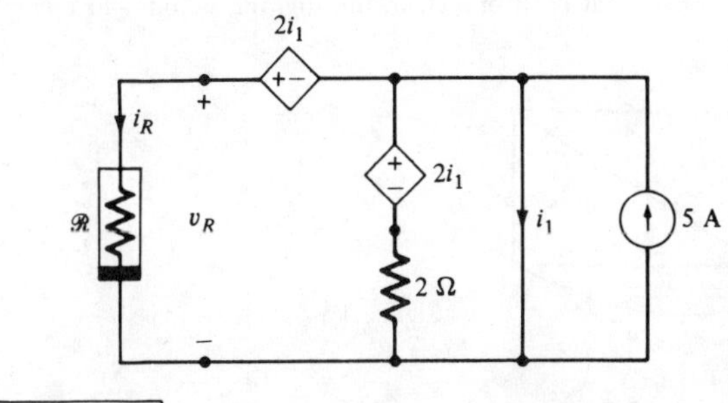

Figure P5.28

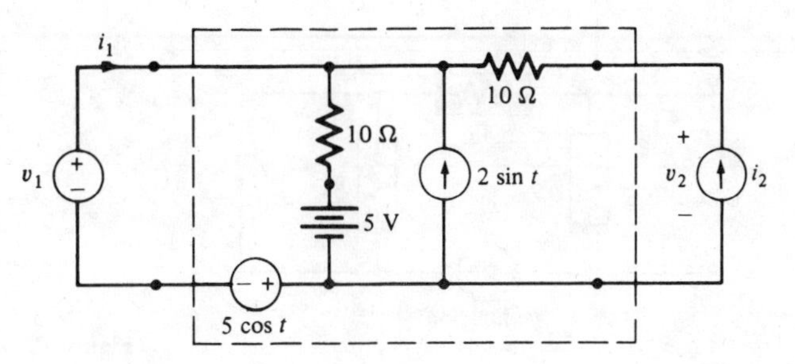

Figure P5.29

Two-port representation theorem

29 (*a*) Find the hybrid 2 representation for the two-port shown in Fig. P5.29.
(*b*) Rearrange the equations from (*a*) into a voltage-controlled representation.
(*c*) Repeat (*b*) for the current-controlled representation.

30 Repeat Prob. 29 for the two-port shown in Fig. P5.30.

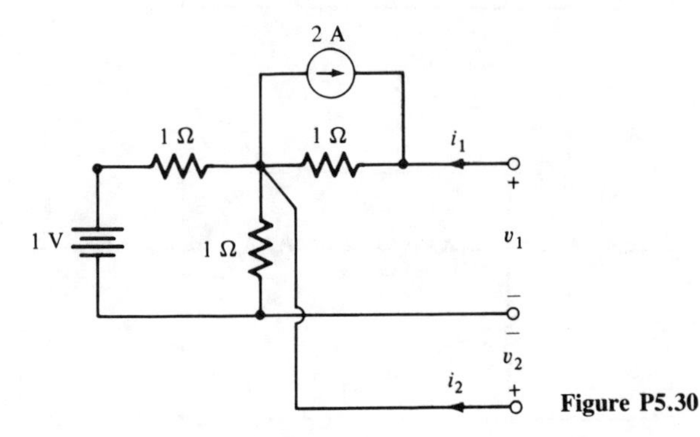

Figure P5.30

Loop-cut-set exclusion property

31 Consider the digraph shown in Fig. P5.31.
(*a*) List all possible loops for this graph.
(*b*) Are there any similarly directed loops in the list above? If not, what can you conclude about each branch of this graph.
(*c*) Verify that each branch in the digraph belongs to a similarly directed cut set.

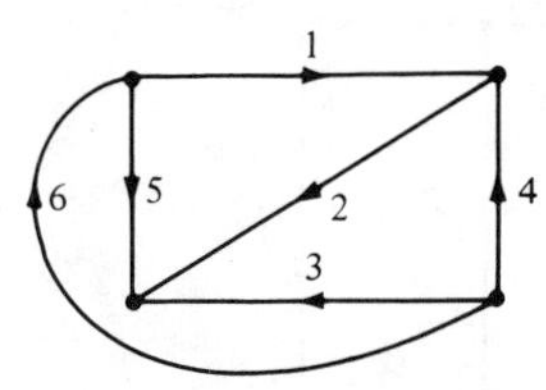

Figure P5.31

Consequences of strict passivity

32 Use the no current-gain property to show that in the circuit shown in Fig. P5.32 with $\mathcal{R}_1$, $\mathcal{R}_2$, and $\mathcal{R}_3$ as strictly passive resistors, no branch current exceeds 3 A.

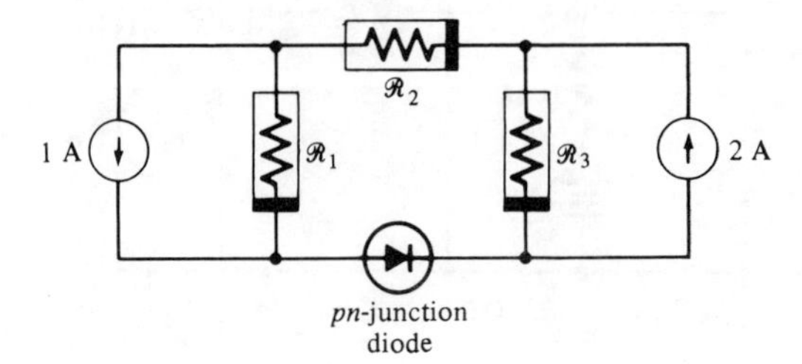

Figure P5.32

CHAPTER

SIX

FIRST-ORDER CIRCUITS

Chapters 2 to 5 have been devoted exclusively to circuits made of resistors and independent sources. The resistors may contain two or more terminals and may be linear or nonlinear, time-varying or time-invariant. We have shown that these resistive circuits are always governed by algebraic equations.

In this chapter, we introduce two new circuit elements, namely, two-terminal capacitors and inductors. We will see that these elements differ from resistors in a fundamental way: They are *lossless*, and therefore *energy* is not dissipated but merely *stored* in these elements.

A circuit is said to be *dynamic* if it includes some capacitor(s) or some inductor(s) or both. In general, dynamic circuits are governed by *differential equations*. In this initial chapter on dynamic circuits, we consider the *simplest* subclass described by only *one* first-order differential equation—hence the name first-order circuits. They include all circuits containing *one* 2-terminal capacitor (or inductor), plus resistors and independent sources.

The important concepts of *initial state*, *equilibrium state*, and *time constant* allow us to find the solution of any first-order linear time-invariant circuit driven by dc sources by inspection (Sec. 3.1). Students should master this material before plunging into the following sections where the inspection method is extended to include *linear switching* circuits in Sec. 4 and *piecewise-linear* circuits in Sec. 5. Here, the important concept of a *dynamic route* plays a crucial role in the analysis of piecewise-linear circuits by inspection.

1 TWO-TERMINAL CAPACITORS AND INDUCTORS

Many devices cannot be modeled accurately using only resistors. In this section, we introduce *capacitors* and *inductors*, which, together with *resistors*, form three basic circuit elements for modeling physical circuits. To emphasize the "dual" character of these two elements, we will often use a two-column format so that each statement on the left is the dual of the one on the right. Once the reader gets used to the idea of duality, he need only read one column while mentally reflecting on the dual statement in the other column.

In elementary circuits courses, the

capacitor is defined by

$$i = C\frac{dv}{dt} \qquad (1.1a)$$

where C is a constant called the *capacitance*.

inductor is defined by

$$v = L\frac{di}{dt} \qquad (1.1b)$$

where L is a constant called the *inductance*.

These definitions are valid only if the elements are *linear* and *time-invariant*. To generalize these definitions to the *nonlinear* and possibly *time-varying* case, it is necessary to invoke the following "dual" circuit variables:

The *charge* $q(t)$ associated with a two-terminal element at any time t is defined by

$$q(t) \triangleq \int_{-\infty}^{t} i(\tau)\,d\tau \qquad (1.2a)$$

where i denotes the branch current.[1] The unit of charge is the *coulomb* (C).

Given the current waveform $i(\cdot)$, the associated charge $q(t)$ can be calculated by integration. Alternately, $q(t)$ can be measured directly by an instrument designed to implement Eq. (1.2a) automatically.

The *flux* $\phi(t)$ associated with a two-terminal element at any time t is defined by

$$\phi(t) \triangleq \int_{-\infty}^{t} v(\tau)\,d\tau \qquad (1.2b)$$

where v denotes the branch voltage.[2] The unit of flux is the *weber* (Wb).

Given the voltage waveform $v(\cdot)$, the associated flux $\phi(t)$ can be calculated by integration. Alternately, $\phi(t)$ can be measured directly by an instrument designed to implement Eq. (1.2b) automatically.

[1] In some devices, we can interpret $q(t)$ as the charge accumulated in a conducting plate (see Example 1a) at time t. In other devices (e.g., see Exercise on page 303) where no simple physical interpretation exists, $q(t)$ may be interpreted as a new variable defined mathematically via Eq. (1.2a).

[2] In some devices, we can interpret $\phi(t)$ as the magnetic flux linking a coil (see Example 1b) at time t. In other devices (e.g., see Exercise on page 303) where no simple physical interpretation exists, $\phi(t)$ may be interpreted as a new variable defined mathematically via Eq. (1.2b).

Let N be a two-terminal element driven by a voltage source $v(t)$ [respectively, current source $i(t)$] and let $i(t)$ [respectively, $v(t)$] denote the corresponding current (respectively, voltage) response. If we plot the locus $\mathscr{L}(v, i)$ of $(v(t), i(t))$ in the v-i plane and obtain a *fixed* curve *independent of the excitation waveforms*, then N can be modeled as a two-terminal *resistor*. If $\mathscr{L}(v, i)$ changes with the excitation waveform, then N does not behave like a resistor and a different *model* must be chosen. In this case, we can calculate the associated charge waveform $q(t)$ using Eq. (1.2*a*), or the flux waveform $\phi(t)$ using Eq. (1.2*b*), and see whether the corresponding locus $\mathscr{L}(q, v)$ of $(q(t), v(t))$ in the q-v plane, or $\mathscr{L}(\phi, i)$ of $(\phi(t), i(t))$ in the ϕ-i plane, is a fixed curve independent of the excitation waveforms.

Exercises

1. Apply at $t = 0$ a voltage source $v(t) = A \sin \omega t$ (in volts) across a 1-F capacitor. (*a*) Calculate the associated *current* $i(t)$, *flux* $\phi(t)$, and *charge* $q(t)$ for $t \geq 0$, using Eqs. (1.2*a*) and (1.2*b*). Assume $\phi(0) = -1$ Wb and $q(0) = 0$ C. (*b*) Sketch the loci $\mathscr{L}(v, i)$, $\mathscr{L}(\phi, i)$, and $\mathscr{L}(q, v)$ in the v-i plane, ϕ-i plane, and q-v plane, respectively, for the following parameters (A is in volts, ω is in radians per second):

A	1	2	1
ω	1	1	2

(*c*) Does it make sense to describe this element by a v-i characteristic? ϕ-i characteristic? q-v characteristic? Explain.

2. Apply at $t = 0$ a current source $i(t) = A \sin \omega t$ (in amperes) across a 1-H inductor. (*a*) Calculate the associated *voltage* $v(t)$, *charge* $q(t)$, and *flux* $\phi(t)$ for $t \geq 0$, using Eqs. (1.2*a*) and (1.2*b*). Assume $q(0) = -1$ C and $\phi(0) = 0$ W. (*b*) Sketch the loci $\mathscr{L}(v, i)$, $\mathscr{L}(q, v)$, and $\mathscr{L}(\phi, i)$ in the v-i plane, q-v plane, and ϕ-i plane, respectively, for the following parameters (A is in amperes, ω is in radians per second):

A	1	2	1
ω	1	1	2

(*c*) Does it make sense to describe this element by a v-i characteristic? q-v characteristic? ϕ-i characteristic? Explain.

1.1 q-v and ϕ-i characteristics

A two-terminal element whose *charge* $q(t)$ and *voltage* $v(t)$ fall on some *fixed* curve in the q-v plane at any time t is called a *time-invariant capacitor*.[3] This curve is called the

A two-terminal element whose *flux* $\phi(t)$ and *current* $i(t)$ fall on some *fixed* curve in the ϕ-i plane at any time t is called a *time-invariant inductor*.[4] This curve is called the

[3] This definition is generalized to that of a *time-varying* capacitor in Sec. 1.2.

[4] This definition is generalized to that of a *time-varying* inductor in Sec. 1.2.

q-v characteristic of the capacitor. It may be represented by the equation[5]

$$f_C(q, v) = 0 \qquad (1.3a)$$

If Eq. (1.3*a*) can be solved for v as a single-valued function of q, namely,

$$v = \hat{v}(q) \qquad (1.4a)$$

the capacitor is said to be *charge-controlled*.

If Eq. (1.3*a*) can be solved for q as a single-valued function of v, namely,

$$q = \hat{q}(v) \qquad (1.5a)$$

the capacitor is said to be *voltage-controlled*.

If the function $\hat{q}(v)$ is differentiable, we can apply the chain rule to express the current entering a *time-invariant voltage-controlled capacitor* in a form similar to Eq. (1.1*a*):[6]

$$i = C(v)\dot{v} \qquad (1.6a)$$

where

$$C(v) \triangleq \frac{d\hat{q}(v)}{dv} \qquad (1.7a)$$

is called the *small-signal capacitance* at the operating point v.

Example 1*a* (Linear time-invariant parallel-plate capacitor) Figure 1.1*a* shows a familiar device made of two flat parallel metal plates sepa-

[5] This equation is also called the *constitutive relation* of the capacitor.

[6] We will henceforth use the notation $\dot{v} \triangleq dv(t)/dt$.

ϕ-i characteristic of the inductor. It may be represented by the equation[7]

$$f_L(\phi, i) = 0 \qquad (1.3b)$$

If Eq. (1.3*b*) can be solved for i as a single-valued function of ϕ, namely,

$$i = \hat{i}(\phi) \qquad (1.4b)$$

the inductor is said to be *flux-controlled*.

If Eq. (1.3*b*) can be solved for ϕ as a single-valued function of i, namely,

$$\phi = \hat{\phi}(i) \qquad (1.5b)$$

the inductor is said to be *current-controlled*.

If the function $\hat{\phi}(i)$ is differentiable, we can apply the chain rule to express the voltage across a *time-invariant current-controlled inductor* in a form similar to Eq. (1.1*b*):[8]

$$v = L(i)\dot{i} \qquad (1.6b)$$

where

$$L(i) \triangleq \frac{d\hat{\phi}(i)}{di} \qquad (1.7b)$$

is called the *small-signal inductance* at the operating point i.

Example 1*b* (Linear time-invariant toroidal inductor) Figure 1.2*a* shows a familiar device made of a conducting wire wound around a toroid

[7] This equation is also called *the constitutive relation* of the inductor.

[8] We will henceforth use the notation $\dot{i} \triangleq di(t)/dt$.

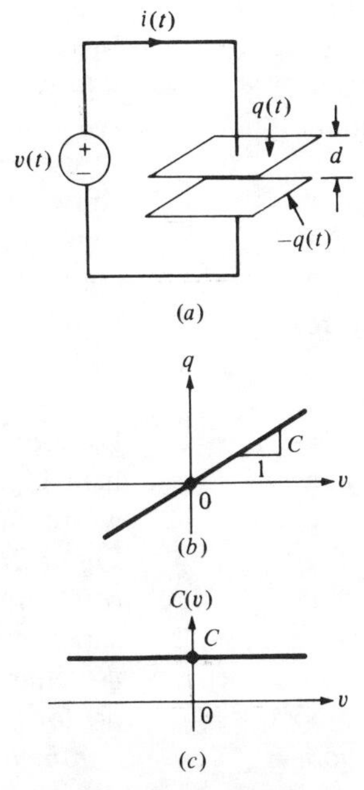

Figure 1.1 Parallel-plate capacitor.

rated in free space by a distance d. When a voltage $v(t) > 0$ is applied, we recall from physics that a charge equal to

$$q(t) = C\,v(t) \qquad (1.8a)$$

is induced at time t on the upper plate, and an equal but opposite charge is induced on the lower plate at time t. The constant of proportionality is given approximately by

$$C = \varepsilon_0 \frac{A}{d} \qquad \text{farad (F)}$$

where $\varepsilon_0 = 8.854 \times 10^{-12}$ F/m is the dielectric constant in free space, A is the plate area in square meters, and d is the separation of the plates in meters.

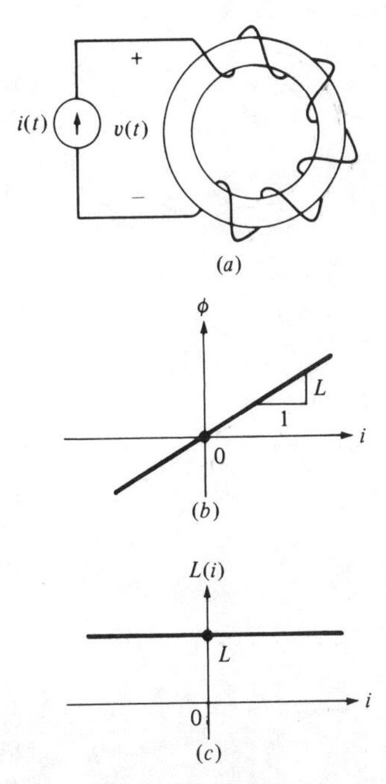

Figure 1.2 Toroidal inductor.

made of a nonmetallic material such as wood. When a current $i(t) > 0$ is applied, we recall from physics that a flux equal to

$$\phi(t) = Li(t) \qquad (1.8b)$$

is induced at time t and circulates around the interior of the toroid. The constant of proportionality is given approximately by

$$L = \mu_0 \frac{N^2 A}{\ell} \qquad \text{henry (H)}$$

where $\mu_0 = 4 \times 10^{-7}$ H/m is the permeability of the wooden core, N is the number of turns, A is the cross-sectional area in square meters, and ℓ is the midcircumference along the toroid in meters.

Equation (1.8*a*) defines the q-v characteristic of a *linear time-invariant capacitor*, namely, a straight line through the origin with slope equal to C, as shown in Fig. 1.1*b*. Its small-signal capacitance $C(v) = C$ is a constant function (Fig. 1.1*c*). Consequently, Eq. (1.6*a*) reduces to Eq. (1.1*a*).

Example 2*a* (Nonlinear time-invariant parallel-plate capacitor) If we fill the space between the two plates in Fig. 1.1*a* with a nonlinear *ferroelectric* material (such as barium titanate), the *measured* q-v characteristic in Fig. 1.3*a* is no longer a straight line. This *nonlinear* behavior is due to the fact that the dielectric constant of *ferroelectric* materials is not a constant—it changes with the applied *electric* field intensity.

Likewise, the small-signal capacitance shown in Fig. 1.3*b* is a *nonlinear* function of v.

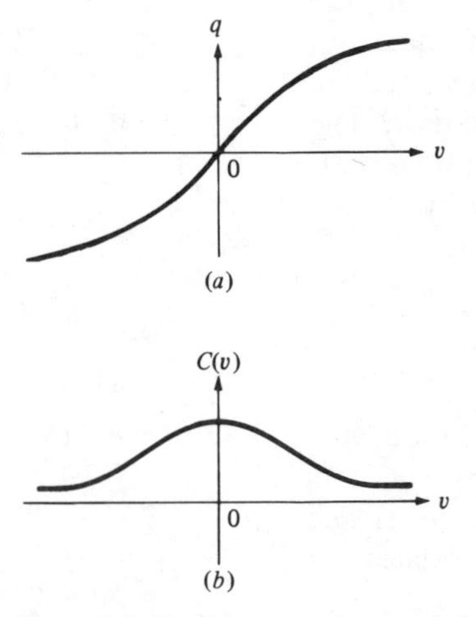

Figure 1.3 Nonlinear q-v characteristic.

Equation (1.8*b*) defines the ϕ-i characteristic of a *linear time-invariant inductor*, namely, a straight line through the origin with slope equal to L, as shown in Fig. 1.2*b*. Its small-signal inductance $L(i) = L$ is a constant function (Fig. 1.2*c*). Consequently, Eq. (1.6*b*) reduces to Eq. (1.1*b*).

Example 2*b* (Nonlinear time-invariant toroidal inductor) If we replace the wooden core in Fig. 1.2*a* with a nonlinear *ferromagnetic* material (such as superpermalloy) the *measured* ϕ-i characteristic in Fig. 1.4*a* is no longer a straight line. This *nonlinear* behavior is due to the fact that the permeability of *ferromagnetic* materials is not a constant—it changes with the applied *magnetic* field intensity.

Likewise, the small-signal inductance shown in Fig. 1.4*b* is a *nonlinear* function of i.

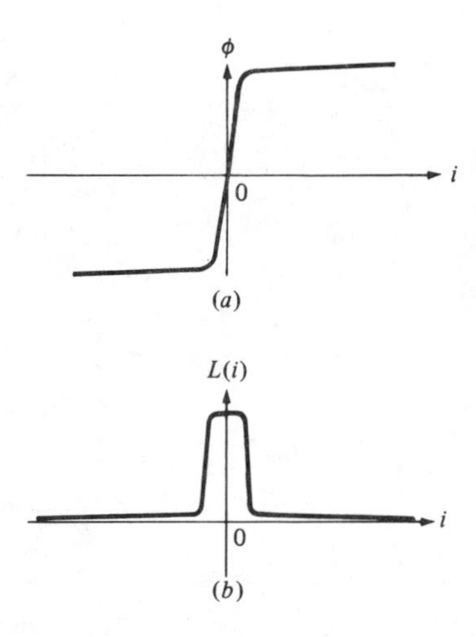

Figure 1.4 Nonlinear ϕ-i characteristic.

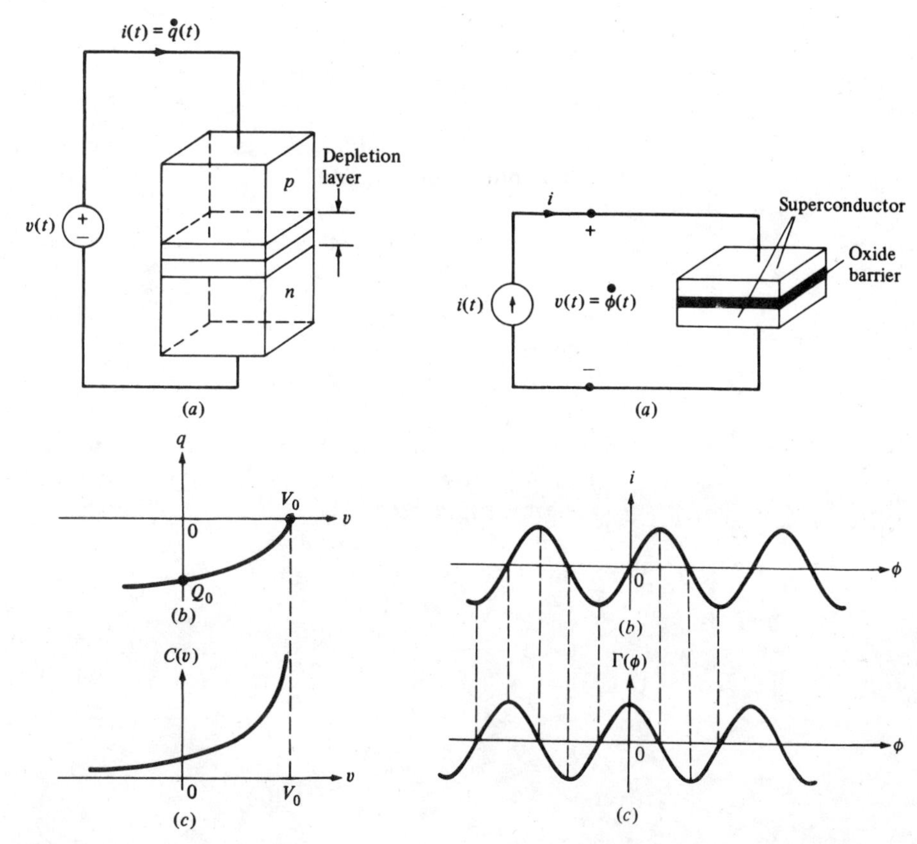

Figure 1.5 q-v characteristic of a varactor diode.

Figure 1.6 ϕ-i characteristic of a Josephson junction.

Example 3a (Varactor diode) The varactor diode[9] shown in Fig. 1.5*a* is a *pn*-junction diode designed specially to take advantage of the depletion layer when operating in reverse bias, i.e., when $v < V_0$ (typically, $0.2V < V_0 < 0.9V$). Semiconductor physics proves that the charge q accumulated on the top layer is equal to

Example 3b (Josephson junction) The Josephson junction[10] shown in Fig. 1.6*a* consists of two superconductors separated by an insulating layer (such as oxide). Superconductor physics proves that the current i varies sinusoidally with ϕ, namely,

[9] Varactor diodes are widely used in many communication circuits. For example, modern radio and TV sets are automatically tuned by applying a suitable dc bias voltage across such a diode.

[10] Josephson proposed this exotic device in 1961 and was awarded the Nobel prize in physics in 1969 for this discovery. The Josephson junction has been used in numerous applications.

$$q = -K(V_0 - v)^{1/2} \triangleq \hat{q}(v) \tag{1.9a}$$

provided $v < V_0$. Here, V_0 is the contact potential and

$$K = \frac{2\varepsilon N_a N_d}{N_a + N_d} A$$

where ε = permittivity of the material, N_a = number of acceptor atoms per cubic centimeter, N_d = number of donor atoms per cubic centimeter, and A = cross-sectional area in square centimeters.

Its small-signal capacitance (Fig. 1.5*c*) is obtained by differentiating Eq. (1.9*a*):

$$C(v) = \tfrac{1}{2} K(V_0 - v)^{-1/2} \qquad v < V_0 \tag{1.10a}$$

Note that unlike the previous examples, the *q-v* characteristic in Fig. 1.5*b* is *not* defined for $v > V_0$. Hence, this capacitor is *not* voltage-controlled for *all* values of $v > V_0$. (For $v > V_0$, the diode becomes forward biased and behaves like a nonlinear resistor.)

$$i = I_0 \sin k\phi \triangleq \hat{i}(\phi) \tag{1.9b}$$

where I_0 is a device parameter and

$$k = 4\pi \frac{e}{h}$$

where e = electron charge and h = Planck's constant.

Note that unlike the previous example, the *ϕ-i* characteristic in Fig. 1.6*b* is *not* current-controlled. Consequently, its small-signal inductance $L(i) \triangleq d\phi(i)/di$ is not uniquely defined.

However, the Josephson junction is *flux-controlled* and has a well-defined slope (Fig. 1.6*c*)

$$\Gamma(\phi) \triangleq \frac{d\hat{i}(\phi)}{d\phi} = kI_0 \cos k\phi \tag{1.10b}$$

We call $\Gamma(\phi)$ the *reciprocal small-signal inductance* since it has the unit of H^{-1}.

REMARK Note that Eqs. (1.9*a*) and (1.9*b*), as well as Eqs. (1.10*a*) and (1.10*b*), are not strictly dual equations because the corresponding variables are not duals of each other. However, if we solved for v in terms of q in Eq. (1.9*a*), we would obtain a dual function $v = \hat{v}(q)$. In this case, the derivative

$$S(q) \triangleq \frac{d\hat{v}(q)}{dq}$$

is called the *reciprocal small-signal capacitance*.

Exercise (*a*) Show that a one-port obtained by connecting port 2 of a *gyrator* (assume unity coefficient) across a k-H linear inductor is equivalent to that of a k-F linear capacitor in the sense that they have identical q-v characteristics. (*b*) Is it possible to give a simple *physical* interpretation of the *charge* associated with this element? (*c*) Generalize the property in (*a*) to the case where the inductor is nonlinear: $\phi = \hat{\phi}(i)$.

Exercise (*a*) Show that a one-port obtained by connecting port 2 of a *gyrator* (assume unity coefficient) across a k-F linear capacitor is equivalent to that of a k-H linear inductor in the sense that they have identical ϕ-i characteristics. (*b*) Is it possible to give a simple *physical* interpretation of the *flux* associated with this element? (*c*) Generalize the property in (*a*) to the case where the capacitor is nonlinear: $q = \hat{q}(v)$.

1.2 Time-Varying Capacitors and Inductors

The examples presented so far are *time-invariant* in the sense that the q-v and ϕ-i characteristics do not change with time.

If the q-v characteristic changes with time, the capacitor is said to be *time-varying*.

For example, suppose we vary the spacing between the parallel-plate capacitor in Fig. 1.1*a*, say by using a motor-driven cam mechanism, so that the capacitance C becomes some *prescribed function of time* $C(t)$. Then Eq. (1.8*a*) becomes

$$q(t) = C(t)v(t) \qquad (1.11a)$$

It follows from Eq. (1.2*a*) that

$$i(t) = C(t)\frac{dv(t)}{dt} + \frac{dC(t)}{dt}v(t) \qquad (1.12a)$$

Note that the current in a *time-varying linear capacitor* differs from Eq. (1.1*a*) *not only* in the replacement of C by $C(t)$, but also in the presence of an extra term.

If the ϕ-i characteristic changes with time, the inductor is said to be *time-varying*.

For example, suppose we vary the number of turns of the winding in Fig. 1.2*a*, say by using a motor-driven sliding contact, so that the inductance L becomes some *prescribed function of time* $L(t)$. Then Eq. (1.8*b*) becomes

$$\phi(t) = L(t)i(t) \qquad (1.11b)$$

It follows from Eq. (1.2*b*) that

$$v(t) = L(t)\frac{di(t)}{dt} + \frac{dL(t)}{dt}i(t) \qquad (1.12b)$$

Note that the voltage in a *time-varying linear inductor* differs from Eq. (1.1*b*) *not only* in the replacement of L by $L(t)$, but also in the presence of an extra term.

To be specific, assume

$$C(t) = 2 + \sin t \qquad (1.13a)$$

then

$$q(t) = (2 + \sin t)v(t) \qquad (1.14a)$$

and

$$i(t) = (2 + \sin t)\frac{dv(t)}{dt} + (\cos t)v(t) \qquad (1.15a)$$

The q-v characteristic of a time-varying linear capacitor consists of a *family* of straight lines, each line valid for a given instant of time. For example, the q-v characteristic of the above time-varying linear capacitor is shown in Fig. 1.7*a*. Its associated small-signal capacitance consists of a family of horizontal lines (Fig. 1.7*b*).

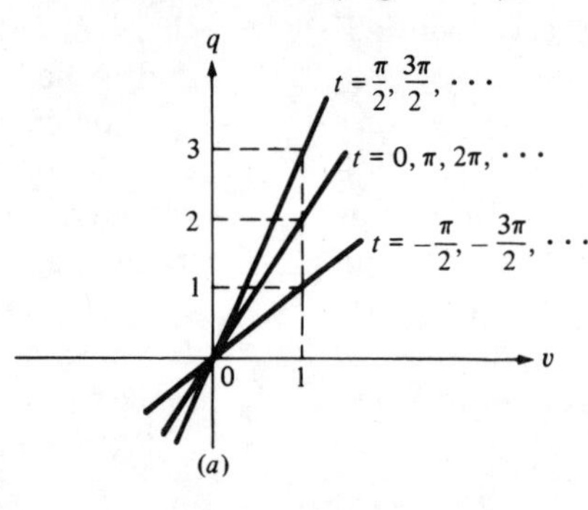

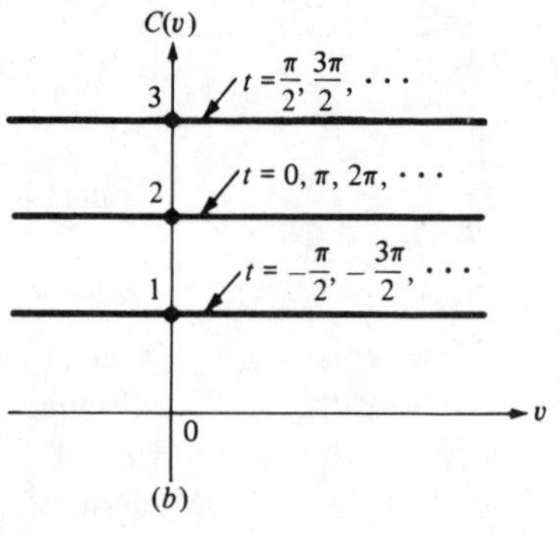

Figure 1.7 Time-varying q-v characteristic of Eq. (1.14*a*).

To be specific, assume

$$L(t) = 2 + \sin t \qquad (1.13b)$$

then

$$\phi(t) = (2 + \sin t)i(t) \qquad (1.14b)$$

and

$$v(t) = (2 + \sin t)\frac{di(t)}{dt} + (\cos t)i(t) \qquad (1.15b)$$

The ϕ-i characteristic of a time-varying linear inductor consists of a *family* of straight lines, each line valid for a given instant of time. For example, the ϕ-i characteristic of the above time-varying linear inductor is shown in Fig. 1.8*b*. Its associated small-signal inductance consists of a family of horizontal lines (Fig. 1.8*b*).

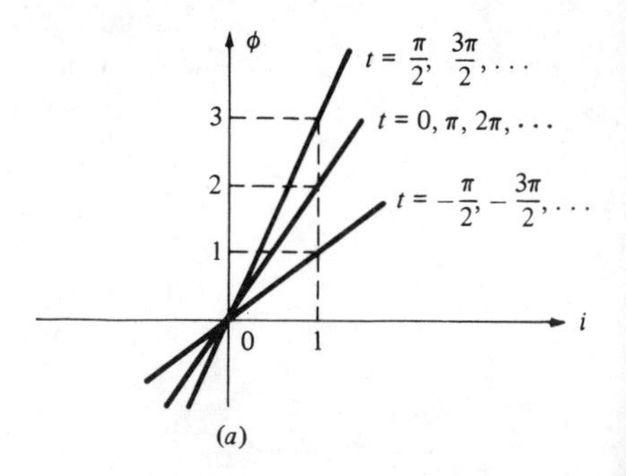

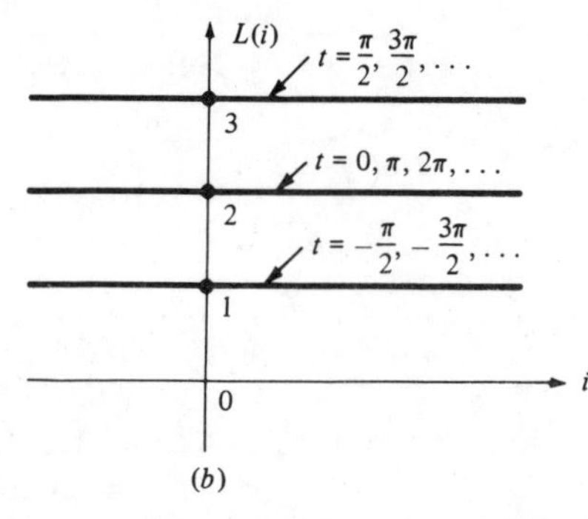

Figure 1.8 Time-varying ϕ-i characterisic of Eq. (1.14*b*).

Time-varying linear capacitors and inductors are useful in the modeling, analysis, and design of many communication circuits (e.g., modulators, demodulators, parametric amplifiers).

In the most general case, a *time-varying nonlinear capacitor* is defined by a family of time-dependent and nonlinear q-v characteristics, namely,

$$f_C(q, v, t) = 0 \quad (1.16a)$$

In the most general case, a *time-varying nonlinear inductor* is defined by a family of time-dependent and nonlinear ϕ-i characteristics, namely,

$$f_L(\phi, i, t) = 0 \quad (1.16b)$$

The two circuit variables used in defining a two-terminal resistor, inductor, or capacitor can be easily remembered with the help of the mnemonic diagram shown in Fig. 1.9. Note that out of the six exhaustive pairings of the four basic variables v, i, q, and ϕ, two are related *by definitions*, namely, $i = \dot{q}$ and $v = \dot{\phi}$. The remaining pairs are constrained by the *constitutive relation* of a two-terminal element, three of which give us the resistor, inductor, and capacitor.[11]

We will use the symbols shown in Fig. 1.9 to denote a nonlinear two-terminal resistor, inductor, or capacitor, respectively. Note that a dark band is included in each symbol in order to distinguish the two terminals. Just as in the case of a *nonbilateral* two-terminal resistor, such distinction is necessary if the q-v or ϕ-i characteristic is *not* odd symmetric. In the special case where the element is *linear*, the v-i, ϕ-i, and q-v characteristics are odd symmetric

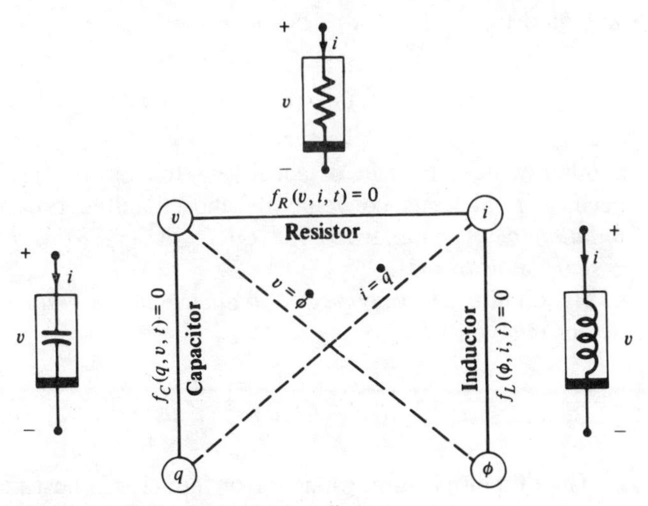

Figure 1.9 Basic circuit element diagram.

[11] A fourth nonlinear two-terminal element called the *memristor* is defined by the remaining relationship between q and ϕ. This circuit element is described in L. O. Chua, "Memristor—The Missing Circuit Element," *IEEE Trans. on Circuit Theory*, vol. 18, pp. 507–519, September 1971.

and hence remain unchanged after the two terminals are interchanged. In this case, we simply delete drawing the enclosing rectangle and revert to the standard symbol for a linear resistor, inductor, and capacitor.

2 BASIC PROPERTIES EXHIBITED BY TIME-INVARIANT CAPACITORS AND INDUCTORS

Capacitors and inductors behave differently from resistors in many ways. The following typical properties illustrate some fundamental differences.[12]

2.1 Memory Property

Suppose we drive the linear capacitor in Fig. 1.1*a* by a current source $i(t)$. The corresponding voltage at any time t is obtained by integrating both sides of Eq. (1.1*a*) from $\tau = -\infty$ to $\tau = t$. Assuming $v(-\infty) = 0$ (i.e., the capacitor has no initial charge when manufactured), we obtain

$$v(t) = \frac{1}{C} \int_{-\infty}^{t} i(\tau)\, d\tau \tag{2.1}$$

Note that unlike the *resistor* voltage which depends on the resistor current only at one instant of time t, the above capacitor voltage depends on the *entire past history* (*i.e.*, $-\infty < \tau < t$) of $i(\tau)$. Hence, *capacitor has memory.*

Now suppose the voltage $v(t_0)$ at some time $t_0 < t$ is given, then Eq. (1.1*a*) integrated from $t = -\infty$ to t becomes

$$v(t) = v(t_0) + \frac{1}{C} \int_{t_0}^{t} i(\tau)\, d\tau \qquad t \geq t_0 \tag{2.2}$$

In other words, instead of specifying the entire past history, we need only specify $v(t)$ at some conveniently chosen initial time t_0. In effect, the *initial condition* $v(t_0)$ summarizes the effect of $i(\tau)$ from $\tau = -\infty$ to $\tau = t_0$ on the present value of $v(t)$.

By duality, it follows that *inductor has memory* and that the inductor current is given by

$$i(t) = i(t_0) + \frac{1}{L} \int_{t_0}^{t} v(\tau)\, d\tau \qquad t \geq t_0 \tag{2.3}$$

The "memory" in a capacitor or inductor is best manifested by the "dual" equivalent circuits shown in Fig. 2.1*a* and *b* which asserts the following:

[12] Unless otherwise specified, all capacitors and inductors are assumed to be *time-invariant* in this book.

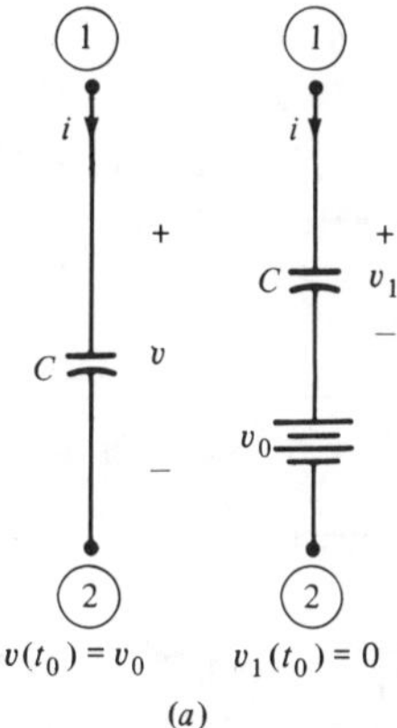

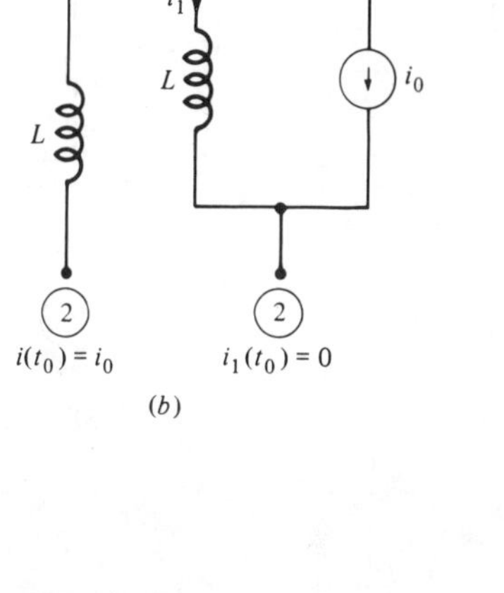

Figure 2.1 Initial condition transformation.

Initial capacitor voltage transformation A linear time-invariant capacitor with an *initial voltage* v_0 is indistinguishable externally from a one-port made of an *initially uncharged* capacitor (having the same capacitance) in series with a battery of v_0 volts.

The circuits shown in Fig. 2.1*a* are equivalent because they are characterized by the same equation, namely, Eq. (2.2).

Initial inductor current transformation A linear time-invariant inductor with an *initial current* i_0 is indistinguishable externally from a one-port made of an inductor (having the same inductance) with *zero initial current* in parallel with a current source of i_0 amperes.

The circuits shown in Fig. 2.1*b* are equivalent because they are characterized by the same equation, namely, Eq. (2.3).

The memory property of capacitors and inductors has been exploited in the design of many practical circuits. For example, consider the "peak detector" circuit shown in Fig. 2.2*a*. Since the ideal diode current vanishes whenever $v_{\text{in}}(t) \leq v_o(t)$, it follows from Eq. (2.2) that at any time t_1, $v_o(t)$ is equal to the *maximum* value of $v_{\text{in}}(t)$ from $t = -\infty$ to $t = t_1$. A typical waveform of $v_o(t)$ and $v_{\text{in}}(t)$ is shown in Fig. 2.2*b*. In practice, this circuit is usually implemented as shown in Fig. 2.2*c*, where the op-amp circuit from Fig. 3.13 of Chap. 4 is used to simulate an ideal diode and where the op-amp buffer from Fig. 2.1 of Chap. 4 is used to avoid output loading effects.

Exercise The switch S in the "track-and-hold" circuit shown in Fig. 2.3 is periodically open and closed every Δt seconds. Sketch $v_o(t)$, and suggest a typical application.

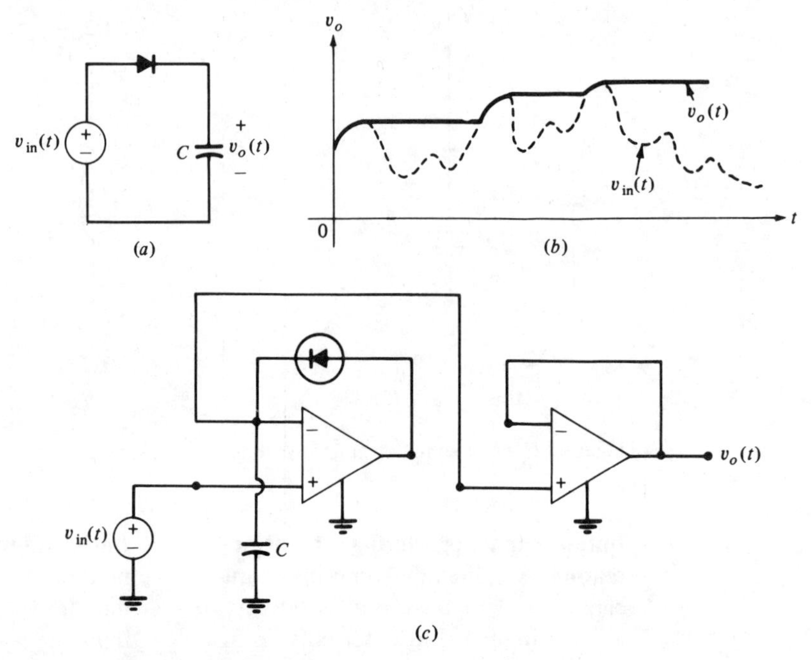

Figure 2.2 A peak detector circuit.

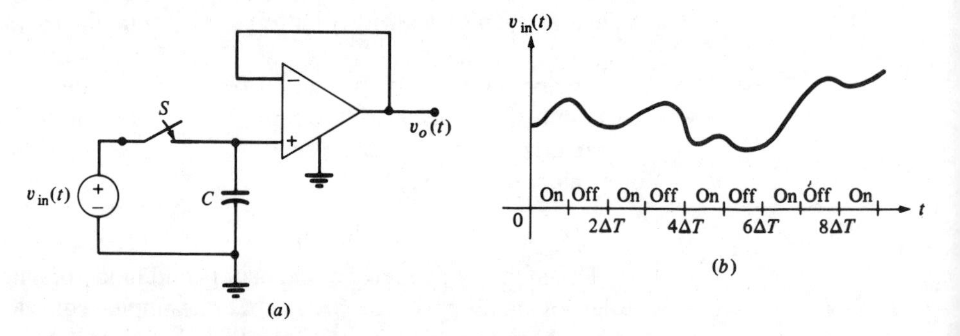

Figure 2.3 A track-and-hold circuit.

Equation (2.2) and Fig. 2.1*a* are valid only when the capacitor is linear. To show that *nonlinear* capacitors also exhibit memory, note that its voltage $v(t)$ depends on the charge and by Eq. (1.2*a*),

$$q(t) = q(t_0) + \int_{t_0}^{t} i(\tau)\, d\tau \qquad t > t_0 \tag{2.4a}$$

Equation (2.3) and Fig. 2.1*b* are valid only when the inductor is linear. To show that *nonlinear* inductors also exhibit memory, note that its current $i(t)$ depends on the flux and by Eq. (1.2*b*),

$$\phi(t) = \phi(t_0) + \int_{t_0}^{t} v(\tau)\, d\tau \qquad t > t_0 \tag{2.4b}$$

which in turn depends on the past history of $i(\tau)$ for $-\infty < \tau < t_0$.

which in turn depends on the past history of $v(\tau)$ for $-\infty < \tau < t_0$.

2.2 Continuity Property

Consider the circuit shown in Fig. 2.4*a*, where the current source is described by the "discontinuous" square wave shown in Fig. 2.4*b*. Assuming that $v_C(0) = 0$ and applying Eq. (2.2), we obtain the "continuous" capacitor voltage waveform shown in Fig. 2.4*c*. This "smoothing" phenomenon turns out to be a general property shared by both capacitor voltages and inductor currents. More precisely, we can state this important property as follows:

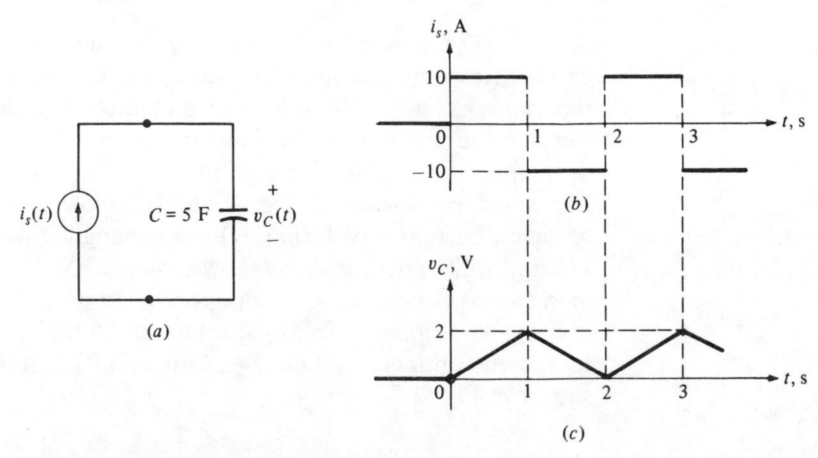

Figure 2.4 The discontinuous capacitor current waveform in (*b*) is smoothed out by the capacitor to produce the continuous voltage waveform in (*c*).

> *Capacitor voltage-inductor current continuity property*
>
> (*a*) If the current waveform $i_C(t)$ in a linear time-invariant *capacitor* remains bounded in a closed interval $[t_a, t_b]$, then the voltage waveform $v_C(t)$ across the capacitor is a *continuous* function in the open interval (t_a, t_b). In particular,[13] for any time T satisfying $t_a < T < t_b$, $v_C(T-) = v_C(T+)$.
>
> (*b*) If the voltage waveform $v_L(t)$ in a linear time-invariant *inductor* remains bounded in a closed interval $[t_a, t_b]$, then the current waveform $i_L(t)$ through the inductor is a *continuous* function in the open interval (t_a, t_b). In particular, for any time T satisfying $t_a < T < t_b$, $i_L(T-) = i_L(T+)$.

[13] We denote the left-hand limit and right-hand limit of a function $f(t)$ at $t = T$ by $f(T-)$ and $f(T+)$, respectively.

PROOF We will prove only (*a*) since (*b*) follows by duality. Substituting $t = T$ and $t = T + dt$ into Eq. (2.2), where $t_a < T < t_b$ and $t_a < T + dt \leq t_b$, and subtracting, we get

$$v_C(T + dt) - v_C(T) = \frac{1}{C} \int_T^{T+dt} i_C(t')\, dt' \tag{2.5}$$

Since $i_C(t)$ is *bounded* in $[t_a, t_b]$, there is a *finite* constant M such that $|i_C(t)| < M$ for all t in $[t_a, t_b]$. It follows that the area under the curve $i_C(t)$ from T to $T + dt$ is at most $M\,dt$ (in absolute value), which tends to zero as $dt \to 0$. Hence Eq. (2.5) implies $v_C(T + dt) \to v_C(T)$ as $dt \to 0$. This means that the waveform $v_C(\cdot)$ is continuous at $t = T$. ∎

REMARK The above continuity property does *not* hold if the capacitor current (respectively, inductor voltage) is *unbounded*. Before we illustrate this remark, let us first give an example showing how a capacitor current can become unbounded—at least in theory.

Suppose we apply a voltage source across a 1-F linear capacitor having the waveform shown in Fig. 2.5*b*. It follows from Eq. (1.1*a*) that the capacitor current waveform $i_C(t)$ is a rectangular pulse with height equal to $1/\Delta$ and width equal to Δ, as shown in Fig. 2.5*c*. Note that the pulse height increases as Δ decreases. It is important to note that the area of this pulse is equal to 1, *independent of* Δ. Now in the limit where $\Delta \to 0$, $v_s(t)$ tends to the discontinuous "unit step" function [henceforth denoted by $1(t)$] shown in Fig. 2.5*d*, i.e.,[14]

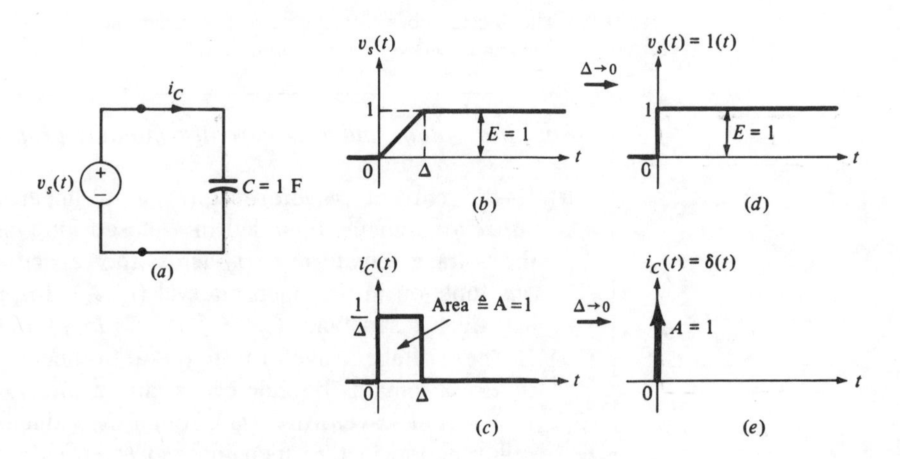

Figure 2.5 Circuit for generating a unit current impulse.

[14] The value of the unit step function $1(t)$ at $t = 0$ does not matter from the physical point of view. However, sometimes it is convenient to define it to be equal to $\frac{1}{2}$ in circuit theory.

$$\lim_{\Delta \to 0} v_s(t) = 1(t) \triangleq \begin{cases} 0 & t<0 \\ 1 & t>0 \end{cases} \tag{2.6}$$

To show that this discontinuity in the capacitor voltage does *not* contradict the preceding continuity property, note that the corresponding capacitor *current* is *unbounded* in this case, namely,

$$\lim_{\Delta \to 0} i_C(t) = \infty \qquad \text{at } t=0 \tag{2.7}$$

Since this waveform is of great importance in engineering analysis, let us pause to study its properties carefully.

Impulse (delta function) As $\Delta \to 0$, the height of the "rectangular" pulse in Fig. 2.5*c* tends to infinity at $t=0$, and to zero elsewhere, while the area under the pulse remains unchanged, i.e., $A=1$. This limiting waveform is called an *impulse* and will be denoted by $\delta(t)$.

More precisely, an unbounded signal $\delta(t)$ is called a *unit impulse*[15] iff it satisfies the following *two* properties:

1.	$\delta(t) \triangleq \begin{cases} \text{singular} & t=0 \\ 0 & t \neq 0 \end{cases}$		(2.8*a*) (2.8*b*)
2.	$\int_{-\varepsilon_1}^{\varepsilon_2} \delta(t)\, dt = 1$	for any $\varepsilon_1 > 0$ and $\varepsilon_2 > 0$	(2.8*c*)

Since the unit impulse is *unbounded*, we will denote it symbolically by a "bold" arrowhead as shown in Fig. 2.5*e*.

Now, if $E \neq 1$ and $C \neq 1$ in Fig. 2.5, the above discussion still holds provided the area A of the impulse is changed from $A=1$ to $A=CE$. Note that this situation can be simulated by connecting an E-V battery across a C-F capacitor *at $t=0$*. The resulting current waveform would then be an impulse with an area equal to $A=CE$.[16]

Now that we have demonstrated how a current impulse can be generated, let us drive the circuit in Fig. 2.4*a* with a current impulse of area $A=10$ applied at $t=5$ s, as shown in Fig. 2.6*a*. It follows from Eq. (2.8) that this impulse can be denoted by

$$i_s(t) = A\delta(t-5) \tag{2.9}$$

[15] In physics, the unit impulse is called a *delta function*. Using the theory of distribution in advanced mathematics, the unit impulse can be rigorously defined as a "generalized" function imbued with most of the standard properties of a function. In particular, most of the time $\delta(t)$ can be manipulated like an ordinary function.

[16] In practice, only a very large (but finite) current pulse is actually observed because all *physical* batteries have a small but nonzero internal resistance (recall Fig. 2.4*a* of Chap. 2). Given the value of R, we will be able to calculate the exact current waveform $i_C(t)$ in Sec. 3.1.

Substituting $i_s(t)$ into Eq. (2.2) with $v(0)=0$ we obtain

$$v_C(t) = \frac{1}{5}\int_0^t A\delta(\tau - 5)\, d\tau \qquad t \geq 0 \tag{2.10}$$

Defining a new dummy variable $x \triangleq \tau - 5$ and using Eq. (2.8), we obtain

$$v_C(t) = \frac{A}{5}\int_{-5}^{t-5} \delta(x)\, dx \qquad t \geq 0$$

$$= \begin{cases} 0 & t<5 \quad \text{[in view of Eq. (2.8}b\text{)]} \\ 2 & t>5 \quad \text{[in view of Eq. (2.8}c\text{)]} \end{cases} \tag{2.11}$$

The resulting capacitor voltage waveform is shown in Fig. 2.6*b*. Note that it is *discontinuous* at $t = 5$ s.

Exercise Prove that whenever the current waveform $i_C(t)$ entering a C-F linear time-invariant capacitor contains an *impulse* of area A at $t = t_0$, the associated capacitor voltage waveform $v_C(t)$ *will change abruptly* at t_0 by an amount equal to A/C.

Exercise Prove that whenever the voltage waveform $v_L(t)$ across an L-H linear time-invariant inductor contains an *impulse* of area A at $t = t_0$, the associated inductor current waveform $i_L(t)$ will change *abruptly* at t_0 by an amount equal to A/L.

2.3 Lossless Property

Since $p(t) = v(t)\, i(t)$ is the *instantaneous power* in *watts* entering a two-terminal element at any time t, the total *energy* $w(t_1, t_2)$ in joules entering the element during any time interval $[t_1, t_2]$ is given by

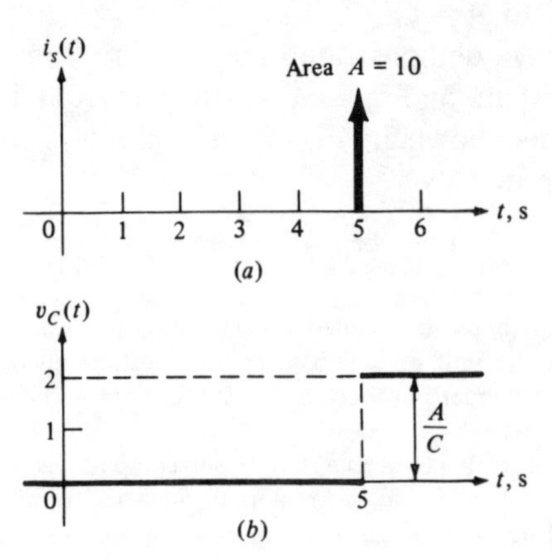

Figure 2.6 The voltage waveform $v_C(t)$ is discontinuous at $t = 5$ s.

$$w(t_1, t_2) = \int_{t_1}^{t_2} v(t)\, i(t)\, dt \qquad \text{joules} \tag{2.12}$$

For example, the total energy $w_R(t_1, t_2)$ entering a linear resistor with resistance $R > 0$ is given by

$$w_R(t_1, t_2) = \int_{t_1}^{t_2} [R\, i(t)]\, i(t)\, dt = R \int_{t_1}^{t_2} i^2(t)\, dt = \frac{1}{R} \int_{t_1}^{t_2} v^2(t)\, dt \tag{2.13}$$

This energy is dissipated in the form of heat and is lost as far as the circuit is concerned. Such an element is therefore said to be *lossy*.

In general, the energy $w(t_1, t_2)$ entering a two-terminal element during $[t_1, t_2]$ depends on the entire voltage waveform $v(t)$ or current waveform $i(t)$ over the *entire interval* $[t_1, t_2]$. For example, if we drive the 10-Ω resistor in Fig. 2.7*a* by the waveforms shown in Fig. 2.7*b* and *c*, respectively, the energy dissipated during the interval $[\frac{1}{4}, \frac{3}{4}]$ is given respectively by

$$w_1(\tfrac{1}{4}, \tfrac{3}{4}) = 10 \int_{1/4}^{3/4} (2 \sin 2\pi t)^2\, dt = 10.00 \text{ joules} \tag{2.14}$$

$$w_2(\tfrac{1}{4}, \tfrac{3}{4}) = 10 \int_{1/4}^{3/4} [-8(t - \tfrac{1}{2})]^2\, dt = 6.67 \text{ joules} \tag{2.15}$$

Note that $w_1(\frac{1}{4}, \frac{3}{4}) \neq w_2(\frac{1}{4}, \frac{3}{4})$ even though the resistor currents $i_1(t)$ and $i_2(t)$, and hence also their voltages $v_1(t)$ and $v_2(t)$, are *identical at the end points*, namely, $i_1(\frac{1}{4}) = i_2(\frac{1}{4}) = 2$ A and $i_1(\frac{3}{4}) = i_2(\frac{3}{4}) = -2$ A.

In sharp contrast to the above typical observations, the following calculation shows that

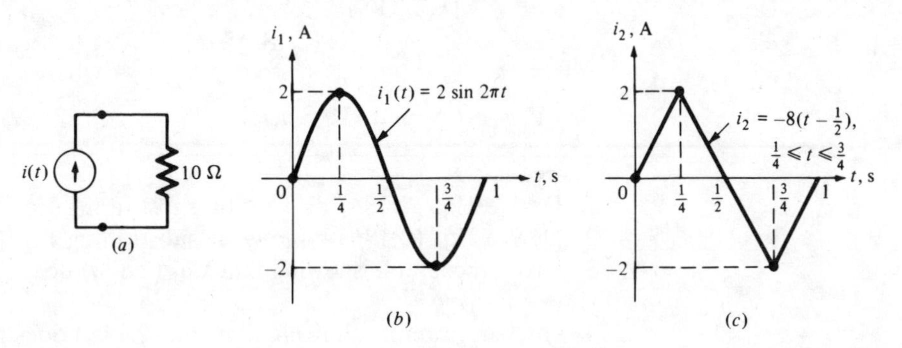

Figure 2.7 Resistor driven by two distinct current waveforms whose values coincide at $t_1 = \frac{1}{4}$ s and $t_2 = \frac{3}{4}$ s.

The energy $w_C(t_1, t_2)$ entering a charge-controlled capacitor during any time interval $[t_1, t_2]$ *is independent of the capacitor voltage or charge waveforms*: It is uniquely determined by the capacitor charge at the end points, namely by $q(t_1)$ and $q(t_2)$. Indeed,

$$w_C(t_1, t_2) = \int_{t_1}^{t_2} \hat{v}(q(t)) \frac{dq(t)}{dt} dt \tag{2.16a}$$

It follows from Eq. (2.16*a*) that

$$w_C(q_1, q_2) = \int_{q_1}^{q_2} \hat{v}(q)\, dq \tag{2.17a}$$

where we switched from t to q as the dummy variable, and $q_1 \triangleq q(t_1)$ and $q_2 \triangleq q(t_2)$.

Example For a C-F linear capacitor, we have $\hat{v}(q) = q/C$ and hence Eq. (2.17*a*) reduces to

$$\begin{aligned} w_C(q_1, q_2) &= \frac{1}{2C}[q_2^2 - q_1^2] \\ &= \tfrac{1}{2} C[V_2^2 - V_1^2] \end{aligned} \tag{2.18a}$$

where

$$V_1 \triangleq v(t_1) \quad \text{and} \quad V_2 \triangleq v(t_2).$$

The energy $w_L(t_1, t_2)$ entering a flux-controlled inductor during any time interval $[t_1, t_2]$ *is independent of the inductor current or flux waveforms*: It is uniquely determined by the inductor flux at the end points, namely, by $\phi(t_1)$ and $\phi(t_2)$. Indeed,

$$w_L(t_1, t_2) = \int_{t_1}^{t_2} \hat{i}(\phi(t)) \frac{d\phi(t)}{dt} dt \tag{2.16b}$$

It follows from Eq. (2.16*b*) that

$$w_L(\phi_1, \phi_2) = \int_{\phi_1}^{\phi_2} \hat{i}(\phi)\, d\phi \tag{2.17b}$$

where we switched from t to ϕ as the dummy varaible, and $\phi_1 \triangleq \phi(t_1)$ and $\phi_2 \triangleq \phi(t_2)$.

Example For an L-H linear inductor, we have $\hat{i}(\phi) = \phi/L$ and hence Eq. (2.17*b*) reduces to:

$$\begin{aligned} w_L(\phi_1, \phi_2) &= \frac{1}{2L}[\phi_2^2 - \phi_1^2] \\ &= \tfrac{1}{2} L[I_2^2 - I_1^2] \end{aligned} \tag{2.18b}$$

where

$$I_1 \triangleq i(t_1) \quad \text{and} \quad I_2 \triangleq i(t_2).$$

Exercises

1. Derive Eq. (2.18*a*) directly by substituting Eq. (1.1*a*) into Eq. (2.12).
2. Derive Eq. (2.18*b*) directly by substituting Eq. (1.1*b*) into Eq. (2.12).
3. Give an example showing that Eq. (2.17*a*) does not hold if the capacitor is *time-varying*.
4. Give an example showing that Eq. (2.17*b*) does not hold if the inductor is *time-varying*.

Equation (2.17*a*) shows that the energy $w_C(t_1, t_2)$ entering a charge-controlled capacitor is equal to the shaded area shown in Fig. 2.8*a*. Any waveform pair $[v(\cdot), q(\cdot)]$ taking the values $[v(t_1), q(t_1)]$ at t_1 and $[v(t_2), q(t_2)]$ at t_2 will give the same $w_C(t_1, t_2)$.

Now suppose $v(t)$ and $q(t)$ are *periodic* with period $T = t_2 - t_1$. Then $q(t_2) = q(t_1 + T) = q(t_1)$, and hence $w_C(t_1, t_2) = 0$. In this case, P_1 and P_2 in Fig. 2.8*a* coincide, thereby resulting in a zero area.

This observation can be summarized as follows: *Under periodic excitation, the total energy entering a charge-controlled capacitor is zero over any period.*

Equation (2.17*b*) shows that the energy $w_L(t_1, t_2)$ entering a flux-controlled inductor is equal to the shaded area shown in Fig. 2.8*b*. Any waveform pair $[i(\cdot), \phi(\cdot)]$ taking the values $[i(t_1), \phi(t_1)]$ at t_1 and $[i(t_2), \phi(t_2)]$ at t_2 will give the same $w_L(t_1, t_2)$.

Now suppose $i(t)$ and $\phi(t)$ are *periodic* with period $T = t_2 - t_1$. Then $\phi(t_2) = \phi(t_1 + T) = \phi(t_1)$, and hence $w_L(t_1, t_2) = 0$. In this case, P_1 and P_2 in Fig. 2.8*b* coincide, thereby resulting in a zero area.

This observation can be summarized as follows: *Under periodic excitation, the total energy entering a flux-controlled inductor is zero over any period.*

It follows from the above observation that the instantaneous power entering any charge-controlled capacitor or flux-controlled inductor is positive only during parts of each cycle, and must necessarily become negative elsewhere in order for the net area over each cycle to cancel out. Hence, unlike resistors, the power entering the capacitor or inductor is *not* dissipated. Rather, energy is stored during parts of each cycle and is "spit" out during the remaining part of the cycle. Such elements are therefore said to be *lossless*.

One immediate consequence of this lossless property is that in a periodic regime where $v(t) = v(t + T)$ and $i(t) = i(t + T)$ for all t, the voltage waveform $v(t)$ and current waveform $i(t)$ associated with any capacitor or inductor must necessarily *cross the time axis at different instants of time*. Otherwise, the integrand in Eq. (2.12) would always be positive, or negative, for all t, thereby implying $w(t_1, t_2) \neq 0$.

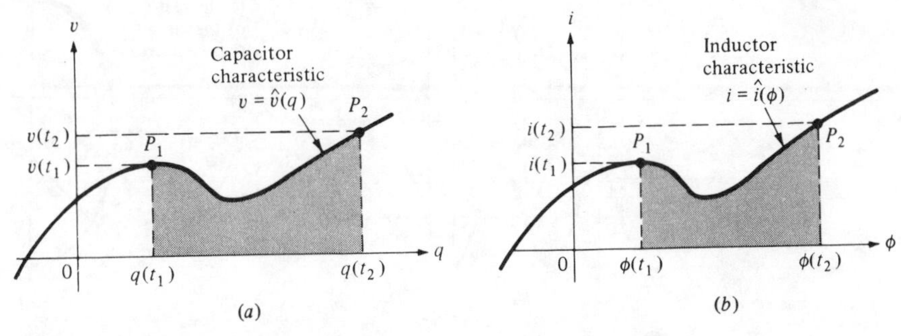

Figure 2.8 Geometric interpretation of $w_C(t_1, t_2)$ and $w_L(t_1, t_2)$.

For a linear capacitor or inductor operating in a sinusoidal steady state, this distinct "zero-crossing property" manifests itself as a 90° *phase shift* between the voltage and current, respectively. For example, if we drive the linear capacitor in Fig. 1.1*a* with a sinusoidal voltage $v(t) = E \sin \omega t$ as shown in Fig. 2.9*a*, then the corresponding current $i(t) = \omega CE \cos \omega t$ *leads* the voltage by 90° as shown in Fig. 2.9*b*. The locus $\mathscr{L}(v, i)$ in the *v*-*i* plane is therefore an ellipse as shown in Fig. 2.9*c*. Note that this locus is *frequency dependent*. Indeed, by adjusting ω from $\omega = 0$ to $\omega = +\infty$, the locus can be made to pass through *any* point lying within the vertical strip $-E < v < E$. Hence, it does not make sense to describe a capacitor, or an inductor, by a characteristic in the *v*-*i* plane.

2.4 Energy Stored in a Linear Time-Invariant Capacitor or Inductor

Consider a *C*-F linear capacitor having an initial voltage $v(t_1) = V$ and an initial charge $q(t_1) = Q = CV$ at $t = t_1$. Let the capacitor be connected to an external circuit, as shown in Fig 2.10*a*, at $t = t_1$. The energy entering the capacitor during $[t_1, t_2]$ is given by Eq. (2.17*a*):

$$w_C(t_1, t_2) = \frac{1}{2C}[q^2(t_2) - Q^2] = \tfrac{1}{2}C[v^2(t_2) - V^2] \qquad (2.19)$$

Note that whenever $q(t_2) < Q$, or $v(t_2) < V$, then $w_C(t_1, t_2) < 0$. This can also be seen in Fig. 2.10*b* where $w_C(t_1, t_2)$ is negative because we are integrating from right (P_1) to left (P_2) in the first quadrant. Note that $w_C(t_1, t_2) < 0$ means energy is actually being spit out of the capacitor and returned to the external circuit $\mathscr{N}$. It follows from Eq. (2.19) and Fig. 2.10*b* that $w_C(t_1, t_2)$ is *most negative* when $q(t_2) = v(t_2) = 0$, whereupon $w_C(t_1, t_2) = -Q^2/2C = -\tfrac{1}{2}CV^2$.

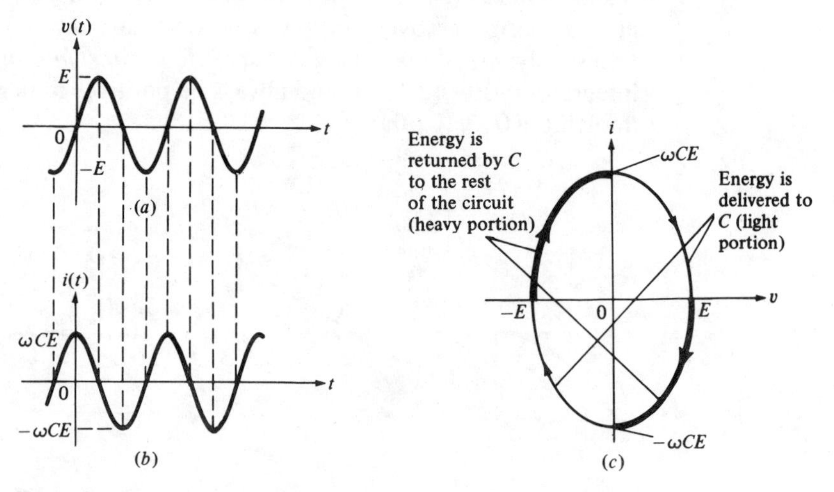

Figure 2.9 Voltage and current waveforms in a linear capacitor.

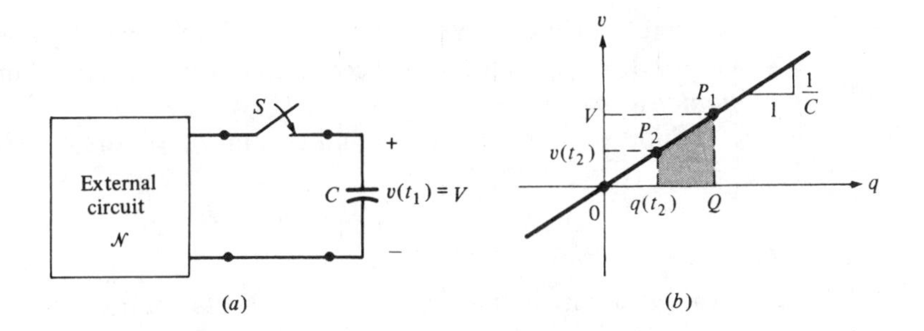

Since this represents the maximum amount of energy that could be *extracted* from the capacitor, it is natural to say that *an energy equal to*

$$\mathscr{E}_C(Q) = \frac{1}{2C} Q^2 = \tfrac{1}{2} CV^2 \tag{2.20}$$

is stored in a *linear capacitor* C having an initial voltage $v(t_1) = V$ or initial charge $q(t_1) = Q = CV$.

By duality, *an energy equal to*

$$\mathscr{E}_L(\phi) = \frac{1}{2L} \phi^2 = \tfrac{1}{2} LI^2 \tag{2.21}$$

is stored in a linear inductor L having an initial current $i(t_1) = I$ or initial flux $\phi(t_1) = \phi = LI$.

2.5 Energy Stored in a Nonlinear Time-Invariant Capacitor or Inductor[17]

Following the same reasoning for the linear case, we define the *energy stored in a nonlinear capacitor or inductor* to be equal *to the magnitude of* the *maximum energy* that can be extracted from the element at a given initial condition. Since the q-v (or ϕ-i) characteristic need not pass through the origin (Fig. 2.11*a*) and

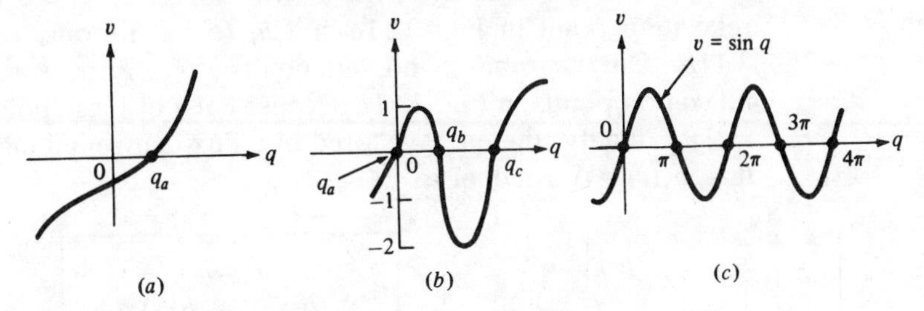

Figure 2.11 Examples of nonlinear q-v characteristics.

[17] May be omitted without loss of continuity.

can have several zero crossings (Fig. 2.11b) or infinitely many zero crossings (Fig. 2.11c), special care is needed to derive a formula for stored energy in the *nonlinear* case.

Suppose it is possible to find a point q_* on the q-v characteristic such that

$$\int_{q_*}^{q} \hat{v}(q')\, dq' \geq 0 \qquad \text{for all } -\infty < q < \infty \tag{2.22}$$

Now given any initial charge $q(t_1) = Q$, the energy entering a charge-controlled capacitor during $[t_1, t_2]$ is given by Eq. (2.17a):

$$w_C(t_1, t_2) = \int_{Q}^{q(t_2)} \hat{v}(q)\, dq$$

$$= \int_{Q}^{q_*} \hat{v}(q)\, dq + \int_{q_*}^{q(t_2)} \hat{v}(q)\, dq \tag{2.23}$$

It follows from Eq. (2.22) that the *first term is negative* while the second term is positive if $q(t_2) \neq q_*$ in Eq. (2.23). Consequently, $w_C(t_1, t_2)$ is most negative when we choose $q(t_2) = q_*$, and the maximum energy that can be extracted is equal to $w_C(t_1, t_2) = \int_Q^{q_*} \hat{v}(q)\, dq < 0$. It follows from the lossless property that *the energy stored in a charge-controlled capacitor having an initial charge* $q(t_1) = Q$ *is equal to*

$$\boxed{\mathscr{E}_C(Q) = \int_{q_*}^{Q} \hat{v}(q)\, dq} \tag{2.24}$$

where q_* is *any* point satisfying Eq. (2.22). Note that $\mathscr{E}_C(Q) \geq 0$ for all Q in view of Eq. (2.22).

Since $\mathscr{E}_C(q_*) = 0$, it follows that no energy is stored when the initial charge is equal to q_* and the capacitor is therefore said to be initially relaxed. Consequently, any point q_* satisfying Eq. (2.22) is called a *relaxation point*.

For the q-v characteristic shown in Fig. 2.11, we find q_a to be the only relaxation point in Fig. 2.11a and q_c to be the only relaxation point in Fig. 2.11b. On the other hand, all points $q = \pm k2\pi$, $k = 0, 1, 2, \ldots$, qualify as relaxation points in Fig. 2.11c because each of these points satisfies Eq. (2.22).

By duality, the energy stored in a flux-controlled inductor having an initial flux $\phi(t_1) = \Phi$ is equal to

$$\boxed{\mathscr{E}_L(\Phi) = \int_{0_*}^{\Phi} \hat{i}(\phi)\, d\phi} \tag{2.25}$$

where ϕ_* is *any* relaxation point, namely,

$$\int_{\phi_*}^{\phi} \hat{i}(\phi)\, d\phi \geq 0 \qquad \text{for all } -\infty < \phi < \infty \tag{2.26}$$

Special case In the special case where the

q-v characteristic passes through the origin and satisfies

$$\int_0^q \hat{v}(q)\, dq \geq 0 \qquad -\infty < q < \infty \tag{2.27a}$$

the origin is a relaxation point and the stored energy is simply given by

$$\mathscr{E}_C(Q) = \int_0^Q \hat{v}(q)\, dq \tag{2.28a}$$

In this case, $\mathscr{E}_C(Q)$ is equal to the net area under the *q-v* characteristic from $q = 0$ to $q = Q$, as shown in Fig. 2.12*a*.

ϕ-*i* characteristic passes through the origin and satisfies

$$\int_0^\phi \hat{i}(\phi)\, d\phi > 0 \qquad -\infty < \phi < \infty \tag{2.27b}$$

the origin is a relaxation point and the stored energy is simply given by

$$\mathscr{E}_L(\Phi) = \int_0^\Phi \hat{i}(\phi)\, d\phi \tag{2.28b}$$

In this case, $\mathscr{E}_L(\Phi)$ is equal to the net area under the ϕ-*i* characteristic from $\phi = 0$ to $\phi = \Phi$, as shown in Fig. 2.12*b*.

REMARK The results presented throughout Sec. 2 are valid only if the capacitors and inductors are time-invariant. When the element is time-varying, additional energy is contributed by an external energy source which causes the time variation.

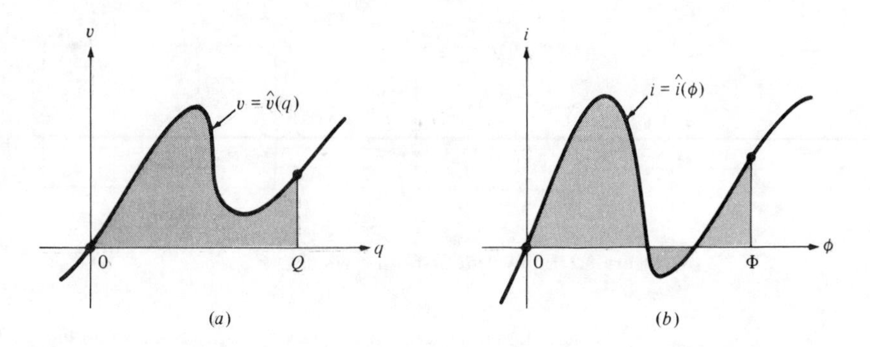

Figure 2.12 The net area under the curve is equal numerically to the stored energy.

Exercises

1. Verify that out of the three zero crossings in Fig. 2.11*b*, only q_C qualifies as a relaxation point.
2. Find all relaxation points associated with the Josephson junction defined earlier by Eq. (1.9*b*).
3. Prove that if a nonlinear capacitor or inductor has more than one relaxation point, then each point will give the *same* stored energy $\mathscr{E}_C(Q)$ or $\mathscr{E}_L(\Phi)$.

3 FIRST-ORDER LINEAR CIRCUITS

Circuits made of *one* capacitor (or one inductor), resistors, and independent sources are called *first-order circuits*. Note that "resistor" is understood in the broad sense: It includes controlled sources, gyrators, ideal transformers, etc.

In this section, we study first-order circuits made of linear time-invariant elements and independent sources. Any such circuit can be redrawn as shown in either Fig. 3.1*a* or *b*, where the one-port N is assumed to include all other elements (e.g., independent sources, resistors, controlled sources, gyrators, ideal transformers, etc.).[18]

Applying the *Thévenin-Norton equivalent one-port theorem* from Chap. 5, we can, in most instances, replace N by the equivalent circuit shown in Fig. 3.2*a* and *b*, respectively.

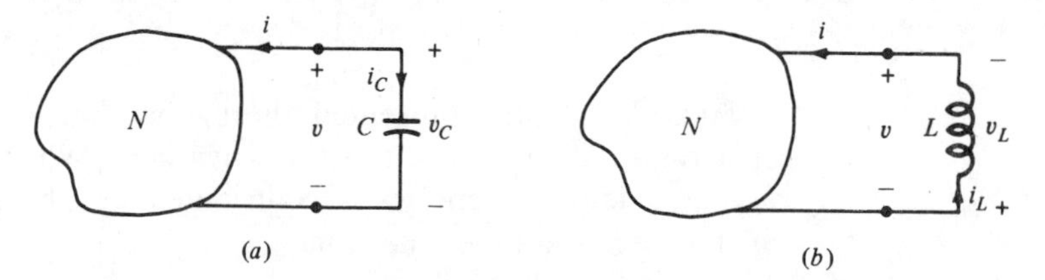

Figure 3.1 (*a*) First-order *RC* circuit. (*b*) First-order *RL* circuit.

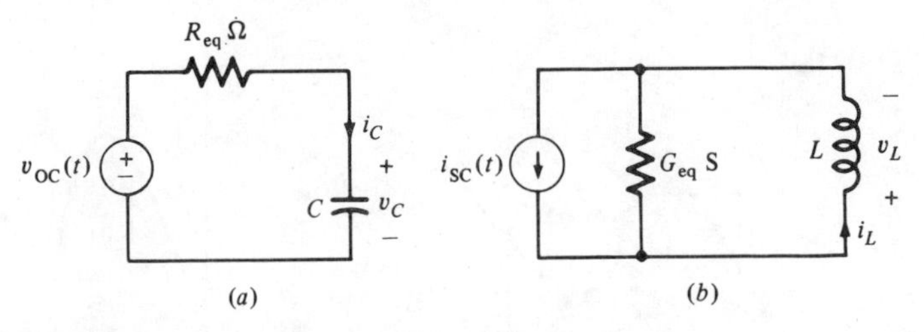

Figure 3.2 Equivalent first-order circuits.

[18] Without loss of generality, we draw v_L and i_L as shown in Fig. 3.1*b* so that $i_L = i$ (the dual of $v_C = v$ in Fig. 3.1*a*). This will guarantee the state equation (3.2*b*) will come out to be the dual of Eq. (3.2*a*).

Applying KVL we obtain

$$R_{eq} i_C + v_C = v_{OC}(t) \quad (3.1a)$$

Substituting $i_C = C\dot{v}_C$ and solving for $\dot{v}_C$, we obtain

$$\boxed{\dot{v}_C = -\frac{v_C}{R_{eq}C} + \frac{v_{OC}(t)}{R_{eq}C}} \quad (3.2a)$$

Applying KCL we obtain

$$G_{eq} v_L + i_L = i_{SC}(t) \quad (3.1b)$$

Substituting $v_L = L\dot{i}_L$ and solving for $\dot{i}_L$, we obtain

$$\boxed{\dot{i}_L = -\frac{i_L}{G_{eq}L} + \frac{i_{SC}(t)}{G_{eq}L}} \quad (3.2b)$$

When written in the above standard form, this *first*-order linear differential equation is called a *state equation* and the variable v_C (respectively, i_L) is called a *state variable*.

Given any *initial condition* $v_C(t_0)$ at any initial time t_0, our objective is to find the solution $v_C(t)$ for all $t \geq t_0$. We will show that $v_C(t)$ depends only on the initial condition $v_C(t_0)$ and the waveform $v_{OC}(\cdot)$ over $[t_0, t]$.

Once the solution $v_C(\cdot)$ is found, we can apply the *substitution theorem* from Chap. 5 and replace the capacitor in Fig. 3.1*a* by a voltage source $v_C(t)$.

Given any *initial condition* $i_L(t_0)$ at any initial time t_0, our objective is to find the solution $i_L(t)$ for all $t \geq t_0$. We will show that $i_L(t)$ depends only on the initial condition $i_L(t_0)$ and the waveform $i_{SC}(\cdot)$ over $[t_0, t]$.

Once the solution $i_L(\cdot)$ is found, we can apply the *substitution theorem* from Chap. 5 and replace the inductor in Fig. 3.1*b* by a current source $i_L(t)$.

The resulting equivalent circuit, being resistive, can then be solved using techniques developed in the preceding chapters.

In Sec. 3.1 we show that the solution of any first-order linear circuit can be found by inspection, provided N contains only *dc* sources. By repeated application of this "inspection method," Sec. 3.2 shows how the solution can be easily found if N contains only *piecewise-constant* sources. This method is then applied in Sec. 3.3 for finding the solution—called the *impulse response*—when the circuit is driven by an *impulse* $\delta(t)$. Finally, Sec. 3.4 gives an explicit integration formula for finding solutions under arbitrary excitations.

3.1 Circuits Driven by DC Sources

When N contains only *dc* sources, $v_{OC}(t) = v_{OC}$ and $i_{SC}(t) = i_{SC}$ are constants in Fig. 3.2 and in Eq. (3.2). Let us rewrite Eqs. (3.2*a*) and (3.2*b*) as follows:

State equation

$$\boxed{\dot{x} = -\frac{x}{\tau} + \frac{x(t_\infty)}{\tau}} \tag{3.3}$$

where

$$x \triangleq v_C$$
$$x(t_\infty) \triangleq v_{OC} \tag{3.4a}$$
$$\tau \triangleq R_{eq}C$$

for the *RC* circuit.

where

$$x \triangleq i_L$$
$$x(t_\infty) \triangleq i_{SC} \tag{3.4b}$$
$$\tau \triangleq G_{eq}L$$

for the *RL* circuit.

Given any initial condition $x = x(t_0)$ at $t = t_0$, Eq. (3.3) has a unique solution[19]

$$\boxed{x(t) - x(t_\infty) = [x(t_0) - x(t_\infty)] \exp \frac{-(t - t_0)}{\tau}} \tag{3.5}$$

which holds for *all* times t, i.e., $-\infty < t < \infty$. To verify that this is indeed the solution, simply substitute Eq. (3.5) into Eq. (3.3) and show that both sides are identical. Observe that at $t = t_0$, both sides of Eq. (3.5) reduce to $x(t_0) - x(t_\infty)$. Note also that the solution given by Eq. (3.5) is valid whether τ is positive or negative.

The solution (3.5) is determined by only three parameters: $x(t_0)$, $x(t_\infty)$, and τ. We call them *initial state*, *equilibrium state*, and *time constant*, respectively. To see why $x(t_\infty)$ is called the equilibrium state, note that if $x(t_0) = x(t_\infty)$, then Eq. (3.3) gives $\dot{x}(t_0) = 0$ and thus $x(t) = x(t_\infty)$ for all t. Hence the circuit remains "motionless," or in equilibrium.

Since the "inspection method" to be developed in this section depends crucially on the ability to sketch the *exponential* waveform quickly, the following properties are extremely useful.

A. Properties of exponential waveforms Depending on whether τ is positive or negative, the exponential waveform in Eq. (3.5) tends either to a constant or to infinity, as the time t tends to infinity. Hence, it is convenient to consider these two cases separately.

$\tau > 0$ (*Stable case*) When $\tau > 0$, Eq. (3.5) shows that $x(t) - x(t_\infty)$, i.e., the distance between the present state and the equilibrium state $x(t_\infty)$, *decreases* exponentially: For all initial states, the solution $x(t)$ is *sucked* into the equilibrium and $|x(t) - x(t_\infty)|$ decreases exponentially with a time constant τ.

[19] We write $x(t_\infty)$ on the left side to make it easier to remember this important formula.

The solution (3.5) for $\tau > 0$ is sketched in Fig. 3.3 for two different initial states $\tilde{x}(t_0)$ and $x(t_0)$ for $t \geq t_0$. Observe that because the time constant τ is positive,

$$x(t) \to x(t_\infty) \qquad \text{as } t \to \infty \tag{3.6}$$

Thus, when $\tau > 0$, we say the equilibrium state $x(t_\infty)$ is *stable* because any initial deviation $x(t_0) - x(t_\infty)$ decays exponentially and $x(t) \to x(t_\infty)$ as $t \to \infty$.

The *exponential* waveforms in Fig. 3.3 can be accurately sketched using the following observations:

1. The *tangent* at $t = t_0$ passes through the point $[t_0, x(t_0)]$ and the point $[t_0 + \tau, x(t_\infty)]$.
2. After one time constant τ, the distance between $x(t)$ and $x(t_\infty)$ decreases approximately by 63 percent of the initial distance $|x(t_0) - x(t_\infty)|$.
3. After five time constants, $x(t)$ practically attains the *steady-state* value $x(t_\infty)$. (Indeed, $e^{-5} \approx 0.007$.)

Example 1 (Op-amp voltage follower: Stable configuration) Consider the op-amp circuit shown in Fig. 3.4*a*. Using the ideal op-amp model, this circuit was analyzed earlier in Sec. 2.2 (Fig. 2.1) of Chap. 4. Assuming the switch is closed at $t = 0$, we found $v_0(t) = v_{\text{in}}(t) = 10\text{ V}$ for $t \geq 0$.

In practice, the output is observed to reach the 10-V solution after a small but finite time. In order to predict the *transient* behavior before the

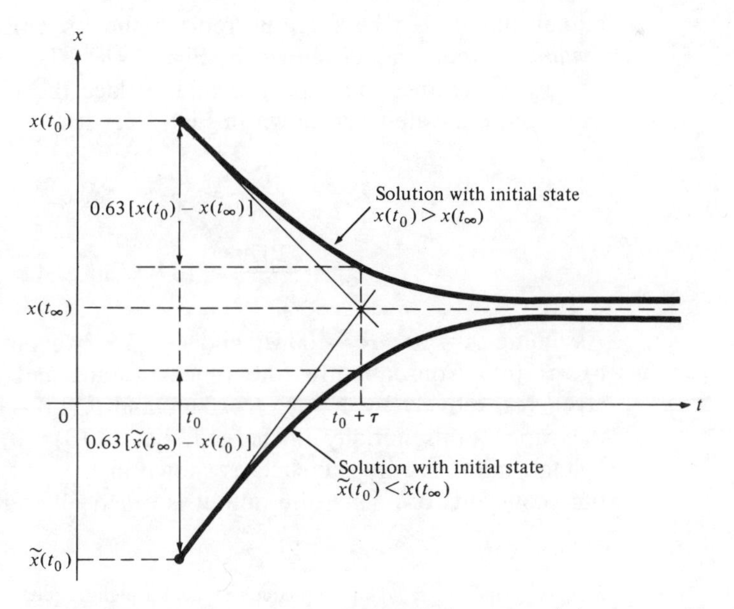

Figure 3.3 The solution tends to the equilibrium state $x(t_\infty)$ as $t \to \infty$ when the time constant τ is *positive*.

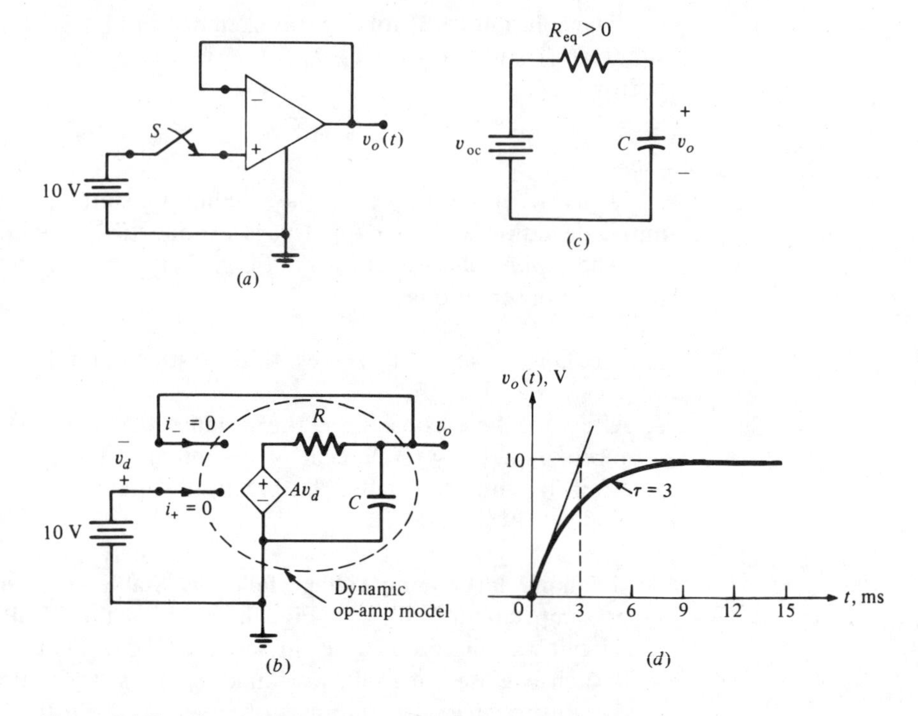

Figure 3.4 Transient behavior of op-amp voltage follower circuit.

equilibrium is reached, let us replace the op amp in Fig. 3.4*a* by the *dynamic* circuit model shown in Fig. 3.4*b*.[20] To analyze this first-order circuit, we extract the capacitor and replace the remaining circuit by its Thévenin equivalent as shown in Fig. 3.4*c*, where

$$R_{eq} = \frac{R}{A+1} \approx \frac{R}{A} \qquad \text{since } A \gg 1 \tag{3.7}$$

$$v_{OC} = \frac{10A}{A+1} \approx 10 \qquad \text{since } A \gg 1 \tag{3.8}$$

Assuming $A = 10^5$, $R = 100\ \Omega$, and $C = 3\ \text{F}$, we obtain $R_{eq} \approx 10^{-3}\ \Omega$ and $v_{OC} \approx 10\ \text{V}$. Consequently, the time constant and equilibrium state are given respectively by $\tau = R_{eq}C = 3\ \text{ms}$ and $v_0(t_\infty) = v_{OC} \approx 10\ \text{V}$. Assuming the capacitor is initially uncharged, i.e., $v_0(0) = 0$, the resulting output voltage can be easily sketched as shown in Fig. 3.4*d*. Note that after five time constants or 15 ms, the output is practically equal to 10 V.

[20] A more realistic *dynamic* op-amp circuit model for high-frequency applications would require several linear capacitors. The one-capacitor model chosen in Fig. 3.4, though not valid in general, does predict the transient behavior correctly for the voltage follower circuit.

$\tau < 0$ ***(Unstable case)*** When $\tau < 0$, Eq. (3.5) shows that the quantity $x(t) - x(t_\infty)$ *increases* exponentially for all initial states, i.e., the solution $x(t)$ *diverges* from the equilibrium, and $x(t) - x(t_\infty)$ *increases* exponentially with a time constant τ.

The solution (3.5) for $\tau < 0$ is sketched in Fig. 3.5 for two different initial states $x(t_0)$ and $\tilde{x}(t_0)$.

Observe that, since the time constant τ is negative, as $t \to \infty$, $x(t) \to \infty$ if $x(t_0) > x(t_\infty)$, and $x(t) \to -\infty$ if $x(t_0) < x(t_\infty)$.

Thus, when $\tau < 0$, we say the equilibrium state $x(t_\infty)$ is *unstable* because any initial deviation $x(t_0) - x(t_\infty)$ *grows* exponentially with time and $|x(t)| \to \infty$ as $t \to \infty$.

However, if we run time *backward*, then

$$x(t) \to x(t_\infty) \qquad \text{as } t \to -\infty \tag{3.9}$$

Consequently, $x(t_\infty)$ can be interpreted as a *virtual equilibrium state*.

The exponential waveform in Fig. 3.5 can be accurately sketched using the following observations:

1. The *tangent* at $t = t_0$ passes through the point $[t_0, x(t_0)]$ and the point $[t_0 - |\tau|, x(t_\infty)]$.
2. At $t = t_0 + |\tau|$, the distance $|x(t_0 + |\tau|) - x(t_\infty)|$ is approximately 1.72 times the initial distance $|x(t_0) - x(t_\infty)|$.

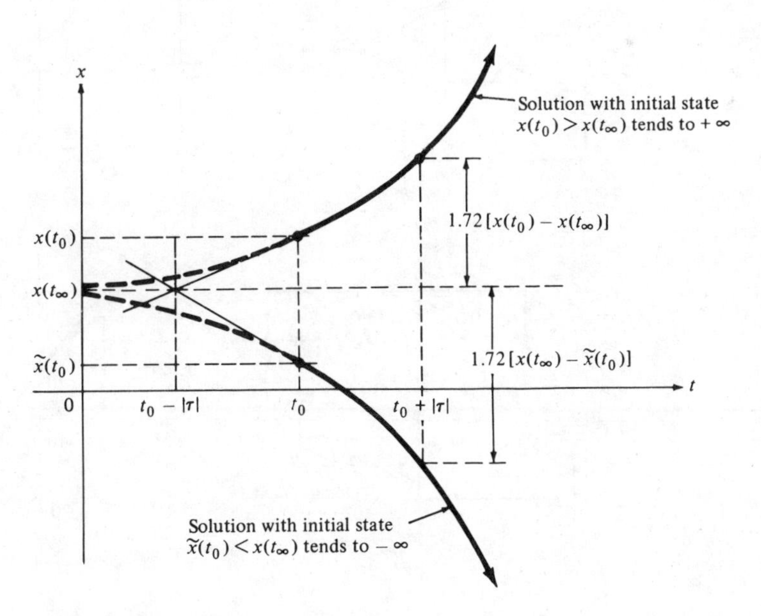

Figure 3.5 The solution tends to the "virtual" equilibrium state $x(t_\infty)$ as $t \to -\infty$ when the time constant τ is negative.

Example 2 (Op-amp voltage follower: Unstable configuration) The op-amp circuit in Fig. 3.6*a* is identical to that of Fig. 3.4*a* except for an interchange between the inverting (−) and the noninverting (+) terminals. Using the ideal op-amp model in the linear region, we would obtain exactly the same answer as before, namely, $v_0 = 10$ V for $t \geq 0$, provided $E_{sat} > 10$ V. Let us see what happens if the op amp is replaced by the dynamic model adopted earlier in Fig. 3.4*b*. The resulting circuit shown in Fig. 3.6*b* resembles that of Fig. 3.4*b* except for an important difference: The polarity of v_d is now reversed. The parameters in the Thévenin equivalent circuit now become

$$R_{eq} = -\frac{R}{A-1} \approx -\frac{R}{A} \qquad \text{since } A \gg 1 \tag{3.10}$$

$$v_{OC} = \frac{10A}{A-1} \approx 10 \qquad \text{since } A \gg 1 \tag{3.11}$$

Assuming the same parameter values as in Example 1, we obtain $R_{eq} \approx -10^{-3}\ \Omega$ and $v_{OC} \approx 10$ V. Consequently, the time constant and equilibrium state are given respectively by $\tau \approx -3$ ms and $v_0(t_\infty) \approx 10$ V. Assuming $v_0(0) = 0$ as in Example 1, the resulting output voltage can be easily sketched as shown in Fig. 3.6*d*.

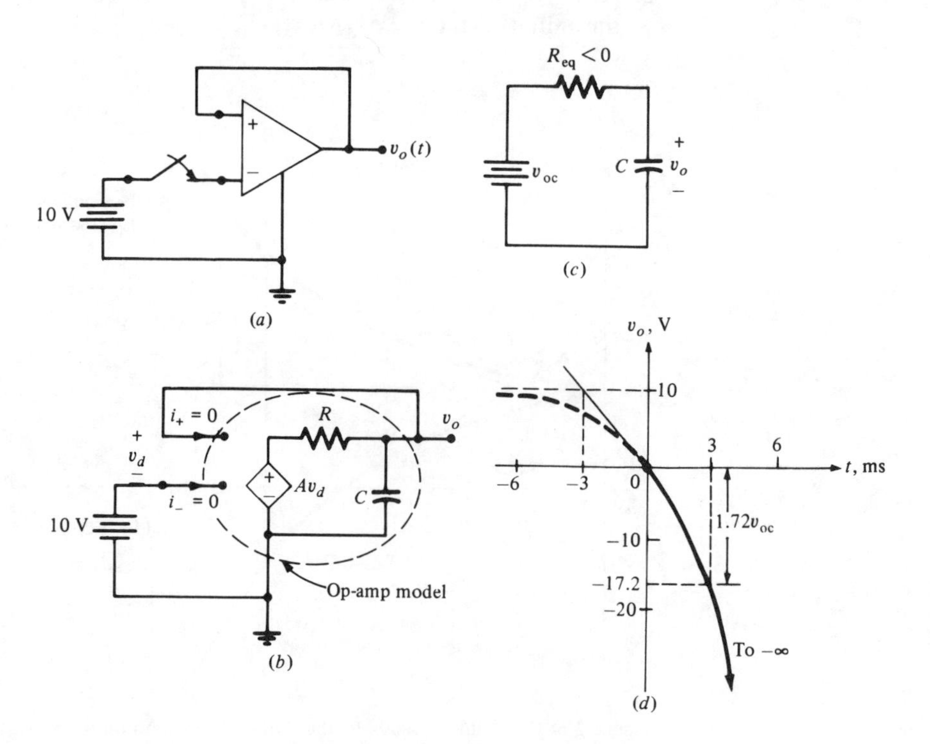

Figure 3.6 Unstable transient behavior of op-amp voltage follower circuit.

Note that the solution differs drastically from that of Fig. 3.4d: It tends to $-\infty$! Of course, in practice, when $v_0(t)$ decreases to $-E_{sat}$, the op-amp negative saturation voltage, the solution would remain constant at $-E_{sat}$. Clearly, this circuit would not function as a voltage follower in practice.

B. Elapsed time formula We will often need to calculate the time interval between two prescribed points on an exponential waveform. For example, to obtain the actual solution waveform for the circuit in Fig. 3.6, we need to calculate the time that elapsed when v_0 decreases from $v_0 = 0$ to $v_0 = -15$ V (assuming $E_{sat} = 15$ V) in Fig. 3.6d.

Given any two points $[(t_j, x(t_j))$ and $(t_k, x(t_k))]$ on an exponential waveform (see, e.g., Figs. 3.3 and 3.5), the time it takes to go from $x(t_j)$ to $x(t_k)$ is given by

Elapsed time formula

$$t_k - t_j = \tau \ln \frac{x(t_j) - x(t_\infty)}{x(t_k) - x(t_\infty)} \tag{3.12}$$

To derive Eq. (3.12), let $t = t_j$ and $t = t_k$ in Eq. (3.5), respectively:

$$x(t_j) - x(t_\infty) = [x(t_0) - x(t_\infty)] \exp \frac{-(t_j - t_0)}{\tau} \tag{3.13}$$

$$x(t_k) - x(t_\infty) = [x(t_0) - x(t_\infty)] \exp \frac{-(t_k - t_0)}{\tau} \tag{3.14}$$

Dividing Eq. (3.13) by Eq. (3.14) and taking the logarithm on both sides, we obtain Eq. (3.12).

REMARK The above derivation does not depend on whether τ is positive or negative.

C. Inspection method (First-order linear time-invariant circuits driven by dc sources) Consider first the first-order RC circuit in Fig. 3.1a where all independent sources inside N are dc sources. Equation (3.5) gives us the voltage waveform across the capacitor, namely,

$$v_C(t) = v_C(t_\infty) + [v_C(t_0) - v_C(t_\infty)] \exp \frac{-(t - t_0)}{\tau} \tag{3.15}$$

Suppose we replace the capacitor by a *voltage source* defined by Eq. (3.15). Assuming the resulting resistive circuit is uniquely solvable, we can apply the *substitution theorem* to conclude that the solution inside N of the resistive circuit is identical to that of the first-order RC circuit.

Let v_{jk} denote the voltage across any pair of nodes, say ⓙ and ⓚ and assume that N contains α independent dc voltage sources $V_{s1}, V_{s2}, \ldots, V_{s\alpha}$ and

β independent dc current sources $I_{s1}, I_{s2}, \ldots, I_{s\beta}$. Applying the *superposition theorem* from Chap. 5, we know the solution $v_{jk}(t)$ is given by an expression of the form

$$v_{jk}(t) = H_0 v_C(t) + \sum_{j=1}^{\alpha} H_j V_{sj} + \sum_{j=1}^{\beta} K_j I_{sj} \tag{3.16}$$

where H_0, H_j, and K_j are *constants* (which depend on element values and circuit configuration). Substituting Eq. (3.15) for $v_C(t)$ in Eq. (3.16) and rearranging terms, we obtain

$$v_{jk}(t) - v_{jk}(t_\infty) = [v_{jk}(t_0) - v_{jk}(t_\infty)] \exp \frac{-(t - t_0)}{\tau} \tag{3.17}$$

where

$$v_{jk}(t_\infty) \triangleq H_0 v_C(t_\infty) + \sum_{j=1}^{\alpha} H_j V_{sj} + \sum_{j=1}^{\beta} K_j I_{sj} \tag{3.18}$$

and

$$v_{jk}(t_0) \triangleq H_0 v_C(t_0) + \sum_{j=1}^{\alpha} H_j V_{sj} + \sum_{j=1}^{\beta} K_j I_{sj} \tag{3.19}$$

Since Eq. (3.17) has exactly the same form as Eq. (3.5), and since nodes ⓙ and ⓚ are arbitrary, we conclude that:

The voltage $v_{jk}(t)$ across any pair of nodes in a first-order RC circuit driven by dc sources is an exponential waveform having the same time constant τ as that of $v_C(t)$.

By the same reasoning, we conclude that:

The current $i_j(t)$ in any branch j of a first-order RC circuit driven by dc sources is an exponential waveform having the same time constant τ as that of $v_C(t)$.

It follows from duality that the voltage $v_{jk}(t)$ across any pair of nodes, or the current $i_j(t)$ in any branch j of a first-order RL circuit driven by dc sources is an exponential waveform having the same time constant τ as that of $i_L(t)$.

The above "exponential solution waveform" property, of course, assumes that the first-order circuit is *not* degenerate, i.e., that it is uniquely solvable and that $0 < |\tau| < \infty$.

It is important to remember that all voltage and current waveforms in a given first-order circuit have the *same time constant* τ as defined in Eq. (3.4).

Moreover, as we approach the *equilibrium*, i.e., when $t \to +\infty$ (if $\tau > 0$) or $t \to -\infty$ (if $\tau < 0$), *the capacitor current and the inductor voltage both tend to zero*. This follows from Figs. 3.3 and 3.5, $i_C = C\dot{v}_C$, and $v_L = L\dot{i}_L$.

Since an exponential waveform is uniquely determined by only three parameters [initial state $x(t_0)$, equilibrium state $x(t_\infty)$, and time constant τ], the following "inspection method" can be used to find the voltage solution $v_{jk}(t)$

across any pair of nodes ⓙ and ⓚ or the current solution $i_j(t)$ in any branch j, in any uniquely solvable linear first-order circuit driven by dc sources:

RC circuit: given $v_C(t_0)$.	*RL circuit*: given $i_L(t_0)$.
1. Replace the *capacitor* by a dc *voltage source* with a terminal voltage equal to $v_C(t_0)$. Label the voltage across node-pair ⓙ, ⓚ as $v_{jk}(t_0)$ and the current i_j as $i_j(t_0)$. Solve the resulting resistive circuit for $v_{jk}(t_0)$ or $i_j(t_0)$.	1. Replace the *inductor* by a dc *current source* with a terminal current equal to $i_L(t_0)$. Label the voltage across node-pair ⓙ, ⓚ as $v_{jk}(t_0)$ and the current i_j as $i_j(t_0)$. Solve the resulting resistive circuit for $v_{jk}(t_0)$ or $i_j(t_0)$.
2. Replace the *capacitor* by any *open circuit*. Label the voltage across node-pair ⓙ, ⓚ as $v_{jk}(t_\infty)$ and the current i_j as $i_j(t_\infty)$. Solve for $v_{jk}(t_\infty)$ or $i_j(t_\infty)$.	2. Replace the *inductor* by a *short circuit*. Label the voltage across node-pair ⓙ, ⓚ as $v_{jk}(t_\infty)$ and the current i_j as $i_j(t_\infty)$. Solve for $v_{jk}(t_\infty)$ or $i_j(t_\infty)$.
3. Find the Thévenin equivalent circuit of N. Calculate the time constant $\tau = R_{eq}C$.	3. Find the Norton equivalent circuit of N. Calculate the time constant $\tau = G_{eq}L$.
4. If $0<\|\tau\|<\infty$, use the above three parameters to sketch the *exponential* solution waveform.	4. If $0<\|\tau\|<\infty$, use the above three parameters to sketch the *exponential* solution waveform.

REMARKS
1. The above inspection method eliminates the usual step of writing the differential equation: It reduces each step to resistive circuit calculations.
2. The above method is valid only if the circuit is uniquely solvable. For example, if the one-port N in Fig. 3.1 does not have a Thévenin *and* Norton equivalent circuit, it is not uniquely solvable.
3. The above method assumes the circuit is *not* degenerate in the sense that $0<|\tau|<\infty$. This means that $R_{eq} \neq 0$ and is finite in Fig. 3.2*a*, and that $G_{eq} \neq 0$ and is finite in Fig. 3.2*b*.

3.2 Circuits Driven by Piecewise-Constant Signals

Consider next the case where the independent sources in N of Fig. 3.1 are piecewise-constant for $t > t_0$. This means that the semi-infinite time interval $t_0 \leq t < \infty$ can be partitioned into *subintervals* $[t_j, t_{j+1})$, $j = 1, 2, \ldots$, such that all sources assume a *constant* value during each subinterval. Hence, we can analyze the circuit as a sequence of first-order circuits driven by dc sources, each one analyzed separately by the inspection method. Since the circuit remains unchanged except for the sources, the *time constant* τ remains unchanged throughout the analysis.

The *initial state* $x(t_0)$ and *equilibrium state* $x(t_\infty)$ will of course vary from one subinterval to another. Although the same procedure holds in the determination of $x(t_\infty)$, one must be careful in calculating the *initial value* at the beginning of each subinterval t_j because at least one source changes its value *discontinuously* at each boundary time t_j between two consecutive subintervals. In general, $x(t_j-) \neq x(t_j+)$, where the $-$ and $+$ denote the *limit* of $x(t)$ as $t \to t_j$ *from the left* and *from the right*, respectively. The *initial value* to be used in the calculation during the subinterval $[t_j, t_{j+1})$ is $x(t_j+)$.

Although in general both $v_{jk}(t)$ and $i_j(t)$ can jump, the "continuity property" in Sec. 2.2 guarantees that in the usual case where the capacitor current (respectively, inductor voltage) waveform is bounded, the capacitor voltage (respectively, inductor current) waveform is a *continuous* function of time and therefore cannot jump. This property is the key to finding the solution by inspection, as illustrated in the following examples.

Example 1 Consider the *RC* circuit shown in Fig. 3.7*a*: $v_s(\cdot)$ is given by Fig. 3.7*a* and $v_C(0) = 0$. Our objective is to find $i_C(t)$, $v_C(t)$, and $v_R(t)$ for

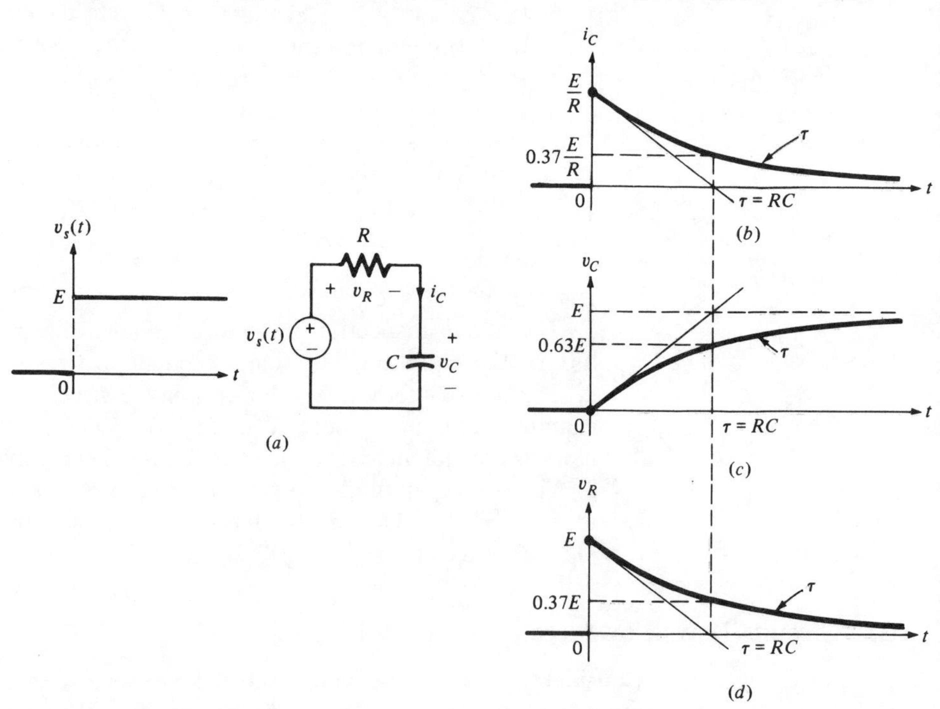

Figure 3.7 Solution waveforms for *RC* circuit. Here, τ denotes the *time constant* of the exponential.

$t \geq 0$ by *inspection*. Since $v_C(0) = 0$ and $v_s(t) = 0$ for $t \leq 0$, it follows that $i_C(t) = v_C(t) = V_R(t) = 0$ for $t \leq 0$.

The solution waveforms for $t > 0$ consists of *exponentials* with a time constant $\tau = RC$. At $t = 0+$, using the continuity property, we have $v_C(0+) = v_C(0-) = 0$. Therefore, $v_R(0+) = v_s(0+) - v_C(0+) = E$ and $i_C(0+) = v_R(0+)/R = E/R$. To find the equilibrium state, we *open* the capacitor and find $i_C(t_\infty) = 0$, $v_C(t_\infty) = E$, and $v_R(t_\infty) = 0$.

These three pieces of information allow us to sketch $i_C(t)$, $v_C(t)$, and $v_R(t)$ for $t \geq 0$ as shown in Fig. 3.7*b*, *c*, and *d*, respectively. Note that $i_C(t) = C\, dv_C(t)/dt$ and $v_R(t) + v_C(t) = E$ for $t \geq 0$, as they should. Observe also that whereas $v_R(t)$ is *discontinuous* at $t = 0$, $v_C(t)$ is *continuous* for all t, as expected.

REMARKS

1. The circuit in Fig. 3.7 is often used to model the situation where a dc voltage source is suddenly connected across a resistive circuit which normally draws a zero-input current. The linear capacitor in this case is used to model the small *parasitic* capacitance between the connecting wires. Without this capacitor, the input voltage would be identical to $v_s(t)$. However, in practice, a "transient" is always observed and the circuit in Fig. 3.7*a* represents a more realistic situation. In this case, the time constant τ gives a measure of how "fast" the circuit can respond to a step input. Such a measure is of crucial importance in the design of high-speed circuits, say in computers, measuring equipment, etc.
2. Since the term *time constant* is meaningful only for *first-order* circuits, a more general measure of "response speed" called the *rise time* is used in specifying practical equipments.

The *rise time* t_r is defined as the time it takes the output waveform to rise from *10 percent* to *90 percent* of the steady-state value after application of a step input.

For first-order circuits, the following simple relationship between t_r and τ follows directly from Eq. (3.12):

Rise time

$$t_r = \tau \ln \frac{0.1E - E}{0.9E - E} = \tau \ln 9 \approx 2.2\tau \tag{3.20}$$

Example 2 Consider the *RL* circuit shown in Fig. 3.8*a*, driven by a periodic square-wave current source in Fig. 3.8*b*. Our objective is to find $i_o(t)$ through the resistor when (*a*) $R = 10\ \text{k}\Omega$, $L = 1\ \text{mH}$ and (*b*) $R = 1\ \text{k}\Omega$, $L = 10\ \text{mH}$.

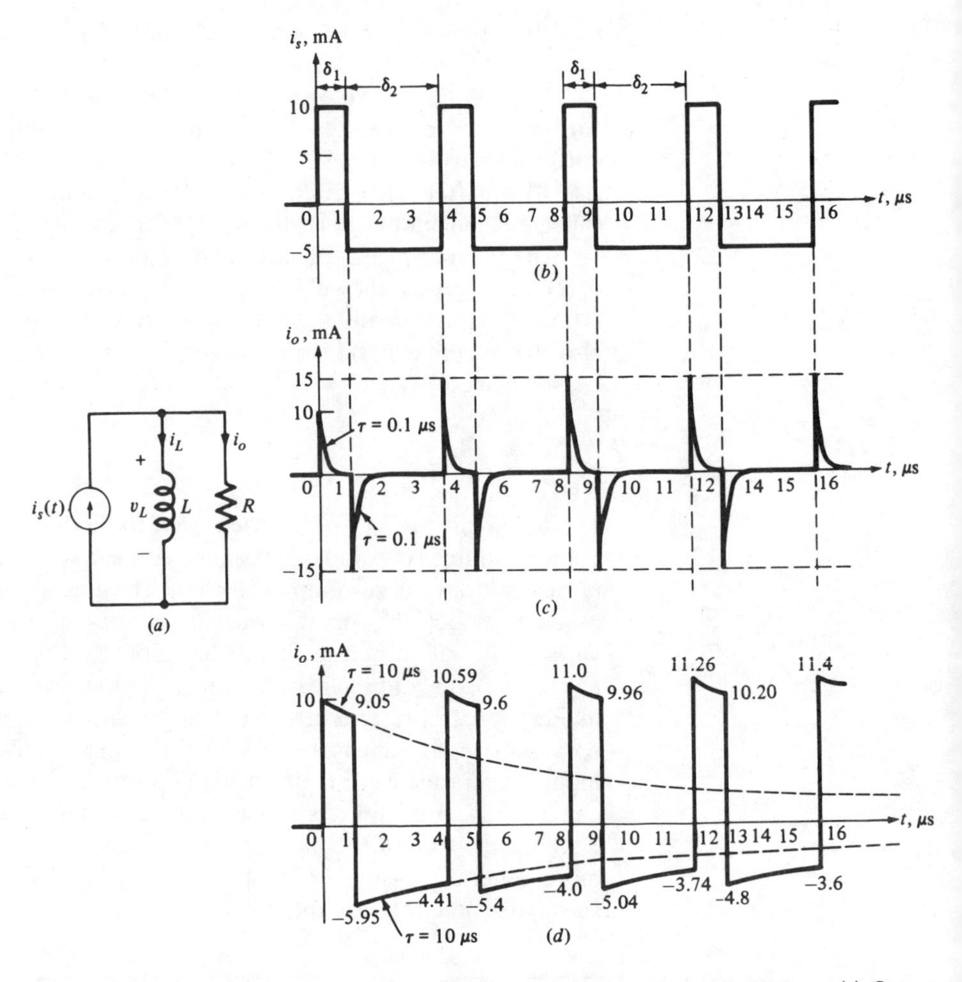

Figure 3.8 (*a*) *RL* circuit. (*b*) Input current waveform with $\delta_1 = 1\ \mu$s and $\delta_2 = 3\ \mu$s. (*c*) Output current waveform when $\tau \ll \delta_1$. (*d*) Output current waveform when $\tau \gg \delta_1$.

(*a*) *Small time constant case*: $\tau = GL = L/R = 0.1\ \mu$s. Since $\tau \ll \delta_1 = 10\tau$, the exponential waveform solution in each subinterval of width δ_1 or δ_2 will have essentially reached its steady state and we only need to calculate $i_o(t)$ over one period. In other words, the solution is periodic for all practical purposes.

Since $i_s(t) = 0$ for $t \leq 0$, the inductor is in equilibrium and can be replaced by a *short circuit* at $t = 0-$ so that $i_L(0+) = i_L(0-) = 0$. Hence $i_o(0+) = i_s(0+) - i_L(0+) = 10 - 0 = 10$ mA.

To find $i_o(t_\infty)$ for the circuit during the subinterval $[0, \delta_1)$, we replace the inductor by a short circuit and obtain $i_L(t_\infty) = 10$ mA and $i_o(t_\infty) = 0$.

At $t=\delta_1=1\ \mu s$, $i_L(\delta_1+)=i_L(\delta_1-)=10\ \text{mA}$. Hence $i_o(\delta_1+)=i_s(\delta_1+)-i_L(\delta_1+)=-5-10=-15\ \text{mA}$. Hence i_o jumps at $t=\delta_1$ from 0 to $-15\ \text{mA}$.

To find $i_o(t_\infty)$ for the circuit during the subinterval $[\delta_1, \delta_1+\delta_2)$, we replace the inductor by a short circuit again and obtain $i_o(t_\infty)=0$.

At $t=\delta_1+\delta_2=4\ \mu s$, $i_o(t)$ jumps again from 0 to 15 mA, and the solution repeats itself thereafter, as shown in Fig. 3.8*c*.

(*b*) *Large time constant case*: $\tau=10\ \mu s$. Since $\tau \gg \delta_1 = 0.1\tau$, the exponential waveform does not have enough time to reach a steady state during each subinterval. Consequently, the solution $i_o(t)$ is *not* periodic and we will have to partition $0 \le t < \infty$ into *infinitely* many subintervals $[0, \delta_1)$, $[\delta_1, \delta_1+\delta_2)$, $[\delta_1+\delta_2, 2\delta_1+\delta_2)$, ... We will see, however, that $i_o(t)$ will tend to a periodic waveform after a few periods.

Starting at $t=0$ as in (*a*), we find $i_o(0+)=10\ \text{mA}$ and $i_o(t_\infty)=0$. The exponential solution is drawn in a solid line during $0 \le t < \delta_1$ and in a dotted line thereafter in Fig. 3.8*d* to emphasize the relative magnitudes of τ and δ_1.

To determine $i_o(\delta_1+)=i_o(1+)$, it is necessary to write the solution $i_o(t)=10\exp(-t/10)$ in order to calculate $i_o(1-)=9.05\ \text{mA}$. This gives us $i_L(1-)=i_s(1-)-i_o(1-)=10-9.05=0.95\ \text{mA}$. Since $i_L(1+)=i_L(1-)=0.95\ \text{mA}$, $i_o(1+)=i_s(1+)-i_L(1+)=-5-0.95=-5.95\ \text{mA}$. Hence $i_o(t)$ jumps from 9.05 to -5.95 mA at $t=1\ \mu s$, as shown in Fig. 3.8*d*.

Again, the exponential solution during $[1, 4)$ has not reached steady state when $i_s(t)$ changes from -5 to 10 mA at $t=4\ \mu s$. To calculate $i_o(t)$ at $t=4+$, it is necessary to write the solution $i_o(t)=-5.95\exp\{-[(t-1)/10]\}$ and obtain $i_o(4-)=-4.41\ \text{mA}$. This gives $i_L(4+)=i_L(4-)=i_s(4-)-i_o(4-)=-5-(-4.41)=-0.59\ \text{mA}$ and $i_o(4+)=i_s(4+)-i_L(4+)=10-(-0.59)=10.59\ \text{mA}$. Hence $i_o(t)$ jumps from -4.41 to 10.59 mA at $t=4\ \mu s$, as shown in Fig. 3.8*d*.

Repeating the above procedure, we find $i_o(t)$ jumps from 9.6 to -5.4 mA at $t=5\ \mu s$, from -4.0 to 11.0 mA at $t=8\ \mu s$, from 9.96 to -5.04 mA at $t=9\ \mu s$, from -3.74 to 11.26 mA at $t=12\ \mu s$, from 10.20 to -4.8 mA at $t=13\ \mu s$, and from -3.6 to 11.4 mA at $t=16\ \mu s$, etc., as shown in Fig. 3.8*d*.

It is clear from Fig. 3.8*d* that $i_o(t)$ is tending toward a periodic waveform. To determine this periodic waveform, note that if we let I_o denote the "peak" value of each "falling" exponential segment in Fig. 3.8*d* (e.g., $I_o=10$, 10.59, 11, 11.26, and 11.4 mA at $t=0$, 4, 8, 12, 16 μs, etc.) then this periodic waveform must satisfy the following *periodicity condition*:

$$I_o \exp\frac{-\delta_1}{\tau} - 15\exp\frac{-\delta_2}{\tau} + 15 = I_o$$

where $\delta_1 = 1\ \mu s$, $\delta_2 = 3\ \mu s$, and $\tau = 10\ \mu s$. The solution of this equation gives one point on the periodic solution, namely, the *peak value*.

Exercise

(*a*) Calculate the peak value I_o from the periodicity condition.
(*b*) Specify the initial inductor current $i_L(0)$ in Fig. 3.8*a* so that the solution $i_o(t)$ is periodic for $t \geq 0$.
(*c*) Sketch this periodic solution.

3.3 Linear Time-Invariant Circuits Driven by an Impulse

Consider the *RC* circuit shown in Fig. 3.9*a* and the *RL* circuit shown in Fig. 3.9*b*. Let the input voltage source $v_s(t)$ and input current source $i_s(t)$ be a square pulse $p_\Delta(t)$ of width Δ and height $1/\Delta$, as shown in Fig. 3.9*c*. Assuming *zero initial* state [i.e., $v_C(0-) = 0$, $i_L(0-) = 0$], the response voltage $v_C(t)$ and current $i_L(t)$ are given by the same waveform shown in Fig. 3.9*d*, where $\tau = RC$ for the *RC* circuit and $\tau = GL$ for the *RL* circuit, and

$$h_\Delta(\Delta) \triangleq \frac{1 - \exp(-\Delta/\tau)}{\Delta} \triangleq \frac{f(\Delta)}{g(\Delta)} \tag{3.21}$$

The *input* and *response* corresponding to $\Delta = 1, \frac{1}{2}$, and $\frac{1}{3}$ s are shown in Fig. 3.9*e* and *f*, respectively. Note that as $\Delta \to 0$, $p_\Delta(t)$ tends to the unit *impulse* shown in Fig. 3.9*g* [recall Eq. (2.8)], namely,

$$\lim_{\Delta \to 0} p_\Delta(t) = \delta(t) \tag{3.22}$$

Note also that the "peak" value $h_\Delta(\Delta)$ of the response waveform in Fig. 3.9*d* increases as Δ decreases. To obtain the limiting value of $h_\Delta(\Delta)$ as $\Delta \to 0$, we apply L'Hospital's rule:

$$\lim_{\Delta \to 0} h_\Delta(\Delta) = \lim_{\Delta \to 0} \frac{f'(\Delta)}{g'(\Delta)} = \lim_{\Delta \to 0} \frac{(1/\tau)\exp(-\Delta/\tau)}{1} = \frac{1}{\tau} \tag{3.23}$$

Hence, the response waveform in Fig. 3.9*f* tends to the exponential waveform

$$h(t) = \begin{cases} \dfrac{1}{\tau} \exp\left(-\dfrac{t}{\tau}\right) & t > 0 \\ 0 & t < 0 \end{cases} \tag{3.24}$$

shown in Fig. 3.9*h*. Using the *unit step function* $1(t)$ defined earlier in Eq. (2.6), we can rewrite Eq. (3.24) as follows:

$$h(t) = \frac{1}{\tau} \exp\left(\frac{-t}{\tau}\right) 1(t) \tag{3.24$'$}$$

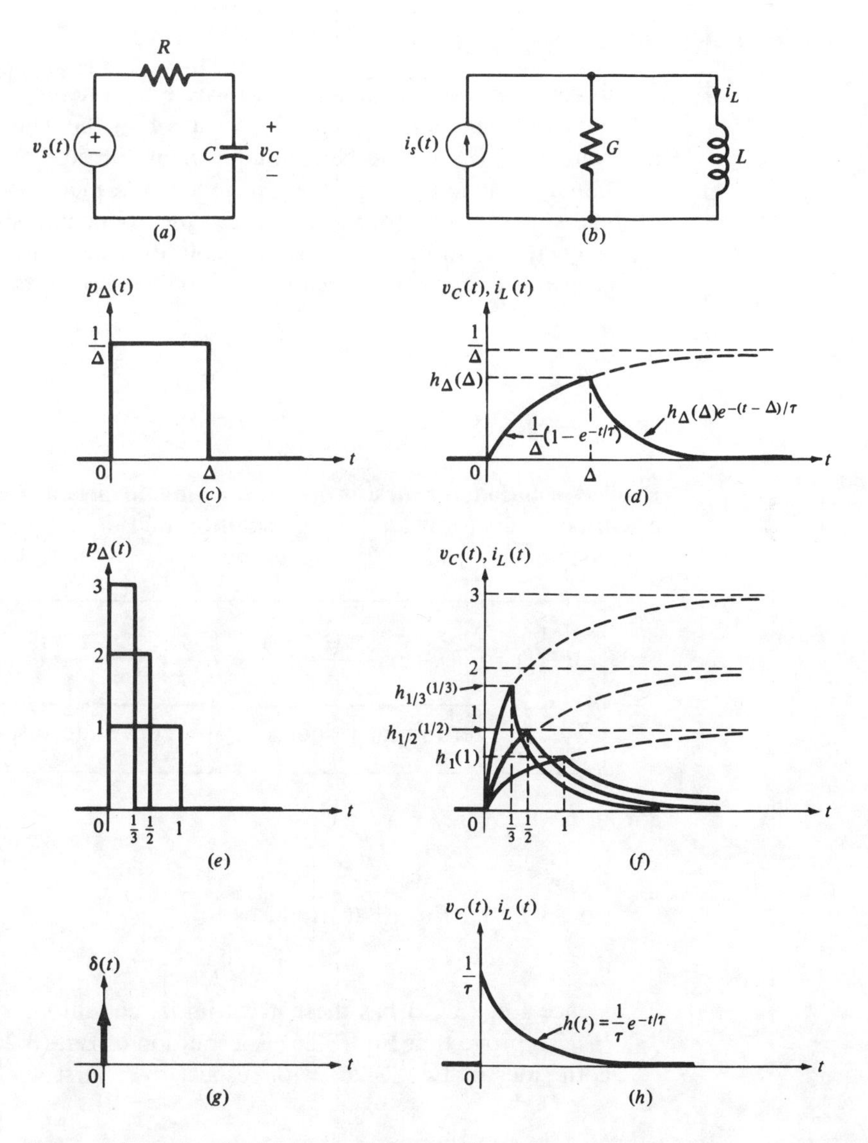

Figure 3.9 As $\Delta \to 0$, the square pulse in (c) tends to the *unit* impulse $\delta(\cdot)$ in (g). The corresponding response tends to the *impulse* response $h(t)$ in (h).

Because $h(t)$ is the response of the circuit when driven by a unit impulse under *zero initial condition*, it is called an *impulse response*. Note that $h(t) = 0$ for $t < 0$.

In Chap. 10, we will show that given the impulse response of any linear time-invariant circuit, we can use it to calculate the response when the circuit is driven by any other input waveform.

3.4 Circuits Driven by Arbitrary Signals

Let us consider now the general case where the one-port N in Fig. 3.1 contains arbitrary independent sources. This means that the Thévenin equivalent voltage source $v_{OC}(t)$, or the Norton equivalent current source $i_{SC}(t)$, in Fig. 3.2 can be *any* function of time, say, in practice, a piecewise-continuous function of time: square wave, triangular wave, synchronization signal of a TV set, etc. Our objective is to derive an explicit solution and draw the consequences.

Consider first the RC circuit in Fig. 3.2*a* whose state equation is

$$\dot{v}_C(t) = -\frac{v_C(t)}{\tau} + \frac{v_{OC}(t)}{\tau} \tag{3.25}$$

where $\tau \triangleq R_{eq}C$.

Explicit solution for first-order linear time-invariant *RC* circuits Given *any* prescribed waveform $v_{OC}(t)$, the solution of Eq. (3.25) corresponding to *any* initial state $v_C(t_0)$ at $t = t_0$ is given by

$$v_C(t) = \underbrace{v_C(t_0) \exp \frac{-(t-t_0)}{\tau}}_{\text{zero-input response}} + \underbrace{\int_{t_0}^{t} \frac{1}{\tau} \exp \frac{-(t-t')}{\tau} v_{OC}(t')dt'}_{\text{zero-state response}} \tag{3.26}$$

for all $t \geq t_0$. Here, $\tau = R_{eq}C$.

PROOF

(*a*) At $t = t_0$, Eq. (3.26) reduces to

$$v_C(t)|_{t=t_0} = v_C(t_0) \tag{3.27}$$

Hence Eq. (3.26) has the correct initial condition.

(*b*) To prove that Eq. (3.26) is a solution of Eq. (3.25), let us differentiate both sides of Eq. (3.26) with respect to t: First we rewrite Eq. (3.26) as

$$v_C(t) = v_C(t_0) \exp \frac{-(t-t_0)}{\tau} + \left(\frac{1}{\tau} \exp \frac{-t}{\tau}\right) \int_{t_0}^{t} \exp \frac{t'}{\tau}\, v_{OC}(t')\, dt' \tag{3.28}$$

Then upon differentiating we obtain for $t > 0$,

$$\dot{v}_C(t) = -\frac{1}{\tau} v_C(t_0) \exp \frac{-(t-t_0)}{\tau} + \left(-\frac{1}{\tau^2} \exp \frac{-t}{\tau}\right) \times \int_{t_0}^{t} \exp \frac{t'}{\tau}\, v_{OC}(t')\, dt' + \left(\frac{1}{\tau} \exp \frac{-t}{\tau}\right)\left[\exp \frac{t}{\tau}\, v_{OC}(t)\right] \tag{3.29}$$

where we used the fundamental theorem of calculus:

$$\frac{d}{dt}\int_0^t f(t')\,dt' = f(t) \qquad \text{if } f(\cdot) \text{ is continuous at time } t$$

Simplifying Eq. (3.29), we obtain

$$\dot{v}_C(t) = -\frac{1}{\tau} v_C(t_0) \exp \frac{-(t-t_0)}{\tau} - \frac{1}{\tau}\left[\int_{t_0}^{t} \frac{1}{\tau} \exp \frac{-(t-t')}{\tau} v_{OC}(t')\,dt'\right] + \frac{1}{\tau} v_{OC}(t)$$

$$= -\frac{v_C(t)}{\tau} + \frac{v_{OC}(t)}{\tau} \tag{3.30}$$

Hence Eq. (3.26) is a solution of Eq. (3.25).

(*c*) From mathematics we learned that the differential equation (3.25) has a unique solution. Hence Eq. (3.26) is indeed the solution. ■

Zero-input response and zero-state response The solution (3.26) consists of two terms. The first term is called the *zero-input response* because when all independent sources in N are set to zero, we have $v_{OC}(t) = 0$ for all times, and $v_C(t)$ reduces to the first term only. The second term is called the *zero-state response* because when the initial state $v_C(t_0) = 0$, $v_C(t)$ reduces to the second term only.

Example Let us find the solution $v_C(t)$ of Fig. 3.7*a* using the above general formula. In this case, we have

$$v_C(t_0) = 0 \qquad t_0 = 0 \qquad \text{and} \qquad v_{OC}(t) = E \qquad t \geq 0$$

Substituting these parameters into Eq. (3.26), we obtain

$$v_C(t) = 0 \times \exp\left[-\frac{(t-0)}{\tau}\right] + \int_0^t \frac{1}{\tau} \exp\left[-\frac{(t-t')}{\tau}\right] \cdot E\,dt'$$

$$= \frac{E}{\tau} \exp\left(-\frac{t}{\tau}\right) \int_0^t \exp\frac{t'}{\tau}\,dt' = \frac{E}{\tau} \exp\left(-\frac{t}{\tau}\right)\left(\exp\frac{t}{\tau} - 1\right)\tau$$

$$= E\left[1 - \exp\left(-\frac{t}{\tau}\right)\right] \qquad t \geq 0 \tag{3.31}$$

which coincides with that shown in Fig. 3.7*c*, as it should.

By duality, we have the following:

Explicit solution for first-order linear time-invariant *RL* circuit Given any prescribed waveform $i_{SC}(t)$, the solution of Eq. (3.26) corresponding to any initial state $i_L(t_0)$ at $t = t_0$ is given by

$$i_L(t) = \underbrace{i_L(t_0) \exp \frac{-(t-t_0)}{\tau}}_{\text{zero-input response}} + \underbrace{\int_{t_0}^{t} \frac{1}{\tau} \exp \frac{-(t-t')}{\tau}\, i_{SC}(t')\, dt'}_{\text{zero-state response}} \tag{3.32}$$

for all $t \geq t_0$. Here, $\tau = G_{eq}L$.

REMARKS

1. In both Eqs. (3.26) and (3.32), the *zero-input* response does *not* depend on the inputs and the *zero-state response* does *not* depend on the initial condition. In both cases, the total response can be interpreted as the superposition of two terms, one due to the initial condition acting alone (with all independent sources set to zero) and the other due to the input acting alone (with the initial condition set to zero).
2. Formulas (3.26) and (3.32) are valid for both $\tau > 0$ and $\tau < 0$. Consider the stable case $\tau > 0$. For values of t' such that $t - t' \gg \tau$, the factor $\exp[-(t-t')/\tau]$ is very small; consequently the values of $v_{OC}(t)$ [respectively, $i_{SC}(t)$] for such times contribute almost nothing to the integral in Eq. (3.26) [respectively, Eq. (3.32)]. In other words, the stable *RC* circuit (respectively, the stable *RL* circuit) has a *fading memory*: Inputs that have occurred many time constants ago have practically no effect at the present time.

   Thus we may say that the time constant τ is a measure of the memory time of the circuit.
3. Using the impulse response $h(t)$ for the *RC* circuit derived earlier in Eq. (3.24), we can rewrite the zero-state response in Eq. (3.26) as follows:

$$\int_{t_0}^{t} h(t-t')\, v_{OC}(t')\, dt' \tag{3.33}$$

Equation (3.33) is an example of a *convolution integral* to be developed in Chap. 10.

4. Once $v_C(t)$ is found using Eq. (3.26), we can replace the capacitor in Fig. 3.2*a* by an independent voltage source described by $v_C(t)$. We can then apply the *substitution theorem* to find the corresponding solution inside N by solving the resulting linear resistive circuit using the methods from the preceding chapters.
5. The *zero-state response* due to a *unit step* input $1(t)$ is called the *step response*, and will be denoted in this book by $s(t)$. The step response for a first-order *RC* (respectively, *RL*) circuit can be found by the *inspection method* in Sec. 3.1C, upon choosing $v_C(0) = 0$ (respectively, $i_L(0) = 0$).

The significance of the *step response* is that for *any* linear time-invariant circuit, the *impulse response* $h(t)$ needed in the convolution integral (6.5) of Chap. 10 can be derived from $s(t)$ (which is usually much easier to derive) via the formula

$$h(t) = \frac{ds(t)}{dt} \tag{3.34}$$

This important relationship is the subject of Exercise 1 in Chap. 10, page 615 [Eq. (4.64)].

The dual remark of course applies to the RL circuit in Fig. 3.2b.

4 FIRST-ORDER LINEAR SWITCHING CIRCUITS

Suppose now that the one-port N in Fig. 3.1 contains one or more *switches*, where the *state* (open or closed) of each switch is specified for all $t \geq t_0$. Typically, a switch may be open over several disjoint time intervals, and closed during the remaining times. Although a switch is a *time-varying* linear resistor, such a linear switching circuit may be analyzed as a sequence of first-order linear time-invariant circuits, each one valid over a time interval where all switches remain in a given state. This class of circuits can therefore be analyzed by the same procedure used in the preceding section. The only difference here is that unlike Sec. 3, the time constant τ will generally vary whenever a switch changes state, as demonstrated in the following example.

Example Consider the RC circuit shown in Fig. 4.1a, where the switch S is assumed to have been open for a long time prior to $t = 0$.

Given that the switch is *closed* at $t = 1$ s and then *reopened* at $t = 2$ s, our objective is to find $v_C(t)$ and $v_o(t)$ for all $t \geq 0$.

Since we are only interested in $v_C(t)$ and $v_o(t)$, let us replace the remaining part of the circuit by its Thévenin equivalent circuit. The result is shown in Fig. 4.1b and c corresponding to the case where S is "open" or "closed," respectively. The corresponding *time constant* is $\tau_2 = 1$ s and $\tau_1 = 0.9$ s, respectively.

Since the switch is initially open and the capacitor is initially in equilibrium, it follows from Fig. 4.1b that $v_C(t) = 6$ V and $v_o(t) = 0$ for $t \leq 1$ s. At $t = 1+$ we change to the equivalent circuit in Fig. 4.1c. Since, by continuity, $v_C(1+) = v_C(1-) = 6$ V, we have $i_C(1+) = (10 - 6)\text{V}/(2 + 1.6)\text{ k}\Omega \approx 1.11$ mA and hence $v_o(1+) = (1.6\text{ k}\Omega)(1.11\text{ mA}) \approx 1.78$ V.

To determine $v_C(t_\infty)$ and $v_o(t_\infty)$ for the equivalent circuit in Fig. 4.1c, we open the capacitor and obtain $v_C(t_\infty) = 0$. The waveforms of $v_C(t)$ and $v_o(t)$ during [1, 2) are drawn as solid lines in Figs. 4.1d and e, respectively. The dotted portion shows the respective waveform if S had been left closed for all $t \geq 1$ s.

Since S is closed at $t = 2$ s, we must write the equation of these two waveforms to calculate $v_C(2-) = 8.68$ V and $v_o(2-) = 0.59$ V.

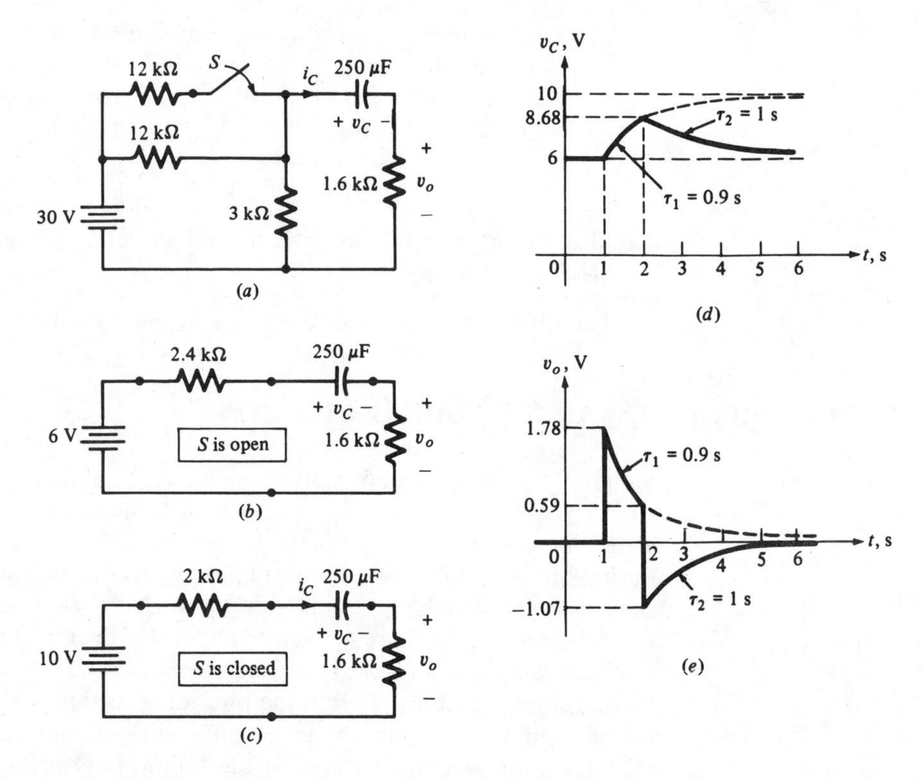

Figure 4.1 An RC switching circuit and the solution waveforms corresponding to the case where S is open during $t<1$ s and $t \geq 2$ s, and closed during $1 \leq t < 2$.

At $t=2+$, we return to the equivalent circuit in Fig. 4.1b. Since $v_C(2+)=v_C(2-)=8.68$ V, we have $i_C(2+)=(6-8.68)\text{V}/(2.4+1.6)\text{ k}\Omega \approx -0.67$ mA and $v_o(2+)=(1.6\text{ k}\Omega)(-0.67\text{ mA}) \approx -1.07$ V.

To determine $v_C(t_\infty)$ and $v_o(t_\infty)$ for the circuit in Fig. 4.1b, we open the capacitor and obtain $v_C(t_\infty)=6$ V and $v_o(t_\infty)=0$. The remaining solution waveforms are therefore as shown in Figs. 4.1d and e, respectively.

5 FIRST-ORDER PIECEWISE-LINEAR CIRCUITS

Consider the first-order circuit shown in Fig. 5.1 where the resistive one-port N may now contain *nonlinear* resistors in addition to linear resistors and dc sources. As before, all resistors and the capacitor are time-invariant. This class of circuits includes many important nonlinear electronic circuits such as multivibrators, relaxation oscillators, time-base generators, etc. In this section, we assume that all nonlinear elements inside N are *piecewise-linear* so that the one-port N is described by a *piecewise-linear driving-point characteristic*.

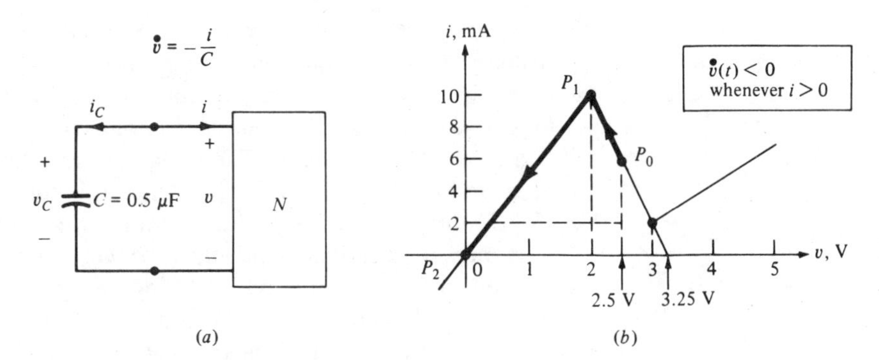

Figure 5.1 (*a*) A piecewise-linear *RC* circuit. (*b*) Driving-point characteristic of *N*.

Our main problem is to find the solution $v_C(t)$ for the *RC* circuit, or $i_L(t)$ for the *RL* circuit, subject to any given initial state. Since the corresponding port variables of *N*, namely, $[v(t), i(t)]$, must fall on the driving-point characteristic of *N*, the evolution of $[v(t), i(t)]$ can be visualized as the motion of a point on the characteristic starting from a given initial point.

5.1 The Dynamic Route

Since the driving-point characteristic is *piecewise-linear*, the solution $[v(t), i(t)]$ can be found by determining first the specific "route" and "direction," henceforth called the *dynamic route*, along the characteristic where the motion actually takes place. Once this route is identified, we can apply the "inspection method" developed in Sec. 3.1 to obtain the solution traversing along *each segment* separately, as illustrated in the following examples.

Example 1 Consider the *RC* circuit shown in Fig. 5.1*a*, where the one-port *N* is described by the voltage-controlled *piecewise-linear* characteristic shown in Fig. 5.1*b*.

Given the initial capacitor voltage $v_C(0) = 2.5$ V, our objective is to find $v_C(t)$ for all $t \geq 0$.

Step 1. Identify the initial point. Since $v(t) = v_C(t)$, for all t, initially $v(0) = v_C(0) = 2.5$ V. Hence the initial point on the driving-point characteristic of *N* is P_0, as shown on Fig. 5.1*b*.

Step 2. Determine the dynamic route. The dynamic route starting from P_0 contains two pieces of information: (*a*) the route traversed and (*b*) the direction of motion. They are determined from the following information:

Key to dynamic route for *RC* circuit

(*a*) The driving-point characteristic of *N*

(*b*) $\dot{v}(t) = -\dfrac{i(t)}{C}$

Since $\dot{v}(t) = -i(t)/C < 0$ whenever $i(t) > 0$, the voltage $v(t)$ *decreases* so long as the associated current $i(t)$ is positive. Hence, for $i(t) > 0$, the dynamic route starting at P_0 must always move *along* the v-i curve *toward the left*, as indicated by the *bold directed* line segments $P_0 \rightarrow P_1$ and $P_1 \rightarrow P_2$ in Fig. 5.1*b*. The dynamic route for this circuit ends at P_2 because at P_2, $i = 0$, so $\dot{v} = 0$. Hence the capacitor is in equilibrium.

Step 3. Obtain the solution for each straight line segment. Replace N by a sequence of *Thévenin equivalent circuits* corresponding to *each* line segment in the dynamic route. Using the method from Sec. 3.1, find a sequence of solutions $v_C(t)$. For this example, the dynamic route $P_0 \rightarrow P_1 \rightarrow P_2$ consists of only two segments. The corresponding equivalent circuits are shown in Fig. 5.2*a* and *b*, respectively.

To obtain $v_C(t)$ for segment $P_0 \rightarrow P_1$, we calculate $\tau = -62.5\ \mu s$, $v_C(0) = 2.5$ V, and $v_C(t_\infty) = 3.25$ V. Since the time constant in this case is *negative*, $v_C(t)$ consists of an "unstable" exponential passing through $v_C(0) = 2.5$ V and tending asymptotically to the "unstable" equilibrium value $v_C(t_\infty) = 3.25$ V as $t \rightarrow -\infty$. This solution is shown in Fig. 5.2*c* from P_0 to P_1. To calculate the time t_1 when $v_C(t) = 2$ V, we apply Eq. (3.12) and obtain

$$t_1 - 0 = 62.5\ \mu s \times \ln\left[\frac{2.5\text{ V} - 3.25\text{ V}}{2\text{ V} - 3.25\text{ V}}\right] = 31.9\ \mu s \qquad (5.1)$$

Applying Eq. (3.5), we can write the solution from P_0 to P_1 analytically as follows (all voltages are in volts):

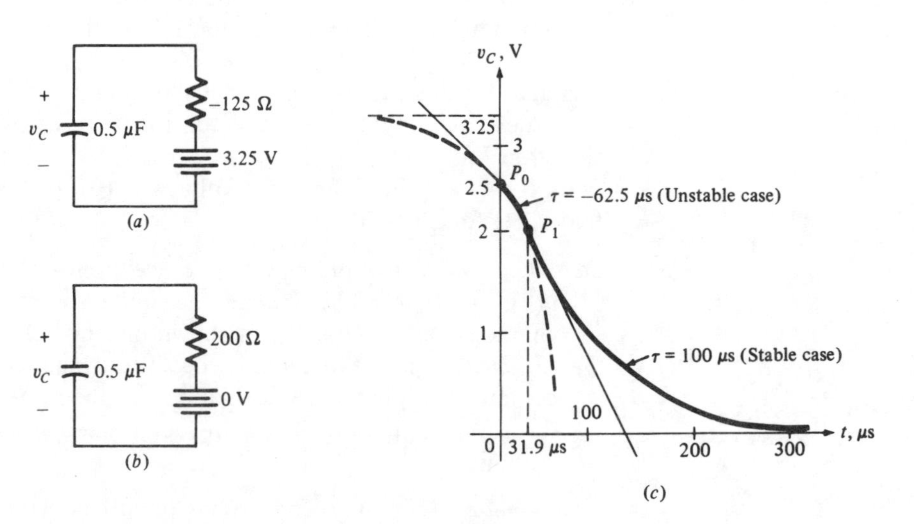

Figure 5.2 (*a*) Equivalent circuit corresponding to $P_0 \rightarrow P_1$. (*b*) Equivalent circuit corresponding to $P_1 \rightarrow P_2$. (*c*) Solution $v_C(t)$.

$$v_C(t) = 3.25 + [2.5 - 3.25] \exp \frac{-t}{62.5\ \mu s}$$

$$= 3.25 - 0.75 \exp \frac{-t}{62.5\ \mu s} \qquad 0 \le t \le 31.9\ \mu s \tag{5.2}$$

To obtain $v_C(t)$ for segment $P_1 \to P_2$, we calculate $\tau_2 = 100\ \mu s$, $v_C(t_0) = 2$ V, $t_0 = 31.9\ \mu s$, and $v_C(t_\infty) = 0$ V. The resulting exponential solution is shown in Fig. 5.2*c*. Applying Eq. (3.5), we can write the solution from P_1 to P_2 analytically as follows:

$$v_C(t) = 2 \exp \frac{-t - 31.9\ \mu s}{100\ \mu s} \qquad t \ge 31.9\ \mu s \tag{5.3}$$

Example 2 Consider the *RL* circuit shown in Fig. 5.3*a*, where *N* is described by the piecewise-linear characteristic shown in Fig. 5.3*b*.

Given the *initial inductor current* $i_L(t_0) = -I_0$, our objective is to find $i_L(t)$ for all $t \ge t_0$. (Note I_0 is the initial current *into* the one-port).

Step 1. Identify initial point. Since $i(t_0) = I_0$, we identify the initial point at P_0 on Fig. 5.3*b*.

Step 2. Determine the dynamic route. The dynamic route starting from P_0 is determined from the following information:

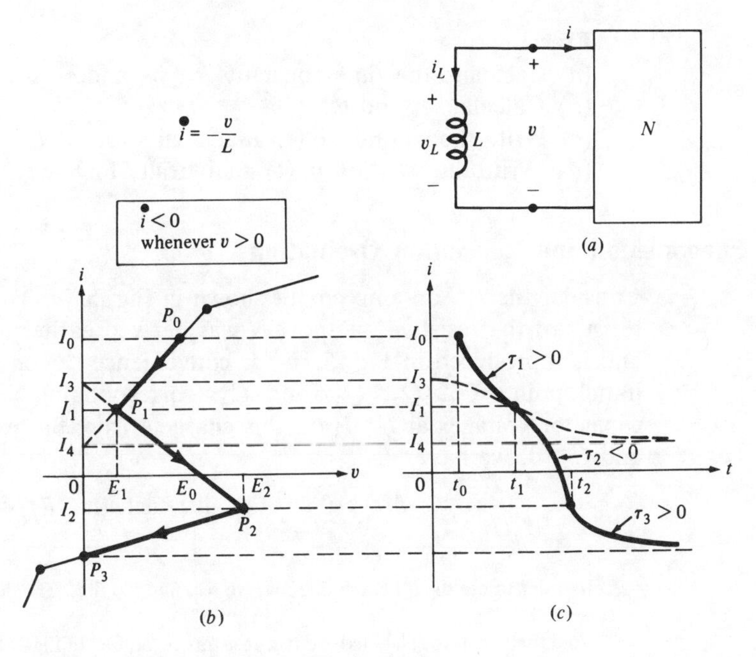

Figure 5.3 A piecewise-linear *RL* circuit.

Key to dynamic route for *RL* circuit

(*a*) The driving-point characteristic of N

(*b*) $\dot{i}(t) = -\dfrac{v(t)}{L}$

Since $\dot{i}(t) = -v(t)/L < 0$ whenever $v(t) > 0$, it follows that the *current* solution $i(t)$ must decrease so long as the associated $v(t)$ is positive.[21] Hence the dynamic route from P_0 must always move downward and consists of three segments $P_0 \rightarrow P_1$, $P_1 \rightarrow P_2$, and $P_2 \rightarrow P_3$ as shown in Fig. 5.3*b*. The dynamic route ends at P_3 because at P_3, $v = 0$ so $\dot{i} = 0$. Hence the inductor is in equilibrium.

Step 3. Replacing N by a sequence of Norton equivalent circuits corresponding to *each* line segment in the dynamic route, we obtain the solution in Fig. 5.3*c* by inspection.

REMARKS

1. After some practice, one can obtain the solution in Figs. 5.2*c* and 5.3*c* *directly* from the dynamic route, i.e., without drawing the Thévenin or Norton equivalent circuits.
2. In the *RC* case, since $\dot{v}(t) = -i(t)/C$, when $\tau > 0$, the dynamic route always terminates upon intersecting the v axis ($i = 0$).
3. In the *RL* case, since $\dot{i}(t) = -v(t)/L$, when $\tau > 0$, the dynamic route always terminates upon intersecting the i axis ($v = 0$).

Exercise

(*a*) Calculate the time constants τ_1, τ_2, and τ_3 in Fig. 5.3*c*.
(*b*) Calculate t_1 and t_2.
(*c*) Write the solution $i_L(t)$ analytically for $t \geq t_0$.
(*d*) Write the solution $v_L(t)$ analytically for $t \geq t_0$.

5.2 Jump Phenomenon and Relaxation Oscillation

Consider the *RC* op-amp circuit shown in Fig. 5.4*a*. The driving-point characteristic of the resistive one-port N was derived earlier in Fig. 3.8*b* of Chap. 4 and is reproduced in Fig. 5.4*b* for convenience.[22] Consider the four different initial points Q_1, Q_2, Q_3, and Q_4 (corresponding to four different initial capacitor voltages at $t = 0$) on this characteristic. Since $\dot{v}(t) = \dot{v}_C(t) = -i(t)/C$ and $C > 0$, we have

$$\dot{v}(t) > 0 \qquad \text{for all } t \text{ such that } i(t) < 0 \tag{5.4a}$$

[21] In order to use the v-i curve directly, we will find $i(t)$ first. The desired solution is then simply $i_L(t) = -i(t)$.

[22] Note that we have relabeled the two resistors R_1 and R_2 in Fig. 3.8*b* of Chap. 4 as R_A and R_B, respectively, in Fig. 5.4*a*. The symbols R_1, R_2, and R_3 in Fig. 5.4 denote the reciprocal slope of segments 1, 2, and 3, respectively, in Fig. 5.4*b*.

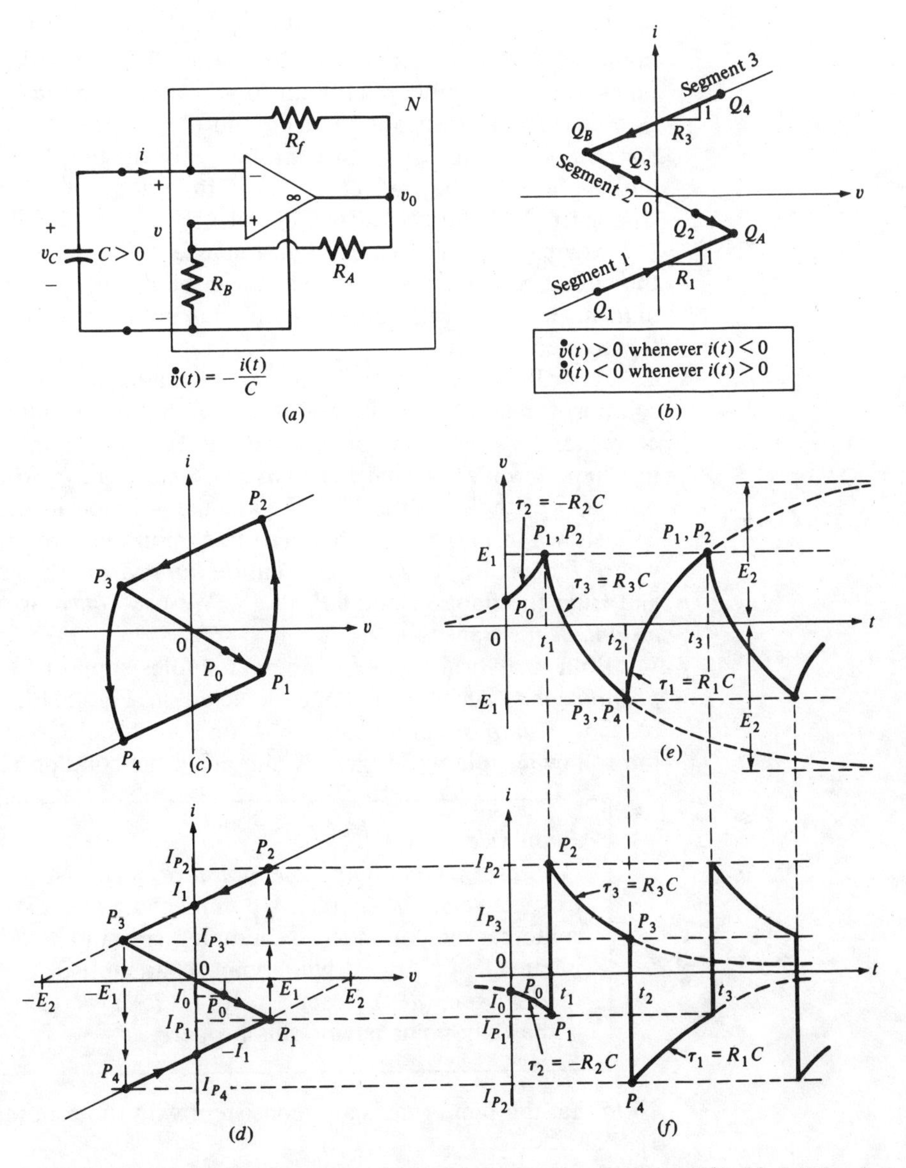

Figure 5.4 (*a*) *RC* op-amp circuit. (*b*) Driving-point characteristic of *N*. (*c*) Solution locus of $(v(t), i(t))$ for the remodeled circuit. (*d*) Dynamic route for the limiting case. (*e*) Voltage waveform $v(t)$. (*f*) Current waveform $i(t)$.

and
$$\mathring{v}(t) < 0 \qquad \text{for all } t \text{ such that } i(t) > 0 \tag{5.4b}$$

Hence the *dynamic route* from any initial point must move *toward the left in the upper half plane*, and *toward the right in the lower half plane*, as indicated by the arrow heads in Fig. 5.4*b*.

Since $i \neq 0$ at the two *breakpoints* Q_A and Q_B, they are *not* equilibrium points of the circuit. It follows from Eq. (3.12) that the amount of time T it takes to go from any initial point to Q_A or Q_B is finite [because $x(t_k) \neq x(t_\infty)$].

Since the arrowheads from Q_1 and Q_2 (or from Q_3 and Q_4) *are oppositely directed*, it is impossible to continue drawing the dynamic route (from any initial point P_0) beyond Q_A or Q_B. In other words, an *impasse* is reached whenever the solution reaches Q_A or Q_B.

Any circuit which exhibits an impasse is the result of poor modeling. For the circuit of Fig. 5.4*a*, the impasse can be resolved by inserting a *small* linear *inductor* in series with the capacitor; this inductor models the inductance L of the connecting wires.

As will be shown in Chap. 7, the remodeled circuit has a well-defined solution for all $t \geq 0$ so long as $L > 0$. A typical solution locus of $(v(t), i(t))$ corresponding to the initial condition at P_0 is shown in Fig. 5.4*c*. Our analysis in Chap. 7 will show that the *transition time* from P_1 to P_2, or from P_3 to P_4, decreases with L. In the limit $L \rightarrow 0$, the solution locus tends to the limiting case shown in Fig. 5.4*d* with a *zero* transition time. In other words in the limit where L decreases to zero, the solution *jumps* from the impasse point P_1 to P_2, and from the impasse point P_3 to P_4. We use *dotted* arrows to emphasize the *instantaneous* transition.

Both analytical and experimental studies support the existence of a *jump phenomenon*, such as the one depicted in Fig. 5.4*d*, whenever a solution reaches an *impasse point* such as P_1 or P_3. This observation allows us to state the following rule which greatly simplifies the solution procedure.

> *Jump rule*
>
> Let Q be an *impasse point* of any first-order RC circuit (respectively, RL circuit). Upon reaching Q at $t = T$, the dynamic route can be continued by jumping (instantaneously) to another point Q' on the driving-point characteristic of N such that $v_C(T+) = v_C(T-)$ [respectively, $i_L(T+) = i_L(T-)$] provided Q' is the only point having this property.

Note that the jump rule is also consistent with the continuity property of v_C, or i_L.

OBSERVATIONS

1. The concepts of an *impasse point* and the *jump rule* are applicable regardless of whether the driving-point characteristic of N is piecewise-linear or not.
2. A first-order RC circuit has at least one impasse point if N is described by a continuous *nonmonotonic* current-controlled driving-point characteristic. The instantaneous transition in this case consists of a *vertical jump* in the v-i plane, assuming i is the vertical axis.

3. A first-order RL circuit has at least one impasse point if N is described by a continuous *nonmonotonic* voltage-controlled driving-point characteristic. The instantaneous transition in this case consists of a *horizontal jump* in the v-i plane, assuming i is the vertical axis.
4. Once the dynamic route is determined, with the help of the jump rule, *for all* $t > t_0$, the solution waveforms of $v(t)$ and $i(t)$ can be determined *by inspection*, as illustrated below.

Example The solution waveforms $v(t)$ and $i(t)$ corresponding to the initial point P_0 in Fig. 5.4c can be found as follows:

Applying the *jump rule* at the two impasse points P_1 and P_3, we obtain the closed dynamic route shown in Fig. 5.4d. This means that the solution waveforms become *periodic* after the short transient time interval from P_0 to P_1. Since the two *vertical* routes occur instantaneously, the *period of oscillation* is equal to the sum of the time it takes to go from P_2 to P_3 and from P_4 to P_1.

Following the same procedure as in the preceding examples, we obtain the voltage waveform $v(\cdot)$ shown in Fig. 5.4e and the current waveform $i(\cdot)$ shown in Fig. 5.4f. As expected, these solution waveforms are periodic and the op-amp circuit functions as an *oscillator*.

Observe that the oscillation waveforms of $v(t)$ and $i(t)$ are far from being sinusoidal. Such oscillators are usually called *relaxation oscillators*.[23]

Exercise
(a) Find the time constants τ_1, τ_2, τ_3, and the time instants t_1, t_2, and t_3 indicated in Fig. 5.4e and f in terms of the element values in Fig. 5.4a. (Assume the ideal op-amp model.)
(b) Use the v_o-vs.-v_i transfer characteristic derived earlier in Fig. 3.8c of Chap. 4 to show that the op-amp output voltage waveform $v_0(\cdot)$ is a *square wave* of period T. Calculate T in terms of the element parameters.

5.3 Triggering a Bistable Circuit (Flip-Flop)

Suppose we replace the capacitor in Fig. 5.4a by the inductor-voltage source combination as shown in Fig. 5.5a. Consider first the case where $v_s(t) \equiv 0$ so that the inductor is directly connected across N. Since $\dot{i}(t) = -v(t)/L$ and $L > 0$, it follows that $di/dt > 0$ whenever $v < 0$ and $di/dt < 0$ whenever $v > 0$. Hence the current i *decreases* in the *right half* v-i plane and *increases* in the *left half* v-i plane, as depicted by the typical dynamic routes in Fig. 5.5b.

Since the equilibrium state of a first-order RL circuit is determined by replacing the inductor by a short circuit, i.e., $v = v_L = 0$, it follows that this

[23] Historically, relaxation oscillators are designed using only two vacuum tubes, or two transistors, such that one device is operating in a "cut-off" or *relaxing* mode, while the other device is operating in an "active" or "saturation" mode.

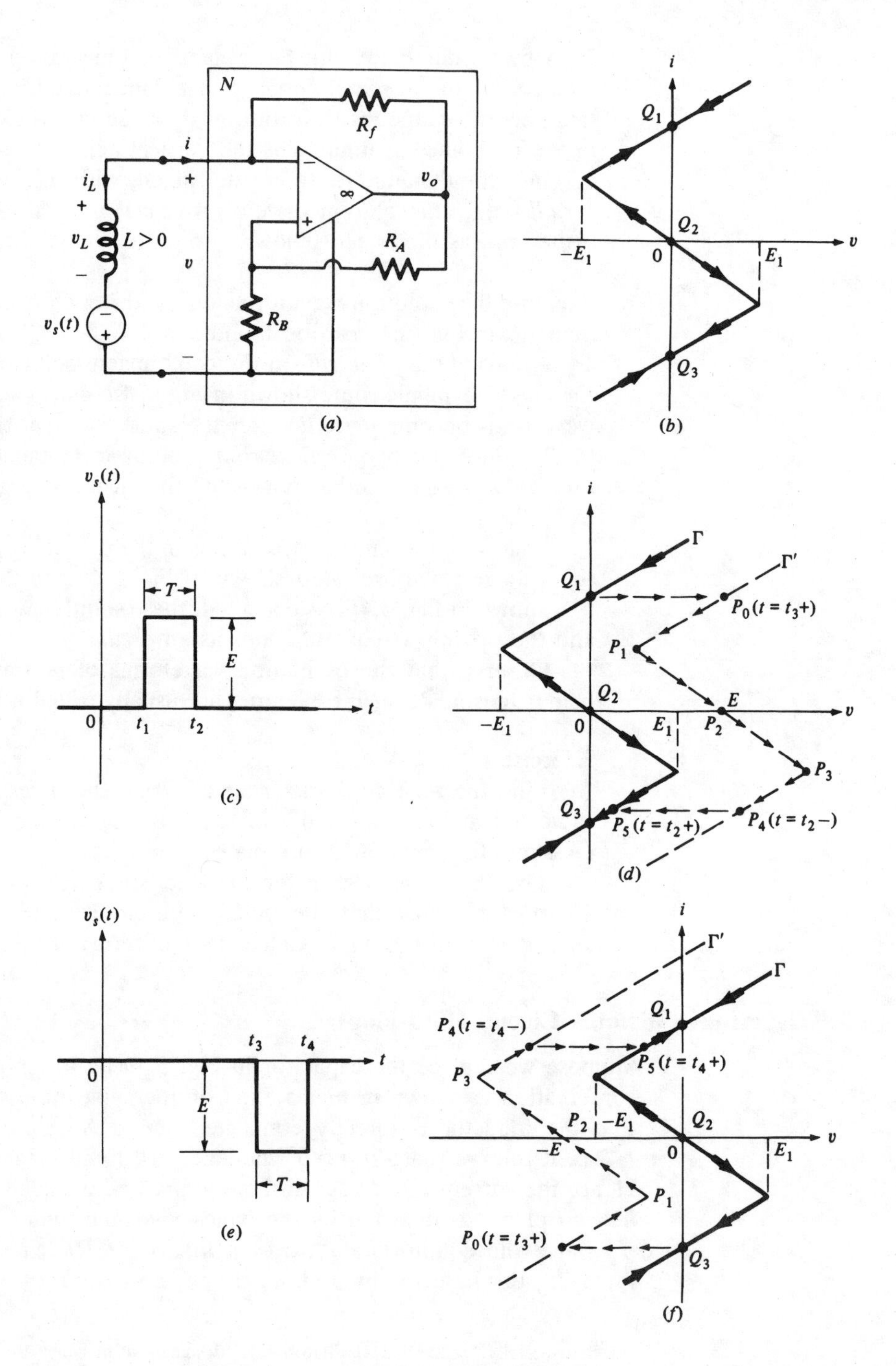

Figure 5.5 A bistable op-amp circuit and the dynamic routes corresponding to two typical triggering signals.

circuit has three *equilibrium points*; namely, Q_1, Q_2, and Q_3. These equilibrium points are the operating points of the associated resistive circuit obtained by short-circuiting the inductor L.

Since the dynamic route in Fig. 5.5*b* either tends to Q_1 or Q_3, but always diverges from Q_2, we say that the equilibrium point Q_2 is *unstable*. Hence even though the associated resistive circuit has three operating points, Q_2 can never be observed in practice—the slightest noise voltage will cause the dynamic route to diverge from Q_2, even if the circuit is operating initially at Q_2.

Whether Q_1 or Q_3 is actually observed depends on the initial condition. Such a circuit is said to be *bistable*.

Bistable circuits (flip-flops) are used extensively in digital computers, where the two stable equilibrium points correspond to the two binary states; say Q_1 denotes "1" and Q_3 denotes "0." In order to perform logic operations, it is essential to switch from Q_1 to Q_3, and vice versa. This is done by using a small *triggering signal*. We will now show how the voltage source in Fig. 5.5*a* can serve as a triggering signal.

Suppose initially the circuit is operating at Q_1. Let us at $t = t_1$ apply a *square pulse* of width $T = t_2 - t_1$ as shown in Fig. 5.5*c*. During the time interval $t_1 < t < t_2$, $v_s(t)$ can be replaced by an E-V battery, so that the inductor sees a translated driving-point characteristic as shown in Fig. 5.5*d* in broken line segments. Let us denote the original and the translated piecewise-linear driving-point characteristics by Γ and Γ' respectively. Then Γ holds over the time intervals $t < t_1$ and $t > t_2$, whereas Γ' holds over the time interval $t_1 < t < t_2$.

Since the inductor current cannot change instantaneously $[i_L(t_1-) = i_L(t_1+)]$, the dynamic route must jump horizontally from Q_1 to P_0 at time $t = t_1$. From P_0, the current i must subsequently decrease so long as $v > 0$. Hence, the dynamic route will be as indicated ($Q_1 \rightarrow P_0 \rightarrow P_1 \rightarrow P_2 \rightarrow P_3 \rightarrow P_4$) in Fig. 5.5*d*. Here, we assume that at time $t = t_2-$, the dynamic route arrives at some point P_4 in the lower half plane. At time $t = t_2+$, Γ' switches back to Γ, and the dynamic route must jump horizontally from P_4 to P_5 at $t = t_2+$. After approximately five time constants, the dynamic route has essentially reached Q_3, and we have succeeded in triggering the circuit from equilibrium point Q_1 to equilibrium point Q_2.

To trigger from Q_3 back to Q_1, simply apply a similar triggering pulse of *opposite* polarity, as shown in Fig. 5.5*e*. The resulting dynamic route is shown in Fig. 5.5*f*.

Triggering criteria The following two conditions must be satisfied by the triggering signal in order to trigger from Q_1 to Q_3, or vice versa.

Minimum pulse width condition If t_2 occurs before the dynamic route in Fig. 5.5*d* (respectively, *f*) crosses the v axis at P_2, the route will jump (horizontally) to a point on Γ in the upper *left half* plane (respectively, lower *right half* plane) and return to Q_1 (respectively, Q_3). Hence, for successful triggering, we must require $T > T_{min}$, where T_{min} is the time it takes to go from P_0 to P_2 in Fig. 5.5*d* or *f*.

Minimum pulse height condition If E is too small such that the breakpoint P_1 on Γ' is located in the left half plane, (respectively, the right half plane), then the dynamic route will also return to Q_1 (respectively, Q_3). Hence, for successful triggering, we must require $E > E_{min}$, where $E_{min} = E_1$.

Exercise

(*a*) Express T_{min} and E_{min} in terms of the circuit parameters.
(*b*) Sketch the solution waveforms of $i(t)$ and $v_o(t)$ for the case when $T = 1.5T_{min}$ and $E = 1.5E_{min}$.
(*c*) Repeat (*b*) for the case where $T = 0.5T_{min}$ and $E = 0.5E_{min}$.

SUMMARY

- A two-terminal element described by a q-v characteristic $f_C(q, v) = 0$ is called a *time-invariant capacitor*.
- In the special case where $q = Cv$, where C is a constant called the *capacitance*, the capacitor is *linear* and *time-invariant*. In this case, it can be described by

$$i = C\frac{dv}{dt}$$

or

$$v(t) = v(t_0) + \frac{1}{C}\int_{t_0}^{t} i(\tau)\, d\tau$$

- A *linear time-varying capacitor* is described by

$$q = C(t)v$$

This implies that

$$i(t) = C(t)\frac{dv(t)}{dt} + \frac{dC(t)}{dt}v(t)$$

requires an additional term compared to the time-invariant case.

- A two-terminal element described by a ϕ-i characteristic $f_L(\phi, i) = 0$ is called a *time-invariant inductor*.
- In the special case where $\phi = Li$, where L is a constant called the *inductance*, the inductor is *linear* and *time-invariant*. In this case, it can be described by

$$v = L\frac{di}{dt}$$

or

$$i(t) = i(t_0) + \frac{1}{L}\int_{t_0}^{t} v(\tau)\, d\tau$$

- A *linear time-varying inductor* is described by

$$\phi = L(t)i$$

This implies that

$$v(t) = L(t)\frac{di(t)}{dt} + \frac{dL(t)}{dt}i(t)$$

requires an additional term compared to the time-invariant case.

- Memory property: The capacitor voltage at any time T depends on the *entire* capacitor current waveform for all $t < T$.
- *Initial capacitor voltage transformation*: A C-F capacitor with an initial voltage $v_C(0)$ is equivalent to a C-F capacitor with *zero* initial voltage in series with a $v_C(0)$-V voltage source.
- *Capacitor voltage continuity property*: For any $t \in (t_1, t_2)$,

$$v_C(t+) = v_C(t-)$$

provided that, for some M,

$$|i_C(t)| \leq M < \infty \qquad \text{for all } t \in [t_1, t_2]$$

- Memory property: The inductor current at any time T depends on the *entire* inductor voltage waveform for all $t < T$.
- *Initial inductor current transformation*: An L-H inductor with an initial current $i_L(0)$ is equivalent to an L-H inductor with *zero* initial current in parallel with an $i_L(0)$-A current source.
- *Inductor current continuity property*: For any $t \in (t_1, t_2)$,

$$i_L(t+) = i_L(t-)$$

provided that, for some M,

$$|v_L(t)| \leq M < \infty \qquad \text{for all } t \in [t_1, t_2]$$

- A *unit step* function $1(t)$ is defined by

$$1(t) \triangleq \begin{cases} 0 & t < 0 \\ 1 & t > 0 \end{cases}$$

- A *unit impulse* (or delta function) $\delta(t)$ is defined by the following two properties:

1. $$\delta(t) \triangleq \begin{cases} \text{singular} & t = 0 \\ 0 & t \neq 0 \end{cases}$$

2. $$\int_{\varepsilon_1}^{\varepsilon_2} \delta(t)\, dt = 1 \qquad \text{for any } \varepsilon_1 < 0 \text{ and } \varepsilon_2 > 0$$

- The zero-state response $h(t)$ to a *unit impulse* $\delta(t)$ is called the *impulse response*.
- The zero-state response $s(t)$ to a *unit step* $1(t)$ is called the *step response*.
- For any linear time-invariant circuit, the impulse response $h(t)$ and the step response $s(t)$ are related by

$$h(t) = \frac{ds(t)}{dt}$$

- *Lossless property*: A time-invariant charge-controlled capacitor cannot dissipate energy. Rather, energy is stored and can be recovered subsequently.

- *Lossless property*: A time-invariant flux-controlled inductor cannot dissipate energy. Rather, energy is stored and can be recovered subsequently.

- The energy w_C entering a time-invariant *charge-controlled* capacitor during $[t_1, t_2]$ depends on the charge at the end points, namely, $q_C(t_1)$ and $q_C(t_2)$. In particular,

$$w_C(q_1, q_2) = \int_{q_1}^{q_2} \hat{v}(q)\, dq$$

where $q_1 \triangleq q_C(t_1)$ and $q_2 \triangleq q_C(t_2)$.

- The energy $\mathscr{E}_C$ stored in a *C*-F capacitor with initial voltage $v_C(0) = V$ is equal to

$$\mathscr{E}_C = \tfrac{1}{2}\, CV^2$$

- q_* is called a *relaxation point* for a time-invariant charge-controlled capacitor iff

$$\int_{q_*}^{q} \hat{v}(q')\, dq' \geq 0 \qquad \text{for all } -\infty < q < \infty$$

The energy $\mathscr{E}_C$ stored in a time-invariant charge-controlled capacitor with initial charge $q(0) = Q$ is equal to

$$\mathscr{E}_C(Q) = \int_{q_*}^{Q} \hat{v}(q)\, dq$$

where q_* is any relaxation point of the capacitor.

- A first-order linear *parallel RC* circuit is described by a state equation

$$\dot{v}_C = -\frac{v_C}{R_{\text{eq}}C} + \frac{v_{\text{OC}}(t)}{R_{\text{eq}}C}$$

where R_{eq} is the Thévenin equivalent resistance and $v_{\text{OC}}(t)$ is the open-circuit voltage of the resistive one-port seen by the capacitor.

- The energy w_L entering a time-invariant *flux-controlled* inductor during $[t_1, t_2]$ depends *only* on the flux at the end points, namely, $\phi_L(t_1)$ and $\phi_L(t_2)$. In particular,

$$w_L(\phi_1, \phi_2) = \int_{\phi_1}^{\phi_2} \hat{i}(\phi)\, d\phi$$

where $\phi_1 \triangleq \phi_L(t_1)$ and $\phi_2 \triangleq \phi_L(t_2)$.

- The energy $\mathscr{E}_L$ stored in an *L*-H inductor with initial current $i_L(0) = I$ is equal to

$$\mathscr{E}_L = \tfrac{1}{2}\, LI^2$$

- ϕ_* is called a *relaxation point* for a time-invariant flux-controlled inductor iff

$$\int_{\phi_*}^{\phi} \hat{i}(\phi')\, d\phi' \geq 0 \qquad \text{for all } -\infty < \phi < \infty$$

The energy $\mathscr{E}_L$ stored in a time-invariant flux-controlled inductor with initial flux $\phi(0) = \Phi$ is equal to

$$\mathscr{E}_L(\Phi) = \int_{\phi_*}^{\Phi} \hat{i}(\phi)\, d\phi$$

where ϕ_* is any relaxation point of the inductor.

- A first-order linear *series RL* circuit is described by a state equation

$$\dot{i}_L = -\frac{i_L}{G_{\text{eq}}L} + \frac{i_{\text{SC}}(t)}{G_{\text{eq}}L}$$

where G_{eq} is the Norton equivalent conductance and $i_{\text{SC}}(t)$ is the short-circuit current of the resistive one-port seen by the inductor.

- Any first-order linear time-invariant circuit driven by dc sources is described by a state equation of the form

$$\dot{x} = -\frac{x}{\tau} + \frac{x(t_\infty)}{\tau}$$

where τ is called the *time constant*, and $x(t_\infty)$ is called the *equilibrium state*.

- $\tau = R_{eq}C$ for an *RC* circuit
- $\tau = G_{eq}L$ for an *RL* circuit

- The solution of the above state equation is always given explicitly by an *exponential* waveform:

$$x(t) - x(t_\infty) = [x(t_0) - x(t_\infty)] \exp \frac{-(t - t_0)}{\tau}$$

for *all time t*.

This solution is uniquely specified by three pieces of information: the *initial state* $x(t_0)$, the *equilibrium state* $x(t_\infty)$, and the *time constant* τ.

- Let $x(t_j)$ and $x(t_k)$ denote any two points on the above *exponential waveform*. The *elapsed time* between t_j and t_k can be calculated explicitly as follows:

$$t_k - t_j = \tau \ln \frac{x(t_j) - x(t_\infty)}{x(t_k) - x(t_\infty)}$$

- The solution of any first-order linear time-invariant circuit driven by dc sources, or by piecewise-constant signals, or circuits containing switches can be obtained *by* inspection (i.e., without writing the state equation): Simply determine the three relevant pieces of information over appropriate time intervals.
- The solution of any first-order *piecewise-linear* circuit can be determined by inspection by drawing the associated *dynamic route*.
- When the dynamic route contains *impasse points*, the capacitor voltage waveform and the inductor current waveform must exhibit one or more instantaneous jumps.

PROBLEMS

Nonlinear capacitors and inductors

1 A varactor diode behaves like a capacitor when $v < V_0$ ($V_0 = 0.5$ V in this case). Its *q-v* characteristic is given by

$$q = 10^{-15}(0.5 - v)^{1/2} \qquad v < 0.5\text{ V}$$

(*a*) Calculate its small-signal capacitance $C(v)$ for $v < 0.5$ V.

(*b*) A voltage $v(t) = -1 + 0.3\cos(2\pi 10^8 t)$ is applied. Obtain an explicit expression for the current through the capacitor.

2 A time-invariant nonlinear inductor has the characteristic shown in Fig. P6.2.

(*a*) Using the graph, estimate the small-signal inductance $L(i)$ when $i = 3$ mA.

(*b*) Repeat (*a*) for $i = 2$ mA.

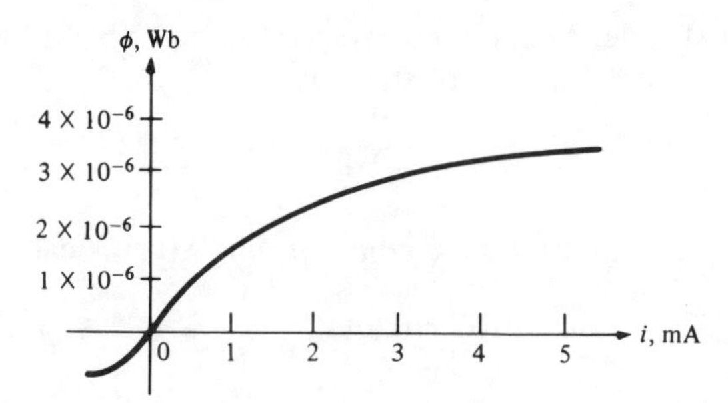

Figure P6.2

Time-varying capacitors

3 A linear-varying capacitor is given by $C(t) = (2 + \sin t)\ \mu\text{F}$; the voltage across the capacitor is $v(t) = 10 \cos (2\pi 10^3 t)$ V. Calculate the current into the capacitor at $t = 0$.

Stored energy

4 A time-invariant inductor has the characteristic shown in Fig. P6.4. Calculate the stored energy in the inductor when $i = 3$ mA.

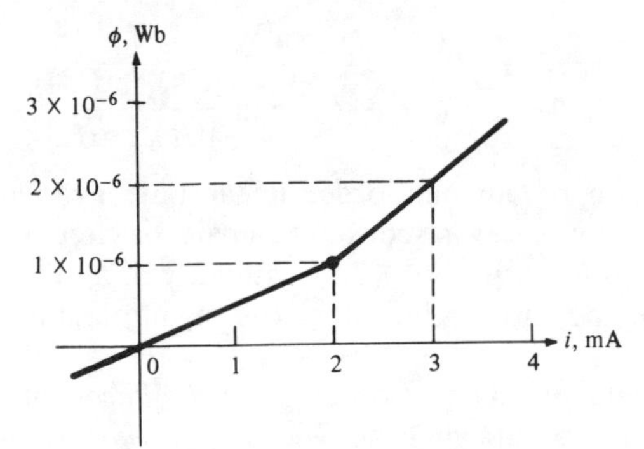

Figure P6.4

Nonlinear capacitors and stored energy

5 Consider a *nonlinear* time-invariant capacitor whose q-v characteristic goes through the origin and is specified by its small-signal capacitance $C(v)$, where $C(v) > 0$ for all v. Show that when the capacitor is charged to v_0 volts, the energy stored is

$$\mathscr{E}_C(v_0) = \int_0^{v_0} C(v) v \, dv$$

Circuit driven by dc sources

6 Consider the circuit shown in Fig. P6.6 with $R_2 = R_3 = 2\ \text{k}\Omega$, $C_1 = 1\ \mu\text{F}$, $L_4 = 1$ mH, $v_1(0-) = 5$ V, $i_4(0-) = 5$ mA, and $v_s = 10$ V. S_1 is closed at $t = 0$.

(*a*) Calculate all voltages and currents at $t = 0+$.

(*b*) Calculate all voltages and currents at $t = \infty$.

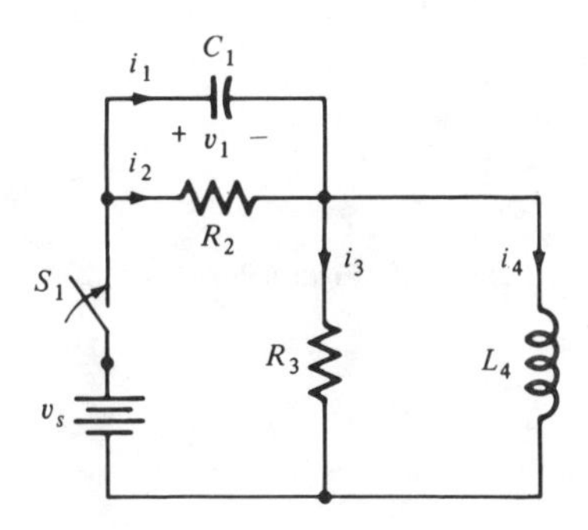

Figure P6.6

7 For the circuit shown in Fig. P6.7 calculate $v_o(t)$ for $t \geq 0$, given $i_L(0) = 2$ A.

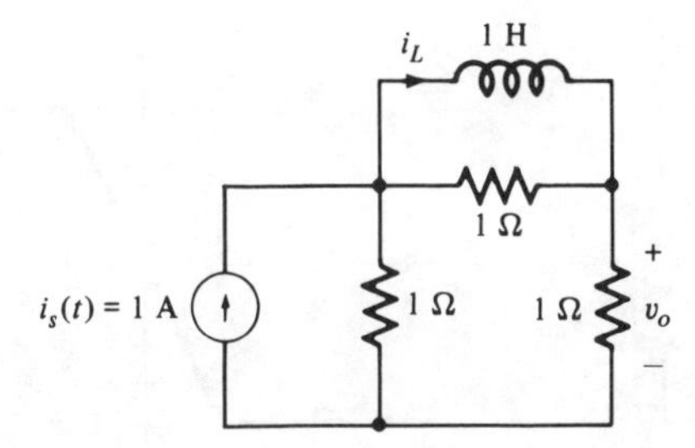

Figure P6.7

8 For the linear time-invariant circuit shown in Fig. P6.8, given $v_s = 10$ V, $R_1 = 2$ kΩ, $R_2 = 2$ kΩ, $C = 1\ \mu$F and $i_C = 5$ mA at $t = 1$ ms,

(*a*) Calculate and sketch $v_R(t)$ for $t \geq 1$ ms.
(*b*) Repeat (*a*) with $R_2 = -1$ kΩ.
(*c*) For part (*a*), what is the elapsed time for which v_R changes from 8 to 6 V?

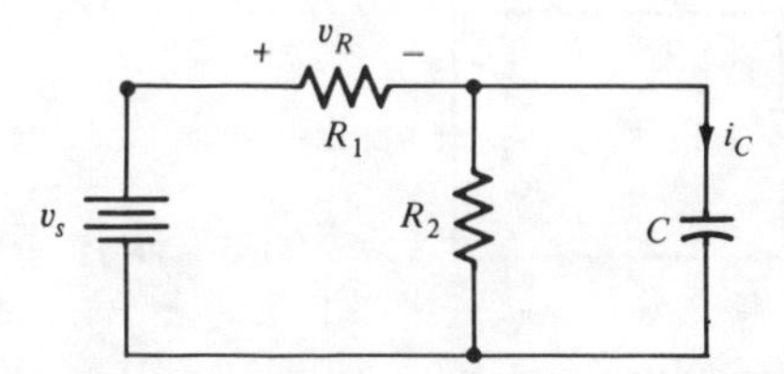

Figure P6.8

Stored energy

9 The circuit shown in Fig. P6.9 is made of linear time-invariant elements. Prior to time 0, the left capacitor is charged to V_0 volts, and the right capacitor is uncharged. The switch is closed at time 0. Calculate the following:

(*a*) The current i for $t \geq 0$.
(*b*) The energy dissipated during the interval $(0, T)$.
(*c*) The limiting values for $t \to \infty$ of (1) the capacitor voltages v_1 and v_2, (2) the current i, and (3) the energy stored in the capacitor and the energy dissipated in the resistor.
(*d*) Is there any relation between these energies? If so, state what it is.
(*e*) What happens when $R \to 0$?

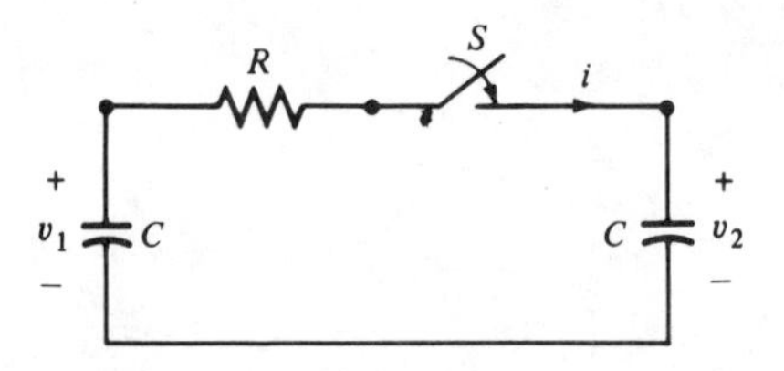

Figure P6.9

Nonlinear first-order circuits

10 Consider the circuit shown in Fig. P6.10 where the inductor is nonlinear and is given by the i-ϕ characteristic shown.

(*a*) Let $i_s(t) = 31(t)$ and $i(0-) = -1$ A. Determine the current $i(t)$ for $t \geq 0$.

(*b*) What is the amount of energy delivered to the inductor from $t = 0$ to $t = \infty$?

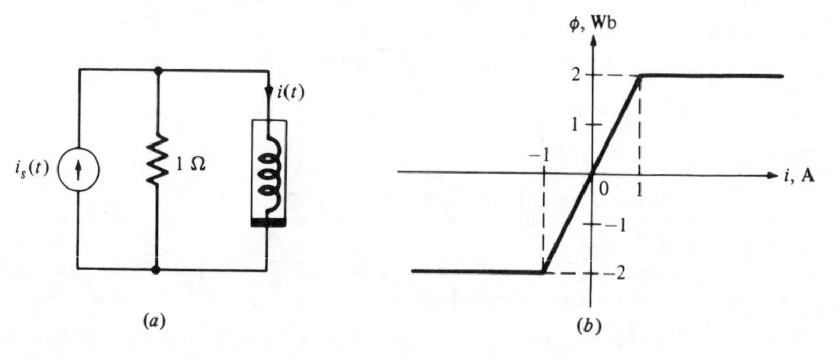

Figure P6.10

11 For the circuit shown in Fig. P6.11, calculate and sketch $v_C(t)$ for $t > 0$. Assume $v_C(0) = 0$.

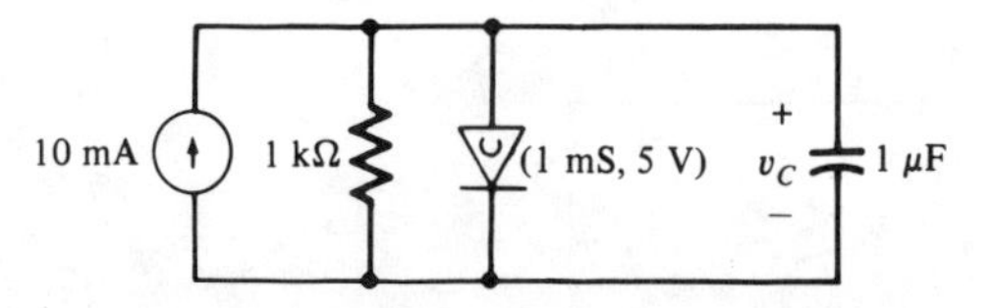

Figure P6.11

Circuits driven by piecewise-constant signals

12 The RC circuit in Fig. P6.12*a* is driven by the stepwise signal shown in Fig. P6.12*b*. Assuming $v_C(0) = 2E_0$, calculate and sketch $i_2(t)$ for $t \geq 0$.

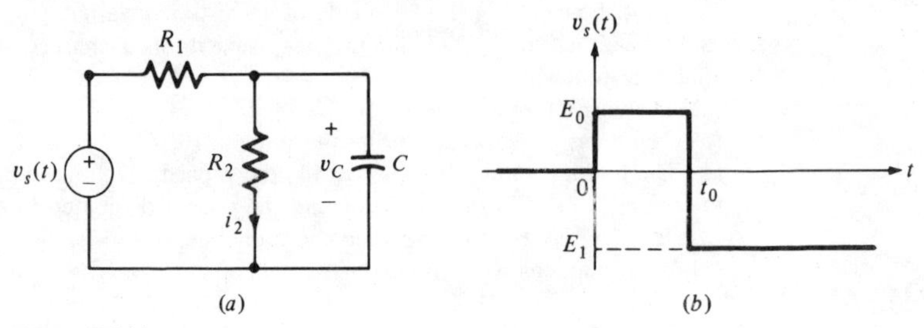

Figure P6.12

13 Calculate and sketch $v(t)$, $t \geq 0$ for the circuit shown in Fig. P6.13*a* with $i_s(t)$ as in Fig. P6.13*b*. Assume $v_C(0) = 0$.

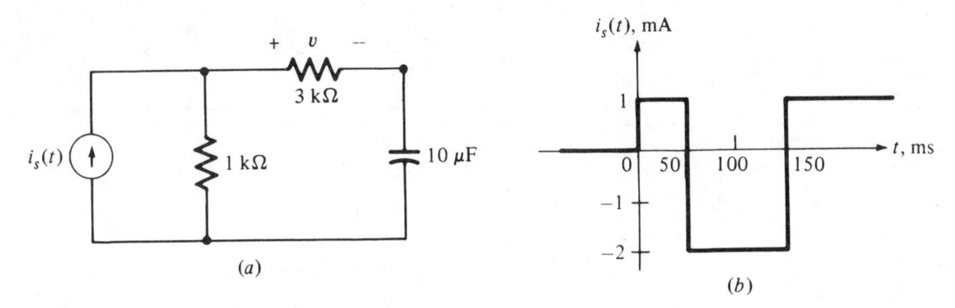

Figure P6.13

14 In the circuit shown in Fig. P6.14*a*, the switch is caused to snap back and forth between the two positions *A* and *B* at regular intervals equal to L/R s. After a large number of cycles, the current becomes periodic as shown in Fig. P6.14*b*. Determine the current levels I_1, and I_2 characterizing this periodic waveform.

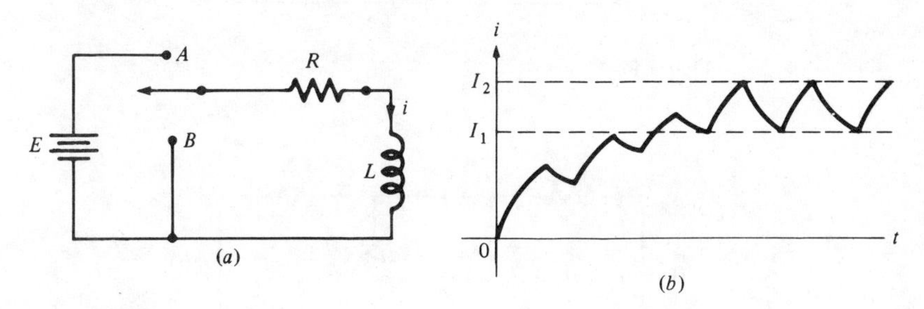

Figure P6.14

Impulse response

15 (*a*) Find the impulse response $v_o = h(t)$ of the *RC* circuit shown in Fig. P6.15*a*.

(*b*) Repeat (*a*) for the *RL* circuit in Fig. P6.15*b*.

(*c*) Repeat (*a*) for the op-amp circuit in Fig. P6.15*c*. Assume an ideal op-amp model in the linear region.

16 Find the impulse response $v_o = h(t)$ of the circuit shown in Fig. P6.7.

Zero-state and zero-input response

17 For the linear time-invariant circuit shown in Fig. P6.17, let $i_0(t)$ be the response.

(*a*) What is the zero-state response due to a unit step input?

(*b*) Determine $i_0(t)$ for $v_s(t) = \delta(t)$.

(*c*) If $v_C(0) = 2$ V, what is the zero-input response?

(*d*) If $v_s(t) = E \cdot 1(t)$ and $v_C(0) = 2$ V, determine *E* such that $i_0(t)$ has no transient.

18 For the linear time-invariant circuit in Fig. P6.7, let v_0 be the output.

(*a*) Write a first-order differential equation in terms of v_0 and the input $i_s(t)$.

(*b*) Find the zero-input response for $i_L(0) = 2$ A.

(*c*) Find the complete response for $i_L(0) = 2$ A and $i_s(t) = 2 \cdot 1(t)$.

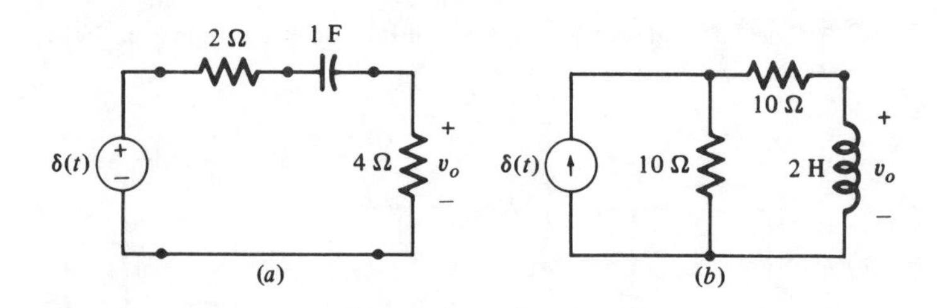

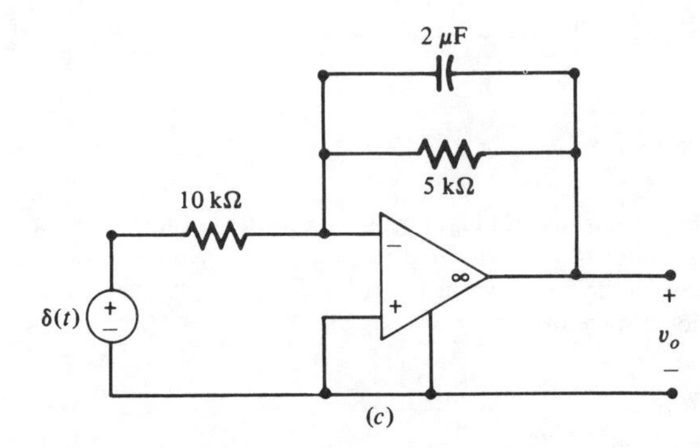

Figure P6.15

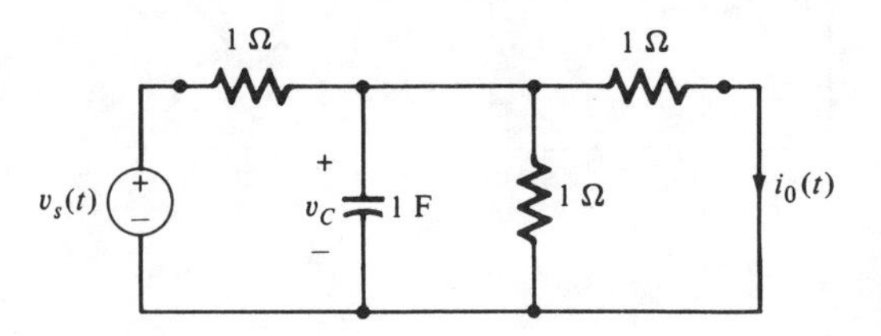

Figure P6.17

19 The zero-state response v_1, v_2, and v_3 of three linear time-invariant one-ports N_1, N_2, and N_3 to the same input current i are shown in Fig. P6.19. Knowing that these one-ports can be described by first-order differential equations, propose a circuit topology and appropriate element values for each case.

Dynamic route

20 Consider the circuit shown in Fig. P6.20*a* where N is described by the v-i characteristic shown in Fig. P6.20*b*.

(*a*) Indicate the dynamic route. Label all equilibrium points and state whether they are stable or unstable.

(*b*) Suppose $v_C(0) = 15$ V. Find and sketch $v_C(t)$ and $i_C(t)$ for $t \geq 0$. Indicate all pertinent information on the sketches.

21 Consider the circuit shown in Fig. P6.21*a* where N is described by the v-i characteristic in Fig. P. 6.21*b*.

(*a*) Indicate the dynamic route. Label all equilibrium points and state whether they are stable or unstable.

(*b*) Suppose $i_L(0) = -20$ mA; calculate and sketch $i(t)$ and $v(t)$ for $t \geq 0$.

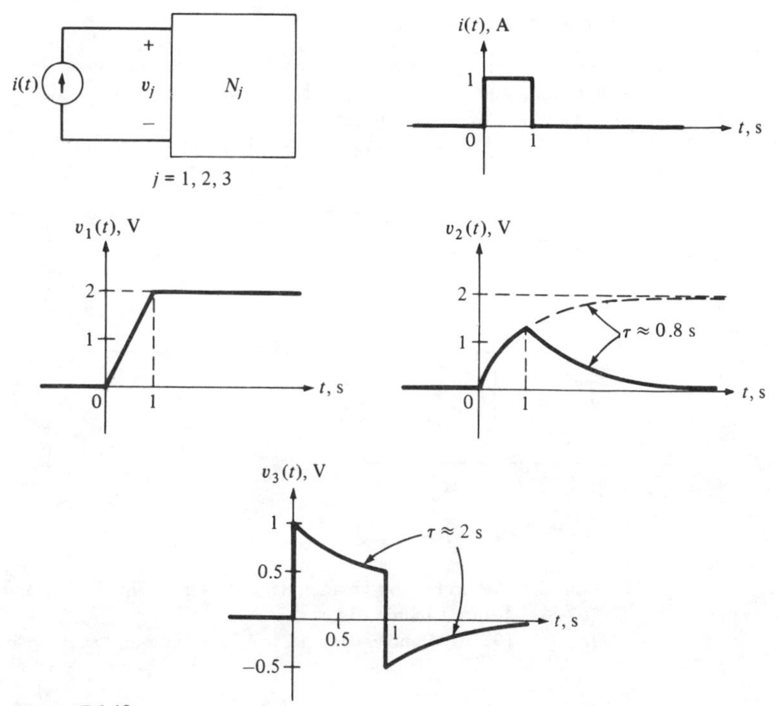

Figure P6.19

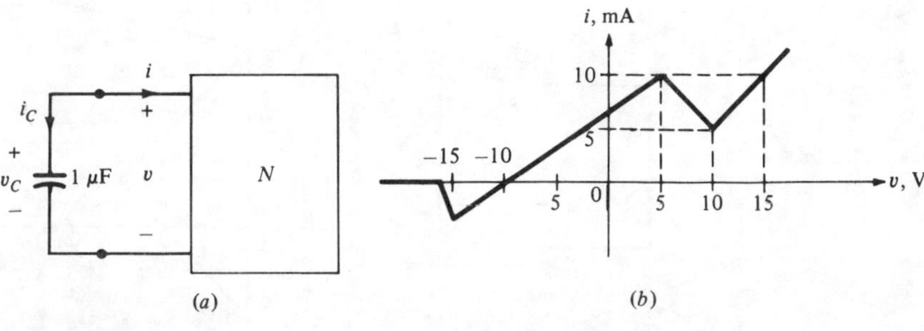

Figure P6.20

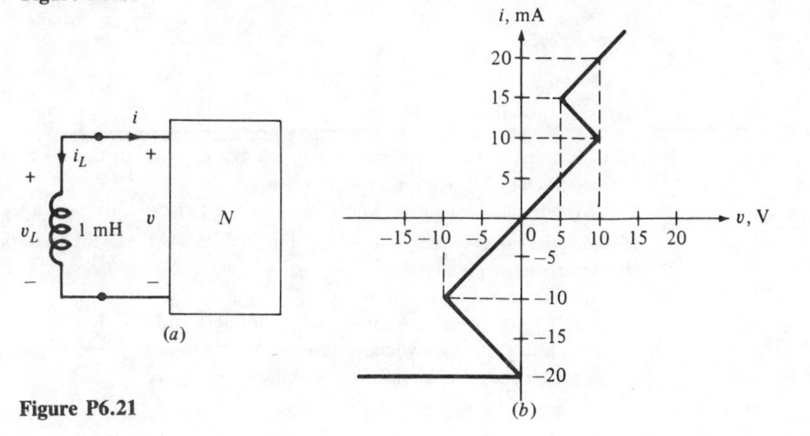

Figure P6.21

Jump phenomenon

22 For the circuit shown in Fig. P6.22*a* with the nonlinear resistor as in Fig. P6.22*b*, find all equilibrium states and classify each as stable or unstable:

(*a*) When switch S is in position 1.

(*b*) When switch S is in position 2.

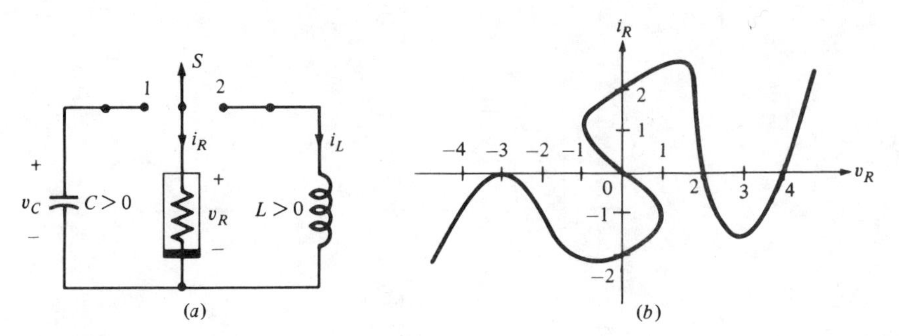

Figure P6.22

23 Consider the circuit and characteristic shown in Fig. P6.23*a* and *b*.

(*a*) Sketch the dynamic route.

(*b*) If $v_C(0)=2$ V and $i_C(0)=-2$ mA, calculate and sketch $i(t)$ and $v(t)$ for $t\geq 0$.

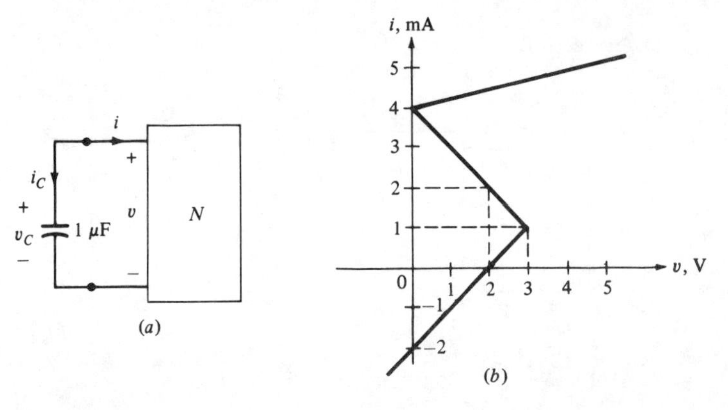

Figure P6.23

Relaxation oscillation

24 Consider the circuit shown in Fig. P6.21*a* along with the v-i characteristic in Fig. P6.24.

(*a*) Sketch the dynamic route.

(*b*) For what range of initial condition $i_L(0)$ does this circuit exhibit relaxation oscillation?

25 Consider the circuit shown in Fig. P6.25*a* where the one-port is described by the v-i characteristic in Fig. P6.25*b*.

(*a*) Sketch the dynamic route.

(*b*) If $i_L(0)=-15$ mA, find and sketch $i(t)$ and $v(t)$ for $t\geq 0$.

(*c*) Determine all switching times.

(*d*) Calculate the period of oscillation.

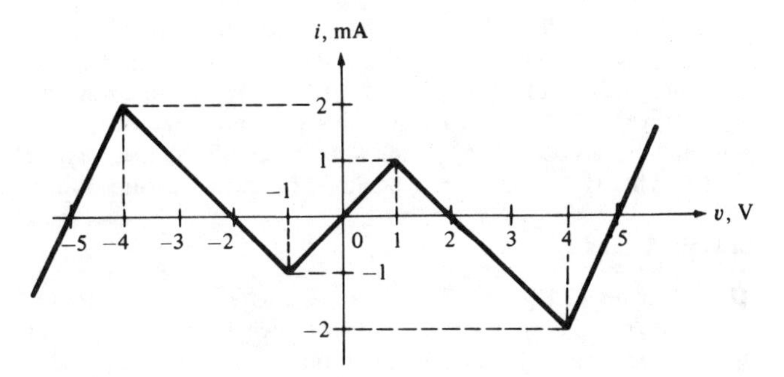

Figure P6.24

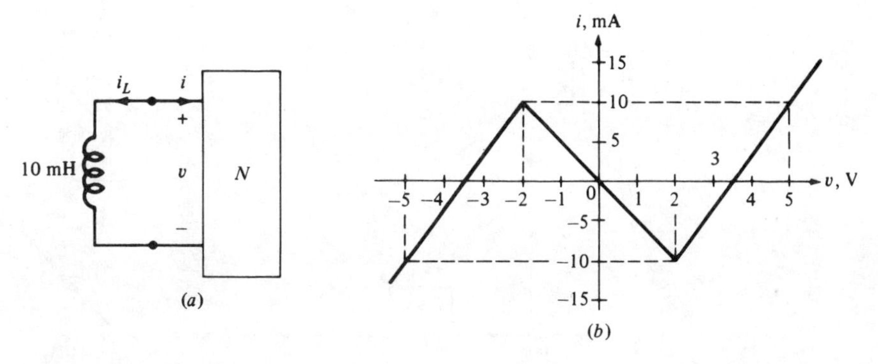

Figure P6.25

26 The circuit shown in Fig. P6.26*a* is to be used as a time-base generator in an oscilloscope. The v-i characteristic of the nonlinear resistor is given in Fig. P6.26*b*.

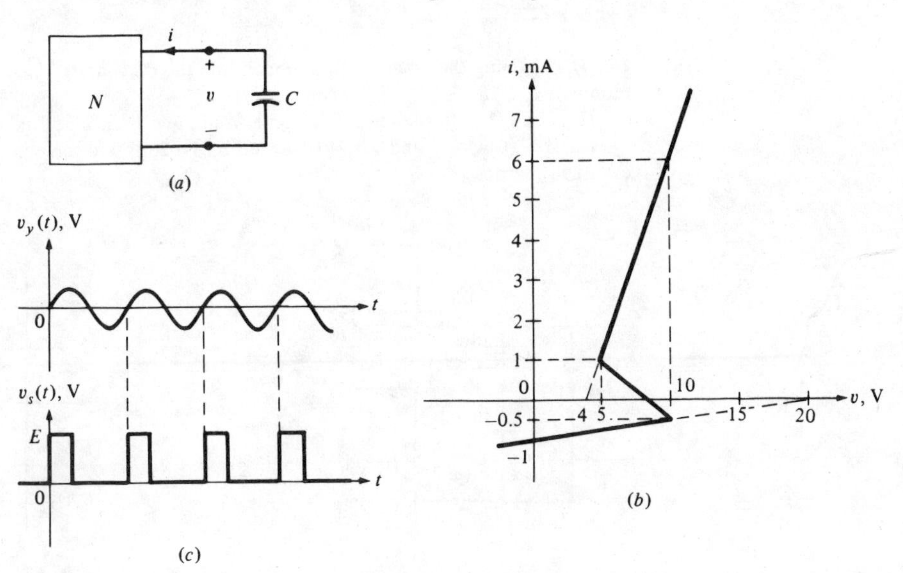

Figure P6.26

(*a*) Find C such that the frequency of oscillation is 1 kHz and sketch the waveform $v(t)$.

(*b*) It is desired to synchronize this time-base generator with a 1.1 kHz sine wave. To accomplish this a pulse train $v_s(t)$ is derived from the sine wave as shown in Fig. P6.26*c*. Indicate how $v_s(t)$ should be applied in order to synchronize $v(t)$ with $v_s(t)$. Assuming that E is of sufficient magnitude to accomplish switching instantaneously, sketch a typical dynamic route.

(*c*) What is the minimum magnitude for E that ensures synchronization?

Bistable circuits

27 The circuit in Fig. P6.27 is to be used as a flip-flop. The v-i characteristic of the nonlinear resistor is given in Fig. P6.27*b*. In order to switch from Q_1 to Q_3 a triggering signal, Fig. P6.27*c*, is applied. Determine the minimum duration of the pulse required for successful switching.

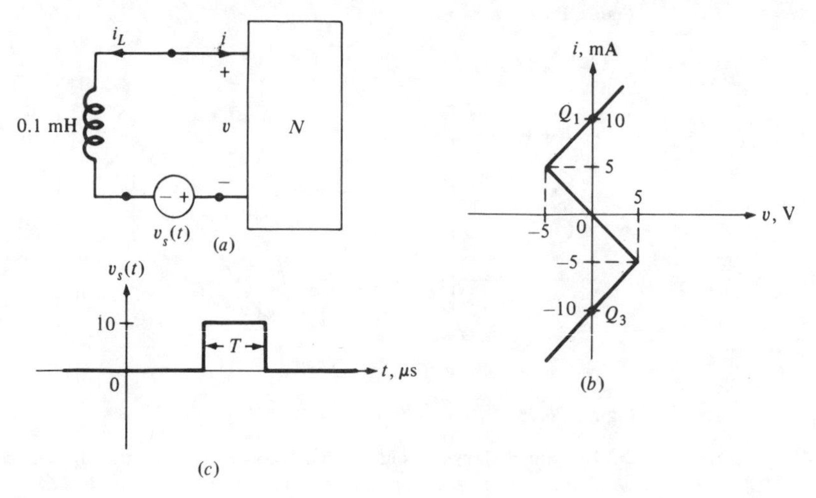

Figure P6.27

28 (*a*) Assuming the ideal op-amp model, find the v-i characteristic of the one-port to the right of terminals L and L'.

(*b*) Let $i_s(t) = 0$. Sketch the dynamic route and indicate all equilibrium points.

(*c*) If $v_C(0) = E_{sat}$ and with $i_s(t)$ as in Fig. P6.28*b*, find the conditions on I and T such that v_C eventually settles at $-E_{sat}$.

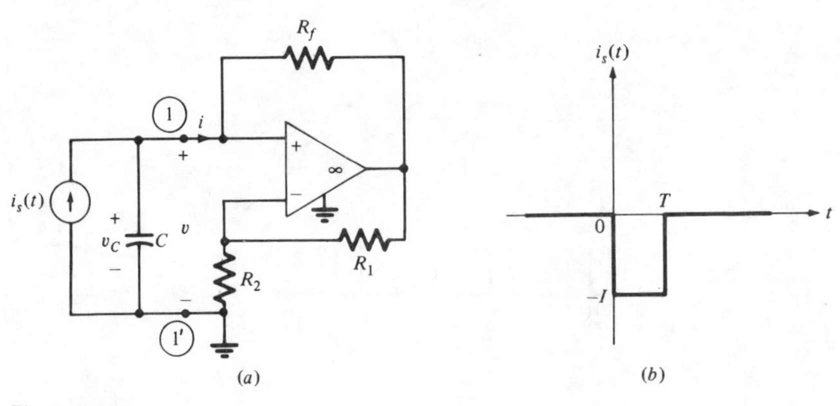

Figure P6.28

CHAPTER

SEVEN

SECOND-ORDER CIRCUITS

Dynamic circuits containing two capacitors, or two inductors, or one capacitor and one inductor, are called *second-order circuits*. This chapter is devoted exclusively to an in-depth study of the behavior and capabilities of both *linear* and *nonlinear* time-invariant second-order circuits. In each case, we first learn how to formulate the equations governing such circuits.

In Sec. 1 we show that the equations governing any second-order circuit made of *linear time-invariant* elements can *always* be written into the following two distinct but equivalent *standard* forms:

Scalar differential equation

$$\ddot{x} + 2\alpha\dot{x} + \omega_0^2 x = u_s(t)$$

State equation

$$\begin{bmatrix} \dot{x}_1 \\ \dot{x}_2 \end{bmatrix} = \begin{bmatrix} a_{11} & a_{12} \\ a_{21} & a_{22} \end{bmatrix} \begin{bmatrix} x_1 \\ x_2 \end{bmatrix} + \begin{bmatrix} u_1(t) \\ u_2(t) \end{bmatrix}$$

or, in vector form, $\dot{\mathbf{x}} = \mathbf{A}\mathbf{x} + \mathbf{u}(t)$.

Although the scalar differential equation is more familiar to students, having seen them already in physics, it is difficult to formulate a simple yet systematic method for writing such an equation for *arbitrary* circuits. On the other hand, it is conceptually almost trivial to write the associated *state equation*. Moreover, once the state equation has been written, it can be easily transformed via *explicit* formulas in Sec. 1.2 into an equivalent scalar differential equation. In other words, except for simple circuits where the scalar differential equation may be written by inspection, it is easier, if not necessary, to formulate the state equation first.

Although equivalent to each other, the solutions of these two standard forms provide different perspectives on the behavior of linear time-invariant second-order circuits. They are therefore studied in great detail in Secs. 2 and 3. While the materials in Sec. 2 are *quantitative* in nature (the solutions are represented analytically), those in Sec. 3 are *qualitative* in the sense that the solutions are represented geometrically by a family of *trajectories* called a *phase portrait*. The concept of a phase portrait is important not only because it is more revealing of the circuit's dynamic behavior, but also because it generalizes to the nonlinear case in Secs. 5 and 6.

When the circuit is nonlinear, it is difficult, if not impossible, to write a corresponding *nonlinear scalar* differential equation. However, any properly modeled second-order circuit can be described by the following standard form:

Nonlinear state equation

$$\dot{x}_1 = f_1(x_1, x_2, t)$$
$$\dot{x}_2 = f_2(x_1, x_2, t)$$

or, in vector form, $\dot{\mathbf{x}} = \mathbf{f}(\mathbf{x}, t)$.

Moreover, all general tools for studying *nonlinear* differential equations are couched in terms of this standard form. It is natural that we begin our study of nonlinear time-invariant circuits in Sec. 4 by learning how to formulate their associated state equations.

Since the solution of *nonlinear* state equations can*not* in general be written in *closed* form, it is of utmost importance to investigate its *qualitative* behavior. This is studied in Sec. 5 for the important subclass of time-invariant circuits containing only dc sources—called *autonomous circuits*. The three fundamental concepts of *equilibrium state*, *trajectory*, and *phase portrait* encountered earlier in Sec. 3 now play an indispensable role in this study. The readers are strongly urged to master these concepts in Sec. 3 before embarking upon their generalizations in Sec. 5.

Finally, the techniques from Sec. 5 are applied in Sec. 6 to carry out a study of electronic oscillators. As a bonus, this analysis also justifies the validity of the *jump rule* which we postulated earlier in Chap. 6 to predict the jump phenomenon.

Remark on notation In this chapter, *double-level* subscripts in $v_{C_1}, v_{C_2}, i_{C_1}, i_{C_2}$ for capacitors (and $v_{L_1}, v_{L_2}, i_{L_1}, i_{L_2}$ for inductors) in the *figures* (e.g., Fig. 1.3) are set *single-level* as $v_{C1}, v_{C2}, i_{C1}, i_{C2}$ (and $v_{L1}, v_{L2}, i_{L1}, i_{L2}$) in the text.

1 EQUATION FORMULATION: LINEAR TIME-INVARIANT CIRCUITS

1.1 Two Standard Forms: $\ddot{x} + 2\alpha\dot{x} + \omega_0^2 x = u_s(t)$ and $\dot{x} = Ax + u(t)$

Let us begin with two dual *linear time-invariant* circuits whose governing equations can be written, by inspection, into the form of a *scalar* differential equation.

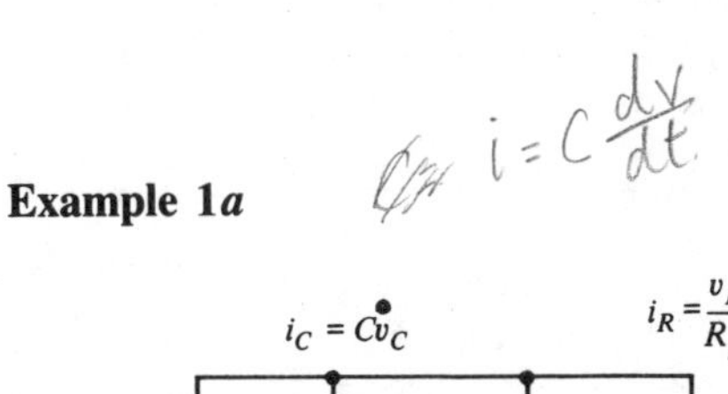

Example 1*a*

Example 1*b*

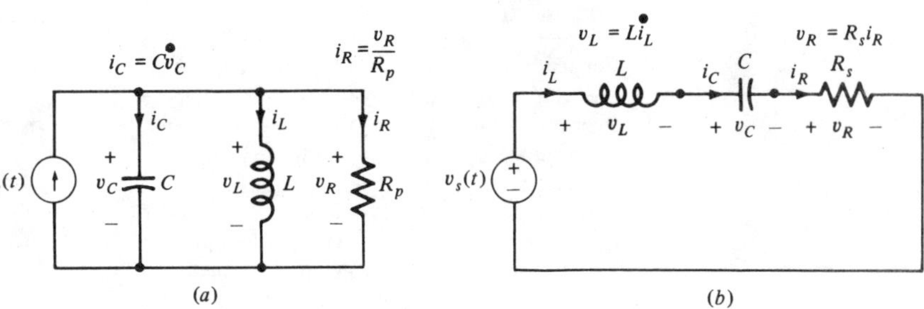

Figure 1.1 (*a*) Parallel linear *RLC* circuit driven by a current source. (*b*) Series linear *RLC* circuit driven by a voltage source.

Applying KCL to the circuit in Fig. 1.1*a*, we obtain

$$C\dot{v}_C + \frac{1}{R_p} v_R + i_L = i_s(t) \tag{1.1a}$$

Substituting $v_C = v_R = L\dot{i}_L$ into Eq. (1.1*a*) and simplifying, we obtain

$$\ddot{i}_L + \frac{1}{R_p C}\dot{i}_L + \frac{1}{LC} i_L = \frac{1}{LC} i_s(t) \tag{1.2a}$$

Applying KVL to the circuit in Fig. 1.1*b*, we obtain

$$L\dot{i}_L + R_s i_R + v_C = v_s(t) \tag{1.1b}$$

Substituting $i_L = i_R = C\dot{v}_C$ into Eq. (1.1*b*) and simplifying, we obtain

$$\ddot{v}_C + \frac{R_s}{L}\dot{v}_C + \frac{1}{LC} v_C = \frac{1}{LC} v_s(t) \tag{1.2b}$$

Note that we could have obtained Eq. (1.2*b*) directly from Eq. (1.2*a*), or vice versa, upon identifying the "dual" entries in Table 1.1.

Mechanical analog It is often instructive to associate simple electric circuits with mechanical analogs. For example, consider the mechanical system in Fig. 1.2*a* in which a block with mass M is tied to the wall by a spring and is driven by a horizontal force $f_s(t)$. Let us model the spring and the friction by the following linear constitutive relations:

$$f_k = Kx \tag{1.3}$$

$$f_B = B\dot{x} \tag{1.4}$$

where K and B are called the *spring constant* and *coefficient of friction*, respectively. Equation (1.3) is usually referred to as Hooke's law, while Eq. (1.4) represents *force* due to friction. Here, x and $v \triangleq \dot{x}$ denote the *displacement* and *velocity* of the block, respectively.

To write the equation of motion for this mechanical system, let us draw the

Table 1.1 Dual variables, parameters, and names

Capacitor voltage v_C	$\leftrightarrow$ Inductor current i_L
Inductor current i_L	$\leftrightarrow$ Capacitor voltage v_C
Resistor voltage v_R	$\leftrightarrow$ Resistor† current i_R
Capacitance C	$\leftrightarrow$ Inductance L
Inductance L	$\leftrightarrow$ Capacitance C
Conductance $G_P = 1/R_P$	$\leftrightarrow$ Resistance $R_s = 1/G_s$
Current source	$\leftrightarrow$ Voltage source
Parallel connection	$\leftrightarrow$ Series connection

†To be precise, we should call this a *conductance* current. Note that whereas the dual of a capacitor is an inductor (which is a different circuit element), or vice versa, the dual of a resistor is also a resistor having the same v-i characteristic except that the voltage and current axis labels are interchanged. In the linear case, therefore, the dual of an R-Ω resistor is an R-S or $(1/R)$-Ω resistor.

associated free-body diagram in Fig. 1.2*b* and note that the net force to accelerate the body is $f_s - f_K - f_B$. It follows from Newton's law that this force is equal to the derivative of the momentum, namely,

$$f_s - f_K - f_B = \frac{d}{dt}(Mv) \tag{1.5}$$

Substituting Eq. (1.4) into Eq. (1.5) and rearranging the equation, we obtain

$$M\dot{v} + Bv + f_K = f_s(t) \tag{1.6}$$

Using Eq. (1.3), we obtain

$$\dot{x} = \frac{1}{K}\dot{f}_K = v \tag{1.7}$$

Substituting Eq. (1.7) for v in Eq. (1.6) and simplifying, we obtain

$$\ddot{f}_K + \frac{B}{M}\dot{f}_K + \frac{K}{M}f_K = \frac{K}{M}f_s(t) \tag{1.8}$$

Figure 1.2 Linear mechanical system and its free-body diagram.

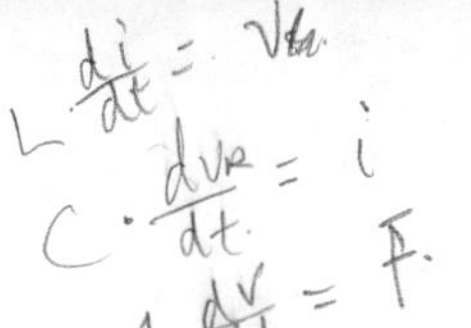

Table 1.2 Electrical analogs of linear mechanical systems

Mechanical	Electrical analog 1	Electrical analog 2
Force f_s	Current i_s	Voltage v_s
Velocity v	Voltage v_R	Current i_R
Spring	Inductor	Capacitor
Friction	Resistor	Resistor
Mass	Capacitor	Inductor

The solution of Eq. (1.8) due to any *initial condition* is well known from physics when $f_s(t) = 0$. We will use this familiar system to provide us with some physical interpretations in Sec. 2.1.

Note that Eq. (1.6) is identical in form to Eqs. (1.1*a*) and (1.1*b*). Likewise, Eq. (1.8) is identical in form to Eqs. (1.2*a*) and (1.2*b*). Consequently, we say the parallel *RLC* circuit in Fig. 1.1*a* is an *electrical* analog of the mechanical system in Fig. 1.2, provided we identify the analogous entries listed in columns 1 and 2 of Table 1.2. Likewise, we say the series *RLC* circuit in Fig. 1.1*b* is an *electrical* analog of the mechanical system in Fig. 1.2, provided we identify the analogous entries listed in columns 1 and 3 of Table 1.2.

The above observations are useful because many terminologies, phenomena, and concepts to be presented in Secs. 1.2 and 1.3 can be given a clearer physical interpretation if we think of them in terms of their mechanical analogs.

Equations (1.2*a*), (1.2*b*), and (1.8) are examples of a *second-order scalar linear differential equation* with constant coefficients, henceforth written in the following standard form:

$$\boxed{\ddot{x} + 2\alpha\dot{x} + \omega_0^2 x = u(t)} \tag{1.9}$$

For reasons that will be obvious soon, α is called the *damping constant* and $\omega_0 \triangleq 2\pi f_0$ is called the (angular) *frequency* in radians per second (f_0 is the frequency in *hertz*).

In Sec. 1.2, we will show that *any* linear time-invariant second-order circuit can be described by an equation of this form. In particular, we will derive explicit formulas which specify α, ω_0, and $u(t)$ in terms of the circuit parameters and independent sources. These formulas, however, require that we first formulate the *state equation* governing the circuit.

We will introduce the standard form of a state equation for a linear time-invariant second-order circuit by an example.

Example 2 Consider the *RC* op-amp circuit shown in Fig. 1.3,[1] where the *ideal* op amp is assumed to be operating in its linear region. For this

[1] This circuit is an example of a *low-pass filter* to be presented in Chap. 14.

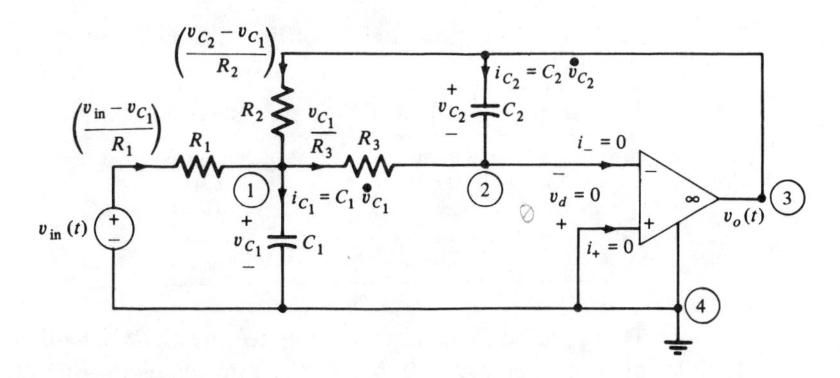

Figure 1.3 Linear time-invariant *RC* op-amp circuit driven by a time-dependent voltage source $v_{\text{in}}(\cdot)$.

circuit, it is easier to write the governing equation in the form of two first-order linear differential equations as follows:

1. Applying KCL at node ① and noting that $v_d = 0$, we obtain

$$C_1\dot{v}_{C1} = \frac{v_{\text{in}} - v_{C1}}{R_1} + \frac{v_{C2} - v_{C1}}{R_2} - \frac{v_{C1}}{R_3} \tag{1.10a}$$

2. Applying KCL at node ②, we obtain

$$C_2\dot{v}_{C2} = -\frac{v_{C1}}{R_3} \tag{1.10b}$$

Equation (1.10) can be recast into the form

$$\begin{bmatrix} \dot{x}_1 \\ \dot{x}_2 \end{bmatrix} = \begin{bmatrix} a_{11} & a_{12} \\ a_{21} & a_{22} \end{bmatrix} \begin{bmatrix} x_1 \\ x_2 \end{bmatrix} + \begin{bmatrix} u_1(t) \\ u_2(t) \end{bmatrix} \tag{1.11}$$

where $x_i \triangleq v_{Ci}$, $\dot{x}_i \triangleq dx_i/dt$, $u_1(t) \triangleq (1/R_1C_1)v_{\text{in}}(t)$, $u_2(t) \triangleq 0$, and

$$\mathbf{A} = \begin{bmatrix} a_{11} & a_{12} \\ a_{21} & a_{22} \end{bmatrix} \triangleq \begin{bmatrix} -\dfrac{1}{C_1}\left(\dfrac{1}{R_1} + \dfrac{1}{R_2} + \dfrac{1}{R_3}\right) & \dfrac{1}{R_2C_1} \\ -\dfrac{1}{R_3C_2} & 0 \end{bmatrix} \tag{1.12}$$

1.2 State Equation and State Variables

The system of two first-order differential equations defined by Eq. (1.11) is called a *state equation* describing the associated circuit. The two variables x_1 and x_2 are called *state variables*.

We will sometimes rewrite Eq. (1.11) into the following compact vector form:

$$\dot{\mathbf{x}}(t) = \mathbf{A}\mathbf{x}(t) + \mathbf{u}(t) \tag{1.13}$$

where $\mathbf{x} \triangleq \begin{bmatrix} x_1 \\ x_2 \end{bmatrix}$ is called the *state vector*, $\mathbf{u}(t) \triangleq \begin{bmatrix} u_1(t) \\ u_2(t) \end{bmatrix}$ is called the *input vector*, and $\mathbf{A}$ is usually called the $\mathbf{A}$ matrix.[2]

Given any initial condition $\mathbf{x}(t_0)$ at $t = t_0$, henceforth called the *initial state*, and the *input* waveform $\mathbf{u}(t)$ for $t \geq t_0$, Eq. (1.13) specifies the *velocity* $\dot{\mathbf{x}}(t)$ at $t = t_0$. This information allows us to calculate numerically the state $\mathbf{x}(t)$ a short time increment Δt later, namely, $\mathbf{x}(t_0 + \Delta t)$. Taking the calculated $\mathbf{x}(t_0 + \Delta t)$ as the new initial state, we can calculate $\mathbf{x}(t_0 + 2\,\Delta t)$. Iterating this process, we can eventually calculate $\mathbf{x}(t)$ for all $t \geq t_0$. Moreover, we can determine the solution $\mathbf{x}(t)$ to any desired accuracy by choosing a sufficiently small Δt. This important property of the state equation is in fact the basis of every computer program for solving state equations. Hence, *the state equation specifies implicitly the future time evolution, or dynamics, of the circuit.*

Another motivation for studying *state equations* is that, unlike the scalar equation from Sec. 1.1, the basic concept of *state equations* and *state variables* is applicable also for *time-varying* and *nonlinear circuits*.

Given the state equation, the following formulas allow us to recast it into an equivalent scalar differential equation.

Transforming a state equation into a second-order scalar differential equation *State equation* (1.11) can be transformed into the *scalar* differential equation

$$\ddot{x}_1 - T\dot{x}_1 + \Delta x_1 = u_a(t) \qquad \text{if } a_{12} \neq 0 \tag{1.14a}$$

or the *scalar* differential equation

$$\ddot{x}_2 - T\dot{x}_2 + \Delta x_2 = u_b(t) \qquad \text{if } a_{21} \neq 0 \tag{1.14b}$$

where[3]

$$T \triangleq (a_{11} + a_{22}) \tag{1.15a}$$

$$\Delta \triangleq a_{11}a_{22} - a_{12}a_{21} \tag{1.15b}$$

$$u_a(t) \triangleq -a_{22}\,u_1(t) + a_{12}\,u_2(t) + \dot{u}_1(t) \tag{1.15c}$$

$$u_b(t) \triangleq a_{21}\,u_1(t) - a_{11}\,u_2(t) + \dot{u}_2(t) \tag{1.15d}$$

PROOF If both a_{12} and a_{21} are zero, then Eq. (1.11) reduces to two *uncoupled* first-order equations and there is no point in writing a second-order equation.

So consider the case where $a_{12} \neq 0$. Hence by the first equation in Eq.

[2] This symbol is *not* related to the reduced incidence matrix in Chap. 1.

[3] In linear algebra, T and Δ are called the *trace* and the *determinant* of $\mathbf{A}$, respectively.

(1.11), x_2 affects the velocity of x_1. Differentiating the first equation in Eq. (1.11) with respect to time, we obtain

$$\ddot{x}_1 = a_{11}\dot{x}_1 + a_{12}\dot{x}_2 + \dot{u}_1(t) \tag{1.16}$$

Substituting $\dot{x}_1$ and $\dot{x}_2$ from Eq. (1.11) into Eq. (1.16) and simplifying, we obtain

$$\ddot{x}_1 = (a_{11}^2 + a_{12}a_{21})x_1 + (a_{11} + a_{22})a_{12}x_2 + [a_{11}\, u_1(t) + a_{12}\, u_2(t)] + \dot{u}_1(t) \tag{1.17}$$

If $a_{12} \neq 0$, x_2 appears in Eq. (1.17). Using the first equation of Eq. (1.11), we obtain

$$a_{12}x_2 = [\dot{x}_1 - a_{11}x_1 - u_1(t)] \tag{1.18}$$

Substituting Eq. (1.18) for x_2 in Eq. (1.17) and simplifying, we obtain Eq. (1.14*a*).

If $a_{21} \neq 0$, a similar proof yields Eq. (1.14*b*). ∎

Example To illustrate the application of Eq. (1.14), let us substitute Eq. (1.12)—which describes the *RC* op-amp circuit of Fig. 1.3—into Eq. (1.15):

$$T = -\frac{1}{C_1}\left(\frac{1}{R_1} + \frac{1}{R_2} + \frac{1}{R_3}\right) \tag{1.19a}$$

$$\Delta = \frac{1}{R_2R_3C_1C_2} \tag{1.19b}$$

It follows from Eq. (1.14*a*) that this circuit is described by the *scalar* differential equation

$$\ddot{x}_1 + \frac{1}{C_1}\left(\frac{1}{R_1} + \frac{1}{R_2} + \frac{1}{R_3}\right)\dot{x}_1 + \frac{1}{R_2R_3C_1C_2}\, x_1 = \dot{u}_1(t) \tag{1.20}$$

Comparing Eq. (1.20) with Eq. (1.9), we obtain

$$\alpha \triangleq \frac{1}{2C_1}\left(\frac{1}{R_1} + \frac{1}{R_2} + \frac{1}{R_3}\right) \qquad \omega_0 \triangleq \frac{1}{\sqrt{R_2R_3C_1C_2}} \tag{1.21}$$

Equation (1.14) allows us to write, trivially, the scalar linear differential equation for *any* linear time-invariant second-order circuit provided we know how to formulate its state equation. This observation provides us with one motivation for our next section. An even more fundamental reason for writing state equations is that the same equation formulation procedure can be easily generalized for *nonlinear* circuits, as will be clear in Sec. 4.

1.3 Linear State Equation Formulation

Any linear time-invariant circuit containing two energy-storage elements can be drawn in one of the three standard configurations shown in Fig. 1.4, where N denotes a *two-port* made of an arbitrary interconnection of two-terminal, multiterminal, and multiport *linear* time-invariant resistors, and independent sources. Our goal is to formulate a system of two first-order differential equations with *capacitor voltages* and *inductor currents* chosen as state variables.

A. Two-capacitor configuration Consider the circuit in Fig. 1.4*a*. Each capacitor current is given by

$$i_{Cj} = C_j \dot{v}_{Cj} = -i_j \qquad j = 1, 2 \tag{1.22}$$

Hence,

$$\dot{v}_{Cj} = -\frac{1}{C_j} i_j \qquad j = 1, 2 \tag{1.23}$$

An inspection of the standard form in Eq. (1.11) shows that Eq. (1.23) is not yet in the form of a state equation because the variables i_1 and i_2 on the right-hand side are *not* state variables. Consequently, it is necessary to express i_1 and i_2 in terms of $v_{C1} = v_1$ and $v_{C2} = v_2$. Since v_1 and v_2 are *independent* variables, this task is equivalent to solving for the port currents i_1 and i_2 when N is terminated by *voltage sources* v_1 and v_2, as shown in Fig. 1.5*a*.

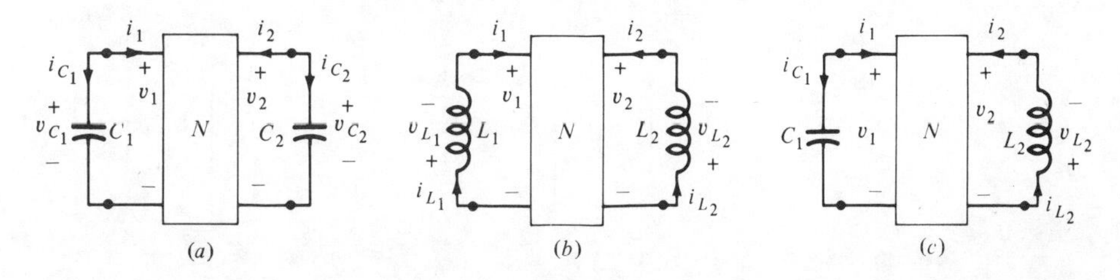

Figure 1.4 Three standard configurations for second-order circuits. The capacitor voltage polarity is chosen so that it coincides with the port voltage. The inductor current direction is chosen so that it coincides with the port current.

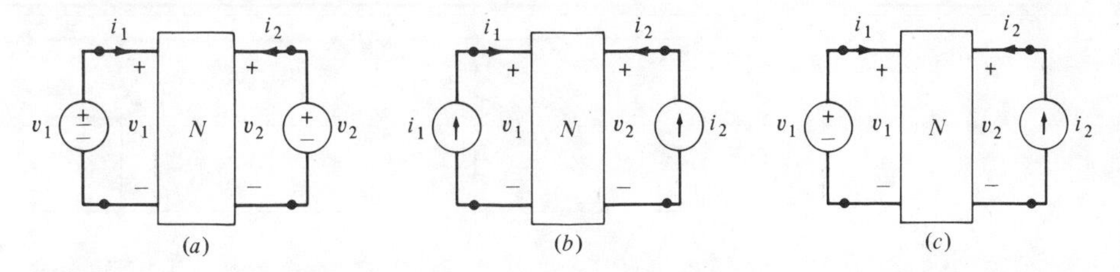

Figure 1.5 Circuits for deriving the three representations needed in the state equation formulation.

Since N contains only linear time-invariant resistors and independent sources, it follows from Eq. (4.41) of the *two-port representation theorem* from Chap. 5 that provided the circuit in Fig. 1.5*a* is *uniquely solvable*, i_1 and i_2 are related to v_1 and v_2 as follows:

$$i_1(t) = g_{11}v_1(t) + g_{12}v_2(t) + i_{s1}(t) \tag{1.24a}$$

$$i_2(t) = g_{21}v_1(t) + g_{22}v_2(t) + i_{s2}(t) \tag{1.24b}$$

where g_{11}, g_{12}, g_{21}, and g_{22} are real constants and $i_{s1}(t)$ and $i_{s2}(t)$ are time functions depending on the *independent* sources inside N.

Substituting Eq. (1.24) in place of i_1 and i_2 in Eq. (1.23), we obtain the following state equation for the two-capacitor configuration:

State equation for two-capacitor configuration

$$\begin{bmatrix} \dot{v}_{C1} \\ \dot{v}_{C2} \end{bmatrix} = \begin{bmatrix} -\dfrac{g_{11}}{C_1} & -\dfrac{g_{12}}{C_1} \\ -\dfrac{g_{21}}{C_2} & -\dfrac{g_{22}}{C_2} \end{bmatrix} \begin{bmatrix} v_{C1} \\ v_{C2} \end{bmatrix} + \begin{bmatrix} -\dfrac{1}{C_1} i_{s1}(t) \\ -\dfrac{1}{C_2} i_{s2}(t) \end{bmatrix} \tag{1.25}$$

Example Let us derive the state equation of the *RC* op-amp circuit in Fig. 1.3 using the above systematic procedure.

The first step is to replace capacitors C_1 and C_2 by voltage sources v_1 and v_2 as shown in Fig. 1.6. This circuit has already been analyzed in Chap. 5 (Fig. 4.13) and the resulting two-port representation is given by Eq. (4.45) of Chap. 5, namely,

$$\begin{bmatrix} i_1 \\ i_2 \end{bmatrix} = \begin{bmatrix} \dfrac{1}{R_1} + \dfrac{1}{R_2} + \dfrac{1}{R_3} & -\dfrac{1}{R_2} \\ \dfrac{1}{R_3} & 0 \end{bmatrix} \begin{bmatrix} v_1 \\ v_2 \end{bmatrix} + \begin{bmatrix} -\dfrac{1}{R_1} v_{\text{in}}(t) \\ 0 \end{bmatrix} \tag{1.26}$$

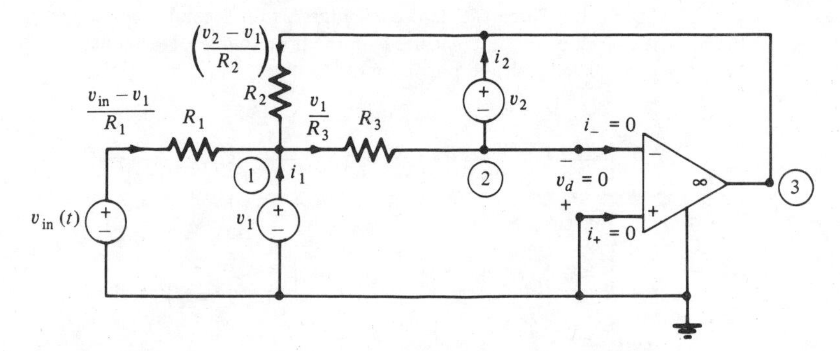

Figure 1.6 The resistive two-port N associated with the *RC* op-amp circuit in Fig. 1.3.

Comparing Eq. (1.26) with Eq. (1.24), we identify the following:

$$g_{11} = \left(\frac{1}{R_1} + \frac{1}{R_2} + \frac{1}{R_3}\right) \qquad g_{12} = -\frac{1}{R_2} \qquad g_{21} = \frac{1}{R_3} \qquad g_{22} = 0$$

$$i_{s1}(t) = -\frac{1}{R_1} v_{\text{in}}(t) \qquad \text{and} \qquad i_{s2}(t) = 0$$

Substituting these coefficients into Eq. (1.25), we obtain the same state equation obtained earlier in Eqs. (1.11) and (1.12), as we should.

B. Two-inductor configuration By a dual procedure, we replace the two inductors in Fig. 1.4*b* by two current sources $i_1 = i_{L1}$ and $i_2 = i_{L2}$ as shown in Fig. 1.5*b*. Assuming the resulting resistive circuit is uniquely solvable, we can express the voltages v_1 and v_2 in the form of Eq. (4.42) from Chap. 5:

$$v_1(t) = r_{11} i_1(t) + r_{12} i_2(t) + v_{s1}(t) \tag{1.27a}$$

$$v_2(t) = r_{21} i_1(t) + r_{22} i_2(t) + v_{s2}(t) \tag{1.27b}$$

The associated state equation is then given by

State equation for two-inductor configuration

$$\begin{bmatrix} \dot{i}_{L1} \\ \dot{i}_{L2} \end{bmatrix} = \begin{bmatrix} \dfrac{-r_{11}}{L_1} & \dfrac{-r_{12}}{L_1} \\ \dfrac{-r_{21}}{L_2} & \dfrac{-r_{22}}{L_2} \end{bmatrix} \begin{bmatrix} i_{L1} \\ i_{L2} \end{bmatrix} + \begin{bmatrix} -\dfrac{1}{L_1} v_{s1}(t) \\ -\dfrac{1}{L_2} v_{s2}(t) \end{bmatrix} \tag{1.28}$$

C. Capacitor-inductor configuration In this case, we replace capacitor C_1 and inductor L_2 in Fig. 1.4*c* by voltage source v_1 and current source i_2, respectively, as shown in Fig. 1.5*c*. Assuming the resulting resistive circuit is uniquely solvable, we can express the current i_1 and the voltage v_2 in the form of Eq. (4.43) from Chap. 5:

$$i_1(t) = h'_{11} v_1(t) + h'_{12} i_2(t) + i_{s1}(t) \tag{1.29a}$$

$$v_2(t) = h'_{21} v_1(t) + h'_{22} i_2(t) + v_{s2}(t) \tag{1.29b}$$

The associated state equation is given by

State equation for capacitor-inductor configuration

$$\begin{bmatrix} \dot{v}_{C1} \\ \dot{i}_{L2} \end{bmatrix} = \begin{bmatrix} \dfrac{-h'_{11}}{C_1} & \dfrac{-h'_{12}}{C_1} \\ \dfrac{-h'_{21}}{L_2} & \dfrac{-h'_{22}}{L_2} \end{bmatrix} \begin{bmatrix} v_{C1} \\ i_{L2} \end{bmatrix} + \begin{bmatrix} -\dfrac{1}{C_1} i_{s1}(t) \\ -\dfrac{1}{L_2} v_{s2}(t) \end{bmatrix} \tag{1.30}$$

Summary:

The state equation of any second-order linear time-invariant circuit as depicted in the general configurations in Fig. 1.4 can be obtained systematically in three steps:

State equation formulation algorithm

Step 1. Replace each capacitor C_j by an independent voltage source v_j and each inductor L_j by an independent current source i_j, as shown in Fig. 1.5.

Step 2. Use any method from *Chap. 5* to solve for the *unknown* port current or port voltage indicated in Fig. 1.5. (Be sure to use associated reference directions.) If the associated resistive circuit is *uniquely solvable*, the equations can always be recast into the required form of Eq. (1.24), (1.27), or (1.29).

Step 3. Write the state equation using Eq. (1.25), (1.28), or (1.30).

REMARKS

1. It follows from the *two-port representation theorem* from Chap. 5 that the coefficients g_{ij}, r_{ij}, and h'_{ij} in Eqs. (1.24), (1.27), and (1.29), respectively, are *real* numbers which depend *only* on the resistors inside N and *not* the independent sources.
2. It follows from the *two-port representation theorem* from Chap. 5 that the *input* vector $\mathbf{u}(t)$ in the state equation (1.13) is a *constant* vector if N contains only *dc* independent sources and is identically zero if N contains no independent sources.

Exercise

(*a*) Use the above general procedure to derive the state equation for the two circuits in Fig. 1.1.

(*b*) Use Eq. (1.14*b*) to transform the state equations from (*a*) into a second-order scalar differential equation.

2 ZERO-INPUT RESPONSE

In this and the next section, we will consider only second-order linear time-invariant circuits containing *no independent* sources. To give "life" to the circuit, we assume the capacitor has some initial voltage $v_C(t_0) \neq 0$ and/or the inductor has some initial current $i_L(t_0) \neq 0$, at some initial time t_0. The corresponding solution for $t \geq t_0$ is, by definition, the *zero-input response*.[4] In

[4] To avoid redundancy, the complementary *zero-state response* defined in Chap. 6 will be presented in its full generality in Chap. 10.

the context of the mechanical analog in Fig. 1.2, this amounts to stretching the spring initially from its equilibrium position and then letting it go at $t = t_0$. The resulting "motion" in this case is the zero-input response of the stretched spring.

We study zero-input response first, and in great detail, not only because it is easier to understand in view of our experience with the mechanical analog, but also because most of the "qualitative features" of zero-input responses are *preserved* even in *nonlinear* circuits, as we will show in Sec. 5.

Since the circuit contains no *independent* sources, it follows from Remark 2 from the preceding section that $i_{s1}(t) = i_{s2}(t) = 0$ in Eq. (1.25), $v_{s1}(t) = v_{s2}(t) = 0$ in Eq. (1.28), and $i_{s1}(t) = v_{s2}(t) = 0$ in Eq. (1.30). Hence, the state equations (1.25), (1.28), and (1.30) reduce to the form

$$\begin{bmatrix} \dot{x}_1 \\ \dot{x}_2 \end{bmatrix} = \begin{bmatrix} a_{11} & a_{12} \\ a_{21} & a_{22} \end{bmatrix} \begin{bmatrix} x_1 \\ x_2 \end{bmatrix} \quad \text{or} \quad \dot{\mathbf{x}} = \mathbf{A}\mathbf{x} \tag{2.1}$$

It follows from Eq. (1.14) that the associated second-order scalar differential equation reduces to the form

$$\ddot{x} + 2\alpha\dot{x} + \omega_0^2 x = 0 \tag{2.2}$$

where

$$2\alpha \triangleq -T = -(a_{11} + a_{22}) \tag{2.3a}$$

$$\omega_0^2 \triangleq \Delta = a_{11}a_{22} - a_{12}a_{21} \tag{2.3b}$$

Equation (2.3) gives the relationships between the damping constant α and the natural frequency ω_0 [defined earlier in Eq. (1.9)] in terms of the coefficients of the **A** matrix and in terms of the two parameters T and Δ. We will see later that whereas the parameters α and ω_0 have nice *physical* interpretations, the parameters T and Δ have natural *geometric* interpretations.

Our objective in this section is to study the *zero-input response* associated with both Eqs. (2.1) and (2.2). We begin (in Sec. 2.1) with Eq. (2.2) because the same material has already been treated in elementary physics. To avoid redundancy, we assume that $\alpha \geq 0$ and $\omega_0^2 > 0$ in Eq. (2.2). The general case for arbitrary T and Δ (and hence arbitrary α and ω_0^2) will be treated in Sec. 2.2 in connection with Eq. (2.1).

2.1 Determining Zero-Input Response from $\ddot{x} + 2\alpha\dot{x} + \omega_0^2 x = 0$; $\alpha \geq 0$ and $\omega_0^2 > 0$

Equation (2.2) is a second-order linear homogeneous differential equation with constant coefficients. The *characteristic polynomial* for this differential equation is

$$p(s) \triangleq s^2 + 2\alpha s + \omega_0^2 \tag{2.4}$$

The zeros of the characteristic polynomial $p(s)$ are called the *natural frequencies* of the circuit; they are[5]

$$\left.\begin{matrix} s_1 \\ s_2 \end{matrix}\right\} = -\alpha \pm \sqrt{\alpha^2 - \omega_0^2} = \begin{cases} -\alpha \pm \alpha_d & \text{if } \alpha > \omega_0 > 0 \\ -\alpha \pm j\omega_d & \text{if } 0 < \alpha < \omega_0 \\ -\alpha & \text{if } \alpha = \omega_0 > 0 \end{cases} \tag{2.5}$$

where

$$\alpha_d \triangleq \sqrt{\alpha^2 - \omega_0^2} \tag{2.6}$$

$$\omega_d \triangleq \sqrt{\omega_0^2 - \alpha^2} \tag{2.7}$$

From an elementary course on differential equations, we have learned that depending on the relative values of α and ω_0, the solution (zero-input response) of Eq. (2.2) can exhibit one of the four qualitatively distinct forms listed in Table 2.1, assuming that $\alpha \geq 0$ and $\omega_0^2 > 0$. These four explicit solutions can be verified by direct substitution into Eq. (2.2). In each of these four cases, the corresponding pair of constants (k_1, k_2), (k, k'), (k, θ), and (k, θ) are uniquely determined by the initial conditions $x(0)$ and $\dot{x}(0)$.

Using Eq. (2.5), the four responses in Table 2.1 can also be classified in terms of the *natural frequencies* s_1 and s_2. Since s_1 and s_2 can be either real or complex, it is instructive to represent each natural frequency as a point in the Re(s) versus the Im(s) plane,[6] henceforth called the s plane. The locations of s_1 and s_2 for each of the four responses are sketched in Table 2.1.

The *overdamped response* in Table 2.1 is a sum of two *damped exponentials* with time constants equal to $1/s_1$ and $1/s_2$, respectively. As α increases, the damping becomes greater and the response will reach zero sooner.

On the other hand, if we keep ω_0^2 fixed and if we decrease α beyond the critical value $\alpha = \omega_0$, we would obtain a "ringing" *underdamped response* consisting of a sine wave whose amplitude decays exponentially.

In the limiting case where $\alpha = 0$, the waveform becomes a pure sinusoid with an angular oscillation frequency equal to ω_0.[7]

In terms of the mechanical system in Fig. 1.2, note that $\alpha = B/2M \geq 0$ is proportional to the coefficient of friction B. Assuming that $f_s(t) = 0$ for all $t \geq 0$ and that the friction is sufficiently large (overdamped case), we expect from experience that given any initial displacement of the block from its equilibrium position, the block will eventually tend to the equilibrium monotonically: The greater the friction, the sooner the block settles to equilibrium.

[5] We assume the readers are familiar with the solutions of linear differential equations. To avoid confusing the standard symbol $i \triangleq \sqrt{-1}$ with the current i, throughout this book, we use j to denote $\sqrt{-1}$.

[6] We denote the real and imaginary part of a complex number z by Re(z) and Im(z), respectively.

[7] It is instructive to glance at Fig. 2.4 now which shows the natural frequencies s_1 and s_2 traversed along a semicircular path as we increase the damping coefficient from $\alpha = 0$ to $\alpha = \omega_0$, where ω_0 is held fixed in this diagram. Note that s_1 and s_2 move *horizontally* in opposite directions as α increases from $\alpha = \omega_0$ to $\alpha = \infty$.

Table 2.1 Zero-input response† of $\ddot{x} + 2\alpha\dot{x} + \omega_0^2 x = 0$; $\alpha \geq 0$ and $\omega_0^2 > 0$

1. *Overdamped Response* ($\alpha > \omega_0 > 0$)

$x(t) = k_1 e^{s_1 t} + k_2 e^{s_2 t}$

s_1 and s_2 are real numbers

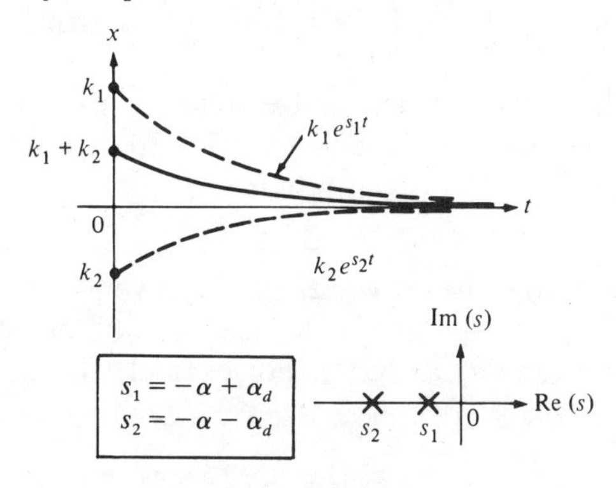

2. *Critically Damped Response* ($\alpha = \omega_0 > 0$)

$x(t) = (k + k't)e^{-\alpha t}$

$s_1 = s_2$ is a real number

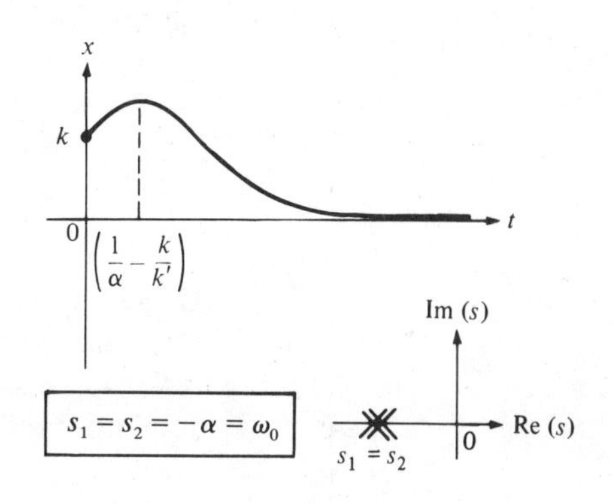

3. *Underdamped Response* ($0 < \alpha < \omega_0$)

$x(t) = ke^{-\alpha t}\cos(\omega_d t - \theta)$

s_1 and s_2 are complex-conjugate numbers

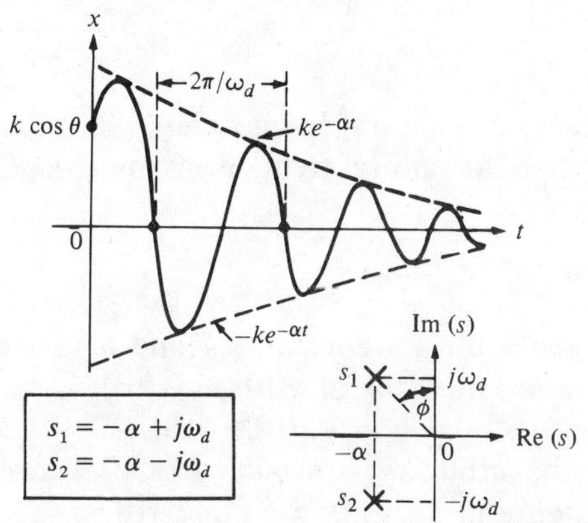

4. *Lossless Response* ($\alpha = 0$, $\omega_0 > 0$)

$x(t) = k\cos(\omega_0 t - \theta)$

s_1 and s_2 are complex-conjugate imaginary numbers

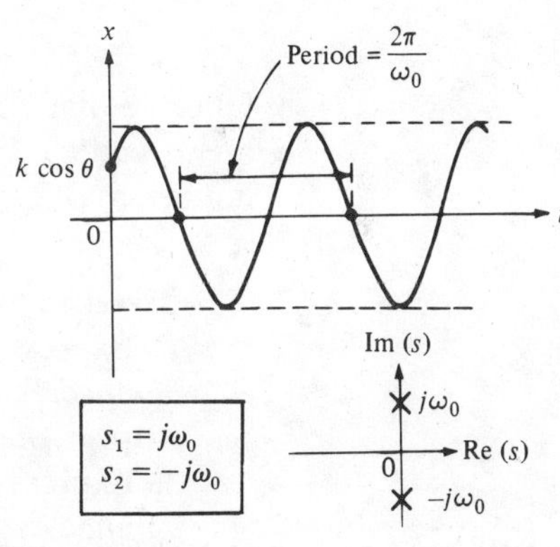

† The solution $x(t) = k_1 e^{s_1 t} + k_2 e^{s_2 t}$ in Table 2.1 is valid for any equation where $s_1 \neq s_2$. In particular, it is valid for the *undamped* and *lossless* case. However, the forms of the solution given in Table 2.1 are more convenient for plotting purposes.

As we decrease the friction (say by lubricating the block), there comes a point beyond which (underdamped case) the block starts swinging back and forth with decreasing amplitude. In the limit where the friction tends to zero (lossless case) we expect the block to *oscillate* with a simple harmonic (sinusoidal) motion of frequency $\omega_0 = \sqrt{M/K}$.

The above physical interpretations justify our calling α and ω_0 the *damping constant* and (angular) *frequency*, respectively.

Throughout this section, we have assumed that both $\alpha \geq 0$ and $\omega_0^2 > 0$. There exist circuits of course where $\alpha < 0$ and/or $\omega_0^2 < 0$. Although we could carry out a similar analysis of Eq. (2.4) for this case, it is more illuminating to use the state equilibrium (2.1), which will be presented in Sec. 2.2.

Exercise Show that the natural frequencies of Eq. (2.2) have negative real parts if and only if $\alpha > 0$ and $\omega_0^2 > 0$. (Note for the circuit of Fig. 1.1, $R > 0$, $L > 0$, and $C > 0 \Rightarrow \alpha > 0$ and $\omega_0^2 > 0$.)

A. Determination of arbitrary constants The two arbitrary constants associated with each zero-input response in Table 2.1 can be determined by setting $t = 0$ in the expression for $x(t)$ and $\dot{x}(t)$. For example, in the *overdamped response* case, we have

$$x(0) = k_1 + k_2 \tag{2.8a}$$

$$\dot{x}(0) = k_1 s_1 + k_2 s_2 \tag{2.8b}$$

Solving Eq. (2.8) for k_1 and k_2, we obtain

$$k_1 = \frac{\dot{x}(0) - x(0)s_2}{s_1 - s_2} \qquad k_2 = \frac{\dot{x}(0) - x(0)s_1}{s_2 - s_1} \tag{2.9}$$

Observe that when both natural frequencies s_1 and s_2 are *real* numbers, so too are k_1 and k_2. When the natural frequencies are *complex* numbers, then $s_2 = \bar{s}_1$ implies that $k_2 = \bar{k}_1$.

Exercises

1. Express the arbitrary constants k and k' associated with the *critically damped response* in terms of $x(0)$ and $\dot{x}(0)$.
Answer: $k = x(0)$ and $k' = \alpha x(0) + \dot{x}(0)$

2. Express the arbitrary constants k and θ associated with the *under-damped response* in terms of $x(0)$ and $\dot{x}(0)$.
Answer: $\theta = \tan^{-1}\left[\dfrac{\alpha}{\omega_d} + \dfrac{\dot{x}(0)}{\omega_d x(0)}\right]$ and $k = x(0) \sec \theta$.

3. Express the arbitrary constants k and θ associated with the *lossless response* in terms of $x(0)$ and $\dot{x}(0)$.
Answer: $\theta = \tan^{-1}\left[\dfrac{\dot{x}(0)}{\omega_0 x(0)}\right]$ and $k = x(0) \sec \theta$.

B. Physical interpretation via parallel *RLC* circuits Consider the parallel *RLC* circuit in Fig. 2.1. Since Fig. 2.1 is obtained by setting $i_s(t) = 0$ in Fig. 1.1*a*, it follows from Eq. (1.9) that this circuit is described by

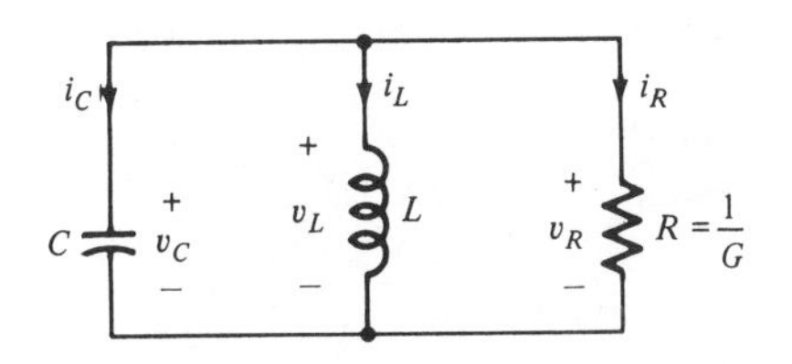

Figure 2.1 Parallel RLC circuit with $R > 0$, $L > 0$, and $C > 0$.

$$\ddot{i}_L + 2\alpha \dot{i}_L + \omega_0^2 i_L = 0 \tag{2.10}$$

where

$$\alpha = \frac{1}{2RC} = \frac{G}{2C} \quad \text{and} \quad \omega_0 = \frac{1}{\sqrt{LC}}$$

At the critical value $\alpha = \omega_0$, or $G/2C = 1/\sqrt{LC}$, we find $G = 2\sqrt{C/L}$. Hence, the zero-input response is as listed in Table 2.2.

In the limiting case where $G = 0$ ($R = \infty$), the circuit reduces to an LC *harmonic oscillator*. In this case, the capacitor voltage $v_C(t) = v_L(t) = L\dot{i}_L$ becomes a pure sinusoid:

$$v_C(t) = -(\omega_0 Lk)\sin(\omega_0 t - \theta) \tag{2.11}$$

The *total instantaneous energy* $w_{LC}(t)$ stored in the inductor and in the capacitor is [recall Eqs. (2.20) and (2.21) from Chap. 6] given by

$$\begin{aligned} w_{LC}(t) &= \tfrac{1}{2}Li_L^2(t) + \tfrac{1}{2}Cv_C^2(t) \\ &= \tfrac{1}{2}Lk^2\cos^2(\omega_0 t - \theta) + \tfrac{1}{2}C(\omega_0 Lk)^2\sin^2(\omega_0 t - \theta) \\ &= \tfrac{1}{2}Lk^2[\cos^2(\omega_0 t - \theta) + \sin^2(\omega_0 t - \theta)] = \tfrac{1}{2}Lk^2 \end{aligned} \tag{2.12}$$

As expected, the stored energy remains constant:

$$w_{LC}(t) = w_{LC}(0) = \tfrac{1}{2}Li_L^2(0) + \tfrac{1}{2}Cv_C^2(0) \tag{2.13}$$

for all $t \geq 0$. Hence, as the energy stored in the magnetic field of the inductor increases, the energy stored in the electric field of the capacitor decreases, and vice versa. This implies that energy is continuously being transferred back and forth from the inductor to the capacitor, without any loss.

Table 2.2 Zero-input response of the circuit in Fig. 2.1

1. *Overdamped response* $\left(\alpha > \omega_0 \Leftrightarrow G > 2\sqrt{\frac{C}{L}}\right)$	2. *Critically damped response* $\left(\alpha = \omega_0 \Leftrightarrow G = 2\sqrt{\frac{C}{L}}\right)$
$i_L(t) = k_1 e^{s_1 t} + k_2 e^{s_2 t}$	$i_L(t) = (k + k't)e^{-\alpha t}$
3. *Underdamped response* $\left(\alpha < 0 \Leftrightarrow G < 2\sqrt{\frac{C}{L}}\right)$	4. *Lossless response* $(\alpha = 0 \Leftrightarrow G = 0)$
$i_L(t) = ke^{-\alpha t}\cos(\omega_d t - \theta)$	$i_L(t) = k\cos(\omega_0 t - \theta)$

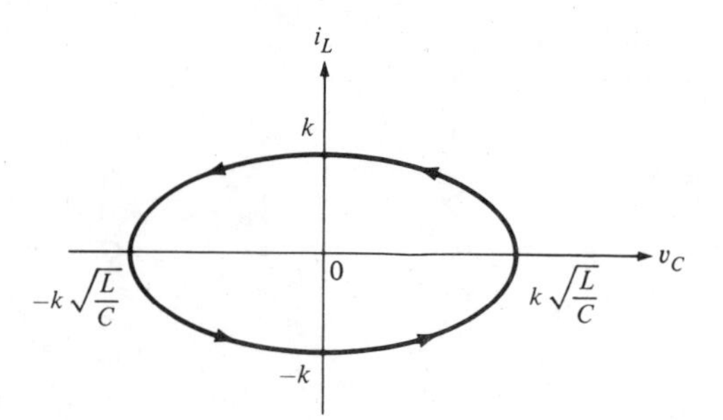

Figure 2.2 v_C-i_L locus in the *lossless* case (drawn with $L > C$).

The above energy swapping mechanism is best depicted by plotting the locus of

$$i_L(t) = k \cos(\omega_0 t - \theta) \tag{2.14a}$$

and[8]

$$v_C(t) = -k\sqrt{\frac{L}{C}} \sin(\omega_0 t - \theta) \tag{2.14b}$$

in the v_C-i_L plane, as shown in Fig. 2.2. To see that this locus is indeed an ellipse, note from Eq. (2.14) that

$$\left(\frac{v_C}{k\sqrt{L/C}}\right)^2 + \left(\frac{i_L}{k}\right)^2 = 1 \tag{2.15}$$

Figure 2.2 shows clearly that the *lossless circuit* oscillates along an elliptical locus whose semi-axes depend on the parameters L and C and the initial state $[v_C(t_0), i_L(t_0)]$.

Now consider the *underdamped* case, where $G > 0$. In this case, during each cycle the resistor dissipates part of the energy being transferred between the inductor and the capacitor. Hence, the total energy left in the magnetic and electric fields gradually diminishes. This translates into an exponential decay of the oscillation.

The above energy dissipation mechanism is best depicted by plotting the locus of

$$i_L(t) = k e^{-\alpha t} \cos(\omega_d t - \theta) \tag{2.16a}$$

and[9]

$$\begin{aligned} v_C(t) &= -kL e^{-\alpha t}[\omega_d \sin(\omega_d t - \theta) + \alpha \cos(\omega_d t - \theta)] \\ &= -kL e^{-\alpha t}[(\omega_0 \cos \phi) \sin(\omega_d t - \theta) - (\omega_0 \sin \phi) \cos(\omega_d t - \theta)] \\ &= -k\sqrt{\frac{L}{C}}\, e^{-\alpha t}[\sin(\omega_d t - \theta - \phi)] \end{aligned} \tag{2.16b}$$

[8] Here, we substitute $\omega_0 L = \sqrt{L/C}$ in Eq. (2.11).

[9] Here we use the relationships [recall Eq. (2.6)] $\omega_d = \omega_0 \cos \phi$ and $\alpha = -\omega_0 \sin \phi$, where ϕ is the angle subtended between the vertical axis and the vector from the origin to s_1 in Table 2.1.

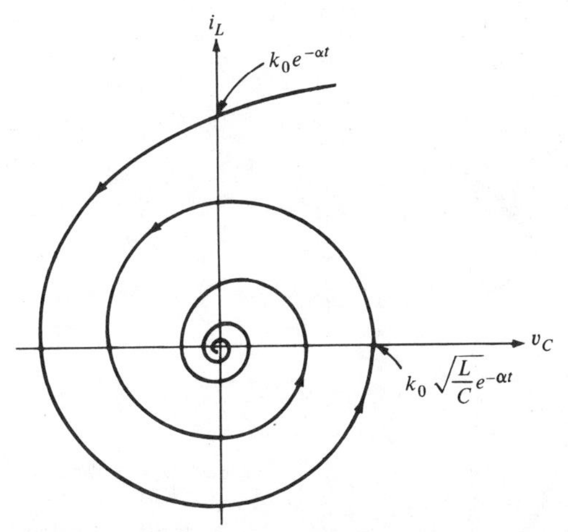

Figure 2.3 v_C-i_L locus in the *underdamped* case (drawn with $L > C$).

in the v_C-i_L plane, as shown in Fig. 2.3. To see that this locus is a shrinking spiral, note from Eq. (2.16) that

$$\left(\frac{v_C}{k_0\sqrt{L/C}\,\mathrm{e}^{-\alpha t}}\right)^2 + \left(\frac{i_L}{k_0 \mathrm{e}^{-\alpha t}}\right)^2 = 1 \tag{2.17}$$

where[10] $\quad k_0 \triangleq k\{1 + \frac{1}{2}[\cos 2(\omega_d t - \theta) - \cos 2(\omega_d t - \theta - \phi)]\}^{1/2}$

We can interpret Eq. (2.17) as an ellipse whose semi-axes shrink continuously at an exponential rate equal to α.

Exercise Derive Eq. (2.17) with k_0 as defined above.

C. Quality factor Table 2.1 shows that the damping constant α determines the rate of exponential decay when $\alpha > 0$. In the *underdamped* case, the frequency of oscillation is given by $\omega_d = \sqrt{\omega_0^2 - \alpha^2}$ rad and *not* ω_0, and hence the amount in which the oscillation amplitude decreases per period depends on both ω_0 and α. This information is often characterized by a single number

$$Q \triangleq \frac{\omega_0}{2\alpha} \tag{2.18}$$

called the *quality factor*, or simply *Q-factor*, of the associated circuit.

Observe that, for fixed ω_0, Q increases as α decreases: in particular, $Q = \infty$ when $\alpha = 0$ (lossless case) and $Q = \frac{1}{2}$ when $\alpha = \omega_0$ (critically damped case). It follows from Table 2.1 that the circuit exhibits an *underdamped response* for $\frac{1}{2} < Q < \infty$ and an *overdamped response* or $0 < Q < \frac{1}{2}$ (assuming $\alpha > 0$). In Fig. 2.4, the values of Q are related to the locations of the natural frequencies s_1 and s_2 for fixed ω_0.

[10] Note that whereas θ in Eq. (2.16) is an arbitrary phase angle whose value is fixed by the *initial condition*, the angle ϕ is fixed by the circuit parameters. These symbols will be used consistently in subsequent examples as well.

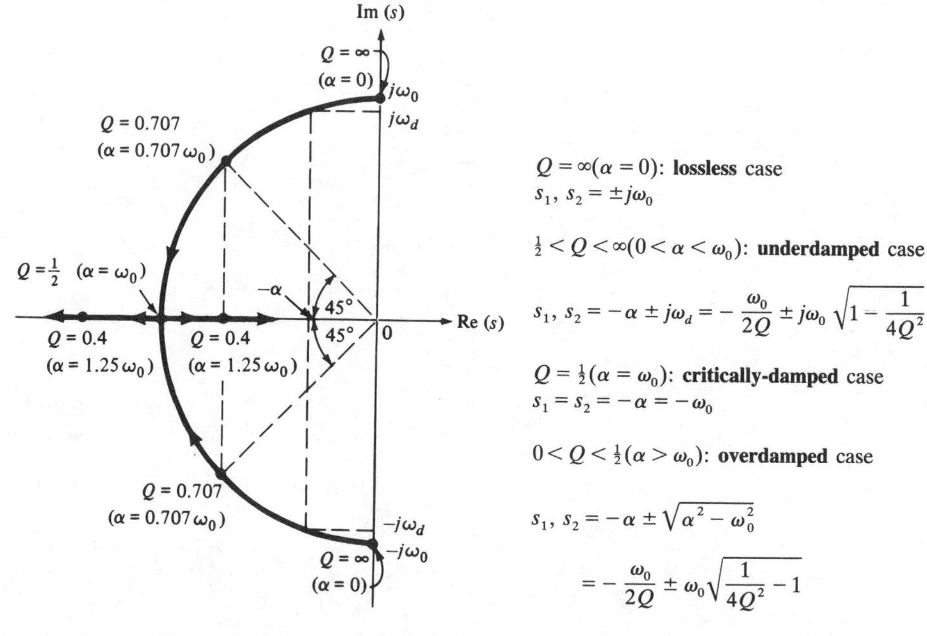

Figure 2.4 Locus of s_1 and s_2 as Q decreases from $Q = \infty$ (as α increases from 0) to $Q = 0$ ($\alpha = \infty$) with ω_0 held fixed.

Example For the parallel *RLC* circuit in Fig. 2.1, we have

$$Q = \frac{\omega_0 C}{G} = \frac{R}{\omega_0 L} = R\sqrt{\frac{C}{L}} = \frac{1}{G}\sqrt{\frac{C}{L}} \tag{2.19}$$

Note that if we fix the values of L and C so that ω_0 is held constant, then Q increases as we increase R. In particular $Q \to \infty$ as $R \to \infty$ (*lossless* case).

In Chap. 9, we will see that the quality factor Q is a useful *figure of merit* for characterizing the selectivity of a *resonant circuit*, of which Fig. 2.1 is a simple example.

Exercise (Interpretation of Q) In practical lumped circuits, a Q on the order of a few hundred is available. In mechanical filters, Q may range from 10^3 to 2×10^4 in the frequency range of 0.1 to 600 kHz. In crystal filters, Q may range from 10^4 to 3×10^5 in the frequency range of 2 to 200 MHz. At microwave frequencies using superconductors, Q on the order of 10^{11} have been achieved.

To get a feeling of the meaning of Q, assume $Q \gg 1$ (hence take $\omega_d \approx \omega_0$) and show that the amplitude of the damped oscillation decreases to 4.3 percent of its initial value after Q periods of $2\pi/\omega_0$ seconds.

2.2 Determining Zero-Input Response from $\dot{\mathbf{x}} = \mathbf{A}\mathbf{x}$; $\Delta \neq \frac{1}{4}T^2$

It is often more revealing to obtain the *zero-input response* by solving the *state equation* (2.1) directly. The main advantage of this approach is that the same procedure can be easily generalized for linear time-invariant circuits containing any number of inductors and capacitors.

The first step in solving Eq. (2.1) is to find the *eigenvalues* λ_1 and λ_2 of the matrix **A**: They are the solutions of the equation[11]

$$\det\begin{bmatrix} a_{11} - \lambda & a_{12} \\ a_{21} & a_{22} - \lambda \end{bmatrix} = \lambda^2 - T\lambda + \Delta = 0 \tag{2.20}$$

where $T \triangleq a_{11} + a_{22}$ and $\Delta \triangleq a_{11}a_{22} - a_{12}a_{21}$ have been defined earlier in Eqs. (1.15*a*) and (1.15*b*). It follows from Eq. (2.3) that the solution of Eq. (2.20) are precisely the *natural frequencies* s_1 and s_2 of the circuit, namely,[12]

$$\lambda_1 = \frac{T}{2} + \sqrt{\frac{1}{4}T^2 - \Delta} = -\alpha + \sqrt{\alpha^2 - \omega_0^2} = s_1 \tag{2.21a}$$

$$\lambda_2 = \frac{T}{2} - \sqrt{\frac{1}{4}T^2 - \Delta} = -\alpha - \sqrt{\alpha^2 - \omega_0^2} = s_2 \tag{2.21b}$$

To emphasize this important identity, we will henceforth use the symbol s_1 and s_2 to denote both *natural frequencies* of the circuit and the *eigenvalues* of the associated **A** matrix. It follows from Eq. (2.21) that s_1 and s_2 are either real numbers or else they are complex conjugates of each other, i.e., $s_2 = \bar{s}_1$. In the special case where $\alpha \geq 0$ and $\omega_0^2 > 0$, s_1 and s_2 can be written in the form of Eq. (2.5). Since we assume $\Delta \neq \frac{1}{4}T^2$ in this section, the *eigenvalues are distinct*.

Our next step in solving Eq. (2.1) is to find the *eigenvectors*

$$\boldsymbol{\eta}_1 = \begin{bmatrix} \eta_{11} \\ \eta_{12} \end{bmatrix} \qquad \text{and} \qquad \boldsymbol{\eta}_2 = \begin{bmatrix} \eta_{21} \\ \eta_{22} \end{bmatrix} \tag{2.22}$$

associated with the *eigenvalues* s_1 and s_2. By definition, $\boldsymbol{\eta}_1$ and $\boldsymbol{\eta}_2$ are any *nonzero* vectors which satisfy

$$\mathbf{A}\boldsymbol{\eta}_i = s_i\boldsymbol{\eta}_i \qquad i = 1, 2 \tag{2.23}$$

Note that the eigenvectors are unique only to within a scaling real or complex coefficient γ. Indeed, if $\boldsymbol{\eta}_i$ is a vector satisfying Eq. (2.22), then so is the "scaled" vector $\gamma\boldsymbol{\eta}_i$ because Eq. (2.23) implies $\mathbf{A}(\gamma\boldsymbol{\eta}_i) = s_i(\gamma\boldsymbol{\eta}_i)$.

[11] The reader should review the concept of *eigenvalues* and *eigenvectors* from linear algebra before reading this section.

[12] Unlike Sec. 2.1, we now allow the general case where both α and ω_0^2 may be negative. For simplicity, we exclude the pathological case $\Delta = \frac{1}{4}T^2$, i.e., $s_1 = s_2$.

Once the eigenvalues s_1 and s_2 and their associated eigenvectors have been found, the solution of the state equation (2.1) can be easily written (for the usual case where $s_1 \neq s_2$). Assuming the real matrix $\mathbf{A}$ has distinct eigenvalues s_1 and s_2, then the *zero-input response of* $\dot{\mathbf{x}} = \mathbf{A}\mathbf{x}$ is given by

$$\mathbf{x}(t) = (k_1 e^{s_1 t})\boldsymbol{\eta}_1 + (k_2 e^{s_2 t})\boldsymbol{\eta}_2 \tag{2.24}$$

where $\boldsymbol{\eta}_1$ and $\boldsymbol{\eta}_2$ are distinct eigenvectors associated with s_1 and s_2, respectively, and k_1 and k_2 are arbitrary constants which depend on the initial state $\mathbf{x}(0)$.

Before we prove Eq. (2.24), it is instructive to consider first an example.

Example: Zero-input response of a parallel *RLC* circuit Let us consider the same circuit studied earlier in Fig. 2.1. Its *state equation* can be written by inspection:

$$\begin{bmatrix} \dot{v}_C \\ \dot{i}_L \end{bmatrix} = \begin{bmatrix} -\dfrac{G}{C} & -\dfrac{1}{C} \\ \dfrac{1}{L} & 0 \end{bmatrix} \begin{bmatrix} v_C \\ i_L \end{bmatrix} \tag{2.25}$$

Substituting

$$\alpha = -\frac{T}{2} = \frac{G}{2C} \qquad \text{and} \qquad \omega_0^2 = \Delta = \frac{1}{LC} \tag{2.26}$$

into Eq. (2.21), we obtain the following natural frequencies:[13]

$$s_1 = -\frac{G}{2C} + \sqrt{\left(\frac{G}{2C}\right)^2 - \frac{1}{LC}} \qquad s_2 = -\frac{G}{2C} - \sqrt{\left(\frac{G}{2C}\right)^2 - \frac{1}{LC}} \tag{2.27}$$

To find the associated eigenvectors, we must find $\boldsymbol{\eta}_i$ which satisfies Eq. (2.23):

$$\begin{bmatrix} -\dfrac{G}{C} & -\dfrac{1}{C} \\ \dfrac{1}{L} & 0 \end{bmatrix} \begin{bmatrix} \eta_{i1} \\ \eta_{i2} \end{bmatrix} = s_i \begin{bmatrix} \eta_{i1} \\ \eta_{i2} \end{bmatrix} \qquad i = 1, 2 \tag{2.28}$$

It follows from Eq. (2.28) that the second component of each eigenvector $\boldsymbol{\eta}_i$ in Eq. (2.22) is related to the first component as follows:

$$\eta_{i2} = \left(\frac{1}{s_i L}\right)\eta_{i1} \qquad i = 1, 2 \tag{2.29}$$

[13] Our standing assumption that the eigenvalues are distinct (i.e., $\Delta \neq \frac{1}{4}T^2$) in this example becomes $G \neq 2\sqrt{C/L}$, i.e., the circuit in Fig. 2.1 is *not* critically damped (recall Table 2.2).

Hence, the eigenvectors associated with Eq. (2.27) are given by

$$\boldsymbol{\eta}_1 = \begin{bmatrix} \gamma_1 \\ \dfrac{\gamma_1}{s_1 L} \end{bmatrix} \qquad \boldsymbol{\eta}_2 = \begin{bmatrix} \gamma_2 \\ \dfrac{\gamma_2}{s_2 L} \end{bmatrix} \tag{2.30}$$

where γ_1 and γ_2 are *arbitrary* real or complex scaling constants which can be chosen to simplify the resulting expressions. In particular, let us choose $\gamma_1 = s_1$ and $\gamma_2 = s_2$ to obtain

$$\boldsymbol{\eta}_1 = \begin{bmatrix} s_1 \\ \dfrac{1}{L} \end{bmatrix} \qquad \boldsymbol{\eta}_2 = \begin{bmatrix} s_2 \\ \dfrac{1}{L} \end{bmatrix} \tag{2.31}$$

Substituting Eqs. (2.27) and (2.31) into Eq. (2.24), we obtain the following zero-input response:

$$\begin{bmatrix} v_C(t) \\ i_L(t) \end{bmatrix} = k_1 e^{s_1 t} \begin{bmatrix} s_1 \\ \dfrac{1}{L} \end{bmatrix} + k_2 e^{s_2 t} \begin{bmatrix} s_2 \\ \dfrac{1}{L} \end{bmatrix} \tag{2.32}$$

Substituting $t = 0$ in Eq. (2.32) and solving for k_1 and k_2, we obtain

$$k_1 = \frac{v_C(0) - s_2 L i_L(0)}{s_1 - s_2} \qquad k_2 = \frac{v_C(0) - s_1 L i_L(0)}{s_2 - s_1} \tag{2.33}$$

Hence, the zero-input response $v_C(t)$, or $i_L(t)$, for the parallel *RLC* circuit in Fig. 2.1 is given by Eqs. (2.32) and (2.33). Note that the constants k_1 and k_2 depend on the *initial state* $(v_C(0), i_L(0))$.

To verify that this solution is identical to the one obtained earlier in Sec. 2.1, let us rewrite $i_L(t)$ in Eq. (2.32) as follows:

$$\begin{aligned} i_L(t) &= \left[\frac{v_C(0) - s_2 L i_L(0)}{s_1 - s_2}\right] \frac{e^{s_1 t}}{L} + \left[\frac{v_C(0) - s_1 L i_L(0)}{s_2 - s_1}\right] \frac{e^{s_2 t}}{L} \\ &= \left[\frac{\dot{i}_L(0) - s_2 i_L(0)}{s_1 - s_2}\right] e^{s_1 t} + \left[\frac{\dot{i}_L(0) - s_1 i_L(0)}{s_2 - s_1}\right] e^{s_2 t} \end{aligned} \tag{2.34}$$

where we have substituted $v_C(0) = L\dot{i}_L(0)$. Note that this is identical to $i_L(t)$ given in Table 2.2 (for the *overdamped response*) with k_1 and k_2 given by Eq. (2.9), as it should.

The above example should make it easier to understand the following proof of Eq. (2.24).

PROOF Differentiating Eq. (2.24) with respect to time, we obtain

$$\dot{\mathbf{x}}(t) = (k_1 e^{s_1 t})(s_1 \boldsymbol{\eta}_1) + (k_2 e^{s_2 t})(s_2 \boldsymbol{\eta}_2) \tag{2.35}$$

Since $\boldsymbol{\eta}_i$ is an eigenvector associated with the eigenvalue s_i, $i = 1, 2$, we have

$$\mathbf{A}\boldsymbol{\eta}_1 = s_1 \boldsymbol{\eta}_1 \qquad \mathbf{A}\boldsymbol{\eta}_2 = s_2 \boldsymbol{\eta}_2 \tag{2.36}$$

Substituting Eq. (2.36) into Eq. (2.35), we obtain

$$\begin{aligned}\dot{\mathbf{x}}(t) &= (k_1 e^{s_1 t})\mathbf{A}\boldsymbol{\eta}_1 + (k_2 e^{s_2 t})\mathbf{A}\boldsymbol{\eta}_2 \\ &= \mathbf{A}[(k_1 e^{s_1 t})\boldsymbol{\eta}_1 + (k_2 e^{s_2 t})\boldsymbol{\eta}_2] = \mathbf{A}\mathbf{x}(t)\end{aligned} \tag{2.37}$$

Hence $\mathbf{x}(t)$ given by Eq. (2.24) is a solution of $\dot{\mathbf{x}} = \mathbf{A}\mathbf{x}$. ■

Exercise Find $v_C(t)$ and $i_L(t)$ for the circuit in Fig. 2.1 corresponding to the following circuit parameters:
(*a*) $G = 8\,\text{S}$, $L = \frac{1}{7}\,\text{H}$, $C = 1\,\text{F}$.
(*b*) $G = 2\,\text{S}$, $L = \frac{1}{5}\,\text{H}$, $C = 1\,\text{F}$.
(*c*) $G = 0\,\text{S}$, $L = 1\,\text{H}$, $C = 1\,\text{F}$.
Specify the *eigenvectors* in each case, and sketch them whenever they are real vectors.

3 QUALITATIVE BEHAVIOR OF $\dot{\mathbf{x}} = \mathbf{A}\mathbf{x}$

Let $[x_1(t), x_2(t)]$ denote the solution of the state equation

$$\begin{bmatrix} \dot{x}_1 \\ \dot{x}_2 \end{bmatrix} = \begin{bmatrix} a_{11} & a_{12} \\ a_{21} & a_{22} \end{bmatrix} \begin{bmatrix} x_1 \\ x_2 \end{bmatrix} \tag{3.1}$$

with an *initial state* $(x_1(t), x_2(t))$. This solution is given by Eq. (2.24) which we reproduce here for convenience:

$$\mathbf{x}(t) = (k_1 e^{s_1 t})\boldsymbol{\eta}_1 + (k_2 e^{s_2 t})\boldsymbol{\eta}_2 \tag{3.2}$$

Here s_1 and s_2 are *eigenvalues* of $\mathbf{A}$, $\boldsymbol{\eta}_1$ and $\boldsymbol{\eta}_2$ are the associated *eigenvectors*, and k_1 and k_2 are constants which are uniquely determined by the initial state.

The locus of the solution waveforms $(x_1(t), x_2(t))$ in the x_1-x_2 plane is called a *trajectory* of the state equation.

For example, the two loci shown earlier in Figs. 2.2 and 2.3 are trajectories of the state equation (2.5) in the v_C-i_L plane.

To investigate the *qualitative behavior* of a circuit, it is more informative to have an *approximate* sketch of a family of trajectories originating from suitably[14] chosen initial states (e.g., a grid of uniformly spaced points) in the x_1-x_2 plane. Such a *family* of trajectories is called a *phase portrait* of the state equation.

The phase portrait is valuable because it reveals all possible dynamic behaviors of the associated circuit in a single picture.

Our objective in this section is to investigate the phase portraits of *second-order* linear time-invariant circuits. We will show that, except for two degenerate cases which we have relegated to Exercises 2 and 3 at the end of this section, such circuits can display only six *qualitatively* distinct phase portraits. In particular, we will identify various *geometric* features which allow us to sketch the associated phase portrait rapidly. We will see in Sec. 5 that these features can be used as building blocks, much like a jigsaw puzzle, for sketching the phase portrait of *nonlinear* circuits. Let us now describe one of these features.

A point (x_{1Q}, x_{2Q}) in the x_1-x_2 plane is called an *equilibrium state* iff $(x_1(0), x_2(0)) = (x_{1Q}, x_{2Q})$ implies $(x_1(t), x_2(t)) = (x_{1Q}, x_{2Q})$ for all $t \geq 0$. Since $(\dot{x}_1(t), \dot{x}_2(t)) = (0, 0)$ for all $t \geq 0$ under this condition, the circuit is said to be in equilibrium whenever its initial state happens to be an equilibrium state.

The equilibrium states associated with Eq. (3.1) can be found by solving the associated *equilibrium equation*

$$\begin{bmatrix} a_{11} & a_{12} \\ a_{21} & a_{22} \end{bmatrix} \begin{bmatrix} x_1 \\ x_2 \end{bmatrix} = \begin{bmatrix} 0 \\ 0 \end{bmatrix} \tag{3.3}$$

It follows from Eq. (3.3) that $(x_1, x_2) = (0, 0)$ is the *only* equilibrium state for a *linear second-order* circuit if and only if

$$\Delta \triangleq a_{11}a_{22} - a_{12}a_{21} \neq 0 \tag{3.4}$$

In the case where $\Delta = 0$, the rows of **A** become *linearly dependent* so that both equations $a_{11}x_1 + a_{12}x_2 = 0$ and $a_{21}x_1 + a_{22}x_2 = 0$ define *the* same *straight line* Γ through the origin. In this case, Eq. (3.3) has *infinitely many* equilibrium states, namely, all points on Γ. Note from Eq. (2.21*b*) that $s_2 = 0$ when $\Delta = 0$.

For purposes of analysis, it is convenient to investigate first the simpler case where both eigenvalues are *real* numbers, namely $\Delta \leq \frac{1}{4}T^2$. This is presented in Sec. 3.1 for all such Δ and T except the following:

Degenerate case 1: $$\Delta = 0 \tag{3.5}$$

Degenerate case 2: $$\Delta = \tfrac{1}{4}T^2 \tag{3.6}$$

Observe from Eq. (2.21) that the matrix **A** has a *zero* eigenvalue in degenerate case 1, and two *equal* real eigenvalues (double root) in degenerate case 2. We exclude these two cases in Sec. 3.1 because they are seldom encountered in practice.

[14] Historically, the x_1-x_2 plane is usually called the *phase plane*. Consequently, the portrait of trajectories on this plane is called a *phase portrait*.

The other case where both eigenvalues are *complex* numbers is investigated in Sec. 3.2.

Exercises

1. Show that if both eigenvalues s_1 and s_2 are real and distinct, then the *trajectory* originating from $\mathbf{x}(0) = k_1\boldsymbol{\eta}_1$ or $\mathbf{x}(0) = k_2\boldsymbol{\eta}_2$ is a *straight line*, where $\boldsymbol{\eta}_i$ is an eigenvector associated with s_i, and k_i is any real constant, $i = 1, 2$.

2. Verify that the following four matrices correspond to *degenerate case 1* ($\Delta = 0$). Sketch the family of trajectories when $\lambda > 0$ and identify all equilibrium states in the x_1-x_2 plane.

$$(a)\begin{bmatrix}\lambda & 0\\ 0 & 0\end{bmatrix} \quad (b)\begin{bmatrix}0 & 0\\ 0 & \lambda\end{bmatrix} \quad (c)\begin{bmatrix}0 & \lambda\\ 0 & 0\end{bmatrix} \quad (d)\begin{bmatrix}0 & 0\\ \lambda & 0\end{bmatrix}$$

3. Verify that the following state equations correspond to *degenerate case 2* ($\Delta = \frac{1}{4}T^2$). Sketch the family of trajectories when $\lambda > 0$, and when $\lambda < 0$, respectively.

$$(a)\begin{bmatrix}\dot{x}_1\\ \dot{x}_2\end{bmatrix} = \begin{bmatrix}\lambda & 0\\ 0 & \lambda\end{bmatrix}\begin{bmatrix}x_1\\ x_2\end{bmatrix} \quad (b)\begin{bmatrix}\dot{x}_1\\ \dot{x}_2\end{bmatrix} = \begin{bmatrix}1 & \lambda\\ 0 & 1\end{bmatrix}\begin{bmatrix}x_1\\ x_2\end{bmatrix}$$

3.1 Two Distinct Real Eigenvalues: $\Delta < \frac{1}{4}T^2$, $\Delta \neq 0$ (equivalently, $\alpha^2 > \omega_0^2$, $\omega_0^2 \neq 0$)

The conditions $\Delta < \frac{1}{4}T^2$ and $\Delta \neq 0$ guarantee that the two eigenvalues s_1 and s_2 are *distinct*, *real*, and *nonzero*. This implies that the associated eigenvectors $\boldsymbol{\eta}_1$ and $\boldsymbol{\eta}_2$ are linearly independent real vectors, as shown in Fig. 3.1.

It follows from Eq. (3.2) that the solution $\mathbf{x}(t)$ due to any initial state not falling on either $\boldsymbol{\eta}_1$ or $\boldsymbol{\eta}_2$ is equal to the sum of two vectors proportional to $\boldsymbol{\eta}_1$ and $\boldsymbol{\eta}_2$, respectively, as depicted in Fig. 3.1. Here, the scaling factors are given by $k_1e^{s_1t}$ and $k_2e^{s_2t}$, respectively. Note that if $k_i < 0$, the solution component $(k_ie^{s_it})\boldsymbol{\eta}_i$ represents a vector along $\boldsymbol{\eta}_i$ but directed in the opposite direction (shown dotted in Fig. 3.1).

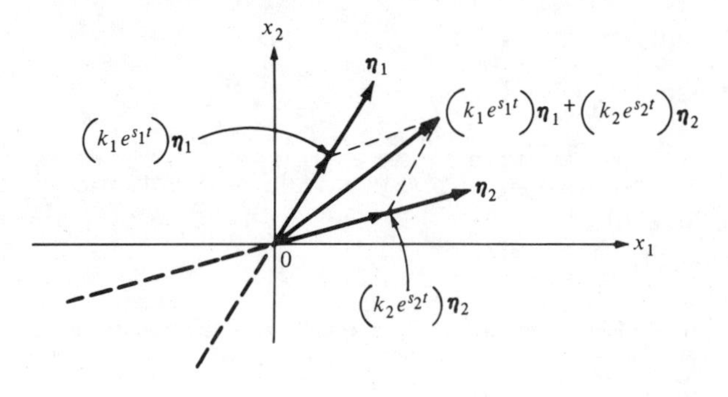

Figure 3.1 Components of solution $x(t)$ along the two linearly independent real eigenvectors $\boldsymbol{\eta}_1$ and $\boldsymbol{\eta}_2$.

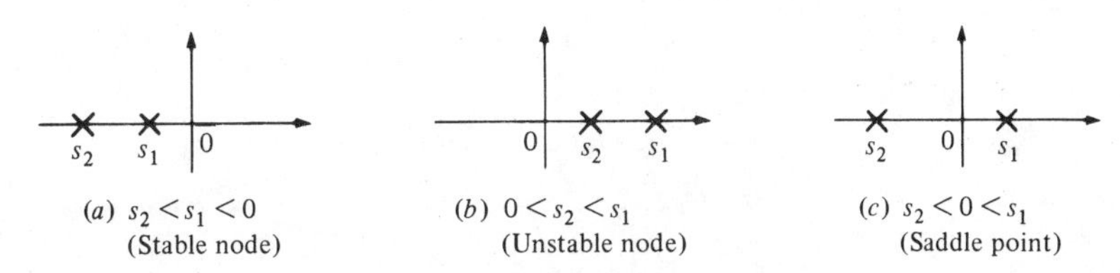

(a) $s_2 < s_1 < 0$ (Stable node)   (b) $0 < s_2 < s_1$ (Unstable node)   (c) $s_2 < 0 < s_1$ (Saddle point)

Figure 3.2 Three qualitatively distinct combinations of two real eigenvalues where $s_2 < s_1$ and $s_i \neq 0$.

The locus of all these vector sums from $t = 0$ to $t = \infty$ gives the *trajectory*: It represents the path followed by $\mathbf{x}(t)$ from $\mathbf{x}(0)$ to $\mathbf{x}(\infty)$.

Without loss of generality, we assume that $s_2 < s_1$ in the solution given by Eq. (3.2). Since $s_i \neq 0$ (by assumption, $\Delta \neq 0$), we need only investigate the three qualitatively distinct combinations depicted in Fig. 3.2.

A. Stable node: Two real negative eigenvalues ($s_2 < s_1 < 0$; equivalently,[15] $\alpha > \omega_0 > 0$) (s_2 s_1 0). Since s_1 and s_2 are negative, both vectors $(k_1 e^{s_1 t})\boldsymbol{\eta}_1$ and $(k_2 e^{s_2 t})\boldsymbol{\eta}_2$ tend to zero exponentially as $t \to \infty$. Moreover, since $s_2 < s_1 < 0$, the second component $(k_2 e^{s_2 t})\boldsymbol{\eta}_2$ in Eq. (3.1) will tend to zero *faster* than the first component $(k_1 e^{s_1 t})\boldsymbol{\eta}_1$. Hence, we call $\boldsymbol{\eta}_2$ the *fast eigenvector* and $\boldsymbol{\eta}_1$ the *slow eigenvector*.

To find the limiting behavior of the trajectories as $t \to \pm\infty$, let us recast Eq. (3.2) as follows:

$$\mathbf{x}(t) = e^{s_2 t}(k_1 e^{(s_1 - s_2)t}\boldsymbol{\eta}_1 + k_2 \boldsymbol{\eta}_2) \tag{3.7}$$

The following properties follow from Eq. (3.7) and $s_2 < s_1 < 0$:

1. As $t \to \infty$, the vector $(k_1 e^{s_1 t})\boldsymbol{\eta}_1$ dominates: Hence, as $t \to \infty$, all trajectories tend to the *origin* and become tangent to $\boldsymbol{\eta}_1$ as they approach the origin $\mathbf{0}$.
2. As $t \to -\infty$, the vector $(k_2 e^{s_2 t})\boldsymbol{\eta}_2$ dominates: Hence, as $t \to -\infty$, all trajectories tend toward the point at infinity and become parallel to $\boldsymbol{\eta}_2$.

These observations allow us to sketch the typical family of trajectories shown in Fig. 3.3. The behavior in this case corresponds to the *overdamped response* in Table 2.1. The equilibrium state $\mathbf{x} = \mathbf{0}$ in this case is called a *stable node*.

Exercise Let $\mathbf{A} = \begin{bmatrix} -3 & 1 \\ 1 & -3 \end{bmatrix}$

(a) Find the parameters T, Δ, α, and ω_0 associated with $\dot{\mathbf{x}} = \mathbf{A}\mathbf{x}$.
(b) Sketch the *slow* and *fast eigenvectors* associated with $\mathbf{A}$.
(c) Sketch the family of trajectories associated with $\dot{\mathbf{x}} = \mathbf{A}\mathbf{x}$.

[15] This follows from the relationships $2\alpha = -(s_1 + s_2)$ and $\omega_0^2 = s_1 s_2$.

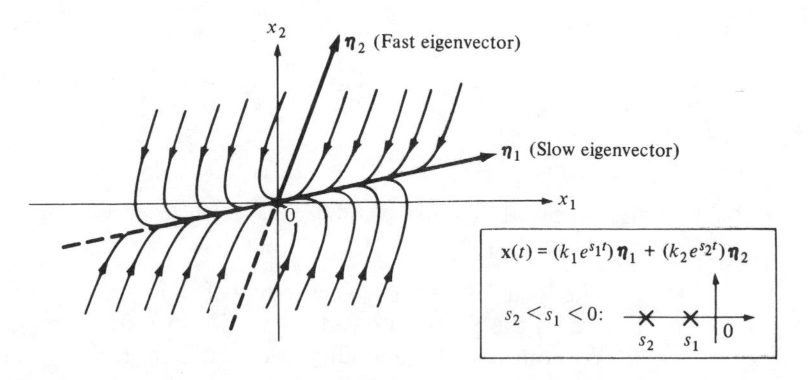

Figure 3.3 Qualitative behavior defining a *stable node* at the origin.

B. Unstable node: Two real positive eigenvalues ($0 < s_2 < s_1$; equivalently, $\alpha < -\omega_0 < 0$) Since s_1 and s_2 are positive, both vectors $(k_1 e^{s_1 t})\boldsymbol{\eta}_1$ and $(k_2 e^{s_2 t})\boldsymbol{\eta}_2$ grow exponentially as t increases; however, they tend to zero at $t \to -\infty$. Moreover, since $0 < s_2 < s_1$, the first component $(k_1 e^{s_1 t})\boldsymbol{\eta}_1$ in Eq. (3.1) will grow *faster* than the second component $(k_2 e^{s_2 t})\boldsymbol{\eta}_2$. Hence, we call $\boldsymbol{\eta}_1$ the *fast eigenvector* and $\boldsymbol{\eta}_2$ the *slow eigenvector* in this case.

The following properties follow from Eq. (3.7) and $0 < s_2 < s_1$:

1. As $t \to \infty$ the vector $(k_1 e^{s_1 t})\boldsymbol{\eta}_1$ dominates: Hence, as $t \to \infty$, all trajectories tend toward the point at infinity and become parallel to $\boldsymbol{\eta}_1$.
2. As $t \to -\infty$, the vector $(k_2 e^{s_2 t})\boldsymbol{\eta}_2$ dominates: Hence, as $t \to -\infty$, the trajectories recede to the origin and become tangent to $\boldsymbol{\eta}_2$ as they approach the origin $\mathbf{0}$.

These observations allow us to sketch the typical family of trajectories shown in Fig. 3.4. The equilibrium state $\mathbf{x} = \mathbf{0}$ in this case is called an *unstable node*.

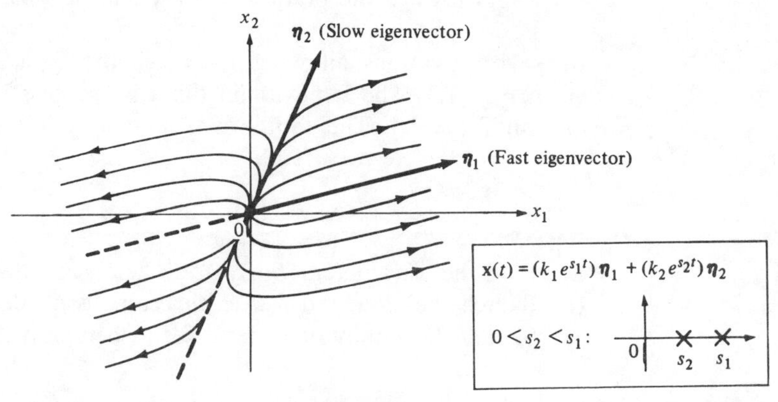

Figure 3.4 Qualitative behavior defining an *unstable node* at the origin (drawn with the same eigenvectors as Fig. 3.3 for comparison).

Exercise Let $\mathbf{A} = \begin{bmatrix} 3 & -1 \\ -1 & 3 \end{bmatrix}$

(*a*) Find the parameters T, Δ, α, and ω_0 associated with $\dot{\mathbf{x}} = \mathbf{A}\mathbf{x}$.
(*b*) Sketch the *slow* and *fast eigenvectors* associated with **A**.
(*c*) Sketch the family of trajectories associated with $\dot{\mathbf{x}} = \mathbf{A}\mathbf{x}$.

C. Saddle point: One negative and one positive eigenvalue ($s_2 < 0 < s_1$; equivalently, $\omega_0^2 < 0$) (s_2 × — 0 — × s_1). In this case, $(k_1 e^{s_1 t})\boldsymbol{\eta}_1 \to \infty$ while $(k_2 e^{s_2 t})\boldsymbol{\eta}_2 \to 0$ as $t \to \infty$. Conversely, $(k_1 e^{s_1 t})\boldsymbol{\eta}_1 \to 0$ while $(k_2 e^{s_2 t})\boldsymbol{\eta}_2 \to \infty$ as $t \to -\infty$. Consequently, the qualitative behavior of the trajectories in this case is as follows:

1. As $t \to \infty$, the vector $(k_1 e^{s_1 t})\boldsymbol{\eta}_1$ dominates: Hence, as $t \to \infty$, all trajectories tend toward the point at infinity and become parallel to $\boldsymbol{\eta}_1$.
2. As $t \to -\infty$, the vector $(k_2 e^{s_2 t})\boldsymbol{\eta}_2$ dominates: Hence, as $t \to -\infty$ the trajectories recede toward the point at infinity and become parallel to $\boldsymbol{\eta}_2$.

These observations allow us to sketch the typical family of trajectories shown in Fig. 3.5. The equilibrium state $\mathbf{x} = \mathbf{0}$ in this case is called a *saddle point*.

Exercise Let $\mathbf{A} = \begin{bmatrix} 1 & -3 \\ -3 & 1 \end{bmatrix}$

(*a*) Find the parameters T, Δ, α, and ω_0 associated with $\dot{\mathbf{x}} = \mathbf{A}\mathbf{x}$.
(*b*) Sketch the slow and fast eigenvectors associated with **A**.
(*c*) Sketch the family of trajectories associated with $\dot{\mathbf{x}} = \mathbf{A}\mathbf{x}$.

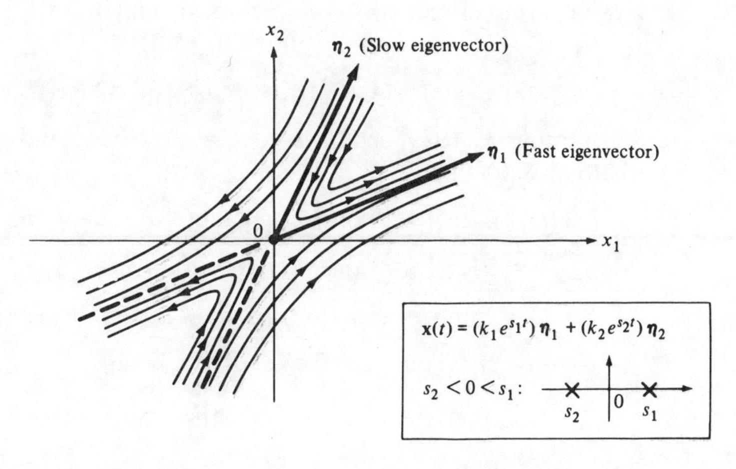

Figure 3.5 Qualitative behavior defining a saddle point at the origin (drawn with the same eigenvectors as Figs. 3.3 and 3.4 for comparison).

3.2 Two Complex-Conjugate Eigenvalues: $\Delta > \frac{1}{4}T^2$ (equivalently, $\alpha^2 < \omega_0^2$)

The condition $\Delta > \frac{1}{4}T^2$ guarantees that the two eigenvalues s_1 and s_2 are *complex-conjugate* numbers ($s_2 = \bar{s}_1$). It follows from Eq. (2.5) that

$$s_1 = -\alpha + j\omega_d \qquad s_2 = -\alpha - j\omega_d \tag{3.8}$$

where α is the *damping constant* and ω_d is defined in Eq. (2.7). Since $\mathbf{A}$ is a *real* matrix, the associated eigenvectors $\boldsymbol{\eta}_1$ and $\boldsymbol{\eta}_2$ can always be written as a pair of *complex-conjugate vectors*, i.e.,

$$\boldsymbol{\eta}_1 = \boldsymbol{\eta}_r + j\boldsymbol{\eta}_i \qquad \boldsymbol{\eta}_2 = \boldsymbol{\eta}_r - j\boldsymbol{\eta}_i = \bar{\boldsymbol{\eta}}_1 \tag{3.9}$$

where $\boldsymbol{\eta}_r$ and $\boldsymbol{\eta}_i$ denote *real* vectors called the *real* and *imaginary* component vectors of $\boldsymbol{\eta}_1$, respectively. For example, if $\boldsymbol{\eta}_1 = \begin{bmatrix} 1 - j2 \\ -3 + j5 \end{bmatrix}$, then

$$\boldsymbol{\eta}_1 = \begin{bmatrix} 1 \\ -3 \end{bmatrix} + j\begin{bmatrix} -2 \\ 5 \end{bmatrix} \quad \text{so} \quad \boldsymbol{\eta}_r = \begin{bmatrix} 1 \\ -3 \end{bmatrix} \quad \text{and} \quad \boldsymbol{\eta}_i = \begin{bmatrix} -2 \\ 5 \end{bmatrix}$$

Exercises

1. Give an example of a *complex* matrix $\mathbf{A}$ such that its eigenvectors cannot be written in the form of Eq. (3.9).
2. Show that if the two eigenvectors associated with Eq. (3.1) are *complex-conjugate vectors* and $\mathbf{A}$ is a *real* matrix, then $k_2 = \bar{k}_1$ for the solution given by Eq. (3.2).

Since $\boldsymbol{\eta}_1$ and $\boldsymbol{\eta}_2$ in Eq. (3.2) are *complex* vectors in this section, they cannot be represented geometrically as in Fig. 3.1. Hence the phase portraits constructed earlier in Figs. 3.3 to 3.5 with the help of $\boldsymbol{\eta}_1$ and $\boldsymbol{\eta}_2$ no longer hold in this section. To obtain an analogous geometric interpretation, we can recast $\mathbf{x}(t)$ in terms of the two *real* vectors $\boldsymbol{\eta}_r$ and $\boldsymbol{\eta}_i$ in Eq. (3.9) by substituting $s_2 = \bar{s}_1$, $k_2 = \bar{k}_1$, and $\boldsymbol{\eta}_2 = \bar{\boldsymbol{\eta}}_1$ into Eq. (3.2):[16]

$$\mathbf{x}(t) = (k_1 e^{s_1 t})\boldsymbol{\eta}_1 + (\bar{k}_1 e^{\bar{s}_1 t})\bar{\boldsymbol{\eta}}_1 = 2\,\mathrm{Re}[(k_1 e^{s_1 t})\boldsymbol{\eta}_1] \tag{3.10}$$

Substituting $k_1 = |k_1| \exp(j\sphericalangle k_1)$, $s_1 = -\alpha + j\omega_d$, and $\boldsymbol{\eta}_1 = \boldsymbol{\eta}_r + j\boldsymbol{\eta}_i$ into Eq. (3.10), we obtain

$$\begin{aligned}
\mathbf{x}(t) &= 2\,\mathrm{Re}\{[|k_1| \exp(j\sphericalangle k_1)][\exp(-\alpha + j\omega_d)t](\boldsymbol{\eta}_r + j\boldsymbol{\eta}_i)\} \\
&= 2\,\mathrm{Re}\{[|k_1| \exp(-\alpha t)][\exp j(\omega_d t + \sphericalangle k_1)](\boldsymbol{\eta}_r + j\boldsymbol{\eta}_i)\} \\
&= 2\,\mathrm{Re}\{[|k_1| \exp(-\alpha t)][\cos(\omega_d t + \sphericalangle k_1) + j\sin(\omega_d t + \sphericalangle k_1)](\boldsymbol{\eta}_r + j\boldsymbol{\eta}_i)\} \\
&= 2[|k_1| \exp(-\alpha t)] \cos(\omega_d t + \sphericalangle k_1)\,\boldsymbol{\eta}_r \\
&\quad - 2[|k_1| \exp(-\alpha t)] \sin(\omega_d t + \sphericalangle k_1)\,\boldsymbol{\eta}_i
\end{aligned} \tag{3.11}$$

[16] Note that $(z + \bar{z}) = (a + jb) + (a - jb) = 2a = 2\,\mathrm{Re}(z)$, where $\mathrm{Re}(z)$ denotes the real part of z. Note also that $\overline{uv} = \bar{u}\bar{v}$ for any complex numbers u and v.

Since $\boldsymbol{\eta}_r$ and $\boldsymbol{\eta}_i$ in Eq. (3.11) are *real* vectors, they can once again be represented geometrically in Fig. 3.6 and the trajectory corresponding to Eq. (3.11) can be sketched as in the previous section. The only difference is that the scale factors multiplying $\boldsymbol{\eta}_r$ and $\boldsymbol{\eta}_i$ now oscillate and change sign every π/ω_d s. Hence, during every other half period, the component solution vectors along $\boldsymbol{\eta}_r$ and $\boldsymbol{\eta}_i$ in Fig. 3.6 would fall along the dotted lines. The locus of all these vector sums from $t=0$ to $t=\infty$ gives the *trajectory* originating from $(x_1(0), x_2(0))$. The family of all such trajectories forms the *phase portrait*.

The *qualitative* behavior of the trajectories in the *complex* eigenvalue case depends on the *real* part of the eigenvalues in Eq. (3.8), namely, the *damping constant* α. There are only three qualitatively distinct combinations, as depicted in Fig. 3.7.

A. Center: Two imaginary eigenvalues (equivalently, $\alpha = 0$, $\omega_d = \omega_0$)

$j\omega_0$, 0, $-j\omega_0$. In this case $s_1 = -s_2 = j\omega_0$ and the solution in Eq. (3.11) reduces to

$$\mathbf{x}(t) = 2|k_1| \cos(\omega_0 t + \measuredangle k_1)\, \boldsymbol{\eta}_r - 2|k_1| \sin(\omega_0 t + \measuredangle k_1)\, \boldsymbol{\eta}_i \tag{3.12}$$

Since the scaling factors along $\boldsymbol{\eta}_r$ and $\boldsymbol{\eta}_i$ are both periodic of the same angular frequency ω_0, the trajectory is closed and returns to the initial point after $T = 2\pi/\omega_0$ s, as shown in Fig. 3.8. We leave it as an *exercise* for the reader to

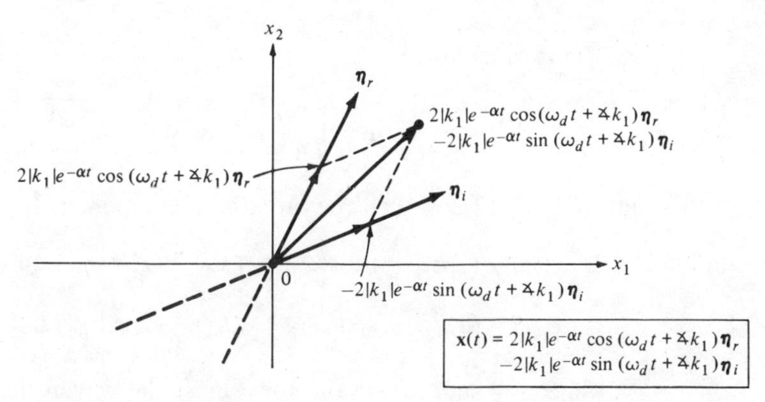

Figure 3.6 In the case $\Delta > \frac{1}{4}T^2$ $(\alpha^2 < \omega_0^2)$, the state vector $\mathbf{x}(t)$ is equal to the sum of two real vectors proportional to $\boldsymbol{\eta}_r$ and $\boldsymbol{\eta}_i$.

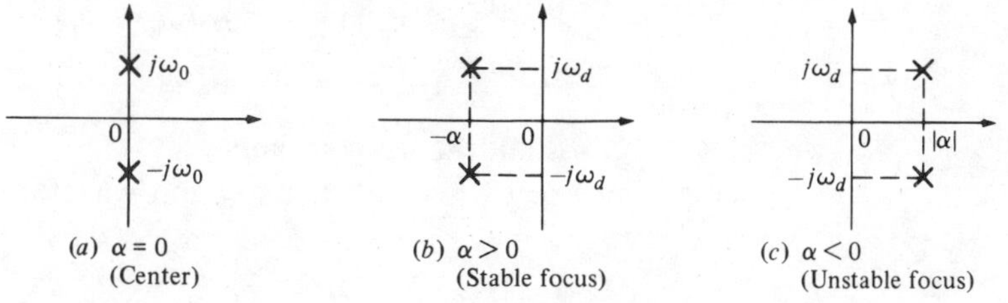

Figure 3.7 Three qualitatively distinct combinations of two complex-conjugate eigenvalues.

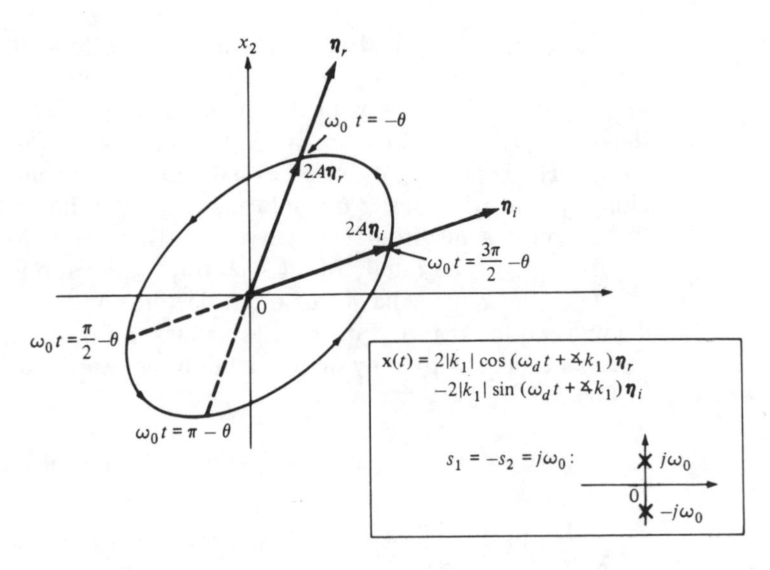

Figure 3.8 Trajectory for the $\alpha = 0$, $\omega_0^2 > 0$ case (center).

show that this trajectory is an *ellipse* which has $\boldsymbol{\eta}_r$ and $\boldsymbol{\eta}_i$ as conjugate directions.[17]

From Eq. (3.12) and for the specific $\boldsymbol{\eta}_r$ and $\boldsymbol{\eta}_i$ drawn in Fig. 3.6, the trajectory is traversed in a counterclockwise direction. The equilibrium state $\mathbf{x} = \mathbf{0}$ in this case is called a *center*.

If we substitute

$$\boldsymbol{\eta}_r = \begin{bmatrix} \boldsymbol{\eta}_{r1} \\ \boldsymbol{\eta}_{r2} \end{bmatrix} \quad \text{and} \quad \boldsymbol{\eta}_i = \begin{bmatrix} \boldsymbol{\eta}_{i1} \\ \boldsymbol{\eta}_{i2} \end{bmatrix} \tag{3.13}$$

into Eq. (3.12), we can rewrite $\mathbf{x}(t)$ in component form as follows:

$$x_1(t) = (2|k_1|\eta_{r1}) \cos(\omega_0 t + \measuredangle k_1) - (2|k_1|\eta_{i1}) \sin(\omega_0 t + \measuredangle k_1) \tag{3.14a}$$

$$x_2(t) = (2|k_1|\eta_{r2}) \cos(\omega_0 t + \measuredangle k_1) - (2|k_1|\eta_{i2}) \sin(\omega_0 t + \measuredangle k_1) \tag{3.14b}$$

Since the sum of two or more sinusoidal waveforms of *the same* frequency ω_0 always results in another sinusoidal waveform of frequency ω_0, we can rewrite Eq. (3.14) as follows:

$$x_1(t) = A_1 \cos(\omega_0 t + \theta_1) \tag{3.15a}$$

$$x_2(t) = A_2 \cos(\omega_0 t + \theta_2) \tag{3.15b}$$

where A_1, A_2, θ_1, and θ_2 are constants.

Comparing Eq. (3.15) with the solution waveforms listed earlier in Table

[17] Figure 2.2 is an example whose conjugate directions coincide with the x_1-x_2 axes.

2.1, we can identify the qualitative behavior of a *center* to be identical to that of a *lossless* response.

Exercises

1. Express the constants A_1, A_2, θ_1, and θ_2 in Eq. (3.15) in terms of $|k_1|$, $\measuredangle k_1$, η_{r1}, η_{r2}, η_{i1}, and η_{i2} in Eq. (3.14).

2. Let $\mathbf{A} = \begin{bmatrix} 0 & -2 \\ 2 & 0 \end{bmatrix}$

(*a*) Find the parameters T, Δ, α, and ω_0 associated with $\dot{\mathbf{x}} = \mathbf{A}\mathbf{x}$.

(*b*) Sketch the real and imaginary component vectors $\boldsymbol{\eta}_r$ and $\boldsymbol{\eta}_i$ associated with the complex eigenvector $\boldsymbol{\eta}_1$ of $\mathbf{A}$.

(*c*) Sketch the family of trajectories associated with $\dot{\mathbf{x}} = \mathbf{A}\mathbf{x}$.

B. Stable focus: Two complex-conjugate eigenvalues with a negative real part;

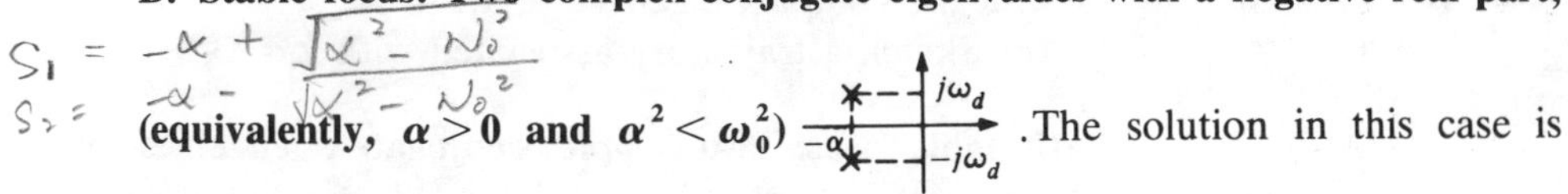

(equivalently, $\alpha > 0$ and $\alpha^2 < \omega_0^2$) . The solution in this case is given by Eq. (3.11) which we rewrite as follows:

$$\mathbf{x}(t) = e^{-\alpha t}[2|k_1| \cos(\omega_d t + \measuredangle k_1)\, \boldsymbol{\eta}_r - 2|k_1| \sin(\omega_d t + \measuredangle k_1)\, \boldsymbol{\eta}_i] \qquad (3.16)$$

Note that this solution differs from that of Eq. (3.12) for the *center* case only in the presence of a weighting factor $e^{-\alpha t}$ which tends exponentially to zero at $t \to \infty$. Hence, all trajectories in this case are logarithmic *spirals* which shrink to the origin as $t \to \infty$, as shown in Fig. 3.9. The equilibrium state $\mathbf{x} = \mathbf{0}$ in this case is called a *stable focus*.[18]

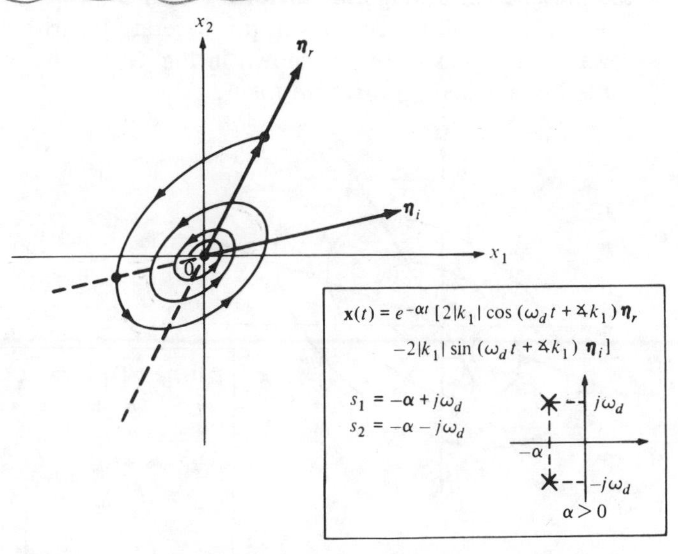

Figure 3.9 Trajectory for the $\alpha > 0$, $\alpha^2 < \omega_0^2$ case (*stable focus*).

[18] We have already seen an example of a spiral trajectory in Fig. 2.3.

We can rewrite Eq. (3.16) in component form as follows:

$$x_1(t) = A_1 e^{-\alpha t} \cos(\omega_d t + \theta_1) \tag{3.17a}$$

$$x_2(t) = A_2 e^{-\alpha t} \cos(\omega_d t + \theta_2) \tag{3.17b}$$

where A_1, A_2, θ_1, and θ_2 are the same constants as in Eq. (3.15).

Comparing Eq. (3.17) with the solution waveforms listed earlier in Table 2.1, we can identify the qualitative behavior of a *stable focus* to be identical to that of an *underdamped response*.

Exercise Let $\mathbf{A} = \begin{bmatrix} -1 & 2 \\ -2 & -1 \end{bmatrix}$

(*a*) Find the parameters T, Δ, α, and ω_0 associated with $\dot{\mathbf{x}} = \mathbf{A}\mathbf{x}$.
(*b*) Sketch the real and imaginary component vectors $\boldsymbol{\eta}_r$ and $\boldsymbol{\eta}_i$ associated with the complex eigenvector $\boldsymbol{\eta}_1$ of $\mathbf{A}$.
(*c*) Sketch a trajectory associated with $\dot{\mathbf{x}} = \mathbf{A}\mathbf{x}$.

C. Unstable focus: Two complex-conjugate eigenvalues with a positive real part (equivalently, $\boldsymbol{\alpha < 0}$ and $\boldsymbol{\alpha^2 < \omega_0^2}$) [diagram: $j\omega_d$, $-j\omega_d$, 0, $|\alpha|$]. The solution in this case is given by Eq. (3.11) which we rewrite as follows:

$$\mathbf{x}(t) = e^{|\alpha|t}[2|k_1|\cos(\omega_d t + \measuredangle k_1)\,\boldsymbol{\eta}_r - 2|k_1|\sin(\omega_d t + \measuredangle k_1)\,\boldsymbol{\eta}_i] \tag{3.18}$$

Note that this solution differs from that of Eq. (3.12) for the *center* case only in the presence of a weighting factor $e^{|\alpha|t}$ which grows exponentially to infinity as $t \to \infty$. Hence, all trajectories in this case are logarithmic spirals which expand toward infinity as $t \to \infty$, as shown in Fig. 3.10. The equilibrium state $\mathbf{x} = \mathbf{0}$ in this case is called an *unstable focus*.

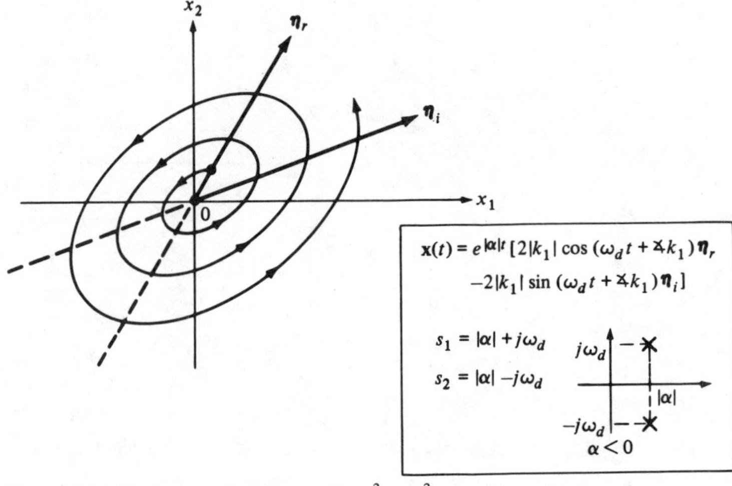

Figure 3.10 Trajectory for the $\alpha < 0$, $\alpha^2 < \omega_0^2$ case (*unstable focus*).

We can rewrite Eq. (3.18) in component form as follows:

$$x_1(t) = A_1 e^{|\alpha|t} \cos(\omega_d t + \theta_1) \tag{3.19a}$$

$$x_2(t) = A_2 e^{|\alpha|t} \cos(\omega_d t + \theta_2) \tag{3.19b}$$

where A_1, A_2, θ_1, and θ_2 are the same constants as in Eq. (3.15).

Exercise Let $\mathbf{A} = \begin{bmatrix} 1 & 2 \\ -2 & 1 \end{bmatrix}$

(*a*) Find the parameters T, Δ, α, and ω_0 associated with $\dot{\mathbf{x}} = \mathbf{A}\mathbf{x}$.
(*b*) Sketch the real and imaginary component vectors $\boldsymbol{\eta}_r$ and $\boldsymbol{\eta}_i$ associated with the complex eigenvector $\boldsymbol{\eta}_1$ of $\mathbf{A}$.
(*c*) Sketch a trajectory associated with $\dot{\mathbf{x}} = \mathbf{A}\mathbf{x}$.

3.3 Summary of Equilibrium State Classification

Given the state equation

$$\dot{\mathbf{x}} = \mathbf{A}\mathbf{x} \tag{3.20}$$

the origin $\mathbf{x} = \mathbf{0}$ is called an *equilibrium state* because $\dot{\mathbf{x}}(t) = \mathbf{0}$ for all $t > 0$ when the initial state is $\mathbf{x}(0) = \mathbf{0}$, and the circuit is therefore in equilibrium. The preceding analysis reveals that the qualitative features of the trajectories associated with Eq. (3.20) are determined by the behavior of trajectories in a small neighborhood of the origin. Except for the two degenerate cases corresponding to $\Delta = 0$ and $\Delta = \frac{1}{4}T^2$, the qualitative behavior near $\mathbf{x} = \mathbf{0}$ can be unambiguously classified into one of six types, namely, a *stable node* (Fig. 3.3), an *unstable node* (Fig. 3.4), a *saddle point* (Fig. 3.5), a *center* (Fig. 3.8), a *stable focus* (Fig. 3.9), and an *unstable focus* (Fig. 3.10).

The precise classification criteria for each of the above six types of equilibrium states has been derived in terms of the *trace* T and *determinant* Δ of the matrix $\mathbf{A}$, and is summarized in Table 3.1 for future reference. Rather than memorizing this table, it is often easier to use the mnemonic diagram shown in Fig. 3.11, which entails the drawing of only the parabola $\Delta = \frac{1}{4}T^2$. Note that the T axis, the Δ axis, and the parabola together partition the T-Δ plane into six regions, each one corresponding to a particular type of equilibrium state.

If the matrix $\mathbf{A}$ is given, it is easier to calculate T and Δ directly from their definitions:

$$T \triangleq a_{11} + a_{22} \qquad \Delta \triangleq a_{11}a_{22} - a_{12}a_{21} \tag{3.21}$$

On the other hand, if the circuit is described by a scalar differential equation

$$\ddot{x} + 2\alpha\dot{x} + \omega_0^2 x = 0 \tag{3.22}$$

then it is easier to calculate T and Δ from the formula

$$T = -2\alpha \qquad \Delta = \omega_0^2 \tag{3.23}$$

In this case, we do not have to derive the state equation.

Table 3.1. Equilibrium state classification for nondegenerate case; i.e., $\Delta \neq 0$ and $\Delta \neq \frac{1}{4}T^2$

$\mathbf{A} = \begin{bmatrix} a_{11} & a_{12} \\ a_{21} & a_{22} \end{bmatrix}$ $T = a_{11} + a_{22} = s_1 + s_2$ $\Delta = a_{11}a_{22} - a_{12}a_{21} = s_1 s_2$	$\det(s\mathbf{l} - \mathbf{A}) = s^2 - \underbrace{(a_{11} + a_{22})}_{T}s + \underbrace{(a_{11}a_{22} - a_{12}a_{21})}_{\Delta}$ *Eigenvalue* s_i: *Eigenvector* $\boldsymbol{\eta}_i$: $s_1 = \frac{T}{2} + \sqrt{\frac{1}{4}T^2 - \Delta}$, $\mathbf{A}\boldsymbol{\eta}_i = s_1\boldsymbol{\eta}_i$ $s_2 = \frac{T}{2} - \sqrt{\frac{1}{4}T^2 - \Delta}$, $\mathbf{A}\boldsymbol{\eta}_2 = s_2\boldsymbol{\eta}_2$ *Solution*: $\mathbf{x}(t) = (k_1 e^{s_1 t})\boldsymbol{\eta}_1 + (k_2 e^{s_2 t})\boldsymbol{\eta}_2$			
Type of equilibrium state	Real eigenvalues		Complex eigenvalues	
	$\Delta < \frac{1}{4}T^2$		$\Delta > \frac{1}{4}T^2$	
Stable node	$T < 0$, $\Delta > 0$	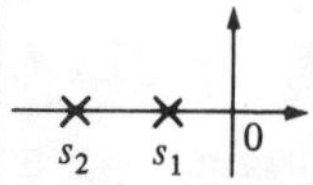 s_2 s_1 0		
Unstable node	$T > 0$, $\Delta > 0$	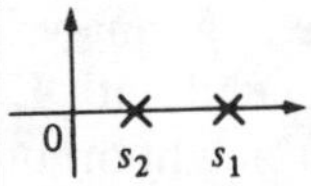 0 s_2 s_1		
Saddle point	$\Delta < 0$	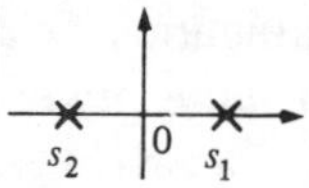 s_2 0 s_1		
Center			$\Delta > 0$, $T = 0$	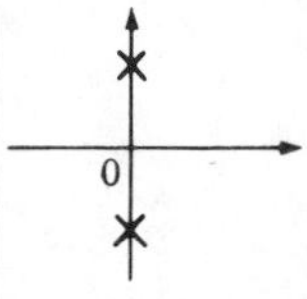 0
Stable focus			$\Delta > 0$, $T < 0$	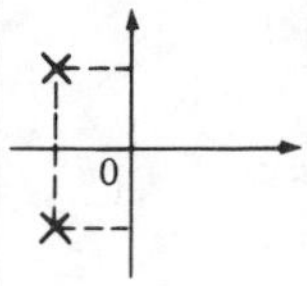 0
Unstable focus			$\Delta > 0$, $T > 0$	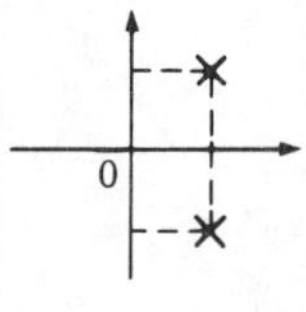 0

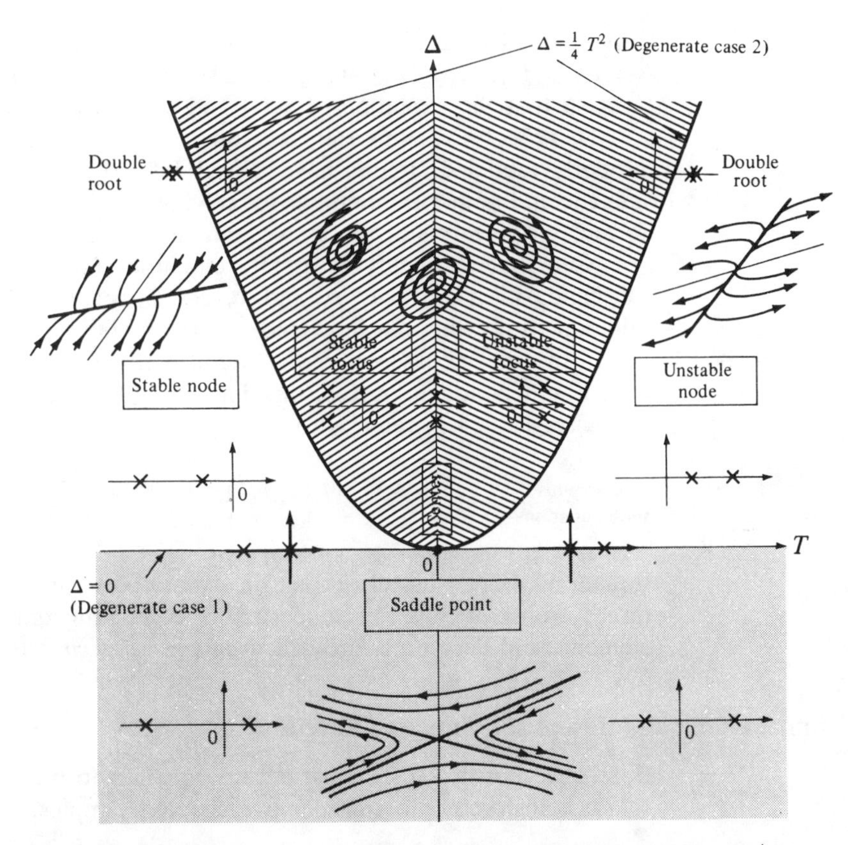

Figure 3.11 Equilibrium state classification diagram (nondegenerate case): $\Delta \triangleq a_{11}a_{22} - a_{12}a_{21} = \omega_0^2$, $T \triangleq a_{11} + a_{22} = \alpha$. A natural-frequency pattern and its associated phase portrait is shown near a *typical* point having this pattern.

In the case where the *natural frequencies* s_1 and s_2 of the circuit, or equivalently the *eigenvalues* λ_1 and λ_2 of $\mathbf{A}$, are given, T and Δ can be calculated from the following relationships:

$$T = s_1 + s_2 = \lambda_1 + \lambda_2 \qquad \Delta = s_1 s_2 = \lambda_1 \lambda_2 \tag{3.24}$$

4 NONLINEAR STATE EQUATION FORMULATION

Our objective in this section is to give a systematic method for formulating the governing equations of a *second-order nonlinear circuit* in the following standard form:

Nonlinear state equation

$$\begin{aligned}\dot{x}_1 &= f_1(x_1, x_2, t)\\ \dot{x}_2 &= f_2(x_1, x_2, t)\end{aligned} \qquad \text{or} \qquad \dot{\mathbf{x}} = \mathbf{f}(\mathbf{x}, t) \tag{4.1}$$

Equation (4.1) is called a *state equation* because given the initial state $\mathbf{x}(t_0) = (x_1(t_0), x_2(t_0))$ at any time t_0, it specifies the *rate of change* of x_1 and x_2 at t_0 so that the evolution of $x_1(t)$ and $x_2(t)$ can be calculated for all $t \geq t_0$.

There are two important reasons for writing state equations. *First*, all properly modeled circuits have a well-defined state equation. *Second*, most analytic and numerical methods for solving *nonlinear* differential equations are formulated in terms of the above standard form.

Note that to qualify as a state equation, $\dot{x}_1$ and $\dot{x}_2$ must appear on the left-hand side, and only x_1 and x_2 can appear on the right-hand side, in addition to possibly the independent time variable t.

In the special case where the circuit contains only time-invariant elements and dc independent sources, the time variable does not appear explicitly. In this case, Eq. (4.1) reduces to

Autonomous state equation

$$\begin{aligned} \dot{x}_1 &= f_1(x_1, x_2) \\ \dot{x}_2 &= f_2(x_1, x_2) \end{aligned} \qquad \text{or} \qquad \dot{\mathbf{x}} = \mathbf{f}(\mathbf{x}) \tag{4.2}$$

Equation (4.2) is usually called an *autonomous* state equation in the literature.[19] Consequently, dynamic circuits containing only linear time-invariant elements and dc sources are called *autonomous circuits*.

4.1 Tunnel Diode and Josephson Junction Circuits

Before we develop a systematic state equation formulation procedure in Sec. 4.2, it is instructive to consider two simple examples.

Example 1 (Tunnel diode circuit) Consider the tunnel diode circuit shown in Fig. 4.1*a*, where the tunnel diode is characterized by $i_R = \hat{i}_R(v_R)$, as shown in Fig. 4.1*b*. Applying KCL at node ①, we obtain

$$C_1 \dot{v}_{C1} + \hat{i}_R(v_R) - i_{L2} = 0 \tag{4.3}$$

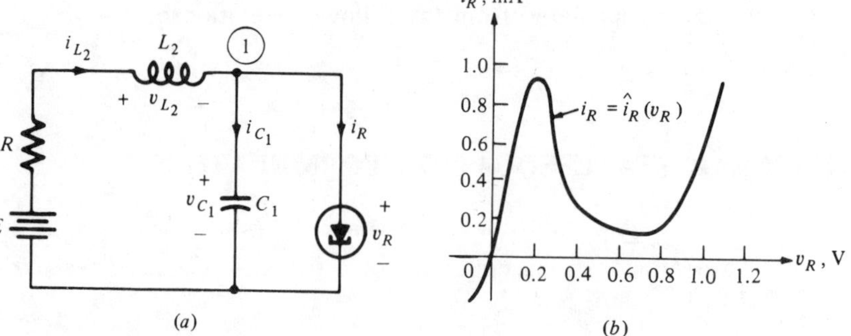

Figure 4.1 (*a*) Tunnel diode circuit. (*b*) Tunnel diode v_R-i_R characteristic.

[19] Historically, the term "autonomous" pertains to the differential equation describing vacuum tube oscillators which generate a periodic waveform without applying any time-dependent source as input to the circuit.

Differentiating both sides with respect to time, we obtain

$$C_1 \ddot{v}_{C1} + g(v_R)\dot{v}_R - \dot{i}_{L2} = 0 \tag{4.4a}$$

where

$$g(v_R) \triangleq \frac{d\hat{i}(v_R)}{dv_R} \tag{4.4b}$$

is the *small-signal conductance* at v_R. Now

$$\dot{i}_{L2} = \frac{v_{L2}}{L_2} = \frac{E - Ri_{L2} - v_{C1}}{L_2} \tag{4.5a}$$

$$= \frac{E - R[C_1\dot{v}_{C1} + \hat{i}_R(v_R)] - v_{C1}}{L_2} \qquad \text{[in view of Eq. (4.3)]} \tag{4.5b}$$

Substituting $v_R = v_{C1}$ and Eq. (4.5*b*) into Eq. (4.4*a*), and simplifying, we obtain the following nonlinear second-order scalar differential equation:

$$\ddot{v}_{C1} + \left[\frac{R}{L_2} + \frac{1}{C_1}\,g(v_{C1})\right]\dot{v}_{C1} - \left[\frac{E - R\hat{i}_R(v_{C1})}{C_1L_2} - \frac{v_{C1}}{C_1L_2}\right] = 0 \tag{4.6}$$

It is generally impossible to solve a nonlinear differential equation (of any order) in *closed form*. However, the solution can be found by efficient *numerical* methods.

To formulate the state equation for the circuit in Fig. 4.1, we can choose v_{C1} and i_{L2} as the state variables and combine Eqs. (4.3) and (4.5*a*) as follows:

$$\dot{v}_{C1} = \frac{1}{C_1}[-\hat{i}_R(v_{C1}) + i_{L2}] \triangleq f_1(v_{C1}, i_{L2}) \tag{4.7a}$$

$$\dot{i}_{L2} = \frac{1}{L_2}[E - Ri_{L2} - v_{C1}] \triangleq f_2(v_{C1}, i_{L2}) \tag{4.7b}$$

Note that the time variable t does not appear explicitly in the state equation (4.7) because the circuit contains only time-invariant elements and a dc source. Hence, Eq. (4.7) is an *autonomous* state equation.

Example 2 (Josephson junction circuit) Consider next the Josephson circuit shown in Fig. 4.2, where the Josephson junction is a time-invariant *nonlinear inductor* described by $i_{L2} = I_0 \sin k\phi_{L2}$. Applying KCL at node ①, we obtain

$$C_1\dot{v}_{C1} + Gv_{C1} + I_0 \sin k\phi_{L2} = i_s(t) \tag{4.8}$$

Substituting

$$v_{C1} = v_{L2} = \dot{\phi}_{L2} \tag{4.9}$$

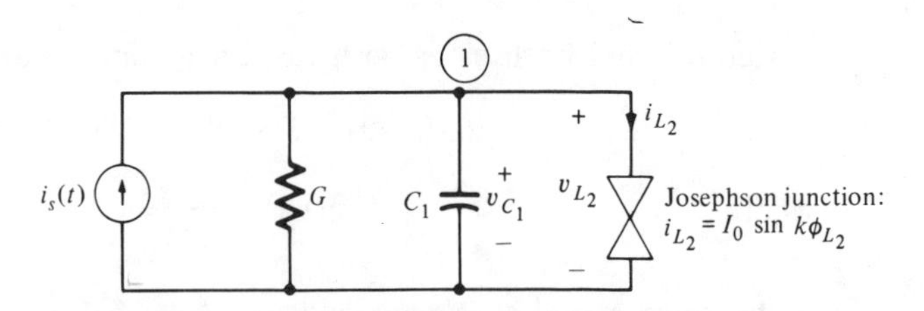

Figure 4.2 A Josephson junction circuit driven by a time-dependent current source.

into Eq. (4.8), we obtain

$$\ddot{\phi}_{L2} + \frac{G}{C_1}\dot{\phi}_{L2} + \frac{I_0}{C_1}\sin k\phi_{L2} = \frac{1}{C_1}\, i_s(t) \tag{4.10}$$

Again Eq. (4.10) is a *nonlinear* scalar differential equation whose solution cannot be found in closed form. Hence, let us formulate the state equation so that its solution can be found numerically.

Note that if we choose v_{C1} and i_{L2} as the state variables as in Example 1, it will be necessary to eliminate the variable ϕ_{L2} from Eq. (4.8). To do this, it is necessary to rewrite Eq. (4.8) into the form

$$\sin k\phi_{L2} = \frac{1}{I_0}[i_s(t) - C_1\dot{v}_{C1} - Gv_{C1}] \triangleq y \tag{4.11}$$

and then solve for $k\phi_{L2}$ as a *single-valued* function of y. But this is impossible because $\sin k\phi_{L2}$ is not a one-to-one function and hence does not possess an inverse: For each value of y, there correspond infinitely many values of $k\phi_{L2}$ (if $|y| \leq 1$).

To overcome this difficulty, let us choose v_{C1} and ϕ_{L2} as the state variables so that we do not have to eliminate ϕ_{L2}. In terms of v_{C1} and ϕ_{L2}, the state equation follows directly from Eqs. (4.8) and (4.9):

$$\dot{v}_{C1} = \frac{1}{C_1}[-Gv_{C1} - I_0 \sin k\phi_{L2} + i_s(t)] \triangleq f_1(v_{C1}, \phi_{L2}, t) \tag{4.12a}$$

$$\dot{\phi}_{L2} = v_{C1} \triangleq f_2(v_{C1}, \phi_{L2}, t) \tag{4.12b}$$

4.2 How to Write Nonlinear State Equations

The state equations (4.7) and (4.12) were both derived by inspection of the circuits. For more complicated circuits, however, the above ad hoc procedure often fails. Fortunately, the state equations can be formulated using the following systematic procedure:

Table 4.1 State equation for second-order nonlinear circuits

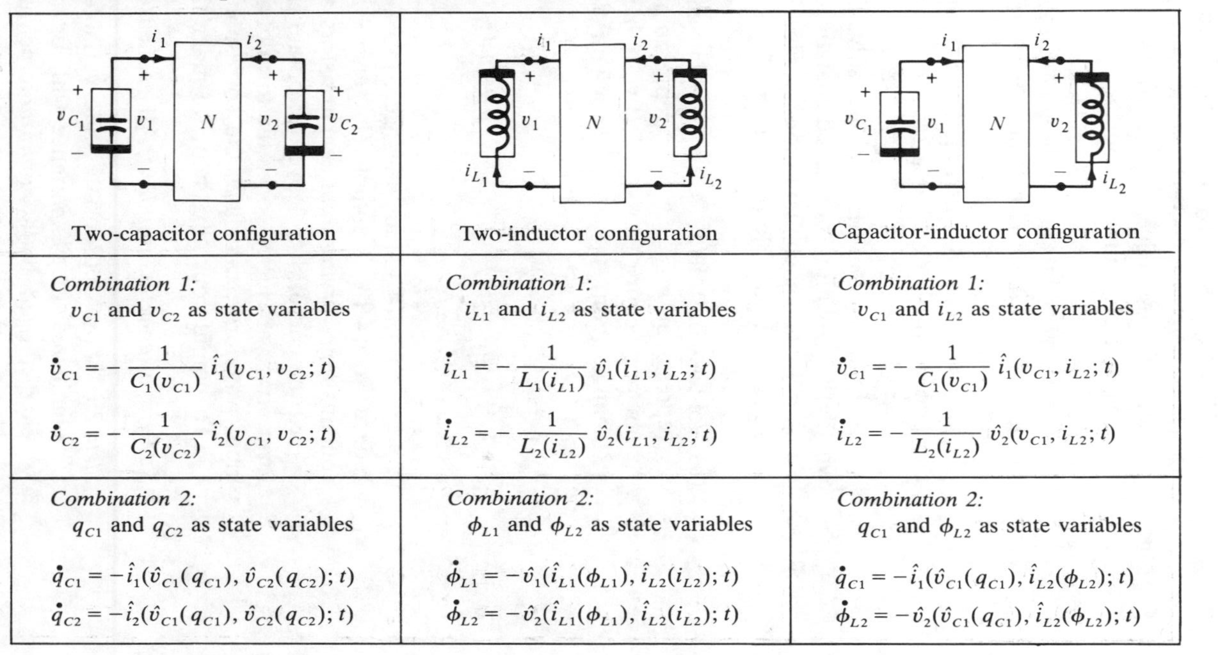

Two-capacitor configuration	Two-inductor configuration	Capacitor-inductor configuration
Combination 1: v_{C1} and v_{C2} as state variables	*Combination 1:* i_{L1} and i_{L2} as state variables	*Combination 1:* v_{C1} and i_{L2} as state variables
$\dot{v}_{C1} = -\dfrac{1}{C_1(v_{C1})}\hat{i}_1(v_{C1}, v_{C2}; t)$	$\dot{i}_{L1} = -\dfrac{1}{L_1(i_{L1})}\hat{v}_1(i_{L1}, i_{L2}; t)$	$\dot{v}_{C1} = -\dfrac{1}{C_1(v_{C1})}\hat{i}_1(v_{C1}, i_{L2}; t)$
$\dot{v}_{C2} = -\dfrac{1}{C_2(v_{C2})}\hat{i}_2(v_{C1}, v_{C2}; t)$	$\dot{i}_{L2} = -\dfrac{1}{L_2(i_{L2})}\hat{v}_2(i_{L1}, i_{L2}; t)$	$\dot{i}_{L2} = -\dfrac{1}{L_2(i_{L2})}\hat{v}_2(v_{C1}, i_{L2}; t)$
Combination 2: q_{C1} and q_{C2} as state variables	*Combination 2:* ϕ_{L1} and ϕ_{L2} as state variables	*Combination 2:* q_{C1} and ϕ_{L2} as state variables
$\dot{q}_{C1} = -\hat{i}_1(\hat{v}_{C1}(q_{C1}), \hat{v}_{C2}(q_{C2}); t)$	$\dot{\phi}_{L1} = -\hat{v}_1(\hat{i}_{L1}(\phi_{L1}), \hat{i}_{L2}(i_{L2}); t)$	$\dot{q}_{C1} = -\hat{i}_1(\hat{v}_{C1}(q_{C1}), \hat{i}_{L2}(\phi_{L2}); t)$
$\dot{q}_{C2} = -\hat{i}_2(\hat{v}_{C1}(q_{C1}), \hat{v}_{C2}(q_{C2}); t)$	$\dot{\phi}_{L2} = -\hat{v}_2(\hat{i}_{L1}(\phi_{L1}), \hat{i}_{L2}(i_{L2}); t)$	$\dot{\phi}_{L2} = -\hat{v}_2(\hat{v}_{C1}(q_{C1}), \hat{i}_{L2}(\phi_{L2}); t)$

Step 1. Extract the energy-storage elements as shown in Table 4.1 for three possible configurations.[20] Drive each port of the *resistive* two-port N by a voltage (respectively, current) source if the port is connected to a capacitor (respectively, inductor), as shown earlier in Fig. 1.5. Solve for the port current (respectively, port voltage) if the port is driven by a voltage (respectively, current) source.

Let us now introduce the following crucial assumption which is always satisfied if the circuit is properly modeled:

Unique solvability assumption:[21] For all input source waveforms, the resistive two-port N in Fig. 1.5 has a *unique* solution for all times.

It follows from the unique solvability assumption that regardless of whether the solution is obtained analytically or numerically, we can express the solution for each configuration in Fig. 1.5 as a *function* of the input port variables, and possibly the time t:

(*a*) Voltage-controlled representation (two-capacitor configuration)	(*b*) Current-controlled representation (two-inductor configuration)	(*c*) Hybrid representation (capacitor-inductor configuration)
$i_1 = \hat{i}_1(v_1, v_2, t)$	$v_1 = \hat{v}_1(i_1, i_2, t)$	$i_1 = \hat{i}_1(v_1, i_2, t)$
$i_2 = \hat{i}_2(v_1, v_2, t)$	$v_2 = \hat{v}_2(i_1, i_2, t)$	$v_2 = \hat{v}_2(v_1, i_2, t)$

(4.13)

Note that Eq. (4.13) includes Eqs. (1.24), (1.27), and (1.29) as special cases. Indeed Eq. (4.13) is completely general and includes even the case where N contains *time-varying* nonlinear resistors and *time-varying* sources.

At the end of Step 1, we will obtain Eq. (4.13) either in explicit analytical form—if we are lucky—or in the form of a computer subroutine which, for any input waveform, calculates the corresponding (unique) output at any time t.

To simplify our next step, we make the following assumption:

Time-invariance assumption: All capacitors and inductors are time-invariant.

Step 2. Choose appropriate state variables so that *all other* variables can be eliminated. Our experience with Example 2 suggests that if an element is characterized by a *nonmonotonic* function $y = f(x)$, then x can*not* be eliminated and hence must be chosen as the state variable. This observation leads to the following method for choosing state variables:

[20] Note that the reference directions in Table 4.1 are chosen, without loss of generality, such that $v_{Cj} = v_j$ and $i_{Lj} = i_j$. The expressions in Table 4.1 are valid *only* under this assumption.

[21] This crucial assumption can often be checked by inspection. See footnote 17 on page 244.

Capacitor

1. If the capacitor is *voltage-controlled*, i.e.,

$$q_{Cj} = \hat{q}_{Cj}(v_{Cj}) \qquad (4.14a)$$

then choose *capacitor voltage* v_{Cj} as the state variable.

2. If the capacitor is *charge-controlled*, i.e.,

$$v_{Cj} = \hat{v}_{Cj}(q_{Cj}) \qquad (4.14b)$$

then choose *capacitor charge* q_{Cj} as the state variable.

Inductor

1. If the inductor is *current-controlled*, i.e.,

$$\phi_{Lj} = \hat{\phi}_{Lj}(i_{Lj}) \qquad (4.15a)$$

then choose *inductor current* i_{Lj} as the state variable.

2. If the inductor is *flux-controlled*, i.e.,

$$i_{Lj} = \hat{i}_{Lj}(\phi_{Lj}) \qquad (4.15b)$$

then choose *inductor flux* ϕ_{Lj} as the state variable.

Since, depending on the element characteristics, one of two distinct state variables must be chosen for each capacitor and each inductor, there are four distinct combinations of state variables for each of the three configurations in Table 4.1; they are listed in Table 4.2. Each choice of state variables will lead to a distinct state equation, as shown in Table 4.1 for the first two combinations.

If we choose v_{C1} and v_{C2} as state variables in the *two-capacitor configuration*, the associated state equation is obtained by substituting column 1 of Eq. (4.13) into

$$i_{Cj} = C_j(v_{Cj})\dot{v}_{Cj} = -i_1 \qquad (4.16a)$$

where[22]

$$C_j(v_{Cj}) \triangleq \frac{d\hat{q}_{Cj}(v_{Cj})}{dv_{Cj}} \qquad (4.16b)$$

and then solving for v_{Cj}.

If we choose i_{L1} and i_{L2} as state variables in the *two-inductor configuration*, the associated state equation is obtained by substituting column 2 of Eq. (4.13) into

$$v_{Lj} = L_j(i_{Lj})\dot{i}_{Lj} = -v_2 \qquad (4.17a)$$

where[23]

$$L_j(i_{Lj}) \triangleq \frac{d\hat{\phi}_{Lj}(i_{Lj})}{di_{Lj}} \qquad (4.17b)$$

and then solving for i_{Lj}.

Table 4.2 Four distinct pairs of state variables

(*a*) Two-capacitor configuration	(*b*) Two-inductor configuration	(*c*) Capacitor-inductor configuration
1. (v_{C1}, v_{C2})	1. (i_{L1}, i_{L2})	1. (v_{C1}, i_{L2})
2. (q_{C1}, q_{C2})	2. (ϕ_{L1}, ϕ_{L2})	2. (q_{C1}, ϕ_{L2})
3. (v_{C1}, q_{C2})	3. (i_{L1}, ϕ_{L2})	3. (v_{C1}, ϕ_{L2})
4. (q_{C1}, v_{C2})	4. (ϕ_{L1}, i_{L2})	4. (q_{C1}, i_{L2})

[22] Recall $C_j(v_{Cj})$ is called the *small-signal capacitance* evaluated at v_{Cj}.

[23] Recall $L_j(i_{Lj})$ is called the *small-signal inductance* evaluated at i_{Lj}.

If we choose q_{C1} and ϕ_{L2} as the state variables in the *capacitor-inductor configuration*, the associated state equation is obtained by substituting $\dot{q}_{C1}$ for i_1, $\dot{\phi}_{L2}$ for v_2, Eq. (4.14*b*) for v_1, and Eq. (4.15*b*) for i_2, in Eq. (4.13).

Example 1 revisited To illustrate Steps 1 and 2 in the preceding state equation formulation method, let us return to Example 1 and derive the state equation (4.7) using this systematic procedure.

Step 1. Extract the capacitor and inductor and redraw the circuit in Fig. 4.1*a* as shown in Fig. 4.3*a*. Next we consider the associated nonlinear resistive two-port N in Fig. 4.3*b* and derive its *hybrid* representation [column 3 of Eq. (4.13)] as follows:

$$i_1 = \hat{i}_R(v_1) - i_2 \triangleq \hat{i}_1(v_1, i_2) \tag{4.18a}$$

$$v_2 = v_1 - E + Ri_2 \triangleq \hat{v}_2(v_1, i_2) \tag{4.18b}$$

Substituting Eq. (4.18*a*) into

$$i_{C1} = C_1\dot{v}_{C1} = -i_1 \tag{4.19a}$$

and solving for $\dot{v}_{C1}$, we obtain

$$\dot{v}_{C1} = -\frac{1}{C_1}[\hat{i}_R(v_{C1}) - i_{L2}] \tag{4.20a}$$

Substituting Eq. (4.18*b*) into

$$v_{L2} = L_2\dot{i}_{L2} = -v_2 \tag{4.19b}$$

and solving for $\dot{i}_{L2}$, we obtain

$$\dot{i}_{L2} = -\frac{1}{L_2}[v_{C1} - E + Ri_{L2}] \tag{4.20b}$$

upon substituting $v_{C1} = v_1$ and $i_{L2} = i_2$. Note that the state equations (4.20) and (4.7) are identical, as expected.

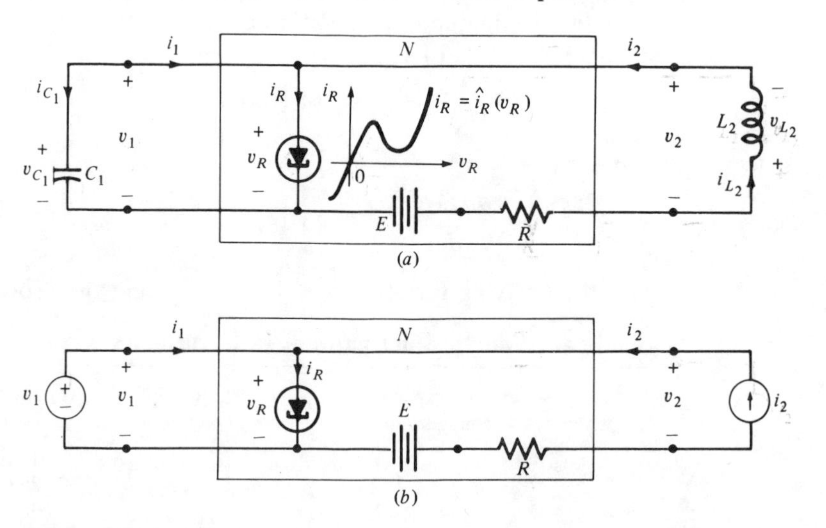

Figure 4.3 (*a*) The tunnel diode circuit in Fig. 4.1*a* is redrawn into the capacitor-inductor configuration of Table 4.1. (*b*) A nonlinear resistive two-port N results upon replacing the capacitor by a voltage source and the inductor by a current source.

Exercises

1. Derive the state equation (4.12) for the Josephson junction circuit in Fig. 4.2 using the preceding systematic method.
2. Complete the entries in Table 4.1 in terms of the remaining combinations of state variables listed in Table 4.2.
3. Consider the circuit shown in Fig. 4.4, where the elements are characterized as follows:

$$\begin{aligned} R_a&: \ i_a = v_a^5 \qquad & C_1&: \ q_{C1} = v_{C1}^3 \\ R_b&: \ v_b = e^{i_b} \qquad & L_2&: \ \phi_{L2} = i_{L2}^{1/3} \end{aligned} \tag{4.21}$$

Formulate the state equation in terms of the following state variables:

(*a*) v_{C1} and i_{L2} (*c*) v_{C1} and ϕ_{L2}

(*b*) q_{C1} and ϕ_{L2} (*d*) q_{C1} and i_{L2}

4. Repeat *Exercise 3* if the capacitor and inductor are described as follows:

$$C_1: \ q_{C1} = v_{C1}^4 \qquad L_2: \ \phi_{L2} = i_{L2}^2 \tag{4.22}$$

5. Repeat *Exercise 3* if the capacitor and inductor are described as follows:

$$C_1: \ v_{C1} = \cos q_{C1} \qquad L_2: \ i_{L2} = \sin \phi_{L2} \tag{4.23}$$

6. Repeat *Exercise 3* if the capacitor and inductor are described as follows:

$$C_1: \ v_{C1} = \cos q_{C1} \qquad L_2: \ \phi_{L2} = i_{L2}^3 \tag{4.24}$$

5 QUALITATIVE BEHAVIOR OF $\dot{\mathbf{x}} = \mathbf{f}(\mathbf{x})$

The concepts of *trajectory*, *phase portrait*, and *equilibrium state* introduced in Sec. 3 are indispensable in the study of autonomous nonlinear circuits. Let us therefore restate these concepts in a *nonlinear* setting.

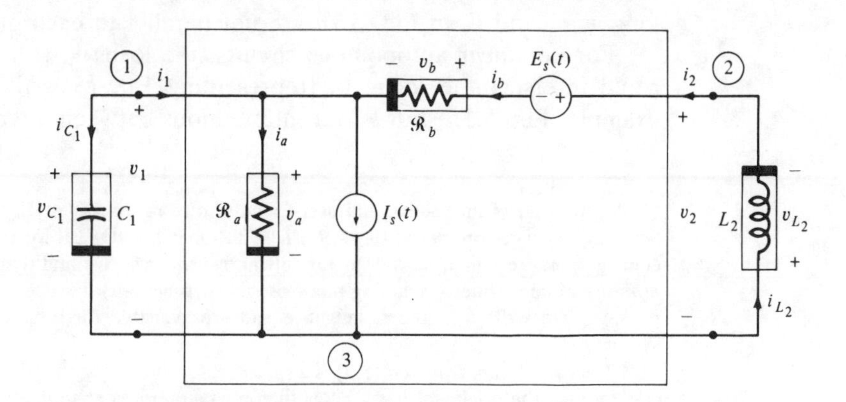

Figure 4.4 Second-order circuit for Exercises 3 to 6.

Given an initial state $\mathbf{x}_0$, the locus in the x_1-x_2 plane of the corresponding solution $(x_1(t), x_2(t))$ of the autonomous state equation (4.2) for all $t \geq 0$ is called a *trajectory* of Eq. (4.2) from $\mathbf{x}_0$.

If we apply the signals $x_1(t)$ and $x_2(t)$ to the horizontal and vertical channels of an oscilloscope, the resulting Lissajous figure traced out in the scope is precisely the associated trajectory. Note that since the *time t is suppressed* in a trajectory, it is not possible to recover the waveforms $x_1(t)$ and $x_2(t)$ associated with a given trajectory. Hence a trajectory gives only the *qualitative* but not quantitative behavior of the associated solution. For example, a closed trajectory tells us the circuit oscillates, whereas a shrinking *spiral* trajectory tells us the circuit rings (a decaying oscillation).

A family of trajectories from a large number of initial states spread (usually uniformly) all over the x_1-x_2 plane is called a *phase portrait* of the autonomous state equation (4.2).

We can photograph the phase portrait of an autonomous second-order circuit with a storage (memory) oscilloscope using different initial conditions. Even easier, we can use a computer to do the job. Since subroutines for solving general nonlinear state equations are widely available, a simple program can be written to calculate a family of trajectories from a set of preprogrammed or user-specified initial states.[24]

Any point $\mathbf{x}_Q = (x_{1Q}, x_{2Q})$ in the x_1-x_2 plane is called an *equilibrium state* of the *autonomous* state equation (4.2) iff the *velocity* $\dot{\mathbf{x}} = (\dot{x}_1, \dot{x}_2)$ at x_Q is *zero*.

An equilibrium state is therefore *any* solution of the following equation:

Equilibrium equation

$$f_1(x_1, x_2) = 0 \tag{5.1a}$$

$$f_2(x_1, x_2) = 0 \tag{5.1b}$$

For *linear* autonomous circuits, the *equilibrium equation* is given by two *linear* equations representing two straight lines Γ_1 and Γ_2, as shown in Fig. 5.1*a*.[25] Any intersection Q between Γ_1 and Γ_2 is an equilibrium state. Note that there is only one equilibrium state so long as $\Delta \triangleq a_{11}a_{22} - a_{12}a_{21} \neq 0$; i.e., so long as Γ_1 and Γ_2 in Fig. 5.1*a* are not parallel to each other.

For *nonlinear* autonomous circuits, the locus Γ_1 [representing Eq. (5.1*a*)] could intersect the locus Γ_2 [representing Eq. (5.1*b*)] at many points. For example, Fig. 5.1*b* shows four intersections between a typical "open" curve Γ_1

[24] A collection of interactive and user-friendly software packages [L. O. Chua and A. C. Deng, Software packages on *NO*NLINEAR *EL*ECTRONICS (NOEL)] for implementing this task on a color graphics terminal, as well as for solving *general* resistive and dynamics circuits which allow *arbitrary* nonlinearities (including piecewise-linear functions) has been developed. These software packages are written in the C language and are available for both UNIX and DOS operating systems.

[25] The *equilibrium equation* (3.3) is a special case where $u_1 = u_2 = 0$. In this case, both Γ_1 and Γ_2 pass through the origin so that $\mathbf{x} = \mathbf{0}$ is the only equilibrium state if $\Delta \neq 0$. In the degenerate case where $\Delta = 0$, Γ_1 coincides with Γ_2 so that there are infinitely many *equilibrium states*.

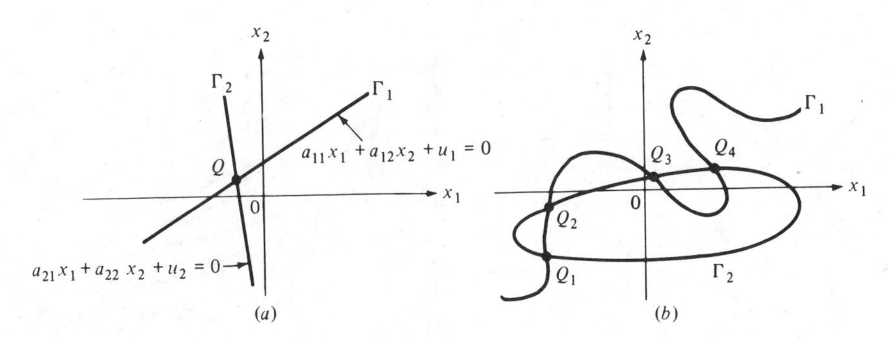

Figure 5.1 (*a*) The intersection between Γ_1 and Γ_2 is the only equilibrium state for a *linear* autonomous circuit with $\Delta \neq 0$. (*b*) The intersections between Γ_1 and Γ_2 are equilibrium states for a *nonlinear* autonomous circuit.

and a "closed" curve Γ_2. In the more general case, each locus Γ_1 and Γ_2 could contain several disconnected branches. In any case, *each* intersection between Γ_1 and Γ_2 is an equilibrium state of the associated *state equation* (4.2).

5.1 Phase Portrait

Our objective in this section is to derive the *qualitative* behaviors of the tunnel-diode circuit in Fig. 4.1 and the Josephson junction circuit in Fig. 4.2 [with $i_s(t) = 0$] by examining their respective phase portraits. We assume the phase portraits are given, as they can be obtained from a canned computer program, such as NOEL (see footnote 24).

Example 1 (Qualitative analysis of tunnel-diode circuit) Consider the tunnel-diode circuit in Fig. 4.1*a* with the following circuit parameters: $E = 1.2\text{ V}$, $R = 1.5\text{ k}\Omega$, $C_1 = 2\text{ pF}$, $L_2 = 5\text{ nH}$. The tunnel-diode characteristic shown in Fig. 4.1*b* is described by the following equation:

$$i_R = \hat{i}_R(v_R) = 17.76v_R - 103.79v_R^2 + 229.62v_R^3 - 226.31v_R^4 + 83.72v_R^5 \text{ mA} \tag{5.2}$$

Substituting the above parameters into Eq. (4.7), we obtain the following autonomous state equation:

$$\dot{v}_{C1} = \frac{i_{L2} - \hat{i}_R(v_{C1})(10^{-3})}{2 \times 10^{-12}} \triangleq f_1(v_{C1}, i_{L2}) \tag{5.3a}$$

$$\dot{i}_{L2} = \frac{1.2 - (1.5 \times 10^3)i_{L2} - v_{C1}}{5 \times 10^{-9}} \triangleq f_2(v_{C1}, i_{L2}) \tag{5.3b}$$

where the function $\hat{i}_R(\cdot)$ is given by Eq. (5.2).

The phase portrait associated with Eq. (5.3) is shown in Fig. 5.2: It shows a family of trajectories, each one emanating from a different initial

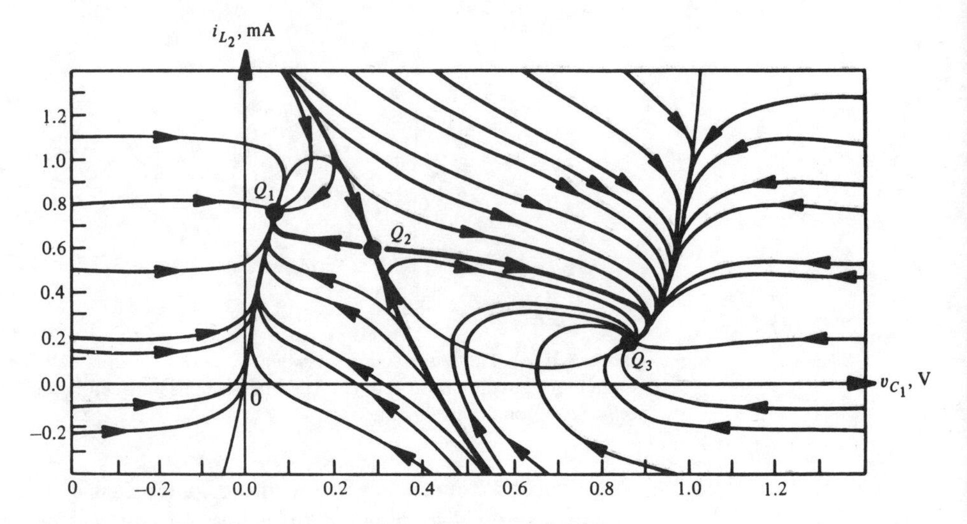

Figure 5.2 Phase portrait associated with state equation (5.3).

state and "flowing" in the direction of the arrow head. We will see that "this picture is worth a thousand words"!

It is often more illuminating to associate the phase portrait with a *vector field* analogous to the iron-filing pattern generated by a static magnetic field in a typical sophomore physics laboratory. Here, each point $P\ (v_{C1}, i_{L2})$ is associated with a vector whose slope is calculated from

$$\left.\frac{di_{L2}}{dv_{C1}}\right|_P = (0.4 \times 10^{-3})\left[\frac{1.2 - (1.5 \times 10^3)i_{L2} - v_{C1}}{i_{L2} - \hat{i}_R(v_{C1})(10^{-3})}\right]\Bigg|_P = \left.\frac{f_2(v_{C1}, i_{L2})}{f_1(v_{C1}, i_{L2})}\right|_P \tag{5.3c}$$

obtained by dividing Eq. (5.3*b*) by Eq. (5.3*a*).

Equation (5.3*c*) uniquely specifies the slope at *any* point P in the v_{C1}-i_{L2} plane. Hence, one can in principle fill up the plane with such a vector drawn at each point. In practice, only a sample of these vectors drawn at uniformly distributed points is sufficient. Given this vector field, any trajectory can be generated by drawing a smooth curve *tangent* to the vector at each point on the curve.

An inspection of the phase portrait in Fig. 5.2 reveals the following important *qualitative* behavior: Except for two rather *special* trajectories which tend to the point Q_2 (0.29 V, 0.6 mA), all other trajectories eventually tend to either the point Q_1 (0.06 V, 0.7 mA) or to the point Q_3 (0.88 V, 0.2 mA).

If we substitute the coordinates of each of these points into Eq. (5.2), we would find

$$\dot{v}_{C1} = \left.\frac{i_{L2} - \hat{i}_R(v_{C1})(10^{-3})}{2 \times 10^{-12}}\right|_{Q_i} = 0 \tag{5.4a}$$

$$\dot{i}_{L2} = \frac{1.2 - (1.5 \times 10^3) i_{L2} - v_{C1}}{5 \times 10^{-9}} \bigg|_{Q_i} = 0 \qquad (5.4b)$$

Hence, Q_1, Q_2, and Q_3 are *equilibrium states* and the circuit is in steady state (i.e., motionless) when operating at these points. Now, when the circuit in Fig. 4.1*a* is in equilibrium, the capacitor becomes an open circuit and the inductor becomes a short circuit. Consequently, $v_{C1} = v_R$ and $i_{L2} = i_R$ and the tunnel-diode v_R-i_R curve in Fig. 4.1*b* can be superimposed on top of the phase portrait, as shown by the lighter curve (without arrowhead) in Fig. 5.2. Note that the equilibrium points Q_1, Q_2, and Q_3 are identical to the operating points of the associated "dc" resistive circuit,[26] namely, the intersection points between the tunnel-diode characteristic and the dc load line drawn with $R = 1.5\ \text{k}\Omega$ and $E = 1.2\ \text{V}$.

The equilibrium states Q_1 and Q_3 in Fig. 5.2 are said to be *asymptotically stable* because all trajectories originating from points in a small neighborhood of Q_1 or Q_3 tend to Q_1 or Q_3 as $t \to \infty$. In contrast, the equilibrium state Q_2 is said to be *unstable* because there exist points arbitrarily close to Q_2 whose trajectories diverge from Q_2 as $t \to \infty$.

Translated into laboratory measurements, the above qualitative property says that depending on the *initial* capacitor voltage and inductor current, we will measure either a steady-state tunnel-diode voltage of 0.06 V (if the trajectory tends to Q_1) or of 0.88 V (if the trajectory tends to Q_3). The voltage of 0.29 V at Q_2 is *never* observed in practice because the ever present *physical noise* would cause the trajectory to diverge from Q_2 even if it were possible to set up the *exact* initial conditions corresponding to Q_2.

If we imagine the two *special* trajectories converging to Q_2 in Fig. 5.2 as a fence, then all trajectories originating from the *left* of the fence will tend to equilibrium state Q_1. All those originating from the *right* will tend to equilibrium state Q_3. Such an imaginary fence is called a *separatrix* because it partitions the v_{C1}-i_{L2} plane into two mutually exclusive regions, each one displaying a similar *qualitative* behavior.

Incidentally, this tunnel-diode circuit has been used as a computer memory, where the equilibrium state Q_1 is associated with the binary state "0" and the equilibrium state Q_3 is associated with the binary state "1." In the context of the phase portrait in Fig. 5.2, triggering from Q_1 to Q_3, or vice versa, consists of applying a small triggering signal of sufficient duration which allows the trajectory to move over to the other side of the separatrix.

The range of v_{C1} and i_{L2} in the phase portrait in Fig 5.2 was chosen so that all essential qualitative features are displayed.[27] The portrait outside this range does not contain any new qualitative features and is therefore omitted.

[26] This important property will be discussed again in more detail in Example 1 of Sec. 5.2.

[27] An interactive computer graphics program, such as NOEL (see footnote 24), can be developed to automatically "zoom" in on any region specified by the user.

The phase portrait in Fig. 5.2 tells us the *global* qualitative behavior of the tunnel-diode circuit. This information is much more valuable than any solution waveform corresponding to a *particular* initial state $\mathbf{x}_0$ because such a solution may not predict how the circuit will behave when the initial state is changed even slightly. For example, a solution with initial state falling on the separatrix would be totally useless if not misleading!

Example 2 (Qualitative analysis of Josephson junction circuit) Consider the autonomous Josephson junction circuit shown in Fig. 5.3, obtained by setting $i_s(t) = 0$ in Fig. 4.2. The state equation for this circuit is obtained by setting $i_s(t) = 0$ in Eq. (4.12):

$$\dot{v}_{C1} = \frac{1}{C_1}(-Gv_{C1} - I_0 \sin k\phi_{L2}) \triangleq f_1(v_{C1}, \phi_{L2}) \tag{5.5a}$$

$$\dot{\phi}_{L2} = v_{C1} \triangleq f_2(v_{C1}, \phi_{L2}) \tag{5.5b}$$

A typical phase portrait of Eq. (5.5) is sketched in Fig. 5.4*a* with ϕ_{L2} chosen as the *horizontal* axis. A computer-generated phase portrait with $C_1 = 1$, $G = 0.25$, $I_0 = 1$, and $k = 1$ is shown in Fig. 5.4*b*. Note that the slope

$$\frac{dv_{C1}}{d\phi_{L2}} = \frac{-Gv_{C1} - I_0 \sin k\phi_{L2}}{C_1 v_{C1}} \tag{5.6}$$

at any point (ϕ_{L2}, v_{C1}) is identical to that at $(\phi_{L2} + 2\pi/k, v_{C1})$. Hence, the phase portrait in Fig. 5.4*a* is *periodic* with a period equal to $2\pi/k$.

To find the equilibrium state, we set Eq. (5.5) to zero and solve for v_{C1} and ϕ_{L2}, namely,

$$v_{C1} = 0 \qquad \phi_{L2} = \pm\frac{n\pi}{k} \qquad n = 0, 1, 2, \ldots \tag{5.7}$$

Hence, there are *infinitely many* equilibrium states; seven of them are identified by solid dots in Fig. 5.4*a*.

Since this phase portrait is $(2\pi/k)$-periodic, it is not necessary to draw the portrait over the infinite extent $-\infty < \phi_{L2} < \infty$ in order to capture all distinct qualitative behaviors of the circuit. Rather, we need only draw the portrait over any vertical strip of width $2\pi/k$. For example, it suffices to

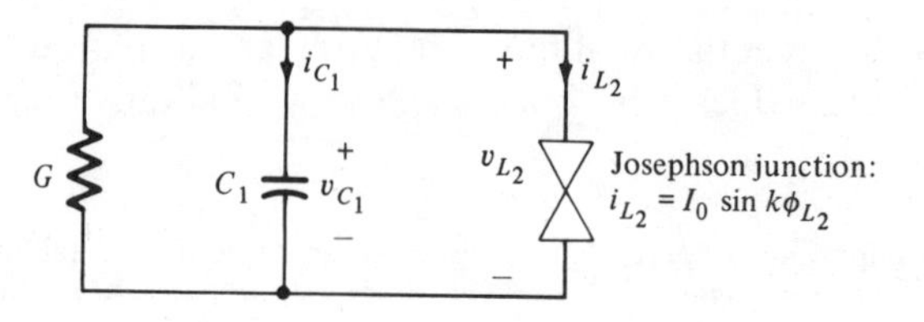

Figure 5.3 A source-free Josephson junction circuit.

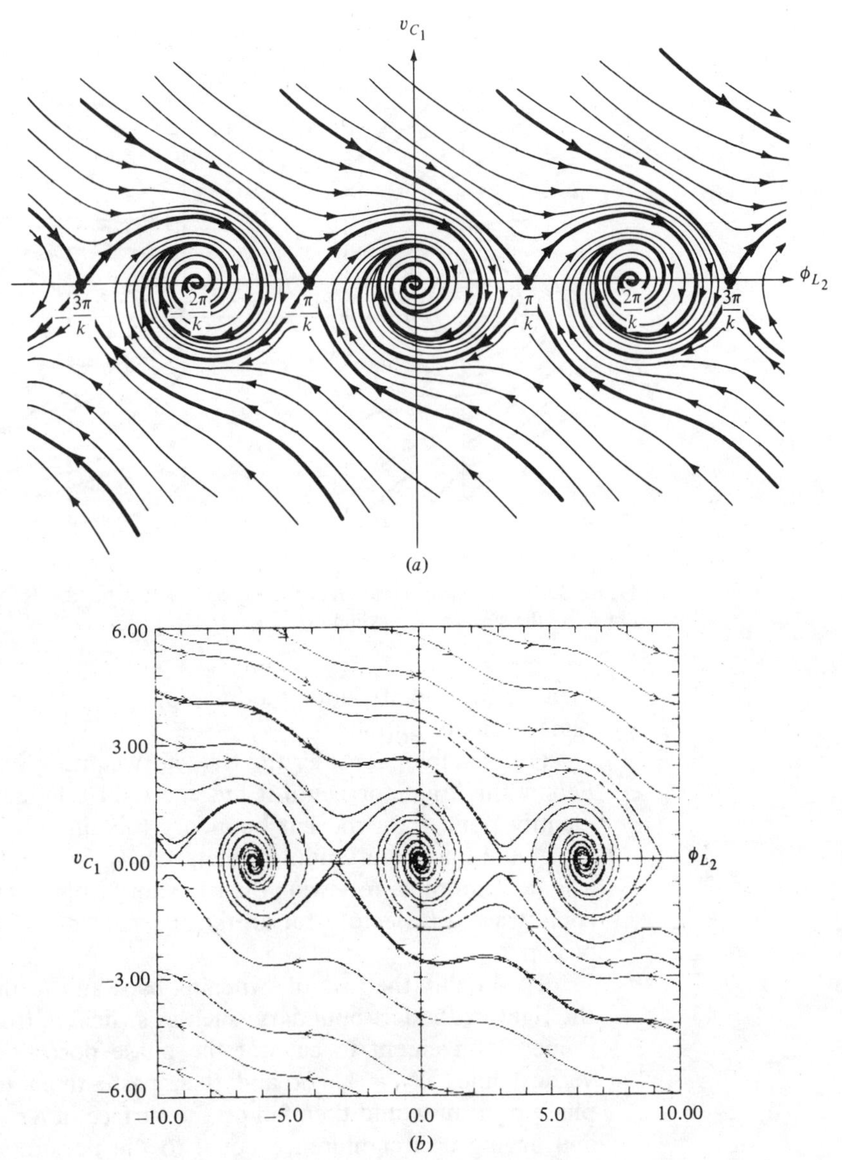

Figure 5.4 (*a*) Typical phase portrait associated with state equation (5.5). (*b*) Computer-generated phase portrait.

sketch the phase portrait over the interval $-\pi/k \leq \phi_{L2} \leq \pi/k$, as shown in Fig. 5.5*a*. The entire phase portrait can be reproduced by using Fig. 5.5*a* as a template.

If we try to follow a trajectory from any initial point on the right-hand margin using the phase portrait in Fig. 5.4*a*, we would eventually run out of paper. A more practical way is to use the phase portrait in Fig. 5.5*a* and

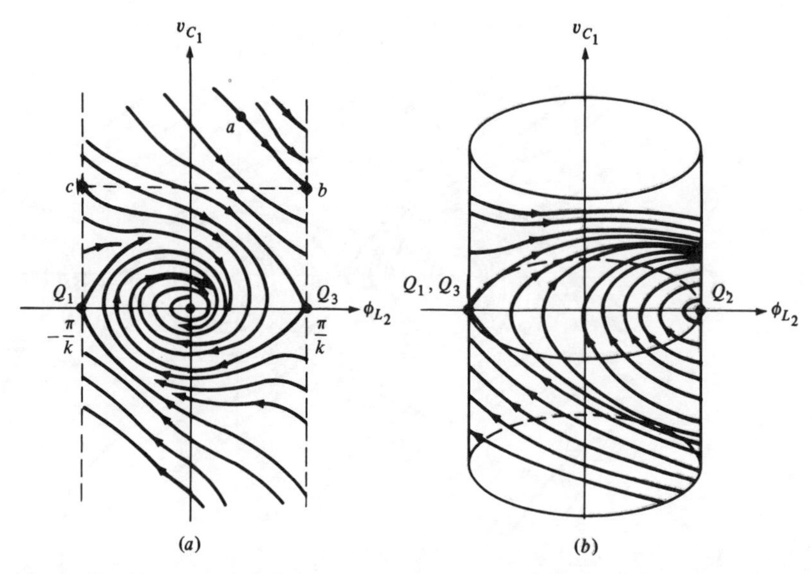

Figure 5.5 (*a*) Phase portrait over the region $-\pi/k \leq \phi_{L2} \leq \pi/k$. (*b*) The same phase portrait drawn on the surface of a cylinder.

then identify points along the lines $\phi_{L2} = -\pi/k$ and $\phi_{L2} = \pi/k$ having the same ordinate v_{C1}.

For example, consider the trajectory starting from point *a* in Fig. 5.5*a*. Follow this trajectory until it hits the right boundary at point *b*. We then identify point *c* on the left boundary (obtained by drawing an imaginary horizontal line from *b* until it intersects $\phi_{L2} = -\pi/k$) to be the same point as *b*, and continue following the trajectory from *c*. Using this algorithm, we could trace a trajectory for all times $-\infty < t < \infty$ without ever running out of paper.

To simplify the task of switching back and forth when a trajectory hits the right or the left boundary, such as switching from point *b* to point *c*, it is more convenient to cut out the phase portrait in Fig. 5.5*a* along the vertical lines $\phi_{L2} = \pm\pi/k$ and then paste them together. The resulting phase portrait would then fall on the surface of a cylinder of infinite extent and having a circumference equal to the period $2\pi/k$, as shown in Fig. 5.5*b*.

When represented in this manner, we say the state equation (5.5) has a "cylindrical state space," instead of the more usual "euclidean" plane. One advantage of working with the cylindrical state space in Fig. 5.5*b* is that we can follow a trajectory continuously for all times *t* without having to switch points periodically as in Fig. 5.5*a*.

In fact, for physical systems which exhibit a "cylindrical" state space, there always exists a "physical" interpretation which suggests that the cylindrical state space is a more natural setting. This is so with the above

Josephson junction circuit provided the flux ϕ_{L2} is interpreted as the quantum "phase difference" in Josephson junction device physics. Since not many students are familiar with this rather subtle physical mechanism, we will now turn to a more mundane example which demonstrates why a cylindrical phase portrait makes good physical sense.

Mechanical analog of autonomous Josephson junction circuit Consider the *simple* pendulum shown in Fig. 5.6*a* where ℓ denotes the length of the rod and m denotes the mass of the bob. For simplicity assume the rod is rigid and has zero mass. Let θ denote the angle subtended by the rod and the vertical axis through the point 0.

The pendulum is free to swing in a vertical plane so that a big initial push would cause the rod to execute more than one complete 360° revolution. Each counterclockwise revolution causes θ to increase by 2π, and each clockwise revolution causes θ to decrease by 2π. Of course, friction will damp out the motion so that the rod will eventually come to rest at its stable equilibrium position $\theta = \pm n2\pi$, where n is the number of complete 360° revolutions executed before coming to rest.

Note that in principle the rod could come to rest at another equilibrium position corresponding to $\theta = \pm(2n+1)\pi$, where n is the number of complete 360° revolutions before coming to rest. However, such an equilibrium position is unstable and is therefore not attainable in practice.

To write the equation of motion describing the simple pendulum, let us assume the damping is proportional to the angular velocity $\dot{\theta}$ with a coefficient of friction equal to μ. Now from physics, we know gravity applies a torque on the pendulum equal to $\tau = -mg\ell \sin\theta$. This is balanced by $I\ddot{\theta}$, where I denotes the moment of inertia (here, $I = m\ell^2$ in view of our assumption on the rod), *and* by a friction torque equal to $\mu\dot{\theta}$, namely, $I\ddot{\theta} + \mu\dot{\theta} = -mg\ell \sin\theta$, or

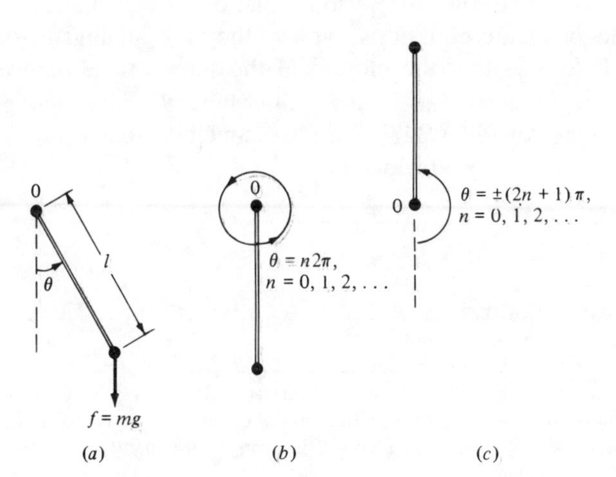

Figure 5.6 (*a*) Pendulum of length ℓ and mass m. (*b*) Pendulum at stable equilibrium position. (*c*) Pendulum at unstable equilibrium position.

$$\ddot{\theta} + \frac{\mu}{I}\dot{\theta} + \frac{mg\ell}{I}\sin\theta = 0 \tag{5.8}$$

Note that Eq. (5.8) is similar in form to the Josephson junction equation (4.10) with $i_s(t) = 0$. To obtain an equivalent state equation for the simple pendulum, let us define

$$x_1 \triangleq \dot{\theta} \tag{5.9a}$$

$$x_2 \triangleq \theta \tag{5.9b}$$

and recast Eq. (5.8) as follows:

Pendulum state equation

$$\dot{x}_1 = -\frac{\mu}{I}x_1 - \frac{mg\ell}{I}\sin x_2 \tag{5.10a}$$

$$\dot{x}_2 = x_1 \tag{5.10b}$$

To show that Eq. (5.10) is analogous to the autonomous Josephson junction state equation (5.5), let us introduce the "normalized" variables

$$y_1 \triangleq kv_{C1} \tag{5.11a}$$

$$y_2 \triangleq k\phi_{L2} \tag{5.11b}$$

Substituting Eq. (5.11) into Eq. (5.5), we obtain the following "normalized" Josephson junction state equation:

Normalized Josephson state equation

$$\dot{y}_1 = -\frac{G}{C}y_1 - \frac{kI_0}{C}\sin y_2 \tag{5.12a}$$

$$\dot{y}_2 = y_1 \tag{5.12b}$$

Clearly, Eq. (5.12) is identical to Eq. (5.10) if we identify x_1 with y_1, x_2 with y_2, μ/I with G/C_1, and $mg\ell/I$ with kI_0/C_1. Since the simple pendulum in Fig. 5.6*a* and the Josephson junction circuit in Fig. 5.3 are described by analogous state equations, we say they are analogs of each other.[28]

It follows that the motion of the simple pendulum is completely specified by the phase portrait in Fig. 5.4, where θ is the analog of $k\phi_{L2}$ and $\dot{\theta}$ is the analog of kv_{C1}. In particular, we conclude that the equilibrium states of the simple pendulum are located at

$$\dot{\theta} = 0$$

$$\theta = \pm n\pi \qquad n = 0, 1, 2, \ldots \tag{5.13}$$

The equilibrium states $\theta = \pm n2\pi$, $n = 0, 1, 2, \ldots$, correspond to *stable*

[28] Interestingly enough, several unrelated physical systems are also modeled by a similar equation. For example, the equation used by power engineers to model the pull-out torque of synchronous motors (swing equation) and the equation used by communication engineers to model "phase-lock loops" are also identical in form. Consequently, the state equation (5.12) is of great practical importance.

equilibrium positions in Fig. 5.6*b*. Likewise, the equilibrium states $\theta = \pm(2n+1)\pi$, $n = 0, 1, 2, \ldots$, correspond to *unstable* equilibrium positions in Fig. 5.6*c*.

Although the pendulum state equation (5.10) has infinitely many equilibrium states, only two are really distinct from a physical point of view: One corresponds to the "stable" equilibrium position in Fig. 5.6*b*; the other corresponds to the "unstable" equilibrium position in Fig. 5.6*c*.

It follows from the above physical interpretation that the *cylindrical* phase portrait in Fig. 5.5*b* is a more "physical" representation than the rectangular phase portrait in Fig. 5.4.

Exercise Let $C = 0$ and $i_s(t) = A \sin \omega t$ in Fig. 4.2, and show that the solution waveform $\phi(t)$ of the resulting first-order state equation can be drawn on the surface of a torus.

5.2 Equilibrium States and Operating Points

In the preceding section, the equilibrium states associated with the autonomous state equation (4.2) are found by solving the *equilibrium equation* (5.1). This assumes implicitly that the state equation (4.2) has been derived so that the functions $f_1(\cdot)$ and $f_2(\cdot)$ are available either in closed form or in the form of a computer subroutine. For complicated circuits, deriving the state equations can be a time-consuming task. Our objective in this section is to offer an alternate method which is often easier. Moreover, we will identify equilibrium states of an *RLC* circuit with the operating points of an associated nonlinear resistive circuit. Using this identification, the "nonphysical" notion of a resistive circuit exhibiting *multiple* operating points in Chaps. 2 to 4 can be given a meaningful physical interpretation.

Let $\mathbf{x}_Q = (x_{1Q}, x_{2Q})$ be an equilibrium state of the autonomous state equation (4.2). Suppose we pick $\mathbf{x}_Q$ as the *initial state* at $t = t_0$, then it follows from Eq. (5.1) that

$$\dot{x}_1(t) = f_1(x_{1Q}, x_{2Q}) = 0 \qquad \text{for all } t \geq t_0 \tag{5.14a}$$

$$\dot{x}_2(t) = f_2(x_{1Q}, x_{2Q}) = 0 \qquad \text{for all } t \geq t_0 \tag{5.14b}$$

Hence, if $\mathbf{x}(t_0) = \mathbf{x}_Q$, then $\mathbf{x}(t) = \mathbf{x}_Q$ for all $t \geq t_0$; thus the circuit is said to be in equilibrium, or in dc steady state.

Let us examine the behavior of a capacitor or an inductor at an equilibrium state:

Capacitor at equilibrium: If $x_j = v_{Cj}$ is the state variable, then

$$\dot{x}_j = \dot{v}_{Cj} = \frac{i_{Cj}}{C_j} = 0 \tag{5.15a}$$

Inductor at equilibrium: If $x_j = i_{Lj}$ is the state variable, then

$$\dot{x}_j = \dot{i}_{Lj} = \frac{v_{Lj}}{L_j} = 0 \tag{5.16a}$$

If $x_j = q_{Cj}$ is the state variable, then

$$\dot{x}_j = \dot{q}_{Cj} = i_{Cj} = 0 \qquad (5.15b)$$

Hence, *regardless of the choice of state variable, the capacitor behaves like an open circuit at any equilibrium state.*

If $x_j = \phi_{Lj}$ is the state variable, then

$$\dot{x}_j = \dot{\phi}_{Lj} = v_{Lj} = 0 \qquad (5.16b)$$

Hence, *regardless of the choice of state variable, the inductor behaves like a short circuit at any equilibrium state.*

Let us now apply the above property to find the equilibrium states of the tunnel-diode and Josephson junction circuits considered earlier *without* writing the state equation.

Example 1 (Equilibrium states of tunnel-diode circuits) Suppose the tunnel-diode circuit in Fig. 4.1*a* is in equilibrium. Then we can replace the capacitor by an open circuit and the inductor by a short circuit as shown in Fig. 5.7*a*. Note that the resulting circuit $\mathcal{N}_Q$ is a *resistive* circuit. Using the load-line method from Chap. 2, we obtain the operating points Q_1 (0.06 V, 0.76 mA), Q_2 (0.29 V, 0.60 mA), and Q_3 (0.88 V, 0.20 mA) shown in Fig. 5.7*b*).

Now, since we earlier chose v_{C1} and i_{L2} as the *state variables*, let us calculate the value of v_{C1} and i_{L2} at each operating point. For the circuit in Fig. 5.7, we find trivially $v_{C1} = v_R$ and $i_{L2} = i_R$. Consequently, for this simple circuit, the *equilibrium states* coincide numerically with the operating points of $\mathcal{N}_Q$, namely,

$$\begin{aligned} &\text{Equilibrium state} \quad Q_1\text{: } v_{C1}(Q_1) = 0.06 \text{ V} \quad i_{L2}(Q_1) = 0.76 \text{ mA} \\ &\text{Equilibrium state} \quad Q_2\text{: } v_{C1}(Q_2) = 0.29 \text{ V} \quad i_{L2}(Q_2) = 0.60 \text{ mA} \\ &\text{Equilibrium state} \quad Q_3\text{: } v_{C1}(Q_3) = 0.88 \text{ V} \quad i_{L2}(Q_3) = 0.20 \text{ mA} \end{aligned} \qquad (5.17)$$

These equilibrium states agree exactly with those obtained earlier by solving Eq. (5.4) directly, as expected.

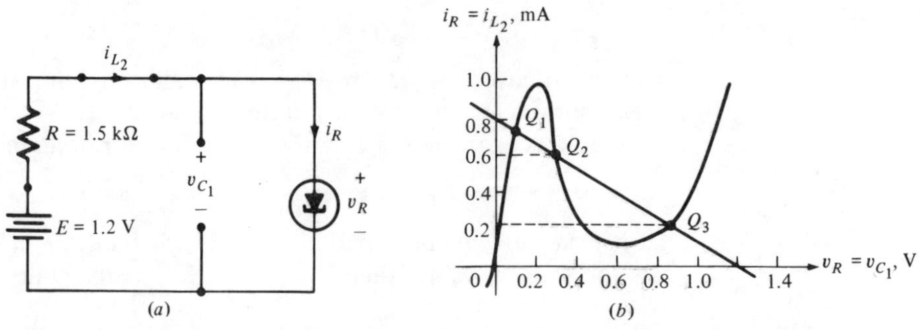

Figure 5.7 (*a*) Resistive circuit $\mathcal{N}_Q$ associated with the tunnel-diode circuit. (*b*) Operating points determined by load-line method.

REMARKS

1. In the above example the coordinate (v_{C1}, i_{L2}) at each equilibrium state Q_j of the dynamic circuit $\mathcal{N}$ coincides exactly with the coordinate (v_R, i_R) at the corresponding operating point of the tunnel diode in the associated resistive circuit $\mathcal{N}_Q$. In this case, there is no ambiguity in labeling both the equilibrium states in Fig. 5.2 and the operating points in Fig. 5.7 by the same symbols Q_1, Q_2, and Q_3.

2. In most circuits, the coordinate at an equilibrium state Q_j of the *dynamic* circuit will not be numerically equal to that of the operating point of the nonlinear resistor, even if the circuit contains only one nonlinear resistor. For example, suppose the capacitor C_1 in the tunnel-diode circuit in Fig. 4.1*a* is connected across the linear resistor R instead of across the tunnel diode. The equilibrium state (v_{C1}, i_{L2}) of the dynamic circuit in this case will be different from the corresponding operating point (v_R, i_R) of the associated resistive circuit because $v_{C1} = -Ri_{L2} \neq v_R$ in this case.

However, for *each* operating point Q_j, we can always calculate the *corresponding* value of v_{C1} and i_{L2}. Hence, as long as we have chosen *capacitor voltages* v_{Cj} and/or *inductor currents* i_{Lj} as state variables, the distinction between equilibrium states and operating points is only superficial. We will summarize this important observation as follows:

Equilibrium state operating point identification property With *capacitor voltage* (respectively, *inductor current*) chosen as state variable, there is a one-to-one correspondence between the *equilibrium states* of an *autonomous* circuit $\mathcal{N}$ and the *operating points* of its associated *resistive* circuit $\mathcal{N}_Q$ obtained by open-circuiting all capacitors (respectively, short-circuiting all inductors). In particular, for each operating point Q_j, we simply calculate the corresponding "open-circuit" capacitor voltage (respectively, "short-circuit" inductor current).

3. In view of the above one-to-one correspondence, we will often abuse notation by labeling equilibrium states and operating points by the same symbols $Q_1, Q_2, \ldots$.

Multiple operating point paradox resolved We have encountered many resistive circuits having *multiple* operating points. For example, the tunnel-diode circuit in Fig. 4.7 of Chap. 2 (which is identical to the circuit in Fig. 5.7) has three operating points for certain values of E and R. This answer seems to contradict thc fact that a single laboratory measurement on the corresponding physical circuit can only give one operating point. We can now resolve this *paradox* as follows:

The tunnel-diode circuit in Fig. 4.7 of Chap. 2 is *not* a realistic model of the *physical* circuit. In any physical circuit, there always exists second-order effects such as fringing electric and magnetic fields, "parasitic" inductances, capacitances, and resistances, due to the connecting wires and device terminals, etc. In circuits having a unique solution these second-order effects can often be neglected without discernible errors. In circuits exhibiting *multiple* solutions, however, some of these parasitic elements can*not* be neglected, no matter how small their values are.

The above cited tunnel-diode circuit is a case in point. By augmenting the tunnel diode with a shunt capacitance and series inductance, we obtain the more realistic circuit model shown in Fig. 4.1*a*. The three operating points in the *resistive* circuit can now be interpreted as *equilibrium states* in the remodeled *dynamic* circuit. Which of these three operating points is actually measured depends on the *stability* of the corresponding equilibrium state as well as on the *initial condition*.

For this circuit, it is clear that the operating point Q_2 is not observable in practice because the slightest physical noise will cause the initial condition to "slip" over to one side of the fence in Fig. 5.2 even if it were possible to set it up there initially.

The other two operating points Q_1 and Q_3, however, are observable. The phase portrait in Fig. 5.2 allows us to determine which point will actually be measured for a given initial state.

Example 2 (Equilibrium states of Josephson junction circuit) Suppose the Josephson junction circuit in Fig. 5.3 is in equilibrium. Then we can replace the capacitor by an open circuit and the inductor by a short circuit as shown in Fig. 5.8*a*. Since the associated resistive circuit $\mathcal{N}_Q$ is linear and source-free, it has a *unique* operating point. In particular, we have $v_{C1} = 0$ and $i_{L2} = 0$.

However, recall that the state variables in the state equation (5.5) consist of v_{C1} and ϕ_{L2}, and not v_{C1} and i_{L2} as in Example 1. Consequently, we must calculate the value of ϕ_{L2} corresponding to $i_{L2} = 0$. This can be obtained by inspection of the ϕ_{L2}-i_{L2} characteristic of the Josephson junction shown in Fig. 5.8*b*, namely,

$$\phi_{L2} = \pm \frac{n}{k} \qquad n = 0, 1, 2, \ldots \tag{5.18}$$

We conclude therefore that the Josephson junction circuit in Fig. 5.3 has *infinitely many* equilibrium states located at $(v_{C1}, \phi_{L2}) = (0, \phi_{L2})$, where ϕ_{L2} is given by Eq. (5.18).

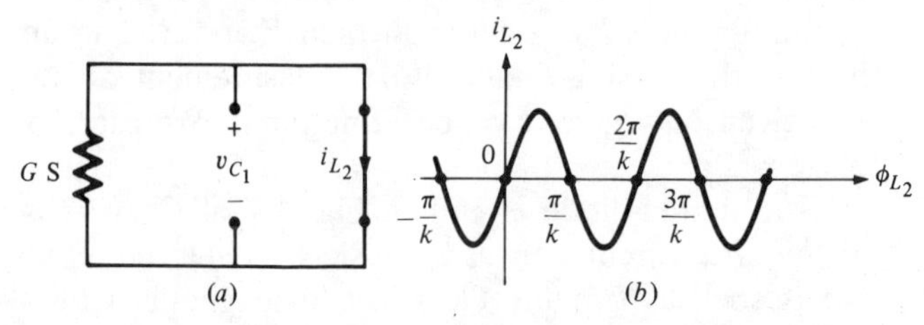

Figure 5.8 (*a*) Resistive circuit $\mathcal{N}_Q$ associated with the Josephson junction circuit. (*b*) When $i_{L2} = 0$, there correspond infinitely many values of ϕ_{L2} (solid dots).

These equilibrium states agree exactly with those obtained by solving state equation (5.5) directly, as expected.

Note that unlike Example 1, when the state variables are not restricted to capacitor voltages and/or inductor currents, there is *no longer* a one-to-one correspondence between the equilibrium states of an autonomous circuit $\mathcal{N}$ and the operating points of its associated resistive circuit $\mathcal{N}_Q$. However, the *first step* for finding the equilibrium states remains the same: Find the operating points of the associated resistive circuit $\mathcal{N}_Q$. Once the "open-circuit" capacitor voltages and/or "short-circuit" inductor currents have been found, the *second step* consists simply of calculating the *charge(s)* q_{Cj} (from the q_{Cj}-v_{Cj} *capacitor* characteristic) and/or the *flux*(es) ϕ_{Lj} (from the ϕ_{Lj}-i_{Lj} inductor characteristic) corresponding to *each* operating point of $\mathcal{N}_Q$. If these characteristics are not strictly monotonic, which is sometimes the case if q_{Cj} and/or ϕ_{Lj} are chosen as state variables, then each operating point of $\mathcal{N}_Q$ would in general give rise to more than one equilibrium state.

In either case, we have shown how to find the equilibrium states without ever writing the state equation.

5.3 Qualitative Behavior Near Equilibrium States

An examination of the phase portraits in Figs. 5.2 and 5.4 reveals that the qualitative behavior of the trajectories in the vicinity of each equilibrium state looks just like those we saw earlier in Sec. 3 for *linear* time-invariant circuits (see Fig. 3.11). For example, in Fig. 5.2, the trajectories near Q_1, Q_2, and Q_3 are similar to those associated with a *stable node*, *saddle point*, and *stable node* respectively. Likewise, the trajectories near $v_{C1} = 0$, $\phi_{L2} = \pm n2\pi/k$, $n = 0, 1, 2, \ldots$, in Fig. 5.4 are similar to those associated with a *stable focus*, whereas those at $v_{C1} = 0$, $\phi_{L2} = \pm(2n+1)\pi/k$ are similar to those associated with a *saddle point*. Our objective in this section is to show that these observations are actually quite general, thereby allowing us to make use of the "linear" techniques in Sec. 3 for predicting the qualitative behavior of autonomous "nonlinear" systems.

Let $Q \triangleq (x_{1Q}, x_{2Q})$ denote any *equilibrium state* of the autonomous state equation (4.2). Expanding the right-hand side of Eq. (4.2) into its Taylor series about the point (x_{1Q}, x_{2Q}), we obtain

$$\dot{x}_1 = f_1(x_{1Q}, x_{2Q}) + a_{11}(x_1 - x_{1Q}) + a_{12}(x_2 - x_{2Q}) + \text{h.o.t.} \tag{5.19a}$$

$$\dot{x}_2 = f_2(x_{1Q}, x_{2Q}) + a_{21}(x_1 - x_{1Q}) + a_{22}(x_2 - x_{2Q}) + \text{h.o.t.} \tag{5.19b}$$

where

$$\begin{aligned} a_{11} &\triangleq \left.\frac{\partial f_1(x_1, x_2)}{\partial x_1}\right|_{\substack{x_1 = x_{1Q} \\ x_2 = x_{2Q}}} & a_{12} &\triangleq \left.\frac{\partial f_1(x_1, x_2)}{\partial x_2}\right|_{\substack{x_1 = x_{1Q} \\ x_2 = x_{2Q}}} \\ a_{21} &\triangleq \left.\frac{\partial f_2(x_1, x_2)}{\partial x_1}\right|_{\substack{x_1 = x_{1Q} \\ x_2 = x_{2Q}}} & a_{22} &\triangleq \left.\frac{\partial f_2(x_1, x_2)}{\partial x_2}\right|_{\substack{x_1 = x_{1Q} \\ x_2 = x_{2Q}}} \end{aligned} \tag{5.20}$$

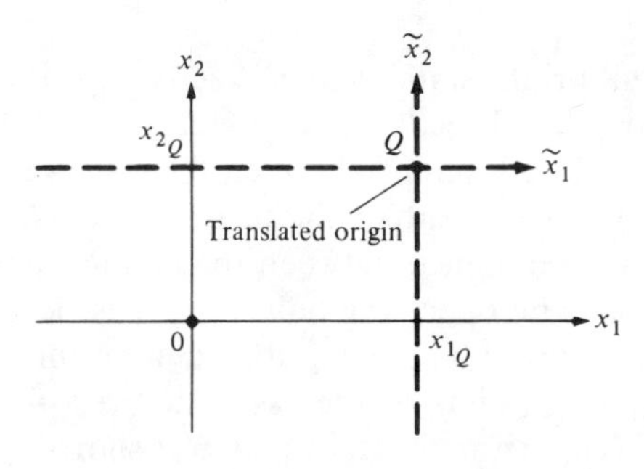

Figure 5.9 Relationship between the original state variables (x_1, x_2) and the translated state variables $(\tilde{x}_1, \tilde{x}_2)$. The same phase portrait is described by either Eq. (4.2) or Eq. (5.23).

denote first partial derivatives evaluated at (x_{1Q}, x_{2Q}), and h.o.t. denotes *higher-order terms* involving $(x_1 - x_{1Q})^n$ and $(x_2 - x_{2Q})^n$, $n = 2, 3, \ldots$.

Since (x_{1Q}, x_{2Q}) is an equilibrium state, we have $f_1(x_{1Q}, x_{2Q}) = 0$ in Eq. (5.19*a*) and $f_2(x_{1Q}, x_{2Q}) = 0$ in Eq. (5.19*b*). Since $(d/dt)(x_1 - x_{1Q}) = \dot{x}_1$ and $(d/dt)(x_2 - x_{2Q}) = \dot{x}_2$, we can rewrite Eq. (5.19) as follows:

$$\frac{d}{dt}(x_1 - x_{1Q}) = a_{11}(x_1 - x_{1Q}) + a_{12}(x_2 - x_{2Q}) + \text{h.o.t.} \tag{5.21a}$$

$$\frac{d}{dt}(x_2 - x_{2Q}) = a_{21}(x_1 - x_{1Q}) + a_{22}(x_2 - x_{2Q}) + \text{h.o.t.} \tag{5.21b}$$

Since we are interested in the trajectories near (x_{1Q}, x_{2Q}), let us define

$$\tilde{x}_1 \triangleq x_1 - x_{1Q} \tag{5.22a}$$

$$\tilde{x}_2 \triangleq x_2 - x_{2Q} \tag{5.22b}$$

and rewrite Eq. (5.21) as follows:

$$\dot{\tilde{x}}_1 = a_{11}\tilde{x}_1 + a_{12}\tilde{x}_2 + \text{h.o.t.} \tag{5.23a}$$

$$\dot{\tilde{x}}_2 = a_{21}\tilde{x}_1 + a_{22}\tilde{x}_2 + \text{h.o.t.} \tag{5.23b}$$

where h.o.t. contains higher-order terms in $\tilde{x}_1$ and $\tilde{x}_2$.

Note that so far we have not made any approximation yet: State equations (4.2) and (5.23) have *identical* phase portraits except that the origin in the phase portrait of Eq. (4.2) has been translated to the equilibrium state (x_{1Q}, x_{2Q}) in the phase portrait of Eq. (5.23). This translation of variables is illustrated in Fig. 5.9.

Linearized approximation Now, if we restrict our attention to a sufficiently small neighborhood of Q such that the higher-order terms (h.o.t.) in Eq. (5.23) become negligible, then we may *approximate* the nonlinear state equation [Eq. (5.23)] by the following linear equation:[29]

Linearized state equation about Q

$$\begin{bmatrix} \dot{\tilde{x}}_1 \\ \dot{\tilde{x}}_2 \end{bmatrix} = \begin{bmatrix} a_{11} & a_{12} \\ a_{21} & a_{22} \end{bmatrix} \begin{bmatrix} \tilde{x}_1 \\ \tilde{x}_2 \end{bmatrix} \tag{5.24}$$

[29] The solution of the *linearized state equation* about the *equilibrium point* Q can be interpreted as the solution obtained by a *small-signal analysis* of the circuit about the dc *operating point* Q.

The *exact* nonlinear state equation (4.2) and the *linearized* state equation (5.24) about the equilibrium point $\mathbf{x}_Q$ are written in vector form as follows for comparison purposes:

$$\dot{\mathbf{x}} = \mathbf{f}(\mathbf{x}) \qquad \text{or} \qquad \dot{\tilde{\mathbf{x}}} = \mathbf{f}(\mathbf{x}_Q + \tilde{\mathbf{x}}) \tag{5.25}$$

$$\dot{\tilde{\mathbf{x}}} = \mathbf{A}\tilde{\mathbf{x}} \tag{5.26}$$

where

$$\mathbf{A} \triangleq \begin{bmatrix} \dfrac{\partial f_1}{\partial x_1} & \dfrac{\partial f_1}{\partial x_2} \\ \dfrac{\partial f_2}{\partial x_1} & \dfrac{\partial f_2}{\partial x_2} \end{bmatrix}_{\mathbf{x}=\mathbf{x}_Q} \tag{5.27}$$

is called the *jacobian matrix* of $\mathbf{f}(\mathbf{x})$ evaluated at the equilibrium state $\mathbf{x}_Q \triangleq (x_{1Q}, x_{2Q})$.

It is reasonable to expect that the trajectories in a small neighborhood of Q of Eq. (5.23) are "close" to those in a small neighborhood of the origin ($\tilde{x}_1 = 0, \tilde{x}_2 = 0$) of Eq. (5.24). This "closeness" property is made precise in the following theorem whose proof can be found in Hartman.[30]

Phase portrait linearization property

Assumption: $f_1(x_1, x_2)$ and $f_2(x_1, x_2)$ have *continuous* first-order partial derivatives in a neighborhood of the equilibrium state (x_{1Q}, x_{2Q}).

Conclusions: If the origin of the linearized state equation is a *stable* (respectively, *unstable*) *node*, a *stable* (respectively, *unstable*) *focus*, or a *saddle point*, then the trajectories in a small neighborhood of (x_{1Q}, x_{2Q}) of the associated *nonlinear* state equation will also behave "like" a *stable* (respectively, *unstable*) *node*, *stable* (respectively, *unstable*) *focus*, or a *saddle point*.[31]

Moreover, in the case of a *stable* (respectively, *unstable*) *node* or a *saddle point*, the slow and fast eigenvectors of $\mathbf{A}$ in Eq. (5.27) determine the *limiting slope* of the trajectories near $\mathbf{x}_Q$ for the *nonlinear* state equation in the same way depicted in Figs. 3.3 to 3.5.

[30] P. Hartman, *Ordinary Differential Equation*, Wiley, New York, 1964. (The generalization of the *phase portrait linearization property* to higher dimensions is usually referred to as *Hartman's theorem*.)

[31] Roughly speaking, this property asserts that if the trajectories of the *nonlinear* state equation [Eq. (4.2)] were drawn on a rubber sheet, then those trajectories restricted to a sufficiently small neighborhood of $\mathbf{x}_Q$ could be made to *coincide* with those of the *linearized* state equation [Eq. (5.24)] by stretching the rubber sheet in an appropriate way.

In view of the above property, we will henceforth call an equilibrium state $\mathbf{x}_Q$ of the *nonlinear* state equation (4.2) a *stable* (respectively, *unstable*) *node*, a *stable* (respectively, *unstable*) *focus*, or a *saddle point* if its *linearized* state equation (5.24) about $\mathbf{x}_Q$ has the same behavior.

REMARKS

1. The above property asserts that the *qualitative* behavior of a *stable* (respectively, *unstable*) *node*, a *stable* (respectively, *unstable*) *focus*, or a *saddle point* of the *nonlinear* state equation is *preserved* in its linearized state equation about $\mathbf{x}_Q$.
2. There exists examples showing that the qualitative behavior describing a *center* in the linearized state equation is *not* preserved in the nonlinear state equation. In this case, the higher-order terms in Eq. (5.23) may either push the trajectories towards the origin, or away from it.

Let us now verify the above property on the phase portraits in Figs. 5.2 and 5.4.

Example 1 (Linearized tunnel-diode state equation) For the tunnel-diode circuit state equation (5.3), we calculate the following parameters for the jacobian matrix **A**:

$$a_{11} \triangleq \frac{\partial f_1}{\partial v_{C1}} = (-0.5 \times 10^9)\,\hat{i}_R'(v_{C1}) \qquad a_{12} \triangleq \frac{\partial f_1}{\partial i_{L2}} = 0.5 \times 10^{12}$$

$$a_{21} \triangleq \frac{\partial f_2}{\partial v_{C1}} = -0.2 \times 10^9 \qquad a_{22} \triangleq \frac{\partial f_2}{\partial i_{L2}} = -0.3 \times 10^{12} \tag{5.28}$$

where $\hat{i}_R'(v_{C1})$ denotes the *slope* of the tunnel-diode characteristic at the equilibrium state Q_1, Q_2, or Q_3.

To classify the equilibrium state Q_j, $j = 1, 2, 3$, we calculate

$$T = a_{11} + a_{22} = (-0.5 \times 10^9)\,\hat{i}_R'(v_{C1}) - 0.3 \times 10^{12} \tag{5.29}$$

$$\Delta = a_{11}a_{22} - a_{12}a_{21} = (0.15 \times 10^{21})\,\hat{i}_R'(v_{C1}) + 0.10 \times 10^{21} \tag{5.30}$$

At $Q_1(v_{C1} = 0.06\text{ V},\ i_{L2} = 0.76\text{ mA})$, we find $\hat{i}_R'(v_{C1}) = 7.6\text{ mS}$, $T = -3.8 \times 10^9 - 0.3 \times 10^{12} \approx -0.3 \times 10^{12}$, and $\Delta = 1.24 \times 10^{21}$. Since $T < 0$ and $0 < \Delta < \frac{1}{4}T^2$, it follows from Fig. 3.11 that Q_1 is a *stable node*. An inspection of the trajectories in the neighborhood of Q_1 in the phase portrait of Fig. 5.2 confirms that Q_1 indeed behaves like a *stable node*. Observe that all trajectories near Q_1 approach Q_1, with the same limiting slope which can be verified to coincide with that of the *slow eigenvector* of **A** evaluated at Q_1.

At $Q_2(v_{C1} = 0.29\text{ V},\ i_{L2} = 0.60\text{ mA})$, we find $\hat{i}_R'(v_{c1}) = -3.55\text{ mS}$, $T \approx -0.3 \times 10^{12}$, and $\Delta = -0.43 \times 10^{21}$. Since $\Delta < 0$, it follows from Fig. 3.11 that Q_2 is a *saddle point*. An inspection of the trajectories in the neighborhood of Q_2 in the phase portrait of Fig. 5.2′ confirms that Q_2 indeed behaves like a saddle point. The slope of the four "bold" trajectories at Q_2 can be verified to coincide, respectively, with that of the *slow* and *fast eigenvectors* of **A** evaluated at Q_2.

At $Q_3(v_{C1} = 0.88\text{ V},\ i_{L2} = 0.20\text{ mA})$, we find $\hat{i}'_R(v_{C1}) = 2.67\text{ mS}$, $T \approx -0.3 \times 10^{12}$, and $\Delta = 20.2 \times 10^{21}$. Since $0 < \Delta < \frac{1}{4}T^2$, it follows from Fig. 3.11 that Q_3 is a *stable node*. An inspection of the trajectories in the neighborhood of Q_3 in the phase portrait of Fig. 5.2 confirms that Q_3 indeed behaves like a *stable node*. Again all trajectories near Q_3 approach Q_3 with the same limiting slope which can be verified to coincide with that of the *slow eigenvector* of **A** evaluated at Q_3.

Example 2 (Linearized Josephson junction state equation) For the Josephson junction circuit state equation (5.5), we calculate the following parameters for the jacobian matrix **A**:

$$
\begin{aligned}
a_{11} &\triangleq \frac{\partial f_1}{\partial v_{C1}} = -\frac{G}{C_1} & a_{12} &\triangleq \frac{\partial f_1}{\partial \phi_{L2}} = \frac{-kI_0}{C_1}\cos k\,\phi_{L2} \\
a_{21} &\triangleq \frac{\partial f_2}{\partial v_{C1}} = 1 & a_{22} &\triangleq \frac{\partial f_2}{\partial \phi_{L2}} = 0
\end{aligned}
\tag{5.31}
$$

To classify the equilibrium state Q_j, $j = 1, 2, \ldots,$ we calculate

$$T = a_{11} + a_{22} = -\frac{G}{C_1} \tag{5.32}$$

$$\Delta = a_{11}a_{22} - a_{12}a_{21} = \frac{kI_0}{C_1}\cos k\,\phi_{L2} \tag{5.33}$$

At $\phi_{L2} = \pm n2\pi/k$, $n = 0, 1, 2, \ldots,$ we find $T = -G/C_1 < 0$ and $\Delta = kI_0/C_1 > 0$. Moreover, for the typical Josephson junction parameters which give the phase portrait in Fig. 5.4, we have $\Delta > \frac{1}{4}T^2$. It follows from Fig. 3.11 that all equilibrium states $(v_{C1}, \phi_{L2}) = (0, \pm n2\pi/k)$, $n = 0, 1, 2, \ldots,$ are *stable foci*. An inspection of the phase portrait in Fig. 5.4 confirms that these equilibrium states indeed behave like a *stable focus*.

At $\phi_{L2} = \pm(2n+1)\pi/k$, $n = 0, 1, 2, \ldots,$ we find $T = -G/C_1$ and $\Delta = -(kI_0/C_1) < 0$. It follows from Fig. 3.11 that all equilibrium states $(v_{C1}, \phi_{L2}) = (0, \pm(2n+1)\pi/k)$, $n = 0, 1, 2, \ldots,$ *are saddle points*. An inspection of the phase portrait in Fig. 5.4 confirms that these equilibrium states indeed behave like *saddle points*.

6 NONLINEAR OSCILLATION

Oscillation is one of the most important and exciting phenomenon that occurs in physical systems (e.g., electronic watch) and in Mother Nature (e.g., planetary motions). In Sec. 5.2 of Chap. 6, we have already seen how a simple first-order op-amp circuit could burst into a relaxation oscillation. Our analysis of this phenomenon depends on a key assumption, namely, the *jump rule*. This rule was decreed by faith in Chap. 6 with a promise for justification in Chap. 7. One of our objectives in this section is to honor the promise.

Every electronic oscillator requires at least two energy-storage elements and at least one *nonlinear* element. We will therefore begin with the simplest nonlinear oscillator circuit, analyze its qualitative behavior via its *phase portrait*, and then examine how the oscillation waveform varies as we tune a parameter, say the inductance. We will then show that as the inductance decreases, the oscillation changes from a nearly "sinusoidal" waveform into a nearly "discontinuous" waveform. In the limit when the inductance tends to zero, the waveform becomes *discontinuous* and we obtain the *jump rule*.

6.1 Basic Negative-Resistance Oscillator

Figure 6.1*a* shows the basic circuit structure of an important class of electronic oscillators. Since both inductor and capacitor are *linear* and passive (i.e., $L>0$ and $C>0$), we claim that the resistive one-port N_R must be *active* (i.e., its driving-point characteristic contains at least some points in the second and/or fourth quadrant of the v-i plane) in order for oscillation to be possible.

To see why N_R must be active, suppose it is *strictly passive* so that $v(t)\, i(t)>0$ for all t; then energy will continually enter N_R, only to be dissipated in the form of heat.[32] This dissipated energy must of course come from the initial energy stored in the capacitor and the inductor. Hence, as time goes on, the total energy stored in the capacitor and inductor will decrease continuously until it becomes completely dissipated as $t\to\infty$. Since the instantaneous energy stored in the capacitor and in the inductor is equal to

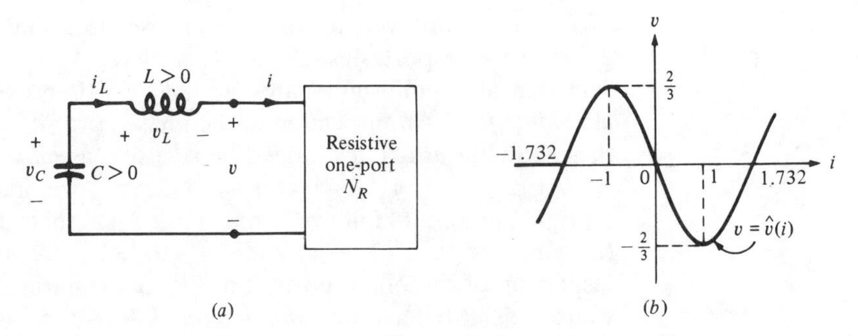

Figure 6.1 (*a*) Basic oscillator circuit. (*b*) Typical nonlinear driving-point characteristic.

[32] Although the circuit would oscillate *in theory* when N_R is a short circuit (passive but not strictly passive), no *oscillation is possible in practice* because the connecting wire always has some small but nonzero resistance.

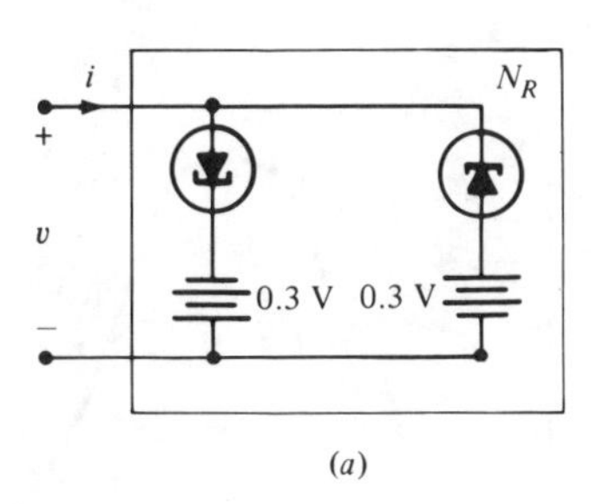

(a)

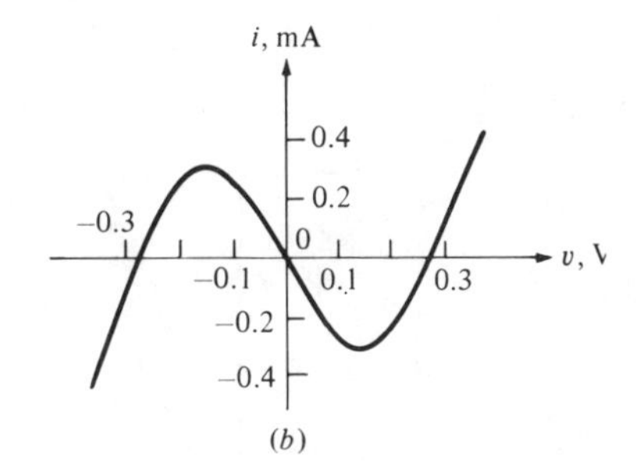

(b)

Figure 6.2 (*a*) A negative-resistance twin-tunnel-diode circuit. (*b*) Typical driving-point characteristic.

$\mathscr{E}_C(t) = \frac{1}{2}Cv_C^2(t)$ and $\mathscr{E}_L(t) = \frac{1}{2}Li_L^2(t)$, respectively [recall Eqs. (2.20) and (2.21) from Chap. 6], it follows that:

$$\text{Total energy} = \tfrac{1}{2}Cv_C^2(t) + \tfrac{1}{2}Li_L^2(t) \to 0 \qquad \text{as } t \to \infty \tag{6.1}$$

Hence, both $v_C(t)$ and $i_L(t)$ must eventually tend to zero and no sustained oscillation is possible.

A typical *active* resistive one-port is described by the three-segment negative-resistance characteristic of the relaxation oscillator in Fig. 5.4*a* of Chap. 6. In general, any *continuous nonmonotonic* current-controlled v-i characteristic described by $v = \hat{v}(i)$ satisfying the conditions

$$\hat{v}(0) = 0 \tag{6.2a}$$

$$\hat{v}'(0) < 0 \tag{6.2b}$$

$$\hat{v}(i) \to \infty \qquad \text{as } i \to \infty \tag{6.2c}$$

$$\hat{v}(i) \to -\infty \qquad \text{as } i \to -\infty \tag{6.2d}$$

would cause the circuit in Fig. 6.1*a* to oscillate.[33]

The op-amp "negative-resistance" driving-point characteristic in Fig. 5.4*b* of Chap. 6 clearly satisfies these conditions. Likewise, the driving-point characteristic described by the "cubic" function

$$v = \hat{v}(i) = \tfrac{1}{3}i^3 - i \tag{6.3}$$

and sketched in Fig. 6.1*b* also satisfies these conditions. Note that here, the current i is chosen as the *horizontal* axis in order that $\hat{v}(i)$ can be plotted as a single-valued function of i.

Indeed, conditions (6.2) are satisfied by many electronic circuits. For example, the typical driving-point characteristic of the twin-tunnel-diode circuit in Fig. 6.2*a* clearly satisfies conditions (6.2). Likewise, the driving-point

[33] This statement can be proved rigorously. We can also prove it for the special limiting case where $L \to 0$ by exactly the same reasoning used in the relaxation oscillator with the help of the *jump rule*.

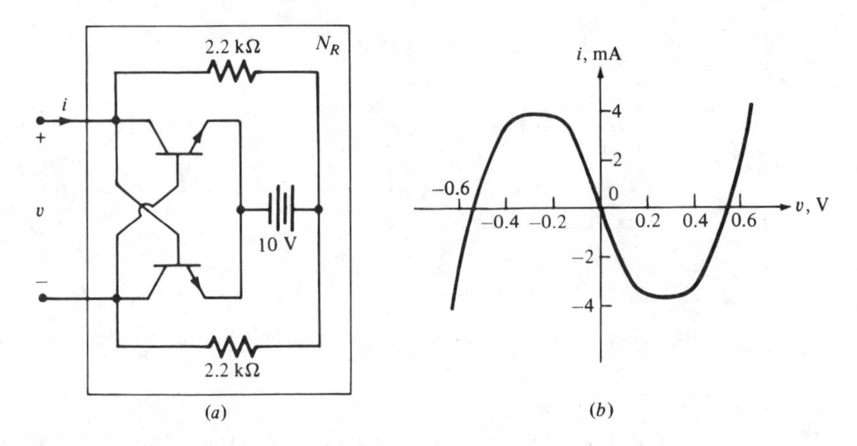

Figure 6.3 (*a*) Twin-transistor circuit. (*b*) Typical driving-point characteristic.

characteristic of the twin-transistor circuit[34] in Fig. 6.3 also satisfies conditions (6.2).

6.2 Physical Mechanisms for Oscillation

For the simple series *RLC* circuit in Fig. 6.1*a*, the following state equation can be obtained by inspection:

Basic oscillator state equation

$$\dot{v}_C = \frac{-i_L}{C} \triangleq f_1(v_C, i_L) \tag{6.4a}$$

$$\dot{i}_L = \frac{v_C - \hat{v}(i_L)}{L} \triangleq f_2(v_C, i_L) \tag{6.4b}$$

Assuming *only* that $\hat{v}(i_L)$ satisfies conditions (6.2) it is possible to derive several general qualitative behaviors for this circuit. Indeed, equating $f_1(\cdot)$ and $f_2(\cdot)$ in Eq. (6.4) to zero and making use of condition (6.2*a*), we conclude that this circuit has a *unique* equilibrium state located at the *origin*:

Equilibrium state

$$v_{CQ} = 0 \qquad i_{LQ} = 0 \tag{6.5}$$

To classify this equilibrium state, we calculate

$$\begin{aligned} a_{11} &= 0 & a_{12} &= -\frac{1}{C} \\ a_{21} &= \frac{1}{L} & a_{22} &= -\frac{\hat{v}'(0)}{L} \end{aligned} \tag{6.6}$$

[34] This circuit is taken from L. A. Rosenthal, "Inductively Tuned Astable Multivibrator," *IEEE Transactions on Circuits and Systems*, vol. CAS-27, pp. 963–964, October 1980.

From these parameters, we calculate

$$T = a_{11} + a_{22} = -\frac{\hat{v}'(0)}{L} \tag{6.7a}$$

$$\Delta = a_{11}a_{22} - a_{12}a_{21} = \frac{1}{LC} \tag{6.7b}$$

Since $\Delta > 0$ and $T > 0$ in view of condition (6.2*b*), it follows from Fig. 3.11 that the origin is an *unstable focus* if

$$\frac{1}{LC} > \frac{1}{4}\left[\frac{-\hat{v}'(0)}{L}\right]^2 \tag{6.8}$$

or equivalently, if

$$|\hat{v}'(0)| < 2\sqrt{\frac{L}{C}} \tag{6.9}$$

Likewise, the origin is an *unstable node* if the inequality sign in Eq. (6.9) is reversed. It follows from Eq. (6.9) that if we tune the value of L while keeping $\hat{v}'(0)$ and C fixed, then the origin will be an *unstable focus* for *large* L and an *unstable node* for small L.

In any case, all trajectories starting near the origin would *diverge* from it and head toward infinity. Indeed, the trajectories would eventually reach infinity if N_R had been a pure *negative linear resistor*. However, since the v-i characteristic of N_R must lie in the first and third quadrants *beyond* a certain *finite* distance from the origin in view of conditions (6.2*c*) and (6.2*d*), N_R becomes *passive* ($vi > 0$) there and must absorb energy from the external world—the capacitor and inductor in this case.

Consequently, the energy initially *supplied* by the "active" N_R (when the trajectory is near the "unstable" origin) to propel the trajectory toward infinity eventually fizzles as N becomes passive and begins to *absorb* energy instead. Therefore, the initial outward motion of the trajectory will be damped out by losses due to power dissipated inside N when the trajectory is sufficiently far out. Soon, the trajectory must "grind to a halt" and start "falling" back toward the origin.

The above scenario is depicted in Fig. 6.4 using the i-v characteristic in Fig. 6.1*b*. Observe that since the circuit has only one equilibrium state, and since it is unstable, there is no point where any trajectory could come to rest. Therefore all trajectories must continue to move at *all times*. Since they cannot stray too far beyond the active region and since no trajectory of any *autonomous* state equation can intersect itself,[35] except at *equilibrium* points, *each*

[35] If a trajectory intersects itself at $(\hat{x}_1, \hat{x}_2)$ at some time $t_0 < \infty$, then its *slope* dx_2/dx_1 at $(\hat{x}_1, \hat{x}_2)$ would have two distinct values. But this is impossible because the vector field $[f_1(x_1, x_2), f_2(x_1, x_2)]$ defines a *unique* slope

$$\frac{dx_2}{dx_1} = \frac{f_2(x_1, x_2)}{f_1(x_1, x_2)}$$

at any point $(\hat{x}_1, \hat{x}_2)$ which is *not* an equilibrium point.

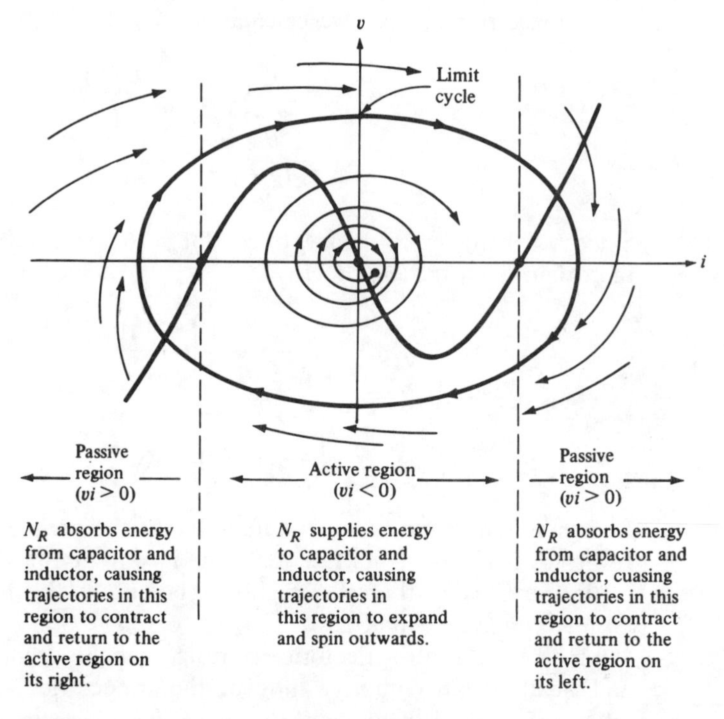

Figure 6.4 Physical mechanism for oscillation.

trajectory must eventually tend toward some limiting orbit,[36] henceforth called a *limit cycle*. One such limit cycle is shown in Fig. 6.4. Clearly, starting from any point on a limit cycle, the trajectory will revolve around the limit cycle periodically. The result is an *oscillator*.

To summarize, we note that the following two *physical mechanisms* are necessary for producing *oscillation* using the circuit structure of Fig. 6.1*a*:

1. An *unstable* equilibrium point which spits out and repels trajectories around it. This in turn requires that N_R must be *active* at least in a small neighborhood around the equilibrium point.
2. A *dissipative* mechanism which restrains the trajectories from running away to infinity. This in turn requires that N_R be *eventually passive*.

   Since any *continuous v-i* characteristic satisfying conditions (6.2*a*) to (6.2*d*) fulfills the above two mechanisms, we have shown via physical reasoning that the basic circuit structure in Fig. 6.1*a* indeed functions as an *oscillator*.

[36] Our physical reasoning does not prove that *all* trajectories must tend to a *unique* limit cycle, even though this is actually the case for the above choice of *v-i* characteristic. Indeed, for more complicated *v-i* characteristics, different trajectories could tend to different limit cycles. For example, see Fig. 6.8.

6.3 Phase Portrait of Typical Oscillators

In this section we will examine the phase portrait of two typical oscillators, namely, the *linear* oscillator and the *Van der Pol oscillator*.

A. Linear oscillator In order to contrast between a linear and a nonlinear oscillator, let us begin with the special case where N_R is a short circuit so that Fig. 6.1*a* reduces to a parallel linear *LC* circuit. Since $\hat{v}(i_L) = 0$, the state equation (6.4) degenerates into

Linear oscillator

$$\dot{v}_C = \frac{-i_L}{C} \tag{6.10a}$$

$$\dot{i}_L = \frac{v_C}{L} \tag{6.10b}$$

The solution of Eq. (6.10) is given by

$$v_C(t) = A\cos(\omega_0 t - \theta) \tag{6.11a}$$

$$i_L(t) = (\omega_0 CA)\sin(\omega_0 t - \theta) \tag{6.11b}$$

where
$$\omega_0 \triangleq \frac{1}{\sqrt{LC}} \tag{6.12}$$

is the oscillation frequency in rad/s, and A and θ are constants dependent on the initial conditions.

Squaring both sides of Eq. (6.11) and adding, we obtain

$$v_C^2(t) + \left[\frac{i_L(t)}{\omega_0 C}\right]^2 = A^2 \tag{6.13}$$

Hence, the phase portrait of the linear *oscillator* consists of a *continuum* of nested *ellipses*, as shown in Fig. 6.5. Each ellipse represents a distinct oscillation corresponding to an *initial condition* lying on the ellipse. Hence, the *amplitude* A of $v_C(t)$ depends on the initial condition.

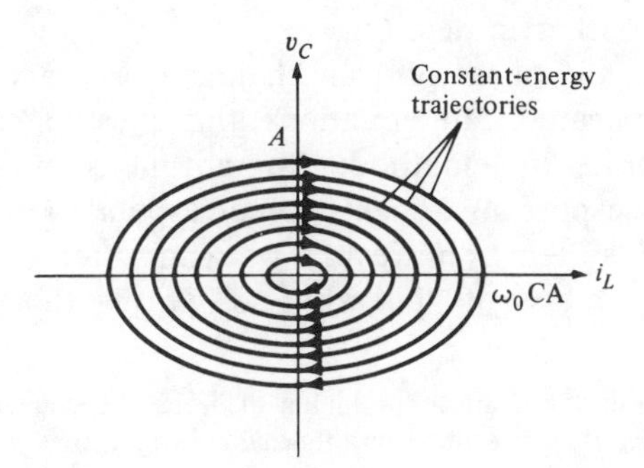

Figure 6.5 Phase portrait of the linear oscillator.

The total energy stored in the capacitor and the inductor at any time t is given by

$$\begin{aligned}\mathscr{E}_C(t) + \mathscr{E}_L(t) &= \tfrac{1}{2}Cv_C^2(t) + \tfrac{1}{2}Li_L^2(t) \\ &= \tfrac{1}{2}CA^2[\cos^2(\omega_0 t - \theta) + (LC)\omega_0^2 \sin^2(\omega_0 t - \theta)] \\ &= \tfrac{1}{2}CA^2 \end{aligned} \tag{6.14}$$

Hence, even though the energy $\mathscr{E}_C(t)$ stored in the capacitor and the energy $\mathscr{E}_L(t)$ stored in the inductor, individually vary as a periodic function of time, the total energy is constant at all times. This is of course expected as both capacitor and inductor are lossless and there are no other elements around to dissipate their stored energy. Consequently, we can interpret each orbit[37] in Fig. 6.5 as a *constant-energy trajectory*, or *level curve* in physics. This property is not true in the *nonlinear* case where the amplitude of oscillation is generally independent of the initial condition.

The oscillation mechanism in the linear oscillator consists of a periodic exchange of energy stored in the capacitor electric field with the energy stored in the inductor magnetic field.

Of course, *linear oscillators exist only on paper*: It is impossible to build a linear oscillator in practice because the resistance in the connecting wires alone will eventually consume whatever energy was initially stored in the capacitor and inductor.

B. Van der Pol oscillator If we choose the "cubic" current-controlled *i-v* characteristic defined in Eq. (6.3), the resulting circuit in Fig. 6.1 is called the *Van der Pol oscillator*. Its state equation is obtained by substituting Eq. (6.3) into Eq. (6.4):

Van der Pol oscillator

$$\dot{v}_C = \frac{-i_L}{C} \tag{6.15a}$$

$$\dot{i}_L = \frac{v_C - (\frac{1}{3}i_L^3 - i_L)}{L} \tag{6.15b}$$

For fixed values of L and C, we could use a computer to generate the phase portrait of Eq. (6.15). The result would of course possess the qualitative features derived earlier in Sec. 6.2.

But how does the phase portrait change as we vary the parameters L and C? In more complicated state equations, this question could only be answered in general by a *brute-force* method: Program the computer to generate a large collection of phase portraits, each one corresponding to a different L and C.

For *physical* systems, however, we can often reduce the number of parameters without loss of generality by writing the equations in terms of

[37] These closed trajectories are a continuum of *orbits*. They are *not* limit cycles because, *by definition*, a *limit cycle* Γ must contain no other *closed* trajectories in a small band around Γ.

dimensionless variables. For the Van der Pol oscillator, let us introduce the following "scaled" time variables:

Dimensionless time

$$\tau \triangleq \frac{1}{\sqrt{LC}}\,t \tag{6.16}$$

Note that since $\sqrt{LC}$ has the unit of time in seconds [in view of Eq. (6.12)], τ is *dimensionless* and will henceforth be called the "dimensionless time."[38]

Observe that

$$\dot{v}_C = \frac{dv_C}{d\tau}\,\frac{d\tau}{dt} = \frac{1}{\sqrt{LC}}\,\frac{dv_C}{d\tau} \tag{6.17a}$$

and

$$\dot{i}_L = \frac{di_L}{d\tau}\,\frac{d\tau}{dt} = \frac{1}{\sqrt{LC}}\,\frac{di_L}{d\tau} \tag{6.17b}$$

Substituting Eq. (6.15) into Eq. (6.17), we obtain the following equivalent state equation in terms of the dimensionless time variable τ:

Van der Pol oscillator (dimensionless form)

$$\frac{dv_C}{d\tau} = -\frac{1}{\varepsilon}\,i_L \tag{6.18a}$$

$$\frac{di_L}{d\tau} = \varepsilon[v_C - (\tfrac{1}{3}\,i_L^3 - i_L)] \tag{6.18b}$$

where

$$\varepsilon \triangleq \sqrt{\frac{C}{L}} \tag{6.19}$$

Observe that Eq. (6.18) now contains only one parameter, ε, as defined by Eq. (6.19). The qualitative behavior of Eq. (6.18) can now be more easily studied as ε is varied by drawing the phase portrait for different values of the *single* parameter ε. Three such phase portraits coresponding to a small, medium, and large value of ε are shown in Fig. 6.6. Note that in each case, all trajectories tend to a *unique limit cycle* (retraced in bold lines in Fig. 6.6).

For small ε (say $\varepsilon < 0.1$), the limit cycle is approximately a smooth *ellipse*, as shown in Fig. 6.6*a* for $\varepsilon = 0.05$. We can derive an approximate solution for Eq. (6.18) for the *small* ε case as follows: Differentiate Eq. (6.18*b*) with respect to τ and then substitute Eq. (6.18*a*) into the resulting expression:

$$\begin{aligned}\frac{d^2 i_L}{d\tau^2} &= \varepsilon\,\frac{dv_C}{d\tau} - \varepsilon(i_L^2 - 1)\\ &= -i_L - \varepsilon(i_L^2 - 1)\,\frac{di_L}{d\tau}\\ &\approx -i_L \qquad (\text{since } \varepsilon \approx 0)\end{aligned} \tag{6.20}$$

[38] The symbol τ here is not related to the *time constant* defined in Chap. 6.

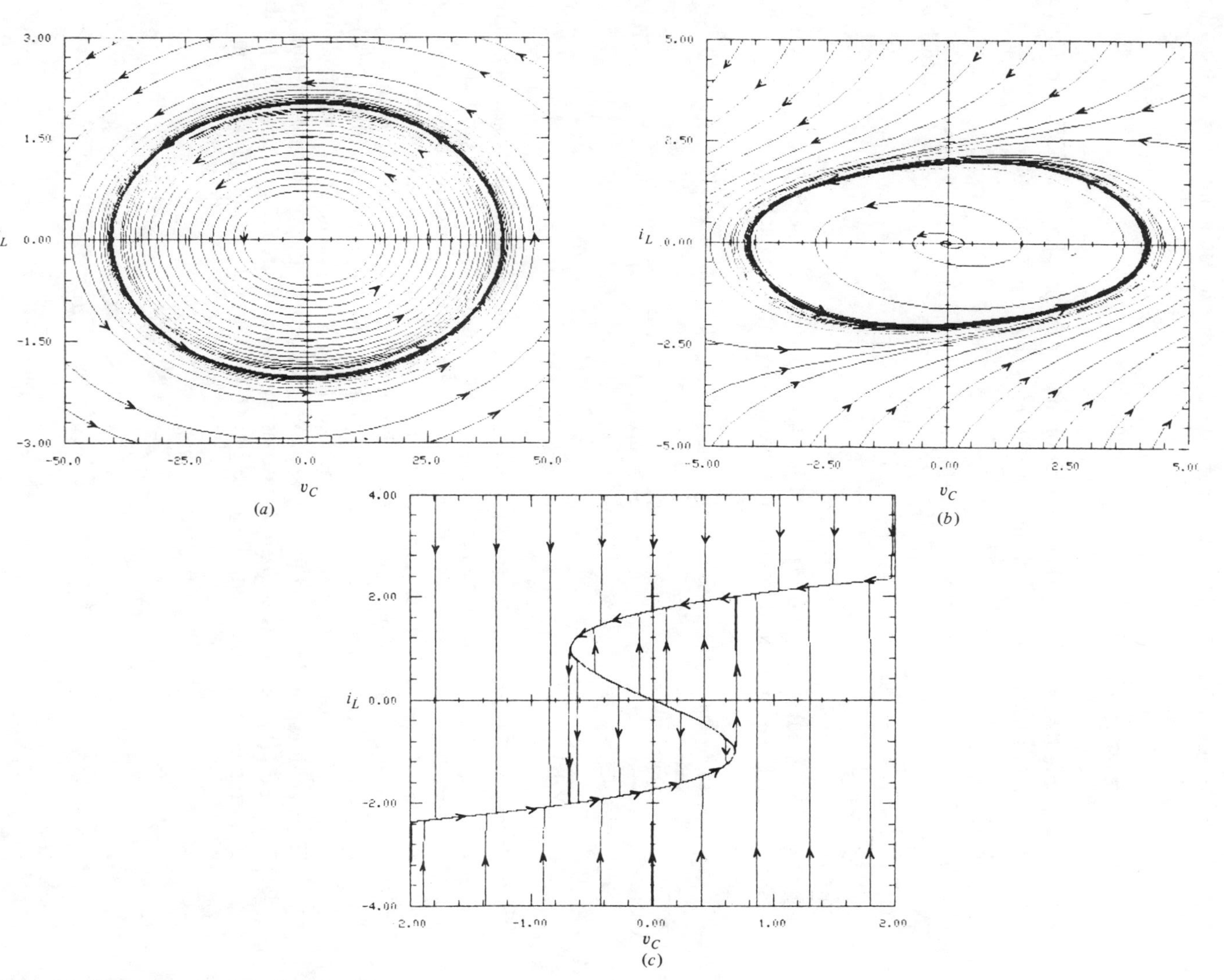

Figure 6.6 Phase portrait of the Van der Pol oscillator for small, medium, and large values of ε: (*a*) $\varepsilon = 0.05$; (*b*) $\varepsilon = 0.5$; and (*c*) $\varepsilon = 30$.

The solution of the approximate equation

$$\frac{d^2 i_L}{d\tau^2} + i_L = 0 \tag{6.21}$$

is

$$i_L = A\cos(t+\theta) \tag{6.22a}$$

It follows from Eq. (6.18*a*) that

$$v_C = \frac{-A}{\varepsilon}\sin(t+\theta) \tag{6.22b}$$

Hence, for small ε, the solution waveforms of both $i_L(t)$ and $v_C(t)$ are nearly sinusoidal, though the amplitude of $v_C(t)$ is $1/\varepsilon$ as large as that of $i_L(t)$.

Note that the above approximate analysis fails to tell us what the amplitude A for $i_L(t)$ is. This is because for the *linear* differential equation (6.21), the amplitude of oscillation depends on the initial condition, as we have seen in Fig. 6.5. In the *Van der Pol oscillator*, it can be proved rigorously that Eq. (6.18) has a *unique* limit cycle and that for *small* ε, $A \approx 2$.

For medium values of ε (say $0.3 < \varepsilon < 0.7$), the "elliptical" limit cycle becomes distorted as shown in Fig. 6.6*b* for $\varepsilon = 0.5$. The solution waveforms $v_C(t)$ and $i_L(t)$ in this case are no longer sinusoidal and do not admit of even an approximate closed form expression.

For *large* ε (say $\varepsilon > 20$), the limit cycle is seen in Fig. 6.6*c* for $\varepsilon = 30$ to cling closely to the curve $v_C = \frac{1}{3} i_L^3 - i_L$ except at the corners, where it becomes nearly vertical. Note that all trajectories in Fig. 6.6*c* are also nearly vertical. The corresponding waveforms for $v_C(t)$ and $i_L(t)$ as obtained by a computer consist of *nearly instantaneous* transition from the lower branch to the upper branch, and vice versa.

We can explain this phenomenon by examining Eq. (6.18) directly. Note that as $\varepsilon \to \infty$, Eq. (6.18) becomes:

$$\frac{dv_C}{d\tau} \to 0 \quad \text{or} \quad v_C \to \text{constant} \tag{6.23a}$$

$$\frac{di_L}{dt} \to \infty \quad \text{or} \quad i_L \to \text{vertical jump} \tag{6.23b}$$

Note from the definition of ε in Eq. (6.19) that if we keep C constant while varying L, then $\varepsilon \to \infty$ is equivalent to $L \to 0$. Hence, the "jump phenomenon" depicted in the phase portrait in Fig. 6.6*c* is completely consistent with that assumed earlier by the *jump rule* in Chap. 6. In the next section, we will return to show that the same "jump phenomenon" in the above Van der Pol oscillator occurs in any negative-resistance oscillator whose nonlinear resistor characteristic $v = \hat{v}(i)$ satisfies conditions (6.2).

REMARK: RELATIONSHIP BETWEEN EQ. (6.18) AND THE VAN DER POL EQUATION If we define $x_1 \triangleq i_L$ and $x_2 \triangleq di_L/d\tau$, the state equation (6.18) can be recast into the following equivalent form:

Van der Pol state equation

$$\frac{dx_1}{d\tau} = x_2 \tag{6.24a}$$

$$\frac{dx_2}{d\tau} = \varepsilon(1 - x_1^2)x_2 - x_1 \tag{6.24b}$$

It follows from the above equation that

$$\begin{aligned}
\frac{d}{d\tau}(x_1^2 + x_2^2) &= 2x_1\dot{x}_1 + 2x_2\dot{x}_2 \\
&= 2x_1(x_2) + 2x_2[\varepsilon(1 - x_1^2)x_2 - x_1] \\
&= (1 - x_1^2)(2\varepsilon x_2^2) \\
&> 0 \text{ if } |x_1| < 1 \\
&< 0 \text{ if } |x_1| > 1
\end{aligned}$$

Since $x_1^2 + x_2^2 = \rho^2$, where ρ is the length of the vector from the origin (in x_1-x_2 plane) to a point (x_1, x_2) on the trajectory of Eq. (6.24), the above inequality asserts that any trajectory originating from inside the strip $|x_1| < 1$ must grow (spiraling outward) with time, whereas any trajectory originating from outside the strip $|x_1| < 1$ must shrink (spiraling inward) toward the origin. This qualitative behavior is identical to the basic oscillation mechanism depicted in Fig. 6.4. Note, however, that unlike the intuitive physical arguments given earlier in Sec. 6.2, our present conclusion is derived rigorously directly from the state equation.

The Van der Pol state equation (6.24) can also be recast into the following equivalent scalar nonlinear second-order differential equation upon differentiating Eq. (6.24*a*) with respect to τ and changing symbol x_1 to x:

Van der Pol equation

$$\ddot{x} + \varepsilon(x^2 - 1)\dot{x} + x = 0 \tag{6.25}$$

Historically, Eq. (6.25) was first extensively studied by the Dutch electrical engineer Van der Pol in the early 1900s and by many mathematicians later, because it represents a good model of the first vacuum tube oscillators.[39] It is Eq. (6.25) that is usually referred to in the literature as the *Van der Pol equation*.

C. Jump phenomenon revisited Let us return to the state equation (6.4) describing the general class of *series* negative-resistance oscillator circuits in Fig. 6.1*a*:

[39] The *Van der Pol equation* has since been found to be a useful model of several other physical and biological systems. For example, the human heartbeat has been modeled by a Van der Pol equation with large ε, thereby establishing a close analogy between the heartbeat mechanism and relaxation *oscillators*, such as that studied in Chap. 6.

$$\dot{v}_C = -\frac{1}{C} i_L \tag{6.4a}$$

$$\dot{i}_L = \frac{1}{L} [v_C - \hat{v}(i_L)] \tag{6.4b}$$

The function $\hat{v}(\cdot)$ representing the nonlinear resistor characteristic can be quite arbitrary except that it satisfies the conditions in Eq. (6.2). This class of course includes the negative-resistance op-amp oscillator circuit analyzed earlier in Chap. 6 (with $L = 0$) by invoking the *jump rule*. Dividing Eq. (6.4*b*) by (6.4*a*), we obtain the *slope*

$$m(P) \triangleq \frac{di_L}{dv_C} = -\frac{C}{L}\left[\frac{v_C - \hat{v}(i_L)}{i_L}\right] \tag{6.26}$$

of the *tangent* vector at any point $P \triangleq (v_C, i_L)$ on a trajectory in the v_C-i_L plane. As $L \to 0$ in Eq. (6.26), the limiting slope

$$|m(P)| = \lim_{L\to 0} \left|\frac{di_L}{dv_C}\right| \to \infty \tag{6.27}$$

provided $v_C \neq \hat{v}(i_L)$. This means that, except for a "narrow band" (whose width shrinks to zero as $L \to 0$) surrounding the nonlinear resistor characteristic $v_C = \hat{v}(i_L)$, all trajectories of Eq. (6.4) consist of *vertical* line segments as shown in Fig. 6.7.

The direction of each line segment can be determined from Eq. (6.4*b*), namely,

$$\frac{di_L}{dt}\begin{cases} >0 & \text{if } v_C > \hat{v}(i_L) \\ <0 & \text{if } v_C < \hat{v}(i_L) \end{cases} \tag{6.28}$$

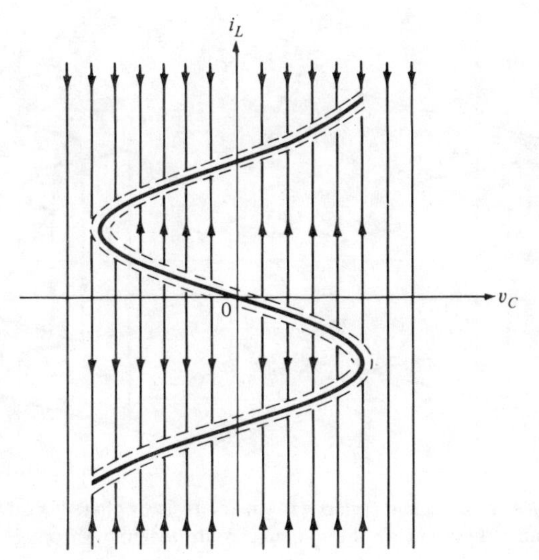

Figure 6.7 Typical phase portrait of series negative-resistance oscillator as $L \to 0$.

Hence, a trajectory through any point P outside of the "dotted band" must go up if P lies to the *right* of the nonlinear characteristic $v_C = \hat{v}(i_L)$ and down otherwise.

To complete our analysis of the jump phenomenon, we must estimate the amount of time it takes a trajectory line segment to go from one branch to another. This is easily found from the velocity along the vertical direction as specified by Eq. (6.4*b*), namely,

$$\lim_{L \to 0} \left| \frac{di_L}{dt} \right| \to \infty \tag{6.29}$$

provided $v_C \neq \hat{v}(i_L)$.

Hence, as $L \to 0$, the time it takes a trajectory to go from one branch of the nonlinear characteristic to another branch tends to zero. In particular, the trajectory through each "impasse point" defined in Chap. 6 must execute a vertical "instantaneous" jump as $L \to 0$. This formally justifies our introduction of the *jump rule* in Chap. 6.

We will close this chapter by applying the *jump rule* to the circuit in Fig. 6.1*a* with $L = 0$ but with the nonlinear resistor characteristic $v = \hat{v}(i)$ chosen as shown in Fig. 6.8*a*. Three dynamic routes associated with this limiting *first-order* circuit are drawn in bold lines in Fig. 6.8*b*. They correspond to three distinct modes of oscillation. To determine which mode is actually observed in practice, it is necessary to reinsert the inductor and examine the phase portrait of the remodeled *second-order circuit*.

These *multimode* oscillations are quite typical in *autonomous* circuits containing nonlinear resistors with *multiple negative-resistance* regions, such as Fig. 6.8*a*. They are even more common in *nonautonomous* circuits, in fact, even in simple second-order *nonautonomous* circuits containing a monotone

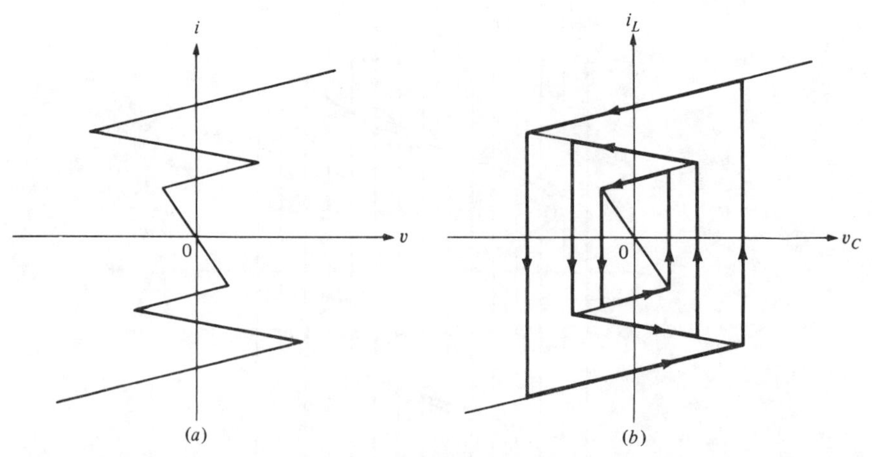

Figure 6.8 (*a*) A nine-segment piecewise-linear resistor characteristic. (*b*) Three distinct periodic solutions obtained by applying the dynamic route technique from Chap. 6 and the jump rule.

increasing nonlinear element.[40] In addition, extremely complicated *nonperiodic* phenomena, generally known as *chaos*, could occur in simple second-order nonautonomous or third-order autonomous circuits.[41–44] The study of these exotic but ubiquitous phenomena is currently an active research area. This yet unsatisfactory state of the art[45] explains why our coverage of *dynamic nonlinear* circuits in this book is restricted to those covered in Chaps. 6 and 7.

Exercise

(*a*) Sketch the phase portrait of the above second-order circuit for very small but finite L.

(*b*) Show among the three *limit cycles* in this phase portrait that only the inner and outer limit cycles can be observed in practice.

(*c*) Specify the regions (called the *basin of attraction*) where points originating within each region would tend to either the inner or the outer limit cycle.

(*d*) Give the analogy between the qualitative behavior of this phase portrait with that obtained earlier for the tunnel-diode circuit in Fig. 5.2.

SUMMARY

- Circuits made of two *energy-storage* elements (two capacitors, two inductors, or one capacitor and one inductor), resistors, and independent sources are called *second-order circuits* because they are described by a second-order *scalar* differential equation or by two first-order scalar differential equations.
- Equations describing a *linear time-invariant* second-order circuit can be cast into the following *equivalent* standard forms:

  Scalar differential equation form: $\ddot{x} + 2\alpha\dot{x} + \omega_0^2 x = u_s(t)$

  Here, α is called the *damping constant* and $\omega_0 \triangleq 2\pi f_0$ is called the (angular) *frequency* in radians per second (f_0 is the frequency in hertz).

[40] M. Hasler and J. Neirynck, *Nonlinear Circuits* (original French edition published by Presses Polytechniques Romandes in 1985) English translation published by Artech, Boston, Mass., 1987.

[41] T. Matsumoto, "A Chaotic Attractor from Chua's Circuit," *IEEE Trans. Circuits and Systems*, vol. CAS-31, pp. 1055–1058, December 1984.

[42] G. Q. Zhong and F. Ayrom, "Experimental Confirmation of Chaos from Chua's Circuit," *International Journal of Circuit Theory and Applications*, vol. 13, pp. 93–98, January 1985.

[43] T. Matsumoto, L. O. Chua, and M. Komuro, "The Double Scroll," *IEEE Trans. Circuits and Systems*, vol. CAS-32, pp. 797–818, August 1985.

[44] L. O. Chua, M. Komuro, and T. Matsumoto, "The Double Scroll Family," *IEEE Trans. Circuits and Systems*, vol. CAS-33, pp. 1072–1118, November 1986.

[45] For a tutorial on the state of the art, see *Proceedings of the IEEE*, Special Section on *Chaotic Systems*, August 1987.

State equation form:
$$\begin{bmatrix} \dot{x}_1 \\ \dot{x}_2 \end{bmatrix} = \begin{bmatrix} a_{11} & a_{12} \\ a_{21} & a_{22} \end{bmatrix} \begin{bmatrix} x_1 \\ x_2 \end{bmatrix} + \begin{bmatrix} u_1(t) \\ u_2(t) \end{bmatrix}$$

or in vector form, $\dot{\mathbf{x}} = \mathbf{A}\mathbf{x} + \mathbf{u}(t)$.

- The *state equation formulation algorithm* given in Sec. 1.3 allows us to write the state equation of any linear time-invariant second-order circuit in a systematic way.
- Once the state equation is obtained, we can transform it into an equivalent second-order scalar differential equation via Eq. (1.14). In particular, we have the following important relationships on the coefficients of these two standard forms:

$$2\alpha = -(a_{11} + a_{22}) = -\text{ trace of } \mathbf{A} \triangleq -T$$
$$\omega_0^2 = a_{11}a_{22} - a_{12}a_{21} = \text{determinant of } \mathbf{A} \triangleq \Delta$$

The *zero-input response* of $\ddot{x} + 2\alpha\dot{x} + \omega_0^2 x = 0$ with $\alpha \geq 0$ and $\omega_0^2 > 0$ can take only four qualitatively distinct forms. They are summarized in Table 2.1. The parameter ranges determining these zero-input responses are (here, $Q \triangleq \omega_0/2\alpha$ is called the *quality factor*):

1. $\alpha > \omega_0 > 0$ (or $\frac{1}{2} < Q < \infty$) $\Rightarrow$ underdamped response
2. $\alpha = \omega_0 > 0$ (or $Q = \frac{1}{2}$) $\Rightarrow$ critically damped response
3. $0 < \alpha < \omega_0$ (or $0 < Q < \frac{1}{2}$) $\Rightarrow$ overdamped response
4. $\alpha = 0$, $\omega_0 > 0$ (or $Q = \infty$) $\Rightarrow$ lossless response

- The *zero-input response* of $\dot{\mathbf{x}} = \mathbf{A}\mathbf{x}$ $(\Delta \neq \frac{1}{4}T^2)$ depends on the *natural frequencies*

$$s_1 = \frac{T}{2} + \sqrt{\tfrac{1}{4}T^2 - \Delta} = -\alpha + \sqrt{\alpha^2 - \omega_0^2}$$

$$s_2 = \frac{T}{2} - \sqrt{\tfrac{1}{4}T^2 - \Delta} = -\alpha - \sqrt{\alpha^2 - \omega_0^2}$$

of $\mathbf{A}$, which are *distinct* if $\Delta \neq \frac{1}{4}T^2$.

- In particular, if s_1 and s_2 are two distinct natural frequencies of $\mathbf{A}$ and if $\boldsymbol{\eta}_1$ and $\boldsymbol{\eta}_2$ are corresponding associated *eigenvectors*, then the zero-input response of $\dot{\mathbf{x}} = \mathbf{A}\mathbf{x}$ is given by

$$\mathbf{x}(t) = (k_1 e^{s_1 t})\boldsymbol{\eta}_1 + (k_2 e^{s_2 t})\boldsymbol{\eta}_2$$

where k_1 and k_2 are arbitrary constants depending on the initial state $\mathbf{x}(0)$. This solution is valid regardless of whether s_1 and s_2 are real or complex, so long as they are distinct.

- Any realistically modeled *second-order nonlinear* circuit can be described by a *nonlinear state* equation

$$\dot{x}_1 = f_1(x_1, x_2, t) \qquad \text{or} \qquad \dot{\mathbf{x}} = \mathbf{f}(\mathbf{x}, t)$$
$$\dot{x}_2 = f_2(x_1, x_2, t)$$

- The time variable t appears explicitly whenever the circuit contains *time-dependent* voltage and/or current sources, or *time-varying* circuit elements. When the time variable is absent, the resulting equation

$$\dot{x}_1 = f_1(x_1, x_2) \qquad \text{or} \qquad \dot{\mathbf{x}} = \mathbf{f}(\mathbf{x})$$
$$\dot{x}_2 = f_2(x_1, x_2)$$

  is called an *autonomous state equation*.
- Systematic methods for formulating nonlinear state equations for second-order nonlinear circuits are summarized in Table 4.1.
- A *trajectory* of an *autonomous* state equation from an initial state $\mathbf{x}_0$ is the *locus* of the corresponding solution $(x_1(t), x_2(t))$ in the x_1-x_2 plane.
- A closed trajectory characterizes a *periodic* solution.
- An *equilibrium state* $\mathbf{x}_0$ of an *autonomous state equation* $\dot{\mathbf{x}} = \mathbf{f}(\mathbf{x})$ is any solution of $\mathbf{f}(\mathbf{x}) = \mathbf{0}$.
- When we choose an equilibrium state to be the initial state, i.e., $\mathbf{x}(0) = \mathbf{x}_Q$, then the solution is $\mathbf{x}(t) = \mathbf{x}_Q$ for all $t \geq 0$. Hence the *circuit is motionless at an equilibrium state* or is in equilibrium.
- An equilibrium state is a *trajectory* consisting of only one point.
- A *phase portrait* of an *autonomous state equation* $\dot{\mathbf{x}} = \mathbf{f}(\mathbf{x})$ consists of a *family* of trajectories originating from a large number of initial states spread (usually uniformly) all over the x_1-x_2 plane or over a particular region R when the portrait outside this region is either uninteresting, or is a periodic repetition of the same portrait within R.
- The qualitative behavior of $\dot{\mathbf{x}} = \mathbf{A}\mathbf{x}$ is completely determined by the phase portrait in the neighborhood of the equilibrium state $\mathbf{x}_Q = \mathbf{0}$.
- Except for the two degenerate cases $\Delta = 0$ and $\Delta = \frac{1}{4}T^2$, there are only six *qualitatively distinct* phase portraits of $\dot{\mathbf{x}} = \mathbf{A}\mathbf{x}$. They are summarized in Table 3.1 and in the mnemonic diagram in Fig. 3.11. The parameters determining the phase portrait in a neighborhood of the origin are summarized as follows (here T and Δ denotes the *trace* and the *determinant* of $\mathbf{A}$, respectively):

$$\left.\begin{array}{ll} 1.\ T<0,\ \Delta>0 & \Rightarrow \text{stable node} \\ 2.\ T>0,\ \Delta>0 & \Rightarrow \text{unstable node} \\ 3.\ \Delta<0 & \Rightarrow \text{saddle point} \end{array}\right\} \Delta < \tfrac{1}{4}T^2$$

$$\left.\begin{array}{ll} 4.\ \Delta>0,\ T=0 & \Rightarrow \text{center} \\ 5.\ \Delta>0,\ T<0 & \Rightarrow \text{stable focus} \\ 6.\ \Delta>0,\ T>0 & \Rightarrow \text{unstable focus} \end{array}\right\} \Delta > \tfrac{1}{4}T^2$$

- An autonomous *nonlinear* state equation $\dot{\mathbf{x}} = \mathbf{f}(\mathbf{x})$ can have *many* equilibrium states. The phase portrait in this case must include the region containing all these equilibrium states (except in the periodic case where a cylindrical surface suffices, as in Fig. 5.5 for the Josephson junction circuit).
- If $\mathbf{f}(\mathbf{x})$ and its first partial derivatives are continuous functions, then the qualitative behavior in the vicinity of *each* equilibrium state $\mathbf{x}_Q$ of $\dot{\mathbf{x}} = \mathbf{f}(\mathbf{x})$ is *similar* to that of the *linearized* equation

$$\begin{bmatrix} \dot{\tilde{x}}_1 \\ \dot{\tilde{x}}_2 \end{bmatrix} = \begin{bmatrix} \dfrac{\partial f_1}{\partial x_1} & \dfrac{\partial f_1}{\partial x_2} \\ \dfrac{\partial f_2}{\partial x_1} & \dfrac{\partial f_2}{\partial x_2} \end{bmatrix}_{\mathbf{x}=\mathbf{x}_Q} \begin{bmatrix} \tilde{x}_1 \\ \tilde{x}_2 \end{bmatrix} \quad \text{or} \quad \dot{\tilde{\mathbf{x}}} = \mathbf{A}\tilde{\mathbf{x}}$$

  associated with five of the six possible qualitatively distinct phase portraits, excluding the *center*, in Table 3.1.
- With the capacitor voltage and inductor current chosen as state variables, each equilibrium state (v_{CQ}, i_{LQ}) is identical to the "open-circuit" capacitor voltage v_C and "short-circuit" inductor current i_L of the associated *resistive* circuit obtained by open-circuiting all capacitors and short-circuiting all inductors. Here, each *equilibrium state* (v_{CQ}, i_{LQ}) can be identified as an *operating point* of the associated resistive circuit.
- The *operating point paradox* is resolved by interpreting each operating point as an equilibrium state of an associated dynamic circuit. Since *nonlinear* circuits can have many equilibrium states, their associated resistive circuit can have many operating points. Which operating point is actually observed in a given laboratory measurement depends on the stability of the equilibrium states *and* on the initial state.
- Every electronic oscillator requires at least two energy-storage elements and at least one *active nonlinear* element. An important class of electronic oscillators can be modeled by a series *RLC* circuit where the inductor and capacitor are linear and the resistor is *nonlinear* and *active* (its *v-i* curve contains a *negative-resistance* region). This class of "*negative-resistance oscillators*," which includes the classic *Van der Pol oscillator* as a special case, can be understood by studying the associated phase portrait.
- When the inductance $L \to 0$ in the negative-resistance oscillator circuit in Fig. 6.1, the phase portrait exhibits instantaneous jumps, thereby justifying the *jump rule* postulated in Chap. 6.

PROBLEMS

Equation formulation

1 Consider the circuit of Fig. 1.3. Let $R_1 = 2\,\Omega$, $R_2 = 4\,\Omega$, and $R_3 = 4\,\Omega$. Replace capacitor C_1 by inductor L_1 of value 2 H and capacitor C_2 by inductor L_2 of value 4 H. Write state equations for the resulting circuit. (Assume ideal op-amp model in the linear region.)

2 (*a*) Write the state equations for the circuits shown in Fig. P7.2.

(*b*) Write a second-order differential equation for each circuit; identify α and ω_0^2 in each case.

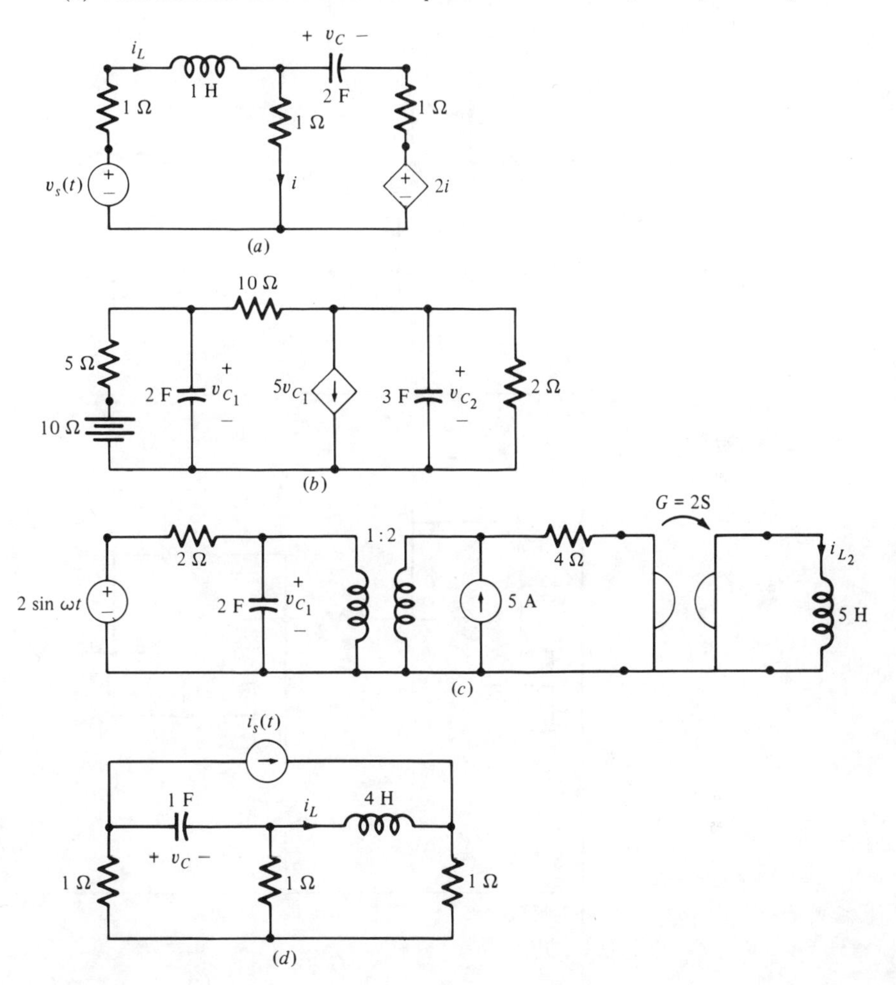

Figure P7.2

3 (*a*) Write the state equations for the circuits shown in Fig. P7.3. Assume that all op amps are ideal and operate in the linear region.

(*b*) For each circuit in (*a*) express v_0 in terms of the state variables and the input.

Equation formulation for circuits containing capacitor loops

4 (*a*) In the loop of linear capacitors shown in Fig. P7.4*a*, C_3 can be replaced by a CCCS, I, as in Fig. P7.4*b*, where

$$I = \frac{C_2 C_3}{\Delta} i_1 - \frac{C_1 C_3}{\Delta} i_2 \qquad \Delta = C_1 C_2 + C_2 C_3 + C_1 C_3$$

Show that these two circuits are equivalent in the sense that they have the same two-port

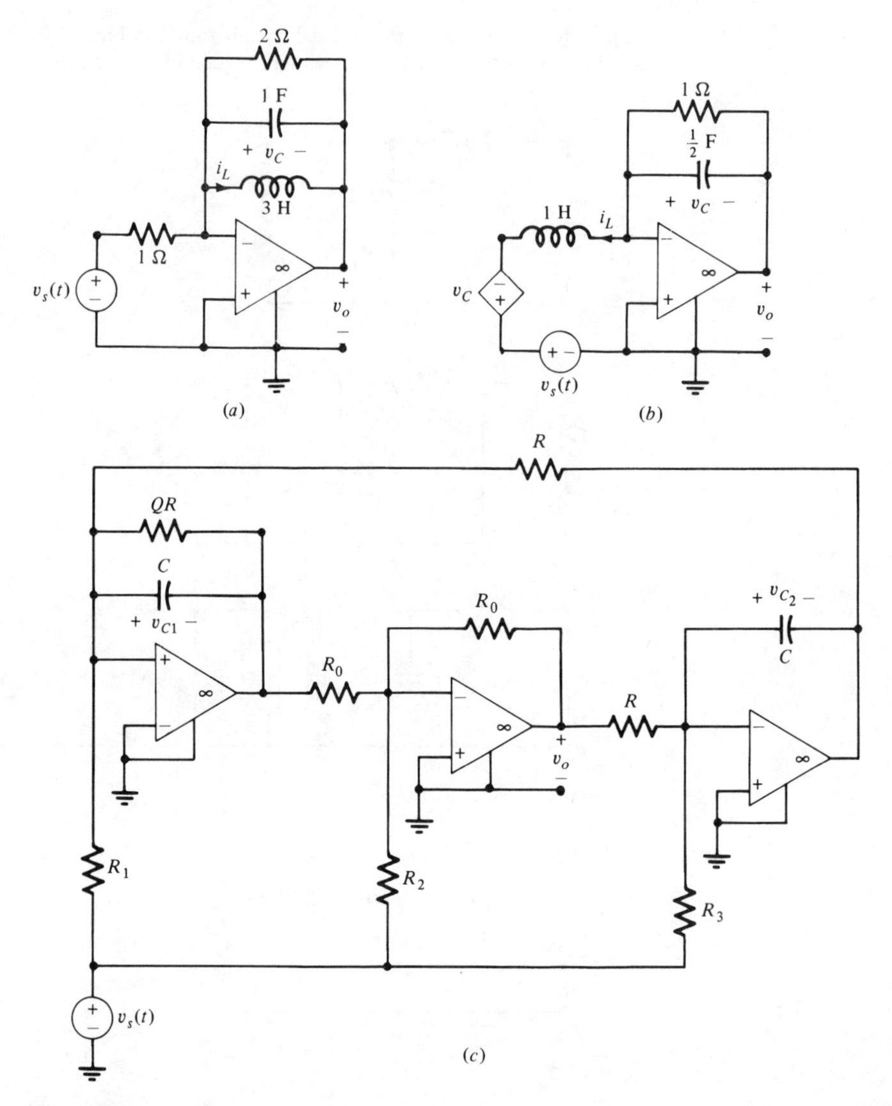

Figure P7.3

representation of the following form:

$$\begin{bmatrix} i_1 \\ i_2 \end{bmatrix} = \begin{bmatrix} C_{11} & C_{12} \\ C_{12} & C_{22} \end{bmatrix} \begin{bmatrix} \dot{v}_1 \\ \dot{v}_2 \end{bmatrix}$$

Note that unlike a two-port resistor, this representation involves $(i_1, i_2; \dot{v}_1, \dot{v}_2)$.

(*b*) The above representation defines the *two-port capacitor* shown in Fig. P7.4*c* where C_{12} is called the coefficient of coupling. Find C_{11}, C_{12}, and C_{22} in terms of C_1, C_2, and C_3 so that this two-port capacitor is equivalent to the one shown in Fig. P7.4*a*.

(*c*) Apply the transformation technique in part (*a*) to the circuit shown in Fig. P7.4*d*. Write the state equations for this equivalent circuit.

(*d*) Derive the state equations directly from Fig. P7.4*d*.

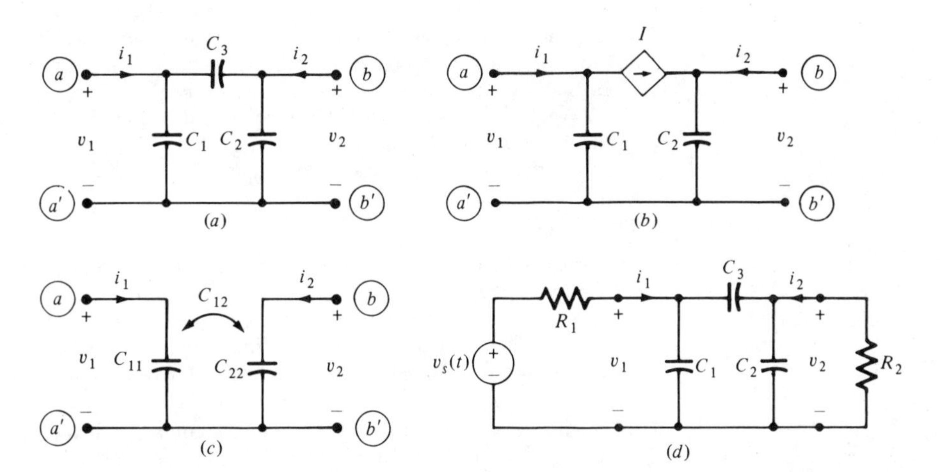

Figure P7.4

Equation formulation for circuits containing inductor cut sets

5 (*a*) In the cut set of inductors shown in Fig. P7.5*a*, L_3 can be replaced by a VCVS, V, as in Fig. P7.5*b* where

$$V = \frac{L_2 L_3}{\Delta} v_1 + \frac{L_1 L_3}{\Delta} v_2 \qquad \Delta = L_1 L_2 + L_2 L_3 + L_1 L_3$$

Show that these two circuits are equivalent in the sense that they have the same two-port representation of the following form:

$$\begin{bmatrix} v_1 \\ v_2 \end{bmatrix} = \begin{bmatrix} L_{11} & L_{12} \\ L_{12} & L_{22} \end{bmatrix} \begin{bmatrix} \dot{i}_1 \\ \dot{i}_2 \end{bmatrix}$$

Note that unlike a two-port resistor, this representation involves $(v_1, v_2; \dot{i}_1, \dot{i}_2)$.

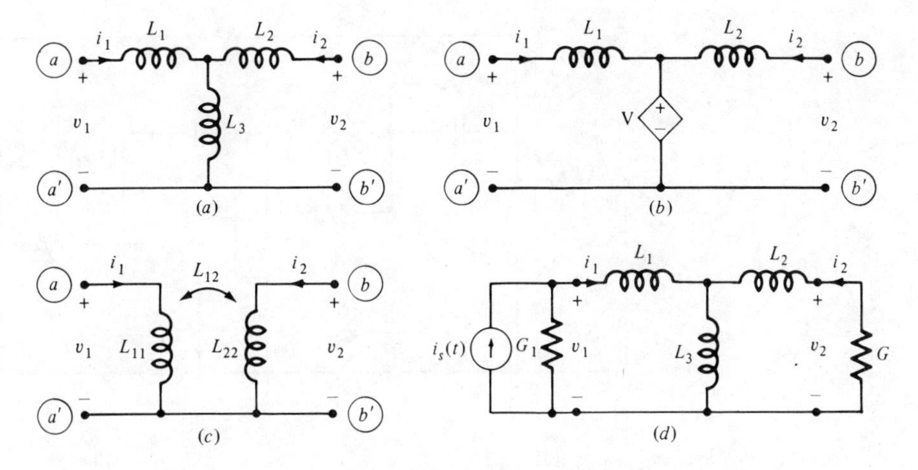

Figure P7.5

(*b*) The above representation defines the *two-port inductor* shown in Fig. P7.5*c* where L_{12} is called the coefficient of coupling. Find L_{11}, L_{12}, and L_{22} in terms of L_1, L_2, and L_3 so that this two-port inductor is equivalent to the one shown in Fig. P7.5*a*.

(*c*) Apply the transformation technique in part (*a*) to the circuit shown in Fig. P7.5*d*. Write the state equations for this equivalent circuit.

(*d*) Derive the state equations directly from Fig. P7.5*d*.

Impulse response

6 The circuit shown in Fig. P7.6 is made of linear time-invariant elements. The voltage v_s is the input, and v_C is the response.

(*a*) Obtain a second-order differential equation in terms of v_C.

(*b*) Calculate the impulse response. *Hint*: Apply Eq. (3.34) of Chap. 6.

(*c*) Calculate the complete response to the input $v_s = 1(t)$ and the initial state $i_L(0) = 2$ A, $v_C(0) = 1$ V.

7 (*a*) Find the step response $v_0 = s(t)$ of the op-amp circuit in Fig. 1.3 of Chap. 7. Let $R_1 = R_2 = 1\ \Omega$, $R_3 = 2\ \Omega$, and $C_1 = C_2 = 2$ F.

(*b*) Find the impulse response $v_0 = h(t)$. *Hint*: Apply Eq. (3.34) of Chap. 6.

8 Consider the op-amp circuit shown in Fig. P7.8. Assume the three op amps are ideal and operate in the linear region.

(*a*) Write a differential equation in terms of v_o.

(*b*) Let $R = 2$ kΩ, $Q = 10$, and $C = 0.01$ μF. Obtain the impulse response $v_o = h(t)$ *Hint*: Apply Eq. (3.34) of Chap. 6.

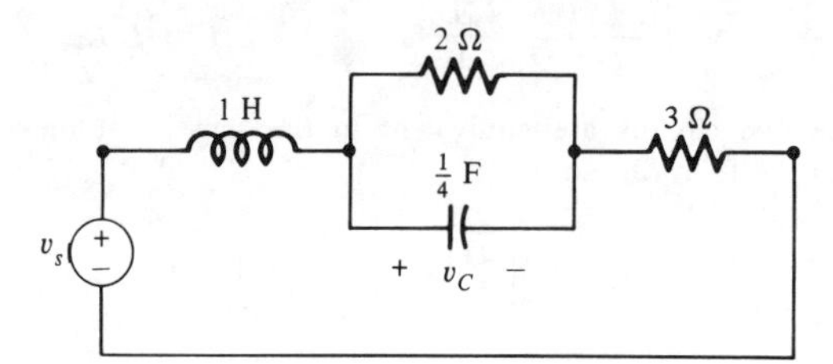

Figure P7.6

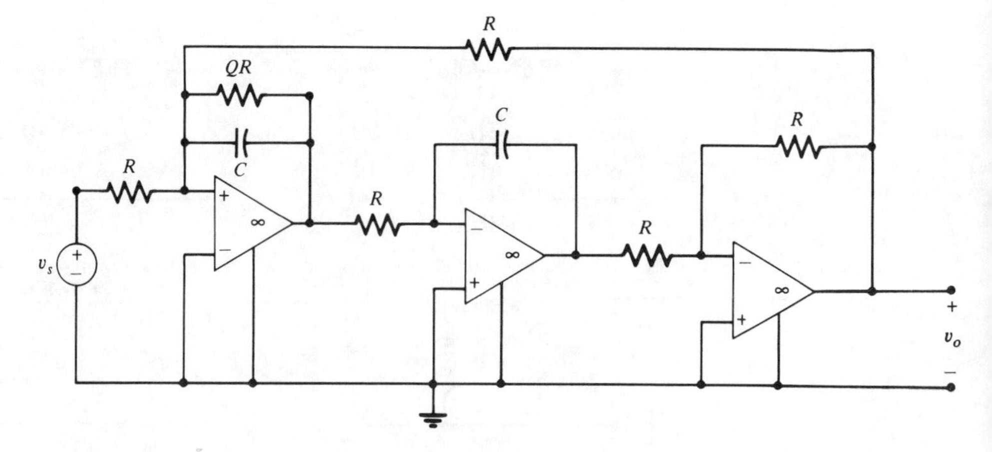

Figure P7.8

Zero-input response

9 Suppose that the natural frequencies of a linear time-invariant circuit consist of one of the following sets:

(i) $s_1 = -2, s_2 = -1$ (iii) $s_1 = j2, s_2 = -j2$
(ii) $s_1 = -2, s_2 = -2$ (iv) $s_1 = -2 + j3, s_2 = -2 - j3$

For each case:

(*a*) Find the corresponding α, ω_0, and Q.
(*b*) Give the general expression for the zero-input response as a real-valued function of time.
(*c*) Suppose $x(0) = 2$, and $\dot{x}(0) = 3$, determine the arbitrary constants for the expressions from (*b*).

10 Let $L = 5$ mH and $C = 20\ \mu$F for the circuits in Fig. 1.1*a* and *b*. Sketch the locus of the natural frequencies s_1 and s_2 as a function of R_p (for Fig. 1.1*a*) and R_s (for Fig. 1.1*b*), respectively.

11 For the circuit shown in Fig. P7.11.

(*a*) Write the state equation.
(*b*) Find the eigenvalues and the eigenvectors.
(*c*) If $v_C(0) = 2$ V and $i_L(0) = 1$ A, determine the zero-input response for $v_C(t)$ and $i_L(t)$.
(*d*) Plot the state trajectory on the v_C-i_L plane.

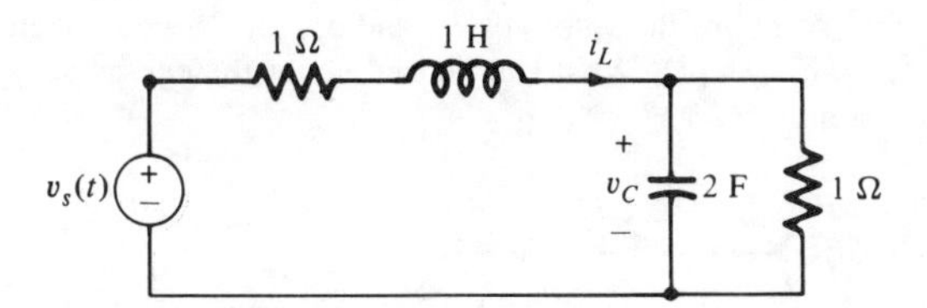

Figure P7.11

Quality factor

12 Given an *RLC* circuit with a Q of 500, how many periods need one wait to have the envelope of the zero-input response reduced to 10 percent, 1 percent, and 0.1 percent of its peak value during the first period? (Give in each case, an answer to the nearest one half period.)

13 Consider two linear time-invariant *RLC* circuits. The first one is a parallel circuit with element values R_p, L, and C, and the second one is a series circuit with element values R_s, L, and C. If the circuits must have the same Q, what is the relation between R_p and R_s? What happens when $Q \to \infty$?

14 Given a linear time-invariant parallel *RLC* circuit with $\omega_0 = 10$ rad/s, $Q = \frac{1}{2}$, and $C = 1$ F, write the differential equation and determine the zero-input response for the voltage v_C across the capacitor. The initial conditions are $v_C(0) = 2$ V and $i_L(0) = 5$ A.

State trajectory

15 Consider the circuit shown in Fig. P7.15. The only data available from measurements taken of this circuit are the time derivatives of the state vector at two different states, namely,

$$\dot{\mathbf{x}} = \begin{bmatrix} -15 \\ 10 \end{bmatrix} \quad \text{at } \mathbf{x} = \begin{bmatrix} 2 \\ 1 \end{bmatrix} \quad \text{and} \quad \dot{\mathbf{x}} = \begin{bmatrix} 3 \\ -5 \end{bmatrix} \quad \text{at } \mathbf{x} = \begin{bmatrix} -1 \\ 1 \end{bmatrix}$$

where

$$\mathbf{x} = \begin{bmatrix} i_L \\ v_C \end{bmatrix}$$

(*a*) Determine the element values R, L, and C.

(*b*) Calculate the derivative of the state vector at

$$\mathbf{x} = \begin{bmatrix} 3 \\ 0 \end{bmatrix}$$

(*c*) Calculate the slope dv_C/di_L of the state trajectory at

$$\mathbf{x} = \begin{bmatrix} 3 \\ 0 \end{bmatrix}$$

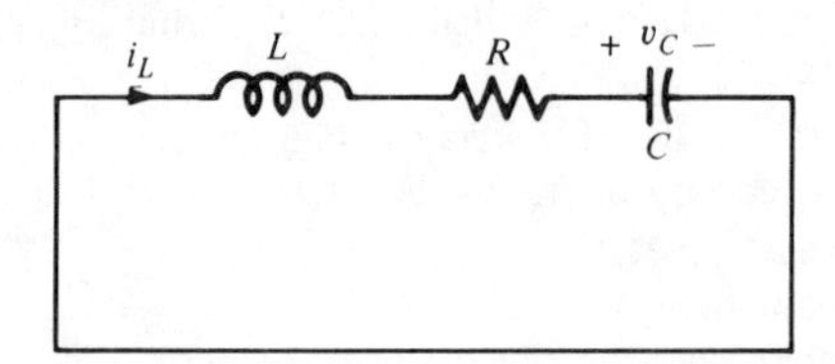

Figure P7.15

16 Consider the *LC* circuit shown in Fig. P7.16. Before time $t = 0$ the switch is open, and the voltages across the capacitors are $v_{C1} = 1\ V$ and $v_{C2} = 4$ V. The switch is closed at time $t = 0$ and remains in this condition for a time interval of 2π s. The switch is opened at $t = 2\pi$ s and remains open thereafter. What are the values of v_{C1} and v_{C2} for $t > 2\pi$ s? Sketch the state trajectory in the i_L-v_C plane ($v_C = v_{C1} + v_{C2}$). What can be said about the energy stored in the circuit before time $t = 0$ and after time $t = 2\pi$ s?

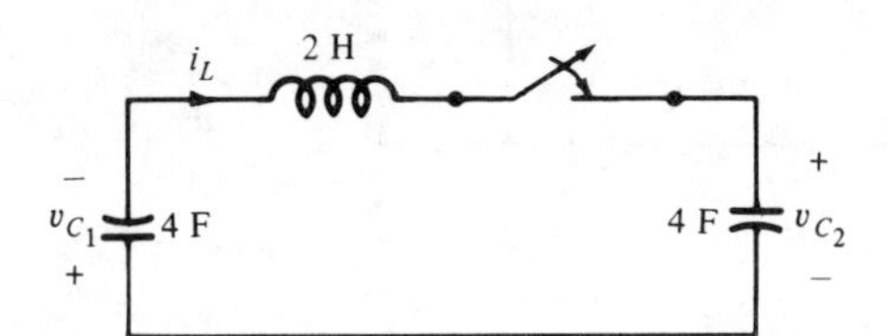

Figure P7.16

Qualitative behavior of $\dot{\mathbf{x}} = \mathbf{A}\mathbf{x}$

17 Sketch a family of trajectories of the state equation $\dot{\mathbf{x}} = \mathbf{A}\mathbf{x}$, where $\mathbf{A}$ is given below. Specify the two eigenvectors if they are real; otherwise specify $\boldsymbol{\eta}_r$ and $\boldsymbol{\eta}_i$. Identify the type of qualitative behavior in each case.

(*a*) $\mathbf{A} = \begin{bmatrix} 0 & 1 \\ -1 & 0 \end{bmatrix}$ (*d*) $\mathbf{A} = \begin{bmatrix} 1 & 1 \\ -1 & 1 \end{bmatrix}$

(*b*) $\mathbf{A} = \begin{bmatrix} 1 & -2 \\ -2 & 1 \end{bmatrix}$ (*e*) $\mathbf{A} = \begin{bmatrix} 2 & 1 \\ 1 & 2 \end{bmatrix}$

(*c*) $\mathbf{A} = \begin{bmatrix} -2 & 1 \\ 1 & -2 \end{bmatrix}$ (*f*) $\mathbf{A} = \begin{bmatrix} -1 & 1 \\ -1 & -1 \end{bmatrix}$

18 For the circuit shown in Fig. P7.18:

(*a*) Write the state equations.
(*b*) Find the equilibrium point.
(*c*) Specify the range of values of *R* such that the equilibrium point is of the following type:
  (i) Stable node
  (ii) Stable focus

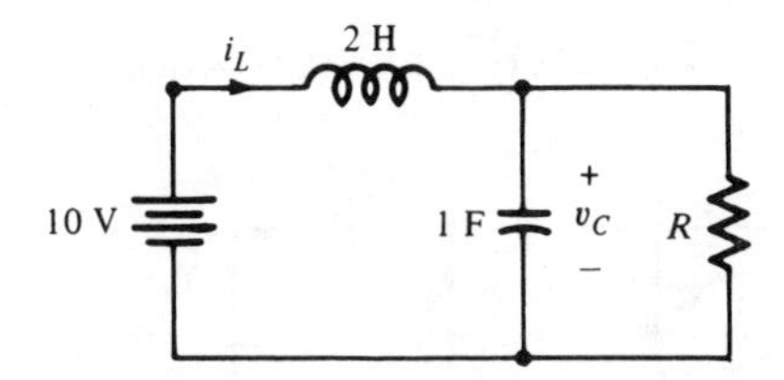

Figure P7.18

19 (*a*) Write the state equations for the circuit in Fig. P7.15.

(*b*) Find its eigenvalues and eigenvectors for each of the following cases:

(i) $R = 0\,\Omega,\ L = 1\,\text{H},\ C = 1\,\text{F}$

(ii) $R = 1\,\Omega,\ L = 1\,\text{H},\ C = 1\,\text{F}$

(iii) $R = -1\,\Omega,\ L = 1\,\text{H},\ C = 1\,\text{F}$

(*c*) For each case in (*b*) sketch a family of trajectories.

(*d*) For each case give one representative sketch of v_C vs. t indicating the trajectory to which it corresponds.

Nonlinear state equations

20 Write the state equations for the circuits shown in Fig. P7.20.

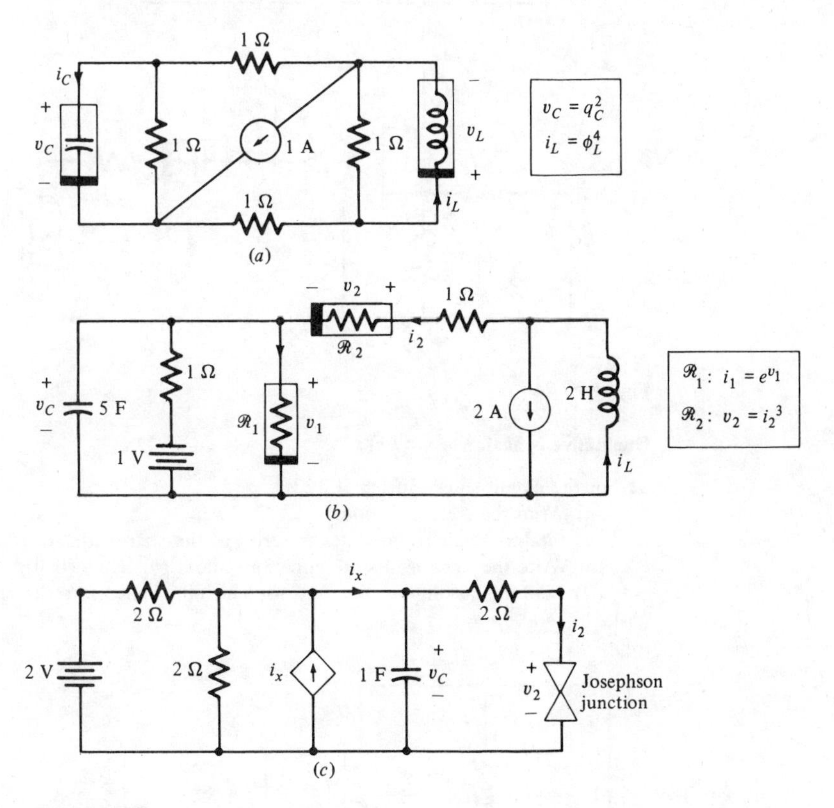

Figure P7.20

21 Repeat Prob. 20 for the circuits shown in Fig. P7.21. Assume the ideal op-amp model operating in the linear region.

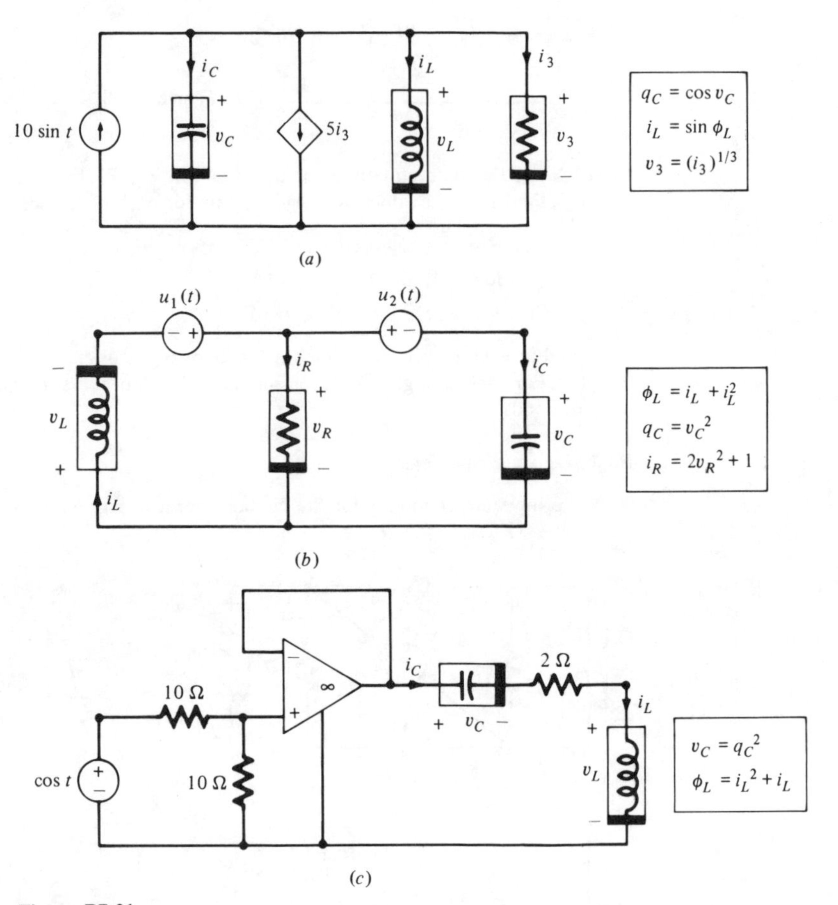

Figure P7.21

Qualitative behavior of $\dot{\mathbf{x}} = \mathbf{f}(\mathbf{x})$

22 For the circuit shown in Fig. P7.22:

(*a*) Write the state equations.

(*b*) Find all equilibrium states in terms of the state variables (v_C, i_L).

(*c*) Write the linearized state equations about $(v_C, i_L) = (0, 0)$.

(*d*) Find the maximum value of C for which the phase portrait near the origin behaves like an unstable node.

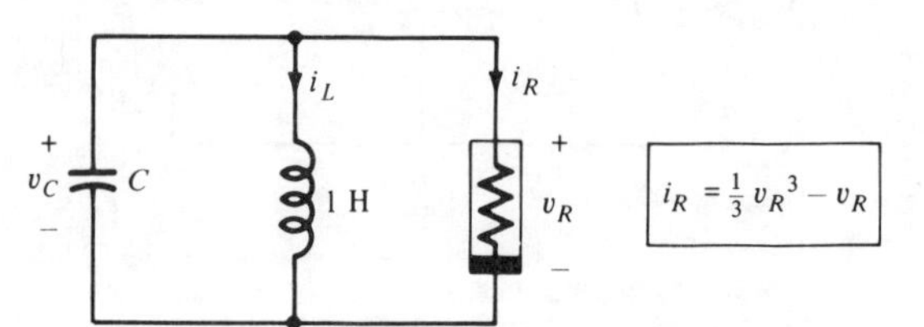

Figure P7.22

23 A nonlinear circuit has the following state equations:

$$\begin{bmatrix} \dot{v}_C \\ \dot{i}_L \end{bmatrix} = \begin{bmatrix} 2v_C^2 + i_L \\ v_C^3 + v_C - i_L \end{bmatrix}$$

(*a*) Find the two equilibrium points.

(*b*) Determine the qualitative behavior of these equilibrium points using the linearized approximation method.

24 A nonlinear circuit is described by the following state equations:

$$\dot{x}_1 = -x_1 + x_1 x_2$$
$$\dot{x}_2 = x_1 + x_2$$

Find the equilibrium points for this circuit and determine their qualitative behavior.

25 Consider the circuit shown in Fig. P7.25*a* where N is described by the v-i characteristic in Fig. P7.25*b*. Find the following:

(*a*) Assuming $R = 0.5\ \Omega$, find and classify each equilibrium point when $E = 2$, 4, and 10 V, respectively.

(*b*) Assuming $E = 10$ V, find and classify each equilibrium point when $R = 1\ \Omega$, $2.5\ \Omega$, and $10\ \Omega$, respectively.

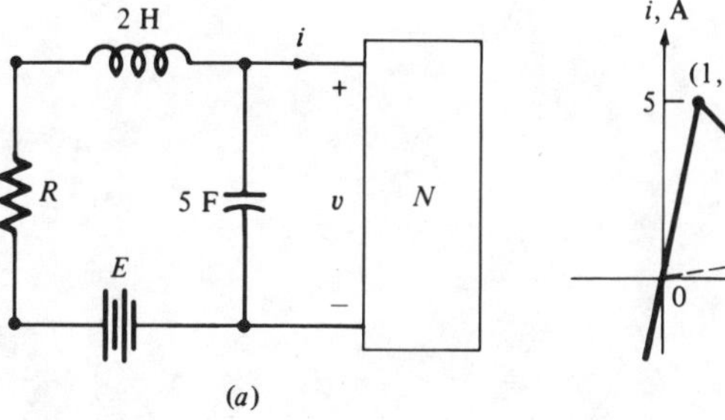

(*a*)

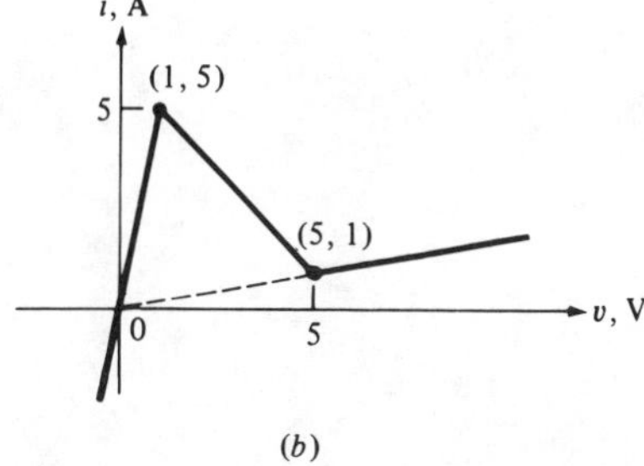

(*b*)

Figure P7.25

26 The nonlinear resistor in Fig. P7.26*a* is described by the v-i characteristic shown in Fig. P7.26*b*.

(*a*) Without writing state equations, find all equilibrium states of this circuit.

(*b*) Calculate the jacobian matrix associated with each equilibrium state and classify it using Fig. 3.11.

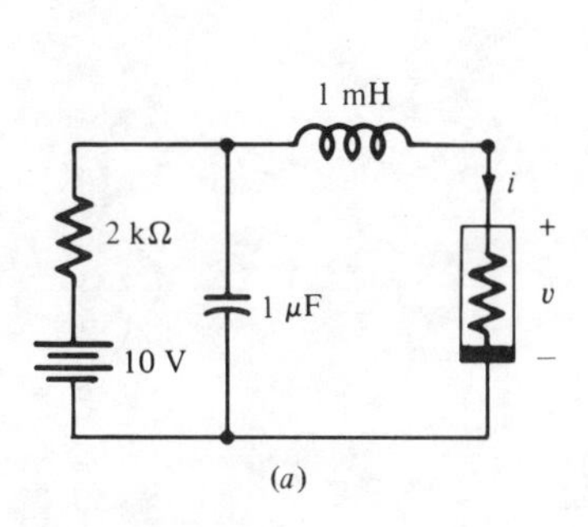

(*a*)

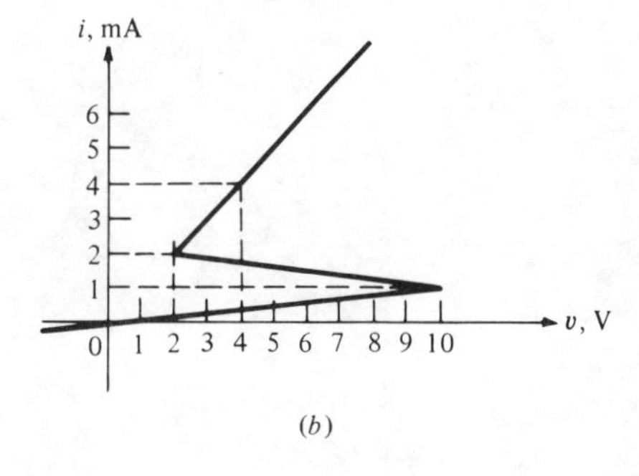

(*b*)

Figure P7.26

Nonlinear oscillation

27 Consider the circuit shown in Fig. 6.1. State which of the following *i-v* characteristics for N_R may cause the circuit to oscillate.

(i) $v = i^n - i$ (n is an odd integer)
(ii) $v = -\sin i$
(iii) $v = -e^i \sin i$
(iv) $v = \cos i - i - 1$

CHAPTER

EIGHT

GENERAL DYNAMIC CIRCUITS

Before plunging into the study of general dynamic circuits, let us develop some perspective and review our coverage. In Chap. 1, we studied the concepts of lumped circuits and Kirchhoff's laws. In Chaps. 2, 3, and 4 we learned about a number of *resistive* elements both two-terminal and multiterminal; we also studied simple circuits and their properties. Then Chap. 5 introduced us to *general resistive circuits*: We learned to write their equations and to derive some of their properties. One key feature of the analysis of *any resistive* circuit is that the equations are *algebraic*. In Chaps. 6 and 7, we encountered *energy-storage* elements, inductors and capacitors, and we studied in detail some simple first-order and second-order circuits. One key feature of the analysis of such circuits was the appearance of *differential equations*. This was our first introduction to the *dynamics* of circuits. Roughly speaking, the present chapter will teach us how to write the equations of any *dynamic circuit* and to derive some of their properties.

But our first task will be to complete our basic set of circuit elements by including coupled inductors. Then, in the next two sections we develop two completely general methods of analysis applicable to *any dynamic circuit*: the tableau analysis and the modified node analysis. These are the methods used by the circuit simulators ASTAP and SPICE. These two methods of analysis will be used repeatedly in the remainder of this book.

In the fourth section, we develop the small-signal analysis of a nonlinear circuit about an operating point. This is followed by two general theorems: the superposition theorem and the substitution theorem. We already encountered these theorems for *resistive* circuits in Chap. 5, we now develop them for general *dynamic* circuits. The last section describes the simplest reliable method for solving the system of algebraic and differential equations of any dynamic circuit on a computer, namely, the backward Euler method.

1 COUPLED INDUCTORS

In Chap. 6 we studied linear and nonlinear two-terminal inductors. Now we study the circuits in which two or more inductors are coupled magnetically. Coupled inductors have many uses in communication circuits and in measuring equipment. Transformers, which are coupled inductors satisfying special design objectives, are of crucial importance in power networks linking the power stations to the users of electric energy. Electric motors and electric generators may also be modeled as time-varying inductors.

Consider a torus made of ferromagnetic material: typically ferrite or thin sheets of special steel. As shown on Fig. 1.1, let us wind on this torus two coils; we thus obtain a two-port. If we drive the first port with a generator so that the current i_1 is *positive* and have the second port open (hence $i_2 = 0$), there will be a strong magnetic field set up in the torus. (This is indicated by the arrow in the figure.) Note that the lines of the magnetic field link the second coil, so that if $i_1(t)$ varies with time, there will be a time-varying flux through that second coil; hence, by Faraday's law, a voltage will be induced and $v_2(t) \neq 0$.

1.1 Linear Time-Invariant Coupled Inductors

A. Characterization If we choose a linear model for the physical device shown in Fig. 1.1, we obtain linear equations relating the currents i_1 and i_2 and the fluxes ϕ_1 and ϕ_2:

$$\begin{aligned}\phi_1 &= L_{11}i_1 + Mi_2 \\ \phi_2 &= Mi_1 + L_{22}i_2\end{aligned} \qquad \text{or} \qquad \boldsymbol{\phi} = \mathbf{L}\,\mathbf{i} \qquad \text{with} \qquad \mathbf{L} \triangleq \begin{bmatrix} L_{11} & M \\ M & L_{22} \end{bmatrix} \tag{1.1}$$

where the ϕ_k's are in webers, the L's are in henrys, and the i's in amperes; $\mathbf{L}$ is called the *inductance matrix*, L_{11} is the *self-inductance* of inductor 1, and M is

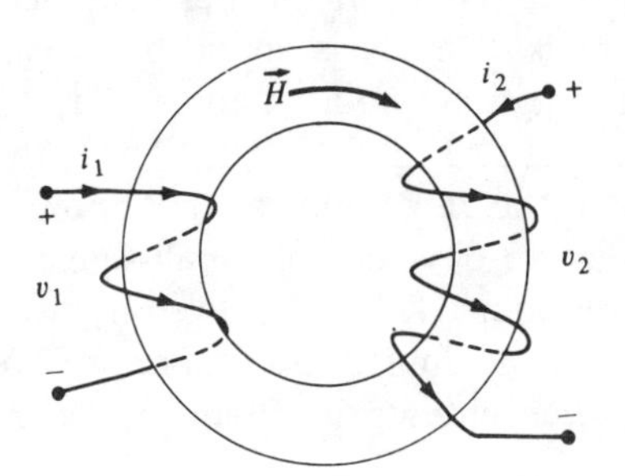

Figure 1.1 Two coupled coils wound on a torus of ferromagnetic material.

called the *mutual inductance* of inductor 1 and inductor 2. In physics, the equality of the off-diagonal terms of the inductance matrix is established by fundamental energy considerations.

Equation (1.1) describes the *ϕ-i characteristic* of the two linear coupled inductors: Eq. (1.1) gives $\boldsymbol{\phi}$ in terms of $\mathbf{i}$, $\boldsymbol{\phi} = \hat{\boldsymbol{\phi}}(\mathbf{i})$, therefore it is a *current-controlled* representation.

Since, for $k = 1, 2$, ϕ_k is the (total) flux through the kth inductor (i.e., due to *both* currents), by Faraday's law, we have

$$\dot{\phi}_k = v_k \qquad k = 1, 2 \tag{1.2}$$

From Eqs. (1.1) and (1.2), we obtain the description of the linear two-port made up of the two inductors shown in Fig. 1.1:

$$\begin{bmatrix} v_1(t) \\ v_2(t) \end{bmatrix} = \begin{bmatrix} L_{11} & M \\ M & L_{22} \end{bmatrix} \begin{bmatrix} \dot{i}_1(t) \\ \dot{i}_2(t) \end{bmatrix} \tag{1.3}$$

or in vector form

$$\mathbf{v}(t) = \mathbf{L}\,\dot{\mathbf{i}}(t) \tag{1.4}$$

The *inductance matrix* $\mathbf{L}$ characterizes the pair of linear time-invariant inductors. Equation (1.4) gives the voltages across the inductors in terms of the rate of change of the inductor currents.

In some cases it is convenient to use the voltages as independent variables; then we solve Eq. (1.3) for $\dot{i}_1$ and $\dot{i}_2$. Thus, provided $\det(\mathbf{L}) \neq 0$, we obtain

$$\begin{bmatrix} \dot{i}_1(t) \\ \dot{i}_2(t) \end{bmatrix} = \begin{bmatrix} \Gamma_{11} & \Gamma_M \\ \Gamma_M & \Gamma_{22} \end{bmatrix} \begin{bmatrix} v_1(t) \\ v_2(t) \end{bmatrix} \qquad \text{or} \qquad \dot{\mathbf{i}}(t) = \boldsymbol{\Gamma}\,\mathbf{v}(t) \tag{1.5}$$

where

$$\Gamma_{11} = \frac{L_{22}}{\det(\mathbf{L})} \qquad \Gamma_{22} = \frac{L_{11}}{\det(\mathbf{L})} \qquad \text{and} \qquad \Gamma_M = \frac{-M}{\det(\mathbf{L})} \tag{1.6}$$

Equation (1.5) is particularly useful in writing node equations. The matrix $\boldsymbol{\Gamma}$ is called the *reciprocal inductance matrix*.

Exercises

1. Verify by direct computation that $\mathbf{L}\boldsymbol{\Gamma} = \mathbf{1}$.

2. You are given a pair of coupled inductors specified by $\mathbf{L}$ as in Eq. (1.1) above. Determine the inductance seen between terminals
(*a*) ① and ② when ① and ② are connected by a short circuit.
(*b*) ① and ② when each pair ① and ②, and ① and ② is connected by a short circuit.
(*c*) ① and ② when ① and ② are connected by a short circuit.

3. Consider the linear time-invariant inductors described by Eq. (1.3) and shown in Fig. 1.2. Suppose we tie nodes ① and ② to a common ground.

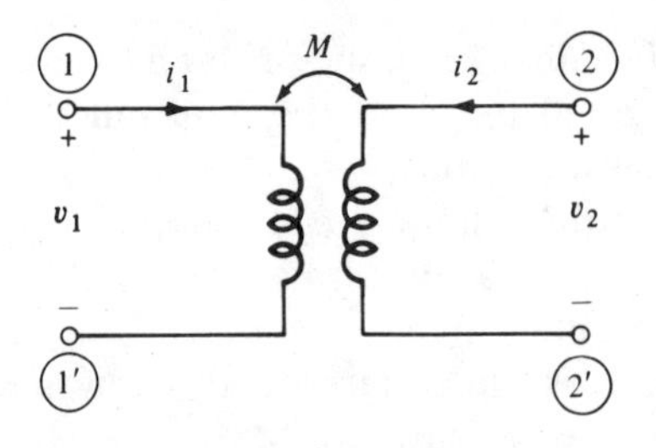

Figure 1.2 Pair of coupled inductors with labeled nodes.

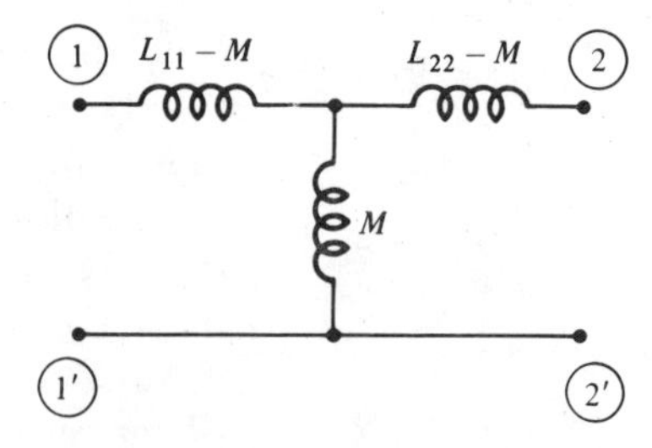

Figure 1.3 This two-port is equivalent to the one shown in Fig. 1.2 provided nodes ① and ② are tied to a common ground.

Show that the resulting two-port is equivalent to the two-port shown in Fig. 1.3: Note that it is made of three *uncoupled* inductors. Show which of these inductors may be negative.

B. Stored energy We are going to calculate the energy stored in the pair of inductors described by Eq. (1.3). We connect two generators at each port so that we can vary the inductor currents i_1 and i_2. We start at $t = 0$ with $i_1(0) = i_2(0) = 0$. No current flows in the inductors; no magnetic field exists in the magnetic material.[1] Hence, at $t = 0$, $\phi_1(0) = \phi_2(0) = 0$ as prescribed by Eq. (1.1). Naturally, this condition is taken to be the zero level for the stored magnetic energy $\mathscr{E}_M$: for $i_1 = i_2 = 0$, $\mathscr{E}_M(0, 0) = 0$.

Now let us increase the currents $i_1(t)$ and $i_2(t)$ so that at some fixed time T, we have $i_1(T) = I_1$, $i_2(T) = I_2$ where the constants I_1 and I_2 have been specified in advance. Let us calculate the energy delivered *by* the generators *to* the two-port over the interval $[0, T]$:

$$\text{Energy delivered} = \int_0^T [v_1(t)\, i_1(t) + v_2(t)\, i_2(t)]\, dt$$

$$= \int_0^T \{L_{11}\, \dot{i}_1(t)\, i_1(t) + M[\dot{i}_1(t)\, i_2(t) + i_1(t)\, \dot{i}_2(t)] + L_{22}\, \dot{i}_2(t)\, i_2(t)\}\, dt \quad (1.7)$$

or

$$\mathscr{E}_M[i_1(T), i_2(T)] \overset{\Delta}{=} \tfrac{1}{2}[L_{11}\, i_1(T)^2 + 2M\, i_1(T)\, i_2(T) + L_{22}\, i_{22}(T)^2] \quad (1.8)$$

where we used elementary calculus and $i_1(0) = i_2(0) = 0$ to go from Eq. (1.7) to Eq. (1.8).

Expression (1.8) gives the energy delivered by the generators to the two coupled inductors, when, at $t = 0$, $i_1(0) = i_2(0) = 0$ and the currents at time T are $i_1(T)$ and $i_2(T)$. This energy is *stored energy*: To see this, consider the following experiment. At $t = 0$, we start with $i_1(0) = i_2(0) = 0$, hence $\mathscr{E}_M(0, 0) = 0$ at $t = 0$. Let us drive the generators so that, at some time $t_1 > 0$,

[1] Since we have a linear model, there is no hysteresis.

$i_1(t_1)$ and $i_2(t_1)$ take the values $i_1(T)$ and $i_2(T)$ as in expression (1.8). Then let us decrease the currents so that at some time $t_2 > t_1$, $i_1(t_2) = i_2(t_2) = 0$. By the calculation above, $\mathscr{E}_M[i_1(t_2), i_2(t_2)] = 0$. Consequently all the energy that was delivered to the inductors at time t_1, namely $\mathscr{E}_M[i_1(T), i_2(T)]$ of Eq. (1.8), has been returned to the generators! This proves that the energy given by expression (1.8) represents *stored energy*. In physics it is proven that the energy is stored in the magnetic field.

The exercise below describes another way of proving that expression (1.8) represents stored energy.

Exercise Consider the coupled inductors shown in Fig. 1.1. Let Eq. (1.3) hold. Suppose that for $t<0$, the currents i_1 and i_2 supplied by external generators (not shown in the figure) are constant and equal to I_1 and I_2, respectively. Now at $t=0$, the generators are switched off and instantaneously replaced by two linear strictly passive resistors R_1 and R_2. Show that the total energy dissipated in the resistors over the time interval $[0, \infty)$, namely,

$$\int_0^\infty [R_1\, i_1(t)^2 + R_2\, i_2(t)^2]\, dt$$

is equal to the energy stored at $t=0$ in the inductors:

$$\tfrac{1}{2}(L_{11}I_1^2 + 2MI_1I_2 + L_{22}I_2^2)$$

(This shows that the total magnetic energy stored in the inductors can be recovered as heat dissipated in the resistors R_1 and R_2.)

In conclusion, we have shown that the expression (1.8) gives the stored magnetic energy when the inductors are traversed by the currents $i_1(T)$ and $i_2(T)$.

For simplicity, let us label these currents by i_1 and i_2; hence we write

$$\mathscr{E}_M(i_1, i_2) = \tfrac{1}{2}(L_{11}i_1^2 + 2Mi_1i_2 + L_{22}i_2^2) \tag{1.9}$$

or, in matrix form,

$$\mathscr{E}_M(i_1, i_2) = \tfrac{1}{2}[i_1 \quad i_2]\begin{bmatrix} L_{11} & M \\ M & L_{22} \end{bmatrix}\begin{bmatrix} i_1 \\ i_2 \end{bmatrix} = \tfrac{1}{2}\mathbf{i}^T\mathbf{L}\,\mathbf{i} \tag{1.10}$$

For obvious physical reasons, $\mathscr{E}_M(i_1, i_2) \geq 0$ for *all* i_1 and i_2. This fact imposes the following restrictions on the inductance matrix $\mathbf{L}$.

Property of the inductance matrix The stored magnetic energy $\mathscr{E}_M(i_1, i_2)$ is *positive* (>0) for all $\mathbf{i} \neq \mathbf{0}$ if and only if

$$L_{11} > 0 \qquad L_{22} > 0 \qquad \text{and} \qquad M^2 < L_{11}L_{22} \tag{1.11}$$

COMMENTS AND INTERPRETATIONS
1. The conditions (1.11) imposed on the matrix **L** may also be written as

$$L_{11} > 0 \quad \text{and} \quad \det(\mathbf{L}) = (L_{11}L_{22} - M^2) > 0 \tag{1.12}$$

It is easy to see that these *two* conditions imply that $L_{22} > 0$. Note that the conditions above are the necessary and sufficient conditions for the real symmetric matrix **L** to be *positive definite*.
2. The mutual inductance M is often expressed in terms of the *coefficient of coupling* k which is defined by

$$k \triangleq \frac{M}{\sqrt{L_{11}L_{22}}}$$

The last inequality (1.12) states that $|k| < 1$, that is, it is impossible to achieve a coefficient of coupling larger or equal to 1. When $k = 0$, we have $M = 0$, that is, there is no magnetic coupling between the inductors. As k increases toward 1, the coupling becomes tighter.
3. To illustrate the meaning of the coefficient of coupling, let us consider the limiting situation where $k = 1$.

Let us rewrite Eq. (1.9) as

$$\mathscr{E}_M(i_1, i_2) = \tfrac{1}{2} L_{11}\left(i_1 + \frac{M}{L_{11}} i_2\right)^2 + \tfrac{1}{2}\left(L_{22} - \frac{M^2}{L_{11}}\right) i_2^2 \tag{1.13}$$

With $k = 1$, for all i_2, the second term in the right-hand side of Eq. (1.13) is *zero*; furthermore, for any $i_2 \neq 0$, if $i_1 = -(M/L_{11})i_2$, the first term is also zero. That is $\mathscr{E}_M = 0$ with $i_1 \neq 0$ and $i_2 \neq 0$: For this to happen, *all* the magnetic field due to i_1 must be *completely canceled* by the magnetic field due to i_2: This is clearly impossible to realize in practice.

PROOF OF THE PROPERTY Consider $\mathscr{E}_M(i_1, i_2)$ as given by Eq. (1.13): If the conditions (1.12) hold, we claim that, for all $\mathbf{i} \neq \mathbf{0}$, $\mathscr{E}_M > 0$. Indeed, if $i_2 = 0$, then $i_1 \neq 0$, and the first term is positive, while the second is zero; if $i_2 \neq 0$, the second term is positive, while the first one is ≥ 0, irrespective of the value of i_1. Hence Eq. (1.12) implies that $\mathscr{E}_M > 0$ for all $\mathbf{i} \neq \mathbf{0}$.

Conversely, suppose that $\mathscr{E}_M(i_1, i_2)$ is > 0 for all $\mathbf{i} \neq \mathbf{0}$. We establish the conditions (1.12) by contradiction. (i) Suppose $L_{11} \leq 0$; then $\mathscr{E}_M(1, 0) = \frac{1}{2} L_{11} \leq 0$, a contradiction. (ii) Interchanging subscripts 1 and 2, we conclude that $L_{22} > 0$. (iii) Suppose $M^2 \geq L_{11}L_{22}$. Then for any $i_2 \neq 0$ and $i_1 \triangleq -(M/L_{11})i_2$, $\mathscr{E}_M \leq 0$ by Eq. (1.13); again a contradiction. Hence the three inequalities of (1.12) must hold. ∎

C. Sign of M The mutual inductance M may be positive or negative. Once the reference directions have been assigned to the inductors, the sign of M is specified by the physical arrangement of the coils. Refer to Fig. 1.4, and let us drive the two inductors by two dc currents $I_1 > 0$ and $I_2 > 0$.

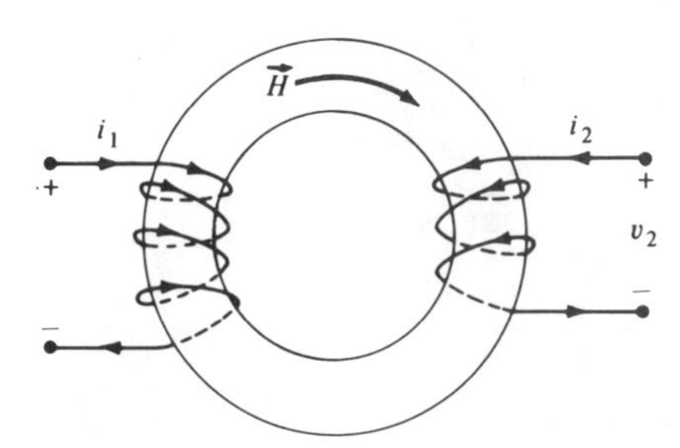

Figure 1.4 Torus with two inductor windings: in this case the mutual inductance M is positive.

By the right-hand rule, the *positive* dc current I_1 creates in the core a magnetic field along the direction of the arrow, and the *positive* dc current I_2 creates in the core a magnetic field in the *same* direction.

Suppose we repeat the experiment with $i_1 = I_1$ and $i_2 = -I_2$, that is, the current in the second inductor now flows in the direction opposite to that of the reference direction shown in Fig. 1.4. By the right-hand rule, the two magnetic fields are now in *opposite* directions.

In the first experiment, there is a larger magnetic field in the core—hence a larger stored energy—than in the second experiment. So (recall that $I_1 > 0$ and $I_2 > 0$)

$$\mathscr{E}(I_1, I_2) > \mathscr{E}(I_1, -I_2)$$

Consequently, using Eq. (1.9) and canceling common terms, we obtain

$$MI_1I_2 > -MI_1I_2$$

and, since $I_1I_2 > 0$, we conclude that

$$M > 0 \tag{1.14}$$

Thus we have shown that, *for the reference directions chosen*, for the circuit shown in Fig. 1.4 the mutual inductance M is positive.

D. More than two coupled inductors It is intuitively clear that if we have more than two linear time-invariant coupled inductors, we obtain equations similar to Eqs. (1.1) to (1.4). For example, if there are *three* inductors, we have

$$\begin{bmatrix} v_1(t) \\ v_2(t) \\ v_3(t) \end{bmatrix} = \begin{bmatrix} L_{11} & M_{12} & M_{13} \\ M_{12} & L_{22} & M_{23} \\ M_{13} & M_{23} & L_{33} \end{bmatrix} \begin{bmatrix} \dot{i}_1(t) \\ \dot{i}_2(t) \\ \dot{i}_3(t) \end{bmatrix} \tag{1.15a}$$

where we used the fact that the inductance matrix is symmetric: M_{12} denotes the mutual inductance between inductor 1 and inductor 2, etc. Note that the mutual inductances M_{12}, M_{23}, and M_{31} need not have the same sign. This is illustrated by the following exercise.

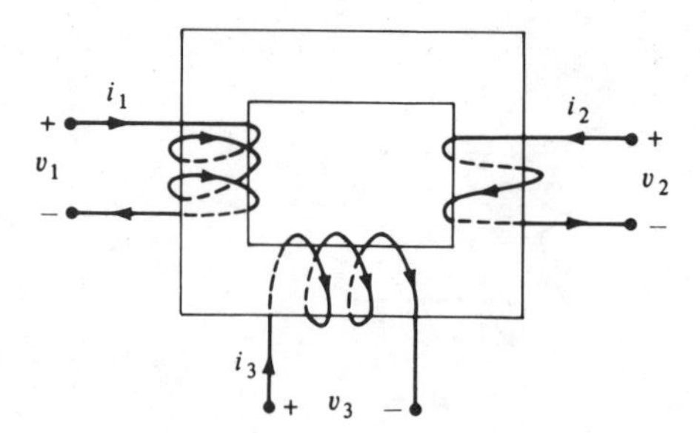

Figure 1.5 Three inductors wound on the same core: the mutual inductances are not all positive.

Exercise Consider the three mutually coupled inductors shown in Fig. 1.5. Show that $M_{12} > 0$, $M_{13} < 0$, and $M_{23} < 0$.

Equation (1.15*a*) is of the form $\mathbf{v}(t) = \mathbf{L}\,\dot{\mathbf{i}}(t)$. As before we can calculate the magnetic energy stored and we find the same formula as Eq. (1.10), namely,

$$\mathscr{E}_M(\mathbf{i}) = \tfrac{1}{2}\mathbf{i}^T\mathbf{L}\,\mathbf{i} \tag{1.15b}$$

E. Relation with ideal transformers It is very useful to know that any pair of coupled linear time-invariant inductors is equivalent to a two-port made up of an ideal transformer and two (uncoupled) inductors L_a and L_m as shown in Fig. 1.6.

The calculations are left as an exercise.

Exercise

(*a*) Calculate the inductance matrix of the two-port shown in Fig. 1.6: More precisely show that we have $\mathbf{v} = \mathbf{L}\,\dot{\mathbf{i}}$ where

$$\mathbf{L} = \begin{bmatrix} L_a + L_m & nL_m \\ nL_m & n^2L_m \end{bmatrix} \tag{1.16}$$

(*b*) Given any pair of coupled inductors specified by Eq. (1.3) show that they are equivalent to the two-port shown in Fig. 1.6 provided

$$n = \frac{L_{22}}{M} \qquad L_m = \frac{M^2}{L_{22}} \qquad L_a = L_{11} - \frac{M^2}{L_{22}} \tag{1.17}$$

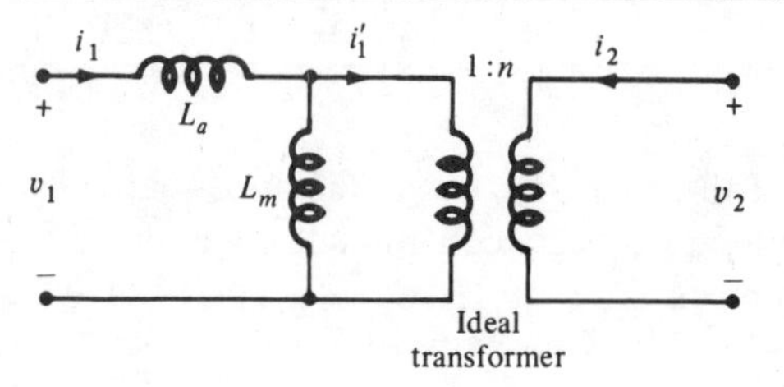

Figure 1.6 A two-port equivalent to a pair of coupled inductors.

The physical interpretation of L_a and L_m is as follows: L_a is the *leakage inductance*, that is, the inductance seen at the first port due to the leakage flux, i.e., the lines of magnetic field that do not link both coils. Indeed as $k^2 \to 1$, $M^2 \to L_{11}L_{22}$, and $L_a \to 0$. L_m is called the *magnetizing inductance*: Its role is to model the magnetic flux common to both coils.

Suppose we wish to build a high-quality transformer. We choose a torus of magnetic material with a very high permeability μ (e.g., ferrite, or super-permalloy, etc.), then we wind tightly on the torus the two coils forming a two-port, say as in Fig. 1.1. Suppose that we were able to find magnetic materials with increasingly high μ: As μ becomes larger and larger (*a*) the leakage flux would become smaller and smaller, hence L_a would become smaller, and (*b*) the common flux would also increase, i.e., L_m would increase. Thus in the limit of $\mu \to \infty$, $L_a \to 0$ and $L_m \to \infty$: Referring to Fig. 1.6 we see that we are left with the ideal transformer!

1.2 Nonlinear Time-Invariant Coupled Inductors

In the *linear* case, the ϕ-i characteristic is described by the linear equations (1.1): The fluxes depend *linearly* on the currents. In the *nonlinear* case, the fluxes will depend in a nonlinear fashion on the currents, hence we write

$$\begin{aligned} \phi_1 &= \hat{\phi}_1(i_1, i_2) \\ \phi_2 &= \hat{\phi}_2(i_1, i_2) \end{aligned} \qquad \text{or} \qquad \boldsymbol{\phi} = \hat{\boldsymbol{\phi}}(\mathbf{i}) \tag{1.18}$$

where the functions $\hat{\phi}_1(\cdot\,,\cdot)$ and $\hat{\phi}_2(\cdot\,,\cdot)$ are scalar-valued functions of two real variables. These equations specify the ϕ-i characterisic of a pair of *current-controlled* nonlinear coupled inductors. Since $\hat{\boldsymbol{\phi}}$ in Eq. (1.18) does not depend explicitly on time, we say that these nonlinear coupled inductors are *time-invariant*.

Suppose that we use these coupled nonlinear inductors in a circuit and that they are traversed by the currents $i_1(\cdot)$ and $i_2(\cdot)$; then, at time t, $\phi_k(t) = \hat{\phi}_k[i_1(t), i_2(t)]$ for $k = 1, 2$. To obtain the port voltages, we must differentiate with respect to time ($v_k = \dot{\phi}_k$), hence, using the chain rule, we obtain

$$\begin{aligned} v_1(t) &= \left.\frac{\partial \hat{\phi}_1}{\partial i_1}\right|_{\mathbf{i}(t)} \cdot \dot{i}_1(t) + \left.\frac{\partial \hat{\phi}_1}{\partial i_2}\right|_{\mathbf{i}(t)} \cdot \dot{i}_2(t) \\ v_2(t) &= \left.\frac{\partial \hat{\phi}_2}{\partial i_1}\right|_{\mathbf{i}(t)} \cdot \dot{i}_1(t) + \left.\frac{\partial \hat{\phi}_2}{\partial i_2}\right|_{\mathbf{i}(t)} \cdot \dot{i}_2(t) \end{aligned} \tag{1.19}$$

where $(\partial \hat{\phi}_j / \partial i_k)|_{\mathbf{i}(t)}$ denotes $(\partial \hat{\phi}_j / \partial i_k)[i_1(t), i_2(t)]$, i.e., all the partial derivatives in Eq. (1.19) are evaluated at $\mathbf{i}(t)$.

Note that Eqs. (1.18) and (1.19) are straightforward generalizations of Eqs. (1.1) and (1.3).

We can rewrite the Eq. (1.19) in matrix form if we use the jacobian matrix of $\hat{\boldsymbol{\phi}}$:

$$\mathbf{v}(t) = \mathbf{J}(\mathbf{i}(t))\,\dot{\mathbf{i}}(t) \tag{1.20}$$

where the (k, ℓ) element of $\mathbf{J}$ is equal to $(\partial\hat{\phi}_k/\partial i_\ell)(\mathbf{i}(t))$, for $k, \ell = 1, 2$. [Recall that we encountered the jacobian matrix in Eq. (3.3) of Chap. 5.]

Thus a pair of *nonlinear time-invariant* inductors is specified by the two scalar-valued functions $\hat{\phi}_1(\cdot\,,\cdot)$ and $\hat{\phi}_2(\cdot\,,\cdot)$, or equivalently, by the vector-valued function $\hat{\boldsymbol{\phi}}(\cdot\,,\cdot)$ as in Eq. (1.18). When we calculate the voltages across the inductors, Eqs. (1.19) and (1.20) show that we need only know $\mathbf{J}(\mathbf{i})$ the jacobian matrix of $\hat{\boldsymbol{\phi}}(\cdot\,,\cdot)$. Thus, as in the two-terminal inductor case, a pair of *nonlinear time-invariant* inductors can be specified by the matrix-valued function $\mathbf{J}(\mathbf{i})$. For reasons that will become clear in Sec. 4, this matrix is called the *small-signal inductance matrix*, and it is denoted by $\mathbf{L}(\mathbf{i})$ or $\mathbf{L}(i_1, i_2)$. Then Eq. (1.19) becomes

$$\begin{aligned} v_1(t) &= L_{11}[i_1(t), i_2(t)]\dot{i}_1(t) + L_{12}[i_1(t), i_2(t)]\dot{i}_2(t) \\ v_2(t) &= L_{21}[i_1(t), i_2(t)]\dot{i}_1(t) + L_{22}[i_1(t), i_2(t)]\dot{i}_2(t) \end{aligned} \tag{1.21}$$

Equation (1.21) may be viewed as defining a two-port inductor whose port voltages are v_1 and v_2 and port currents are i_1 and i_2.

2 TABLEAU ANALYSIS

With the coupled inductors introduced in Sec. 1, we are now in possession of all the circuit elements that we'll need for the remainder of this book. In this section and the next we shall develop *two completely general* methods for writing down the equations for *any dynamic circuit*. In the present section we describe the tableau analysis: In other words, we shall extend to general *dynamic* circuits the method described in Chap. 5, Sec. 2 for *resistive* circuits.

Tableau analysis is a powerful analytical tool for studying circuits: Its power lies in that it separates completely Kirchhoff's laws from the branch equations. Tableau analysis turns out to be an excellent tool for deriving properties of circuits.

As in Chap. 5, we use $\mathbf{e}$, $\mathbf{v}$, $\mathbf{i}$ as variables; we write first KCL ($\mathbf{Ai} = \mathbf{0}$) then KVL ($\mathbf{v} - \mathbf{A}^T\mathbf{e} = \mathbf{0}$) and finally the branch equations. In writing the branch equations for capacitors or inductors we *always* write them in *differential equation form*, *never* in integral form. We consider first linear circuits.

2.1 Linear Dynamic Circuits

A. Linear time-invariant circuits We start with an example.

Example Consider the linear circuit $\mathcal{N}$ shown in Fig. 2.1: Each node has been assigned a number, and each circuit element has been assigned a current reference direction.

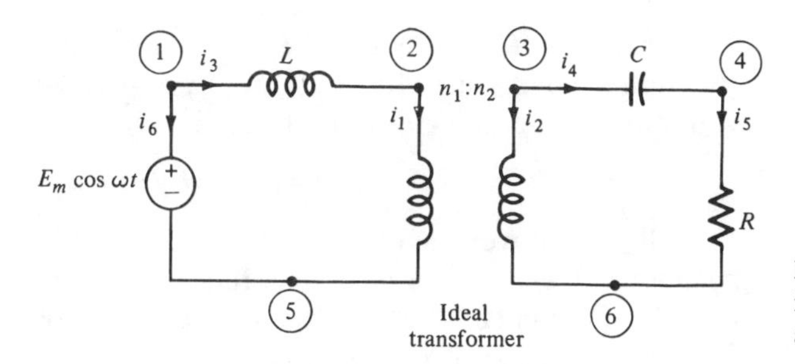

Figure 2.1 Linear circuit which includes an ideal transformer, an inductor, and a capacitor.

As in Chap. 5, let us draw the digraph of this circuit and, by hinging nodes ⑤ and ⑥, we obtain a *connected* digraph of five nodes and six branches, as shown in Fig. 2.2.

In order to alleviate the notation, we suppress the explicit dependence on time in the variables **e**, **v**, and **i**: For example, we write i_6 instead instead of $i_6(t)$.

$$\text{KCL:}\quad \begin{array}{lcccccccl} ① & & & i_3 & & & +i_6 & = 0 \\ ② & i_1 & & -i_3 & & & & = 0 \\ ③ & & i_2 & & +i_4 & & & = 0 \\ ④ & & & & -i_4 & +i_5 & & = 0 \end{array} \Bigg\} \qquad \mathbf{A}\,\mathbf{i}(t) = \mathbf{0} \qquad (2.1)$$

$$\text{KVL:}\quad \left.\begin{array}{l} v_1 = e_2 \\ v_2 = e_3 \\ v_3 = e_1 - e_2 \\ v_4 = \qquad\qquad e_3 - e_4 \\ v_5 = \qquad\qquad\qquad e_4 \\ v_6 = e_1 \end{array}\right\} \qquad \mathbf{v}(t) = \mathbf{A}^T\mathbf{e}(t) \qquad (2.2)$$

Branch equations:

$$\begin{array}{lr} n_2 v_1 - n_1 v_2 \qquad\qquad\qquad\qquad = 0 & (2.3a) \\ n_1 i_1 + n_2 i_2 = 0 & (2.3b) \\ v_3 \qquad - L\dot{i}_3 = 0 & (2.3c) \\ C\dot{v}_4 \qquad - i_4 = 0 & (2.3d) \\ v_5 \qquad - R i_5 = 0 & (2.3e) \\ v_6 \qquad = E_m \cos \omega t & (2.3f) \end{array}$$

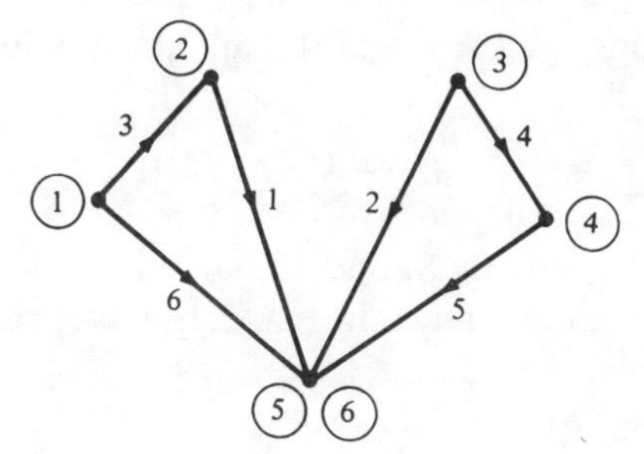

Figure 2.2 Digraph of the circuit shown in Fig. 2.1.

Since Eqs. (2.3*c*) *and* (2.3*d*) are differential equations, we must add the initial conditions: Indeed, they are part of the description of the condition of the inductor and capacitor:

$$i_3(0) = I_L \qquad v_4(0) = V_C \tag{2.3g}$$

All the above equations (2.1), (2.2), and (2.3) are *linear algebraic equations except* for the *inductor* branch equation (2.3*c*) and the *capacitor* branch equation (2.3*d*). This is a general property of the tableau equations for *linear* circuits: They are a mixture of linear algebraic equations and linear differential equations. The latter are the branch equations of inductors and capacitors.

The branch equations (2.3) may be written in matrix form [compare with Eq. (2.4) of Chap. 5]

$$\underbrace{\begin{bmatrix} 0 & & & & & \\ & 0 & & & \mathbf{0} & \\ & & 0 & & & \\ & & & C & & \\ & & & & 0 & \\ & \mathbf{0} & & & & 0 \end{bmatrix}}_{\mathbf{M}_0} \begin{bmatrix} \dot{v}_1 \\ \dot{v}_2 \\ \dot{v}_3 \\ \dot{v}_4 \\ \dot{v}_5 \\ \dot{v}_6 \end{bmatrix} + \underbrace{\begin{bmatrix} n_2 & -n_1 & 0 & 0 & 0 & 0 \\ 0 & 0 & 0 & 0 & 0 & 0 \\ 0 & 0 & 1 & 0 & 0 & 0 \\ 0 & 0 & 0 & 0 & 0 & 0 \\ 0 & 0 & 0 & 0 & 1 & 0 \\ 0 & 0 & 0 & 0 & 0 & 1 \end{bmatrix}}_{\mathbf{M}_1} \begin{bmatrix} v_1 \\ v_2 \\ v_3 \\ v_4 \\ v_5 \\ v_6 \end{bmatrix}$$

$$+ \underbrace{\begin{bmatrix} 0 & & & & & \\ & 0 & & & \mathbf{0} & \\ & & -l & & & \\ & & & 0 & & \\ & \mathbf{0} & & & 0 & \\ & & & & & 0 \end{bmatrix}}_{\mathbf{N}_0} \begin{bmatrix} \dot{i}_1 \\ \dot{i}_2 \\ \dot{i}_3 \\ \dot{i}_4 \\ \dot{i}_5 \\ \dot{i}_6 \end{bmatrix} + \underbrace{\begin{bmatrix} 0 & 0 & 0 & 0 & 0 & 0 \\ n_1 & n_2 & 0 & 0 & 0 & 0 \\ 0 & 0 & 0 & 0 & 0 & 0 \\ 0 & 0 & 0 & -1 & 0 & 0 \\ 0 & 0 & 0 & 0 & -R & 0 \\ 0 & 0 & 0 & 0 & 0 & 0 \end{bmatrix}}_{\mathbf{N}_1} \begin{bmatrix} i_1 \\ i_2 \\ i_3 \\ i_4 \\ i_5 \\ i_6 \end{bmatrix} = \underbrace{\begin{bmatrix} 0 \\ 0 \\ 0 \\ 0 \\ 0 \\ E_m \cos \omega t \end{bmatrix}}_{\mathbf{u}_s(t)} \tag{2.4}$$

Note that the $b \times b$ matrices $\mathbf{M}_0$, $\mathbf{M}_1$, $\mathbf{N}_0$, $\mathbf{N}_1$ are sparse and have *real constant* elements. If we use D to denote the differential operator d/dt, we may write Eq. (2.4) as

$$(\mathbf{M}_0 D + \mathbf{M}_1)\, \mathbf{v}(t) + (\mathbf{N}_0 D + \mathbf{N}_1)\, \mathbf{i}(t) = \mathbf{u}_s(t) \tag{2.5a}$$

To these equations, we must add the initial conditions; in the present case we must specify

$$v_4(0) = V_c \qquad i_3(0) = I_L \tag{2.5b}$$

B. Linear time-varying circuit Suppose that the inductor, the capacitor, and the resistor of the circuit $\mathcal{N}$, shown in Fig. 2.1 above, become time-varying per the following relations:

$$\phi(t) = L(t)\, i_3(t) \qquad q(t) = C(t)\, v_4(t) \qquad v_5(t) = R(t)\, i_5(t) \tag{2.6}$$

where the functions $L(\cdot)$, $C(\cdot)$, and $R(\cdot)$ are part of the specification of these elements. Call $\mathcal{N}_T$ the resulting *linear time-varying* circuit.

Thus the tableau equations of the circuit $\mathcal{N}_T$ consist of the Eqs. (2.1), (2.2), and (2.3) except that the branch equations (2.3*c*), (2.3*d*), and (2.3*e*) now read

$$v_3(t) - \dot{L}_3(t)\, i_3(t) - L_3(t)\, \dot{i}_3(t) = 0 \tag{2.7a}$$

$$\dot{C}_4(t)\, v_4(t) + C_4(t)\, \dot{v}_4(t) - i_4(t) = 0 \tag{2.7b}$$

$$v_5(t) - R_5(t)\, i_5(t) = 0 \tag{2.7c}$$

Since the functions $L_3(\cdot)$ and $C_4(\cdot)$, are known, so are their derivatives $\dot{L}_3(\cdot)$ and $\dot{C}_4(\cdot)$; thus Eqs. (2.7*a*) and (2.7*b*) are *linear* differential equations with time-varying coefficients.

The examples above suggest an algorithm for writing the tableau equations of any linear dynamic circuit.

Algorithm: Tableau equations for linear dynamic circuits

Data: • Circuit diagram with nodes numbered from ① to ⓝ and with current reference directions.

• Branch equation for *each* element of the circuit.

Step 1. Choose a datum node, say, node ⓝ and draw a *connected* digraph. (This may require hinging some nodes.)

Step 2. Write KCL for $n-1$ nodes except datum node

$$\mathbf{A}\, \mathbf{i}(t) = \mathbf{0} \tag{2.8}$$

Step 3. Write KVL for the b branches of the circuit

$$\mathbf{v}(t) - \mathbf{A}^T \mathbf{e}(t) = \mathbf{0} \tag{2.9}$$

Step 4. Write the b branch equations of the circuit:

If the circuit is time-invariant, they read

$$(\mathbf{M}_0 D + \mathbf{M}_1)\, \mathbf{v}(t) + (\mathbf{N}_0 D + \mathbf{N}_1)\, \mathbf{i}(t) = \mathbf{u}_s(t) \tag{2.10a}$$

where $\mathbf{M}_0$, $\mathbf{M}_1$, $\mathbf{N}_0$, $\mathbf{N}_1$ are (sparse) $b \times b$ matrices with constant real elements, D denotes the operator d/dt, and $\mathbf{u}_s(t)$ denotes the waveforms of the *independent sources*.

If the circuit is time-varying, they read

$$[\mathbf{M}_0(t)\, D + \mathbf{M}_1(t)]\, \mathbf{v}(t) + [\mathbf{N}_0(t)\, D + \mathbf{N}_1(t)]\, \mathbf{i}(t) = \mathbf{u}_s(t) \tag{2.10b}$$

where $\mathbf{M}_0(t)$, $\mathbf{M}_1(t)$, $\mathbf{N}_0(t)$, $\mathbf{N}_1(t)$ are (sparse) $b \times b$ matrices whose elements are *known* functions of time.

In matrix form the tableau equations of a linear dynamic circuit read:

Linear dynamic circuits

$$\begin{bmatrix} \mathbf{0} & \mathbf{0} & \mathbf{A} \\ -\mathbf{A}^T & \mathbf{1} & \mathbf{0} \\ \mathbf{0} & \mathbf{M}_0 D + \mathbf{M}_1 & \mathbf{N}_0 D + \mathbf{N}_1 \end{bmatrix} \begin{bmatrix} \mathbf{e}(t) \\ \mathbf{v}(t) \\ \mathbf{i}(t) \end{bmatrix} = \begin{bmatrix} \mathbf{0} \\ \mathbf{0} \\ \mathbf{u}_s(t) \end{bmatrix} \tag{2.11}$$

where if the circuit is *time-varying*, $\mathbf{M}_0, \mathbf{M}_1, \mathbf{N}_0, \mathbf{N}_1$ are *known* functions of time. To these equations we must append the initial conditions: the initial capacitor voltages and the initial inductor currents.

The tableau equations (2.11) form a system of *linear* coupled *algebraic* and *differential* equations: There are $2b + n - 1$ equations in $2b + n - 1$ variables (b branch voltages, b branch currents, and $n - 1$ node voltages).

2.2 Nonlinear Dynamic Circuits

We have studied simple nonlinear dynamic circuits in Chap. 6, Sec. 5 and Chap. 7, Sec. 4. Our task is now to show that with tableau analysis we can write the equations of *any* nonlinear dynamic circuit.

Example Consider the *nonlinear time-invariant* circuit $\mathcal{N}$ shown in Fig. 2.3: Each node has been assigned a node number and each branch a current reference direction. The nonlinear inductor is specified by its small-signal inductance $L(\cdot)$; the nonlinear resistor is voltage-controlled and is specified by its characteristic $\hat{i}_6(\cdot)$; finally, the op amp is specified by the *two-port* model shown in Fig. 2.4. The VCVS of the circuit shown in Fig. 2.4*a* has a characteristic given by

$$f_0(v_d) = \frac{A}{2}|v_d + \varepsilon| - \frac{A}{2}|v_d - \varepsilon| \tag{2.12}$$

where $A\varepsilon = E_{\text{sat}}$. The graph of $f_0(\cdot)$ is shown in Fig. 2.4*b*. Note that f_0 is

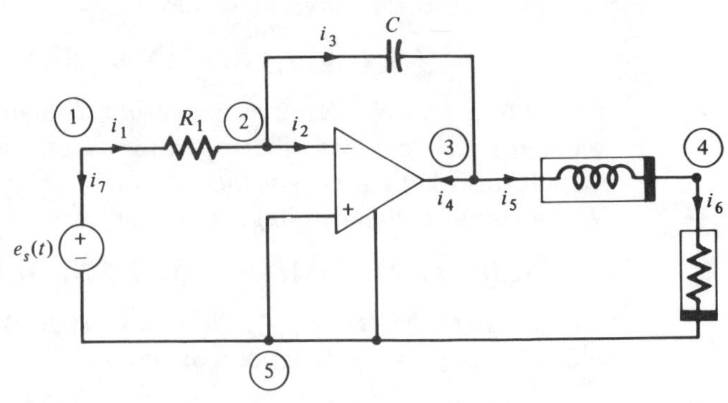

Figure 2.3 Nonlinear time-invariant circuit $\mathcal{N}$, where the nonlinear inductance is specified by its small-signal inductance $L(\cdot)$.

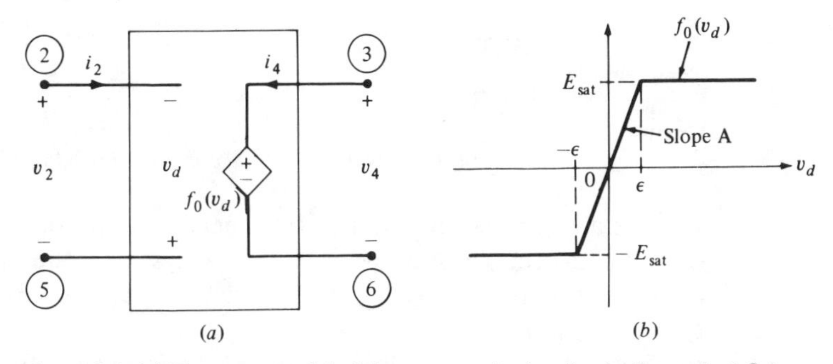

Figure 2.4 (*a*) Two-port model of the op amp: in the circuit $\mathcal{N}$, terminal ⑥ is connected to node ⑤. (*b*) Graph of the nonlinear characteristic f_0 of the VCVS in the op amp model.

an *odd* function: $f_0(v_d) = -f_0(-v_d)$, for all v_d. The two equations specifying the nonlinear two-port model of Fig. 2.4*a* are

$$i_2 = 0 \qquad v_4 = f_0(v_d) = f_0(-v_2) \tag{2.13}$$

where we used the fact that $v_d = -v_2$, see Fig. 2.4*a*.

By inspection we can write down Kirchhoff's laws for $\mathcal{N}$

$$\mathbf{A}\,\mathbf{i}(t) = \mathbf{0} \qquad \mathbf{v}(t) - \mathbf{A}^T\,\mathbf{e}(t) = \mathbf{0} \tag{2.14}$$

(using the suitable reduced incidence matrix **A**) and the branch equations of $\mathcal{N}$

$$v_1 - R_1 i_1 = 0 \tag{2.15a}$$

$$i_2 = 0 \tag{2.15b}$$

$$C\dot{v}_3 - i_3 = 0 \tag{2.15c}$$

$$v_4 - f_0(-v_2) = 0 \tag{2.15d}$$

$$L(i_5)\dot{i}_5 - v_5 = 0 \tag{2.15e}$$

$$i_6 - \hat{i}_6(v_6) = 0 \tag{2.15f}$$

$$v_7 = e_s(t) \tag{2.15g}$$

In Eqs. (2.15), we know the constants R_1 and C and the functions $f_0(\cdot)$, $L(\cdot)$, $\hat{i}_6(\cdot)$, and $e_s(\cdot)$; the unknown functions are $\mathbf{e}(\cdot)$, $\mathbf{v}(\cdot)$, and $\mathbf{i}(\cdot)$.

Equations (2.14) and (2.15) are the tableau equations of circuit $\mathcal{N}$ of Fig. 2.3.

Exercise Suppose we change the model of the op amp and take $\varepsilon = 0$ (see Fig. 2.4*b*): The steeply rising segment of Fig. 2.4*b* becomes vertical. Show how the tableau equations change.

It is clear that this procedure allows us to write the tableau equations of *any nonlinear dynamic circuit*. The resulting equations will always have the form:

Nonlinear dynamic circuits

$$\begin{aligned} &\text{KCL:} && \mathbf{A}\,\mathbf{i}(t) = \mathbf{0} \\ &\text{KVL:} && \mathbf{v}(t) - \mathbf{A}^T\,\mathbf{e}(t) = \mathbf{0} \\ &\text{Branch equations:} && \mathbf{h}(\dot{\mathbf{v}}(t), \mathbf{v}(t), \dot{\mathbf{i}}(t), \mathbf{i}(t), t) = \mathbf{0} \end{aligned} \tag{2.16}$$

For a *connected* digraph of b branches and n nodes, the tableau equations (2.16) constitute a system of $2b + n - 1$ scalar equations in the $2b + n - 1$ unknown functions $e_j(\cdot)$, $j = 1, 2, \ldots, n-1$, $v_k(\cdot)$, $k = 1, 2, \ldots, b$, and $i_\ell(\cdot)$, $\ell = 1, 2, \ldots, b$.

The tableau equations are *coupled algebraic* and *differential* equations: They always include the $b + n - 1$ *linear* equations originating from Kirchhoff's laws. The $\dot{v}_k(\cdot)$ are contributed by the capacitors; the $\dot{i}_l(\cdot)$ are contributed by the inductors. The function $\mathbf{h}$ depends explicitly on time when some *independent* sources are not constant or when some elements are *time-varying*, or both.

REMARK In the derivation of the tableau equation (2.16) we considered only a nonlinear inductor specified by its small-signal inductance $L(i)$. The dual case would be a nonlinear capacitor specified by its small-signal capacitance $C(v)$. In each case, the branch equations read

$$\dot{q}(t) = i_C(t) = C(v)\,\dot{v}(t) \qquad \dot{\phi}(t) = v_L(t) = L(i)\,\dot{i}(t) \tag{2.17}$$

Suppose now that the capacitor is specified by a charge-controlled representation or that the inductor is specified by a flux-controlled representation:

$$v_C = \hat{v}(q) \qquad i_L = \hat{i}(\phi) \tag{2.18}$$

If we use the chain rule as before, we are stuck because q and ϕ appear as arguments in $\hat{v}'(q)$ and $\hat{i}'(\phi)$. The remedy is to use q and ϕ as additional variables and to describe the elements by two equations

$$\begin{aligned} v_C &= \hat{v}(q) & \qquad i_L &= \hat{i}(\phi) \\ \dot{q} &= i_C & \dot{\phi} &= v_L \end{aligned} \tag{2.19}$$

3 MODIFIED NODE ANALYSIS

In Chap. 5, Sec. 1 we studied node analysis for resistive circuits: For any resistive circuit made up of *voltage-controlled* resistors we can write the node equations by inspection. The modified node analysis is based on node analysis but is suitably modified so that it can be used on *any* dynamic circuit.[2] The goal of the modified node analysis is to obtain a set of coupled *algebraic* and

[2] Modified node analysis (usually abbreviated MNA in the literature) is the second general method for analyzing *any* dynamic circuit.

differential equations. Consequently to specify a linear time-invariant inductor we use the *differential* equation

$$v(t) = L \frac{di}{dt} \tag{3.1}$$

rather than the *integral* equation

$$i(t) = i(t_0) + \frac{1}{L} \int_{t_0}^{t} v(t')\, dt'$$

The underlying ideas of modified node analysis are (*a*) write node equations using node voltages as variables, and (*b*) whenever an element is encountered that is *not* voltage-controlled, introduce in the node equation the corresponding branch current as a new variable and add, as a new equation, the branch equation of that element. The result is a system of equations where the unknowns are the *node voltages* and some *selected branch currents*.

The equations of the modified node analysis (MNA) can be written down by inspection. The number of equations is always smaller than that of tableau analysis. But since the MNA equations contain information about the interconnection as well as about the nature of the branches, the equations of MNA do not have the conceptual clarity of the tableau equations. Many circuit analysis programs use MNA, SPICE in particular.

Example (MNA for a linear time-invariant circuit) The circuit $\mathcal{N}$ shown in Fig. 3.1 includes an independent voltage source, a pair of coupled inductors, two resistors, and a capacitor. We have $b = 6$ and $n = 4$. In writing the node equation for node ①, since the independent voltage source is *not* voltage-controlled, we insert the branch current i_6. In considering nodes ② and ③, we introduce the inductor currents i_1 and i_2. We append these three suitably modified node equations with the branch equations of the voltage source and of the two (coupled) inductors. The result is

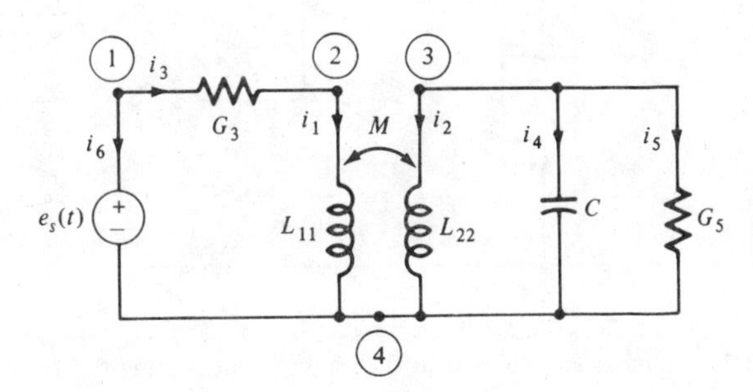

Figure 3.1 Linear circuit used in the example illustrating the MNA analysis.

$$
\begin{array}{ll}
\text{Node equations:} & \left\{\begin{array}{ll} G_3e_1 - G_3e_2 & + i_6 = 0 \\ -G_3e_1 + G_3e_2 + i_1 & = 0 \\ C\dot{e}_3 + G_5e_3 + i_2 & = 0 \end{array}\right. \\
\text{Coupled inductors:} & \left\{\begin{array}{ll} -e_2 + L_{11}\dot{i}_1 + M\dot{i}_2 & = 0 \\ -e_3 + M\dot{i}_1 + L_{22}\dot{i}_2 & = 0 \end{array}\right. \\
\text{Voltage source:} & e_1 \qquad = e_s(t)
\end{array}
\tag{3.2}
$$

MNA gives six equations in the node voltages e_1, e_2, and e_3 and in the selected currents i_1, i_2, and i_6.

Equations (3.2) form a set of coupled algebraic and differential equations.

If we had written tableau equations, we would have had $2b + n - 1 =$ 15 equations.

Exercises

1. Suppose we replace the inductors in the circuit $\mathcal{N}$ of Fig. 3.1 by an ideal transformer; show how the MNA equations must be modified.
2. Suppose that the resistor G_3 and the capacitor C become time varying—hence specified by the functions $G_3(\cdot)$ and $C(\cdot)$, respectively. Show how the MNA equations must be modified.

Let us show that the basic idea of MNA works quite easily for nonlinear circuits.

Example (MNA for a nonlinear circuit) Let us use a circuit analogous to the one used in Sec. 2.2 above. The nonlinear circuit is shown in Fig. 3.2. The nonlinear capacitor is specified by its small-signal capacitance $C(\cdot)$, the nonlinear inductance by its small-signal inductance $L(\cdot)$, and the nonlinear resistor is *current-controlled* and specified by its characteristic $\hat{v}_6(\cdot)$.

For the op amp we use the two-port model shown in Fig. 3.3. The equations of the modified node analysis are

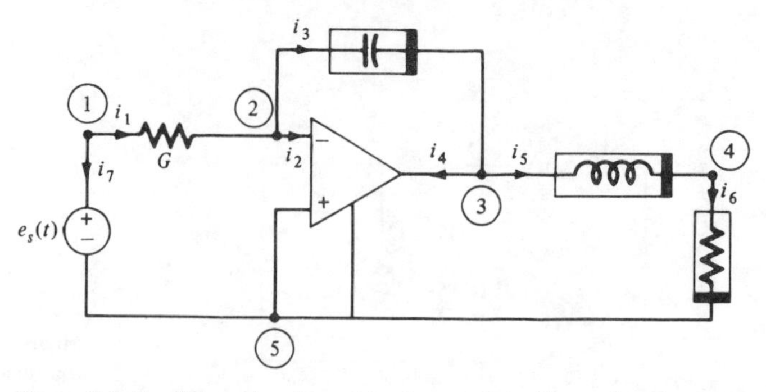

Figure 3.2 Nonlinear circuit used in the example illustrating the MNA analysis.

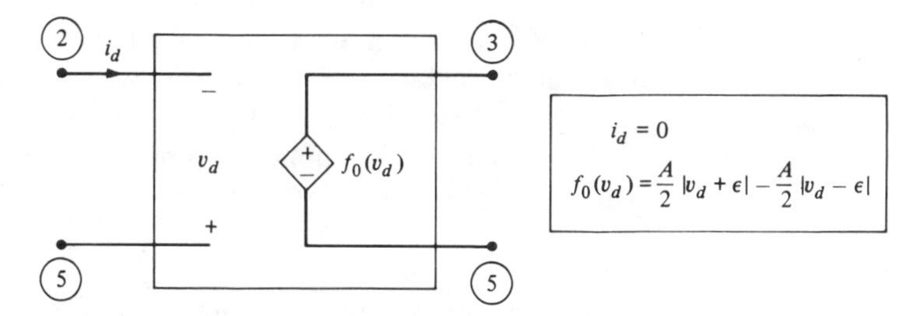

Figure 3.3 Model for the op amp of Fig. 3.2.

$$
\begin{array}{ll}
\textbf{Node equations} & \left\{\begin{array}{l}
Ge_1 - Ge_2 \qquad\qquad\qquad\qquad\qquad\qquad\qquad\qquad\qquad + i_7 = 0 \\
-Ge_1 + Ge_2 + C(e_2 - e_3)\,\dot{e}_2 - C(e_2 - e_3)\,\dot{e}_3 + i_2 = 0 \\
\qquad - C(e_2 - e_3)\,\dot{e}_2 + C(e_2 - e_3)\,\dot{e}_3 + i_4 + i_5 = 0 \\
\qquad\qquad\qquad\qquad\qquad\qquad\qquad - i_5 + i_6 = 0
\end{array}\right. \\
\textbf{Op amp} & \left\{\begin{array}{l}
i_2 = 0 \\
-f_0(-e_2) + e_3 = 0
\end{array}\right. \\
\textbf{NL inductor} & -e_3 + e_4 + L(i_5)\,\dot{i}_5 = 0 \\
\textbf{NL resistor} & e_4 - \hat{v}_6(i_6) = 0 \\
\textbf{Voltage source} & e_1 = e_s
\end{array}
\tag{3.3}
$$

Equations (3.3) constitute a set of nine coupled algebraic and differential equations in nine unknown functions: the four node voltages $e_1(\cdot)$, $e_2(\cdot)$, $e_3(\cdot)$, $e_4(\cdot)$ and the five selected currents $i_2(\cdot)$, $i_4(\cdot)$, $i_5(\cdot)$, $i_6(\cdot)$, $i_7(\cdot)$.

Note that the variable i_4, the op-amp output current, appears only in the third node equation: This node equation is thus a recipe for calculating i_4 once e_2, e_3, and i_5 are known. Also if i_4 is not required, the third node equation may be dropped from (3.3).

These two examples show how easy it is to write down the MNA equations for any circuit.

Algorithm: For writing the MNA equations of any dynamic circuit

Data:
- Circuit diagram with assigned node numbering (from ① to ⓝ), and assigned current reference directions.
- Branch equation(s) for *each* element of the circuit.

Step 1. Choose a datum node, say, node ⓝ, and draw a *connected* digraph. (This may require hinging some nodes.)

Step 2. For $k = 1, 2, \ldots, n-1$, write KCL for node ⓚ using the *node-to-datum voltages* as variables, keeping in mind that (*a*) if one or more *inductors* are connected to node ⓚ, then the branch current of that inductor is entered in the node equation *and* the branch equation of the inductor is appended to the $n-1$ node equation; (b) if one or more branches which are *not voltage-controlled* are connected to node ⓚ, then the corresponding branch current is entered in the node equation *and* the corresponding branch equation is appended to the $n-1$ node equations.

REMARKS

1. In MNA, the variables used are the $n-1$ node voltages and the *selected* currents, that is, all inductor currents and the currents in all branches that are *not voltage-controlled*.
2. MNA is the second completely general method of circuit analysis. In general, MNA requires far less equations than tableau analysis, consequently it is the preferred method for circuit analysis programs (SPICE) and for hand analysis.

4 SMALL-SIGNAL ANALYSIS

The concept of small-signal equivalent circuits is the basis of an extremely important *approximation* technique which is used in communications circuits, measurement systems, and power systems. The method reduces the analysis of a nonlinear dynamic circuit to that of a *nonlinear resistive* circuit, then to that of a *linear dynamic* circuit. In Chap. 7, Sec. 5 we've seen a second-order circuit example of this technique.

The goal of this section is to state and justify the algorithm which delivers the small-signal equivalent circuit of any *nonlinear* time-invariant dynamic circuit about a fixed operating point.

In order to avoid complicated notations, let us reason on an example. This example is chosen so that it includes most of the analyses required for obtaining a small-signal equivalent circuit.

Example Consider the *nonlinear time-invariant* circuit $\mathcal{N}$ shown in Fig. 4.1. Note that it is driven by a dc voltage source E_s and a time-varying voltage source $e_s(\cdot)$. We shall refer to $e_s(\cdot)$ as the *ac source*. The goal of the approximate analysis technique is to take advantage of the fact that $e_s(\cdot)$ is small[3] [in the sense that, for all $t \geq 0$, the values $|e_s(t)|$ are small]. The circuit $\mathcal{N}$ includes a linear resistor R, a linear capacitor C, and a linear inductor L, a *nonlinear* VCCS specified by its characteristic $f_0(\cdot)$, a *nonlinear* current-controlled *inductor* specified by $\hat{\phi}_6(\cdot)$, a *nonlinear* voltage-controlled *capacitor* specified by $\hat{q}_7(\cdot)$ and a *nonlinear* voltage-controlled *resistor* specified by $\hat{i}_2(\cdot)$.

[3] The meaning of the word *small* will be made clear later on: Roughly, $e_s(\cdot)$ must be "small" enough so that some higher-order terms of an expression are negligible.

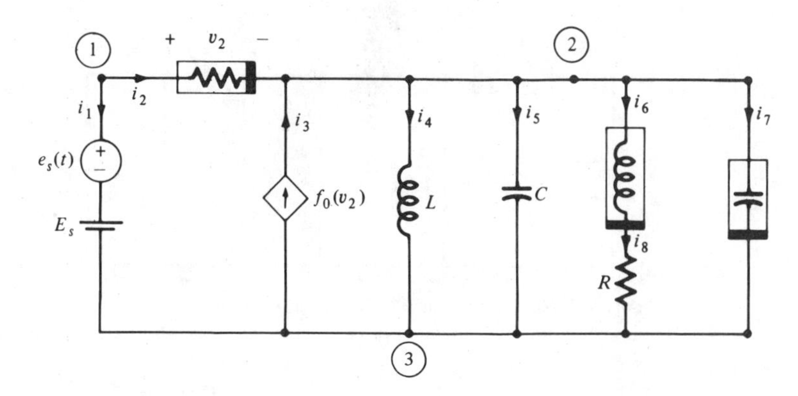

Figure 4.1 Nonlinear time-invariant circuit $\mathcal{N}$ driven by the dc source E_s and the ac source $e_s(\cdot)$.

The tableau equations of the circuit $\mathcal{N}$ can be written in the form

Tableau equations of $\mathcal{N}$

KCL: $$\mathbf{A}\,\mathbf{i}(t) = \mathbf{0} \tag{4.1a}$$

KVL: $$\mathbf{A}^T\,\mathbf{e}(t) - \mathbf{v}(t) = \mathbf{0} \tag{4.1b}$$

Branch equations: $$\mathbf{f}(\dot{\mathbf{v}}(t), \mathbf{v}(t), \dot{\mathbf{i}}(t), \mathbf{i}(t)) = \mathbf{u}_s(t) \tag{4.1c}$$

where[4] the column vector $\mathbf{u}_s(t)$ bookkeeps the contribution of the *independent* sources: $E_s + e_s(t)$. Except for the independent source $e_s(t)$, all the elements are *time-invariant*: consequently, in Eq. (4.1*c*), the function $\mathbf{f}$ *does not depend explicitly on time*.

In order to derive approximate equations representing $\mathcal{N}$ we proceed in three steps:

Step 1. Calculate the DC Operating Point Q, i.e., $\mathbf{E}_Q$, $\mathbf{V}_Q$, $\mathbf{I}_Q$

Set the ac source $e_s(\cdot)$ to zero, turn on the dc source, and call $\mathbf{E}_Q$, $\mathbf{V}_Q$, $\mathbf{I}_Q$ the resulting dc steady state. The corresponding tableau equations read

Tableau equations for computing the dc operating point

$$\mathbf{A}\mathbf{I}_Q = \mathbf{0} \tag{4.2a}$$

$$\mathbf{A}^T\,\mathbf{E}_Q - \mathbf{V}_Q = \mathbf{0} \tag{4.2b}$$

$$\mathbf{f}(\mathbf{0}, \mathbf{V}_Q, \mathbf{0}, \mathbf{I}_Q) = \mathbf{U}_s \tag{4.2c}$$

where $\mathbf{U}_s$ denotes the contribution of the dc source E_s. Since $\mathbf{V}_Q$ and $\mathbf{I}_Q$ are *constant* vectors, $\dot{\mathbf{V}}_Q = \mathbf{0}$ and $\dot{\mathbf{I}}_Q = \mathbf{0}$, and hence these are two *zero* vectors in the argument of $\mathbf{f}$ in Eq. (4.2*c*). In fact the scalar equations of (4.2*c*) read

[4] Equation (4.1*c*) is written in a slightly different form than Eq. (2.16). We do so to emphasize the fact that the $\mathbf{f}$ in (4.1*c*) does not depend explicitly on time.

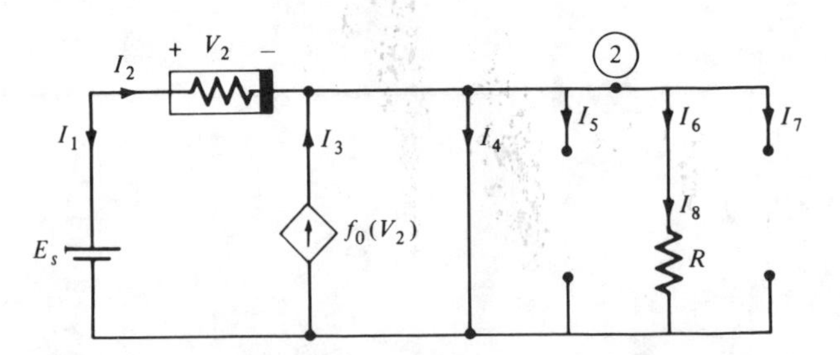

Figure 4.2 Nonlinear resistive circuit whose solution $V_1, V_2, \ldots, I_1, I_2, \ldots$ specifies the operating point Q. Note that inductors have been replaced by short circuits and capacitors by open circuits.

Branch equations for the dc operating point

$$V_1 = E_s \tag{4.3a}$$

$$\hat{i}_2(V_2) - I_2 = 0 \tag{4.3b}$$

$$-f_0(V_2) + I_3 = 0 \tag{4.3c}$$

$$V_4 = 0 \quad \left(\text{because } \frac{dI_4}{dt} = 0\right) \tag{4.3d}$$

$$I_5 = 0 \quad \left(\text{because } \frac{dV_5}{dt} = 0\right) \tag{4.3e}$$

$$V_6 = 0 \quad \left(\text{because } \frac{dI_6}{dt} = 0\right) \tag{4.3f}$$

$$I_7 = 0 \quad \left(\text{because } \frac{dV_7}{dt} = 0\right) \tag{4.3g}$$

$$V_8 - RI_8 = 0 \tag{4.3h}$$

From Eqs. (4.3), we see that to calculate the dc operating point, (*a*) we replace each inductor by a short circuit ($V_4 = V_6 = 0$); (*b*) we replace each capacitor by an open circuit ($I_5 = I_7 = 0$); (*c*) we solve the resulting *nonlinear resistive circuit*. (See Chap. 5, Sec. 3.) This resistive circuit is shown in Fig. 4.2.

In the following, we assume that $\mathbf{E}_Q$, $\mathbf{V}_Q$, and $\mathbf{I}_Q$ are known.[5]

Step 2. Change of Variables

The idea is to use the fact that the ac source is small, and consequently[6] the actual node voltages $\mathbf{e}(t)$ will be close to $\mathbf{E}_Q$, $\mathbf{v}(t)$ will be close to $\mathbf{V}_Q$, and $\mathbf{i}(t)$ will be close to $\mathbf{I}_Q$: So we write

$$\mathbf{e}(t) = \mathbf{E}_Q + \tilde{\mathbf{e}}(t) \tag{4.4a}$$

[5] If Eqs. (4.2) have several solutions, we choose one and stick to it.

[6] This "consequently" assumes some stability property in the neighborhood of Q. More on this later.

$$\mathbf{v}(t) = \mathbf{V}_Q + \tilde{\mathbf{v}}(t) \tag{4.4b}$$

$$\mathbf{i}(t) = \mathbf{I}_Q + \tilde{\mathbf{i}}(t) \tag{4.4c}$$

The point is that $\tilde{\mathbf{e}}(t)$, $\tilde{\mathbf{v}}(t)$, and $\tilde{\mathbf{i}}(t)$ are *small* deviations from the operating point $\mathbf{E}_Q$, $\mathbf{V}_Q$, and $\mathbf{I}_Q$, respectively. If we substitute the $\mathbf{e}(t)$, $\mathbf{v}(t)$, $\mathbf{i}(t)$ given by Eqs. (4.4) into the KCL equation (4.1a) and the KVL equation (4.1b) and take into account (4.2a) and (4.2b), we obtain

$$\mathbf{A}\,\tilde{\mathbf{i}}(t) = \mathbf{0} \tag{4.5a}$$

$$\mathbf{A}^T\,\tilde{\mathbf{e}}(t) - \tilde{\mathbf{v}}(t) = \mathbf{0} \tag{4.5b}$$

Note that Eqs. (4.5a) and (4.5b) are *exact* (no approximation is involved!).

We could perform the same substitution in (4.1c) and use (4.2c) to obtain

$$f\big(\dot{\tilde{\mathbf{v}}}(t), \mathbf{V}_Q + \tilde{\mathbf{v}}(t), \dot{\tilde{\mathbf{i}}}(t), \mathbf{I}_Q + \tilde{i}(t)\big) - f\big(\mathbf{0}, \mathbf{V}_Q, \mathbf{0}, \mathbf{I}_Q\big) = \mathbf{u}_s(t) - \mathbf{U}_s = \tilde{\mathbf{u}}_s(t) \tag{4.5c}$$

However, it is more instructive to proceed by considering one branch equation at a time, because Eq. (4.5c) is still a nonlinear equation and we would like to linearize it (by the use of Taylor's theorem) in order to take advantage of the fact that $\tilde{\mathbf{e}}(t)$, $\tilde{\mathbf{v}}(t)$, and $\tilde{\mathbf{i}}(t)$ are small.

Step 3. Obtain Approximate Branch Equations

We consider successively the resistors, inductors, and capacitors: The final result will be obtained by using a Taylor expansion and dropping the higher-order terms. The result is a set of *approximate* linear time-invariant equations relating $\tilde{\mathbf{v}}(t)$, $\tilde{\mathbf{i}}(t)$, and the ac source. These will be the branch equations of *the small-signal equivalent circuit about the operating point* Q.

Resistors

For the *nonlinear* resistor we have:

$$i_2(t) = \hat{i}_2(v_2(t))$$

Substituting $I_2 + \tilde{i}_2(t)$ for $i_2(t)$, etc., we obtain

$$\begin{aligned} i_2(t) &= I_2 + \tilde{i}_2(t) \\ &= \hat{i}_2(V_2 + \tilde{v}_2(t)) \\ &= \hat{i}_2(V_2) + \hat{i}_2'(V_2)\,\tilde{v}_2(t) \\ &\quad + \text{h.o.t.} \end{aligned}$$

For the *linear* resistor we have:

$$v_8(t) = R_8 i_8(t)$$

Substituting

$$V_8 + \tilde{v}_8(t) = R_8(I_8 + \tilde{i}_8(t))$$

and using Eq. (4.3h) we obtain the *exact* equation

where we expanded $\tilde{i}_2(V_2 + \tilde{v}(t))$ in Taylor series and used "h.o.t." to denote the higher-order terms. Now if $\tilde{v}_2(t)$ is small, we may neglect the h.o.t., take Eq. (4.3*b*) into account, and obtain the *approximate* equation[7]

$$\tilde{i}_2(t) = \hat{i}_2'(V_2)\,\tilde{v}_2(t) \quad (4.6b)$$

$$\tilde{v}_8(t) = R_8\,\tilde{i}_8(t) \quad (4.6h)$$

Remark. Since V_2, the dc voltage calculated in Step 1 is *constant*, Eq. (4.6*b*) is the equation of a *linear time-invariant resistor* with conductance $\hat{i}_2'(V_2)$, the slope of the resistor characteristic at its operating point.

Controlled source

$$i_3(t) = f_0(v_2(t))$$

hence

$$I_3 + \tilde{i}_3(t) = f_0(V_2 + \tilde{v}_2(t))$$

Expanding in Taylor series, using (4.3*c*), and dropping the h.o.t. we obtain successively:

$$I_3 + \tilde{i}_3(t) = f_0(V_2) + f_0'(V_2)\tilde{v}_2(t) + \text{h.o.t.}$$

$$\tilde{i}_3(t) = f_0'(V_2)\,\tilde{v}_2(t) \quad (4.6c)$$

This is the equation of a *linear time-invariant* VCCS.

Inductors

We obtain successively for the *nonlinear* inductor

$$\phi_6(t) = \hat{\phi}_6(i_6(t)) = \hat{\phi}_6(I_6 + \tilde{i}_6(t))$$

Hence

$$v_6(t) = \frac{d}{dt}\,[\hat{\phi}_6(I_6 + \tilde{i}_6(t))] = \hat{\phi}_6'[I_6 + \tilde{i}_6(t)] \cdot \dot{\tilde{i}}_6(t)$$

We obtain successively for the *linear* inductor

$$\phi_4(t) = L i_4(t)$$

Hence

$$\tilde{v}_4(t) = \dot{\phi}_4(t) = \frac{d}{dt}\,L(I_4 + \tilde{i}_4(t)) = L\dot{\tilde{i}}_4(t)$$

[7] The equations below are labeled (4.6*b*), (4.6*h*), . . . in order to display their connection to equations (4.3*b*), (4.3*h*), . . . , resp.

where we used the chain rule. Expanding in Taylor series gives

$$v_6(t) = \hat{\phi}_6'(I_6) \cdot \dot{\tilde{i}}_6(t) + \hat{\phi}_6''(I_6)\, \tilde{i}_6(t)\, \dot{\tilde{i}}_6(t) + \cdots$$

Hence dropping the higher-order terms and using Eq. (4.3*f*) we get the *approximate* equation

$$\tilde{v}_6(t) = \hat{\phi}_6'(I_6)\dot{\tilde{i}}_6(t) \qquad (4.6f)$$

Thus we have the *exact* equation

$$\tilde{v}_4(t) = L\dot{\tilde{i}}_4(t) \qquad (4.6d)$$

Equation (4.6*f*) represents a *linear time-invariant* inductor whose inductance $\hat{\phi}_6'(I_6)$ is the slope of the original nonlinear inductor characteristic at its operating point (I_6, Φ_6).

Capacitors The calculation for the capacitors is the dual of that of the inductors:

$$q_7(t) = \hat{q}_7(v_7(t)) = \hat{q}_7(V_7 + \tilde{v}_7(t))$$

so, by the chain rule,

$$\tilde{i}_7(t) = \hat{q}_7'(V_7 + \tilde{v}_7(t)) \cdot \dot{\tilde{v}}_7(t)$$

and after expanding, dropping the h.o.t. and using Eq. (4.3*g*) we have the *approximate* equation

$$\tilde{i}_7(t) = \hat{q}_7'(V_7) \cdot \dot{\tilde{v}}_7(t) \qquad (4.6g)$$

Again the constant $\hat{q}'(V_7)$ is the slope of the nonlinear capacitor characteristic at the operating point V_7.

$$q_5(t) = Cv_5(t)$$

$$i_5(t) = \tilde{i}_5(t) = C\frac{d}{dt}(V_5 + \tilde{v}_5(t))$$

$$\tilde{i}_5(t) = C\dot{\tilde{v}}_5(t) \qquad (4.6e)$$

Independent sources

$$v_1(t) = V_1 + \tilde{v}_1(t) = E_s + e_s(t)$$

Using Eq. (4.3*a*) we get

$$\tilde{v}_1(t) = e_s(t) \qquad (4.6a)$$

Collecting the resulting eight branch equations, we have

$$\tilde{v}_1(t) = e_s(t) \tag{4.6a}$$

$$\tilde{i}_2(t) - \hat{i}_2'(V_2)\,\tilde{v}_2(t) = 0 \tag{4.6b}$$

$$\tilde{i}_3(t) - f_0'(V_2)\,\tilde{v}_2(t) = 0 \tag{4.6c}$$

$$\tilde{v}_4(t) - L\,\dot{\tilde{i}}_4(t) = 0 \tag{4.6d}$$

$$\tilde{i}_5(t) - C\,\dot{\tilde{v}}_5(t) = 0 \tag{4.6e}$$

$$\tilde{v}_6(t) - \hat{\phi}_6'(I_6)\,\dot{\tilde{i}}_6(t) = 0 \tag{4.6f}$$

$$\tilde{i}_7(t) - \hat{q}_7'(V_7)\,\dot{\tilde{v}}_7(t) = 0 \tag{4.6g}$$

$$\tilde{v}_8(t) - R\,\tilde{i}_8(t) = 0 \tag{4.6h}$$

These equations are interpreted to be the branch equations of one independent voltage source and seven *linear time-invariant* elements. Let us abbreviate these equations in the form

$$(\mathbf{M}_{0Q}D + \mathbf{M}_{1Q})\,\tilde{\mathbf{v}}(t) + (\mathbf{N}_{0Q}D + \mathbf{N}_{1Q})\,\tilde{\mathbf{i}}(t) = \tilde{\mathbf{u}}_s(t) \tag{4.7}$$

where the *constant matrices* $\mathbf{M}_{0Q}$, $\mathbf{M}_{1Q}$, $\mathbf{N}_{0Q}$, $\mathbf{N}_{1Q}$ are directly read from Eqs. (4.6) and $\tilde{\mathbf{u}}_s(t)$ is the column vector of ac sources in Eqs. (4.6). Equation (4.7) has precisely the same form as Eq. (2.10*a*) of the tableau analysis.

Equations (4.5*a*), (4.5*b*), and (4.6) represent the *linear time-invariant* circuit shown in Fig. 4.3.

Exercise We are given a pair of nonlinear coupled inductors (because we drive the ferrite into saturation):

$$\phi_1 = \hat{\phi}_1(i_1, i_2)$$

$$\phi_2 = \hat{\phi}_2(i_1, i_2)$$

Suppose that the dc operating point Q is specified by the inductor currents I_1 and I_2.

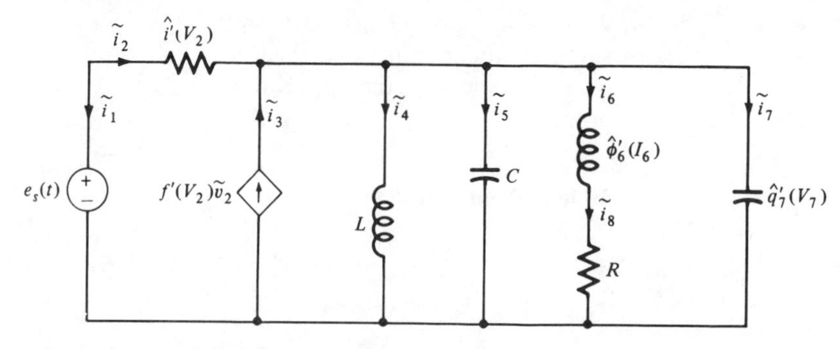

Figure 4.3 The linear time-invariant circuit described by Eqs. (4.5*a*), (4.5*b*), and (4.6). It is the small-signal equivalent circuit of $\mathcal{N}$ about the operating point $(\mathbf{V}_Q, \mathbf{I}_Q)$.

(*a*) Show that the small-signal equations are

$$\tilde{v}_1(t) = \left.\frac{\partial \hat{\phi}_1}{\partial i_1}\right|_{(I_1, I_2)} \cdot \dot{\tilde{i}}_1(t) + \left.\frac{\partial \hat{\phi}_1}{\partial i_2}\right|_{(I_1, I_2)} \cdot \dot{\tilde{i}}_2(t)$$

$$\tilde{v}_2(t) = \left.\frac{\partial \hat{\phi}_2}{\partial i_1}\right|_{(I_1, I_2)} \cdot \dot{\tilde{i}}_1(t) + \left.\frac{\partial \hat{\phi}_2}{\partial i_2}\right|_{(I_1, I_2)} \cdot \dot{\tilde{i}}_2(t)$$

(*b*) What is the self-inductance of the first inductor; what is the mutual inductance? (Note that by adjusting the dc currents I_1 and I_2, we may modify each of these quantities!)

Conclusion

If we collect Eqs. (4.5*a*), (4.5*b*), and (4.7), we get

Tableau equation of small-signal equivalent circuit

KCL: $$\mathbf{A}\,\tilde{\mathbf{i}}(t) = \mathbf{0} \tag{4.8a}$$

KVL: $$\mathbf{A}^T\,\tilde{\mathbf{e}}(t) - \tilde{\mathbf{v}}(t) = \mathbf{0} \tag{4.8b}$$

$$(\mathbf{M}_{0Q}D + \mathbf{M}_{1Q})\,\tilde{\mathbf{v}}(t) + (\mathbf{N}_{0Q}D + \mathbf{N}_{1Q})\,\tilde{\mathbf{i}}(t) = \tilde{\mathbf{u}}_s(t) \tag{4.8c}$$

Examining Eqs. (4.8) we conclude that

1. They are the tableau equations of a *linear time-invariant* circuit with node voltages $\tilde{\mathbf{e}}(t)$, branch voltages $\tilde{\mathbf{v}}(t)$, branch currents $\tilde{\mathbf{i}}(t)$, and independent sources $\tilde{\mathbf{u}}_s(t)$: We denote this circuit by $\mathscr{L}_Q$ and call it the *small-signal equivalent circuit about the operating point Q of $\mathcal{N}$.*
2. This circuit has the *same digraph* as the original *nonlinear time-invariant* circuit $\mathcal{N}$.

If we examine carefully the steps used to obtain the equations of $\mathscr{L}_Q$, we see that the method is perfectly general and that it will apply to any time-invariant circuit $\mathcal{N}$. Since the concept of small-signal equivalent circuits is so important, we specify the procedure in detail by the following algorithm.

Algorithm to obtain the small-signal equivalent circuit $\mathscr{L}_Q$ of $\mathcal{N}$.

Data:
- Circuit diagram of the *nonlinear time-invariant* circuit $\mathcal{N}$ driven by dc sources and ac sources, with nodes numbering from ① to ⓝ and with current reference directions.
- Branch equation(s) of each element of $\mathcal{N}$.

First, we determine the operating point Q.

Step 1. Set all ac *independent* sources to zero.

Step 2. Replace all inductors by short circuits and all capacitors by open circuits.

Step 3. Solve the resulting *resistive* circuit (which is now driven by dc sources only!). Call Q the resulting operating point specified by the solution $(\mathbf{V}_Q, \mathbf{I}_Q)$.[8]

Second, we determine $\mathscr{L}_Q$.

Step 4. In $\mathcal{N}$, set all dc *independent* sources to zero.

Step 5. Leave all *linear* elements alone.

Step 6. Replace every *nonlinear* element by its (linear) small-signal equivalent circuit about the operating point found in Step 3 [see Eq. (4.6) above]. The resulting *linear time-invariant circuit* is $\mathscr{L}_Q$, the small-signal equivalent circuit of $\mathcal{N}$ about the operating point Q.

Example Consider the following circuit shown in Fig. 4.4 (values are normalized for ease of calculation): We are given

$$L(i_1) = 2i_1 - 0.1i_1^3 \qquad C_3(v_3) = \exp(v_3)$$
$$\hat{i}(v_2) = v_2 + v_2^3 \qquad I_s = 1\text{ A}$$

First we apply Steps 1, 2, and 3 to determine the operating points: We have to solve the nonlinear resistive circuit shown in Fig. 4.5. So

$$I_1 - 0.2I_1 = 1\text{ A} \qquad \text{hence } I_1 = 1.25\text{ A}$$
$$V_2 + V_2^3 = 0.25 \qquad \text{hence } V_2 = 0.23673\text{ V}$$
$$V_3 = -V_2 = 0.23673\text{ V}$$

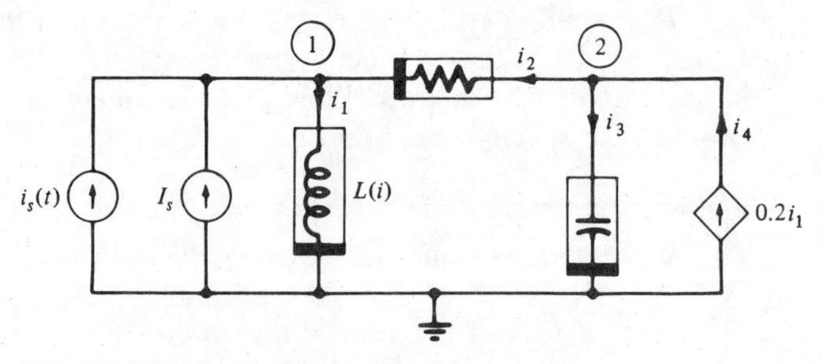

Figure 4.4 Nonlinear circuit used in the example.

[8] If there are several operating points Q, we choose the one of interest and study the dynamics of the circuit about that operating point.

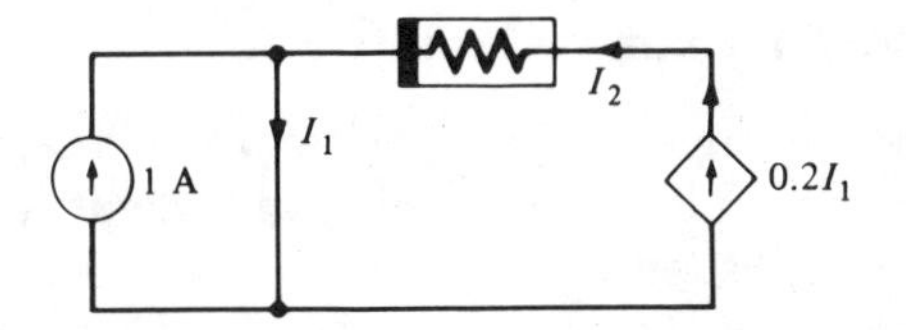

Figure 4.5 The dc equivalent circuit.

Applying Steps 4, 5, and 6 we obtain

$$L(I_1) = 2 \times 1.25 - 0.1 \times (1.25)^3 = 2.3047 \text{ H}$$

$$\hat{i}'(V_2) = 1 + 3V_2^2 = 1.1681 \text{ S}$$

$$C_3 = 1.2671 \text{ F}$$

and the small-signal equivalent circuit is shown in Fig. 4.6.

Summary

The small-signal analysis of a *nonlinear time-invariant* circuit $\mathcal{N}$ consists of four steps:

1. Determine the dc operating point Q: $(\mathbf{V}_Q, \mathbf{I}_Q)$.
2. Determine $\mathcal{L}_Q$, the small-signal equivalent circuit of $\mathcal{N}$ about the operating point Q.
3. Solve $\mathcal{L}_Q$ to obtain the branch voltages $\tilde{v}(\cdot)$ and the branch currents $\tilde{i}(\cdot)$ of $\mathcal{L}_Q$.
4. Obtain the actual (approximate) branch voltages and branch currents of $\mathcal{N}$ by the equations

$$\mathbf{v}(t) = \mathbf{V}_Q + \tilde{\mathbf{v}}(t) \qquad \text{and} \qquad \mathbf{i}(t) = \mathbf{I}_Q + \tilde{\mathbf{i}}(t) \tag{4.9}$$

5 GENERAL PROPERTIES OF DYNAMIC CIRCUITS

In our general study of *resistive* circuits we established the superposition theorem (Chap. 5, Sec. 4.1) and the substitution theorem (Chap. 5, Sec. 5.1). We propose to prove that these theorems (suitably modified, of course) still hold for general *dynamic* circuits.

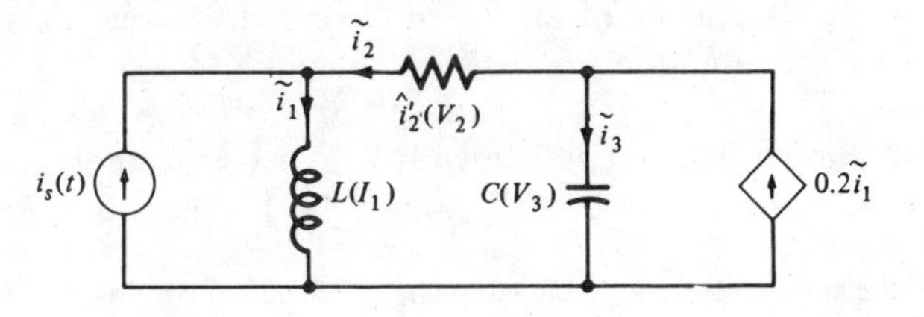

Figure 4.6 The small-signal equivalent circuit.

5.1 Superposition Theorem for Linear Dynamic Circuits

Roughly speaking the superposition theorem says that the *zero-state* response of a *linear dynamic* circuit due to several independent sources is equal to the sum of the zero-state responses due to each source acting alone. Clearly, this is a very valuable theorem: Indeed, consider the design of a high-fidelity system for an orchestra. Consider the response to, say, a clarinet. To guarantee a high-fidelity output when several notes are played together, by the superposition theorem we need consider only one note played at a time, and, even further, only the fundamental and then each harmonic driving the system alone. Thus to ensure high-fidelity reproduction we need only[9] assure ourselves that each tone be reproduced faithfully. This is the benefit of the superposition theorem. It is such a useful theorem that it is used almost unconsciously by engineers who design circuits for communication systems, measuring systems, etc.

Superposition theorem for linear time-varying dynamic circuits Let $\mathcal{N}$ be a *linear, possibly time-varying, uniquely solvable dynamic* circuit. Let $\mathcal{N}$ be driven from $t=0$ on by α *independent voltage sources* $v_{s1}(\cdot), v_{s2}(\cdot), \ldots, v_{s\alpha}(\cdot)$ and β *independent current sources* $i_{s1}(\cdot), i_{s2}(\cdot), \ldots, i_{s\beta}(\cdot)$.

Let $y(\cdot)$ be the *zero-state response* of $\mathcal{N}$ due to these $\alpha+\beta$ independent sources: $y(t)$ may be the jth node voltage $e_j(t)$, or the pth branch voltage $v_p(t)$, or the qth branch current $i_q(t)$. For $k=1,2,\ldots,\alpha$, let $y_{vk}(\cdot)$ be the *zero-state response* due to $v_{sk}(\cdot)$ *alone*, and, for $\ell = 1,2,\ldots,\beta$, let $y_{k\ell}(\cdot)$ be the *zero-state response* due to $i_{s\ell}(\cdot)$ *alone*. Then, for all $t \geq 0$,

$$y(t) = \sum_{k=1}^{\alpha} y_{vk}(t) + \sum_{\ell=1}^{\beta} y_{i\ell}(t) \tag{5.1}$$

Circuit interpretations

1. Consider the first sum in Eq. (5.1): The term $y_{vk}(t)$ is the zero-state response of $\mathcal{N}$ observed at y, when all *independent* sources except $v_{sk}(\cdot)$ are set to zero (i.e., voltage sources replaced by short circuits and current sources replaced by open circuits).
2. Keep in mind that since the circuit $\mathcal{N}$ is *dynamic*, the response at time t, $y(t)$, depends on the initial conditions (here all set to zero) and, in general, on all the values of all the independent sources during $[0, t]$. Contrast this with the situation for resistive circuits.

Example The circuit shown in Fig. 5.1 is a simple model for a communication system: the current source $i_s(t)$ injects the audio signal, the modu-

[9] Of course, in practice, we need to check that we do not drive the transistors out of their linear operating range.

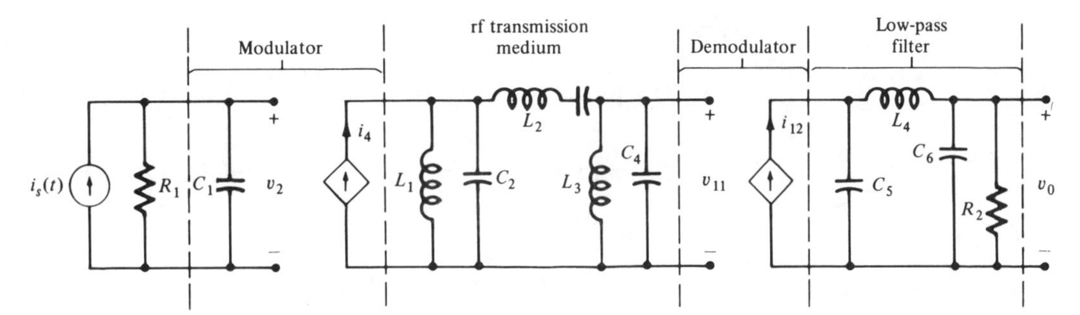

Figure 5.1 The circuit $\mathcal{N}$ is a linear time-varying circuit which models a communication system.

lator transforms the voltage $v_2(\cdot)$ into a radio-frequency current suitable for transmission over airwaves, cables, etc. The demodulator brings back the radio-frequency signal to audio frequency. The low-pass filter blocks the remaining radio-frequency signal from the output v_0: All elements R_1, R_2, $C_1, \ldots, C_6$, $L_1, \ldots, L_4$ are *linear* and *time-invariant*, but the controlled sources are *linear time-varying*:

$$i_4(t) = g_m \cdot (\cos \omega_c t) \cdot v_3(t) \tag{5.2}$$

$$i_{12}(t) = g_m \cdot (\cos \omega_c t) \cdot v_{11}(t) \tag{5.3}$$

where g_m is a constant, a characteristic of the modulator, and, in both cases, the sinusoid $\cos \omega_c t$ is supplied by a local oscillator. Thus the circuit $\mathcal{N}$ shown in Fig. 5.1 is a *linear time-varying* circuit.

For example, if we want to assure ourselves that it transmits properly any signal combining frequencies, say from 16 Hz to 18 kHz, by the superposition theorem, we need only check that it does so for a few representative sinusoidal signals in the desired range. Thus an apparently complicated problem has been reduced to a series of simple problems.

Proof of superposition theorem. Let $\mathcal{N}$ be a uniquely solvable, linear, time-varying circuit. For simplicity, let it be driven by one voltage source $v_{s1}(\cdot)$ and one current source $i_{s1}(\cdot)$. We shall use tableau analysis to obtain the zero-state response when (*a*) $v_{s1}(\cdot)$ is on and $i_{s1}(\cdot)$ is set to zero, and (*b*) $v_{s1}(\cdot)$ is set to zero and $i_{s1}(\cdot)$ is on.

$v_{s1}(\cdot)$ *is on*

$$\mathbf{A}\,\mathbf{i}(t) = \mathbf{0}$$

$$\mathbf{v}(t) - \mathbf{A}^T\,\mathbf{e}(t) = \mathbf{0}$$

$$[\mathbf{M}_0(t)\,D + \mathbf{M}_1(t)]\,\mathbf{v}(t) + [\mathbf{N}_0(t)\,D + \mathbf{N}_1(t)]\,\mathbf{i}(t) = \mathbf{u}_{s1}(t) \tag{5.4a}$$

$i_{s1}(\cdot)$ *is on*

$$\mathbf{A}\,\mathbf{i}(t) = \mathbf{0}$$

$$\mathbf{v}(t) - \mathbf{A}^T\,\mathbf{e}(t) = \mathbf{0}$$

$$[\mathbf{M}_0(t)\,D + \mathbf{M}_1(t)]\,\mathbf{v}(t) + [\mathbf{N}_0(t)\,D + \mathbf{N}_1(t)]\,\mathbf{i}(t) = \mathbf{u}_{s2}(t) \tag{5.4b}$$

where

$$\mathbf{u}_{s1}(t) = [0, 0, \ldots, 0, v_{s1}(t), 0]^T$$

Initial conditions:

$$\mathbf{v}_C(0) = \mathbf{i}_L(0) = \mathbf{0}$$

where

$$\mathbf{u}_{s2}(t) = [0, 0, \ldots, 0, 0, i_{s1}(t)]^T \tag{5.5}$$

Initial conditions:

$$\mathbf{v}_C(0) = \mathbf{i}_L(0) = \mathbf{0} \tag{5.6}$$

where $\mathbf{v}_C(t)$ [$\mathbf{i}_L(t)$, respectively] denotes the vector of all *capacitor voltages* (*inductor currents*, respectively). The tableau Eqs. (5.4*a*) and (5.4*b*) are identical except for the forcing term $\mathbf{u}_{s1}(t)$ and $\mathbf{u}_{s2}(t)$, see Eq. (5.5). Note also that the initial conditions are identical [see Eq. (5.6)].

Since $\mathcal{N}$ is uniquely solvable, call

$$[\mathbf{e}^{(1)}(\cdot), \mathbf{v}^{(1)}(\cdot), \mathbf{i}^{(1)}(\cdot)] \quad \text{and} \quad [\mathbf{e}^{(2)}(\cdot), \mathbf{v}^{(2)}(\cdot), \mathbf{i}^{(2)}(\cdot)] \tag{5.7}$$

their respective solutions: In view of Eq. (5.6), these solutions are *zero-state responses*.

By direct computation, we shall verify that the waveforms

$$[\mathbf{e}^{(1)}(\cdot) + \mathbf{e}^{(2)}(\cdot), \mathbf{v}^{(1)}(\cdot) + \mathbf{v}^{(2)}(\cdot), \mathbf{i}^{(1)}(\cdot) + \mathbf{i}^{(2)}(\cdot)] \tag{5.8}$$

are the unique *zero-state* solution of $\mathcal{N}$ when it is driven by $v_{s1}(\cdot)$ *and* $i_{s1}(\cdot)$. By (5.4*a*) and (5.4*b*) we have

$$\begin{aligned} \mathbf{0} &= \mathbf{A}\,\mathbf{i}^{(1)}(t) + \mathbf{A}\,\mathbf{i}^{(2)}(t) \\ &= \mathbf{A} \cdot [\mathbf{i}^{(1)}(t) + \mathbf{i}^{(2)}(t)] \end{aligned} \tag{5.9}$$

Adding the next equations of (5.4), we obtain

$$\begin{aligned} \mathbf{0} &= [\mathbf{v}^{(1)} - \mathbf{A}^T\,\mathbf{e}^{(1)}(t)] + [\mathbf{v}^{(2)}(t) - \mathbf{A}^T\,\mathbf{e}^{(2)}(t)] \\ &= [\mathbf{v}^{(1)}(t) + \mathbf{v}^{(2)}(t)] - \mathbf{A}^T[\mathbf{e}^{(1)}(t) + \mathbf{e}^{(2)}(t)] \end{aligned} \tag{5.10}$$

So Eqs. (5.9) and (5.10) show that the waveforms (5.8) satisfy the required KCL and KVL.

Prior to adding the last equations (5.4), consider this sum:

$$\begin{aligned} &[\mathbf{M}_0(t)\,D + \mathbf{M}_1(t)]\,\mathbf{v}^{(1)}(t) + [\mathbf{M}_0(t)\,D + \mathbf{M}_1(t)]\mathbf{v}^{(2)}(t) \\ &\quad = \mathbf{M}_0(t)\,\dot{\mathbf{v}}^{(1)}(t) + \mathbf{M}_1(t)\,\mathbf{v}^{(1)}(t) + \mathbf{M}_0(t)\,\dot{\mathbf{v}}^{(2)}(t) + \mathbf{M}_1(t)\,\mathbf{v}^{(2)}(t) \\ &\quad = \mathbf{M}_0(t)[\dot{\mathbf{v}}^{(1)}(t) + \dot{\mathbf{v}}^{(2)}(t)] + \mathbf{M}_1(t)[\mathbf{v}^{(1)}(t) + \mathbf{v}^{(2)}(t)] \\ &\quad = [\mathbf{M}_0(t)\,D + \mathbf{M}_1(t)][\mathbf{v}^{(1)}(t) + \mathbf{v}^{(2)}(t)] \end{aligned} \tag{5.11}$$

Now if we add the last equations of (5.4*a*) and (5.4*b*) we obtain

$$\begin{aligned} &[\mathbf{M}_0(t)\,D + \mathbf{M}_1(t)]\,\mathbf{v}^{(1)}(t) + [\mathbf{N}_0(t)\,D + \mathbf{N}_1(t)]\,\mathbf{i}^{(1)}(t) + [\mathbf{M}_0(t)\,D \\ &\qquad + \mathbf{M}_1(t)]\,\mathbf{v}^{(2)}(t) + [\mathbf{N}_0(t)\,D + \mathbf{N}_1(t)]\,\mathbf{i}^{(2)}(t) = \mathbf{u}_{s1}(t) + \mathbf{u}_{s2}(t) \end{aligned} \tag{5.12}$$

Collecting terms and using what we learned in obtaining Eq. (5.11), we get

$$[\mathbf{M}_0(t)\,D + \mathbf{M}_1(t)][\mathbf{v}^{(1)}(t) + \mathbf{v}^{(2)}(t)]$$
$$+ [\mathbf{N}_0(t)\,D + \mathbf{N}_1(t)][\mathbf{i}^{(1)}(t) + \mathbf{i}^{(2)}(t)] = \mathbf{u}_{s1}(t) + \mathbf{u}_{s2}(t) \quad (5.13)$$

Also note that from Eq. (5.6)

$$\mathbf{v}^{(1)}(0) + \mathbf{v}^{(2)}(0) = \mathbf{0} = \mathbf{i}^{(1)}(0) + \mathbf{i}^{(2)}(0) \quad (5.14)$$

and from Eq. (5.5)

$$\mathbf{u}_{s1}(t) + \mathbf{u}_{s2}(t) = [0, 0, \ldots, 0, v_{s1}(t), i_{s1}(t)]^T \quad (5.15)$$

Now Eqs. (5.9), (5.10), (5.13) with (5.14) and (5.15) show that the waveforms $[\mathbf{e}^{(1)}(\cdot) + \mathbf{e}^{(2)}(\cdot),\ \mathbf{v}^{(1)}(\cdot) + \mathbf{v}^{(2)}(\cdot),\ \mathbf{i}^{(1)}(\cdot) + \mathbf{i}^{(2)}(\cdot)]$ constitute the unique solution of $\mathcal{N}$, starting from zero initial conditions and driven by $v_{s1}(\cdot)$ and $i_{s1}(\cdot)$. Hence superposition is established. ∎

In general, the superposition theorem is not valid if $\mathcal{N}$ includes nonlinear elements. It is this feature of nonlinear circuits that makes their behavior so hard to understand: We can no longer break up inputs into a sum of simple terms and study successively the response to each term. In general, the sum of these responses bears little or no relation to the circuit response when all sources are acting simultaneously.

Exercise (To emphasize that the superposition theorem involves *only zero-state* responses) Consider the linear time-invariant parallel *RC* circuit shown in Fig. 5.2: The output under consideration is $y = v_C$. For simplicity, let $R = 1\,\Omega$ and $C = 1\,\text{F}$.
(i) $i_s(t) = 1$ A for $t \geq 0$ and $v_C(0) = 1$ V
(ii) $i_s(t) = 3$ A for $t \geq 0$ and $v_C(0) = 1$ V
(*a*) Calculate the *complete* response $y(t)$ in case (i) and (ii).
(*b*) Show that if $y_s(t)$ denotes the complete response for $i_s(t) = 4$ A for $t \geq 0$ and $v_C(0) = 1$ V, then $y_s(t)$ is *not* the sum of the responses calculated in (*a*).
(*c*) Explain your result.

Corollary to the superposition theorem Let $\mathcal{N}$ be a *linear*, possibly *time-varying*, *uniquely solvable* dynamic circuit. Let $\mathcal{N}$ start, at time t_0, from a given set of initial conditions $i_{L1}(t_0), i_{L2}(t_0), \ldots$, and $v_{C1}(t_0), v_{C2}(t_0), \ldots$, and let $\mathcal{N}$ be driven by a given set of independent sources. Choose any circuit variable as a "response" and call it y_c. Let

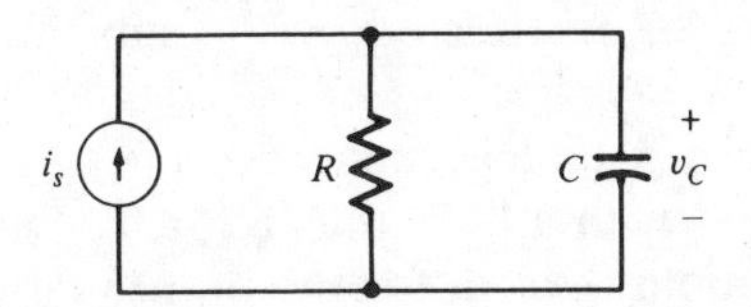

Figure 5.2 Parallel *RC* circuit.

$y_{zs}(\cdot)$ denote the *zero-state response* of $\mathcal{N}$ due to the given independent sources
$y_{zi}(\cdot)$ denote the *zero-input response* of $\mathcal{N}$ due to the given initial conditions

Then

$$y_c(t) = y_{zi}(t) + y_{zs}(t) \qquad \text{for all } t \geq t_0 \tag{5.16}$$

Intuitively, we write

For any *linear, time-varying dynamic* circuit

$$\begin{pmatrix}\text{Complete}\\ \text{response}\end{pmatrix} = \begin{pmatrix}\text{Zero-input}\\ \text{response}\end{pmatrix} + \begin{pmatrix}\text{Zero-state}\\ \text{response}\end{pmatrix}$$

Exercise Prove the corollary. (Hint: Replace the initial conditions by appropriate *independent* sources as in Fig. 2.1 of Chap. 6.)

5.2 Substitution Theorem for Dynamic Circuits

Many engineering problems reduce to the analysis of two 1-ports connected to each other. The substitution theorem simplifies the analysis of such cases. Here we generalize the substitution theorem for *resistive* circuits (see Chap. 5, Sec. 5.1) to general *dynamic* circuits.

Substitution theorem Let $\mathcal{N}_c$ be the circuit resulting from the connection of two *nonlinear time-varying* dynamic one-ports N and N' as shown in Fig. 5.3. It is assumed that, in the circuit $\mathcal{N}_c$, the *only* interaction between the one-ports N and N' occurs as a result of the wires connecting the nodes ① and ①′ and connecting the nodes ⓝ and ⓝ′: More precisely, it is assumed that there is *no* mechanical, thermal, magnetic, optical, etc., interaction between any element of N and any element of N'. Suppose that for the given independent sources and for the specified initial conditions, $\mathcal{N}_c$ has a *unique* solution: In particular, the current $\hat{i}_1(\cdot)$ and the port voltage $\hat{e}_1(\cdot)$.

Call $\mathcal{N}_v$, the circuit obtained from $\mathcal{N}_c$ by replacing N' by the voltage source $\hat{e}_1(\cdot)$ (see Fig. 5.3*b*). Call $\mathcal{N}_i$, the circuit obtained from $\mathcal{N}_c$ by replacing N' by the current source $\hat{i}_1(\cdot)$, as shown in Fig. 5.3*b*.

If $\mathcal{N}_v$ is uniquely solvable, then all the branch voltages and all the branch currents of N are unaffected by the substitution of N' by $\hat{e}_1(\cdot)$.

If $\mathcal{N}_i$ is uniquely solvable, then all the branch currents and all the branch voltages of N are unaffected by the substitution of N' by $\hat{i}_1(\cdot)$.

PROOF Let us perform a tableau analysis of $\mathcal{N}_c$: We label the node and branch variables of N' with a "′", thus $\mathbf{e}'$, $\mathbf{v}'$, $\mathbf{i}'$. Let us write KCL, KVL, and the branch equations, except for the branch connecting ① and ①′. For

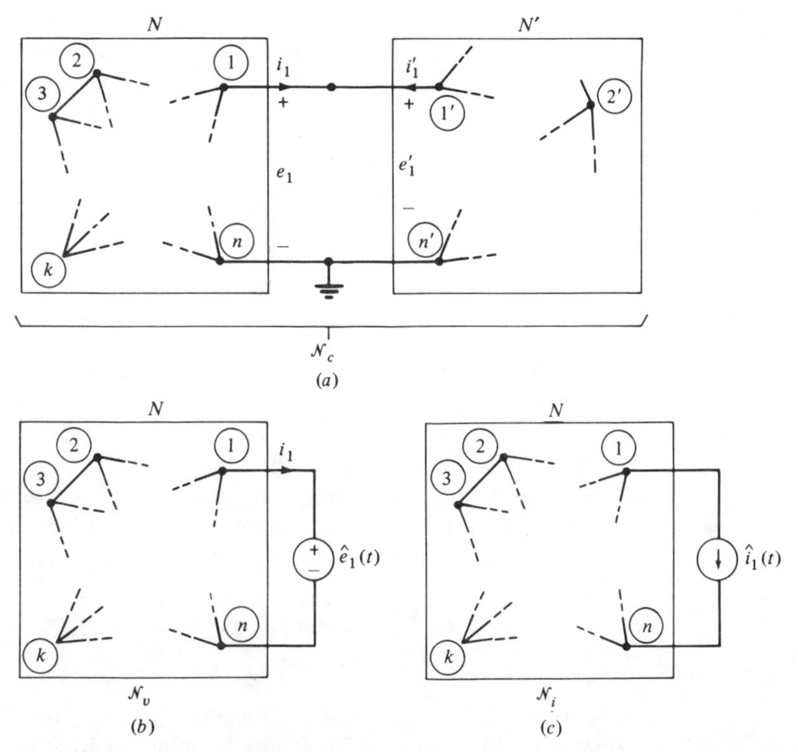

Figure 5.3 (*a*) The nonlinear circuit $\mathcal{N}_c$ consists of the interconnection of the one-port N and the one-port N'. (*b*) $\mathcal{N}_v$ consists of the one-port N terminated by the independent voltage source $\hat{e}_1(t)$; $\hat{e}_1(t)$ is the actual voltage appearing across ① and ⓝ in $\mathcal{N}_c$. (*c*) $\mathcal{N}_i$ consists of the one-port N terminated by the independent current source $\hat{i}_1(t)$.

this purpose, let **A** denote the reduced incidence matrix of N to which we have added a branch from node ① to node ⓝ (the reference direction of that branch is from ① to ⓝ). Similarly **A**′ denotes the reduced incidence matrix of N' to which we added a branch from ①′ to ⓝ′. The currents in these added branches are called $i_1(t)$ and $i'_1(t)$, respectively.

$$\mathbf{A}\,\mathbf{i}(t) = \mathbf{0} \qquad \mathbf{A}'\,\mathbf{i}'(t) = \mathbf{0} \tag{5.17}$$

$$\mathbf{v}(t) - \mathbf{A}^T\,\mathbf{e}(t) = \mathbf{0} \qquad \mathbf{v}'(t) - \mathbf{A}'^T\,\mathbf{e}'(t) = \mathbf{0} \tag{5.18}$$

$$\mathbf{h}(\dot{\mathbf{v}}, \mathbf{v}, \dot{\mathbf{i}}, \mathbf{i}, t) = \mathbf{0} \qquad \mathbf{h}'(\dot{\mathbf{v}}', \mathbf{v}', \dot{\mathbf{i}}', \mathbf{i}', t) = \mathbf{0} \tag{5.19}$$

To these equations, we must add the connection equations:

$$e_1(t) = e'_1(t) \qquad \text{and} \qquad i_1(t) = -i'_1(t) \tag{5.20}$$

By assumption, together with the specified initial conditions, the Eqs. (5.17), (5.18), (5.19), and (5.20) have a unique solution which we denote by

$$\hat{\mathbf{e}}(\cdot), \hat{\mathbf{v}}(\cdot), \hat{\mathbf{i}}(\cdot), \hat{\mathbf{e}}'(\cdot), \hat{\mathbf{v}}'(\cdot), \hat{\mathbf{i}}'(\cdot) \tag{5.21}$$

Since $\mathcal{N}_v$ is obtained from $\mathcal{N}_c$, by replacing the one-port N' by the voltage source $\hat{\mathbf{e}}_1(\cdot)$ (see Fig. 5.3*b*), the tableau equations of $\mathcal{N}_v$ are

$$\mathbf{A}\,\mathbf{i}(t) = \mathbf{0} \tag{5.22}$$

$$\mathbf{v}(t) - \mathbf{A}^T\,\mathbf{e}(t) = \mathbf{0} \tag{5.23}$$

$$\mathbf{h}(\dot{\mathbf{v}}, \mathbf{v}, \dot{\mathbf{i}}, \mathbf{i}, t) = \mathbf{0} \tag{5.24}$$

and

$$e_1(t) - \hat{e}_1(t) = 0 \tag{5.25}$$

where the last equation is the branch equation of the voltage source. It is obvious that since Eqs. (5.22), (5.23), and (5.24) are identical with those on the left-hand side of (5.17), (5.18), and (5.19), they are satisfied by the solution $\hat{\mathbf{e}}(\cdot)$, $\hat{\mathbf{v}}(\cdot)$, $\hat{\mathbf{i}}(\cdot)$. The same is true for Eq. (5.25). Thus, $\hat{\mathbf{e}}(\cdot)$, $\hat{\mathbf{v}}(\cdot)$, $\hat{\mathbf{i}}(\cdot)$ constitute *the* (unique, by assumption) solution of $\mathcal{N}_v$.

Therefore we have shown that all branch voltages and all branch currents of N are unaffected by the substitution of N' by the voltage source $\hat{\mathbf{e}}_1(\cdot)$.

The second assertion of the theorem is proved in a similar manner. ∎

Extension of the substitution theorem Let $\mathcal{N}_c$ be the circuit obtained by connecting the two *nonlinear time-varying dynamic* $(\nu+1)$-terminal circuits $\mathcal{N}$ and $\mathcal{N}'$ as shown in Fig. 5.4. Assume that there is no interaction between $\mathcal{N}$ and $\mathcal{N}'$ except through the wires connecting them. Suppose that, for the given independent sources and for the specified initial conditions, $\mathcal{N}_c$ has a *unique* solution: In particular, the terminal currents $\hat{i}_1(\cdot), \hat{i}_2(\cdot), \ldots, \hat{i}_\nu(\cdot)$ and the node voltages $\hat{e}_1(\cdot), \hat{e}_2(\cdot), \ldots, \hat{e}_\nu(\cdot)$.

Call $\mathcal{N}_v$ the circuit obtained from $\mathcal{N}_c$ by replacing $\mathcal{N}'$ by the independent voltage sources $\hat{e}_1(\cdot), \hat{e}_2(\cdot), \ldots, \hat{e}_\nu(\cdot)$. *If* $\mathcal{N}_v$ is *uniquely solvable*, *then* all the branch voltages and all the branch currents of $\mathcal{N}$ are unaffected by the substitution of $\mathcal{N}'$ by the voltage sources $\hat{e}_1(\cdot), \ldots, \hat{e}_\nu(\cdot)$.

Call $\mathcal{N}_i$ the circuit obtained from $\mathcal{N}_c$ by replacing $\mathcal{N}'$ by independent current sources $\hat{i}_1(\cdot), \hat{i}_2(\cdot), \ldots, \hat{i}_\nu(\cdot)$. *If* $\mathcal{N}_i$ is *uniquely solvable*, *then* all the branch voltages and all the branch currents of $\mathcal{N}$ are unaffected by the substitution of $\mathcal{N}'$ by the independent current sources $\hat{i}_1(\cdot), \ldots, \hat{i}_\nu(\cdot)$.

The proof follows exactly the same pattern as the one above, and is therefore omitted.

6 NUMERICAL SOLUTION OF CIRCUIT EQUATIONS

In this section we describe the simplest numerical methods for solving the circuit equations that result from either tableau analysis, node analysis, modified node analysis (MNA), or state equations.

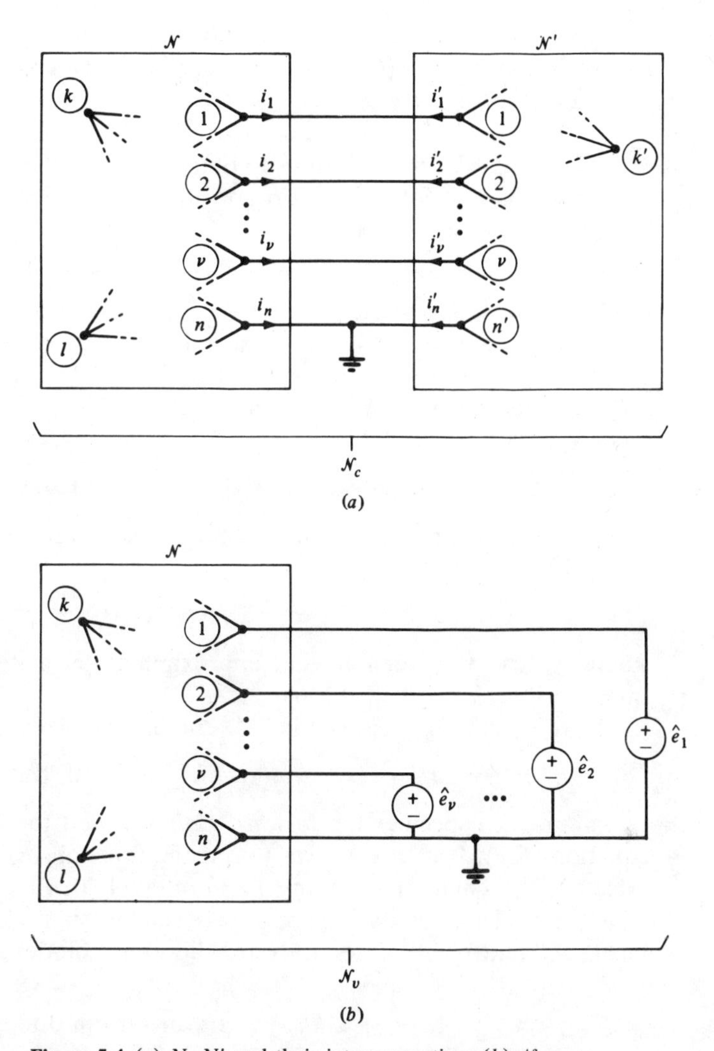

Figure 5.4 (*a*) N, N' and their interconnection. (*b*) $\mathcal{N}_\nu$.

6.1 The Forward Euler Method

To start out let us consider the calculation of the solution of differential equations. The simplest and most intuitive method for solving differential equations is the *forward Euler method*. To ease the exposition suppose that the equations of our circuit are written as[10]

$$\dot{\mathbf{x}}(t) = \mathbf{f}(\mathbf{x}(t), t) \tag{6.1}$$

[10] Our purpose here is to illustrate an idea of how one might proceed to solve a differential equation numerically. Table 4.1 of Chap. 7 shows how to obtain equations of the form (6.1).

Thus, for any time t, if we know the circuit variables, we know $\mathbf{x}(t)$ and we can calculate $\dot{\mathbf{x}}(t)$ by Eq. (6.1), that is, Eq. (6.1) gives us the rate of change (at time t) of the circuit variables.

Exercise 1 Use node analysis to obtain the circuit equations of the circuit $\mathcal{N}$ (shown in Fig. 6.1) in the form of Eq. (6.1): The nonlinear resistor has the characteristic: $i = \hat{i}(v)$.

To obtain a numerical solution of Eq. (6.1), starting at time $t=0$ from $\mathbf{x}(0) = \mathbf{x}_0$ where $\mathbf{x}_0$ is the set of initial conditions, we reason as follows: Suppose we consider a *small* time interval of length h; if h is small enough, then, over the interval from 0 to h, the derivative $\dot{\mathbf{x}}(t)$ will be *approximately constant*, namely,

$$\dot{\mathbf{x}}(t) \cong \dot{\mathbf{x}}(0) = \mathbf{f}(\mathbf{x}_0, 0) \qquad \text{for } 0 \le t \le h \tag{6.2}$$

Since, with this approximation, the function $\mathbf{x}(t)$ varies with a constant velocity $\mathbf{f}(\mathbf{x}(0), 0)$, we have at time $t = h$

$$\mathbf{x}_1 = \mathbf{x}_0 + h\,\mathbf{f}(\mathbf{x}_0, 0) \tag{6.3}$$

where we use $\mathbf{x}_1$ to denote the (approximate) calculated value of $\mathbf{x}(h)$. The small positive h is called the *step size.*

If we repeat this step, we have (assuming that we keep the same step size),

$$\mathbf{x}_{k+1} = \mathbf{x}_k + h\,\mathbf{f}(\mathbf{x}_k, kh) \qquad k = 0, 1, 2, \ldots \tag{6.4}$$

Equation (6.4) specifies the *forward Euler method* for solving the differential equation (6.1). It can be shown that if the function $\mathbf{f}$ is smooth enough, if we calculate the solution over any *fixed* interval $[0, T]$ using Eq. (6.4), if we repeat the calculation using successively smaller step sizes, then, as $h \to 0$, the *calculated* values will *all* tend toward the *exact* solution, i.e., as $h \to 0$,

$$\mathbf{x}_k \to \mathbf{x}(kh) \qquad \text{for all } k \text{ from 0 to } \frac{T}{h}$$

where $\mathbf{x}(kh)$ denotes the *exact* solution from Eq. (6.1) at time $t = kh$.

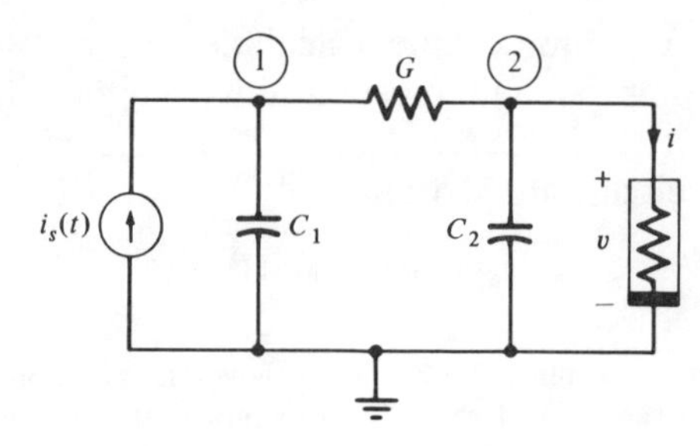

Figure 6.1 Nonlinear circuit for Exercise 1.

From an engineering point of view, we ask how small h must be. What if h is not chosen small enough? It turns out that even if the circuit is stable, if h is taken too large, the algorithm (6.4) *will always produce a sequence* $\mathbf{x}_k$ *that blows up*! This is illustrated in Exercise 2.

Exercise 2 Consider a linear time-invariant circuit with $G = 1$ S and $C = 1$ F (see Fig. 6.2). It is represented by $\dot{x} = -x$. Hence $x(t) = e^{-t}x_0$: The solution always goes to zero exponentially. The forward Euler method gives

$$x_{k+1} = x_k + hf(x_k) = x_k - hx_k = (1-h)x_k \tag{6.5}$$

(*a*) Assume $x_0 > 0$. Let $0 < h < 1$; show that the sequence of calculated values, $x_0, x_1, x_2, \ldots$, decays exponentially to zero, [i.e., it resembles the *exact* solution $x(t) = x_0 \exp(-t)$].
(*b*) Show that if $1 < h < 2$, the sequence of calculated values, $x_0, x_1, \ldots$, oscillates, but tends to zero as $k \to \infty$. [Thus the calculated values have successively positive and negative values; whereas $x_0 \exp(-t)$ is *always* positive.]
(*c*) Show that if $h > 2$, then the sequence oscillates and blows up as $k \to \infty$, i.e., as k increases, the successive values $|x_k|$ become arbitrarily large.

COMMENT This exercise shows that even though the circuit has a solution that tends to zero exponentially, the *solution* calculated by the forward Euler method may oscillate and even blow up if h is taken too large.

This fundamental defect of the forward Euler method leads us to the more useful backward Euler method.

6.2 The Backward Euler Method

The two basic steps behind the forward Euler method were: For $k = 0, 1, 2, \ldots$

1. *Assume* $\dot{\mathbf{x}}(t)$ *constant* over the time interval $(kh, (k+1)h)$.
2. *Approximate* $\dot{\mathbf{x}}(t)$ by its value at the *beginning* of the interval, namely,

$$\dot{\mathbf{x}}(t) = \mathbf{f}(\mathbf{x}_k, kh) \qquad \text{for } t \in (kh, (k+1)h) \tag{6.6}$$

The backward Euler method repeats Step 1 but modifies Step 2 as follows:

2*b*. *Approximate* $\dot{\mathbf{x}}(t)$ by its value at the *end* of the interval, namely,

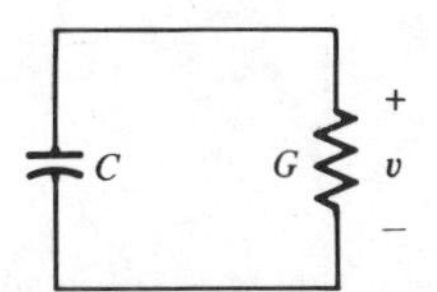

Figure 6.2 With C and G normalized to 1, the equation of the circuit reduces to $\dot{v} + v = 0$.

$$\dot{\mathbf{x}}(t) = \mathbf{f}[\mathbf{x}_{k+1}, (k+1)h] \qquad \text{for } t \in (kh, (k+1)h) \tag{6.7}$$

The backward Euler method works as follows: We use Eq. (6.7) to calculate $\dot{\mathbf{x}}(t)$ in the interval $(kh, (k+1)h)$: as a result of this approximation we obtain for $\mathbf{x}_{k+1}$ the expression

$$\mathbf{x}_{k+1} = \mathbf{x}_k + h\mathbf{f}[\mathbf{x}_{k+1}, (k+1)h] \qquad k = 0, 1, 2, \ldots \tag{6.8}$$

Note that, in Eq. (6.8), $\mathbf{x}_k$ and h are known as well as the function $\mathbf{f}$: Hence Eq. (6.8) is a set of *nonlinear algebraic equations that we have to solve for* $\mathbf{x}_{k+1}$.

The following exercise will show the benefits of this extra complication, namely, at each step of the algorithm we have to solve an algebraic equation.

Exercise Considering the same circuit as in Exercise 2 of Sec. 6.1, we have $\dot{x} = -x$. Now Eq. (6.8) gives

$$x_{k+1} = x_k - hx_{k+1}$$

or

$$x_{k+1} = \left(\frac{1}{1+h}\right)x_k \qquad k = 0, 1, 2, \ldots$$

Show that, *for all* $h > 0$, the sequence x_k *does not oscillate* and *tends to zero*. (Hence, the sequence of calculated values, $x_0, x_1, x_2, \ldots$, looks like the exact solution e^{-t} at the points $0, h, \ldots, kh, \ldots$.)

It can be shown that, for a *stable* linear circuit, if h is taken sufficiently small, the backward Euler method will give an accurate solution. If h is taken too large, the calculated solution will lose accuracy but it will never blow up! This inherent stability makes the backward Euler method very attractive.

Example Consider the first-order linear time-invariant circuit shown in Fig. 6.3. Let us write the node equation for node ①, divide by C, and collect the derivative in the left-hand side; we obtain

$$\dot{e}(t) = -\frac{G}{C}\, e(t) + \frac{1}{C}\, i_s(t) \tag{6.9}$$

Let the initial capacitor voltage be V_0. Equation (6.9) has the same form as Eq. (6.1), so using the backward Euler approximation for $\dot{x}(t)$ given by

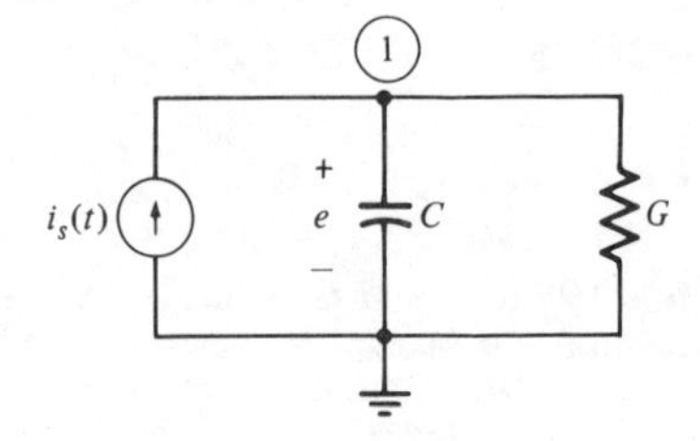

Figure 6.3 First-order circuit used in the example of the backward Euler method.

Eq. (6.8), we finally obtain

$$e_{k+1} = e_k - h\frac{G}{C}e_{k+1} + \frac{h}{C}i_s((k+1)h) \tag{6.10}$$

and $e_0 = V_0$. Since we know G, C, h, e_k, and $i_s(\cdot)$, Eq. (6.10) is an *algebraic equation in the unknown* e_{k+1}. Equation (6.10) is precisely Eq. (6.8) applied to the circuit of Fig. 6.3.

Equation (6.10) has a useful circuit interpretation. An easy calculation verifies that Eq. (6.10) is the node equation of the *resistive* circuit shown in Fig. 6.4. Note that the current source $i_s(\cdot)$ and the resistor (with conductance G) are left intact but that the *linear capacitor* C of Fig. 6.3 is replaced by a linear resistor of conductance C/h and an *independent voltage source* e_k.

6.3 The Backward Euler Method Applied to Circuit Equations

Suppose we have a *nonlinear time-varying* circuit: Then if we use the tableau analysis or if we use the modified node analysis, we end up with equations of the form

$$\mathbf{f}(\dot{\mathbf{x}}(t), \mathbf{x}(t), t) = \mathbf{0} \tag{6.11}$$

where $\mathbf{x}(t) = [\mathbf{e}(t), \mathbf{v}(t), \mathbf{i}(t)]^T$ if we use tableau analysis or $\mathbf{x}(t) = [\mathbf{e}(t)$, selected branch currents$]^T$ if we use MNA.

The backward Euler method assumes that $\dot{\mathbf{x}}(t)$ is constant over the interval $(kh, (k+1)h)$, and takes it to be given by

$$\dot{\mathbf{x}}(t) = \frac{\mathbf{x}_{k+1} - \mathbf{x}_k}{h} \tag{6.12}$$

where h is the chosen step size, $\mathbf{x}_k$ is known, and $\mathbf{x}_{k+1}$ is to be calculated.

Since we calculate $\dot{\mathbf{x}}(t)$ at the *end* of the interval, we use Eq. (6.12) in Eq. (6.11) to obtain

$$\mathbf{f}\left(\frac{\mathbf{x}_{k+1} - \mathbf{x}_k}{h}, \mathbf{x}_{k+1}, (k+1)h\right) = \mathbf{0} \tag{6.13}$$

This is a set of *nonlinear algebraic equations* in the unknown $\mathbf{x}_{k+1}$.

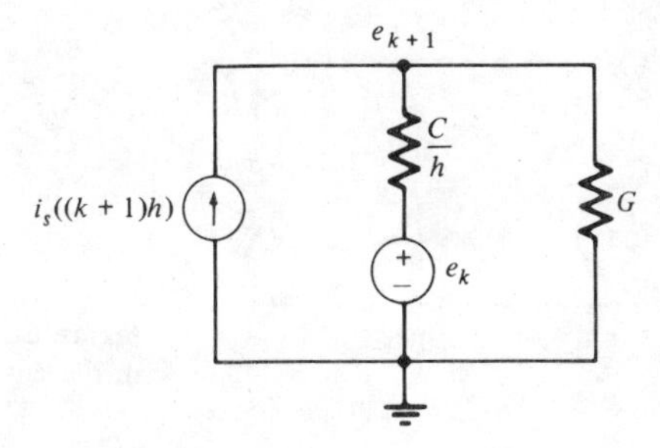

Figure 6.4 Resistive circuit whose node voltage e_{k+1} is the voltage obtained by the backward Euler method. The resistors have conductances C/h and G, respectively.

Example Let $\mathcal{N}$ be the nonlinear time-invariant circuit shown in Fig. 6.5: The nonlinear capacitor is specified by its small-signal capacitance $C(\cdot)$ and the nonlinear resistor $\mathcal{R}$ by its characteristic, $i = \hat{i}(v)$. The constants C_1 and G are also given. The node equations of $\mathcal{N}$ read

$$C_1\dot{e}_1(t) + Ge_1(t) - Ge_2(t) = i_s(t) \tag{6.14}$$

$$-Ge_1(t) + Ge_2(t) + C(e_2(t))\dot{e}_2(t) + \hat{i}(e_2(t)) = 0 \tag{6.15}$$

Note that these node equations are of the form of Eq. (6.11). To obtain the equation (6.13) we replace

$$\dot{e}_1(t) \text{ by } \frac{e_{1,k+1} - e_{1,k}}{h} \qquad \text{and} \qquad \dot{e}_2(t) \text{ by } \frac{e_{2,k+1} - e_{2,k}}{h} \tag{6.16}$$

hence

$$\begin{aligned} &\frac{C_1}{h}[e_{1,k+1} - e_{1,k}] + Ge_{1,k+1} - Ge_{2,k+1} = i_s((k+1)h) \\ -Ge_{1,k+1} + Ge_{2,k+1} + C(e_{2,k+1})&\frac{1}{h}[e_{2,k+1} - e_{2,k}] + \hat{i}(e_{2,k+1}) = 0 \end{aligned} \tag{6.17}$$

Note that the constants C_1, G, and h are known, the functions $C(\cdot)$, $i_s(\cdot)$, and $\hat{i}(\cdot)$ are known, and $\mathbf{e}_k = (e_{1,k}, e_{2,k})$ is known. Hence, Eq. (6.17) is a set of two *nonlinear algebraic equations* in the unknown $\mathbf{e}_{k+1} = (e_{1,k+1}, e_{2,k+1})$. Examination of Eq. (6.17) shows that it represents the *nonlinear resistive* circuit shown in Fig. 6.6.

Let us compare the given circuit $\mathcal{N}$ (of Fig. 6.5) and the circuit representation of Eq. (6.17) shown in Fig. 6.6. The linear resistor (of conductance G), the nonlinear resistor $\mathcal{R}$, and the independent current source $i_s(\cdot)$ are left intact. The *nonlinear capacitor* of $\mathcal{N}$ [characterized by $C(\cdot)$] is replaced by the *nonlinear resistor* $\mathcal{R}_2$ characterized by its voltage-controlled characteristic

$$i_2 = \frac{1}{h}C(v_2)(v_2 - e_{2,k})$$

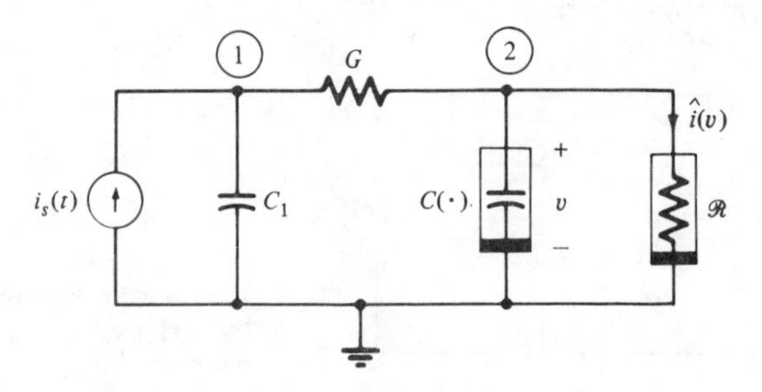

Figure 6.5 The nonlinear circuit used in the example.

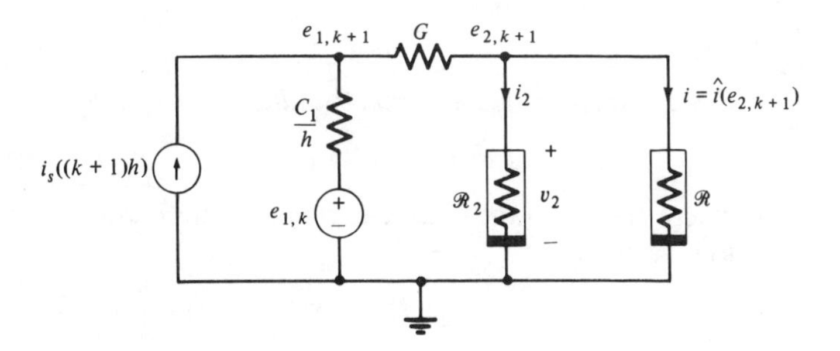

Figure 6.6 Circuit representation of Eq. (6.17).

Exercise Consider the nonlinear time-invariant circuit shown in Fig. 6.7. The nonlinear inductor is specified by its small-signal inductance $L_2(\cdot)$ and the nonlinear resistor is specified by its characteristic $i_3 = \hat{i}_3(v_3)$ where $\hat{i}_3(\cdot)$ is a given function. The capacitance C is also given.
(a) Write the MNA for the circuit.
(b) Write the algebraic equations required to solve the equations of (a) by the backward Euler method.
(c) Draw the resistive circuit that the equations in (b) represent: Specify precisely each element of this circuit.

Summary To integrate the circuit equations (6.11) by the backward Euler method, we must, at each step, solve the nonlinear algebraic Eq. (6.13). This is done by applying the Newton-Raphson algorithm (see Chap. 5, Sec. 3). If $\mathbf{x}_k$ is used as an initial guess, experience shows that one or two iterations are usually sufficient to obtain an adequate solution of (6.13).

Circuit simulators usually use more sophisticated numerical methods for solving the circuit equations. The discussion above shows why engineers are willing to incur the cost of solving at each step a *nonlinear* resistive circuit. Also once you understand the backward Euler method, the more sophisticated methods are easy to learn.

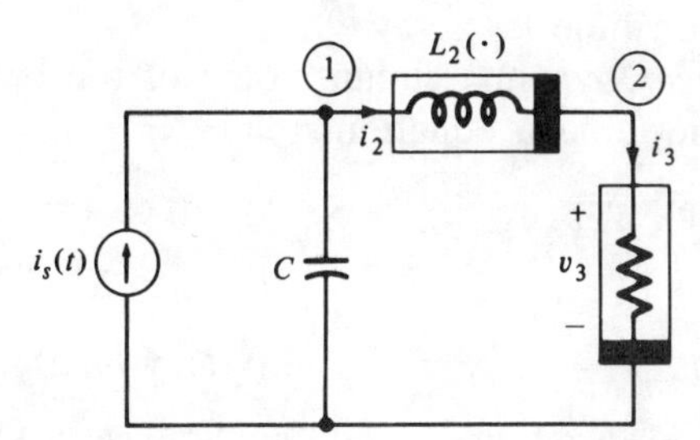

Figure 6.7 Nonlinear circuit for the exercise.

SUMMARY

- Linear time-invariant *coupled inductors* are described by

$$\boldsymbol{\phi}(t) = \mathbf{L}\,\mathbf{i}(t) \qquad \text{or} \qquad \mathbf{v}(t) = \mathbf{L}\,\dot{\mathbf{i}}(t)$$

where the constant matrix $\mathbf{L}$ is called the *inductance matrix*. If we solve for $\dot{\mathbf{i}}(t)$ we obtain

$$\dot{\mathbf{i}}(t) = \boldsymbol{\Gamma}\,\mathbf{v}(t) \qquad \text{where } \boldsymbol{\Gamma} = \mathbf{L}^{-1}$$

where the constant matrix $\boldsymbol{\Gamma}$ is called the *reciprocal inductance matrix*.

- The *magnetic energy stored* in a pair of linear time-invariant coupled inductors traversed by the currents i_1 and i_2 is given by

$$\mathscr{E}_M(i_1, i_2) = \tfrac{1}{2}\,(L_{11}i_1^2 + 2Mi_1i_2 + L_{22}i_2^2) = \tfrac{1}{2}\,\mathbf{i}^T\mathbf{L}\,\mathbf{i}$$

- $\mathscr{E}_M > 0$ for all $\mathbf{i} \neq \mathbf{0}$ if and only if $L_{11} > 0$, $L_{22} > 0$, $M^2 < L_{11}L_{22}$, or equivalently $\mathbf{L}$ is positive definite.
- $k = M/\sqrt{L_1L_2}$ is called the *coefficient of coupling*. For $\mathbf{L}$ positive definite, $|k| < 1$.
- Nonlinear time-invariant inductors are described by

$$\boldsymbol{\phi} = \hat{\boldsymbol{\phi}}(\mathbf{i})$$

and

$$\mathbf{v}(t) = \mathbf{J}(\mathbf{i}(t)) \cdot \dot{\mathbf{i}}(t)$$

where $\mathbf{J}(\mathbf{i}(t))$ is the jacobian matrix of $\hat{\boldsymbol{\phi}}$ evaluated at $\mathbf{i}(t)$. $\mathbf{J}(\mathbf{i})$ is also called the *small-signal inductance matrix*; it is then denoted $\mathbf{L}(\mathbf{i})$.

- Let $\mathcal{N}$ be a *linear time-invariant* dynamic circuit. The tableau equations consist of $2b + n - 1$ coupled algebraic and differential equations

KCL ($n - 1$ equations): $\qquad \mathbf{A}\,\mathbf{i}(t) = \mathbf{0}$

KVL (b equations): $\qquad -\mathbf{A}^T\,\mathbf{e}(t) + \mathbf{v}(t) = \mathbf{0}$

Branch equations (b equations):

$$(\mathbf{M}_0D + \mathbf{M}_1)\,\mathbf{v}(t) + (\mathbf{N}_0D + \mathbf{N}_1)\,\mathbf{i}(t) = \mathbf{u}_s(t)$$

where $\mathbf{A}$, $\mathbf{M}_0$, $\mathbf{M}_1$, $\mathbf{N}_0$, $\mathbf{N}_1$ are constant matrices, D denotes the operator d/dt, and $\mathbf{u}_s(t)$ represents the *independent* sources.

- If $\mathcal{N}$ is *linear* but *time-varying*, then the matrices $\mathbf{M}_0$, $\mathbf{M}_1$, $\mathbf{N}_0$, and $\mathbf{N}_1$ will in general depend on time.
- If $\mathcal{N}$ is a *nonlinear* dynamic circuit, some of the branch equations will be nonlinear and the tableau equations read

KCL ($n - 1$ equations): $\qquad \mathbf{A}\,\mathbf{i}(t) = \mathbf{0}$

KVL (b equations): $\qquad -\mathbf{A}^T\,\mathbf{e}(t) + \mathbf{v}(t) = \mathbf{0}$

Branch equations: $\qquad \mathbf{h}(\dot{\mathbf{v}}(t), \mathbf{v}(t), \dot{\mathbf{i}}(t), \mathbf{i}(t), t) = \mathbf{0}$

- The variables used in modified node analysis (MNA) are the *node voltages* and the *selected currents*. One writes the node equations (for all nodes except the datum node) with two provisos. First, whenever an inductor is encountered, the inductor current is entered in the node equation and the inductor branch equation is appended to the node equations. Second, whenever a branch which is not voltage-controlled is encountered, the branch current is entered in the node equation and the branch equation of that element is appended to the node equations.

- The tableau analysis and the MNA are two completely general methods of analysis. MNA is very efficient and easy to program.
- To perform the *small-signal analysis* of a *nonlinear time-invariant dynamic* circuit $\mathcal{N}$ driven by dc and "small" ac independent sources, we proceed in two steps: First, solve the nonlinear *resistive* circuit obtained from $\mathcal{N}$ by (*a*) setting to zero all ac independent sources, (*b*) replacing all inductors by short circuits and all capacitors by open circuits. Call Q the calculated operating point; it is specified by $(V_{Q1}, V_{Q2}, \ldots; I_{Q1}, I_{Q2}, \ldots)$. We obtain $\mathcal{L}_Q$, the small-signal equivalent circuit of $\mathcal{N}$ about the operating point Q, as follows: Set all dc independent sources to zero and replace all *nonlinear* elements by their small-signal equivalent circuits about the operating point Q. The resulting circuit $\mathcal{L}_Q$ is *linear*, *time-invariant*, and driven by the "small" ac independent sources. For each branch, the *approximate* branch voltage and *approximate* branch current are given by

$$v_k(t) \cong V_{Qk} + \tilde{v}_k(t) \qquad i_k(t) \cong I_{Qk} + \tilde{i}_k(t)$$

where $\tilde{v}_k(\cdot)$ and $\tilde{i}_k(\cdot)$ are the solutions of $\mathcal{L}_Q$.

- *Superposition theorem*: Let $\mathcal{N}$ be any *uniquely solvable linear time-varying* dynamic circuit. The *zero-state response* of $\mathcal{N}$ to a number of independent voltage sources and independent current sources is equal to the sum of the *zero-state responses* due to each source acting alone.
- *Substitution theorem*: Let N and N' be two *nonlinear time-varying* dynamic one-ports connected together. Let $\hat{e}_1(\cdot)$ and $\hat{i}_1(\cdot)$ be the observed port voltage and current. Let $\mathcal{N}_v$ ($\mathcal{N}_i$, respectively) be the circuit obtained by replacing N' by the voltage-source $\hat{e}_1(\cdot)$ [the current source $i_1(\cdot)$, respectively]. Let $\mathcal{N}_c$ be the circuit obtained when N and N' are connected together. If the three circuits, $\mathcal{N}_c$, $\mathcal{N}_v$, and $\mathcal{N}_i$, are uniquely solvable, then all branch currents and all branch voltages of N are unaffected by the substitution of N' by $\hat{e}_1(\cdot)$ [or $\hat{i}_1(\cdot)$, respectively]. (See Fig. 5.3.)
- Let the tableau equations or the modified node analysis equations read $\mathbf{f}(\dot{\mathbf{x}}(t), \mathbf{x}(t), t) = \mathbf{0}$. In the *backward Euler* method, first we approximate $\dot{\mathbf{x}}(t)$ by $(\mathbf{x}_{k+1} - \mathbf{x}_k)/h$—where h is the *step size*—and second, we calculate $\mathbf{x}_{k+1}$ at the end of the time interval $[kh, (k+1)h]$; namely, at each step, we solve the *nonlinear algebraic* equation

$$\mathbf{f}\left(\frac{\mathbf{x}_{k+1} - \mathbf{x}_k}{h}, \mathbf{x}_{k+1}, (k+1)h\right) = \mathbf{0}$$

for $\mathbf{x}_{k+1}$. We start from a specified $\mathbf{x}_0$ at time 0 and proceed successively for $k = 0, 1, 2, \ldots$.

PROBLEMS

Coupled inductors

1 A pair of coupled inductors has (for the reference directions shown in Fig. P8.1*a*) the inductance matrix

$$\mathbf{L} = \begin{bmatrix} 4 & -3 \\ -3 & 6 \end{bmatrix}$$

Find the equivalent inductance of the four connections shown in Fig. P8.1*b* to *e*.

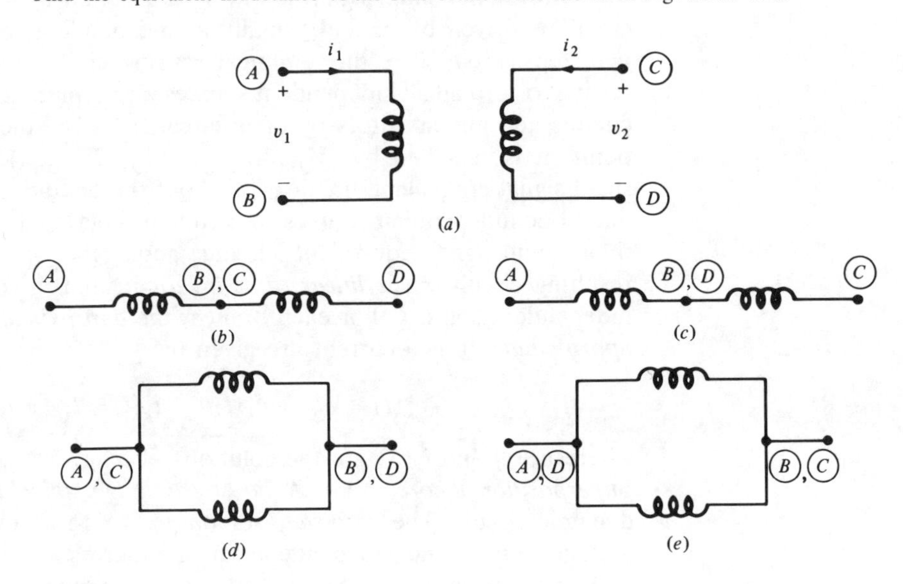

Figure P8.1

2 (*a*) Obtain the inductance matrix for the two-port shown in Fig. P8.2.

(*b*) Find conditions on n_1, n_2, L_a, and L_b such that this two-port is equivalent to the one described by Eqs. (1.3).

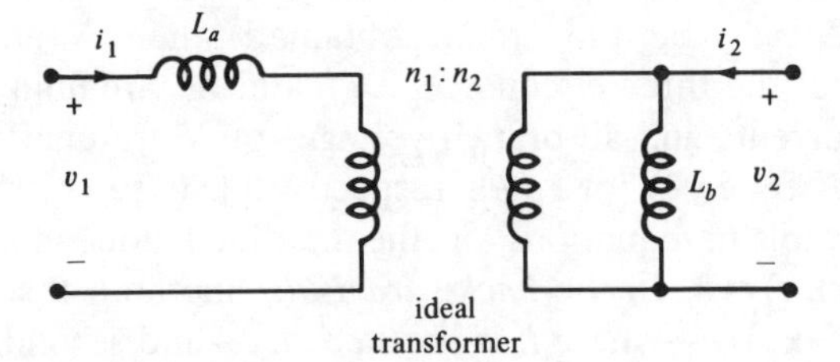

Figure P8.2

3 Consider the circuit shown in Fig. P8.3. The three coupled inductors are defined by the following inductance matrix:

$$\mathbf{L} = \begin{bmatrix} 5 & 2 & 1 \\ 2 & 4 & -1 \\ 1 & -1 & 2 \end{bmatrix}$$

Assume that i_s is a constant source and $i_1 = 2$ A, $i_2 = 1$ A, and $i_3 = -3$ A. What is the energy stored in the inductors?

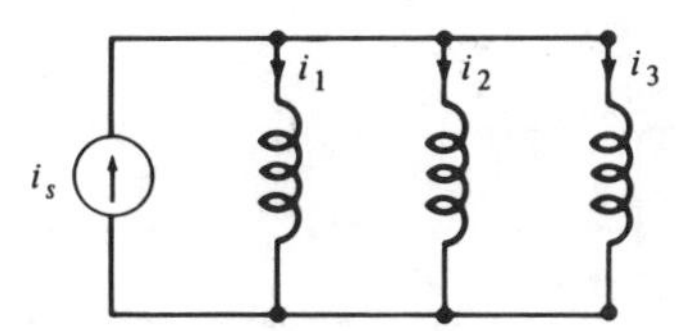

Figure P8.3

Tableau equations

4 Write the tableau equations for the linear time-invariant circuits shown in Fig. P8.4.

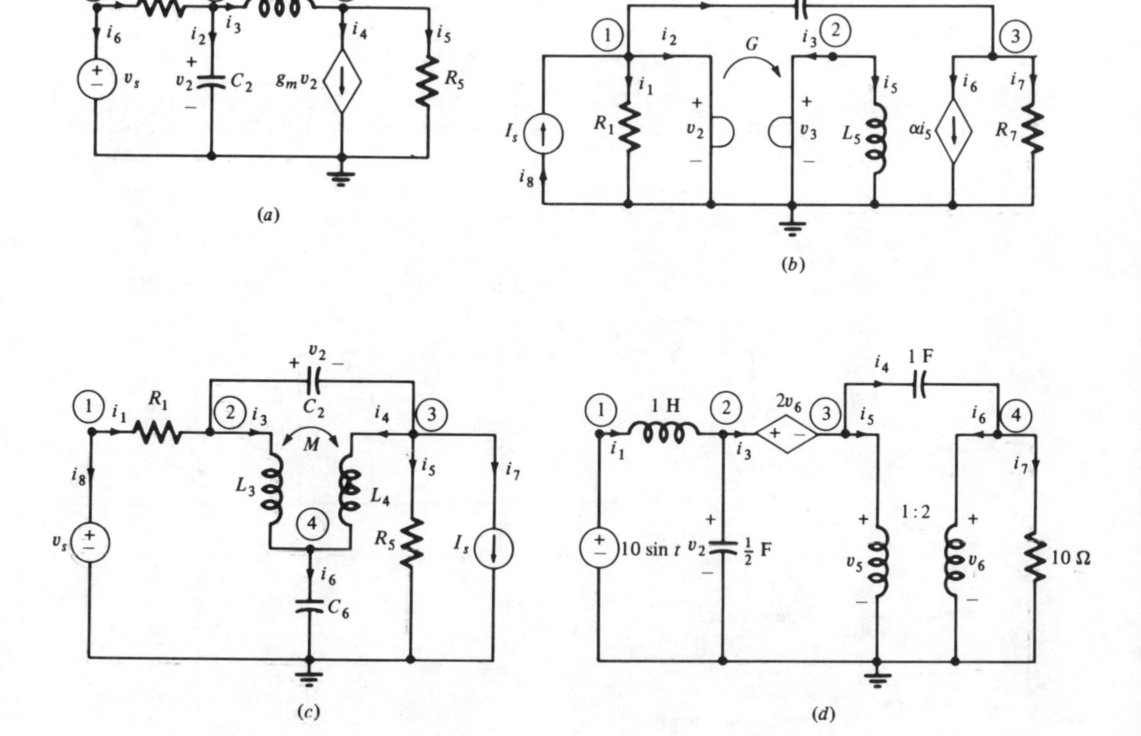

Figure P8.4

5 Write the tableau equations for the linear time-varying circuits shown in Fig. P8.5.

6 Write the tableau equations for the nonlinear dynamic circuits shown in Fig. P8.6. Assume the transistor is modeled by the Ebers-Moll equations and the op amp is modeled by Fig. 2.6.

MNA

7 Write the modified node equations for the circuits shown in Fig. P8.4.

8 Write the modified node equations for the circuits shown in Fig. P8.5.

9 Repeat Prob. 6 for modified node equations.

10 Write the modified node equations for the circuit shown in Fig. 4.1.

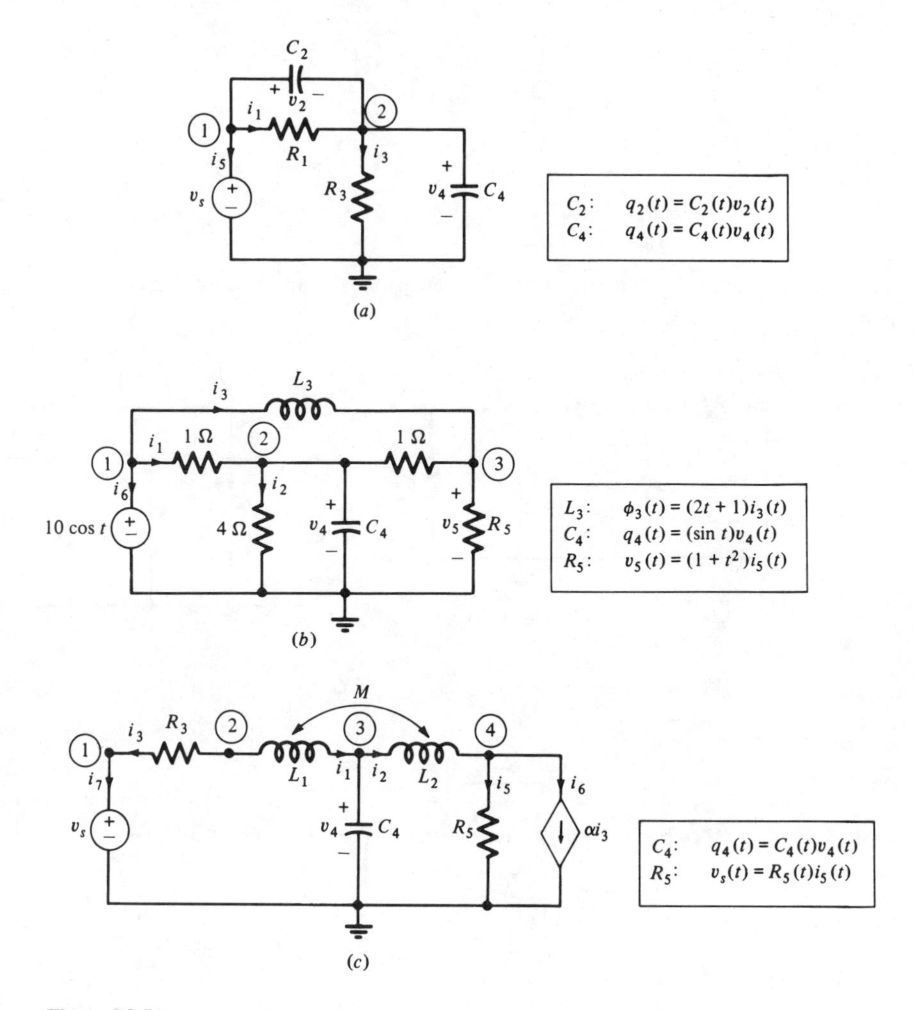

Figure P8.5

Small-signal analysis

11 Consider the nonlinear dynamic circuit shown in Fig. P8.11.
(a) Find the operating point $(\mathbf{V}_Q, \mathbf{I}_Q)$.
(b) Construct the associated small-signal equivalent circuit about this operating point.
(c) Write the modified node equations for the circuit obtained in (b).

12 Repeat Prob. 11 for the circuit shown in Fig. P8.12.

13 For the circuits shown in Fig. P8.13:
(a) Find the operating point $(\mathbf{V}_Q, \mathbf{I}_Q)$.
(b) Draw the small-signal equivalent circuit about this operating point.

14 (a) Use the Newton-Raphson algorithm to obtain the operating point for the circuit shown in Fig. P8.14.
(b) Draw the small-signal equivalent circuit about this operating point.

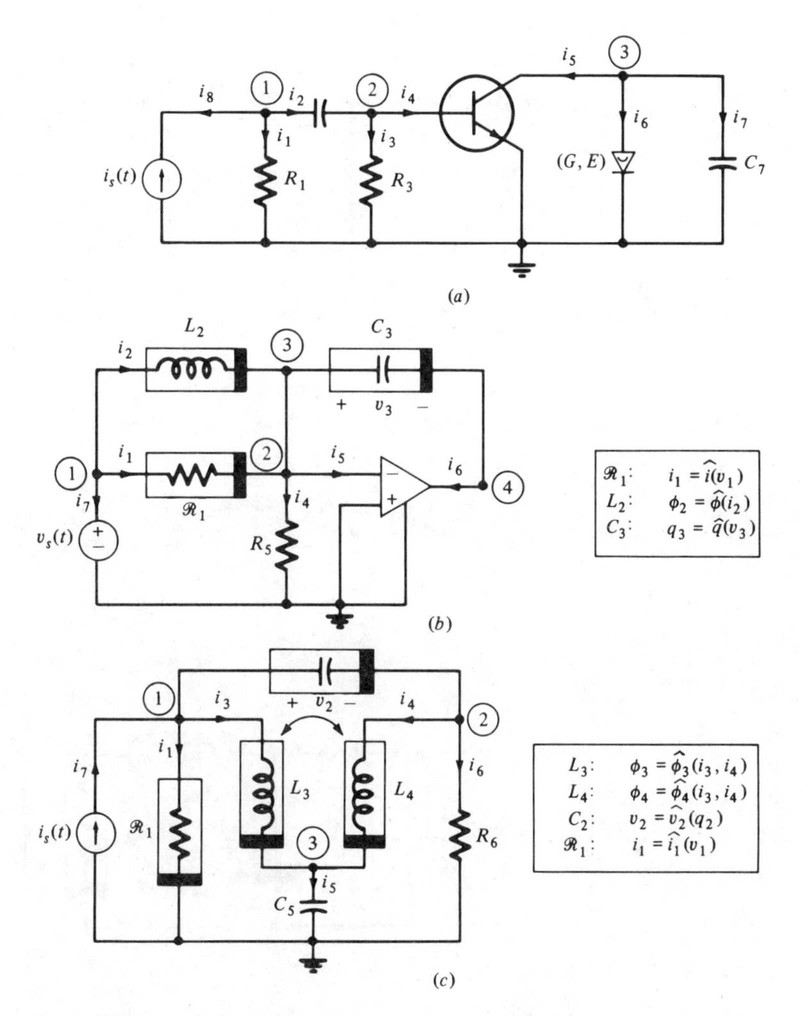

Figure P8.6

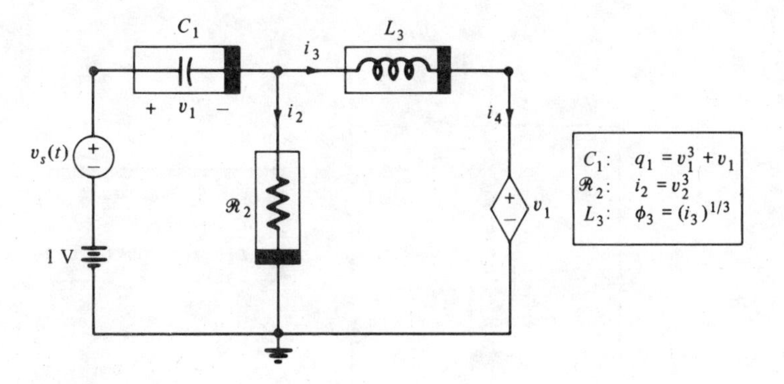

Figure P8.11

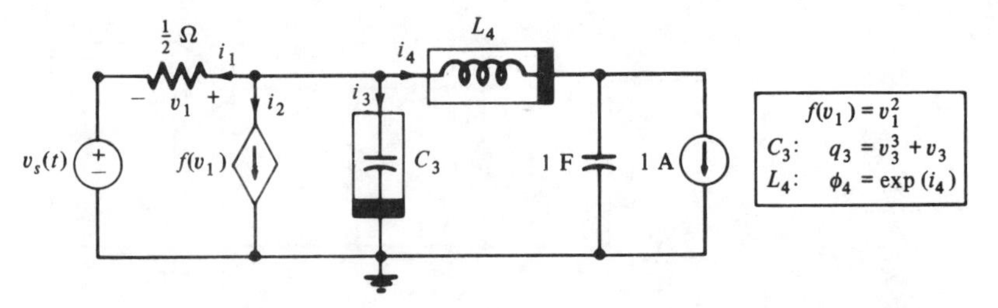

Figure P8.12

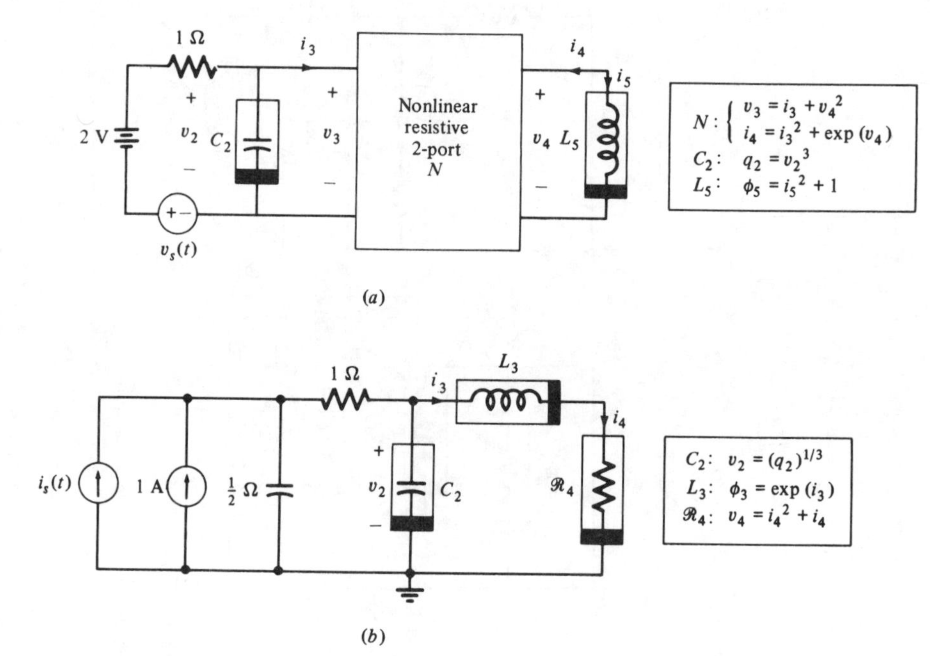

Figure P8.13

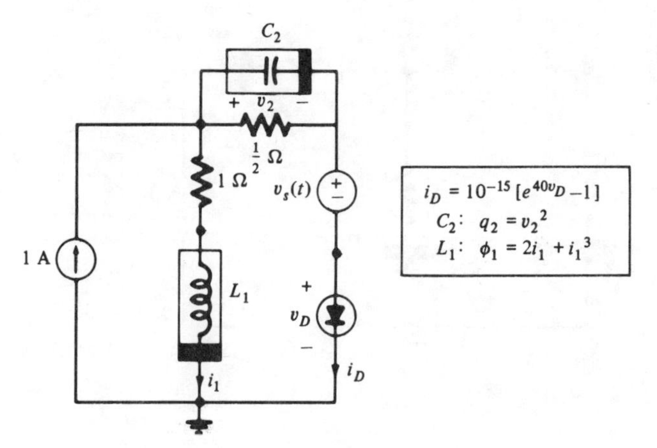

Figure P8.14

Superposition, zero-state response

15 The zero-state response v of a linear time-invariant RC circuit to a unit step of current is given by

$$v(t) = 2(1 - e^{-t})1(t)$$

Calculate the zero-state response to a current having the waveform shown in Fig. P8.15.

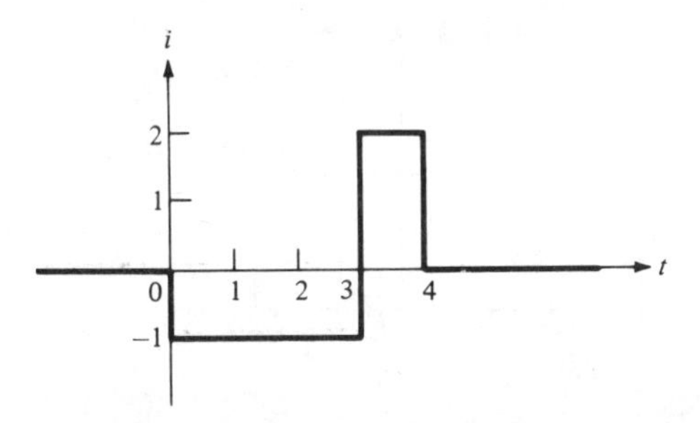

Figure P8.15

16 Consider a first-order linear time-invariant RC circuit and call v_c the capacitor voltage. Let $u_s(t)$ be the input and $y(t)$ be the response. The following measurements have been made:

$$u_s(t) = i_{s1}(t) \qquad y_1(t) = (\tfrac{3}{2}e^{-2t} + \tfrac{1}{2})1(t)$$
$$v_c(0) = V_0$$

$$u_s(t) = i_{s2}(t) \qquad y_2(t) = (3e^{-2t})1(t)$$
$$v_c(0) = V_0$$

$$u_s(t) = i_{s1}(t) + i_{s2}(t) \qquad y_3(t) = (\tfrac{1}{2}e^{-2t} + \tfrac{1}{2})1(t)$$
$$v_c(0) = 0$$

Determine the zero-state response to $i_{s1}(t)$.

Numerical methods

17 For the first-order differential equation $\dot{x} = \lambda x$, $\lambda < 0$, using step size $h > 0$, determine the region of stability.

(*a*) The forward Euler method is used.
(*b*) The backward Euler method is used.

In each case give a graphic interpretation as a set in the h-λ plane.

18 For differential equations and their solutions given below, evaluate the solutions at $t = 0$ to obtain the initial condition and then integrate over several steps with $h = 0.1$ using

(*a*) The forward Euler method
(*b*) The backward Euler method

(i) $\dot{x} = -xt$ Solution: $x = \exp\left(-\dfrac{t^2}{2}\right)$

(ii) $\dot{x} = t^2 - tx + 1$ $x = t + \exp\left(-\dfrac{t^2}{2}\right)$

(iii) $\dot{x} = x - 2\sin t$ $x = \sin t + \cos t$

Compare the numerical results with the exact solution.

19 For the nonlinear time-invariant circuit in Fig. P8.19:

(*a*) Write the MNA equations.

(*b*) Write the algebraic equations required to solve the equations of (*a*) by the backward Euler method.

(*c*) Draw the resistive circuit representing the equations in (*b*).

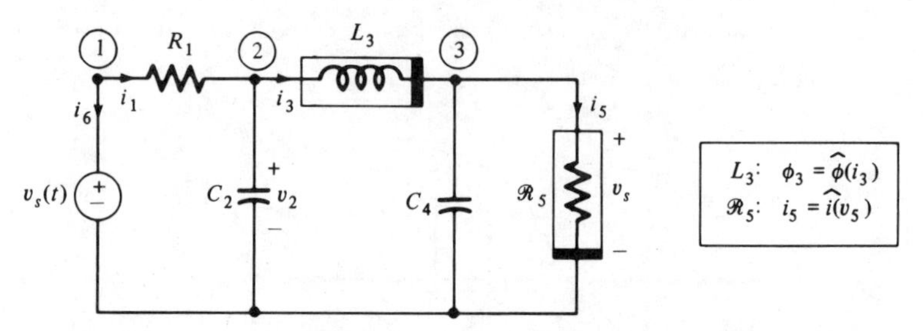

Figure P8.19

20 For the circuits of Figs. P8.11 and P8.12 set the dc sources to zero and construct the backward Euler equivalent circuits.

CHAPTER

NINE

SINUSOIDAL STEADY-STATE ANALYSIS

INTRODUCTION

In this chapter we consider exclusively *linear time-invariant* circuits and we concentrate our study on their sinusoidal steady-state behavior, that is, their behavior when they are driven by one or more sinusoidal sources at some frequency ω and when, after all "transients" have died down, all currents and voltages are sinusoidal at frequency ω.

Our first task is to systematically develop the technique of phasor analysis: The idea is to associate with each sine wave (of voltage or current) a complex number called the *phasor*. This technique is very important in engineering for three reasons: (*a*) many circuits essentially are operating in the sinusoidal steady state; (*b*) the technique is extremely efficient, hence has wide applicability (electric circuits, control systems, quantum electronics, electromagnetics, etc.); (*c*) as we shall learn later, if we know the response of a linear time-invariant circuit to a sinusoidal input of *any* frequency, then we can calculate efficiently its response to any signal (by, for example, fast Fourier transforms).

The technique of phasor analysis leads naturally to the important concept of network functions: In Sec. 3 we study their properties and how they are used to calculate efficiently the circuit response. Next we turn to the analysis of power and energy in the sinusoidal steady state: This topic is of great importance in communication, measurements, and, of course, in power networks. The maximum power transfer theorem is a key tool in many designs. The final section gives a brief and self-contained treatment of three-phase circuits: We establish why economics dictates their use, we explain the concepts behind the electric generators, and finally we show how simply they can be analyzed.

This chapter has a somewhat narrow focus but the concepts and techniques it teaches are fundamental to engineering. In particular, the concept of network functions will be generalized in Chap. 10 and used to solve many problems in Chap. 11.

0 REVIEW OF COMPLEX NUMBERS

Let us summarize some useful facts about complex numbers. Let z be a complex number: z is specified by its *real part*, $x = \text{Re}(z)$, and its *imaginary part*, $y = \text{Im}(z)$:

$$z = x + jy$$

where $j = \sqrt{-1}$. The representation given above suggests associating, with the complex number z, the point (x, y) in the complex plane (see Fig. 0.1). Thus $x + jy$ is called the *rectangular representation* of z. Similarly we may use *polar coordinates*:

$$|z| = \sqrt{x^2 + y^2} \qquad \text{and } \measuredangle z \text{ is specified by} \qquad \begin{cases} x = |z| \cos \measuredangle z \\ y = |z| \sin \measuredangle z \end{cases}$$

We write

$$z = |z|\, e^{j\measuredangle z} \qquad \text{or} \qquad z = |z| \exp(j\measuredangle z)\,.$$

where $|z|$ is called the *magnitude* of z and $\measuredangle z$ is called the *phase* of z. Referring to Fig. 0.1, it is clear that $|z| \exp(j\measuredangle z)$ should be called the polar *representation* of z.

Two complex numbers, say, $z_1 = x_1 + jy_1$ and $z_2 = x_2 + jy_2$ are by definition *equal* if and only if $x_1 = x_2$ and $y_1 = y_2$. Thus *one* equality between complex numbers is equivalent to *two* equalities between real numbers.

REMARKS

1. The equations above imply that, if $x \neq 0$, $y/x = \tan \measuredangle z$ or equivalently $\tan^{-1}(y/x) = \measuredangle z$. In order to find the phase of z, one must consider both $\text{Re}(z) = x = |z| \cos \measuredangle z$ and $\text{Im}(z) = y = |z| \sin \measuredangle z$. Knowing the real part

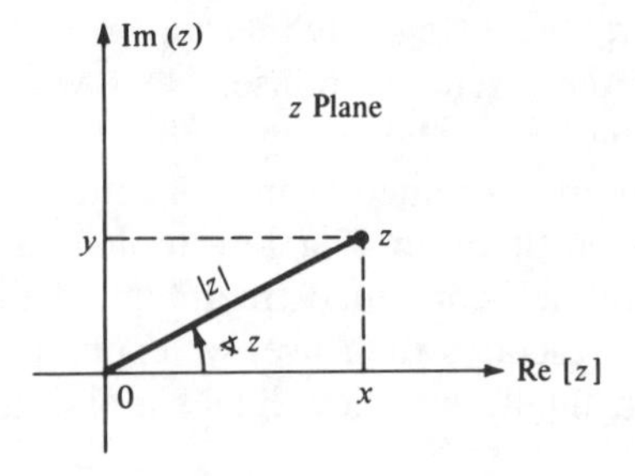

Figure 0.1 The complex z plane: the point z has cartesian coordinates $\text{Re}(z)$, $\text{Im}(z)$ and polar coordinates $|z|$, $\measuredangle z$.

and the imaginary part of z determines the quadrant in which z is located: For the first quadrant, $x \geq 0$ and $y \geq 0$, hence $0 \leq \measuredangle z \leq 90°$; for the second quadrant, $x \leq 0$ and $y \geq 0$, hence $90° \leq \measuredangle z \leq 180°$; etc.

Blind use of the formula $\measuredangle z = \tan^{-1}(y/x)$ leads to wrong answers:

$$z = 1 + j2 \qquad \text{hence } \measuredangle z = 63.43° \qquad \left(\text{note that } \frac{y}{x} = 2\right)$$

$$z = -1 - j2 \qquad \text{hence } \measuredangle z = -116.565° \qquad \left(\text{note that } \frac{y}{x} = 2\right)$$

2. Given $z = x + jy$, the magnitude $|z|$ is uniquely defined. *Not so for the phase* $\measuredangle z$: This is a consequence of the 2π-periodicity of the functions cos and sin. For example, if $z = 1 + j2$, then $|z| = 2.236$ and $\measuredangle z = 63.435°$ or $63.435° - 360°$, or $63.435° + 360°$, etc. For this reason, in Sec. 3.1 below, we shall use the phrase "the phase may be chosen so that. . . ."
3. Note that if $|z| = 0$, the phase is not defined.

OPERATIONS WITH COMPLEX NUMBERS

The usual operations are defined as follows: Let z_1 and z_2 be any two complex numbers:

$$z_1 = x_1 + jy_1 = |z_1| \exp(j\measuredangle z_1) \qquad z_2 = x_2 + jy_2 = |z_2| \exp(j\measuredangle z_2)$$

Addition:

$$z_1 + z_2 = (x_1 + jy_1) + (x_2 + jy_2) = (x_1 + x_2) + j(y_1 + y_2)$$

Multiplication: Using $j^2 = -1$,

$$z_1 z_2 = (x_1 + jy_1)(x_2 + jy_2) = x_1x_2 - y_1y_2 + j(x_1y_2 + x_2y_1)$$

or in polar representations

$$z_1 z_2 = |z_1| \exp(j\measuredangle z_1) \cdot |z_2| \exp(j\measuredangle z_2) = |z_1| \cdot |z_2| \exp j(\measuredangle z_1 + \measuredangle z_2)$$

Complex conjugate: Given $z = x + jy$, the complex number $x - jy = \bar{z}$ is called the *complex conjugate* of z. Note that

$$\bar{z} = x - jy = |z| \exp(-j\measuredangle z)$$

and $$z + \bar{z} = 2x \qquad z - \bar{z} = 2jy \qquad z\bar{z} = |z|^2 = x^2 + y^2$$

Division:

$$\frac{z_1}{z_2} = \frac{x_1 + jy_1}{x_2 + jy_2} = \frac{z_1\bar{z}_2}{z_2\bar{z}_2} = \frac{(x_1x_2 + y_1y_2) + j(-x_1y_2 + x_2y_1)}{x_2^2 + y_2^2}$$

$$\frac{z_1}{z_2} = \frac{|z_1|}{|z_2|} \exp j(\measuredangle z_1 - \measuredangle z_2)$$

REMARK We know that real numbers form what algebraists call a *field*. It is easy to show that with the operations defined above, the complex numbers also obey the axioms of field. As a consequence, *all the algebraic manipulations that we learned for real numbers apply to complex numbers*. For example: solution of linear algebraic equations, gaussian elimination, properties of determinants and Cramer's rule, computations with eigenvalues and eigenvectors, etc.

Exercises

1. Use the definitions above to prove that

(*a*) $\overline{z_1 + z_2} = \bar{z}_1 + \bar{z}_2$

(*b*) $\overline{z_1 z_2} = \bar{z}_1 \cdot \bar{z}_2$

(*c*) $\overline{\left(\dfrac{z_1}{z_2}\right)} = \dfrac{\bar{z}_1}{\bar{z}_2}$

(*d*) For any *real* numbers a_1, a_2

$$\operatorname{Re}(a_1 z_1 + a_2 z_2) = a_1 \operatorname{Re}(z_1) + a_2 \operatorname{Re}(z_2)$$

(*e*) For any integer $k \geq 1$, $\overline{z^k} = (\bar{z})^k = |z|^k \exp k(-j\measuredangle z)$

2. Let a_0, a_1, a_2 be *real* numbers, and let z be a complex number; then

$$\overline{a_2 z^2 + a_1 z + a_0} = a_2 \bar{z}^2 + a_1 \bar{z} + a_0$$

3. Let $z_1, z_2, \ldots, z_5$ be any complex numbers; show that

$$\left| \frac{z_1 \cdot z_2}{z_3 \cdot z_4 \cdot z_5} \right| = \frac{|z_1| \cdot |z_2|}{|z_3| \cdot |z_4| \cdot |z_5|}$$

and
$$\measuredangle \frac{z_1 \cdot z_2}{z_3 \cdot z_4 \cdot z_5} = (\measuredangle z_1 + \measuredangle z_2) - (\measuredangle z_3 + \measuredangle z_4 + \measuredangle z_5)$$

4. Choose a complex number $z = x + jy$ in the first quadrant. Draw the x and y axes, and plot the points z, $-z$, $\bar{z}$, $-\bar{z}$.

(*a*) Describe the symmetry between z and $-z$; z and $\bar{z}$; z and $-\bar{z}$.

(*b*) Let z be fixed with $\operatorname{Re} z > 0$. Show that for all s in the right-hand plane $[\operatorname{Re}(s) \geq 0]$, $\left| \dfrac{s - z}{s + \bar{z}} \right| \leq 1$.

1 PHASORS AND SINUSOIDAL SOLUTIONS

1.1 Sinusoids and Phasors

A *sinusoid* of angular frequency ω (rad/s) is by definition a function of the form

$$A_m \cos(\omega t + \phi) \tag{1.1}$$

where the *amplitude* A_m, the *phase* ϕ, and the *frequency* ω are real *constants*. The amplitude A_m is always taken to be *positive*. The frequency ω is measured in radians per second. The *period* is $T = 2\pi/\omega$ in seconds.

If the frequency is measured in hertz and labeled f, then the sinusoid becomes $A_m \cos(2\pi f t + \phi)$ and $T = 1/f$.

To the sinusoid above we associate a complex number A called the *phasor* (of that sinusoid) according to the rule

$$A \triangleq A_m \, e^{j\phi} \tag{1.2}$$

The crucial point is that the sinusoid specified by Eq. (1.1) defines *uniquely* the phasor A by Eq. (1.2).

Conversely, *knowing the frequency* ω, the phasor A specifies *uniquely* the sinusoid by the formula

$$\text{Re}[A \, e^{j\omega t}] = \text{Re}[A_m \, e^{j(\omega t + \phi)}] = A_m \cos(\omega t + \phi) \tag{1.3}$$

In summary, there is a one-to-one correspondence between sinusoids (at frequency ω) and phasors. This correspondence is illustrated by

$$\begin{array}{ll} \textit{Sinusoid} & \textit{Phasor} \\ \left.\begin{array}{l} A_m \cos(\omega t + \phi) \\ = (A_m \cos \phi) \cos \omega t \\ + (-A_m \sin \phi) \sin \omega t \end{array}\right\} \Leftrightarrow & \left\{\begin{array}{l} A = A_m \, e^{j\phi} \\ = (A_m \cos \phi) + j(A_m \sin \phi) \end{array}\right. \end{array} \tag{1.4}$$

Equivalence (1.4) says that

$$A_m \cos(\omega t + \phi) = \text{Re}[A] \cos \omega t - \text{Im}[A] \sin \omega t \tag{1.5}$$

1.2 Three Lemmas

The use of phasors in the analysis of linear time-invariant circuits in sinusoidal steady state becomes completely obvious once the following three lemmas are thoroughly understood.

Lemma 1 (Uniqueness) Two sinusoids are equal if and only if they are represented by the same phasor; symbolically, for all t,

$$\text{Re}(A \, e^{j\omega t}) = \text{Re}(B \, e^{j\omega t}) \Leftrightarrow A = B \tag{1.6}$$

PROOF ($\Leftarrow$) By assumption $A = B$. Consequently, for all t,

$$A \, e^{j\omega t} = B \, e^{j\omega t} \qquad \text{and} \qquad \text{Re}(A \, e^{j\omega t}) = \text{Re}(B \, e^{j\omega t})$$

($\Rightarrow$) By assumption for all t,

$$\text{Re}(A \, e^{j\omega t}) = \text{Re}(B \, e^{j\omega t}) \tag{1.7}$$

In particular, for $t = 0$, Eq. (1.7) gives

$$\text{Re}(A) = \text{Re}(B) \tag{1.8}$$

Similarly, for $t_0 = \pi/2\omega$, $\exp j\omega t_0 = \exp j(\pi/2) = j$ and $\mathrm{Re}(Aj) = -\mathrm{Im}(A)$; hence (1.7) gives

$$\mathrm{Im}(A) = \mathrm{Im}(B) \tag{1.9}$$

Hence from (1.8) and (1.9), we obtain successively

$$A = \mathrm{Re}(A) + j\,\mathrm{Im}(A) = \mathrm{Re}(B) + j\,\mathrm{Im}(B) = B$$

Lemma 2 (Linearity) The phasor representing a linear combination of sinusoids (with *real* coefficients) is equal to the *same* linear combination of the phasors representing the individual sinusoids.

Symbolically, let the sinusoids be

$$x_1(t) = \mathrm{Re}[A_1\, e^{j\omega t}] \qquad \text{and} \qquad x_2(t) = \mathrm{Re}[A_2\, e^{j\omega t}]$$

Thus the phasor A_1 represents the sinusoid $x_1(t)$ and the phasor A_2 represents $x_2(t)$.

Let a_1 and a_2 be *any two real numbers*; then the sinusoid $a_1\, x_1(t) + a_2\, x_2(t)$ is represented by the phasor $a_1A_1 + a_2A_2$.

Proof We verify the assertion by computation:

$$a_1\, x_1(t) + a_2\, x_2(t) = a_1\, \mathrm{Re}[A_1\, e^{j\omega t}] + a_2\, \mathrm{Re}[A_2\, e^{j\omega t}] \tag{1.10}$$

Now a_1 and a_2 are *real* numbers, hence for any complex numbers z_1 and z_2,

$$a_i\, \mathrm{Re}[z_i] = \mathrm{Re}[a_i z_i] \qquad i = 1, 2 \tag{1.11}$$

and

$$a_1\, \mathrm{Re}[z_1] + a_2\, \mathrm{Re}[z_2] = \mathrm{Re}[a_1 z_1 + a_2 z_2]$$

Now applying this fact to Eq. (1.10) we have

$$a_1\, \mathrm{Re}[A_1\, e^{j\omega t}] + a_2\, \mathrm{Re}[A_2\, e^{j\omega t}] = \mathrm{Re}[(a_1A_1 + a_2A_2)\, e^{j\omega t}] \tag{1.12}$$

Thus (1.10) and (1.12) yield

$$a_1x_1(t) + a_2x_2(t) = \mathrm{Re}[(a_1A_1 + a_2A_2)\, e^{j\omega t}] \tag{1.13}$$

that is, the sinusoid $a_1x_1(t) + a_2x_2(t)$ is represented by the phasor $a_1A_1 + a_2A_2$.

The proof is easily extended to a linear combination (with *real* coefficients) of n sinusoids:

$$\sum_{k=1}^{n} a_k x_k(t) \qquad \text{is represented by the phasor} \qquad \sum_{k=1}^{n} a_k A_k$$

Exercise Show by example that if a_i is *not* a real number but z_i is a complex number, then Eq. (1.11) is false.

COMMENT ON NOTATION Since a phasor, say A, is a complex number, we may represent it in rectangular coordinates and in polar coordinates; hence the notations

$$A = \text{Re}(A) + j\,\text{Im}(A) = A_m \exp(i\sphericalangle A) = |A| \exp(j\sphericalangle A)$$

The magnitude of A is denoted by A_m or by $|A|$, as in calculus. Note that we always have $A_m = |A| \geq 0$.

Lemma 3 (Differentiation rule) A is the phasor of a given sinusoid $A_m \cos(\omega t + \sphericalangle A)$ if and only if $j\omega A$ is the phasor of its derivative, $\frac{d}{dt}[A_m \cos(\omega t + \sphericalangle A)]$.

Symbolically,

$$\text{Re}[j\omega A\, e^{j\omega t}] = \frac{d}{dt}[\text{Re}(A\, e^{j\omega t})] \tag{1.14}$$

COMMENT It is convenient to think of Eq. (1.14) as stating that the linear operators Re and $\frac{d}{dt}$ commute:

$$\text{Re}\left[\frac{d}{dt}(A\, e^{j\omega t})\right] = \text{Re}[j\omega A\, e^{j\omega t}] = \frac{d}{dt}[\text{Re}(A\, e^{j\omega t})]$$

PROOF: BY CALCULATION. We write $A = A_m \exp(j\sphericalangle A)$.

$$\frac{d}{dt}[\text{Re}(A \exp(j\omega t))]$$

$$= \frac{d}{dt}[\text{Re}(A_m \exp j(\omega t + \sphericalangle A))] \qquad \text{where } A_m \text{ is real}$$

$$= \frac{d}{dt}[A_m \cos(\omega t + \sphericalangle A)] \qquad \text{since} \qquad \exp jx = \cos x + j \sin x$$

$$= -A_m \omega \sin(\omega t + \sphericalangle A)$$

$$= \text{Re}[j\omega A_m \exp j(\omega t + \sphericalangle A)]$$

$$= \text{Re}[j\omega A \exp j\omega t]$$

Application: Sum of sinusoids at the same frequency ω Consider the sum

$$A_{1m} \cos(\omega t + \sphericalangle A_1) + A_{2m} \cos(\omega t + \sphericalangle A_2) + \frac{d}{dt}[A_{3m} \cos(\omega t + \sphericalangle A_3)] \tag{1.15}$$

Clearly, after performing the differentiation, this expression reduces to a single sinusoid of frequency ω: We could calculate it by using trigonometric formulas; however, this approach gets very complicated.

So let us use the phasor rules that we just learned. To be specific let

$$A_1 \triangleq A_{1m} \exp(j\sphericalangle A_1) = 12 \exp(j23°) = 11.046 + j4.689$$
$$A_2 \triangleq A_{2m} \exp(j\sphericalangle A_2) = 7 \exp(-j57°) = 3.812 - j5.871$$
$$A_3 \triangleq A_{3m} \exp(j\sphericalangle A_3) = 0.2 \exp(j\sphericalangle 71°) = 0.06511 + j0.18910$$

and let $\omega = 377$ rad/s.

Using the differentiation rule, the last term of the sum (1.15) is represented by the phasor

$$j\omega A_3 = 377.0 \times 0.2 \exp(j\sphericalangle 161°) = -71.292 + j24.548$$

Thus

$$A_1 + A_2 + j\omega A_3 = -56.434 + j23.366 = 61.080 \exp(j157°.51) \triangleq A_s$$

and the resulting sinusoid is

$$\text{Re}[A_s \, e^{j\omega t}] = 61.080 \cos(377t + 157°.51)$$

1.3 Example of Sinusoidal Steady-State Solution

The differential equation

$$\ddot{i}_L(t) + 2\alpha \dot{i}_L(t) + \omega_0^2 i_L(t) = \omega_0^2 i_s(t) \tag{1.16}$$

governs the behavior of the linear time-invariant parallel *RLC* circuit shown in Fig. 1.1. Note that since we assume $R > 0$, $L > 0$, and $C > 0$, $2\alpha \triangleq 1/RC$ and $\omega_0^2 \triangleq 1/LC$ are both *positive*. Let the source current be sinusoidal, namely,

$$i_s(t) = \text{Re}[I_s \, e^{j\omega t}] = I_{sm} \cos(\omega t + \sphericalangle I_s) \tag{1.17}$$

where I_s is a phasor with $I_s = I_{sm} \exp(j\sphericalangle I_s)$. We want to find the sinusoidal particular solution of Eq. (1.16) for $i_s(t)$ given by Eq. (1.17). We try the solution $\text{Re}(I_L \, e^{j\omega t})$ where the complex number I_L is the yet-undetermined phasor which specifies this particular sinusoidal solution. Substituting into Eq. (1.16), we obtain: For all t,

$$\frac{d^2}{dt^2}[\text{Re}(I_L \, e^{j\omega t})] + 2\alpha \frac{d}{dt}[\text{Re}(I_L \, e^{j\omega t})] + \omega_0^2 \, \text{Re}(I_L \, e^{j\omega t}) = \omega_0^2 \, \text{Re}(I_s \, e^{j\omega t}) \tag{1.18}$$

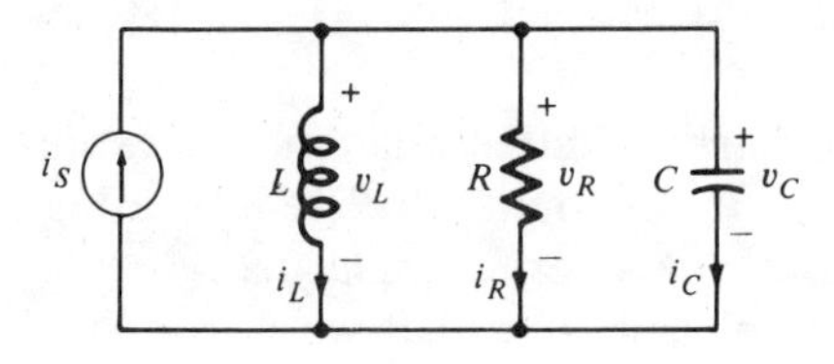

Figure 1.1 Linear *RLC* circuit studied in the example.

1. Using (three times) the *differentiation rule* we get

$$\mathrm{Re}[(j\omega)^2 I_L \,e^{j\omega t}] + 2\alpha\, \mathrm{Re}[(j\omega) I_L \,e^{j\omega t}] + \omega_0^2\, \mathrm{Re}(I_L \,e^{j\omega t}) = \omega_0^2\, \mathrm{Re}(I_s \,e^{j\omega t})$$

2. Using the *linearity lemma* we obtain (since α and ω_0^2 are real),

$$\mathrm{Re}\{[(j\omega)^2 + 2\alpha(j\omega) + \omega_0^2] I_L \,e^{j\omega t}\} = \omega_0^2\, \mathrm{Re}(I_s \,e^{j\omega t})$$

3. Using uniqueness, we obtain an algebraic equation for I_L

$$[(j\omega)^2 + 2\alpha(j\omega) + \omega_0^2] I_L = \omega_0^2 I_s$$

Hence

$$I_L = \frac{\omega_0^2 I_s}{(\omega_0^2 - \omega^2) + 2\alpha j\omega} \triangleq I_{Lm} \exp(j\theta_L + j\measuredangle I_s) \tag{1.19}$$

with

$$I_{Lm} = \frac{\omega_0^2}{\sqrt{(\omega_0^2 - \omega^2)^2 + (2\alpha\omega)^2}}\, I_{sm} \qquad \theta_L = \measuredangle\left[\frac{\omega_0^2}{(\omega_0^2 - \omega^2) + 2\alpha j\omega}\right]$$

$$= -\tan^{-1} \frac{2\alpha\omega}{\omega_0^2 - \omega^2}$$

The sinusoidal solution is then

$$i_{Lp}(t) = \frac{\omega_0^2 I_{sm}}{\sqrt{(\omega_0^2 - \omega^2)^2 + (2\alpha\omega)^2}} \cos(\omega t + \measuredangle I_s + \theta_L) \tag{1.20}$$

where the subscript p reminds us that i_{Lp} is the sinusoidal *particular* solution.

The *physical meaning* of this particular solution is the following: Since R, L, and C are positive constants, it follows that $\alpha > 0$ and $\omega_0^2 > 0$. Consequently the two natural frequencies s_1 and s_2 of the circuit, i.e., the zeros of its characteristic polynomial $\chi(s) = s^2 + 2\alpha s + \omega_0^2$, have *negative* real parts. Therefore, *any* solution of Eq. (1.16) starting, at *any* t_0, from *any* initial condition, has the form

$$i_L(t) = k_1 \exp s_1(t - t_0) + k_2 \exp s_2(t - t_0) + i_{Lp}(t) \tag{1.21}$$

where the constants k_1 and k_2 depend on the initial conditions and $i_{Lp}(t)$ is the particular solution defined in Eq. (1.20). [In case $s_1 = s_2$, Eq. (1.21) must be modified appropriately.] In any case, since $\mathrm{Re}(s_1) < 0$ and $\mathrm{Re}(s_2) < 0$, as $t \to x$,

$$i_L(t) \longrightarrow i_{Lp}(t)$$

This particular solution is called *the sinusoidal steady-state solution of the circuit*. The difference between the complete solution $i_L(t)$, given by Eq. (1.21), and the particular solution $i_{Lp}(t)$, given by Eq. (1.20), is called the *transient* response.

REMARK In Chap. 10, Sec. 5 we shall prove a generalization of this property: For any *linear time-invariant* circuit, if all its natural frequencies have *negative* real parts, then for *any* initial conditions and for *any* set of independent sources, each one sinusoidal at the *same* frequency ω, *all currents* and *all voltages* will tend exponentially as $t \to \infty$ to sinusoidal waveforms at frequency ω. When that situation occurs the circuit is said to be in the *sinusoidal steady state*. Note that the sinusoidal steady state does *not* depend on the initial conditions.

Exercise Consider a linear time-invariant *RLC* circuit as in Fig. 1.1; let its normalized element values[1] be $C = 1$, $L = (25.01)^{-1}$, $R = 5$, and $I_s = 3\exp(j\pi/6)$.
(*a*) Calculate the *sinusoidal steady-state solution* for $\omega = 1$, $\omega = 5$, and $\omega = 24$. (Express it each time as a real-valued function of time.)
(*b*) For $\omega = 5$ rad/s calculate the *complete* solution for $i_L(0) = 1$, $v_C(0) = 0$.
(*c*) For $\omega = 5$ rad/s obtain initial conditions at $t = 0$ so that in Eq. (1.21), with $t_0 = 0$, $k_1 = k_2 = 0$ (i.e., for that solution there is no transient).

2 PHASOR FORMULATION OF CIRCUIT EQUATIONS

Consider the connected *linear time-invariant* circuit $\mathcal{N}$ driven by some sinusoidal sources all at frequency ω. We say that the circuit $\mathcal{N}$ is *in the sinusoidal steady state at frequency* ω iff *all* branch voltages, *all* branch currents, and *all* node voltages are *sinusoids at the same frequency* ω.

If $\mathcal{N}$ is in the sinusoidal steady state at frequency ω, then all currents and voltages can be described by phasors. We shall prove that an analysis of circuit $\mathcal{N}$ in the sinusoidal steady state reduces to solving linear *algebraic* equations with *complex* coefficients.

2.1 Kirchhoff's Laws

A. Kirchhoff's current law: KCL Let the linear time-invariant circuit $\mathcal{N}$ be in the sinusoidal steady state at frequency ω. (See Fig. 2.1.) Consider node ①: KCL applied to node ① reads:

$$i_1(t) + i_2(t) - i_3(t) = 0 \qquad \text{for all } t \tag{2.1}$$

For $k = 1, 2, 3, \ldots, b$, let I_k be the phasor representing the sinusoid $i_k(t)$. Thus (2.1) gives: For all t,

$$\mathrm{Re}(I_1 e^{j\omega t}) + \mathrm{Re}(I_2 e^{j\omega t}) - \mathrm{Re}(I_3 e^{j\omega t}) = 0$$

[1]See Sec. 3.5 below.

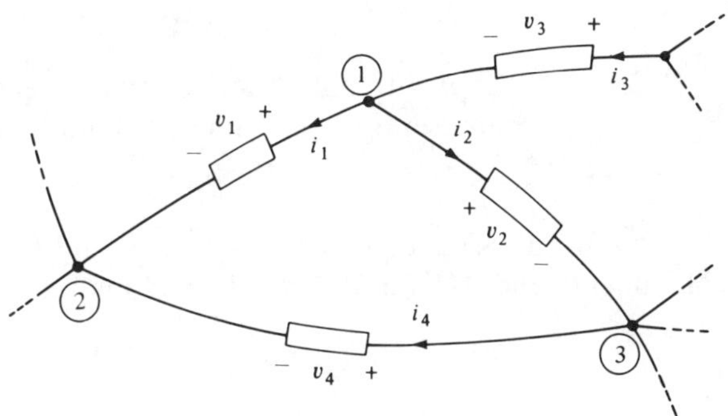

Figure 2.1 The circuit $\mathcal{N}$.

Hence using *linearity* and *uniqueness* we obtain

$$I_1 + I_2 - I_3 = 0 \tag{2.2}$$

Since the reasoning is quite general, we can state the following conclusion.

Conclusion In the sinusoidal steady state, any node equation can be written down directly in terms of phasors.

More generally, for any connected circuit $\mathcal{N}$, KCL reads

$$\mathbf{A}\,\mathbf{i}(t) = \mathbf{0} \qquad \text{for all } t \tag{2.3}$$

where $\mathbf{A}$ is the $(n-1) \times b$ reduced incidence matrix of the circuit $\mathcal{N}$. Since the elements of $\mathbf{A}$ are real, we may repeat the reasoning above and show that we may write KCL directly in terms of phasors:

$$\mathbf{AI} = \mathbf{0} \tag{2.4}$$

where $\mathbf{I}$ is the b-dimensional column vector whose elements are the branch current phasors $I_1, I_2, \ldots, I_b$ representing the sinusoidal branch currents $i_1(t), i_2(t), \ldots, i_b(t)$.

> REMARK In the KCL equation $\mathbf{AI} = \mathbf{0}$, $\mathbf{A}$ is an $(n-1) \times b$ matrix of *real* numbers and $\mathbf{I}$ is a b-vector with *complex* components.

B. Kirchhoff's voltage law: KVL Consider now KVL applied to the closed node sequence ①–②–③–① of Fig. 2.1:

$$v_1(t) - v_2(t) - v_4(t) = 0 \qquad \text{for all } t \tag{2.5}$$

Since $v_k(t) = \text{Re}[V_k\, e^{j\omega t}]$, using linearity and uniqueness in Eq. (2.5) we obtain

$$V_1 - V_2 - V_4 = 0 \tag{2.6}$$

In general, the KVL which reads

$$\mathbf{v}(t) = \mathbf{A}^T\,\mathbf{e}(t) \qquad \text{for all } t$$

becomes in terms of phasors

$$\mathbf{V} = \mathbf{A}^T\mathbf{E} \tag{2.7}$$

where $\mathbf{E}$ denotes the $(n-1)$-dimensional column vector made of the phasors $E_1, E_2, \ldots, E_{n-1}$ representing the sinusoidal node-to-datum voltages $e_1(t)$, $e_2(t), \ldots, e_{n-1}(t)$.

REMARK In the KVL equation $\mathbf{V} = \mathbf{A}^T\mathbf{E}$, $\mathbf{V}$ and $\mathbf{E}$ are vectors with *complex* components and $\mathbf{A}^T$ is a $b \times (n-1)$ *real* matrix.

2.2 Branch Equations

Straightforward application of the three lemmas to the time domain from the branch equations (also called constitutive relations) yields the following table:

Resistor:	$v(t) = R\, i(t)$	$V = RI$	(2.8)
Inductor:	$v(t) = L\,\dfrac{di}{dt}$	$V = j\omega LI$	(2.9)
Capacitor:	$i(t) = C\,\dfrac{dv}{dt}$	$I = j\omega CV$	(2.10)
VCVS:	$v_3(t) = \mu v_1(t)$	$V_3 = \mu V_1$	(2.11*a*)
VCCS:	$i_4(t) = g_m\, v_5(t)$	$I_4 = g_m V_5$	(2.11*b*)
CCVS:	$v_6(t) = r_m\, i_5(t)$	$V_6 = r_m I_5$	(2.11*c*)
CCCS:	$i_8(t) = \alpha\, i_7(t)$	$I_8 = \alpha I_7$	(2.11*d*)
Gyrator:	$i_9(t) = G\, v_{10}(t)$ $i_{10}(t) = -G\, v_9(t)$	$I_9 = GV_{10}$ $I_{10} = -GV_9$	(2.11*e*)
Ideal transformer	$v_1(t) = \dfrac{1}{n}\, v_2(t)$ $i_1(t) = -n\, i_2(t)$	$V_1 = \dfrac{1}{n} V_2$ $I_1 = -nI_2$	(2.12)

The expressions R, $j\omega L$, and $1/j\omega C$ are the *impedances* at frequency ω of the circuit elements R, L, and C, respectively; $1/R$, $1/j\omega L$, and $j\omega C$ are the corresponding *admittances*; μ is a *voltage gain*; α is a *current gain*; g_m is a *transfer conductance*; and r_m is a *transfer resistance*.

The crucial point is that, in terms of phasors, the branch equations become *algebraic* equations with *complex* coefficients: In Eqs. (2.8) to (2.12), the V's and the I's representing phasors are complex numbers; the factors $j\omega L$ and $j\omega C$ are complex, in fact purely imaginary.

REMARKS ON PHASOR DIAGRAMS

1. To the sinusoid $x(t) = A_m \cos(\omega t + \measuredangle A)$ is associated the phasor $A \triangleq A_m \exp(j\measuredangle A)$. We may plot the complex number A in the complex plane (see Fig. 2.2) as a vector from the origin to the point $A = A_m \exp(j\measuredangle A)$. To visualize the sinusoid $x(t)$: (*a*) we imagine the vector rotating counterclockwise at the angular velocity of ω rad/s, namely, we consider $A\,e^{j\omega t}$ as t increases; (*b*) we project orthogonally on the x axis the tip of the vector to obtain $x(t)$.

2. The branch equations (2.8) to (2.12) may be graphically represented by plotting the vectors representing the V's and the I's. For example, we may imagine these pairs of vectors (V, I) rotating at ω rad/s and, by projection on the x axis, giving the sinusoids $v(t)$ and $i(t)$.

Since, as in Fig. 2.2, it is common to visualize the phasors as rotating *counterclockwise*, referring to Fig. 2.3, we say "the inductor current phasor I_L *lags* the inductor voltage phasor V_L by 90°" and "the capacitor current phasor I_C *leads* the capacitor voltage phasor V_C by 90°."

2.3 The Concept of Impedance and Admittance

Consider the circuit shown in Fig. 2.4*a*: The one-port N is formed by an arbitrary interconnection of *linear time-invariant* elements; the input is a sinusoidal current source at frequency ω given by

$$i_s(t) = \text{Re}[I_s\,e^{j\omega t}] = |I_s| \cos(\omega t + \measuredangle I_s) \tag{2.13}$$

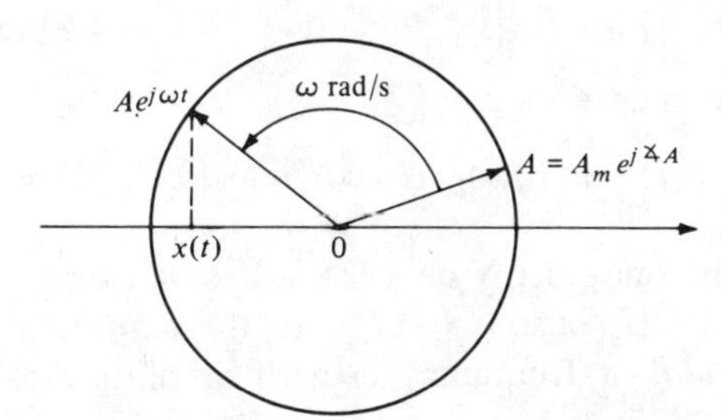

Figure 2.2 The sinusoid $x(t) = A_m \cos(\omega t + \measuredangle A)$ may be viewed as generated by the projection of the tip of the "rotating phasor" $A\,e^{j\omega t}$.

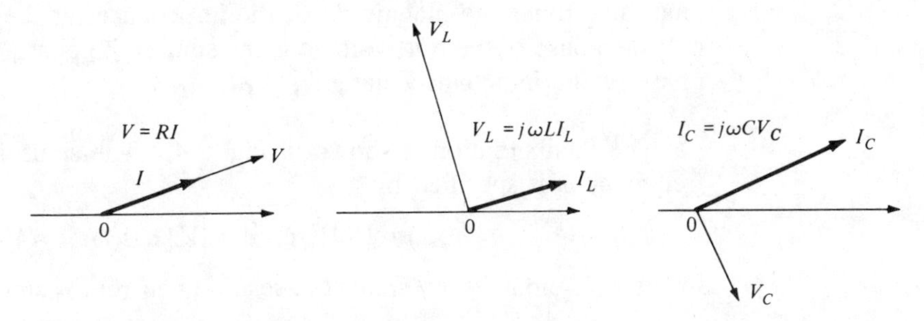

Figure 2.3 Graphical representation of phasor relations.

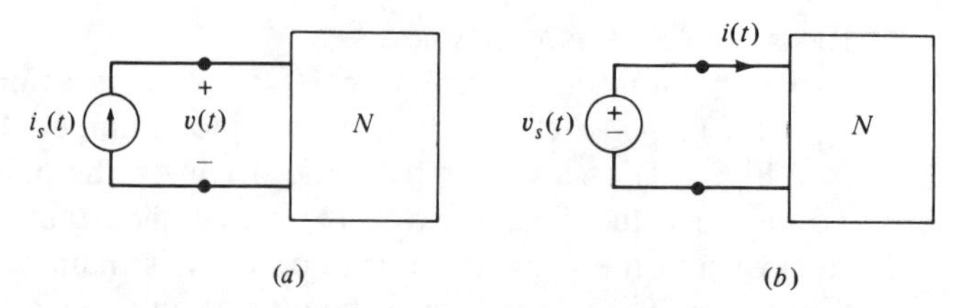

Figure 2.4 (*a*) The sinusoidal current source $i_s(t) = \text{Re}(I_s\, e^{j\omega t})$ is driving the one-port N. (*b*) The sinusoidal voltage source $v_s(t) = \text{Re}(V_s\, e^{j\omega t})$ is driving the one-port N.

Assume that N is in the sinusoidal steady state; then the port voltage is a sinusoid at frequency ω represented by a phasor V; thus

$$v(t) = \text{Re}[V\, e^{j\omega t}] = |V| \cos(\omega t + \measuredangle V) \tag{2.14}$$

We define the *driving-point impedance* of the one-port N at the frequency ω to be the ratio of the *port-voltage phasor* V and the *input-current phasor* I_s, that is,

$$Z(j\omega) \triangleq \frac{V}{I_s}$$

or

$$V = Z(j\omega) I_s \tag{2.15}$$

We explicitly indicate the dependence of Z on the frequency of the source. From (2.15), we deduce

$$|V| = |Z(j\omega)| \cdot |I_s| \quad \text{and} \quad \measuredangle V = \measuredangle Z(j\omega) + \measuredangle I_s \tag{2.16}$$

Hence (2.14) becomes

$$v(t) = |Z(j\omega)| \cdot |I_s| \cos[\omega t + \measuredangle Z(j\omega) + \measuredangle I_s] \tag{2.17}$$

Conclusion Let the one-port N have $Z(j\omega)$ as driving-point impedance, and let its input current be $|I_s| \cos(\omega t + \measuredangle I_s)$. In the sinusoidal steady state, its port voltage is *sinusoidal* at frequency ω with amplitude $|Z(j\omega)| \cdot |I_s|$ and phase $\measuredangle Z(j\omega) + \measuredangle I_s$.

Thus the amplitude of the port voltage is the product of the current amplitude times the magnitude of the impedance (at the appropriate frequency); the phase of the port voltage is the sum of the phase of the current and the phase of the impedance, namely, $\measuredangle Z(j\omega)$.

The dual situation is shown in Fig. 2.4b: The sinusoidal voltage source at frequency ω is specified by

$$v_s(t) = \text{Re}[V_s\, e^{j\omega t}] = |V_s| \cos(\omega t + \measuredangle V_s)$$

In the sinusoidal steady state, the resulting current is sinusoidal and written as

$$i(t) = \text{Re}[I\, e^{j\omega t}] = |I| \cos(\omega t + \measuredangle I)$$

We define the *driving-point admittance* of the one-port N at the frequency ω to be the ratio of the *port-current phasor* I and the *port-voltage phasor* V, that is,

$$Y(j\omega) = \frac{I}{V_s} \qquad \text{or equivalently} \qquad I = Y(j\omega)\,V_s \tag{2.18}$$

Following the same reasoning as the one used to derive Eq. (2.17) from Eqs. (2.14) and (2.15), we easily obtain

$$i(t) = |Y(j\omega)| \cdot |V_s| \cos[\omega t + \measuredangle Y(j\omega) + \measuredangle V_s] \tag{2.19}$$

As expected Eq. (2.19) is precisely the dual of Eq. (2.17).

REMARK It is intuitively obvious that since Z is the ratio of the port-voltage phasor V to the port-current phasor I, $Z = V/I$, and since $Y \triangleq I/V$, we have, for all frequencies,

$$Z(j\omega) = \frac{1}{Y(j\omega)} \tag{2.20}$$

This useful relation can be derived rigorously by using the substitution theorem.

Application Consider the series connection of two circuit elements or of two 1-ports. Let them be specified by their impedances $Z_1(j\omega)$ and $Z_2(j\omega)$. Suppose the circuit shown in Fig. 2.5*a* is in the sinusoidal steady state. Let us calculate the driving-point impedance Z of the one-port. By inspection of Fig. 2.5*a*, we see that KVL gives $V = V_1 + V_2$ and KCL gives $I = I_1 = I_2$. Hence

$$Z = \frac{V}{I} = \frac{V_1}{I} + \frac{V_2}{I} = Z_1 + Z_2$$

Thus

Impedances in series

$$Z = Z_1 + Z_2 \tag{2.21}$$

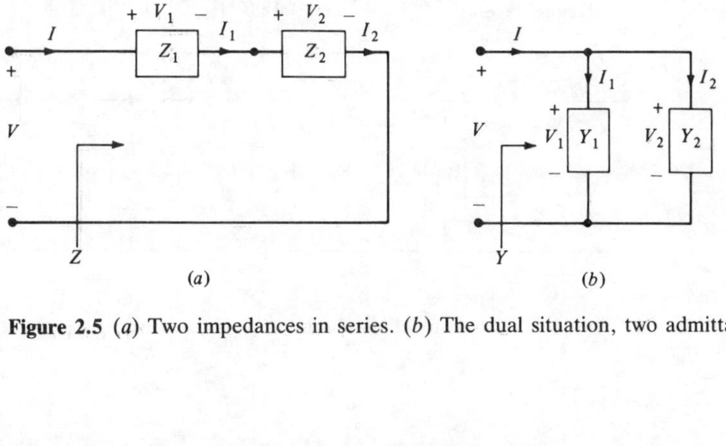

Figure 2.5 (*a*) Two impedances in series. (*b*) The dual situation, two admittances in parallel.

For the dual situation shown in Fig. 2.5*b*, KVL gives $V = V_1 = V_2$ and KCL gives $I = I_1 + I_2$. Hence the driving-point admittance $Y(j\omega)$ of the resulting one-port is given by

$$Y = \frac{I}{V} = \frac{I_1 + I_2}{V} = Y_1 + Y_2$$

Thus

Admittances in parallel

$$Y = Y_1 + Y_2 \tag{2.22}$$

Since we know how to calculate the *impedance* of *series* connections and the *admittance* of *parallel* connections, it is easy to calculate the driving-point impedance or admittance of series-parallel circuits.

Exercise Calculate the driving-point impedance and admittance of the one-ports of Fig. 2.6

2.4 Tableau Equations

Let $\mathcal{N}$ be a *connected linear time-invariant* circuit (with b branches and n nodes) driven by independent sources whose waveforms are sinusoidal at the *same* frequency ω rad/s. We assume that $\mathcal{N}$ is in the sinusoidal steady state: Hence the $n-1$ node voltages are represented by the vector $\mathbf{E} = (E_1, \ldots, E_{n-1})^T$ where E_k is the *phasor* representing the kth-node voltage; similarly $\mathbf{V} = (V_1, V_2, \ldots, V_b)^T$ and $\mathbf{I} = (I_1, I_2, \ldots, I_b)^T$ represent, respectively, the branch-voltage *phasors* and branch-current *phasors*.

The tableau equations consist of KCL, KVL, and the branch equations. We have already obtained them: See Eqs. (2.4), (2.7), and (2.8) to (2.12). Hence we obtain a system of $(2b + n - 1)$ *linear* algebraic equations with *complex* coefficients:

Tableau equations for the sinusoidal steady state at frequency ω

$$\begin{bmatrix} \mathbf{0} & \mathbf{0} & \mathbf{A} \\ -\mathbf{A}^T & \mathbf{I} & \mathbf{0} \\ \mathbf{0} & \mathbf{M}(j\omega) & \mathbf{N}(j\omega) \end{bmatrix} \begin{bmatrix} \mathbf{E} \\ \mathbf{V} \\ \mathbf{I} \end{bmatrix} = \begin{bmatrix} \mathbf{0} \\ \mathbf{0} \\ \mathbf{U}_s \end{bmatrix} \tag{2.23}$$

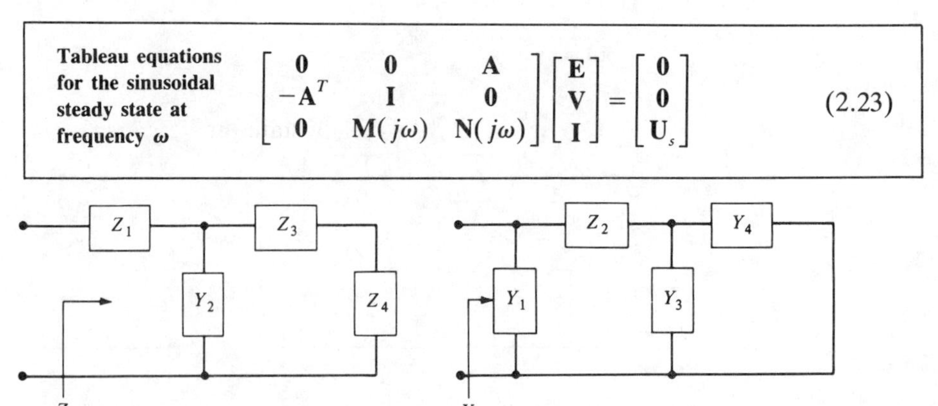

Figure 2.6 Series-parallel circuits used in the exercise.

The b-dimensional vector $\mathbf{U}_s$ lists the phasors representing the sinusoidal waveforms of the *independent* sources.

Note that the branch equations, namely $\mathbf{MV} + \mathbf{NI} = \mathbf{U}_s$, can always be written so that $\mathbf{M}(j\omega)$ and/or $\mathbf{N}(j\omega)$ are matrices of *polynomials* in $j\omega$ (i.e., we may avoid the presence of terms like $1/j\omega$).

Example Consider the linear time-invariant circuit $\mathcal{N}$ shown in Fig. 2.7. Since Kirchhoff's laws are straightforward, let us concentrate on the branch equations so that we can see the structure of $\mathbf{M}(j\omega)$, $\mathbf{N}(j\omega)$, and $\mathbf{U}_s$. Writing successively the branch equations of the elements we obtain

$$\underbrace{\begin{bmatrix} 0 & 0 & 0 & 0 & 0 & 0 \\ 0 & -G & 0 & 0 & 0 & 0 \\ 0 & 0 & 1 & 0 & 0 & 0 \\ 0 & 0 & 0 & -j\omega C & 0 & 0 \\ 0 & 0 & 0 & 0 & 0 & 0 \\ 0 & -\alpha & 0 & 0 & 0 & 1 \end{bmatrix}}_{\mathbf{M}(j\omega)} \begin{bmatrix} V_1 \\ V_2 \\ V_3 \\ V_4 \\ V_5 \\ V_6 \end{bmatrix} + \underbrace{\begin{bmatrix} -1 & 0 & 0 & 0 & 0 & 0 \\ 0 & 1 & 0 & 0 & 0 & 0 \\ 0 & 0 & -j\omega L & 0 & 0 & 0 \\ 0 & 0 & 0 & 1 & 0 & 0 \\ 0 & 0 & -\beta & 0 & 1 & 0 \\ 0 & 0 & 0 & 0 & 0 & 0 \end{bmatrix}}_{\mathbf{N}(j\omega)} \begin{bmatrix} I_1 \\ I_2 \\ I_3 \\ I_4 \\ I_5 \\ I_6 \end{bmatrix} = \begin{bmatrix} I_s \\ 0 \\ 0 \\ 0 \\ 0 \\ 0 \end{bmatrix} \tag{2.24}$$

REMARKS

1. Every two-terminal element contributes only to the diagonal of $\mathbf{M}$ and/or $\mathbf{N}$; controlled sources, representing coupling between different branches, contribute off-diagonal elements. Coupled inductors will also contribute off-diagonal elements: This will be caused by the inductive coupling between two-terminal inductors.
2. For any linear time-invariant circuit made of elements listed in the table of Sec. 2.2, the tableau equations may always be written so that thc tableau matrix $\mathbf{T}(j\omega)$ is a *polynomial* matrix with *real* coefficients, namely,

$$\mathbf{T}(j\omega) = \mathbf{T}_0 + \mathbf{T}_1 j\omega \tag{2.25}$$

where $\mathbf{T}_0$ and $\mathbf{T}_1$ are $(2b + n - 1) \times (2b + n - 1)$ matrices with *real coefficients*.
3. The sinusoidal steady state (at frequency ω) of circuit $\mathcal{N}$ is well-defined and unique if and only if $\det[\mathbf{T}(j\omega)] \neq 0$. This follows immediately from Cramer's rule.

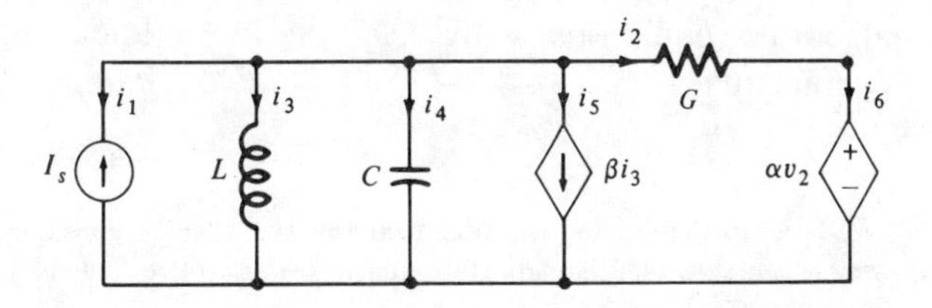

Figure 2.7 Example used to illustrate the tableau analysis.

Notation In the sinusoidal steady state we will write the tableau equations as follows:

$$\mathbf{T}(j\omega)\mathbf{W}(j\omega) = \mathbf{B} \tag{2.26}$$

where **B** is the constant vector whose nonzero elements are the phasors representing the *independent* sources; $\mathbf{W}(j\omega)$ is the $(2b + n - 1)$-column vector which represents $(\mathbf{E}, \mathbf{V}, \mathbf{I})^T$.

2.5 Sinusoidal Steady-State Analysis

A. Analogy with resistive circuit analysis Let $\mathcal{N}_R$ be a linear time-invariant *resistive* circuit with a connected graph having n nodes and b branches. Suppose that we, first, replace a number of resistors of $\mathcal{N}_R$ by inductors or capacitors, and second, drive the resulting circuit by sinusoidal sources *all operating at the same frequency* ω. Assume that the resulting circuit is in the sinusoidal steady state and call the circuit $\mathcal{N}_\omega$.[2]

Let us compare the tableau equations for $\mathcal{N}_R$ and for $\mathcal{N}_\omega$

Linear time-invariant resistive circuit $\mathcal{N}_R$ [see Eq. (2.4) of Chap. 5]

$$\begin{bmatrix} \mathbf{0} & \mathbf{0} & \mathbf{A} \\ -\mathbf{A}^T & \mathbf{I} & \mathbf{0} \\ \mathbf{0} & \mathbf{M} & \mathbf{N} \end{bmatrix} \begin{bmatrix} \mathbf{e}(t) \\ \mathbf{v}(t) \\ \mathbf{i}(t) \end{bmatrix} = \begin{bmatrix} \mathbf{0} \\ \mathbf{0} \\ \mathbf{u}_s(t) \end{bmatrix} \tag{2.27a}$$

Notes:

(1) $\mathbf{e}(\cdot)$, $\mathbf{v}(\cdot)$, $\mathbf{i}(\cdot)$, and $\mathbf{u}_s(\cdot)$ are vector-valued functions of time.

(2) The tableau matrix **T** has *real* elements.

(3) $\mathcal{N}_R$ is completely described by Eq. (2.27*a*), i.e., a set of linear algebraic equations with *real* coefficients.

Linear time-invariant circuit $\mathcal{N}_\omega$ *operating in the sinusoidal steady state* [see Eq. (2.23)]

$$\begin{bmatrix} \mathbf{0} & \mathbf{0} & \mathbf{A} \\ -\mathbf{A}^T & \mathbf{I} & \mathbf{0} \\ \mathbf{0} & \mathbf{M}(j\omega) & \mathbf{N}(j\omega) \end{bmatrix} \begin{bmatrix} \mathbf{E} \\ \mathbf{V} \\ \mathbf{I} \end{bmatrix} = \begin{bmatrix} \mathbf{0} \\ \mathbf{0} \\ \mathbf{U}_s \end{bmatrix} \tag{2.27b}$$

Notes:

(1) **E**, **V**, **I**, and $\mathbf{U}_s$ are vectors whose components are *phasors*; each one representing a sinusoidal waveform, e.g., $e_k(t) = |E_k| \cos(\omega t + \measuredangle E_k)$.

(2) The tableau matrix $\mathbf{T}(j\omega)$ has *complex* elements in its bottom b rows.

(3) $\mathcal{N}_\omega$ is completely described by Eq. (2.27*b*), i.e., a set of linear algebraic equations with *complex* coefficients.

[2] We label this circuit $\mathcal{N}_\omega$ in order to emphasize that we consider its sinusoidal steady state *at frequency* ω: In sinusoidal steady-state analysis, the tableau matrix $\mathbf{T}(j\omega)$ is a function of ω.

Conclusions

1. The substitution theorem holds for $\mathcal{N}_\omega$: Indeed the analysis of Chap. 5, Sec. 5.1 holds almost word for word.
2. The superposition theorem holds for $\mathcal{N}_\omega$: Provided $\det[\mathbf{T}(j\omega)] \neq 0$, the sinusoidal steady state (at frequency ω) due to several independent sources (at frequency ω) is equal to the sum of the sinusoidal steady states due to each independent source acting alone. (See Chap. 5, Sec. 4.1.)
3. Thévenin-Norton equivalent: For example, if the driving-point characteristic of $\mathcal{N}_\omega$ at terminals ①, ①′ is current-controlled, then the resulting one-port may be replaced by the Thévenin circuit shown in Fig. 2.8 where V_{oc} is the phasor representing the open-circuit voltage at ①, ①′ and Z_{eq} is the impedance of $\mathcal{N}_{\omega 0}$ seen at ①, ①′. (See Chap. 5, Sec. 4.2.)

B. Example of node analysis in the sinusoidal steady state From the discussion above, it is clear that the sinusoidal steady-state analysis is almost identical with the analysis of linear time-invariant *resistive* circuits: In sinusoidal steady-state analysis, the variables are *phasors*, the circuit elements are represented by their impedances, admittances, etc., and, as in the resistive case, Cramer's rule plays a major role. Consequently any analysis method for *linear resistive* circuits extends automatically to sinusoidal steady-state analysis.

For example, let us perform a *node analysis* on the op-amp circuit shown in Fig. 2.9. We assume that it is in the sinusoidal steady state: Let V_i, V_0, E_1, and E_2 denote the phasors representing the corresponding sinusoidal voltages. We use the ideal op-amp model ($A = \infty$); hence $V_d = 0$, and consequently (see Fig 2.9) $E_2 = V_0$.

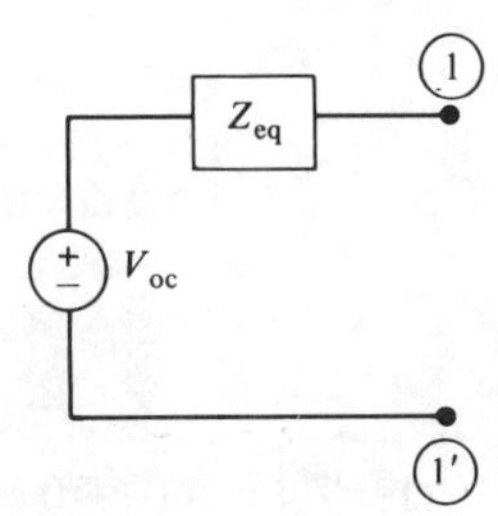

Figure 2.8 Thévenin equivalent circuit.

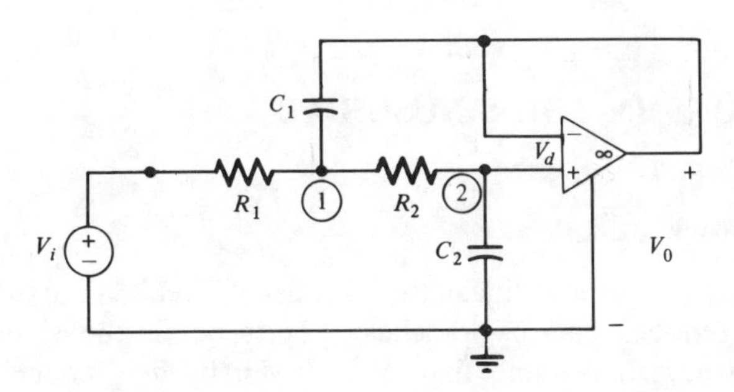

Figure 2.9 Op-amp circuit realizing a second-order filter.

Now write the node equations for nodes ① and ②:

$$
\begin{aligned}
&\text{①:} \qquad \frac{1}{R_1}(E_1 - V_i) + j\omega C_1(E_1 - V_0) + \frac{1}{R_2}(E_1 - V_0) = 0 \\
&\text{②:} \qquad \frac{1}{R_2}(V_0 - E_1) + j\omega C_2 V_0 = 0
\end{aligned}
\tag{2.28}
$$

Hence we obtain successively

$$E_1 = (1 + j\omega R_2 C_2)V_0$$

and

$$V_i = \left(1 + \frac{R_1}{R_2} + j\omega R_1 C_1\right)E_1 - \left(\frac{R_1}{R_2} + j\omega R_1 C_1\right)V_0$$

And finally

$$\frac{V_0(j\omega)}{V_i} = \frac{1}{1 + j\omega(R_1 + R_2)C_2 + (j\omega)^2 R_1 R_2 C_1 C_2} \tag{2.29}$$

This formula gives, for all ω, the ratio of the output-voltage phasor $V_0(j\omega)$ to the input-voltage phasor, V_i.

Example In Eq. (2.29) let $R_1 = R_2 = 1\,\Omega$ and $C_1 = C_2 = 1\,\text{F}$; then

$$\frac{V_0(j\omega)}{V_i} = \frac{1}{1 - \omega^2 + j2\omega}$$

Thus if $v_i(t) = 3\cos t = \text{Re}[3e^{jt}]$, then

$$v_0(t) = \text{Re}\left[\frac{1}{2j}\,3e^{jt}\right] = 1.5\sin t$$

If $v_i(t) = 3\cos 10t$, then

$$v_0(t) = \text{Re}\left[\frac{1}{1 - 100 + j20}\,3e^{j10t}\right] = 0.02970\cos(10t - 168.58°)$$

3 NETWORK FUNCTIONS IN THE SINUSOIDAL STEADY STATE

3.1 The Concept of Network Function

Consider a general linear time-invariant circuit $\mathcal{N}$; assume it is connected and has n nodes and b branches. Let it be in the sinusoidal steady state at frequency ω. Assume that $\mathcal{N}$ is driven by *one* independent source, say, the sinusoidal current source represented by the phasor I_s. Hence the right-hand side of the tableau equations, Eq. (2.23), has only one nonzero element, namely, I_s. Suppose we want to calculate the node voltage E_k and consider the dependence of the phasor E_k on ω (while the phasor I_s is kept constant throughout). Assuming that $\det[\mathbf{T}(j\omega)] \neq 0$ for all ω's of interest, we obtain from Cramer's rule

$$E_k(j\omega) = \frac{\text{cofactor of } \mathbf{T}(j\omega)}{\det[\mathbf{T}(j\omega)]} I_s \tag{3.1}$$

where the cofactor is the one associated with the kth column and with the row in which we find I_s. Equation (3.1) implies that $E_k(j\omega)/I_s$ is the ratio of two polynomials in $j\omega$ with *real* coefficients. Indeed, every element of $\mathbf{T}(j\omega)$ is a *real* constant or a *real* number times $j\omega$; hence $\det[\mathbf{T}(j\omega)]$ and any of its cofactors is a polynomial in $j\omega$ with *real* coefficients.

Furthermore, Eq. (3.1) shows that the ratio $E_k(j\omega)/I_s$ is a function of $j\omega$ which depends *only on the circuit* $\mathcal{N}$ and *not on* I_s: In the present case, it is called the *transfer impedance from* I_s *to* E_k.

More generally, we have for $k = 2, 3, \ldots, b$, *transfer impedances* from I_s to V_k (the kth-branch voltage) and *current gains* from I_s to I_k (the kth-branch current) (see Fig. 3.1). In case the current source is connected from the datum node, (node ①'), to node ① as shown in Fig. 3.1, then the ratio $E_1(j\omega)/I_s$ is called the *driving-point impedance of* $\mathcal{N}$ *seen at port* ①, ①'.

Dually, if we drive the circuit $\mathcal{N}$ with a voltage source E_s, we will have *voltage* transfer functions from E_s to E_k or to V_k, and *transfer admittances* from E_s to I_k. If I_1 is the current through the source E_s, its ratio I_1/E_s is called the *driving-point admittance of* $\mathcal{N}$.

Collectively, these functions are called *network functions*: At the frequency ω, the network function evaluated at ω gives the ratio of the output phasor to the input phasor.

We have proven an important result:

Theorem *If* $\det[\mathbf{T}(\mathbf{j}\omega)] \neq 0$, *then any network function of* $\mathcal{N}$ *is well defined and is the ratio of two polynomials in* $j\omega$ *with real coefficients.*

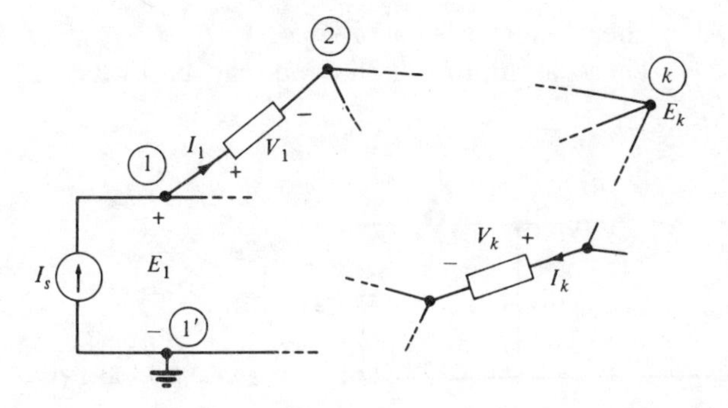

Figure 3.1 Circuit for illustrating the definition of network functions.

Another way of expressing this fact is to say: Any network function of $\mathcal{N}$ is a *real rational* function. Indeed a *rational function* in $j\omega$ is defined to be the ratio of polynomials in $j\omega$, and the adjective *real* means that all coefficients in the rational function are real numbers.

Examples

1. Refer to the example of node analysis at the end of the previous section (Chap. 9, Sec. 2), in particular, the op-amp circuit of Fig. 2.9. The calculations performed in that example gave the voltage transfer function from the source V_i to the output voltage V_0 [see Eq. (2.29)].
2. Consider the ladder circuit shown in Fig. 3.2. Determine the driving-point impedance V_i/I_1 and the voltage transfer function V_0/V_i.

We could write, say, node equations and upon solving them obtain the required network functions. However, since the circuit is a ladder, let us use the technique of Chap. 2, Sec. 2.3. That is, we assume we know V_0 and successively compute V_i and I_1. By inspection

$$V_3 = V_0 + R_2 j\omega C_1 V_0 = (1 + j\omega C_1 R_2)V_0$$

$$V_i = V_3 + R_4(j\omega C_3 V_3 + j\omega C_1 V_0)$$

Using the previous result, this reduces to

$$V_i = [1 + j\omega(R_2C_1 + R_4C_3 + R_4C_1) + (j\omega)^2 R_2C_1R_4C_3]V_0$$

Hence the voltage transfer function from V_1 to V_0 is

$$\frac{V_0}{V_i} = [1 + j\omega(R_2C_1 + R_4C_3 + R_4C_1) + (j\omega)^2 R_2C_1R_4C_3]^{-1} \tag{3.2}$$

To obtain I_1, note that

$$R_4 I_1 = V_i - V_3 = [j\omega R_4(C_3 + C_1) + (j\omega)^2 R_2C_1R_4C_3]V_0$$

Hence the driving-point impedance is given by

$$\frac{V_i}{I_1} = R_4 \frac{1 + j\omega(R_2C_1 + R_4C_3 + R_4C_1) + (j\omega)^2 R_2C_1R_4C_3}{j\omega[R_4(C_3 + C_1) + j\omega R_2C_1R_4C_3]} \tag{3.3}$$

As a check, note that as $\omega \to \infty$, $V_i/I_1 \to R_4$; as $\omega \to 0$, $V_i/I_1 \cong 1/j\omega(C_3 + C_1)$: These asymptotic behaviors can easily be checked by inspection of Fig. 3.2.

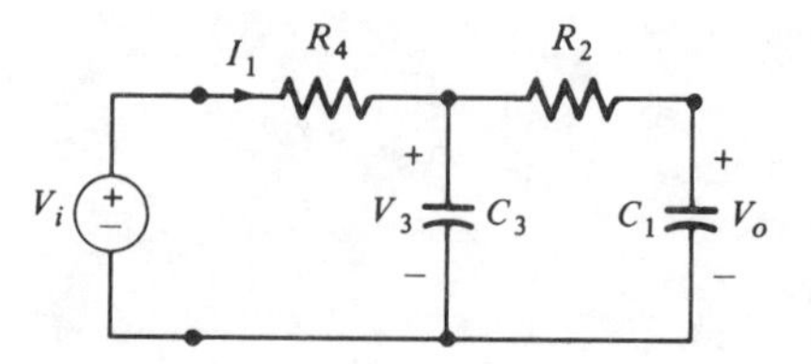

Figure 3.2 Ladder circuit for Example 2.

3. The purpose of this example is to show that some analyses turn out to be quite simple. (We assume ideal op amps throughout, and we assume that they all operate in their linear range.) The op-amp circuit shown in Fig. 3.3 is called the *Akerberg-Mossberg biquad*. This circuit has eight nodes. By the ideal op-amp assumption we have $E_3 = E_5 = E_6 = 0$. We note that $E_7 = V_i$ hence is known. Consequently, we need only write the equations of nodes ③, ⑤, and ⑥ in order to solve for E_1, E_2, and E_4 (do keep in mind that $E_3 = E_5 = E_6 = 0$):

$$
\begin{aligned}
&③: \quad -\left(\frac{1}{QR} + j\omega C\right)E_1 - \frac{1}{R}E_2 && = \frac{k}{R}V_i \\
&⑤: \quad -\frac{1}{R}E_1 \qquad\qquad -j\omega C E_4 && = 0 \\
&⑥: \quad -\frac{E_2}{r} - \frac{E_4}{r} && = 0
\end{aligned}
\tag{3.4}
$$

So, $E_4 = -E_2$ and we are left with two equations in two unknowns. Letting $\omega_0 \overset{\Delta}{=} 1/RC$, we obtain

$$\frac{E_2}{V_i} = -\frac{k}{1 + (1/Q)j(\omega/\omega_0) + j(\omega/\omega_0)^2} \tag{3.5}$$

and since $E_1 = -j\omega RCE_4 = j\omega RCE_2$

$$\frac{E_1}{V_i} = -\frac{kj(\omega/\omega_0)}{1 + (1/Q)j(\omega/\omega_0) + j(\omega/\omega_0)^2} \tag{3.6}$$

This network function has, except for the constant factor, the same dependence on ω as the parallel resonant circuit (see Eq. (4.15*a*), Sec. 4, below).

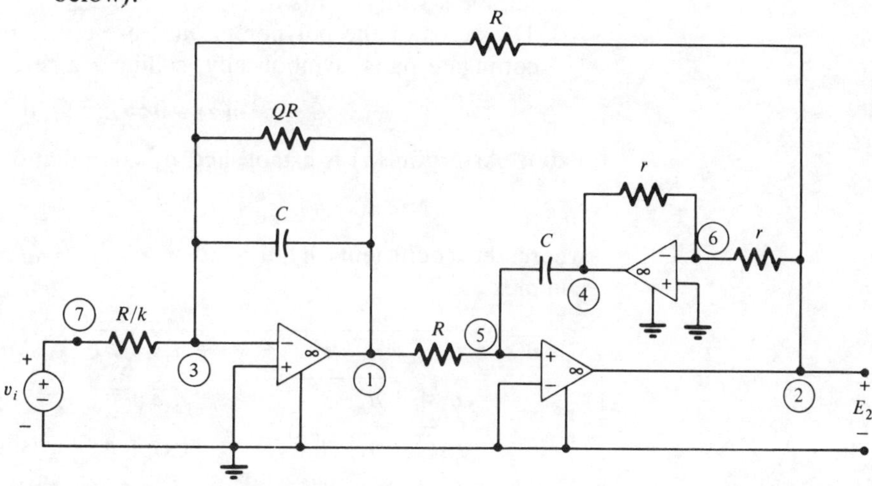

Figure 3.3 The Akerberg-Mossberg biquad.

If we were to write the node equations for nodes ①, ②, and ④, we would have to introduce in each node equation a new variable: the output current of the op amp. Thus these three node equations are merely recipes for computing the op-amp output currents.

REMARK ON STABILITY Examples 1 and 3 above are op-amp circuits, hence they are potentially unstable, i.e., they may have natural frequencies in the open right-half plane. In principle we should have first checked that *all* natural frequencies of these two circuits have negative real parts so that, as in Eq. (1.21) above, the transient goes to zero exponentially. A general treatment of this question is the subject of Chap. 10, Sec. 5.

3.2 Symmetry Property of Network Functions

To be specific, let $Z_k(j\omega) = E_k(j\omega)/I_s$ be the network function under consideration. Let us represent $Z_k(j\omega)$ in polar form

$$Z_k(j\omega) = |Z_k(j\omega)| \exp[j\sphericalangle Z_k(j\omega)] \tag{3.7}$$

We are going to show that its magnitude function $|Z_k(j\omega)|$ is an *even* function of ω and that its phase function $\sphericalangle Z_k(j\omega)$ *may always be chosen to be an odd function of* ω.

As a preliminary step let us establish an important property of polynomials with real coefficients. (Recall that by the theorem above, any network function is the ratio of two such polynomials.)

Lemma Let $n(s)$ be a polynomial with *real* coefficients in the complex variable $s = \sigma + j\omega$; then

(*a*) The complex conjugate of $n(s)$ is $n(\bar{s})$; symbolically, for all complex numbers s, $\overline{n(s)} = n(\bar{s})$.

(*b*) The zeros of the polynomial $n(s)$ are either real or occur in complex conjugate pairs; symbolically, calling z a zero of the polynomial $n(s)$,

$$n(z) = 0 \Leftrightarrow n(\bar{z}) = 0$$

PROOF Assertion (*a*) is established by computation. Let

$$n(s) = n_k s^k + n_{k-1}s^{k-1} + \cdots + n_1 s + n_0$$

where the coefficients $n_k, n_{k-1}, \ldots, n_0$ are *real* numbers. Then for any complex s,

$$\begin{aligned}
\overline{n(s)} &= \overline{n_k s^k + n_{k-1}s^{k-1} + \cdots + n_1 s + n_0} \\
&= \overline{n_k s^k} + \overline{n_{k-1}s^{k-1}} + \cdots + \overline{n_1 s} + \overline{n_0} \\
&= n_k \overline{s^k} + n_{k-1}\overline{s^{k-1}} + \cdots + n_1 \bar{s} + n_0 \qquad \text{(since the } n_i\text{'s are } \textit{real}) \\
&= n_k \bar{s}^k + n_{k-1}\bar{s}^{(k-1)} + \cdots + n_1 \bar{s} + n_0 \\
&= n(\bar{s})
\end{aligned}$$

Thus assertion (*a*) of the lemma is established.

To prove assertion (*b*), we observe that

$$n(z) = 0 \Leftrightarrow \overline{n(z)} = \bar{0} = 0$$

Now by assertion (*a*), $\overline{n(z)} = n(\bar{z})$, hence $n(z) = 0 \Leftrightarrow n(\bar{z}) = 0$. ∎

Now we are ready to use this lemma to state and prove the symmetry property of network functions.

Symmetry property The *magnitude* of any network function is an *even* function of ω, and its *phase* may always be chosen to be an *odd* function of ω. Symbolically, calling $H(j\omega)$ the *network function*, we have by choosing an appropiate phase,

$$|H(j\omega)| = |H(-j\omega)| \text{ and } \measuredangle H(j\omega) = -\measuredangle H(-j\omega) \qquad \text{for all real } \omega\text{'s} \tag{3.8}$$

PROOF Call $H(j\omega)$ the network function of interest. We have to show that (*a*) the function $|H(j\omega)|$ is *even*, and (*b*) that we may choose the phase so that the function

$$\measuredangle H(j\omega) \qquad \text{is } odd \tag{3.9}$$

Equivalently, for all ω real,

$$|H(j\omega)| = |H(-j\omega)| \qquad \text{and} \qquad \measuredangle H(j\omega) = -\measuredangle H(-j\omega)$$

Call $n(j\omega)$ and $d(j\omega)$ the numerator polynomial and the denominator polynomial of $H(j\omega)$, i.e., $H(j\omega) = n(j\omega)/d(j\omega)$. Then, we obtain successively,

$$\begin{aligned} \overline{H(j\omega)} &= \overline{\left[\frac{n(j\omega)}{d(j\omega)}\right]} = \frac{\overline{n(j\omega)}}{\overline{d(j\omega)}} \\ &= \frac{n(-j\omega)}{d(-j\omega)} \qquad \text{(since } \overline{j\omega} = -j\omega \text{ and using the lemma above)} \\ &= H(-j\omega) \end{aligned}$$

Since, for all real ω's, $\overline{H(j\omega)} = H(-j\omega)$, we have

$$|H(j\omega)| = |H(-j\omega)|$$

and $\measuredangle H(-j\omega)$, the phase of $H(j\omega)$, can always be chosen so that[3]

$$-\measuredangle H(j\omega) = \measuredangle H(-j\omega)$$

∎

REMARK This symmetry property is a great labor saving device: Indeed if we know $|H(j\omega)|$ and $\measuredangle H(j\omega)$ for positive frequencies, we know them for negative frequencies.

[3]The phase of $\bar{z}$ may always be chosen to be $-\measuredangle z$.

Example Consider the driving-point admittance of the circuit shown in Fig. 3.4. It is easy to see that

$$Y(j\omega) = \frac{j\omega - 1}{j\omega + 1}$$

[As a check, for $\omega = 0$, $Y(0) = -1$; for $\omega = \infty$, $Y(\infty) = 1$.] Now, for any ω real,

$$|Y(j\omega)| = \left|\frac{j\omega - 1}{j\omega + 1}\right| = \frac{\sqrt{1+\omega^2}}{\sqrt{1+\omega^2}} = 1$$

Consider now the phase for $\omega > 0$, e.g., for $\omega = 0.1$

$$j\omega - 1 = 1.005\,\mathrm{e}^{j174.3°} \quad \text{and} \quad j\omega + 1 = 1.005\,\mathrm{e}^{j5.170°}$$

so

$$\measuredangle Y(j0.1) = 168.6°$$

Suppose that we have calculated the functions $|Y(j\omega)|$ and $\measuredangle Y(j\omega)$ for $\omega > 0$, then we do not need to calculate them for $\omega < 0$, since the function $|Y(j\omega)|$ is *even* and the phase $\measuredangle Y(j\omega)$ can always be chosen so that it is an *odd* function of ω. The resulting graph of $\measuredangle Y(j\omega)$ is shown in Fig. 3.5.

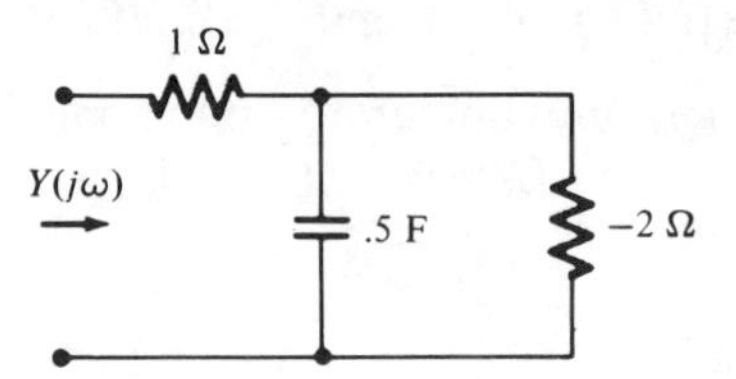

Figure 3.4 Linear circuit whose driving-point admittance $Y(j\omega)$ has magnitude 1 for all ω.

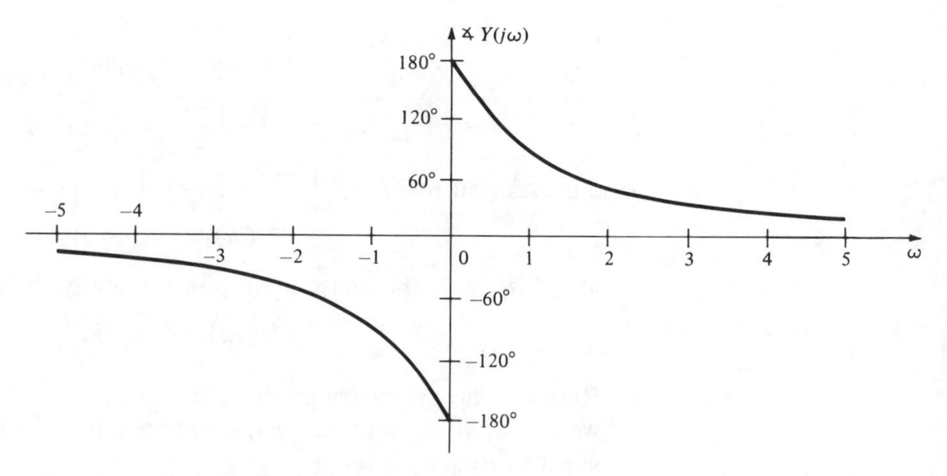

Figure 3.5 $\measuredangle Y(j\omega)$ for the circuit shown in Fig. 3.4.

3.3 Network Functions and Sinusoidal Waveforms

Consider a linear time-invariant circuit $\mathcal{N}$ in the sinusoidal steady state of frequency ω. Assume that it has only one independent source, the voltage source E_s. Let V_k be the output phasor and $H(j\omega)$ be the network function from E_s to V_k; then the phasors V_k and E_s are related by

$$V_k = H(j\omega)E_s \tag{3.10}$$

Hence, in the time domain, using $v_k(t) = \text{Re}[V_k\, e^{j\omega t}]$, we have

$$v_k(t) = |V_k| \cos(\omega t + \sphericalangle V_k)$$

$$v_k(t) = |H(j\omega)| \cdot |E_s| \cos[\omega t + \sphericalangle H(j\omega) + \sphericalangle E_s] \tag{3.10a}$$

Thus the *output sine wave* has an amplitude equal to $|H(j\omega)|$ times $|E_s|$, the amplitude of the *input* sine wave; furthermore, the *output sine wave* is shifted in phase by $\sphericalangle H(j\omega)$ with respect to the *input* sine wave.

These facts can also be visualized by writing Eq. (3.10) in terms of its polar representation

$$|V_k| = |H(j\omega)| \cdot |E_s| \qquad \text{and} \qquad \sphericalangle V_k = \sphericalangle H(j\omega) + \sphericalangle E_s \tag{3.11}$$

COMMENT The two equations (3.11) are of great engineering importance: Indeed there are precise and reliable sinusoidal generators and it is easy to make precise measurements in the sinusoidal steady state. Thus by comparing the amplitude of the input, namely $|E_s|$, with the amplitude of the output, namely $|V_k|$, we obtain the amplitude of the network function at frequency ω: $|H(j\omega)|$. By measuring the phase difference between the output sinusoid and the input sinusoid, namely $\sphericalangle V_k - \sphericalangle E_s$, we obtain the phase of the network function at frequency ω: $\sphericalangle H(j\omega)$.

Thus for any exponentially stable[4] linear time-invariant circuit, we may, in principle,[5] *measure directly* any of its network functions.

3.4 Superposition of Sinusoidal Steady States

If a given linear time-invariant circuit $\mathcal{N}$ is driven by a number of independent sources all operating at the *same* frequency ω, then Eq. (2.27*b*) gives the tableau equations: By Cramer's rule, the sinusoidal steady state due to all these sources operating simultaneously is the sum of the sinusoidal steady states that each source would create if it were acting alone.

Let $\mathcal{N}$ be a linear time-invariant circuit which is driven by two sinusoidal independent sources *operating at two different frequencies* (see Fig. 3.6). The

[4]This concept will be discussed in detail in Chap. 10, Sec. 4.3. By definition, "$\mathcal{N}$ is exponentially stable" means that all its natural frequencies have negative real parts.

[5]We say "in principle," because there may be practical limitations: some nodes are not accessible, or the measuring equipment perturbs the circuit, etc.

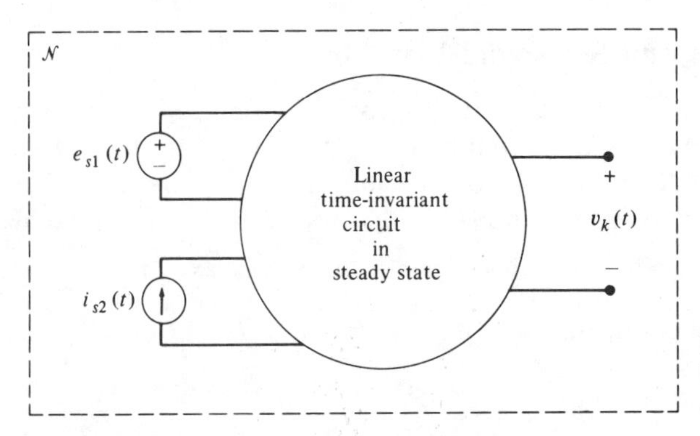

Figure 3.6 The circuit $\mathcal{N}$ is driven by the independent sources $e_{s1}(t)$ and $i_{s2}(t)$.

voltage source is specified by phasor E_{s1} and operates at frequency ω_1: Hence

$$e_{s1}(t) = |E_{s1}| \cos(\omega_1 t + \measuredangle E_{s1}) \tag{3.12}$$

The current source is specified by phasor I_{s2} and operates at frequency ω_2: Hence

$$i_{s2}(t) = |I_{s2}| \cos(\omega_2 t + \measuredangle I_{s2}) \tag{3.13}$$

Recall that $\omega_2 \neq \omega_1$.

We shall prove in Chap. 10, Sec. 5 that these two sinusoidal sources will establish a steady state which is the superposition of the steady states that each source would establish if each one operated alone.

Suppose that the response of interest is the voltage across the kth branch of $\mathcal{N}$. Let $H_{k1}(j\omega)$ be the voltage transfer function of $\mathcal{N}$ from E_{s1} to V_k; then the voltage across branch k due to E_{s1} acting alone is

$$\begin{aligned} v_k^{(1)}(t) &= \text{Re}[H_{k1}(j\omega_1)E_{s1}\,\text{e}^{j\omega_1 t}] \\ &= |H_{k1}(j\omega_1)| \cdot |E_{s1}| \cos[\omega_1 t + \measuredangle E_{s1} + \measuredangle H_{k1}(j\omega_1)] \end{aligned} \tag{3.14}$$

Note that the network function H_{k1} is evaluated at the frequency of the source e_{s1}. Similarly, let $H_{k2}(j\omega)$ be the transfer impedance from I_{s2} to V_k; then the voltage across branch k due to I_{s2} acting alone is

$$\begin{aligned} v_k^{(2)}(t) &= \text{Re}[H_{k2}(j\omega_2)I_{s2}\,\text{e}^{j\omega_2 t}] \\ &= |H_{k2}(j\omega_2)| \cdot |I_{s2}| \cos[\omega_2 t + \measuredangle I_{s2} + \measuredangle H_{k2}(j\omega_2)] \end{aligned} \tag{3.15}$$

By the superposition theorem, the resulting *steady state* is the superposition of two sinusoids—given by Eqs. (3.14) and (3.15)—of different frequencies, different amplitudes, and different phases. This steady state is no longer sinusoidal, since $\omega_1 \neq \omega_2$.

COMMENT The steady state need not even be periodic! The simple case is where $\omega_1 = r\omega_2$ where r is a *rational* number; then the two sinusoids (3.14) and (3.15) have a common period. If $\omega_1 = r\omega_2$ but r is not a rational number (e.g., $\sqrt{2}$, π, ln 2, . . .), then the steady state is *not* periodic: It is called *almost periodic*.

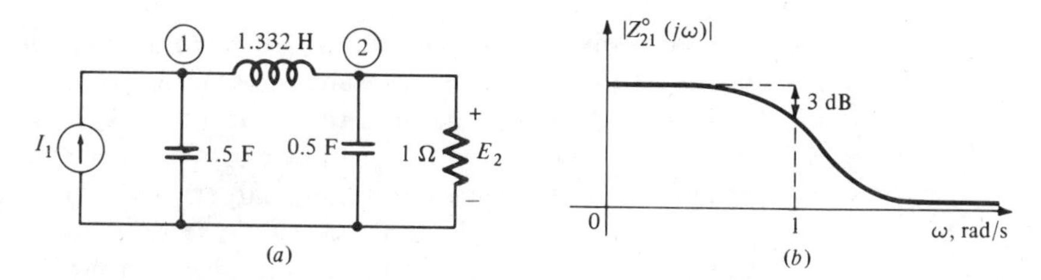

Figure 3.7 (*a*) Normalized linear circuit $\mathcal{N}_0$. (*b*) The magnitude of the transfer impedance $Z^0_{21} = E_2/I_1$ is plotted versus ω.

3.5 Impedance and Frequency Normalization

The idea of normalization is very simple and intuitive. In order not to get involved in complicated notations we consider a simple example.

Suppose that we have in our files a third-order Butterworth low-pass filter: It is driven by a current source I_1, its output voltage is E_2, and its transfer impedance $Z^0_{21}(j\omega)$ relates the input I_1 to the output E_2 (see Fig. 3.7*a*). Call this normalized circuit $\mathcal{N}_0$. Figure 3.7*b* shows that the normalized circuit $\mathcal{N}_0$ has a transfer impedance $Z^0_{21}(j\omega)$ which is close to 1 for $\omega \ll 1$, which dips down 3 dB at $\omega = 1$, and which decreases monotonically to zero as ω increases.[6]

Suppose we need a filter at an impedance level of 1200 Ω (i.e., the load resistor should be 1200 Ω rather than 1 Ω) and that the 3-dB frequency should be $3.8\text{ kHz} = 2\pi \times 3.8 \times 10^3\text{ rad/s} = 23.88 \times 10^3\text{ rad/s}$.

Thus for this case we have an impedance normalization factor given by

$$r \stackrel{\Delta}{=} \frac{\text{desired impedance level}}{\text{impedance level of normalized design}} = \frac{1200\ \Omega}{1\ \Omega} = 1200 \tag{3.16}$$

and a frequency normalization factor ω_n given by

$$\omega_n \stackrel{\Delta}{=} \frac{\text{desired typical frequency}}{\text{typical frequency of normalized design}} = \frac{23.88 \times 10^3\text{ rad/s}}{1\text{ rad/s}} = 23.88 \times 10^3 \tag{3.17}$$

[6]The abbreviation dB stands for *decibel*. Voltages and currents can be expressed in decibels according to the formula

$$\text{Voltage}\Big|_{\text{in decibels}} = 20 \log \text{voltage}\Big|_{\text{in volts}}$$

(and similarly for currents). Any network function H, being a ratio of currents and/or voltages is also expressible in decibels as follows:

$$|H(j\omega)|\Big|_{\text{in decibels}} = 20 \log |H(j\omega)|$$

Since in the present case, $H(j\omega_0) = 1$, at ω_0 the network functions is 0 dB and 0°. Since $20 \log(1/\sqrt{2}) \approx -3$, if for some frequency ω_1, $|H(j\omega_1)|$ is -3 dB, it means that

$$\frac{|H(j\omega_1)|}{|H(j\omega_0)|} = \frac{1}{\sqrt{2}} = 0.707$$

Intuitively, it is clear that (*a*) if we multiply *all impedances* by r (this implies also dividing all admittances by r), and (*b*) if we divide every resulting L, M, and C by ω_n, then (1) the new design will have an impedance level r times larger than the normalized design; (2) since the current source I_1 is left unchanged, all voltages are r times larger; and (3) at frequency ω_n rad/s, all impedances of the new design are r times those of the normalized design at frequency 1 rad/s. It is then clear that the transfer impedance Z_{21} of the new design is related to that of the normalized design by

$$Z_{21}(j\omega) = rZ_{21}^0\left(j\,\frac{\omega}{\omega_n}\right) \qquad \text{for all } \omega \tag{3.18}$$

A little thought shows that this procedure is quite general.

Exercise Use tableau analysis to demonstrate that the procedure of the example above is correct. {Hint: Consider only *one independent current source* and separate its branch equation from the rest. Show that if $[\mathbf{E}(j\omega), \mathbf{V}(j\omega), \mathbf{I}(j\omega)]$ is the solution of the normalized circuit $\mathcal{N}_0$, then

$$\left[r\mathbf{E}\left(j\,\frac{\omega}{\omega_n}\right), r\mathbf{V}\left(j\,\frac{\omega}{\omega_n}\right), \mathbf{I}\left(j\,\frac{\omega}{\omega_n}\right)\right]$$

is the solution of the new circuit.}

Procedure Define r and ω_n by Eqs. (3.16) and (3.17). Let R_0 (L_0, C_0, g_{m0}, r_{m0}, α_0 and μ_0) be a typical resistance (inductance, capacitance, transconductance, transimpedance, current gain, and voltage gain) of the *normalized* design $\mathcal{N}_0$, then for the new circuit we have

$$R = rR_0 \qquad g_m = \frac{1}{r}\,g_{m0}$$

$$L = \frac{r}{\omega_n}\,L_0 \qquad r_m = rr_{m0}$$

$$C = \frac{1}{r\omega_n}\,C_0 \qquad \alpha = \alpha_0 \qquad \mu = \mu_0$$

Normalization is also used to simplify calculations and to reduce roundoff errors. For example, in designing high-frequency communication circuits, it is convenient to use volts (V), milliamperes (mA), and nanoseconds (ns) as units: Using $v = Ri$, $V = L\,di/dt$, and $i = C\,dv/dt$, it is easy to see that this amounts to using the kilohm (kΩ) as unit of resistance, the microhenry (μH) as unit of inductance, and the picofarad (pF) as unit of capacitance.

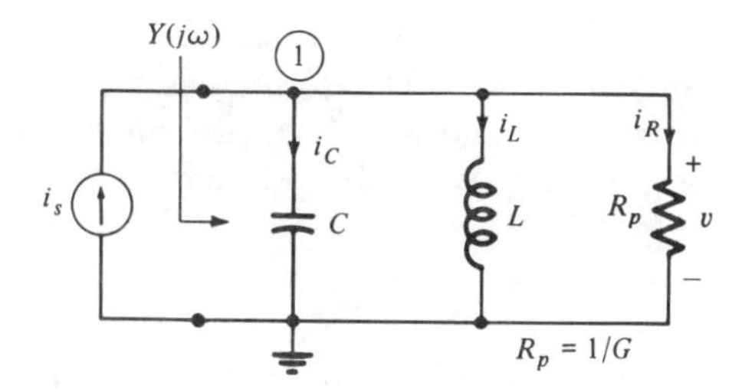

Figure 4.1 Resonant circuit.

4 RESONANT CIRCUIT

Resonant circuits are of great engineering importance: First they are used in measuring equipment, communication circuits, dc to dc converters, etc.; second, a resonant (electric) circuit is an example of the general phenomenon of resonance; and third, it is our first opportunity to exhibit the connection between time-domain behavior of circuits and their frequency-domain behavior.

We consider throughout this section the parallel resonant[7] circuit shown in Fig. 4.1. We assume that $R_p > 0$, $L > 0$, and $C > 0$. (We write R_p to recall that R_p is in *parallel* with L and C.)

4.1 Time-Domain Analysis

From Chap. 7, we know that KCL applied to node ① gives the differential equation of the circuit

$$LC\ddot{i}_L + GL\dot{i}_L + i_L = i_s(t) \tag{4.1}$$

Using the standard notations

$$\omega_0^2 \triangleq \frac{1}{LC} \qquad 2\alpha \triangleq \frac{G}{C} = \frac{1}{R_pC} \qquad Q \triangleq \frac{\omega_0}{2\alpha} = \frac{\omega_0 C}{G} = \frac{R_p}{\sqrt{L/C}} \tag{4.2}$$

the circuit differential equation becomes

$$\ddot{i}_L + 2\alpha\dot{i}_L + \omega_0^2 i_L = \omega_0^2 i_s(t) \tag{4.3}$$

or

$$\ddot{i}_L + \frac{\omega_0}{Q}\dot{i}_L + \omega_0^2 i_L = \omega_0^2 i_s(t) \tag{4.4}$$

We also know that the natural frequencies are

$$\left.\begin{matrix} s_1 \\ s_2 \end{matrix}\right\} = -\alpha \pm \sqrt{\alpha^2 - \omega_0^2} = -\frac{\omega_0}{2Q} \pm j\omega_0\sqrt{1 - \frac{1}{4Q^2}} \tag{4.5}$$

[7] As an exercise, the reader should use the following developments to derive, by duality, the theory of the series resonant circuit.

and are displayed in Fig. 4.2. For L and C fixed (hence $\omega_0 = 1/\sqrt{LC}$ is constant), as R_p increases to ∞, Q also increases to infinity. Now, for $Q > \frac{1}{2}$, the natural frequencies s_1 and $s_2 = \bar{s}_1$ move on the semicircle of radius ω_0 centered at the origin; as $Q \to \infty$, $s_1 \to j\omega_0$ and $\bar{s}_1 \to -j\omega_0$.

4.2 Frequency-Domain Analysis

A. Behavior of $Y(j\omega)$ Let $Y(j\omega)$ be the driving-point admittance of the resonant circuit (see Fig. 4.1): Since the elements R_p, L, and C are in parallel, we obtain by inspection

$$Y(j\omega) = G + j\omega C + \frac{1}{j\omega L} = G + j\left(\omega C - \frac{1}{\omega L}\right) \tag{4.6}$$

So the real part of $Y(j\omega)$, i.e., G, is *constant*; in contrast, its imaginary part, $B(\omega) \overset{\Delta}{=} \omega C - 1/\omega L$, varies from $-\infty$ to $+\infty$ as ω varies from 0 to ∞, as shown in Fig. 4.3. Now if for each $\omega > 0$, we plot the complex number $Y(j\omega) = \text{Re}[Y(j\omega)] + j\,\text{Im}[Y(j\omega)] = G + jB(\omega)$ in the Y plane and vary ω from 0 to ∞, we obtain a vertical straight line of abscissa G (see Fig. 4.3). For $\omega > 0$, very small, $|Y(j\omega)|$ is very large because at very low frequencies the inductor is almost a short circuit; similarly for $\omega > 0$, very large, $|Y(j\omega)|$ is again very large because the capacitor is almost a short circuit. Analytically from Eq. (4.6)

$$Y(j\omega) \cong \frac{1}{j\omega L} = -j\,\frac{1}{\omega L} \qquad \text{for } 0 < \omega \ll \omega_0 \tag{4.7a}$$

$$Y(j\omega) \cong j\omega C \qquad \text{for } \omega \gg \omega_0 \tag{4.7b}$$

and at resonance, $\omega = \omega_0$,

$$Y(j\omega_0) = G \tag{4.7c}$$

B. Behavior of $Z(j\omega)$ Let $Z(j\omega)$ be the driving-point *impedance* of the resonant circuit; hence

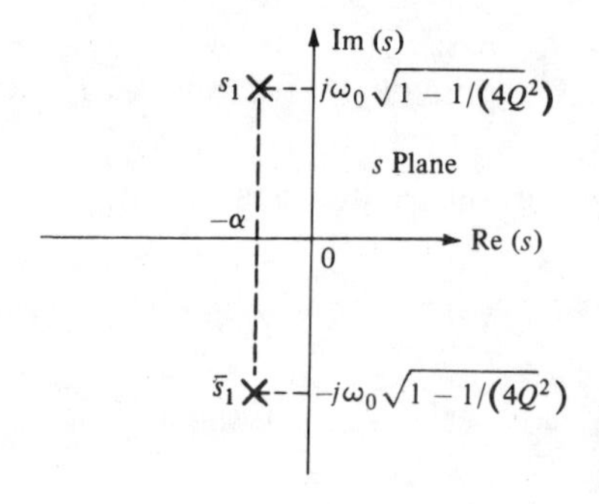

Figure 4.2 Location of natural frequencies of the circuit of Fig. 4.1.

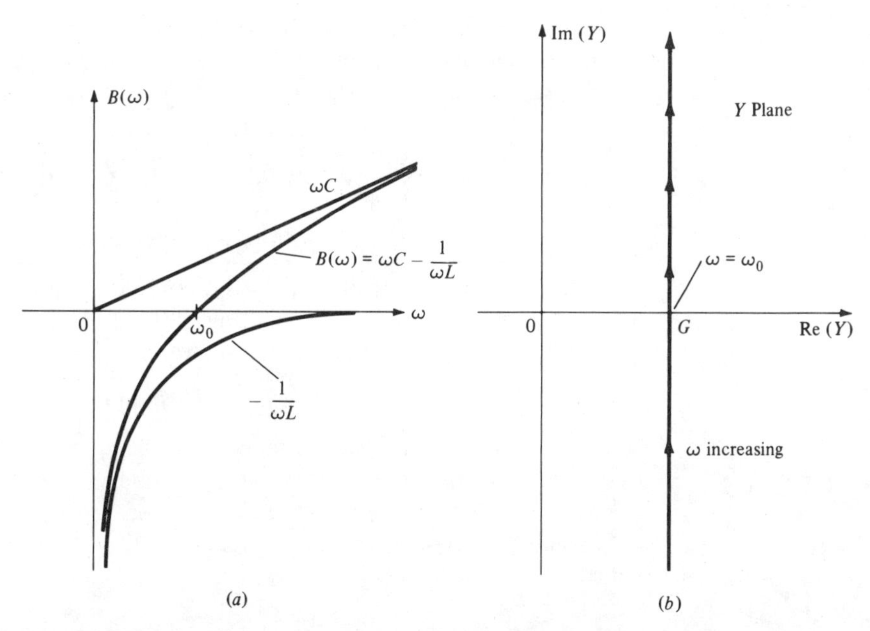

Figure 4.3 (*a*) The imaginary part of $Y(j\omega)$ for the resonant circuit of Fig. 4.1. (*b*) The locus of $Y(j\omega) = \text{Re}[Y(j\omega)] + j\,\text{Im}[Y(j\omega)]$ shown on the complex plane Y versus frequency for the circuit of Fig. 4.1.

$$Z(j\omega) = \frac{1}{Y(j\omega)} = \frac{1}{G + j(\omega C - 1/\omega L)} \tag{4.8}$$

and, using Eq. (4.2),

$$Z(j\omega) = \frac{R_p}{1 + jQ[(\omega/\omega_0) - (\omega_0/\omega)]}$$

In particular if I_s denotes the phasor representing the sinusoid $i_s(t)$, then, in the sinusoidal steady state, the output voltage v is also sinusoidal and its phasor is given by

$$V = Z(j\omega)I_s \tag{4.9}$$

So if we imagine driving the resonant circuit with a *constant amplitude* sine wave and with a slowly increasing frequency, the plot of $|Z(j\omega)|$ and of $\measuredangle Z(j\omega)$ versus ω will give a picture of the successive amplitudes and phases of the output voltage $V(\cdot)$.

Now by algebra

$$Z(j\omega) = \frac{1}{Y(j\omega)} = \frac{1}{G + jB(\omega)} = \frac{G - jB(\omega)}{G^2 + B(\omega)^2}$$

hence

$$Z(j\omega) = \frac{G}{G^2 + B(\omega)^2} + j\,\frac{-B(\omega)}{G^2 + B(\omega)^2} \tag{4.10}$$

Exercise Use algebra to show that as ω increases from 0 to ∞, $Z(j\omega)$ moves clockwise around the circle of radius $1/2G = R_p/2$ and of center $(R_p/2, 0)$, shown in Fig. 4.4. [Hint: Verify that, for all $\omega > 0$,

$$\left\{\text{Re}[Z(j\omega)] - \frac{1}{2G}\right\}^2 + \{\text{Im}[Z(j\omega)]\}^2 = \left(\frac{1}{2G}\right)^2$$

Check the direction of motion as ω increases from 0 to ∞ by Eqs. (4.7).]

The locus of $Z(j\omega)$ plotted in the complex plane as ω varies from $-\infty$ to $+\infty$ is shown in Fig. 4.4. The resulting plots of $|H(j\omega)|$ and $\measuredangle H(j\omega)$ are given in Fig. 4.5, where, for normalization, we set

$$Z(j\omega) = H(j\omega) \cdot R_p \tag{4.11}$$

Note that $|Z(j\omega)|$ reaches a maximum at $\omega = \omega_0 = 1/\sqrt{LC}$, the *resonant frequency*.

C. Circuit at resonance For the circuit of Fig. 4.6, at resonance, $Y(j\omega_0) = G$, hence

$$Z(j\omega_0) = \frac{1}{G} = R_p \tag{4.12}$$

Thus, at resonance $V = R_p I_s$ and $I_s = I_R$, that is, the source current appears to go through the resistor. Now, defining Q as usual by $Q = \omega_0 C R_p = R_p/\omega_0 L = R_p/\sqrt{L/C}$, we have

$$I_C = j\omega_0 CV = j\omega_0 C R_p I_s = jQI_s \tag{4.13}$$

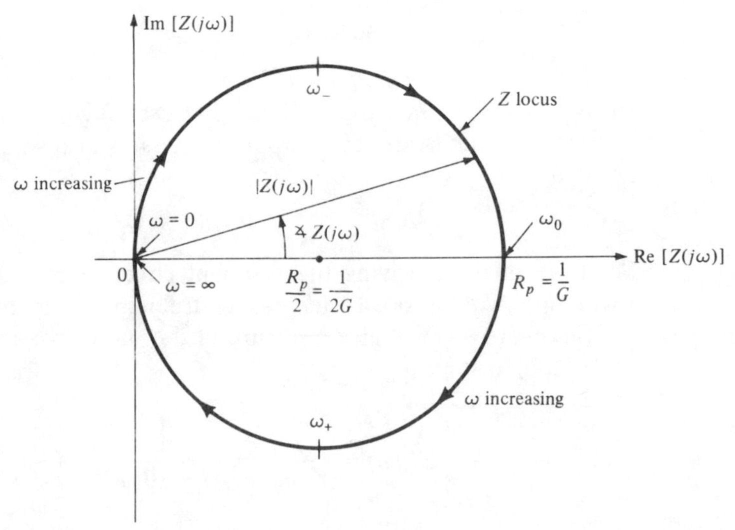

Figure 4.4 The locus of $Z(j\omega) = \text{Re}[Z(j\omega)] + j\,\text{Im}[Z(j\omega)]$ is a circle: its center is $(1/2G, 0)$ and its radius is $1/2G$.

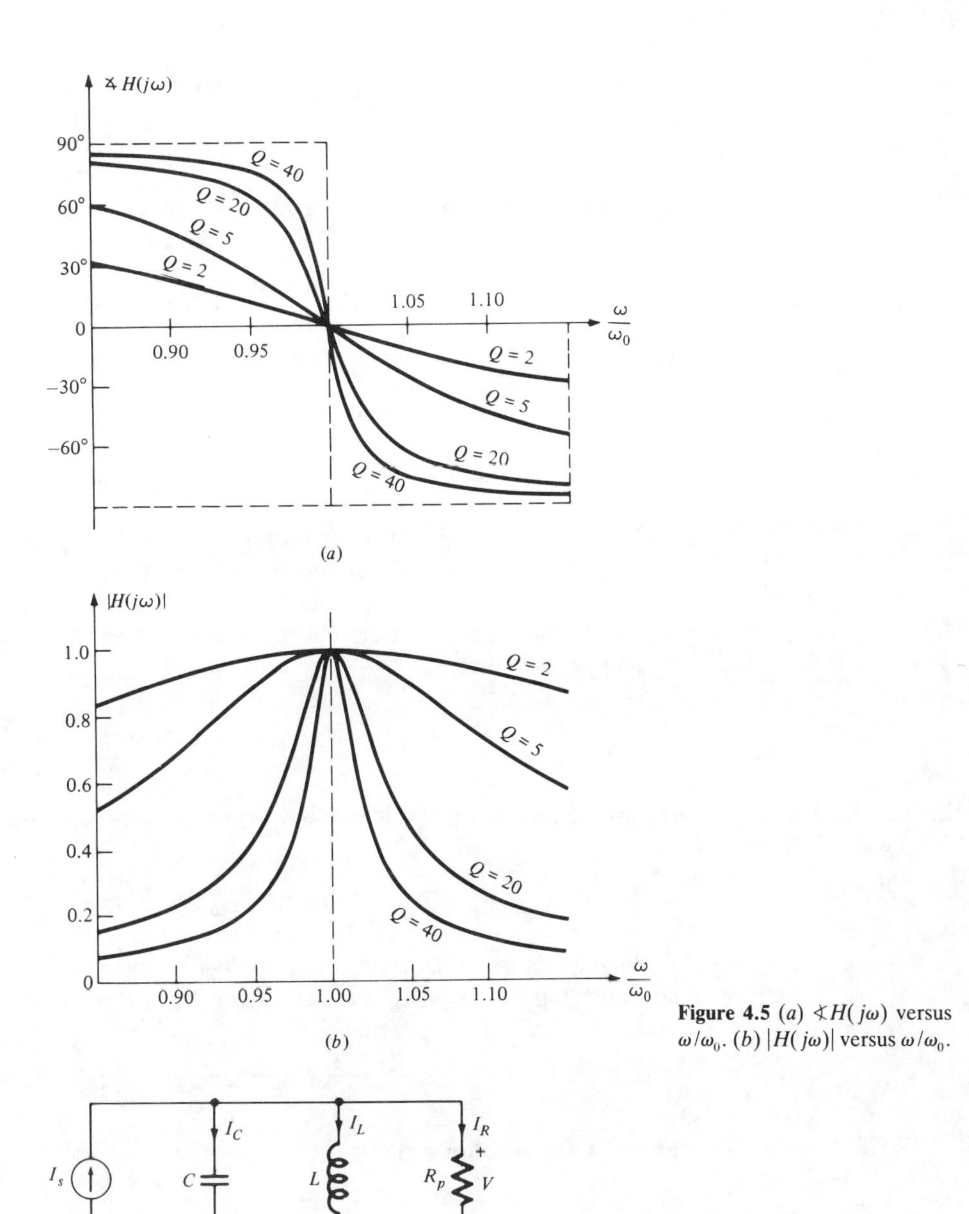

Figure 4.5 (*a*) $\measuredangle H(j\omega)$ versus ω/ω_0. (*b*) $|H(j\omega)|$ versus ω/ω_0.

Figure 4.6 Resonant circuit.

$$I_L = \frac{V}{j\omega_0 L} = -j\,\frac{R_p}{\omega_0 L}\,I_s = -jQI_s \tag{4.14}$$

Thus everything happens as if there is a circulating current through L and C—note that, at resonance, $I_C = -I_L = jQI_s$—and this circulating current is Q *times larger than the source current* I_s!

Exercise In the above setup, calculate the peak energy stored in L and in C at resonance.

D. Behavior near resonance We have

$$V = Z(j\omega)\, I_s = \frac{1}{1 + jQ[(\omega/\omega_0) - (\omega_0/\omega)]} R_p I_s \tag{4.15a}$$

In order to emphasize the dependence of $Z(j\omega)$ on Q and ω, we write $Z(j\omega) = H(j\omega)\, R_p$ where

$$H(j\omega) = \frac{1}{1 + jQ[(\omega/\omega_0) - (\omega_0/\omega)]} \tag{4.15b}$$

As seen from Eq. (4.15a) or from Fig. 4.4, $|Z(j\omega)|$ reaches a maximum equal to R_p at frequency ω_0. As ω decreases (increases, respectively) monotonically from ω_0, $|Z(j\omega)|$ decreases (decreases, respectively) also monotonically. Consequently there are two frequencies ω_- and ω_+ between 0 and ω_0 and between ω_0 and ∞ such that

$$|Z(j\omega_-)| = |Z(j\omega_+)| = \frac{1}{\sqrt{2}}\,|Z(j\omega_0)| = \frac{R_p}{\sqrt{2}} \tag{4.16}$$

By inspection of Eq. (4.15), this will occur when ω satisfies the equation

$$Q\left(\frac{\omega}{\omega_0} - \frac{\omega_0}{\omega}\right) = \pm 1 \tag{4.17}$$

Reducing this to a quadratic equation in ω/ω_0, we see that the solutions ω_- and ω_+ satisfying $0 < \omega_- < \omega_0 < \omega_+$ are given by

$$\frac{\omega_\mp}{\omega_0} = \sqrt{1 + \frac{1}{4Q^2}} \mp \frac{1}{2Q} \tag{4.18}$$

From (4.18) it follows that

$$\omega_+ - \omega_- = \frac{\omega_0}{Q} \qquad \text{and} \qquad \omega_-\omega_+ = \omega_0^2 \tag{4.19}$$

The frequencies ω_- and ω_+ are called the *3-dB frequencies* and $\omega_+ - \omega_-$ is the *3-dB bandwidth* (in rad/s). (See Fig. 4.7.) Since $\omega_+ - \omega_- = \omega_0/Q$, for fixed ω_0, the larger Q is, the smaller the 3-dB bandwidth is, the sharper the resonance curve $|Z(j\omega)|$ is, and the more abrupt the change in the slope of the phase curve $\measuredangle Z(j\omega)$ at resonance is. The change in the slope of the resonance curve as Q increases is illustrated by Fig. 4.5a and 4.5b.

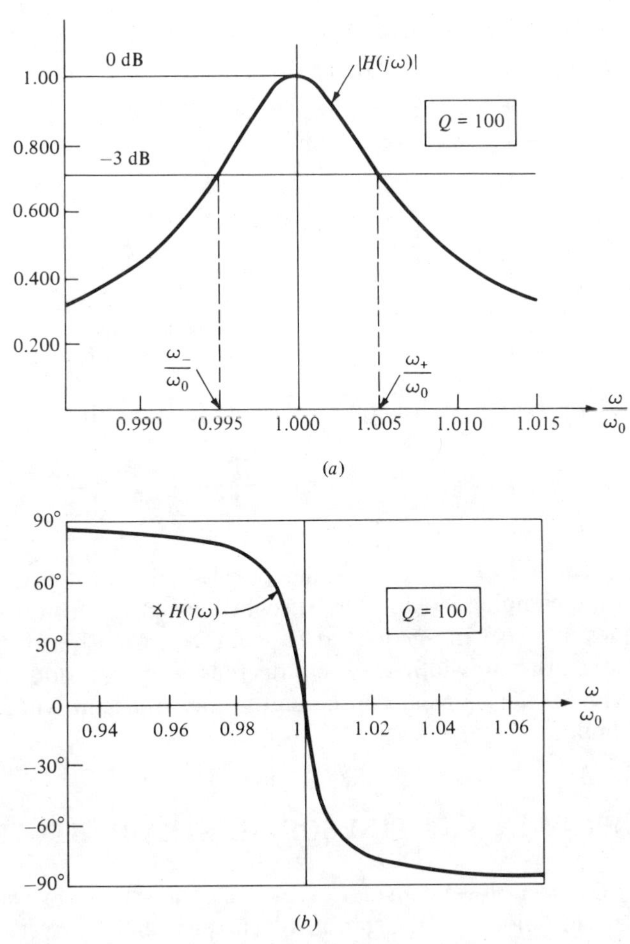

Figure 4.7 Resonance curves (*a*) $|H(j\omega)|$ and (*b*) $\sphericalangle H(j\omega)$.

REMARK

$$Z(j\omega_-) = \frac{R_p}{\sqrt{2}}\,e^{+j45^\circ} \qquad Z(j\omega_+) = \frac{R_p}{\sqrt{2}}\,e^{-j45^\circ} \qquad \text{(see Fig. 4.4)}$$

High-Q Approximation If $Q \gg 1$, we may neglect $1/Q^2$; furthermore, since $\sqrt{1+x^2} \cong 1 + (x^2/2) - (x^4/8) + \cdots$ we obtain the following useful approximate expressions

$$\omega_- \cong \omega_0\left(1 - \frac{1}{2Q}\right) \qquad \omega_+ \cong \omega_0\left(1 + \frac{1}{2Q}\right) \tag{4.20}$$

In both expressions (4.20), the first neglected term is $\omega_0/(8Q^2)$.

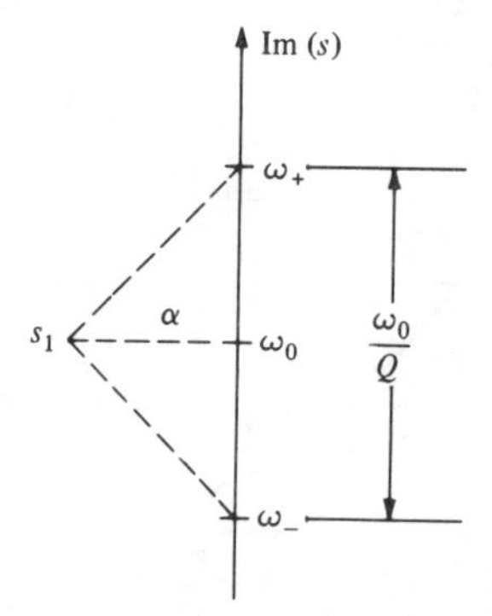

Figure 4.8 For high Q, the natural frequency s_1 defines the center frequency and $2a = \omega_0/Q$ is the 3-dB bandwidth.

Exercise Use Taylor expansions to show that

$$\frac{\omega_{\pm}}{\omega_0} = 1 \pm \frac{1}{2Q} + \frac{1}{8Q^2} - \frac{1}{128Q^4} + \cdots$$

Refer to Fig. 4.8. Assume $Q \gg 1$, and from the natural frequency s_1 draw in the complex s plane two straight lines at 45° from the horizontal; these two lines intersect the $j\omega$ axis at $j\omega_+$ and $j\omega_-$. Clearly, for fixed ω_0 as Q increases, the 3-dB bandwidth $\omega_+ - \omega_-$ decreases to zero and the natural frequency s_1 tends to the $j\omega$ axis; consequently, any transient will experience less and less damping.

5 POWER AND ENERGY IN THE SINUSOIDAL STEADY STATE

In this section we consider exclusively linear time-invariant circuits operating in the sinusoidal steady state: All currents and all voltages are assumed to be sinusoidal at the same frequency ω.

We shall use repeatedly the following well-known formulas:

$$(\cos x)^2 = \tfrac{1}{2}(1 + \cos 2x) \tag{5.1}$$

$$\cos(x + y) = \cos x \cos y - \sin x \sin y \tag{5.2}$$

$$2 \cos x \cos y = \cos(x + y) + \cos(x - y) \tag{5.3}$$

5.1 Instantaneous Power and Average Power

A. Examples

1. Consider a *resistor* R traversed by a current $i(t)$ specified by the phasor $I = I_m \exp(j\measuredangle I)$. The power delivered by the generator to the resistor is—recall that $i(t) = I_m \cos(\omega t + \measuredangle I)$—

$$p(t) = v(t)\, i(t) = RI_m \cos(\omega t + \measuredangle I) I_m \cos(\omega t + \measuredangle I)$$

hence, using Eq. (5.1), we obtain

$$p(t) = \tfrac{1}{2} RI_m^2[1 + \cos 2(\omega t + \sphericalangle I)] \tag{5.4}$$

Note that the power oscillates *twice* per period $T = 2\pi/\omega$ between 0 and $R_m I_m^2$. The *average power over one period* T is given by

$$P_{\text{ave}} = \frac{1}{T}\int_0^T p(t)\, dt = \tfrac{1}{2} RI_m^2 \tag{5.5}$$

2. Replace the resistor R by a *capacitor* C. Let the voltage across C be specified by the phasor $V = V_m \exp(j\sphericalangle V)$. Hence the current through C is given by the phasor $I = j\omega CV$ and

$$\begin{aligned} i(t) &= \text{Re}[\, j\omega CV_m \exp(j\sphericalangle V) \exp(j\omega t)] \\ &= \omega CV_m \cos\left(\omega t + \sphericalangle V + \frac{\pi}{2}\right) \end{aligned}$$

So

$$\begin{aligned} p(t) = v(t)\, i(t) &= \omega CV_m^2 \cos(\omega t + \sphericalangle V) \cos\left(\omega t + \sphericalangle V + \frac{\pi}{2}\right) \\ &= \tfrac{1}{2}\, \omega CV_m^2 \cos 2\left(\omega t + \sphericalangle V + \frac{\pi}{4}\right) \end{aligned} \tag{5.6}$$

Note that the power into the capacitor oscillates twice per period between $-\tfrac{1}{2}\omega CV_m^2$ and $\tfrac{1}{2}\omega CV_m^2$. The average power over one period is zero [see Eq. (5.6)] as we expect since no energy is dissipated and $\tfrac{1}{2}Cv(0)^2$ (the initially stored energy) equals $\tfrac{1}{2}Cv(T)^2$ (the finally stored energy) because $v(\cdot)$ is sinusoidal with period T.

3. If we had an *inductor* L, we would obtain (by duality!)

$$p(t) = \tfrac{1}{2}\, \omega LI_m^2 \cos 2\left(\omega t + \sphericalangle I + \frac{\pi}{4}\right) \tag{5.7}$$

Since capacitors and inductors are lossless elements and do not dissipate energy, as Eqs. (5.6) and (5.7) show, the *average power* delivered to C and to L is zero.

Consider now the energy stored in L: At time t, it is

$$\tfrac{1}{2}Li(t)^2 = \tfrac{1}{2}LI_m[\cos(\omega t + \sphericalangle I)]^2$$

Note that it is always nonnegative and that the *average energy stored* in L is given by

$$\tfrac{1}{4}LI_m^2$$

Of course, the dual formula holds for capacitors.

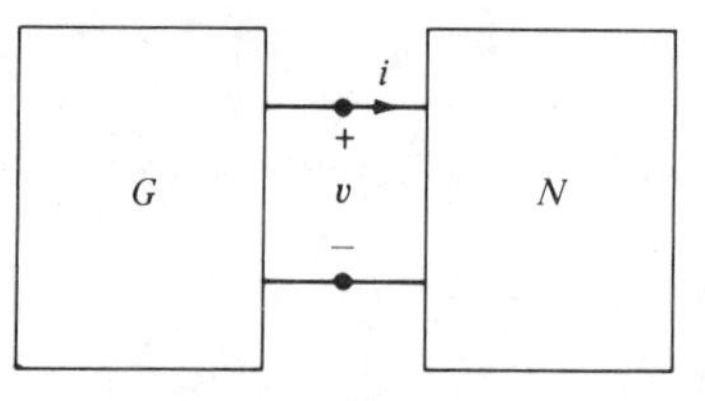

Figure 5.1 The generator G drives, in the sinusoidal steady state, the linear one-port N.

REMARK ON EFFECTIVE VALUES In the equations above, we expressed the current as

$$i(t) = I_m \cos(\omega t + \sphericalangle I)$$

that is, I_m is the *amplitude* of the sinusoid $i(t)$; as a consequence, we got factors of $\frac{1}{2}$ in expression for the average power, Eq. (5.5). In order to get rid of this factor $\frac{1}{2}$, power engineers use *effective* values, or *RMS values*—RMS stands for "root mean square"—define,[8]

$$I_{\text{eff}} \triangleq \frac{I_m}{\sqrt{2}} \qquad V_{\text{eff}} \triangleq \frac{V_m}{\sqrt{2}} \tag{5.8}$$

then, since $V_m = RI_m$, we have

$$P_{\text{ave}} = RI_{\text{eff}}^2 \qquad \text{and} \qquad P_{\text{ave}} = \frac{V_{\text{eff}}^2}{R} \tag{5.9}$$

These two formulas are identical with those for the dc case.

B. Power into a one-port Consider a sinusoidal generator G driving the linear time-invariant one-port N shown in Fig. 5.1. Assume that the circuit shown in Fig. 5.1 is in the sinusoidal steady state. Call V and I the phasors representing the port variables v and i. Then *the power delivered*, at time t, *by the generator G to the one-port N* is given by

$$\begin{aligned} p(t) &= v(t)\, i(t) \\ &= V_m \cos(\omega t + \sphericalangle V) I_m \cos(\omega t + \sphericalangle I) \end{aligned}$$

so, by (5.3),

$$p(t) = \tfrac{1}{2} V_m I_m \cos(\sphericalangle V - \sphericalangle I) + \tfrac{1}{2} V_m I_m \cos(2\omega t + \sphericalangle V + \sphericalangle I) \tag{5.10}$$

The current $i(\cdot)$, voltage $v(\cdot)$, and power $p(\cdot)$ are plotted in Fig. 5.2.

Call *average power* the average over one period ($T = 2\pi/\omega$) of $p(t)$:

$$P_{\text{ave}} \triangleq \frac{1}{T} \int_0^T p(t)\, dt \tag{5.11}$$

[8] For example, in our homes, the power company supplies us with what they call 110 volts ac: it is 110 volts RMS. Thus the actual voltage is $\sqrt{2}\ 110 \cos(2\pi 60 t + \phi)$.

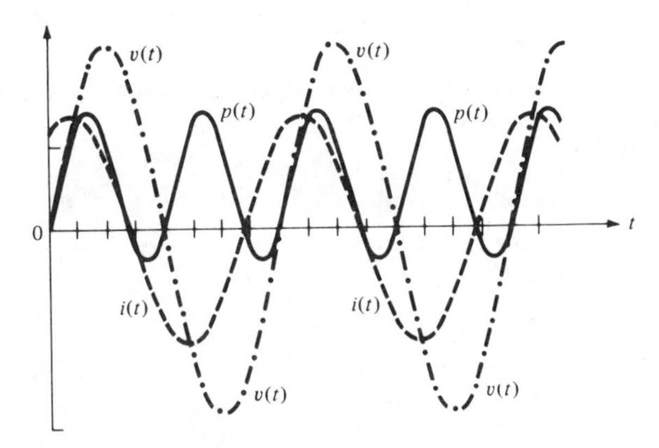

Figure 5.2 Plots of the port voltage $v(t)$, port current $i(t)$, and the power $p(t)$ delivered to N. Note that for some time intervals $p(t)$ is negative!

From Eq. (5.10) we obtain

$$P_{\text{ave}} = \tfrac{1}{2} V_m I_m \cos(\measuredangle V - \measuredangle I) \tag{5.12}$$

REMARKS

1. The average power depends not only on the *magnitude* of the sinusoids $v(\cdot)$ and $i(\cdot)$ but also on their *phase difference* $\measuredangle V - \measuredangle I$. The cosine factor $\cos(\measuredangle V - \measuredangle I)$ is of extreme engineering importance and is called the *power factor*.
2. Since $V = ZI$, where Z is the driving-point impedance of N,

$$\measuredangle V - \measuredangle I = \measuredangle Z \tag{5.13}$$

Consequently

$$P_{\text{ave}} \geq 0 \qquad \text{if and only if} \qquad |\measuredangle Z| \leq 90° \tag{5.14}$$

3. Note the presence of the oscillating term in Eq. (5.10): Whenever $|\cos(\measuredangle V - \measuredangle I)| < 1$, Eq. (5.10) shows that the instantaneous power $p(t)$ *changes sign four times per period*. (See Fig. 5.2.)
4. In Eq. (5.12), the average power is expressed in terms of the current and the voltage; if we use Z and Y, we obtain

$$P_{\text{ave}} = \tfrac{1}{2} V_m I_m \cos(\measuredangle Z) = \tfrac{1}{2} I_m^2 \operatorname{Re}(Z) \tag{5.15}$$

and dually

$$P_{\text{ave}} = \tfrac{1}{2} V_m I_m \cos(\measuredangle Y) = \tfrac{1}{2} V_m^2 \operatorname{Re}(Y) \tag{5.16}$$

5.2 Complex Power

A. Definition For the circuit of Fig. 5.1 operating in the sinusoidal steady state, we define the complex number P by[9]

[9] Power engineers use the letter S for complex power. In circuit theory we need S for other uses.

$$P \stackrel{\Delta}{=} \tfrac{1}{2}\mathbf{V}\bar{\mathbf{I}} \tag{5.17}$$

P is called the *complex power* delivered *by* the generator G *to* the one-port N. Obviously

$$P = \underbrace{\frac{V_m I_m}{2}\cos(\measuredangle V - \measuredangle I)}_{P_{\text{ave}}} + \underbrace{j\,\frac{V_m I_m}{2}\sin(\measuredangle V - \measuredangle I)}_{jQ} \tag{5.18}$$

Q is called the *reactive power*[10] delivered to the one-port N by the generator.

The interpretation of the reactive power is the following: Consider Z via its admittance $Y = 1/Z$; then split Y into its real part G and its imaginary part $B: Y = G + jB$. Next in the second term of Eq. (5.10) write

$$\cos(2\omega t + \measuredangle V + \measuredangle I) = \cos[(2\omega t + 2\measuredangle I) + (\measuredangle V - \measuredangle I)] \tag{5.19}$$

and use Eq. (5.2); then Eq. (5.10) becomes

$$p(t) = \frac{V_m I_m}{2}\cos(\measuredangle V - \measuredangle I)[1 + \cos 2(\omega t + \measuredangle I)] - \frac{V_m I_m}{2}\sin(\measuredangle V - \measuredangle I)\sin 2(\omega t + \measuredangle I) \tag{5.20}$$

The first term of Eq. (5.20) represents the power absorbed by G, the real part of $Y = 1/Z$; note that it is always ≥ 0, if $G \geq 0$, or equivalently, if $|\measuredangle V - \measuredangle I| \leq 90°$. The second term of Eq. (5.20) has a zero average (it is purely oscillating); it represents power going back and forth through the reactive part of Y, and its magnitude is equal to $|Q|$.

B. Conservation of complex power In Chap. 1 we showed that KCL and KVL imply that energy is conserved in *any* electric circuit. There is a similar theorem for linear time-invariant circuits operating *in the sinusoidal steady state*: It says that complex power is conserved. More precisely we have the theorem:

> **Theorem** Consider a *linear time-invariant* circuit driven by a number of *independent* sources, all sinusoidal at the *same* frequency ω. We assume that the circuit is *in the sinusoidal steady state*. Then the sum of the complex power delivered by each independent source to the circuit is equal to the sum of the complex power absorbed by each element of the circuit.

Thus not only does it say that the average power is conserved but also that the reactive power is conserved. Thus if some element absorbs an amount Q of reactive power then that amount must be supplied to the circuit by either one or more independent sources or by one or more elements. This fact is of crucial importance in power engineering.

[10] The reactive power Q has nothing to do with the quality factor Q of a resonant circuit. Common usage dictates that the reactive power be labeled Q.

PROOF For simplicity assume that there is only one independent source, a current source (see Fig. 5.3). Assign associated reference directions to all branches of the circuit and call $V_1, V_2, \ldots, V_b$ the branch-voltage phasors and $I_1, I_2, \ldots, I_b$ the branch-current phasors. Of course, the V_k's obey KVL and the I_k's obey KCL (see Sec. 2.1 above). In particular, we have $\mathbf{AI} = \mathbf{0}$. Since the reduced incidence matrix $\mathbf{A}$ has *real* elements, if we take the complex conjugate of that equation, we obtain

$$\mathbf{A\bar{I}} = \mathbf{0} \tag{5.20a}$$

Thus the $\bar{I}_k$'s also satisfy KCL! Consequently by Tellegen's theorem

$$\sum_{k=1}^{b} V_k \bar{I}_k = 0$$

hence

$$-\tfrac{1}{2} V_1 \bar{I}_1 = \sum_{k=2}^{b} \tfrac{1}{2} V_k \bar{I}_k \tag{5.21}$$

The term on the left is the complex power delivered *by* the current source *to* the circuit, and the sum on the right is the sum of the complex power absorbed by each branch of the circuit.

The generalization to the case where there are many independent sources is immediate. ■

5.3 Maximum Power Transfer Theorem

Consider a one-port driving a load Z_L. Assume that the circuit is in the sinusoidal steady state at frequency ω. The problem is to determine the load impedance Z_L so that the average power received by the load is maximum. Such a problem arises in radar design: The radar antenna picks up a signal which must be amplified. The problem is to choose the input impedance of the amplifier (Z_L) so that it receives maximum average power.

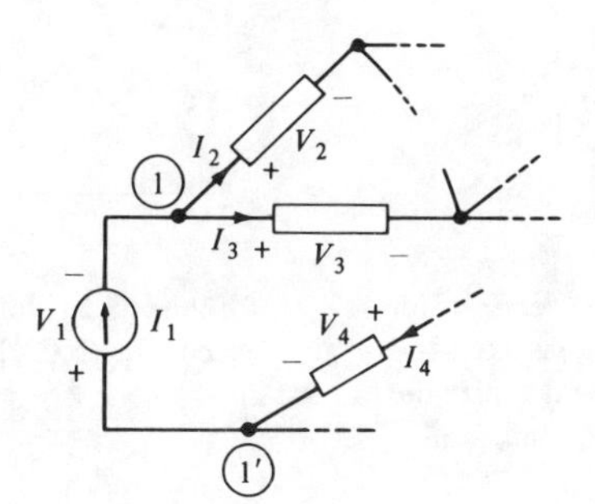

Figure 5.3 This circuit driven by the single source I_1 is used in establishing the conservation of complex power.

A. Analysis Let us model the given one-port by its Thévenin equivalent circuit (see Fig. 5.4). We consider E_G, the phasor representing the sinusoidal source, and $Z_G = R_G + jX_G$ as *given*. For simplicity, we use a time origin so that E_G is *real*, i.e., $E_G = E_{Gm}$, the magnitude of the sinusoidal source voltage.

The conservation of complex power implies the conservation of average power. Hence, the average power into the load, $P_{L\text{ave}}$, is equal to the average power delivered by the source minus the average power received by Z_G. Thus calling $I_L = I_{Lm} \exp(j\measuredangle I_L)$ the yet undetermined load current, we obtain

$$P_{L\text{ave}} = \tfrac{1}{2}\,\text{Re}[E_G \bar{I}_L] - \tfrac{1}{2}R_G|I_L|^2$$

or

$$P_{L\text{ave}} = \tfrac{1}{2}E_G I_{Lm}\cos(\measuredangle I_L) - \tfrac{1}{2}R_G I_{Lm}^2 \tag{5.22}$$

Since E_G and R_G are given, we see that $P_{L\text{ave}}$ is a function of two real variables: $\measuredangle I_L$ and I_{Lm}. Note that I_{Lm}, being the *magnitude* of the current, is necessarily nonnegative: $I_{Lm} \geq 0$. Consequently, to maximize $P_{L\text{ave}}$ we must have $\cos(\measuredangle I_L) = 1$, equivalently,

$$\measuredangle I_L = 0 \tag{5.23}$$

With $\cos(\measuredangle I_L) = 1$, Eq. (5.22) shows that, for the problem to make sense, R_G *must be positive*. Indeed, if R_G is <0, for large (and necessarily positive) I_{Lm}, $P_{L\text{ave}}$ would become arbitrarily large! Hence, for a realistic problem we must have $R_G > 0$.

With $\cos(\measuredangle I_L) = 1$ and $R_G > 0$, Eq. (5.22) shows that the curve $P_{L\text{ave}}$ versus I_{Lm} is a parabola (see Fig. 5.5). To maximize $P_{L\text{ave}}$ we calculate its derivatives with respect to I_{Lm}:

$$\frac{\partial P_{L\text{ave}}}{\partial I_{Lm}} = \tfrac{1}{2}E_G - R_G I_{Lm} \tag{5.24}$$

$$\frac{\partial^2 P_{L\text{ave}}}{\partial I_{Lm}^2} = -R_G < 0 \tag{5.25}$$

From Eqs. (5.23) and (5.24) the maximizing I_L, denoted by $I_L^0 = I_{Lm}^0 \exp(j\measuredangle I_L^0)$, is given by

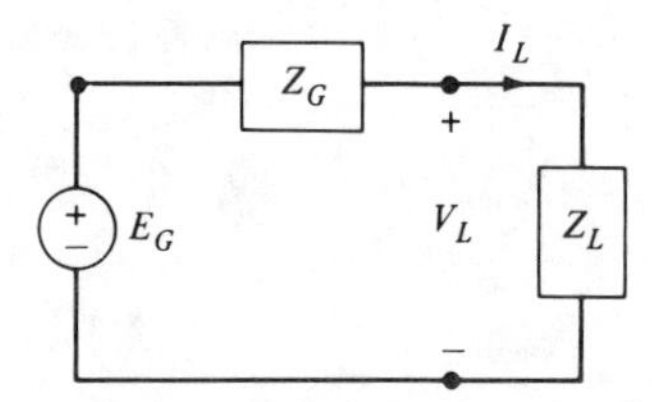

Figure 5.4 In the maximum power transfer theorem, the generator (E_G, Z_G) is fixed and the load impedance is varied to maximize the *average power* delivered to the load.

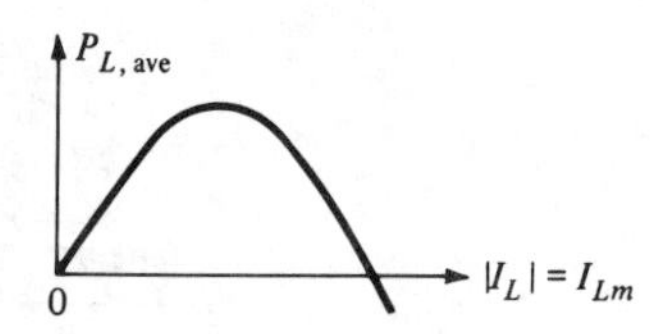

Figure 5.5 Plot of $P_{L\text{ave}}$ versus $|I_L| = I_{Lm}$ [see Eq. (5.22)].

$$I_{Lm}^0 = \frac{1}{2}\frac{E_G}{R_G} \qquad \text{and} \qquad \sphericalangle I_L^0 = 0 \tag{5.26}$$

Equations (5.26) show that the loop impedance in Fig. 5.4 is *real* and equal to $2R_G$. Hence the *optimum load impedance* Z_L^0 is

$$Z_L^0 = R_G - jX_G \qquad \text{or, equivalently,} \qquad Z_L^0 = \bar{Z}_G \tag{5.27}$$

It is easily checked that[11]

$$P_{L\text{ave}}^0 = \frac{|E_G|^2}{8R_G} \tag{5.28}$$

Note that $P_{L\text{ave}}$ as a function of I_{Lm} and $\sphericalangle I_L$ has a *unique* maximum at $(I_{Lm}^0, \sphericalangle I_L^0 = 0)$ [see Eqs. (5.22) and (5.24)].

B. Conclusions We have derived the following result:

Maximum power transfer theorem We are given a one-port operating in the sinusoidal steady state at frequency ω and driving a load impedance Z_L (see Fig. 5.4). This one-port is specified by its Thévenin equivalent $(E_G, Z_G = R_G + jX_G)$, where $R_G > 0$. The load impedance Z_L will receive from the one-port a maximum average power if and only if

$$Z_L = \bar{Z}_G \tag{5.29}$$

In that case the average power delivered to the load is given by (5.28).

REMARKS
1. When $Z_L = \bar{Z}_G$, we say that the load is *matched to the one-port*.
2. When $Z_L = \bar{Z}_G$, since $R_L = R_G$, only 50 percent of the energy of the source goes into the load! Still $Z_L = \bar{Z}_G$ is the best we can do if Z_G is *given* and *not* under our control.
3. The engineer who designs alternators never uses matched loads: His alternator receives power from, say, the turbine (mechanical power input = torque × angular velocity), and he wishes that most of that power flows out of the alternator as electric energy. Therefore he designs his alternator so that R_G is as small as possible. For him Z_G is *not given*: he designs Z_G! Big alternators of power companies have efficiencies that are as high as 99 percent.

Exercise Suppose that you are in the situation specified in the maximum power transfer theorem, but now you are required to have a *real* load impedance R_L. Find the optimal R_L when $R_G = 0$ (Answer: $R_L = |X_G|$).

[11]Note that $|E_G| = E_{Gm}$ represents the *peak* value of the source voltage. If $|E_G|$ is in RMS volts, the denominator should be $4R_G$.

5.4 Average Power Due to Several Sinusoidal Inputs

We all know that instantaneous power does not obey superposition: For example, consider a resistor R driven by a voltage source $v_s(t) = V_0 + V_m \cos \omega t$. Let us view the source v_s as the sum of a dc source V_0 and a sinusoidal source $V_m \cos \omega t$. If V_0 acts alone, the power dissipated in R is V_0^2/R; if $V_m \cos \omega t$ acts alone, the power dissipated in R is $V_m^2(\cos \omega t)^2/R$. Now the power dissipated when $v_s(t) = V_0 + V_m \cos \omega t$ drives the resistor R is given by

$$\frac{V_0^2 + 2V_0V_m \cos \omega t + V_m^2(\cos \omega t)^2}{R} \tag{5.30}$$

This expression is not the sum of the two previous ones. That is what is meant by saying that "instantaneous power does not obey superposition."

Note that if we average expression (5.30) over the period $2\pi/\omega = T$, then the *average* is the sum of the averages of the two previous expressions. This fact is true in general as we shall see.

Suppose that as shown in Fig. 5.6 we drive a linear time-invariant one-port N by a voltage source $v_s(t)$ given by

$$v_s(t) = V_{m1} \cos(\omega_1 t + \sphericalangle V_1) + V_{m2} \cos(\omega_2 t + \sphericalangle V_2) \tag{5.31}$$

$v_s(\cdot)$ is the sum of two sinusoids of *different* frequencies. To be specific let $\omega_1 > \omega_2$. Assume that the steady state has been reached. Thus the port current $i(\cdot)$ (see Fig. 5.6) is given by

$$i(t) = I_{m1} \cos(\omega_1 t + \sphericalangle I_1) + I_{m2} \cos(\omega_2 t + \sphericalangle I_2) \tag{5.32}$$

where, for $k = 1, 2$

$$I_k = I_{mk} \exp(j\sphericalangle I_k) = Y(j\omega_k)V_{mk} \exp(j\sphericalangle V_k) \tag{5.33}$$

The (instantaneous) power $p(t)$ delivered *by* the voltage source v_s *to* the one-port N is given by $v_s(t)\, i(t)$, and, in view of Eqs. (5.31) and (5.32), this is a sum of four terms, each one a product of cosines. Using standard formulas we obtain

$$p(t) = v_s(t)\, i(t)$$

$$p(t) = \frac{V_{m1}I_{m1}}{2} \cos(\sphericalangle V_1 - \sphericalangle I_1) + \frac{V_{m2}I_{m2}}{2} \cos(\sphericalangle V_2 - \sphericalangle I_2)$$

$$+ \frac{V_{m1}I_{m1}}{2} \cos(2\omega_1 + \sphericalangle V_1 + \sphericalangle I_1) + \frac{V_{m2}I_{m2}}{2} \cos(2\omega_2 t + \sphericalangle V_2 + \sphericalangle I_2)$$

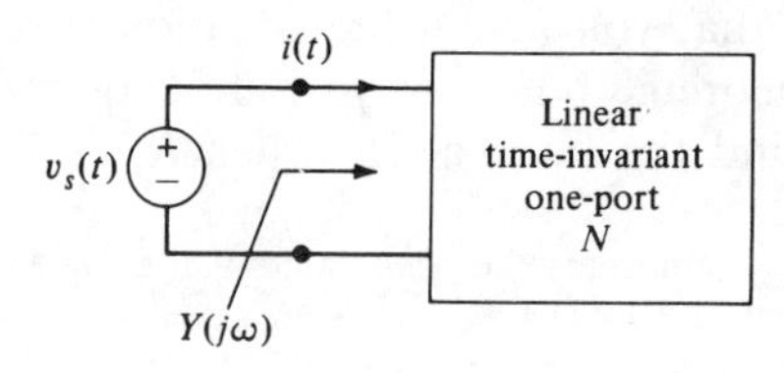

Figure 5.6 Here the source $v_s(t)$ produces the sum of two sinusoids at *different* frequencies.

$$+\frac{V_{1m}I_{2m}}{2}\cos[(\omega_1+\omega_2)t+\sphericalangle V_1+\sphericalangle I_2]$$

$$+\frac{V_{2m}I_{1m}}{2}\cos[(\omega_1+\omega_2)t+\sphericalangle V_2+\sphericalangle I_1]$$

$$+\frac{V_{1m}I_{2m}}{2}\cos[(\omega_1-\omega_2)t+\sphericalangle V_1-\sphericalangle I_2]$$

$$+\frac{V_{2m}I_{1m}}{2}\cos[(\omega_1-\omega_2)t+\sphericalangle I_1-\sphericalangle V_2] \tag{5.34}$$

The first four terms of expression (5.34) give the sum of the (instantaneous) powers that would be delivered by V_1 and V_2 *if they were acting alone.*

Let us *assume* that the periods $T_1 = 2\pi/\omega_1$ and $T_2 = 2\pi/\omega_2$ of the voltage source are harmonically related, that is, there is a "common period" T_c such that

$$T_c = n_1 T_1 = n_2 T_2$$

where n_1 and n_2 are suitable *positive integers.*[12] Consider now the power *averaged over* T_c:

$$P_{\text{ave}} \triangleq \frac{1}{T_c}\int_0^{T_c} p(t)\,dt \tag{5.35}$$

Note that the sinusoids of frequencies—here $k = 1, 2$—

$$\left.\begin{matrix}\omega_k\\ \omega_1+\omega_2\\ \omega_1-\omega_2\end{matrix}\right\} \quad \text{have periods} \quad \left\{\begin{matrix} T_k = \dfrac{T_c}{n_k}\\[2ex] T_+ = \dfrac{T_c}{n_1+n_2}\\[2ex] T_- = \dfrac{T_c}{n_1-n_2}\end{matrix}\right\} \text{respectively}$$

Consequently, the interval $[0, T_c]$ has an integral number of periods of each of the *six* sinusoids in expression (5.34). Therefore each sinusoid will have a zero contribution to the average defined in Eq. (5.35). Thus

$$P_{\text{ave}} \triangleq \frac{1}{T_c}\int_0^{T_c} p(t)\,dt = \frac{V_{m1}I_{m1}}{2}\cos(\sphericalangle V_1-\sphericalangle I_1)+\frac{V_{m2}I_{m2}}{2}\cos(\sphericalangle V_2-\sphericalangle I_2) \tag{5.36}$$

[12] If there is no such "common period," we must define P_{ave} by

$$P_{\text{ave}} = \lim_{T\to\infty}\frac{1}{T}\int_0^T p(t)\,dt$$

The calculations are long but straightforward. The conclusion is the same.

Conclusion The *average power* delivered *by* $v_s(\cdot)$ *to* the one-port N is equal to the sum of the average powers that each individual sinusoidal waveform would deliver to N if it were acting alone.

COMMENT This is a very important result that simplifies many calculations in circuits, communications systems, and in noise theory.

Exercises

1. Show that the conclusion above *does not hold* if $\omega_1 = \omega_2$.
2. Suppose that a linear time-invariant one-port N is driven by a current source i_s and that $i_s(\cdot)$ is periodic with period $T = 2\pi/\omega$. More precisely, assume that

$$i_s(t) = \sum_{k=0}^{N} I_{mk} \cos(k\omega t + \measuredangle I_k)$$

Let the driving-point impedance $Z(j\omega)$ of the one-port N be a known function of ω. Assume that the steady state has been reached. Calculate the *average power* delivered by i_s to N in terms of the I_{mk}'s, $\measuredangle I_k$'s, and $Z(j\omega)$.

5.5 Driving-Point Impedance, Stored Energy, and Dissipated Power

Consider a linear time-invariant circuit $\mathcal{N}$ driven by one sinusoidal independent current source of frequency ω rad/s (see Fig. 5.7). Assume that the circuit $\mathcal{N}$ is in the sinusoidal steady state. Assume that $\mathcal{N}$ includes only *passive* elements: resistors, capacitors, inductors, ideal transformers, gyrators. (In other words, no active elements such as negative resistors and controlled sources are allowed.)

Let

$$P_{\text{ave}} = \text{average power dissipated in the resistors of } \mathcal{N} \tag{5.37a}$$

$$\mathscr{E}_M = \text{average } magnetic \text{ energy stored in } \mathcal{N} \tag{5.37b}$$

$$\mathscr{E}_E = \text{average } electric \text{ energy stored in } \mathcal{N} \tag{5.37c}$$

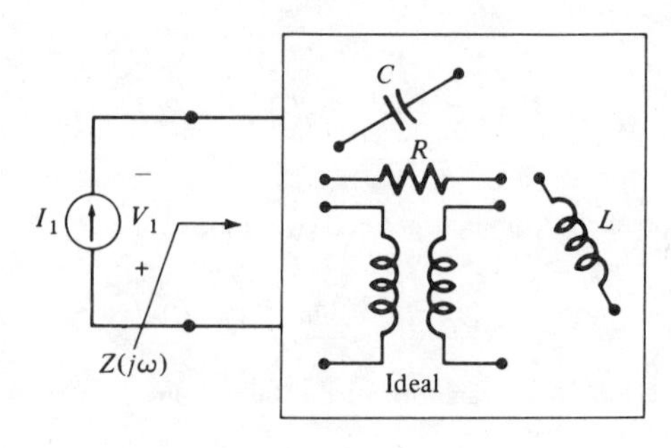

Figure 5.7 $Z(j\omega)$ is the driving-point impedance of a linear time-invariant circuit made of R's, L's, C's, and ideal transformers.

then we claim that

$$Z(j\omega) = \frac{2P_{\text{ave}} + 4j\omega[(\mathscr{E}_M(\omega) - \mathscr{E}_E(\omega)]}{|I_1|^2} \tag{5.38}$$

where I_1 is the phasor representing the input current.

COMMENTS

1. As expected $\text{Re}[Z(j\omega)]$ is $2P_{\text{ave}}/|I_1|^2$: The larger the dissipation, the larger the real part of the input impedance. [See also Eq. (5.15) above.]
2. If $\mathcal{N}$ is purely *reactive* (only L's and C's, no resistors), then $Z(j\omega)$ is purely imaginary; furthermore $\text{Im}[Z(j\omega)] \stackrel{\Delta}{=} X(j\omega)$ is positive or negative according to whether $\mathscr{E}_M(\omega) > \mathscr{E}_E(\omega)$ or vice versa. [Recall the series RL, $Z = R + j\omega L$, or the series RC circuit, $Z = R - j(1/\omega C)$.]
3. Equation (5.38) implies the following facts:

(*a*) If $\mathcal{N}$ consists only of *passive resistors* and/or *passive inductors*—often called a passive RL circuit—then

$$0 \leq \measuredangle Z(j\omega) \leq 90° \qquad \text{for all real } \omega\text{'s}$$

(*b*) If $\mathcal{N}$ consists only of *passive resistors* and/or *passive capacitors*—called a passive RC circuit—then

$$-90° \leq \measuredangle Z(j\omega) \leq 0 \qquad \text{for all real } \omega\text{'s}$$

(*c*) In the general case, if $\mathcal{N}$ has only *passive* R's, L's, C's, transformers, and gyrators, then

$$-90° \leq \measuredangle Z(j\omega) \leq +90° \qquad \text{for all real } \omega\text{'s}$$

or equivalently

$$\text{Re}[Z(j\omega)] \geq 0 \qquad \text{for all real } \omega\text{'s}$$

4. Similar conclusions hold for $Y(j\omega)$, by duality.

PROOF OF EQ. (5.38) Let us use throughout associated reference directions. By Tellegen's theorem we see that the complex power P delivered by the source is given by

$$\begin{aligned} P &= -V_1\bar{I}_1 = Z_{\text{in}}(j\omega)I_1\bar{I}_1 \\ &= Z_{\text{in}}(j\omega)\cdot|I_1|^2 \\ &= \sum_{\text{Res}} R_\rho|I_\rho|^2 + \sum_{\text{Ind}} j\omega L_\lambda|I_\lambda|^2 - \sum_{\text{Cap}} j\omega C_\gamma|V_\gamma|^2 \end{aligned} \tag{5.39}$$

where[13] the three sums are extended, respectively, over all resistors, inductors, and capacitors. Ideal transformers and gyrators being non-energic elements need not be included because their contribution is zero.

Now the *average* power dissipated in the resistor R_ρ is given by $\frac{1}{2}R_\rho|I_\rho|^2$: Hence the first sum is $2P_{\text{ave}}$ [see Eq. (5.37*a*)]. The average energy stored in inductor L_λ is given by $L_\lambda|I_\lambda|^2/4$, and the average energy stored in capacitor C_γ is given by $C_\gamma|V_\gamma|^2/4$. Formula (5.38) follows from these observations and Eq. (5.39). ■

Exercise Consider the setup shown in Fig. 5.7: Let $\mathcal{N}$ be as specified above except that it contains *no resistors* (hence $P_{\text{ave}} = 0$ and $Z = jX$). We say that $\mathcal{N}$ is an *LC circuit* and $X(\omega)$ is its *input reactance*. Show that

$$\frac{dX}{d\omega} = \frac{4[\mathscr{E}_M(\omega) + \mathscr{E}_E(\omega)]}{|I_1|^2} \geq \frac{|X(\omega)|}{\omega} = \frac{4|\mathscr{E}_M(\omega) - \mathscr{E}_E(\omega)|}{|I_1|^2} \tag{5.40}$$

Hint: The second equality in Eq. (5.40) comes directly from Eq. (5.38). To derive the first equality of Eq. (5.40), imagine successive sinusoidal steady states caused by the current $|I_1| \cos(\omega t + \sphericalangle I_1)$ where I_1 is *constant* and ω increases from experiment to experiment. Consequently the V_k's and I_k's become functions of ω. Apply Tellegen to $(\bar{I}_k, dV_k/d\omega)$ and then to $(dI_k/d\omega, \bar{V}_k)$.]

6 THREE-PHASE CIRCUITS[14]

The purpose of this section is to show why three-phase generators and three-phase transmission lines are used in power circuits. Then we show how the analysis of a balanced three-phase circuit reduces to that of a one-phase circuit.

6.1 General Considerations

The purpose of this subsection is to explain why most transmission lines that we see in the countryside are (*a*) high voltage, (*b*) three-phase (i.e., have three wires), and (*c*) are driven by ac generators (as opposed to dc generators).

Alternating current is more convenient than dc current because with ac it is easy to step up and step down the ac voltage with transformers. Furthermore

[13]
$$V_\lambda \bar{I}_\lambda = j\omega L_\lambda I_\lambda \bar{I}_\lambda = j\omega L_\lambda |I_\lambda|^2$$
$$V_\gamma \bar{I}_\gamma = V_\gamma(\overline{j\omega C_\gamma V_\gamma}) = -j\omega C_\gamma |V_\gamma|^2$$

[14]*References*:

A. E. Fitzgerald, C. Kingsley, Jr., and A. Kusko, *Electric Machinery*, 3d Edition, McGraw-Hill, New York, 1971.

O. I. Elgerd, *Electric Energy Systems*, McGraw-Hill, New York, 1971.

transformers are very efficient at 60 Hz and require practically no maintenance. Typically, in a power station the output voltage of the ac generator is in the range of 11 to 30 kV. It is then stepped up into hundreds of kilovolts for long distance transmission; then it is stepped down for the factories and homes.

High voltage is used in power transmission because the average power loss in a line[15] of impedance $R + jX$ is $W_L = \frac{1}{2}RI_m^2$; the average power transmitted is $W = \frac{1}{2}V_m I_m \cos(\measuredangle V - \measuredangle I)$. Hence, for a *given* transmitted power W, the power lost in heat in the transmission line is easily obtained by eliminating I_m

$$W_L = \frac{2RW^2}{V_m^2 \cos^2(\measuredangle V - \measuredangle I)} \qquad (6.1)^{16}$$

Thus for a given line (hence R is fixed) and for a given power to be transmitted (hence W is fixed), the power lost is reduced by using a voltage V_m as large as possible (typically up to 765 kV) and by keeping the power factor $\cos(\measuredangle V - \measuredangle I)$ close to 1.

In practice *ac generators are easier to build* than dc machines mainly for two reasons: (*a*) high-voltage high-current windings are on the stator; (*b*) the voltage induced in the stator is naturally oscillatory, and by pole shaping and/or winding design one can make the induced voltage almost sinusoidal.

Finally, *three-phase circuits* are used for engineering and economic reasons: (*a*) under a balanced load, *the torque on the generator is constant*, hence no vibrations. (This is far better than the multicylinder internal combustion engine or the Wankel rotary engine.) (*b*) With three-phase ac it is easy to create *a rotating magnetic* field (hence cheap induction motors); (*c*) with three-phase ac *one saves aluminum* on transmission lines: Under a balanced load, we will show that we need only three wires instead of six. These last three observations will be discussed in the following subsections.

6.2 Elementary One-Phase Generator

Figure 6.1 shows the cross section of an elementary *two-pole one-phase* ac generator. The rotor is a two-pole magnet: a permanent magnet in small ac generators and a magnet created by a large dc current in the case of high-power generators. The stator is slotted and contains the armature winding; for simplicity, the armature winding is shown as two conductors a and a' in Fig. 6.1. The dotted lines show the lines of the magnetic field.

The rotor position is specified by the angle θ. A turbine drives the rotor in the counterclockwise direction at a constant angular velocity of ω rad/s, typically 3600 r/min.

[15]We are considering a single-phase line: V and I represent the voltage and the current at the input of the line.

[16]The factor $\frac{1}{2}$ in the formula $W_L = \frac{1}{2}RI_m^2$ and hence the factor 2 in Eq. (6.1) is due to the fact that *we* use throughout *peak values* for sinusoidally varying quantities. (Power engineers always use RMS values, therefore, *they* write $W_L = RI_{\text{RMS}}^2$.)

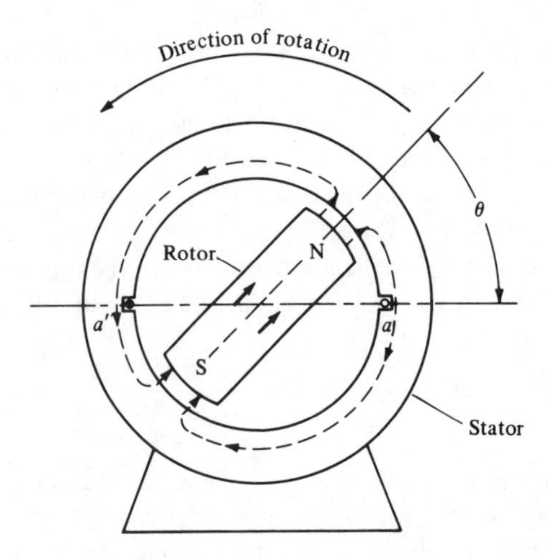

Figure 6.1 Elementary two-pole one-phase ac generator.

By shaping the rotor pole face and with proper design of the windings, the flux through the armature coil can be made to obey the law

$$\phi_a = \Phi_m \sin \theta \qquad (\text{where } \Phi_m \text{ is a constant}) \tag{6.2}$$

Suppose the rotor rotates at a constant velocity of

$$3600 \text{ r/min} = 60 \text{ r/s} = 2\pi \cdot 60 \text{ rad/s} \cong 377 \text{ rad/s}$$

Then, as a function of time

$$\phi_a(t) = \Phi_m \sin \omega t \qquad (\omega = 377 \text{ rad/s}) \tag{6.3}$$

By Faraday's law, $e = d\phi/dt$, the induced voltage in the coil aa' is then

$$e_a(t) = \omega \Phi_m \cos \omega t \tag{6.4}$$

or with

$$E_p \triangleq \omega \Phi_m \tag{6.5}$$

$$e_a(t) = E_p \cos \omega t \qquad \begin{cases} E_p = \textit{peak} \text{ phase voltage} \\ \omega = \text{angular frequency} \end{cases} \tag{6.6}$$

where E_p and ω are constant. This ac generator is called *synchronous* because the angular frequency of the ac voltage is *equal* to the angular velocity of the rotor.

REMARKS ON POWER AND TORQUE If the stator winding a–a' is connected to a load of impedance Z_L, the current in sinusoidal steady state will be (see Fig. 6.2)

$$i_a(t) = \frac{E_p}{|Z_L|} \cos(\omega t - \measuredangle Z_L) \tag{6.7}$$

The instantaneous power delivered by the generator to the load is (see Fig. 6.2)

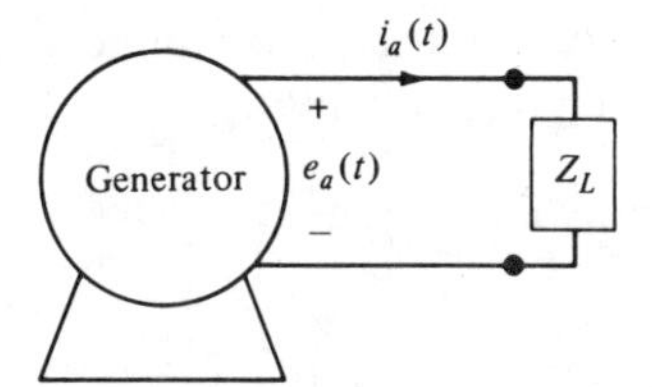

Figure 6.2 The one-phase ac generator drives the load Z_L.

$$p_a(t) = e_a(t)\, i_a(t) = \frac{E_p^2}{2|Z_L|}\,[\cos(\measuredangle Z_L) + \cos(2\omega t - \measuredangle Z_L)] \qquad (6.8)$$

By conservation of energy (neglecting internal losses inside the generator), the electric power delivered *to* Z_L is equal to the mechanical power delivered *by* the turbine *to* the generator; hence

$$p_a(t) = T(t)\,\omega \qquad (6.9)$$

where T is the applied torque and ω is the angular velocity of the rotor. From Eqs. (6.8) and (6.9), if the angular velocity ω is assumed constant, it is clear that the torque will be a constant term plus a sinusoidal term at the angular frequency 2ω. Hence single-phase ac generators have pulsating torques: At high power levels they would generate lots of vibration.

6.3 Elementary Three-Phase Generator

Figure 6.3 shows the cross section of an elementary *two-pole* three-phase ac generator. The rotor is a two-pole magnet. The stator contains three armature windings: a–a', b–b', c–c'. Winding b–b' is identical to winding a–a' except for a rotation of $120° = 2\pi/3$ rad in the direction of motion of the rotor. Similarly, winding c–c' is identical to winding a–a' except for a rotation of

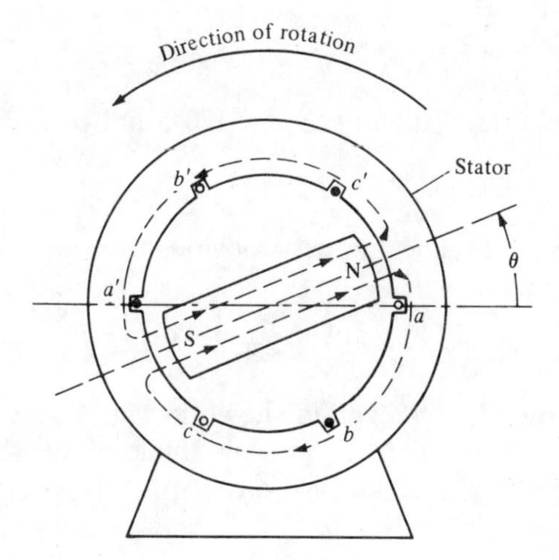

Figure 6.3 Elementary two-pole three-phase generator.

$240° = 4\pi/3$ rad. As a function of the rotor angle θ, the fluxes through the windings a–a', b–b', c–c' are, respectively,

$$\begin{aligned} \phi_a(\theta) &= \Phi_m \sin \theta \\ \phi_b(\theta) &= \Phi_m \sin\left(\theta - \frac{2\pi}{3}\right) \\ \phi_c(\theta) &= \Phi_m \sin\left(\theta - \frac{4\pi}{3}\right) \end{aligned} \tag{6.10}$$

If the rotor rotates at the constant angular velocity ω rad/s, then as functions of time, the fluxes and the induced voltages are

$$\left.\begin{aligned} \phi_a(t) &= \Phi_m \sin \omega t \\ \phi_b(t) &= \Phi_m \sin\left(\omega t - \frac{2\pi}{3}\right) \\ \phi_c(t) &= \Phi_m \sin\left(\omega t - \frac{4\pi}{3}\right) \end{aligned}\right\} \Rightarrow \left\{\begin{aligned} e_a(t) &= E_p \cos \omega t \\ e_b(t) &= E_p \cos\left(\omega t - \frac{2\pi}{3}\right) \\ e_c(t) &= E_p \cos\left(\omega t - \frac{4\pi}{3}\right) \end{aligned}\right. \tag{6.11}$$

where again $E_p \triangleq \omega\Phi_m$. In terms of phasors, we have (denoting phasors with a superscript "$\tilde{}$")

$$\begin{aligned} \tilde{E}_a &= E_p + j0 = E_p \qquad [E_p \text{ is real as the amplitude of } e_a(t) \text{ in (6.11)}] \\ \tilde{E}_b &= E_p \exp\left(-j\frac{2\pi}{3}\right) \\ \tilde{E}_c &= E_p \exp\left(-j\frac{4\pi}{3}\right) \end{aligned} \tag{6.12}$$

Thus phasor $\tilde{E}_b$ "lags" phasor $\tilde{E}_a$ by 120°, and phasor $\tilde{E}_c$ lags phasor $\tilde{E}_a$ by 240°.

REMARK $\tilde{E}_a + \tilde{E}_b + \tilde{E}_c = 0$ as is obvious from the equation

$$1 + \exp\left(-j\frac{2\pi}{3}\right) + \exp\left(-j\frac{4\pi}{3}\right) = 0$$

and from the phasor diagram shown in Fig. 6.4.

Since, by convention, we can think of phasors as rotating in the counterclockwise direction, we say that $\tilde{E}_b$ "lags" $\tilde{E}_a$ by 120° and $\tilde{E}_c$ "lags" $\tilde{E}_b$ by 120°.

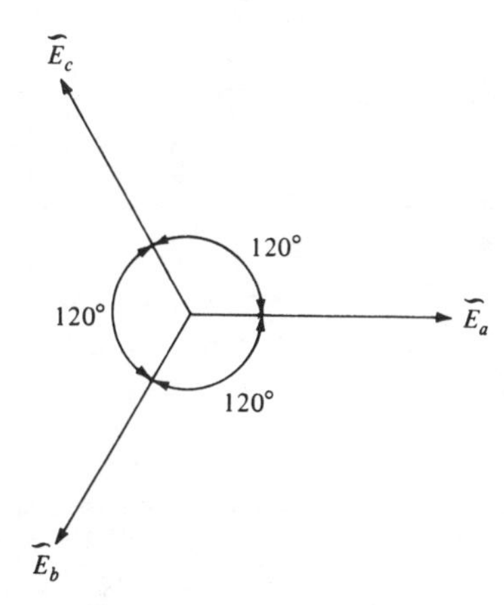

Figure 6.4 Plot of the phasors $\tilde{E}_a$, $\tilde{E}_b$, and $\tilde{E}_c$, which represent the voltages generated in the three-phase generator.

6.4 Three-Phase Generator Under Balanced Load

Suppose we have a generator which produces three-phase ac voltages as specified by Eq. (6.12). Thus the generator has three pairs of terminals as shown in Fig. 6.5. If we connect *identical* loads Z_L to each phase (i.e., the phases are "balanced"), then

$$\tilde{I}_a = \frac{\tilde{E}_a}{Z_L} \qquad \tilde{I}_b = \frac{\tilde{E}_b}{Z_L} \qquad \tilde{I}_c = \frac{\tilde{E}_c}{Z_L} \tag{6.13}$$

and, by Eq. (6.12)

$$\tilde{I}_a + \tilde{I}_b + \tilde{I}_c = 0 \tag{6.14}$$

Furthermore, we may tie one terminal of each phase, i.e., armature winding, to a common terminal *n*. This suggests redrawing the circuit as in Fig. 6.6. Clearly, because of Eq. (6.14), there is no need to put down the wire connecting node ⓝ to node ⓝ' since it carries no current!

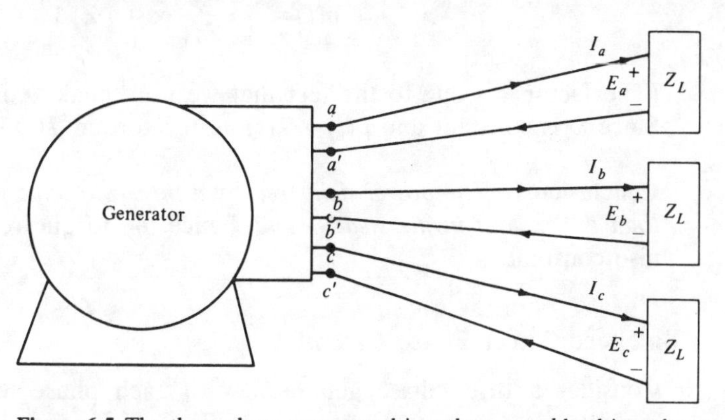

Figure 6.5 The three-phase generator drives three *equal* load impedances Z_L.

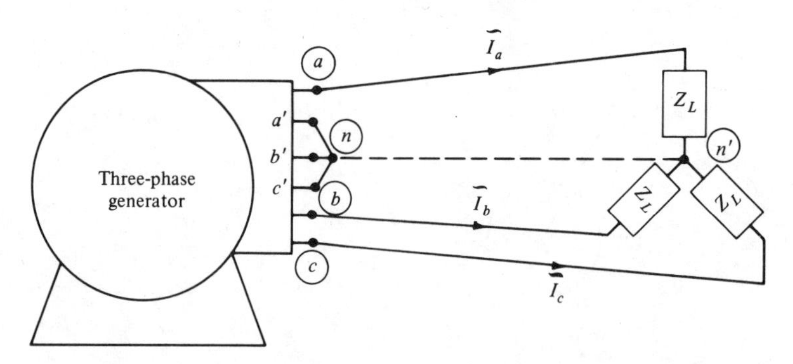

Figure 6.6 The three-phase generator drives the three load impedances Z_L which are connected in the Y configuration. Because the circuit is balanced, the wire connecting ⓝ to ⓝ' can be dispensed with.

Conclusion I For *balanced three-phase electric power transmission* instead of *six wires* as in Fig. 6.5 we need only *three wires* as shown in Fig. 6.6.

Let us now calculate the instantaneous power delivered *by* the generator *to* the loads shown in Fig. 6.6. Using Eqs. (6.12) and (6.13), we obtain

$$p_a(t) = e_a(t)\, i_a(t) = \frac{1}{2}\frac{E_p^2}{|Z_L|}\,[\cos(\sphericalangle Z_L) + \cos(2\omega t - \sphericalangle Z_L)]$$

$$p_b(t) = e_b(t)\, i_b(t) = \frac{1}{2}\frac{E_p^2}{|Z_L|}\left[\cos(\sphericalangle Z_L) + \cos\left(2\omega t - \sphericalangle Z_L - \frac{4\pi}{3}\right)\right]$$

$$p_c(t) = e_c(t)\, i_c(t) = \frac{1}{2}\frac{E_p^2}{|Z_L|}\left[\cos(\sphericalangle Z_L) + \cos\left(2\omega t - \sphericalangle Z_L - \frac{8\pi}{3}\right)\right]$$

It is easy to check by direct calculation that the total power $p_a + p_b + p_c$ is *constant* and equal to

$$p(t) = \frac{3}{2}\frac{E_p^2}{|Z_L|}\cos(\sphericalangle Z_L) \tag{6.15}$$

(The factor $\frac{1}{2}$ is due to the fact that we use "peak" voltages in the formula.) Since ω is constant and $p(t) = T(t)\,\omega$, the torque $T(t)$ is also constant!

Conclusion II *The power delivered by a three-phase ac generator to a balanced load is constant in the steady state*; hence the torque required for the rotor is also constant.

6.5 Analysis of a Balanced Three-Phase Circuit

Consider a three-phase generator with each phase terminated in the load impedance Z_L as shown in Fig. 6.7. Call Z_ℓ the impedance of the line

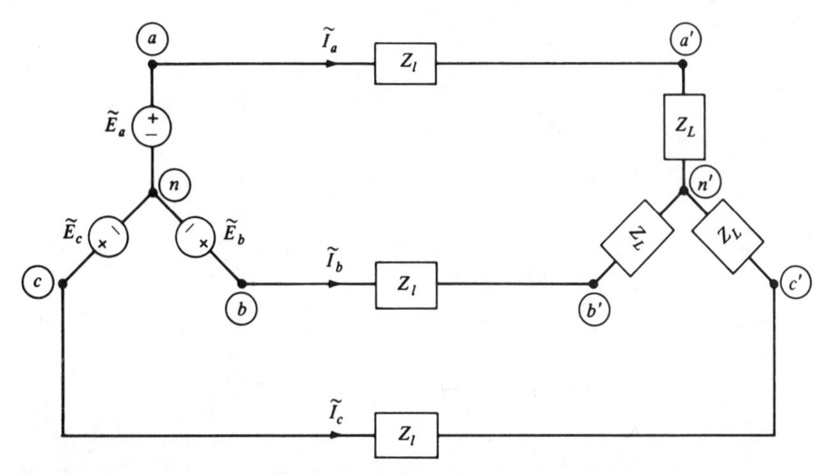

Figure 6.7 Equivalent circuit for the physical circuit shown in Fig. 6.6.

connecting each phase to the load. We wish to show that the analysis of this balanced three-phase circuit reduces to that of a one-phase circuit.

Let

$$Y \triangleq \frac{1}{Z_\ell + Z_L} \tag{6.16}$$

and let node ⓝ be the datum node; then the node equation of ⓝ' reads (using phasors)

$$Y(\tilde{E}_{n'} - \tilde{E}_a) + Y(\tilde{E}_{n'} - \tilde{E}_b) + Y(\tilde{E}_{n'} - \tilde{E}_c) = 0 \tag{6.17}$$

Since $\tilde{E}_a + \tilde{E}_b + \tilde{E}_c = 0$, Eq. (6.17) shows that $\tilde{E}_{n'} = 0$; thus, *node* ⓝ' *is at the same potential as* ⓝ, namely, *zero*. Writing KVL around the node sequence ⓝ–ⓐ–ⓐ'–ⓝ'–ⓝ, we obtain

$$\tilde{E}_a = (Z_L + Z_\ell)\tilde{I}_a \tag{6.18}$$

Hence, with Eq. (6.12), we obtain

$$\tilde{E}_b = (Z_\ell + Z_L)\tilde{I}_b \qquad \tilde{E}_c = (Z_\ell + Z_L)\tilde{I}_c$$

Consequently

$$\tilde{I}_b = \tilde{I}_a \exp\frac{-j2\pi}{3} \qquad \tilde{I}_c = \tilde{I}_a \exp\frac{-j4\pi}{3}$$

Conclusion There is a complete symmetry between the three phases: Once we solve one, we solve them all.

Exercise This exercise suggests the fact that three-phase currents produce easily rotating magnetic fields. These form the basis of both the asynchronous motor and the synchronous motor.

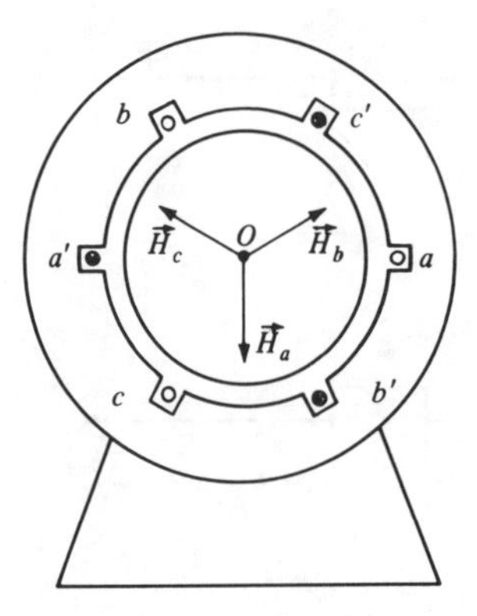

Figure 6.8 A balanced set of three sinusoidal currents into the windings a–a', b–b', and c–c' produce a total magnetic field $\vec{H}$ that is rotating at angular velocity ω.

Suppose the three windings a–a', b–b', and c–c' of a stator built as the one in Fig. 6.8 are traversed by three balanced ac currents i_a, i_b, and i_c. The reference directions are shown in Fig. 6.8. Suppose that the rotor is symmetric. Let $\vec{H}_a(t)$ be the magnetic field vector produced by $i_a(t)$ at the point 0 at the center of the rotor. $\vec{H}_b(t)$ and $\vec{H}_c(t)$ are similarly defined. Neglect saturation (i.e., use superposition) and show that the total field $\vec{H}(t) = \vec{H}_a(t) + \vec{H}_b(t) + \vec{H}_c(t)$ is a rotating field. [Hint: Choose the time origin so that $i_a(t) = I_m \cos \omega t$; hence $\vec{H}_a(t) = \vec{H}_{ma} \cos \omega t, \ldots$.]

(A thorough analysis would require analysis of the fields in the air gap. This is the subject of courses on electric machinery.)

SUMMARY

Phasors

- The sinusoid $A_m \cos(\omega t + \phi)$ is represented by the phasor $A = A_m e^{j\phi}$. Conversely, the phasor A defines the sinusoid by

$$\text{Re}[A\,e^{j\omega t}] = \text{Re}[A_m\,e^{j\phi}\,e^{j\omega t}] = A_m \cos(\omega t + \phi)$$

- Any sinusoid defines the corresponding phasor uniquely, and given the frequency ω, the phasor defines the sinusoid uniquely.
- Let $x_1(t)$, $x_2(t)$, and $x_3(t)$ be sinusoids at the *same* frequency represented by the phasors A_1, A_2, and A_3. If a_1, a_2, and a_3 are *real* numbers, then the sinusoid $a_1\,x_1(t) + a_2\,x_2(t) + a_3\,x_3(t)$ is represented by the phasor $a_1A_1 + a_2A_2 + a_3A_3$.
- If sinusoid $a(t)$ is represented by phasor A, then sinusoid $\dfrac{da}{dt}$ is represented by $j\omega A$.

Circuit Equations in Terms of Phasors

- In terms of phasors, the branch equations read: $V = RI$, $V = j\omega LI$, $I = j\omega CV$, $I = g_m V$, etc.
- The tableau equation reads

$$\begin{bmatrix} \mathbf{0} & \mathbf{0} & \mathbf{A} \\ -\mathbf{A}^T & \mathbf{1} & \mathbf{0} \\ \mathbf{0} & \mathbf{M}(j\omega) & \mathbf{N}(j\omega) \end{bmatrix} \begin{bmatrix} \mathbf{E} \\ \mathbf{V} \\ \mathbf{I} \end{bmatrix} = \begin{bmatrix} \mathbf{0} \\ \mathbf{0} \\ \mathbf{U}_s \end{bmatrix}$$

Network Functions

- Let the linear time-invariant circuit be driven by the *sinusoidal* source I_s, and let it be in the sinusoidal steady state. Any circuit variable, say V_k, is related to the phasor I_s by

$$V_k = H(j\omega)\, I_s$$

where the *network function* $H(j\omega)$ is a ratio of two polynomials in $j\omega$ with *real* coefficients: $H(j\omega) = n(j\omega)/d(j\omega)$.
- For all ω real, $|H(j\omega)| = |H(-j\omega)|$ and, by proper choice of the phase,

$$\measuredangle H(j\omega) = -\measuredangle H(-j\omega)$$

- From the phasor equation $V_k = H(j\omega)\, E_s$, we have the time-domain equation

$$v_k(t) = |H(j\omega)| \cdot |E_s| \cos[\omega t + \measuredangle E_s + \measuredangle H(j\omega)]$$

- If a linear time-invariant circuit is driven by several independent sinusoidal sources *at different frequencies*, then the steady-state waveforms that exist in the circuit are the sum of the *sinusoidal steady states* due to each source acting *alone*.
- *Normalization*: Let $r \triangleq$ (desired impedance)/(normalized impedance), and let $\omega_n \triangleq$ (desired typical frequency)/(normalized typical frequency). If we denote by the subscript 0 the normalized element values, then

$$R = rR_0 \qquad L = \frac{r}{\omega_n} L_0 \qquad C = \frac{1}{r\omega_n} C_0 \qquad g_m = \frac{1}{r} g_{m0} \qquad r_m = rr_{m0}$$

Resonant Circuit

Let $\mathcal{N}$ be the parallel RLC circuit driven by a sinusoidal source I_s at frequency ω. The driving-point admittance of $\mathcal{N}$ is

$$Y(j\omega) = \frac{1}{R_p} + j\omega C + \frac{1}{j\omega L}$$

As ω goes from $-\infty$ to $+\infty$, $Y(j\omega)$ traces a vertical line in the Y plane of abscissa $1/R_p$. $|Y(j\omega)|$ reaches a minimum of $1/R_p$ at $\omega_0 = 1/\sqrt{LC}$. Correspondingly

$$Z(j\omega) = \frac{1}{Y(j\omega)} = \frac{R_p}{1 + jQ[(\omega/\omega_0) - (\omega_0/\omega)]}$$

and $Z(j\omega)$ traces a circle of center $(R_p/2, 0)$ and diameter R_p. For $Q \gg 1$, the 3-dB frequencies are given by $\omega_\pm = \omega_0(1 \pm 1/2Q)$.

Power

- In sinusoidal steady state, the *average power* into a one-port is given by

$$P_{\text{ave}} = \tfrac{1}{2}|V| \cdot |I| \cos(\measuredangle V - \measuredangle I) = \tfrac{1}{2}|V|^2 \operatorname{Re}[Y] = \tfrac{1}{2}|I|^2 \operatorname{Re}[Z]$$

- The complex power delivered to a one-port is given by

$$P \triangleq \tfrac{1}{2}V\bar{I} = P_{\text{ave}} + jQ$$

  where Q is called the *reactive power*.
- In a linear time-invariant circuit operating in the sinusoidal steady state, *complex power is conserved*; equivalently, using associated reference directions for all branches of the circuit,

$$\sum_{k=1}^{b} V_k \bar{I}_k = 0$$

- Let a *given* sinusoidal generator, specified by its Thévenin equivalent E_G, $Z_G = R_G + jX_G$ with $R_G > 0$, drive a one-port with driving-point impedance Z_L. In the sinusoidal steady state, the one-port will receive a maximum average power if and only if $Z_L = \bar{Z}_G$. Then $P_{\text{ave}} = |E_G|^2/8R_G$.
- Suppose that $v_s(\cdot)$ is a sum of sinusoidal waveforms with different frequencies, that $v_s(\cdot)$ drives a linear time-invariant circuit $\mathcal{N}$, and that the circuit is in the steady state; then the *average power* delivered *by* $v_s(\cdot)$ *to* the circuit $\mathcal{N}$ is equal to the sum of the *average powers* that each sinusoidal waveform would deliver to $\mathcal{N}$ if it were acting alone.
- Power transmission is almost universally carried by three-phase circuits because (*a*) under a balanced load a three-phase generator needs a *constant* torque, (*b*) three-phase circuits create easily *rotating magnetic fields*, and (*c*) three-phase transmission lines require only *three* wires instead of *six*.
- By symmetry, the analysis of a balanced three-phase circuit reduces to that of a one-phase circuit.

PROBLEMS

Complex numbers

1 Evaluate the following complex quantities and express the answers in terms of both the polar and the rectangular coordinates:

(*a*) $\dfrac{(1+j)(1+j2)}{j5(1-j)}$

(*b*) $2e^{j30°} - e^{-j45°}$

(*c*) $\left(\dfrac{1-j}{1+2j}\right)e^{j45°}$

Phasor representation

2 (*a*) Determine the phasors which represent the following real-valued time functions.

(i) $10\cos\left(2\pi + \dfrac{\pi}{6}\right) + 5\sin 2t$

(ii) $\sin\left(3t - \dfrac{\pi}{3}\right) + \cos\left(3t + \dfrac{\pi}{3}\right)$

(iii) $\cos t + \cos\left(t + \dfrac{\pi}{3}\right) + \cos\left(t + \dfrac{\pi}{3}\right)$

(iv) $\cos t + \cos\left(t + \dfrac{2\pi}{3}\right) + \cos\left(t + \dfrac{4\pi}{3}\right)$

(*b*) If the complex quantities in Prob. 9.1 above are phasors, determine the corresponding real-valued time function in each case. Assume ω as the angular frequency.

Sinusoidal steady-state solution

3 For the circuits shown in Fig. P9.3,
(*a*) Write a second-order differential equation in terms of v_0.
(*b*) Using phasors obtain the sinusoidal steady-state solution for $v_0(t)$ and $i_L(t)$.

4 Find the complete solution of the following differential equations. Indicate whether the steady-state solution exists for each case.

(*a*) $\ddot{x} + \dot{x} + x = \left(\dfrac{d}{dt} + 1\right)\cos 2t$

$x(0^-) = 1 \quad \dot{x}(0^-) = -1$

(*b*) $\ddot{x} + x = \cos t$

$x(0^-) = 1 \quad \dot{x}(0^-) = 0$

(*c*) $\ddot{x} - 3\dot{x} + 2x = \sin 2t$

$x(0^-) = 1 \quad \dot{x}(0^-) = 1$

Phasor diagram

5 The circuit shown in Fig. P9.5 is in sinusoidal steady state. Assuming $v_c(t) = \cos 2t$,
(*a*) Construct a phasor diagram showing all voltages and currents indicated.
(*b*) Find the sinusoidal steady-state voltage $e_1(t)$.

Impedance and admittance

6 Given, for all t,

$$v(t) = 50\sin\left(10t + \frac{\pi}{4}\right)$$

$$i(t) = 400\cos\left(10t + \frac{\pi}{6}\right)$$

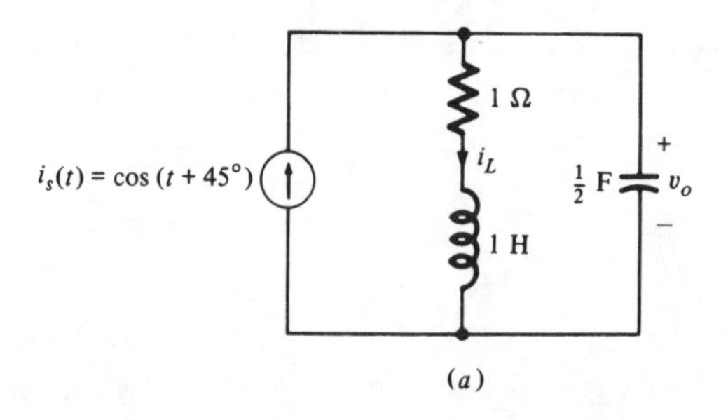

(a)

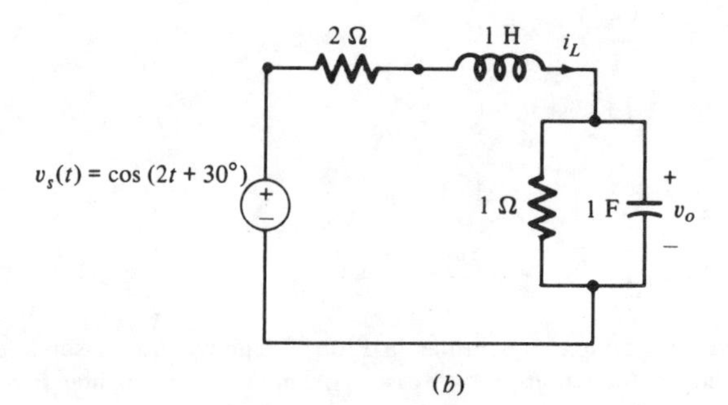

(b)

Figure P9.3

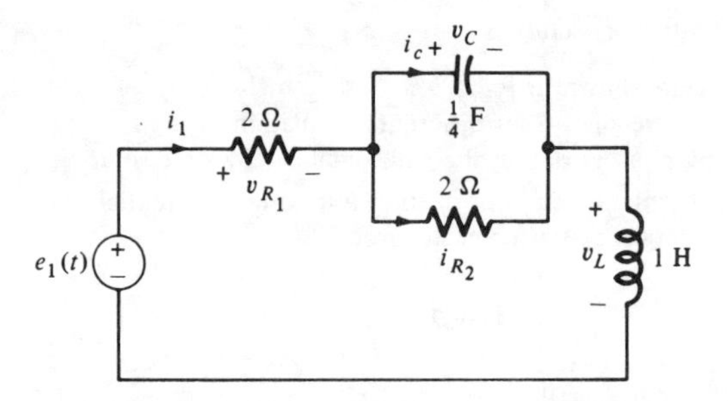

Figure P9.5

find suitable elements of the linear, time-invariant circuit shown in Fig. P9.6, and indicate their values in ohms, henrys, or farads.

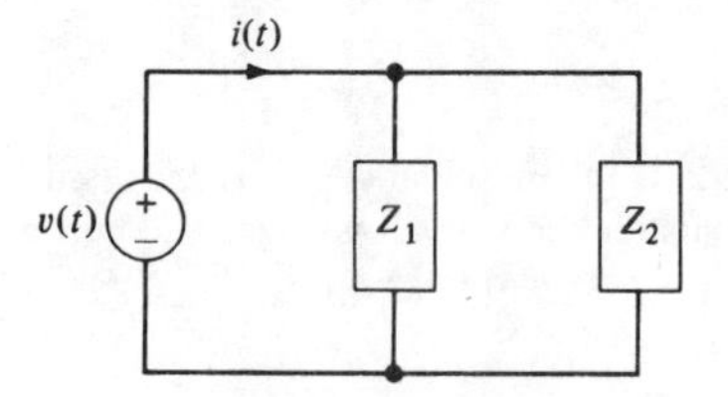

Figure P9.6

7 Determine the driving-point impedance $Z(j\omega)$ of the circuits shown in Fig. P9.7.

8 For the circuit shown in Fig. P9.8,

(a) Find the driving-point impedance $Z(j\omega)$.

(b) Calculate the value of the impedance for $\omega = 0$ and $\omega = 1$ rad/s. Express the impedance as a magnitude and an angle.

(c) Explain by physical reasoning the value of the impedance for $\omega = 0$ and $\omega = \infty$.

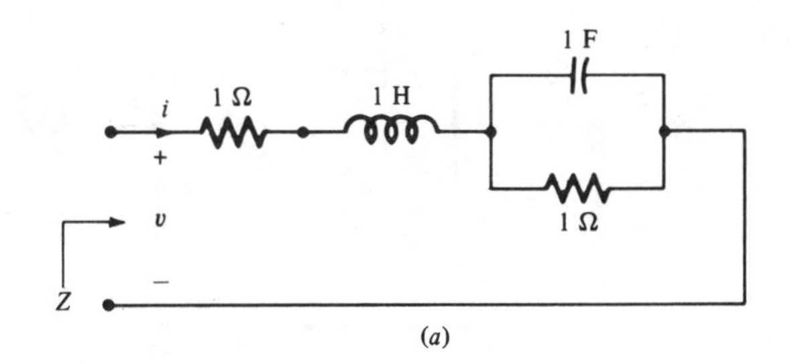

(a)

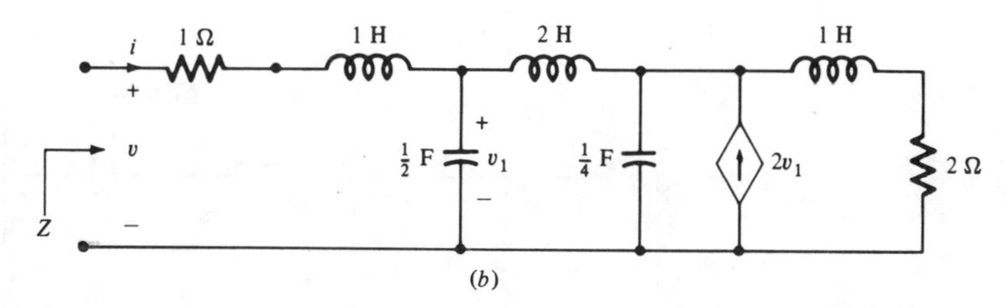

(b)

Figure P9.7

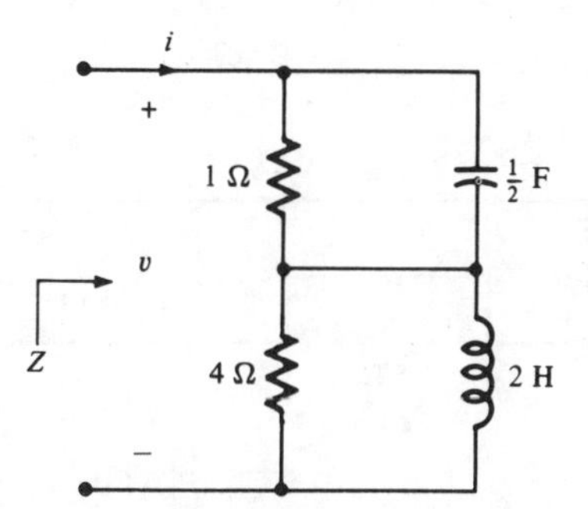

Figure P9.8

Tableau equations

9 Assuming sinusoidal steady state, write the tableau equations for the circuits shown in Fig. P8.4.

Node analysis

10 Write by inspection the node equations for the circuits shown in Fig. P9.10. Assume sinusoidal steady state.

Superposition theorem

11 If a current source $i_s(t) = 1 + \cos t + \cos 2t$ is applied to the one-port in Fig. P9.7a, determine the steady-state port voltage.

12 For the circuit in Fig. P9.12, calculate the steady-state voltage v as a function of time.

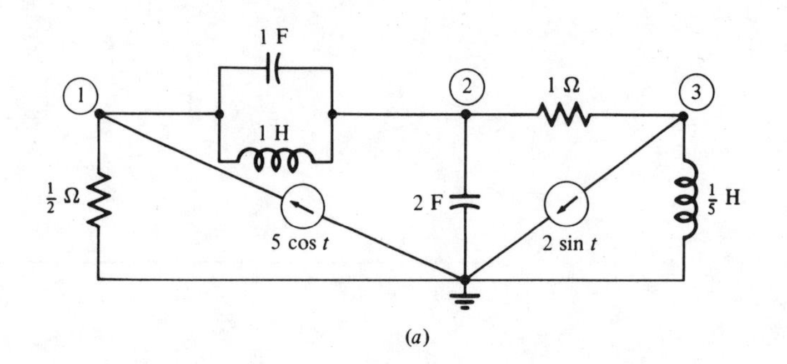

(a)

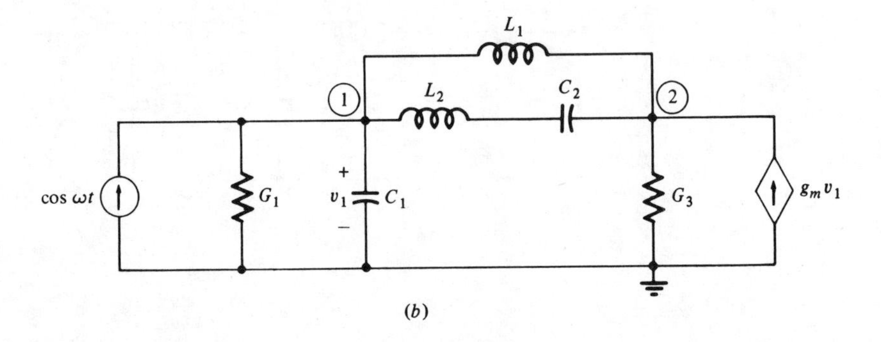

(b)

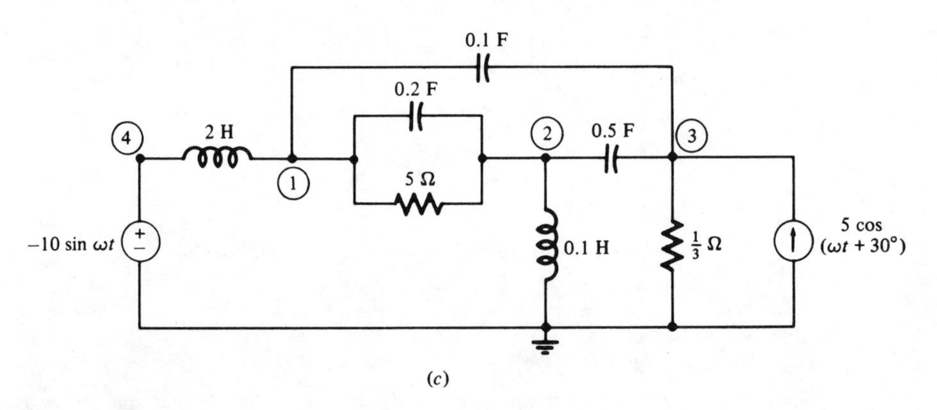

(c)

Figure P9.10

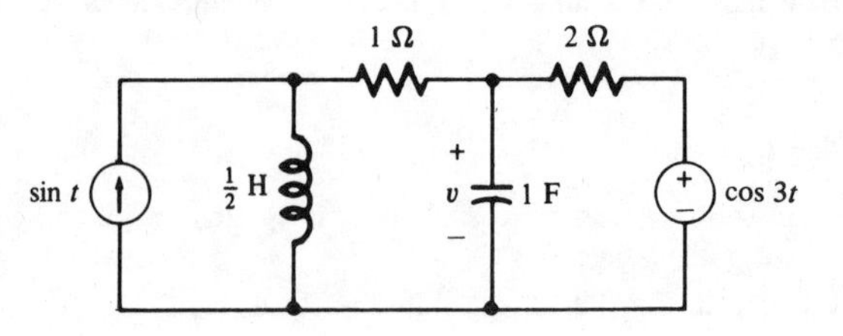

Figure P9.12

Thévenin-Norton

13 The circuit shown in Fig. P9.13 is in the sinusoidal steady state, $e_s(t) = 9\cos 10t$ and $i_s(t) = 2\cos[10t - (\pi/3)]$. For the circuit to the left of the terminals ① and ①′, obtain

(*a*) The Thévenin equivalent circuit.

(*b*) The Norton equivalent circuit.

(*c*) Calculate v for $R = 1\ \Omega$ and $R = 10\ \Omega$ (express your answer as a real-valued function of time).

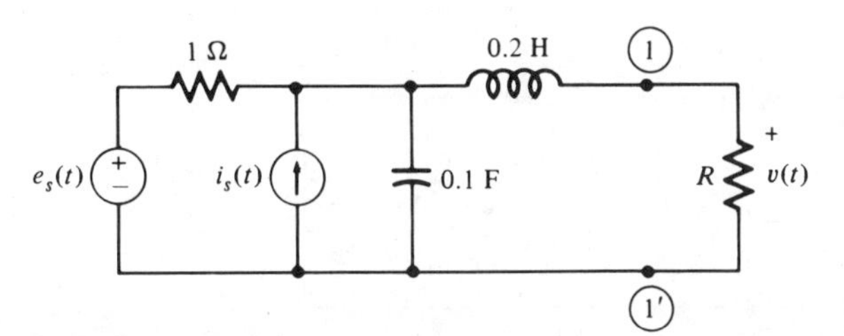

Figure P9.13

14 The linear time-invariant circuit shown in Fig. P9.14 is in the steady state. In order to determine the steady-state inductor current i, use Thévenin's theorem to

(*a*) Determine the open-circuit voltage v_{oc} at terminals ① and ①′ when the inductor is open-circuited.

(*b*) Determine Z_{eq}, the equivalent impedance which is faced by the inductor.

(*c*) Determine the steady-state current i.

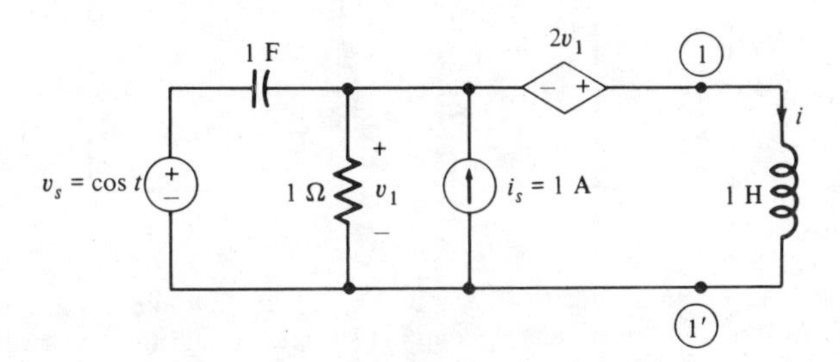

Figure P9.14

15 Using Thévenin's theorem obtain the steady-state current i as a real-valued function of time for the circuit shown in Fig. P9.15.

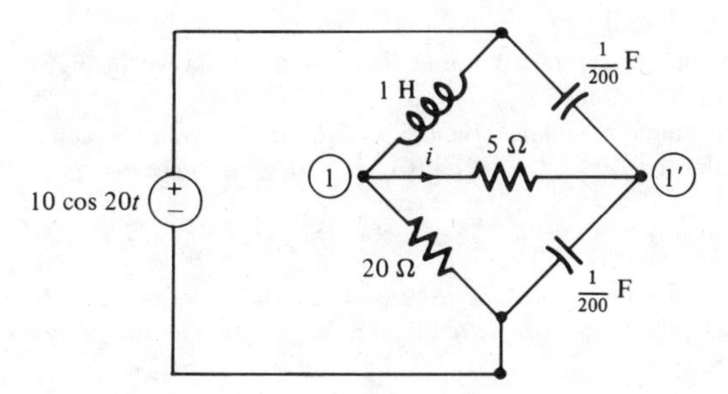

Figure P9.15

Network functions

16 For the ladder circuit shown in Fig. P9.16,

(*a*) Determine the driving-point admittance $Y(j\omega)$.

(*b*) Calculate the steady-state current $i_1(t)$ due to the sinusoidal voltage source $e_s(t) = 2\cos 2t$.

(*c*) Determine the transfer admittance $Y_{21}(j\omega) = I_2/E_s$ where I_2 and E_s are the phasors representing the sinusoidal current i_2 and sinusoidal voltage e_s, respectively.

(*d*) Calculate the steady-state current i_2.

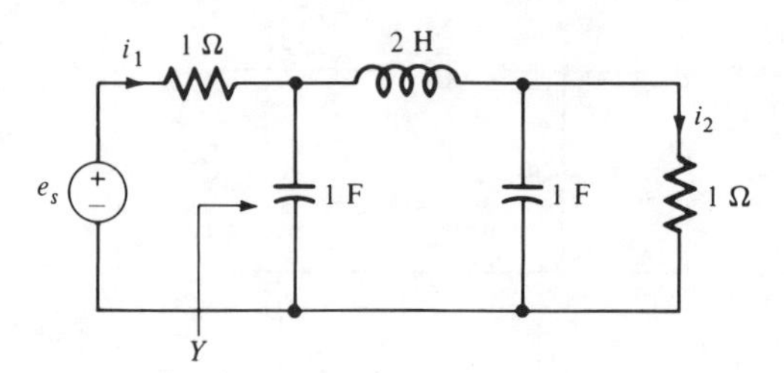

Figure P9.16

17 Consider the coupled circuit shown in Fig. P9.17. Determine:

(*a*) The driving-point impedance V_1/I_1.

(*b*) The transfer impedance V_2/I_1.

(*c*) The transfer voltage ratio V_2/V_1.

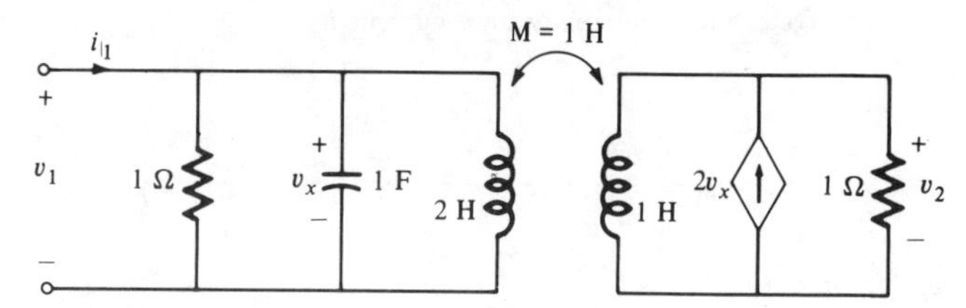

Figure P9.17

18 For the op-amp circuits shown in Fig. P9.18 find the voltage transfer function $H(j\omega) = V_0(j\omega)/V_i(j\omega)$. Assume ideal op-amp model operating in the linear region.

Resonant circuits

19 For the resonant circuit in Fig. P9.19,

(*a*) Calculate the resonant frequency ω_0 and the value of Q.

(*b*) Calculate the driving-point impedance $Z(j\omega)$.

(*c*) Plot $|Z(j\omega)|$ and $\measuredangle Z(j\omega)$ versus ω/ω_0.

20 The resonance curve of a parallel *RLC* circuit is shown in Fig. P9.20.

(*a*) Find R, L, and C.

(*b*) The same resonance behavior is desired around a center frequency of 20 kHz. The maximum value of $|Z(j\omega)|$ is to be 0.1 MΩ. Find the new values of R, L, and C.

Energy and power

21 The circuit in Fig. P9.21 is in the sinusoidal steady state. If $e_s(t) = 10\cos\omega t$ V, find the instantaneous power into the circuit and the instantaneous energy stored in the inductor as a function of time.

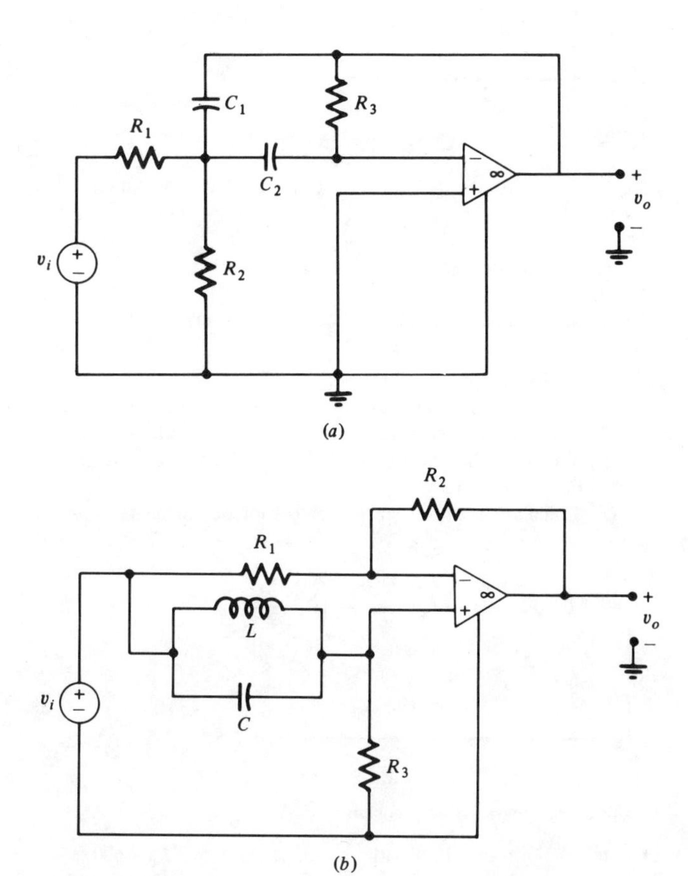

Figure P9.18

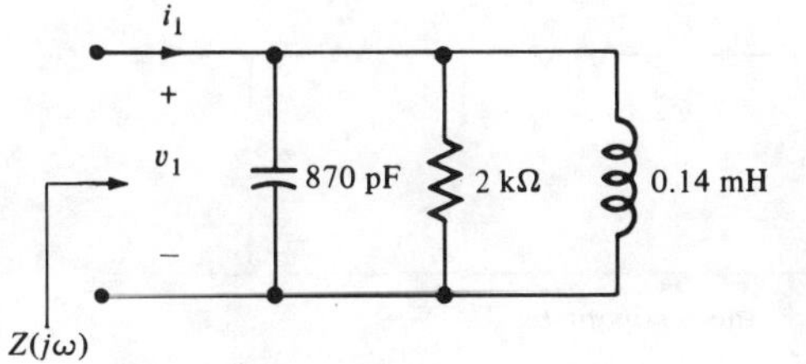

Figure P9.19

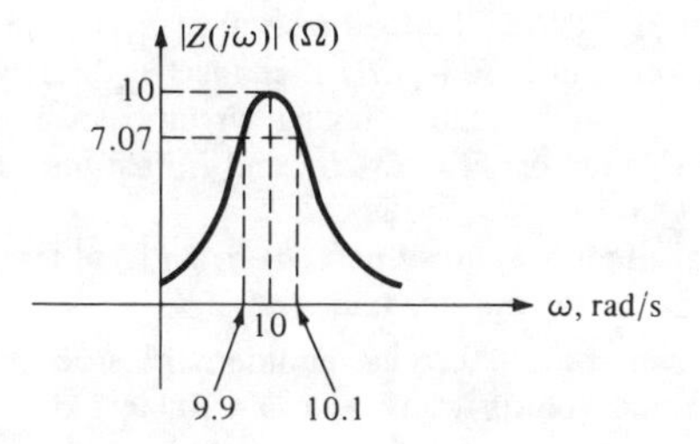

Figure P9.20

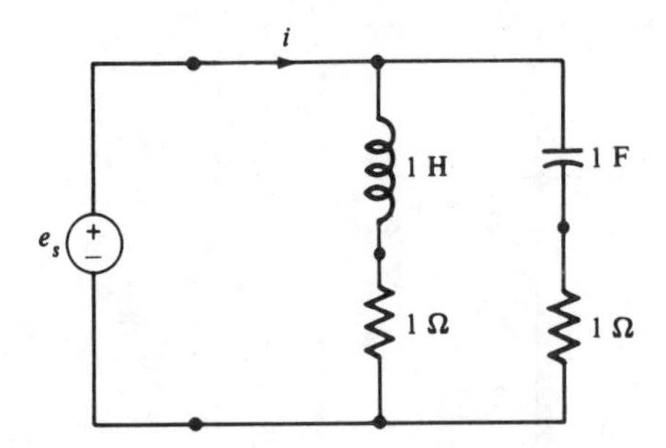

Figure P9.21

22 For the circuit shown in Fig. P9.22,

(*a*) Calculate the sinusoidal steady-state response i to $e_s = \sin \omega t$ V for values of $\omega = 2$, 2.02, and 2.04 rad/s. Give each result as a real-valued function of time.

(*b*) Calculate the energy stored in the capacitor and in the inductor as functions of time for $\omega = 2$, 2.02, and 2.04 rad/s.

(*c*) Calculate the average power dissipated in the resistor for $\omega = 2$, 2.02, and 2.04 rad/s.

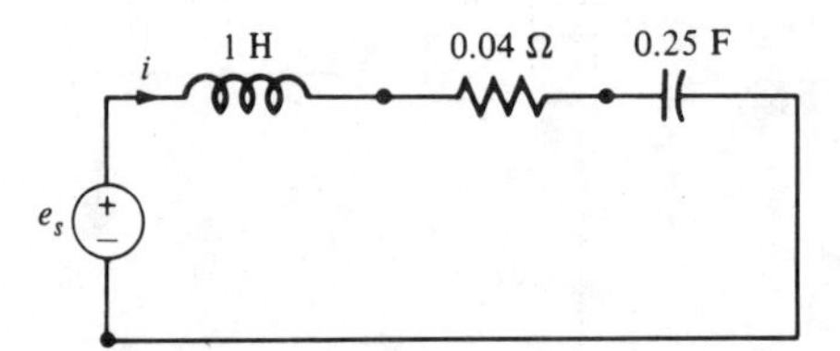

Figure P9.22

Maximum power transfer theorem

23 A telephone transmitter with an output resistance $R_0 = 600\ \Omega$ is to feed a transmission line that can be modeled by an infinite ladder of resistances as in Fig. P9.23. Specify R for maximum power transfer.

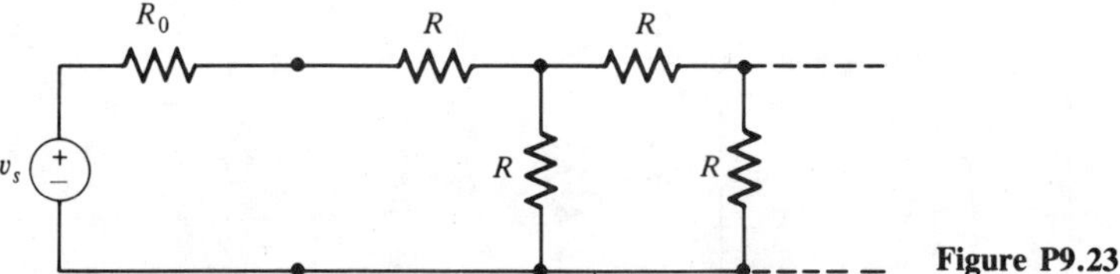

Figure P9.23

24 A load Z is supplied from two energy sources as in Fig. P9.24. Specify the value of Z to absorb maximum average power, and find the average power absorbed by the specified value of Z.

25 For the circuit shown in Fig. P9.25, the load resistance R_L is equal to $R_G/2$.

(*a*) Determine the power transferred to R_L if connected directly to the generator.

(*b*) To increase power transfer, the coupling circuit shown is used as an "impedance-matching" device. Find the relation between L_1, L_2, and L_3 that must be satisfied if power transfer is to be maximum.

(*c*) Suppose you replace the two-port of part (*b*) by an ideal transformer; find the turns ratio to achieve maximum average power transfer into R_L.

26 The signal generator shown in Fig. P9.26 has an internal resistance $R_G = 500\ \Omega$. Specify values of L and C to provide maximum power transfer at $\omega = 10{,}000$ rad/s to a load $R_L = 5\ \Omega$.

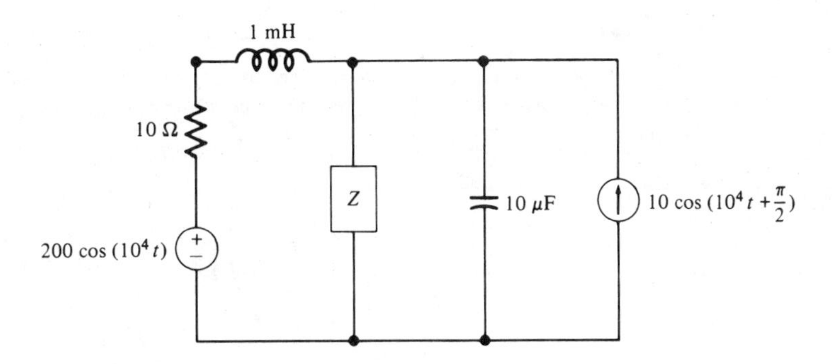

Figure P9.24

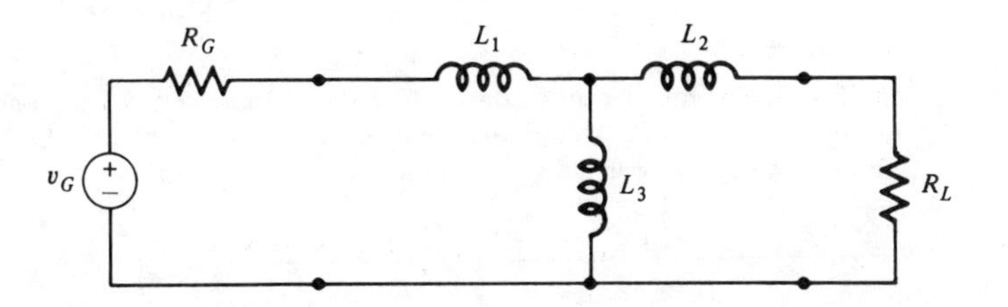

Figure P9.25

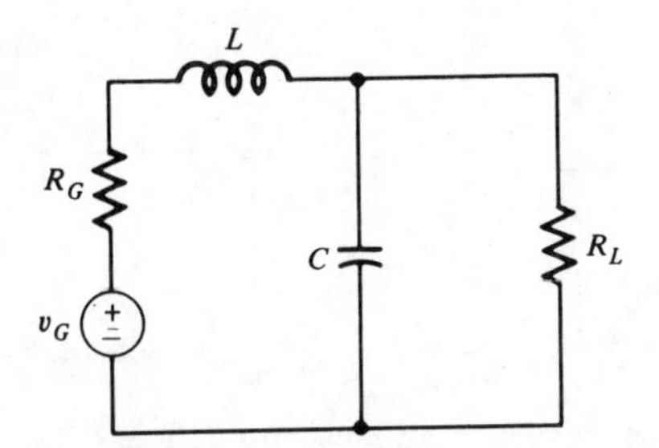

Figure P9.26

Properties of *RLC* circuits

27 Your assistant measures the driving-point impedance (or admittance) at a fixed frequency ω_0 of a number of linear time-invariant circuits made of passive elements. In each case, state whether or not you have any reasons to believe the validity of these measurements (in ohms or siemens).

(*a*) *RC* circuit: $Z = 5 + j2$
(*b*) *RL* circuit: $Z = 5 - j7$
(*c*) *RLC* circuit: $Y = 2 - j3$
(*d*) *LC* circuit: $Z = 2 + j3$
(*e*) *RLC* circuit: $Z = -5 - j19$
(*f*) *RLC* circuit: $Z = -j7$

Whenever you accept a measurement as plausible, assume $\omega_0 = 1$ rad/s and give a linear time-invariant passive circuit which has the specified network function.

Three-phase circuits

28 A load consisting of three identical impedances $Z = 10\measuredangle -45° \ \Omega$ connected in Δ is connected to a three-phase 220-V source (Fig. P9.28). Determine the line currents I_a, I_b and I_c, and the currents through each impedance Z.

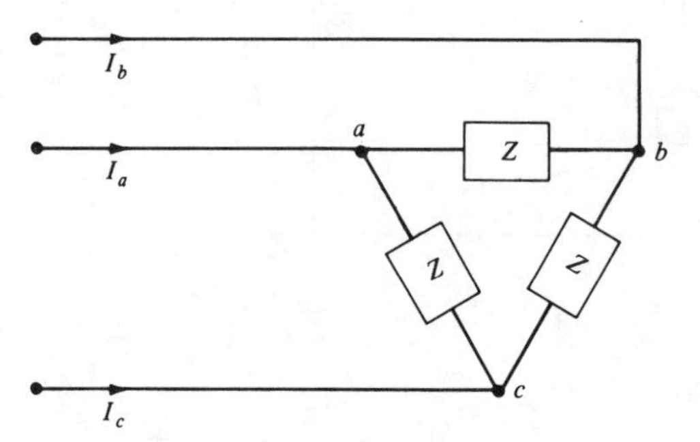

Figure P9.28

29 Three equal impedances of $Z = 10\measuredangle 45° \ \Omega$ are connected in Y across a three-phase 220-V supply as shown in Fig. P9.29. Determine the phase voltages V_{ac}, V_{bc}, the line currents I_a, I_b and I_c. Find the total average power delivered to the three impedances.

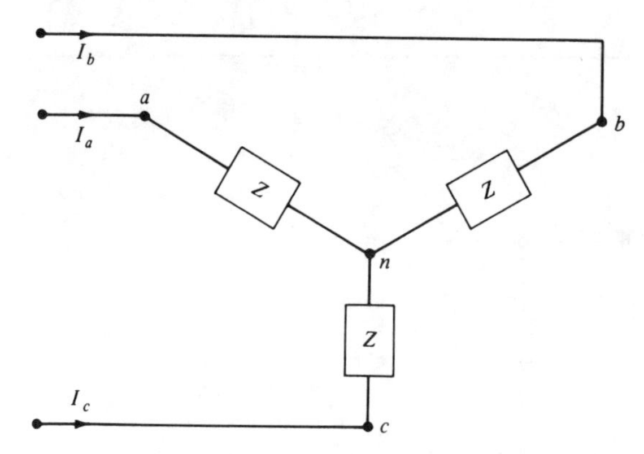

Figure P9.29

CHAPTER

TEN

LINEAR TIME-INVARIANT CIRCUITS

INTRODUCTION

In the preceding chapter we studied linear time-invariant circuits in the *sinusoidal steady state*, and our main tool was phasor analysis. In the present chapter we also study linear time-invariant circuits, but we do it now under general excitation. We will encounter a number of basic concepts and basic properties of such circuits that are indispensable to the solution of many engineering problems.

The first three sections are devoted to a brief statement on the most useful properties of Laplace transforms. The Laplace transform is a generalization of the phasor concept: The phasor A is the *complex number* associated with the sinusoid $|A|\cos(\omega t + \measuredangle A)$. Similarly, the *Laplace transform* associates a *complex-valued function* of s, say $F(s)$, to a given *time function*, $f(t)$, defined on $[0, \infty)$. The Laplace transform plays a crucial role in relating the *time-domain behavior* to the *frequency-domain behavior* of linear time-invariant circuits. A detailed tabulation of the properties of the Laplace transform is given in the summary section at the end of this chapter.

Section 4, first, obtains the tableau and MNA equations in terms of Laplace transforms; second, it relates the natural frequencies of a circuit to its zero-input response; and third, it relates the network function to the impulse response.

Section 5 exhibits the mechanism whereby the sinusoidal steady state is independent of the initial conditions.

Section 6 studies in detail the operation of convolution, and explains its importance in engineering.

All the concepts and results of this chapter will be used repeatedly in the remainder of this book.

Review of Rational Functions

By definition, a *real rational function* is the ratio of two *polynomials* with *real* coefficients. For example,

$$H(s) = \frac{n(s)}{d(s)} = \frac{s+1}{(s+2)[(s+1)^2+3^2]}$$

It is of crucial importance to note that

$$\frac{s}{s+1} = \frac{s(s-1)}{(s+1)(s-1)} = \frac{s(s-1)(s+3)^2}{(s+1)(s-1)(s+3)^2}$$

These three expressions represent the *same* rational *function*, in precisely the same manner as the expressions $\frac{1}{2}$, 0.5, $\frac{16}{32}$ represent the same point on the real number line. Recall that a/b is equal to c/d if and only if $ad = bc$.

For obvious reasons, it is efficient to cancel any common factors between the polynomials $n(s)$ and $d(s)$; *after* cancellation of all common factors, the polynomials $n(s)$ and $d(s)$ are called *coprime*. Given

$$H(s) = \frac{n(s)}{d(s)}$$

and given that $n(s)$ and $d(s)$ are *coprime*, z_i is called a *zero* of $H(s)$ iff $n(z_i) = 0$ [by coprimeness, $d(z_i) \neq 0$, hence $H(z_i) = 0$]; p_i is called a *pole* of $H(s)$ iff $d(p_i) = 0$ [by coprimeness $n(p_i) \neq 0$], hence as $s \to p_i$, $H(s) \to \infty$.

1 DEFINITION OF THE LAPLACE TRANSFORM

1.1 Definition

Throughout this chapter, the variable s will be a *complex* variable and its real and imaginary part will be denoted by σ and ω, respectively; thus

$$s = \sigma + j\omega \qquad \sigma \text{ real and } \omega \text{ real} \tag{1.1}$$

We view s as a point in the complex plane: σ is its abscissa and ω is its ordinate.

Suppose we are given a function of time $f(t)$ defined on $[0, \infty)$; we then form the product $f(t)\,e^{-st}$ and integrate from $0-$ to $+\infty$, thus[1]

$$\boxed{F(s) \triangleq \int_{0-}^{\infty} f(t)\,e^{-st}\,dt} \tag{1.2}$$

[1] In the integral (1.2), it is understood that σ is taken sufficiently large and positive so that $e^{-\sigma t}$ decays fast enough to make the integral converge, in other words, so that the area under the curve $|f(t)|\,e^{-\sigma t}$ is finite.

In the integral above, t is the integration variable and it is integrated between the limits $0-$ and ∞; hence the integral depends only on the time function $f(\cdot)$ and on the value of s.

The function $F(s)$ defined in Eq. (1.2) is called the *Laplace transform* of the time function $f(t)$; s is called the *complex frequency*.

REMARKS

1. The lower limit of integration is chosen to be $0-$, so that whenever $f(t)$ includes an impulse at the origin, it is included in the interval of integration. [See Example 2 below, Eq. (1.15).]
2. In some calculations, $f(t)$ will be complex-valued [e.g., $\exp(\alpha + j\beta)t$]; then the integral (1.2) is easily reduced to a linear combination of integrals of real-valued functions. Wc have

$$e^{st} = e^{\sigma t + j\omega t} = (e^{\sigma t} \cos \omega t) + j(e^{\sigma t} \sin \omega t)$$

Let us split $f(t)$ into its real and its imaginary part, viz,

$$f(t) = f_r(t) + j\, f_i(t)$$

then, it is easy to see that Eq. (1.2) reduces to

$$F(s) = \int_{0-}^{\infty} [f_r(t)\, e^{-\sigma t} \cos \omega t + f_i(t)\, e^{-\sigma t} \sin \omega t]\, dt$$

$$+ j \int_{0-}^{\infty} [-f_r(t)\, e^{-\sigma t} \sin \omega t + f_i(t)\, e^{-\sigma t} \cos \omega t]\, dt$$

Thus, conceptually, there is no difficulty with considering a complex-valued $f(t)$ together with complex values of s.

Notations

1. The operation of taking the Laplace transform is denoted by $\mathscr{L}$, thus we write

$$F(s) = \mathscr{L}\{f(t)\} \tag{1.3}$$

As will be discussed later, the operation of going back to the time-domain function $f(t)$ from its Laplace transform is denoted by

$$f(t) = \mathscr{L}^{-1}\{F(s)\}$$

2. If we take the Laplace transform of a voltage $v(t)$ or a current $i(t)$, we denote them by $V(s)$ and $I(s)$, respectively, thus we use *capital letters* to denote Laplace transforms.

Relation to Fourier transforms The Laplace transform may be viewed as a special case of the Fourier transform, namely, when *$f(t) = 0$, for all $t < 0$.* This added restriction on the class of time functions, allows us to let s take values far into the right-half plane and in doing so, allows us to consider problems with exponentially increasing time functions (as in unstable circuits!).

Remark on Laplace transformable functions Not all functions have a Laplace transform: For example, e^{t^2} grows so fast as $t \to \infty$ that there is no value of s for which the integral in Eq. (1.2) is finite. Any function $f(t)$, for which there exists an s such that the integral in Eq. (1.2) is finite, is called *Laplace transformable*. By direct calculation, any function $f(t)$ for which there are constants $M > 0$ and $c > 0$ such that

$$|f(t)| \leq M\,e^{ct} \qquad \text{for all } t > 0 \tag{1.4}$$

is Laplace transformable. Without comment, *we shall always assume* that the functions that we encounter are Laplace transformable.

Exercise Let $f(t)$ satisfy Eq. (1.4). Show that if s is chosen *real* and $s > c$, then

$$|F(s)| \leq M \int_0^\infty e^{(c-s)t}\,dt \leq \frac{M}{s-c}$$

1.2 Two Important Examples

Example 1 Let $f(t) = e^{at}$, with a real (positive or negative). Substituting into Eq. (1.2) we obtain successively

$$F(s) = \int_{0-}^\infty e^{at}\,e^{-st}\,dt = \int_{0-}^\infty e^{-(s-a)t}\,dt$$

$$= \frac{e^{-(s-a)t}}{-(s-a)}\bigg|_{0-}^{\infty} \tag{1.5}$$

Now, if s is real with $s > a$, then $s - a > 0$ and $e^{-(s-a)t} \to 0$ as $t \to \infty$. Similarly, if s is complex with $\sigma > a$, then $\sigma - a > 0$ and

$$|\exp[-(s-a)t]| = \exp[-\text{Re}(s-a)\cdot t] = \exp[-(\sigma - a)t] \to 0$$

as $t \to \infty$. Therefore from Eq. (1.5) we have, for s real and $s > a$,

$$F(s) = \frac{1}{s-a} \tag{1.6}$$

and we write

$$\boxed{\mathscr{L}\{e^{at}\} = \frac{1}{s-a}} \tag{1.7}$$

REMARKS

1. Now let a and s be complex; then Eq. (1.7) still holds. Indeed, for an s complex, by definition of the exponential function,

$$e^{st} = 1 + st + \frac{1}{2!}(st)^2 + \cdots + \frac{1}{n!}(st)^n + \cdots \tag{1.8}$$

where the series converges absolutely for *all* complex numbers s and for all real t. Differentiating term by term we can easily check that

$$\frac{d}{dt}(e^{st}) = s\, e^{st} \tag{1.9}$$

Hence, for s and a complex

$$\frac{d}{dt}\left[\frac{e^{-(s-a)t}}{-(s-a)}\right] = e^{-(s-a)t}$$

Therefore the derivation of Eq. (1.7) holds for the case where s and a are complex. In order that $e^{-(s-a)t} \to 0$ as $t \to \infty$, we must take $\text{Re}(s) \triangleq \sigma > \text{Re}(a)$.

2. Strictly speaking, $F(s)$, defined by the integral in Eq. (1.5), is defined only for $\text{Re}(s) > \text{Re}(a)$. However the resulting expression (1.7) makes sense for all s in the complex plane, except for $s = a$. Consequently we consider $F(s)$ to be well-defined by Eq. (1.7), for all $s \neq a$. This extension can be validated by a process called "analytic continuation."

3. Consider $F(s) = 1/(s-a)$. Note that $s = a$ is a *pole* of $F(s)$ since for $s = a$, $d(s) = 0$; as $s \to a$, $s - a \to 0$ and $|F(s)| \to \infty$. We say $F(s)$ "blows up" at $s = a$ and that the rational function $1/(s-a)$ has a *pole* at $s = a$.

For a real, if $a > 0$, $e^{at} \to \infty$ as $t \to \infty$; then the pole of $F(s)$ is at $s = a$ in the *open right*-half plane; if $a < 0$, $e^{at} \to 0$ as $t \to \infty$; then the pole of $F(s)$ is at $s = a$ in the *open left*-half plane. For $a < 0$, $-a \triangleq 1/T$ where T is called the *time constant*.

This relation between pole position in the s plane and rates of exponential increase or decrease as time increases is of crucial importance in engineering.

Exercises

1. Use Eq. (1.7) to conclude that

$$\mathcal{L}\{1(t)\} = \frac{1}{s} \qquad \text{and} \qquad \mathcal{L}\{e^{j\omega t}\} = \frac{1}{s - j\omega} \tag{1.10}$$

2. Consider Eq. (1.7)

$$\int_{0-}^{\infty} e^{at}\, e^{-st}\, dt = \frac{1}{s-a} \tag{1.11}$$

and view both sides of the equation as functions of a. Differentiate Eq. (1.11) with respect to a to obtain successively

$$\mathscr{L}\{t\,e^{at}\} = \frac{1}{(s-a)^2} \tag{1.12}$$

$$\mathscr{L}\{t^2\,e^{at}\} = \frac{2}{(s-a)^3} \tag{1.13}$$

Use induction to show that, for any positive integer n,

$$\mathscr{L}\{t^n\,e^{at}\} = \frac{n!}{(s-a)^{n+1}} \tag{1.14}$$

3. Let $\mathscr{L}\{f(t)\} = F(s)$. Show that $\mathscr{L}\{t\,f(t)\} = -d\,F(s)/ds$. Give an expression for $\mathscr{L}\{t^n\,f(t)\}$.

Example 2 Let $f(t) = \delta(t)$, that is, $f(t)$ is a *unit impulse* at the origin.

$$\boxed{\mathscr{L}\{\delta(t)\} = \int_{0-}^{\infty} \delta(t)\,e^{-st}\,dt = 1} \tag{1.15}$$

To justify Eq. (1.15), let us approximate $\delta(t)$ by the unit area rectangular pulse $p_\Delta(t)$ where

$$p_\Delta(t) = \begin{cases} \dfrac{1}{\Delta} & \text{for } 0 \le t \le \Delta \\ 0 & \text{elsewhere} \end{cases}$$

Using p_Δ in the definition of the Laplace transform, Eq. (1.2), we obtain successively

$$\int_0^\infty p_\Delta(t)\,e^{-st}\,dt = \int_0^\Delta \frac{1}{\Delta}\,e^{-st}\,dt = \frac{e^{-st}}{-s\Delta}\bigg|_0^\Delta = \frac{1-e^{-s\Delta}}{s\Delta}$$

Thus

$$\mathscr{L}\{p_\Delta(t)\} = \frac{1-e^{-s\Delta}}{s\Delta} \tag{1.16}$$

Now let $\Delta \to 0$, then $p_\Delta(t) \to \delta(t)$ and $\mathscr{L}\{p_\Delta(t)\} \to \mathscr{L}\{\delta(t)\}$, so by Eq. (1.16)

$$\begin{aligned}\mathscr{L}\{\delta(t)\} &= \lim_{\Delta\to 0} \frac{1-e^{-s\Delta}}{s\Delta} \\ &= \lim_{\Delta\to 0} \frac{1-(1-s\Delta+s^2\Delta^2/2-\cdots)}{s\Delta} \\ &= 1\end{aligned}$$

Exercises

1. Let A and T be positive constants. Consider the rectangular pulse of amplitude A and duration T:

$$f(t) = A[1(t) - 1(t - T)] \tag{1.17}$$

(a) Sketch $f(t)$ versus t.
(b) Show that

$$F(s) = A\,\frac{1 - e^{-sT}}{s} \tag{1.18}$$

2. Let T be constant and $T > 0$. Consider the unit impulse occurring at time T, namely $\delta(t - T)$. Show that

$$\mathscr{L}\{\delta(t - T)\} = e^{-sT} \tag{1.19}$$

2 FOUR BASIC PROPERTIES OF LAPLACE TRANSFORMS

The following four properties are basic to the use of Laplace transforms in circuits, control, and communications.

2.1 Uniqueness Property

The Laplace transform $\mathscr{L}$ defined in Eq. (1.2) establishes a *one-to-one correspondence* between the time function $f(t)$, defined on $[0, \infty)$, and its Laplace transform $F(s)$.[2]

This is a deep theorem of mathematical analysis, which we do not prove, but which is *extremely useful*: it allows us to transform a time-domain problem into a frequency-domain problem, solve it in the frequency domain, and then go back to the time-domain solution. The uniqueness guarantees that the procedure gives *the* solution of the original problem.

Also if we have found, by hook or by crook, that some function $f(t)$ has some given $F(s)$ as its Laplace transform, then, by the uniqueness property, we know that $f(t)$ is *the* time function $\mathscr{L}^{-1}\{F(s)\}$.

2.2 Linearity Property

Let $\mathscr{L}\{f_k(t)\} = F_k(s)$ for $k = 1, 2$; then for any *constants* (real or complex) c_1 and c_2

$$\begin{aligned}\mathscr{L}\{c_1\, f_1(t) + c_2\, f_2(t)\} &= c_1\, \mathscr{L}\{f_1(t)\} + c_2\, \mathscr{L}\{f_2(t)\}\\ &= c_1\, F_1(s) + c_2\, F_2(s)\end{aligned} \tag{2.1}$$

[2] To be technical: if $\mathscr{L}\{g(t)\} = \mathscr{L}\{f(t)\} = F(s)$, then $\int_0^\infty |f(t) - g(t)|\, dt = 0$, i.e., from an engineering point of view, f and g are considered to be the *same* function.

PROOF The derivation is based on the linearity properties of the integral:

$$\mathcal{L}\{c_1 f_1(t) + c_2 f_2(t)\} = \int_{0-}^{\infty} [c_1 f_1(t) + c_2 f_2(t)]\, e^{-st}\, dt \qquad \text{(by definition)}$$

$$= c_1 \int_{0-}^{\infty} f_1(t)\, e^{-st}\, dt$$

$$+ c_2 \int_{0-}^{\infty} f_2(t)\, e^{-st}\, dt \qquad \text{(because } c_1 \text{ and } c_2 \text{ are } \textit{constants}\text{)}$$

$$= c_1 F_1(s) + c_2 F_2(s) \qquad \text{(by definition)}$$

■

This linearity property is an extremely useful work-saving device.

Example Let the k_α's and p_α's be *constants* (possibly complex); we'll show that

$$\mathcal{L}^{-1}\left\{\sum_{\alpha=1}^{a} \frac{k_\alpha}{s - p_\alpha}\right\} = \sum_{\alpha=1}^{a} k_\alpha\, e^{p_\alpha t} \tag{2.2}$$

Indeed, apply the linearity property to the left-hand side and note that the k_α's are constants,

$$\mathcal{L}^{-1}\left\{\sum_{\alpha=1}^{a} \frac{k_\alpha}{s - p_\alpha}\right\} = \sum_{\alpha=1}^{a} \mathcal{L}^{-1}\left\{\frac{k_\alpha}{s - p_\alpha}\right\} = \sum_{\alpha=1}^{a} k_\alpha \mathcal{L}^{-1}\left\{\frac{1}{s - p_\alpha}\right\}$$

and the right-hand side of Eq. (2.2) follows from Eq. (1.7).

Exercise Use $e^{j\omega_0 t} = \cos \omega_0 t + j \sin \omega_0 t$ to prove the following, and in each case sketch the time function and the location of the pole(s) of $F(s)$.

(*a*)
$$\mathcal{L}\{\cos \omega_0 t\} = \frac{s}{s^2 + \omega_0^2} \qquad \mathcal{L}\{\sin \omega_0 t\} = \frac{\omega_0}{s^2 + \omega_0^2} \tag{2.3}$$

(*b*)
$$\mathcal{L}\{e^{\alpha t} \cos \omega_0 t\} = \frac{s - \alpha}{(s - \alpha)^2 + \omega_0^2} \qquad \mathcal{L}\{e^{\alpha t} \sin \omega_0 t\} = \frac{\omega_0}{(s - \alpha)^2 + \omega_0^2} \tag{2.4}$$

(*c*) Let k be complex and a and b be real:

$$\mathcal{L}^{-1}\left\{\frac{k}{s - (a + jb)} + \frac{\bar{k}}{s - (a - jb)}\right\} = 2|k|\, e^{at} \cos(bt + \measuredangle k) \tag{2.5}$$

2.3 Differentiation Rule

We have already seen that it is useful to generalize the concept of differentiation from calculus to the idea of differentiating "in the distribution sense":[3] In particular we already know that

$$\frac{d}{dt}1(t) = \delta(t) \tag{2.6}$$

A. Property and proof

Property Let $\mathscr{L}\{f(t)\} = F(s)$. Let $d\,f(t)/dt$ denote the derivative of $f(t)$ in the distribution sense, then

$$\boxed{\mathscr{L}\left\{\frac{d}{dt}f(t)\right\} = s\,F(s) - f(0-)} \tag{2.7}$$

and
$$\mathscr{L}\left\{\frac{d^2}{dt^2}f(t)\right\} = s^2\,F(s) - s\,f(0-) - \dot{f}(0-) \tag{2.8}$$

PROOF Use integration by parts:

$$\int_{0-}^{\infty} \underbrace{\dot{f}(t)}_{\dot{u}}\,\underbrace{e^{-st}}_{v}\,dt = \underbrace{f(t)}_{u}\,\underbrace{e^{-st}}_{v}\Big|_{0-}^{\infty} - \int_{0-}^{\infty}\underbrace{f(t)}_{u}\,\underbrace{(-s\,e^{-st})}_{\dot{v}}\,dt \tag{2.9}$$

$$= -f(0-) + s\int_{0-}^{\infty} f(t)\,e^{-st}\,dt \tag{2.10}$$

$$= s\,F(s) - f(0-) \tag{2.11}$$

where, to obtain Eq. (2.10), note that we always take Re(s) sufficiently large so that $f(t)\,e^{-st} \to 0$ as $t \to \infty$; also note that s is a constant, when we integrate with respect to t, so it can be factored out of the second integral in Eq. (2.9). So Eq. (2.7) is established.

Equation (2.8) is obtained by applying Eq. (2.7) to $\dot{f}(t)$. ■

[3] We say "differentiating in the distribution sense" in order to emphasize that whenever we differentiate a function which has a jump [like $1(t)$ in Eq. (2.9) below], we must include the corresponding impulse in the derivative.

In general, if at $t = t_0$, $f(t)$ jumps from $f(t_0-)$ to $f(t_0+)$, then the derivative includes the "impulses" term

$$[f(t_0+) - f(t_0-)]\,\delta(t - t_0)$$

i.e., an impulse at time t_0 of area $f(t_0+) - f(t_0-)$.

Examples

1. By Eq. (1.10), we have $\mathscr{L}\{1(t)\} = 1/s$. Using Eqs. (2.6) and (2.7), and $1(0-) = 0$, we find that

$$\mathscr{L}\{\delta(t)\} = 1$$

as it should from our direct computation in Example 2 of Sec. 1.2, Eq. (1.15) in particular.

2. Let $f(t) = 1(t)\cos\omega t$. Note that $f(0-) = 0$ and $f(0+) = 1$; furthermore

$$\dot{f}(t) = \delta(t) - 1(t)\omega \sin \omega t \tag{2.12}$$

Thus, by linearity,

$$\mathscr{L}\{\dot{f}(t)\} = 1 - \frac{\omega^2}{s^2 + \omega^2} = \frac{s^2}{s^2 + \omega^2} \tag{2.13}$$

Using the differentiation rule (2.7) and Eq. (2.3), we obtain

$$\mathscr{L}\{\dot{f}(t)\} = s\,\frac{s}{s^2 + \omega^2} - 0 = \frac{s^2}{s^2 + \omega^2} \tag{2.14}$$

and the answer checks with Eq. (2.13) above.

Note that if, in Eq. (2.12), we did *not* differentiate in the distribution sense, Eq. (2.14) would not agree with Eq. (2.13).

3. (Calculation of the impulse response.) Let us calculate the impulse response $v_c(t) = h(t)$ of the linear time-invariant circuit shown in Fig. 2.1. By definition of the impulse response, KCL applied to ① gives

$$C\,\dot{h}(t) + G\,h(t) = \delta(t) \qquad h(0-) = 0 \tag{2.15}$$

This differential equation with the initial condition completely specifies the impulse response $h(t)$. Let us take the Laplace transform of both sides of Eq. (2.15):

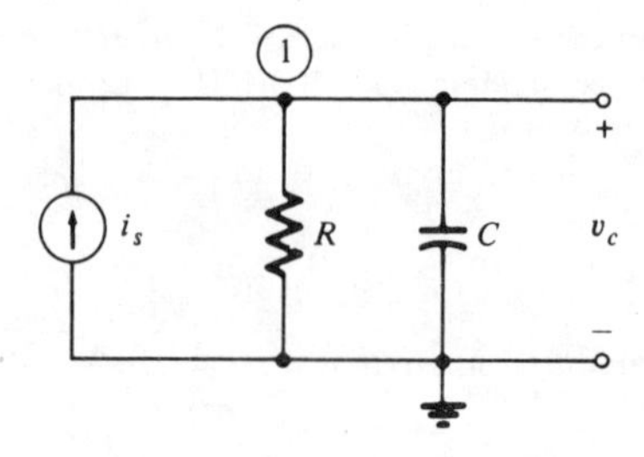

Figure 2.1 The input is $i_s(t) = \delta(t)$, and the response is $v_C(t)$.

$$\mathscr{L}\{C\dot{h} + Gh\} = \mathscr{L}\{\delta(t)\} = 1$$
$$= C\mathscr{L}\{\dot{h}\} + G\mathscr{L}\{h\} \qquad \text{(using linearity)}$$
$$= Cs\,H(s) - C\,h(0-) + G\,H(s) \qquad \text{(using the differentiation rule)}$$

So

$$H(s) = \frac{1}{Cs + G} = \frac{R}{1 + RCs} \qquad \text{and} \qquad h(t) = 1(t)\,\frac{1}{C}\,e^{-t/RC}$$

where we used Eq. (1.7) and linearity.

Exercise Use the Laplace transform to calculate the impulse response of the *RLC* circuit shown in Fig. 4.1 of Chap. 9. [Assume $Q > \frac{1}{2}$; the current source $i_s(t)$ is the input and the output is $v(t)$ as shown in Fig. 4.1.]

2.4 Integration Rule

Property Let $\mathscr{L}\{f(t)\} = F(s)$; then

$$\mathscr{L}\left\{\int_{0-}^{t} f(t')\,dt'\right\} = \frac{1}{s}\,F(s) \tag{2.16}$$

PROOF The proof is very simple: Use integration by parts (again!).

$$\int_{0-}^{\infty}\underbrace{\left[\int_{0-}^{t} f(t')\,dt'\right]}_{u}\underbrace{e^{-st}}_{\dot{v}}\,dt = \underbrace{\left[\int_{0-}^{t} f(t')\,dt'\right]}_{u}\underbrace{\frac{e^{-st}}{-s}}_{v}\Bigg|_{0}^{\infty} - \int_{0-}^{\infty}\underbrace{f(t)}_{\dot{u}}\,\underbrace{\frac{e^{-st}}{-s}}_{v}\,dt$$

$$= 0 + \frac{1}{s}\int_{0-}^{\infty} f(t)\,e^{-st}\,dt = \frac{1}{s}\,F(s)$$

where again we take s to have $\text{Re}(s)$ sufficiently large so that $f(t)\,e^{-st} \to 0$ as $t \to \infty$. ∎

Exercise From the definition of the unit impulse $\delta(t)$ we know that $\int_{0-}^{t} \delta(t')\,dt' = 1(t)$. Use this fact and the integration rule to obtain successively (*a*) $\mathscr{L}\{1(t)\}$, (*b*) $\mathscr{L}\{t\,1(t)\}$, and (*c*) $\mathscr{L}\left[\frac{t^n}{n!}\,1(t)\right] = \frac{1}{s^{n+1}}$.

2.5 Laplace Transform Rules and Phasor Calculation Rules

It is of crucial importance to note that *provided all time functions are 0 at $t = 0-$* (equivalently, all initial conditions are zero at $t = 0-$) *the rules for manipulating phasors and the rules for manipulating Laplace transforms are identical*, except for replacing $j\omega$ by s.

This is displayed in detail in Table 2.1.

Table 2.1 Phasor Rules and Laplace Transform Rules

Phasors	Laplace transform
Uniqueness: 1–1 correspondence between sinusoid $a(t) = \text{Re}[A\, e^{j\omega t}]$ and phasor A.	*Uniqueness*: 1–1 correspondence between time function $f(t)$ and its Laplace transform $F(s) = \int_{0-}^{\infty} f(t)\, e^{-st}\, dt$.
Linearity: Given any two sinusoids (at the *same* frequency ω), $a_1(t)$ and $a_2(t)$, and any two *real* constants c_1 and c_2: $$\text{Phasor of } [c_1\, a_1(t) + c_2\, a_2(t)] = c_1 A_1 + c_2 A_2$$ or $$\text{Re}(c_1 A_1\, e^{j\omega t} + c_2 A_2\, e^{j\omega t}) = c_1\, \text{Re}(A_1\, e^{j\omega t}) + c_2\, \text{Re}(A_2\, e^{j\omega t})$$	*Linearity*: Given any two time functions, $f_1(t)$ and $f_2(t)$, and any two (possibly complex) *constants* c_1 and c_2: $$\mathcal{L}\{c_1\, f_1(t) + c_2\, f_2(t)\} = c_1\, \mathcal{L}\{f_1(t)\} + c_2\, \mathcal{L}\{f_2(t)\} = c_1\, F_1(s) + c_2\, F_2(s)$$
Differentiation: Let $a(t) \triangleq$ sinusoid at frequency ω. $$\text{Phasor of } \left[\frac{da(t)}{dt}\right] = j\omega \cdot \text{phasor of } [a(t)]$$ or $$\frac{d}{dt} \underbrace{\text{Re}(A\, e^{j\omega t})}_{a(t)} = \text{Re}(j\omega A\, e^{j\omega t})$$	*Differentiation*: **Assuming** $f(0-) = 0$, $$\mathcal{L}\{\dot{f}(t)\} = s\, \mathcal{L}\{f(t)\} = s\, F(s)$$
Integration: Let $b(t)$ be the sinusoid which is an antiderivative of the sinusoid $a(t)$; then in terms of phasors we have $$B = \frac{1}{j\omega} A$$	*Integration*: $$\mathcal{L}\left\{\int_{0-}^{t} f(t')\, dt'\right\} = \frac{1}{s}\, \mathcal{L}\{f(t)\} = \frac{1}{s}\, F(s)$$

Example Consider the *RLC* circuit shown in Fig. 4.1 of Chap. 9, Sec. 4 and described by Eq. (4.3), namely,

$$\ddot{i}_L(t) + 2\alpha\, \dot{i}_L(t) + \omega_0^2\, i_L(t) = \omega_0^2\, i_s(t) \tag{2.17}$$

(*a*) *Phasor calculation*. Assume $i_s(t) = \text{Re}[I_s\, e^{j\omega t}]$, and seek only the sinusoidal steady-state response described by the *phasor* I_L: Using the rules of phasor analysis we obtain

$$[(\omega_0^2 - \omega^2) + 2\alpha j\omega]I_L = \omega_0^2 I_s \tag{2.18}$$

equivalently,

$$I_L = \frac{\omega_0^2}{(\omega_0^2 - \omega^2) + 2\alpha j\omega}\, I_s \tag{2.19}$$

(*b*) *Laplace transform calculation* (*with zero initial conditions*). The differential equation (2.17) gives

$$\mathcal{L}\{\ddot{i}_L(t) + 2\alpha\, \dot{i}_L(t) + \omega_0^2\, i_L(t)\} = \mathcal{L}\{\omega_0^2\, i_s(t)\} \tag{2.20}$$

and by linearity,

$$\mathcal{L}\{\ddot{i}_L(t)\} + 2\alpha\mathcal{L}\{\dot{i}_L(t)\} + \omega_0^2\mathcal{L}\{i_L(t)\} = \omega_0^2\mathcal{L}\{i_s(t)\}$$

and by the differentiation rule used three times (note the zero initial conditions!),

$$(s^2 + 2\alpha s + \omega_0^2)\, I_L(s) = \omega_0^2\, I_s(s)$$

or equivalently,

$$I_L(s) = \frac{\omega_0^2}{s^2 + 2\alpha s + \omega_0^2}\, I_s(s) \tag{2.21}$$

The phasor equation (2.19) and the Laplace transform equation (2.21) have exactly the *same form* except for $j\omega$ being replaced by s: this is a consequence of the fact that all initial conditions are zero (see Table 2.1).

These two equations have *different meanings*. The phasor equation (2.19) relates the given *phasor* I_s to the *phasor* I_L which represents the *sinusoidal steady-state response* of the circuit. The Laplace transform Eq. (2.21) relates the given Laplace transform $I_s(s)$ to the *function* $I_L(s)$ which represents the Laplace transform of the *zero-state response* of the circuit to the input $i_s(t)$. [Note the generality, $i_s(t)$ may be *any* Laplace transformable function!]

Exercise Show that if the initial conditions were not set to zero, then instead of Eq. (2.21) we would have obtained

$$I_s(s) = \frac{\omega_0^2}{s^2 + 2\alpha s + \omega_0^2}\, I_s(s) + \frac{(s + 2\alpha)i_L(0-) + \dot{i}_L(0-)}{s^2 + 2\alpha s + \omega_0^2} \tag{2.22}$$

3 PARTIAL FRACTION EXPANSIONS

The analysis of a circuit by Laplace transforms yields the transform of the output variable [e.g., see Eq. (2.21)]. The next step is to go from the Laplace transform back to the time function, or as engineers say, from the frequency domain to the time domain. An extremely useful technique is the partial fraction expansion.

Suppose we are given a Laplace transform $F_0(s)$ which is a rational function $n_0(s)/d_0(s)$, where $n_0(s)$ and $d_0(s)$ are *polynomials* with real coefficients. We further assume that $n_0(s)$ and $d_0(s)$ are *coprime*, that is, any nontrivial common factor has been canceled out.

3.1 Reduction Step

If the degree of n_0 is larger or equal to the degree of d_0, we divide the polynomial $n_0(s)$ by $d_0(s)$ to obtain the *quotient polynomial* $q(s)$ and the *remainder polynomial* $r(s)$: Hence

$$n_0(s) = q(s)\, d_0(s) + r(s) \qquad \text{degree } (r) < \text{degree } (d_0) \tag{3.1}$$

Hence

$$F_0(s) = \frac{n_0(s)}{d_0(s)} = q(s) + \frac{r(s)}{d_0(s)} \tag{3.2}$$

and, by linearity,

$$\mathscr{L}^{-1}\{F_0(s)\} = \mathscr{L}^{-1}\{q(s)\} + \mathscr{L}^{-1}\left\{\frac{r(s)}{d_0(s)}\right\} \tag{3.3}$$

For example,

$$\frac{2s^2+8s+7}{(s+1)(s+3)} = 2 + \frac{1}{(s+1)(s+3)}$$

hence

$$\mathscr{L}^{-1}\left\{\frac{2s^2+8s+7}{(s+1)(s+3)}\right\} = 2\delta(t) + \mathscr{L}^{-1}\left\{\frac{1}{(s+1)(s+3)}\right\} \tag{3.4}$$

Exercise Consider the polynomial $q(s) = q_0 + q_1 s$, and show that

$$\mathscr{L}^{-1}[q_0 + q_1 s] = q_0\,\delta(t) + q_1\,\dot{\delta}(t) \tag{3.5}$$

where $\dot{\delta}(t)$ is the derivative of $\delta(t)$.

As a consequence of the reduction step, we need only consider the case where

$$F(s) = \frac{n(s)}{d(s)} \qquad \begin{cases} n(s) \text{ and } d(s) \text{ are real polynomials in } s \\ n(s) \text{ and } d(s) \text{ are } \textit{coprime} \\ \text{degree } (n) < \text{degree } (d) \end{cases} \tag{3.6}$$

3.2 Simple Poles

The rational function $F(s)$ specified by Eq. (3.6) is said to have *simple poles* iff the denominator polynomial $d(s)$ has simple zeros. Call these poles $p_1, p_2, \ldots, p_m$.[4] We know from algebra that $F(s)$ may be written as

$$F(s) = \frac{n(s)}{d(s)} = \frac{n(s)}{(s-p_1)(s-p_2)\cdots(s-p_m)} = \sum_{\alpha=1}^{m} \frac{k_\alpha}{s-p_\alpha} \qquad \text{for } s \neq p_\alpha \tag{3.7}$$

that is, both sides of the last equality in Eq. (3.7) are equal for all $s \neq p_\alpha$, $\alpha = 1, 2, \ldots, m$. The number k_α is called the *residue* of $F(s)$ at the pole p_α.

If the values of s are very close to p_α but different from p_α, $s - p_\alpha$ is very small. Consequently, by Eq. (3.7), $F(s)$ is very large: In the neighborhood of any one of its poles, $F(s)$ "blows up."

To obtain k_1, multiply Eq. (3.7) by $(s - p_1)$ and obtain

$$(s-p_1)\,F(s) = \frac{(s-p_1)\,n(s)}{(s-p_1)(s-p_2)\cdots(s-p_m)} = k_1 + \sum_{\alpha=2}^{m} (s-p_1)\frac{k_\alpha}{(s-p_\alpha)} \tag{3.8}$$

Now let $s \to p_1$. Then the last sum in Eq. (3.8) tends to zero; thus

$$k_1 = \lim_{s \to p_1} (s-p_1)\,F(s) = \frac{n(p_1)}{(p_1-p_2)(p_1-p_3)\cdots(p_1-p_2)} \tag{3.9}$$

and, in general, for $\alpha = 1, 2, \ldots, m$,

$$k_\alpha = \lim_{s \to p_\alpha} (s-p_\alpha)\,F(s) = \frac{n(p_\alpha)}{(p_\alpha-p_1)\cdots(p_\alpha-p_m)} \tag{3.10}$$

$\uparrow$ factor $(p_\alpha - p_\alpha)$ is missing

Example Let $F(s) = \dfrac{1}{(s+1)(s+3)}$; hence $p_1 = -1$ and $p_2 = -3$. By Eq. (3.9), we have

$$k_1 = \lim_{s \to -1} (s+1)\,F(s) = \lim_{s \to -1} \frac{1}{s+3} = 0.5$$

[4] Some of these poles may be *complex* numbers. Since $d(s)$ has *real* coefficients, such complex zeros of $d(s)$ necessarily occur in complex-conjugate pairs. Thus if $F(s)$ has complex poles, they occur in complex-conjugate pairs.

Similarly $k_2 = -0.5$. Hence

$$f(t) = (0.5\,e^{-t} - 0.5\,e^{-3t})1(t)$$

We insert the unit step factor to remind ourselves that $f(t) = 0$ for $t < 0$.

Exercise Let $F(s)$ be given by Eq. (3.7). Use Eq. (3.10) to show that if $p_2 = \bar{p}_1$, then the residue k_2 of the pole p_2 satisfies $k_2 = \bar{k}_1$.

REMARKS

1. From Eq. (3.7)

$$f(t) \triangleq \mathcal{L}^{-1}\{F(s)\} = \left(\sum_{\alpha=1}^{m} k_\alpha\, e^{p_\alpha t}\right) 1(t) \tag{3.11}$$

Thus, if some pole, say p_i, is in the *open right-half plane* [i.e., $\mathrm{Re}(p_i) > 0$], it contributes, in the time domain, a *growing exponential*, and the farther out p_i is into the right-half plane, the faster this exponential growth is. Similarly if some pole p_j is in the *open left-half plane* [i.e., $\mathrm{Re}(p_j) < 0$], it contributes a *decaying exponential*, and the farther out p_j is into the left-half plane, the faster this *exponential decay* is.

2. For $\alpha = 1, 2, \ldots, m$, $n(p_\alpha) \neq 0$ because $n(s)$ and $d(s)$ are coprime. Hence by Eq. (3.10) $k_\alpha \neq 0$, for all α's.

3. If there is a pair of complex-conjugate poles p_1 and $\bar{p}_1$, the corresponding residues are, respectively, k_1 and $\bar{k}_1$: Hence (using the notations $k_1 = |k_1|\, e^{j\measuredangle k_1}$ and $p_1 = p_{1r} + jp_{1i}$) we obtain

$$\mathcal{L}^{-1}\left\{\frac{k_1}{s - p_1} + \frac{\bar{k}_1}{s - \bar{p}_1}\right\} = (k_1\, e^{p_1 t} + \bar{k}_1\, e^{\bar{p}_1 t})\, 1(t) = 2\,\mathrm{Re}(k_1\, e^{p_1 t}) \cdot 1(t)$$

$$= [2|k_1|\, e^{p_{1r} t} \cos(p_{1i} t + \measuredangle k_1)]\, 1(t) \tag{3.12}$$

3.3 Multiple Poles

Suppose that the rational function $F(s)$ has a double pole at p_1 and simple poles at $p_2, p_3, \ldots, p_m$; equivalently, $d(s)$ has a double zero at p_1 and simple zeros at $p_2, p_3, \ldots, p_m$; equivalently,

$$d(s) = (s - p_1)^2 (s - p_2) \cdots (s - p_m)$$

Then we know from algebra that there are constants $k_{11}, k_{12}, k_2, \ldots, k_m$ such that

$$F(s) = \frac{n(s)}{d(s)} = \frac{k_{11}}{(s - p_1)^2} + \frac{k_{12}}{s - p_1} + \sum_{\alpha=2}^{m} \frac{k_\alpha}{s - p_\alpha} \tag{3.13}$$

From Eqs. (1.11) and (1.12),

$$f(t) = \mathscr{L}^{-1}\{F(s)\} = \left(k_{11}t\,\mathrm{e}^{p_1 t} + k_{12}\,\mathrm{e}^{p_1 t} + \sum_{\alpha=2}^{m} k_\alpha\,\mathrm{e}^{p_\alpha t}\right)1(t) \tag{3.14}$$

To obtain the constants $k_{11}, k_{12}, k_2, \ldots, k_\alpha$ we proceed as follows:

1. Multiply $F(s)$ by $(s - p_1)^2$; then let $s \to p_1$

$$k_{11} = \lim_{s \to p_1} (s - p_1)^2 F(s) = \frac{n(p_1)}{(p_1 - p_2)(p_1 - p_3)\cdots(p_1 - p_m)}$$

2. Differentiate with respect to s the expression $(s - p_1)^2 F(s)$ and let $s \to p_1$

$$k_{12} = \lim_{s \to p_1} \frac{d}{ds}\left[(s - p_1)^2 F(s)\right]$$

3. The remaining k_α's are obtained as before [see Eq. (3.10)].

COMMENT Equations (3.7) and (3.11)—and Eqs. (3.13) and (3.14)—exhibit the close relation between the locations of the poles of $F(s)$ *in the s plane* with the exponential behavior *in time* of the corresponding terms of $f(t)$. This connection has important consequences in engineering design.

Example Let $F(s) = \dfrac{s+2}{(s+1)^2(s+3)}$; thus $p_1 = -1$, $p_2 = -3$

$$k_{11} = \lim_{s \to -1} \frac{s+2}{s+3} = \frac{1}{2}$$

$$k_{12} = \lim_{s \to -1} \frac{d}{ds}\,\frac{s+2}{s+3} = \lim_{s \to -1} \frac{1}{(s+3)^2} = \frac{1}{4}$$

$$k_2 = \frac{-1}{4}$$

Hence $$f(t) = (0.5t\,\mathrm{e}^{-t} + 0.25\,\mathrm{e}^{-t} - 0.25\,\mathrm{e}^{-3t})\,1(t)$$

3.4 Example: Circuit Analysis Using Laplace Transform

This example illustrates the use of most of the rules of Laplace transforms. Let the elements be normalized and have values shown in Fig. 3.1. Consider the linear time-invariant second-order circuit shown in Fig. 3.1. Let

$$e(t) = 10 \sin 5t \qquad \text{hence } E(s) = \frac{50}{s^2 + 5^2}$$

$i_L(0-) = 2$ A, and $v_C(0-) = 0$. We wish to calculate the current $i(t)$ defined in

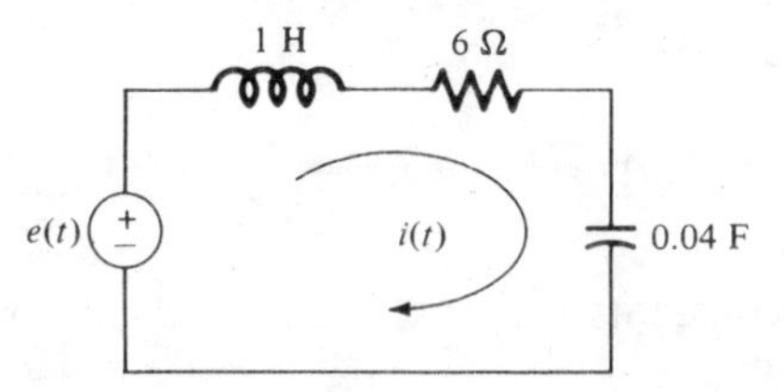

Figure 3.1 Second-order linear time-invariant circuit to be analyzed.

Fig. 3.1. Using the loop current $i(t)$ as variable, KVL gives

$$L\dot{i}(t) + Ri(t) + \frac{1}{C}\int_{0-}^{t} i(t')\, dt' = e(t) \qquad i_L(0-) = 2\text{ A}$$

Taking Laplace transforms of both sides and using the properties of the transform, we obtain

$$\left(Ls + R + \frac{1}{C}\frac{1}{s}\right) I(s) = E(s) + Li_L(0-)$$

hence $I(s) = \dfrac{s}{(s+3)^2 + 4^2} E(s) + \dfrac{2s}{(s+3)^2 + 4^2} = \dfrac{s}{(s+3)^2 + 4^2} \dfrac{2s^2 + 100}{s^2 + 25}$

Using partial fractions we obtain

$$I(s) = \frac{k_1}{s + 3 - j4} + \frac{\bar{k}_1}{s + 3 + j4} + \frac{k_2}{s - j5} + \frac{\bar{k}_2}{s + j5}$$

Performing the computations we obtain

$$\begin{aligned} k_1 &= (s + 3 - j4)\, I(s)\big|_{s=-3+j4} = \left[\frac{s(2s^2 + 100)}{(s + 3 + j4)(s^2 + 25)}\right]\Bigg|_{s=-3+j4} \\ &= 2.052\measuredangle 60.83^\circ \\ k_2 &= (s - j5)\, I(s)\big|_{s=j5} \\ &= 0.8333\measuredangle -90^\circ \end{aligned}$$

So, for $t \geq 0$

$$\begin{aligned} i(t) &= 2|k_1|\, e^{-3t} \cos(4t + \measuredangle k_1) + 2|k_2| \cos(5t + \measuredangle k_2) \\ &= 4.104\, e^{-3t} \cos(4t + 60.83^\circ) + 1.667 \cos(5t - 90^\circ) \end{aligned}$$

This current includes a transient term and a sinusoidal steady-state term: After about five time constants—five thirds of a second in the present case—the transient term is negligible and the circuit is essentially in sinusoidal steady state.

4 PROPERTIES OF LINEAR TIME-INVARIANT CIRCUITS

In this section we study the general properties of *linear time-invariant circuits*. We consider a general *linear time-invariant* circuit $\mathcal{N}$ which is *connected*, has *b branches*, and *n nodes*.

First, we develop, both in the frequency domain and in the time domain, the tableau equations and the modified node analysis.

Next, we use these formulations to study the *zero-input* and the *zero-state* response of such circuits. We obtain general definitions and characterizations for *natural frequencies*. We generalize the concept of *network functions* and establish their properties. Finally, we establish their connection to the time-domain behavior of the circuit.

4.1 Tableau Equations in the Frequency Domain

Given *a uniquely solvable linear time-invariant* circuit $\mathcal{N}$, we are going to obtain its tableau equations in terms of the Laplace transformed variables $\mathbf{E}(s) = \mathcal{L}\{\mathbf{e}(t)\}$, $\mathbf{V}(s) = \mathcal{L}\{\mathbf{v}(t)\}$, and $\mathbf{I}(s) = \mathcal{L}\{\mathbf{i}(t)\}$.

Let us consider the same example as in Sec. 2.1 of Chap. 8. We reproduce the linear time-invariant circuit $\mathcal{N}$ in Fig. 4.1.

A. Laplace transform of Kirchhoff's laws Kirchhoff's laws applied to node ① read

$$i_3(t) + i_6(t) = 0 \tag{4.1}$$

Taking the Laplace transform of this equation and using the linearity property, we obtain

$$I_3(s) + I_6(s) = 0 \tag{4.2}$$

This example shows that the transcription of the KCL from the time domain to the frequency domain is straightforward: In each KCL equation, we replace $i_k(t)$ by $I_k(s)$.

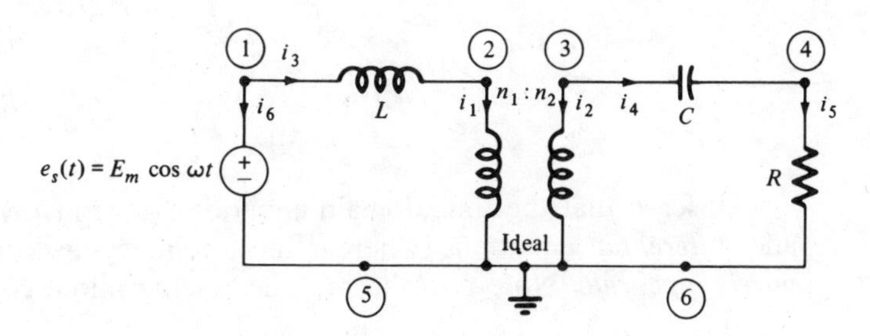

Figure 4.1 Linear time-invariant circuit under study.

Consider now the first equation of KVL for the circuit $\mathcal{N}$:

$$v_1(t) = e_2(t)$$

Taking Laplace transforms we obtain

$$V_1(s) = E_2(s)$$

So, in each KVL equation, we replace $v_k(t)$ by $V_k(s)$ and $e_j(t)$ by $E_j(s)$.

In summary, Kirchhoff's law equations read

$$\begin{array}{cc} \textit{Time domain} & \textit{Frequency domain} \\ \mathbf{A}\,\mathbf{i}(t) = \mathbf{0} & \mathbf{A}\,\mathbf{I}(s) = \mathbf{0} \\ \mathbf{A}^T\,\mathbf{e}(t) - \mathbf{v}(t) = \mathbf{0} & \mathbf{A}^T\,\mathbf{E}(s) - \mathbf{V}(s) = \mathbf{0} \end{array} \tag{4.3}$$

where $\mathbf{E}(s) = (n-1)$-vector whose components are the Laplace transforms of the node voltages
$\mathbf{V}(s) = b$-vector whose components are the Laplace transforms of the branch voltages
$\mathbf{I}(s) = b$-vector whose components are the Laplace transforms of the branch currents

B. Laplace transform of the branch equations In the time domain, the branch equations read

$$\begin{aligned} n_2\,v_1(t) - n_1\,v_2(t) &= 0 \\ n_1\,i_1(t) + n_2\,i_2(t) &= 0 \\ v_3(t) - L\,\dot{i}_3(t) &= 0 \\ C\,\dot{v}_4(t) - i_4(t) &= 0 \\ v_5(t) - R_5\,i_5(t) &= 0 \\ v_6(t) &= E_m \cos \omega t \end{aligned} \tag{4.4}$$

We are also given the initial conditions $v_4(0-)$ and $i_3(0-)$. Let us take the Laplace transform of each of these equations: As above we use the *linearity property*—since the circuit is time-invariant where n_1, n_2, L, C, and R_5 are constants—but we also use the *differentiation rule*:

$$\begin{aligned} n_2\,V_1(s) - n_1\,V_2(s) &= 0 \\ n_1\,I_1(s) + n_2\,I_2(s) &= 0 \\ V_3(s) - Ls\,I_3(s) &= -L\,i_3(0-) \\ Cs\,V_4(s) - I_4(s) &= C\,v_4(0-) \\ V_5(s) - R_5\,I_5(s) &= 0 \\ V_6(s) &= E_m \frac{s}{s^2+\omega^2} \end{aligned} \tag{4.5}$$

Observe that the time-domain equations (4.4) involve *algebraic* equations and *differential* equations, whereas the frequency-domain equations (4.5) are *purely algebraic*. Note also the presence of the initial conditions.

C. The tableau equations in the frequency domain In summary, the tableau equations read in the time domain [see Chap. 8, Eq. (2.11)]

$$\underbrace{\begin{bmatrix} \mathbf{0} & \mathbf{0} & \mathbf{A} \\ -\mathbf{A}^T & \mathbf{1} & \mathbf{0} \\ \mathbf{0} & \mathbf{M}_0 D + \mathbf{M}_1 & \mathbf{N}_0 D + \mathbf{N}_1 \end{bmatrix}}_{\mathbf{T}(D)} \underbrace{\begin{bmatrix} \mathbf{e}(t) \\ \mathbf{v}(t) \\ \mathbf{i}(t) \end{bmatrix}}_{\mathbf{w}(t)} = \begin{bmatrix} \mathbf{0} \\ \mathbf{0} \\ \mathbf{u}_s(t) \end{bmatrix} \tag{4.6}$$

where the $b \times b$ matrices $\mathbf{M}_0$, $\mathbf{M}_1$, $\mathbf{N}_0$, and $\mathbf{N}_1$ are constant (since $\mathcal{N}$ is linear *time-invariant*) and D denotes the differentiation operator d/dt.

In the frequency domain we have

Tableau equation in frequency domain

$$\underbrace{\begin{bmatrix} \mathbf{0} & \mathbf{0} & \mathbf{A} \\ -\mathbf{A}^T & \mathbf{1} & \mathbf{0} \\ \mathbf{0} & \mathbf{M}_0 s + \mathbf{M}_1 & \mathbf{N}_0 s + \mathbf{N}_1 \end{bmatrix}}_{\mathbf{T}(s)} \underbrace{\begin{bmatrix} \mathbf{E}(s) \\ \mathbf{V}(s) \\ \mathbf{I}(s) \end{bmatrix}}_{\mathbf{W}(s)} = \begin{bmatrix} \mathbf{0} \\ \mathbf{0} \\ \mathbf{U}_s(s) + \mathbf{U}^i \end{bmatrix} \tag{4.7}$$

Note that the equations (4.7) form a system of $2b + (n - 1)$ *algebraic* equations in the unknowns $\mathbf{E}(s)$, $\mathbf{V}(s)$, and $\mathbf{I}(s)$. The matrix of the system, $\mathbf{T}(s)$, has *constants* or *polynomials* of degree one as elements: Consequently we obtain the following:

Property: $\det[\mathbf{T}(s)]$ is a *polynomial* in s with *real coefficients*.

Note that

$$\mathbf{U}^i = \mathbf{M}_0\, \mathbf{v}_c(0-) + \mathbf{N}_0\, \mathbf{i}_L(0-) \tag{4.8}$$

is a b-column vector which keeps track of the initial conditions.

Exercise Show that the term $C\, v_4(0-)$ $[-L\, i_3(0-)$, respectively] in Eq. (4.5) may be interpreted as representing a current source $C\, v_4(0-)\, \delta(t)$ in parallel with the uncharged capacitor C [voltage source $-L\, i_3(0-)\, \delta(t)$ in series with the unenergized inductor L, respectively].

D. Two properties of linear time-invariant circuits

Unique solvability property Consider an arbitrary *linear time-invariant* circuit $\mathcal{N}$. Let us write its tableau equations, say as in Eq. (4.7) above. Since the circuit $\mathcal{N}$ is specified, we know the constant matrices $\mathbf{A}$, $\mathbf{M}_0$, $\mathbf{M}_1$, $\mathbf{N}_0$, and $\mathbf{N}_1$; since the independent sources and the initial conditions are given, we know the b-vector $\mathbf{U}_s(s)$ and the constant b-vector $\mathbf{U}^i$. Hence Eq. (4.7) is a system of *linear algebraic* equations in the unknown vector $[\mathbf{E}(s), \mathbf{V}(s), \mathbf{I}(s)]$, and the matrix of the system is the matrix $\mathbf{T}(s)$. Therefore by Cramer's rule we conclude that

Property (Unique solvability) The linear time-invariant circuit $\mathcal{N}$ is uniquely solvable if and only if

$$\det[\mathbf{T}(s)] \not\equiv 0 \tag{4.9}$$

The expression "det[$\mathbf{T}(s)$] $\not\equiv 0$" expresses the fact that the real polynomial det[$T(s)$] is *not identically zero*. (Recall that any real polynomial of degree $n \geq 1$ has exactly n zeros, counting multiplicities: Some of these zeros may be complex.)

REMARK The tableau equations (4.6) or (4.7) give a *global view* of the circuit behavior. The solution of these equations say, $\mathbf{w}(t) \triangleq [\mathbf{e}(t), \mathbf{v}(t), \mathbf{i}(t)]$, or in the frequency domain $\mathbf{W}(s) = [\mathbf{E}(s), \mathbf{V}(s), \mathbf{I}(s)]$, lists *all* the node voltages, *all* the branch voltages, and *all* the branch currents. It is convenient to give a common name to these $2b + n - 1$ variables: We call them *the circuit variables*.

The tableau equations (4.7) yield immediately a superposition type result which simplifies greatly the analysis of any linear time-invariant circuit.

Superposition property Given any uniquely solvable *linear time-invariant* circuit $\mathcal{N}$, every node voltage $e_k(t)$, every branch voltage $v_k(t)$, and every branch current $i_k(t)$ is the sum of two responses: the corresponding *zero-state response* (due to all the inputs alone) and the corresponding *zero-input response* (due to all the initial conditions alone).

PROOF From Eq. (4.6) we obtain, by multiplying on the left by $\mathbf{T}(s)^{-1}$,

$$\mathbf{W}(s) = \underbrace{\mathbf{T}(s)^{-1}\begin{bmatrix} \mathbf{0} \\ \mathbf{0} \\ \mathbf{U}_s(s) \end{bmatrix}}_{\substack{\text{Due to the inputs} \\ \text{(with the initial} \\ \text{conditions set} \\ \text{to zero)}}} + \underbrace{\mathbf{T}(s)^{-1}\begin{bmatrix} \mathbf{0} \\ \mathbf{0} \\ \mathbf{U}_i \end{bmatrix}}_{\substack{\text{Due to the initial} \\ \text{conditions (with all} \\ \text{the inputs set} \\ \text{to zero)}}} \tag{4.10}$$

For example, the kth component of $\mathbf{W}(s)$ is $\mathbf{E}_k(s) = \mathcal{L}\{e_k(t)\}$, so Eq. (4.10) shows that $\mathbf{E}_k(s)$ is the sum of two terms; one *due to all the inputs* and the other *due to all the initial conditions*. Going back to the time domain, we see that *every circuit variable* of $\mathcal{N}$ is the sum of two terms which, using previous definitions, we may write

$$\begin{pmatrix} \text{Complete} \\ \text{response} \end{pmatrix} = \begin{pmatrix} \text{zero-state} \\ \text{response} \end{pmatrix} + \begin{pmatrix} \text{zero-input} \\ \text{response} \end{pmatrix} \tag{4.11}$$

■

4.2 Modified Node Analysis in the Frequency Domain

Given a *uniquely solvable linear time-invariant* circuit $\mathcal{N}$, we are going to obtain its modified node analysis (MNA) equations in the frequency domain in terms of the Laplace transformed variables $\mathbf{E}(s) = \mathcal{L}\{\mathbf{e}(t)\}$ and $\mathbf{I}_2(s) = \mathcal{L}\{\mathbf{i}_2(t)\}$, where $\mathbf{i}_2(t)$ denotes the selected branch currents of MNA.

Consider the simple linear time-invariant circuit in Fig. 4.2. MNA gives

$$\begin{array}{llllll}
① & G_2\,e_1(t) & & +\,i_3(t) & & = I_m \cos \omega t \\
② & & C_5\,\dot{e}_2(t) + G_6\,e_2(t) & & +i_4(t) & = 0 \\
& e_1(t) & & -L_3\,\dot{i}_3(t) & & = 0 \quad \text{(inductor)} \\
& -\alpha\,e_1(t) & +\,e_2(t) & & & = 0 \quad \text{(VCVS)}
\end{array} \tag{4.12}$$

Note that we have two selected currents: i_3 the current through the inductor and i_4 the current through the VCVS.

Let us take the Laplace transform of the four equations (4.12): Using linearity and the differentiation rule we obtain

$$\begin{array}{lllll}
G_2\,E_1(s) & & +I_3(s) & & = I_m \dfrac{s}{s^2+\omega^2} \\
& (C_5 s + G_6)\,E_2(s) & & +I_4(s) & = C_5\,e_2(0-) \\
E_1(s) & & -sL_3\,I_3(s) & & = -L_3\,i_3(0-) \\
-\alpha\,E_1(s) & +E_2(s) & & & = 0
\end{array} \tag{4.13}$$

The MNA equations in the time domain were algebraic and differential equations; in the frequency domain, they are purely algebraic. Equation (4.13) is a set of four linear algebraic equations in four unknowns $E_1(s)$, $E_2(s)$, $I_3(s)$, and $I_4(s)$. In matrix form Eq. (4.13) reads

$$\begin{bmatrix} G_2 & 0 & 1 & 0 \\ 0 & C_5 s + G_6 & 0 & 1 \\ 1 & 0 & -sL_3 & 0 \\ -\alpha & 1 & 0 & 0 \end{bmatrix} \begin{bmatrix} E_1(s) \\ E_2(s) \\ I_3(s) \\ I_4(s) \end{bmatrix} = \begin{bmatrix} I_m \dfrac{s}{s^2+\omega^2} \\ C_5\,e_2(0-) \\ -L_3\,i_3(0-) \\ 0 \end{bmatrix} \tag{4.14}$$

It is clear that this analysis is applicable to any linear time-invariant circuit and that the resulting frequency-domain equations are in the form

MNA equations in frequency domain

$$\Big[\mathbf{P}(s)\Big] \begin{bmatrix} \mathbf{E}(s) \\ \mathbf{I}_2(s) \end{bmatrix} = \Big[\mathbf{U}(s)\Big] \tag{4.15}$$

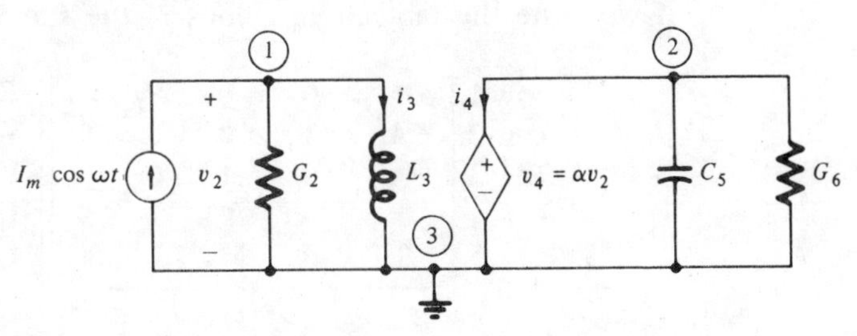

Figure 4.2 Linear time-invariant circuit to be analyzed by MNA.

where the elements of the matrix $\mathbf{P}(s)$ are either real *constants* or real *polynomials of degree 1*; $\mathbf{E}(s)$ is the $n-1$ vector of the Laplace transformed node voltages $e_1(t), \ldots, e_{n-1}(t)$; $\mathbf{I}_2(s)$ is the vector of the Laplace transformed selected branch currents; and the vector $\mathbf{U}(s)$ includes the contributions of the independent sources and of the initial conditions.

REMARK ON UNIQUE SOLVABILITY In the guided exercise that follows, we show that if we analyze a given linear time-invariant circuit $\mathcal{N}$ by the tableau method [hence obtain, in particular, the matrix $\mathbf{T}(s)$] and by the MNA [hence obtain in particular, the matrix $\mathbf{P}(s)$], then we have

$$\det[\mathbf{T}(s)] = k \det[\mathbf{P}(s)] \tag{4.16}$$

where k is a *nonzero constant*.

Consequently by Cramer's rule applied to Eq. (4.15) and by the unique solvability property of the previous subsection we have

$$\left\{\begin{matrix} \mathcal{N} \text{ is} \\ \text{uniquely} \\ \text{solvable} \end{matrix}\right\} \Leftrightarrow \det[\mathbf{T}(s)] \not\equiv 0 \Leftrightarrow \det[\mathbf{P}(s)] \not\equiv 0 \tag{4.17}$$

Guided exercise Obtain the MNA equations from the tableau equations in order to prove Eq. (4.17).

Given $\mathcal{N}$, suppose we write the tableau equations, keeping in mind that we want to extract from them the MNA equations. Therefore we partition all branches into two sets: first, the voltage-controlled branches which are *not* inductors, and second, the remaining branches. We denote by $\mathbf{V}_1$, $\mathbf{I}_1$ and $\mathbf{V}_2$, $\mathbf{I}_2$ the Laplace transforms of these branch voltage vectors and branch current vectors, respectively. Since $\mathbf{V}_1$ and $\mathbf{I}_1$ are associated with voltage-controlled branches which are not inductors, we have the following branch equations:

$$-\mathbf{Y}_1\mathbf{V}_1 + \mathbf{I}_1 = \mathbf{I}_{s1}$$

where $\mathbf{Y}_1$ is the branch admittance matrix of these elements and the nonzero elements of $\mathbf{I}_{s1}$ represent the independent current sources of $\mathcal{N}$ and initial conditions. We also partition the reduced incidence matrix $\mathbf{A}$ into $[\mathbf{A}_1 \vdots \mathbf{A}_2]$.

If we write the tableau equations in the standard manner we obtain

$$\begin{matrix} \text{Row 1} \\ \text{Row 2} \\ \text{Row 3} \\ \text{Row 4} \\ \text{Row 5} \end{matrix} \underbrace{\begin{bmatrix} \mathbf{0} & \mathbf{0} & \mathbf{0} & \mathbf{A}_1 & \mathbf{A}_2 \\ -\mathbf{A}_1^T & \mathbf{1} & \mathbf{0} & \mathbf{0} & \mathbf{0} \\ -\mathbf{A}_2^T & \mathbf{0} & \mathbf{1} & \mathbf{0} & \mathbf{0} \\ \mathbf{0} & -\mathbf{Y}_1 & \mathbf{0} & \mathbf{1} & \mathbf{0} \\ \mathbf{0} & \mathbf{0} & \mathbf{M}_2 & \mathbf{0} & \mathbf{N}_2 \end{bmatrix}}_{\mathbf{T}(s)} \begin{bmatrix} \mathbf{E} \\ \mathbf{V}_1 \\ \mathbf{V}_2 \\ \mathbf{I}_1 \\ \mathbf{I}_2 \end{bmatrix} = \begin{bmatrix} \mathbf{0} \\ \mathbf{0} \\ \mathbf{0} \\ \mathbf{I}_{s1} \\ \mathbf{U}_{s2} \end{bmatrix}$$

$\mathbf{U}_{s2}$ bookkeeps the independent voltage sources and initial conditions. The matrices $\mathbf{Y}_1$, $\mathbf{M}_2$, and $\mathbf{N}_2$ have elements that are real polynomials in s of at most degree 1.

Step 1. (Purpose: to have $\mathbf{E}$ and $\mathbf{I}_2$ be listed last.) Interchange row 1 and row 4, and then interchange column 1 and column 4.

Step 2. (Purpose: to eliminate $\mathbf{I}_1$ from all but the first row.) Perform

$$\text{row } 1 \leftarrow \text{row } 1 + \mathbf{Y}_1 \cdot \text{row } 2$$

$$\text{row } 4 \leftarrow \text{row } 4 - \mathbf{A}_1 \cdot \text{row } 1$$

Step 3. (Purpose: to eliminate $\mathbf{V}_2$ from row 5.)

$$\text{row } 5 \leftarrow \text{row } 5 - \mathbf{M}_2 \cdot \text{row } 3$$

The result is

$$\underbrace{\begin{bmatrix} \mathbf{1} & \mathbf{0} & \mathbf{0} & -\mathbf{Y}_1\mathbf{A}_1^T & \mathbf{0} \\ \mathbf{0} & \mathbf{1} & \mathbf{0} & -\mathbf{A}_1^T & \mathbf{0} \\ \mathbf{0} & \mathbf{0} & \mathbf{1} & -\mathbf{A}_2^T & \mathbf{0} \\ \mathbf{0} & \mathbf{0} & \mathbf{0} & \mathbf{A}_1\mathbf{Y}_1\mathbf{A}_1^T & \mathbf{A}_2 \\ \mathbf{0} & \mathbf{0} & \mathbf{0} & \mathbf{M}_2\mathbf{A}_2^T & \mathbf{N}_2 \end{bmatrix}}_{\mathbf{T}_2(s)} \begin{bmatrix} \mathbf{I}_1 \\ \mathbf{V}_1 \\ \mathbf{V}_2 \\ \mathbf{E} \\ \mathbf{I}_2 \end{bmatrix} = \begin{bmatrix} \mathbf{I}_{s1} \\ \mathbf{0} \\ \mathbf{0} \\ -\mathbf{A}_1\mathbf{I}_{s1} \\ \mathbf{U}_{s1} \end{bmatrix} \tag{4.18}$$

It is clear that (a) the bottom two equations are the MNA equations, hence

$$\mathbf{P}(s) = \begin{bmatrix} \mathbf{A}_1\mathbf{Y}_1\mathbf{A}_1^T & \mathbf{A}_2 \\ \mathbf{M}_2\mathbf{A}_2^T & \mathbf{N}_2 \end{bmatrix} \tag{4.19}$$

and (b) once $\mathbf{E}$ and $\mathbf{I}_2$ have been obtained, backsubstitution into the first three equations gives $\mathbf{I}_1$, $\mathbf{V}_1$, and $\mathbf{V}_2$.

Now, by inspection,

$$\det[\mathbf{T}_2(s)] = \det[\mathbf{P}(s)]$$

where the matrix $\mathbf{T}_2(s)$ is obtained from $\mathbf{T}(s)$ by the row and column operations of the three steps above. From the properties of determinants, we see that

$$\det[\mathbf{T}_2(s)] = \det[\mathbf{T}(s)] \tag{4.20}$$

So in the present case we find that $k = 1$. If we had written, for the MNA, the equations in a different order or had labeled the variables in a different order, then the constant k of Eq. (4.16) could be -1.

4.3 Zero-Input Response and Natural Frequencies

We are going to generalize the concept of natural frequencies that we defined in Chap. 7, Sec. 2.1.

A. Characteristic polynomial and natural frequencies Consider a *linear time-invariant* circuit $\mathcal{N}$. Assume that (1) $\mathcal{N}$ is *uniquely solvable* and (2) $\mathcal{N}$ has *no independent sources*.

Consequently all the voltages and currents of $\mathcal{N}$ are caused by the initial conditions, i.e., by the energy initially stored in the inductors and capacitors. In other words, we are considering now only the *zero-input response of* $\mathcal{N}$.

Consider the differential equations (4.6); since $\mathbf{u}_s(t) = 0$, they reduce to the *homogeneous* system of *algebraic* and *differential* equations

$$\mathbf{T}(D)\,\mathbf{w}(t) = \mathbf{0} \tag{4.21}$$

where $\mathbf{T}(D)$ is a matrix of real polynomials in $D = d/dt$ and $\mathbf{w}(t)$ is the column vector made up of $\mathbf{e}(t)$, $\mathbf{v}(t)$, and $\mathbf{i}(t)$, namely, of the circuit variables.

Let us find a solution of Eq. (4.21) of the form

$$\mathbf{w}(t) = \mathbf{w}_0\, e^{\lambda t} \tag{4.21a}$$

where both the *constant nonzero* vector $\mathbf{w}_0$ and the *number* λ will be determined later. Note that, since $\mathbf{w}_0$ is *constant*,

$$D\mathbf{w}_0\, e^{\lambda t} = \frac{d}{dt}(\mathbf{w}_0\, e^{\lambda t}) = \lambda \mathbf{w}_0\, e^{\lambda t}$$

Hence substituting $\mathbf{w}_0\, e^{\lambda t}$ into Eq. (4.21)

$$\mathbf{T}(D)\,\mathbf{w}_0\, e^{\lambda t} = \mathbf{T}(\lambda)\,\mathbf{w}_0\, e^{\lambda t} = \mathbf{0} \tag{4.22}$$

Now for any number λ and any $t \geq 0$, $e^{\lambda t} \neq 0$, and hence Eq. (4.22) is equivalent to

$$\mathbf{T}(\lambda)\,\mathbf{w}_0 = \mathbf{0} \tag{4.23}$$

This is a linear *homogeneous* system of *algebraic* equations: By Cramer's rule, Eq. (4.23) has a nonzero solution $\mathbf{w}_0$ if and only if

$$\det[\mathbf{T}(\lambda)] = 0 \tag{4.24}$$

Since every element of $\mathbf{T}(\lambda)$ is a polynomial in λ (with real coefficients and at most of degree 1), $\det[\mathbf{T}(\lambda)]$ is a polynomial in λ with real coefficients. The polynomial

$$\chi(\lambda) \triangleq \det[\mathbf{T}(\lambda)] \tag{4.25}$$

is called the *characteristic polynomial of the circuit* $\mathcal{N}$. Any zero of $\chi(\lambda)$ is called *a natural frequency of* $\mathcal{N}$ (sometimes also called, by abuse of language, an eigenvalue of $\mathcal{N}$).

REMARKS

1. If we had analyzed the linear time-invariant circuit $\mathcal{N}$ by MNA the natural frequencies would be defined as the zeros of the polynomial

$$\det[\mathbf{P}(\lambda)] = 0 \tag{4.26}$$

In Eq. (4.16), we noted that $\det[\mathbf{T}(\lambda)] = k \det[\mathbf{P}(\lambda)]$, for some *nonzero constant* k. Hence as expected, the tableau analysis and the MNA yield the *same list* of natural frequencies, with the *same* multiplicities.

2. We know that $\det[\mathbf{T}(\lambda)]$ is a *polynomial* in λ with *real* coefficients [see Eq. (4.7)]. Consequently, the natural frequencies of $\mathcal{N}$ are either real or occur in complex-conjugate pairs.

B. Physical interpretation of natural frequencies The natural frequencies just defined are based on the analysis of $\mathcal{N}$ by the tableau equations; let us show that these natural frequencies depend only on $\mathcal{N}$. We establish this by showing that a natural frequency is an intrinsic physical property of the circuit.

Natural frequency property Consider a *uniquely solvable linear time-invariant* circuit $\mathcal{N}$ *with no independent sources*. Then the circuit $\mathcal{N}$ will exhibit, for some initial conditions, a solution described by

$$\mathbf{w}(t) = \mathbf{w}_0\, e^{\lambda t} = \begin{bmatrix} \mathbf{e}_0 \\ \mathbf{v}_0 \\ \mathbf{i}_0 \end{bmatrix} e^{\lambda t} \quad \begin{array}{l}\text{(where } \mathbf{e}_0, \mathbf{v}_0, \mathbf{i}_0, \text{ and } \mathbf{w}_0 \text{ are} \\ \text{constant vectors of suitable} \\ \text{dimensions, with } \mathbf{w}_0 \neq \mathbf{0})\end{array} \tag{4.27}$$

if and only if

$$\lambda \text{ is a natural frequency of } \mathcal{N} \tag{4.28}$$

When a linear time-invariant circuit $\mathcal{N}$ exhibits a purely exponential zero-input response of the form

$$\mathbf{w}(t) = \begin{bmatrix} \mathbf{e}(t) \\ \mathbf{v}(t) \\ \mathbf{i}(t) \end{bmatrix} = \mathbf{w}_0\, e^{\lambda t}$$

(where $\mathbf{w}_0$ and λ are constant), we say that *the mode* $\mathbf{w}_0\, e^{\lambda t}$ *of* $\mathcal{N}$ has been set up, or that $\mathcal{N}$ *is in the mode* $\mathbf{w}_0\, e^{\lambda t}$. Note that if only that mode has been excited, all the circuit variables behave exponentially: Each circuit variable is proportional to the same exponential $e^{\lambda t}$.

PROOF OF THE NATURAL FREQUENCY PROPERTY

$\Rightarrow$ We proved it above when we showed that if there is a solution of the form of Eq. (4.27)—which is a repeat of Eq. (4.21*a*)—then $\det[\mathbf{T}(\lambda)] = 0$, that is, Eq. (4.23) holds for some $\mathbf{w}_0 \neq \mathbf{0}$.

$\Leftarrow$ Let λ_i be *any* natural frequency of $\mathcal{N}$, that is, λ_i is *any* zero of $\det[\mathbf{T}(\lambda)]$.

Consider now the system of linear algebraic equations:

$$\mathbf{T}(\lambda_i)\mathbf{z} = \mathbf{0} \tag{4.29}$$

where $\mathbf{z}$ is an unknown *constant vector* of dimension $2b + (n-1)$. The linear homogeneous Eq. (4.29) has nonzero solutions because by assumption $\det[\mathbf{T}(\lambda_i)] = 0$. That is, there is a nonzero vector $\mathbf{z}$ which satisfies Eq. (4.29) and consequently, in view of our previous calculations, $\mathbf{w}(t) = \mathbf{z}\exp(\lambda_i t)$ is a solution of the circuit equations (4.21). ∎

REMARK Knowing the constant vector $\mathbf{z}$, a nonzero solution of Eq. (4.29), the corresponding initial conditions are obtained by picking out of the components of $\mathbf{z}$ those that correspond to *capacitor voltages* and *inductor currents*.

Example Considering the linear time-invariant circuit $\mathcal{N}$ shown in Fig. 4.3 (*a*) *Let us find the natural frequencies of* $\mathcal{N}$. Since all the elements of $\mathcal{N}$ are voltage-controlled, MNA reduces to node analysis. The node equations read

$$\underbrace{\begin{bmatrix} s+1.1 & -0.1 \\ -32.1 & s+1.1 \end{bmatrix}}_{\mathbf{P}(s)} \begin{bmatrix} E_1(s) \\ E_2(s) \end{bmatrix} = \begin{bmatrix} v_1(0-) \\ v_2(0-) \end{bmatrix} \tag{4.30}$$

Now

$$\det[\mathbf{P}(s)] = (s+1.1)^2 - 0.1 \times 32.1 = s^2 + 2.2s - 2$$

The natural frequencies are 0.6916 and -2.891. Thus there is an exponentially growing unstable mode and an exponentially stable mode (-2.891): in other words, the origin of the (e_1, e_2) plane is a saddle point (see Chap. 7, Fig. 3.5. (*b*) *Let us find initial conditions so that the unstable mode is excited*. Thus we seek an initial state which lies on the unstable eigenvector of the saddle point. We have to solve the following equation for $\mathbf{w}_0$,

$$\mathbf{P}(0.6916)\mathbf{w}_0 = \mathbf{0}$$

i.e.,

$$\begin{bmatrix} 1.7916 & -0.1 \\ -32.1 & 1.7916 \end{bmatrix} \begin{bmatrix} w_1 \\ w_2 \end{bmatrix} = \begin{bmatrix} 0 \\ 0 \end{bmatrix} \tag{4.31}$$

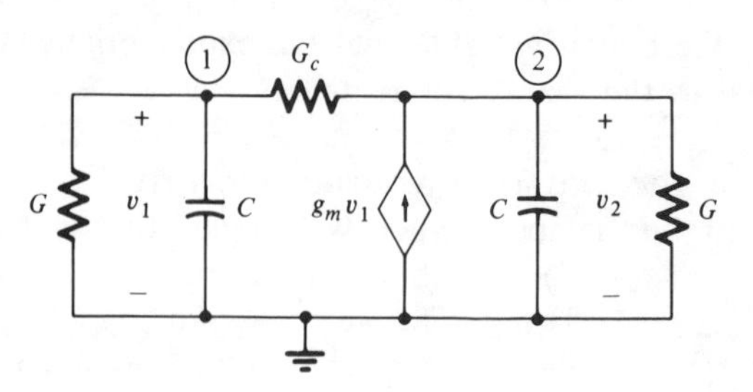

Figure 4.3 Let $G = 1\,\mathrm{S}$, $G_C = 0.1\,\mathrm{S}$, $C = 1\,\mathrm{F}$, and $g_m = 32.0\,\mathrm{S}$.

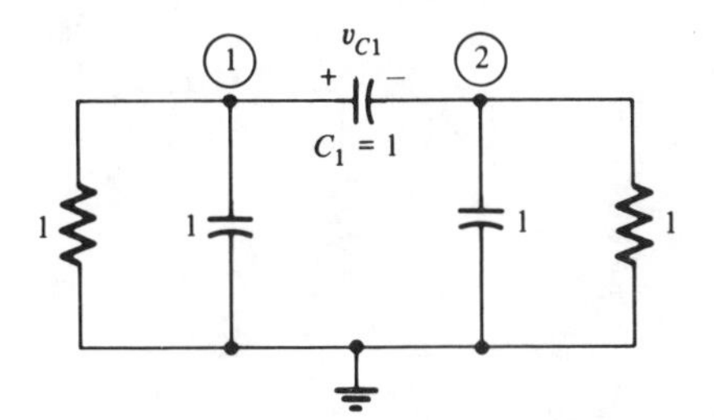

Figure 4.4 Linear time-invariant circuit whose natural frequencies are -1 and $-\frac{1}{3}$ rad/s.

Since det[$\mathbf{P}(0.6916)$] = 0, the second equation is linearly dependent on the first, hence it may be deleted. From the first equation we see that

$$w_1 = 1 \qquad w_2 = 17.916$$

is a solution. [Note that all solutions of Eq. (4.31) are multiples of that one.] Consequently, the initial conditions

$$e_1(0) = 1 \text{ V}$$

$$e_2(0) = 17.916 \text{ V}$$

will give the zero-input response

$$\begin{bmatrix} e_1(t) \\ e_2(t) \end{bmatrix} = \begin{bmatrix} 1\,\mathrm{e}^{0.6916t} \\ 17.916\,\mathrm{e}^{0.6916t} \end{bmatrix} 1(t) \tag{4.32}$$

REMARKS

1. If we view this problem from a *Laplace transform* point of view—see Eq. (4.7) with $\mathbf{U}_s(s) = \mathbf{0}$—we see that the initial conditions just obtained correspond to an initial-value vector $\mathbf{U}^i$ in Eq. (4.7) with the property that the partial fraction expansion of the vector

$$\mathbf{w}(s) = \mathbf{T}(s)^{-1} \begin{bmatrix} \mathbf{0} \\ \mathbf{0} \\ \mathbf{U}^i \end{bmatrix} \tag{4.33}$$

has only components of the form $k/(s - \lambda_j)$. All the other residues are zero as a result of the special choice of $\mathbf{U}^i$ obtained from the solution $\mathbf{z}$ of Eq. (4.29) above.

2. Examples show that for some natural frequencies, *some of the components of* $\mathbf{w}_0$ *may be zero*.[5] Since the natural frequencies are properties of $\mathcal{N}$, we may calculate them using any method of analysis. For example, consider the circuit shown in Fig. 4.4. To make calculations easy, we normalize all elements to have a value of 1 (ohms and farads). The node equations give

[5] Here we consider $\mathbf{w}(t) = \mathbf{w}_0 \exp \lambda t$, the solution of the tableau equations [see Eqs. (4.21) and (4.21a) above].

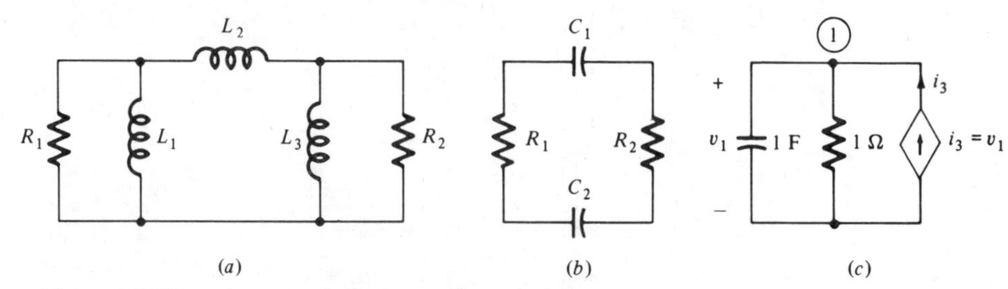

Figure 4.5 These three circuits have a zero natural frequency.

$$\begin{aligned} 2\dot{e}_1 + e_1 \qquad\quad - \dot{e}_2 &= 0 \\ -\dot{e}_1 \qquad + 2\dot{e}_2 + e_2 &= 0 \end{aligned} \tag{4.34}$$

Adding and subtracting those equations gives

$$\begin{aligned} \frac{d}{dt}(e_1 + e_2) + (e_1 + e_2) &= 0 \\ 3\frac{d}{dt}(e_1 - e_2) + (e_1 - e_2) &= 0 \end{aligned} \tag{4.35}$$

Since $v_{C1} = e_1 - e_2$, the last equation reads

$$\dot{v}_{C1} + \tfrac{1}{3} v_{C1} = 0 \tag{4.36}$$

that is, *the only solution* for $v_{C1}(t)$ is $v_{C1}(t) = v_{C1}(0)\,\mathrm{e}^{-t/3}$. Since

$$e_1 = \tfrac{1}{2}[(e_1 + e_2) + (e_1 - e_2)] \tag{4.37}$$

the two equations (4.35) show that

$$e_1(t) = \tfrac{1}{2}[e_1(0) + e_2(0)]\,\mathrm{e}^{-t} + \tfrac{1}{2} v_{C1}(0)\,\mathrm{e}^{-t/3} \tag{4.38}$$

Thus both natural frequencies -1 and $-\frac{1}{3}$ may be *observed* at $e_1(t)$, the node voltage of node ①, whereas only the natural frequency $-\frac{1}{3}$ is *observable* at $v_{C1}(t)$, the voltage across capacitor v_{C1}.

Thus we can say that the mode associated with the natural frequency $-\frac{1}{3}$ is *observable* as a voltage or a current at any of the five elements of the circuit; but the mode associated with the natural frequency -1 is *not observable* at $v_{C1}(t)$, the voltage across the capacitor C_1.

3. (*Zero natural frequencies*)[6] Note that, for $\lambda = 0$, $\mathrm{e}^{\lambda t} = 1$ for all t: hence a zero natural frequency is associated with a *constant current* and/or a *constant voltage*. This occurs, for example, when there is a *loop of inductors*[7] or, dually, *a cut set of capacitors*. (See Fig. 4.5*a* and *b*.) The

[6] This remark may be skipped without loss of continuity.

[7] This property is used in particle accelerators where, to save energy, the magnet coils are cooled to become superconducting, i.e., have precisely zero resistance.

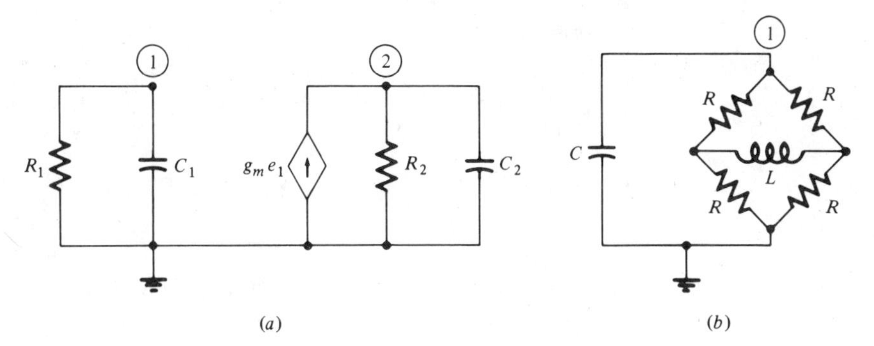

Figure 4.6 These two circuits have some natural frequencies that do not depend on some elements.

circuit shown in Fig. 4.5*c* also has zero as a natural frequency: Indeed the node equation for node ① reads

$$\dot{v}_1 + v_1 - v_1 = \dot{v}_1 = 0 \tag{4.39}$$

So for any initial condition $v_1(0)$, the voltage $v_1(t)$ is constant and a dc current flows through the resistor.

4. The natural frequencies of the circuit $\mathcal{N}$ are properties of the *whole circuit* $\mathcal{N}$; indeed they are the zeros of $\chi(s) = \det[\mathbf{T}(s)]$.[8] Consequently, in general, any change in any element value of $\mathcal{N}$ will change all the natural frequencies. We say "in general" because when certain symmetry or certain topology or certain numerical relations hold, some natural frequencies do not depend on the values of *some* elements.

Example For the circuit of Fig. 4.6*a*, the natural frequency $-1/R_1C_1$ of e_1 does not depend on R_2C_2 and g_m. For the circuit of Fig. 4.6*b*, the natural frequency $-1/RC$ of e_1 does not depend on L (because the bridge is balanced).

C. The zero-input response Consider a *uniquely solvable linear time-invariant* circuit $\mathcal{N}$ which has no independent sources. Consider its tableau equations. The vector $\mathbf{w}(t)$ lists all the circuit variables, that is, all the node voltages, all the branch voltages, and all the branch currents of $\mathcal{N}$. For some arbitrary initial conditions specified by $\mathbf{U}^i$ and zero independent sources, the tableau equations read:

$$\mathbf{T}(s)\,\mathbf{W}(s) = \begin{bmatrix} \mathbf{0} \\ \mathbf{0} \\ \mathbf{U}^i \end{bmatrix} \tag{4.40}$$

where $\mathbf{T}(s) = \mathbf{T}_0 + s\mathbf{T}_1$ and the matrices $\mathbf{T}_0$, $\mathbf{T}_1$, and $\mathbf{U}^i$ are real and constant.

[8] In the case of modified node analysis, they are the zeros of $\det[\mathbf{P}(s)]$.

1. Suppose that the characteristic polynomial of $\mathcal{N}$, $\chi(s) \triangleq \det[\mathbf{T}(s)]$, has *simple zeros*: $\lambda_1, \lambda_2, \ldots, \lambda_\nu$. (Equivalently, $\mathcal{N}$ has ν simple natural frequencies.) Applying Cramer's rule to Eq. (4.40) we obtain, after partial fraction expansion,

$$\mathbf{W}(s) = \frac{\mathbf{k}_1}{s-\lambda_1} + \frac{\mathbf{k}_2}{s-\lambda_2} + \cdots + \frac{\mathbf{k}_\nu}{s-\lambda_\nu} \tag{4.41}$$

hence

$$\mathbf{w}(t) = \sum_{\alpha=1}^{\nu} \mathbf{k}_\alpha \, e^{\lambda_\alpha t} \tag{4.42}$$

Thus, for any arbitrary initial condition, each circuit variable of $\mathcal{N}$ is a linear combination of exponentials of the form $e^{\lambda_\alpha t}$, where each λ_α is a natural frequency of $\mathcal{N}$. Of course, if we consider a particular component of $\mathbf{w}(t)$, say $e_k(t)$, some of the exponentials may be absent (because the corresponding weighting coefficient is zero!).

Equation (4.42) shows a decomposition of the zero-input response as a sum of *modes* with, respectively, natural frequencies λ_α, $\alpha = 1, 2, \ldots, \nu$.

2.[9] Suppose that λ_1 is a *second-order* zero of $\det[\mathbf{T}(s)]$ and that the remaining $\nu - 2$ zeros are simple. Then it can be shown that there are two possibilities, either

(*a*)
$$\mathbf{w}(t) = \mathbf{w}_1 \, e^{\lambda_1 t} + \mathbf{w}_1' \, e^{\lambda_1 t} + \sum_{\alpha=2}^{\nu-1} \mathbf{w}_\alpha \, e^{\lambda_\alpha t} \tag{4.43}$$

(where $\mathbf{w}_1$ and $\mathbf{w}_1'$ are linearly independent vectors) or

(*b*)
$$w(t) = \mathbf{w}_1 \, e^{\lambda_1 t} + \mathbf{w}_1'' t \, e^{\lambda_1 t} + \sum_{\alpha=2}^{\nu-1} \mathbf{w}_\alpha \, e^{\lambda_\alpha t} \tag{4.44}$$

(where $\mathbf{w}_1$ and $\mathbf{w}_1''$ are not necessarily linearly independent vectors: the functions $e^{\lambda_1 t}$ and $t\,e^{\lambda_1 t}$ are linearly independent).

Note that associated with the *second-order* zeros of $\det[\mathbf{T}(s)]$, there are *two* terms and that, in both cases, there is a total of ν terms; this makes sense since ν is the degree of $\det[\mathbf{T}(s)]$.

3. In general, associated with an mth-order zero of $\det[\mathbf{T}(s)]$ there are m terms which may be pure exponentials or polynomials times exponentials (and the polynomial is at most of degree $m - 1$).

These results lead to a very important engineering conclusion.

> If *all the natural frequencies* of the linear time-invariant circuit $\mathcal{N}$ have negative (<0) real parts, then, *for all initial conditions*, the zero-input response $\mathbf{w}(t) \to \mathbf{0}$ *exponentially*, i.e., *all* the circuit variables of $\mathcal{N}$ go to zero exponentially as $t \to \infty$.

(4.45)

The result follows immediately from Eqs. (4.42), (4.43), and (4.44).

[9] This brief discussion of multiple zeros may be skipped without loss of continuity.

Exercise Let $\alpha > 0$ and m be an integer > 0. Show that $f(t) = t^m \mathrm{e}^{-\alpha t} \to 0$ as $t \to \infty$. [*Hint*: Take the log of $f(t)$.]

We say that a linear time-invariant circuit $\mathcal{N}$ is *exponentially stable* iff all its natural frequencies have negative (<0) real parts.

The purpose of the following example is to illustrate the geometric meaning of Eq. (4.42). To simplify our equations we'll use node analysis.

Example Consider the linear time-invariant circuit shown in Fig. 4.7. Since for this circuit all elements are voltage-controlled and include no inductors, we can write the node equations by inspection:

$$\begin{bmatrix} 3s+1 & -s & 0 \\ -4-s & 3s+1 & 0 \\ 0 & -4 & s+1 \end{bmatrix} \begin{bmatrix} E_1(s) \\ E_2(s) \\ E_3(s) \end{bmatrix} = \begin{bmatrix} 3e_1(0-) - e_2(0-) \\ 3e_2(0-) - e_1(0-) \\ e_3(0-) \end{bmatrix} \tag{4.46}$$

It is easy to see that

$$\det[\mathbf{P}(s)] = (8s^2 + 2s + 1)(s+1) \tag{4.47}$$

Hence the three natural frequencies are

$$\lambda_{1,2} = \frac{-1 \pm j\sqrt{7}}{8} \qquad \lambda_3 = -1 \tag{4.48}$$

Since $\mathrm{Re}(\lambda_i) \le -\frac{1}{8}$, for $i = 1, 2$ and $\mathrm{Re}(\lambda_3) = -1$, the circuit of Fig. 4.7 is exponentially stable. (*a*) Consider the *real* natural frequency $\lambda_3 = -1$. To set up the corresponding mode we solve

$$\mathbf{P}(-1)\,\mathbf{w}_3 = \mathbf{0} \tag{4.49}$$

for a constant vector $\mathbf{w}_3$. Clearly for any nonzero real k, $\mathbf{w}_3 = (0, 0, k)^T$ is a solution. The corresponding mode is

$$\mathbf{e}(t) = \begin{bmatrix} e_1(t) \\ e_2(t) \\ e_3(t) \end{bmatrix} = \begin{bmatrix} 0 \\ 0 \\ k \end{bmatrix} \mathrm{e}^{-t} \tag{4.50}$$

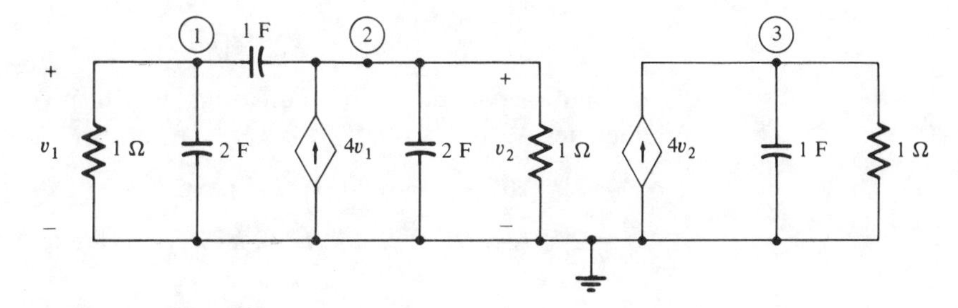

Figure 4.7 Third-order circuit whose nodes are to be determined.

The constant k is equal to the initial capacitor voltage $e_3(0)$. (*b*) Consider now *the pair of complex-conjugate natural frequencies* λ_1, λ_2, so $\overline{\lambda_1} = \lambda_2$. To set up the mode we solve

$$\mathbf{P}(\lambda_1)\,\mathbf{w}_1 = \mathbf{0} \tag{4.51}$$

for the constant vector $\mathbf{w}_1$. Now λ_1 is complex, so the vector $\mathbf{w}_1$ will also be complex. In fact it is easy to see that

$$\mathbf{w}_1 = \begin{bmatrix} 1 \\ 2 - j\sqrt{7} \\ 3.5 - j4.5\sqrt{7} \end{bmatrix}$$

is a solution. Formally, $\mathbf{w}_1\,e^{\lambda_1 t}$ is a solution of the differential equations of the circuit. Of course, it has *no* physical meaning since λ_1 and $\mathbf{w}_1$ are both complex. Consider the second natural frequency λ_2 and its corresponding $\mathbf{w}_2$: $\mathbf{w}_2$ is a nonzero solution of

$$\mathbf{P}(\lambda_2)\mathbf{w}_2 = \mathbf{0} \tag{4.52}$$

Now $\lambda_2 = \overline{\lambda_1}$; furthermore if we take the complex conjugate of *every* element of the numerical matrix $\mathbf{P}(\lambda_1)$ and if we note that $\mathbf{P}(s)$ is a polynomial matrix with *real* coefficients, we have

$$\overline{\mathbf{P}(\lambda_1)} = \mathbf{P}(\overline{\lambda}_1) = \mathbf{P}(\lambda_2)$$

Therefore if we take the complex conjugate of Eq. (4.51), we obtain successively

$$\overline{\mathbf{P}(\lambda_1)\mathbf{w}_1} = \overline{\mathbf{P}(\lambda_1)}\,\overline{\mathbf{w}}_1 = \mathbf{P}(\overline{\lambda}_1)\,\overline{\mathbf{w}}_1 = \mathbf{P}(\lambda_2)\,\overline{\mathbf{w}}_1 = \mathbf{0}$$

Thus $\overline{\mathbf{w}}_1$ is a solution of $\mathbf{P}(\lambda_2)\,\mathbf{w}_2 = \mathbf{0}$. Thus associated to the natural frequency $\lambda_2 = \overline{\lambda}_1$ we have a solution $\overline{\mathbf{w}}_1\,e^{\overline{\lambda}_1 t}$, which is precisely the complex conjugate of the preceding one.

It turns out that associated with the *pair* of complex-conjugate natural frequencies λ_1 and $\lambda_2 = \overline{\lambda}_1$, we have a family of solutions that depend on *two* real independent parameters. Let $k = |k| \exp j\sphericalangle k$ be some nonzero complex number, then

$$\begin{aligned} \mathbf{e}(t) &= k\mathbf{w}_1\,e^{\lambda_1 t} + \overline{k}\mathbf{w}_2\,e^{\lambda_2 t} \\ &= 2\,\mathrm{Re}[k\mathbf{w}_1\,e^{\lambda_1 t}] \end{aligned} \tag{4.53}$$

is a solution of the node equations which involves only the natural frequencies λ_1 and $\overline{\lambda}_1$. Let $\lambda_1 = \lambda_{1r} + j\lambda_{1i}$ and $\mathbf{w}_1 = \mathbf{w}_r + j\mathbf{w}_i$; then

$$\begin{aligned} \mathbf{e}(t) &= 2|k|\,\mathrm{Re}[(\mathbf{w}_r + j\mathbf{w}_i)\,(\exp \lambda_{1r}t)\exp j(\lambda_{1i}t + \sphericalangle k)] \\ &= 2|k|\,(\exp \lambda_{1r}t)\cos(\lambda_{1i}t + \sphericalangle k)\,\mathbf{w}_r - 2|k|\,(\exp \lambda_{1r}t)\sin(\lambda_{1i}t + \sphericalangle k)\,\mathbf{w}_i \end{aligned} \tag{4.54}$$

Thus any initial condition of the form

$$\mathbf{e}(0) = 2\,\mathrm{Re}(k\mathbf{w}_1) = 2|k|\cos(\measuredangle k)\mathbf{w}_r - 2|k|\sin(\measuredangle k)\mathbf{w}_i$$

will result in the solution described by Eq. (4.54), that is, a solution that involves exclusively the natural frequencies λ_1 and $\bar{\lambda}_1$. For that solution the vector $\mathbf{e}(t)$ moves in the plane spanned by the *real vectors* $\mathbf{w}_r$ and $\mathbf{w}_i$.

Note that we have two degrees of freedom, $|k|$ and $\measuredangle k$, in choosing these initial conditions; consequently, we can specify any point in that plane as an initial point of the solution $\mathbf{e}(t)$.

The physical meaning of the solutions (4.50) and (4.54) can be established as follows: (*a*) Let the circuit of Fig. 4.7 start with initial conditions $e_1(0) = e_2(0) = 0$ V and $e_3(0) = 5$ V; then Eq. (4.50) shows that the solution is an exponential $e_1(t) = e_2(t) = 0$ V for $t \geq 0$ and $e_3(t) = 5\,\mathrm{e}^{-t}$ V for $t \geq 0$. The mode $\mathbf{w}_3\,\mathrm{e}^{-t}$ of the circuit has been set up. (*b*) Choose now as initial conditions for the same circuit the real part of the vector $\mathbf{w}_1$ [equivalently take $k = \frac{1}{2}$ in Eq. (4.53)], namely, $e_1(0) = 1$ V, $e_2(0) = 2$ V, and $e_3(0) = 3.5$ V. By Eq. (4.54) we have

$$e_1(t) = \mathrm{e}^{-t/8}\cos\left(\frac{\sqrt{7}}{8}t\right)$$

$$e_2(t) = 2\,\mathrm{e}^{-t/8}\cos\left(\frac{\sqrt{7}}{8}t\right) + \sqrt{7}\,\mathrm{e}^{-t/8}\sin\left(\frac{\sqrt{7}}{8}t\right)$$

$$e_3(t) = 3.5\,\mathrm{e}^{-t/8}\cos\left(\frac{\sqrt{7}}{8}t\right) + 4.5\cdot\sqrt{7}\,\mathrm{e}^{-t/8}\sin\left(\frac{\sqrt{7}}{8}t\right)$$

Thus only the mode associated with $\lambda_1 = (-1 + j\sqrt{7})/8$ and $\bar{\lambda}_1$ has been excited by these initial conditions.

Exercise Calculate the resulting solution caused by the initial condition $k_r\mathbf{w}_r + k_i\mathbf{w}_i$ (k_r and k_i are any real numbers). [*Hint*: $\mathbf{w}_r = \frac{1}{2}(\mathbf{w}_1 + \mathbf{w}_2)$, $\mathbf{w}_i = \frac{1}{2j}(\mathbf{w}_1 - \mathbf{w}_2)$.]

4.4 Zero-State Response, Network Functions, and Impulse Response

In the preceding chapter we considered exclusively circuits in the sinusoidal steady state at some arbitrary frequency ω; in Sec. 3 of Chap. 9 we defined the network function $H(j\omega)$ as the ratio of the *response phasor* to the *input phasor*. In the present section, we generalize the concept of a network function so that it is applicable with *any input waveform*: The price of that generalization is that we must restrict ourselves exclusively to *zero-state responses*, i.e., we consider arbitrary inputs but all initial conditions are set to zero.

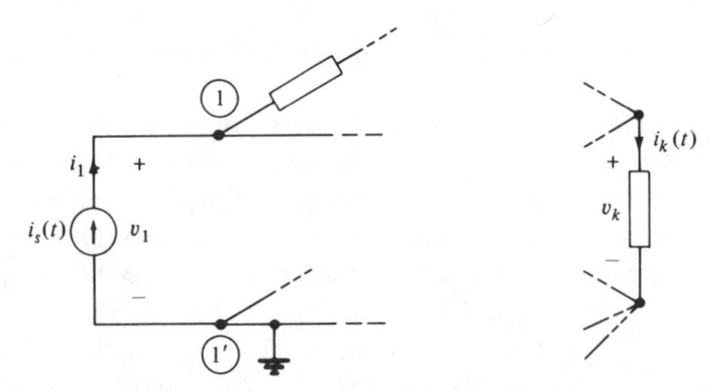

Figure 4.8 General circuit under consideration: the input is i_s, the output is v_k or i_k.

The benefit of this generalization is that it will allow us to establish in this and the next chapter many different connections between the time-domain and the frequency-domain behavior of circuits.

In this section we'll study some properties of network functions and their relation to the corresponding impulse response.

A. Network functions Consider a *uniquely solvable linear time-invariant* circuit $\mathcal{N}$. Suppose that it is driven by *only one independent current source* $i_s(t)$, as shown in Fig. 4.8. Suppose that we are interested in calculating $i_k(t)$, the current in the kth branch due to $i_s(t)$ alone, that is, we set *all the initial conditions to zero*. We may rephrase the problem as: Calculate the *zero-state response* $i_k(t)$ resulting from the input $i_s(t)$.

Let us write the tableau equations using Laplace transform. Using Eq. (4.7), noting that $\mathbf{U}^i = \mathbf{0}$, we obtain

$$\mathbf{T}(s)\,\mathbf{W}(s) = \begin{bmatrix} \mathbf{0} \\ \mathbf{0} \\ \mathbf{U}_s(s) \end{bmatrix} \tag{4.55}$$

By unique solvability $\det[\mathbf{T}(s)] \neq 0$, so Cramer's rule gives

$$I_k(s) = \frac{\text{cofactor of } \mathbf{T}(s)}{\det[\mathbf{T}(s)]}\, I_s(s) \tag{4.56}$$

where the cofactor is the one associated with the $[b + (n-1) + k]$th column and with the $(b + n)$th row.[10] Equation (4.56) relates the Laplace transform of the *zero-state response* $i_k(t)$ to the Laplace transform of the input $i_s(t)$. Observe that the ratio of $I_k(s)$ to $I_s(s)$ defined in Eq. (4.56) depends only on the *circuit* $\mathcal{N}$: It does *not* depend on the nature of the function $i_s(t)$. Here the ratio $I_k(s)/I_s(s)$ is called the current transfer function; it is a *network function*. As before, we will usually denote a network function by $H(s)$.

Note that the presently defined concept of network function is *much more general* than the one defined for the sinusoidal steady state: Indeed it "works" for *any* input waveform $i_s(\cdot)$.

[10] These are, respectively, the column and the row corresponding to i_k and i_s.

From now on we adopt the following *general definition* of a *network function*:

$$\begin{pmatrix}\text{Network} \\ \text{function}\end{pmatrix} \triangleq \frac{\mathcal{L}\{\text{zero-state response}\}}{\mathcal{L}\{\text{input}\}}$$

Recalling that $\mathbf{T}(s)$ may be written as $\mathbf{T}_0 + s\mathbf{T}_1$ where $\mathbf{T}_0$ and $\mathbf{T}_1$ are matrices with constant real elements, we conclude that any expression of the form

$$H(s) = \frac{\text{cofactor of } \mathbf{T}(s)}{\det[\mathbf{T}(s)]} \tag{4.57}$$

is a *rational* function with *real coefficients*. Note that the denominator in Eq. (4.57) is the characteristic polynomial of the circuit $\mathcal{N}$.

Call $c(s)$ a real polynomial containing all the common factors of the numerator of Eq. (4.57) and of its denominator.[11] Then we may factor out the polynomial $c(s)$ from the numerator and the denominator of $H(s)$ and write

$$H(s) = \frac{n(s)\,c(s)}{d(s)\,c(s)} = \frac{n(s)}{d(s)} \tag{4.58}$$

where the polynomial $d(s)$ is a factor of the characteristic polynomial $\det[T(s)]$; furthermore, $n(s)$ and $d(s)$ are *real* polynomials and are *coprime*. Therefore all the zeros of $H(s)$—i.e., all the z_i such that $n(z_i) = 0$—and all its poles—i.e., all the p_i such that $d(p_i) = 0$—are either *real* or occur in *complex-conjugate pairs*. From Eqs. (4.57) and (4.58) we see that if p_i is a pole of $H(s)$, then $d(p_i) = 0$ and, consequently, $\det[\mathbf{T}(p_i)] = 0$; that is, p_i is necessarily a natural frequency of $\mathcal{N}$.

We may summarize this discussion by stating the following property:

Properties of network functions. Let $\mathcal{N}$ be a *uniquely solvable linear time-invariant circuit*, then

(i) Any network function of $\mathcal{N}$ is the ratio of two polynomials with *real* coefficients, hence its zeros and its poles occur in complex-conjugate pairs:

(ii) $$\mathcal{L}\begin{Bmatrix}\text{Zero-state} \\ \text{response}\end{Bmatrix} = \begin{Bmatrix}\text{network} \\ \text{function}\end{Bmatrix} \cdot \mathcal{L}\{\text{input}\}$$

(iii) The poles of any network function are natural frequencies of $\mathcal{N}$.

(4.59)

[11] In other words, the polynomial $c(s)$ is a greatest common divisor of the numerator and denominator of $H(s)$ [in Eq. (4.57)].

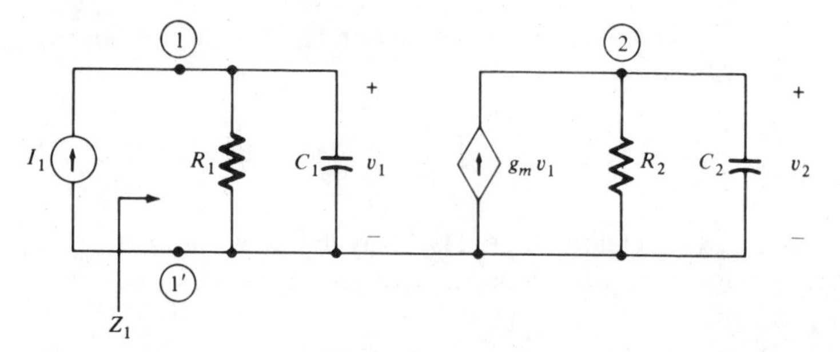

Figure 4.9 Circuit used to illustrate the concepts of natural frequencies and poles of network functions.

REMARKS

1. This concept of network functions is much more general than the one originating from the sinusoidal steady-state theory. Of course if we set $s = j\omega$ in the present network function, we obtain the one we would have calculated from the sinusoidal steady-state analysis. This is a consequence of the conclusions of Sec. 2.5; Table 2.1 in particular. (We shall encounter this fact again in Sec. 5, below.)

2. Statement (ii) in Property (4.59) above implies a very useful superposition property: Let us consider statement (ii) applied to the circuit of Fig. 4.8 above. We assume that the input $i_s(t)$ is the sum of a few simple waveforms $i_{s1}(t), \ldots, i_{sn}(t)$; then, since the network function is independent of the input, by Eq. (4.59), we have

$$V_k(s) = \sum_{j=1}^{n} H(s)\, I_{sj}(s)$$

B. Remark on cancellations Statement (iii) of Property (4.59) says that if p_i is a pole of a network function, then p_i is necessarily a natural frequency. In the calculation of the polynomials n and d in Eq. (4.58), factors common to the cofactor and det[$\mathbf{T}(s)$] must be canceled out; consequently any natural frequency associated with these canceled factors is *not* necessarily[12] a pole of that network function. In other words, *every pole of $H(s)$ is a natural frequency*, but, *for a given $H(s)$, some natural frequencies of $\mathcal{N}$ may not appear as poles of $H(s)$*.

Exercise For the circuit shown in Fig. 4.9, (*a*) calculate by MNA its natural frequencies, (*b*) calculate the poles of the driving-point impedance $Z_1(s)$ seen at the port ①, ①', (*c*) calculate the poles of $Z_{21}(s) \stackrel{\Delta}{=} V_2(s)/I_1(s)$.

[12] It may happen [see Eq. (4.58)] that $c(s)$ and $d(s)$ have a common zero p_k. In that case p_k is a pole of H: The order of p_k as a pole of H is smaller than the order of p_k as a zero of det[$\mathbf{T}(s)$].

Physical interpretation of cancellations Consider the three linear time-invariant circuits shown in Fig. 4.10. The three circuits $\mathcal{N}_a$, $\mathcal{N}_b$, $\mathcal{N}_c$ differ only by the location or absence of the voltage-controlled current source. In each case the input is the current source i_s and the output is the voltage v_0; therefore we are interested in the network function

$$H(s) = \frac{V_0(s)}{I_s(s)}$$

It can easily be seen that each circuit has two natural frequencies -1 and 1; furthermore, their network functions are equal and

$$H_a(s) = H_b(s) = H_c(s) = \frac{1}{s+1}$$

because the common factor $s-1$ cancels out from the numerator and denominator when we calculate the H's by Eq. (4.57).

For $\mathcal{N}_a$, the unstable mode e^t is *excitable* by the input i_s but is *not observable* at the output v_0.

For $\mathcal{N}_b$, the unstable mode e^t is *not excitable* by the input i_s but is *observable* at the output v_0.

For $\mathcal{N}_c$, the unstable mode e^t is *neither excitable* by the input i_s *nor observable at* the output v_0.

It can be shown that these are the only possible physical mechanisms whereby a cancellation may occur in Eq. (4.57).

C. Impulse response and network functions By definition, the impulse response is the zero-state response to a unit impulse. Suppose that in Eq. (4.56), we have $i_s(t) = \delta(t)$; then the response $i_k(t)$ becomes the corresponding impulse response, and since $\mathcal{L}\{\delta(t)\} = 1$, we may conclude that

$$\boxed{\mathcal{L}\begin{Bmatrix}\text{Impulse}\\ \text{response}\end{Bmatrix} = \begin{Bmatrix}\text{network}\\ \text{function}\end{Bmatrix}} \tag{4.60}$$

This is a crucial relation because it relates the *frequency-domain behavior*, $H(s)$, to the *time-domain behavior* of $\mathcal{N}$. For example, if the network function $H(s)$ has simple poles $p_1, p_2, \ldots, p_\nu$, and if[13]

$$H(s) = \sum_{\alpha=1}^{\nu} \frac{k_\alpha}{s - p_\alpha} \tag{4.61}$$

[13] Note that some natural frequencies of $\mathcal{N}$ may be missing from the list $(p_1, p_2, \ldots, p_\nu)$.

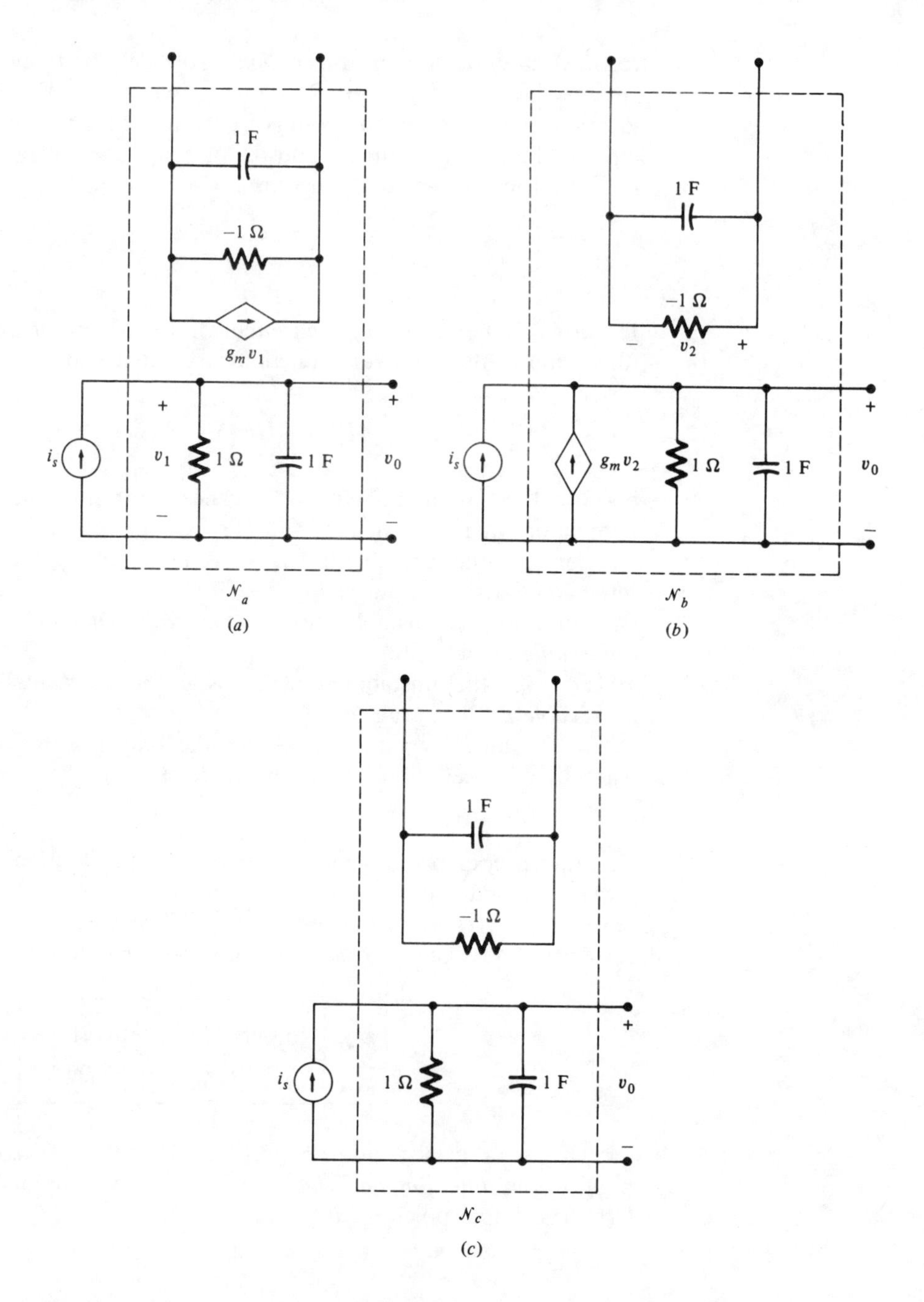

Figure 4.10 The three linear time-invariant circuits $\mathcal{N}_a$, $\mathcal{N}_b$, $\mathcal{N}_c$ have the same pair of natural frequencies $\lambda_1 = -1$ and $\lambda_2 = 1$. They also have a common network function

$$H_a(s) = H_b(s) = H_c(s) = \frac{V_0(s)}{I_s(s)} = \frac{1}{s+1}$$

then, the impulse response is given by

$$h(t) = \left(\sum_{\alpha=1}^{\nu} k_\alpha \, e^{p_\alpha t} \right) 1(t) \tag{4.62}$$

Note that we inserted the unit step factor $1(t)$ to emphasize the fact that $h(t) = 0$ for $t < 0$: Indeed in calculating the impulse response, the input is $\delta(t)$, and, for $t < 0$, the input is identically zero and all initial conditions at time $0-$ are zero.

Exercises

1. Consider a linear *time-invariant* circuit $\mathcal{N}$ as above. For some given choice of input and output, call $h(t)$ and $s(t)$ its impulse response and its step response, respectively. Prove that

(*a*)
$$\mathcal{L}\{s(t)\} = \frac{1}{s} H(s) \tag{4.63}$$

(*b*)
$$h(t) = \frac{ds(t)}{dt} \tag{4.64}$$

(Note that the differentiation d/dt is understood in the distribution sense.)

2. In Exercise 1, Eq. (4.64) is derived using Laplace transforms and the concept of network functions. It is important to know that *Eq. (4.64) does not hold for linear time-varying circuits*. Consider the circuit shown in Fig. 4.11: $R(t)$ is a time-varying resistor with a resistance $R(t)$ where $R(t) = 1\,\Omega$ for $t < 1$ and $R(t) = \infty$ for $t > 1$ s. Here i_s is the input and v_0 is the response. (*a*) Calculate $h(t)$ and $s(t)$ from the circuit equations. (*b*) Show that for $t > 1$, Eq. (4.64) does not hold!

5 THE FUNDAMENTAL THEOREM OF THE SINUSOIDAL STEADY STATE

Consider a uniquely solvable linear time-invariant circuit $\mathcal{N}$. We have seen that if *all* its natural frequencies have *negative* real parts, (equivalently, lie in the open left-half plane), then (*a*) for all initial conditions, its zero-input responses

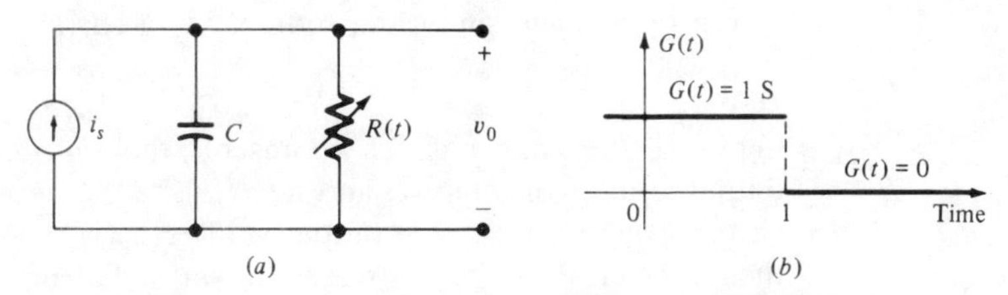

Figure 4.11 Example of a linear circuit with a time-varying resistor.

go to zero exponentially [see Conclusion (4.45)]; (*b*) any of its impulse responses go to zero exponentially [see Eq. (4.62)]; and (*c*) any of its transfer functions have only poles in the open left-half plane [see Property (4.59)]. It is therefore convenient to give a name to such circuits. A uniquely solvable linear time-invariant circuit is called *exponentially stable* iff *all* its natural frequencies have negative (<0) real parts. Clearly, most linear circuits used in practice are exponentially stable.

The following theorem exhibits the rich connection that exists between the concept of a network function defined in terms of Laplace transforms and the concept defined in terms of the sinusoidal steady state. Furthermore it justifies a considerable number of measuring procedures.

> *Fundamental theorem of the sinusoidal steady state*. Consider any *linear time-invariant* circuit $\mathcal{N}$. Assume that $\mathcal{N}$ is *uniquely solvable*. Let all the independent sources driving $\mathcal{N}$ be *sinusoidal* at the *same* frequency ω. If $\mathcal{N}$ is *exponentially stable*, then, for all initial conditions,
>
> (i) As $t \to \infty$, *all branch voltages* and *all branch currents* tend to a *unique* sinusoidal steady-state at frequency ω,
> (ii) This sinusoidal steady state is, of course, easily obtained by phasor analysis.

REMARK This theorem is conceptually very important: Indeed it shows that for any uniquely solvable linear time-invariant exponentially stable circuit $\mathcal{N}$, for any sinusoidal drive at frequency ω, *there is one and only one sinusoidal steady state* which is independent of the initial conditions. That justifies the phrase "*The* sinusoidal steady state of $\mathcal{N}$. . . ."

For simplicity we prove only a special case; this special case is very instructive because it demonstrates again that the network function defined on the basis of the sinusoidal steady-state analysis is identical with that obtained from Laplace transforms.

Special case Let $\mathcal{N}$ be a uniquely solvable linear time-invariant circuit driven by a single sinusoidal source, say a current source,

$$i_s(t) = \text{Re}[\hat{I}\, e^{j\omega t}] \tag{5.1}$$

here $\hat{I}$ is the *phasor* which represents the *sinusoid* $i_s(t)$. Let v_0 be the output voltage of interest, and let $H(s)$ be the network function from the current source $I_s(s)$ to the output voltage $V_0(s)$. If $\mathcal{N}$ is *exponentially stable* then the zero-state response to the sinusoidal input $i_s(t)$ tends to a sine wave of the same frequency:

$$v_0(t) \to \text{Re}[H(j\omega)\,\hat{I}\,e^{j\omega t}] \qquad \text{as } t \to \infty \tag{5.2}$$

REMARK The purpose of this remark is to show that once the *special case* is established, the fundamental theorem is essentially proved.

The special case considers only *one input* and only the *zero-state* response, i.e., all initial conditions are set to zero. We are going to show that these restrictions do not affect the conclusion.

(*a*) Suppose that the initial conditions are *not* zero: By the superposition property, Eq. (4.11), we have

$$\begin{pmatrix}\text{Complete}\\ \text{response}\end{pmatrix} = \begin{pmatrix}\text{zero-state}\\ \text{response}\end{pmatrix} + \begin{pmatrix}\text{zero-input}\\ \text{response}\end{pmatrix} \tag{5.3}$$

Assuming that $\mathcal{N}$ has simple natural frequencies $\lambda_1, \lambda_2, \ldots, \lambda_\nu$, then the zero-input response is of the form

$$\sum_{\alpha=1}^{n} k_\alpha\, e^{\lambda_\alpha t} \tag{5.4}$$

where the k_α's depend on the initial conditions. Now since $\mathcal{N}$ is exponentially stable, $\text{Re}(\lambda_\alpha) < 0$ for all α's, and hence the sum (5.4) goes to zero as $t \to \infty$. Hence the generalization to arbitrary initial conditions does not affect the conclusion.

(*b*) Suppose that $\mathcal{N}$ is driven by *several* independent *sinusoidal* sources at the *same* frequency ω: Then using Eq. (4.10) it can be shown that the zero-state response due to all sinusoidal sources acting together is the sum of the zero-state response of each source acting alone. But the special case proves that each one of these contributions tends to the sine wave predicted by phasor analysis. Therefore, as $t \to \infty$, the zero-state response due to all the sources acting together tends to the sum of the sine waves predicted by phasor analysis.

PROOF OF THE SPECIAL CASE. Note that

$$i_s(t) = \text{Re}(\hat{I}\,e^{j\omega t}) = \tfrac{1}{2}\hat{I}\,e^{j\omega t} + \tfrac{1}{2}\overline{\hat{I}\,e^{j\omega t}}$$

where the second term is the complex conjugate of the first. Taking Laplace transforms we obtain

$$I_s(s) = \frac{1}{2}\,\frac{\hat{I}}{s - j\omega} + \frac{1}{2}\,\frac{\bar{\hat{I}}}{s + j\omega}$$

and $V_0(s)$, the Laplace transform of the zero-state response, is given by

$$V_0(s) = \frac{1}{2}\,H(s)\,\frac{\hat{I}}{s - j\omega} + \frac{1}{2}\,H(s)\,\frac{\bar{\hat{I}}}{s + j\omega} \tag{5.5}$$

Let us evaluate the time function corresponding to the first term in Eq. (5.5). For simplicity, assume that $H(s)$ has *simple* poles, say, $p_1, p_2, \ldots, p_\nu$. Since $\mathcal{N}$ is exponentially stable, for $i = 1, 2, \ldots, \nu$, $\text{Re}(p_i) < 0$; consequently, $j\omega \neq p_i$ for all i and the two terms in Eq. (5.5) have *simple* poles. Let us perform a partial fraction expansion on the first term: For $i = 1, 2, \ldots, \nu$, call k_i its residue at p_i and note that its residue at $j\omega$ is given by

$$\lim_{s \to j\omega} \left[\cancel{(s - j\omega)}\, H(s) \, \frac{1}{2} \, \frac{\hat{I}}{\cancel{s - j\omega}} \right] = \frac{1}{2} \, H(j\omega) \, \hat{I} \tag{5.6}$$

Therefore, the first term contributes

$$\sum_{i=1}^{\nu} k_i \, e^{p_i t} + \tfrac{1}{2} H(j\omega) \, \hat{I} \, e^{j\omega t} \qquad \text{for all } t \geq 0 \tag{5.7}$$

Now, in view of the exponential stability of $\mathcal{N}$, for all i's, $\exp(p_i t) \to 0$ as $t \to \infty$; hence, as $t \to \infty$, the first term in Eq. (5.3) tends to

$$\tfrac{1}{2} H(j\omega) \, \hat{I} \, e^{j\omega t} \tag{5.8}$$

The same reasoning applied to the second term of Eq. (5.5) shows that its contribution tends to

$$\tfrac{1}{2} H(-j\omega) \, \bar{\hat{I}} \, e^{-j\omega t} = \tfrac{1}{2} \overline{H(j\omega) \, \hat{I} \, e^{j\omega t}} \qquad \text{as } t \to \infty \tag{5.9}$$

Therefore, taking into account Eqs. (5.8) and (5.9) we conclude from Eq. (5.5) that, as $t \to \infty$, we have

$$v_0(t) \to \text{Re}[H(j\omega) \, \hat{I} \, e^{j\omega t}]$$

■

REMARK Suppose that for some natural frequency, say λ_1, we have $\text{Re}(\lambda_1) > 0$ and that, for all other natural frequencies, $\text{Re}(\lambda_\alpha) < 0$; then, in view of the natural frequency property [see Eq. (4.27)], there will always be initial conditions which will cause the zero-input response $\mathbf{w}(t)$ to grow exponentially as $t \to \infty$. In fact one can prove that almost all initial conditions will cause this exponential growth of $\mathbf{w}(t)$. So *a single natural frequency in the open right-half plane makes the linear circuit* $\mathcal{N}$ *exponentially unstable.*

Exercise Let the circuit $\mathcal{N}$ be linear, time-invariant, and exponentially stable; let its natural frequencies be $\lambda_1, \lambda_2, \ldots, \lambda_\nu$. Suppose that we drive it with a voltage source $v_s(t)$ where

$$v_s(t) = V_i \, e^{s_i t}$$

where V_i and s_i are constants (possibly complex). Let the output be the current i_o and call $H(s)$ the transfer admittance from v_s to i_o. Show that irrespective of the initial conditions, if $\text{Re}(\lambda_\alpha) < \text{Re}(s_i)$ for all α, then

$$i_o(t) \to H(s_o) V_i \, e^{s_i t} \qquad \text{as } t \to \infty$$

6 CONVOLUTION

Convolution is an extremely important concept that permeates engineering and physics. In circuit theory and in communications, its importance lies in that (*a*) it ties the time-domain description to the frequency-domain description and (*b*) it is the clearest consequence of the properties of *linearity* and *time-invariance*.

6.1 Engineering Interpretation of Convolution

A. The convolution operation Let $f_1(t)$ and $f_2(t)$ be functions defined on $[0, \infty)$, and let us take them to *be equal to zero for* $t < 0$: The *convolution of the time functions* f_1 *and* f_2 is a new time function denoted by $(f_1 * f_2)(t)$ and defined for all t by[14]

$$(f_1 * f_2)(t) = \int_{0-}^{t+} f_1(t-\tau)\, f_2(\tau)\, d\tau \tag{6.1}$$

Note that since $f_2(\tau) = 0$ for all $\tau < 0$, we have

$$(f_1 * f_2)(t) = 0 \qquad \text{for all } t < 0 \tag{6.2}$$

Example Let us calculate $f(\cdot)$, the convolution of $1(t)\, \mathrm{e}^{at}$, with $1(t)\, \mathrm{e}^{bt}$. By definition

$$f(t) \triangleq \int_0^t \mathrm{e}^{a(t-\tau)}\, \mathrm{e}^{b\tau}\, d\tau = \mathrm{e}^{at} \int_0^t \mathrm{e}^{(b-a)\tau}\, d\tau$$

$$= \mathrm{e}^{at} \left. \frac{\mathrm{e}^{(b-a)\tau}}{b-a} \right|_0^t = \mathrm{e}^{at} \left[\frac{\mathrm{e}^{(b-a)t} - 1}{b-a} \right]$$

$$= \frac{\mathrm{e}^{bt} - \mathrm{e}^{at}}{b-a} \tag{6.3}$$

Corollary The convolution is a commutative operation:[15] More precisely, let the functions f_1 and f_2 be defined as above, then, for all t,

$$(f_1 * f_2)(t) = (f_2 * f_1)(t) \tag{6.4}$$

Exercise Prove Eq. (6.4) directly by performing in Eq. (6.1) the change of variable $t' \triangleq t - \tau$.

[14] The limits of integration are purposefully chosen to be $0-$ and $t+$ so that if either $f_1(t)$ or $f_2(t)$ had an impulse at the origin, it would be included in the integration process.

[15] Like addition and multiplication of real or complex numbers.

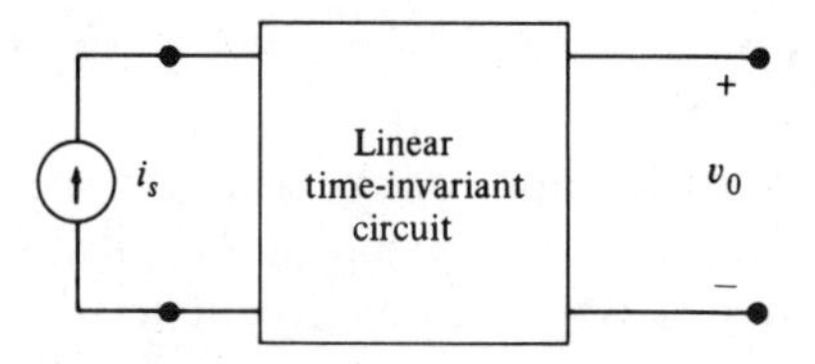

Figure 6.1 The circuit $\mathcal{N}$ is driven by the current source i_s and v_0 denotes its zero-state response.

B. The convolution integral Consider a linear time-invariant circuit $\mathcal{N}$ driven by a current source $i_s(t)$; let $v_0(t)$ be the output voltage of interest (see Fig. 6.1). Suppose we are interested in the *zero-state response* $v_0(t)$ of $\mathcal{N}$ to $i_s(t)$. Then,[16] for all $t > 0$,

$$\boxed{v_0(t) = \int_0^t h(t-\tau)\, i_s(\tau)\, d\tau} \tag{6.5}$$

Equation (6.5) shows that the value of v_0 at time t is an integral—versus τ—over the interval $[0, t]$ of the product $h(t-\tau)\, i_s(\tau)$; $v_0(\cdot)$ is the convolution of the *impulse response* $h(\cdot)$ with the *input* $i_s(\cdot)$.

This is a very important observation: For *linear time-invariant circuits* we have in the time domain

$$\boxed{\begin{Bmatrix}\text{Zero-state}\\ \text{response}\end{Bmatrix} = \begin{Bmatrix}\text{impulse}\\ \text{response}\end{Bmatrix} * \{\text{input}\}} \tag{6.6}$$

Now we are going to derive Eq. (6.5) and interpret it graphically.

C. Proof of Eq. (6.5) based on linearity and time-invariance Our goal is to derive the convolution integral (6.5) on the basis of the *time-invariance* of the circuit and the *linearity* of the *zero-state* response with respect to the input. As in Fig. 6.1, we consider a *linear time-invariant* circuit $\mathcal{N}$ driven by a single independent source: a current source $i_s(\cdot)$. The response under consideration is the output voltage $v_0(\cdot)$.

Two preliminary observations First, suppose we decompose the input waveform $i_s(\cdot)$ as a sum of say $N+1$ terms:

$$i_s(t) = \sum_{k=0}^{N} i_{sk}(t) \tag{6.7}$$

then we may calculate the zero-state response at time t, $v_0(t)$, as the sum of the responses due to each of the $i_{sk}(\cdot)$ acting alone. Second, for any $\Delta > 0$, we define the rectangular pulse p_Δ shown in Fig. 6.2*a*: Equivalently, we have

[16] For simplicity we assume that i_s and h do not have impulses at the origin, and hence we drop the $0-$ and $t+$ and replace them by 0 and t, respectively.

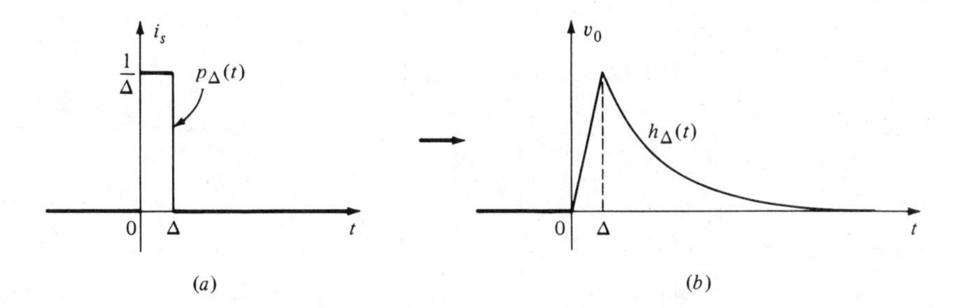

Figure 6.2 Input pulse p_Δ and corresponding zero-state response h_Δ.

$$p_\Delta(t) = \frac{1}{\Delta}[1(t) - 1(t - \Delta)] \tag{6.8}$$

Note that p_Δ has a unit area and that as $\Delta \to 0$, $p_\Delta(t) \to \delta(t)$. Call $h_\Delta(t)$ the *zero-state response* of $\mathcal{N}$ due to p_Δ. (See Fig. 6.2.) Suppose now that we *delay* the pulse p_Δ by τ_k seconds, with $\tau_k > 0$, and that we multiply it by a *constant* a_k, then, by linearity and time-invariance, the zero-state response of $\mathcal{N}$ to $a_k p_\Delta(t - \tau_k)$ is $a_k h_\Delta(t - \tau_k)$ as shown in Fig. 6.3.

Analysis In the following, t will denote the *fixed* observation time and we assume $t > 0$: We want to calculate the zero-state response v_0 at time t, namely, the voltage $v_0(t)$. The input current $i_s(\cdot)$ is applied at time 0. We use τ to denote any time between 0 and t.

Step 1. Consider the waveform $i_s(\cdot)$ for $0 \le \tau \le t$. Let us approximate it by, first, dividing $[0, t]$ in $N + 1$ equal intervals of length Δ where

$$\Delta \triangleq \frac{t}{N+1}$$

and second, over each successive interval, say (τ_k, τ_{k+1}), we replace the

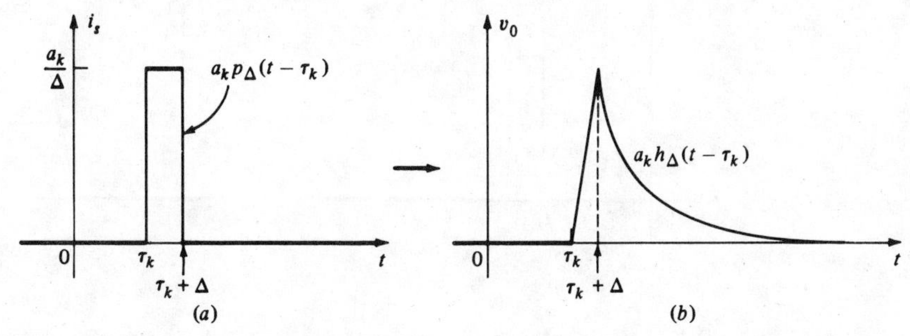

Figure 6.3 (*a*) The input pulse $p_\Delta(t)$ of Fig. 6.2 is delayed by τ_k seconds and multiplied by a_k. (*b*) Correspondingly, the zero-state response is multiplied by a_k and delayed by τ_k.

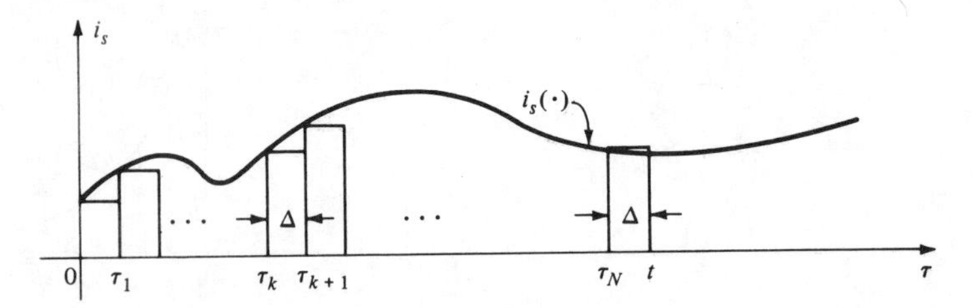

Figure 6.4 The input $i_s(\cdot)$ and its approximation as a sum of pulses of length Δ.

curve $i_s(\tau)$ by a *constant* $i_s(\tau_k)$ (See Fig. 6.4.) Since, for $\tau_k < \tau < \tau_k + \Delta = \tau_{k+1}$,

$$\Delta p_\Delta(\tau - \tau_k) = 1 \tag{6.9}$$

and since $p_\Delta(\tau - \tau_k)$ is equal to zero outside this interval, this piecewise-constant approximation of $i_s(\tau)$ is given by

$$i_s(\tau) \cong \sum_{k=0}^{N} i_s(\tau_k)\,\Delta p_\Delta(\tau - \tau_k) \qquad \text{for } 0 \le \tau \le t \tag{6.10}$$

Step 2. We now use the linearity and time-invariance of $\mathcal{N}$ to calculate the zero-state response of $\mathcal{N}$ to the *approximate* input given by Eq. (6.10). Recalling the second observation above and Fig. 6.3, we see that

$$v_0(t) \cong \sum_{k=0}^{N} i_s(\tau_k)\,\Delta h_\Delta(t - \tau_k) \tag{6.11}$$

Intuitively, it is clear that to the "pulse" of *area* $i_s(\tau_k)\,\Delta$, *applied at time* τ_k, the circuit $\mathcal{N}$ gives a response, *at time t*, given by $i_s(\tau_k)\,\Delta h_\Delta(t - \tau_k)$, since the time interval $t - \tau_k$ separates the "observation time" t from the "hit time" τ_k. (See Fig. 6.5.)

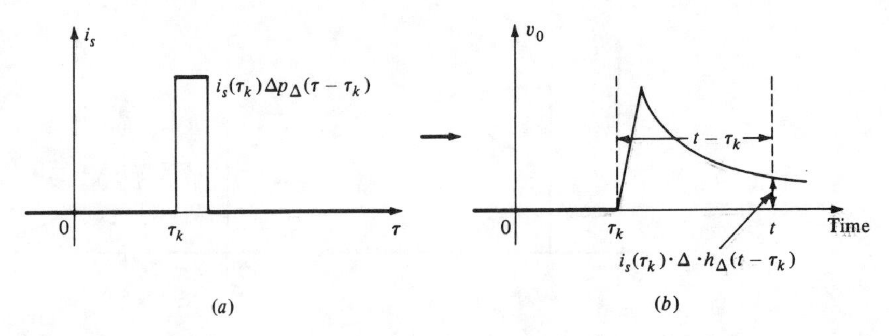

Figure 6.5 The pulse is delayed by τ_k seconds, and the corresponding zero-state response is also delayed by τ_k seconds.

Step 3. Equation (6.11) gives an approximation for $v_0(t)$. From calculus, we know that assuming that the input $i_s(\cdot)$ and the impulse response $h(\cdot)$ are well-behaved (say, piecewise-continuous), then *as Δ decreases to zero*, (consequently, $N \to \infty$),

(*a*) The piecewise-constant approximation [see Eq. (6.10) and Fig. 6.4] tends to the actual input $i_s(\cdot)$.

(*b*) The zero-state response to the piecewise-constant approximation becomes the actual zero-state response to $i_s(\cdot)$, i.e., the sum in Eq. (6.11) tends to the true value $v_0(t)$.

(*c*) $h_\Delta \to h$, the impulse response, since, as $\Delta \to 0$, $p_\Delta(t) \to \delta(t)$ [see Fig. 6.2].

(*d*) The sum in Eq. (6.11) becomes an integral.

Taking these four facts into consideration we see that Eq. (6.11) becomes

$$v_0(t) = \int_0^t h(t-\tau)\, i_s(\tau)\, d\tau \tag{6.12}$$

Thus we have shown that the convolution integral in (6.12) gives the *zero-state response* of $\mathcal{N}$ at time t to the input $i_s(\cdot)$ applied at time 0. ■

D. Graphic interpretation: Flip and drag We want to interpret graphically the convolution integral. Note that, from Eq. (6.12), if we consider several different values of t, the interval of integration $[0, t]$ changes and the factor $h(t-\tau)$ changes also; the function $i_s(\tau)$ does *not* change.

Consider the graph of $h(\tau)$ versus τ (see Fig. 6.6*a*). From this graphic data, the graph of $h(-\tau)$ versus τ is given by Fig. 6.6*b*: In other words, the graph is

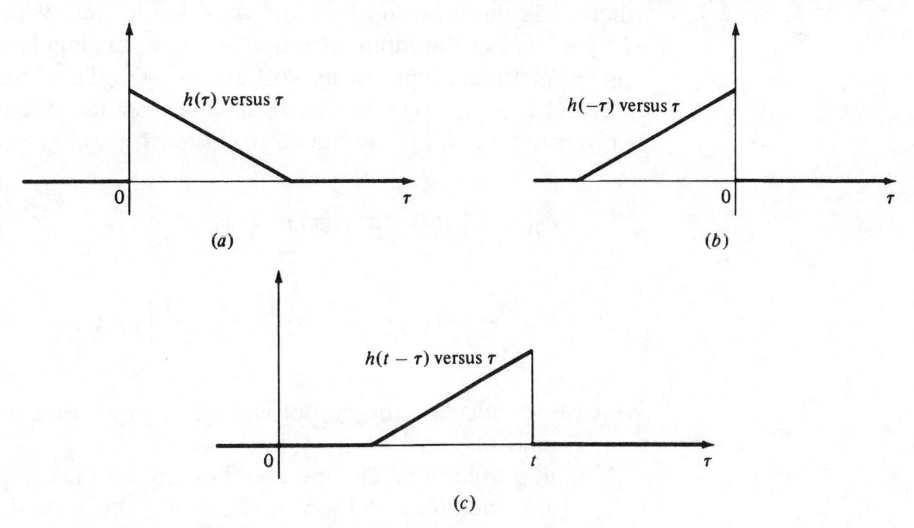

Figure 6.6 (*a*) The given impulse response $h(\cdot)$; (*b*) h is flipped. (*c*) h is first flipped [see (*b*)] and then dragged by t seconds.

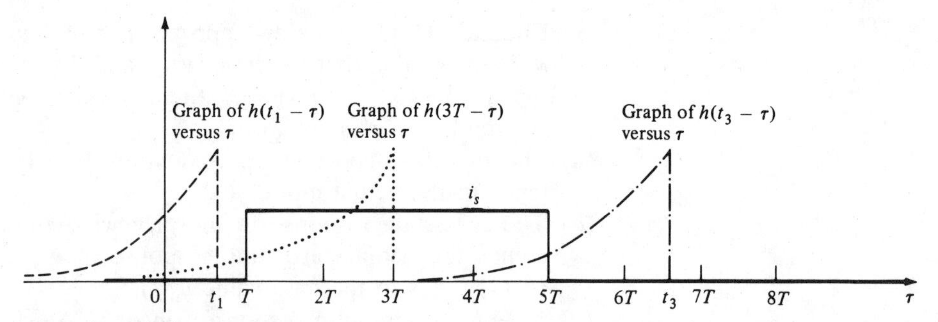

Figure 6.7 Calculating for times t_1, $t_2 = 3T$, and t_3, the response to the rectangular pulse applied at $t = T$ and turned off at $t = 5T$.

flipped around the vertical axis. Now, for a fixed $t > 0$, let us plot $h(t - \tau)$ versus τ: A little thought shows that the answer is that shown in Fig. 6.6*c*. In other words, to go from $h(\tau)$ versus τ to $h(t - \tau)$ versus τ, we must, first, *flip* the graph of $h(\cdot)$ around the vertical axis, and, second, *drag* it *to the right* by t seconds. We call this process "flip and drag."[17]

Thus the calculation of $v_0(t)$ in Eq. (6.12) amounts to (*i*) plotting $i_s(\tau)$ versus τ, (*ii*) plotting $h(t - \tau)$ versus τ, (*iii*) performing the product, and (*iv*) integrating the product from $\tau = 0$ to $\tau = t$.

E. Example Suppose that we have a linear amplifier driven by a current source $i_s(\cdot)$ and that the impulse response of the amplifier is

$$h(t) = 1(t)\,\frac{A}{T}\,\exp\left[-\left(\frac{t}{T}\right)\right] \tag{6.13}$$

where T is the time constant and A is the dc gain of the amplifier [verify that $H(0) = A$]. Let the input current $i_s(\cdot)$ be a rectangular pulse of amplitude I_m, applied at time T, and turned off at time $5T$: Thus, for all t outside the time interval $[T, 5T]$, $i_s(t) = 0$. The zero-state response due to that pulse of current is given for any $t \geq 0$, by the convolution integral

$$\begin{aligned} v_0(t) &= \int_0^t h(t - \tau)\, i_s(\tau)\, d\tau \\ &= \int_0^t 1(t - \tau)\,\frac{A}{T}\,\exp\left[\frac{-(t - \tau)}{T}\right] I_m[1(\tau - T) - 1(\tau - 5T)]\, d\tau \end{aligned} \tag{6.14}$$

We wish to calculate the output voltage $v_0(t)$ at some arbitrary time $t_1 < T$, at $t_2 = 3T$, and at some arbitrary time $t_3 > 5T$.

Think graphically: The integrand of Eq. (6.14) is a product of two factors: $h(t - \tau)$ depends on t and is zero for $\tau > t$; the second factor, $i_s(\tau)$, does not depend on t, is equal to I_m on $[T, 5T]$, and is zero elsewhere (see Fig. 6.7).

[17] We thank Professor Alan Willson for this creative label.

For any $t_1 < T$, the input current $i_s(\cdot)$ has not yet been turned on, and hence intuitively $v_0(t_1) = 0$. From the convolution integral (6.14), $h(t_1 - \tau)$ is zero for all $\tau > t_1$, but the second factor is zero on $[0, t_1]$ because $t_1 < T$; hence the product is identically zero, and Eq. (6.14) also gives $v_0(t) = 0$ for all $t_1 < T$.

Now, intuitively $v_0(3T) \neq 0$ because the current has been on for $2T$; from Eq. (6.14) and Fig. 6.7, the first factor $h(3T - \tau)$ is positive on $[0, 3T]$ and $i_s(\tau) = I_m$ on $[T, 5T]$ (see Fig. 6.7) so we expect $v_0(3T) \neq 0$. In fact, from Eq. (6.14) we obtain

$$v_0(3T) = \int_T^{3T} \frac{A}{T} \exp\left[\frac{-(3T - \tau)}{T}\right] I_m \, d\tau \tag{6.15}$$

and with $3T - \tau = \tau'$,

$$v_0(3T) = \int_0^{2T} \frac{AI_m}{T} \exp\left(\frac{-\tau'}{T}\right) d\tau' = AI_m(1 - e^{-2}) \tag{6.16}$$

Now, for any $t_3 > 5T$, we intuitively expect $v_0(t_3) \to 0$ as t_3 increases. In Eq. (6.14) we note that $i_s(\tau) = 0$ for $\tau > 5T$ and for $\tau < T$, and hence

$$v_0(t_3) = \int_T^{5T} \frac{A}{T} \exp\left[\frac{-(t_3 - \tau)}{T}\right] I_m \, d\tau$$

and with $\tau' = t_3 - \tau$

$$\begin{aligned} v_0(t_3) &= \int_{t_3-5T}^{t_3-T} \frac{AI_m}{T} \exp\left(\frac{-\tau'}{T}\right) d\tau' \\ &= AI_m(1 - e^{-4}) \exp\left[\frac{-(t_3 - 5T)}{T}\right] \end{aligned} \tag{6.17}$$

As expected $v(t_3) \to 0$ exponentially as $t_3 \to \infty$.

The final response $v_0(t)$ for $t \geq 0$ is shown in Fig. 6.8.

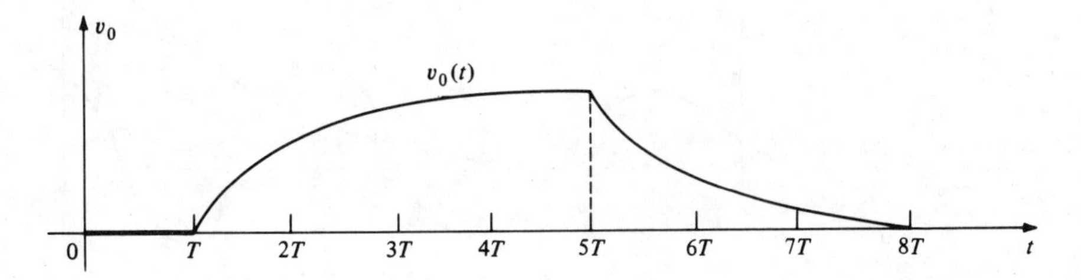

Figure 6.8 Zero-state response v_0 of the amplifier (6.13) driven by the rectangular pulse shown in Fig. 6.7.

F. Memory time of a circuit We know that for most circuits, the output, say $v_0(\cdot)$, does not depend on inputs that have occurred far in the past: Roughly speaking, such circuits have a finite-time memory. The convolution integral shows that this memory time is roughly the time beyond which the impulse response becomes negligible.

To see this, consider the calculation of $v_0(t) = (h * i_s)(t)$: So we plot $i_s(\tau)$ versus τ, $h(t-\tau)$ versus τ, take the product, and integrate. (See Fig. 6.9.) From Fig. 6.9, for $\tau \leq 3$, the product $h(t-\tau)\, i_s(\tau)$ is negligible, so consequently, in this case, the memory time of the circuit is about 5 s: Any input applied before $t-5$ is "forgotten" at the output at time t.

6.2 The Convolution Theorem

Let $f_1(t)$ and $f_2(t)$ have $F_1(s)$ and $F_2(s)$ as Laplace transforms. We assume that for $i = 1, 2$, $f_i(t) = 0$ for $t < 0$. The convolution of f_1 and f_2 is defined by Eq. (6.1) and its Laplace transform is given by

$$\mathscr{L}\{(f_1 * f_2)(t)\} = F_1(s)\, F_2(s) \tag{6.18}$$

Thus, the operation of convolution in the time domain is equivalent to multiplication in the frequency domain. It is this property that makes the Laplace and Fourier transforms so useful in engineering.

Equation (6.18) has a right-hand side which does not depend on the order of F_1 and F_2 since $F_1(s)\, F_2(s) = F_2(s)\, F_1(s)$; consequently we see again that the convolution operation is commutative.

PROOF OF THE CONVOLUTION THEOREM Let us calculate $\mathscr{L}\{(f_1 * f_2)(t)\}$: By definition,

$$J \triangleq \mathscr{L}\{(f_1 * f_2)(t)\} = \int_{0-}^{\infty} \left[\int_{0-}^{t+} f_1(t-\tau)\, f_2(\tau)\, d\tau \right] e^{-st}\, dt. \tag{6.19}$$

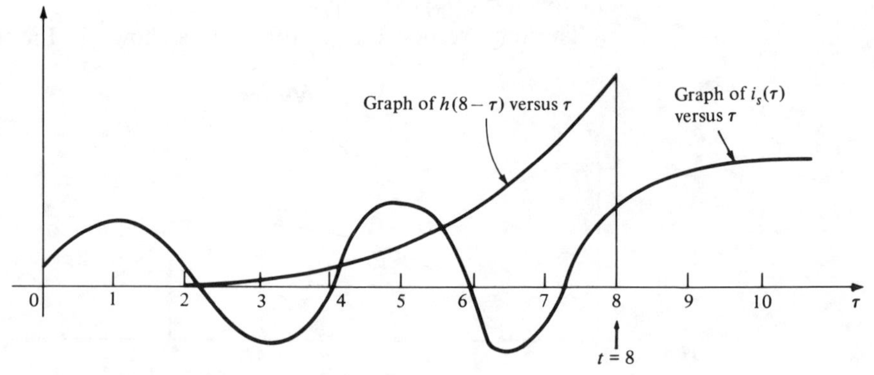

Figure 6.9 The two functions appearing in the integral of (6.12) are plotted versus τ. Clearly, the values of $i_s(\tau)$ for $\tau < 3$ hardly affect $v_0(t)$ at $t = 8$.

where Re(s) is chosen sufficiently large so that the integral converges. Since $f_1(t) = 0$ for all $t < 0$, we have $f(t - \tau) = 0$ for all $\tau > t$; hence, in the internal integral, we may replace the upper limit $t+$ by ∞. We thus obtain

$$J = \int_{0-}^{\infty} \left[\int_{0-}^{\infty} f_1(t - \tau)\, f_2(\tau)\, d\tau \right] \mathrm{e}^{-s\tau}\, d\tau$$

Now let us replace e^{-st} by $\mathrm{e}^{-s(t-\tau)}\,\mathrm{e}^{-s\tau}$ and then interchange the order of integration:

$$J = \int_{0-}^{\infty} \left[\int_{0-}^{\infty} f_1(t - \tau)\, \mathrm{e}^{-s(t-\tau)}\, d\tau \right] f_2(\tau)\, \mathrm{e}^{-s\tau}\, d\tau$$

Now in the inside integral, let us put $t' \stackrel{\Delta}{=} t - \tau$ so

$$dt' = dt \qquad \text{and} \qquad \begin{cases} t = 0- & \text{gives } t' = -\tau \\ t = \infty & \text{gives } t' = \infty \end{cases}$$

so we get

$$J = \int_{0-}^{\infty} \left[\int_{-\tau}^{\infty} f_1(t')\, \mathrm{e}^{-st'}\, dt' \right] f_2(\tau)\, \mathrm{e}^{-s\tau}\, d\tau$$

Since $f_1(t') = 0$ for all $t' < 0$, we may replace the lower limit of integration $-\tau$ by $0-$; consequently the bracket becomes $F_1(s)$, and, since $F_1(s)$ is independent of τ, it may be pulled out of the integral, so we have

$$J = F_1(s) \int_{0-}^{\infty} f_2(\tau)\, \mathrm{e}^{-s\tau}\, d\tau = F_1(s)\, F_2(s)$$ ■

6.3 The Sinusoidal Steady State Analyzed by Convolution[18]

In Sec. 5 we considered an exponentially stable circuit $\mathcal{N}$, driven by sinusoidal sources and we established that *all* circuit variables tend to the sinusoidal steady state predicted by phasor analysis. We are going to use the convolution theorem to analyze this problem when (*a*) the circuit *starts from the zero-state*, (*b*) *only one independent source* drives the circuit, and (*c*) only *one circuit variable* is under study. More precisely we are going to prove the following:

Theorem Let $\mathcal{N}$ be a uniquely solvable linear time-invariant circuit. Let it be driven by a *sinusoidal* current source, say $i_s(t) = \mathrm{Re}[\hat{I}\,\mathrm{e}^{j\omega t}]$, and let us observe the output voltage $v_0(t)$ (see Fig. 6.10). If $\mathcal{N}$ is exponentially stable, then the zero-state response $v_0(t)$ satisfies

$$v_0(t) = \mathrm{Re}[H(j\omega)\, \hat{I}\, \mathrm{e}^{j\omega t}] - \mathrm{Re}\left[\int_t^{\infty} h(\tau)\, \mathrm{e}^{-j\omega t}\, d\tau \cdot \hat{I}\, \mathrm{e}^{j\omega t} \right] \tag{6.20}$$

[18] This subsection may be skipped without loss of continuity.

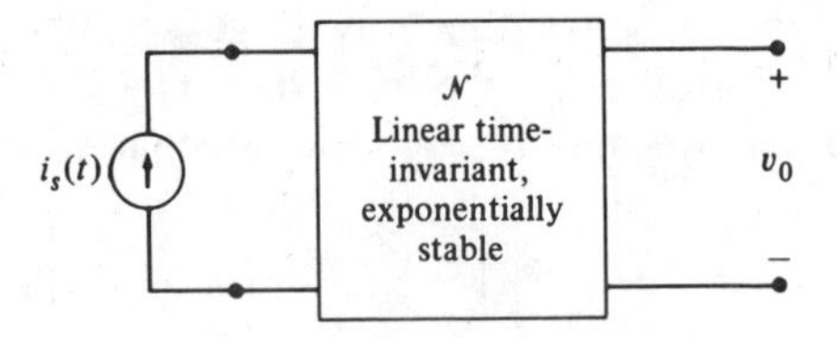

Figure 6.10 $\mathcal{N}$ is driven by i_s, and the corresponding zero-state response v_0 is observed.

COMMENT Since $\mathcal{N}$ is exponentially stable, the impulse response $h(\tau) \to 0$ exponentially as $\tau \to \infty$; hence, in Eq. (6.20), the integral goes to zero as $t \to \infty$. Thus Eq. (6.20) implies that

$$v_0(t) \to \text{Re}[H(j\omega)\hat{I}\,e^{j\omega t}] \qquad \text{as } t \to \infty \tag{6.21}$$

as expected from Eq. (5.2).

PROOF The zero-state response $v_0(t)$ due to the input $i_s(\cdot)$ is given by

$$v_0(t) = \int_0^t h(t-\tau')\, i_s(\tau')\, d\tau' = \int_0^t h(\tau)\, i_s(t-\tau)\, d\tau \tag{6.22}$$

where to obtain the last equation we performed the change of variable $\tau = t - \tau'$. Now expressing $i_s(t-\tau)$ in terms of the phasor $\hat{I}$ and noting that $h(\tau)$ is real for all τ, we obtain

$$\begin{aligned} v_0(t) &= \int_0^t h(\tau)\, \text{Re}[\hat{I}\, e^{j\omega(t-\tau)}]\, d\tau \\ &= \text{Re}\Big[\underbrace{\int_0^t h(\tau)\, e^{-j\omega\tau}\, d\tau}_{\triangleq\, \Gamma(t)} \cdot \hat{I}\, e^{j\omega t}\Big] \end{aligned} \tag{6.23}$$

But
$$\begin{aligned} \Gamma(t) &= \int_0^t h(\tau)\, e^{-j\omega\tau}\, d\tau \\ &= \int_0^\infty h(\tau)\, e^{-j\omega\tau}\, d\tau - \int_t^\infty h(\tau)\, e^{-j\omega\tau}\, d\tau \end{aligned} \tag{6.24}$$

Since $h(\tau) \to 0$ exponentially, the first integral is well-defined and is equal to $H(j\omega)$—recall the definition of $H(s)$—furthermore, the second integral $\to 0$ as $t \to \infty$ because (*a*) $h(\tau)$ decays exponentially and (*b*) for all τ and for all ω

$$|e^{-j\omega\tau}| = 1 \tag{6.25}$$

These conclusions and Eq. (6.24) inserted in Eq. (6.23) yield Eq. (6.20).

∎

REMARKS

1. Equation (6.20) exhibits how fast the zero-state response $v_0(t)$ reaches the steady state. It shows that what counts is the time t at which the area under the curve $|h(\tau)|$ over the interval $[t, \infty)$ becomes negligible: Indeed, in view of Eq. (6.25), the second term of Eq. (6.24) is bounded by

$$\left| \int_t^\infty h(\tau)\, e^{-j\omega\tau}\, d\tau \right| \leq \int_t^\infty |h(\tau)|\, d\tau \tag{6.26}$$

This again shows us how convolution gives us the memory time of the circuit.

2. The proof is quite general: It does not use the fact that $H(s)$ is a real rational function. It merely uses the fact that the right-hand side of Eq. (6.26) goes to zero as $t \to \infty$. Since for *distributed* (i.e., *not* lumped) circuits, the network functions are, in general, not rational, the derivation above applies to this broader class of circuits.

The following exercise is a good example of the connection between time and frequency domain.

Exercise

(*a*) Verify by time-domain calculations that, for any given fixed $T > 0$ and for any *continuous* function $f(\cdot)$,

$$f(t) * \delta(t - T) = f(t - T)$$

(*b*) Show that, for any fixed $T > 0$,

$$\mathscr{L}\{f(t - T)\} = F(s)\, e^{-sT}$$

(*c*) Show that if a communication channel had $H(s) = e^{-sT}$ as a network function[19] relating its input, $f(t)$, to its output, $f_0(t)$, then (i) $f_0(t)$ is a *perfect* replica of the input except for a delay of T seconds and (ii) find the gain and phase curve of $H(s)$. {Note that $-\dfrac{d}{d\omega}[\measuredangle H(j\omega)]$ is equal to the delay T.}

SUMMARY

- *Laplace transform*: To a time function $f(t)$ defined on $[0, \infty)$, the Laplace transform associates a function $F(s)$ of the complex frequency s

$$F(s) \triangleq \int_{0-}^\infty f(t)\, e^{-st}\, dt$$

where $\text{Re}(s)$ is taken sufficiently large so that the integral converges.

[19] Since e^{-sT} is not a rational function, this communication channel is an example of a *distributed* circuit.

1. *Uniqueness*: $f(t)$ has only one Laplace transform $F(s)$ and, conversely, there is only one time function that has $F(s)$ as a Laplace transform.
2. *Linearity*: For all *constants* c_1 and c_2, real or complex,

$$\mathscr{L}\{c_1 f_1(t) + c_2 f_2(t)\} = c_1 \mathscr{L}\{f_1(t)\} + c_2 \mathscr{L}\{f_2(t)\}$$

3. *Differentiation*:

$$\mathscr{L}\left\{\frac{df}{dt}\right\} = s\,F(s) - f(0-)$$

{If f has a "jump" at t_0, $\frac{df}{dt}$ includes the impulse

$$[f(t_0+) - f(t_0-)]\delta(t - t_0)\}$$

4. *Integration*:

$$\mathscr{L}\left\{\int_{0-}^{t} f(t')\,dt'\right\} = \frac{1}{s}\,\mathscr{L}\{f(t)\}$$

5. *Convolution*: The convolution of f_1 and f_2 is denoted by $f_1 * f_2$ and is defined by

$$(f_1 * f_2)(t) \triangleq \int_{0-}^{t+} f_1(t - \tau)\, f_2(\tau)\, d\tau$$

then

$$\mathscr{L}\{f_1 * f_2\} = F_1(s)\, F_2(s)$$

- *Partial fraction expansion*: Let $H(s) = n(s)/d(s)$ be a rational function with *real* coefficients, with $n(s)$ and $d(s)$ *coprime*. Let p_1 be a simple pole of $H(s)$; then p_1 contributes the terms

$$\frac{k_1}{s - p_1} \text{ to } H(s) \qquad \text{and} \qquad k_1\, e^{p_1 t} \text{ to } h(t)$$

Let p_2 and $\bar{p}_2$ be a pair of complex-conjugate simple poles of $H(s)$; then they contribute the terms

$$\frac{k_2}{s - p_2} + \frac{\bar{k}_2}{s - \bar{p}_2} \qquad \text{and} \qquad 2|k_2|\, e^{p_{2r} t} \cos(p_{2i} t + \measuredangle k_2)$$

(Here $p_2 \triangleq p_{2r} + jp_{2i}$.) Let p_3 be a second-order pole of $H(s)$; then p_3 contributes

$$\frac{k_{31}}{(s - p_3)^2} + \frac{k_{32}}{s - p_3} \qquad \text{and} \qquad (k_{31} t + k_{32})\, e^{p_3 t}$$

Laplace transforms of elementary functions

$f(t)$	$F(s) \triangleq \int_{0-}^{\infty} f(t)\,e^{-st}\,dt$	
$\delta(t)$	1	
$\delta^{(n)}(t)$	s^n	$(n = 1, 2, \ldots)$
$1(t)$	$\frac{1}{s}$	
$\frac{t^n}{n!}$	$\frac{1}{s^{n+1}}$	$(n = 1, 2, \ldots)$
e^{-at} $\begin{pmatrix} a \text{ real or} \\ \text{complex} \end{pmatrix}$	$\frac{1}{s+a}$	
$\frac{t^n}{n!} e^{-at}$ $\begin{pmatrix} a \text{ real or} \\ \text{complex} \end{pmatrix}$	$\frac{1}{(s+a)^{n+1}}$	$(n = 1, 2, \ldots)$
$e^{-\alpha t} \cos \beta t$	$\frac{s+\alpha}{(s+\alpha)^2 + \beta^2}$	
$e^{-\alpha t} \sin \beta t$	$\frac{\beta}{(s+\alpha)^2 + \beta^2}$	
$2\lvert k\rvert e^{-\alpha t} \cos(\beta t + \measuredangle k)$	$\frac{k}{s+\alpha-j\beta} + \frac{\bar{k}}{s+\alpha+j\beta}$	

- *Tableau equations*: In the frequency domain the tableau equations of any linear time-invariant circuit $\mathcal{N}$ have the form

$$\underbrace{\begin{bmatrix} \mathbf{0} & \mathbf{0} & \mathbf{A} \\ -\mathbf{A}^T & \mathbf{1} & \mathbf{0} \\ \mathbf{0} & \mathbf{M}_0 s + \mathbf{M}_1 & \mathbf{N}_0 s + \mathbf{N}_1 \end{bmatrix}}_{\mathbf{T}(s)} \underbrace{\begin{bmatrix} \mathbf{E}(s) \\ \mathbf{V}(s) \\ \mathbf{I}(s) \end{bmatrix}}_{\mathbf{W}(s)} = \begin{bmatrix} \mathbf{0} \\ \mathbf{0} \\ \mathbf{U}_s(s) + \mathbf{U}^i \end{bmatrix}$$

where $\mathbf{M}_0$, $\mathbf{M}_1$, $\mathbf{N}_0$, $\mathbf{N}_1$ are *constant* matrices, $\mathbf{U}_s(s)$ represents the *independent* sources, and $\mathbf{U}^i$ is a *constant* vector, the contribution of the initial conditions, $\mathbf{v}_C(0-)$ and $\mathbf{i}_L(0-)$.
- The linear time-invariant circuit $\mathcal{N}$ is *uniquely solvable* if and only if $\det[\mathbf{T}(s)] \not\equiv 0$.
- *Modified node analysis*: In the frequency domain, for any linear time-invariant circuit $\mathcal{N}$, the modified node analysis (MNA) leads to equations of the form

$$\begin{bmatrix} \mathbf{P}(s) \end{bmatrix} \begin{bmatrix} \mathbf{E}(s) \\ \mathbf{I}_2(s) \end{bmatrix} = \begin{bmatrix} \mathbf{U}(s) \end{bmatrix}$$

where the elements of $\mathbf{P}(s)$ are either *real constants* or *real polynomials of degree 1*; $\mathbf{I}_2(s)$ is the Laplace transform of the chosen currents.

- *Circuit properties*: For any *uniquely solvable linear time-invariant* circuit $\mathcal{N}$, the following thirteen properties hold:

1. $$\begin{Bmatrix} \text{complete} \\ \text{response} \end{Bmatrix} = \begin{Bmatrix} \text{zero-input} \\ \text{response} \end{Bmatrix} + \begin{Bmatrix} \text{zero-state} \\ \text{response} \end{Bmatrix}$$

2. The characteristic polynomial of $\mathcal{N}$ has *real* coefficients and

$$\chi(s) = \det[\mathbf{T}(s)] = k \det[\mathbf{P}(s)]$$

where k is a *nonzero constant* [hence the zeros of $\chi(s)$ are either real or occur in complex-conjugate pairs].

3. Any zero of $\chi(s)$ is called a *natural frequency* of the circuit $\mathcal{N}$.
4. For any natural frequency λ there are initial conditions producing a *purely exponential zero-input response*

$$\mathbf{w}(t) = \begin{bmatrix} \mathbf{e}_0 \\ \mathbf{v}_0 \\ \mathbf{i}_0 \end{bmatrix} e^{\lambda t} = \mathbf{w}_0 \, e^{\lambda t}$$

[this zero-input response $\mathbf{w}(t) = \mathbf{w}_0 \, e^{\lambda t}$ is called the *mode associated with* λ]; the constant vector $\mathbf{w}_0$ satisfies $\mathbf{T}(\lambda)\,\mathbf{w}_0 = \mathbf{0}$.

5. If $\mathcal{N}$ is *exponentially stable* (equivalently, if *all* natural frequencies of $\mathcal{N}$ have *negative* real parts), then, *for all initial conditions*, the zero-input response

$$\mathbf{w}(t) \to \mathbf{0} \qquad \text{exponentially as } t \to \infty$$

6. Network functions are defined by the equation

$$\mathcal{L}\begin{Bmatrix} \text{zero-state} \\ \text{response} \end{Bmatrix} = \begin{Bmatrix} \text{network} \\ \text{function} \end{Bmatrix} \cdot \mathcal{L}\{\text{input}\}$$

7. *Any* network function is of the form

$$H(s) = \frac{n(s)}{d(s)}$$

where $n(s)$ and $d(s)$ are *real* polynomials in s and any pole of $H(s)$ is a natural frequency.

8. $$\mathcal{L}\begin{Bmatrix} \text{impulse} \\ \text{response} \end{Bmatrix} = \begin{Bmatrix} \text{network} \\ \text{function} \end{Bmatrix}$$

9. The poles and zeros of $H(s)$ may be real or complex; if they are complex, they occur in complex-conjugate pairs.

10. For any network function $H(s)$, for all real ω's,

$$|H(j\omega)| = |H(-j\omega)|$$

and the phase can be chosen so that

$$\measuredangle H(j\omega) = -\measuredangle H(-j\omega)$$

i.e., the function $|H(j\omega)|$ is even and $\measuredangle H(j\omega)$ is odd.

11. If $\mathcal{N}$ is *exponentially stable* and is driven by sinusoidal independent sources at the *same* frequency ω, then, for *all initial conditions*, *all circuit variables* tend to a *unique sinusoidal steady state* at frequency ω.
12. In the time domain,

$$\left\{\begin{array}{c}\text{zero state}\\ \text{response}\end{array}\right\} = \left\{\begin{array}{c}\text{impulse}\\ \text{response}\end{array}\right\} * \text{input}$$

For example, $$v_0(t) = \int_0^t h(t-\tau)i_s(\tau)\,d\tau$$

where the integral is calculated by the *flip and drag* technique. The convolution is a consequence of *linearity* and *time-invariance*.

13. The derivative of the step response is equal to the impulse response.

PROBLEMS

Calculating Laplace transforms

1 Find the Laplace transform of the following time functions:

(*a*) $3\,e^{-2t} + e^{-t}$
(*b*) $1 - 2\,e^{-2t} + 4\,e^{-t}\cos(3t)$
(*c*) $t^3 + t^2 + t + 1$
(*d*) $\sinh(at)$
(*e*) $\sum_{n=1}^{\infty} \frac{1}{n^2} e^{-nt}$

Properties of Laplace transforms

2 Given that $\mathcal{L}\{f(t)\} = F(s)$ and that a is a positive constant prove that

(*a*) $\mathcal{L}\{e^{-at} f(t)\} = F(s+a)$

$$\mathcal{L}\{f(at)\} = \left(\frac{1}{a}\right) F\left(\frac{s}{a}\right)$$

(*b*) Let T be a positive constant, assume that $f(t) = 0$ for $t < 0$, and show that

$$\mathcal{L}\{f(t-T)\,1(t-T)\} = e^{-Ts} F(s)$$

3 Using the results of Prob. 2, find the Laplace transforms of the following time functions:

(*a*) $e^{-t} + 1(t-1)\,e^{-(t-1)} + \delta(t-2)$
(*b*) $e^{-at}\sin(\omega_0 t)\,1(t)$
(*c*) $\sinh(at)[1(t) - 1(t-1)]$

4 Find the Laplace transforms of the time functions shown in Fig. P10.4.

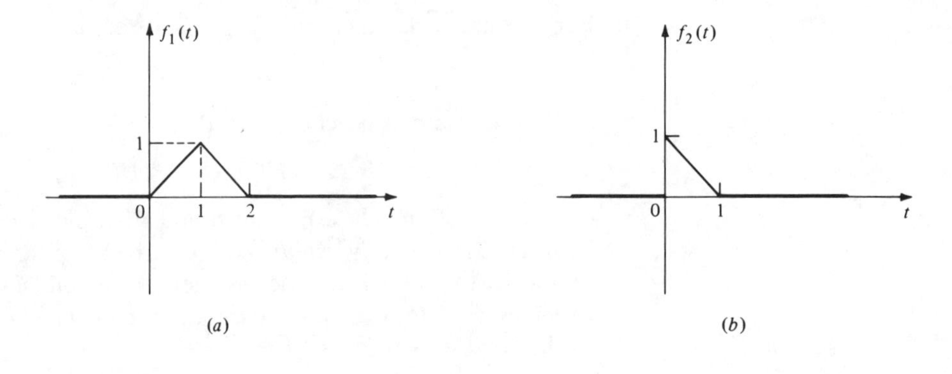

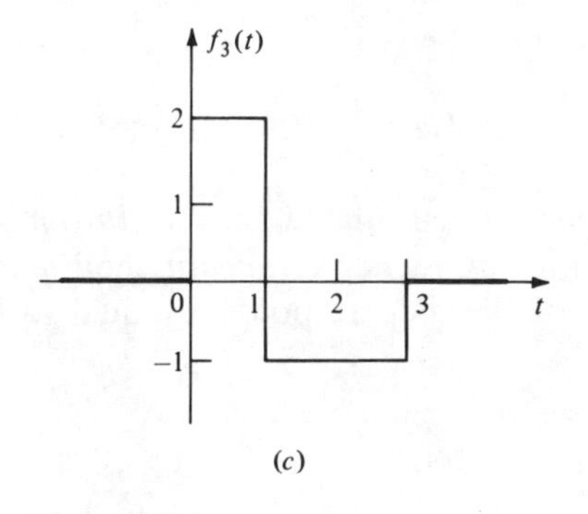

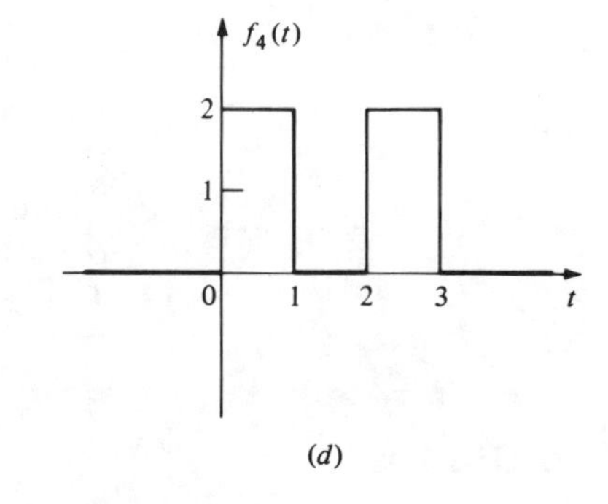

Figure P10.4

Laplace transform of periodic functions

5 (*a*) Let $f(t)$ be a Laplace transformable periodic function with period T. Show that

$$F(s) = \frac{F_T(s)}{1 - e^{-Ts}}$$

where $F_T(s)$ is the Laplace transform of $f(t)$ over the period $[0, T]$.

(*b*) Find the Laplace transform of the periodic function shown in Fig. P10.5.

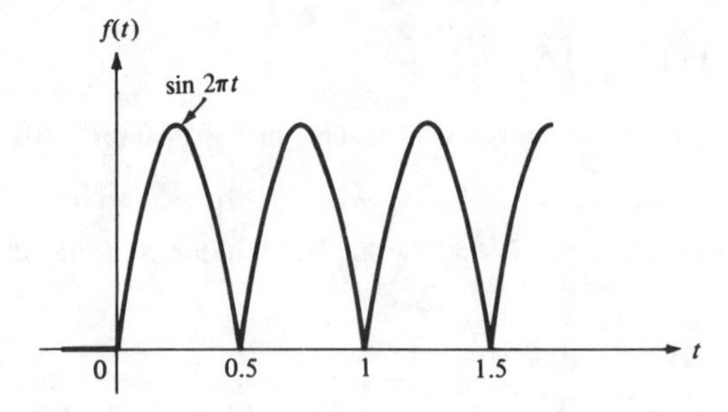

Figure P10.5

Partial fraction expansion

6 Using the partial fraction expansion find the inverse Laplace transform of the following functions

(*a*) $\dfrac{s^2+6s+8}{s^2+4s+3}\,e^{-sT}$ (*f*) $\dfrac{(s+1)(s+3)}{s(s+2)(s+5)}$

(*b*) $\dfrac{s^2+s}{s^3+2s^2+s+2}$ (*g*) $\dfrac{s(s^2+2)}{(s^2+1)(s^2+3)}$

(*c*) $\dfrac{e^{-s}}{s^3+2s^2+2s+1}$ (*h*) $\dfrac{1}{s^2(s+1)^2(s+2)}$

(*d*) $\dfrac{1}{s^4+3s^3+4s^2+3s+1}$ (*i*) $\dfrac{s^3+1}{s^2+2s+2}$

(*e*) $\dfrac{1-e^{-4s}}{5s^2}$ (*j*) $\dfrac{s\,e^{-s}+2s^2+9}{s(s^2+9)}$

7 Solve the following differential equations using Laplace transform methods:

(*a*) $\ddot{x}+4x=0$ where $x(0)=1$ $\dot{x}(0)=0$

(*b*) $\ddot{x}+2\dot{x}+x=21(t)$ where $x(0)=1$ $\dot{x}(0)=1$

(*c*) $\ddot{x}+\dot{x}=t^2+2t$ where $x(0)=4$ $\dot{x}(0)=-2$

(*d*) $2\,\dot{x}(t)+4\,x(t)+\dot{y}(t)-y(t)=0$ where $x(0)=0$ $\dot{x}(0)=2$
$\dot{x}(t)+2\,x(t)+\dot{y}(t)+y(t)=0$ $y(0)=1$ $\dot{y}(0)=-3$

Complete response

8 (*a*) The circuit shown in Fig. P10.8 is linear and time-invariant. Write integrodifferential equations for the circuit using node-to-datum voltages v_1 and v_2 as variables and initial conditions $v_1(0-)=\gamma$, $i(0-)=\rho$.

(*b*) Determine $V_2(s)$, where $V_2(s)=\mathcal{L}\{v_2(t)\}$.

(*c*) Let $v_s(t)=1(t)\sin t$, $\gamma=-2$ V, and $\rho=1$ A. Calculate $v_2(t)$ for $t\geq 0$. Write the answer as the sum of the zero-input response and the zero-state response.

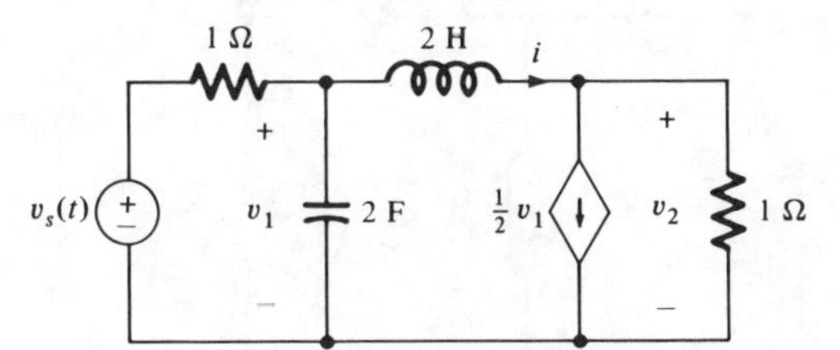

Figure P10.8

9 The circuit shown in Fig. P10.9 is in the steady state with the switch *S* closed. The switch *S* is opened at $t=0$. Find the currents $i_{L1}(t)$ and $i_{L2}(t)$ and the voltage $v(t)$ for $t\geq 0$. Relate the values of i_{L1} and i_{L2} at 0+ to their values at 0−. Explain physically.

Tableau equations

10 Write the tableau equations in the frequency domain for the circuits shown in Fig. P10.10.

11 Write the tableau equations in the frequency domain for the circuits shown in Fig. P8.4.

12 Write the modified node equations for the circuits shown in Fig. P10.10.

13 For the circuit in Fig. P10.10*c* let $v_1(0-)=2$ V and $i_2(0-)=3$ A. Determine $i_3(t)$.

14 Write the modified node equations for the circuits shown in Fig. P8.4.

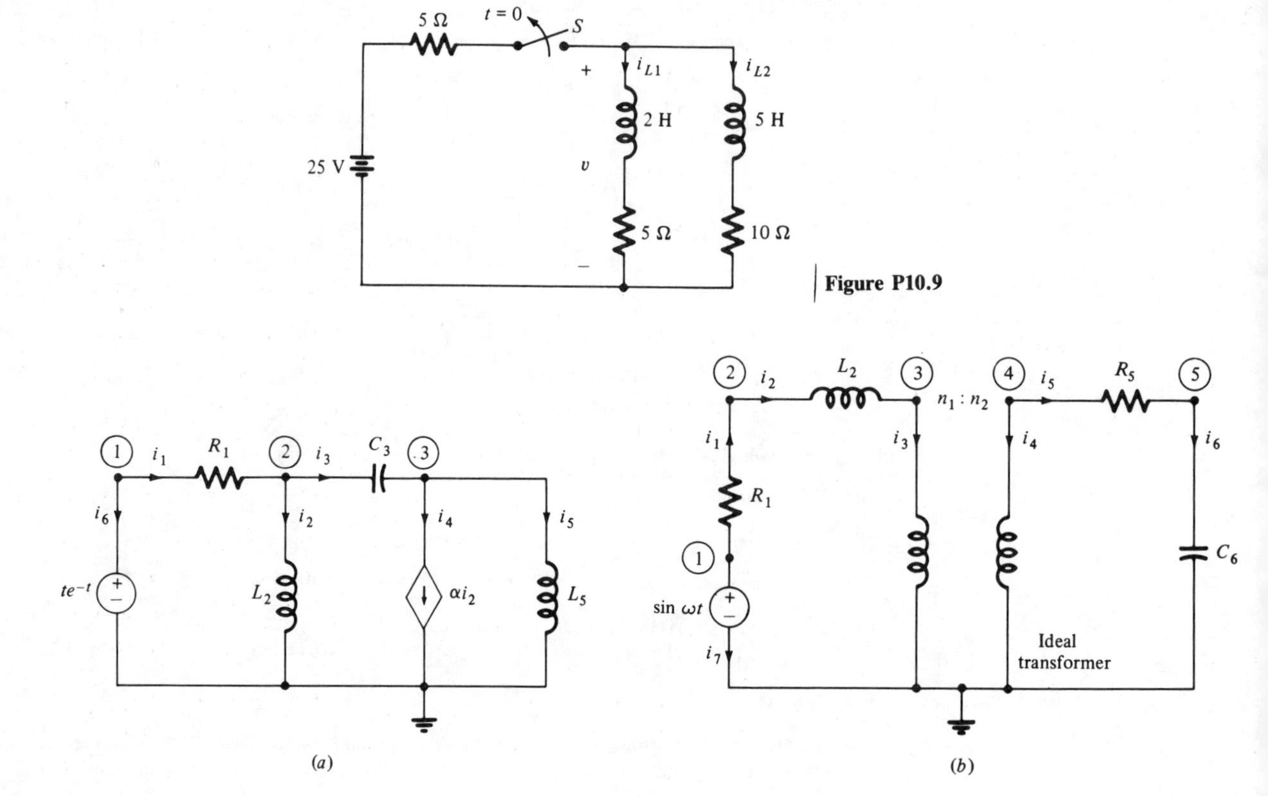

Figure P10.9

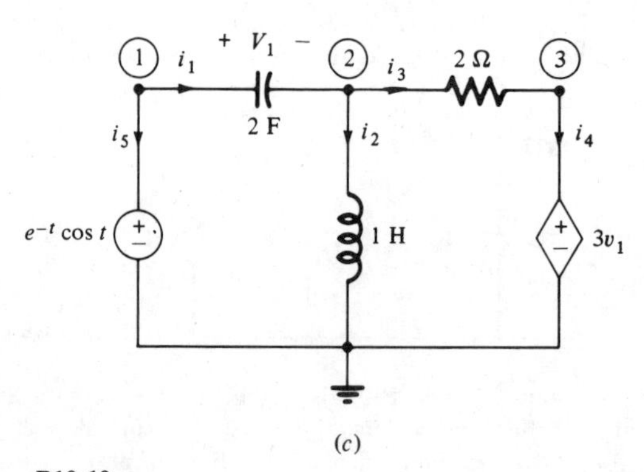

Figure P10.10

Natural frequencies

15 Find the natural frequencies of the circuits shown in Fig. P10.15.

16 Consider the circuit in Fig. 4.6*a*. Let $R_1 = 1\,\Omega$, $R_2 = 2\,\Omega$, $C_1 = C_2 = 1\,\text{F}$, and $g_m = 1\,\text{S}$.

(*a*) Find the natural frequencies.

(*b*) Show that only one of the natural frequencies is observable at e_1.

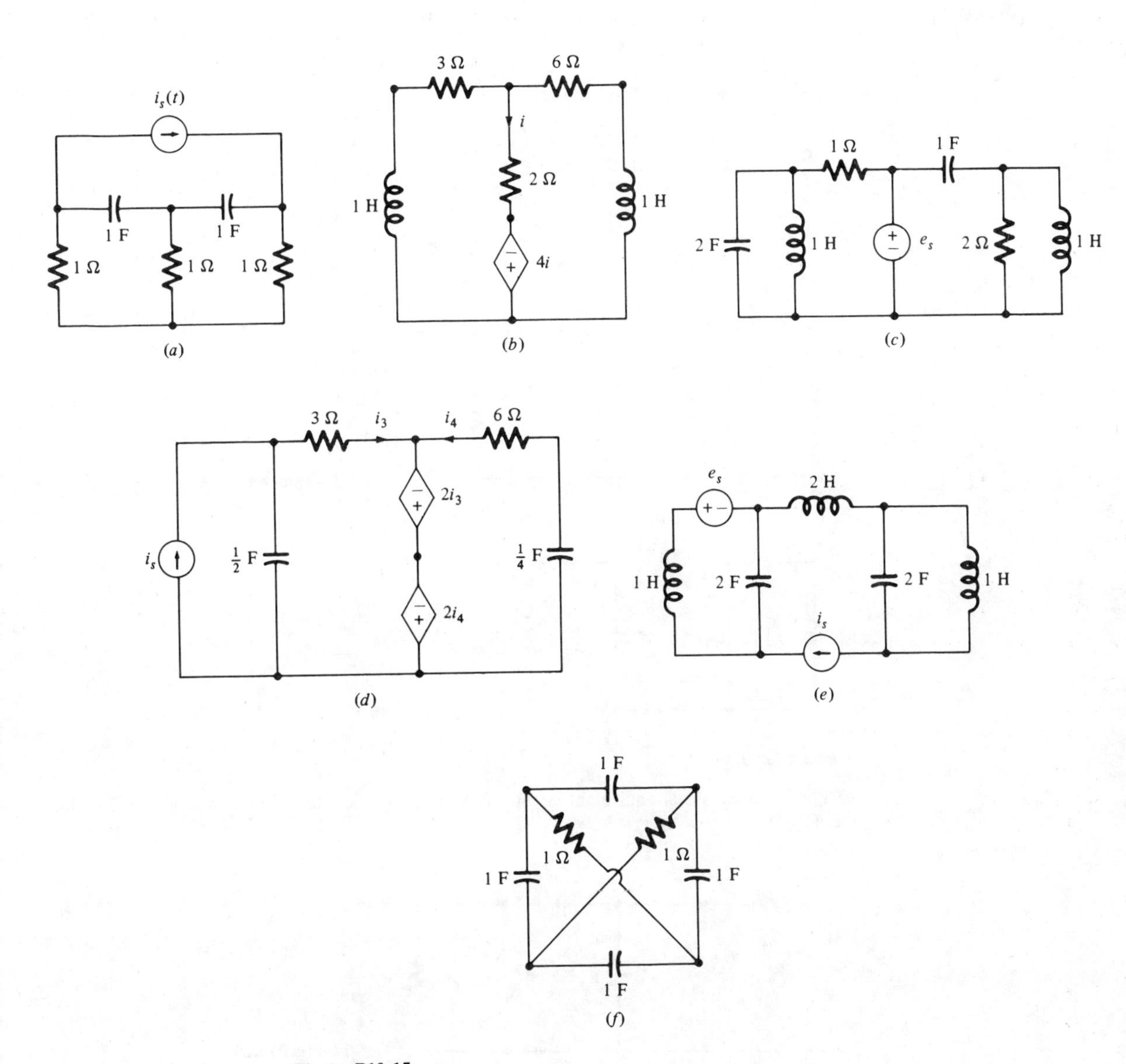

Figure P10.15

17 For the circuit shown in Fig. P10.17:

(*a*) Find the natural frequencies.

(*b*) Choose initial conditions to set up each mode.

18 Find the initial conditions to set up each mode of the circuit shown in Fig. P10.18.

19 The circuit shown in Fig. P10.19 is linear and time-invariant.

(*a*) Find its natural frequencies.

(*b*) Find an initial state such that only the smallest natural frequency is excited.

(*c*) Find how to locate (at $t = 0$) 1 J of energy in the circuit so that the resulting (zero-input) response includes only the largest natural frequency. (Specify the required initial voltage on each capacitor.)

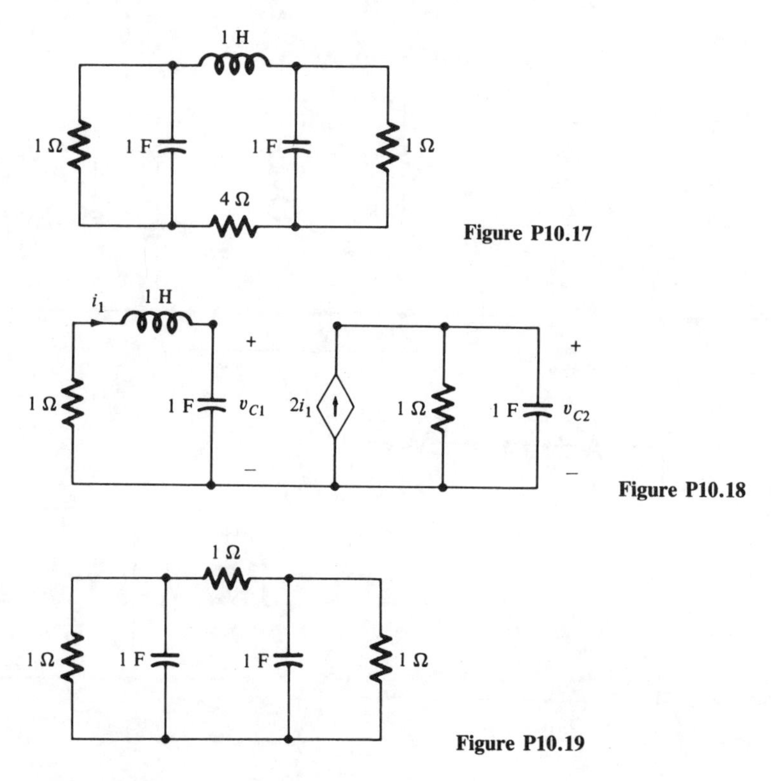

Figure P10.17

Figure P10.18

Figure P10.19

Network functions

20 For the small-signal equivalent circuit of the transistor amplifier shown in Fig. P10.20, obtain the network function $H(s) = V_2/V_s$.

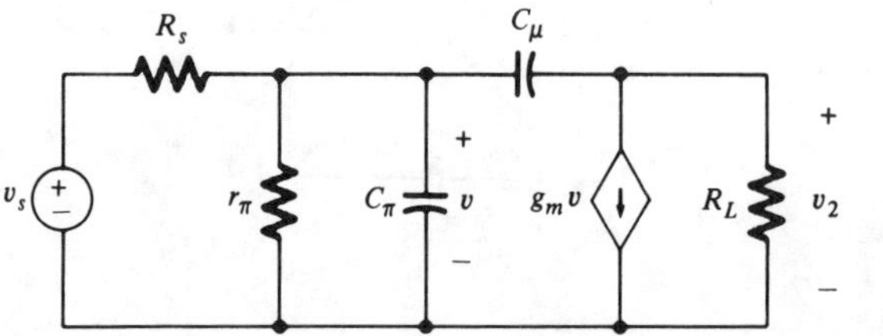

Figure P10.20

21 For the op-amp circuits shown in Fig. P10.21, obtain the network functions, $H(s) = V_0/V_i$. Assume ideal op-amp models operating in the linear region.

22 Consider the linear time-invariant one-port N shown in Fig. P10.22. Call $Y(s)$ and $Z(s)$ its driving-point impedance and driving-point admittance, respectively.

(*a*) Suppose the terminals ① and ①′ are short-circuited and that we observe the current i through the short circuit. Is it true that if s_1 is a pole of $Y(s)$, then for some initial state we may observe $i(t) = K_1 \exp s_1 t$? Explain.

(*b*) Suppose that the terminals ① and ①′ are open-circuited and that we observe the voltage v across the port. Is it true that if s_2 is a pole of $Z(s)$, then for some initial state we may observe $v(t) = K_2 \exp s_2 t$? Explain.

(*c*) Work out the details for the circuit shown in Fig. P10.22*c*.

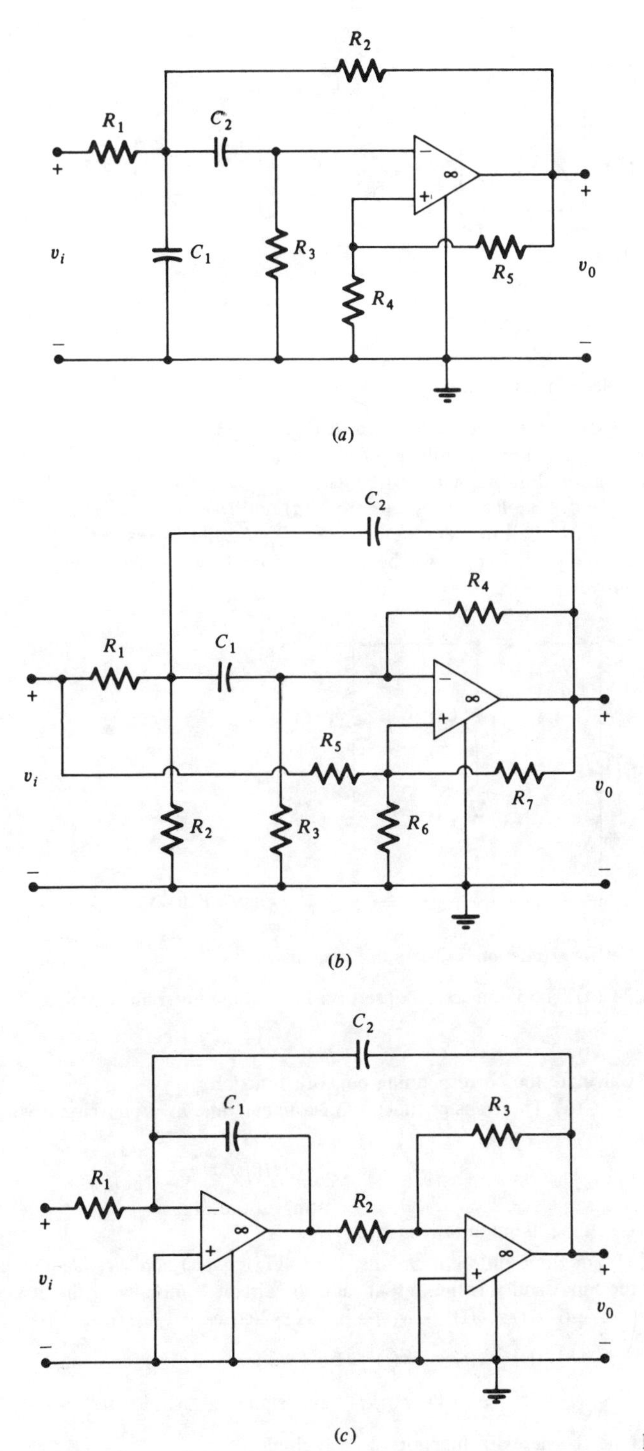

Figure P10.21

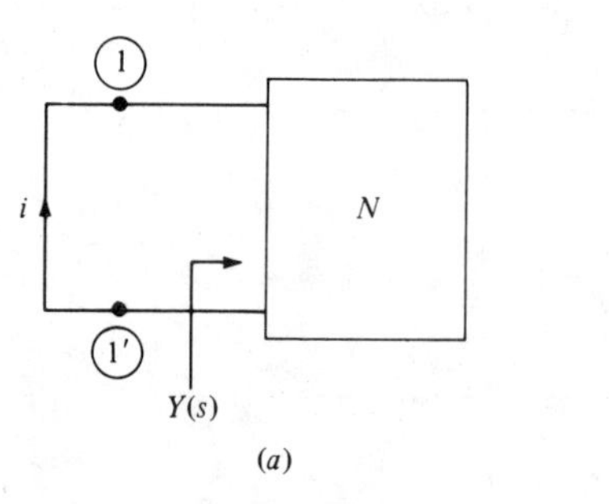

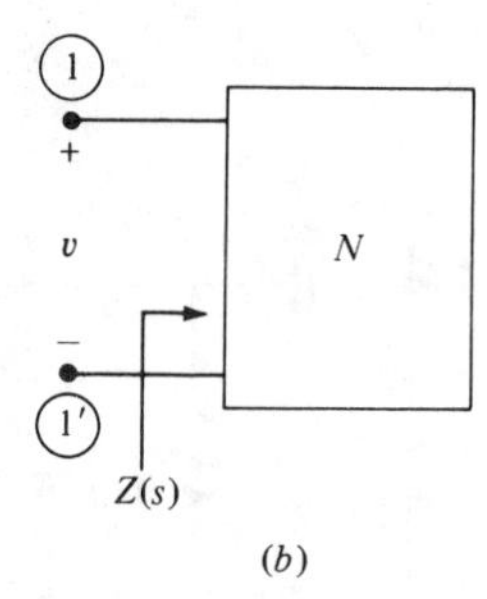

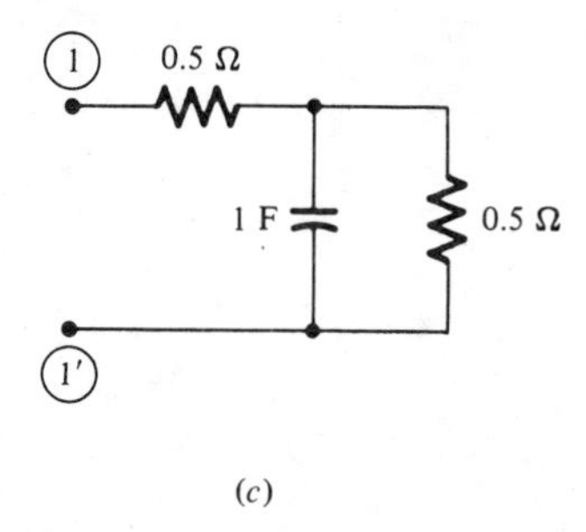

Figure P10.22

Pole-zero cancellation

23 Consider the circuit shown in Fig. P10.23.

(*a*) Find the driving-point impedance.

(*b*) Find its natural frequencies.

(*c*) Explain the results from (*a*) and (*b*).

(*d*) Call the four elements R_1, R_2, L, and C, respectively. Under what condition(s) will the driving-point impedance be independent of frequency? Give a physical interpretation for your answer.

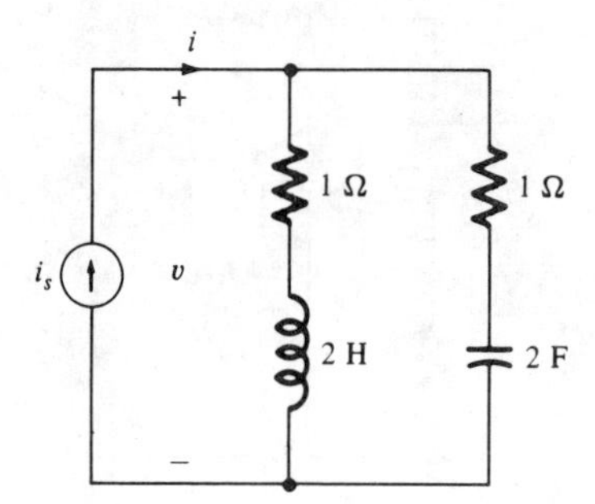

Figure P10.23

Network function and impulse response

24 (*a*) The impulse response of a linear time-invariant circuit is

$$h(t) = (e^{-t} + 2\,e^{-2t})\,1(t)$$

Calculate the corresponding network function.

(*b*) The transfer function of a linear time-invariant circuit is

$$H(j\omega) = \frac{1}{1 + j(\omega/\omega_0)}$$

Calculate its impulse response.

25 For three different circuits, the following $x(t)$ and $y(t)$ combinations are obtained, where x is the input and y is the output, and the circuit is initially in the zero state.

(*a*) $x(t) = \delta(t)$ $\quad y(t) = [\sinh t + 2(\cosh t - 1)]\,1(t)$

(*b*) $x(t) = 1(t)$ $\quad y(t) = (2\,e^{-t} + \sin 2t - 2)\,1(t)$

(*c*) $x(t) = e^{-t}1(t)$ $\quad y(t) = (e^{-t}\sin t - e^{-2t}\cos 3t)\,1(t)$

Find the network function of each circuit.

Zero-state response

26 For each case below, $H(s)$ is the network function, and $e(\cdot)$ is the input, calculate the corresponding zero-state response.

$$(a)\ H(s) = \frac{s+1}{s^2+5s+6}$$
$$e(t) = 31(t)\sin 5t$$
$$(b)\ H(s) = \frac{s+1}{s^2+s-2}$$
$$e(t) = (1+t)\,1(t)$$
$$(c)\ H(s) = \frac{s}{s^2+25}$$
$$e(t) = 1(t)\cos 5t$$

27 For the circuit shown in Fig. P10.27, let e_s and i_3 be the input and response of the circuit, respectively.

(*a*) Find the network function $H(s) = I_3/E_3$.
(*b*) Find the zero-state response when $e_s(t) = 3\,e^{-t}\cos 6t$.
(*c*) Find the steady-state response when $e_s(t) = 2 + \cos 2t$.

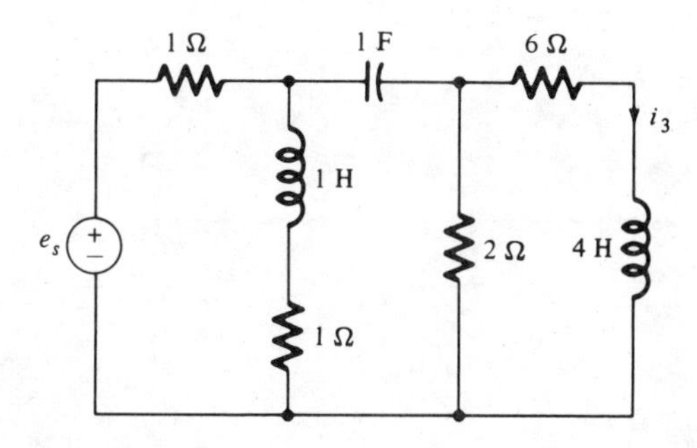

Figure P10.27

Convolution

28 Calculate and sketch the function that results from convolving each pair of functions shown in Fig. P10.28.

29 (*a*) Calculate the Laplace transforms of $f_1(t) = 1(t)\,e^{-at}$ and $f_2(t) = 1(t)\,e^{-bt}\cos\omega t$, where a, b, and ω are constants.

(*b*) Calculate the convolution of f_1 and f_2.
(*c*) Calculate the Laplace transform of this convolution. Is it what you expect?

30 For the same f_1 and f_2 as in the preceding problem,

(*a*) Evaluate the convolution integral of f_1 and df_2/dt.
(*b*) Repeat (*a*) for df_1/dt and f_2.
(*c*) Verify these calculations using Laplace transforms.

31 If the impulse response of a linear time-invariant circuit is given as

$$h(t) = \begin{cases} 2\,e^{-t} & 0 \le t < 3 \\ 0 & t \ge 3 \end{cases}$$

find the zero-state response to the circuit due to an input

$$i_s(t) = \begin{cases} 4 & 0 \le t < 2 \\ 0 & t \ge 2 \end{cases}$$

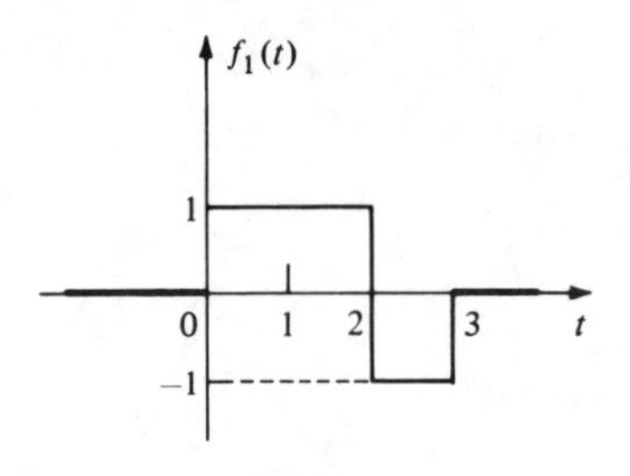

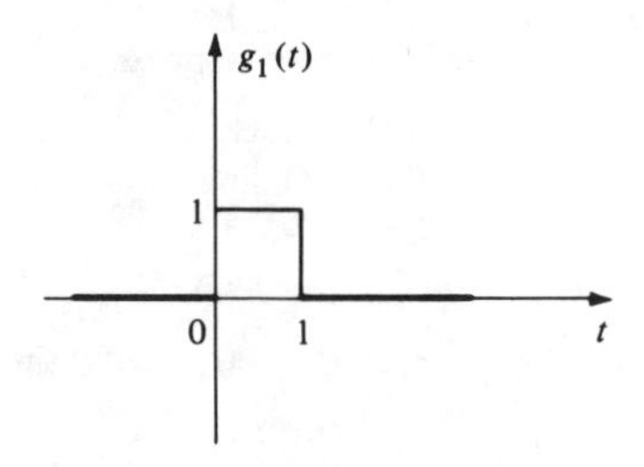

(a)

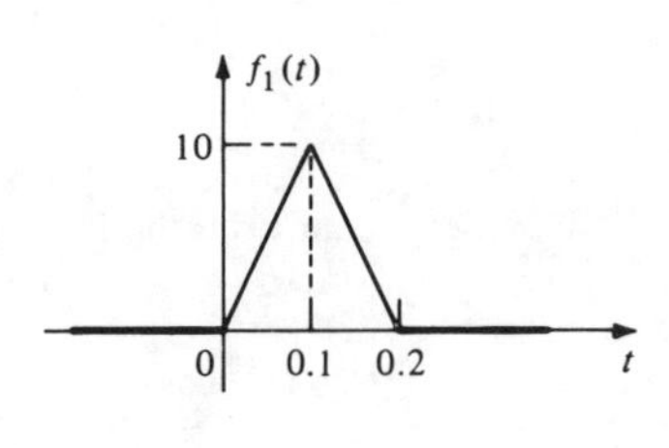

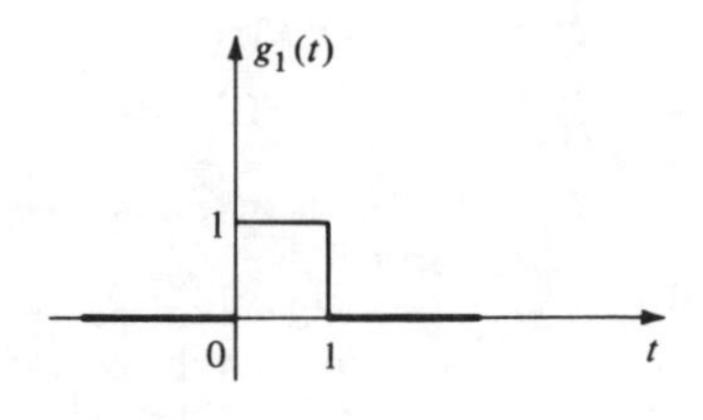

(b)

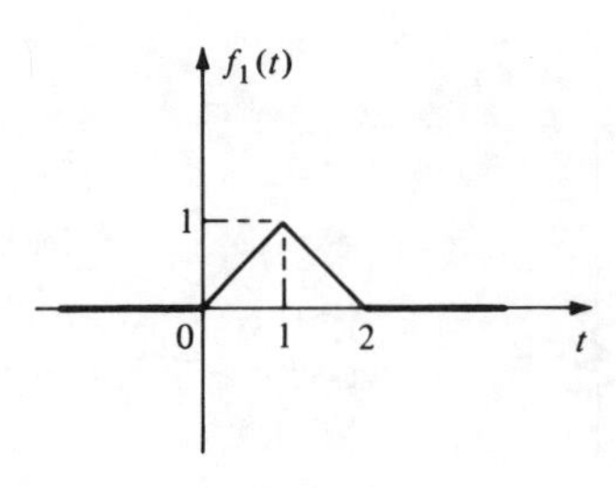

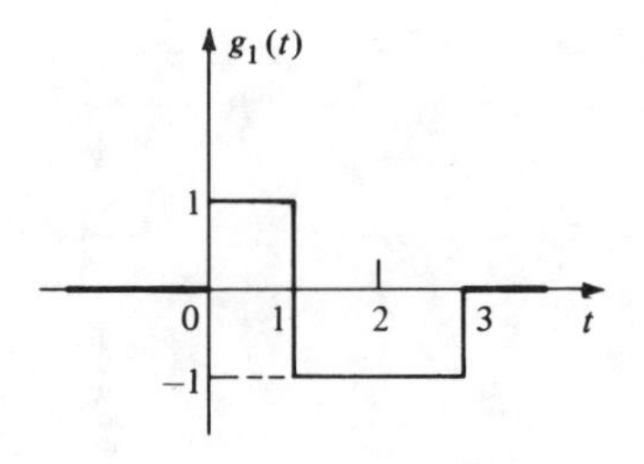

(c)

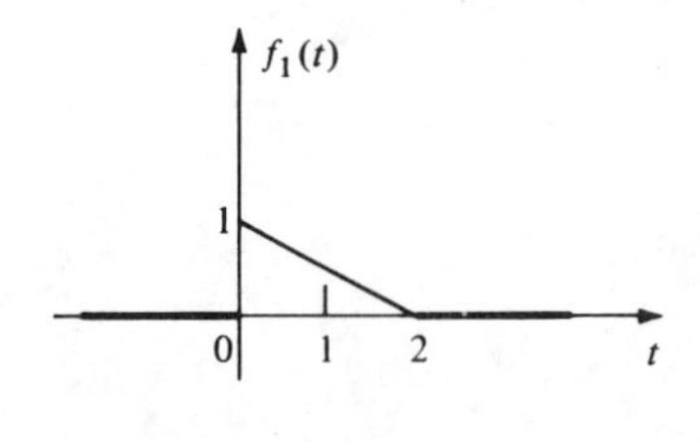

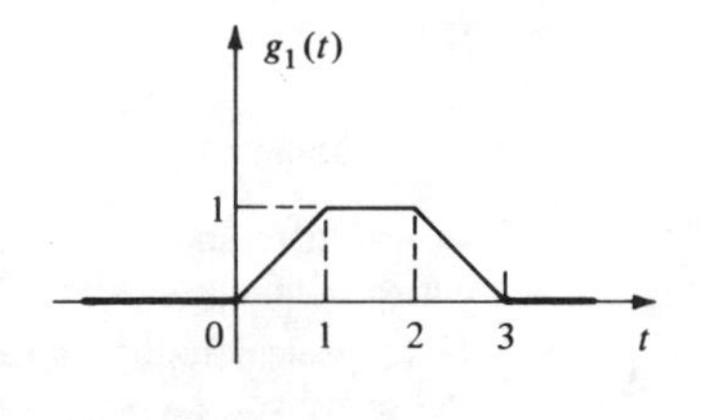

(d)

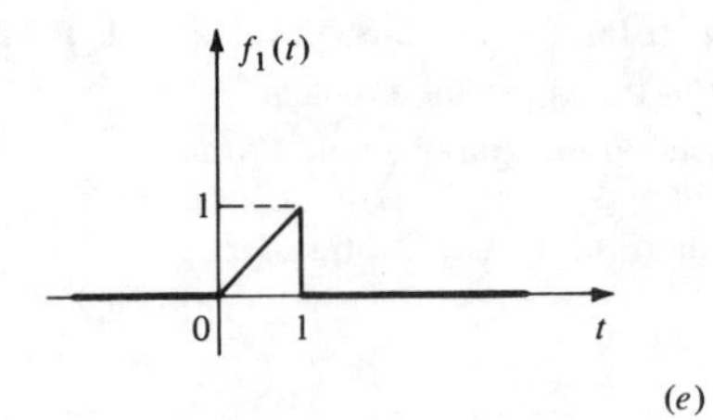

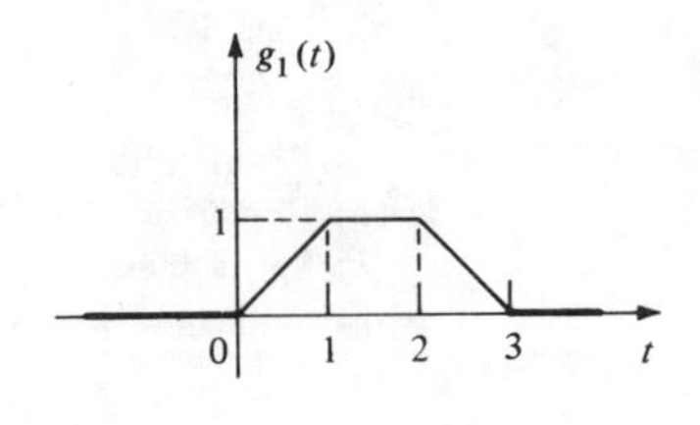

(e)

$f_1(t) = 1(t) - 1(t - 2\pi)$ $\qquad$ $g_1(t) = (\sin 2\pi t)\, 1(t)$

(f)

Figure P10.28

32 A linear time-invariant circuit has impulse response $h(\cdot)$ as shown in Fig. P10.32*a*.
(*a*) Find the step response $s(\cdot)$.
(*b*) Find the zero-state response due to the waveform shown in Fig. P10.32*b*.

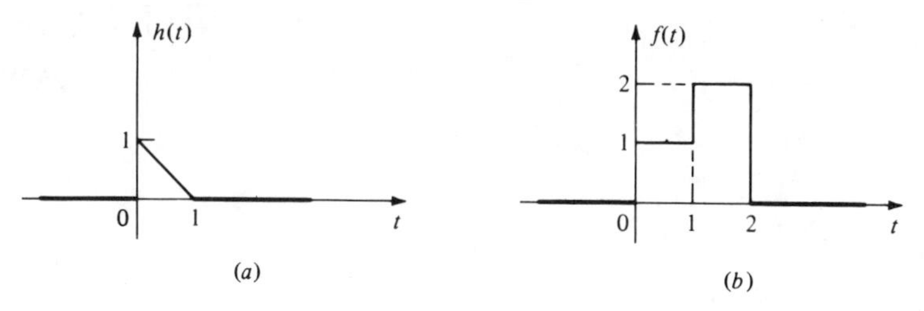

Figure P10.32

Symmetry

33 Figure P10.33 shows a uniquely solvable linear time-invariant circuit made of R's, L's, C's, ideal op amps, and VCCS's. The dashed vertical line is an axis of symmetry: The left part, $\mathscr{L}$, is precisely the mirror image of the right part, $\mathscr{R}$. In particular, to any node ⓘ of $\mathscr{L}$ corresponds its mirror image ⓘ' of $\mathscr{R}$; also, for any pair of nodes ⓘ, ⓚ joined by a capacitor C in $\mathscr{L}$ there is a mirror-image pair of nodes ⓘ' and ⓚ' joined by a capacitor C in $\mathscr{R}$, etc. The four horizontal connections cutting the axis of symmetry include a resistor with resistance $R > 0$: The points a, b, and c are precisely in the middle of the resistor (so that there is $R/2\ \Omega$ both to the left of a as well as to the right of a, etc.). A student asserts that one can obtain the list of the natural frequencies of the given circuit as follows:

1. Cut the top three horizontal wires at points a, b, and c, and calculate the natural frequencies of the resulting "open-circuited" $\mathscr{L}$.
2. Connect, by a short wire, points a, b, and c to the ground d, and calculate the natural frequencies of the resulting "short-circuited" $\mathscr{L}$.
3. The desired list is the union of the lists found in 1 and 2 above.

Do you believe the student? If yes, use node analysis to prove him right; if no, propose a counter example.

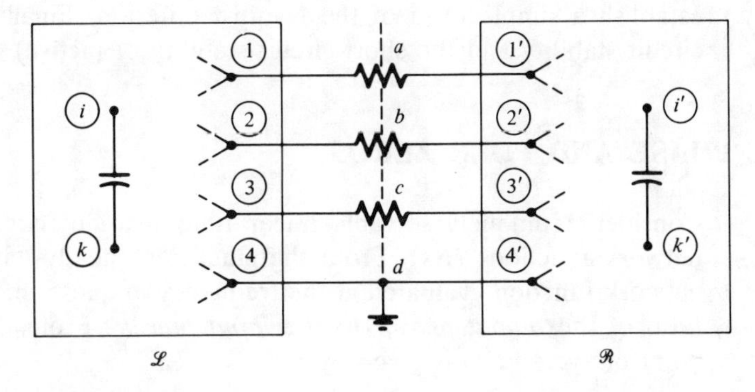

Figure P10.33

CHAPTER

ELEVEN

NETWORK FUNCTIONS AND STABILITY

INTRODUCTION

In this chapter we deepen our study of linear time-invariant circuits. Throughout this chapter our basic approach is the network function. We study the relation between the magnitude and phase curves and the pole and zero pattern of the network function. We demonstrate that a zero of a network function, which was defined purely algebraically, has in fact the physical meaning of being a zero of transmission. In Sec. 3, we demonstrate the use of network functions to study the stability of linear time-invariant circuits: We establish a simple form of the Nyquist criterion. Finally we study the open-circuit stability and the short-circuit stability of (active) one-ports.

1 MAGNITUDE PHASE AND POLE ZEROS

Consider a uniquely solvable linear time-invariant circuit $\mathcal{N}$ and one of its network functions $H(s)$. From the sinusoidal steady-state point of view, the network function evaluated at the frequency in question, namely $H(j\omega)$, is the ratio of the *output phasor* to the *input phasor*. From the Laplace transform point of view, $H(s)$ is given by

$$H(s) = \frac{\mathcal{L}\{\text{zero-state response}\}}{\mathcal{L}\{\text{input}\}} \tag{1.1}$$

We know that $H(s)$ is a *real* rational function [Chap. 10, Property (4.59)]

$$H(s) = \frac{n(s)}{d(s)} \tag{1.2}$$

and, with the polynomials $n(s)$ and $d(s)$ *coprime*, the zeros of $n(s)$ are the *zeros of* $H(s)$ and the zeros of $d(s)$ are the *poles of* $H(s)$. Finally, the impulse response $h(t)$ is given by

$$h(t) = \mathscr{L}^{-1}\{H(s)\} \tag{1.3}$$

For ω real, let us write the complex number $H(j\omega)$ in polar form:

$$H(j\omega) = |H(j\omega)| \exp j\sphericalangle H(j\omega) \tag{1.4}$$

Consider the curve of $|H(j\omega)|$ and $\sphericalangle H(j\omega)$ versus ω say from $\omega = 0$ to $\omega = \infty$. These curves are called the *magnitude* and *phase* versus frequency curves of the network function $H(j\omega)$. These two curves allow us to immediately visualize how the circuit will behave in the sinusoidal steady state *at any frequency* [Chap. 9, Eq. (3.10*a*)]. That is why these magnitude and phase curves are so often used in engineering design and in writing specifications.

For a few simple network functions we are going to study the relation between the magnitude and phase curves of a network function and the location of its poles and zeros. We shall also study the relation between the poles and zeros and the impulse response.

1.1 First-Order Circuits

A. Analysis Consider the linear time-invariant circuit of Fig. 1.1. Let $i_0(t)$ be the output variable. Since KCL gives

$$CR\,\dot{i}_0(t) + i_0(t) = i_s(t)$$

we have, assuming zero initial conditions,

$$H(s) \triangleq \frac{I_0(s)}{I_s(s)} = \frac{1}{RCs + 1} \tag{1.5}$$

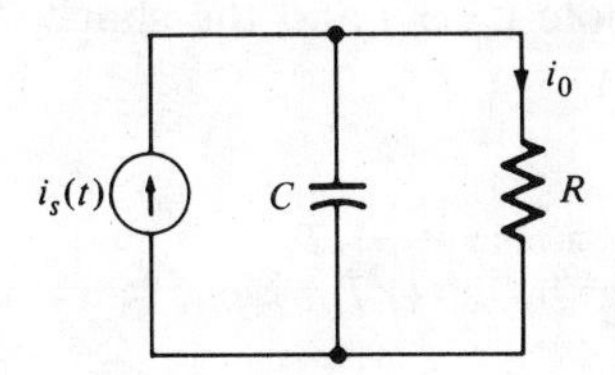

Figure 1.1 First-order circuit.

Let

$$T \triangleq RC \qquad \text{and} \qquad \omega_c \triangleq \frac{1}{RC} = \frac{1}{T} \tag{1.6}$$

ω_c is called the *corner frequency* and T is the *time constant*.[1] We rewrite Eq. (1.5) as

$$H(s) = \frac{\omega_c}{s + \omega_c} = \frac{1}{1 + s/\omega_c} \tag{1.7}$$

This rational function $H(s)$ has *no* (finite)[2] *zero* and has *one pole* at $s = -\omega_c$.

B. Magnitude and phase curves From Eq. (1.7)

$$|H(j\omega)| = \frac{1}{|1 + j(\omega/\omega_c)|} = \frac{1}{\sqrt{1 + (\omega/\omega_c)^2}} \tag{1.8}$$

The curve of $H(j\omega)$ versus ω is given in Fig. 1.2b: $|H(0)| = 1$ and $|H(j\omega)|$ decreases monotonically as ω increases.

By inspection from Eq. (1.7), the phase starts at 0° for $\omega = 0$, decreases monotonically as ω increases, and tends to $-90°$ as $\omega \to \infty$:

$$\measuredangle H(j\omega) = -\tan^{-1} \frac{\omega}{\omega_c} \tag{1.9}$$

From the pole and zero point of view, $H(j\omega)$ can also be written as

$$H(j\omega) = \frac{\omega_c}{j\omega - (-\omega_c)} \qquad \text{so} \qquad \begin{cases} |H(j\omega)| = \dfrac{\omega_c}{|j\omega - (-\omega_c)|} & (1.10) \\ \measuredangle H(j\omega) = -\measuredangle[j\omega - (-\omega_c)] & (1.11) \end{cases}$$

Refer to Fig. 1.2a: The denominator of Eq. (1.10) is the length of the vector from the pole $-\omega_c$ to the point $j\omega$, i.e., the length of the vector PQ; $\measuredangle[j\omega - (-\omega_c)]$ is the angle of the same vector with the σ axis of the s plane, i.e., the angle OPQ.

Figures 1.2b and c show the magnitude and phase plotted versus ω and obtained from Eqs. (1.8) and (1.9).

The important point to note is that the consideration of Fig. 1.2a explains the shape of *both* curves $|H(j\omega)|$ and $\measuredangle H(j\omega)$: We have found a connection between the location of the pole $(-\omega_c)$ and the shape of the magnitude curve and phase curve.

[1] Note that $H(s)$ of Eq. (1.5) has a pole at $-1/T$.

[2] Some authors say that $H(s)$ given by Eq. (1.7) has one zero at infinity: This is equivalent to saying that, as $|s| \to \infty$, $H(s) \to 0$.

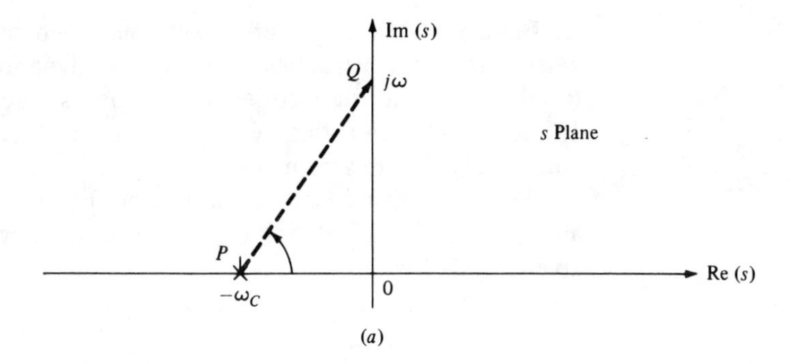

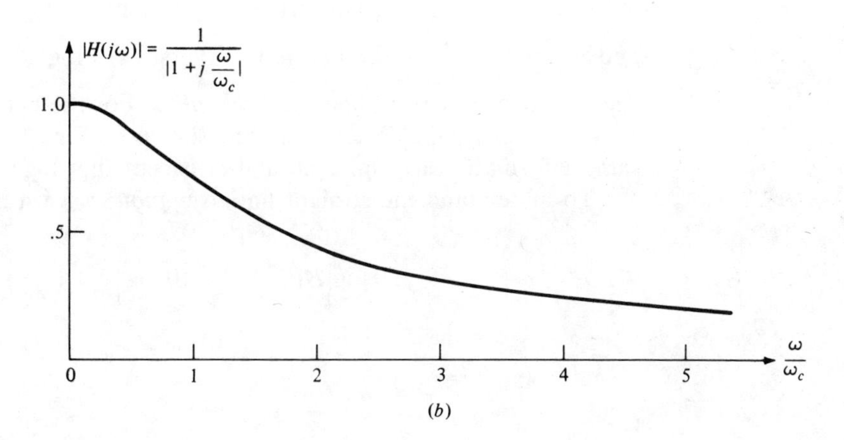

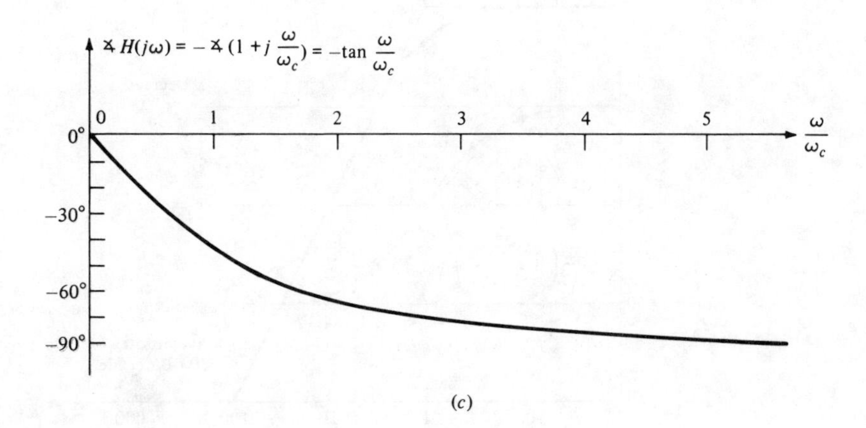

Figure 1.2 (*a*) The pole-zero plot of $H(s) = 1/[1 + s/\omega_c]$; $H(s)$ has a pole at $-\omega_c$. (*b*) The magnitude curve. (*c*) The phase curve.

C. Bode plots In engineering systems, such as a cable TV distribution network, the circuit variables cover a large dynamic range, say from tens of volts to millivolts, and the frequency also covers an even larger dynamic range from 10^2 to 10^8 Hz. For this reason, engineers like to use logarithmic scales for amplitudes and for frequency.

For the magnitude $|H(j\omega)|$, the commonly used unit is the decibel, abbreviated dB. By definition, at the frequency ω_1, the amplitude $|H(j\omega_1)|$ is given in dB by

$$20 \log |H(j\omega_1)|$$

For example, if $|H(j\omega_1)| = 23$, then $20 \log |H(j\omega_1)| = 27.235$ dB. Thus, for convenience, we plot

$$20 \log |H(j\omega)| \qquad \text{versus} \qquad \log \omega$$

and

$$\measuredangle H(j\omega) \qquad \text{versus} \qquad \log \omega$$

These two curves are called the *Bode plots*. For the network function $H(j\omega) = 1/[1 + j(\omega/\omega_c)]$, the Bode plots are shown in Fig. 1.3. Figure 1.3 gives the same information as Fig. 1.2*b* and *c*, except that logarithmic scales are used.

To understand the straight line *asymptotes* shown in Fig. 1.3 note that

$$20 \log |H(j\omega)| = -10 \log \left[1 + \left(\frac{\omega}{\omega_c}\right)^2\right] \tag{1.12}$$

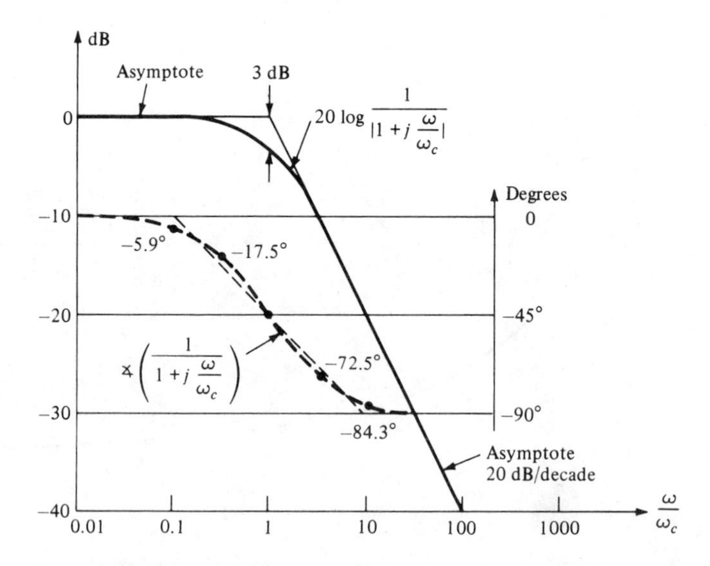

Figure 1.3 Bode plots of $H(j\omega) = 1/[1 + j(\omega/\omega_c)]$: $20 \log[1/|1 + j(\omega/\omega_c)|]$ versus $\log(\omega/\omega_c)$ and $\measuredangle\{1/[1 + j(\omega/\omega_c)]\}$ versus $\log(\omega/\omega_c)$.

Now if $\omega \gg \omega_c$, the square term dominates the 1 in the brackets, and hence

$$\text{for } \omega \gg \omega_c, \qquad 20 \log|H(j\omega)| \cong -10 \log\left(\frac{\omega}{\omega_c}\right)^2$$

$$= -20 \log \omega + 20 \log \omega_c \tag{1.13}$$

Thus, for $\omega \gg \omega_c$, the magnitude graph becomes *linear in* $\log \omega$, i.e., a straight line when plotted versus $\log \omega$; the slope of this asymptote is -20 dB/decade.[3]

Next if $\omega \ll \omega_c$, the argument of the log on the right-hand side of Eq. (1.12) is approximately one: Thus, since $\log 1 = 0$,

$$20 \log|H(j\omega)| \to 0 \qquad \text{as } \omega \to 0 \tag{1.14}$$

and, for $\omega \ll \omega_c$, using Taylor series about $\omega/\omega_c = 0$, we obtain

$$20 \log|H(j\omega)| \cong -4.4329\left(\frac{\omega}{\omega_c}\right)^2 \tag{1.15}$$

Thus as ω decreases to zero, $20 \log|H(j\omega)|$ tends to zero by Eq. (1.14) and furthermore its slope becomes more and more horizontal; consequently we have a *horizontal asymptote at the ordinate of 0 dB*.

The horizontal asymptote and the descending asymptote intersect at the point (1, 0 dB), thus ω_c is called *the corner frequency*. (See Fig. 1.3.)

Example To illustrate the use of these asymptotes, let us consider a slightly more complicated network function, say,

$$H_1(j\omega) = \frac{1 + j(\omega/\omega_z)}{[1 + j(\omega/\omega_1)][1 + j(\omega/\omega_2)]} \tag{1.16}$$

The dB versus log frequency plot of $H_1(j\omega)$ is easily obtained from the graphs given in Fig. 1.3 by simple addition:

$$\log|H_1(j\omega)| = \log\left|1 + j\frac{\omega}{\omega_z}\right| - \log\left|1 + j\frac{\omega}{\omega_1}\right| - \log\left|1 + j\frac{\omega}{\omega_2}\right| \tag{1.17}$$

Here $\omega_z = 0.8$, $\omega_1 = 10$, and $\omega_2 = 25$, as shown in Fig. 1.4. Hence we have one *rising* asymptote with a corner frequency of ω_z and two *descending* asymptotes with corner frequencies of ω_1 and ω_2. The asymptotes of $20 \log|H_1(j\omega)|$ are easily obtained by adding the asymptotes of the individual terms in Eq. (1.17). The process is illustrated in Fig. 1.4.

[3] If two frequencies ω_1 and ω_2 are such that $10\omega_1 = \omega_2$, we say that the frequencies ω_1 and ω_2 are a *decade apart*.

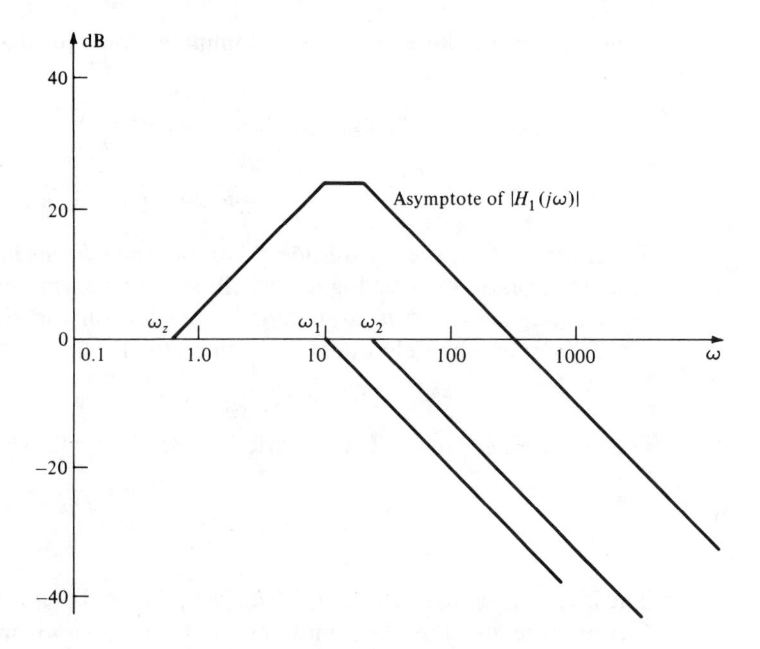

Figure 1.4 Asymptotes of $H_1(j\omega)$ given by Eq (1.17).

From the asymptotes of $|H_1(j\omega)|$ sketched in Fig. 1.4, one can use the curve of Fig. 1.3 to obtain a more accurate picture of the behavior of $|H_1(j\omega)|$ versus ω.

D. Nyquist plot of $H(j\omega) = 1/[1 + j(\omega/\omega_c)]$ Now, for each ω from 0 to ∞, let us plot the complex number $H(j\omega)$ in the $H(j\omega)$ plane: We obtain a curve whose parametric equations are

$$x = \text{Re}[H(j\omega)] \qquad y = \text{Im}[H(j\omega)] \qquad \text{for } \omega \geq 0 \tag{1.18}$$

Let us rewrite Eq. (1.7):

$$H(j\omega) = \frac{1}{1 + j(\omega/\omega_c)} \tag{1.19}$$

Consequently, for $\omega = 0$, $H(0) = 1$,

$$H(j\omega) \cong 1 - j\frac{\omega}{\omega_c} \qquad \text{for } \left|\frac{\omega}{\omega_c}\right| \ll 1 \tag{1.20}$$

$$H(\infty) = 0 \qquad \text{for } \omega = \infty$$

$$H(j\omega) \cong \frac{1}{j(\omega/\omega_c)} = -j\frac{\omega_c}{\omega} \quad \text{for } \left|\frac{\omega}{\omega_c}\right| \gg 1 \tag{1.21}$$

In fact, the graph of $H(j\omega)$ for ω increasing from 0 to ∞ is the half circle shown in Fig. 1.5.

Since, for all real ω, $H(-j\omega) = \overline{H(j\omega)}$, the curve for $\omega < 0$ is simply the mirror image of the preceding one with respect to the real axis. Consider now the graph as ω increases from $-\infty$ through 0 to $+\infty$: The resulting *closed* curve is called the *Nyquist plot of* $H(j\omega)$. Figure 1.5 shows the Nyquist plot for $H(j\omega) = [1 + j(\omega/\omega_c)]^{-1}$. Note the location of $H(j\omega_c)$ and $H(-j\omega_c)$.

Exercise Prove that the Nyquist plot of Fig. 1.5 is a circle of center (0.5, 0) and radius 0.5.

E. Step and impulse response of $H(s) = 1/(1 + s/\omega_c)$ From Eq. (1.7), we easily calculate the impulse response $h(t)$ and the step response $s(t)$: with $T \stackrel{\Delta}{=} 1/\omega_c$

$$h(t) = 1(t)\, \omega_c\, e^{-\omega_c t} = 1(t)\, \frac{1}{T}\, e^{-t/T} \tag{1.22a}$$

$$s(t) = 1(t)\, (1 - e^{-\omega_c t}) = 1(t)(1 - e^{-t/T}) \tag{1.22b}$$

We explicitly included the unit step factor $1(t)$ to remind ourselves that $h(t) = s(t) = 0$ for all $t < 0$.

The important point is to see the connection between the *pole location*—ω_c is the distance of the pole to the $j\omega$ axis—the *corner frequency* ω_c, the *rise time* of $s(t)$, and the *time constant* T. Recall also that $5T$ is the "memory" time of the circuit. A common definition of *rise time* is the following: It is the time required for the step response $s(t)$ to go from 10 to 90 percent of its final value $s(\infty)$ (see Fig. 1.6). For first-order circuits, the rise time has been calculated in Eq. (3.2) of Chap. 6 to be $2.2T$.

In summary, for the first-order network function $\omega_c/(s + \omega_c)$ we can state that as ω_c becomes large, we have a *large bandwidth*, a *small rise time*, and a *short memory*.

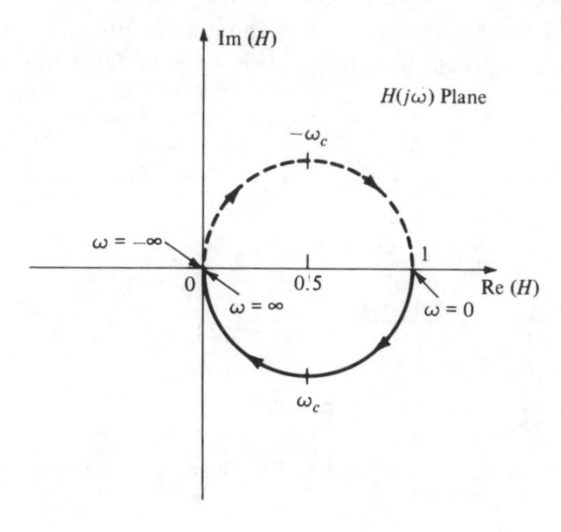

Figure 1.5 Nyquist plot of network function $H(j\omega) = [1 + j(\omega/\omega_c)]^{-1}$.

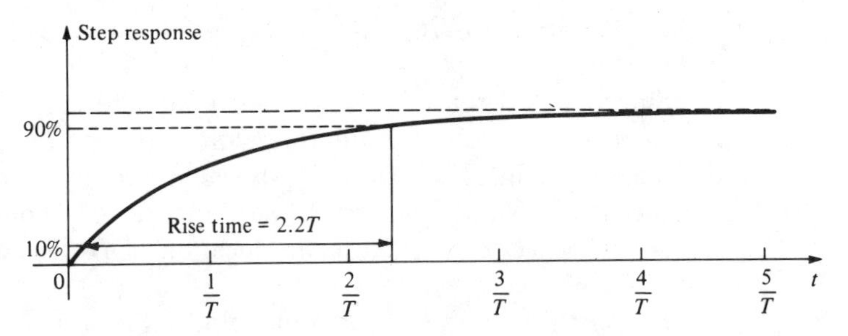

Figure 1.6 Step response of $H(s) = 1 + (s/\omega_c)$: the 10 to 90 percent rise time is shown to be $2.2T$.

1.2 Second-Order Circuits

Consider the familiar second-order circuit, and, as before, let $v_C(t)$ be the output variable; from Fig. 1.7 we have

$$H_1(s) = \frac{V_C(s)}{I_s(s)} = \frac{1}{C}\,\frac{s}{s^2 + s/R_pC + 1/LC} \tag{1.23}$$

Using the standard notations

$$\omega_0 = \frac{1}{\sqrt{LC}} \qquad \alpha = \frac{1}{2R_pC} \qquad Q = \frac{\omega_0}{2\alpha} = \frac{R_p}{\sqrt{L/C}} \qquad \omega_d = \sqrt{\omega_0^2 - \alpha^2} \tag{1.24}$$

we write, as in Eq. (4.15) of Chap. 9,

$$H_1(j\omega) = \frac{R_p}{1 + jQ(\omega/\omega_0 - \omega_0/\omega)} = R_p\,H(j\omega) \tag{1.25}$$

$H(j\omega)$ is the "normalized" gain of the circuit: R_p has been factored out of $H_1(j\omega)$. We have already plotted the magnitude and phase of $H(j\omega)$: Recall Chap. 9, Sec. 4, Figs. 4.3 to 4.5. For $Q > \frac{1}{2}$, $H(s)$ has two complex-conjugate poles

$$p_1 = -\alpha + j\omega_d = -\alpha + j\omega_0\sqrt{1 - \frac{1}{4Q^2}} \qquad p_2 = \bar{p}_1 = -\alpha - j\omega_d \tag{1.26}$$

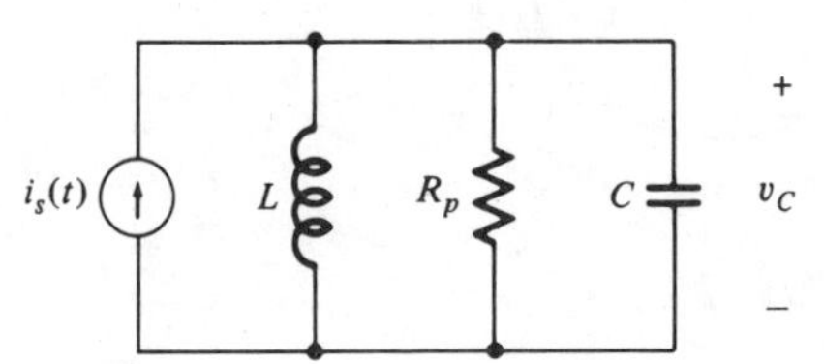

Figure 1.7 Second-order circuit with $R_p > 0$, $L > 0$, and $C > 0$.

These poles are shown in Fig. 1.8. Using this notation, Eq. (1.25) becomes

$$H(j\omega) = \frac{1}{R_pC} \frac{j\omega}{(j\omega - p_1)(j\omega - \bar{p}_1)} \tag{1.27}$$

The magnitude and phase of $H(j\omega)$ are then

$$|H(j\omega)| = \frac{1}{R_pC} \frac{|j\omega|}{|j\omega - p_1||j\omega - \bar{p}_1|} \tag{1.28}$$

and

$$\measuredangle H(j\omega) = 90° - \measuredangle(j\omega - p_1) - \measuredangle(j\omega - \bar{p}_1) \tag{1.29}$$

When Q is equal to 5 or larger, the pole-zero pattern of $H(s)$ explains easily the curves of $|H(j\omega)|$ and of $\measuredangle H(j\omega)$ versus ω. Figure 1.8 shows the pole-zero pattern for $Q = 2.27$. (This modest Q was chosen for ease of drafting.) From the study of Fig. 1.8 and Eqs. (1.28) and (1.29) we obtain the following:

1. $H(0) = 0$.
2. For $\omega > 0$ but $\omega \ll \omega_d$, $\measuredangle H(j\omega) \cong 90°$: Indeed, the angles of $(j\omega - p_1)$ and $(j\omega - \bar{p}_1)$ approximately cancel out.
3. As ω increases from zero to ω_d, $|H(j\omega)|$ starts at 0 and increases monotonically to reach a maximum for $\omega \cong \omega_d$: Indeed the numerator $|j\omega|$ increases linearly with ω and the denominator factor $|j\omega - p_1|$ decreases to reach a minimum at $\omega = \omega_d$. The other denominator factor $|j\omega - \bar{p}_1|$ increases, but its relative increase is not as drastic as the relative decrease of $|j\omega - p_1|$.
4. At $\omega = \omega_d$ [see Eq. (1.27)], the normalized network function is given by

$$\begin{aligned} |H(j\omega_d)| &\cong \frac{1}{R_pC} \frac{\omega_d}{\alpha \cdot 2\omega_d} = \frac{1}{2\alpha R_pC} = 1 \\ \measuredangle H(j\omega_d) &\cong (90°) - 0 - (90°) = 0 \end{aligned} \tag{1.30}$$

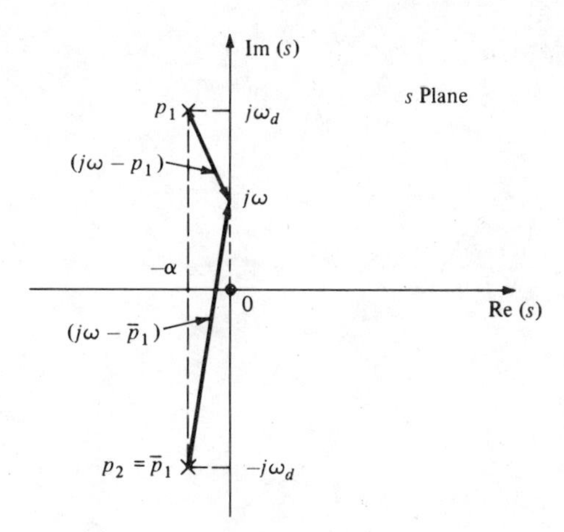

Figure 1.8 The pole-zero pattern of the network function of the resonant circuit of Fig. 1.1.

5. For $\omega > \omega_d$ and $\omega \to \infty$, $|H(j\omega)|$ decreases monotonically, for $\omega \gg \omega_d$ [see Eq. (1.27) and recall that $\mathrm{Im}(p_1) = j\omega_d$],

$$|H(j\omega)| \cong \frac{1}{R_p C} \frac{1}{j\omega} \qquad \sphericalangle H(j\omega) \cong -90° \tag{1.31}$$

Thus we have shown that the magnitudes and phases of $j\omega$, $j\omega - p_1$, and $j\omega - \bar{p}_1$ explain completely the curves of $|H(j\omega)|$ and $\sphericalangle H(j\omega)$ of Fig. 4.5 of Chap. 9: The higher the Q, the closer are the poles p_1 and $\bar{p}_1$ to the $j\omega$ axis; consequently the peak around ω_0 is more pronounced and the 3-dB bandwidth decreases as Q increases; similarly, as Q increases, the phase changes more abruptly in the neighborhood of ω_0.

The Nyquist plot of $H_1(j\omega)$ turns out to be a circle as in the previous case (see Fig. 4.4 of Chap. 9).

The pole locations also play a key role in determining the characteristics of the impulse response. A straightforward calculation shows that the impulse response of $H_1(s)$ is given by

$$h(t) = 1(t)\sqrt{\frac{L}{C}}\,\frac{\omega_0^2}{\omega_d}\,\mathrm{e}^{-\alpha t}\cos(\omega_d t + \phi) \tag{1.32}$$

where ϕ is such that $\tan\phi = \alpha/\omega_0$. Thus the distance α of the pole from the $j\omega$ axis controls the decay of the impulse response and the ordinate of the pole, ω_d, controls the frequency of the oscillatory decay. (See Fig. 1.9.)

The point of this discussion is to show that the pole-zero pattern of $H(s)$ explains its *frequency-domain behavior*—i.e., the curves $|H(j\omega)|$ and $\sphericalangle H(j\omega)$—as well as its *time-domain behavior*—i.e., the impulse response and through the impulse response, via convolution, any zero-state response.

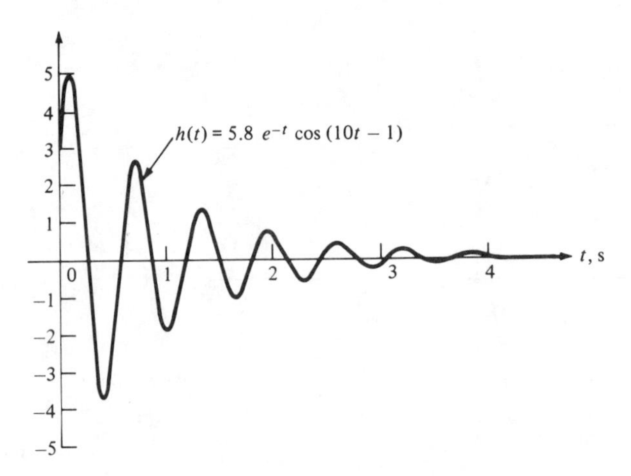

Figure 1.9 Impulse response associated with $H(s)$ for $\alpha = 1$ and $\omega_d = 10$.

1.3 General Case

Engineers frequently use the pole-zero pattern to derive and interpret properties of linear time-invariant circuits. These considerations are based on the following calculations: Consider, for simplicity, the case where in Eq. (1.2), the polynomials n and d are of degrees 2 and 3, respectively, and both have simple zeros, say,

$$H(s) = \frac{n(s)}{d(s)} = k \frac{(s - z)(s - \bar{z})}{(s - p_1)(s - p_2)(s - \bar{p}_2)} \tag{1.33}$$

where z and $\bar{z}$ are the *zeros of* $H(s)$: a pair of complex-conjugate zeros z and $\bar{z}$

p_1, p_2, and $\bar{p}_2$ are the *poles* of $H(s)$: one real pole p_1 and a pair of complex-conjugate poles p_2 and $\bar{p}_2$

k is a real scaling constant (note that k *may* be negative!)

Setting $s = j\omega$ in Eq. (1.33) and taking absolute values, we obtain

$$|H(j\omega)| = |k| \frac{|j\omega - z||j\omega - \bar{z}|}{|j\omega - p_1||j\omega - p_2||j\omega - \bar{p}_2|} \tag{1.34}$$

and, taking the angles, we obtain

$$\begin{aligned}\measuredangle H(j\omega) = \measuredangle k + \measuredangle(j\omega - z) + \measuredangle(j\omega - \bar{z}) \\ -[\measuredangle(j\omega - p_1) + \measuredangle(j\omega - p_2) + \measuredangle(j\omega - \bar{p}_2)]\end{aligned} \tag{1.35}$$

If we write $p_2 \triangleq -\alpha + j\omega_d$, we easily obtain the impulse response from Eq. (1.33) by partial fraction expansion:

$$h(t) = 1(t)\, 2|k_2|\, e^{-\alpha t} \cos(\omega_d t + \measuredangle k_2) + k_1 e^{p_1 t} \tag{1.36}$$

where the residues k_2 and k_1 are given by [here we use Eq. (3.9) of Chap. 10]

$$k_1 = k \frac{(p_1 - z)(p_1 - \bar{z})}{(p_1 - p_2)(p_1 - \bar{p}_2)} \tag{1.37}$$

$$k_2 = k \frac{(p_2 - z)(p_2 - \bar{z})}{(p_2 - p_1)(p_2 - \bar{p}_2)} \tag{1.38}$$

From these expressions and the pole-zero pattern shown in Fig. 1.10, we draw the following conclusions.

CONCLUSIONS

1. In the neighborhood of a *pole close to the $j\omega$ axis*, we expect the magnitude to have a local maximum and the phase to vary rapidly. Also the impulse response will exhibit oscillations that will decay and oscillate according to the location of this pole. [Recall $2|k|\, e^{-\alpha t} \cos(\omega_d t + \measuredangle k)$.]
2. In the neighborhood of a *zero* close to the $j\omega$ axis, the magnitude of $H(j\omega)$ will be small and the phase will vary rapidly.

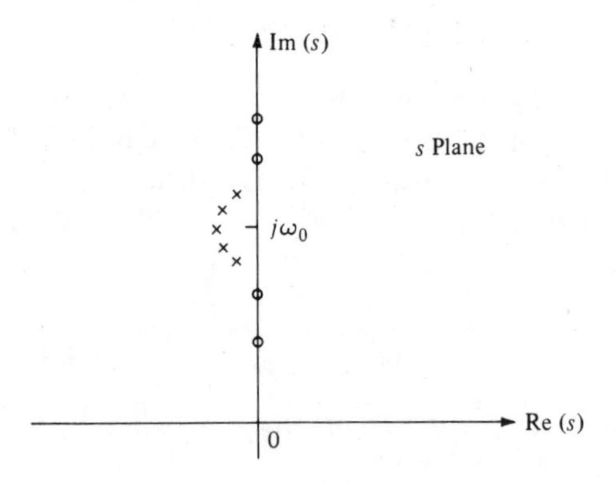

Figure 1.10 Upper half-plane pole and zero pattern for a filter whose band center is ω_0 rad/s.

These considerations have many uses in design. For example, suppose we want to design a filter that passes frequencies close to a center frequency ω_0 and stops frequencies away from ω_0; thus $H(j\omega)$ must be large when ω is close to ω_0 and small at other frequencies. A typical pole-zero pattern is shown in Fig. 1.10: The poles are clustered around ω_0 and the zeros are located on the $j\omega$ axis near those frequencies where we want to have $H(j\omega)$ small. Efficient filter designs have pole-zero patterns of that type.

Physical interpretation of $j\omega$ axis zeros Some circuits have the property that, by inspection, one may detect the presence of $j\omega$ axis zeros. To illustrate this point consider the linear time-invariant circuit shown in Fig. 1.11: All the L's, the C's, and the R's are positive. Call Z_1 the impedance of the series resonant circuit L_1C_1 and call ω_1 its resonant frequency; then $Z_1(j\omega_1)=0$. Call Z_2 the impedance of the parallel resonant circuit L_2C_2 and call ω_2 its resonant frequency; then $Z_2(j\omega_2)=\infty$, or, more precisely, $Y_2(j\omega_2)=0$.

We study the network function Z_{21} from *input* I_s to *output* V_2, the voltage across the load resistor R.

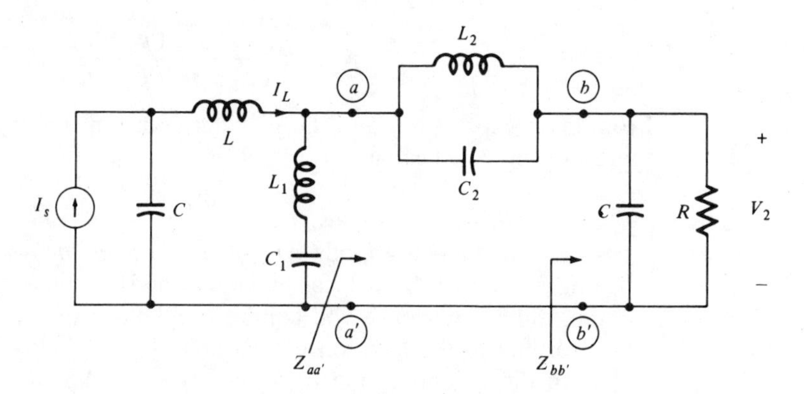

Figure 1.11 Ladder circuit which allows the determination of its zeros by inspection.

Claim 1 $Z_{21}(j\omega_1) = 0$, i.e., the transfer impedance Z_{21} has a zero at ω_1, the series resonant frequency of L_1C_1.

Call $Z_{aa'}(j\omega)$ the driving-point impedance of the circuit that lies to the right of terminals ⓐ, ⓐ'. If $\omega_2 = \omega_1$, then, since the impedance at frequency ω_2 of the resonant circuit L_2C_2 is infinite, $Z_{aa'}(j\omega_1) = Z_{aa'}(j\omega_2) = \infty$. Hence at frequency ω_1 all the current will go through the resonant circuit L_1C_1 and no current will flow in R. Consequently $Z_{21}(j\omega_1) = 0$. If $\omega_2 \neq \omega_1$, we claim that $Z_{aa'}(j\omega_1) \neq 0$: Indeed, in the sinusoidal steady state at frequency ω_1, the positive resistor R is traversed by a *nonzero* current. Hence energy is dissipated, and $\text{Re}[Z_{aa'}(j\omega_1)] > 0$. So in either case, whether $\omega_2 = \omega_1$ or $\omega_2 \neq \omega_1$, $Z_{aa'}(j\omega_1) \neq 0$.

So, at frequency ω_1, the current I_L sees a *zero* impedance $Z_1(j\omega_1) = 0$ in parallel with a *nonzero* impedance $Z_{aa'}(j\omega_1)$: Hence all the current goes through Z_1 and *no current* flows at terminal ⓐ. Consequently, at frequency ω_1, $V_2 = 0$. Hence $Z_{21}(j\omega_1) = 0$.

Roughly speaking, we can say that, at frequency ω_1, *the series resonant circuit L_1C_1 shorts out the remainder of the circuit* from the driving current source I_s.

Claim 2 $Z_{21}(j\omega_2) = 0$, i.e., the transfer impedance Z_{21} has a zero at ω_2, the parallel resonant frequency of L_2C_2.

If $\omega_1 = \omega_2$, then $Z_1(j\omega_2) = 0$ and $Z_2(j\omega_2) = \infty$; hence, repeating the reasoning above, we have $Z_{21}(j\omega_2) = 0$.

If $\omega_2 \neq \omega_1$, then $Z_1(j\omega_2) \neq 0$. Let $Z_{bb'}(j\omega)$ denote the impedance of the circuit that lies to the right of terminals ⓑ, ⓑ'; then, by inspection, $Z_{bb'}(j\omega_2) \neq 0$ and is *finite*. Consequently, since

$$Z_{aa'}(j\omega_2) = Z_2(j\omega_2) + Z_{bb'}(j\omega_2)$$

we conclude that $Z_{aa'}(j\omega_2) = \infty$. Thus at frequency ω_2 the current I_L sees a *finite* impedance $Z_1(j\omega_2)$ in parallel with an *infinite* impedance $Z_{aa'}(j\omega_2)$. Therefore no current flows at terminal ⓐ; consequently at frequency ω_2, $V_2 = 0$. Hence $Z_{21}(j\omega_2) = 0$.

If the circuit of Fig. 1.11 were designed as a low-pass filter, the graph of $Z_{21}(j\omega)$ would typically look like the curve shown in Fig. 1.12.

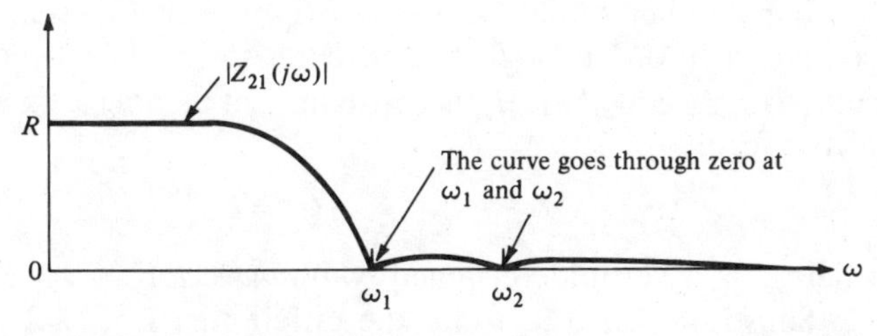

Figure 1.12 Sketch of transfer function of $Z_{21}(j\omega)$: the zeros at ω_1 and ω_2 force $Z_{21}(j\omega)$ to be small at the desired frequencies.

2 ZEROS OF NETWORK FUNCTIONS

Let the linear time-invariant circuit $\mathcal{N}$ be exponentially stable, assume that its network function $H(s) \triangleq V_0/I_s$ has a zero at $j\omega_z$ on the $j\omega$ axis, then if we drive the circuit with a sinusoid at that particular frequency ω_z *and* we wait for the sinusoidal steady state to get established, then the output will be zero. This is easily proven by noting that the sinusoidal steady-state output is given by

$$v_0(t) = \mathrm{Re}[H(j\omega_z)I_m\, \mathrm{e}^{j\omega_z t}] = 0 \tag{2.1}$$

because $H(j\omega_z) = 0$ [here I_m denotes the phasor representing the sinusoidal current $i_s(t)$]. Thus we may say that *a zero of $H(s)$ on the $j\omega$ axis blocks transmission at that frequency*.

We are going to show that the same is true for any other zero, provided we do the experiment appropriately.

The analysis is particularly easy if we use state variables. Suppose that the given linear time-invariant circuit $\mathcal{N}$ is described by its state equations and suppose that they are of the form:

$$\begin{aligned} \dot{\mathbf{x}} &= \mathbf{A}\mathbf{x} + \mathbf{b}\, i_s(t) \\ v_0(t) &= \mathbf{c}\mathbf{x}(t) \end{aligned} \tag{2.2}$$

where $i_s(t)$ is the scalar *input*, due to a current source, and $v_0(t)$ is the (scalar) *output voltage*. The $n \times n$ matrix $\mathbf{A}$ and the n-vectors $\mathbf{b}$ and $\mathbf{c}$ are constant; $\mathbf{b}$ is a column vector, $\mathbf{c}$ is a row vector. Note that we make no assumptions whatsoever as to the locations of the natural frequencies of $\mathcal{N}$.

The network function relating $V_0(s)$ to $I_s(s)$ is

$$H(s) = \frac{V_0(s)}{I_s(s)} = \mathbf{c}(s\mathbf{1} - \mathbf{A})^{-1}\mathbf{b} \tag{2.3}$$

Let z be a zero of the network function $H(s)$, that is, $H(z) = 0$. More precisely,

$$H(z) = \mathbf{c}(z\mathbf{1} - \mathbf{A})^{-1}\mathbf{b} = 0 \tag{2.4}$$

Note that the zero z of the network function $H(s)$ may lie *anywhere* in the s plane.

For simplicity think of z as a real number. (The calculations are valid for z complex.) In order to show that there is a *zero of transmission* at $s = z$, let us set $i_s(t) = I_m\, \mathrm{e}^{zt}$, where I_m is constant. Let us now seek a solution for Eq. (2.2) of the form

$$\mathbf{x}(t) = \boldsymbol{\xi}\, \mathrm{e}^{zt} \tag{2.5}$$

where $\boldsymbol{\xi}$ is a yet undetermined *constant vector*.

Substitute Eq. (2.5) into the differential Eq. (2.2); then

$$z\boldsymbol{\xi}\, \mathrm{e}^{zt} = \mathbf{A}\boldsymbol{\xi}\, \mathrm{e}^{zt} + \mathbf{b}I_m\, \mathrm{e}^{zt} \tag{2.6}$$

Since $e^{zt} \neq 0$ for all t, Eq. (2.6) is equivalent to

$$\boldsymbol{\xi} = (z\mathbf{1} - \mathbf{A})^{-1}\mathbf{b}I_m \tag{2.7}$$

The corresponding solution of Eq. (2.2) is [see Eq. (2.5)]

$$\mathbf{x}(t) = (z\mathbf{1} - \mathbf{A})^{-1}\mathbf{b}I_m\, e^{zt} \tag{2.8}$$

and by Eq. (2.2), the output is given by

$$\begin{aligned} v_0(t) &= \mathbf{c}(z\mathbf{1} - \mathbf{A})^{-1}\mathbf{b}I_m\, e^{zt} \\ &= 0 \qquad \text{for all } t \geq 0 \qquad \text{by Eq. (2.4)} \end{aligned} \tag{2.9}$$

These calculations allow us to state the *blocking property of zeros*.

Let z be *any* zero of the network function $H(s)$, let the circuit be set up at $t = 0$ with the initial condition $\mathbf{x}(0) = \boldsymbol{\xi} = (z\mathbf{1} - \mathbf{A})^{-1}\mathbf{b}I_m$, and let it be driven, for $t \geq 0$, by the exponential $i_s(t) = I_m\, e^{zt}$—here I_m gives the amplitude of the exponential. Then the circuit oscillates according to $\mathbf{x}(t) = \boldsymbol{\xi}\, e^{zt}$, but

$$v_0(t) = 0 \qquad \text{for all } t \geq 0 \tag{2.10}$$

that is, the exponential oscillation of the input $i_s(t)$ is *blocked* out from the output $v_0(t)$, for all $t \geq 0$.

It is important to note that if z were *not* a zero of $H(s)$, then, for the *same* input $i_s(t) = I_m\, e^{zt}$ and the *same* initial state, the output voltage would be

$$v_0(t) = \mathbf{c}(z\mathbf{1} - \mathbf{A})^{-1}\mathbf{b}I_m\, e^{zt} \neq 0 \qquad \text{for all } t \geq 0 \tag{2.11}$$

Exercise 1 Show that for the circuit described by Eq. (2.2), the network function relating the output voltage $V_0(s)$ to the input current $I_s(s)$ is given by Eq. (2.3).

Example In order to obtain the voltage gain V_2/V_1 of the op-amp circuit shown in Fig. 2.1 we write the two op-amp branch equations and the node equations for nodes ③ and ④:

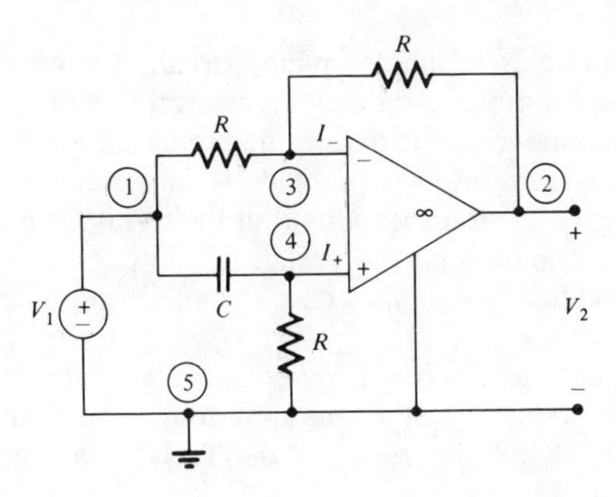

Figure 2.1 Example of a circuit with a transmission zero.

$$I_+ = I_- = 0$$

$$E_3 - E_4 = 0$$

Node ③: $$-GV_2 + 2GE_3 - GV_1 = 0$$

Node ④: $$(G + sC)E_3 - sCV_1 = 0$$

where in the last equation we used $G = 1/R$ and the fact that $E_4 = E_3$. Eliminating E_3 from the last two equations we obtain

$$(G + sC)V_2 = -(G - sC)V_1 \tag{2.12}$$

Consequently, the desired network function is

$$H(s) = \frac{V_2}{V_1} = -\frac{G - sC}{G + sC}$$

Note that $H(s)$ has a zero at G/C and a pole at $-G/C$. Consequently,

$$|H(j\omega)| = 1 \qquad \text{for all real } \omega\text{'s}$$

Such a circuit is called *an all-pass circuit* because it passes all sinusoidal signals for *any* frequency without any attenuation whatsoever.

Exercises

1. Calculate the phase of $H(j\omega)$. [*Answer*: $180° - 2\tan^{-1}(\omega RC)$.]
2. To illustrate the blocking property of the zero, let $G = 1$ S and $C = 1$ F in the circuit of Fig. 2.1.
(*a*) Obtain from Eq. (2.12) the differential equation relating the response $v_2(t)$ to the input $v_1(t)$. (*Answer*: $\dot{v}_2 + v_2 = \dot{v}_1 - v_1$.)
(*b*) If $v_1(t) = 1(t)\,e^t$, obtain the appropriate initial condition $v_2(0-)$ such that $v_2(t) = 0$ for all $t > 0$.

3 NYQUIST CRITERION

In this section we use a simple op-amp circuit to motivate the Nyquist criterion. We then state and prove an elementary form of the Nyquist criterion. In the following section we use it to consider some new circuit problems.

The Nyquist criterion is particularly important for two reasons: first, not only does it give a test for stability, but the test is formulated in such a way that suggestions for improving the design are apparent; second, the test relies on easily obtainable experimental data, namely, sinusoidal steady-state measurements.

Recall that a *uniquely solvable linear time-invariant circuit* $\mathcal{N}$ is said to be *exponentially stable* iff all its natural frequencies have negative real parts. {Equivalently iff all the zeros of $\det[\mathbf{T}(s)]$ or of $\det[\mathbf{P}(s)]$ are in the open

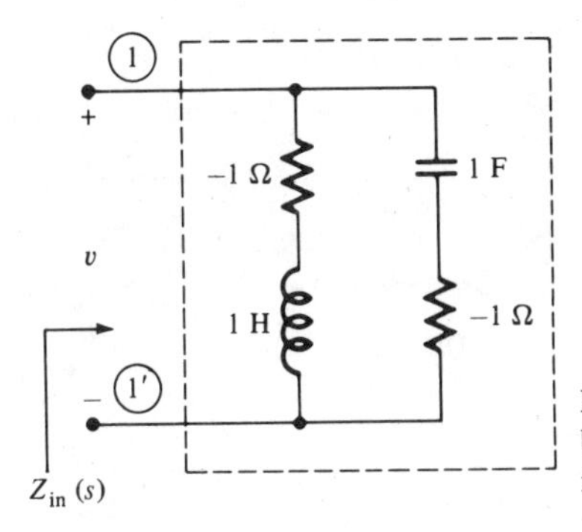

Figure 3.1 Circuit used to demonstrate the difference between "the circuit $\mathcal{N}$ is exp. stable" and "the driving-point impedance $Z_{in}(s)$ is exp. stable."

LHP.}[4] We shall say that a *rational network function* $H(s) = n(s)/d(s)$ is *exponentially stable* (abbreviated exp. stable) iff (a) it is bounded at infinity and (b) H has all its poles in the open LHP. Note that (a) is equivalent to "degree of $n \leq$ degree of d"; and, assuming that the polynomials n and d are coprime, (b) is equivalent to "the zeros of $d(s)$ have negative real parts." Recall also that all zeros of the polynomial $d(s)$ are natural frequencies of the circuit $\mathcal{N}$.

Exercise Consider the circuit $\mathcal{N}$ shown in Fig. 3.1.
(a) Find all the natural frequencies of $\mathcal{N}$ with port ① ①′ open-circuited.
(b) Calculate $Z_{in}(s)$, and show that Z_{in} is exponentially stable.
(c) Discuss your answers to (a) and (b).
(d) Apply a current source $i_s(t) = 1(t)$ across terminals ①, ①′. Show that the *zero-state* response $v(t) = -1(t)$. Show that the unstable mode e^t is *observable* as a voltage across the capacitor, but is *not observable* across the series combination of the capacitor and resistor.

The point of this exercise is to emphasize the fact that $Z_{in}(s)$ is exponentially stable—i.e., the driving-point impedance of the *one-port* shown in Fig. 3.1 is exp. stable—does not imply that the *circuit* shown in Fig. 3.1 is exp. stable: All the poles of Z_{in} are in the open LHP, but some natural frequencies of $\mathcal{N}$ are in the open RHP.

3.1 Example

Consider the op-amp circuit shown in Fig. 3.2a. Let us use the ideal op-amp model and assume that the op amp is operating in its linear range. Thus we have a linear time-invariant circuit and

$$E_2(s) = -\frac{R_f}{R_1} E_s(s) \tag{3.1}$$

where $E_2(s)$ is the Laplace transform of the *zero-state* response to E_s.

[4] LHP (RHP) denotes the left-half plane (right-half plane, respectively). Open LHP denotes all points of the complex plane $\sigma + j\omega$ such that $\sigma < 0$.

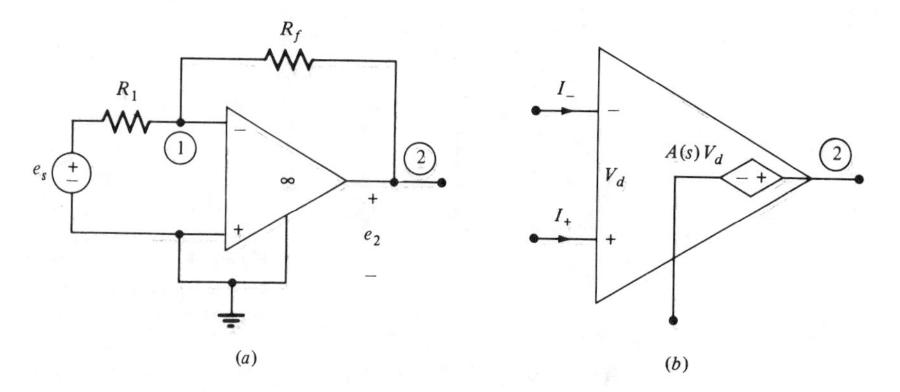

Figure 3.2 (*a*) The circuit under consideration. (*b*) The more sophisticated op-amp model used in the study of stability.

Suppose that we introduce a high frequency model for the op amp, namely, one that models the decreasing op-amp gain as the frequency ω increases. As before we assume for the op amp that $i_- = i_+ = 0$, but we assume a finite frequency-dependent gain for the VCVS (see Fig. 3.2*b*): More precisely, we assume that

$$A(s) = A_0 \frac{p_1' p_2' p_3'}{(s + p_1')(s + p_2')(s + p_3')} \tag{3.2}$$

where A_0 is *large* and *positive* (typically about 10^5), and the p_i''s are *real* and *positive*. Thus the poles of $A(s)$ are located at $-p_1', -p_2', -p_3'$ on the negative real axis: Thus the transfer function $A(s)$ is exponentially stable.

Let us redraw the circuit of Fig. 3.2(*a*) using the model of Fig. 3.2(*b*). We obtain the circuit shown in Fig. 3.3. Using Laplace transforms and assuming zero initial conditions, we write the node equation for node ① as

$$\frac{1}{R_1}[E_1(s) - E_s(s)] + \{E_1(s) - [-A(s)\, E_1(s)]\}\frac{1}{R_f} = 0 \tag{3.3}$$

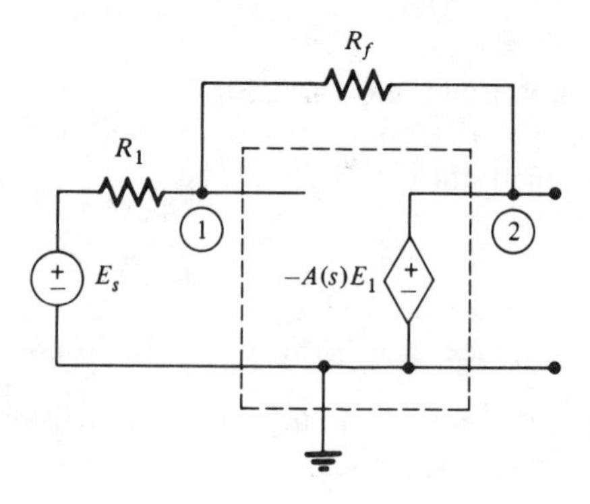

Figure 3.3 More sophisticated model for the circuit of Fig. 3.2.

Multiplying both sides by $R_1R_f/(R_1 + R_f)$, we obtain

$$\left[1 + \frac{R_1}{R_1 + R_f} A(s)\right] E_1(s) = \frac{R_f}{R_1 + R_f} E_s(s) \tag{3.4}$$

Now $E_2(s) = -A(s)\,E_1(s)$; hence the voltage gain is

$$E_2(s) = -\frac{R_f}{R_1} \frac{[R_1/(R_1 + R_f)]A(s)}{1 + [R_1/(R_1 + R_f)]A(s)} E_s(s) \tag{3.5}$$

As a check: let $A \to \infty$, and Eq. (3.5) reduces to Eq. (3.1) as it should. Let

$$k \triangleq \frac{R_1}{R_1 + R_f} \tag{3.6}$$

This constant factor k may be adjusted by changing either R_1 or R_f. In fact we can show that, in some cases, if k is too large, the circuit becomes unstable (because some natural frequencies are in the open right-half plane)!

Equation (3.5) may be rewritten as

$$\frac{E_2}{E_s} = -\frac{R_f}{R_1} \frac{k\,A(s)}{1 + kA(s)} = -\frac{R_f}{R_1} \frac{A(s)}{1/k + A(s)} \tag{3.7}$$

For future reference we set

$$G(s, k) \triangleq \frac{A(s)}{1/k + A(s)} \tag{3.8}$$

$G(s, k)$ is often written as

$$G(s, k) = \frac{k\,A(s)}{1 + k\,A(s)} \tag{3.9}$$

REMARKS[5]

1. The expression $-k\,A(s)$ may be interpreted as a "*loop gain*." Consider Fig. 3.3, and suppose that we cut the input lead to the op amp just to the right of node ①; we thus create a new node ⓡ to which we apply a test voltage E_t as shown in Fig. 3.4. By inspection we have

$$E_1 = -\frac{R_1}{R_1 + R_f} A(s)\,E_t = -k\,A(s)\,E_t$$

So "$-k\,A(s)$" may be thought of as the "voltage gain around the loop" or more succinctly, the "loop gain."

[5] May be skipped without loss of continuity.

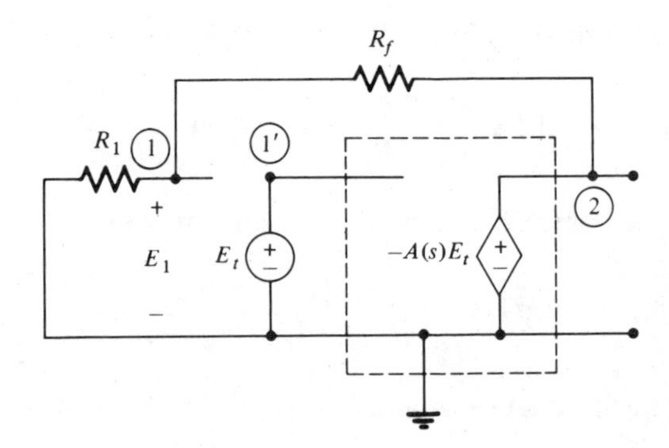

Figure 3.4 Circuit diagram used to illustrate the concept of "loop gain."

2. Equation (3.9) is very important because it describes the behavior of many engineering systems. We derived Eq. (3.9) by calculating the voltage gain of the op-amp circuit of Fig. 3.3. In later courses, you'll learn that this equation characterizes feedback amplifiers and control systems (for example the control system that points the radar antenna toward the plane, steers the robot arm to screw a nut in an automatic assembly machine, keeps constant the output voltage of a power supply in spite of large changes in the load, etc.). In communication and control applications, one draws a *block diagram* as shown in Fig. 3.5: The scalar variable $u(t)$ is called the *input*; $y(t)$ is called the *output*. The + and − signs indicate that $e(t) = u(t) - y(t)$, hence $e(t)$ is called the *error*: It measures the difference between the desired input and the actual output. Roughly speaking, the control system tries at all times to keep the output $y(t)$ close to the input $u(t)$.

The block diagram of Fig. 3.5*a* shows that the error $e(t)$ is fed into a "constant gain block" k whose output is $k\,e(t)$. $P(s)$ denotes the transfer function[6] relating the input $k\,e(t)$ to the output $y(t)$; thus—considering throughout zero-state responses—$Y(s) = P(s)\,k\,E(s)$, where $Y(s)$ denotes the Laplace transform of $y(t)$.

To obtain the relation between the output $Y(s)$ and the system input $U(s)$, we start by writing the summing node equation

$$E(s) = U(s) - Y(s) = U(s) - k\,P(s)\,E(s)$$

equivalently,

$$E(s) = \frac{1}{1 + k\,P(s)}\,U(s) \tag{3.10}$$

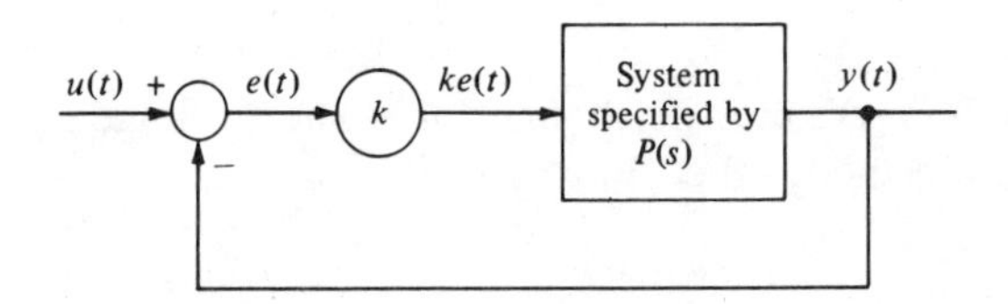

Figure 3.5 Elementary block diagram of a control system. Its closed-loop transfer function is given by Eq. (3.11).

[6] In control, the term "transfer function" is used instead of "network function."

Recalling that $Y(s) = P(s)\, k\, E(s)$, we obtain, using (3.10),

$$Y(s) = \frac{k\,P(s)}{1 + k\,P(s)}\,U(s) \tag{3.11}$$

This equation is precisely of the same form as Eq. (3.9).

3.2 Stability Analysis

We want to determine all the values of the real number k for which the network function

$$G(s, k) = \frac{A(s)}{1/k + A(s)} = \frac{k\,A(s)}{1 + k\,A(s)} \tag{3.12}$$

is exp. stable.

Example Let $A(s)$ be given by Eq. (3.2) with $A_0 = 3 \times 10^5$, $p_1' = 30$, $p_2' = p_3' = 10^5$ rad/s. Writing $G(s, k)$ as a ratio of polynomials, the denominator polynomial is found to be

$$d_G(s, k) = (s + p_1')(s + p_2')(s + p_3') + kA_0 p_1' p_2' p_3'$$

For k very small—more precisely, for $k \ll 10^{-7}$—the last term

$$kA_0 p_1' p_2' p_3' \triangleq k\alpha^3$$

is negligible compared to $p_1' p_2' p_3'$. Consequently, for such small k's,

$$d_G(s, k) \cong (s + p_1')(s + p_2')(s + p_3')$$

Hence, for such small k's, the poles of $G(s, k)$ are close to $-p_1'$, $-p_2'$, and $-p_3'$, the poles of $A(s)$: Hence $G(s, k)$ is exp. stable. Figure 3.6 shows how the poles of $G(s, k)$ vary as k increases from these small values. In particular it shows that increasing k beyond a certain point makes the circuit *unstable*. This phenomenon is very well known to engineers: It occurs in many different fields. If you turn up the gain too high, then the system will become unstable and, for example, go into oscillation or saturation, or burn a component.

Exercise [This is a guided exercise to show why as k increases, the polynomial $d_G(s, k)$ (defined above) has two zeros in the open RHP.] Consider $d_G(s, k)$ and consider values of k such that $k \gg 1$. Let $d_G(s, k) = s^3 + a_1 s^2 + a_2 s + (3 \times 10^{11} + k\alpha^3)$.

(*a*) Show that for k of the order of 1, with a small relative error, the constant term of $d_G(s, k)$ may be taken to be $k\alpha^3$. Verify that $\alpha \cong 4.5 \times 10^5$, $a_1 \cong 2 \times 10^5$, and $a_2 \cong 10^{10}$.

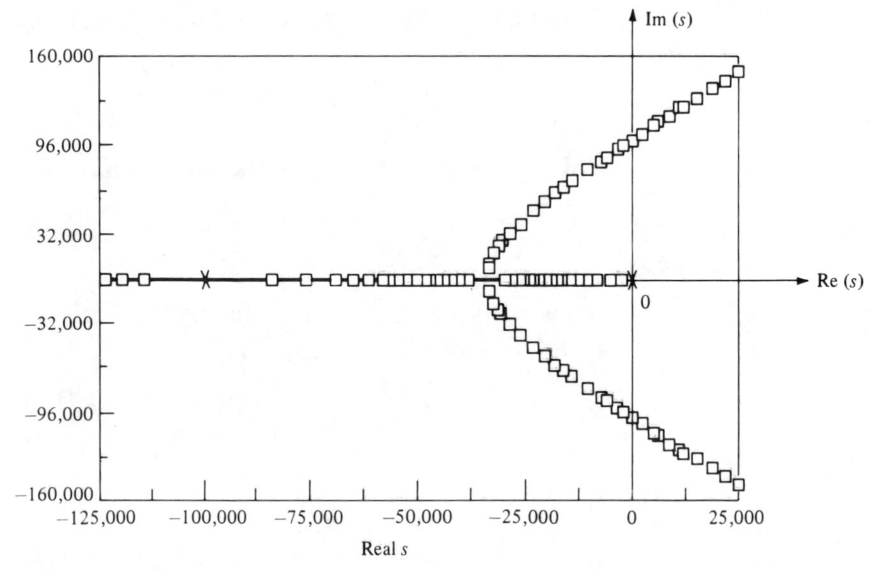

Figure 3.6 The poles of $G(s, k)$ for $k \geq 0$ at $k = 0$, the poles of $G(s, k)$ are indicated by the crosses; at $k = 0.0889$, the three poles of $G(s, k)$ are $0 \pm j100{,}030$ and $-200{,}030$. For larger k's, $G(s, k)$ is unstable: it has two poles in the open right-half plane.

(*b*) Write $d_G(s, k) = 0$ as follows:

$$s^3 = -k\alpha^3 - (a_1 s^2 + a_2 s - 3 \times 10^{11})$$

and show that if, for k very large, you neglect the terms in the parentheses, there are three zeros

$$s_\ell = \alpha k^{1/3} \exp j\left(\frac{\pi}{3} + \ell\,\frac{2\pi}{3}\right) \qquad \ell = 0, 1, 2$$

This shows that for $k \gg 1$, the polynomial $d_G(s, k)$ has two zeros in the open RHP. Hence $G(s, k)$ is unstable.

Since the results of our analysis will be quite general let us broaden the class of $A(s)$ under consideration: We assume that

$$\text{(A1)}\quad A(s) = \frac{n_A(s)}{d_A(s)} \qquad \text{where} \quad \begin{cases} n_A(s),\ d_A(s) \text{ are polynomials} \\ \text{with } \textit{real} \text{ coefficients;} \\ n_A \text{ and } d_A \text{ are } \textit{coprime}. \end{cases} \tag{3.13}$$

(A2) $A(s) \to 0$ as $|s| \to \infty$ (3.14)
(equivalently, the degree of polynomial $d_A >$ degree of polynomial n_A)

(A3) $A(s)$ is *exp. stable* (3.15)
[equivalently, $A(s)$ has no *poles* in the closed RHP, equivalently, all the *zeros* of $d_A(s)$ are in the open LHP]

Substituting Eq. (3.13) into Eq. (3.12) we see that

$$G(s, k) = \frac{k\, n_A(s)}{d_A(s) + k\, n_A(s)} = \frac{n_A(s)/d_A(s)}{1/k + n_A(s)/d_A(s)} \tag{3.16}$$

Note that if $|k| \ll 1$, then the zeros of $d_A(s) + k\, n_A(s)$ are very close to those of $d_A(s)$, and consequently, for $|k|$ very small, $G(s, k)$ is exp. stable.

Now, note that (i) since by assumption polynomials n_A and d_A are coprime, for all constants k, the polynomials n_A and $d_A + kn_A$ are also coprime; consequently, by Eq. (3.16), the poles of $G(s, k)$ are precisely the zeros of the polynomial $d_A + kn_A$. (ii) By assumption (3.15), $d_A(s) \neq 0$ for all s in the closed RHP, and consequently, for all $k \neq 0$, $1/k + n_A(s)/d_A(s)$ has no poles in the closed right-half plane and has the same zeros in the closed RHP as the polynomial $d_A + kn_A$. (iii) By assumption (3.14), $G(s, k) \to 0$ as $s \to \infty$.

With these observations in mind we state our conclusions in the following form:

Property: Given that assumptions (3.13), (3.14), and (3.15) hold, the voltage gain $G(s, k)$ is *exp. stable*

$$\Leftrightarrow \quad d_A(s) + k\, n_A(s) \text{ has } \textit{no zeros} \text{ in the closed RHP} \tag{3.17}$$

$$\Leftrightarrow \quad \frac{1}{k} + \frac{n_A(s)}{d_A(s)} \text{ has } \textit{no zeros} \text{ in the closed RHP} \tag{3.18}$$

$$\Leftrightarrow \quad \frac{1}{k} + A(s) \text{ has } \textit{no zeros} \text{ in the closed RHP} \tag{3.19}$$

So the exponential stability of $G(s, k)$ is equivalent to guaranteeing that the rational function $1/k + n_A(s)/d_A(s) = 1/k + A(s)$ has *no zeros in the closed RHP*. This is the idea that we are going to exploit.

3.3 Graphic Interpretation and the Key Theorem

Let assumptions (3.13), (3.14), and (3.15) hold, and let $k \neq 0$. Let us reason as follows:

(*a*) Consider Fig. 3.7*a* where the contour *OMNPO* is drawn in the s plane. It is customary to call it the *D contour* and to traverse it in the order *O–M–N–P–O*, i.e., in a *clockwise* manner. The radius R of the half circle *MNP* is taken so large that, by assumption (3.14), for all s on the half circle *MNP*, $A(s) \cong 0$: consequently

$$\frac{1}{k} + A(s) \cong \frac{1}{k} \qquad \text{for all } s \text{ on the half circle } MNP \tag{3.20}$$

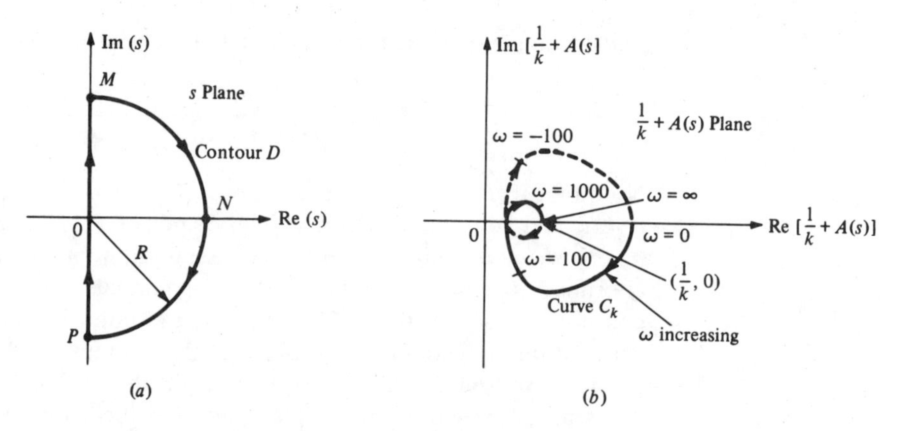

Figure 3.7 (*a*) The contour D in the s plane. (*b*) The curve C_k which is the image of the contour D mapped by $1/k + A(s)$.

and, for all $|s| > R$ with s in the RHP, $1/k + A(s) \neq 0$ [because $1/k + A(s)$ is very close to the *nonzero* number $1/k$].

(*b*) For each successive s on the contour D, let us calculate the complex number $1/k + A(s)$; next, let us plot this complex number in the $[1/k + A(s)]$ plane of Fig. 3.7*b*. Proceeding in this way, the D contour of the s plane is mapped into a *closed oriented* curve C_k in the $[1/k + A(s)]$ plane.[7]

Note that, by assumption (3.14), $A(s) \to 0$ as $|s| \to \infty$, hence the image of the half circle MNP of radius R reduces, for large R, to a small arc near the point $(1/k, 0)$ on the real axis.

We now state the key theorem:

Key theorem Let assumptions (3.13), (3.14), and (3.15) hold, and let $k \neq 0$; then

$$G(s, k) = \frac{A(s)}{(1/k) + A(s)} \quad \text{is exp. stable} \tag{3.21}$$

$$\Leftrightarrow \qquad \frac{1}{k} + A(s) \text{ has no zeros in the closed RHP} \tag{3.22}$$

$\Leftrightarrow$ The closed oriented curve C_k *does not go through nor encircle the origin* (0, 0) of the $\left[\dfrac{1}{k} + A(s)\right]$ plane (3.23)

Proof We know by the property above that condition (3.21) is equivalent to condition (3.22). We prove now that condition (3.22) is equivalent to condition (3.23) by establishing two claims.

[7] This curve is labeled C_k to remind us that it depends on k.

Claim 1 The closed oriented curve C_k goes through the origin (3.24)

$$\Leftrightarrow \qquad \frac{1}{k} + A(s) \text{ has at least one } j\omega \text{ axis zero} \tag{3.25}$$

PROOF OF CLAIM 1

$\Leftarrow$ Call $j\omega_1$ one of the $j\omega$ axis zeros of $1/k + A(s)$. Then as s moves up the D contour and reaches $j\omega_1$, the corresponding point $1/k + A(s)$ traces a curve that goes through the origin $(0, 0)$ since $1/k + A(j\omega_1) = 0$. Thus we have shown that condition (3.25) implies condition (3.24).

$\Rightarrow$ C_k goes through the origin, by assumption. Thus, for some s on the contour D, $1/k + A(s) = 0$. Now, for all s on MNP, the half circle of radius R, $A(s) \cong 0$ by assumption (3.14), and hence $1/k + A(s) \neq 0$ for all such s. Therefore, the point s on the contour D for which $1/k + A(s) = 0$ must be on the $j\omega$ axis, i.e., there is some ω_1 such that $1/k + A(j\omega_1) = 0$. Thus we have shown that condition (3.24) implies condition (3.25).

Claim 2 Suppose now that C_k does not go through the origin; then we claim that

$$C_k \text{ encircles the origin}$$

$$\Leftrightarrow \quad \frac{1}{k} + A(s) \text{ has one or more zeros in the open RHP.}$$

We prove this claim in a number of small steps:

(*a*) By assumption, for all real ω's, $1/k + A(j\omega) \neq 0$; furthermore, by choosing the radius R large enough, $1/k + A(s) \neq 0$, for all s on D. Consequently, for all s on D, $\measuredangle[1/k + A(s)]$ is well-defined and varies continuously as s moves around the contour D.

(*b*) Simple geometry shows that the curve C_k encircles the origin if and only if there is a net change in the phase $\measuredangle[1/k + A(s)]$ as s goes around the closed contour D.

(*c*) Let $p_1, p_2, \ldots$ and $z_1, z_2, \ldots$ be the poles and the zeros of $1/k + A(s)$, then

$$\frac{1}{k} + A(s) = k_0 \frac{\prod_i (s - z_i)}{\prod_j (s - p_j)} \tag{3.26}$$

and, using elementary properties of complex numbers, we obtain

$$\measuredangle\left[\frac{1}{k} + A(s)\right] = \measuredangle k_0 + \sum_i \measuredangle(s - z_i) - \sum_j \measuredangle(s - p_j) \tag{3.27}$$

A typical example of a pole-zero constellation is shown in Fig. 3.8.

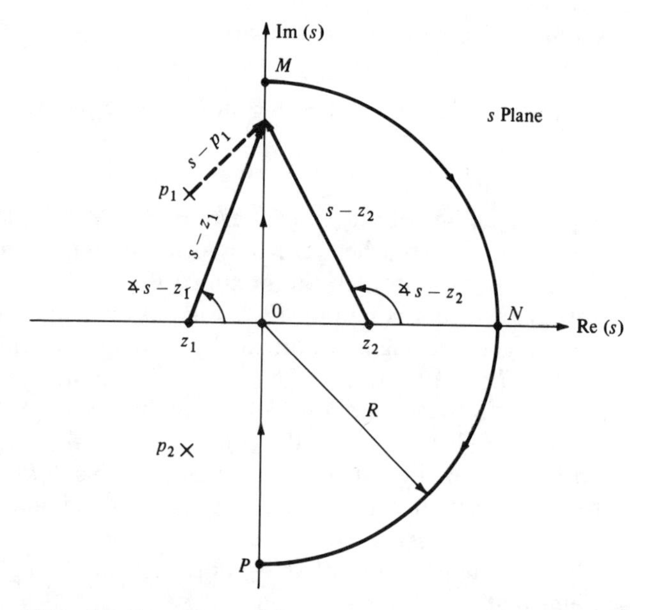

Figure 3.8 Shows that only the open-RHP zeros of $1/k + A(s)$ contribute a *net change* of phase as the point s goes around the contour D.

(*d*) By assumption (3.15), for all j's, $\text{Re}(p_j) < 0$. Hence, as s goes around the contour D, the net change in each of the angles $\measuredangle(s - p_j)$ is *zero*. (See Fig. 3.8.)

(*e*) If some zero, say z_1, is in the open LHP [i.e., $\text{Re}(z_1) < 0$], then, as s goes around the contour D, the net change in the angle $\measuredangle(s - z_1)$ is *zero*.

(*f*) If some zero, say z_2, is in the open RHP [i.e., $\text{Re}(z_2) > 0$], then, as s goes around the contour D, the net change in $\measuredangle(s - z_2)$ is -2π rad $= -360°$; this corresponds to one turn *clockwise* around the origin, and, in terms of C_k, to one *clockwise* encirclement of the origin. (See Fig. 3.9.)

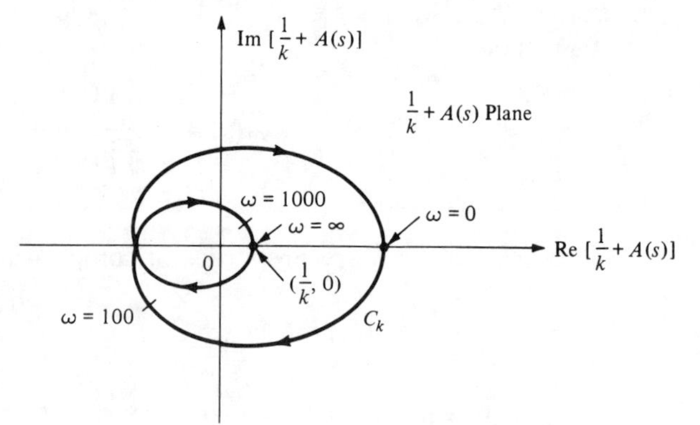

Figure 3.9 Here the curve C_k encircles the origin twice (clockwise).

(*g*) From (*d*), (*e*), and (*f*), we see that only open-RHP zeros of $1/k + A(s)$ contribute to a net change of phase and hence cause an encirclement of the origin by C_k. Hence Claim 2 is established.

Back to the theorem Since Claim 1 and Claim 2 hold, we see that

$$\left.\begin{array}{ll} & C_k \text{ does not go through nor encircle the origin} \\ \Leftrightarrow & \dfrac{1}{k} + A(s) \text{ has no zeros in the closed RHP} \end{array}\right\} \qquad (3.28)$$

Thus the equivalence between conditions (3.22) and (3.23) is established and the key theorem is proved. ■

REMARKS

1. The reasoning above shows that, provided assumptions (3.13), (3.14), and (3.15) hold, the closed oriented curve C_k encircles the origin n_z times in the *clockwise* direction if and only if $1/k + A(s)$ has n_z zeros in the open RHP. (In case of multiple zeros, each zero is counted according to its multiplicity.)
2. A more detailed reasoning would show that if $A(s)$ had n_p poles in the open RHP, then $G(s, k)$ is exponentially stable if and only if C_k encircles the origin n_p times *counterclockwise*. A careful study of this more general situation, including the case where $A(s)$ has poles on the $j\omega$ axis, is an important part of courses on feedback systems.

3.4 Nyquist Criterion

The key theorem above gives a graphic technique for testing whether $1/k + A(s)$ has zeros in the closed RHP. In design problems, the constant k is a design parameter, and feedback theory teaches that k should be taken as large as possible. Thus, the designer wants to consider many values of k, but the key theorem suggests plotting a curve C_k for each k!

Any astute student will suggest that (*a*) we map the contour D by $A(s)$—call the resulting curve C_A because it depends only on the network function $A(s)$; (*b*) we note that C_A is obtained from C_k by moving the curve C_k by $1/k$ to the left. Consequently

$$\begin{array}{ll} & C_k \text{ goes through (encircles, respectively) the origin } (0, 0) \\ \Leftrightarrow & C_A \text{ goes through (encircles, respectively) the point } \left(-\dfrac{1}{k}, 0\right) \end{array} \qquad (3.29)$$

C_A is, by definition, the image of the contour D by $A(s)$ and is called *the Nyquist plot of* $A(s)$. The contour D and the Nyquist plot of $A(s)$ are shown in Fig. 3.10.

With this in mind we can state an elementary form of the Nyquist criterion. The truth of the criterion follows immediately from the key theorem above and statement (3.29).

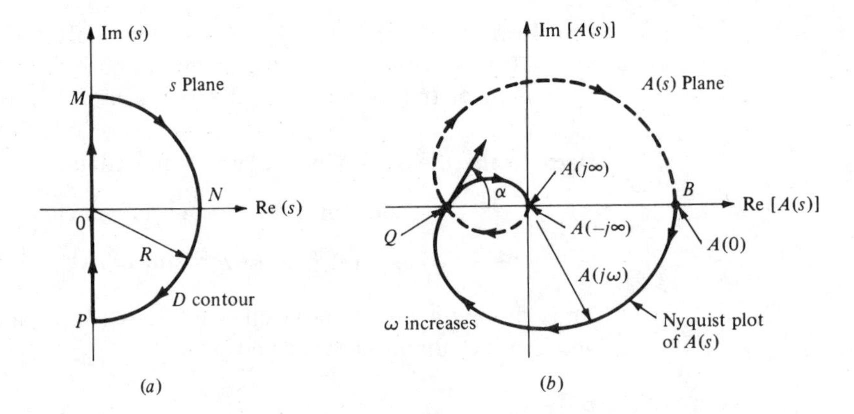

Figure 3.10 (*a*) The *D* contour: note the half circle of radius *R*, where *R* is taken very large. (*b*) The Nyquist plot of $A(s)$.

Nyquist criterion: Given a network function of the form

$$G(s, k) = \frac{A(s)}{(1/k) + A(s)} \qquad \text{with } k \neq 0 \tag{3.30}$$

where

$$A(s) = \frac{n_A(s)}{d_A(s)} \qquad \text{(a ratio of two } \textit{coprime} \text{ polynomials)} \tag{3.31}$$

such that

$$A(s) \to 0 \text{ as } |s| \to \infty \tag{3.32}$$

and such that

$A(s)$ is *exponentially stable* [equivalently, in view of statement (3.31), all the zeros of $d_A(s)$ are in the open LHP] (3.33)

then

$G(s, k)$ is exp. stable
$\Leftrightarrow$ the Nyquist plot of $A(s)$ [i.e., the curve obtained by mapping the *D* contour by $A(s)$ on the $A(s)$ plane] *does not go through nor encircle the point* $\left(-\dfrac{1}{k}, 0\right)$.

REMARKS

1. Let $A(0)=5$; equivalently let the coordinates of the point B in Fig. 3.10*b* be $(5, 0)$. Let the coordinates of the point Q be $(-3, 0)$. Then, for the data of Fig. 3.9*b*, $G(s)$ is exponentially stable if and only if

$$\text{either} \qquad -\frac{1}{k} < -3 \qquad \text{i.e., } k < \tfrac{1}{3}$$

$$\text{or} \qquad -\frac{1}{k} > 5 \qquad \text{i.e., } k > -\tfrac{1}{5} = -0.2$$

So $G(s, k)$ is exp. stable if and only if $-0.2 < k < 0.333$. $G(s, k)$ has two poles in the RHP if $k > 0.333$, and $G(s)$ has one pole in the RHP if $k < -0.2$.

2. Since, for all real ω, $A(-j\omega) = \overline{A(j\omega)}$, there is no need to plot the Nyquist diagram for $\omega < 0$. Of course, we must take the whole diagram from $\omega = -\infty$ to $\omega = \infty$ into account as we count the number of encirclements.

3. Obviously one can obtain the Nyquist plot of $A(s)$ by calculation, but it is extremely important to note that one can also obtain the Nyquist plot *directly by measurements*: By measuring the loop gain. *This is very important because the resulting plot is free from modeling errors.*

4. As ω increases from zero, the values of $A(j\omega)$ go through a considerable dynamical range: 10^4 or 10^5 is not unusual. For this reason rather than plot the point specified by the polar coordinates $[|A(j\omega)|, \measuredangle A(j\omega)]$, one plots $[\log|A(j\omega)|, \measuredangle A(j\omega)]$; or one plots *Bode diagrams*, namely, $20 \log|A(j\omega)|$ versus $\log \omega$ and $\measuredangle A(j\omega)$ versus $\log \omega$. Clearly this stretching of the radial distance does not affect encirclements.

5. We assumed that $A(s) \to 0$ as $|s| \to \infty$, because it is quite a realistic assumption when one considers amplifiers or control systems. A careful examination of the proof of the key theorem and that of the Nyquist criterion will show that the only fact that was used was that as $|s| \to \infty$, $A(s)$ tended to a constant (zero in the discussion above) so that the image of the half circle of radius R of the D contour under the map $A(s)$ reduced in the limit to the point $(0, 0)$. Consequently, the key theorem and the Nyquist criterion still hold if as $|s| \to \infty$, $A(s)$ tends to a *constant* $A(\infty)$. Of course the Nyquist plot will then end up at $A(\infty)$ rather than at the origin as in Fig. 3.10*b*.

3.5 The Nyquist Plot and Root Loci

We now show that the Nyquist plot gives information on the location of some poles of $G(s, k)$.

We consider some $A(s)$ satisfying the three assumptions (3.30) to (3.32) of the Nyquist criterion and the corresponding $G(s, k)$ given by Eq. (3.30). Assume that the Nyquist plot of $A(s)$ is that shown in Fig. 3.10*b*.

For k small and positive, $G(s, k)$ is *exp. stable*; however, as k increases, the point $(-1/k, 0)$ is encircled twice (clockwise) by the Nyquist plot. Hence $G(s, k)$ is *unstable*. (In fact it has *two* unstable poles, see Remark 1 at the end of Sec. 3.3, above.)

Equations (3.17) and (3.18) show that the poles of $G(s, k)$ are precisely the zeros of the polynomial $d_A(s) + k\, n_A(s)$ or, equivalently, precisely the zeros of the rational function

$$f(s, k) \triangleq \frac{1}{k} + A(s) \tag{3.34}$$

Consequently, (*a*) for k small and positive, all the zeros of $d_A(s) + k\, n_A(s)$ [hence all the poles of $G(s, k)$] are close to the zeros of $d_A(s)$ [hence are in the open LPH by assumption (3.33)]; (*b*) as k reaches a critical value, say k_c, the point $(-1/k_c, 0)$ is on the Nyquist plot, say at $A(j\omega_c)$;[8] as k increases beyond k_c, the point $(-1/k, 0)$ is encircled twice clockwise, and hence $G(s, k)$ has two poles in the *open RHP*.

Now it can be shown that assumption (3.31) implies that, as long as the zeros of the rational function $f(s, k)$ are simple, each zero of $f(s, k)$ is an analytic function of k. So if we call s_1 the upper-half plane zero of $f(s, k)$ that crosses into the right-half plane when k increases through k_c, the function $s_1(k)$ can be expanded in Taylor series. Consequently, for k close to k_c, we have

$$f(s_1(k), k) = 0 \qquad \text{for } all\ k \text{ close to } k_c \tag{3.35}$$

We also have $s_1(k_c) = j\omega_c$ since, by definition of k_c and $j\omega_c$,

$$f(j\omega_c, k_c) = \frac{1}{k_c} + A(j\omega_c) = 0$$

Equation (3.35) implies that the derivative of the left-hand side (LHS) with respect to k is equal to zero for all k close to k_c; in particular, it is zero at $k = k_c$. Using the chain rule, we obtain

$$\left.\frac{\partial f}{\partial s}\right|_{(j\omega_c, k_c)} \left.\frac{ds_1}{dk}\right|_{k_c} + \left.\frac{\partial f}{\partial k}\right|_{(j\omega_c, k_c)} = 0 \tag{3.36}$$

Now, recalling Eq. (3.34), we have successively

$$\left.\frac{\partial f}{\partial s}\right|_{(j\omega_c, k_c)} = \left.\frac{\partial A}{\partial s}\right|_{(j\omega_c, k_c)} = \left.\frac{\partial A(j\omega)}{\partial (j\omega)}\right|_{j\omega_c} = \frac{1}{j} \left.\frac{\partial A(j\omega)}{\partial \omega}\right|_{\omega_c}$$

Note that the Nyquist plot is the curve giving the complex number $A(j\omega)$ in terms of ω. So $\left.\dfrac{\partial A(j\omega)}{\partial \omega}\right|_{\omega_c} \triangleq a$ is the "velocity" vector at $\omega = \omega_c$, i.e., at the point Q on the Nyquist plot of Fig. 3.10*b*; a is a complex number, and $\measuredangle a$ is the angle α shown in Fig. 3.10*b*.

[8] Equivalently, the point Q in Fig. 3.10*b* is represented by the complex number $A(j\omega_c)$.

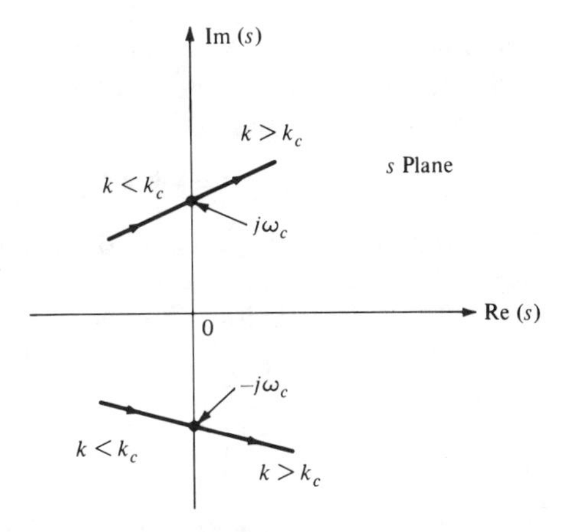

Figure 3.11 The trajectory of two poles of $G(s, k)$ as the parameter k increases through the critical value k_c.

Clearly, from the definition of $f(s, k)$, $\left.\dfrac{\partial f}{\partial k}\right|_{k_c} = -\dfrac{1}{k_c^2}$, so, by Eq. (3.36)

$$\left.\frac{ds_1}{dk}\right|_{k_c} = j\,\frac{1}{k_c^2 a} = \frac{e^{j\pi/2}}{k_c^2|a|}\exp(-j\sphericalangle a)$$

$$= \frac{1}{k_c^2 \cdot |a|}\exp\left[j\left(\frac{\pi}{2} - \alpha\right)\right] \qquad (3.37)$$

Thus, if, as in Fig. 3.11, we plot, in the s plane, the trajectory of s_1 as a function of k[9] we have (a) $s_1(k_c) = j\omega_c$ and (b) Eq. (3.37), the tangent to the trajectory of s_1, at the point $j\omega_c$, makes an angle of $\pi/2 - \alpha$ with the real axis. Thus we have shown that the Nyquist plot gives information on the motion of the poles of $G(s, k)$ that become unstable as k goes through k_c.

3.6 The Nyquist Criterion in Terms of the Bode Plot

In many applications, the magnitude of $A(j\omega)$, $|A(j\omega)|$, and the phase $\sphericalangle A(j\omega)$ *decrease monotonically* as ω increases. This is the case for the op amp. As a result of this simple behavior of $A(j\omega)$ it is easy to draw the conclusions of the Nyquist criterion on the basis of the *Bode plot* of $A(j\omega)$.

Problem Given that $A(s)$ is exponentially stable and that as $\omega \to \infty$, $|A(j\omega)|$ and $\sphericalangle A(j\omega)$ decrease monotonically to zero, determine conditions under which

$$G(s, k) = \frac{A(s)}{(1/k) + A(s)} \qquad \text{with } k \text{ positive}$$

is exponentially stable.

[9] The trajectory of s_1 as a function of k is usually called a *root locus*.

Typically the *Nyquist plot* of such a gain $A(s)$ has the general shape shown in Fig. 3.12*a*. Note that ω_k is defined as the frequency such that $|A(j\omega_k)| = 1/k$ and that ω_{180} is defined by

$$\measuredangle A(j\omega_{180}) = -180° \tag{3.38}$$

Figure 3.12*b* and *c* show the Bode plots of $A(j\omega)$, namely, the graph of $20 \log|A(j\omega)|$ versus $\log \omega$ and $\measuredangle A(j\omega)$ versus $\log \omega$. In view of the simple behavior of $A(j\omega)$ versus ω, it is obvious that we can state the following:

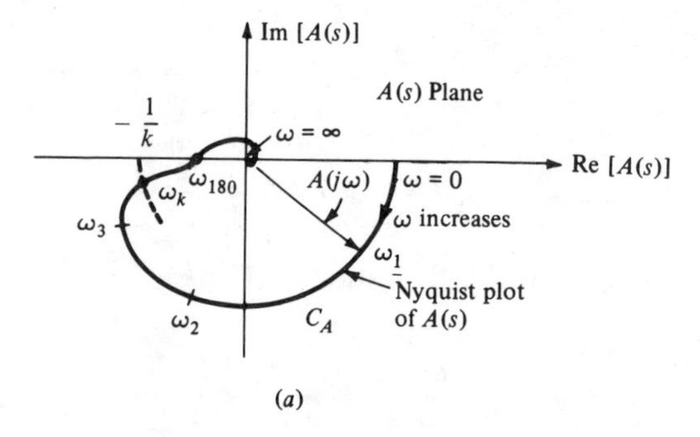

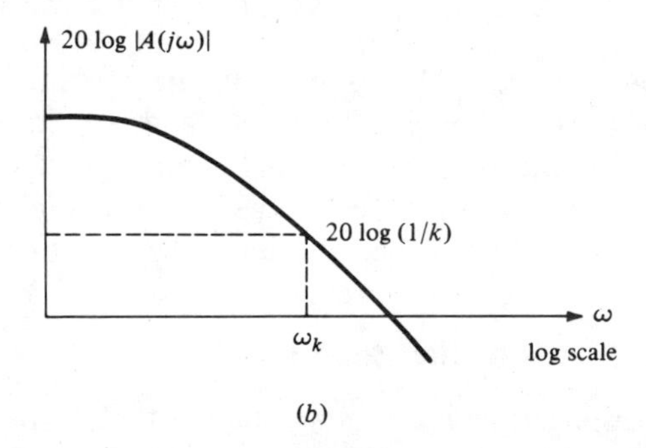

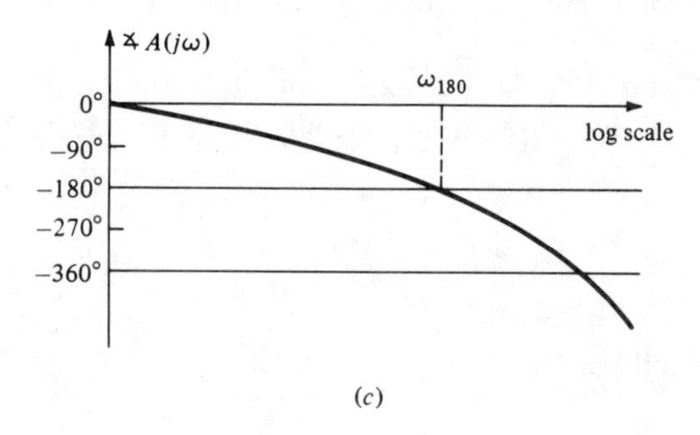

Figure 3.12 The Nyquist plot C_A and the Bode plots of $A(s)$: the definition of the frequency ω_{180} is shown graphically in (c); the frequency ω_k is defined in (b).

Nyquist criterion in terms of Bode plot: Given

$$G(s, k) \triangleq \frac{A(s)}{(1/k) + A(s)} \quad \text{with } k > 0$$

where $A(s) = \dfrac{n_A(s)}{d_A(s)}$ (n_A and d_A are coprime polynomials) (3.39)

$A(j\omega) \to 0$ as $\omega \to \infty$ (3.40)

$|A(j\omega)|$ and $\measuredangle A(j\omega)$ *decrease monotonically* as ω increases (3.41)

$A(s)$ is *exponentially stable* (3.42)

call ω_k the frequency such that $|A(j\omega_k)| = 1/k$, and let ω_{180} be defined by $\measuredangle A(j\omega_{180}) = -180°$; then

$$G(s, k) \text{ is exponentially stable} \Leftrightarrow \omega_k < \omega_{180}$$

This criterion is the basis of the compensation techniques used in the design of op amps and of many circuits in measuring apparatus.

Exercises

1. Consider the op-amp circuit of Fig. 3.2*a* (of Sec. 3.1, above) where $A(s)$ is given by Eq. (3.2). Let $A_0 = 3 \times 10^5$, $p_1' = 30$ rad/s, $p_2' = p_3' = 10^5$ rad/s.
(*a*) Show that if $k = R_1/(R_1 + R_f) = 0.1$, the circuit is unstable. [To do this, obtain the Bode plot. (*Hint*: $\omega_{180} \cong 10^5$.)]
(*b*) Determine the range of values of $k > 0$ for which the circuit is stable.
2. [The purpose of this exercise is to emphasize the importance of the requirement that $|A(j\omega)|$ and $\measuredangle A(j\omega)$ decrease monotonically.] Draw a Nyquist plot for some $A(s)$ of your choice where
(*a*) $A(s)$ satisfies assumptions (3.39), (3.40), and (3.42); and
(*b*) for some k of your choice, $\omega_k > \omega_{180}$ and $G(s, k)$ is exp. stable; explain your assertions carefully.

4 OPEN-CIRCUIT STABILITY AND SHORT-CIRCUIT STABILITY OF ONE-PORTS

If a linear time-invariant one-port includes some active elements—e.g., a negative resistor, or some controlled sources—then it may be unstable, for example, when it is open-circuited or when it is short-circuited. In this section we explore this concept of one-port stability and we derive necessary and sufficient conditions for open-circuit stability and for short-circuit stability of one-ports.

4.1 An Illustrative Example

A linear time-invariant circuit made of *passive* elements is necessarily stable (i.e., it does not have any mode that increases exponentially with time). If the circuit includes op amps or controlled sources or negative resistors, then it may become unstable, since these elements may pump energy into the circuit and, possibly, cause instability. Recall that the natural frequencies of a circuit are, in general, properties of the *whole* circuit: Hence if we, say, short out a pair of nodes in a stable circuit, we will change the natural frequencies and, possibly, cause instability in the process. The reverse situation can also occur.

Here is an example where the one-port is *open-circuit unstable* but *short-circuit stable*. Consider the linear time-invariant circuit $\mathcal{N}_0$ shown in Fig. 4.1 where for ease of calculation we normalized most element values to 1: In the circuit $\mathcal{N}_0$, the port ①, ①' is left *open-circuited*. Since $g_{m1} = 1$ S and $g_{m2} = 2$ S, the node equations read

$$\begin{bmatrix} 1 & -2 \\ -1 & 1+s \end{bmatrix} \begin{bmatrix} E_1(s) \\ E_2(s) \end{bmatrix} = \begin{bmatrix} 0 \\ e_2(0-) \end{bmatrix} \tag{4.1}$$

The characteristic polynomial of the circuit $\mathcal{N}_0$ is

$$\det[P(s)] = s - 1 \tag{4.2}$$

The *circuit* $\mathcal{N}_0$ is unstable since it has $+1$ as a natural frequency.[10]

Now if we *short circuit* the input port, then we obtain a *new circuit* which we call $\mathcal{N}_s$. For $\mathcal{N}_s$, $e_1(t) = 0$, for all t, and $E_1(s) = 0$; also the Eqs. (4.1) are no

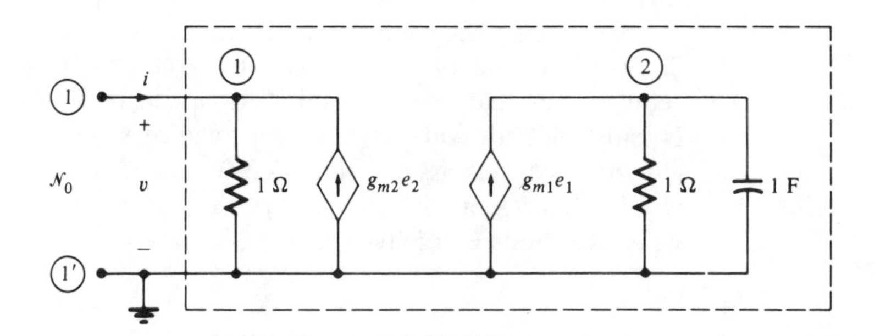

Figure 4.1 Linear time-invariant circuit $\mathcal{N}_0$: with $g_{m2} = 2$ S and $g_{m1} = 1$ S, it is unstable. If we short the input port, we have a new circuit $\mathcal{N}_s$ which is exponentially stable.

[10] Note that the unstable mode proportional to e^t is observable as an exponentially increasing port voltage across ①, ①'.

longer valid. By inspection, it is obvious that the circuit $\mathcal{N}_s$ is exp. stable: Indeed its natural frequency is $s = -1$. Thus Fig. 4.1 shows a one-port that is *open-circuit unstable* but *short-circuit stable*. It is intuitively clear that, by continuity, if we connect between terminals ① and ①′ a sufficiently small linear resistor, the circuit will still be stable, but if the resistor is sufficiently large, the circuit will become unstable.

Exercises

1. Show that the driving-point impedance of the one-port shown in Fig. 4.1 is

$$Z_{\text{in}}(s) = \frac{s+1}{s-1} \tag{4.3}$$

2. Change some of the element values of the circuit of Fig. 4.1 to make it *open-circuit stable* but *short-circuit unstable*.
3. Let N be a one-port made of linear time-invariant elements. Assume that *all its elements are passive*:
(*a*) Use energy arguments to show that its *driving-point impedance* $Z(s)$ cannot have any poles in the open right-half plane, i.e., with a *positive* real part.
(*b*) Show, by example, that such a $Z(s)$ may have poles or zeros on the $j\omega$ axis.
(*c*) Use energy arguments to show that $Z(s)$ cannot have poles of order larger than 1 on the $j\omega$ axis.
(*d*) Use duality to show that the poles and zeros of $Y(s)$ satisfy the properties stated in (*a*), (*b*), and (*c*).
(*e*) Allow now the one-port N to include controlled-sources: construct examples to show that (*a*), (*b*), (*c*), and (*d*) no longer hold.

4.2 Short-Circuit Stability and Open-Circuit Stability Defined

Let us make more precise the concept of *short-circuit stability* when it is applied to a linear time-invariant *one-port*. We start with an important remark.

REMARK ON UNSTABLE HIDDEN MODES Recall the discussion in Sec. 4.4B of Chap. 10: Given a specific network function, say $H(s) = V_0(s)/I_s(s)$, not all the natural frequencies of the linear time-invariant circuit $\mathcal{N}$ will in general be poles of $H(s)$. Some natural frequencies may be canceled out from the ratio of Eq. (4.57) of Chap. 10. We also noted that a natural frequency λ may be canceled either because the input $i_s(\cdot)$ does not *excite* the mode λ or because the mode λ is not *observable* at the output $v_0(\cdot)$, or both. (See the example of Fig. 4.10 in Chap. 10.) If the circuit $\mathcal{N}$ contains active elements (typically controlled sources, op amps, negative resistors, etc.), then some natural frequencies may have the property that $\text{Re}(\lambda) \geq 0$, i.e.,

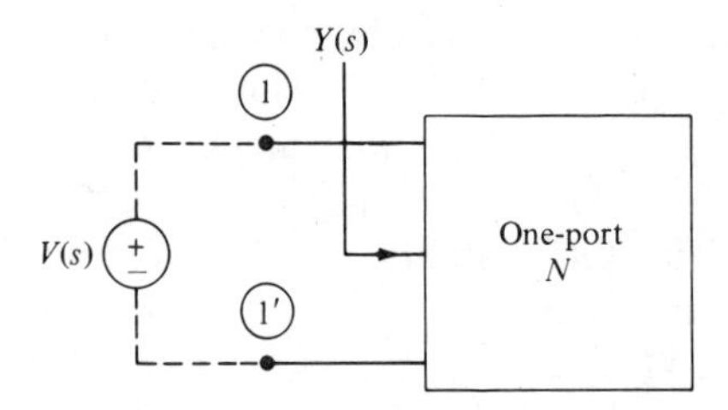

Figure 4.2 The voltage source $V(s)$ is used to test the one-port N for short-circuit stability.

$e^{\lambda t}$ is not exponentially decreasing as $t \to \infty$. Any mode associated with such a natural frequency will be called an *unstable mode*. Thus it may happen that the circuit $\mathcal{N}$ is *unstable* and that some or all the unstable natural frequencies cancel out from the network function $H(s)$:[11] In that case we say that, from the point of view of $H(s) = V_0(s)/I_s(s)$, the circuit $\mathcal{N}$ has *unstable hidden modes*. In that case, the *circuit* $\mathcal{N}$ is *not* exponentially stable—since it has unstable modes—but the *network function* $H(s) = V_0(s)/I_s(s)$ *may* be exponentially stable. This observation leads us to make a very important assumption:

Assumption Whenever a linear time-invariant one-port N is specified by its driving-point impedance $Z(s)$ [or admittance $Y(s)$] *it is always implicitly assumed that the circuit* $\mathcal{N}$ *inside the one-port has no unstable hidden modes.*

Consider a linear time-invariant one-port N specified by its driving-point admittance $Y(s)$. If we connect to its port terminals ① and ①′ an independent *voltage source* $V(s)$ (see Fig. 4.2) then the natural frequencies of the one-port terminated by $V(s)$ are the *same* as its natural frequencies when it is terminated by a short circuit. Therefore it is legitimate to use such a *voltage source* to test *short-circuit stability*.

Let us now use the voltage source to apply a *voltage pulse* during the bounded interval $[0, T]$, that is, $v(t)$ is nonzero but arbitrary during the time interval $[0, T]$ and $v(t) = 0$ for all $t > T$; the time T is positive and finite, but otherwise arbitrary. We say that the one-port N is *short-circuit stable* iff, *for all such voltage pulses*, the resulting port current $i(t)$ tends exponentially to zero as $t \to \infty$.[12]

Consider now the dual situation: The linear time-invariant one-port N is specified by its driving-point impedance $Z(s)$. Now we'll use a current source to test *open-circuit stability*.

Let the current source apply a current pulse during the time interval $[0, T]$; as before, T is positive and finite and $i(t) = 0$ for all $t > T$. We say that the one-port N is *open-circuit stable* iff, for all such pulses of current, the resulting port voltage $v(t)$ tends exponentially to zero as $t \to \infty$.

[11] As examples, consider the circuits shown in Fig. 4.10 in Chap. 10 and the circuit of Fig. 3.1, above.

[12] More precisely, $i(t)$ is the zero-state response due to the voltage pulse.

In terms of the definitions just given we may state a useful stability test.

> *Stability Test*: A linear time-invariant one-port N is *short-circuit stable* if and only if all the poles of $Y(s)$ are in the open LHP.
> Dually, a linear time-invariant one-port N is *open-circuit stable* if and only if all the poles of $Z(s)$ are in the open LHP. (4.4)

PROOF Let us prove it for the open-circuit case.

$\Rightarrow$ By assumption, the one-port is open-circuit stable. Let $I(s)$ be the Laplace transform of the current pulse, then the Laplace transform of the port voltage $v(t)$ is given by

$$V(s) = Z(s)\,I(s) \tag{4.5}$$

Indeed, the one-port starts by assumption from the zero state and is driven *only* by the current source $I(s)$. Since

$$I(s) = \int_0^T \mathrm{e}^{-st}\,i(t)\,dt \tag{4.6}$$

and since T is finite, $I(s)$ is well-behaved for *all* values of s in the complex plane—technically speaking it is an entire function—hence, by Eq. (4.5), the only poles of $V(s)$ are contributed by the poles of $Z(s)$. Since for $t \to \infty$, $v(t) \to 0$ exponentially, all the poles of $Z(s)$ must be in the open LHP, (equivalently, have negative real parts).

$\Leftarrow$ By assumption $Z(s)$ has all its poles in the open left-half plane. Let $h(t) = \mathcal{L}^{-1}\{Z(s)\}$; physically, the impulse response $h(t)$ is the port voltage due to a unit impulse of current applied to the one-port. Since $Z(s)$ has all its poles in the open LHP, $h(t) \to 0$ exponentially as $t \to \infty$. Also, as illustrated in Fig. 4.3, for *some* $k > 0$ and *some* $\alpha > 0$,

$$|h(t)| \le k\,\mathrm{e}^{-\alpha t} \qquad \text{for all } t > 0 \tag{4.7}$$

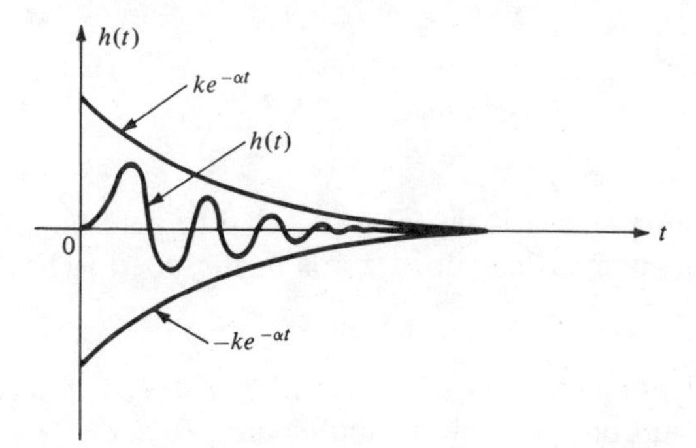

Figure 4.3 These graphs illustrate the fact that for $t \ge 0$, $|h(t)| \le k\,\mathrm{e}^{-\alpha t}$.

Now the port voltage $v(t)$ in response to the pulse of current $i(t)$ is given by the convolution of $h(t)$ and $i(t)$: For any $t \geq 0$, we have

$$v(t) = \int_{0-}^{t} h(t-t')\, i(t')\, dt' \tag{4.8}$$

Since $i(t') = 0$ for all $t' > T$, we have, for $t > T$,

$$|v(t)| = \left| \int_{0-}^{T} h(t-t')\, i(t')\, dt' \right| \tag{4.9}$$

$$\leq \int_{0-}^{T} |h(t-t')| \cdot |i(t')|\, dt' \tag{4.10}$$

$$\leq k \int_{0-}^{T} e^{-\alpha(t-t')} |i(t')|\, dt' \tag{4.11}$$

$$\leq k\, e^{-\alpha(t-T)} \int_{0-}^{T} |i(t')|\, dt' \tag{4.12}$$

where, to obtain the last line, we noted that, for $0 \leq t' < T < t$ and equivalently, $t - t' \geq t - T$ we have

$$\exp[-\alpha(t-t')] \leq \exp[-\alpha(t-T)] \tag{4.13}$$

Since the integral in Eq. (4.12) is a finite *constant* independent of t, the inequality (4.12) shows that, *for any such current pulse*, $v(t) \to 0$ exponentially as $t \to \infty$. Hence the one-port is open-circuit stable. ■

Exercise Given the following impedances or admittances, determine whether the corresponding one-port is short-circuit stable and/or open-circuit stable.

(*a*) $Z(s) = \dfrac{s-1}{s(s+1)}$ (*c*) $Y(s) = \dfrac{s+3}{s^3 - 3s + 2}$

(*b*) $Z(s) = \dfrac{s^2 + 3s - 2}{s^3 + 5s^2 + 7s + 1}$ (*d*) $Y(s) = \dfrac{s^2 + 4s + 9}{s^3 + 5s^2 + 3s + 2}$

4.3 Remark on Polynomials

In simple situations it is useful to be able to determine by inspection whether or not a polynomial has all its zeros in the open LHP. Here are a few useful facts.

1. The polynomial $a_0 s^2 + a_1 s + a_2$, with $a_0 > 0$, has all its zeros in the open LHP if and only if both a_1 and a_2 are *positive*.

2. The polynomial $a_0s^3 + a_1s^2 + a_2s + a_3$, with $a_0 > 0$, has all its zeros in the open LHP if and only if $a_1 > 0$, $a_2 > 0$, $a_3 > 0$, and $a_1a_2 > a_0a_3$.
3. If the polynomial $a_0s^n + a_1s^{n-1} + \cdots + a_{n-1}s + a_n$, with $a_0 > 0$, has one or more *negative* coefficient, then it has at least one zero in the *open* RHP.
4. There exist general tests on the coefficients to determine whether all its zeros are in the open LHP.[13] They are algebraically complicated however. If one has an easy way to draw Nyquist plots, test the rational function

$$F(s) \triangleq \frac{a_0s^n + a_1s^{n-1} + \cdots + a_{n-1}s + a_n}{(s+\alpha)^{n+1}}$$

for any zeros in the closed RHP (here α must be *positive*, and chosen so that the graph has a convenient size). Using the Nyquist criterion, we see that the Nyquist diagram of $F(s)$ goes through or encircles the origin if and only if the polynomial $a_0s^n + a_1s^{n-1} + \cdots + a_{n-1}s + a_n$ has one or more zeros with nonnegative real parts.

5 STABILITY OF A ONE-PORT TERMINATED BY A RESISTOR

In this section we consider exclusively linear time-invariant one-ports. If we terminate a one-port by a linear time-invariant resistor R or better yet by another one-port it may happen that the resulting circuit is unstable. We shall assume that the given one-port is specified by its driving-point impedance $Z_1(s)$. We consider the case where it is terminated by a resistor R.

A. Analysis Suppose that we are given a linear time-invariant one-port N_1 specified by its driving-point impedance $Z_1(s)$. Let $Z_1(s)$ be given by

$$Z_1(s) = \frac{n(s)}{d(s)} \tag{5.1}$$

where the polynomials $n(s)$ and $d(s)$ are *coprime*. Let us now terminate the one-port N_1 by the linear time-invariant resistor R. In order to investigate the stability of the series connection of $Z_1(s)$ and R we use a test voltage source $e_s(t)$. (See Fig. 5.1.) Call $\mathcal{N}_R$ the resulting circuit.

Since $Z_1(s)$ relates the Laplace transform of the *zero-state response* to that of the input and in view of the rules of Laplace transforms (Chap. 10, Sec. 2, Table 2.1), the port voltage $v(t)$ and the port current $i(t)$ are related by the *differential equation*

$$d(D)\,v(t) - n(D)\,i(t) = 0 \tag{5.2}$$

[13] This test is called the *Routh test*. See, for example, T. E. Fortmann and K. L. Hitz, *An Introduction to Linear Control Systems*, Marcel Dekker, Inc., New York, 1977.

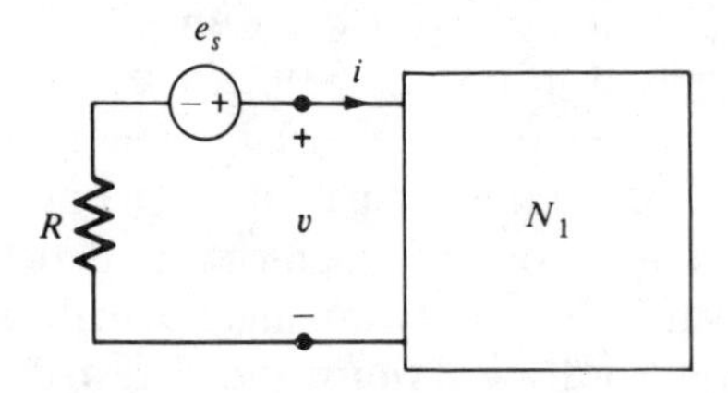

Figure 5.1 The circuit $\mathcal{N}_R$ is the series connection of the one-port N_1, the resistor R, and the test voltage source $e_s(t)$.

where $D \triangleq d/dt$. The resistor imposes the following constraint

$$v(t) + R\,i(t) = e_s(t) \tag{5.3}$$

Equations (5.2) and (5.3) describe the dynamics of the circuit $\mathcal{N}_R$ in terms of the port voltage $v(t)$ and the port current $i(t)$. These equations form a system of algebraic differential equations where $v(t)$ and $i(t)$ are the unknowns, $e_s(t)$ is the known forcing function, and all initial conditions are zero.

The *characteristic polynomial* of the circuit $\mathcal{N}_R$ is

$$\chi(s) = R\,d(s) + n(s) \tag{5.4}$$

Thus λ is a natural frequency of the circuit $\mathcal{N}_R$ shown in Fig. 5.1 if and only if $\chi(\lambda) = 0$.

Now since $n(s)$ and $d(s)$ are coprime, $\chi(s) \neq 0$ at every zero of $d(s)$. Hence every zero of $\chi(s)$ is a zero of $\chi(s)/d(s)$, and vice versa. Thus λ *is a natural frequency of the circuit* $\mathcal{N}_R$ *if and only if*

$$Z_1(\lambda) + R = 0 \tag{5.5}$$

REMARKS

1. Note that no assumption was required concerning the stability (either open-circuit or short-circuit) of $Z_1(s)$.
2. The conclusion (5.5) could also have been obtained by reasoning as follows: The test voltage source $e_s(t)$ sees an impedance $Z(s) + R$ across its terminals, therefore the natural frequencies of $i(t)$ are given by Eq. (5.5) since

$$I(s) = \frac{E_s(s)}{Z_1(s) + R} \tag{5.6}$$

Since $v(t) = e_s(t) - R\,i(t)$ (where R is a *nonzero constant*) and since $e_s(\cdot)$ is a voltage pulse, the voltage $v(t)$ tends to zero as $t \to \infty$ if and only if $i(t)$ does the same. Hence the zeros of $Z_1(s) + R$ indeed constitute the natural frequencies of the circuit $\mathcal{N}_R$ of Fig. 5.1.

B. Nyquist-type test assuming $Z_1(s)$ open-circuit stable Let us now use Nyquist techniques to determine *all* those values of R for which

$$\text{the zeros of } [Z_1(s) + R] \text{ are in the open LHP} \tag{5.7}$$

Let us *assume* now that $(a) Z_1(s)$ is *open-circuit stable*, equivalently that the poles of $Z_1(s)$ are in the open LHP, and (b) that $Z_1(s)$ tends to a constant as $|s| \to \infty$. [Thus we rule out a behavior like $Z_1(s) \cong Ls$ for large $|s|$.]

Clearly $Z_1(s) + R$ is analogous to $1/k + A(s)$ in Sec. 3 above: Indeed in both cases, R and $1/k$ are *real* constants, and both $Z(s)$ and $A(s)$ have their poles in the open LHP. Hence $-R$ will play the role of $-(1/k)$.

To be specific, suppose that the Nyquist plot of $Z_1(s)$ is the one shown in Fig. 5.2. Note the values of the impedance at the Nyquist plot intersects the real axis. The Nyquist plot does not encircle the point $(-R, 0)$ if and only if either

$$-R < -500\,\Omega \qquad \text{equivalently,} \qquad R > 500\,\Omega$$

or $$-200\,\Omega < -R < 0 \qquad \text{equivalently,} \qquad 200\,\Omega > R > 0$$

or $$1\,\text{k}\Omega < -R \qquad \text{equivalently,} \qquad R < -1\,\text{k}\Omega$$

Furthermore, for *all other* values of R, the circuit of Fig. 5.1 is not exponentially stable. More precisely, for any R satisfying $200\,\Omega < R < 500\,\Omega$, there are two clockwise encirclements of $(-R, 0)$ and hence $Z_1(s) + R$ has two zeros in the open right-half plane; for any R satisfying $0 < R < 1\,\text{k}\Omega$, there is one clockwise encirclement of $(-R, 0)$ and hence $Z_1(s) + R$ has one zero in the open right-half plane.

REMARK The value of this test is that
(a) It gives global insight: It gives *all* the values of R for which the circuit $\mathcal{N}_R$ is stable, it gives an idea of how close one is from instability, and finally, it gives the nature of the instability.
(b) This test can be performed directly on the basis of *measurements* of $Z_1(j\omega)$ for a number of frequencies.

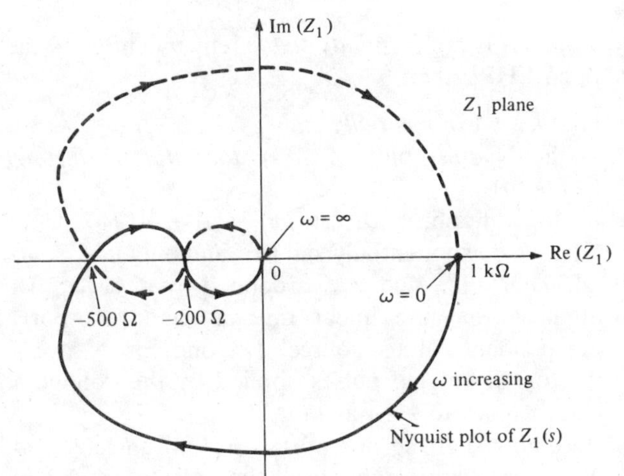

Figure 5.2 Nyquist plot of the driving-point impedance $Z_1(s)$.

SUMMARY

- Let the network function

$$H(s) = \frac{n(s)}{d(s)} \qquad [n(s), d(s) \text{ coprime}]$$

have $z_1, z_2, \bar{z}_2$ as zeros and $p_1, p_2, p_3, \bar{p}_3$ as poles; then, for any frequency ω [which is not a pole of $H(s)$],

$$|H(j\omega)| = |k| \frac{|j\omega - z_1| \cdot |j\omega - z_2| \cdot |j\omega - \bar{z}_2|}{|j\omega - p_1| \cdot |j\omega - p_2| \cdot |j\omega - p_3| \cdot |j\omega - \bar{p}_3|}$$

and

$$\begin{aligned}\measuredangle H(j\omega) = \measuredangle k &+ [\measuredangle(j\omega - z_1) + \measuredangle(j\omega - z_2) + \measuredangle(j\omega - \bar{z}_2)] \\ &- [\measuredangle(j\omega - p_1) + \measuredangle(j\omega - p_2) + \measuredangle(j\omega - p_3) + \measuredangle(j\omega - \bar{p}_3)]\end{aligned}$$

- Let z be a zero of $H(s)$; then for an input $I_m\, e^{zt}$ there are initial conditions $\mathbf{x}_0$ such that if the circuit is started from $\mathbf{x}_0$ and driven by $I_m\, e^{zt}$, the output is equal to *zero for all* $t \geq 0$.
- *Nyquist test.* Given the network function

$$G(s, k) = \frac{k\, A(s)}{1 + k\, A(s)} = \frac{A(s)}{(1/k) + A(s)}$$

where k is a *nonzero constant,*

$$A(s) = \frac{n_A(s)}{d_A(s)} \qquad [\text{with } n_A(s) \text{ and } d_A(s) \text{ coprime}]$$

$$A(s) \to 0 \qquad \text{as } |s| \to \infty$$

and $A(s)$ is *exponentially stable* [equivalently, all the zeros of $d_A(s)$ are in the open LHP], then

$G(s, k)$ is exponentially stable
$\Leftrightarrow$ the Nyquist plot of $A(s)$ *does not go through nor encircle the point* $(-1/k, 0)$.

- If, in addition, both $|A(j\omega)|$ and $\measuredangle A(j\omega)$ decrease monotonically, then $G(s, k)$ is exponentially stable if and only if $\omega_k < \omega_{180}$, where ω_k is defined by $|A(j\omega_k)| = 1/k$ and ω_{180} is defined by $\measuredangle A(j\omega_{180}) = -180°$.
- Suppose that the linear time-invariant one-port N is terminated by an independent voltage source. The one-port N is said to be *short-circuit stable* iff, for all voltage pulses applied by the voltage source, the resulting port current goes to zero as $t \to \infty$.
- Assuming that the one-port N has no unstable hidden modes, the one-port N is *short-circuit stable* $\Leftrightarrow$ all the poles of $Y(s)$ are in the open LHP.

- The open-circuit stability definition and test are the dual of the above.
- Let N be a linear time-invariant one-port whose driving-point impedance $Z_1(s)$ is *exponentially stable*, and where $Z_1(s) \to$ constant as $|s| \to \infty$. Then, the one-port N terminated by a linear resistor R is *exponentially stable* $\Leftrightarrow$ the Nyquist plot of $Z_1(s)$ *does not go through nor encircle* the point $(-R, 0)$.

PROBLEMS

Poles and zeros

1 Find the driving-point impedances of the linear time-invariant circuits shown in Fig. P11.1, and plot their poles and zeros on the complex plane.

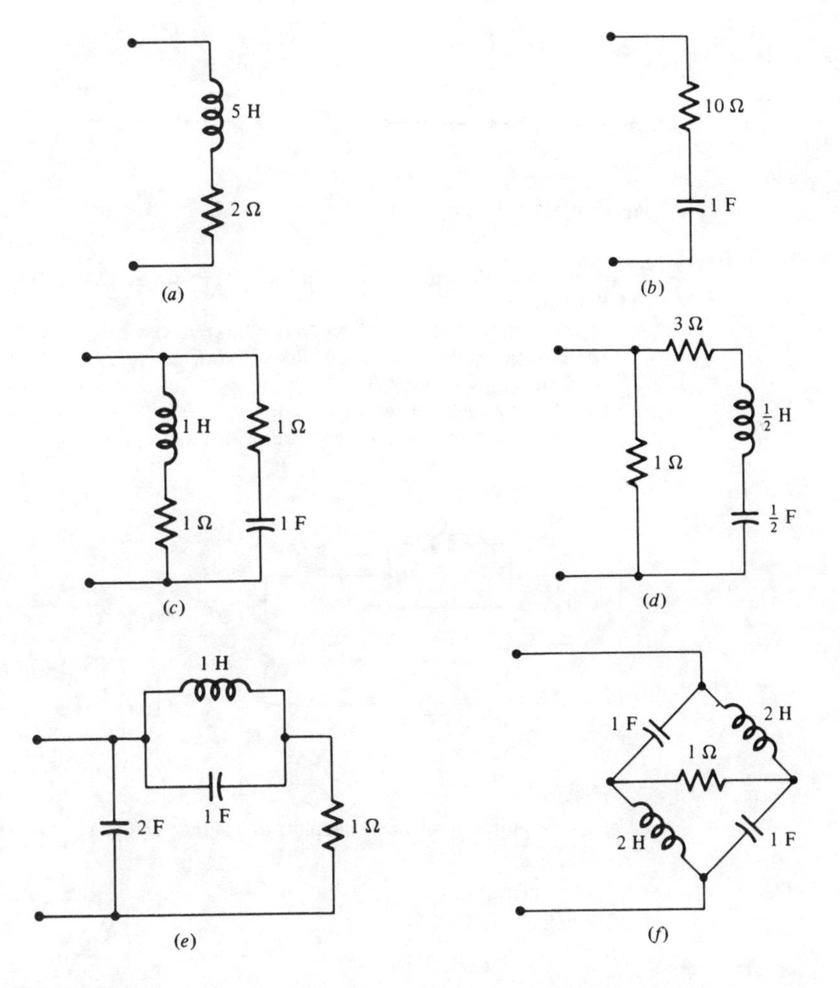

Figure P11.1

Magnitude and phase plots

2 In Fig. P11.2*a* and *b* the element values are in ohms and farads.

(*a*) Show that the network function of a lag compensator (a common control system filter), $H(s)$, can be realized as (i) the voltage transfer function V_0/V_i of the circuit in Fig. P11.2*a* or (ii) the driving-point impedance of the circuit in Fig. P11.2*b*, where

$$H(s) = \frac{1}{\alpha}\frac{s+\omega_1}{s+\omega_1/\alpha} \qquad \alpha > 1 \text{ and } \omega_1 > 0$$

(*b*) Plot the poles and zeros of $H(s)$ on the complex plane.

(*c*) Using asymptotes, draw the Bode plots (magnitude and phase) for $H(s)$.

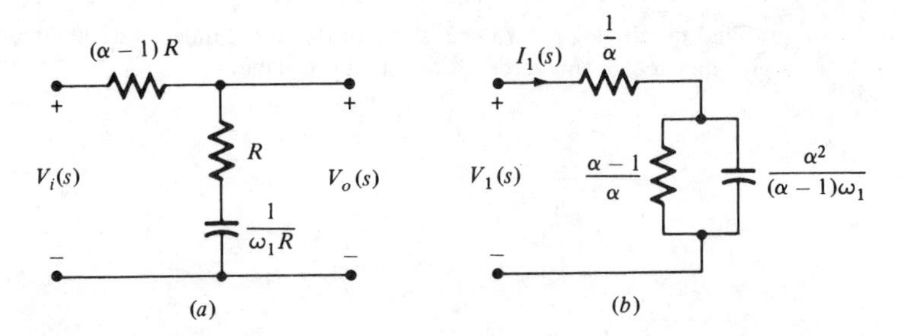

Figure P11.2

3 Consider the filter circuit shown in Fig. P11.3, where $C_1 = 1.73$ F, $C_2 = C_3 = 0.27$ F, $L = 1$ H, and $R = 1\ \Omega$.

(*a*) Find the network function $H(s) = V_2(s)/I_1(s)$.

(*b*) Plot the poles and zeros of this network function.

(*c*) Plot $|H(j\omega)|$ versus ω.

(*d*) Plot $\measuredangle H(j\omega)$ versus ω.

(*e*) Determine the impulse response of this filter.

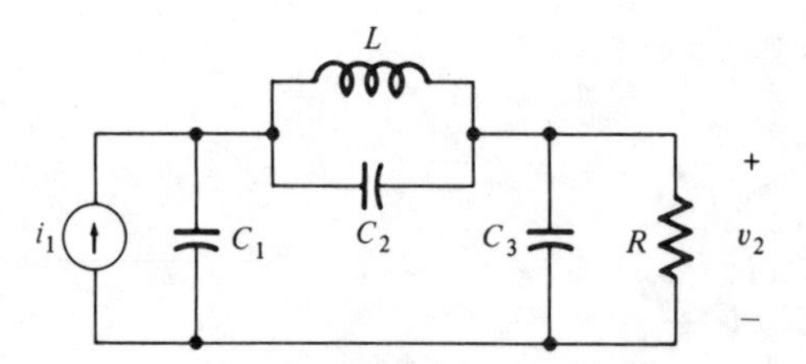

Figure P11.3

Bode plots

4 Using asymptotes, draw the Bode plots (magnitude and phase) for each of the following network functions:

(*a*) $H(s) = \dfrac{5(s+1)}{s+5}$

(*b*) $H(s) = \dfrac{(s+1)(s+3)}{(s+\frac{1}{2})(s+6)}$

(c) $H(s) = \dfrac{20}{s(s+2)(s+5)}$

(d) $H(s) = \dfrac{72}{(s+1)(s+3)^2}$

Nyquist plot

5 Sketch the Nyquist plot for each of the following network functions:

(a) $H(s) = \dfrac{1}{(s+1)(s+2)}$

(b) $H(s) = \dfrac{1}{(s+1)(s+3)(s+4)}$

(c) $H(s) = \dfrac{3(s+1)}{(s+2)(s+5)}$

(d) $H(s) = \dfrac{s+4}{(s+2)(s^2+s+2)}$

Network function and graphic evaluation

6 The transfer function $H(s)$ has the pole-zero plot shown in Fig. P11.6. Given $|H(j2)| = 7.7$ and $0 < \measuredangle H(j2) < \pi$.

(a) Evaluate graphically $H(j4)$.

(b) Give $|H(j4)|$ and $\measuredangle H(j4)$ in decibels and radians, respectively.

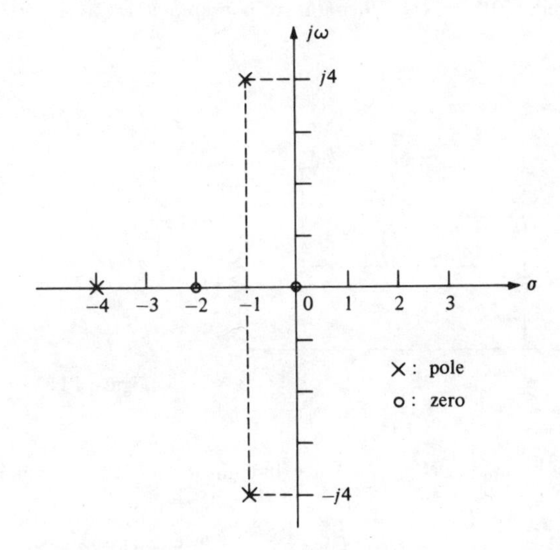

Figure P11.6

Frequency and time response

7 The magnitude plot of a network function is sketched in Fig. P11.7. The function $H(s)$ is assumed to be rational. The sketch is not necessarily to scale, but the following specifics should be noted:

$$|H(j\omega)| \text{ is an even function of } \omega$$

$$|H(0)| = 1 \qquad |H(j1)| = 0 \qquad \text{and} \qquad |H(j\infty)| = 0$$

(*a*) In order to be compatible with the above data, what is the minimum number of zeros that $H(s)$ can have? What is the minimum number of poles? Find an $H(s)$ compatible with the above data.

(*b*) Suppose that the network function $H(s)$ has the minimum number of zeros and simple poles at $s = -1, -1 \pm j$ and has no other poles. Suppose that the zero-state response $y(t)$ due to some imput $x(t)$ and corresponding to this $H(s)$ is

$$y(t) = (1 - \cos t)\, e^{-t} 1(t)$$

Find the input $x(t)$.

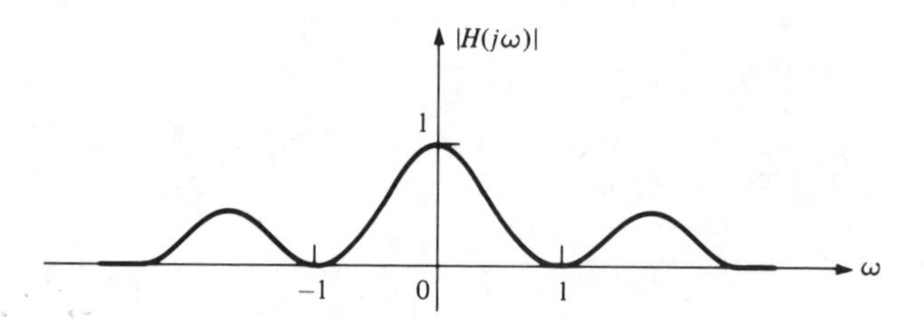

Figure P11.7

Zeros of network functions

8 Consider the circuit shown in Fig. P11.8. Since the parallel-tuned circuit resonates at $\omega = \frac{1}{3}$ rad/s, the network function $H(s) = E_2(s)/I(s)$ has a zero at $\omega = \frac{1}{3}$ rad/s. Show that if $e_1(0) = 0$, $v_2(0) = 0$, and $i_L(0) = -\frac{3}{2}$ A, then the response to $i(t) = 1(t)\cos(t/3)$ is, for all $t \geq 0$, $e_1(t) = 3/2 \sin(t/3)$ and $e_2(t) = 0$.

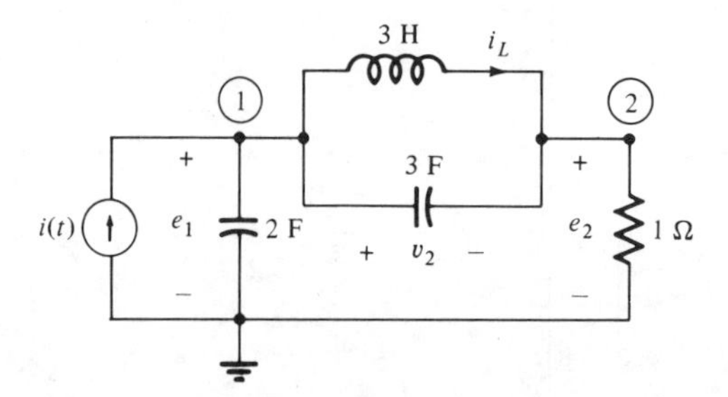

Figure P11.8

9 The network function $H(s)$ of a linear time-invariant circuit is assumed to be a proper rational function. Given

$$H(j\omega) = 0 \qquad \text{at } \omega = 2 \text{ rad/s}$$

and the impulse response

$$h(t) = [k_1 e^{-t} + |k_2| e^{-t} \cos(t + \measuredangle k_2)]\, 1(t)$$

determine $H(s)$ to within a constant.

Nyquist criterion

10 Which of the following network functions satisfy conditions (3.31) to (3.33) of the Nyquist criterion?

(*a*) $H(s) = \dfrac{s^2 + 4}{(s+1)(s+3)}$

(*b*) $H(s) = \dfrac{(s+1)}{(s+2)(s^2 + 5s + 4)}$

(*c*) $H(s) = \dfrac{s^2 + 2s + 1}{(s+2)(s^2 + 3s - 2)}$

(*d*) $H(s) = \dfrac{s - 2}{(s^2 + s + 1)(s+3)}$

(*e*) $H(s) = \dfrac{s + 1}{(s^2 + 4)(s+3)}$

11 For the following specifications of $A(s)$ use the Nyquist criterion to determine the range of values of k for which $G(s, k)$ is stable.

(*a*) $A(s) = \dfrac{1}{(s+1)(s^2 + 3s + 2)}$

(*b*) $A(s) = \dfrac{3}{(s+2)(s^2 + s + 2)}$

(*c*) $A(s) = \dfrac{1}{(s+1)^4}$

12 The Nyquist plots for three specifications of $A(s)$ are shown in Fig. P11.12. In each case $A(s)$ is assumed to be exponentially stable. Use the Nyquist criterion to find the range of values of k for which the closed-loop feedback system is exponentially stable.

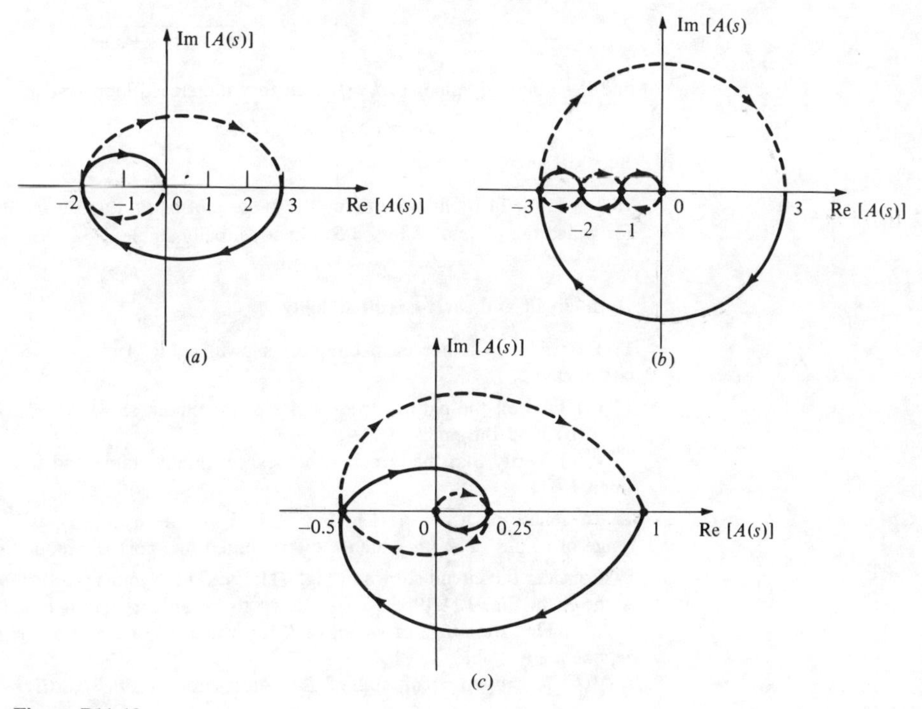

Figure P11.12

13 Consider the circuit shown in Fig. P11.13 where the op amp is modeled as in Fig. 3.2*b*.

(*a*) Show that:

$$\frac{V_0}{V_i} = -\frac{R_f}{R_1}\frac{A(s)}{A(s)+k}$$

where

$$k = \frac{R_1R_2 + R_1R_f + R_2R_f}{R_1R_2}$$

(*b*) Let $R_1 = R_2 = 1\text{ k}\Omega$ and $A(s) = 10^5 \dfrac{6}{(s+1)(s+2)(s+3)}$, and find the range of values of $R_f > 0$ for which V_0/V_i is exponentially stable.

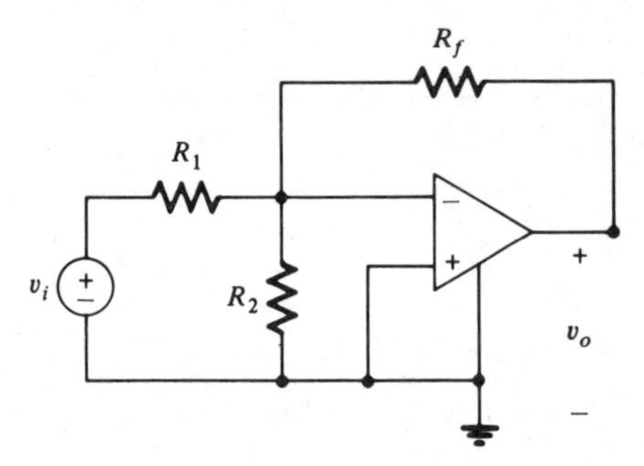

Figure P11.13

14 Consider the control system shown in Fig. 3.5. Given

$$P(s) = \frac{1}{(s+1)(s^2+s+1)}$$

find the range of values of $k > 0$ such that the closed-loop system is exponentially stable.

Bode design

15 Do Prob. 14 by first obtaining the Bode plot of $P(s)$ and using it to determine the values of k for which the system of Fig. 3.5 is exponentially stable.

Open-circuit and short-circuit stability

16 Determine whether each one-port shown in Fig. P11.16 is short-circuit stable and/or open-circuit stable.

17 (*a*) Find all the natural frequencies of the circuit shown in Fig. P11.17.

(*b*) Find the poles of $Z_{\text{in}}(s)$.

(*c*) Verify that the circuit is not exponentially stable and that $Z_{\text{in}}(s)$ has all its poles in the open *LHP*.

18 Terminate each one-port in Fig. P11.16 by a resistance R as in Fig. 5.1, and determine the range of values of R for which the terminated one-port is exponentially stable.

19 Consider the circuit shown in Fig. P11.19*a*. The Nyquist plot for $I(s)$ when $R = 1\ \Omega$ and $V(s) = 1$ is shown in Fig. P11.19b. $I(s)$ is known to be an exponentially stable proper rational function.

(*a*) Find the range of values of R for which the one-port to the right of terminals ① and ①′ is exponentially stable.

(*b*) Is Z open-circuit stable? Is it short-circuit stable? Justify your answers.

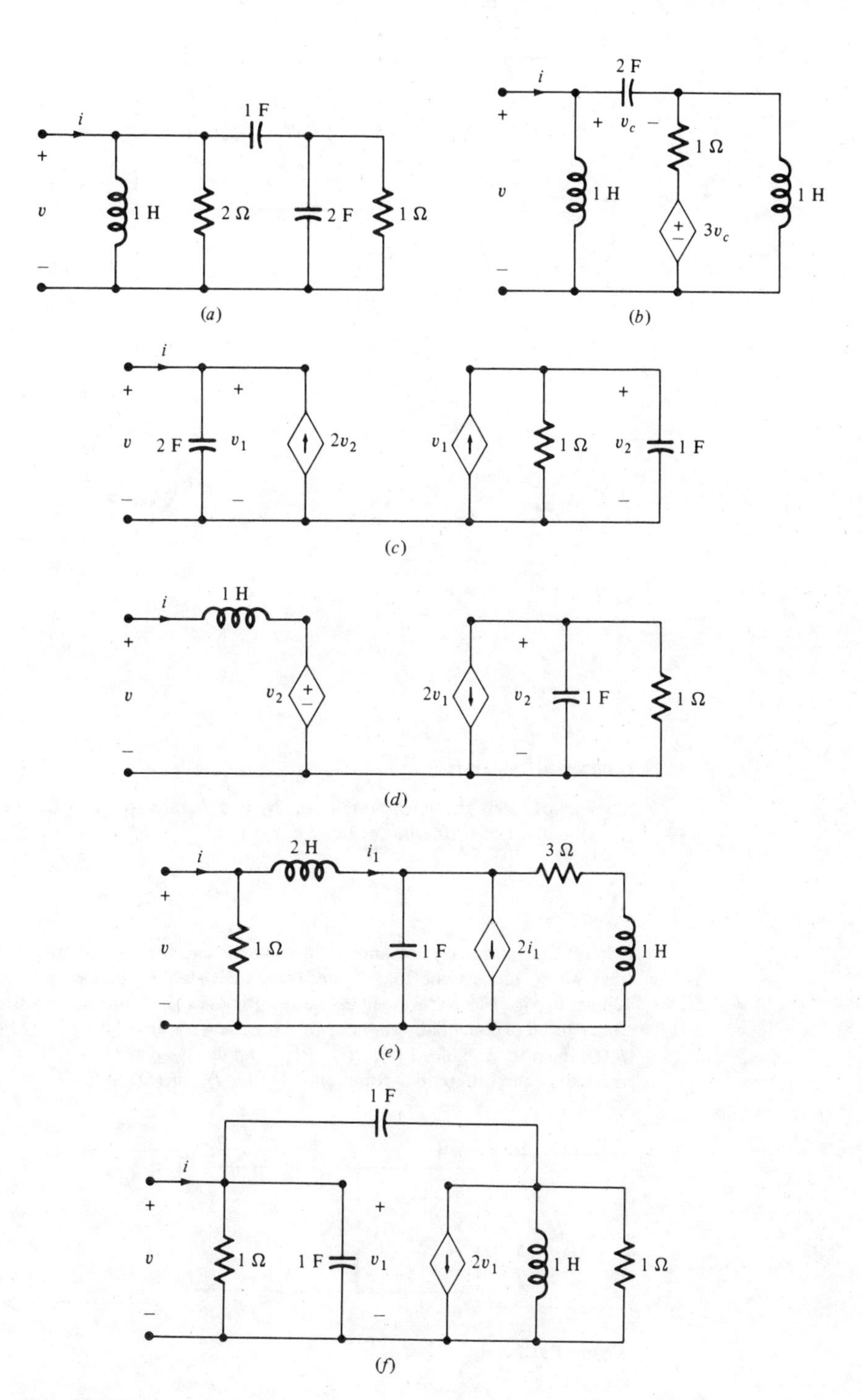

Figure P11.16

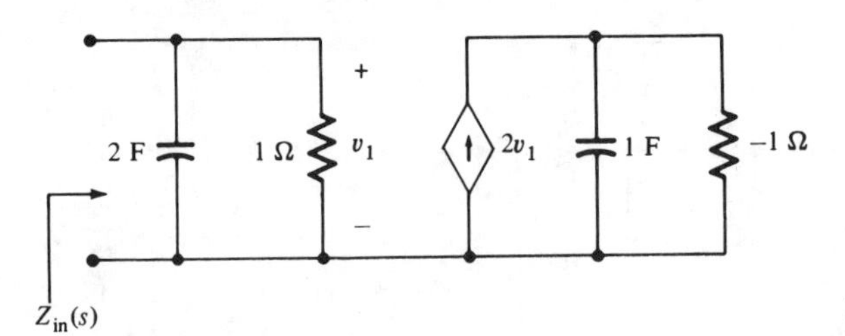

Figure P11.17

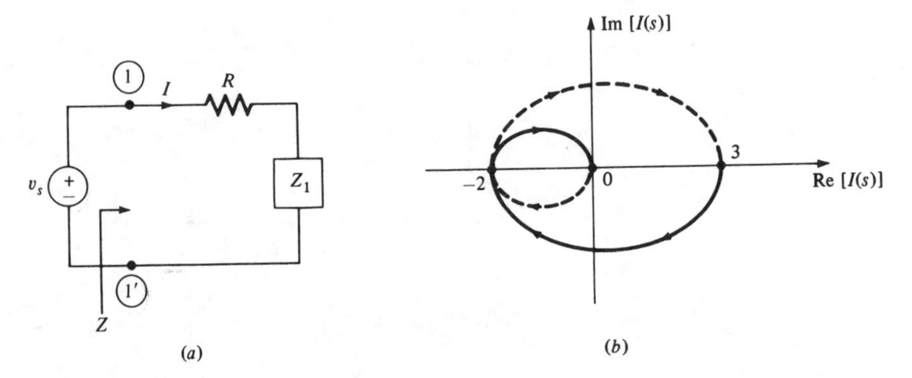

Figure P11.19

Connected two-ports

20 Suppose that the linear one-ports N_1 and N_2 are specified by their driving-point impedance $Z_1(s)$ and $Z_2(s)$; furthermore, let, for $k = 1, 2$,

$$Z_k(s) = \frac{n_k(s)}{d_k(s)}$$

where $n_k(s)$ and $d_k(s)$ are known polynomials and for $k = 1, 2$, the pair (n_k, d_k) is coprime. Note that we do not assume that Z_1 and/or Z_2 are stable. Suppose that N_1 and N_2 are connected as shown in Fig. P11.20. We call $\mathcal{N}$ the resulting circuit. Find the necessary and sufficient conditions for $\mathcal{N}$ to be exponentially stable. More precisely for any short pulse of current $i_s(t)$ and/or voltage $v_s(t)$, the port variables $i_1(t)$, $i_2(t)$, $v_1(t)$, and $v_2(t) \to 0$ as $t \to \infty$. [*Hint*: Note that $v_1(t)$ and $i_1(t)$ are related by the differential equation $d_1(D)\, v_1(t) = n_1(D)\, i_1(t)$.]

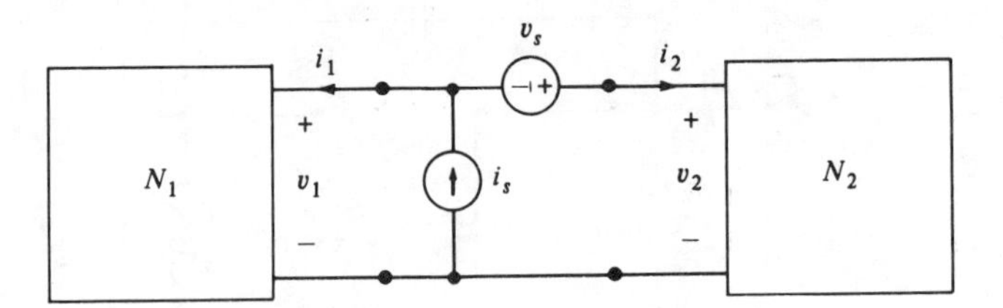

Figure P11.20

CHAPTER

TWELVE

CIRCUIT TOPOLOGY AND GENERAL CIRCUIT ANALYSIS

Kirchhoff's laws are the fundamental postulates of lumped-circuit theory. They are valid regardless of the nature of the circuit elements. It is natural therefore to separate the equations of Kirchhoff's laws from that of the element characteristics as in tableau analysis. The former is prescribed by the topology of the circuit, that is, the way circuit elements are interconnected.

In Chap. 1 we demonstrated the usefulness of a digraph in formulating Kirchhoff's equations. One key concept introduced there was the datum node of a circuit, which has been used throughout the first eleven chapters in analysis. Thus, in KVL, we express branch voltages in terms of node-to-datum voltages, and in KCL we have used the node formulation almost exclusively. This is true in tableau analysis, node analysis, and MNA. There exist other methods of circuit analysis which are sometimes more convenient and useful. By further exploring circuit topology, we can generalize the approaches used based on nodes of a graph to that on cut sets and loops. This will be the major aim of this chapter.

In Sec. 1 we introduce two elementary source transformations which allow us to move sources in a circuit without changing the performance of the circuit. This is useful in deriving general equations for circuit analysis. In Sec. 2 we revisit Kirchhoff's laws and introduce the cut-set and loop matrices which represent generalizations of the incidence matrix of a digraph. Section 3 deals with a fundamental theorem of a digraph in terms of a tree. This then leads to two fundamental matrices associated with a tree: the fundamental cut-set matrix **Q** and the fundamental loop matrix **B**. Section 4 deals with the cut-set tableau equation and the loop tableau equation. In Sec. 5 we introduce cut-set analysis and loop analysis and derive the cut-set equations and loop equations from tableau equations for linear time-invariant circuits. We conclude the chapter with the presentation of a simple method of writing state equations by using the concept of a proper tree.

1 SOURCE TRANSFORMATIONS

In a general discussion of circuit analysis we assume that the number and the location of independent sources are arbitrary as long as Kirchhoff's laws are not violated. For example, a cut set which consists only of independent current sources must satisfy the condition that the sum of the currents is identically zero at all times. A similar constraint exists for all independent voltage sources which form a loop. In many ways, independent sources play quite a different role from the rest of the circuit elements. As we have seen in practical circuit analysis, there exists in many instances only a single independent source in the entire circuit, and the source contribution is usually placed on the right side of circuit equations because it serves as the *input* to the circuit. In contrast, the contributions of dependent sources are always put on the left side because they model interactions between circuit variables. The definition of a linear circuit also gives a special role to independent sources. Recall that a linear circuit contains only linear circuit elements with the exception of independent sources.

Recall that tableau analysis and MNA are completely general; they allow independent voltage and current sources to appear anywhere in the circuit. In node analysis, for example, in linear time-invariant circuits, independent voltage sources are not allowed unless the voltage source is in series with a two-terminal element and can be transformed into its Norton equivalent as shown in Fig. 1.1. This is also true in cut-set analysis to be discussed in this chapter. A dual statement can also be made in terms of independent current sources in loop analysis. Unfortunately, independent sources can appear anywhere in a circuit, and in order to be able to still use these various circuit analysis methods we need to introduce two other source transformations. As we will see, it is always possible by using these transformations to move the independent sources in a circuit without changing the performance of the circuit.

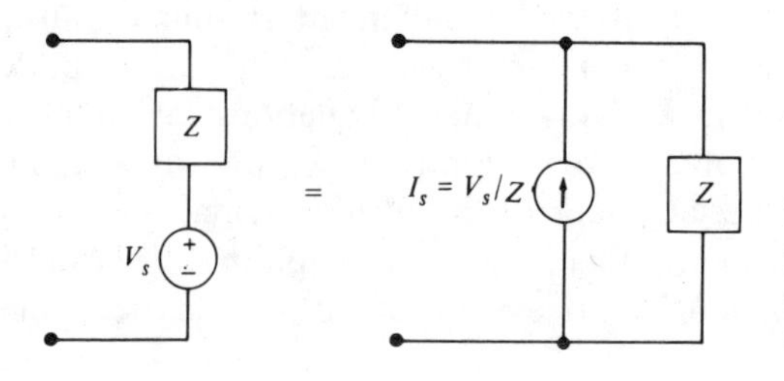

Figure 1.1 A series connection of an independent voltage source and a two-terminal element with impedance Z and its Norton equivalent.

1.1 The v_s-Shift Property

The v_s-shift property is illustrated in Fig. 1.2. In Fig. 1.2*a* an independent voltage source v_s is connected between two nodes ⓪ and ⓪' which belong to two arbitrary circuits $\mathcal{N}$ and $\mathcal{N}'$, respectively. Note that the cut set $\mathscr{C}$ which partitions the circuit into two parts contains, in addition to the source v_s, the branches connecting nodes ①, ①'; ②, ②'; . . . ; ⓚ, ⓚ'. In Fig. 1.2*b* we show an equivalent circuit after the source has been shifted. By *equivalent*, we mean the following: *The v_s-shift property* asserts that the corresponding branch voltages and currents of Fig. 1.2*a* and Fig. 1.2*b* inside $\mathcal{N}$ and $\mathcal{N}'$ are equal. Note that the polarity of the sources in Fig. 1.2*b* after the shift is the reverse of the voltage source in Fig. 1.2*a*.

REMARKS

1. Note that in the equivalent circuit in Fig. 1.2*b*, the branch connecting ⓪ and ⓪' is a short circuit. In other branches in the cut set $\mathscr{C}$, we inserted v_s by means of a *pliers entry*, i.e., cutting the wire by a plier and connecting v_s at the cut. This terminology, "pliers entry" together with its dual term "soldering iron entry," which will be discussed later, represents a convenient and intuitively suggestive way of describing how a source is

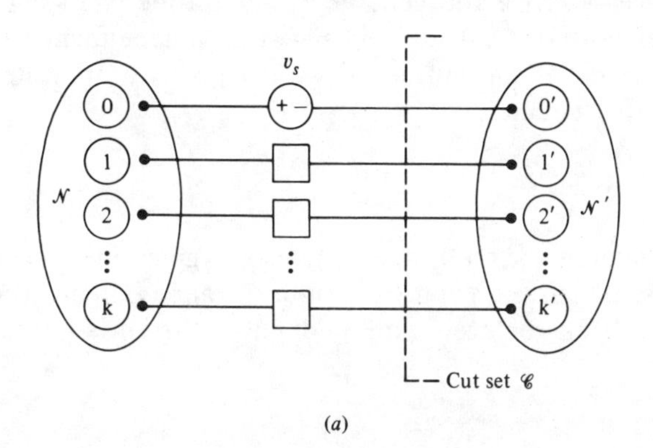

(*a*)

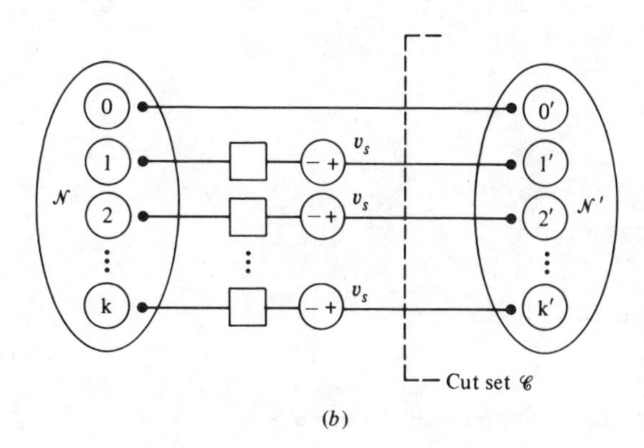

(*b*)

Figure 1.2 Illustrating the v_s-shift property. (*a*) Original circuit. (*b*) Equivalent circuit.

"entered" into a circuit. This concept is particularly useful in explaining the reciprocity theorem in Chap. 13.

2. When the waveform of an independent voltage source is set to zero, the source is reduced to a short circuit. Therefore, without sources, the two circuits in Fig. 1.2*a* and *b* are clearly identical.

3. A special but most useful case of the v_s-shift property is shown in Fig. 1.3 where all branches in the cut set $\mathscr{C}$ share a common node ⓪′. In this case we may think of the transformation as pushing v_s through the node ⓪′ from the source branch to all other branches which are incident to the node. Note that after the transformation, the two nodes ⓪ and ⓪′ coalesce into a single node ⓪″ in Fig. 1.3*b*.

The validity of the v_s-shift property in Fig. 1.3 is easily seen by checking pertinent Kirchhoff's equations for the two circuits. First, for all KVL equations for loops containing the v_s-branch in Fig. 1.3*a*, there exist identical equations for corresponding loops in Fig. 1.3*b* passing through the node ⓪″. Next, the sum of the two KCL equations for nodes ⓪ and ⓪′ in Fig. 1.3*a* is equivalent to the KCL equation for node ⓪″ in Fig. 1.3*b*. Thus the two circuits have identical equations; consequently they have the same performance.

Exercise Prove the general v_s-shift property in Fig. 1.2. *Hint*: Enter by pliers entry at cut set $\mathscr{C}$ in Fig. 1.2*a* independent voltage sources v_s to all branches of the cut set $\mathscr{C}$ with a polarity opposite to that of v_s in the original circuit.

1.2 The i_s-Shift Property

The statement for the i_s-shift property is given in terms of Fig. 1.4. Note that the current source i_s connecting nodes ⓪ and ⓚ in an arbitrary circuit $\mathscr{N}$ of Fig. 1.4*a* has been replaced with k identical current sources i_s in Fig. 1.4*b*, each

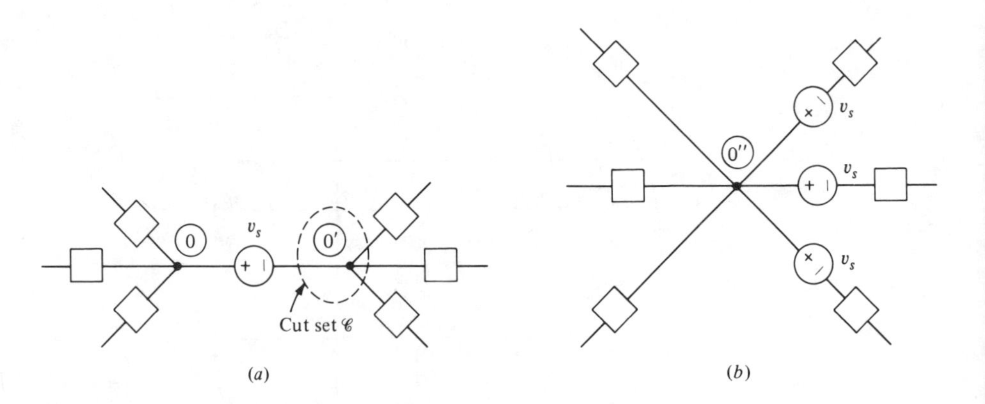

Figure 1.3 The v_s-shift property through a node.

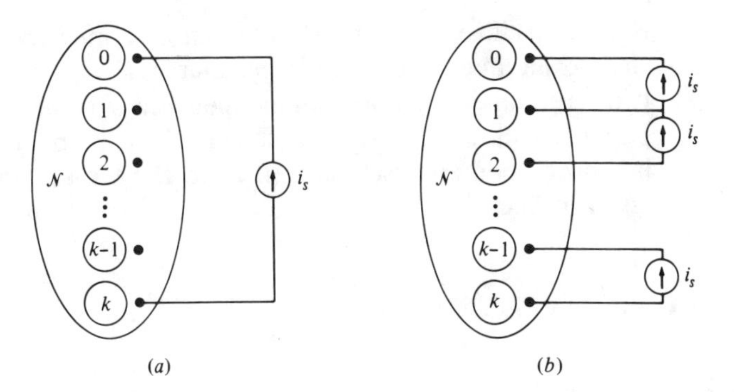

Figure 1.4 Illustrating the i_s-shift property. (*a*) Original circuit. (*b*) Equivalent circuit.

connecting two successive nodes in the node sequence ⓪-①-②- · · · -ⓚ. The two circuits are equivalent. By *equivalent*, we mean the following: The *i_s-shift property* asserts that the corresponding branch voltages and currents of Fig. 1.4*a* and *b* inside $\mathcal{N}$ are equal.

The proof of this equivalence is again straightforward. We only need to examine KCL at the $k+1$ nodes of the two circuits. Obviously, KCL equations at nodes ⓪ and ⓚ for the two circuits are the same. The same is true for node ⓙ, $j = 1, 2, \ldots, k-1$, because in Fig. 1.4*b* there always exists a current source i_s entering a node and another current source i_s leaving the same node.

REMARK The i_s-shift property holds for *any closed* node sequence which includes the current source i_s.

Exercises

1. Demonstrate that the source shift properties remain valid for controlled sources provided that the controlling variable is not affected in the shift.
2. In the circuit in Fig. 1.5, perform source transformations to obtain the Thévenin and Norton equivalent circuits of the one-port.

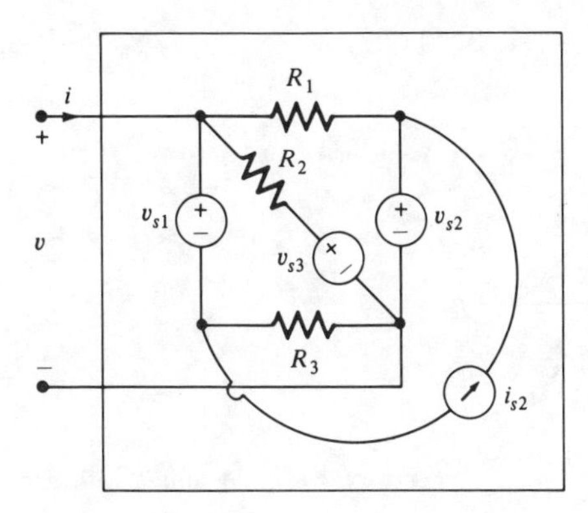

Figure 1.5 Exercise on source transformation.

REMARK With source transformations, we may assume that for a general circuit, an independent voltage source always appears in series with a two-terminal element and an independent current source always appears in parallel with a two-terminal element. This helps us in the general formulation of node analysis, as well as in the cut-set and loop analyses to be given in Sec. 5.

2 KIRCHHOFF'S LAWS REVISITED

2.1 KCL Equations Based on Cut Sets

In Chap. 1 we introduced the matrix formulation of KCL by means of the incidence matrix $\mathbf{A}_a$ of the digraph $\mathcal{G}$ which is associated with a circuit. We assume that $\mathcal{G}$ is connected. Recall $\mathbf{A}_a$ has n rows representing the n nodes of $\mathcal{G}$ and b columns representing the b branches. Thus KCL for the n nodes are written in terms of $\mathbf{A}_a$ and the branch current vector $\mathbf{i}$:

$$\mathbf{A}_a\mathbf{i} = \mathbf{0} \tag{2.1}$$

A key property observed in Chap. 1 is that $\mathbf{A}_a$ has rank $n-1$, and $\mathbf{A}$, the reduced incidence matrix, is obtained from $\mathbf{A}_a$ by deleting the row which corresponds to the chosen datum node. Thus we can write KCL equations on the node basis as a set of $n-1$ linearly independent equations

$$\mathbf{Ai} = \mathbf{0} \tag{2.2}$$

The reduced incidence matrix $\mathbf{A}$ appears in both the tableau and the node analysis and plays a key role so far in circuit analysis.

Let us now generalize the above to write KCL equations based on cut sets. We will use the example from Chap. 1, Fig. 6.1. The digraph is redrawn in Fig. 2.1. We first write *all distinct* cut sets of $\mathcal{G}$ as follows.

$$\mathscr{C}_1 = \{1, 2, 6\} \qquad \mathscr{C}_4 = \{4, 5, 6\}$$
$$\mathscr{C}_2 = \{1, 3, 4\} \qquad \mathscr{C}_5 = \{2, 3, 4, 6\}$$
$$\mathscr{C}_3 = \{2, 3, 5\} \qquad \mathscr{C}_6 = \{1, 3, 5, 6\}$$

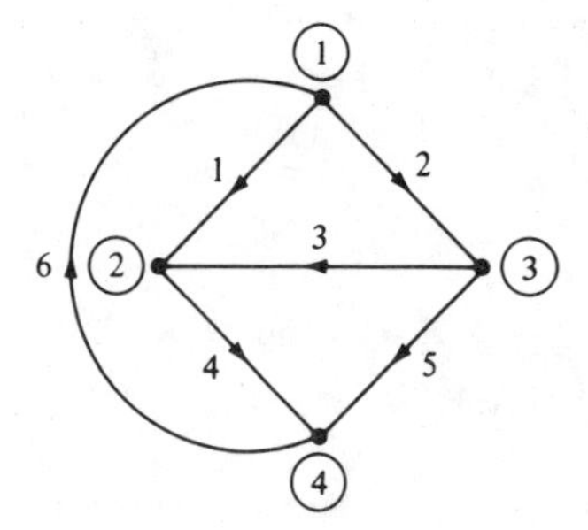

Figure 2.1 A digraph $\mathcal{G}$ with $b = 6$ and $n = 4$.

We next write the KCL equations based on the above cut sets in matrix form, choosing arbitrarily a positive reference direction for each cut set:

$$\begin{bmatrix} 1 & 1 & 0 & 0 & 0 & -1 \\ -1 & 0 & -1 & 1 & 0 & 0 \\ 0 & -1 & 1 & 0 & 1 & 0 \\ 0 & 0 & 0 & -1 & -1 & 1 \\ 0 & 1 & -1 & 1 & 0 & -1 \\ 1 & 0 & 1 & 0 & 1 & -1 \end{bmatrix} \begin{bmatrix} i_1 \\ i_2 \\ i_3 \\ i_4 \\ i_5 \\ i_6 \end{bmatrix} = \begin{bmatrix} 0 \\ 0 \\ 0 \\ 0 \\ 0 \\ 0 \end{bmatrix} \tag{2.3a}$$

or

$$\mathbf{Q}_a \mathbf{i} = \mathbf{0} \tag{2.3b}$$

where $\mathbf{Q}_a$ is called the *cut-set matrix*. Note that the first four rows are identical with $\mathbf{A}_a$ of Chap. 1, Eq. (6.4) since the cut sets $\mathscr{C}_1$, $\mathscr{C}_2$, $\mathscr{C}_3$, and $\mathscr{C}_4$ are made up of the branches which are incident with nodes ①, ②, ③, and ④, respectively. Since cut-set equations are linear combinations of node equations, the set of six equations is linearly dependent and the matrix $\mathbf{Q}_a$ has a rank $n-1=3$. Naturally, we would like to know how, if we write KCL equations based on cut sets of a digraph, we would choose a set which is maximal and linearly independent. By *maximal* we mean the largest set, and from Chap. 1 we know that we need $n-1$ equations. A simple and intuitive way to guarantee such an independent set is to choose $n-1$ cut sets successively in such a way that each new cut set brings in at least a new branch and no branch in $\mathscr{G}$ is left out. The proof of the statement is left as an exercise.

Thus after we have chosen such a set and have deleted the superfluous equations from Eq. (2.3), we obtain

$$\mathbf{Q}_R \mathbf{i} = \mathbf{0} \tag{2.4}$$

The matrix $\mathbf{Q}_R$ has $n-1$ rows and b columns and is called the *reduced cut-set matrix*. It plays the same role as the reduced incidence matrix in writing KCL equations based on nodes instead of cut sets. As a matter of fact, the reduced incidence matrix $\mathbf{A}$ represents a special case of the reduced cut-set matrix $\mathbf{Q}_R$.

Exercise For the diagraph in Fig. 2.2, find all distinct cut sets of $\mathscr{G}$ and write the equation $\mathbf{Q}_a\mathbf{i} = \mathbf{0}$. Determine two reduced cut-set matrices which are not reduced incidence matrices of $\mathscr{G}$.

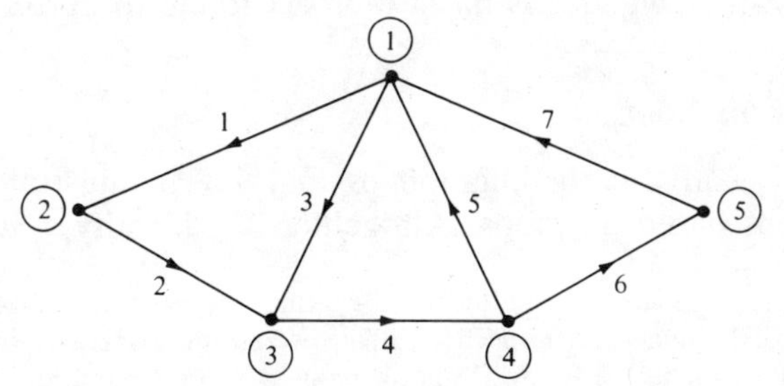

Figure 2.2 A digraph $\mathscr{G}$ with $b=7$ and $n=5$.

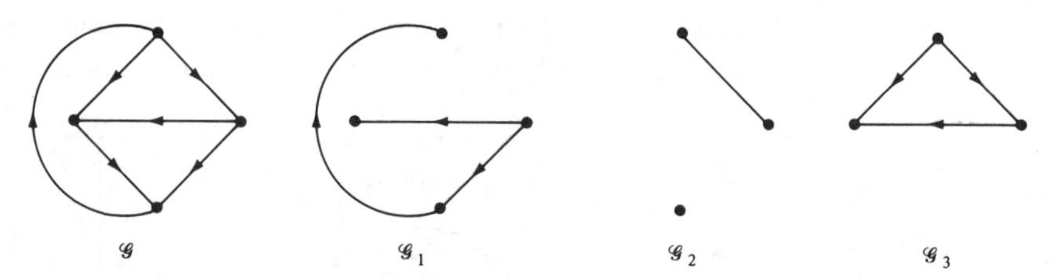

Figure 2.3 Digraph $\mathcal{G}$ and its three subgraphs.

2.2 Graph, Subgraph, and Loop

In Chap. 1 we introduced the concept of a loop via a closed node sequence. To generalize our discussion of KVL from Chap. 1, we need to deal with circuit graphs in a more formal way. A *graph* $\mathcal{G}$ is defined to be a set $\mathcal{V}$ of nodes and a set $\mathcal{B}$ of branches with a prescribed branch-node incidence relation, i.e., each branch is incident with two nodes in $\mathcal{V}$.[1] Symbolically, we write

$$\mathcal{G} = (\mathcal{V}, \mathcal{B}) \tag{2.5}$$

Thus

$$\mathcal{G}_1 = (\mathcal{V}_1, \mathcal{B}_1) \tag{2.6}$$

is called a *subgraph* of $\mathcal{G}$ iff $\mathcal{G}_1$ itself is a graph, $\mathcal{V}_1$ is a subset of $\mathcal{V}$, and $\mathcal{B}_1$ is a subset of $\mathcal{B}$.

Recall that a digraph is a graph whose branches are oriented. In circuit theory we deal with digraphs exclusively.

Example The digraph $\mathcal{G}$ of Fig. 2.1, together with three of its subgraphs, is shown in Fig. 2.3. While we have assumed that $\mathcal{G}$ is always connected, a subgraph need not be. The subgraph $\mathcal{G}_1$ contains all the nodes of $\mathcal{G}$, but only three out of the original six branches. The subgraph $\mathcal{G}_3$ contains three nodes and three branches which form a loop.

Given a connected digraph $\mathcal{G}$, a *loop* $\mathcal{L}$ is defined to be a connected subgraph of $\mathcal{G}$ in which precisely two branches are incident with each node. Figure 2.4 illustrates the definition of a loop in terms of the digraph $\mathcal{G}$. Note that a self-loop which contains precisely one node and one branch as shown in Fig. 2.5 is not a loop according to our definition above.

2.3 KVL Equations Based on Loops

Let us again use the digraph in Fig. 2.1 to illustrate the writing of KVL equations based on loops. Altogether, we identify seven distinct loops for $\mathcal{G}$

[1] In the literature on graph theory, the terms "vertices" and "edges" are often used instead of "nodes" and "branches," respectively. Thus we use the symbol $\mathcal{V}$ for vertices in order not to be confused with $\mathcal{N}$ for nodes, which has been used as the symbol for a circuit.

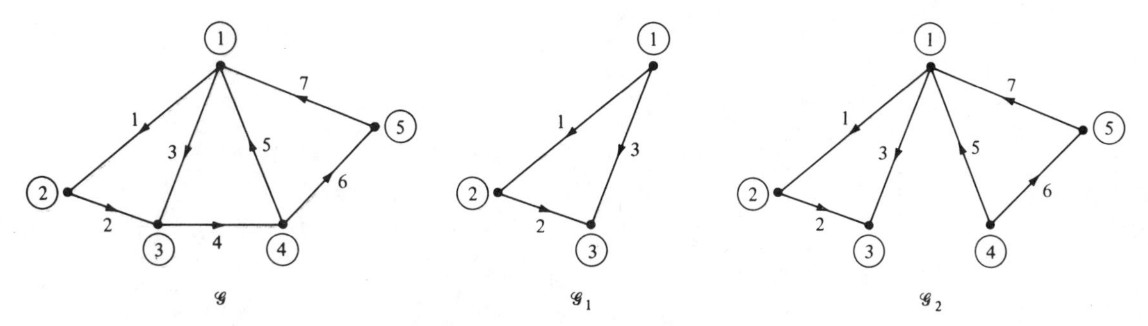

Figure 2.4 $\mathcal{G}_1$ is a loop. $\mathcal{G}_2$ violates the definition at node 1 and therefore is not a loop.

Figure 2.5 A self-loop.

(shown in Fig. 2.6). Their KVL equations are given below with an arbitrarily chosen positive reference direction for each loop:

$$\begin{bmatrix} 1 & -1 & -1 & 0 & 0 & 0 \\ 0 & 0 & 1 & 1 & -1 & 0 \\ 1 & 0 & 0 & 1 & 0 & 1 \\ 0 & 1 & 0 & 0 & 1 & 1 \\ 1 & -1 & 0 & 1 & -1 & 0 \\ 1 & 0 & -1 & 0 & 1 & 1 \\ 0 & 1 & 1 & 1 & 0 & 1 \end{bmatrix} \begin{bmatrix} v_1 \\ v_2 \\ v_3 \\ v_4 \\ v_5 \\ v_6 \end{bmatrix} = \begin{bmatrix} 0 \\ 0 \\ 0 \\ 0 \\ 0 \\ 0 \\ 0 \end{bmatrix} \tag{2.7a}$$

or

$$\mathbf{B}_a \mathbf{v} = \mathbf{0} \tag{2.7b}$$

where $\mathbf{v}$ is the branch voltage vector and $\mathbf{B}_a$ is called the *loop matrix* of the digraph $\mathcal{G}$.

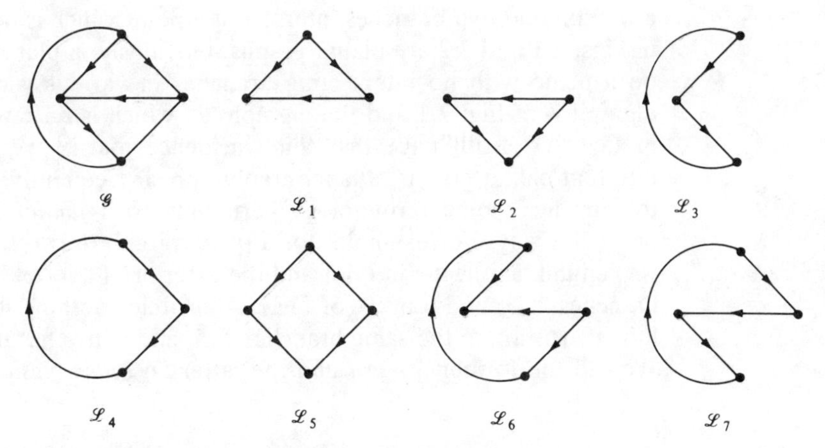

Figure 2.6 The seven distinct loops of $\mathcal{G}$ of Fig. 2.1.

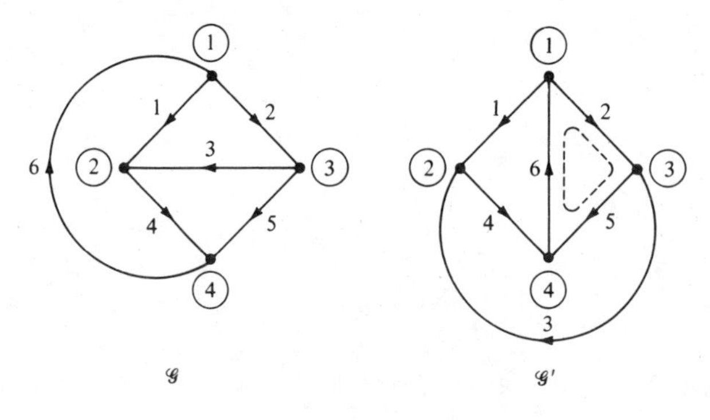

Figure 2.7 The graph $\mathscr{G}$ redrawn as another planar graph $\mathscr{G}'$. The loop consisting of branches 2, 5, and 6 is an outer mesh in $\mathscr{G}$ but a mesh in $\mathscr{G}'$.

Clearly the set of equations in (2.7a) is linearly dependent. Like the cut-set situation, we would like to know what represents a *maximal* and linearly independent subset, i.e., the largest set of linearly independent equations. Stated in another way, this subset represents a minimum set of loops which contain all the information pertaining to KVL equations for the circuit. In Sec. 3 we will prove we need $\ell = b - (n-1)$ loops and give a specific method of choosing an independent set. For the present example, $\ell = 3$, and $\mathscr{L}_1$, $\mathscr{L}_2$, and $\mathscr{L}_3$ represent such a maximal and linearly independent set. Thus, if we delete the superfluous equations corresponding to $\mathscr{L}_4$, $\mathscr{L}_5$, $\mathscr{L}_6$, and $\mathscr{L}_7$ from Eq. (2.7), we obtain

$$\mathbf{B}_R \mathbf{v} = \mathbf{0} \tag{2.8}$$

where $\mathbf{B}_R$ is called the *reduced loop matrix*. It has $\ell = b - (n-1)$ rows and b columns and may be considered as the dual of the reduced cut-set matrix.

Planar graph A *planar graph* is a graph which can be drawn on a plane in such a way that no two branches intersect at a point which is not a node. The graphs in Figs. 2.1 and 2.2 are planar graphs. Obviously, a planar graph can be drawn on a plane with no intersecting branches in various ways. For example, the digraph $\mathscr{G}$ of Fig. 2.1 and the digraph $\mathscr{G}'$, which is redrawn from $\mathscr{G}$, are shown in Fig. 2.7. Both have the same incidence matrix, yet they appear as two different planar graphs.[2] Planar graphs appear frequently in circuits. It is useful to introduce some terminology pertaining to a *planar graph* as *drawn* in a particular way. For example, for a planar graph drawn specifically on a plane, we can talk about the interior and the exterior of a loop. The loop consisting of branches 2, 5, and 6 in $\mathscr{G}'$ of Fig. 2.7 encircles nothing in its interior, and the loop consisting of the same branches 2, 5, and 6 in $\mathscr{G}$ has nothing in its exterior. We call the former a *mesh* and the latter an *outer mesh*.

[2] By relabeling branches and nodes we can, however, deduce one from the other.

Exercises

1. Show that a connected planar graph with no self-loops has $\ell = b - (n - 1)$ meshes.
2. Prove that the ℓ KVL equations based on ℓ meshes of a connected planar graph form a maximal and linearly independent set.
3. Consider the graph in Fig. 2.2; find all distinct loops and write the equation $\mathbf{B}_a \mathbf{v} = \mathbf{0}$. Determine the reduced loop matrix which represents the meshes of the planar graph as drawn in Fig. 2.2.

3 CUT SETS AND LOOPS BASED ON A TREE

3.1 Tree

One of the most important concepts in graph theory is the tree of a graph. It turns out that the concept of a tree also plays a central role in circuit analysis. It will be shown in this section that by picking a tree of $\mathcal{G}$ we can introduce a unique reduced cut-set matrix and a unique reduced loop matrix, called the fundamental cut-set matrix and the fundamental loop matrix, respectively, associated with a tree and thereby establish the basis for cut-set analysis and loop analysis.

A *tree* $\mathcal{T}$ of a connected digraph $\mathcal{G}$ is a subgraph which satisfies the following three fundamental properties: (*a*) it is connected, (*b*) it contains all the nodes of $\mathcal{G}$, and (*c*) it has no loops.

A digraph has many trees. For the digraph $\mathcal{G}$ in Fig. 2.1, numerous distinct trees exist. In Fig. 3.1, we show $\mathcal{G}$ together with four different trees.

Exercises

1. Determine the remaining trees of $\mathcal{G}$ not shown in Fig. 3.1.
2. $\mathcal{G}$ is called a *complete graph* or a *clique* iff every node of $\mathcal{G}$ is connected to every other node of $\mathcal{G}$ by a branch. Show that for a complete graph there exist n^{n-2} distinct trees.

Given a connected digraph $\mathcal{G}$ and a tree $\mathcal{T}$, we can partition the branches of $\mathcal{G}$ into two disjoint sets: those which belong to $\mathcal{T}$, called *tree branches* or, in short, *twigs*, and those which do not belong to $\mathcal{T}$, called *links* (*some authors call them chords and others call them cotree branches*).

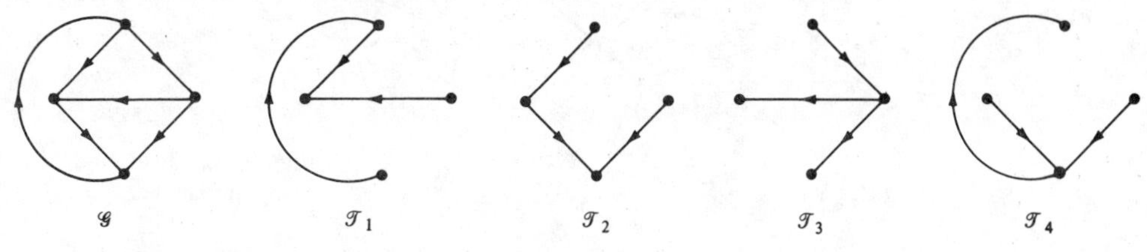

Figure 3.1 Four distinct trees of the digraph $\mathcal{G}$.

Fundamental theorem of graphs Given a connected digraph $\mathcal{G}$ with n nodes and b branches and a tree $\mathcal{T}$ of $\mathcal{G}$.

1. There is a unique path (disregard the branch orientation) along the tree between any pairs of nodes.
2. There are $n-1$ twigs and $\ell = b-(n-1)$ links.
3. Every twig of $\mathcal{T}$ together with some links defines a unique cut set, called the *fundamental cut set associated with the twig.*
4. Every link of $\mathcal{T}$ and the unique path on the tree between its two nodes constitute a unique loop, called the *fundamental loop associated with the link.*

PROOF

1. Since the digraph is connected, there exists a path between any two nodes. If there were two paths between node ① and node ②, the two paths would form a loop, which violates property (c) of a tree.
2. Let $\mathcal{T}$ be a tree of $\mathcal{G}$, then it is a subgraph which contains n nodes according to property (b). Furthermore, due to property (a), starting from any node, if we trace along the tree, we can reach all the nodes of $\mathcal{G}$. Each time we reach a new node, we uncover a twig. Thus the total number of twigs must be equal to the total number of nodes on the tree minus one. Since there are altogether b branches, the number of links is equal to $\ell = b-(n-1)$.
3. Consider the twig b_1 of an arbitrary digraph $\mathcal{G}$ shown in Fig. 3.2. Let us first remove b_1 from $\mathcal{T}$. What remains of $\mathcal{T}$ is then two separate parts, say $\mathcal{T}_1$ and $\mathcal{T}_2$. Since every link connects a node of $\mathcal{T}$ to another node of $\mathcal{T}$, let us consider only those links which connect a node of $\mathcal{T}_1$ to a node of $\mathcal{T}_2$. It is easily seen that this particular set of links together with the twig b_1 constitutes a unique cut set.
4. Consider a link ℓ_1 between two nodes ① and ②. By 1, there is a unique path from ① to ② on the tree. This path together with the link ℓ_1 forms a loop, which is unique. There cannot be any other loop since a tree has no loops. ∎

Exercise List all fundamental cut sets and fundamental loops of the digraph $\mathcal{G}$ associated with the tree $\mathcal{T}_1$ in Fig. 3.1.

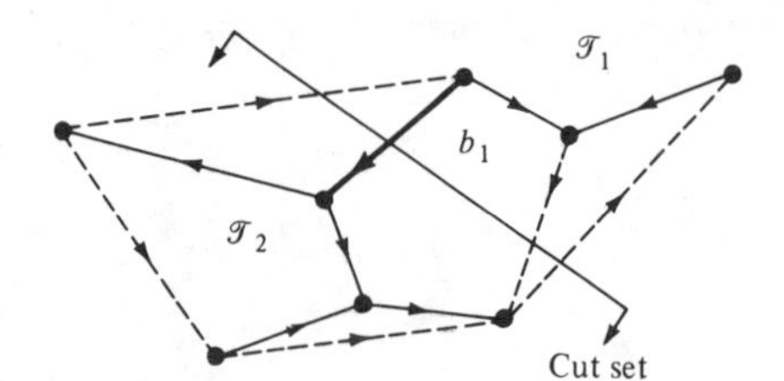

Figure 3.2 Illustration of the properties of a fundamental cut set.

3.2 The Fundamental Cut-Set Matrix Associated with a Tree

Consider a connected digraph $\mathcal{G}$ with b branches and n nodes. Pick a tree $\mathcal{T}$. There are $n-1$ twigs. From the fundamental theorem of graphs, we know that each twig defines a unique cut set. We can therefore write $n-1$ KCL equations from the $n-1$ fundamental cut sets. Clearly this set of $n-1$ equations is linearly independent because each equation contains one and only one twig current.

Examples

1. **(KCL equations based on fundamental cut sets associated with a tree)** Let us consider the digraph shown in Fig. 3.3, where $b=9$ and $n=6$; thus $\ell=4$. A tree $\mathcal{T}$ is picked as shown. There are $n-1=5$ twigs. The five fundamental cut sets are marked with dashed lines. Each fundamental cut set is given an orientation defined by the direction of the twig. KCL for the five fundamental cut sets specifies the following equations:

Cut set 1: $$i_1 - i_6 = 0$$

Cut set 2: $$i_2 - i_6 + i_8 + i_9 = 0$$

Cut set 3: $$i_3 + i_7 + i_8 + i_9 = 0$$

Cut set 4: $$i_4 - i_7 = 0$$

Cut set 5: $$i_5 + i_9 = 0$$

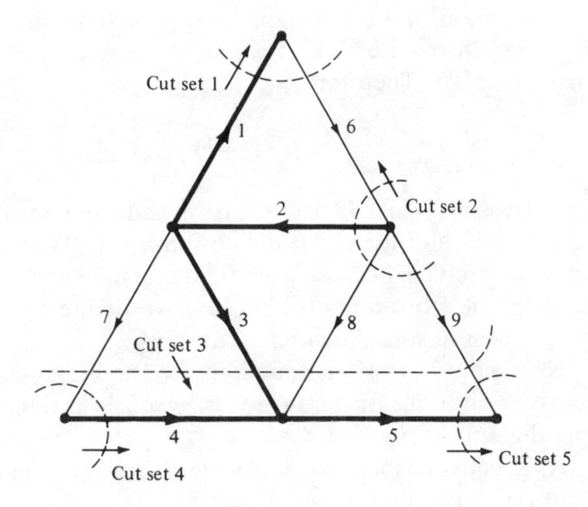

Figure 3.3 Example illustrating the use of fundamental cut sets in writing KCL equations.

In matrix form, we have

$$n-1 \text{ Cut sets} \underbrace{\begin{bmatrix} 1 & 0 & 0 & 0 & 0 & -1 & 0 & 0 & 0 \\ 0 & 1 & 0 & 0 & 0 & -1 & 0 & 1 & 1 \\ 0 & 0 & 1 & 0 & 0 & 0 & 1 & 1 & 1 \\ 0 & 0 & 0 & 1 & 0 & 0 & -1 & 0 & 0 \\ 0 & 0 & 0 & 0 & 1 & 0 & 0 & 0 & 1 \end{bmatrix}}_{n-1 \text{ twigs} \quad\quad \ell \text{ links}} \begin{bmatrix} i_1 \\ i_2 \\ i_3 \\ \cdot \\ \cdot \\ \cdot \\ i_9 \end{bmatrix} = \begin{bmatrix} 0 \\ 0 \\ 0 \\ 0 \\ 0 \end{bmatrix} \tag{3.1}$$

In general, we can write KCL equations based on the fundamental cut sets of a tree:

$$\mathbf{Qi} = \mathbf{0} \tag{3.2}$$

where $\mathbf{Q}$ is an $(n-1) \times b$ matrix called the *fundamental cut-set matrix* associated with a tree $\mathcal{T}$. Its jkth element is defined as follows:

$$q_{jk} = \begin{cases} 1 & \text{if branch } k \text{ belongs to cut set } j \text{ and has the same direction} \\ -1 & \text{if branch } k \text{ belongs to cut set } j \text{ and has the opposite direction} \\ 0 & \text{if branch } k \text{ does not belong to cut set } j \end{cases}$$

Here, the direction of each cut set j is defined to be the direction of its associated twig. We can write the fundamental cut-set matrix associated with the tree $\mathcal{T}$ in a convenient form if we label the b branches as follows: the twigs first, i.e., $1, 2, \ldots, n-1$, and the links next, that is $n, n+1, \ldots, b$. Therefore,

$$\mathbf{Q} = [\mathbf{1}_{n-1}, \mathbf{Q}_\ell] \tag{3.3}$$

where $\mathbf{Q}_\ell$ is a submatrix of $n-1$ rows and ℓ columns made of 0, 1, and -1 and $\mathbf{1}_{n-1}$ is a unit matrix of dimension $n-1$. The reason that there is the unit matrix in $\mathbf{Q}$ is because each fundamental cut set contains one and only one twig, and, furthermore, because we define the orientation of the cut set according to the direction of the twig.

Next we turn to KVL equations. We note that each branch voltage can be written in terms of the twig voltages. Let us first illustrate this with the same digraph in Fig. 3.3.

2. **(KVL equations using twig voltages)** Let the twig voltages be labeled as $v_{t1}, v_{t2}, \ldots, v_{t5}$, that is, we write

$$v_1 = v_{t1}$$
$$v_2 = v_{t2}$$
$$v_3 = v_{t3}$$
$$v_4 = v_{t4}$$
$$v_5 = v_{t5}$$

Then, using KVL for the fundamental loops defined by the four links, we have

$$\begin{aligned} v_6 &= -v_1 - v_2 &&= -v_{t1} - v_{t2} \\ v_7 &= v_3 - v_4 &&= v_{t3} - v_{t4} \\ v_8 &= v_2 + v_3 &&= v_{t2} + v_{t3} \\ v_9 &= v_2 + v_3 + v_5 &&= v_{t2} + v_{t3} + v_{t5} \end{aligned}$$

Combining the above two sets of equations, we have, in matrix form,

$$\begin{bmatrix} v_1 \\ v_2 \\ \\ \cdot \\ \cdot \\ \cdot \\ \\ \\ v_9 \end{bmatrix} = \underbrace{\begin{bmatrix} 1 & 0 & 0 & 0 & 0 \\ 0 & 1 & 0 & 0 & 0 \\ 0 & 0 & 1 & 0 & 0 \\ 0 & 0 & 0 & 1 & 0 \\ 0 & 0 & 0 & 0 & 1 \\ -1 & -1 & 0 & 0 & 0 \\ 0 & 0 & 1 & -1 & 0 \\ 0 & 1 & 1 & 0 & 0 \\ 0 & 1 & 1 & 0 & 1 \end{bmatrix}}_{n-1} \begin{bmatrix} v_{t1} \\ v_{t2} \\ v_{t3} \\ v_{t4} \\ v_{t5} \end{bmatrix} \tag{3.4}$$

In general, the KVL equations are of the form

$$\mathbf{v} = \mathbf{Q}^T \mathbf{v}_t \tag{3.5}$$

where $\mathbf{v}_t$ is the twig voltage vector and $\mathbf{Q}^T$ is the transpose of the fundamental cut-set matrix associated with $\mathcal{T}$. To prove that Eq. (3.5) is valid, we first assume that $\mathbf{v} = \mathbf{D}\mathbf{v}_t$ where $\mathbf{D}$ is a $b \times (n-1)$ matrix expressing the linear combinations. This says that the branch voltages v_k, $k = 1, 2, \ldots, b$, are written in terms of the twig voltages v_{tj}, $j = 1, 2, \ldots, n-1$, by using KVL. Thus, we have

$$v_k = \sum_{j=1}^{n=1} d_{kj} v_{tj}$$

where d_{kj} is given by the following:

$$d_{kj} = \begin{cases} 1 & \text{if branch } k \text{ is in cut set } j \text{ and their reference directions are the same} \\ -1 & \text{if branch } k \text{ is in cut set } j \text{ and their reference directions are opposite} \\ 0 & \text{if branch } k \text{ is not in cut set } j \end{cases}$$

Comparing d_{kj} with q_{jk} of the fundamental cut-set matrix $\mathbf{Q}$, we find that $\mathbf{D}$ is indeed the transpose of $\mathbf{Q}$. Thus $\mathbf{D} = \mathbf{Q}^T$ and Eq. (3.5) is proven. The result of Example 2 can be checked with that of Example 1 to verify this conclusion.

We summarize the results obtained in this subsection based on the fundamental cut-set matrix associated with a tree. We consider a circuit with a *connected* digraph $\mathscr{G}$ of n nodes and b branches; we choose a tree $\mathscr{T}$.

> Kirchhoff's equations based on the fundamental cut-set matrix $\mathbf{Q}$ of $(n-1)$ rows and b columns:
>
> $$\text{KCL:} \quad \mathbf{Qi} = \mathbf{0}$$
> $$\text{KVL:} \quad \mathbf{v} = \mathbf{Q}^T\mathbf{v}_t \qquad (3.6)$$
>
> If twigs are labeled first from 1 to $n-1$, then
>
> $$\mathbf{Q} = [\mathbf{1}_{n-1}, \mathbf{Q}_\ell]$$

REMARKS

1. The two equations in (3.6) are generalizations of the following familiar Kirchhoff's equations based on nodes and the reduced incidence matrix $\mathbf{A}$:

$$\text{KCL:} \quad \mathbf{Ai} = \mathbf{0}$$
$$\text{KVL:} \quad \mathbf{v} = \mathbf{A}^T\mathbf{e} \qquad (3.7)$$

2. Obviously, the fundamental cut-set matrix $\mathbf{Q}$ associated with a tree represents a special case of the reduced cut-set matrix $\mathbf{Q}_R$ introduced in Sec. 2. While for many digraphs, the reduced incidence matrix $\mathbf{A}$ is a special case of the fundamental cut-set matrix, i.e., a tree can be chosen such that the $\mathbf{Q}$ obtained is identical with the reduced incidence matrix for a particular datum node. This is not always possible as in the graph in Fig. 3.3.

Exercise Determine the conditions on the digraph $\mathscr{G}$ under which the reduced incidence matrix represents a special case of the fundamental cut-set matrix.

3.3 The Fundamental Loop Matrix Associated with a Tree

The treatment of this subsection is the dual of that based on the fundamental cut sets. Instead of emphasizing the twig voltage variables, we consider the link currents. We use KCL to express all branch currents in terms of the ℓ link currents. We use KVL to write ℓ equations for the ℓ fundamental loops defined by the ℓ links. Let us again illustrate the above ideas by the same digraph in Fig. 3.3.

Example 3 (KVL equations based on fundamental loops associated with a tree) The digraph is redrawn in Fig. 3.4 to depict the fundamental loops. Again, we choose the same tree; thus there are four links. Since each link defines uniquely a fundamental loop, the four fundamental loops can be used to write four linearly independent KVL equations:

Loop 1: $$v_6 + v_1 + v_2 = 0$$

Loop 2: $$v_7 - v_3 + v_4 = 0$$

Loop 3: $$v_8 - v_2 - v_3 = 0$$

Loop 4: $$v_9 - v_2 - v_3 - v_5 = 0$$

In matrix form, we have

$$\ell \text{ loops} \underbrace{\begin{bmatrix} 1 & 1 & 0 & 0 & 0 \\ 0 & 0 & -1 & 1 & 0 \\ 0 & -1 & -1 & 0 & 0 \\ 0 & -1 & -1 & 0 & -1 \end{bmatrix}}_{n-1 \text{ twigs}} \!\!\! \underbrace{\begin{matrix} 1 & 0 & 0 & 0 \\ 0 & 1 & 0 & 0 \\ 0 & 0 & 1 & 0 \\ 0 & 0 & 0 & 1 \end{matrix}}_{\ell \text{ links}} \begin{bmatrix} v_1 \\ v_2 \\ v_3 \\ \cdot \\ \cdot \\ \cdot \\ v_9 \end{bmatrix} = \begin{bmatrix} 0 \\ 0 \\ 0 \\ 0 \end{bmatrix} \tag{3.8}$$

In general, we can write KVL equations based on the fundamental loops as

$$\mathbf{Bv} = \mathbf{0} \tag{3.9}$$

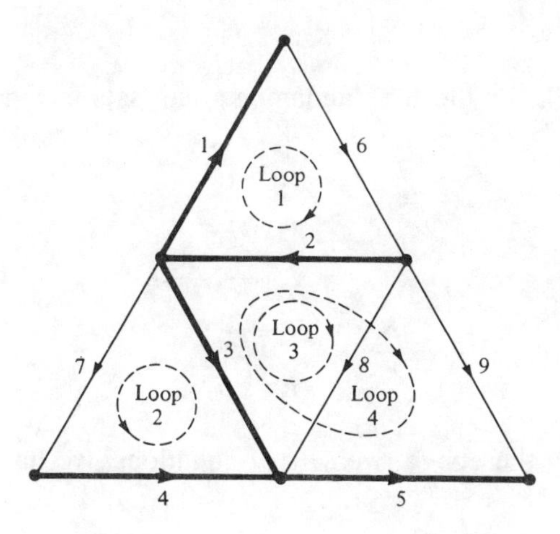

Figure 3.4 Example illustrating the use of fundamental loops associated with tree $\mathcal{T}$ in writing KVL equations.

where **B** is an $\ell \times b$ matrix called the *fundamental loop matrix* associated with the tree $\mathcal{T}$. Its *jk*th element is defined as follows:

$$b_{jk} = \begin{cases} 1 & \text{if branch } k \text{ is in loop } j \text{ and their reference directions are the same} \\ -1 & \text{if branch } k \text{ is in loop } j \text{ and their reference directions are opposite} \\ 0 & \text{if branch } k \text{ is not in loop } j \end{cases}$$

Here, the reference direction of each loop j is defined by the direction of its associated link. Using the same labeling scheme as before, i.e., the twigs are labeled first from 1 to $n-1$ and the links next from n to b, we can express **B** in the form

$$\mathbf{B} = [\mathbf{B}_t, \mathbf{1}_\ell] \tag{3.10}$$

where $\mathbf{1}_\ell$ is a unit matrix of dimension ℓ and $\mathbf{B}_t$ is a submatrix of ℓ rows and $n-1$ columns made of 0, 1, and -1. Again, the unit matrix in **B** is due to the fact that each fundamental loop contains one and only one link and the reference directions for the fundamental loops are the same as that of the links.

Next we turn to KCL. We note that each branch current can be expressed in terms of the ℓ link currents. We illustrate this with the following example.

Example 4 (KCL equations using link currents) For the digraph in Fig. 3.4, let the link currents be $i_{\ell 1}$, $i_{\ell 2}$, $i_{\ell 3}$, and $i_{\ell 4}$; then we can identify the four branch currents as:

$$i_6 = i_{\ell 1}$$
$$i_7 = i_{\ell 2}$$
$$i_8 = i_{\ell 3}$$
$$i_9 = i_{\ell 4}$$

Using KCL for the five fundamental cut sets in Fig. 3.3, we have

$$i_1 = i_{\ell 1}$$
$$i_2 = i_{\ell 1} - i_{\ell 3} - i_{\ell 4}$$
$$i_3 = -i_{\ell 2} - i_{\ell 3} - i_{\ell 4}$$
$$i_4 = i_{\ell 2}$$
$$i_5 = -i_{\ell 4}$$

Combining the above two sets of equations, we have

$$\begin{bmatrix} i_1 \\ i_2 \\ \\ \vdots \\ \\ \\ \\ \\ i_9 \end{bmatrix} = \underbrace{\begin{bmatrix} 1 & 0 & 0 & 0 \\ 1 & 0 & -1 & -1 \\ 0 & -1 & -1 & -1 \\ 0 & 1 & 0 & 0 \\ 0 & 0 & 0 & -1 \\ 1 & 0 & 0 & 0 \\ 0 & 1 & 0 & 0 \\ 0 & 0 & 1 & 0 \\ 0 & 0 & 0 & 1 \end{bmatrix}}_{\ell} \begin{bmatrix} i_{\ell 1} \\ i_{\ell 2} \\ i_{\ell 3} \\ i_{\ell 4} \end{bmatrix} \tag{3.11}$$

In general, the KCL equations are of the form

$$\mathbf{i} = \mathbf{B}^T \mathbf{i}_\ell \tag{3.12}$$

where $\mathbf{i}_\ell$ is the link current vector and $\mathbf{B}^T$ is the transpose of the fundamental loop matrix associated with the tree $\mathcal{T}$. Equation (3.12) is the dual of Eq. (3.5). The proof of the validity of Eq. (3.12) is left as an exercise.

We summarize the result of this subsection based on the fundamental loop matrix associated with a tree:

> Kirchhoff's equations based on the fundamental loop matrix $\mathbf{B}$ of ℓ rows and b columns
>
> KVL: $\mathbf{Bv} = \mathbf{0}$
>
> KCL: $\mathbf{i} = \mathbf{B}^T \mathbf{i}_\ell$ (3.13)
>
> If twigs are numbered first from 1 to $n - 1$,
>
> $$\mathbf{B} = [\mathbf{B}_t, \mathbf{1}_\ell]$$

REMARKS

1. The fundamental loop matrix $\mathbf{B}$ associated with a tree $\mathcal{T}$ is obviously a special case of the reduced loop matrix $\mathbf{B}_R$. In deriving the KVL equations $\mathbf{Bv} = \mathbf{0}$ based on a tree, we have demonstrated that $\mathbf{B}$ is of rank $\ell = b - (n - 1)$. Thus the number of a maximal and linearly independent set of KVL equations based on loops is equal to ℓ, i.e., $\mathbf{B}_R$ is of rank ℓ.
2. We have mentioned duality in deriving the fundamental loop matrix $\mathbf{B}$ from the fundamental cut-set matrix $\mathbf{Q}$. It should be clear to the reader by now that there indeed exist many dual terms. We summarize some of these in Table 3.1.

Table 3.1 Dual terms in loop and cut-set analysis

Loop analysis	Cut-set analysis
Link	Twig
Fundamental loop	Fundamental cut set
Link current, $\mathbf{i}_\ell$	Twig voltage, $\mathbf{v}_t$
Fundamental loop matrix, **B**	Fundamental cut-set matrix, **Q**

3.4 Relation between Q and B

Given a connected digraph $\mathscr{G}$, we pick a tree $\mathscr{T}$. The branches can be partitioned into twigs and links. The twigs define a unique set of fundamental cut sets, and the links define a unique set of fundamental loops. The fundamental cut-set matrix **Q** gives the topological relation between the branches and the fundamental cut sets, while the fundamental loop matrix **B** gives the topological relation between the branches and the fundamental loops. Since **B** and **Q** come from the same tree, naturally, we expect that there exists a relation between the two matrices **Q** and **B**. The following theorem gives the precise relation.

Theorem Let **Q** and **B** be the fundamental cut-set matrix and the fundamental loop matrix, respectively, of a connected digraph $\mathscr{G}$ for a specified tree $\mathscr{T}$; then

$$\mathbf{B}\mathbf{Q}^T = \mathbf{0} \tag{3.14}$$

Proof We have, for an arbitrary twig voltage vector $\mathbf{v}_t = [v_{t1}, v_{t2}, \ldots, v_{t(n-1)}]^T$,

$$\mathbf{v} = \mathbf{Q}^T \mathbf{v}_t \tag{3.15}$$

This says that the b-vector **v** is expressed by KVL in terms of linear combinations of v_{tk}'s by the matrix $\mathbf{Q}^T$. Next, we have, by KVL, a set of linear constraints on the b-vector **v**, given by

$$\mathbf{B}\mathbf{v} = \mathbf{0} \tag{3.16}$$

Thus, premultiplying Eq. (3.15) by **B** and using Eq. (3.16), we obtain

$$\mathbf{B}\mathbf{Q}^T\mathbf{v}_t = \mathbf{0} \qquad \text{for all } \mathbf{v}_t \tag{3.17}$$

Thus,

$$\mathbf{B}\mathbf{Q}^T = \mathbf{0}$$

Note that it is important to emphasize that Eq. (3.17) must be valid *for all* $\mathbf{v}_t$ in order to obtain our conclusion (3.14). ∎

Remarks

1. Consider the submatrices $\mathbf{Q}_\ell$ and $\mathbf{B}_t$ in Eqs. (3.3) and (3.10), respectively. We can derive a relation between the two submatrices $\mathbf{Q}_\ell$ and $\mathbf{B}_t$. Since

$$\mathbf{Q} = [\mathbf{1}_{n-1}, \mathbf{Q}_\ell]$$

and

$$\mathbf{B} = [\mathbf{B}_t, \mathbf{1}_\ell]$$

$$\mathbf{B}\mathbf{Q}^T = [\mathbf{B}_t, \mathbf{1}_\ell]\begin{bmatrix} \mathbf{1}_{n-1} \\ \mathbf{Q}_\ell^T \end{bmatrix}$$

$$= \mathbf{B}_t + \mathbf{Q}_\ell^T = \mathbf{0}$$

We obtain the identities

$$\mathbf{B}_t = -\mathbf{Q}_\ell^T \qquad \text{and} \qquad \mathbf{B}_t^T = -\mathbf{Q}_\ell \tag{3.18}$$

We can check the above identities for Examples 1 and 3 in Eqs. (3.1) and (3.8).

2. Obviously, by taking the transpose of Eq. (3.14), we obtain

$$\mathbf{Q}\mathbf{B}^T = \mathbf{0} \tag{3.19}$$

3. Either Eq. (3.14) or (3.19) can be interpreted as an alternative form of Tellegen's theorem. Premultiplying Eq. (3.19) by $\mathbf{v}_t^T$ and then postmultiplying by $\mathbf{i}_\ell$, we obtain

$$\mathbf{v}_t^T\mathbf{Q}\mathbf{B}^T\mathbf{i}_\ell = 0 \tag{3.20}$$

Using Eqs. (3.15) and (3.12), we reduce (3.20) to

$$\mathbf{v}^T\mathbf{i} = 0$$

which is Tellegen's theorem for a digraph $\mathcal{G}$.

4 TABLEAU ANALYSIS

In tableau analysis, we separate the equations due to Kirchhoff's laws from that of the branch characteristic. The former represents the topological description of the digraph and, as we have learned, can be expressed in terms of the reduced incidence matrix **A**, the fundamental cut-set matrix **Q**, or the fundamental loop matrix **B**. Table 4.1 summarizes the three sets of equations given by Eqs. (3.6), (3.7), and (3.13).

Table 4.1 Summary of Kirchhoff's equations based on nodes, fundamental cut sets, and fundamental loops

	Matrix of the digraph		
	A	**Q**	**B**
KCL	$\mathbf{A}\mathbf{i} = \mathbf{0}$	$\mathbf{Q}\mathbf{i} = \mathbf{0}$	$\mathbf{i} - \mathbf{B}^T\mathbf{i}_\ell = \mathbf{0}$
KVL	$\mathbf{v} - \mathbf{A}^T\mathbf{e} = \mathbf{0}$	$\mathbf{v} - \mathbf{Q}^T\mathbf{v}_t = \mathbf{0}$	$\mathbf{B}\mathbf{v} = \mathbf{0}$

In addition to the branch voltage vector $\mathbf{v}$ and the branch current vector $\mathbf{i}$, the equations in Table 4.1 contain the node-to-datum voltage vector $\mathbf{e}$, the twig voltage vector $\mathbf{v}_t$, and the link current vector $\mathbf{i}_\ell$. As we know, the node-to-datum voltage $\mathbf{e}$ is the *basic variable* used in node analysis. As we will learn in the next section, the twig voltage $\mathbf{v}_t$ is the basic variable used for the cut-set analysis and the link current $\mathbf{i}_\ell$ is the basic variable used for the loop analysis.

The second set of equations in tableau analysis is the branch equations relating the branch voltage vector $\mathbf{v}$ and the branch current vector $\mathbf{i}$. In Chap. 8 we used the equation

$$\mathbf{h}(\mathbf{v}, \mathbf{i}, \dot{\mathbf{v}}, \dot{\mathbf{i}}, t) = \mathbf{0} \tag{4.1}$$

to represent the branch equation for a general nonlinear dynamic circuit. For linear time-invariant dynamic circuits, the equations can be written in the form

$$(\mathbf{M}_0 D + \mathbf{M}_1)\mathbf{v} + (\mathbf{N}_0 D + \mathbf{N}_1)\mathbf{i} = \mathbf{u}_s(t) \tag{4.2}$$

where $\mathbf{M}_0, \mathbf{N}_0, \mathbf{M}_1$, and $\mathbf{N}_1$ are constant $b \times b$ matrices, D is the differentiation operator d/dt, and $\mathbf{u}_s(t)$ is the input vector due to both independent voltage and current sources.

If we combine Kirchhoff's equations in Table 4.1 and the branch equation in (4.1) or (4.2), we have three forms of tableau equations. They are referred to as the *node tableau equations*, *cut-set tableau equations*, and *loop tableau equations*.

Let us specialize our discussion to linear time-invariant circuits. We will use the *frequency domain* and write tableau equations in terms of the Laplace transform variables: $\mathbf{V}$, $\mathbf{I}$, $\mathbf{V}_t$, and $\mathbf{I}_\ell$ corresponding to $\mathbf{v}$, $\mathbf{i}$, $\mathbf{v}_t$, and $\mathbf{i}_\ell$ in the time domain. The equations in the frequency domain based on cut sets become

KCL: $$\mathbf{Q}\mathbf{I}(s) = \mathbf{0}$$

KVL: $$\mathbf{V}(s) - \mathbf{Q}^T\mathbf{V}_t(s) = \mathbf{0} \tag{4.3}$$

Branches: $$\mathbf{M}(s)\mathbf{V}(s) + \mathbf{N}(s)\mathbf{I}(s) = \mathbf{U}_s(s) + \mathbf{U}^i \tag{4.4}$$

where $\mathbf{M}(s) \triangleq \mathbf{M}_0 s + \mathbf{M}_1$, $\mathbf{N}(s) \triangleq \mathbf{N}_0 s + \mathbf{N}_1$, $\mathbf{U}_s(s)$ is the Laplace transform of the source vector $\mathbf{u}_s(t)$, and $\mathbf{U}^i$ represents contributions of initial conditions on the capacitors and inductors.

Equations (4.3) and (4.4) can be combined and put into matrix form:

Cut set tableau equations

$$\begin{bmatrix} \mathbf{0} & \mathbf{0} & \mathbf{Q} \\ -\mathbf{Q}^T & \mathbf{1}_b & \mathbf{0} \\ \mathbf{0} & \mathbf{M} & \mathbf{N} \end{bmatrix} \begin{bmatrix} \mathbf{V}_t \\ \mathbf{V} \\ \mathbf{I} \end{bmatrix} = \begin{bmatrix} \mathbf{0} \\ \mathbf{0} \\ \mathbf{U}_s + \mathbf{U}^i \end{bmatrix} \tag{4.5}$$

where for clarity the dependence of $\mathbf{V}_t$, $\mathbf{V}$, $\mathbf{I}$, $\mathbf{M}$, $\mathbf{N}$, and $\mathbf{U}_s$ on the variable s has been suppressed. Note that the cut-set tableau equation above is similar to the familiar node tableau equation. The differences are that the node-to-datum voltage vector $\mathbf{E}(s)$ has been replaced by the twig voltage vector $\mathbf{V}_t(s)$ and that the reduced incidence matrix $\mathbf{A}$ has been replaced by the fundamental cut-set matrix $\mathbf{Q}$. There are altogether $2b + n - 1$ scalar equations in Eq. (4.5).

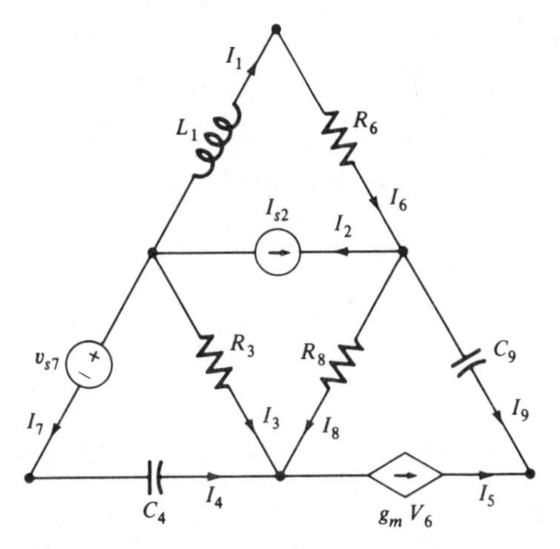

Figure 4.1 Example illustrating loop tableau analysis.

If we write Kirchhoff's equations based on loops, we have instead of Eq. (4.3)

$$\begin{aligned} &\text{KCL:} && \mathbf{I} - \mathbf{B}^T\mathbf{I}_\ell = \mathbf{0} \\ &\text{KVL:} && \mathbf{BV} = \mathbf{0} \end{aligned} \tag{4.6}$$

Combining Eqs. (4.4) and (4.6) we have the loop tableau equations for linear time-invariant circuits in the frequency domain:

Loop tableau equations

$$\begin{bmatrix} -\mathbf{B}^T & \mathbf{0} & \mathbf{1}_b \\ \mathbf{0} & \mathbf{B} & \mathbf{0} \\ \mathbf{0} & \mathbf{M} & \mathbf{N} \end{bmatrix} \begin{bmatrix} \mathbf{I}_\ell \\ \mathbf{V} \\ \mathbf{I} \end{bmatrix} = \begin{bmatrix} \mathbf{0} \\ \mathbf{0} \\ \mathbf{U}_s + \mathbf{U}^i \end{bmatrix} \tag{4.7}$$

The total number of scalar equations here is $2b + \ell$.

Example Consider the circuit shown in Fig. 4.1. We choose the same tree as in Sec. 3 and number the branches like before. In the frequency domain, the branch equations are

$$\begin{aligned} V_1 - sL_1I_1 &= 0 + V_1^i \\ -I_2 &= I_{s2} \\ V_3 - R_3I_3 &= 0 \\ sC_4V_4 - I_4 &= 0 + I_4^i \\ g_mV_6 - I_5 &= 0 \\ V_6 - R_6I_6 &= 0 \\ V_7 &= V_{s7} \\ V_8 - RI_8 &= 0 \\ sC_9V_9 - I_9 &= 0 + I_9^i \end{aligned}$$

where V_1^i, I_4^i, and I_9^i represent initial conditions, pertaining to L_1, C_4, and C_9, respectively.[3] The loop tableau equation in matrix form is

[3] For example, $V_1^i = L_1i_1(0-)$ is the initial flux in the inductor L_1, and $I_4^i = C_4v_4(0-)$ is the initial charge on the capacitor C_4.

$$
\left[\begin{array}{c:c:c}
\begin{array}{cccc}
-1 & 0 & 0 & 0 \\
-1 & 0 & 1 & 1 \\
0 & 1 & 1 & 1 \\
0 & -1 & 0 & 0 \\
0 & 0 & 0 & 1 \\
-1 & 0 & 0 & 0 \\
0 & -1 & 0 & 0 \\
0 & 0 & -1 & 0 \\
0 & 0 & 0 & -1
\end{array} & \mathbf{0} & \mathbf{1} \\
\hdashline
\mathbf{0} &
\begin{array}{ccccccccc}
1 & 1 & 0 & 0 & 0 & 1 & 0 & 0 & 0 \\
0 & 0 & -1 & 1 & 0 & 0 & 1 & 0 & 0 \\
0 & -1 & -1 & 0 & 0 & 0 & 0 & 1 & 0 \\
0 & -1 & -1 & 0 & -1 & 0 & 0 & 0 & 1
\end{array} & \mathbf{0} \\
\hdashline
\mathbf{0} &
\begin{array}{ccccccccc}
1 & & & & & & & & \\
 & 0 & & & & & & & \\
 & & 1 & & & & & & \\
 & & & sC_4 & & & & & \\
 & & & & 0 & g_m & & & \\
 & & & & & 1 & & & \\
 & & & & & & 1 & & \\
 & & & & & & & 1 & \\
 & & & & & & & & sC_9
\end{array} &
\begin{array}{ccccccccc}
-sL_1 & & & & & & & & \\
 & -1 & & & & & & & \\
 & & -R_3 & & & & & & \\
 & & & -1 & & & & & \\
 & & & & -1 & & & & \\
 & & & & & -R_6 & & & \\
 & & & & & & 0 & & \\
 & & & & & & & -R_8 & \\
 & & & & & & & & -1
\end{array}
\end{array}\right]
\left[\begin{array}{c}
\mathbf{I}_{\ell 1} \\ \mathbf{I}_{\ell 2} \\ \mathbf{I}_{\ell 3} \\ \mathbf{I}_{\ell 4} \\
\hdashline
\mathbf{V}_1 \\ \mathbf{V}_2 \\ \mathbf{V}_3 \\ \mathbf{V}_4 \\ \mathbf{V}_5 \\ \mathbf{V}_6 \\ \mathbf{V}_7 \\ \mathbf{V}_8 \\ \mathbf{V}_9 \\
\hdashline
\mathbf{I}_1 \\ \mathbf{I}_2 \\ \mathbf{I}_3 \\ \mathbf{I}_4 \\ \mathbf{I}_5 \\ \mathbf{I}_6 \\ \mathbf{I}_7 \\ \mathbf{I}_8 \\ \mathbf{I}_9
\end{array}\right]
=
\left[\begin{array}{c:c}
\multicolumn{2}{c}{\mathbf{0}} \\
\hdashline
\multicolumn{2}{c}{\mathbf{0}} \\
\hdashline
 & \mathbf{V}_1^i \\
\mathbf{I}_{s2} & \\
0 & \\
 & \mathbf{I}_4^i \\
0 & \\
0 & \\
\mathbf{V}_{s7} & \\
0 & \\
 & \mathbf{I}_9^i
\end{array}\right]
$$

REMARK One important property of the tableau equations is that there always exists a large number of zeros in the equations. Thus tableau equations are often referred to as *sparse tableau equations*. The word "sparse" refers to the fact that the number of nonzero elements in the tableau matrix is small in comparison to the total number of entries in the matrix: $(2b + \ell)^2$ in the loop tableau equations and $(2b + n - 1)^2$ in the node or cut-set tableau equations.

In numerical computation, there exist many efficient sparse matrix algorithms which are crucial in the analysis of large circuits. As a matter of fact, while the solution of say m linear simultaneous algebraic equations takes on the order of $m^3/3$ computations (addition and multiplication) by gaussian elimination, it is found experimentally that it takes approximately $m^{1.4}$ computations for a sparse system by using an efficient sparse matrix algorithm. Because of that we are not very concerned with the size of the total number of equations.

5 LINEAR TIME-INVARIANT CIRCUIT ANALYSIS

In this section we consider only linear time-invariant circuits and derive the cut-set equations and loop equations from the tableau equations. Also, mesh analysis as a special case of loop analysis will be illustrated.

5.1 Cut-Set Analysis

Like node analysis the cut-set analysis is restricted to circuit elements which are *voltage-controlled*. Thus we assume that any voltage source, if present, is in series with a two-terminal element. For convenience in the derivation we also assume that any current source is in parallel with a two-terminal element. If the assumptions are not valid, source transformations as discussed in Sec. 1 may be performed first.

A. Branch equations The branch equation (4.4) can be put in terms of the branch admittance matrix, $\mathbf{Y}_b$. It is useful then to partition the elements into three groups representing resistive, capacitive, and inductive elements. Then, suppose we label the circuit elements by listing the resistors first, capacitors next, and inductors last; then we have

$$\mathbf{V} = \begin{bmatrix} \mathbf{V}_R \\ \mathbf{V}_C \\ \mathbf{V}_L \end{bmatrix} \quad \text{and} \quad \mathbf{I} = \begin{bmatrix} \mathbf{I}_R \\ \mathbf{I}_C \\ \mathbf{I}_L \end{bmatrix} \tag{5.1}$$

as the branch voltage and current vectors in the frequency domain.

We assumed that all the elements are voltage-controlled; thus the branch equations of the resistors can be written as

$$\mathbf{G}\mathbf{V}_R - \mathbf{I}_R = \mathbf{I}_{sR} \tag{5.2}$$

where **G** is the conductance matrix. The resistive elements allowed are two-terminal resistors, VCCS, and gyrators; hence **G** is not necessarily diagonal. A typical branch as shown in Fig. 5.1, which includes an independent voltage source in series with a resistor, has for its branch equation

$$G_k V_k - I_k = G_k V_{sk} \tag{5.3}$$

For the capacitive elements, we have

$$s\mathbf{C}\mathbf{V}_C - \mathbf{I}_C = \mathbf{I}_{sC} + \mathbf{I}_C^i \tag{5.4}$$

where **C** is a diagonal capacitance matrix, $\mathbf{I}_{sC}$ represents the current source vector associated with the capacitors, and $\mathbf{I}_C^i$ represents the initial condition current vector. A typical branch, which includes an independent current source in parallel with a capacitor as shown in Fig. 5.2, has for its branch equation

$$sC_k V_k - I_k = I_{sk} + C_k v_k(0-) \tag{5.5}$$

For the inductive elements, we have

$$\mathbf{V}_L - s\mathbf{L}\mathbf{I}_L = \mathbf{V}_{sL} + \mathbf{V}_L^i \tag{5.6}$$

where **L** is a symmetric inductance matrix which is assumed to be nonsingular (otherwise it is not voltage-controlled). Thus $\mathbf{\Gamma} = \mathbf{L}^{-1}$ exists, and $\mathbf{\Gamma}$ is called the *reciprocal inductance matrix*. Equation (5.6) can be written as

$$\frac{1}{s}\mathbf{\Gamma}\mathbf{V}_L - \mathbf{I}_L = \mathbf{I}_{sL} + \mathbf{I}_L^i \tag{5.7}$$

An example of a coupled inductor with an independent voltage source is shown in Fig. 5.3. The branch equations in the time domain are

$$\begin{bmatrix} v_1 \\ v_2 \end{bmatrix} = D \begin{bmatrix} L_1 & M \\ M & L_2 \end{bmatrix} \begin{bmatrix} i_1 \\ i_2 \end{bmatrix} + \begin{bmatrix} v_{s1} \\ 0 \end{bmatrix}$$

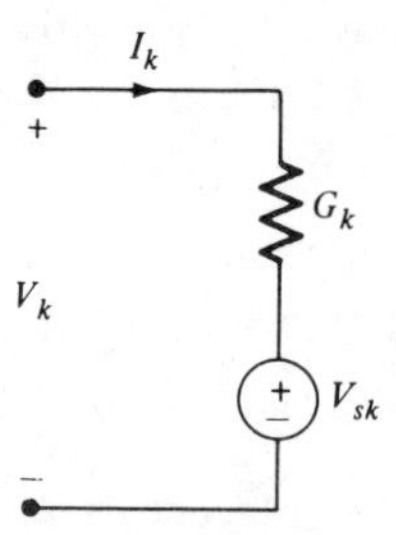

Figure 5.1 A resistive branch which contains a voltage source in series.

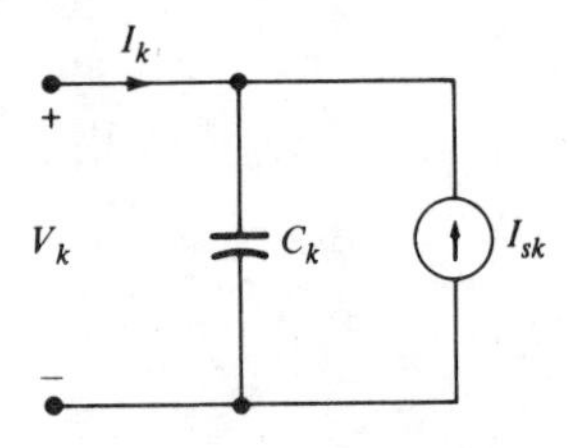

Figure 5.2 A capacitive branch with initial $v_k(0-)$ which contains a current source in parallel.

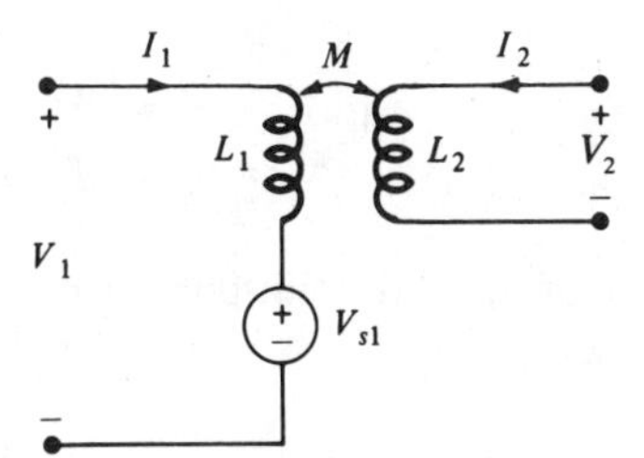

Figure 5.3 A coupled-inductor example.

Taking the Laplace transform of the above, we obtain

$$\begin{bmatrix} V_1 \\ V_2 \end{bmatrix} = s\mathbf{L}\begin{bmatrix} I_1 \\ I_2 \end{bmatrix} + \begin{bmatrix} V_{s1} \\ 0 \end{bmatrix} - \mathbf{L}\begin{bmatrix} i_1(0-) \\ i_2(0-) \end{bmatrix}$$

Premultiplying by $(1/s)\mathbf{L}^{-1}$, we obtain the branch equation in the form of Eq. (5.7):

$$\frac{1}{s}\boldsymbol{\Gamma}\begin{bmatrix} V_1 \\ V_2 \end{bmatrix} - \begin{bmatrix} I_1 \\ I_2 \end{bmatrix} = \frac{1}{s}\boldsymbol{\Gamma}\begin{bmatrix} V_{s1} \\ 0 \end{bmatrix} - \frac{1}{s}\begin{bmatrix} i_1(0-) \\ i_2(0-) \end{bmatrix}$$

Combining Eqs. (5.2), (5.4), and (5.6), we have

$$\begin{bmatrix} \mathbf{G} & \mathbf{0} & \mathbf{0} \\ \mathbf{0} & s\mathbf{C} & \mathbf{0} \\ \mathbf{0} & \mathbf{0} & \mathbf{1} \end{bmatrix}\begin{bmatrix} \mathbf{V}_R \\ \mathbf{V}_C \\ \mathbf{V}_L \end{bmatrix} - \begin{bmatrix} \mathbf{1} & \mathbf{0} & \mathbf{0} \\ \mathbf{0} & \mathbf{1} & \mathbf{0} \\ \mathbf{0} & \mathbf{0} & s\mathbf{L} \end{bmatrix}\begin{bmatrix} \mathbf{I}_R \\ \mathbf{I}_C \\ \mathbf{I}_L \end{bmatrix} = \begin{bmatrix} \mathbf{I}_{sR} \\ \mathbf{I}_{sC} \\ \mathbf{V}_{sL} \end{bmatrix} + \begin{bmatrix} \mathbf{0} \\ \mathbf{I}_C^i \\ \mathbf{V}_L^i \end{bmatrix} \tag{5.8}$$

which is of the form of Eq. (4.4) with

$$\mathbf{M} = \begin{bmatrix} \mathbf{G} & \mathbf{0} & \mathbf{0} \\ \mathbf{0} & s\mathbf{C} & \mathbf{0} \\ \mathbf{0} & \mathbf{0} & \mathbf{1} \end{bmatrix} \quad \text{and} \quad \mathbf{N} = -\begin{bmatrix} \mathbf{1} & \mathbf{0} & \mathbf{0} \\ \mathbf{0} & \mathbf{1} & \mathbf{0} \\ \mathbf{0} & \mathbf{0} & s\mathbf{L} \end{bmatrix} \tag{5.9}$$

Since $\mathbf{L}$ is nonsingular, $\mathbf{N}^{-1}$ exists, and we have from Eq. (5.9)

$$-\mathbf{N}^{-1}\mathbf{M} = \begin{bmatrix} \mathbf{G} & \mathbf{0} & \mathbf{0} \\ \mathbf{0} & s\mathbf{C} & \mathbf{0} \\ \mathbf{0} & \mathbf{0} & \frac{1}{s}\boldsymbol{\Gamma} \end{bmatrix} \triangleq \mathbf{Y}_b \tag{5.10}$$

where $\mathbf{Y}_b$ is the branch admittance matrix. The branch equation can be written in terms of $\mathbf{Y}_b$ by combining Eqs. (5.1), (5.2), (5.4), (5.7), and (5.10) as

$$\mathbf{Y}_b\mathbf{V} - \mathbf{I} = \mathbf{I}_s + \mathbf{I}^i \tag{5.11}$$

where

$$\mathbf{I}_s = \begin{bmatrix} \mathbf{I}_{sR} \\ \mathbf{I}_{sC} \\ \mathbf{I}_{sL} \end{bmatrix} \quad \text{and} \quad \mathbf{I}^i = \begin{bmatrix} \mathbf{0} \\ \mathbf{I}_C^i \\ \mathbf{I}_L^i \end{bmatrix} \tag{5.12}$$

are the current source vector and the initial current vector of the energy-storage elements, respectively.

B. Cut-set equations We next use the branch equation, Eq. (5.11), and combine it with the Kirchhoff equations, Eq. (4.3). We eliminate the branch voltage vector **V** and the branch current vector **I** to obtain

$$\mathbf{Q}\mathbf{Y}_b\mathbf{Q}^T\mathbf{V}_t = \mathbf{Q}\mathbf{I}_s + \mathbf{Q}\mathbf{I}^i \tag{5.13}$$

The *cut-set equation* is

$$\mathbf{Y}_q\mathbf{V}_t = \mathbf{I}_{sq} + \mathbf{I}_q^i \tag{5.14}$$

where

$$\mathbf{Y}_q \triangleq \mathbf{Q}\mathbf{Y}_b\mathbf{Q}^T \tag{5.15a}$$

$$\mathbf{I}_{sq} \triangleq \mathbf{Q}\mathbf{I}_s \tag{5.15b}$$

and

$$\mathbf{I}_q^i \triangleq \mathbf{Q}\mathbf{I}^i \tag{5.15c}$$

$\mathbf{Y}_q$ is an $(n-1) \times (n-1)$ matrix called the *cut-set admittance matrix*, $\mathbf{I}_{sq}$ is the cut-set current source vector of dimension $n-1$, and $\mathbf{I}_q^i$ is the cut-set current vector due to initial conditions. The basic variable for cut-set analysis (based on the fundamental cut set associated with a tree) is the twig voltage vector $\mathbf{V}_t$.

Example (Cut-set analysis) Let us consider the circuit shown in Fig. 5.4, where branches 2 and 4 are parallel combinations of sources and passive elements. A tree is picked to include branches 1, 2, and 3. The digraph which depicts the tree and the fundamental cut sets is shown in Fig. 5.5. The fundamental cut-set matrix is

$$\mathbf{Q} = \begin{bmatrix} 1 & 0 & 0 & 1 & 0 \\ 0 & 1 & 0 & 1 & 1 \\ 0 & 0 & 1 & 0 & -1 \end{bmatrix} \tag{5.16}$$

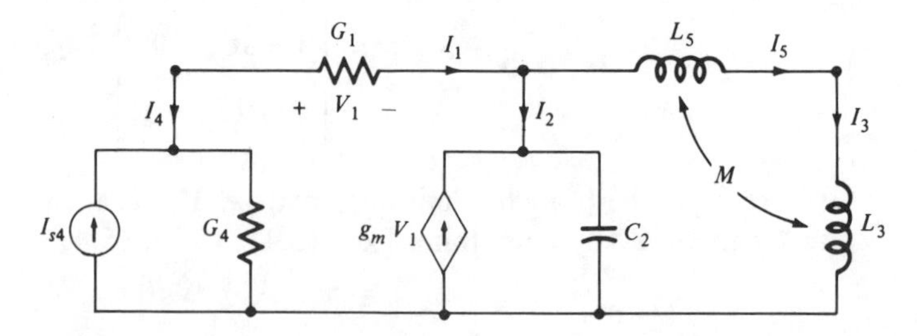

Figure 5.4 Example illustrating cut-set analysis.

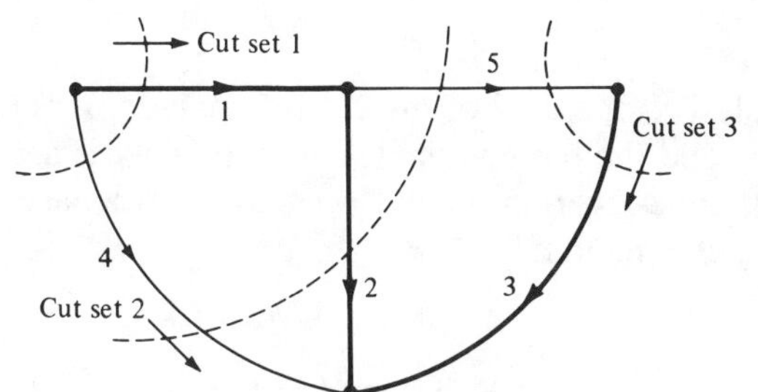

Figure 5.5 Digraph for circuit shown in Fig. 5.4 with tree branches 1, 2, and 3.

For simplicity, we assume all initial conditions are zero. The branch admittance matrix and the branch current source vector are

$$\mathbf{Y}_b = \begin{bmatrix} G_1 & 0 & 0 & 0 & 0 \\ -g_m & sC_2 & 0 & 0 & 0 \\ 0 & 0 & \dfrac{\Gamma_3}{s} & 0 & \dfrac{\Gamma_m}{s} \\ 0 & 0 & 0 & G_4 & 0 \\ 0 & 0 & \dfrac{\Gamma_m}{s} & 0 & \dfrac{\Gamma_5}{s} \end{bmatrix} \qquad \mathbf{I}_s = \begin{bmatrix} 0 \\ 0 \\ 0 \\ I_{s4} \\ 0 \end{bmatrix} \tag{5.17}$$

where

$$\begin{bmatrix} \Gamma_3 & \Gamma_m \\ \Gamma_m & \Gamma_5 \end{bmatrix} \triangleq \begin{bmatrix} L_3 & M \\ M & L_5 \end{bmatrix}^{-1}$$

To obtain the cut-set equations, we use Eqs. (5.15a) and (5.15b). Thus

$$\mathbf{Y}_q = \mathbf{Q}\mathbf{Y}_b\mathbf{Q}^T = \begin{bmatrix} G_1 + G_4 & G_4 & 0 \\ G_4 - g_m & G_4 + \dfrac{1}{s}\Gamma_5 + sC_2 & -\dfrac{\Gamma_5}{s} + \dfrac{\Gamma_m}{s} \\ 0 & -\dfrac{\Gamma_5}{s} + \dfrac{\Gamma_m}{s} & \dfrac{1}{s}(\Gamma_3 + \Gamma_5 + 2\Gamma_m) \end{bmatrix} \tag{5.18}$$

and

$$\mathbf{I}_{sq} = \mathbf{Q}\mathbf{I}_s = \begin{bmatrix} I_{s4} \\ I_{s4} \\ 0 \end{bmatrix} \tag{5.19}$$

The cut-set equation is therefore

$$\mathbf{Y}_q \begin{bmatrix} V_{t1} \\ V_{t2} \\ V_{t3} \end{bmatrix} = \begin{bmatrix} I_{s4} \\ I_{s4} \\ 0 \end{bmatrix} \tag{5.20}$$

5.2 Loop Analysis

Loop analysis is the dual of the cut-set analysis. It is based on the loop tableau equations, and the basic variables in loop analysis are the link currents, $\mathbf{i}_\ell$. For any linear time-invariant circuit we may use Laplace transforms and write the *loop tableau equations*:

$$\begin{aligned} \mathbf{BV}(s) &= \mathbf{0} \\ \mathbf{I}(s) - \mathbf{B}^T\mathbf{I}_\ell(s) &= \mathbf{0} \end{aligned} \tag{5.21}$$

$$\mathbf{M}(s)\mathbf{V}(s) + \mathbf{N}(s)\mathbf{I}(s) = \mathbf{U}_s(s) + \mathbf{U}^i \tag{5.22}$$

The restriction imposed in loop analysis is that all circuit elements must be *current-controlled*. Using a dual approach to the cut-set analysis, we write the branch equation in terms of the *branch impedance matrix* $\mathbf{Z}_b$:

$$\mathbf{V} - \mathbf{Z}_b\mathbf{I} = \mathbf{V}_s + \mathbf{V}^i \tag{5.23}$$

Note that Eq. (5.23) is the dual of Eq. (5.11). In Eq. (5.23) $\mathbf{V}_s$ is the branch voltage source vector and $\mathbf{V}^i$ represents voltage initial conditions for the energy-storage elements. The branch impedance matrix $\mathbf{Z}_b$ is the inverse of the branch admittance matrix $\mathbf{Y}_b$:

$$\mathbf{Z}_b = \mathbf{Y}_b^{-1} \triangleq -\mathbf{M}^{-1}\mathbf{N} = \begin{bmatrix} \mathbf{R} & \mathbf{0} & \mathbf{0} \\ \mathbf{0} & \frac{1}{s}\mathbf{S} & \mathbf{0} \\ \mathbf{0} & \mathbf{0} & s\mathbf{L} \end{bmatrix} \tag{5.24}$$

where $\mathbf{R} \triangleq \mathbf{G}^{-1}$ is the *branch resistance matrix* and $\mathbf{S} \triangleq \mathbf{C}^{-1}$ is called the *branch elastance matrix*.

In loop analysis, an independent current source is not allowed unless it is in parallel with a two-terminal element, i.e., it can be converted to a voltage source. Similarly, in general, the following coupling elements are not allowed: VCVS, CCCS, VCCS, and ideal transformers.

Combining Eqs. (5.21) and (5.23) and eliminating the branch voltage vector V, we obtain the equations for loop analysis in the frequency domain:

$$\mathbf{Z}_\ell\mathbf{I}_\ell = \mathbf{V}_{s\ell} + \mathbf{V}_\ell^i \tag{5.25}$$

where

$$\mathbf{Z}_\ell \triangleq \mathbf{BZ}_b\mathbf{B}^T \tag{5.25a}$$

$$\mathbf{V}_{s\ell} \triangleq -\mathbf{BV}_s \tag{5.25b}$$

$$\mathbf{V}_\ell^i \triangleq -\mathbf{BV}^i \tag{5.25c}$$

are, respectively, the *loop impedance matrix* of dimension $\ell \times \ell$, the loop voltage source vector of dimension ℓ, and the voltage vector due to the initial conditions. Equation (5.25) is the *loop equation* associated with a tree $\mathcal{T}$, and the basic variable is the link current $\mathbf{I}_\ell$.

Example (Loop analysis) Let us again use the example in Fig. 5.4 and show that it is possible to use transformation of sources to write the loop equations. The circuit after transformation is shown in Fig. 5.6*a*, and its digraph is shown in Fig. 5.6*b*. First, let us demonstrate that the VCCS $g_m V_1$ in Fig. 5.4 can be converted to a CCVS. The branch equation for branch 2 in the original circuit is

$$sC_2V_2 - g_mV_1 - I_2 = 0 \tag{5.26}$$

Dividing by sC_2 and setting $V_1 = I_1/G_1$, we obtain

$$V_2 - \frac{g_mI_1}{G_1sC_2} - \frac{I_2}{sC_2} = 0 \tag{5.27}$$

In loop analysis, the basic variables are the two link currents $I_{\ell 1}$ and $I_{\ell 2}$. The fundamental loop matrix is defined by the two fundamental loops shown in Fig. 5.6*b*:

$$\mathbf{B} = \begin{bmatrix} -1 & -1 & 0 & 1 & 0 \\ 0 & -1 & 1 & 0 & 1 \end{bmatrix} \tag{5.28}$$

The branch equations can be written by a similar approach, as in the previous example, or directly from Fig. 5.4. In addition to Eq. (5.27) for branch 2, they are

$$V_1 - \frac{1}{G_1} I_1 = 0$$

$$\begin{bmatrix} V_3 \\ V_5 \end{bmatrix} - s \begin{bmatrix} L_3 & M \\ M & L_5 \end{bmatrix} \begin{bmatrix} I_3 \\ I_5 \end{bmatrix} = \mathbf{0}$$

and

$$V_4 - \frac{1}{G_4} I_4 = \frac{I_{s4}}{G_4}$$

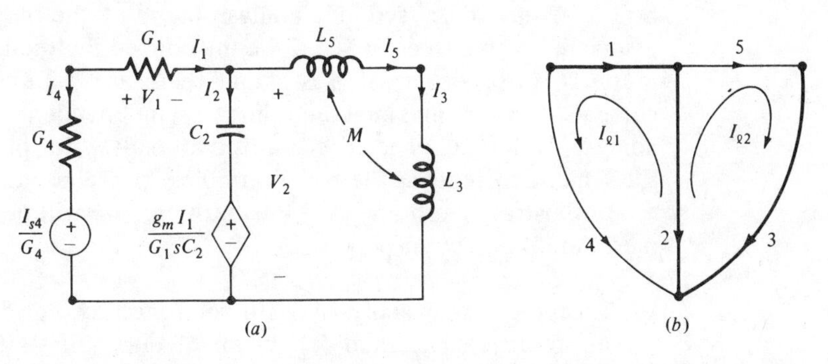

Figure 5.6 Loop analysis: circuit of Fig. 5.3 and its digraph depicting the fundamental loops.

The branch equation is therefore

$$\mathbf{V} - \mathbf{Z}_b\mathbf{I} = \mathbf{V}_s \tag{5.29}$$

or

$$\begin{bmatrix} V_1 \\ V_2 \\ V_3 \\ V_4 \\ V_5 \end{bmatrix} - \begin{bmatrix} \frac{1}{G_1} & 0 & 0 & 0 & 0 \\ \frac{g_m}{G_1 s C_2} & \frac{1}{sC_2} & 0 & 0 & 0 \\ 0 & 0 & sL_3 & 0 & sM \\ 0 & 0 & 0 & \frac{1}{G_4} & 0 \\ 0 & 0 & sM & 0 & sL_5 \end{bmatrix} \begin{bmatrix} I_1 \\ I_2 \\ I_3 \\ I_4 \\ I_5 \end{bmatrix} = \begin{bmatrix} 0 \\ 0 \\ 0 \\ \frac{I_{s4}}{G_4} \\ 0 \end{bmatrix} \tag{5.30}$$

Next we obtain the loop impedance matrix, the loop voltage source vector, and the loop equation:

$$\mathbf{Z}_\ell = \mathbf{B}\mathbf{Z}_b\mathbf{B}^T = \begin{bmatrix} \frac{1}{G_1} + \frac{1}{G_4} + \frac{g_m}{G_1 s C_2} + \frac{1}{sC_2} & \frac{1}{sC_2} \\ \frac{g_m}{G_1 s C_2} + \frac{1}{sC_2} & \frac{1}{sC_2} + s(L_3 + L_5 + 2M) \end{bmatrix} \tag{5.31}$$

$$\mathbf{V}_{s\ell} = \mathbf{B}\mathbf{V}_s = \begin{bmatrix} -\frac{I_{s4}}{G_4} \\ 0 \end{bmatrix} \tag{5.32}$$

and

$$\mathbf{Z}_\ell \begin{bmatrix} I_{\ell 1} \\ I_{\ell 2} \end{bmatrix} = \mathbf{V}_{s\ell} \tag{5.33}$$

5.3 Mesh Analysis

So far we have discussed loop analysis based on the fundamental loop matrix **B** associated with a tree. In Sec. 2 we introduced the reduced loop matrix $\mathbf{B}_R$ of a graph $\mathcal{G}$. In particular, if $\mathcal{G}$ is planar there are $\ell = b - (n-1)$ meshes whose currents form a maximal and linearly independent set. The name "mesh analysis" refers to a special class of loop analysis for planar graphs, which uses the ℓ mesh currents as the basic variables. Let us use the same example in Fig. 5.4 to illustrate mesh analysis. Furthermore, we will demonstrate how to write mesh equations by inspection.

Example (Mesh analysis) With reference to Fig. 5.6, we choose the two mesh currents $I_{\ell 1}$ and $I_{\ell 2}$ shown in Fig. 5.6*b* as the basic variables. To obtain the two mesh equations we need only to write two KVL equations for the two meshes directly in terms of $I_{\ell 1}$ and $I_{\ell 2}$. Thus for mesh 1,

$$\left(\frac{1}{G_4}+\frac{1}{G_1}\right)I_{\ell 1}+\frac{1}{sC_2}(I_{\ell 1}+I_{\ell 2})-\frac{g_m I_1}{G_1 sC_2}=-\frac{I_{s4}}{G_4}$$

For mesh 2,

$$\frac{1}{sC_2}(I_{\ell 1}+I_{\ell 2})-\frac{g_m I_1}{G_1 sC_2}+s(L_3+L_5+2M)I_{\ell 2}=0$$

Since the branch current I_1 is equal to $-I_{\ell 1}$ (see Fig. 5.6*b*), after rearranging the terms in the above two equations, we obtain the following mesh equations:

$$\begin{bmatrix} \dfrac{1}{G_1}+\dfrac{1}{G_4}+\dfrac{1}{sC_2}+\dfrac{g_m}{G_1 sC_2} & \dfrac{1}{sC_2} \\ \dfrac{g_m}{G_1 sC_2}+\dfrac{1}{sC_2} & \dfrac{1}{sC_2}+s(L_3+L_5+2M) \end{bmatrix}\begin{bmatrix} I_{\ell 1} \\ I_{\ell 2} \end{bmatrix}=\begin{bmatrix} -\dfrac{I_{s4}}{G_4} \\ 0 \end{bmatrix}$$

which is seen to be the same as Eqs. (5.31) to (5.33) of the previous section. The two meshes here happen to be the same as the two fundamental loops in the previous section.

REMARK The mesh impedance matrix and the node admittance matrix in Chap. 5 have dual properties. In particular, if a circuit consists of two-terminal elements only (thus there exists no couplings between elements), the mesh impedance matrix is symmetric. The following rules can be used in writing the *mesh impedance matrix* by inspection:

Let
$$\mathbf{Z}_\ell=\begin{bmatrix} z_{11} & z_{12} & \cdots & z_{1\ell} \\ z_{21} & z_{22} & \cdots & z_{2\ell} \\ \cdots & \cdots & \cdots & \cdots \\ z_{\ell 1} & z_{\ell 2} & \cdots & z_{\ell\ell} \end{bmatrix}$$

1. z_{kk} is the sum of the impedances of the elements in mesh k.
2. For $i \neq k$, $z_{ik}=z_{ki}$ is either the sum or the negative sum of the impedances of elements which are common to the two meshes. If in the element common to the two meshes the reference directions of the two meshes are the same, z_{ik} is the sum; otherwise, z_{ik} is the negative sum.

6 STATE EQUATIONS

State equations play an important role in the study of the dynamic behavior of a circuit. In Chap. 7 we discussed methods of writing state equations for second-order circuits. In this section we will present a useful method for writing state equations based on the "proper tree" method. The method is applicable to linear and nonlinear circuits, time-varying and time-invariant circuits. For simplicity we illustrate it with the following linear time-invariant circuit. The purpose is to write the following state and output equations in a systematic way.

$$\dot{\mathbf{x}}(t) = \mathbf{A}\mathbf{x}(t) + \mathbf{b}u(t) \tag{6.1}$$

$$y(t) = \mathbf{c}^T\mathbf{x}(t) + du(t) \tag{6.2}$$

where $\mathbf{x}$ is the state vector, u is the input, and y is the output. $\mathbf{A}$ is a constant matrix, $\mathbf{b}$ and $\mathbf{c}$ are two constant vectors, and d is a scalar.

Example (State equation) Consider the linear time-invariant circuit shown in Fig. 6.1*a*. It contains two capacitors C_1 and C_2 and one inductor L. We choose the capacitor voltage v_{C1} and v_{C2}, and the inductor current i_L as the state variables. Therefore the state is given by

$$\mathbf{x} = \begin{bmatrix} v_{C1} \\ v_{C2} \\ i_L \end{bmatrix} \tag{6.3}$$

which is of dimension three and thus equal to the number of energy-storage elements in the circuit. The question is whether we can always write the state equations in the form of Eq. (6.1) using $\mathbf{x}$ in Eq. (6.3) as the state vector. Recall, from Chap. 7, that we may use KCL to write the capacitor current $C\,\dfrac{dv_C}{dt}$ in terms of other currents and use KVL to write the inductor voltage $L\,\dfrac{di_L}{dt}$ in terms of other voltages. Furthermore, we must be able to express all the current and voltage variables in the equations in terms of *only* the state variables and the inputs.

It turns out that the tree concept and the fundamental theorem of a digraph introduced earlier in the chapter provide an algorithm which works well for a subclass of linear time-invariant circuits. The method consists of four steps and we will illustrate these with the circuit in Fig. 6.1*a*.

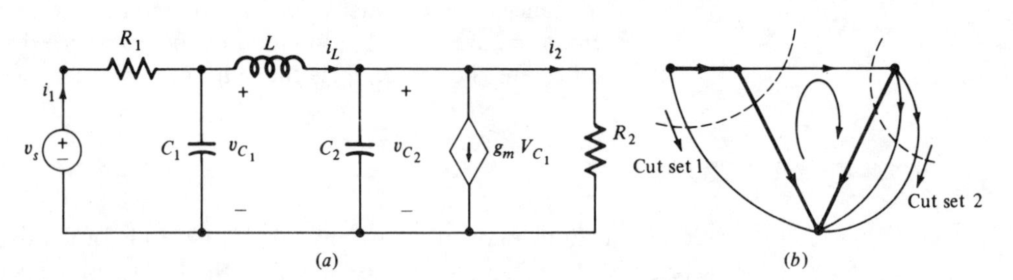

Figure 6.1 Example which illustrates writing state equations using a proper tree: (*a*) Circuit; (*b*) digraph.

Step 1. Pick a tree of the digraph of the circuit, which includes all the capacitors and no inductors. Such a tree is called a *proper tree*.[4]

The proper tree picked for the example is shown by the heavier lines in the digraph of Fig. 6.1*b*. It includes the two capacitors C_1 and C_2 and the resistor R_1 as twigs.

Step 2. The twig capacitor voltages and the link inductor currents are chosen as state variables. In this example, the state variables are v_{C1}, v_{C2}, and i_L.

Step 3. For each fundamental cut set defined by the twig capacitor, write KCL equations. For our circuit, we have

$$C_1 \frac{dv_{C1}}{dt} = i_1 - i_L = \frac{v_s - v_{C1}}{R_1} - i_L \tag{6.4}$$

$$C_2 \frac{dv_{C2}}{dt} = i_L - g_m v_{C1} - i_2 = i_L - g_m v_{C1} - \frac{v_{C2}}{R_2} \tag{6.5}$$

Note that we must express the right-hand side of the equations in terms of the state variables and/or inputs. In this example, the currents i_1 and i_2 have been rewritten in terms of v_{C1}, v_s, and v_{C2}.

Step 4. For each fundamental loop defined by the link inductor, write KVL equations. For our circuit, we have

$$L \frac{di_L}{dt} = v_{C1} - v_{C2} \tag{6.6}$$

Again, we must make sure that the right-hand side of equations is expressed in terms of state variables and/or inputs.

The resulting state equation in matrix form is simply obtained by combining Eqs. (6.4), (6.5), and (6.6) and dividing through the equations by the appropriate C's and L's:

$$\begin{bmatrix} \dfrac{dv_{C1}}{dt} \\ \dfrac{dv_{C2}}{dt} \\ \dfrac{di_L}{dt} \end{bmatrix} = \begin{bmatrix} -\dfrac{1}{R_1 C_1} & 0 & -\dfrac{1}{C_1} \\ -\dfrac{g_m}{C_2} & -\dfrac{1}{C_2 R_2} & \dfrac{1}{C_2} \\ \dfrac{1}{L} & -\dfrac{1}{L} & 0 \end{bmatrix} \begin{bmatrix} v_{C1} \\ v_{C2} \\ i_L \end{bmatrix} + \begin{bmatrix} \dfrac{1}{R_1 C_1} \\ 0 \\ 0 \end{bmatrix} v_s \tag{6.7}$$

Note that the equation is in the form of Eq. (6.1), with $u = v_s$ as the input.

[4] Note that if a circuit has a capacitor loop or an inductor cut set, a proper tree does not exist. Thus the method fails.

If, in the circuit, we choose the response or output variable as the current i_1, we can express the output in terms of the input and the state variables as

$$i_1 = \frac{v_s - v_{C1}}{R_1}$$

$$= \begin{bmatrix} -\frac{1}{R_1} & 0 & 0 \end{bmatrix} \begin{bmatrix} v_{C1} \\ v_{C2} \\ i_L \end{bmatrix} + \frac{1}{R_1} v_s \tag{6.8}$$

This is in the form of Eq. (6.2), with $y = i_1$ as the single output. Note that the output variable is given in terms of the input v_s and the state vector $\mathbf{x}$.

REMARKS

1. From the above example it is clear that there are certain restrictions on the circuit for which the above systematic procedure can be used to obtain the state equations. From the first step of the proposed procedure we can immediately conclude two necessary and sufficient conditions for which a proper tree always exists for circuits with two-terminal elements only:
(*a*) Capacitors in the circuit must not form loops.
(*b*) Inductors in the circuit must not form cut sets.

Condition (*a*) is obvious, for otherwise there does not exist a tree which includes *all* the capacitors in the circuit. Similarly, condition (*b*) is also clear because if there were inductors which form a cut set, one of the inductors would have to be a twig.

2. In Step 2 we chose the capacitor voltages and the inductor current as state variables. We can also use the capacitor charges and the inductor fluxes as state variables. Let $\mathbf{x} = [q_1, q_2, \phi]^T$ be the state vector where

$$q_1 = C_1 v_{C1} \qquad q_2 = C_2 v_{C2} \qquad \text{and} \qquad \phi = L i_L \tag{6.9}$$

We obtain, from Eqs. (6.4), (6.5), and (6.6)

$$\begin{bmatrix} \dfrac{dq_1}{dt} \\ \dfrac{dq_2}{dt} \\ \dfrac{d\phi}{dt} \end{bmatrix} = \begin{bmatrix} -\dfrac{1}{R_1 C_1} & 0 & -\dfrac{1}{L} \\ \dfrac{-g_m}{C_1} & -\dfrac{1}{R_2 C_2} & +\dfrac{1}{L} \\ \dfrac{1}{C_1} & \dfrac{1}{C_2} & 0 \end{bmatrix} \begin{bmatrix} q_1 \\ q_2 \\ \phi \end{bmatrix} + \begin{bmatrix} \dfrac{1}{R_1} \\ 0 \\ 0 \end{bmatrix} v_s \tag{6.10}$$

and

$$i_1 = \frac{v_s - v_{C1}}{R_1}$$

$$= \begin{bmatrix} -\frac{1}{R_1 C_1} & 0 & 0 \end{bmatrix} \begin{bmatrix} q_1 \\ q_2 \\ \phi \end{bmatrix} + \frac{1}{R_1} v_s \tag{6.11}$$

which are the state and output equations using the charges and flux as state variables.

Exercise In the circuit shown in Fig. 6.1, let the inductor be nonlinear, specified by $\phi_L = \hat{\phi}_L(i_L)$, and the capacitor C_2 be nonlinear, specified by $v_{C2} = \hat{v}_{C2}(q_2)$. Using the same tree shown in Fig. 6.1*b*, write the state equations for the circuit.

SUMMARY

- The v_s-shift property and i_s-shift property can be used in any lumped circuit and are useful to move the independent sources in a circuit such that every independent voltage source is in series with a two-terminal element and every independent current source is in parallel with a two-terminal element.
- KCL equations based on all cut sets in a digraph $\mathcal{G}$ are given by the cut-set matrix:

$$\mathbf{Q}_a\mathbf{i} = \mathbf{0}$$

- The reduced cut-set matrix $\mathbf{Q}_R$ is an $(n-1) \times b$ matrix of rank $n-1$. Thus KCL on the cut-set basis is given by the following linearly independent set of equations:

$$\mathbf{Q}_R\mathbf{i} = \mathbf{0}$$

- A loop is defined as a connected subgraph in which precisely two branches are incident with each node.
- KVL equations based on all loops in a digraph $\mathcal{G}$ are given by the loop matrix $\mathbf{B}_a$:

$$\mathbf{B}_a\mathbf{v} = \mathbf{0}$$

- The reduced loop matrix $\mathbf{B}_R$ is an $\ell \times b$ matrix of rank $\ell = b - (n-1)$. Thus KVL on the loop basis is given by the following linearly independent set of equations:

$$\mathbf{B}_R\mathbf{v} = \mathbf{0}$$

- A mesh is a loop of a planar graph drawn on a plane, which encircles nothing in its interior.
- A tree is a connected subgraph of $\mathcal{G}$ which contains all nodes of $\mathcal{G}$ but has no loops.
- The branches of $\mathcal{G}$ can be partitioned into two disjoint sets of branches, those belonging to a tree, called twigs, and those not belonging to a tree, called links. In a connected graph of b branches and n nodes, there are $n-1$ twigs and $\ell = b - (n-1)$ links.
- A fundamental cut set associated with a tree is a cut set which contains one and only one twig. A fundamental loop associated with a tree is a loop which contains one and only one link.

- The fundamental cut-set matrix associated with a tree is a reduced cut-set matrix which consists of all fundamental cut sets defined by the twigs of a tree. The fundamental loop matrix associated with a tree is a reduced loop matrix which consists of all fundamental loops defined by the links of a tree.
- Table 4.1 summarizes Kirchhoff's laws based on nodes, cut sets, and loops in terms of, respectively, the reduced incidence matrix **A**, the fundamental cut-set matrix **Q**, and the fundamental loop matrix **B** associated with a tree.
- The appropriate Kirchhoff's equations above together with the branch equations constitute the node tableau equations, the cut-set tableau equations, and the loop tableau equations.
- In cut-set analysis the basic circuit variables are the $n-1$ twig voltages. For a linear time-invariant circuit, the cut-set equations written in the frequency domain are

$$\mathbf{Y}_q\mathbf{V}_t = \mathbf{I}_{sq} + \mathbf{I}_q^i$$

where
$$\mathbf{Y}_q = \mathbf{Q}\mathbf{Y}_b\mathbf{Q}^T$$

- In loop analysis the basic circuit variables are the ℓ link currents. For a linear time-invariant circuit, the loop equations written in the frequency domain are

$$\mathbf{Z}_\ell\mathbf{I}_\ell = \mathbf{V}_{s\ell} + \mathbf{V}_\ell^i$$

where
$$\mathbf{Z}_\ell = \mathbf{B}\mathbf{Z}_b\mathbf{B}^T$$

- Cut-set analysis is the dual of loop analysis.
- A proper tree is a tree which includes all capacitors in a circuit but none of the inductors. Not all circuits have a proper tree. If the circuit $\mathcal{N}$ has no capacitor loops and no inductor cut sets, a "proper tree" can be used to write the state equations of $\mathcal{N}$.

PROBLEMS

Source transformations

1 (*a*) Use the v_s-shift property to transform the dc voltage source in Fig. P12.1*a* into dc current sources.

(*b*) Use the i_s-shift property to transform the dc current source in Fig. P12.1*b* into dc voltage sources.

2 In the circuit in Fig. P12.2, perform source transformations to obtain the Thévenin and Norton equivalent circuits of the one-port.

Trees, cut sets, and loops

3 For the digraph shown in Fig. P12.3 and for the tree indicated:

(*a*) Indicate all the fundamental loops and the fundamental cut sets.

(*b*) Write all the fundamental loop and cut-set equations.

(*c*) Can you find a tree such that all its fundamental loops correspond to meshes?

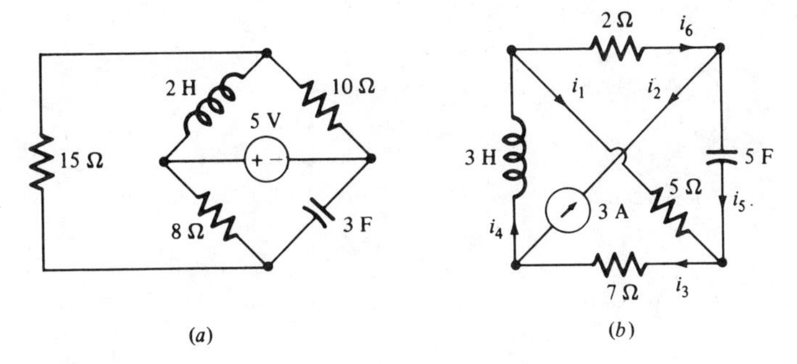

Figure P12.1

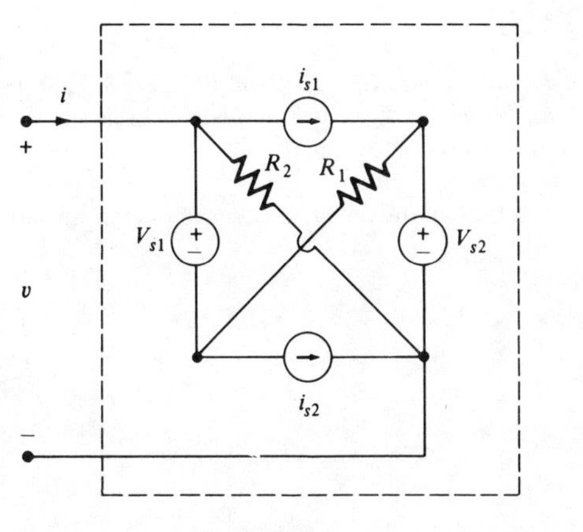

Figure P12.2

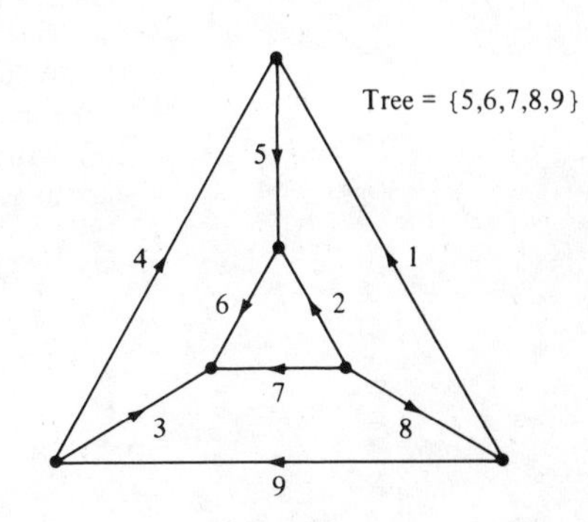

Figure P12.3

Fundamental cut-set and loop matrices

4 For the circuit in Fig. P12.1*b*:

(*a*) List all trees associated with the circuit.

(*b*) Write the fundamental loop matrix **B** associated with the tree branches $\{i_4, i_5, i_6\}$ in the circuit.

(*c*) Write the fundamental cut-set matrix **Q** associated with the same tree in (*b*).

5 For a given connected digraph and for a fixed tree, the fundamental loop matrix is given by

$$\mathbf{B} = \begin{bmatrix} 1 & 0 & 0 & 1 & 0 & 0 \\ 0 & 1 & 0 & 0 & 0 & -1 \\ 0 & 0 & 1 & 1 & -1 & -1 \end{bmatrix}$$

(*a*) Write, by inspection, the fundamental cut-set matrix which corresponds to the same tree.

(*b*) Draw the digraph.

Cut-set analysis

6 For the circuit shown in Fig. P12.6, with the branch currents defined, choose branches 2, 4 and 5 as twigs of a tree. Let $i_1(0) = \rho_1$, $i_3(0) = \rho_3$, and $v_4(0) = \gamma$. Write the cut-set equations in the frequency domain (*s* domain).

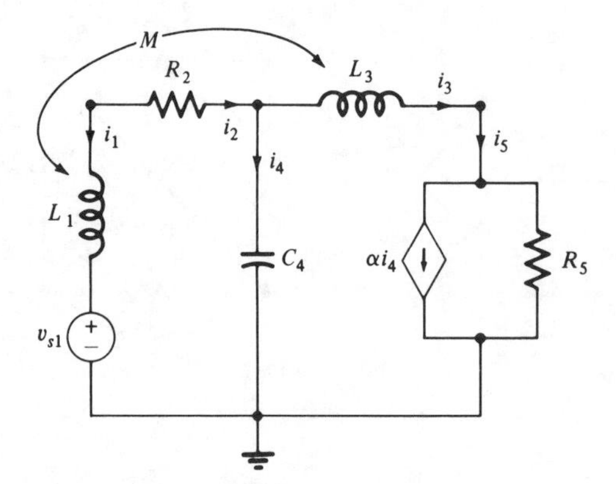

Figure P12.6

7 The linear time-invariant circuit of Fig. P12.7*a*, having a digraph shown in Fig. P12.7*b*, is in the sinusoidal steady state. From the digraph a tree is picked as shown in Fig. P12.7*c*.

(*a*) Write the fundamental cut-set matrix **Q**.

(*b*) Calculate the cut-set admittance matrix $\mathbf{Y}_q$

(*c*) Write the cut-set equations in the frequency domain in terms of the cut-set voltage and the current source:

$$\mathbf{Y}_q \mathbf{V}_t = \mathbf{I}_{sq}$$

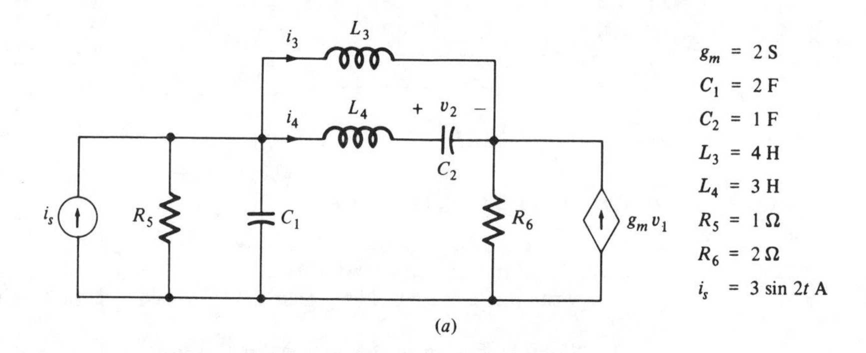

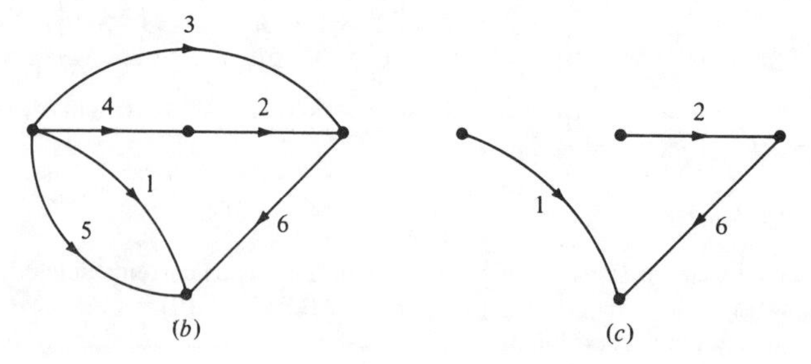

Figure P12.7

8 The linear time-invariant circuit shown in Fig. P12.8 is in the sinusoidal steady state. It originates from delay-line designs. The coupling between inductors is specified by the reciprocal inductance matrix

$$\boldsymbol{\Gamma} = \begin{bmatrix} \Gamma_0 & \Gamma_1 & \Gamma_2 & 0 \\ \Gamma_1 & \Gamma_0 & \Gamma_1 & \Gamma_2 \\ \Gamma_2 & \Gamma_1 & \Gamma_0 & \Gamma_1 \\ 0 & \Gamma_2 & \Gamma_1 & \Gamma_0 \end{bmatrix}$$

Pick a tree such that the corresponding cut-set equations are easy to write. Write these cut-set equations in matrix form $\mathbf{Y}_q \mathbf{V}_t = \mathbf{I}_{sq}$

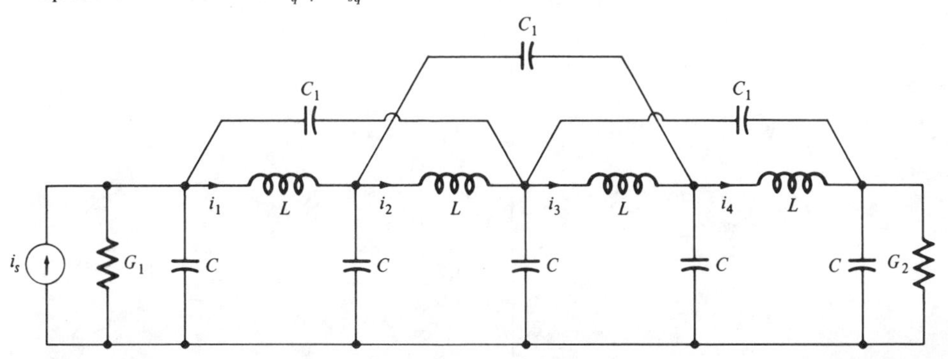

Figure P12.8

Loop analysis

9 For the same circuit and the same tree as in Prob. 6, write the loop equations in the frequency domain.

10 Consider the linear time-invariant circuit shown in Fig. P12.7*a*, suppose it is in the sinusoidal steady state, and let the tree be shown in Fig. P12.7*c*.

(*a*) Write the fundamental loop matrix $\mathbf{B}$.

(*b*) Calculate the loop impedance matrix $\mathbf{Z}_\ell$.

(*c*) Write the loop equations in terms of voltage and current phasors, that is, $\mathbf{Z}_\ell \mathbf{I}_\ell = \mathbf{V}_{s\ell}$.

11 Assume that the linear time-invariant circuit of Fig. P12.7 is in the sinusoidal steady state. Write the fundamental loop equations for the tree indicated by the shortcut method. (First assume that the dependent current source is independent, and introduce its dependence in the last step.)

12 The linear time-invariant circuit shown in Fig. P12.12 is in the sinusoidal steady state. For the reference directions indicated on the inductors, the inductance matrix is

$$\begin{bmatrix} 4 & 2 & 1 \\ 2 & 4 & 2 \\ 1 & 2 & 4 \end{bmatrix}$$

Write the fundamental loop equations for a tree of your choice.

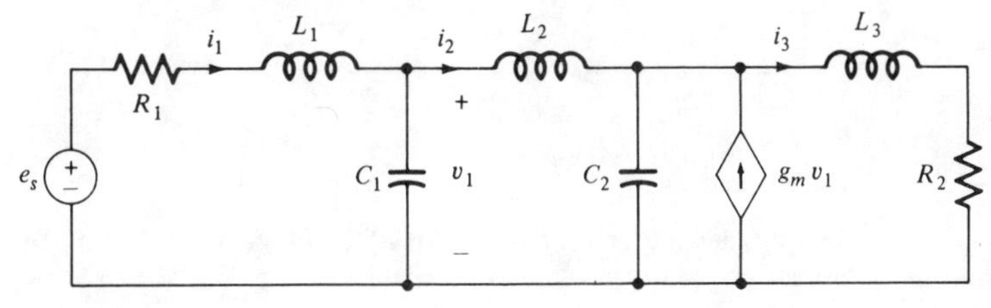

Figure P12.12

Mesh analysis

13 The linear time-invariant circuit shown in Fig. P12.13 represents the Maxwell bridge, an instrument used to measure the inductance L_x and resistance R_x of a linear time-invariant physical inductor. The measurement is made with the circuit in the sinusoidal steady state by adjusting R_1 and C until $v_A = v_B$ (then the bridge is said to be "balanced"). It is important to note that when balance is achieved, the current through the detector D is zero regardless of the value of the detector impedance. Show that when the bridge is balanced,

$$L_x = R_2 R_3 C \qquad R_x = \frac{R_2 R_3}{R_1}$$

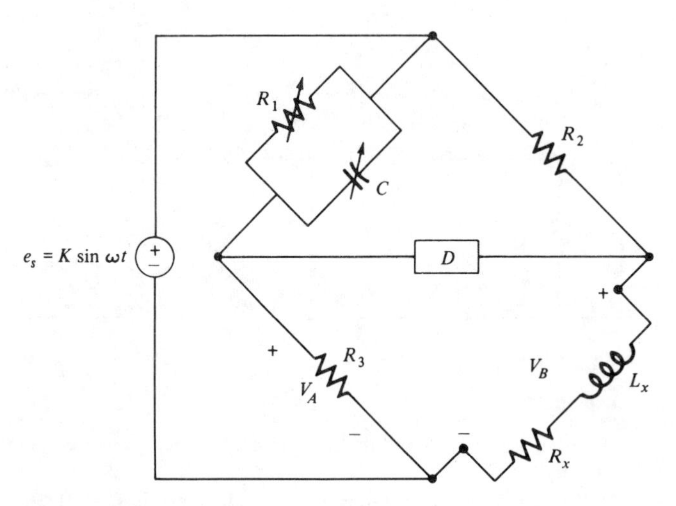

Figure P12.13

14 Consider the circuit shown in Fig. P12.14.
(*a*) Write the mesh equations using the charges q_1 and q_2 as variables.
(*b*) Determine the network function $Q_2(s)/V_s(s)$.

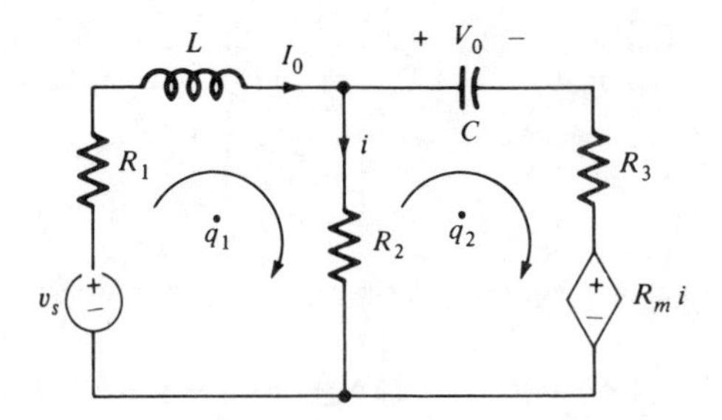

Figure P12.14

State equations

15 Consider the linear time-invariant circuit shown in Fig. P12.15. Give the state equations by using v_4, i_2, and i_3 as the state variables.

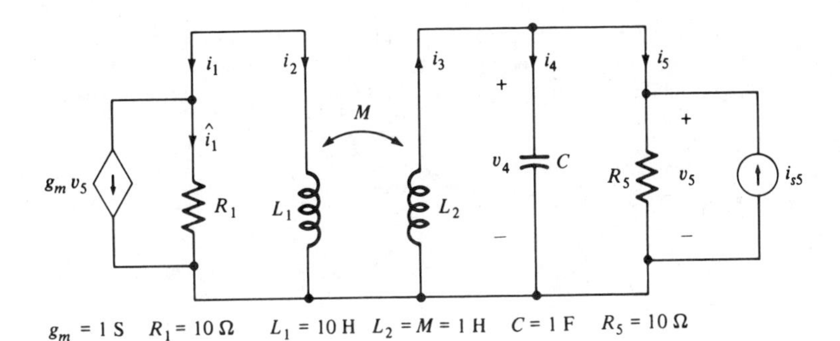

Figure P12.15

16 Write the state equation of the circuit shown in Fig. P12.16 for the following situations:

(*a*) All nonsource elements are linear and time-invariant (use C_1, C_2, C_3, L_4, and L_5).

(*b*) All nonsource elements (except the resistors R_1, R_3) are linear and time-varying [these elements have characteristics specified by the functions of time $C_1(\cdot)$, $C_2(\cdot)$, $C_3(\cdot)$, $L_4(\cdot)$, and $L_5(\cdot)$].

(*c*) All nonsource elements (except the resistors R_1, R_3) are nonlinear and time-invariant [the element characteristics are $v_1 = f_1(q_1)$, $v_2 = f_2(q_2)$, $v_3 = f_3(q_3)$, $i_4 = f_4(\phi_4)$, and $i_5 = f_5(\phi_5)$].

(*d*) All nonsource elements (except the resistors R_1, R_3) are nonlinear and time-varying, and

$$v_1 = f_1(q_1, t), \ldots, i_5 = f_5(\phi_5, t)$$

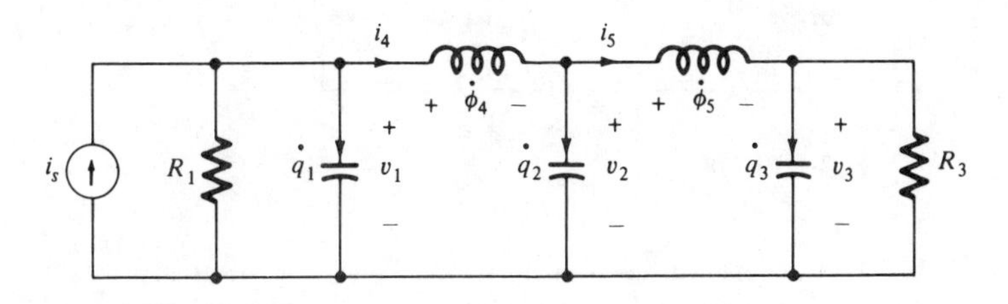

Figure P12.16

17 Consider the two versions (linear and nonlinear) of the time-invariant circuit shown in Fig. P12.17. For each version write a set of state equations using ϕ_1, ϕ_2, ϕ_3, q_4, and q_5 as state variables. The initial conditions are $i_1(0) = 2$ A, $i_2(0) = 0$, $i_3(0) = 0$, $v_4(0) = 4$ V, $v_5(0) = 6$ V, and the source is $i_s = 5 \cos t$ amperes.

Linear version	Nonlinear version
$L_1 = 1$ H	$i_1 = f_1(\phi_1)$
$L_2 = 2$ H	$i_2 = f_2(\phi_2)$
$L_3 = 5$ H	$i_3 = f_3(\phi_3)$
$C_4 = 1$ F	$v_4 = f_4(q_4)$
$C_5 = 3$ F	$v_5 = f_5(q_5)$
$R_6 = 2\ \Omega$	$R_6 = 2\ \Omega$
$R_7 = 3\ \Omega$	$R_7 = 3\ \Omega$

Figure P12.17

18 Obtain for each of the nonlinear time-invariant circuits shown in Fig. P12.18 a set of state equations. Proceed in a systematic way, i.e., pick a proper tree and write fundamental cut-set and fundamental loop equations. Perform whatever algebra is necessary, and finally indicate the initial state.

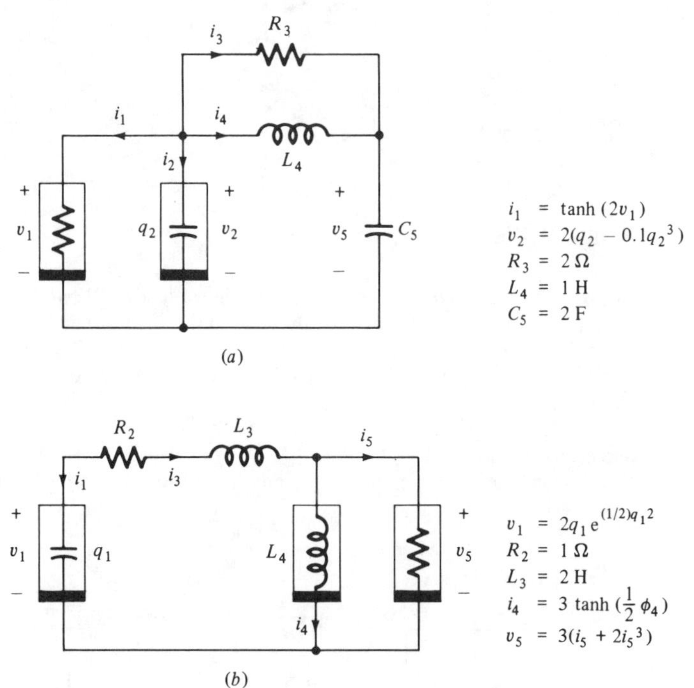

Figure P12.18

19 The nonlinear elements of the time-invariant circuit shown in Fig. P12.19 are specified by the following relations: $i_2 = f(v_2)$ and $q_1 = q_1(v_1)$. Knowing that the remaining components are linear, write two differential equations with v_1 and v_2 as state variables.

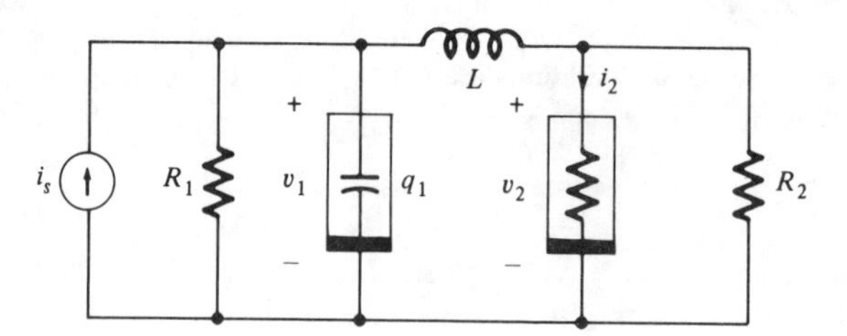

Figure P12.19

Tree, cut-set, loop, and state equations

20 (*a*) Determine all possible trees of the digraph shown in Fig. P12.20*a*.

(*b*) For the tree $\{1, 2, 3\}$, write down the fundamental cut-set matrix and the fundamental loop matrix.

(*c*) Consider the circuit shown in Fig. P12.20*b*. Write the cut-set equations in the frequency domain using the same tree.

(*d*) Write the loop equations in the frequency domain using the same tree.

(*e*) Write the state equations and the output equation with the current i_5 as the output.

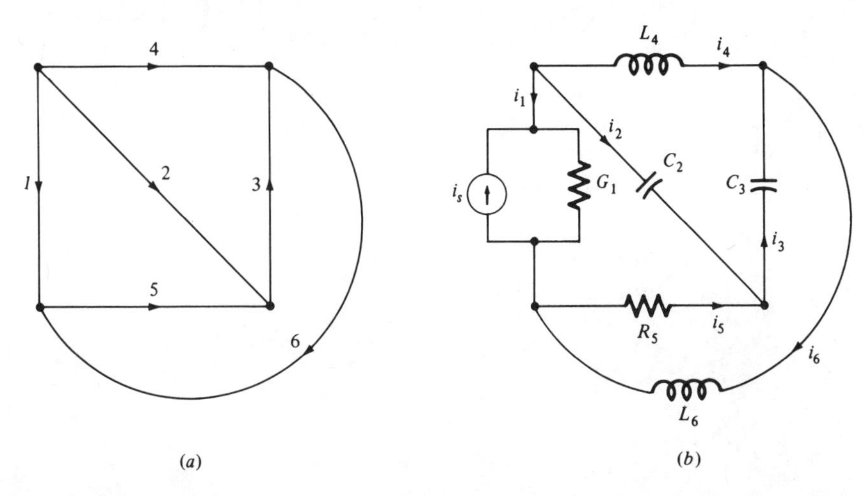

Figure P12.20

CHAPTER

THIRTEEN

TWO-PORTS, MULTIPORTS, AND RECIPROCITY

The concept of a port was first introduced in Chap. 1. Recall that a port can be created from a circuit by accessing two leads to a pair of nodes of the circuit. Thus a one-port can be viewed as a black box which has *one pair of terminals accessible* from the outside. By definition, the only circuit variables needed for the characterization of a one-port are the port voltage $v(t)$ and the port current $i(t)$ as shown in Fig. 0.1. The term "black box" is to signify that we need not know what is inside of the box.

A two-port or an n-port can be similarly created by accessing two or n *pairs of terminals* from a circuit. The important notion of a port is that the current entering from one terminal of a port is identical to the current leaving from the other terminal of the same port. In Fig. 0.2a we show a two-port as a black box with two pairs of external terminals. Thus we have two port voltages and two port currents as circuit variables. In contrast, a four-terminal circuit shown in Fig. 0.2b has three independent terminal currents and three independent terminal-to-datum voltages.

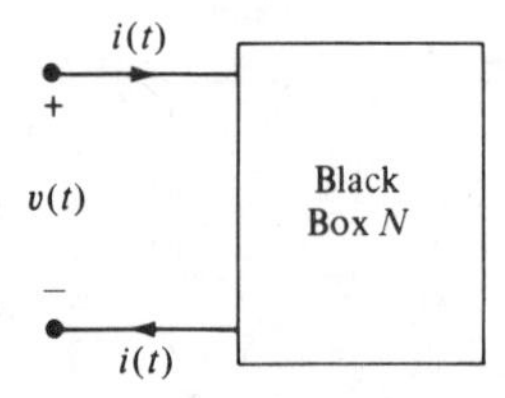

Figure 0.1 A one-port with port voltage and port current specified in the associated reference directions.

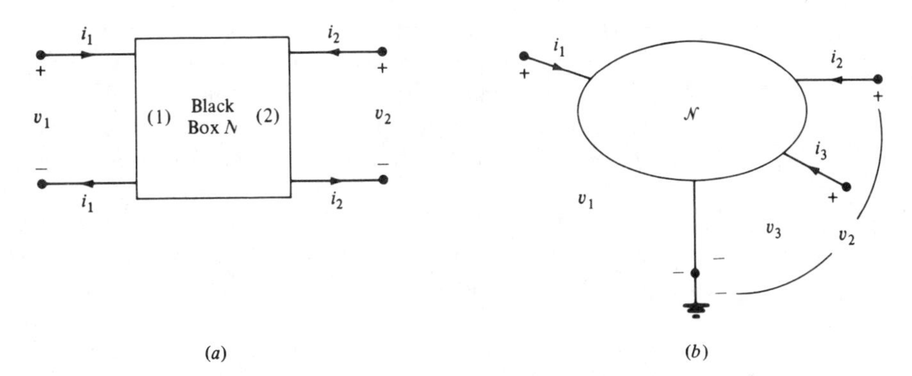

Figure 0.2 (*a*) A two-port; (*b*) a four-terminal circuit.

Two-ports and *n*-ports are useful in device characterization as illustrated in Chap. 3 for resistive elements and in Chap. 8 for coupled inductors. It was pointed out in Chap. 3 that for two-ports there exist six distinct representations relating two port variables to the other two. One aim of this chapter is to demonstrate the usefulness of a general implicit representation of two-ports and to show how each of the six explicit representations can be derived from the implicit representation and how they are interrelated.

After a brief review of the port characterization of different classes of circuits in Sec. 1, we shall restrict our discussion to *linear time-invariant circuits* and use the frequency domain throughout. This is partly because of the fact that general nonlinear two-ports which contain resistors, inductors, and capacitors are difficult to handle. But, more important, a "port" is such a powerful concept in linear time-invariant circuits that it is vital in the design considerations of many electronic and communication systems. We call a *two-port* (respectively, *n*-port) *linear time-invariant* iff it contains only linear time-invariant elements and independent sources.

A feedback amplifier used in the transoceanic submarine cable system can be described in terms of the interconnection of subcircuits shown in Fig. 0.3, where N_g represents the generator, N_ℓ the load, N_μ the amplifier circuit, N_β the feedback circuit, N_1 the input coupling circuit, and N_2 the output coupling circuit. Note that N_g and N_ℓ are one-ports, N_μ and N_β are two-ports, and N_1 and N_2 are three-ports. For design purposes, the specifications of the overall circuit such as gain, bandwidth, impedances, sensitivity, and stability

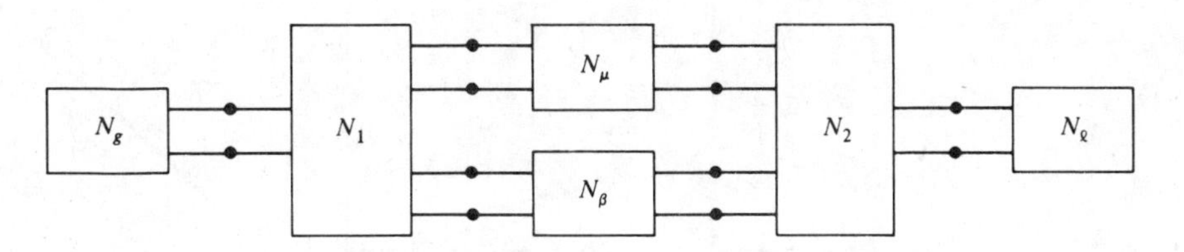

Figure 0.3 A feedback amplifier configuration.

must be translated first into separate specifications of the subcircuits in terms of their port characterizations. Therefore we need to know how port characterizations are obtained from the circuit in each black box and how to analyze a circuit which is formed by the interconnection of subcircuits specified in terms of ports. These are the subjects to be discussed in Secs. 2 and 3.

Section 4 deals with multiport and multiterminal circuits. The indefinite admittance representation of an n-terminal circuit will be introduced. Section 5 deals with the soldering iron entry and the pliers entry. They pertain to how an independent source is connected to a circuit and how a voltage output or a current output is measured. The chapter concludes with an important theorem which holds for a special class of linear time-invariant circuits, the reciprocity theorem. The theorem can be stated explicitly in terms of two-port specifications.

1 REVIEW OF PORT CHARACTERIZATIONS

1.1 One-Port

Resistive one-port Let us consider the one-port shown in Fig. 0.1. If, inside the black box, there are only resistive elements, including independent sources, the one-port is characterized by its driving-point characteristic plotted on the v-i plane or in terms of an implicit function in the form

$$h(v(t), i(t), t) = 0 \tag{1.1}$$

If the resistive circuit is linear time-invariant, the one-port can be represented in general by its Thévenin or Norton equivalent circuit shown in Fig. 1.1. Equation (1.1) can be reduced to the explicit form of voltage v in terms of the current i, the open-circuit voltage v_{oc}, and the equivalent resistance R_{eq}:

$$v = R_{eq}i + v_{oc} \tag{1.2a}$$

or to its dual

$$i = G_{eq}v + i_{sc} \tag{1.2b}$$

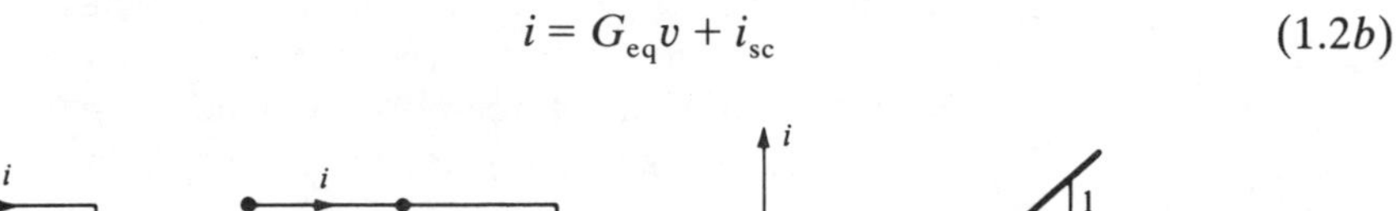

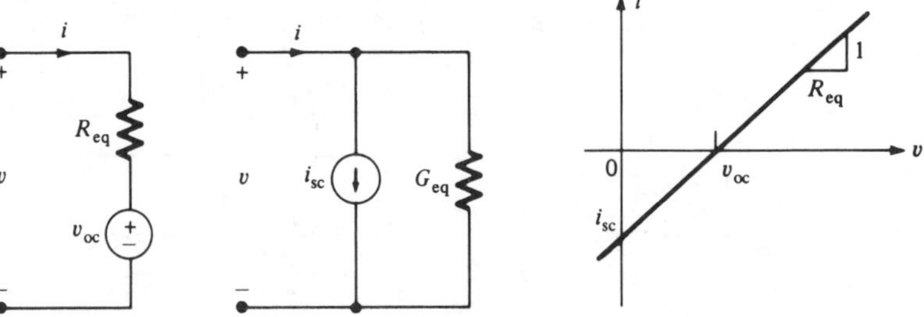

Figure 1.1 Thévenin-Norton equivalent-circuit representations of a linear one-port.

where $$R_{eq} = \frac{1}{G_{eq}} = \frac{-v_{oc}}{i_{sc}} \tag{1.2c}$$

REMARKS

1. Thévenin and Norton's theorem gives added significance to the port characterization of linear time-invariant circuits. It is remarkable that any linear time-invariant resistive one-port can be represented by a single source and an equivalent resistance.
2. Equation (1.2*a*) holds for any linear time-invariant resistive one-port which is current-controlled. Dually, Eq. (1.2*b*) holds for any linear time-invariant resistive one-port which is voltage-controlled.

One-port with resistors, capacitors, and inductors For a general nonlinear one-port with resistors, capacitors, and inductors one possible representation is in terms of the state equations:

$$\dot{\mathbf{x}}(t) = \mathbf{f}(\mathbf{x}(t), u(t)) \qquad \mathbf{x}(0) = \mathbf{x}_0 \tag{1.3a}$$

$$y(t) = \mathbf{g}(\mathbf{x}(t), u(t)) \tag{1.3b}$$

where $\mathbf{x}(\cdot)$ is the state vector, $u(\cdot)$ represents either the current or the voltage input, and $y(\cdot)$ is then the voltage or the current output, respectively. Unfortunately, it is, in general, not possible to find an explicit relation between the port voltage and port current.

Another possible representation of such a general one-port is in terms of $v(t)$, $i(t)$, and their derivatives, together with some initial conditions. For example, the simple one-port shown in Fig. 1.2 has a characterization:

$$i(t) - G\,v(t) - \frac{d\hat{q}}{dv}(v)\dot{v}(t) = 0$$

and $v(0) = V_0$ is the initial voltage. This, however, cannot be generalized easily. As a matter of fact, it is rather difficult and often impossible to obtain the port characterization in explicit analytic form for most nonlinear one-ports with resistors, inductors, and capacitors.

Linear time-invariant one-port Let us specialize in *linear time-invariant circuits*. Assuming that we are interested in either the sinusoidal steady state or the

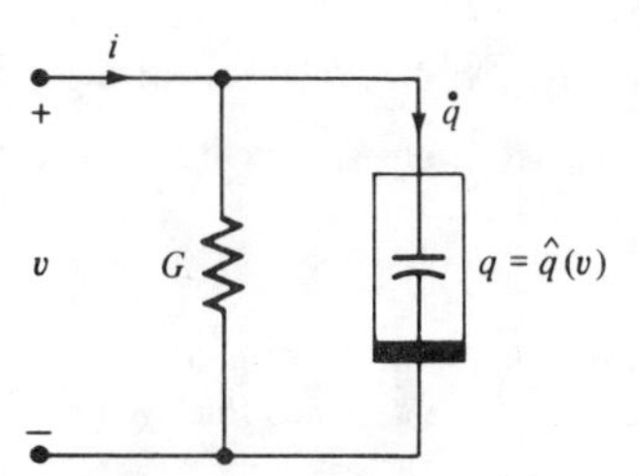

Figure 1.2 A one-port which contains a nonlinear capacitor.

zero-state response, we may use the Thévenin and Norton equivalent circuits in the frequency domain to characterize the one-port:[1]

$$V(s) = Z_{eq}(s)I(s) + V_{oc}(s) \tag{1.4a}$$

or

$$I(s) = Y_{eq}(s)V(s) + I_{sc}(s) \tag{1.4b}$$

where

$$Z_{eq}(s) = \frac{1}{Y_{eq}(s)} = \frac{-V_{oc}(s)}{I_{sc}(s)} \tag{1.4c}$$

In Eqs. (1.4*a*) to (1.4*c*) V, I, V_{oc}, and I_{sc} represent either the phasors or the Laplace transforms of the voltages or currents. Thus any two of the three functions $V_{oc}(s)$, $Z_{eq}(s)$, and $I_{sc}(s)$ in the frequency domain specify completely a linear time-invariant one-port.

1.2 Two-Ports

The general resistive two-port has a characterization in the form of an implicit function:

$$\mathbf{h}(\mathbf{v}(t), \mathbf{i}(t), t) = \mathbf{0} \tag{1.5a}$$

or

$$\begin{aligned} h_1(v_1(t), v_2(t), i_1(t), i_2(t), t) &= 0 \\ h_2(v_1(t), v_2(t), i_1(t), i_2(t), t) &= 0 \end{aligned} \tag{1.5b}$$

where v_1 and v_2 are the port voltages and i_1 and i_2 are the port currents. Recall that there are six possible explicit representations of two of the four variables v_1, v_2, i_1, and i_2 in terms of the remaining two. We have illustrated in Chap. 3 in some detail the current-controlled, the voltage-controlled, and the hybrid representations of resistive two-ports.

Like for a one-port, it is in general difficult to obtain the port characterization of a general nonlinear two-port with resistors, inductors, and capacitors. In Chap. 8 we have seen the characterization of a nonlinear coupled inductor. If the inductor is current-controlled, we have

$$\boldsymbol{\phi} = \hat{\boldsymbol{\phi}}(\mathbf{i}) \tag{1.6a}$$

or

$$\phi_1 = \hat{\phi}_1(i_1, i_2) \tag{1.6b}$$

$$\phi_2 = \hat{\phi}_2(i_1, i_2) \tag{1.6c}$$

where $\boldsymbol{\phi} = \begin{bmatrix} \phi_1 \\ \phi_2 \end{bmatrix}$ is the port flux vector. In small-signal analysis, we may write the above equations in terms of the port voltage vector:

[1] Again, it should be pointed out that Eq. (1.4*a*) holds for any linear time-invariant one-port which is current-controlled. Dually, Eq. (1.4*b*) holds for any linear time-invariant one-port which is voltage-controlled.

$$\mathbf{v} = \frac{d\boldsymbol{\phi}}{dt} = \left.\frac{\partial \hat{\boldsymbol{\phi}}}{\partial \mathbf{i}}\right|_Q \frac{d\mathbf{i}}{dt} \tag{1.7a}$$

or

$$\begin{aligned} v_1 &= \left.\frac{\partial \hat{\phi}_1}{\partial i_1}\right|_Q \frac{di_1}{dt} + \left.\frac{\partial \hat{\phi}_1}{\partial i_2}\right|_Q \frac{di_2}{dt} \\ v_2 &= \left.\frac{\partial \hat{\phi}_2}{\partial i_1}\right|_Q \frac{di_1}{dt} + \left.\frac{\partial \hat{\phi}_2}{\partial i_2}\right|_Q \frac{di_2}{dt} \end{aligned} \tag{1.7b}$$

Exercise Use the Laplace transform to derive the impedance matrix of a coupled inductor from Eq. (1.7).

Linear time-invariant two-ports We now discuss *linear time-invariant circuits* and consider either the sinusoidal steady state or the zero-state response. We will use frequency-domain characterization with voltages and currents representing either phasors or Laplace transforms of waveforms. The Thévenin and Norton equivalent representations for linear resistive two-ports discussed in Chap. 5, Sec. 4 can be easily generalized to linear time-invariant two-ports containing independent sources. For example, the impedance matrix representation is the vector version of Eq. (1.4*a*), and is of the form

$$\mathbf{V} = \mathbf{Z}\mathbf{I} + \mathbf{V}_{oc} \tag{1.8a}$$

or

$$\begin{aligned} V_1 &= z_{11}I_1 + z_{12}I_2 + V_{oc1} \\ V_2 &= z_{21}I_1 + z_{22}I_2 + V_{oc2} \, . \end{aligned} \tag{1.8b}$$

where $\mathbf{Z}$ is a 2×2 matrix whose elements are real rational functions of s. The equivalent circuit is shown in Fig. 1.3. The discussion of two-port representation in the *frequency domain* is the subject of Sec. 2.

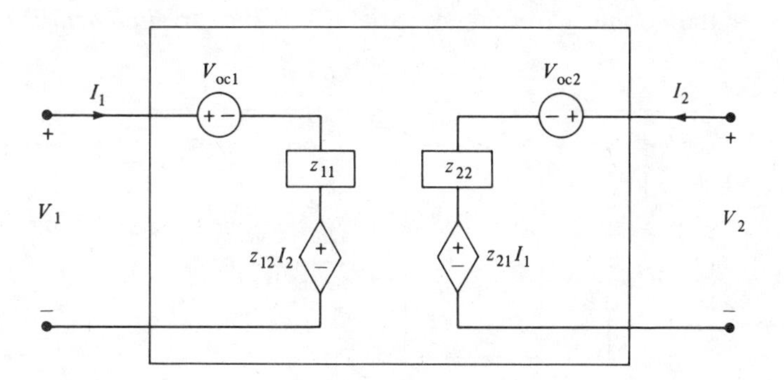

Figure 1.3 Linear time-invariant two-port representation in terms of impedances and two independent voltage sources.

2 LINEAR TIME-INVARIANT TWO-PORTS

Without loss of generality, we may assume that the linear time-invariant two-ports to be studied contain no independent sources. This is because by using the superposition theorem, the effects of any independent sources can be separately accounted for. This can also be seen in the two-port representation in Fig. 1.3. Therefore let us consider the circuit shown in Fig. 2.1, where a linear time-invariant two-port without independent sources is connected to two arbitrary linear time-invariant one-ports, each represented by its Thévenin equivalent circuit (or Norton equivalent circuit). Let us use the frequency-domain characterization and consider zero-state responses only. Thus V_1 and V_2 are the Laplace transforms of the port voltages and I_1 and I_2 are the Laplace transforms of the port currents of the two-port defined with the conventional associated reference directions.

The equations for the two one-ports are

$$V_1 + Z_1 I_1 = V_{s1} \tag{2.1a}$$

$$V_2 + Z_2 I_2 = V_{s2} \tag{2.1b}$$

These two equations may be considered as the boundary conditions imposed on the two-port.

If the linear time-invariant circuit in Fig. 2.1 is *uniquely solvable*, then, by eliminating all the circuit variables except V_1, V_2, I_1, and I_2, we obtain *two linearly independent* equations which give the general representation of the two-port. The two linearly independent equations are of the implicit form in terms of $\mathbf{V}$ and $\mathbf{I}$:

$$\mathbf{MV} + \mathbf{NI} = \mathbf{0} \tag{2.2}$$

or

$$\begin{aligned} m_{11}V_1 + m_{12}V_2 + n_{11}I_1 + n_{12}I_2 &= 0 \\ m_{21}V_1 + m_{22}V_2 + n_{21}I_1 + n_{22}I_2 &= 0 \end{aligned} \tag{2.3}$$

where $\mathbf{M} = [m_{ij}]$ and $\mathbf{N} = [n_{ij}]$ are 2×2 matrices of real rational functions of the complex frequency variable s. They are *independent* of the impedances Z_1

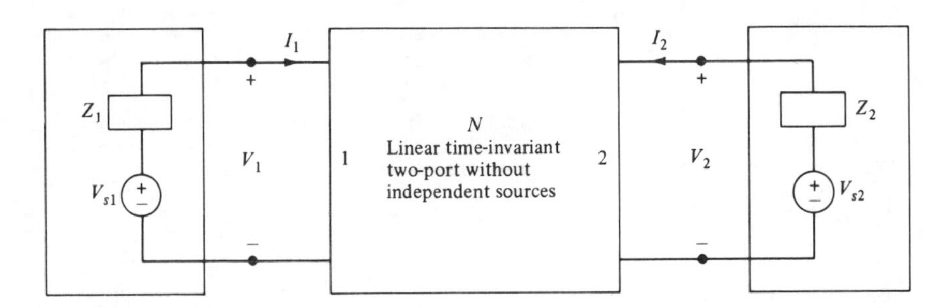

Figure 2.1 A two-port connected to two one-ports represented by their Thévenin equivalents.

and Z_2 and the sources V_{s1} and V_{s2} of the two 1-ports. We use the notation **M** and **N** to convey the similarity of the port characterization of the two-port and the branch equations in tableau analysis. In fact, another way to derive Eq. (2.2) is to connect each port of N to either a voltage source or a current source, write the tableau equation, and then eliminate all variables except the port voltage vector **V** and the port current vector **I**. This approach shows clearly that **M** and **N** in Eq. (2.2) depend only on the circuit elements *inside* N, since the above procedure does not call for inserting Z_1 and Z_2, as in Fig. 2.1. Note that the right-hand sides of Eqs. (2.2) and (2.3) are zero because the two-port is assumed to have no internal independent sources.

Exercise Consider the circuit in Fig. 2.2.
(*a*) Write node equations and solve for V_1 and V_2 in terms of the current sources I_{s1} and I_{s2}.
(*b*) Obtain the implicit representation of the two-port, i.e., Eq. (2.3), by eliminating I_{s1} and I_{s2}.
(*c*) Simplify the equations obtained, and demonstrate that the equations above are independent of G_1 and G_2.

The six explicit representations of a linear resistive two-port introduced in Chap. 3 can be generalized to a linear time-invariant two-port in the frequency domain. Furthermore, it will be demonstrated in this section that all six representations can be derived from Eq. (2.2) or (2.3). In essence, an explicit two-port representation exists for a given pair of port variables iff Eq. (2.2) can be solved for these variables, or equivalently, iff the 2×2 submatrix of $[\mathbf{M}, \mathbf{N}]$, which represent coefficients pertaining to the two variables, is nonsingular. Thus Eq. (2.2) or (2.3) is the most general representation of a linear time-invariant two-port without independent sources.

2.1 The Impedance and Admittance Matrices

For this subsection we are interested in obtaining an explicit relation between the two port voltages V_1 and V_2 and the two port currents I_1 and I_2. If **M** in Eq.

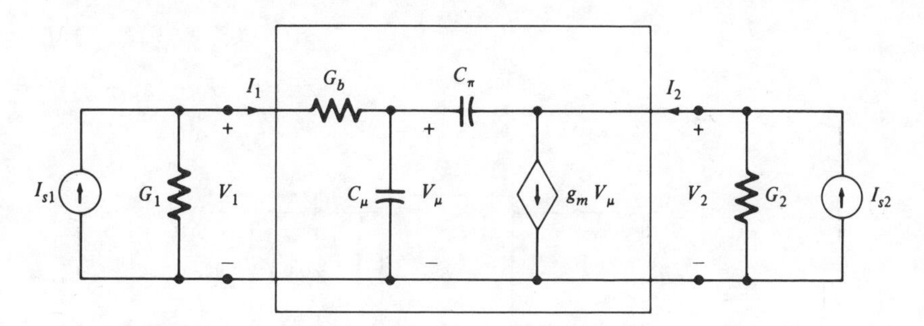

Figure 2.2 Exercise illustrating equations of a linear time-invariant two-port.

(2.2) is nonsingular, we may write $\mathbf{V}$ in terms of $\mathbf{I}$ in Eq. (2.2) as

$$\begin{aligned}\mathbf{V} &= -\mathbf{M}^{-1}\mathbf{N}\mathbf{I} \\ &= \mathbf{Z}\mathbf{I}\end{aligned} \tag{2.4}$$

where $\mathbf{Z}$ is called the *open-circuit impedance matrix of the two-port*. In scalar form, we have

$$V_1 = z_{11}I_1 + z_{12}I_2 \tag{2.5a}$$

$$V_2 = z_{21}I_1 + z_{22}I_2 \tag{2.5b}$$

where z_{11} and z_{22} are called the *open-circuit driving-point impedances* and z_{12} and z_{21} are called the *open-circuit transfer impedances* of the two-port. These are obviously the generalization of the open-circuit driving-point and transfer resistances of a resistive two-port discussed in Chap. 3. Using the substitution theorem on the two 1-ports in Fig. 2.1, we may interpret the meanings of the z_{ij}'s here by considering two current sources as *inputs* to the two-port as shown in Fig. 2.3. Then Eqs. (2.5a) and (2.5b) are simply the superposition of the voltage *responses* at the two-ports due to the two independent current sources. The two driving-point impedances are interpreted as

$$z_{11} = \frac{V_1}{I_1}\bigg|_{I_2=0} \qquad z_{22} = \frac{V_2}{I_2}\bigg|_{I_1=0} \tag{2.6a}$$

and the two transfer impedances are interpreted as

$$z_{21} = \frac{V_2}{I_1}\bigg|_{I_2=0} \qquad z_{12} = \frac{V_1}{I_2}\bigg|_{I_1=0} \tag{2.6b}$$

The imposed conditions $I_1 = 0$ and $I_2 = 0$ explain why we call them open-circuit impedances.

Example 1 Consider the two-port shown in Fig. 2.4. The impedance matrix can be obtained by three methods. First, by writing the two loop equations using KVL, we obtain

$$\begin{aligned} V_1 - (z_a + z_c)I_1 - z_cI_2 = 0 \\ V_2 - (z_c + r_m)I_1 - (z_b + z_c)I_2 = 0 \end{aligned} \tag{2.7}$$

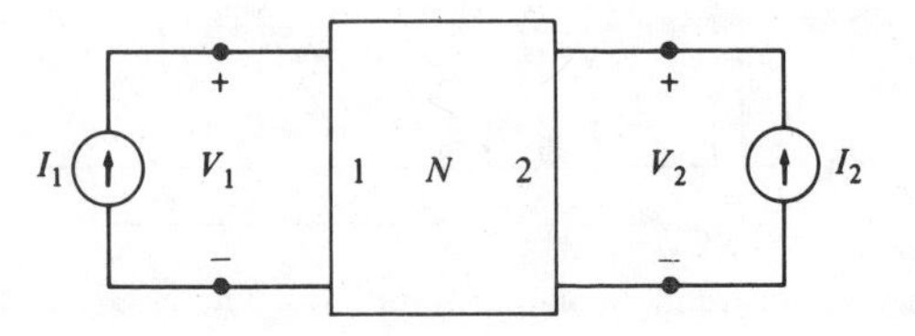

Figure 2.3 A two-port with two current sources as input.

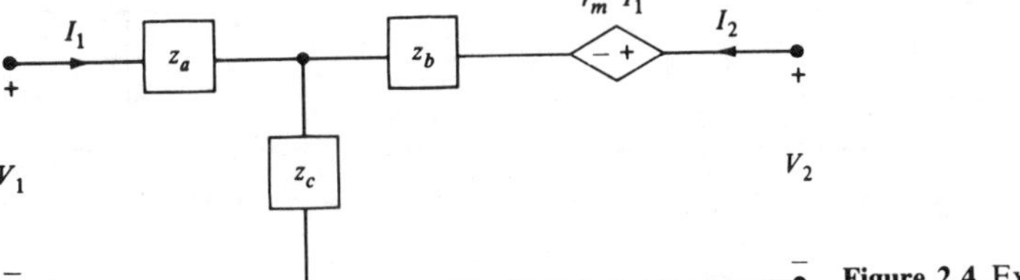

Figure 2.4 Example 1: A T circuit.

Note that these are already in the form of Eq. (2.2), with

$$\mathbf{M} = \begin{bmatrix} 1 & 0 \\ 0 & 1 \end{bmatrix} \quad \text{and} \quad \mathbf{N} = -\begin{bmatrix} z_a + z_c & z_c \\ z_c + r_m & z_b + z_c \end{bmatrix}$$

Thus, since **M** is nonsingular, we obtain from Eq. (2.4).

$$\mathbf{Z} = -\mathbf{M}^{-1}\mathbf{N} = \begin{bmatrix} z_a + z_c & z_c \\ z_c + r_m & z_b + z_c \end{bmatrix}$$

Second, we may apply two current sources $I_{s1} = I_1$ and $I_{s2} = I_2$ to the two-port and use the superposition theorem to calculate V_1 and V_2, thereby obtaining the impedance matrix.

Third, we may calculate each z_{ij} by using the formulas in Eq. (2.6). Using Eq. (2.6*a*) first, we obtain immediately

$$z_{11} = z_a + z_c \qquad z_{22} = z_b + z_c$$

and next, using Eq. (2.6*b*), we obtain

$$z_{21} = z_c + r_m \qquad z_{12} = z_c$$

Note that, if $r_m = 0$, then $z_{21} = z_{12} = z_c$ and the open-circuit impedance matrix is symmetric.

The short-circuited admittance matrix A dual analysis can be given to derive the short-circuit admittance matrix. If, in the original representation in Eq. (2.2), **N** is nonsingular, we may solve explicitly the port currents in terms of the port voltages,

$$\mathbf{I} = -\mathbf{N}^{-1}\mathbf{M}\mathbf{V} = \mathbf{Y}\mathbf{V} \tag{2.8}$$

where **Y** is called the *short-circuit admittance matrix of the two-port*. Clearly, if **Y** is nonsingular, then

$$\mathbf{Z} = \mathbf{Y}^{-1} \tag{2.9}$$

In degenerate situations, for example the two-ports in Fig. 2.5, only one of the two representations exists.

Equation (2.8) written in scalar form becomes

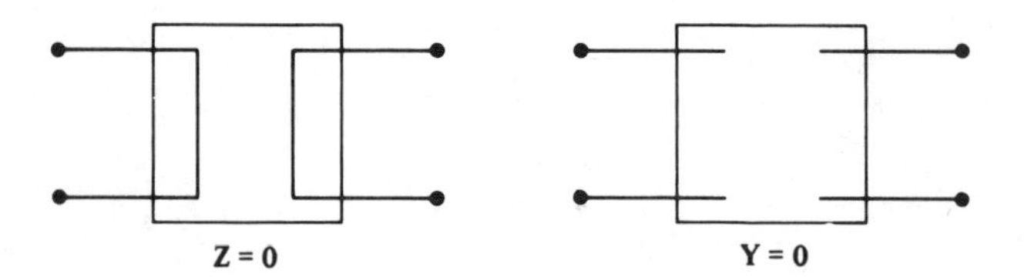

Figure 2.5 Degenerate two-ports.

$$I_1 = y_{11}V_1 + y_{12}V_2 \tag{2.10a}$$

$$I_2 = y_{21}V_1 + y_{22}V_2 \tag{2.10b}$$

where y_{11} and y_{22} are called the *short-circuit driving-point admittances* and y_{12} and y_{21} are called the *short-circuit transfer admittances* of the two-port. These terms can be interpreted by the two-port shown in Fig. 2.6 where the two inputs are two voltage sources applied at the two ports.

Following a dual argument, we have

$$y_{11} = \left.\frac{I_1}{V_1}\right|_{V_2=0} \qquad y_{22} = \left.\frac{I_2}{V_2}\right|_{V_1=0} \tag{2.11a}$$

$$y_{21} = \left.\frac{I_2}{V_1}\right|_{V_2=0} \qquad y_{12} = \left.\frac{I_1}{V_2}\right|_{V_1=0} \tag{2.11b}$$

The imposed conditions $V_1 = 0$ and $V_2 = 0$ are the reason that we call these short-circuit admittances.

Example 2 Consider the two-port shown in Fig. 2.7. We can obtain the admittance matrix by applying two voltage sources as inputs and determin-

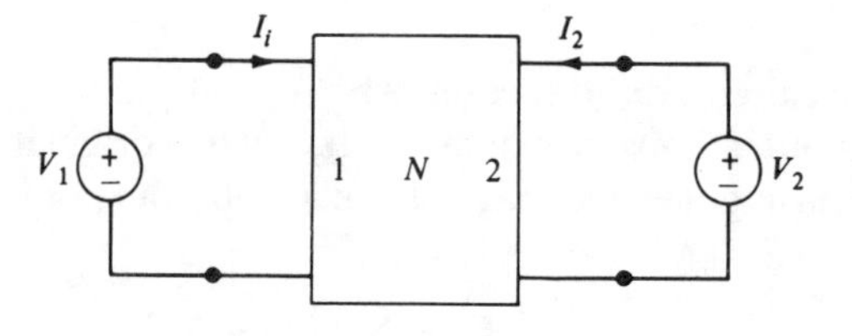

Figure 2.6 A two-port with voltage sources as input.

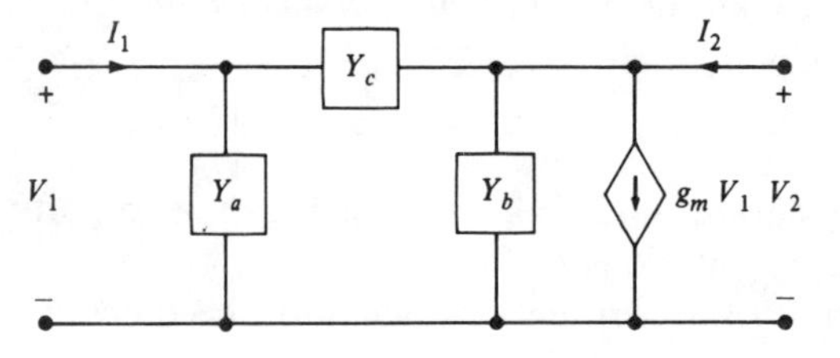

Figure 2.7 Example 2: A π circuit.

ing the two port currents. Or, we may use the formulas in Eqs. (2.11a) and (2.11b). Thus

$$y_{11} = Y_a + Y_c \qquad y_{22} = Y_b + Y_c$$
$$y_{21} = g_m - Y_c \qquad y_{12} = -Y_c$$

Again, if $g_m = 0$, $y_{21} = y_{12}$ and the short-circuit admittance matrix becomes symmetric.

Exercise Determine the impedance and admittance matrices of the two-ports shown in Fig. 2.8 if they exist.

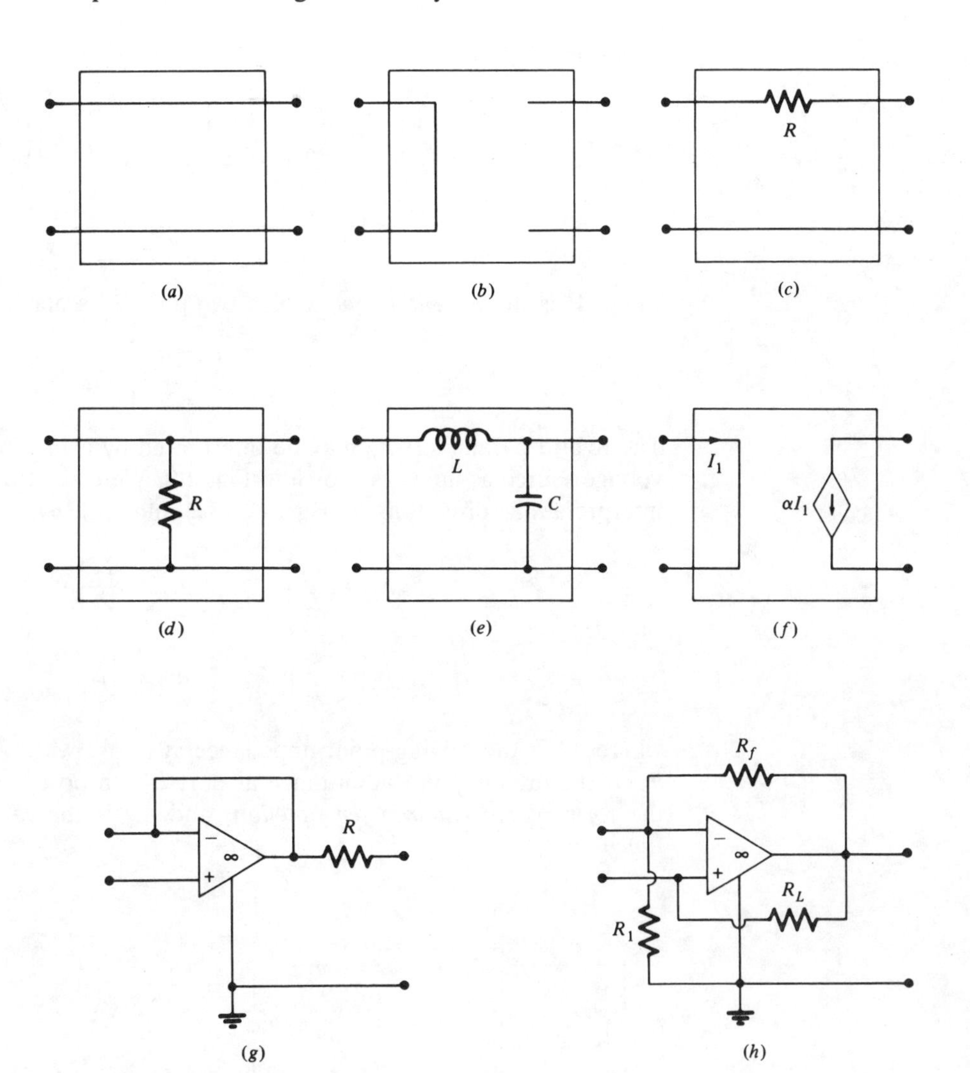

Figure 2.8 Exercises on two-ports.

2.2 Hybrid Matrices

As we learned in Chap. 3 there exists other ways of representing a two-port. With the two port voltages V_1 and V_2 and two port currents I_1 and I_2 in Eq. (2.2) or (2.3) as port variables, there exist four other possibilities in expressing explicitly two variables in terms of the other two. The two hybrid representations express V_1 and I_2 in terms of I_1 and V_2 and the converse. Therefore, with reference to Eq. (2.3), if the submatrix

$$\begin{bmatrix} m_{11} & n_{12} \\ m_{21} & n_{22} \end{bmatrix}$$

is nonsingular, then

$$\begin{bmatrix} V_1 \\ I_2 \end{bmatrix} = -\begin{bmatrix} m_{11} & n_{12} \\ m_{21} & n_{22} \end{bmatrix}^{-1} \begin{bmatrix} n_{11} & m_{12} \\ n_{21} & m_{22} \end{bmatrix} \begin{bmatrix} I_1 \\ V_2 \end{bmatrix}$$

$$= \mathbf{H} \begin{bmatrix} I_1 \\ V_2 \end{bmatrix} \tag{2.12}$$

where $\mathbf{H}$ is the *hybrid 1 matrix* of a two-port. In scalar form, we have

$$V_1 = h_{11}I_1 + h_{12}V_2 \tag{2.13a}$$

$$I_2 = h_{21}I_1 + h_{22}V_2 \tag{2.13b}$$

The hybrid parameters h_{ij} may be interpreted by using a current source and a voltage source as input as shown in Fig. 2.9. We may also obtain the following interpretations directly from Eqs. (2.13*a*) and (2.13*b*):

$$h_{11} = \left.\frac{V_1}{I_1}\right|_{V_2=0} \qquad h_{22} = \left.\frac{I_2}{V_2}\right|_{I_1=0} \tag{2.14a}$$

$$h_{21} = \left.\frac{I_2}{I_1}\right|_{V_2=0} \qquad h_{12} = \left.\frac{V_1}{V_2}\right|_{I_1=0} \tag{2.14b}$$

where h_{11} is the driving-point impedance at port 1 with port 2 short-circuited, h_{22} is the driving-point admittance at port 2 with port 1 open-circuited, h_{21} is the *forward current transfer function*, and h_{12} is the *reverse voltage transfer function*.

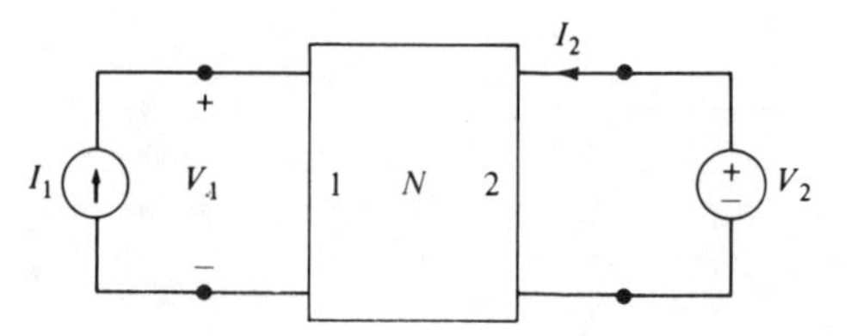

Figure 2.9 Interpreting hybrid parameters of a two-port.

A dual treatment can be given to the *hybrid 2 matrix* defined by

$$\begin{bmatrix} I_1 \\ V_2 \end{bmatrix} = \mathbf{H}' \begin{bmatrix} V_1 \\ I_2 \end{bmatrix} \tag{2.15}$$

Clearly, $\mathbf{H}' = \mathbf{H}^{-1}$ if $\mathbf{H}$ is nonsingular.

Examples

1. An ideal transformer is defined by

$$V_1 - nV_2 = 0 \tag{2.16a}$$

$$nI_1 + I_2 = 0 \tag{2.16b}$$

These two equations are in the implicit form of a general two-port given by Eq. (2.2) or (2.3). Comparing Eq. (2.16) with Eq. (2.2), we find

$$m_{11} = 1 \qquad m_{12} = -n \qquad n_{11} = n_{12} = 0$$

$$m_{21} = m_{22} = 0 \qquad n_{21} = n \qquad n_{22} = 1$$

Clearly, matrices $\mathbf{M}$ and $\mathbf{N}$ are singular, and thus neither $\mathbf{Z}$ nor $\mathbf{Y}$ exists. However, from Eq. (2.12),

$$\mathbf{H} = -\begin{bmatrix} 1 & 0 \\ 0 & 1 \end{bmatrix}^{-1} \begin{bmatrix} 0 & -n \\ n & 0 \end{bmatrix} = \begin{bmatrix} 0 & n \\ -n & 0 \end{bmatrix}$$

and the hybrid 2 matrix is

$$\mathbf{H}' = \mathbf{H}^{-1} = \begin{bmatrix} 0 & -\dfrac{1}{n} \\ \dfrac{1}{n} & 0 \end{bmatrix}$$

Of course, these matrices can be obtained directly from Eqs. (2.14) or Eqs. (2.16); however, the above derivation depicts the relationship between the implicit representation of a two-port and the various explicit representations.

2. Given the impedance matrix $\mathbf{Z}$ of a two-port, determine the hybrid 1 matrix, i.e., express the elements in $\mathbf{H}$ in terms of the open-circuit impedances of $\mathbf{Z}$.

With $\mathbf{Z}$ given, we can rewrite Eqs. (2.5*a*) and (2.5*b*) in the implicit form of Eq. (2.3):

$$\begin{aligned} V_1 - z_{11}I_1 - z_{12}I_2 &= 0 \\ V_2 - z_{21}I_1 - z_{22}I_2 &= 0 \end{aligned} \tag{2.17}$$

Therefore, identifying m_{ij}'s and n_{ij}'s in Eqs. (2.3) with the above, we obtain

$$\begin{array}{llll} m_{11}=1 & m_{12}=0 & n_{11}=-z_{11} & n_{12}=-z_{12} \\ m_{21}=0 & m_{22}=1 & n_{21}=-z_{21} & n_{22}=-z_{22} \end{array}$$

Using Eq. (2.12), we have

$$\mathbf{H}=-\begin{bmatrix}1 & -z_{12}\\ 0 & -z_{22}\end{bmatrix}^{-1}\begin{bmatrix}-z_{11} & 0\\ -z_{21} & 1\end{bmatrix}=\begin{bmatrix}z_{11}-\dfrac{z_{12}z_{21}}{z_{22}} & \dfrac{z_{12}}{z_{22}}\\ -\dfrac{z_{21}}{z_{22}} & \dfrac{1}{z_{22}}\end{bmatrix} \tag{2.18}$$

Alternatively, we may obtain the hybrid matrix directly from Eq. (2.17). In order to obtain **H** we must express V_1 and I_2 in terms of I_1 and V_2; hence we must solve Eq. (2.17) for V_1 and I_2. This is possible since the determinant of the coefficient submatrix for V_1 and I_2 in Eq. (2.17) is

$$\det\begin{bmatrix}1 & -z_{12}\\ 0 & -z_{22}\end{bmatrix}=-z_{22}$$

Hence Eq. (2.18) is obtained.

Exercise Determine the **H** and **H**′, when they exist, of the two-ports shown in Fig. 2.8.

2.3 The Transmission Matrices

The remaining two explicit representations are given by the following two sets of equations:

$$\begin{bmatrix}V_1\\ I_1\end{bmatrix}=\mathbf{T}\begin{bmatrix}V_2\\ -I_2\end{bmatrix} \tag{2.19}$$

and

$$\begin{bmatrix}V_2\\ -I_2\end{bmatrix}=\mathbf{T}'\begin{bmatrix}V_1\\ I_1\end{bmatrix} \tag{2.20}$$

Or, in scalar form

$$V_1=t_{11}V_2-t_{12}I_2 \tag{2.21a}$$

$$I_1=t_{21}V_2-t_{22}I_2 \tag{2.21b}$$

and

$$V_2=t'_{11}V_1+t'_{12}I_1 \tag{2.22a}$$

$$-I_2=t'_{21}V_1+t'_{22}I_1 \tag{2.22b}$$

As seen from these equations, the *forward transmission matrix* **T** expresses the voltage and current at port 1 in terms of the voltage and current at port 2; while the *backward transmission matrix* **T**′ expresses the inverse relation, i.e., $\mathbf{T}'=\mathbf{T}^{-1}$. The reason that a minus sign is associated with I_2 in Eqs. (2.19) and (2.20) is partly historical. It is due to the fact that, in transmission network study, port 2 is usually considered as the output port and port 1 is the input port; and we think in terms of the signal going into port 1 and coming out of port 2. This implies that a generator is connected to port 1 and a load is connected to port 2 so that $-I_2$ denotes the current *entering* the load.

The elements in the transmission matrices in Eqs. (2.21) and (2.22) are related to the open-circuit impedances, the short-circuit admittances, and the hybrid elements by the implicit equations (2.2). In Table 2.1, we give a complete tabulation of the formulas relating the six different representations.

Exercise Determine the transmission matrices of the two-ports shown in Fig. 2.8 if they exist.

Table 2.1 Conversion table for the six representations of a linear two-port.† Note $\Delta_z = \det(\mathbf{Z})$, $\Delta_y = \det(\mathbf{Y})$, $\Delta_h = \det(\mathbf{H})$, $\Delta_h' = \det(\mathbf{H}')$, $\Delta_t = \det(\mathbf{T})$, and $\Delta_t' = \det(\mathbf{T}')$.

	Z		**Y**		**H**		**H′**		**T**		**T′**	
Z	z_{11}	z_{12}	$\frac{y_{22}}{\Delta_y}$	$\frac{-y_{12}}{\Delta_y}$	$\frac{\Delta_h}{h_{22}}$	$\frac{h_{12}}{h_{22}}$	$\frac{1}{h'_{11}}$	$\frac{-h_{12}}{h'_{11}}$	$\frac{t_{11}}{t_{21}}$	$\frac{\Delta_t}{t_{21}}$	$\frac{t'_{22}}{t'_{21}}$	$\frac{1}{t'_{21}}$
	z_{21}	z_{22}	$\frac{-y_{21}}{\Delta_y}$	$\frac{y_{11}}{\Delta_y}$	$\frac{-h_{21}}{h_{22}}$	$\frac{1}{h_{22}}$	$\frac{h'_{21}}{h'_{11}}$	$\frac{\Delta'_h}{h'_{11}}$	$\frac{1}{t_{21}}$	$\frac{t_{22}}{t_{21}}$	$\frac{\Delta'_t}{t'_{21}}$	$\frac{t'_{11}}{t'_{21}}$
Y	$\frac{z_{22}}{\Delta_z}$	$\frac{-z_{12}}{\Delta_z}$	y_{11}	y_{12}	$\frac{1}{h_{11}}$	$\frac{-h_{12}}{h_{11}}$	$\frac{\Delta'_h}{h'_{22}}$	$\frac{h'_{12}}{h_{22}}$	$\frac{t_{22}}{t_{12}}$	$\frac{-\Delta_t}{t_{12}}$	$\frac{t'_{11}}{t'_{12}}$	$\frac{-1}{t'_{12}}$
	$\frac{-z_{21}}{\Delta_z}$	$\frac{z_{11}}{\Delta_z}$	y_{21}	y_{22}	$\frac{h_{21}}{h_{11}}$	$\frac{\Delta_h}{h_{11}}$	$\frac{-h'_{21}}{h'_{22}}$	$\frac{1}{h'_{22}}$	$\frac{-1}{t_{12}}$	$\frac{t_{11}}{t_{12}}$	$\frac{-\Delta'_t}{t'_{12}}$	$\frac{t'_{22}}{t'_{12}}$
H	$\frac{\Delta_z}{z_{22}}$	$\frac{z_{12}}{z_{22}}$	$\frac{1}{y_{11}}$	$\frac{-y_{12}}{y_{11}}$	h_{11}	h_{12}	$\frac{h'_{22}}{\Delta'_h}$	$\frac{h'_{12}}{\Delta'_h}$	$\frac{t_{12}}{t_{22}}$	$\frac{\Delta_t}{t_{22}}$	$\frac{t'_{12}}{t'_{11}}$	$\frac{1}{t'_{11}}$
	$-\frac{z_{21}}{z_{22}}$	$\frac{1}{z_{22}}$	$\frac{y_{21}}{y_{11}}$	$\frac{\Delta_y}{y_{11}}$	h_{21}	h_{22}	$\frac{h'_{21}}{\Delta'_h}$	$\frac{h'_{11}}{\Delta'_h}$	$\frac{-1}{t_{22}}$	$\frac{t_{21}}{t_{22}}$	$\frac{-\Delta'_t}{t'_{11}}$	$\frac{t'_{21}}{t'_{11}}$
H′	$\frac{1}{z_{11}}$	$\frac{-z_{12}}{z_{11}}$	$\frac{\Delta_y}{y_{22}}$	$\frac{y_{12}}{y_{22}}$	$\frac{h_{22}}{\Delta_h}$	$\frac{-h_{12}}{\Delta_h}$	h'_{11}	h'_{12}	$\frac{t_{21}}{t_{11}}$	$\frac{-\Delta_t}{t_{11}}$	$\frac{t'_{21}}{t'_{22}}$	$\frac{-1}{t'_{22}}$
	$\frac{z_{21}}{z_{11}}$	$\frac{\Delta_z}{z_{11}}$	$\frac{-y_{21}}{y_{22}}$	$\frac{1}{y_{22}}$	$\frac{-h_{21}}{\Delta_h}$	$\frac{h_{11}}{\Delta_h}$	h'_{21}	h'_{22}	$\frac{1}{t_{11}}$	$\frac{t_{12}}{t_{11}}$	$\frac{\Delta'_t}{t'_{12}}$	$\frac{t'_{12}}{t'_{22}}$
T	$\frac{z_{11}}{z_{21}}$	$\frac{\Delta_z}{z_{21}}$	$\frac{-y_{22}}{y_{21}}$	$\frac{-1}{y_{21}}$	$\frac{-\Delta_h}{h_{21}}$	$\frac{-h_{11}}{h_{21}}$	$\frac{1}{h'_{21}}$	$\frac{h'_{22}}{h'_{21}}$	t_{11}	t_{12}	$\frac{t'_{22}}{\Delta'_t}$	$\frac{t'_{12}}{\Delta'_t}$
	$\frac{1}{z_{21}}$	$\frac{z_{22}}{z_{21}}$	$\frac{-\Delta_y}{y_{21}}$	$\frac{-y_{11}}{y_{21}}$	$\frac{-h_{22}}{h_{21}}$	$\frac{-1}{h_{21}}$	$\frac{h'_{11}}{h'_{21}}$	$\frac{h'_{\Delta}}{h'_{21}}$	t_{21}	t_{22}	$\frac{t'_{21}}{\Delta'_t}$	$\frac{t'_{11}}{\Delta'_t}$
T′	$\frac{z_{22}}{z_{12}}$	$\frac{\Delta_z}{z_{12}}$	$\frac{-y_{11}}{y_{12}}$	$-\frac{1}{y_{12}}$	$\frac{1}{h_{12}}$	$\frac{h_{11}}{h_{12}}$	$\frac{-\Delta'_h}{h'_{12}}$	$\frac{-h'_{22}}{h'_{12}}$	$\frac{t_{22}}{\Delta_t}$	$\frac{t_{12}}{\Delta_t}$	t'_{11}	t'_{12}
	$\frac{1}{z_{12}}$	$\frac{z_{11}}{z_{12}}$	$\frac{-\Delta_y}{y_{12}}$	$\frac{-y_{22}}{y_{12}}$	$\frac{h_{22}}{h_{12}}$	$\frac{\Delta_h}{h_{12}}$	$\frac{-h'_{11}}{h'_{12}}$	$\frac{-1}{h'_{12}}$	$\frac{t_{21}}{\Delta_t}$	$\frac{t_{11}}{\Delta_t}$	t'_{21}	t'_{22}

† All matrices in the same row are equal.

3 TERMINATED AND INTERCONNECTED TWO-PORTS

3.1 Terminated Two-Ports

In applications, a two-port is usually terminated by two 1-ports as shown in Fig. 2.1. Thus it is important to know the properties of the terminated two-port in relation to the various characterizations of a two-port. In Fig. 2.1, the one-port connected to port 1 of the two-port represents the generator and the one-port connected to port 2 represents the load. Thus, we may assume $V_{s2} = 0$. Under this situation, the two network functions of particular interest are the driving-point impedance at port 1, Z_{11}, and the voltage transfer function, H_V:

$$Z_{11} = \frac{V_1}{I_1} \qquad H_V = \frac{V_2}{V_{s1}}$$

The circuit is redrawn in Fig. 3.1. Clearly, Z_{11} will depend not only on $\mathbf{Z}$, but also on the load Z_2. The four pertinent equations are, for the generator and the load,

$$V_1 + Z_1 I_1 = V_{s1} \tag{3.1a}$$

$$V_2 + Z_2 I_2 = 0 \tag{3.1b}$$

and for the two-port using the open-circuit impedances,

$$V_1 - z_{11} I_1 - z_{12} I_2 = 0 \tag{3.2a}$$

$$V_2 - z_{21} I_1 - z_{22} I_2 = 0 \tag{3.2b}$$

The driving-point impedance Z_{11} with port 2 terminated by Z_2 can be obtained by substituting $V_2 = -Z_2 I_2$ into Eq. (3.2) and eliminating I_2:

$$Z_{11} = \frac{V_1}{I_1} = z_{11} - \frac{z_{12} z_{21}}{z_{22} + Z_2} \tag{3.3}$$

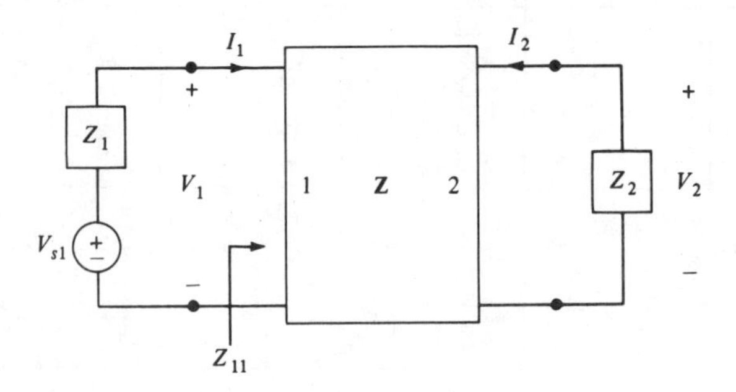

Figure 3.1 Terminated two-port in terms of **Z**.

The voltage transfer function H_V can be obtained by solving for V_2 in the four equations in (3.1) and (3.2):

$$H_V = \frac{V_2}{V_{s1}} = \frac{z_{21}Z_2}{(z_{11} + Z_1)(z_{22} + Z_2) - z_{12}z_{21}} \tag{3.4}$$

By using a dual approach, we can derive expressions in terms of the short-circuit admittances. For the circuit in Fig. 3.2, we have the driving-point admittance Y_{11} when port 2 is terminated by Y_2,

$$Y_{11} = \frac{I_1}{V_1} = y_{11} - \frac{y_{12}y_{21}}{y_{22} + Y_2} \tag{3.5}$$

and the current transfer function

$$H_I = \frac{I_2}{I_{s1}} = \frac{y_{21}Y_2}{(y_{11} + Y_1)(y_{22} + Y_2) - y_{21}y_{21}} \tag{3.6}$$

Exercises

1. Prove that, in terms of the hybrid 1 representation, Z_{11} is given by

$$Z_{11} = h_{11} - \frac{h_{12}h_{21}}{h_{22} + Y_2} \tag{3.7}$$

and, in terms of the hybrid 2 representation, Y_{11} is given by

$$Y_{11} = h'_{11} - \frac{h'_{12}h'_{21}}{h'_{22} + Z_2} \tag{3.8}$$

2. Prove that, in terms of the elements of the transmission matrix, Z_{11} is given by

$$Z_{11} = \frac{t_{11}Z_2 + t_{12}}{t_{21}Z_2 + t_{22}} \tag{3.9}$$

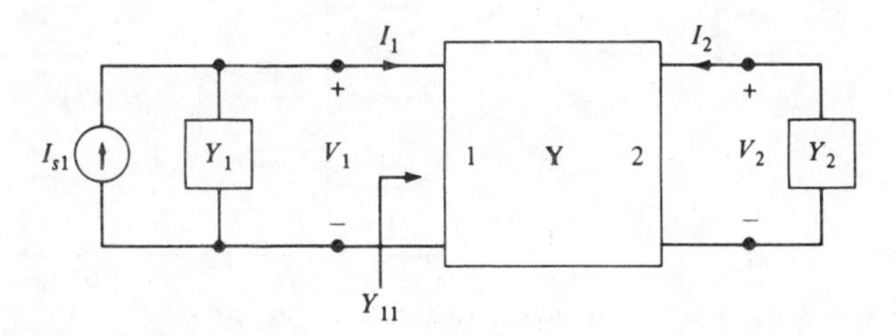

Figure 3.2 Terminated two-port in terms of **Y**.

3.2 Interconnected Two-Ports

Consider the feedback amplifier circuit shown in Fig. 3.3 where an active two-port N_μ representing the amplifier stage is connected in parallel with a passive feedback two-port N_β. The interconnected two-port N is terminated by two 1-ports representing the generator and the load. In an electronic circuits course we usually analyze a feedback amplifier by considering the active device as a *unilateral* two-port which provides one-way amplification and a passive two-port which feeds back the amplified signal to the input of the active device. This method although adequate for analyzing the behavior of a simple feedback amplifier is an approximation which may not be good enough for a stability study of a multistage feedback amplifier. We need to know the effect of the internal feedback of the active device and the forward transmission provided by the passive feedback network. To do that, we use the port-interconnection model in Fig. 3.3 for analysis.

First, we characterize the active two-port N_μ and the passive two-port N_β by their admittance matrices $\mathbf{Y}_\mu$ and $\mathbf{Y}_\beta$, respectively. The parallel connection imposes a relation between the currents and voltages for N_μ, N_β, and the resulting two-port N. Specifically,

$$V_{\mu 1} = V_{\beta 1} = V_1 \qquad V_{\mu 2} = V_{\beta 2} = V_2$$

$$I_{\mu 1} + I_{\beta 1} = I_1 \qquad I_{\mu 2} + I_{\beta 2} = I_2$$

Or, in vector form,

$$\mathbf{V}_\mu = \mathbf{V}_\beta = \mathbf{V}$$

$$\mathbf{I}_\mu + \mathbf{I}_\beta = \mathbf{I}$$

Since, for the two-port N, the admittance matrix $\mathbf{Y}$ relates $\mathbf{I}$ and $\mathbf{V}$ by

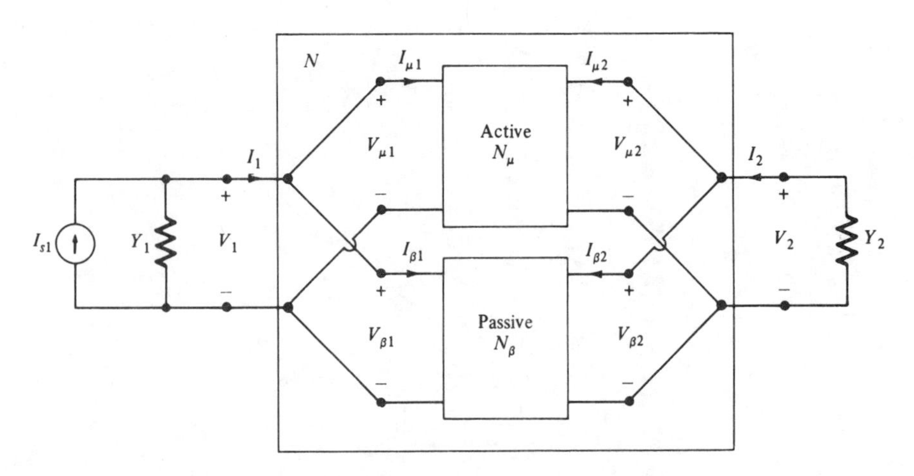

Figure 3.3 A feedback amplifier circuit.

$$\mathbf{I} = \mathbf{Y}\mathbf{V}$$

we obtain immediately

$$\mathbf{Y} = \mathbf{Y}_\mu + \mathbf{Y}_\beta = \begin{bmatrix} y_{11\mu} + y_{11\beta} & y_{12\mu} + y_{12\beta} \\ y_{21\mu} + y_{21\beta} & y_{22\mu} + y_{22\beta} \end{bmatrix}$$

The terminated two-port therefore has the current transfer function given by Eq. (3.6) which turns out to be

$$H_I = \frac{I_2}{I_{s1}} = \frac{Y_2(y_{21\mu} + y_{21\beta})}{(Y_1 + y_{11\mu} + y_{11\beta})(Y_2 + y_{22\mu} + y_{22\beta}) - (y_{21\mu} + y_{21\beta})(y_{12\mu} + y_{12\beta})}$$

Exercises

1. The circuit shown in Fig. 3.4 is called a *twin-T circuit*. Determine the admittance matrix of the two-port, and obtain the driving-point impedance Z_{11} when port 2 is terminated by a resistor R. (*a*) Solve the problem using node analysis. (*b*) Solve the problem viewing the twin-T circuit as the parallel connection of two T circuits.
2. The circuit shown in Fig. 3.5*a* is called a *bridged-T circuit*. Determine the admittance matrix of the two-port and obtain the driving-point impedance Z_{11} when port 2 is terminated by a resistor R. (If $Z_a Z_b = R^2$, what is Z_{11}? The circuit is then called a *constant-resistance circuit*.)

The two-port in Fig. 3.5*a* may be viewed as a series connection of two two-ports shown in Fig. 3.5*b*. Thus an alternate method to determine the characteristic of the two-port is to use the impedance matrices of the two 2-ports $\mathbf{Z}_a$ and $\mathbf{Z}_b$. Since

$$I_1 = I_{1a} = I_{1b} \qquad I_2 = I_{2a} = I_{2b}$$
$$V_1 = V_{1a} + V_{1b} \qquad V_2 = V_{2a} + V_{2b}$$

or, in vector form

$$\mathbf{I} = \mathbf{I}_a = \mathbf{I}_b$$
$$\mathbf{V} = \mathbf{V}_a + \mathbf{V}_b$$

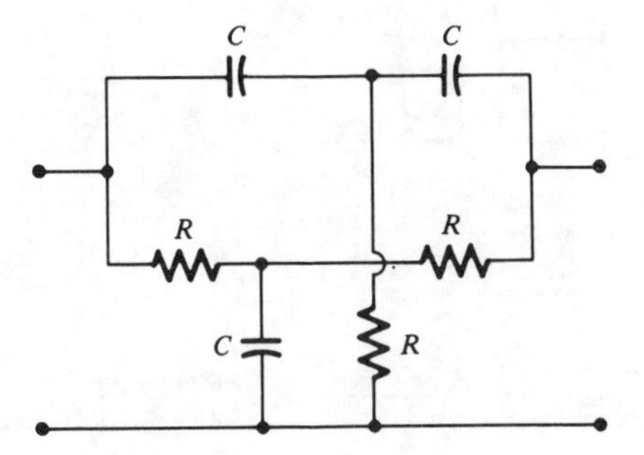

Figure 3.4 A twin-T circuit.

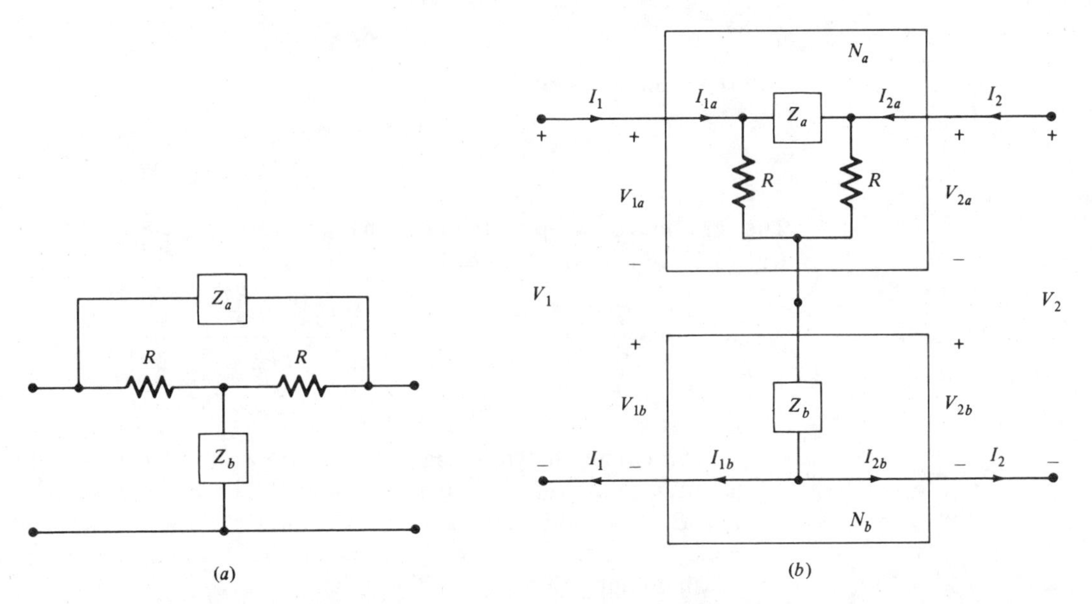

Figure 3.5 (*a*) A bridged-T circuit. (*b*) The same circuit viewed as series connection of two 2-ports.

the impedance matrix of the resulting two-port is given by the sum of the impedance matrices of the individual two-ports. Thus

$$\mathbf{Z} = \mathbf{Z}_a + \mathbf{Z}_b$$

REMARKS

1. In the circuit shown in Fig. 3.6, two 2-ports are connected in series by means of two 1/1 ratio ideal transformers. The purpose of using ideal transformers for this interconnection is to ensure that the port currents maintain the conditions $I_{1a} = I_{1b}$ and $I_{2a} = I_{2b}$ in making the series interconnection. Without the use of ideal transformers, the direct connection may short out certain elements; and as a result the current entering a port is no longer equal to the current leaving the same port, thus violating the

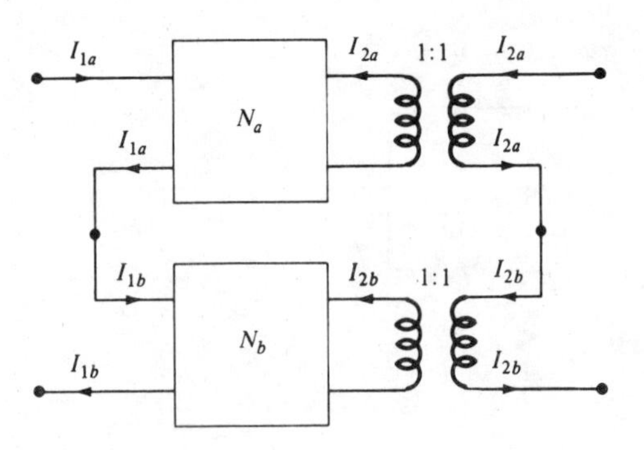

Figure 3.6 Two 2-ports connected in series by means of ideal transformers.

fundamental concept of a port. An example of such a case is shown in Fig. 3.7. It is seen that elements A and B are short-circuited by such a connection.

2. If two 2-ports are connected in such a way that at port 1 they are in series and at port 2 they are in parallel as shown in Fig. 3.8, then we may use the hybrid matrix. The resulting two-port has a hybrid matrix which is equal to the sum of the two individual hybrid matrices.

$$\mathbf{H} = \mathbf{H}_a + \mathbf{H}_b$$

Cascade connection In Fig. 3.9 we show the cascade connection of three 2-ports. The transmission matrix characterization is particularly useful in analyzing cascade connections. Let the transmission matrix for each two-port be designated by $\mathbf{T}^{(1)}$, $\mathbf{T}^{(2)}$, and $\mathbf{T}^{(3)}$. Then it is easy to show that the resulting two-port obtained by the cascade connection has a transmission matrix $\mathbf{T}$, which is equal to the product $\mathbf{T}^{(1)}\mathbf{T}^{(2)}\mathbf{T}^{(3)}$. With reference to Fig. 3.9, let

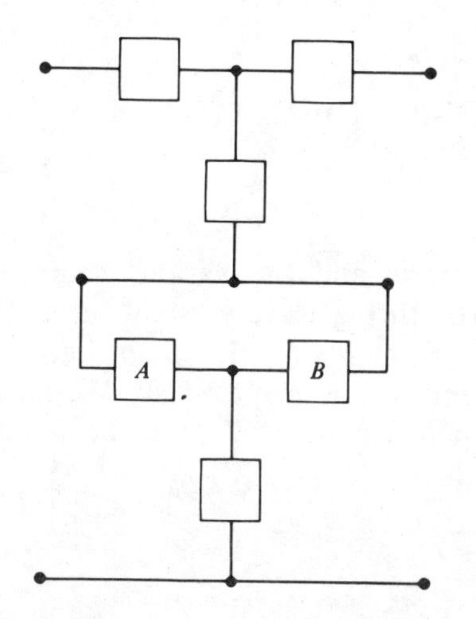

Figure 3.7 Incorrect series connection of two 2-ports.

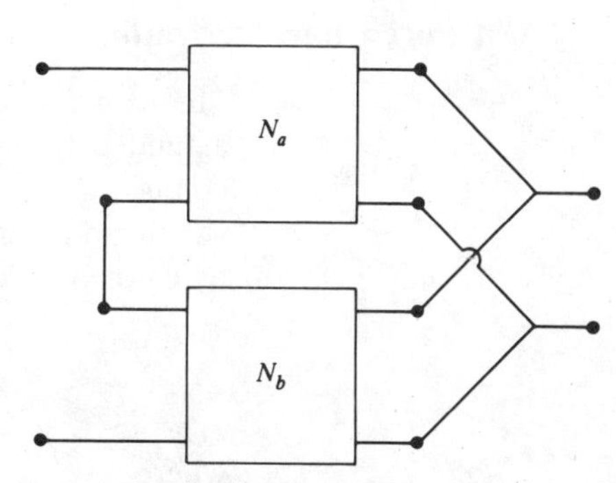

Figure 3.8 Series-parallel connection of two 2-ports.

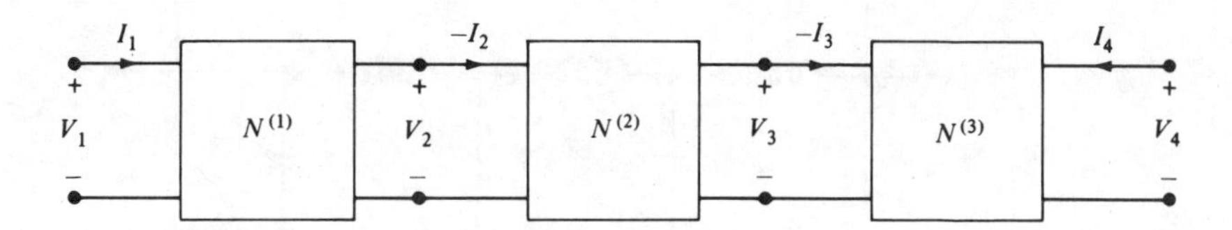

Figure 3.9 Cascade connection of three 2-ports.

$$\begin{bmatrix} V_1 \\ I_1 \end{bmatrix} = \mathbf{T} \begin{bmatrix} V_4 \\ -I_4 \end{bmatrix}$$

Since

$$\begin{bmatrix} V_1 \\ I_1 \end{bmatrix} = \mathbf{T}^{(1)} \begin{bmatrix} V_2 \\ -I_2 \end{bmatrix}$$

$$\begin{bmatrix} V_2 \\ -I_2 \end{bmatrix} = \mathbf{T}^{(2)} \begin{bmatrix} V_3 \\ -I_3 \end{bmatrix}$$

and

$$\begin{bmatrix} V_3 \\ -I_3 \end{bmatrix} = \mathbf{T}^{(3)} \begin{bmatrix} V_4 \\ -I_4 \end{bmatrix}$$

we obtain, by successive substitution,

$$\mathbf{T} = \mathbf{T}^{(1)}\mathbf{T}^{(2)}\mathbf{T}^{(3)} \tag{3.10}$$

This special property is used in the analysis and design of transmission networks and systems.

Exercise Using the property of Eq. (3.10), derive the transmission matrix of the ladder two-port shown in Fig. 3.10. If the ladder is terminated at port 2 by a resistance R, what is the driving-point impedance Z_{11}?

4 MULTIPORTS AND MULTITERMINAL CIRCUITS

4.1 *n*-Port Characterization

The concept of two-ports and the various characterizations can be easily extended to multiports. In Fig. 4.1, we show an n-port together with its port voltages V_k and port currents I_k, $k = 1, \ldots, n$. Let $\mathbf{V}$ and $\mathbf{I}$ be the port voltage vector and port current vector, respectively. We may characterize the n-port by its open-circuit impedance matrix $\mathbf{Z}$ or its short-circuit admittance matrix $\mathbf{Y}$:

$$\mathbf{V} = \mathbf{Z}\mathbf{I} \tag{4.1a}$$

$$\mathbf{I} = \mathbf{Y}\mathbf{V} \tag{4.1b}$$

where

$$\mathbf{Y} = \mathbf{Z}^{-1} \tag{4.1c}$$

Let us illustrate the physical significance of the short-circuit admittances y_{ik} where y_{ik} is an element of the short-circuit admittance matrix $\mathbf{Y}$.

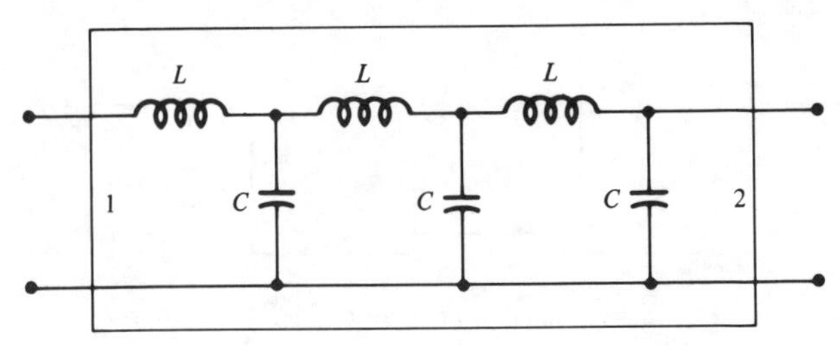

Figure 3.10 A ladder two-port.

In scalar equation form,

$$
\begin{aligned}
I_1 &= y_{11}V_1 + y_{12}V_2 + \cdots + y_{1k}V_k + \cdots + y_{1n}V_n \\
&\vdots \\
I_i &= y_{i1}V_1 + y_{i2}V_2 + \cdots + y_{ik}V_k + \cdots + y_{in}V_n \\
&\vdots \\
I_n &= y_{n1}V_1 + y_{n2}V_2 + \cdots + y_{nk}V_k + \cdots + y_{nn}V_n
\end{aligned}
\tag{4.2}
$$

Clearly,

$$
y_{ik} = \left.\frac{I_i}{V_k}\right|_{V_j=0} \qquad j = 1, 2, \ldots, n \quad \text{and} \quad j \neq k \tag{4.3}
$$

Therefore y_{ik} may be interpreted as the short-circuit transfer admittance from port k to port i with all ports other than the kth port short-circuited. This is illustrated in Fig. 4.2. Thus if $i = k$, y_{kk} is the short-circuit driving-point admittance.

A dual interpretation can be given for elements of the open-circuit impedance matrix of an n-port.

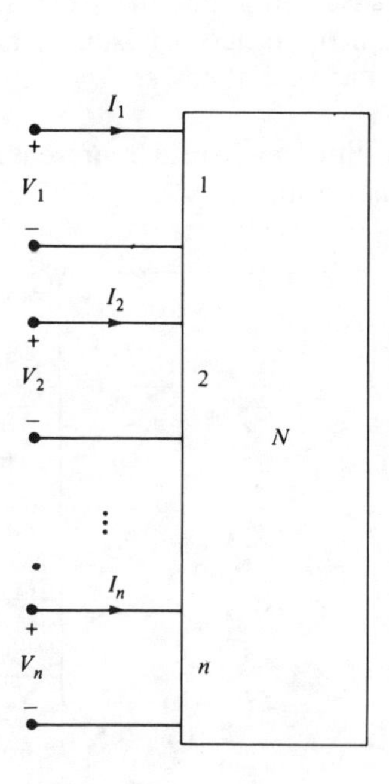

Figure 4.1 An n-port.

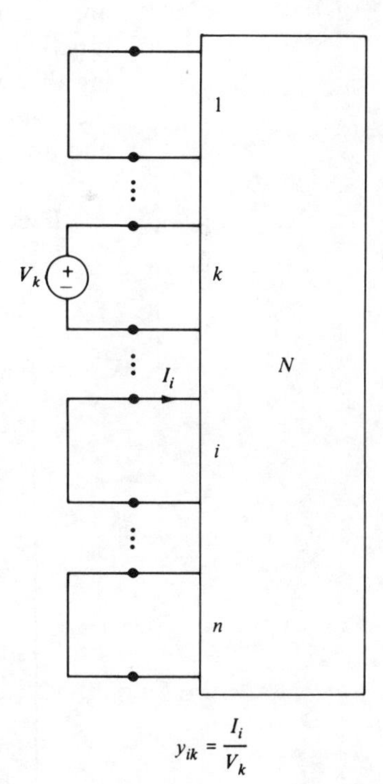

Figure 4.2 Illustrating the short-circuit admittance y_{ik} of an n-port from port k to port i.

For the hybrid matrix, since it expresses a relation between a mixture of currents and voltages, we need to partition the n-ports into two sets. In Fig. 4.3 we show an n-port which has n_1 current ports and n_2 voltage ports. Let

$$\mathbf{V}_1 = \begin{bmatrix} V_1 \\ V_2 \\ \vdots \\ V_{n_1} \end{bmatrix} \qquad \mathbf{V}_2 = \begin{bmatrix} V_{n_1+1} \\ V_{n_1+2} \\ \vdots \\ V_{n_1+n_2} \end{bmatrix} \qquad \mathbf{I}_1 = \begin{bmatrix} I_1 \\ I_2 \\ \vdots \\ I_{n_1} \end{bmatrix} \qquad \mathbf{I}_2 = \begin{bmatrix} I_{n_1+1} \\ I_{n_1+2} \\ \vdots \\ I_{n_1+n_2} \end{bmatrix} \tag{4.4}$$

Then, we may write the hybrid equations in the following form:

$$\begin{bmatrix} \mathbf{V}_1 \\ \mathbf{I}_2 \end{bmatrix} = \begin{bmatrix} \mathbf{H}_{11} & \mathbf{H}_{12} \\ \mathbf{H}_{21} & \mathbf{H}_{22} \end{bmatrix} \begin{bmatrix} \mathbf{I}_1 \\ \mathbf{V}_2 \end{bmatrix} \tag{4.5}$$

The hybrid matrix $\mathbf{H}$ may be interpreted in terms of n inputs applied to the n-port consisting of n_1 current sources and n_2 voltage sources as shown in Fig. 4.3. Under a mixture of current and voltage source inputs, the meaning of the various submatrices can be given as follows: $\mathbf{H}_{11}$ is the impedance matrix of the n_1-port with $\mathbf{V}_2 = \mathbf{0}$, that is, the n_2 voltage ports are short-circuited. $\mathbf{H}_{22}$ is then the admittance matrix of the n_2-port with $\mathbf{I}_1 = \mathbf{0}$, that is, the n_1 current ports are open-circuited. Similarly, $\mathbf{H}_{12}$ has elements which are voltage transfer functions and $\mathbf{H}_{21}$ has elements which are current transfer functions. Comparing this with the impedance and admittance representations, we conclude that the hybrid representation offers more flexibility and includes the impedance and admittance matrices as special cases.

Exercise Give four different hybrid representations of the three-winding ideal transformer shown in Fig. 4.4.

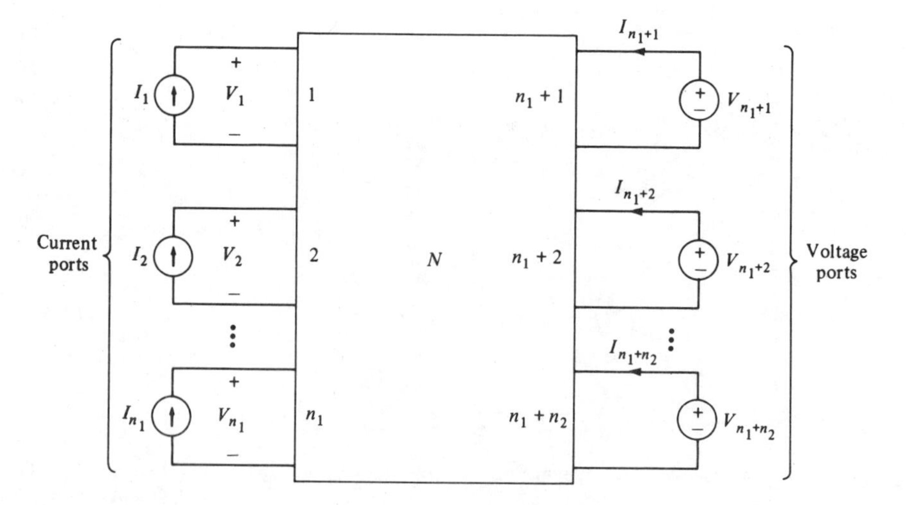

Figure 4.3 An n-port which exhibits the hybrid representation.

4.2 The Indefinite Admittance Matrix

In this section we will study a special admittance matrix useful for characterizing an n-terminal device or circuit. Again, we assume that there are no internal independent sources. We denote a particular terminal as the datum terminal, thus creating a grounded $(n-1)$-port with ports $(1, n)$, $(2, n)$, . . . , $(n-1, n)$ as shown in Fig. 4.5. Note that the nth terminal is shared by the $(n-1)$-ports. The $(n-1)$-port so created has $n-1$ port voltages defined by the $n-1$ terminal-to-datum voltages. The $n-1$ port currents $I_1, I_2, \ldots, I_{n-1}$ must add up to the current $-I_n$ shown in the figure coming out of the nth terminal. Therefore if we consider the admittance matrix representation of the $(n-1)$-port, we have

$$\mathbf{I} = \mathbf{Y}\mathbf{V} \tag{4.6a}$$

or

$$\begin{bmatrix} I_1 \\ I_2 \\ \vdots \\ I_{n-1} \end{bmatrix} = \begin{bmatrix} y_{11} & y_{12} \cdots y_{1,n-1} \\ y_{21} & y_{22} \cdots y_{2,n-1} \\ \cdots & \cdots \\ y_{n-1,1} & y_{n-1,2} \cdots y_{n-1,n-1} \end{bmatrix} \begin{bmatrix} V_1 \\ V_2 \\ \vdots \\ V_{n-1} \end{bmatrix} \tag{4.6b}$$

We assume that $\det \mathbf{Y} \neq 0$.

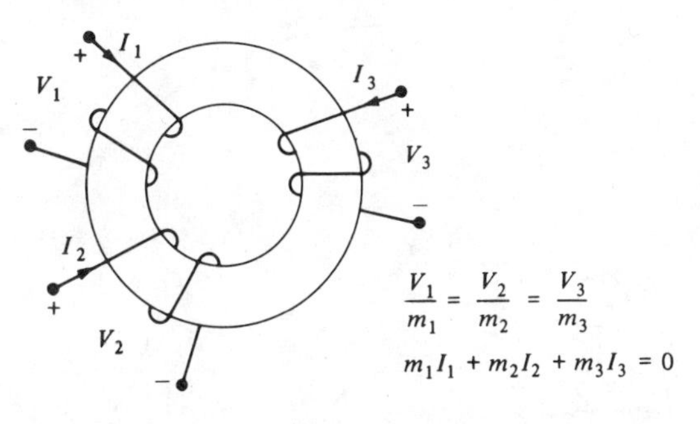

Figure 4.4 A three-winding ideal transformer.

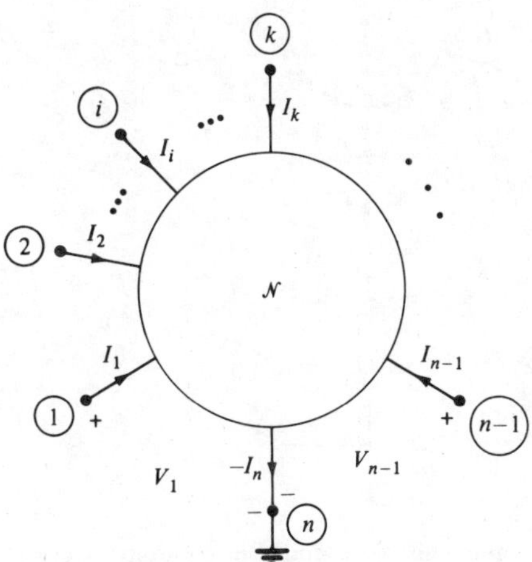

Figure 4.5 An n-terminal circuit with the nth terminal chosen as the datum terminal.

A problem of practical interest is to determine the relation of two admittance matrices of the same n-terminal circuit with respect to a different datum terminal.

Suppose the kth terminal instead of the nth terminal is used as the datum terminal; then for the same circuit, we have a different $(n-1)$-port. We are interested in obtaining the admittance matrix of the new $(n-1)$-port from the original one. It turns out that there exists a simple relation between the two. We will derive below the relation by introducing the concept of the *indefinite admittance matrix*.

In order to treat the n terminals of the circuit without prejudice as to which one is chosen as the datum, we consider the circuit shown in Fig. 4.6 where the n-terminal circuit $\mathcal{N}$ is imbedded in an arbitrary external circuit $\mathcal{N}'$. We let the datum node be marked as shown in the external circuit and the n voltages $E_1, E_2, \ldots, E_n$ be defined from the n terminals of $\mathcal{N}$ to the datum node. We assume that the complete circuit is uniquely solvable. Thus we may use the substitution theorem to substitute the external circuit by n independent voltage sources as shown in Fig. 4.7. This implies that (a) the currents entering $\mathcal{N}$ are the same as before, namely, $I_1, I_2, \ldots, I_n$, and (b) by the superposition theorem, we may write

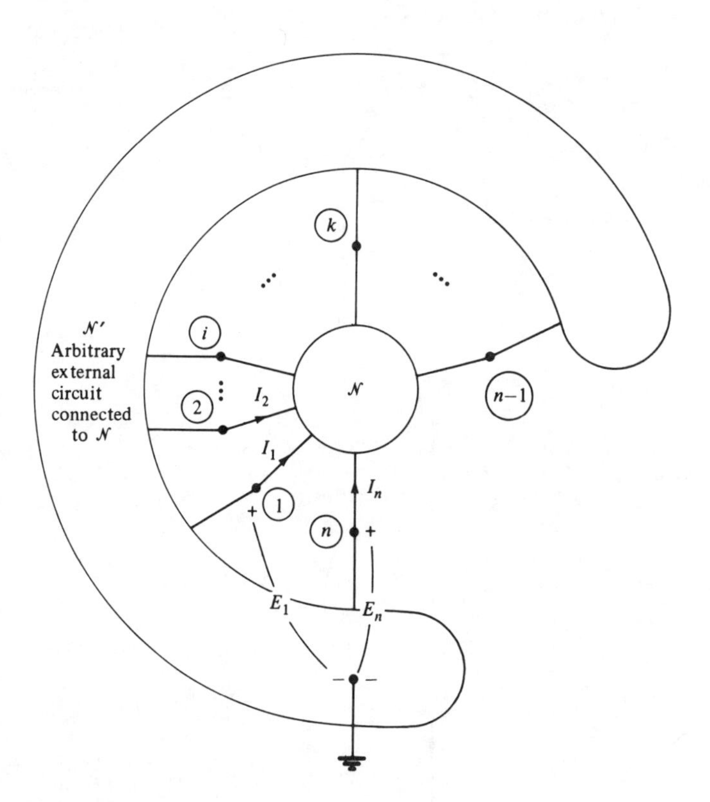

Figure 4.6 An n-terminal circuit $\mathcal{N}$ imbedded in an arbitrary circuit $\mathcal{N}'$.

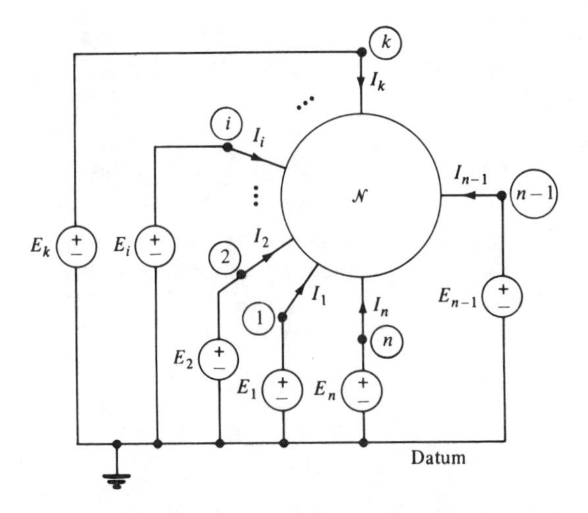

Figure 4.7 Substituted circuit with n voltage sources.

$$\mathbf{I} = \tilde{\mathbf{Y}}\mathbf{E} \tag{4.7a}$$

or, in scalar form

$$\begin{aligned} I_1 &= \tilde{y}_{11}E_1 + \tilde{y}_{12}E_2 + \cdots + \tilde{y}_{1n}E_n \\ I_2 &= \tilde{y}_{21}E_1 + \tilde{y}_{22}E_2 + \cdots + \tilde{y}_{2n}E_n \\ &\vdots \\ I_n &= \tilde{y}_{n1}E_1 + \tilde{y}_{n2}E_2 + \cdots + \tilde{y}_{nn}E_n \end{aligned} \tag{4.7b}$$

We call $\tilde{\mathbf{Y}}$ in Eq. (4.7) the *indefinite admittance matrix* of the n-terminal circuit $\mathcal{N}$. The indefinite admittance matrix is a singular matrix with the following useful *properties*:

1. Any column sum of $\tilde{\mathbf{Y}}$ is identically zero, i.e., for any k,

$$\sum_{i=1}^{n} \tilde{y}_{ik} = 0 \tag{4.8}$$

2. Any row sum of $\tilde{\mathbf{Y}}$ is identically zero, i.e., for any i,

$$\sum_{k=1}^{n} \tilde{y}_{ik} = 0 \tag{4.9}$$

3. If we consider the n-terminal circuit $\mathcal{N}$ by itself, by grounding any terminal, say the kth terminal, the resulting $(n-1)$-port defined with the kth terminal as the datum terminal has a short-circuit admittance matrix $\mathbf{Y}_k$ which can be obtained from $\tilde{\mathbf{Y}}$ by deleting the kth row and the kth column. Note that $\mathbf{Y}$ is a principal submatrix of $\tilde{\mathbf{Y}}_k$. It is a nonsingular $(n-1) \times (n-1)$ matrix.

PROOF

Property 1. Refer to Fig. 4.7: Let $E_k \neq 0$ and $E_j = 0, j \neq k$; then from Eq. (4.7), we obtain

$$\sum_{i=1}^{n} I_i = \sum_{i=1}^{n} \tilde{y}_{ik} E_k$$

This situation corresponds to that of having only the kth node connected to a voltage source E_k while all other nodes are short-circuited. By taking the cut set defined by the n currents entering $\mathcal{N}$, we have, from KCL, $\Sigma_{i=1}^{n} I_i = 0$; therefore with $E_k \neq 0$,

$$\sum_{i=1}^{n} \tilde{y}_{ik} = 0$$

Property 2. Next, we set $E_k = E$ for all k; the circuit in Fig. 4.7 can be redrawn as shown in Fig. 4.8. Since there is no independent source within $\mathcal{N}$, the only independent source in the circuit is the voltage source E shown, which is connected to the datum node. By taking a cut set which consists of the single branch, i.e., the voltage source E, the current through it must be zero; therefore we may cut the branch. Thus the remaining circuit contains no independent sources, and hence all currents $I_1, I_2, \ldots, I_n$ are zero. From Eq. (4.7b), since $E_k = E$, for all $k = 1, 2, \ldots, n$, the row sum for each row is therefore zero.

Property 3. By grounding any terminal say the kth terminal, we set $E_k = 0$ and obtain an $(n-1)$-port with port voltages E_j, $j \neq k$, defined from terminal ⓙ to the datum terminal ⓚ. Thus deleting the kth equation in (4.7b), we have $n-1$ equations in $n-1$ variables. These equations give precisely the $n-1$ port currents of the $(n-1)$-port in terms of the $n-1$ port voltages. ∎

The significance of these properties can be illustrated with an example.

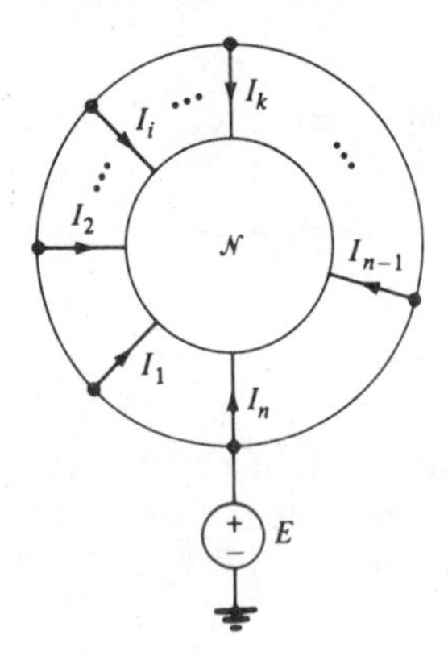

Figure 4.8 An equivalent representation when all voltage sources are the same, that is, $E_k = E$ for all k.

Example Consider the π circuit of Fig. 2.7 which is redrawn in Fig. 4.9*a*. The admittance matrix was found:

$$\mathbf{Y} = \begin{bmatrix} Y_a + Y_c & -Y_c \\ g_m - Y_c & Y_b + Y_c \end{bmatrix}$$

Using properties 1 and 2, we can construct the indefinite admittance matrix for the three-terminal circuit $\tilde{\mathbf{Y}}$ by appending a row and a column to $\mathbf{Y}$:

$$\tilde{\mathbf{Y}} = \begin{bmatrix} Y_a + Y_c & -Y_c & -Y_a \\ g_m - Y_c & Y_b + Y_c & -g_m - Y_b \\ -g_m - Y_a & -Y_b & Y_a + Y_b + g_m \end{bmatrix}$$

Clearly, $\tilde{\mathbf{Y}}$ is singular. By property 3, the admittance matrix for the two-port with terminal ② grounded as shown in Fig. 4.9*b* is simply obtained by deleting the second row and the second column of $\tilde{\mathbf{Y}}$:

$$\mathbf{Y}_2 = \begin{bmatrix} Y_a + Y_c & -Y_a \\ -g_m - Y_a & Y_a + Y_b + g_m \end{bmatrix}$$

Exercise Given the small-signal equivalent circuit of a common-emitter transistor amplifier shown in Fig. 4.10, determine the admittance matrix of (*a*) the common-emitter circuit, (*b*) the common-base circuit, and (*c*) the common-collector circuit.

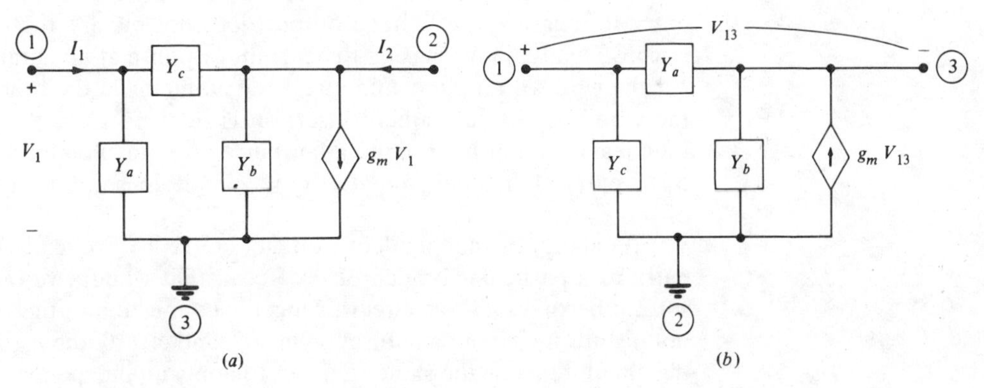

Figure 4.9 A π circuit: (*a*) with ③ as the datum terminal and (*b*) with ② as the datum terminal.

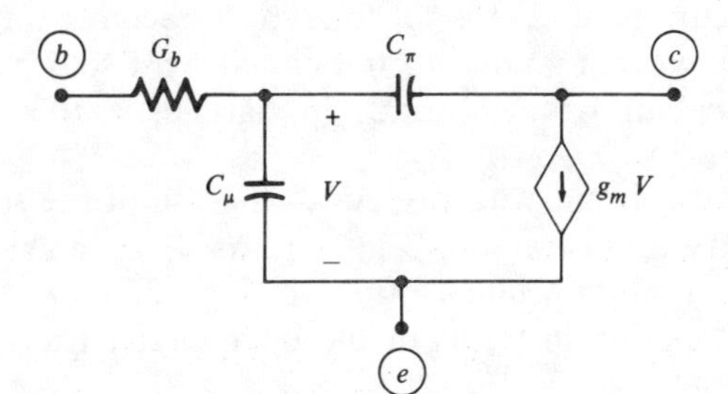

Figure 4.10 Common-emitter transistor small-signal equivalent circuit.

5 SOLDERING IRON ENTRY AND PLIERS ENTRY

We alluded to the subject of *entry* in the previous chapter when we discussed general circuit analysis. Throughout our study of linear circuit theory it is useful to consider separately a circuit which contains no independent sources and the role independent sources play when they are connected to a circuit. In defining linear circuits, we specifically state that we allow linear circuit elements *and* independent sources. In discussing natural frequencies of a linear time-invariant circuit, we set all *independent* sources in the circuit to zero. In defining network functions, we consider the response due to a particular independent source. Similarly, in introducing reciprocity in the next section, we will deal with a circuit that has no independent source and will apply a voltage source or a current source as input. Of course, these sources must be entered into the circuit in an appropriate way.

In measuring the voltage response and the current response, we use an ideal voltmeter and ammeter. An ideal voltmeter has by definition infinite impedance (it draws no current), and an ideal ammeter has by definition infinite admittance (i.e., it has no voltage drop). In measuring the voltage response between two arbitrary nodes, we connect the two terminals of a voltmeter to the two nodes by using the *soldering iron entry*. This means that we solder one node of the circuit to one terminal of the voltmeter and we solder the other node of the circuit to the other terminal of the voltmeter. Because an ideal voltmeter has infinite impedance, using the soldering iron entry to measure the voltage output does not disturb the current flow in the circuit. Dually, in measuring the current response at any branch of a circuit, we cut the wire with a pliers and insert the ammeter in the branch. The cutting of the wire and the subsequent insertion is referred to as the *pliers entry*. Since the ideal ammeter has infinite admittance or, equivalently, zero impedance, the pliers entry of an ideal ammeter into a circuit does not disturb the performance of the circuit.

To apply an independent voltage source to a circuit $\mathcal{N}$, we use a pliers entry to a particular branch of $\mathcal{N}$. Recall that when a voltage source is set to zero, it becomes a short circuit. Thus the connection of the voltage source does not disturb the circuit's natural behavior, and clearly the zero-input response of the circuit remains the same. Dually, to apply an independent current source to a circuit $\mathcal{N}$, we use a soldering iron entry to two nodes of the circuit $\mathcal{N}$. Recall that when a current source is set to zero, it becomes an open circuit. Thus the insertion of a current source to the circuit by a soldering iron entry does not disturb the circuit behavior and, in particular, the zero-input response is unaffected.

The above concept with respect to the use of the soldering iron entry and the pliers entry is illustrated in Fig. 5.1. As seen, in each case, a two-port N is created by an appropriate entry. In Fig. 5.1*b* we may characterize the input-output relation in terms of the open-circuit transfer impedance

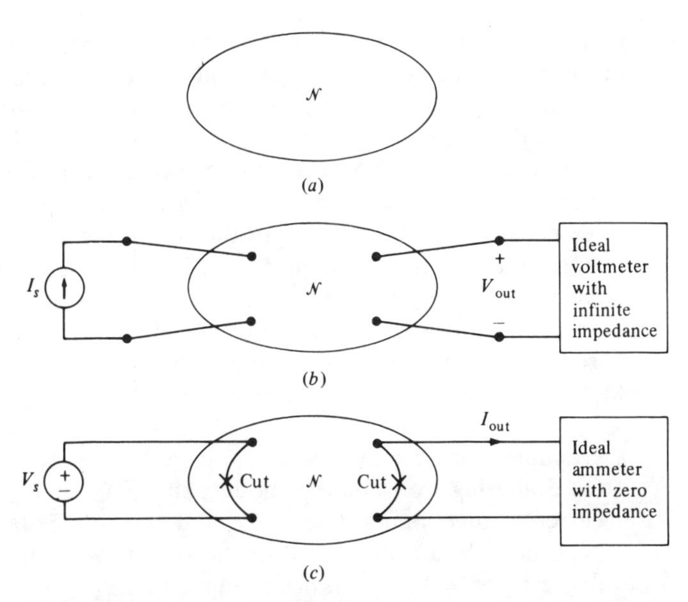

Figure 5.1 Illustrating the soldering iron and the pliers entries: (*a*) Given circuit $\mathcal{N}$ with no independent sources. (*b*) Soldering iron entry: the current source is applied to $\mathcal{N}$ at a pair of nodes by soldering iron entry. The voltage output is measured by an ideal voltmeter by the soldering iron entry. (*c*) Pliers entry: the voltage source is applied to $\mathcal{N}$ at a branch by cutting the branch (pliers entry) and inserting at the cut with the voltage source. The current output is measured by an ideal ammeter with zero impedance by the pliers entry.

$$z_{21} = \frac{V_{out}}{I_s}$$

of the two-port. Dually, in Fig. 5.1*c* we may characterize the input-output relation in terms of the short-circuit transfer admittance

$$y_{21} = \frac{I_{out}}{V_s}$$

of the two-port.

6 THE RECIPROCITY THEOREM

The reciprocity theorem appears in various fields of science and engineering: physics, mechanics, acoustics, electromagnetic waves, and electric circuits. Roughly speaking, it deals with the symmetric role played by the input and output of a physical system. In electric circuits, reciprocity holds for a subclass of linear and nonlinear circuits. In this study we will consider only linear time-invariant circuits: Specifically, the subclass consists of linear time-invariant two-terminal elements (resistors, capacitors, and inductors), coupled inductors, and ideal transformers only. We will denote by $\mathcal{N}_r$ any circuit belonging to this subclass and call it a *reciprocal circuit*. Note that controlled sources and gyrators are ruled out.

In addition, it is important to emphasize that the *reciprocity theorem* holds for the *zero-state response* and *sinusoidal steady-state response* only. Thus, it is convenient to state the theorem in the frequency domain in terms of network functions. Furthermore, since it deals with input-output relations, the theorem can be stated in terms of two-port representations: the impedance, the admittance, the hybrids, and the two transmission matrices. Before we state the theorem let us introduce an example to illustrate reciprocity by way of the soldering iron entry and the pliers entry.

Example Consider the reciprocal circuit $\mathcal{N}_r$ shown in Fig. 6.1*a*.
(*a*) Soldering iron entry at node pairs ①-④ and ②-④. We consider two current source inputs, I_{s1} and I_{s2}, and two voltage responses, V_1 and V_2. In particular we are interested in the two network functions: transfer impedance $z_{21} = V_2/V_{s1}$ and transfer impedance $z_{12} = V_1/I_{s2}$. Writing node equations for nodes ①, ②, and ③ in the frequency domain, we have

$$\begin{bmatrix} C_1 s + \dfrac{1}{sL_1} + \dfrac{1}{R} & -\dfrac{1}{R} & -\dfrac{1}{sL_1} \\ -\dfrac{1}{R} & C_2 s + \dfrac{1}{sL_2} + \dfrac{1}{R} & -\dfrac{1}{sL_2} \\ -\dfrac{1}{sL_1} & -\dfrac{1}{sL_2} & C_3 s + \dfrac{1}{sL_1} + \dfrac{1}{sL_2} \end{bmatrix} \begin{bmatrix} V_1 \\ V_2 \\ V_3 \end{bmatrix} = \begin{bmatrix} I_{s1} \\ I_{s2} \\ 0 \end{bmatrix} \tag{6.1}$$

Note that the node admittance matrix is *symmetric* ($y_{ij}^n = y_{ji}^n$ where y_{ij}^n is the ijth element of the node admittance matrix $\mathbf{Y}_n$ and its ijth cofactor is denoted by Δ_{ij}^n). First, with $I_{s2} = 0$, solving for V_2, we obtain

$$z_{21} = \frac{V_2}{I_{s1}} = \frac{\Delta_{21}^n}{\det(\mathbf{Y}_n)} \tag{6.2}$$

Next, with $I_{s1} = 0$, solving for V_1, we obtain

$$z_{12} = \frac{V_1}{I_{s2}} = \frac{\Delta_{12}^n}{\det(\mathbf{Y}_n)} \tag{6.3}$$

Since $\mathbf{Y}_n$ is symmetric, $\Delta_{12}^n = \Delta_{21}^n$, and consequently $z_{21} = z_{12}$. Thus we may conclude that for a linear time-invariant circuit $\mathcal{N}_r$ which contains two-terminal elements only, the transfer impedances $z_{12} = V_1/I_{s2} = z_{21} = V_2/I_{s1}$, where I_{s1} and I_{s2} are independent current sources applied to $\mathcal{N}_r$ using the

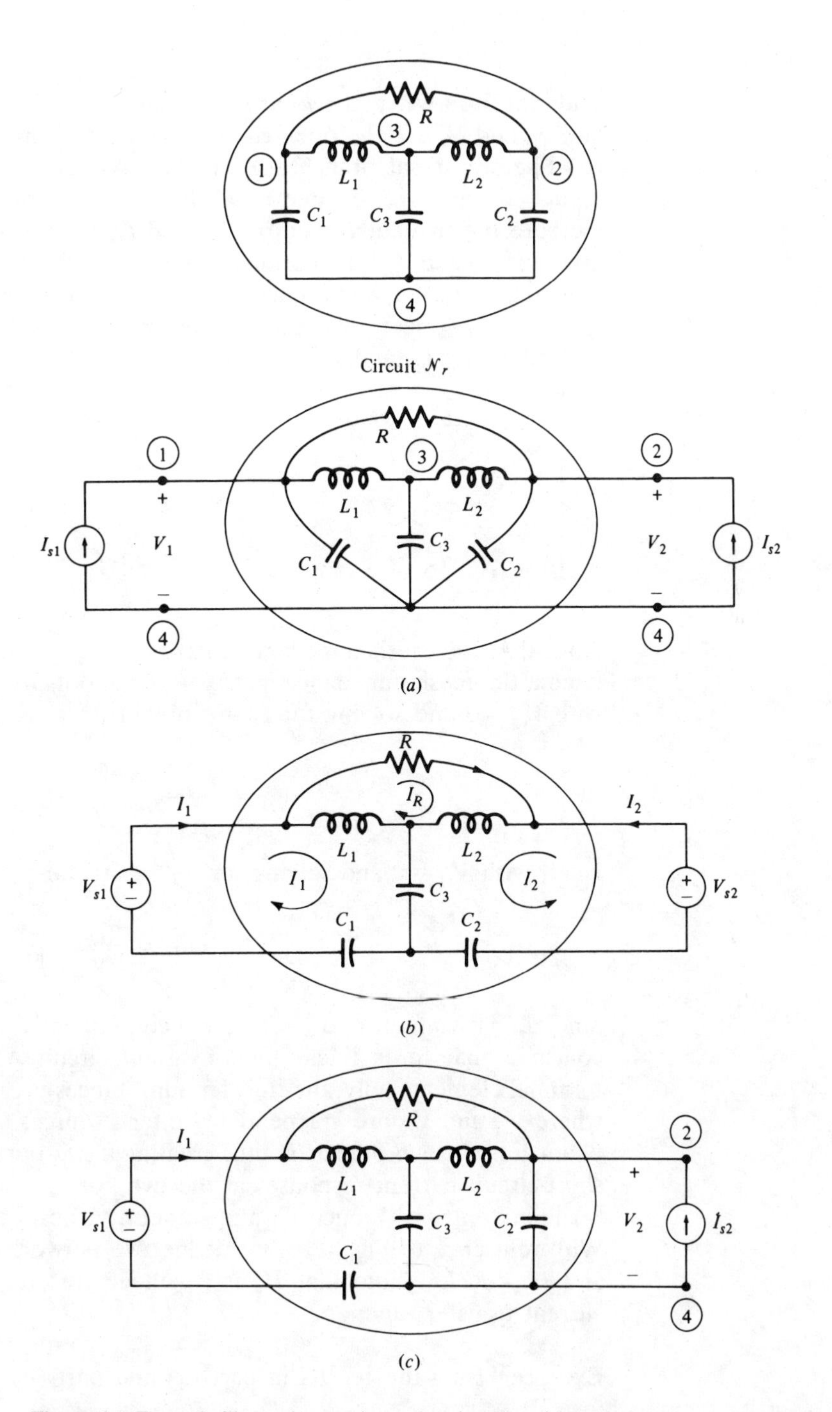

Figure 6.1 Example illustrating entries and reciprocity: (*a*) Soldering iron entry at node pairs ①-④ and ②-④. (*b*) Pliers entry at branches C_1 and C_2. (*c*) Pliers entry at branch C_1 and soldering iron entry at node pair ②-④.

soldering iron entry at two arbitrary node pairs, thus creating a two-port, and V_1 and V_2 are the open-circuit voltage response at the two ports.

(*b*) Pliers entry at branches C_1 and C_2. We consider two voltage source inputs, V_{s1} and V_{s2}, created by a pliers entry at branches C_1 and C_2. We measure the two current outputs I_1 and I_2. (Note the reference directions used for I_1 and I_2.) In particular we are interested in the two network functions: transfer admittance $y_{21} = I_2/V_{s1}$ and transfer admittance $y_{12} = I_1/V_{s2}$. Using mesh analysis for the circuit and writing mesh equations defined by I_1, I_2, and I_R, we have

$$\begin{bmatrix} \frac{1}{sC_1} + \frac{1}{sC_3} + sL_1 & -\frac{1}{sC_3} & -sL_1 \\ -\frac{1}{sC_3} & \frac{1}{sC_2} + \frac{1}{sC_3} + sL_2 & sL_2 \\ -sL_1 & sL_2 & sL_1 + sL_2 + R \end{bmatrix} \begin{bmatrix} I_1 \\ I_2 \\ I_R \end{bmatrix} = \begin{bmatrix} V_{s1} \\ V_{s2} \\ 0 \end{bmatrix} \tag{6.4}$$

Note that the mesh impedance matrix in Eq. (6.4) is symmetric. Let us denote the mesh impedance matrix by $\mathbf{Z}_m$ and its ijth cofactor by Δ_{ij}^m. First, with $V_{s2} = 0$ and solving for I_2, we obtain

$$y_{21} = \frac{I_2}{V_{s1}} = \frac{\Delta_{21}^m}{\det(\mathbf{Z}_m)} \tag{6.5}$$

Next, with $V_{s1} = 0$ and solving for I_1, we obtain

$$y_{12} = \frac{I_1}{V_{s2}} = \frac{\Delta_{12}^m}{\det(\mathbf{Z}_m)} \tag{6.6}$$

Since $\mathbf{Z}_m$ is symmetric, $\Delta_{12}^m = \Delta_{21}^m$, and consequently $y_{12} = y_{21}$. Thus we may conclude that for a linear time-invariant circuit $\mathcal{N}_r$ which contains two-terminal elements only, the transfer admittances $y_{12} = I_1/V_{s2} = y_{21} = I_2/V_{s1}$, where V_{s1} and V_{s2} are independent voltage sources applied to $\mathcal{N}_r$ using the pliers entry at two branches, thus creating a two-port, and I_1 and I_2 are the short-circuit current responses at the two-port.

(*c*) Pliers entry at branch C_1 and soldering iron entry at node pair ②-④. With reference to Fig. 6.1*c*, we define two network functions $H_V = V_2/V_{s1}$ and $H_I = I_1/I_{s2}$. Note that H_V is a voltage transfer function and H_I is a current transfer function.

Exercise Using the results in part (*a*) and part (*b*) show that $H_V = -H_I$.

REMARK If a circuit contains one VCCS, then $\mathbf{Y}_n$ is not symmetric, and hence, in general, for such circuits, $z_{21} \neq z_{12}$. If a circuit contains one CCVS, then $\mathbf{Z}_m$ is not symmetric, and hence, in general, for such circuits, $y_{21} \neq y_{12}$.

Reciprocity theorem We have introduced reciprocity by way of an example. In this section we will state the theorem in terms of general linear time-invariant reciprocal circuits. First, we define a *reciprocal two-port* as a two-port which contains only linear time-invariant two-terminal elements (resistors, capacitors, inductors), coupled inductors, and ideal transformers. We will denote it by N_r, and we will state the reciprocity theorem in terms of an arbitrary reciprocal two-port N_r. The statements are given in terms of transfer network functions defined in Fig. 6.2 in the frequency domain using the impedance, admittance, and two hybrid representations.

Reciprocity theorem For a reciprocal two-port N_r, the following relationship holds for each associated two-port representation which exists:

$$z_{21}(s) = z_{12}(s) \tag{6.7a}$$

$$y_{21}(s) = y_{12}(s) \tag{6.7b}$$

$$h_{21}(s) = -h_{12}(s) \tag{6.7c}$$

$$h'_{21}(s) = -h'_{12}(s) \tag{6.7d}$$

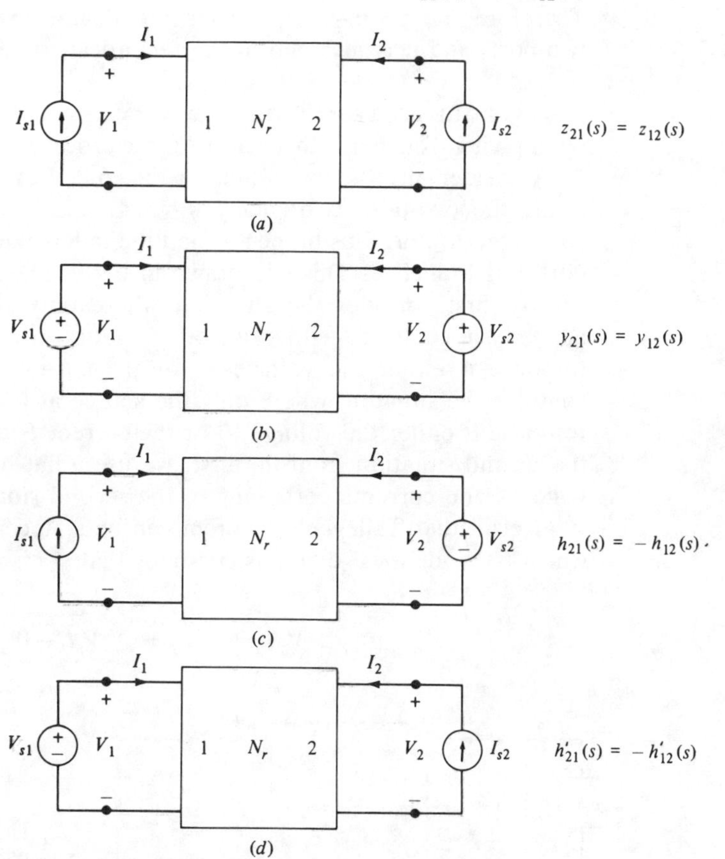

Figure 6.2 Reciprocity theorem in terms of two-port representations.

REMARKS

1. Using Table 2.1, we can easily show that, in terms of the two transmission matrices **T** and **T**′ of a two-port, the reciprocity theorem states that

$$\det(\mathbf{T}) = \det(\mathbf{T}') = 1$$

2. The defining equation for a gyrator is

$$I_1 = GV_2$$
$$I_2 = -GV_1$$

Therefore $y_{21} = -G$ and $y_{12} = G$. Clearly, a gyrator is not a reciprocal two-port; indeed its admittance matrix is skew symmetric. Such two-ports are said to be *antireciprocal*.

3. The defining equation for an ideal transformer is

$$V_1 = nV_2$$
$$I_2 = -nI_1$$

Therefore $h_{21} = -n = -h_{12}$. Thus an ideal transformer is a reciprocal two-port, and it is included in the reciprocal circuit $\mathcal{N}_r$.

PROOF OF THE RECIPROCITY THEOREM We can prove the reciprocity theorem by applying Tellegen's theorem to a reciprocal two-port.

Consider any of the circuits in Fig. 6.2. Let us denote the branches inside the two-port as branch j, $j = 3, 4, \ldots, b$, the independent source connected to port 1 as branch α, and the independent source connected to port 2 as branch β. This is redrawn in Fig. 6.3.

We first consider the situation where only the source at port 1 is applied; it could be either a voltage source or a current source. The response is either the voltage V_β or the current I_β at port 2. We next consider the situation where only the source at port 2 is applied, and the response is either the voltage V_α or the current I_α at port 1. To distinguish the second situation from the first, we use a hat notation to designate all voltages and currents pertaining to the second situation.

Recall that Tellegen's theorem can be stated in terms of two circuits which have identical digraphs. It states that

$$V_\alpha \hat{I}_\alpha + V_\beta \hat{I}_\beta + \sum_{j=3}^{b} V_j \hat{I}_j = 0 \tag{6.8}$$

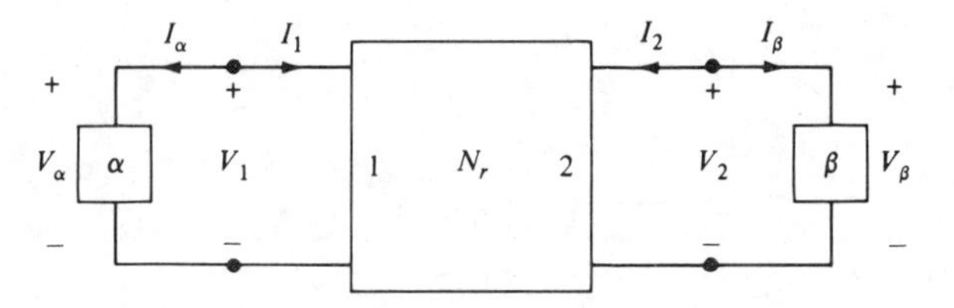

Figure 6.3 Proof of reciprocity theorem using Tellegen's theorem.

and
$$\hat{V}_\alpha I_\alpha + \hat{V}_\beta I_\beta + \sum_{j=3}^{b} \hat{V}_j I_j = 0 \tag{6.9}$$

since, as usual, we use associated reference directions throughout.

Let us consider a typical term $V_j\hat{I}_j$ in Eq. (6.8) and assume that it represents the contribution of a two-terminal element with impedance Z_j. Thus

$$V_j\hat{I}_j = Z_j I_j \hat{I}_j$$

Since
$$Z_j\hat{I}_j = \hat{V}_j$$

we have
$$V_j\hat{I}_j = \hat{V}_j I_j \tag{6.10}$$

For a coupled inductor with two branches j and k, let the branch characteristic be

$$V_j = sL_{jj}I_j + sL_{jk}I_k$$
$$V_k = sL_{kj}I_j + sL_{kk}I_k$$

where L_{jj} and L_{kk} are self-inductances and $L_{jk} = L_{kj}$ is the mutual inductance. Therefore the two typical terms in Eq. (6.8) are

$$V_j\hat{I}_j + V_k\hat{I}_k = sL_{jj}I_j\hat{I}_j + sL_{jk}I_k\hat{I}_j + sL_{kj}I_j\hat{I}_k + sL_{kk}I_k\hat{I}_k \tag{6.11}$$

The key property of a lossless coupled inductor is

$$L_{jk} = L_{kj}$$

Using the above and

$$\hat{V}_j = sL_{jj}\hat{I}_j + sL_{jk}\hat{I}_k$$
$$\hat{V}_k = sL_{kj}\hat{I}_j + sL_{kk}\hat{I}_k$$

we can reduce Eq. (6.11) to

$$V_j\hat{I}_j + V_k\hat{I}_k = \hat{V}_j I_j + \hat{V}_k I_k \tag{6.12}$$

Similarly, for a two-winding ideal transformer[2] with two branches j and k, the branch equations are

$$V_j = nV_k$$

$$I_j = -\frac{1}{n} I_k$$

We have

[2] For a multiwinding transformer, the derivation is similar.

$$V_j\hat{I}_j + V_k\hat{I}_k = nV_k\left(-\frac{1}{n}\hat{I}_k\right) + V_k\hat{I}_k$$

$$= 0$$

$$= \hat{V}_j I_j + \hat{V}_k I_k \tag{6.13}$$

From Eqs. (6.10), (6.12), and (6.13) for two-terminal elements, coupled inductors, and ideal transformers, respectively, we conclude that

$$\sum_{j=3}^{b} V_j\hat{I}_j = \sum_{j=3}^{b} \hat{V}_j I_j \tag{6.14}$$

Therefore, from Eqs. (6.8) and (6.9) we obtain the key equation depicting the α and β branches which represent either an input source or an output source:

$$V_\alpha\hat{I}_\alpha + V_\beta\hat{I}_\beta = \hat{V}_\alpha I_\alpha + \hat{V}_\beta I_\beta \tag{6.15}$$

With the above equation, we only need to identify the α and β branches in Fig. 6.3 with the particular source and response, in Fig. 6.2. For example, in Fig. 6.2*a*, with port 2 open, $I_\alpha = -I_{s1} = -I_1$ and $I_\beta = -I_{s2} = 0$. Similarly, with port 1 open, $\hat{I}_\alpha = -\hat{I}_{s1} = 0$ and $\hat{I}_\beta = -\hat{I}_{s2} = -I_2$. We obtain from Eq. (615)

$$\frac{V_\beta}{-I_{s1}} = \frac{\hat{V}_\alpha}{-\hat{I}_{s2}}$$

Since $\dfrac{V_2}{-I_1} = -z_{21}$ and $\dfrac{\hat{V}_1}{-\hat{I}_2} = -z_{12}$, we have $z_{21} = z_{12}$. For the circuit in Fig. 6.2*c*, with port 2 short-circuited, $I_\alpha = -I_{s1} = -I_1$ and $V_\beta = V_{s2} = 0$, and with port 1 short-circuited, $\hat{I}_\alpha = -\hat{I}_{s1} = 0$ and $\hat{V}_\beta = \hat{V}_{s2} = \hat{V}_2$. We obtain from Eq. (6.15)

$$\frac{I_\beta}{I_{s1}} = \frac{\hat{V}_\alpha}{\hat{V}_{s2}}$$

Since $I_2/I_1 = h_{21}$ and $\dfrac{\hat{V}_1}{-\hat{V}_2} = -h_{12}$, we have $h_{21} = -h_{12}$. The proof for Eqs. (6.7*b*) and (6.7*d*) are left as an exercise. ∎

Remarks

1. The reciprocity theorem stated above for two-ports can be generalized to reciprocal *n*-ports. The property applies to any pair of ports *j* and *k*.
2. The statements of the reciprocity theorem are given in the frequency domain in terms of the transfer functions. In the time domain, the statement which corresponds to $z_{21} = z_{12}$ can be stated as follows: Consider

a reciprocal two-port N_r, and let us first apply a current source input at port 1 with a current waveform $i_{s1}(\cdot)$. We measure the zero-state response at port 2, $v_2(\cdot)$. Let us next apply a current source input to port 2 with a current waveform $\hat{i}_{s2}(\cdot)$ and measure the zero-state response at port 1, $\hat{v}_1(\cdot)$. The *reciprocity theorem* states that if $i_{s1}(t) = \hat{i}_{s2}(t)$ for all t, then $v_2(t) = \hat{v}_1(t)$ for all t.

Exercise Prove that if a two-port contains linear *time-varying two-terminal resistors* only, the reciprocity theorem stated in the time domain above still holds.

SUMMARY

- A port can be created from a circuit by accessing two leads to a pair of nodes of the circuit. As far as the port is concerned, it is characterized in terms of its port voltage and its port current.
- A linear time-invariant two-port with no independent sources can be represented by the general implicit representation $\mathbf{M}(s)\mathbf{V}(s) + \mathbf{N}(s)\mathbf{I}(s) = \mathbf{0}$.
- All six explicit two-port representations can be derived from the implicit representation. The existence of a particular representation depends on the nonsingular property of the corresponding 2×2 submatrix of $[\mathbf{M}, \mathbf{N}]$.
- The open-circuit impedance matrix, the short-circuit admittance matrix, and the hybrid matrices of a linear time-invariant two-port represent extensions and generalizations of the open-circuit resistance matrix, the short-circuit conductance matrix, and the hybrid matrices of a resistive two-port discussed in Chap. 3. For linear time-invariant two-ports, the elements of these matrices are real rational functions of the complex frequency variable s.
- Two-ports can be interconnected without affecting their properties provided that each port maintains the port condition, that is, at each port the current entering one terminal is the same as the current leaving the other terminal of that port.
- The indefinite admittance matrix of an n-terminal circuit is an $n \times n$ singular matrix $\hat{\mathbf{Y}}$ whose $(n-1) \times (n-1)$ principal submatrices are the admittance matrices $\mathbf{Y}_k$ of the grounded $(n-1)$-port defined with respect to a common datum terminal ⓚ: $\mathbf{Y}_k$ is obtained from $\mathbf{Y}$ by deleting the kth row and kth column.
- A soldering iron entry refers to connecting an independent source or a measuring device to a circuit by soldering its two terminals to two nodes of the circuit. We always use the soldering iron entry for connecting current sources and voltmeters to a circuit.
- A pliers entry refers to connecting an independent source or a measuring device to a circuit by cutting a branch in a circuit with pliers and subsequently by inserting it into the circuit. We always use the pliers entry for connecting voltage sources and ammeters to a circuit.

- A reciprocal circuit $\mathcal{N}_r$ is a circuit which contains only linear time-invariant two-terminal elements, coupled inductors, and ideal transformers.
- A reciprocal two-port is a two-port which contains only the elements which are allowed in a reciprocal circuit or, more generally, a two-port which satisfies the reciprocity theorem.
- Reciprocity holds for the zero-state response and sinusoidal steady-state response of linear time-invariant reciprocal circuits only.
- The reciprocity theorem can be stated most conveniently in terms of network functions of a reciprocal two-port. It states that

$$z_{12}(s) = z_{21}(s)$$
$$y_{12}(s) = y_{21}(s)$$
$$h_{12}(s) = -h_{21}(s)$$
$$h'_{12}(s) = -h'_{21}(s)$$

and
$$\det[\mathbf{T}(s)] = \det[\mathbf{T}'(s)] = 1$$

PROBLEMS

Linear time-invariant two-ports

1 Design a loss pad as shown in Fig. P13.1.

(*a*) Find R_1 and R_2 such that, when the two-port is terminated by R_o,

$$\frac{V_1}{V_2} = \alpha \qquad (\alpha > 1) \qquad \text{and} \qquad R_{\text{in}} = R_o$$

(*b*) Find the **T** matrix of the resulting two-port.

(*c*) Verify that α and $\begin{bmatrix} R_o \\ 1 \end{bmatrix}$ are, respectively, an eigenvalue and the corresponding eigenvector for **T**.

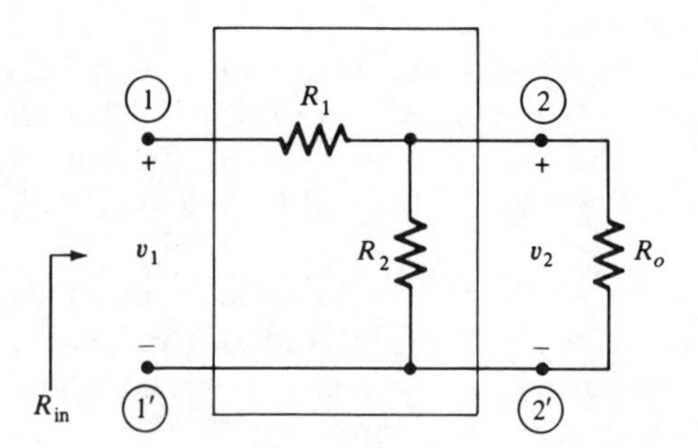

Figure P13.1

Y, H, and T representations

2 Given the two-port shown in Fig. P13.2, determine the short-circuit admittance matrix **Y**, the hybrid matrix **H**, and the transmission matrix **T**.

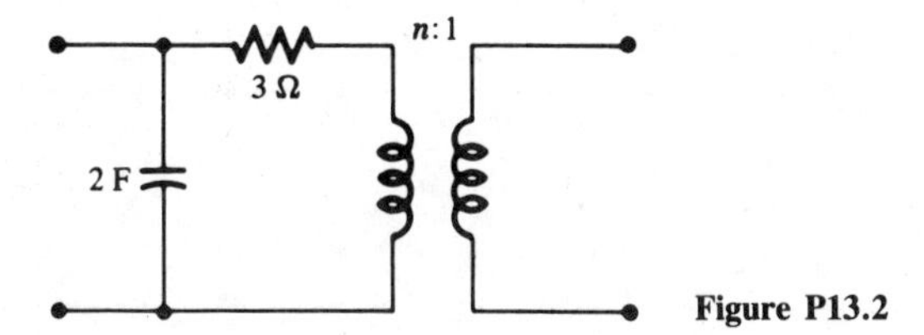

Figure P13.2

Y and Z representations

3 (*a*) Obtain the two-port admittance matrix of the transistor circuit model shown in Fig. P13.3*a* (the ports are *BE* and *CE*) in terms of the frequency ω.

(*b*) Calculate the two-port impedance matrix of the two-port shown in Fig. P13.3*b*, between ports *AE* and *DE*.

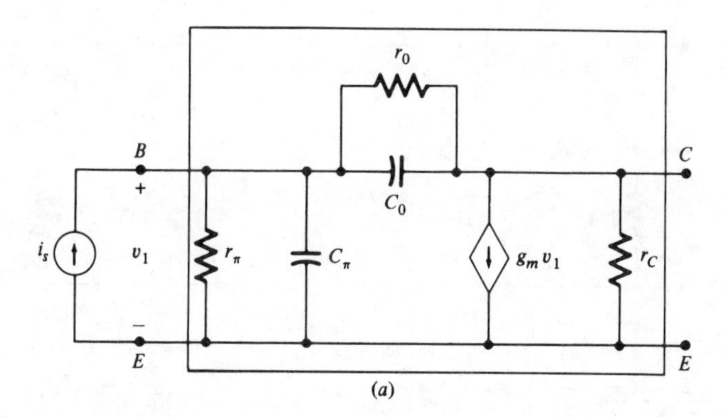

(*a*)

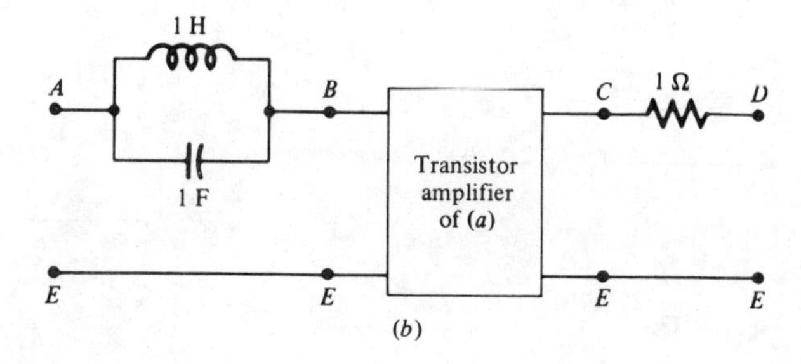

(*b*)

Figure P13.3

Implicit representation

4 Consider the linear time-invariant two-ports shown in Fig. P13.4. For each of these two-ports, write the implicit equations characterizing the two-port. Determine, if they exist:

(*a*) The impedance matrix, **Z**

(*b*) The admittance matrix, **Y**

(*c*) The hybrid matrix, **H**

(*d*) The transmission matrix, **T**

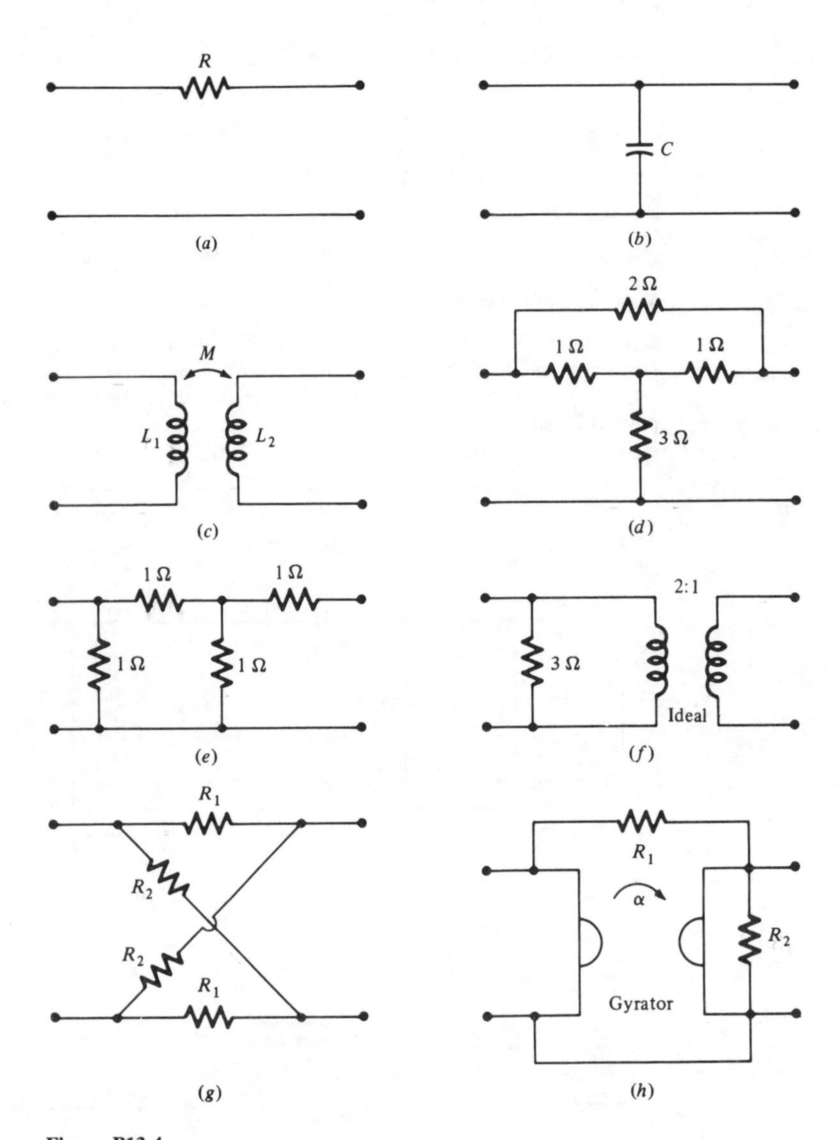

Figure P13.4

Step response and Y of a two-port

5 Consider a linear time-invariant two-port (shown in Fig. P13.5). Zero-state responses to a step-function current source at port 1 are measured in two cases: (*a*) when port 2 is short-circuited and (*b*) when port 2 is terminated by a 4-Ω resistor. From the measurements given below, determine the short-circuit admittance matrix $\mathbf{Y}(s)$ of the two-port.

$$v_{1a} = \tfrac{2}{3}\,1(t)[1 - e^{-3t/2}] \qquad v_{1b} = \tfrac{6}{7}\,1(t)[1 - e^{-7t/6}]$$

$$i_{2a} = \tfrac{1}{2}\,1(t)[1 - e^{-3t/2}] \qquad i_{2b} = \tfrac{1}{14}\,1(t)[1 - e^{-7t/6}]$$

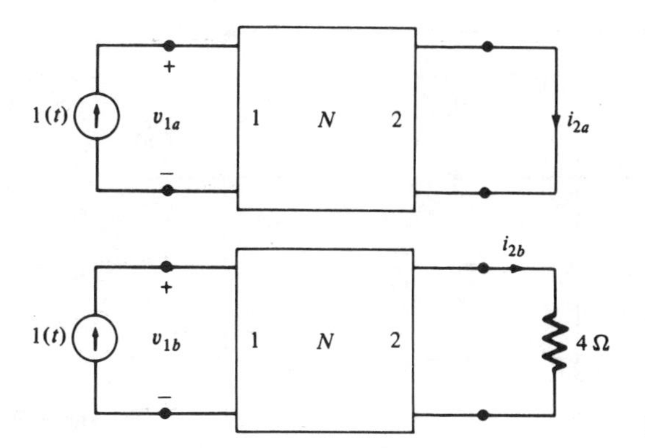

Figure P13.5

T representation

6 Assume the two-ports shown in Fig. P13.6 are in the sinusoidal steady state at frequency ω; calculate the transmission matrix of each two-port.

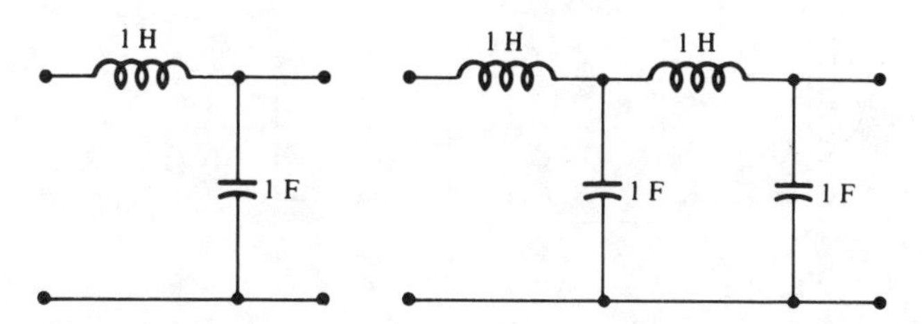

Figure P13.6

Interconnected two-ports

7 Prove that a two-port obtained by the cascade connection of two ideal gyrators is equivalent to an ideal transformer. Give the turns ratio of the ideal transformer in terms of the gyration ratio of gyrators.

8 Given the two-port N shown in Fig. P13.8, determine the short-circuit admittance matrix in the frequency domain. The coupled inductor is specified by its inductance matrix

$$\mathbf{L} = \begin{bmatrix} L_1 & M \\ M & L_2 \end{bmatrix}$$

Natural frequencies and reciprocity

9 The two-port shown in Fig. P13.9 contains linear time-invariant resistors, capacitors, and inductors only. For the connection shown in Fig. P13.9*a*, if the voltage source v_s is a unit impulse, the impulse responses for $t \geq 0$ are

$$v_1(t) = \delta(t) + e^{-t} + e^{-2t}$$

$$v_2(t) = 2\,e^{-t} - e^{-2t}$$

(*a*) What do you know about the natural frequencies of the network?

(*b*) Consider the connection shown in Fig. P13.9*b*. The voltage source v_s has been set to zero, and a current source has been connected to the output port. Is it possible to determine the steady-state voltages v_1' and/or v_2'? If so, determine v_1' and/or v_2'. If not, give your reasoning.

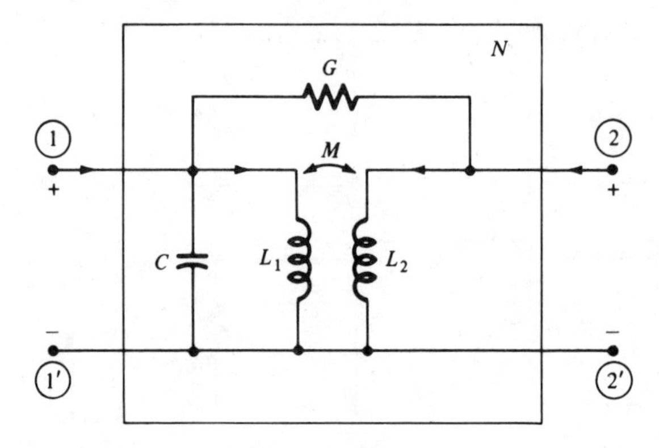

Figure P13.8

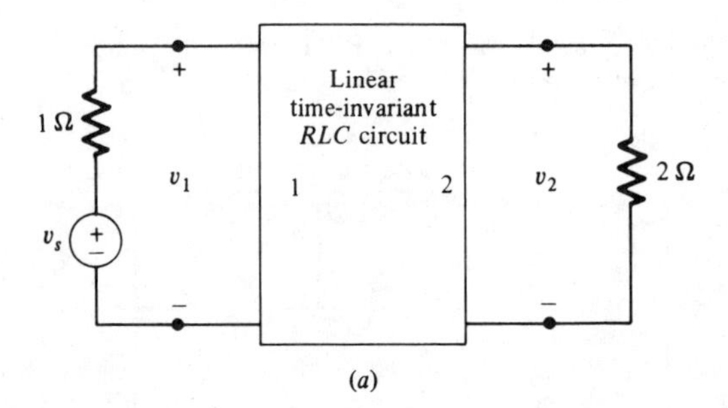

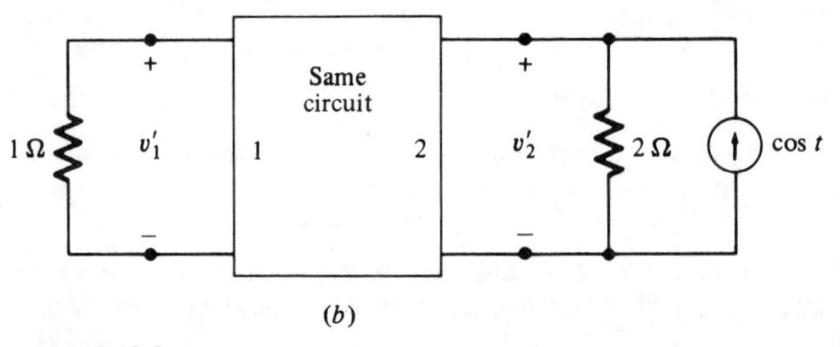

Figure P13.9

Reciprocity

10 The two-port N shown in Fig. P13.10 is made up of linear time-invariant resistors, inductors, and capacitors. We observe the following zero-state response (see Fig. P13.10a):

Input—current source: $i_1(t) = 1(t)$

Response—short-circuit current: $i_2(t) = f(t)$

where $f(\cdot)$ is a given waveform. Find, in terms of $f(\cdot)$, the zero-state response under the following conditions (see Fig. P13.10b):

Input—voltage source: $\hat{v}_2(t) = \delta(t)$

Response—open-circuit voltage: $\hat{v}_1(t) = ?$

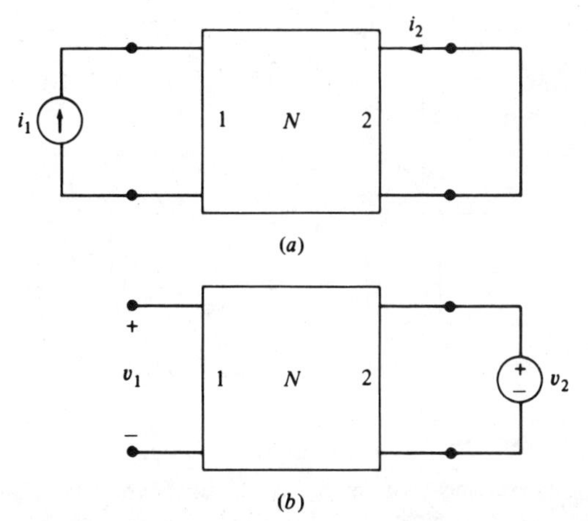

Figure P13.10

11 The two-port N shown in Fig. P13.11 is made up of linear time-invariant resistors, capacitors, inductors, and transformers. The following zero-state response is given:

Input—current source: $i_1(t) = \delta(t)$

Response—open-circuit voltage: $v_2(t) = 3\,e^{-t} + 5\,e^{-t}\cos(500t - 30°)$

(*a*) What do you know about the natural frequencies of the circuit?
(*b*) If $\hat{i}_2$ is a unit step, find the corresponding zero-state response $\hat{v}_1$.
(*c*) If $\hat{i}_2(t) = \cos 500t$, find the sinusoidal steady-state voltage $\hat{v}_1$ as a real-valued function of time.

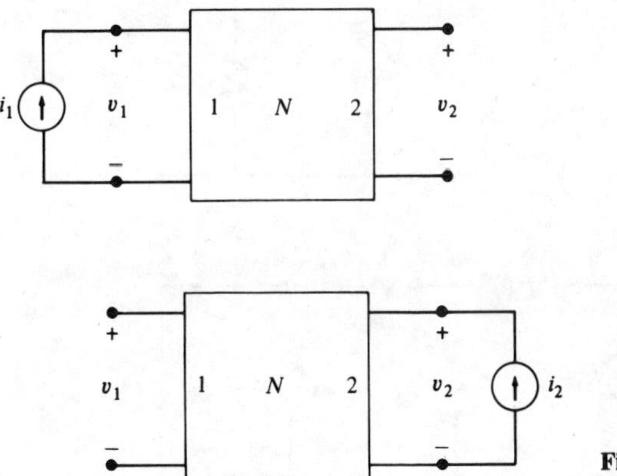

Figure P13.11

Indefinite admittance matrix

12 For the grounded three-port circuit shown in Fig. P13.12, given the following relationship:

$$\begin{bmatrix} v_1 \\ v_3 \end{bmatrix} = \begin{bmatrix} 1 & -1 \\ -2 & 3 \end{bmatrix} \begin{bmatrix} i_1 \\ i_3 \end{bmatrix} \quad \text{when } v_2 = 0$$

find the driving-point impedance of port 2, when $v_1 = 0$ and $i_3 = 0$.

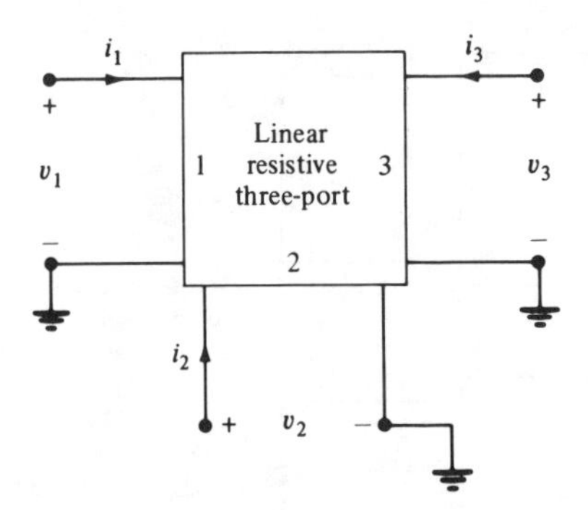

Figure P13.12

Reciprocity and superposition

13 Two sets of measurements are made on a resistive circuit consisting of one known resistor and four unknown resistors, as indicated in Fig. P13.13. In the first measurement we have $i_1 = 0.6i_s$ and $i'_i = 0.3i_s$, as shown in Fig. P13.13*a*; in the second measurement, we have $i_2 = 0.2i_s$ and $i'_2 = 0.5i_s$, as shown in Fig. P13.13*b*.

(*a*) Use the reciprocity theorem to calculate the resistance R_1.

(*b*) Consider the configuration of sources shown in Fig. P13.13*c*, where k has been adjusted so that no voltage appears across R_3 (that is, $i_3 = i'_3$). Use the superposition theorem to determine this value of k.

(*c*) From the value of k obtained above, calculate $i_3(=i'_3)$ in terms of i_s, and hence determine the resistances R_2 and R_4.

(*d*) Determine R_3, using either measurement.

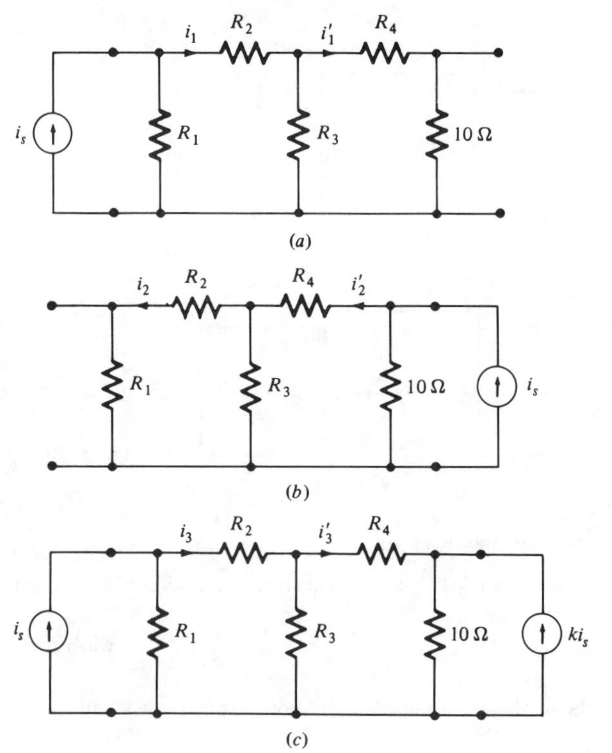

Figure P13.13

14 Two sets of measurements are made on a resistive circuit consisting of one known resistor and four unknown resistors, as indicated in Fig. P13.14*a* and *b*. Given that the resistance $R_5 = 10\ \Omega$, $V_1 = 0.9E$, $V_1' = 0.5E$, $V_2 = 0.3E$, and $V_2' = 0.5E$.

(*a*) Use the reciprocity theorem to determine R_1.

(*b*) Consider the circuit in the configuration shown in Fig. P13.14*c*. Use the superposition theorem to calculate the value of k for which no current flows through R_3 (i.e., for which $V_3 = V_3'$).

(*c*) Determine the values of R_2, R_3, and R_4. (*Hint*: Use the condition set up in Fig. P13.14*c* to calculate R_2 and R_4; then go back to either of the two earlier sets of measurements to find R_3.)

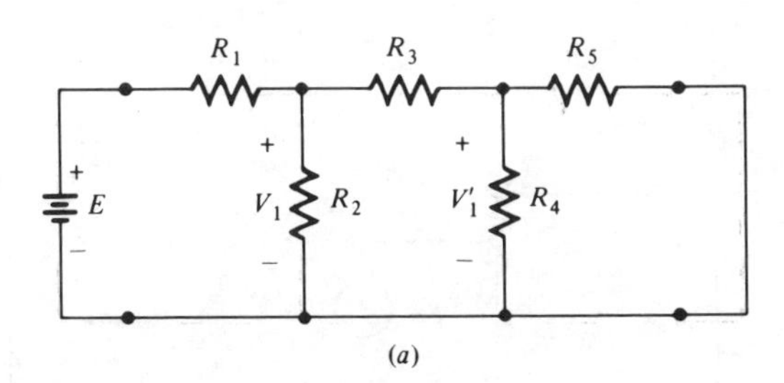

(*a*)

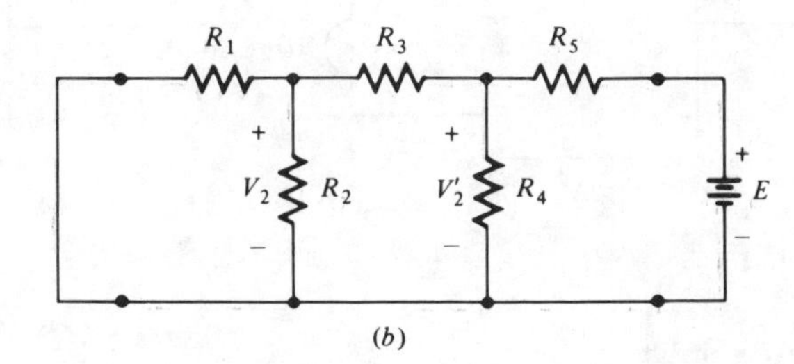

(*b*)

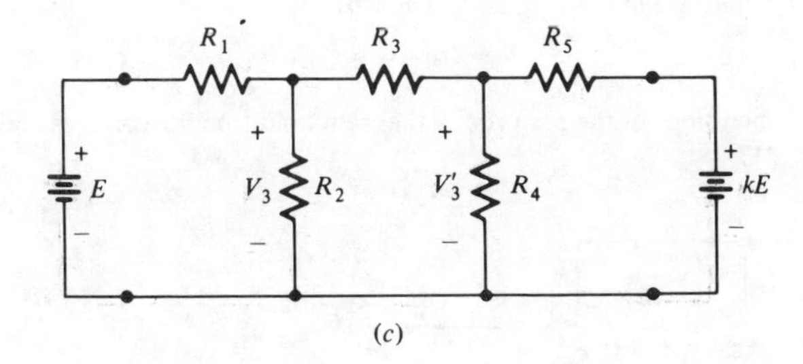

(*c*)

Figure P13.14

Reciprocity

15 The information is known for the linear time-invariant reciprocal three-port shown in Fig. P13.15*a*.

(*a*) For the same three-port N_R, if the inputs are $\hat{v}_2$ and $\hat{i}_3$ as shown in Fig. P13.15*b*, what is the zero-state response $\hat{v}_1(t)$?

(*b*) For the same three-port N_R, if the input is a voltage source v_1^* as shown in Fig. P13.15*c*, can you determine $v_1^*(t)$, $i_2^*(t)$, or $v_3^*(t)$? If so, determine the ones you can, given $v_1^*(t) = \delta(t)$.

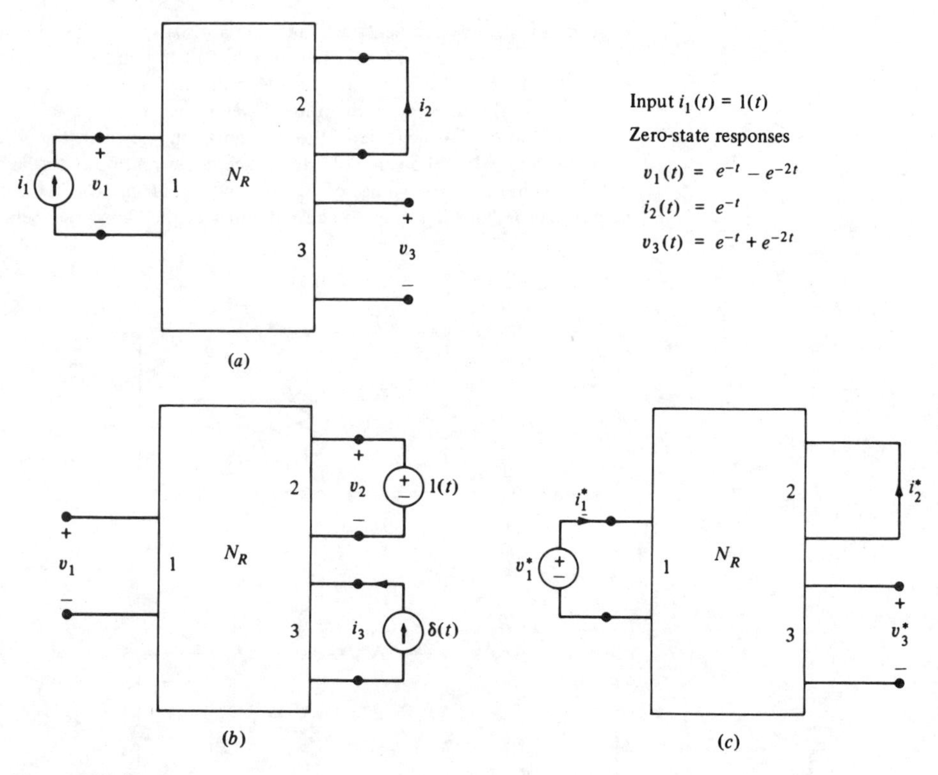

Figure P13.15

Time-varying circuit

16 Consider the *RC* circuit shown in Fig. P13.16. The sliding contact *S* moves so that the resistance between *S* and ground is, for all *t*, given by

$$R(t) = 0.5 + 0.1 \cos 2t \quad \Omega$$

Do the conclusions of the reciprocity theorem hold true for voltages and current at the ports $\{1, 1'\}$ and $\{2, 2'\}$?

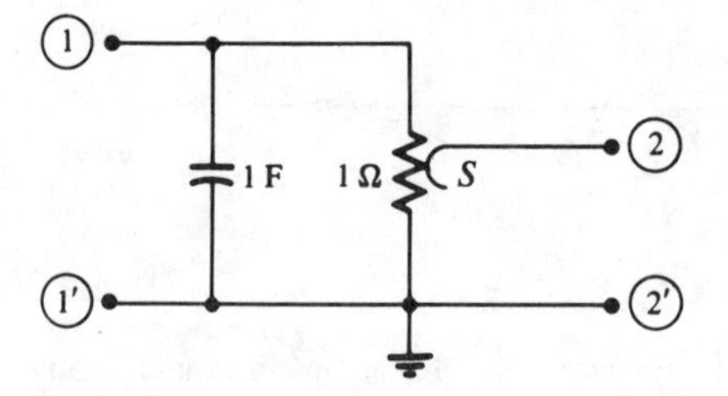

Figure P13.16

Reciprocity

17 Consider a two-port *N* made up of elements allowed by the reciprocity theorem. Consider the following transfer functions $H_f(s)$ and $H_b(s)$:

$$H_f(s) \triangleq \frac{V_2(s)}{I_s(s)} \qquad \text{as shown in Fig. P13.17}a$$

$$H_b(s) \triangleq \frac{V_1(s)}{I_s(s)} \qquad \text{as shown in Fig. P13.17}b$$

A student asserts: obviously, from reciprocity,

$$H_f(s) = H_b(s)$$

(*a*) Do you agree? If so, prove that the student is right. (The two-port N has the six representations.)

(*b*) If you don't agree, is there any additional condition(s) on N that will make the student's claim correct?

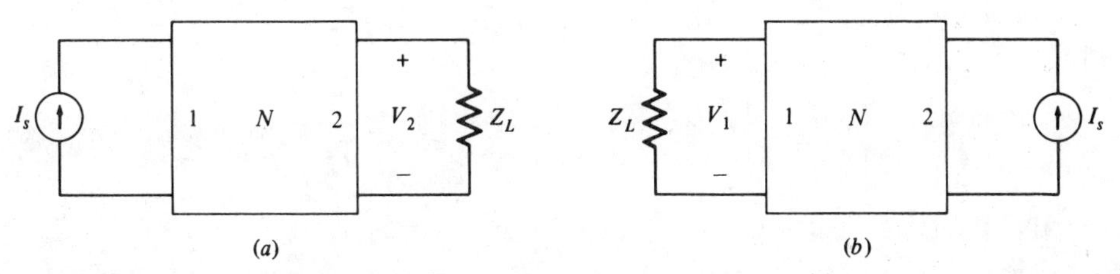

Figure P13.17

CHAPTER

FOURTEEN

DESIGN AND SENSITIVITY

INTRODUCTION

This final chapter provides a glimpse at two topics of great practical interest, namely, designing a circuit having a prescribed specification and analyzing the sensitivity of the resulting circuit to variations in circuit parameters.

In Sec. 1, we present a simple technique for designing *low-pass filters* described by a normalized *Butterworth function* using *RC* op-amp circuits. The concepts of *magnitude*, *frequency*, and *impedance scaling* are then used to rescale the normalized design to fit any specified *gain*, *cutoff frequency*, and *impedance level*.

The concept of *network sensitivity* is defined in Sec. 2, and two explicit formulas for calculating network sensitivity functions are derived: They are expressed in terms of various network functions and in terms of the node admittance matrix. These formulas are useful for hand calculation. For large networks, however, the computer is the only viable tool. Consequently we close this chapter with a computationally efficient algorithm for calculating sensitivity via the adjoint equation.

1 SIMPLE LOW-PASS FILTER DESIGN

The *low-pass filter* is an indispensable "two-port" circuit component found in virtually every communication, measurement, and control system. Its function is to suppress all high-frequency components of a signal beyond some *cutoff frequency* ω_c while allowing the lower frequencies to pass through the filter unattenuated. Hence the magnitude response of an *ideal low-pass filter* is defined by the "brickwall" characteristic in Fig. 1.1*a*, where for simplicity we

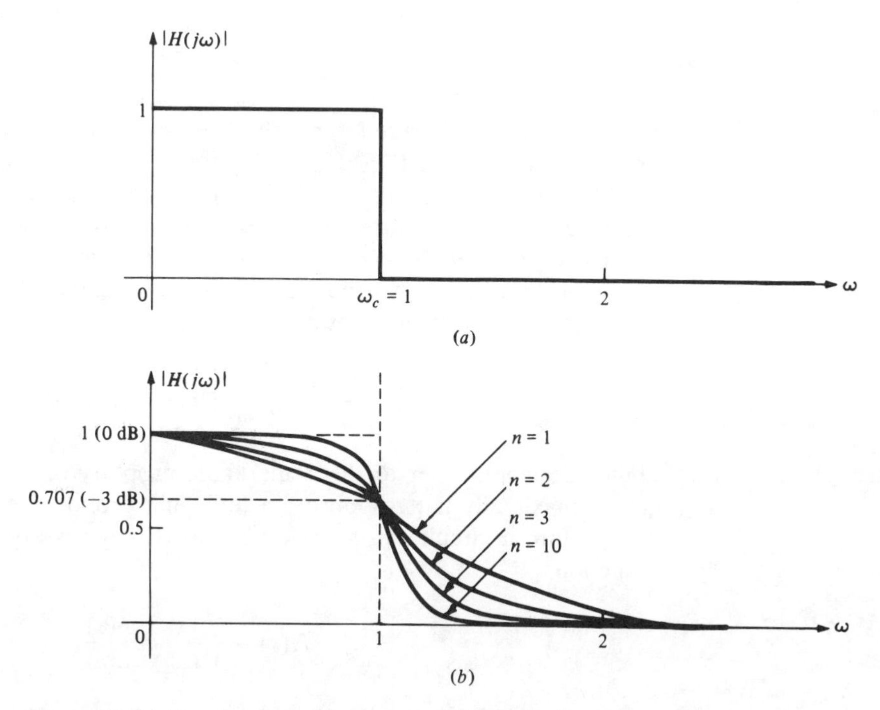

Figure 1.1 (*a*) Ideal low-pass characteristic. (*b*) magnitude response of Butterworth functions.

have normalized both the magnitude $|H(0)|$ and the cutoff frequency ω_c to unity.[1]

Since the transfer function of any linear time-invariant lumped circuit is always a rational function of $s = j\omega$ (recall Chap. 11), its magnitude function $|H(j\omega)|$ is a *continuous* function of ω, except when $s = j\omega$ is a pole of $H(s)$. Consequently, the ideal low-pass characteristic in Fig. 1.1*a* can*not* be exactly realized by a linear time-invariant *lumped* circuit.[2]

1.1 The Butterworth Approximation

Although the ideal low-pass filter is not physically realizable, it is possible to design a physical circuit which *approximates* the ideal characteristics to within any prescribed accuracy. The theory for designing such circuits is beyond the scope of this text. Here, we will whet the reader's appetite by introducing the following simple approximation first proposed by Butterworth:[3]

[1] Both $|H(0)|$ and ω_c can be easily renormalized via the impedance and frequency renormalization technique introduced in Chap. 9.

[2] In fact, it can be proved that no *physical* circuit, lumped or distributed, can realize this ideal characteristic exactly.

[3] S. Butterworth, "On the Theory of Filter Amplifiers," *Wireless Engineers*, vol. 7, pp. 536–541, 1930.

$$|H(j\omega)| = \frac{1}{\sqrt{1+\omega^{2n}}} \tag{1.1}$$

Equation (1.1) is called the nth-order *Butterworth function*; its magnitude response is sketched in Fig. 1.1b for $n=1$, 2, 3, and 10. Observe that $|H(0)|=1$ and $|H(j1)|=0.707$. In terms of the *decibel scale*, i.e., $20\log|H(j\omega)|$, the Butterworth magnitude response starts from 0 dB at dc and drops down *monotonically* by 3 dB (for all n) at the *cutoff frequency* $\omega=1$. Observe that as $n\to\infty$, the Butterworth approximation approaches the ideal low-pass characteristic. Observe also that

$$\left.\frac{d^k|H(j\omega)|}{d\omega^k}\right|_{\omega=0} = 0 \qquad \text{for all } k<2n \tag{1.2}$$

Filter designers refer to this remarkable property of the Butterworth function as a "maximally flat response," a term first proposed by Landon.[4]

Before a circuit can be synthesized, it is necessary to derive the transfer function

$$H(s) = \frac{V_0(s)}{V_i(s)} \triangleq \frac{n(s)}{d(s)} \tag{1.3}$$

associated with Eq. (1.1). Since

$$|H(j\omega)|^2 = \frac{1}{1+\omega^{2n}} = \frac{n(j\omega)n(-j\omega)}{d(j\omega)d(-j\omega)} \tag{1.4}$$

we immediately obtain

$$n(s)=1 \tag{1.5}$$

$$d(s)d(-s) = 1+(-1)^n s^{2n} \tag{1.6}$$

In order to identify the denominator $d(s)$ of $H(s)$ from Eq. (1.6), let us consider some specific values of n:

1. $n=1$: In this case, Eq. (1.6) can be factored as follows:

$$d(s)d(-s) = 1-s^2 = (1+s)(1-s) \tag{1.7}$$

Since the roots of $d(s)=0$ are poles of $H(s)$, we choose $d(s)=s+1$ so that $H(s)$ has only *left-half plane* poles.[5] The associated transfer function is

[4] V. D. Landon, "Cascade Amplifier with Maximal Flatness," *RCA Rev.*, vol. 5, pp. 347–362, 1941.

[5] We could have chosen $d(s)=s-1$ instead so that $s=1$ is a pole of $H(s)$ in the right-half plane. In this case, however, the circuit is *exponentially unstable* and useless as a filter.

$$H_1(s) = \frac{1}{s+1} \tag{1.8}$$

2. $n = 2$
$$d(s)d(-s) = 1 + s^4 \tag{1.9}$$

To factor Eq. (1.9) as in Eq. (1.7), we must find the four complex roots of $1 + s^4 = 0$. Let $s = \rho\, e^{j\theta}$ denote any such root so that $(\rho\, e^{j\theta})^4 = -1$. This can be recast into the following equivalent forms:

$$\rho^4\, e^{j4\theta} = 1 \cdot e^{\pm j\pi} = 1 \cdot e^{\pm j3\pi} \tag{1.10}$$

It follows from Eq. (1.10) that $\rho = 1$, $\theta = \pm\pi/4$ and $\pm 3\pi/4$. These roots are uniformly distributed along the circle of radius 1, as shown in Fig. 1.2*a*. In order for $H(s)$ to be *exponentially stable*, it is necessary to choose the left-half plane roots for $d(s)$, namely,

$$\begin{aligned} d(s) &= \left(s - \exp j\,\frac{3\pi}{4}\right)\left[s - \exp\left(-j\,\frac{3\pi}{4}\right)\right] \\ &= (s + 0.707 - j0.707)(s + 0.707 + j0.707) \end{aligned} \tag{1.11}$$

The associated transfer function is

$$H_2(s) = \frac{1}{s^2 + 1.414s + 1} \tag{1.12}$$

3. $n = 3$
$$d(s)d(-s) = 1 - s^6 \tag{1.13}$$

To factor Eq. (1.13) we must find the six complex roots of $1 - s^6 = 0$. Let $s = \rho\, e^{j\theta}$ denote any such root so that $(\rho\, e^{j\theta})^6 = 1$. This can be recast as follows:

$$\rho^6\, e^{j6\theta} = 1 \cdot e^{\pm j2\pi} = 1 \cdot e^{\pm j4\pi} = 1 \cdot e^{\pm j6\pi} \tag{1.14}$$

It follows from Eq. (1.14) that $\rho = 1$, $\theta = \pm\pi/3$, $\pm 2\pi/3$, and $\pm\pi$. These roots are uniformly distributed along the circle of radius 1, as shown in Fig. 1.2*b*. Choosing only left-half plane roots for $d(s)$, we obtain

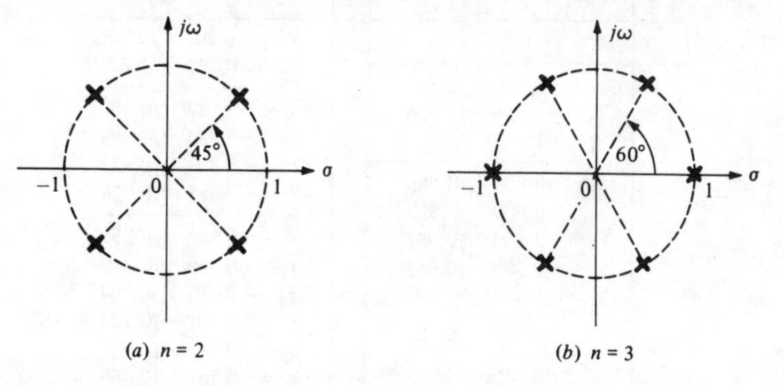

Figure 1.2 Roots of $d(s)d(-s) = 0$ are symmetrical about the $j\omega$ axis.

$$d(s) = \left(s - \exp j\,\frac{2\pi}{3}\right)\left[s - \exp\left(-j\,\frac{2\pi}{3}\right)\right](s - \exp j\pi)$$

$$= (s + 0.5 - j0.866)(s + 0.5 + j0.866)(s + 1) \tag{1.15}$$

The associated transfer function is

$$H_3(s) = \frac{1}{s^3 + 2s^2 + 2s + 1} \tag{1.16}$$

General case Following the same procedure as above, we can identify $H(s)$ for any order n as follows:

$n = odd$: $d(s)d(-s) = 1 - s^{2n}$ The roots of $1 - s^{2n} = 0$ are the $2n$ roots of $+1$: They are uniformly distributed around the unit circle beginning with the *real* root at $s = 1 + j0$. The angle between any two successive roots is equal to π/n, as shown in Fig. 1.2*b* for $n = 3$. These roots are symmetrical about the $j\omega$ axis, and the poles of $H(s)$ correspond to those on the left-half plane. For future reference, the location of the poles of $H(s)$ for $n = 1$, 3, and 5 are listed in Table 1.1.

$n = even$: $d(s)d(-s) = 1 + s^{2n}$ The roots of $1 + s^{2n} = 0$ are the $2n$ roots of -1: They are uniformly distributed around the unit circle beginning with the *complex* root at $s = \cos \pi/2n + j \sin \pi/2n$. The angle between any two successive roots is equal to π/n as shown in Fig. 1.2*a* for $n = 2$. These roots are also symmetrical about the $j\omega$ axis, and the poles of $H(s)$ correspond to those on the left-half plane. For future reference, the location of the poles of $H(s)$ for $n = 2$, 4, and 6 are listed in Table 1.1.

Table 1.1 Pole loctions of the Butterworth function of order *n*.

n	Odd n	n	Even n
1	$s = -1 + j0$	2	$s_1 = -0.707 + j0.707$ $s_2 = -0.707 - j0.707$
3	$s_1 = -0.500 + j0.866$ $s_2 = -0.500 - j0.866$ $s_3 = -1 + j0$	4	$s_1 = -0.924 + j0.383$ $s_2 = -0.924 - j0.383$ $s_3 = -0.383 + j0.924$ $s_4 = -0.383 - j0.924$
5	$s_1 = -0.809 + j0.588$ $s_2 = -0.809 - j0.588$ $s_3 = -0.309 + j0.951$ $s_4 = -0.309 - j0.951$ $s_5 = -1 + j0$	6	$s_1 = -0.966 + j0.259$ $s_2 = -0.966 - j0.259$ $s_3 = -0.707 + j0.707$ $s_4 = -0.707 - j0.707$ $s_5 = -0.259 + j0.966$ $s_6 = -0.259 - j0.966$

Exercise

(*a*) Show that the polynomial $d(s)$ for the nth-order Butterworth transfer function consists of factors of the form $s^2 + 2(\cos\phi_k)s + 1$ and, for $n =$ odd integer, also the factor $s + 1$. Specify the angle ϕ_k geometrically.

(*b*) Use this formula to construct a table where $d(s)$ is factored into first-order and/or quadratic terms for $n \leq 6$.

1.2 Synthesis of All-Pole Transfer Functions

The transfer function associated with the nth-order Butterworth function is given by

$$H(s) = \frac{1}{s^n + b_{n-1}s^{n-1} + \cdots + b_1 s + b_0} \tag{1.17}$$

where the coefficients $b_0, b_1, \ldots, b_{n-1}$ are listed in Table 1.2 for $n = 1, 2, \ldots, 6$. Since $H(s)$ in Eq. (1.17) has no zeros, such transfer functions are called *all-pole functions*.

The Butterworth function is the simplest transfer function capable of approximating the ideal low-pass magnitude characteristic to any degree of accuracy. There exist several more sophisticated classes of transfer functions which can achieve a better approximation for a given order n, but their derivation is too complicated to be included in our introductory treatment. Suffice it to say, however, that they all assume the special form of Eq. (1.17) (i.e., all-pole functions) but with coefficients different from those listed in Table 1.2. These coefficients are listed in numerous filter design handbooks for different classes of approximating functions, including the Butterworth function, up to a fairly high order n, say $n = 12$. Hence, if one can design a circuit having any *prescribed* all-pole function, then one can immediately design many classes of low-pass filters by referring to a standard table for the appropriate coefficients $b_0, b_1, b_2, \ldots, b_n$. Our objective in this section is to develop a simple and widely used procedure for designing all-pole function circuits. Our method is applicable for designing any network function in the form of Eq. (1.17).

Table 1.2 Coefficients of $d(s)$ of $H(s)$

n	b_0	b_1	b_2	b_3	b_4	b_5
1	1.000					
2	1.000	1.414				
3	1.000	2.000	2.000			
4	1.000	2.613	3.414	2.613		
5	1.000	3.236	5.236	5.236	3.236	
6	1.000	3.864	7.764	9.142	7.464	3.864

The *first* step is to find the poles of $H(s)$ and decompose Eq. (1.17) into the form

$$H(s) \triangleq \frac{V_0(s)}{V_i(s)} = H_0 \cdot H_1(s) \cdot H_2(s) \cdot \cdots \cdot H_n(s) \tag{1.18}$$

where

$$H_k(s) = \frac{\sigma_k}{s + \sigma_k} \qquad \sigma_k > 0 \tag{1.19}$$

for each *real* pole $s = \sigma_k$, or where

$$H_k(s) = \frac{\beta_k}{s^2 + \alpha_k s + \beta_k} \qquad \alpha_k > 0 \text{ and } \beta_k > 0 \tag{1.20}$$

for each pair of *complex-conjugate* poles $s = \sigma_k \pm j\omega_k$, and

$$H_0 = H(0) \tag{1.21}$$

is a *constant* equal to the *gain* at dc.

For example, if we choose $H(s)$ to be the nth-order Butterworth low-pass function, then

$$H(s) = \left[\frac{\sigma_1}{s + \sigma_1}\right]\left[\frac{\beta_1}{s^2 + \alpha_1 s + \beta_1}\right]\cdots\left[\frac{\beta_m}{s^2 + \alpha_m s + \beta_m}\right] \qquad m = \frac{n-1}{2} \tag{1.22}$$

when n is *odd*, and

$$H(s) = \left[\frac{\beta_1}{s^2 + \alpha_1 s + \beta_1}\right]\left[\frac{\beta_2}{s^2 + \alpha_2 s + \beta_2}\right]\cdots\left[\frac{\beta_m}{s^2 + \alpha_m s + \beta_m}\right] \qquad m = \frac{n}{2} \tag{1.23}$$

when n is *even*. The coefficients $\sigma_1, \alpha_1, \beta_1, \alpha_2, \beta_2, \ldots,$ etc., can be calculated using Table 1.1. In particular, $\sigma_1 = 1$ and $\beta_k = 1$ for all k. Note that $H_0 = 1$ in view of Eq. (1.17) and Table 1.2 (since $b_0 = 1$).

Equation (1.18) can be synthesized by the cascade circuit configuration shown in Fig. 1.3 where each two-port labeled $H_k(s)$ is described by either Eq. (1.19) or (1.20). The rightmost two-port H_0 denotes a VCVS with a controlling

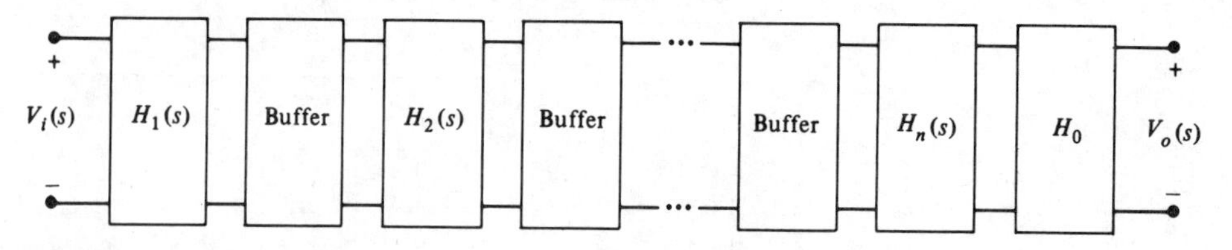

Figure 1.3 Cascade realization of $H(s) = H_0 H_1(s) H_2(s) \cdots H_n(s)$.

coefficient (gain) equal to H_0. The *buffers* (recall Chap. 4) sandwiched between the two-ports guarantee that there is no loading effect so that the overall transfer function is just the product of the individual components. Any circuit valid over the desired frequency range can be used. For low-frequency applications, the op-amp *voltage follower* in Fig. 2.1 of Chap. 4 is a favorite choice.

The *second* step is to design a circuit for each component transfer function [Eq. (1.19) or (1.20)]. Many such circuits can be designed: Some use only *passive* R, L, C elements, while others make use of op amps or other *active* elements. Since inductors are bulky (especially at low frequencies) and are incompatible with modern integrated circuit technology, RC op-amp circuits are widely used for low-frequency applications (say below 100 kHz).

First-order building block To synthesize Eq. (1.19) consider the simple op-amp circuit shown in Fig. 1.4. Its transfer function is given by (assuming an ideal op-amp model):

$$H(s) = \frac{1/RC}{s + 1/RC} \tag{1.24}$$

Comparing this with Eq. (1.19), we identify $\sigma_k = 1/RC$. For normalization purposes (see Sec. 1.3), it is convenient to choose $R = 1\,\Omega$. Hence, Eq. (1.19) is realized by choosing

$$R = 1\,\Omega \qquad C = \frac{1}{\sigma_k}\ \text{F} \tag{1.25}$$

Observe that the Thévenin equivalent circuit seen across the output terminals in Fig. 1.4 is simply an independent voltage source $V_0 = H(s)V_i(s)$ with *zero* internal impedance (assuming an ideal op-amp model). Consequently, no buffer need be connected between $H_1(s)$ and $H_2(s)$ in Fig. 1.3. Indeed, the right portion of the circuit in Fig. 1.4 is a voltage follower, and hence an additional buffer would be superfluous.

Second-order building block To synthesize Eq. (1.20), consider the op-amp circuit shown in Fig. 1.5. Its transfer function is given by (assuming an ideal op-amp model):

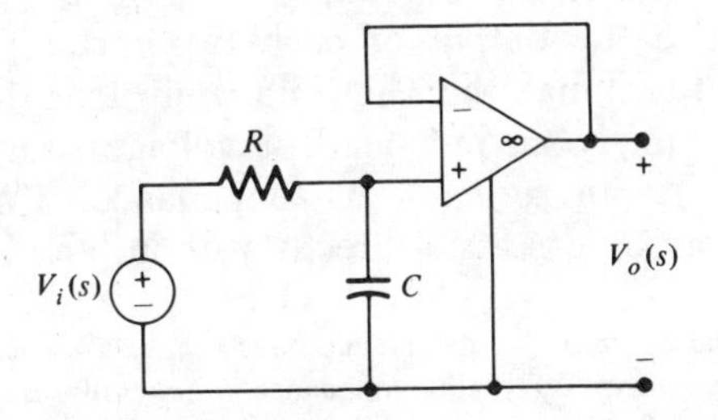

Figure 1.4 *All-pole circuit* having one real pole $s = -1/RC$.

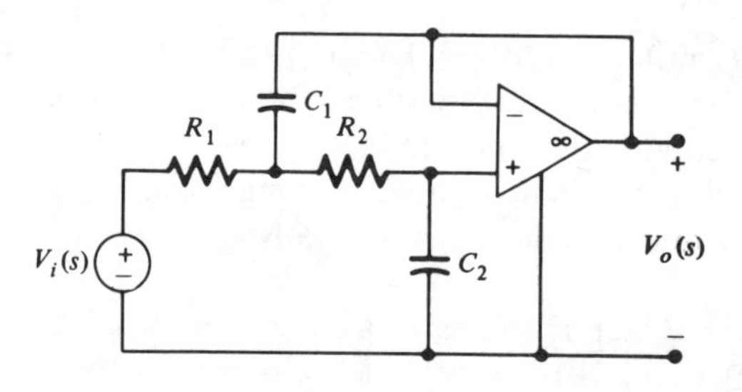

Figure 1.5 *All-pole circuit* having a pair of complex-conjugate poles.

$$H(s) = \frac{1}{s^2(R_1R_2C_1C_2) + s(R_1 + R_2)C_2 + 1} \tag{1.26}$$

Comparing this with Eq. (1.20), we identify

$$\alpha_k = \frac{R_1 + R_2}{R_1R_2C_1} \qquad \beta_k = \frac{1}{R_1R_2C_1C_2} \tag{1.27}$$

Since there are four parameters in Fig. 1.5 but only two independent equations to satisfy the prescribed data α_k and β_k, we can assign any convenient values to R_1 and R_2 and then calculate C_1 and C_2 from Eq. (1.27). For normalization purposes, let us choose $R_1 = 1$ and $R_2 = 1$. The result is

$$\begin{aligned} R_1 &= 1\,\Omega \qquad R_2 = 1\,\Omega \\ C_1 &= \frac{2}{\alpha_k}\,\text{F} \quad C_2 = \frac{\alpha_k}{2\beta_k}\,\text{F} \end{aligned} \tag{1.28}$$

Example Design a third-order Butterworth low-pass filter. Using the data in Table 1.1 for $n = 3$, we decompose the desired function [Eq. (1.16)] as follows:

$$H(s) = \left[\frac{1}{s+1}\right]\left[\frac{1}{s^2 + s + 1}\right] \tag{1.29}$$

Substituting $\sigma_1 = 1$ into Eq. (1.25) and $\alpha_2 = 1$ and $\beta_2 = 1$ into Eq. (1.28), we obtain $C_1 = 1$ F in Fig. 1.4 and $C_1 = 2$ F and $C_2 = 0.5$ F in Fig. 1.5. The resulting circuit is shown in Fig. 1.6.

The above procedure can be repeated to design a Butterworth low-pass filter of any order n: When n is odd, $H_1(s)$ in Fig. 1.3 is realized by Fig. 1.4, whereas $H_k(s)$ is realized by Fig. 1.5 for $k = 2, 3, \ldots, (n-1)/2$. When n is even, $H_k(s)$ is realized by Fig. 1.5 for $n = 1, 2, \ldots, n/2$. Moreover, since the output of each two-port $H_k(s)$ is taken from the op-amp output terminal, the Thévenin equivalent circuit seen to the left of each stage is simply an independent voltage source (assuming an ideal op-amp model) with zero internal impedance.[6] Consequently, these op-amp circuits can be cascaded directly without any buffers!

[6] Although practical op-amp circuits actually have a *nonzero* Thévenin equivalent impedance (called the *output impedance*) $Z_0(s)$, this impedance is generally much smaller compared to the *input impedance* of the following stages and is therefore negligible.

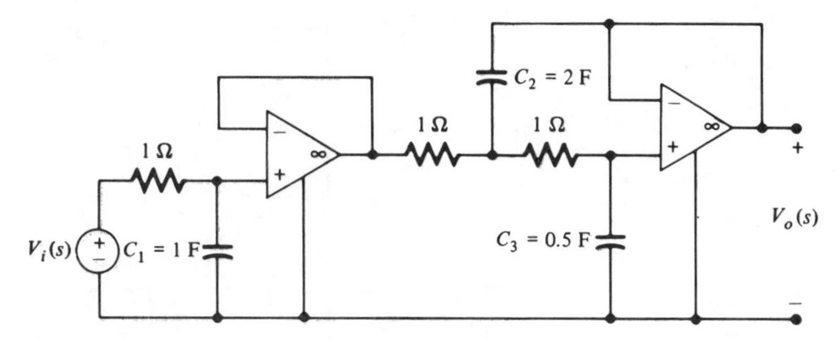

Figure 1.6 Normalized third-order Butterworth low-pass filter.

Exercises

1. Design a normalized fourth-order Butterworth low-pass filter using only two op amps.
2. Design a normalized fifth-order Butterworth low-pass filter using only three op amps.

Third-order building block Since an *odd*-order all-pole function always has one *real* pole, the circuit in Fig. 1.6 will always be present (with different parameters) in the first two stages of a multistage realization. It is desirable therefore to develop a separate third-order circuit using *only* one op amp. This circuit is obtained by deleting the voltage follower in Fig. 1.6 and is shown in Fig. 1.7. Deleting the op amp, of course, introduces serious *loading effects* in this case, and the transfer function of the circuit in Fig. 1.7 will be *different* from that of Fig. 1.6 if we keep all parameters unchanged, i.e., $C_1 = 1\,\text{F}$, $C_2 = 2\,\text{F}$, and $C_3 = 0.5\,\text{F}$ as in Fig. 1.6. In order to obtain the same transfer function, we must rederive the values for C_1, C_2, and C_3. The transfer function in this case is given by (assuming an ideal op-amp model):

$$H(s) = \frac{1}{s^3(C_1C_2C_3) + s^2[2(C_1 + C_2)C_3] + s(C_1 + 3C_3) + 1} \tag{1.30}$$

Hence, for *any prescribed* third-order all-pole function

$$H(s) = \frac{b_0}{s^3 + b_2s^2 + b_1s + b_0} \tag{1.31}$$

we can identify

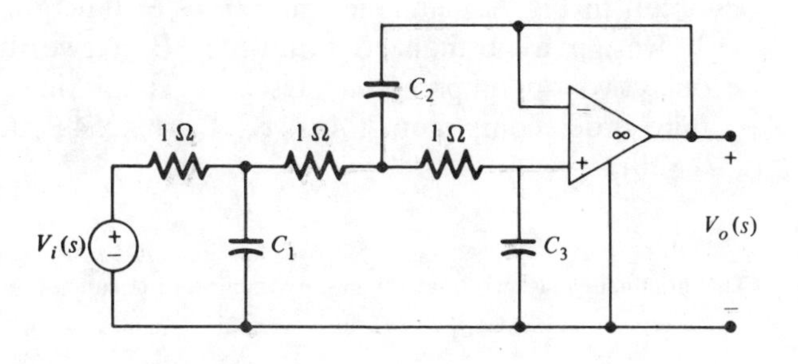

Figure 1.7 Normalized third-order circuit equivalent to that of Fig. 1.6 provided C_1, C_2, and C_3 are calculated from Eqs. (1.32) to (1.34).

$$b_0 = \frac{1}{C_1 C_2 C_3} \tag{1.32}$$

$$b_1 = \frac{C_1 + 3C_3}{C_1 C_2 C_3} \tag{1.33}$$

$$b_2 = \frac{2(C_1 + C_2)}{C_1 C_2} \tag{1.34}$$

The values of C_1, C_2, and C_3 can therefore be found by solving the above system of three *nonlinear* equations. Of course, to be practical, these solutions must be positive, assuming that they exist. Now since the two circuits in Figs. 1.6 and 1.7 differ only by a voltage follower and since C_1, C_2, and C_3 in Fig. 1.6 are always positive so long as all poles of Eq. (1.31) are in the left-half plane, it is reasonable to *assume* that the corresponding capacitances in Fig. 1.7 will also be positive.[7] This assumption will be confirmed by actual calculations for Butterworth filters of order three in the following example, and of order five in the following exercise.

Example Design a third-order Butterworth low-pass filter using the one-op-amp circuit in Fig. 1.7.

From the third-order Butterworth transfer function in Eq. (1.16), we identify $b_0 = 1$, $b_1 = 2$, and $b_2 = 2$. Substituting these parameters into Eqs. (1.32) to (1.34) and solving for C_1, C_2, and C_3 by the Newton-Raphson method (or by any other method), we obtain

$$C_1 = 1.392\text{ F} \qquad C_2 = 3.546\text{ F} \qquad \text{and} \qquad C_3 = 0.2024\text{ F} \tag{1.35}$$

Note that these values are quite different from those in Fig. 1.6, thereby verifying our suspicion that the loading effect introduced by the portion of the circuit to the right of C_1 in Fig. 1.7 can*not* be neglected.

Exercises

1. (*a*) Derive the transfer function in Eq. (1.30).
(*b*) Substitute Eq. (1.35) into Eqs. (1.32) to (1.34), and verify that the circuit in Fig. 1.7 has the same transfer function as that of Fig. 1.6.
2. Design a normalized fifth-order Butterworth low-pass filter using only two op amps. *Hint*: Use the circuit in Fig. 1.7 to realize the third-order component (*Answer*: $C_1 = 1.354$ F, $C_2 = 1.753$ F, and $C_3 = 0.4214$ F).

[7] This argument may not hold for more complicated circuits.

1.3 Renormalization

The low-pass filters designed in the preceding section are called *normalized* filters because the dc magnitude $|H(0)|$ and the 3-dB frequency ω_c (called the *cutoff frequency* in the literature) are both normalized to unity. In practice the specified magnitude and cutoff frequency would in general differ from unity. Our objective in this section is to show how the circuit parameters in the normalized filter can be rescaled to meet the desired specification. Here, we will make use of the concept of *impedance* and *frequency scaling* introduced earlier in Chap. 9.

A. Magnitude scaling Observe that the rightmost two-port in Fig. 1.3 is absent in the preceding low-pass filters because $H_0 = |H(0)| = 1$. To obtain any prescribed $H_0 \neq 1$, we simply add a VCVS with a dc gain equal to H_0 to the normalized filter as in Fig. 1.3. In practice, any voltage amplifier with a high input impedance and a low output impedance will suffice provided the gain is constant over the operating frequency range. For example, the noninverting op-amp circuit in Fig. 1.4 of Chap. 4 can be used for this purpose.

B. Frequency scaling To transform the 3-dB cutoff frequency of the low-pass filter from the normalized value $\omega_c = 1$ rad/s to $\omega_c = \omega_0$ rad/s (see Fig. 1.8), simply divide all capacitances in the circuit by ω_0. This follows because the impedance of each rescaled capacitor at $\omega = \omega_0$ is identical to that of the original (unscaled) capacitor at $\omega = 1$, namely,

$$Z'_{c_k}(j\omega_0) = \frac{1}{j\omega_0(C_k/\omega_0)} = Z_{c_k}(j1) \tag{1.36}$$

Since the remaining circuit elements (resistors and *ideal* op amps) do not depend on frequency, all calculations leading to $H(s)$ remain unchanged provided all element impedances are evaluated at $\omega = \omega_0$.

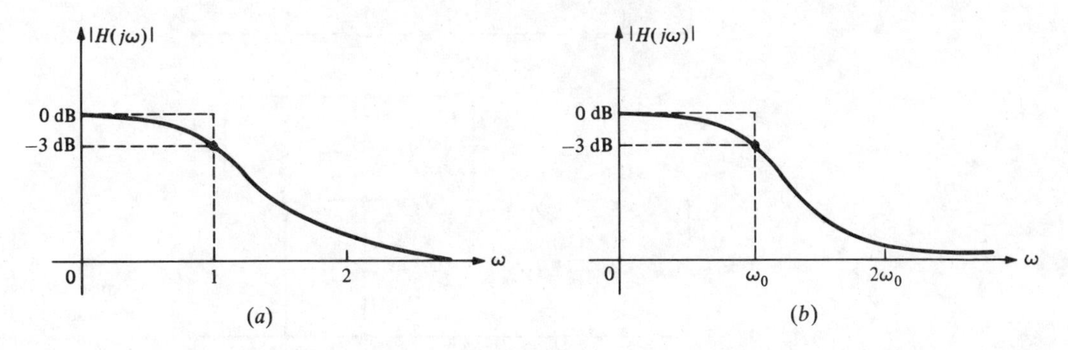

Figure 1.8 Rescaling $|H(j\omega)|$ from $\omega_c = 1$ in (*a*) to $\omega_c = \omega_0$ in (*b*).

Example Rescale the third-order Butterworth filter in Fig. 1.7 (with $C_1 = 1.392$ F, $C_2 = 3.546$ F, and $C_3 = 0.2024$ F) so that the 3-dB cutoff frequency is at 10 kHz.

Since 10 kHz is equivalent to $\omega_0 = 2\pi(10^4)$ rad/s, we must divide C_1, C_2, and C_3 by 62,800 to obtain

$$C_1 = \frac{1.392}{62{,}800} = 22\ \mu\text{F} \qquad C_2 = \frac{3.546}{62{,}800} = 56\ \mu\text{F} \qquad C_3 = \frac{0.2024}{62{,}800} = 3\ \mu\text{F} \tag{1.37}$$

C. Impedance scaling Although the renormalized circuit in Fig. 1.9*a* meets our specification, the element values are clearly impractical: The resistances are too small, whereas the capacitances are too large. Observe however that if we *divide* all capacitances in a linear time-invariant *RC* circuit by any factor Z_0 while *multiplying* all resistances by the same factor Z_0, the resulting transfer function would remain unchanged. Hence, if we choose $Z_0 = 1000$, we would obtain the *equivalent* renormalized circuit in Fig. 1.9*b*. Observe that the element values are now quite practical.

Exercises

1. Give the formulas for simultaneously scaling the frequency to ω_0 and all element impedances by a factor Z_0.
2. Renormalize the third-order Butterworth low-pass filter in Fig. 1.6 to have a 3-dB cutoff frequency at 1 kHz. Rescale the elements so that no capacitance values exceed 0.01 μF.

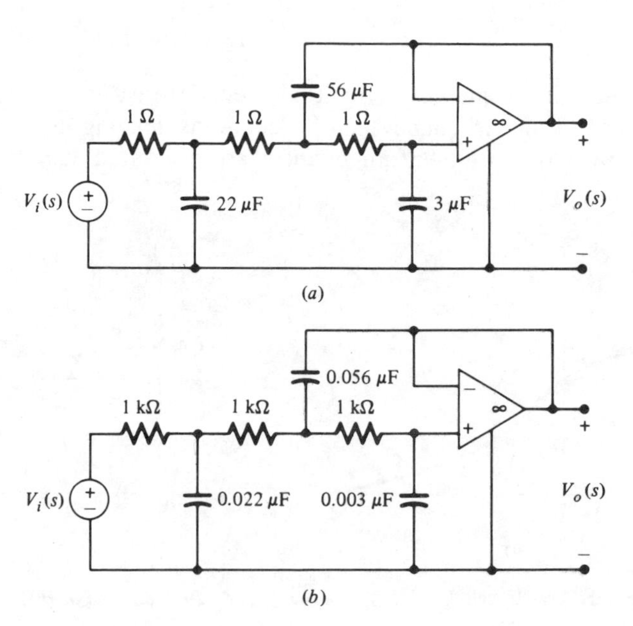

Figure 1.9 Two equivalent renormalized third-order low-pass filters with a 3-db cutoff frequency at 10 kHz.

2 SENSITIVITY ANALYSIS

No physical circuit performs exactly according to design in view of several practical considerations such as tolerance of circuit elements, environmental effects (temperature, humidity, radiation, etc.), and aging. A good design must therefore include an analysis of the effects of element parameter variations on the circuit *performance* $\mathcal{P}$ of interest. For example, $\mathcal{P}$ may be the *transfer function* $|H(j\omega)|$, the *natural frequency* ω_0, the *quality factor* Q, etc., of a linear time-invariant circuit. To estimate the effect of a *small* change in a parameter x_i on $\mathcal{P}$, which in general depends on various parameters $x_1, x_2, \ldots, x_m$, we can calculate its partial derivative $\partial\mathcal{P}/\partial x_i$. For linear time-invariant circuits, it is often more convenient and revealing to calculate the related quantity

$$S_{x_i}^{\mathcal{P}} \triangleq \lim_{\Delta x_i \to 0} \frac{\Delta\mathcal{P}/\mathcal{P}}{\Delta x_i/x_i} \tag{2.1a}$$

$$= \frac{x_i}{\mathcal{P}} \frac{\partial\mathcal{P}}{\partial x_i} \tag{2.1b}$$

$$= \frac{\partial \ln \mathcal{P}}{\partial \ln x_i} \tag{2.1c}$$

where $S_{x_i}^{\mathcal{P}}$ is called the *relative sensitivity* of $\mathcal{P}$ *with respect to the parameter* x_i, or simply *sensitivity*.

Examples

1. Let the property of interest $\mathcal{P}$ be the voltage *transfer function* of a linear voltage divider, namely,

$$T(R_1, R_2) = \frac{R_2}{R_1 + R_2} \tag{2.2}$$

To calculate $S_{R_1}^T$ and $S_{R_2}^T$, we make use of Eq. (2.1*b*):

$$S_{R_1}^T = \frac{R_1}{T} \frac{-R_2}{(R_1 + R_2)^2} = \frac{-R_1}{R_1 + R_2} \tag{2.3a}$$

$$S_{R_2}^T = \frac{R_2}{T} \frac{R_1}{(R_1 + R_2)^2} = \frac{R_1}{R_1 + R_2} \tag{2.3b}$$

Assuming a nominal design value of $R_1 = 2\ \Omega$ and $R_2 = 1\ \Omega$, we obtain $S_{R_1}^T = -\frac{2}{3}$ and $S_{R_2}^T = \frac{2}{3}$.

To illustrate one application of these two sensitivity numbers, suppose the maximum allowable variations in $T(R_1, R_2)$ is ± 1.5 percent and the problem is to determine the tolerance of the two resistors. Equation (2.1*a*) implies that

$$\frac{\Delta T}{T} \approx S_{R_1}^T \frac{\Delta R_1}{R_1} + S_{R_2}^T \frac{\Delta R_2}{R_2} = -\frac{2}{3}\frac{\Delta R_1}{R_1} + \frac{2}{3}\frac{\Delta R_2}{R_2} \tag{2.4}$$

If we choose $|\Delta R_1/R_1| < 0.01$ and $|\Delta R_2/R_2| < 0.01$, then the worst case occurs when ΔR_1 and ΔR_2 have opposite signs, namely, when

$$\left|\frac{\Delta T}{T}\right| = \tfrac{2}{3}(0.01 + 0.01) = 1.3 \text{ percent} \tag{2.5}$$

Hence, it is necessary to use 1 percent resistors in order to attain the prescribed tolerance of ± 1.5 percent. The next most accurate resistors available commercially at 5 percent tolerance will not meet the above specification.

2. Let $\mathcal{P}$ be the *magnitude* of the transfer function of the first-order building block in Fig. 1.4, namely,

$$|H(j\omega)| = \left|\frac{1/RC}{j\omega + 1/RC}\right| = \frac{1}{\sqrt{1 + (\omega RC)^2}} \tag{2.6}$$

Using Eq. (2.1*b*), we obtain

$$S_C^{|H|} = \frac{C}{|H|}\{-\tfrac{1}{2}(2\omega^2 R^2 C)[1 + (\omega RC)^2]^{-3/2}\} = \frac{-(\omega RC)^2}{1 + (\omega RC)^2} \tag{2.7}$$

Note that the sensitivity in this case depends on the input frequency ω.

3. Let $\mathcal{P}$ denote the *natural frequency* ω_0 and *quality factor* Q, respectively, of the transfer function

$$H(s) = \frac{1}{s^2 + (\omega_0/Q)s + \omega_0^2} \tag{2.8}$$

of the second-order building block in Fig. 1.5, namely,

$$\omega_0 \triangleq \frac{1}{\sqrt{R_1 R_2 C_1 C_2}} \tag{2.9}$$

and

$$Q \triangleq \frac{(R_1 + R_2)C_2}{\sqrt{R_1 R_2 C_1 C_2}} \tag{2.10}$$

To calculate the sensitivity of ω_0 with respect to C_1, it is easier to use Eq. (2.1*c*):

$$S_{C_1}^{\omega_0} = \frac{\partial \ln \omega_0}{\partial \ln C_1} = \frac{\partial}{\partial \ln C_1}[-\tfrac{1}{2}\ln(R_1 R_2 C_2) - \tfrac{1}{2}\ln C_1] = -\tfrac{1}{2} \tag{2.11}$$

Similarly, the sensitivity of Q with respect to C_2 is

$$S_{C_2}^{Q} = \frac{\partial \ln Q}{\partial \ln C_2} = \frac{\partial}{\partial \ln C_2}\left[\ln(R_1 + R_2) + \ln C_2 - \tfrac{1}{2}\ln(R_1 R_2 C_1) - \tfrac{1}{2}\ln C_2\right]$$

$$= \tfrac{1}{2} \tag{2.12}$$

2.1 Explicit Sensitivity Formulas

Although the relative sensitivity $S_{x_i}^{\mathcal{P}}$ can always be derived directly from definition (2.1), it is often simpler to make use of explicit formulas, to be derived below.

A. In terms of network function components Let $\mathcal{P}$ denote the voltage transfer function

$$H(j\omega) \triangleq \frac{V_2(j\omega)}{V_1(j\omega)} \tag{2.13}$$

of the linear time-invariant circuit shown in Fig. 2.1*a*, where Z denotes an impedance to be perturbed (i.e., $x_i \triangleq Z$) and N denotes a three-port described by its open-circuit impedance matrix $\mathbf{Z}$, namely,

$$V_1 = Z_{11}I_1 + Z_{12}I_2 + Z_{13}I_3 \tag{2.14}$$

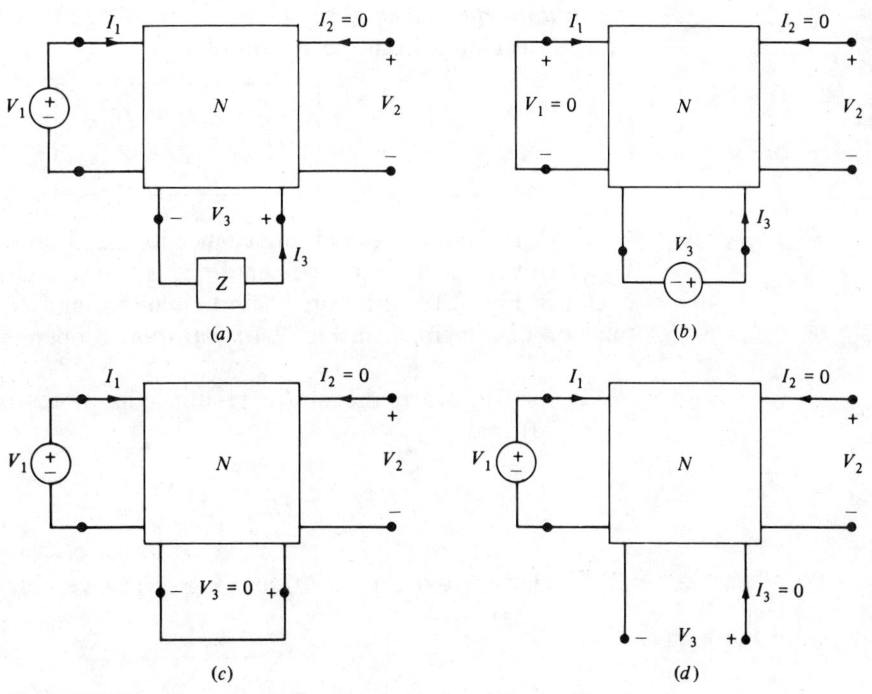

Figure 2.1 The voltage transfer function sensitivity of the circuit in (*a*) relative to Z can be calculated from $Z_3 \triangleq V_3/I_3$ in (*b*), $H_0 \triangleq V_2/V_1$ in (*c*), and $H_\infty \triangleq V_2/V_1$ in (*d*).

$$V_2 = Z_{21}I_1 + Z_{22}I_2 + Z_{23}I_3 \tag{2.15}$$

$$V_3 = Z_{31}I_1 + Z_{32}I_2 + Z_{33}I_3 \tag{2.16}$$

Substituting $I_2 = 0$ and $V_3 = -ZI_3$ into Eqs. (2.14) to (2.16), we obtain

$$V_1 = Z_{11}I_1 + Z_{13}I_3 \tag{2.17}$$

$$V_2 = Z_{21}I_1 + Z_{23}I_3 \tag{2.18}$$

$$I_3 = -\left(\frac{Z_{31}}{Z_{33} + Z}\right) I_1 \tag{2.19}$$

Substituting I_3 from Eq. (2.19) into Eqs. (2.17) to (2.18) and dividing the resulting expressions, we obtain

$$H \triangleq \frac{V_2}{V_1} = \frac{(Z_{21}Z_{33} - Z_{23}Z_{31}) + Z_{21}Z}{(Z_{11}Z_{33} - Z_{13}Z_{31}) + Z_{11}Z} = \frac{A_0 + A_1 Z}{B_0 + B_1 Z} \tag{2.20}$$

where $\quad A_0 \triangleq Z_{21}Z_{33} - Z_{23}Z_{31} \qquad A_1 \triangleq Z_{21}$

$\quad B_0 \triangleq Z_{11}Z_{33} - Z_{13}Z_{31} \qquad$ and $\qquad B_1 \triangleq Z_{11}$

are independent of Z.

Physical interpretations of Eq. (2.20) The voltage transfer function of Fig. 2.1*a* can be calculated from the relationship

$$\boxed{H = \frac{Z_3 H_0 + H_\infty Z}{Z_3 + Z}} \tag{2.21}$$

where Z_3 is the driving-point impedance looking in from port 3 of the circuit in Fig. 2.1*b* with port 1 short-circuited, H_0 is the voltage transfer function of the circuit in Fig. 2.1*c* with port 3 short-circuited, and H_∞ is the voltage transfer function of the circuit in Fig. 2.1*d* with port 3 open-circuited.

PROOF To prove the above relationship, note from Fig. 2.1*c* and Eq. (2.20) that

$$H_0 = \left.\frac{V_2}{V_1}\right|_{Z=0} = \frac{A_0}{B_0} \tag{2.22}$$

Similarly, from Eq. (2.20) and Fig. 2.1*d*, we obtain

$$H_\infty = \left.\frac{V_2}{V_1}\right|_{Z=\infty} = \frac{A_1}{B_1} \tag{2.23}$$

Substituting $A_0 = B_0 H_0$ from Eq. (2.22) and $A_1 = B_1 H_\infty$ from Eq. (2.23) into Eq. (2.20), we obtain

$$H = \frac{B_0 H_0 + B_1 H_\infty Z}{B_0 + B_1 Z} = \frac{(B_0/B_1) H_0 + H_\infty Z}{(B_0/B_1) + Z} \tag{2.24}$$

To interpret B_0/B_1, set $V_1 = 0$ and $I_2 = 0$ in Eq. (2.14) (see Fig. 2.1*b*):

$$I_1 = -\frac{Z_{13}}{Z_{11}} I_3 \tag{2.25}$$

Substituting Eq. (2.25) and $I_2 = 0$ into Eq. (2.16), we obtain

$$V_3 = \left(Z_{33} - \frac{Z_{13} Z_{31}}{Z_{11}} \right) I_3 = Z_3 I_3 \tag{2.26}$$

where
$$Z_3 \triangleq \left. \frac{V_3}{I_3} \right|_{V_1 = 0,\, I_2 = 0} = \frac{Z_{11} Z_{33} - Z_{13} Z_{31}}{Z_{11}} = \frac{B_0}{B_1} \tag{2.27}$$

Substituting Eq. (2.27) into Eq. (2.24), we obtain Eq. (2.21). ∎

Taking the partial derivative of Eq. (2.21) with respect to Z, we obtain

$$\frac{\partial H}{\partial Z} = Z_3 \left[\frac{H_\infty - H_0}{(Z_3 + Z)^2} \right] \tag{2.28}$$

It follows from definition (2.1*b*) that

$$\boxed{S_Z^H = \frac{Z Z_3}{Z + Z_3} \, \frac{H_\infty - H_0}{Z_3 H_0 + H_\infty Z}} \tag{2.29}$$

Observe that S_Z^H is uniquely determined from the *impedance* Z, the *driving-point impedance* Z_3 in Fig. 2.1*b*, the *voltage transfer function* H_0 in Fig. 2.1*c*, and the *voltage transfer function* H_∞ in Fig. 2.1*d*.

Exercise Prove that the sensitivity S_Z^H can be found from the relationship

$$\boxed{S_Z^H = \frac{1}{H} H_a H_b} \tag{2.30}$$

where
$$H_a \triangleq \left. \frac{V_3}{V_1} \right|_{I_2 = 0} \tag{2.31}$$

denotes the voltage transfer function in Fig. 2.2*a*, and

$$H_b \triangleq \left. \frac{V_2}{V_0} \right|_{V_1 = 0,\, I_2 = 0} \tag{2.32}$$

denotes the voltage transfer function in Fig. 2.2*b*.

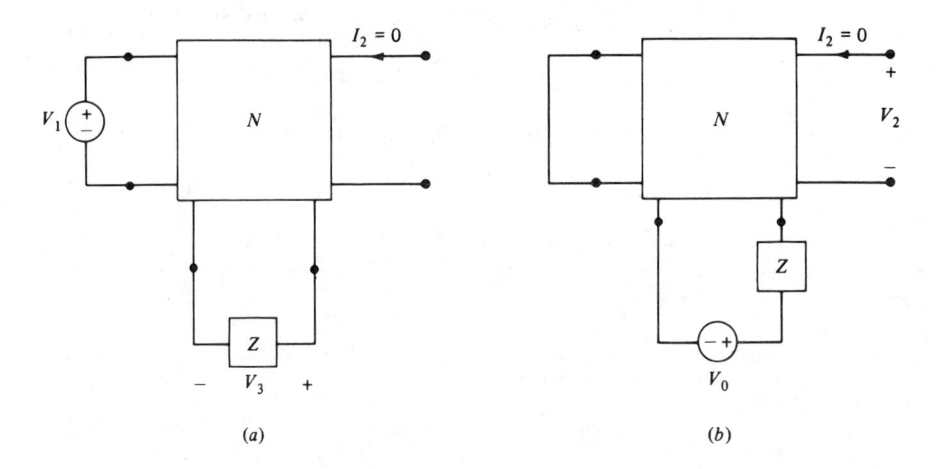

Figure 2.2 Circuit interpretations of H_a and H_b.

B. In terms of the node admittance matrix Recall that the *node equation* of a linear time-invariant circuit in sinusoidal steady state is given by

$$\mathbf{YE} = \mathbf{I}_s \tag{2.33}$$

where $\mathbf{E}$ denotes the *node-to-datum voltage phasor*, $\mathbf{I}_s$ denotes the *equivalent current source phasor*, and

$$\mathbf{Y} \triangleq \mathbf{A}\mathbf{Y}_b\mathbf{A}^T \tag{2.34}$$

denotes the *node admittance matrix* whose elements are phasors. Suppose the branch admittance matrix $\mathbf{Y}_b$ undergoes a small change $\delta\mathbf{Y}_b$ due to a small change in one or more of the circuit parameters (e.g., R, L, C, g_m, μ, etc.). Let $\mathbf{Y} + \delta\mathbf{Y}$ denote the new admittance matrix and let $\mathbf{E} + \delta\mathbf{E}$ be the new node voltage phasor. Since the current source phasor $\mathbf{I}_s$ remains unchanged, we have

$$(\mathbf{Y} + \delta\mathbf{Y})(\mathbf{E} + \delta\mathbf{E}) = \mathbf{I}_s \tag{2.35}$$

Expanding Eq. (2.35) we obtain

$$\mathbf{YE} + \mathbf{Y}\cdot\delta\mathbf{E} + \delta\mathbf{Y}\cdot\mathbf{E} + \delta\mathbf{Y}\cdot\delta\mathbf{E} = \mathbf{I}_s \tag{2.36}$$

Substituting Eq. (2.33) into Eq. (2.36) and neglecting the second-order term $\delta\mathbf{Y}\cdot\delta\mathbf{E}$, we obtain

$$\mathbf{Y}\cdot\delta\mathbf{E} = -\delta\mathbf{Y}\cdot\mathbf{E} \tag{2.37}$$

or

$$\delta\mathbf{E} = -[\mathbf{Y}^{-1}][\delta\mathbf{Y}]\mathbf{E} \tag{2.38}$$

Equation (2.38) gives, within first-order terms, the change $\delta\mathbf{E}$ in the node voltage in terms of the change $\delta\mathbf{Y}$ in the node admittance matrix. To express $\delta\mathbf{Y}$ in terms of the change $\delta\mathbf{Y}_b$ of the *branch* admittance matrix $\mathbf{Y}_b$, substitute $\delta\mathbf{Y} = \mathbf{A}\cdot\delta\mathbf{Y}_b\cdot\mathbf{A}^T$ into Eq. (2.38) and make use of KVL ($\mathbf{V} = \mathbf{A}^T\mathbf{E}$) to obtain

$$\delta \mathbf{E} = -[\mathbf{Y}^{-1}][\mathbf{A}][\delta \mathbf{Y}_b]\mathbf{V} \tag{2.39}$$

Equation (2.39) allows us to calculate the first-order change $\delta\mathbf{E}$ in terms of the *branch* voltage phasor $\mathbf{V}$ of the *unperturbed circuit*. Since we neglected second-order terms in Eq. (2.36), Eq. (2.39) gives the approximate change in $\delta\mathbf{E}$ due to a small change in one or more of the circuit parameters. Now suppose *only one* parameter x changes by a small amount δx. Then all elements in $\delta\mathbf{Y}_b$ are zero except the one corresponding to x. Dividing both sides of Eq. (2.39) by δx and letting $\delta x \to 0$, we obtain the following explicit formula:

$$\boxed{\frac{\partial \mathbf{E}}{\partial x} = -\mathbf{Y}^{-1}\mathbf{A}\left(\frac{\partial}{\partial x}\mathbf{Y}_b\right)\mathbf{V}} \tag{2.40}$$

This formula allows one to calculate the partial derivative of the *node* voltage phasor in terms of the partial derivative of the branch admittance matrix $\mathbf{Y}_b$. Note that whereas Eq. (2.39) gives only an approximate change in E, Eq. (2.40) gives the *exact* derivative of each node voltage phasor E_i with respect to x. The expression $\frac{\partial}{\partial x}\mathbf{Y}_b$ in Eq. (2.40) is a square matrix all of whose elements are zero except the element involving the circuit parameter x. For example, if x denotes the conductance $G \triangleq 1/R$ corresponding to element kk in $\mathbf{Y}_b$, then $Y_{kk} = G$ and $\frac{\partial}{\partial x} Y_{kk} = 1$. On the other hand, if x denotes the capacitance C (respectively, reciprocal inductance $\Gamma \triangleq 1/L$) then $Y_{kk} = j\omega C$ [respectively, $Y_{kk} = (1/j\omega)\Gamma$] and $\partial Y_{kk}/\partial x = j\omega$ (respectively, $1/j\omega$).

Substituting each component of $\partial\mathbf{E}/\partial x$ from Eq. (2.40) into definition (2.1*b*), we obtain the following *explicit* sensitivity formula for E_i:

$$\boxed{S_x^{E_i} = \frac{x}{E_i}\cdot \text{row } i \text{ of } \left[-\mathbf{Y}^{-1}\mathbf{A}\left(\frac{\partial}{\partial x}\mathbf{Y}_b\right)\mathbf{V}\right]} \tag{2.41}$$

This formula allows one to calculate the sensitivity of *any* node voltage E_i in terms of the branch voltage phasor $\mathbf{V}$ of the *unperturbed circuit* and the partial derivative of the branch admittance matrix $\mathbf{Y}_b$. Let us illustrate the application of this explicit formula with a numerical example. To simplify the arithmetic we pick a resistive circuit, but formula (2.41) applies to any linear time-invariant dynamic circuit in sinusoidal steady state.

Example Consider the circuit shown in Fig. 2.3 where $G_1 = 2\text{ S}$, $G_2 = 4\text{ S}$, $G_3 = 1\text{ S}$, and $g_m = 2\text{ S}$. By inspection, we obtain:

$$\mathbf{A} = \begin{bmatrix} 1 & 0 & 1 & 0 \\ 0 & 1 & -1 & 1 \end{bmatrix} \qquad \mathbf{Y} = \begin{bmatrix} 3 & -1 \\ 1 & 5 \end{bmatrix} \qquad \mathbf{I}_s = \begin{bmatrix} 4 \\ 0 \end{bmatrix} \tag{2.42}$$

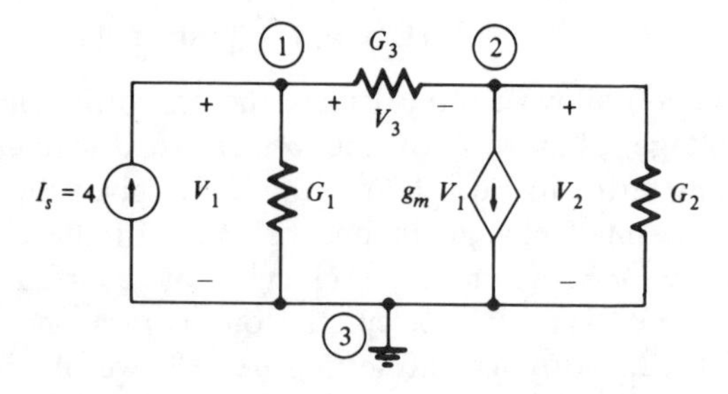

Figure 2.3 Resistive circuit for illustrating sensitivity formula. Here $G_1 = 2$ S, $G_2 = 4$ S, $G_3 = 1$ S, and $g_m = 2$ S.

The node and branch voltage solutions are given by

$$\begin{bmatrix} E_1 \\ E_2 \end{bmatrix} = \begin{bmatrix} 3 & -1 \\ 1 & 5 \end{bmatrix}^{-1} \begin{bmatrix} 4 \\ 0 \end{bmatrix} = \begin{bmatrix} 1.25 \\ -0.25 \end{bmatrix}$$

$$\begin{bmatrix} V_1 \\ V_2 \\ V_3 \\ V_4 \end{bmatrix} = \begin{bmatrix} 1 & 0 \\ 0 & 1 \\ 1 & -1 \\ 0 & 1 \end{bmatrix} \begin{bmatrix} 1.25 \\ -0.25 \end{bmatrix} = \begin{bmatrix} 1.25 \\ -0.25 \\ 1.5 \\ -0.25 \end{bmatrix} \tag{2.43}$$

To calculate the sensitivity of V_1 and V_2 with respect to G_1, we let $x = G_1$ and obtain

$$\frac{\partial}{\partial x} \mathbf{Y}_b = \frac{\partial}{\partial G_1} \begin{bmatrix} G_1 & 0 & 0 & 0 \\ 0 & G_2 & 0 & 0 \\ 0 & 0 & G_3 & 0 \\ g_m & 0 & 0 & 0 \end{bmatrix} = \begin{bmatrix} 1 & 0 & 0 & 0 \\ 0 & 0 & 0 & 0 \\ 0 & 0 & 0 & 0 \\ 0 & 0 & 0 & 0 \end{bmatrix} \tag{2.44}$$

Substituting Eqs. (2.43) to (2.44) into Eq. (2.40) and noting that $V_1 = E_1$ and $V_2 = E_2$, we obtain

$$\begin{bmatrix} \dfrac{\partial V_1}{\partial G_1} \\ \dfrac{\partial V_2}{\partial G_1} \end{bmatrix} = -\begin{bmatrix} 3 & -1 \\ 1 & 5 \end{bmatrix}^{-1} \begin{bmatrix} 1 & 0 & 1 & 0 \\ 0 & 1 & -1 & 1 \end{bmatrix} \begin{bmatrix} 1 & 0 & 0 & 0 \\ 0 & 0 & 0 & 0 \\ 0 & 0 & 0 & 0 \\ 0 & 0 & 0 & 0 \end{bmatrix} \begin{bmatrix} 1.25 \\ -0.25 \\ 1.5 \\ -0.25 \end{bmatrix}$$

$$= \begin{bmatrix} -\frac{25}{64} \\ \frac{5}{64} \end{bmatrix} \tag{2.45}$$

Substituting Eq. (2.45) into Eq. (2.41), we obtain

$$S_{G_1}^{V_1} = \frac{G_1}{V_1} \text{[row 1 of Eq. (2.45)]} = \frac{2}{1.25}\left(\frac{-25}{64}\right) = -\frac{40}{64} \tag{2.46}$$

$$S_{G_1}^{V_2} = \frac{G_1}{V_2} \text{[row 2 of Eq. (2.45)]} = \frac{2}{-0.25}\left(\frac{5}{64}\right) = -\frac{40}{64} \tag{2.47}$$

On the other hand, if we choose $x = g_m$ instead of R_1, then

$$\frac{\partial}{\partial x}\mathbf{Y}_b = \frac{\partial}{\partial g_m}\begin{bmatrix} G_1 & 0 & 0 & 0 \\ 0 & G_2 & 0 & 0 \\ 0 & 0 & G_3 & 0 \\ g_m & 0 & 0 & 0 \end{bmatrix} = \begin{bmatrix} 0 & 0 & 0 & 0 \\ 0 & 0 & 0 & 0 \\ 0 & 0 & 0 & 0 \\ 1 & 0 & 0 & 0 \end{bmatrix} \tag{2.48}$$

Repeating the above calculation we obtain

$$\begin{bmatrix} \dfrac{\partial V_1}{\partial g_m} \\ \dfrac{\partial V_2}{\partial g_m} \end{bmatrix} = \begin{bmatrix} 3 & -1 \\ 1 & 5 \end{bmatrix}^{-1} \begin{bmatrix} 1 & 0 & 1 & 0 \\ 0 & 1 & -1 & 1 \end{bmatrix} \begin{bmatrix} 0 & 0 & 0 & 0 \\ 0 & 0 & 0 & 0 \\ 0 & 0 & 0 & 0 \\ 1 & 0 & 0 & 0 \end{bmatrix} \begin{bmatrix} 1.25 \\ -0.25 \\ 1.5 \\ -0.25 \end{bmatrix} = \begin{bmatrix} -\frac{5}{64} \\ -\frac{15}{64} \end{bmatrix} \tag{2.49}$$

and

$$S_{g_m}^{V_1} = \frac{g_m}{V_1}[\text{row 1 of Eq. (2.49)}] = \frac{2}{1.25}\left(\frac{-5}{64}\right) = -\frac{1}{8} \tag{2.50}$$

$$S_{g_m}^{V_2} = \frac{g_m}{V_2}[\text{row 2 of Eq. (2.49)}] = \frac{2}{-0.25}\left(-\frac{15}{64}\right) = -\frac{30}{16} \tag{2.51}$$

Let us summarize our algorithm as follows:

Algorithm of sensitivity calculation for small circuits (*say*, $n < 4$) *via Eq.* (*2.41*). Given: Linear time-invariant circuit $\mathcal{N}$ made of voltage-controlled elements and independent current sources.

1. Formulate the *reduced incidence matrix* **A** and the node admittance matrix $\mathbf{Y}(j\omega)$ for the *unperturbed circuit* $\mathcal{N}$ at the frequency ω of interest.
2. Calculate the *inverse* matrix $\mathbf{Y}^{-1}(j\omega)$.
3. Calculate the *node voltage phasor* $\mathbf{E}(j\omega)$ using:

$$\mathbf{E}(j\omega) = \mathbf{Y}^{-1}(j\omega)\mathbf{I}_s(j\omega) \tag{2.52}$$

4. Calculate the *branch voltage phasor* $\mathbf{V}(j\omega)$ using

$$\mathbf{V}(j\omega) = \mathbf{A}^T\mathbf{E}(j\omega) \tag{2.53}$$

5. Formulate the matrix

$$\mathbf{M} \triangleq \frac{\partial}{\partial x}\mathbf{Y}_b \tag{2.54}$$

for the *perturbed circuit* where x is the parameter being perturbed. Since all elements of **M** are zero except the element associated with x, no calculation is involved.

6. Calculate $S_x^{E_i}$ from Eq. (2.41).

REMARKS

1. The above calculation must be repeated for *each* frequency ω of interest.
2. Although it is more efficient to calculate $\mathbf{E}(j\omega)$ directly from the node equation (2.33) (say, by Gaussian elimination or LU decomposition), we calculate $\mathbf{Y}^{-1}(j\omega)$ first because the inverse is needed *twice* in this algorithm, namely, in Eqs. (2.52) and (2.41). Moreover, for small circuits (say, $n<4$), $\mathbf{Y}^{-1}(j\omega)$ can be found in explicit symbolic form so that calculation for other frequencies can be obtained more efficiently by direct substitution.
3. For small circuits we need not formulate $\mathbf{A}$ but can rather calculate $\mathbf{Y}(j\omega)$ and

$$\mathbf{A}\left(\frac{\partial}{\partial x}\mathbf{Y}_b\right)\mathbf{V}=\mathbf{A}\left(\frac{\partial}{\partial x}\mathbf{Y}_b\mathbf{A}^T\right)\mathbf{E}=\frac{\partial}{\partial x}\mathbf{Y}(j\omega)\mathbf{E}(j\omega) \tag{2.55}$$

directly by inspection.
4. We can calculate the sensitivity of *any* voltage $V_{jk}\triangleq E_j-E_k$ as follows:

$$S_x^{V_{jk}}=\frac{1}{E_j-E_k}(E_jS_x^{E_j}-E_kS_x^{E_k}) \tag{2.56}$$

5. For the grounded two-port in Fig. 2.4, we can calculate the sensitivity of the network function.

$$H(j\omega)\triangleq\left.\frac{V_2(j\omega)}{I_1(j\omega)}\right|_{I_2=0}=\left.V_2(j\omega)\right|_{I_1=1,\ I_2=0} \tag{2.57}$$

by calculating

$$S_x^H=S_x^{V_2} \tag{2.58}$$

Exercises

1. Calculate the sensitivities $S_{G_2}^{V_3}$ and $S_{G_3}^{V_2}$ for the resistive circuit in Fig. 2.3.
2. Replace G_1 in Fig. 2.3 by a capacitance $C=5\text{ F}$ and replace I_s by a sinusoidal source $I_s=\sin\omega t$ and calculate $S_C^{V_1}$ and $S_{g_m}^{V_2}$.
3. Replace G_3 in Fig. 2.3 by an inductance $L=2\text{ H}$ and replace I_s by a sinusoidal source $I_s=\sin\omega t$ and calculate $S_\Gamma^{V_1}$ and $S_{g_m}^{V_2}$ where $\Gamma\triangleq 1/L$.

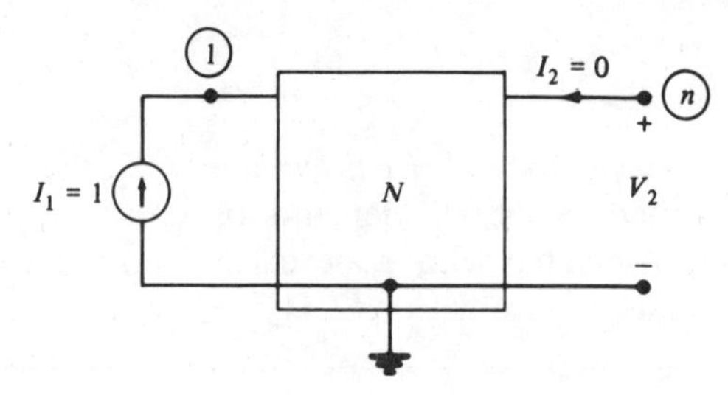

Figure 2.4 A special but common configuration where S_x^H can be calculated by means of Eq. (2.41).

4. Replace G_1 and G_2 in Fig. 2.3 by C and L, respectively, and replace I_s by a sinusoidal source $I_s = \sin \omega t$ and calculate $S_{g_m}^H$ via Eq. (2.58), where

$$H = \frac{V_2(j\omega)}{V_1(j\omega)}$$

2.2 Calculating Sensitivity Via the Adjoint Equation

The preceding algorithm requires that we calculate the *inverse* of matrix $\mathbf{Y}(j\omega)$ for each frequency of interest. This method is computationally expensive for large circuits (say, $n > 100$), especially when the calculation is to be repeated at many frequencies. In this section we will develop a more efficient numerical algorithm based on the *adjoint equation*. Although this approach is applicable for any linear time-invariant circuit, we will restrict our discussion to the simple circuit configuration in Fig. 2.4 so that the main idea can be more easily grasped.

Consider first the problem of calculating the *first-order* change $\delta\mathbf{E}$ (obtained by neglecting second-order terms) of the circuit in Fig. 2.4 due to a small percentage change $\Delta R/R$, $\Delta C/C$, $\Delta g_m/g_m$, etc., of one or more circuit parameters. This has been derived earlier in Eq. (2.38), namely,

$$\delta\mathbf{E} = -[\mathbf{Y}^{-1}][\delta\mathbf{Y}]\mathbf{E} \tag{2.59}$$

where $\mathbf{E}$ denotes the node voltage phasor solution (at the specified frequency ω) of the *unperturbed circuit* $\mathcal{N}$ and $\delta\mathbf{Y}$ denotes the *perturbation* in the *node admittance matrix*.

Suppose we wish to calculate the change $H(j\omega)$ of the network function defined by Eq. (2.57) of Fig. 2.4. If we label the input and output nodes of this circuit by ① and ⓝ, respectively, and assume $I_1 = 1$, then

$$\delta H(j\omega) = \delta E_n(j\omega) \tag{2.60}$$

where $E_n(j\omega)$ is the last element of $\mathbf{E}$, evaluated at the frequency ω. It follows from Eq. (2.59) that

$$\delta E_n(j\omega) = -\boxed{(\text{last row of } \mathbf{Y}^{-1})}\,\boxed{\delta\mathbf{Y}}\,\boxed{\mathbf{E}} \tag{2.61}$$

Instead of calculating the row vector

$$\rho_n \triangleq \text{last row of } \mathbf{Y}^{-1} \tag{2.62}$$

directly (i.e., by calculating $\mathbf{Y}^{-1}$), write the equation

Adjoint equation

$$\boxed{\mathbf{Y}^T\mathbf{E}^a = \mathbf{I}_s^a} \tag{2.63}$$

where $\mathbf{Y}^T$ is simply the *transpose* of $\mathbf{Y}$ and $\mathbf{E}^a$ and $\mathbf{I}_s^a$ are two yet unspecified phasor vectors, henceforth called the *adjoint node voltage vector* and the *adjoint current source vector*, respectively.

Equation (2.63) is called the *adjoint equation* associated with the original node equation

$$\mathbf{Y}\mathbf{E} = \mathbf{I}_s \tag{2.64}$$

A. Physical interpretation of the adjoint equation We can interpret the *adjoint equation* (2.63) as the node equation of a related circuit $\mathcal{N}^a$—henceforth called the *adjoint circuit*—whose node admittance matrix $\mathbf{Y}^a \triangleq \mathbf{Y}^T$ is the *transpose* of that of the original circuit $\mathcal{N}$. Here, $\mathbf{E}^a$ denotes the node voltage phasor solution of the adjoint circuit $\mathcal{N}^a$ and $\mathbf{I}_s^a$ denotes the net current phasor due to the independent current sources in $\mathcal{N}^a$.

For reasons that will be obvious soon, to calculate δE_n via Eq. (2.61), it is convenient to choose[8]

$$\mathbf{I}_s^a = \begin{bmatrix} 0 \\ 0 \\ \vdots \\ 0 \\ -1 \end{bmatrix} \tag{2.65}$$

Physically, this means that we apply a *unit* current source between the output node ⓝ and the ground node with a direction *leaving* node ⓝ.

Example Consider the circuit in Fig. 2.3. Its node equation is given by

$$\begin{bmatrix} 3 & -1 \\ 1 & 5 \end{bmatrix} \begin{bmatrix} E_1 \\ E_2 \end{bmatrix} = \begin{bmatrix} 4 \\ 0 \end{bmatrix} \tag{2.66}$$

The adjoint equation associated with this circuit and the assumed *adjoint current source* vector in Eq. (2.65) is, *by definition*, given by

$$\begin{bmatrix} 3 & 1 \\ -1 & 5 \end{bmatrix} \begin{bmatrix} E_1^a \\ E_2^a \end{bmatrix} = \begin{bmatrix} 0 \\ -1 \end{bmatrix} \tag{2.67}$$

We can interpret Eq. (2.67) as the node equation of the adjoint circuit shown in Fig. 2.5, where all variables have a superscript a to distinguish them from the corresponding variables in the associated circuit $\mathcal{N}$ in Fig. 2.3. Observe that whereas all linear two-terminal elements in $\mathcal{N}$ and $\mathcal{N}^\alpha$ are identical, the two-ports (VCCS in this case) are not. Observe that if we interchange the input and output ports of the VCCS in $\mathcal{N}$ in Fig. 2.3, we would obtain the corresponding VCCS in Fig. 2.5.

A simple procedure can be given to construct the *adjoint circuit* $\mathcal{N}^a$ associated with any given circuit $\mathcal{N}$. For our purpose in this section,

[8] To calculate δE_k, move -1 in Eq. (2.65) to the kth position.

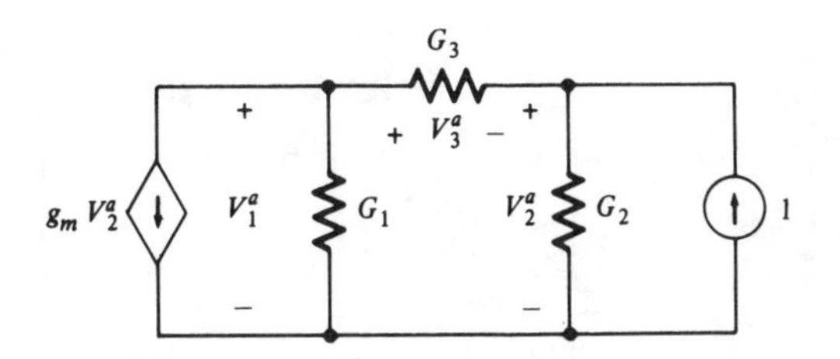

Figure 2.5 Adjoint circuit $\mathcal{N}^a$ associated with that of Fig. 2.3. Here, $G_1 = 2$ S, $G_2 = 4$ S, $G_3 = 1$ S, and $g_m = 2$ S.

however, it is *unnecessary* to draw the adjoint circuit because it is easier to write the *adjoint equation* (2.63) directly from the node equation of the *unperturbed circuit* $\mathcal{N}$.

B. Calculating δE_n in terms of E and E^a The solution of the *adjoint equation* (2.63) can be written in the form

$$\mathbf{E}^a = (\mathbf{Y}^{-1})^T \mathbf{I}_s^a \tag{2.68}$$

where we have made use of the matrix identity $(\mathbf{Y}^T)^{-1} = (\mathbf{Y}^{-1})^T$. Substituting $\mathbf{I}_s^a$ from Eq. (2.65) into Eq. (2.68) and taking the transpose of both sides, we obtain

$$(\mathbf{E}^a)^T = [0\,0 \cdots 0\, -1] \boxed{\mathbf{Y}^{-1}} = -\boxed{\text{last row of } \mathbf{Y}^{-1}} \tag{2.69}$$

Substituting Eq. (2.69) into Eq. (2.61), we obtain

$$\delta E_n = [\mathbf{E}^a]^T [\delta \mathbf{Y}] \mathbf{E} \tag{2.70}$$

Equation (2.70) shows that we can calculate δE_n due to the *first-order* changes in one or more (possibly all) elements in the circuit $\mathcal{N}$ by performing *only two* analyses as follows:

> Algorithm of sensitivity calculation via adjoint equation:
>
> 1. Solve for the *node voltage phasor* $\mathbf{E}$ at frequency ω of the *unperturbed circuit* $\mathcal{N}$.
> 2. Solve for the node voltage phasor $\mathbf{E}^a$ at the same frequency ω of the *adjoint equation* (2.63) where $\mathbf{I}_s^a$ is given by Eq. (2.65).
> 3. Form the *perturbation matrix* $\delta \mathbf{Y}$ at frequency ω, and calculate δE_n from Eq. (2.70).

Example Consider the circuit shown in Fig. 2.6. Its node equation is, by inspection, given by:

$$\begin{bmatrix} G_1 + j\omega(C_1 + C_3) & -j\omega C_3 \\ -j\omega C_3 - g_m & \dfrac{1}{j\omega L} + G_2 + j\omega(C_2 + C_3) \end{bmatrix} \begin{bmatrix} E_1 \\ E_2 \end{bmatrix} = \begin{bmatrix} 1 \\ 0 \end{bmatrix} \tag{2.71}$$

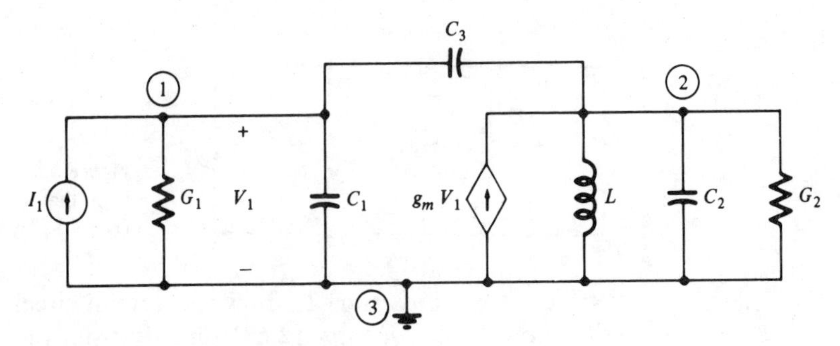

Figure 2.6 Unperturbed circuit $\mathcal{N}$.

The associated *adjoint equation* is given by:

$$\begin{bmatrix} G_1 + j\omega(C_1 + C_3) & -j\omega C_3 - g_m \\ -j\omega C_3 & \dfrac{1}{j\omega L} + G_2 + j\omega(C_2 + C_3) \end{bmatrix} \begin{bmatrix} E_1^a \\ E_2^a \end{bmatrix} = \begin{bmatrix} 0 \\ -1 \end{bmatrix} \tag{2.72}$$

Steps 1 and 2 above consist of solving Eq. (2.71) for $\mathbf{E}$ and Eq. (2.72) for $\mathbf{E}^a$ at the *same* frequency ω.

Now suppose C_1, C_2, C_3, L, and g_m change by a small amount:

$$\begin{gathered} C_1 \to C_1 + \delta C_1 \qquad C_2 \to C_2 + \delta C_2 \qquad C_3 \to C_3 + \delta C_3 \\ L \to L + \delta L \qquad \text{and} \qquad g_m \to g_m + \delta g_m \end{gathered} \tag{2.73}$$

Then the node admittance matrix $\mathbf{Y}$ in Eq. (2.71) changes to $\mathbf{Y} + \delta\mathbf{Y}$ where

$$\delta\mathbf{Y} = \begin{bmatrix} j\omega(\delta C_1 + \delta C_3) & -j\omega\delta C_3 \\ -j\omega\delta C_3 - \delta g_m & -\dfrac{\delta L}{j\omega L^2} + j\omega(\delta C_2 + \delta C_3) \end{bmatrix} \tag{2.74}$$

Observe that since the admittance of the inductance is given by $1/j\omega L$, the *first-order* change in $\mathbf{Y}$ due to a change δL is given by $-\delta L/j\omega L^2$.

Substituting $\mathbf{E}$, $\mathbf{E}^a$, and $\delta\mathbf{Y}$ from Eq. (2.74) into Eq.(2.70), we obtain

$$\delta H(j\omega) = \delta E_2(j\omega) = \begin{bmatrix} E_1^a & E_2^a \end{bmatrix} \begin{bmatrix} \delta\mathbf{Y} \end{bmatrix} \begin{bmatrix} E_1 \\ E_2 \end{bmatrix} \tag{2.75}$$

$$\begin{aligned} &= j\omega(\delta C_1 + \delta C_3)E_1 E_1^a - j\omega\delta C_3 E_2 E_1^a \\ &\quad - j\omega(\delta C_3 - \delta g_m)E_1 E_2^a - \frac{\delta L}{j\omega L^2} \\ &\quad + j\omega(\delta C_2 + \delta C_3)E_2 E_2^a \end{aligned} \tag{2.76}$$

From Eq. (2.76) we can calculate the following partial derivatives:

$$\frac{\partial H}{\partial C_1} = j\omega E_1 E_1^a \qquad \frac{\partial H}{\partial C_2} = j\omega E_2 E_2^a$$

$$\frac{\partial H}{\partial C_3} = j\omega(E_1 E_1^a - E_2 E_1^a - E_1 E_2^a + E_2 E_2^a)$$

$$\frac{\partial H}{\partial L} = \frac{-1}{j\omega L^2}$$

$$\frac{\partial H}{\partial g_m} = j\omega E_1 E_2^a \tag{2.77}$$

It follows from definition (2.1b) and Eq. (2.77) that

$$S_{C_1}^H = \frac{j\omega C_1 E_1 E_1^a}{H(j\omega)} \qquad S_{C_2}^H = \frac{j\omega C_2 E_2 E_2^a}{H(j\omega)}$$

$$S_{C_3}^H = \frac{j\omega C_3[E_1 E_1^a - E_2 E_1^a - E_1 E_2^a + E_2 E_2^a]}{H(j\omega)} \tag{2.78}$$

$$S_L^H = \frac{-1}{j\omega L\, H(j\omega)} \qquad S_{g_m}^H = \frac{j\omega g_m E_1 E_2^a}{H(j\omega)}$$

Observe that whereas $\delta H(j\omega)$ calculated from Eq. (2.76) represents only a *first-order* (hence approximate) change in $H(j\omega)$, the sensitivity of $H(j\omega)$ with respect to C_1, C_2, C_3, L, and g_m as calculated in Eq. (2.78) is *exact* because the partial derivatives (obtained by letting $\delta x \to 0$) in Eq. (2.77) are exact.

REMARKS

1. The node voltage phasor $\mathbf{E}$ in Step 1 can be calculated by any convenient method, including by computer simulation programs, such as SPICE.
2. The node voltage phasor $\mathbf{E}^a$ in Step 2 can also be calculated by a computer simulation program using the *adjoint circuit* $\mathcal{N}^a$. This is one instance in which it would be necessary to construct the adjoint circuit of $\mathcal{N}$. Hence the adjoint circuit is useful when an efficient computer program for calculating the adjoint equation is not available.
3. The perturbation matrix $\delta\mathbf{Y}$ in the above example is calculated by inspection since the circuit is small. For large circuits, $\delta\mathbf{Y}$ can be calculated automatically by a computer program using an algorithm much more efficient than a brute force calculation of

$$\delta\mathbf{Y} = \mathbf{A}(\delta\mathbf{Y}_b)\mathbf{A}^T \tag{2.79}$$

4. The most efficient method for calculating $\mathbf{E}$ and $\mathbf{E}^a$ is via the *LU decomposition method*.

C. LU decomposition method for solving E and $\mathbf{E}^a$

1. Perform an **LU** decomposition of **Y**:

$$\mathbf{Y} = \mathbf{LU} = \begin{bmatrix} \diagdown & \mathbf{0} \\ \mathbf{L} & \diagdown \end{bmatrix} \begin{bmatrix} \diagdown & \mathbf{U} \\ \mathbf{0} & \diagdown \end{bmatrix} \tag{2.80}$$

where **L** is a lower triangular matrix and **U** is an upper triangular matrix. This step requires approximately $n^3/3$ operations (multiplications and divisions).

2. Solve the node equation

$$(\mathbf{LU})\mathbf{E} = \mathbf{I}_s \tag{2.81}$$

for **E** in two steps.

(*a*) Forward substitution: $\mathbf{Lx} = \mathbf{I}_s$

$$\begin{bmatrix} \diagdown & \mathbf{0} \\ \mathbf{L} & \diagdown \end{bmatrix} \begin{bmatrix} x_1 \\ x_2 \\ \vdots \\ x_n \end{bmatrix} = \begin{bmatrix} 1 \\ 0 \\ \vdots \\ 0 \end{bmatrix} \tag{2.82}$$

(*b*) Backward substitution: $\mathbf{UE} = \mathbf{x}$

$$\begin{bmatrix} \diagdown & \mathbf{U} \\ \mathbf{0} & \diagdown \end{bmatrix} \begin{bmatrix} E_1 \\ E_2 \\ \vdots \\ E_n \end{bmatrix} = \begin{bmatrix} x_1 \\ x_2 \\ \vdots \\ x_n \end{bmatrix} \tag{2.83}$$

where **x** is the solution from Eq. (2.82).
These two steps require approximately n^2 operations (multiplications and divisions).

3. Solve the adjoint equation

$$(\mathbf{U}^T\mathbf{L}^T)\mathbf{E}^a = \mathbf{I}_s^a \tag{2.84}$$

for $\mathbf{E}^a$ in two steps.

(*a*) Forward substitution: $\mathbf{U}^T\mathbf{x}^a = \mathbf{I}_s^a$

$$\begin{bmatrix} \diagdown & \mathbf{0} \\ \mathbf{U}^T & \diagdown \end{bmatrix} \begin{bmatrix} x_1^a \\ x_2^a \\ \vdots \\ x_n^a \end{bmatrix} = \begin{bmatrix} 0 \\ 0 \\ \vdots \\ 0 \\ -1 \end{bmatrix} \Rightarrow \begin{bmatrix} x_1^a \\ x_2^a \\ \vdots \\ x_n^a \end{bmatrix} = \begin{bmatrix} 0 \\ 0 \\ \vdots \\ 0 \\ -u_{nn}^{-1} \end{bmatrix} \tag{2.85}$$

where u_{nn} denotes the nnth element of **U**. This step requires only one division.

(*b*) Backward substitution: $\mathbf{L}^T\mathbf{E}^a = \mathbf{x}^a$

$$\begin{bmatrix} \diagdown & \mathbf{L}^T \\ \mathbf{0} & \diagdown \end{bmatrix} \begin{bmatrix} E_1^a \\ E_2^a \\ \vdots \\ E_n^a \end{bmatrix} = \begin{bmatrix} 0 \\ 0 \\ \vdots \\ 0 \\ -u_{nn}^{-1} \end{bmatrix} \tag{2.86}$$

This step requires approximately $n^2/2$ operations (multiplications and divisions).

It follows from the above estimates that the two phasors $\mathbf{E}$ and $\mathbf{E}^a$ can be calculated using approximately $n^3/3 + \frac{3}{2}n^2$ multiplications and divisions. This is significantly smaller (for large n) than that of calculating the *inverse* of the matrix $\mathbf{Y}$.

Exercises

1. Let Z denote the driving-point impedance seen by the current source I_s in Fig. 2.3. Calculate the sensitivity $S_{g_m}^Z$.
2. Calculate the sensitivity $S_{g_m}^{Z(j\omega)}$ for the circuit in Fig. 2.6 where $Z(j\omega)$ denotes the driving-point impedance seen by the current source I_1. Assume $G_1 = 1\,\text{S}$, $G_2 = 2\,\text{S}$, $C_1 = 1\,\text{F}$, $C_2 = 2\,\text{F}$, $C_3 = 3\,\text{F}$, $L = 1\,\text{H}$, and $g_m = 5\,\text{S}$.
3. Given $S_x^{Z_{11}}$ and $S_x^{Z_{21}}$, calculate the sensitivity S_x^H for the circuit configuration in Fig. 2.4, where

$$Z_{11} \triangleq \left.\frac{V_1}{I_1}\right|_{I_2=0} \qquad Z_{21} \triangleq \left.\frac{V_2}{I_1}\right|_{I_2=0} \qquad H \triangleq \left.\frac{V_2}{V_1}\right|_{I_2=0} \tag{2.87}$$

SUMMARY

- The *ideal low-pass characteristic* is not physically realizable.
- The *ideal low-pass magnitude characteristic* can be approximated to any desired degree of accuracy by a *Butterworth function*

$$|H(j\omega)| = \frac{1}{\sqrt{1 + \omega^{2n}}}$$

by choosing a sufficiently large integer n, called the *order* of the filter.
- An nth-order low-pass filter can be described by an nth-order *all-pole function*

$$H(s) = \frac{H_0}{s^n + b_{n-1}s^{n-1} + \cdots + b_1 s + b_0}$$

where $H_0 = 1$ and $b_0, b_1, \ldots, b_{n-1}$ are listed in Table 1.1 for $n = 1, 2, \ldots, 6$.
- Every *odd*-order all-pole function can be factored into the form

$$H(s) = H_0\left[\frac{\sigma_1}{s+\sigma_1}\right]\left[\frac{\beta_1}{s^2+\alpha_1 s+\beta_1}\right]\cdots\left[\frac{\beta_m}{s^2+\alpha_m s+\beta_m}\right]$$

where $m = \frac{1}{2}(n-1)$ for an nth-order function.

- Every *even*-order all-pole function can be factored into the form

$$H(s) = H_0\left[\frac{\beta_1}{s^2+\alpha_1 s+\beta_1}\right]\left[\frac{\beta_2}{s^2+\alpha_2 s+\beta_2}\right]\cdots\left[\frac{\beta_m}{s^2+\alpha_m s+\beta_m}\right]$$

where $m = n/2$ for an nth order function.

- Every all-pole function $H(s)$ can be synthesized by alternately cascading two-ports and buffers using the configuration shown in Fig. 1.3.
- Every *first-order* two-port described by

$$H(s) = \frac{\sigma_1}{s+\sigma_1}$$

can be synthesized by the op-amp circuit shown in Fig. 1.4.

- Every *second-order* two-port described by

$$H(s) = \frac{\beta}{s^2+\alpha s+\beta}$$

can be synthesized by the op-amp circuit shown in Fig. 1.5.

- Every *third-order* two-port described by

$$H(s) = \frac{\gamma}{s^3+\alpha s^2+\beta s+\gamma}$$

can be synthesized by the op-amp circuit shown in Fig. 1.6.

- The *sensitivity* of $\mathscr{P}$ relative to a parameter x is given by

$$S_x^{\mathscr{P}} = \frac{x}{\mathscr{P}}\frac{\partial \mathscr{P}}{\partial x} = \frac{\partial \ln \mathscr{P}}{\partial \ln x}$$

- The *sensitivity* of a voltage transfer function

$$H(j\omega) \triangleq \left.\frac{V_2(j\omega)}{V_1(j\omega)}\right|_{I_2=0}$$

relative to an impedance $Z(j\omega)$ is given by

$$S_Z^H = \left[\frac{ZZ_3}{Z+Z_3}\right]\left[\frac{H_\infty - H_0}{Z_3H_0 + H_\infty Z}\right]$$

where Z_3 is the driving-point impedance in Fig. 2.1*b*, H_0 is the voltage transfer function in Fig. 2.1*c*, and H_∞ is the voltage transfer function in Fig. 2.1*d*.

- For a grounded two-port driven by a current source and containing only voltage-controlled linear time-invariant elements,

$$S_x^{E_i} = \frac{x}{E_i} \cdot \text{row } i \text{ of } \left[-\mathbf{Y}^{-1}\mathbf{A}\left(\frac{\partial}{\partial x}\mathbf{Y}_b\right)\mathbf{V} \right]$$

- The change in the output voltage δE_n in a grounded two-port driven by a current source can be calculated to the first order from

$$\delta E_n = [\mathbf{E}^a]^T[\delta \mathbf{Y}]\mathbf{E}$$

where $\mathbf{E}$ is the *node voltage phasor* solution of the *unperturbed circuit* $\mathcal{N}$, $\mathbf{E}^a$ is the *node voltage phasor* solution of the adjoint circuit $\mathcal{N}^a$, and $\delta\mathbf{Y}$ is the *perturbation* in the *node admittance matrix* $\mathbf{Y}$, all of which are calculated at the source frequency ω. The two phasors can be efficiently calculated via the **LU** decomposition algorithm.

PROBLEMS

Butterworth approximation

1 Design a Butterworth filter with the magnitude characteristic 3 dB down from its maximum value (which is at zero frequency) at the edge of the passband $\omega = 1$ rad/s and at least 15 dB down at all frequencies above $\omega = 2$ rad/s.

2 Design a Butterworth filter to realize the following transfer function:

$$|H(j\omega)| \geq 0.9|H(0)| \qquad \omega \leq \omega_a$$

$$|H(j\omega)| \leq 0.1|H(0)| \qquad \omega \geq \omega_b$$

where $\omega_a = 0.75\omega_c$, $\omega_b = 2\omega_c$, and $|H(j\omega_c)| = \frac{1}{\sqrt{2}}|H(0)|$.

3 The generalized Butterworth function is defined as

$$|H(j\omega)| = \frac{1}{\sqrt{1 + \varepsilon^2\omega^{2n}}}$$

where ε is a constant.

(*a*) Find $|H(0)|$ and $|H(j1)|$.

(*b*) Show that

$$\frac{d^k}{d\omega^k}|H(j\omega)|_{\omega=0} = 0 \qquad \text{for all } k < 2n$$

4 (*a*) Given

$$H(s) = \frac{s+2}{b_4s^4 + b_3s^3 + b_2s^2 + b_1s + b_o}$$

find the values for b_i, $i = 0, 1, \ldots, 4$ to make $H(s)$ realizable as a Butterworth filter.

(*b*) Given

$$H(s) = \frac{1}{s+a} - \frac{1}{s+b}$$

choose a and b to realize $H(s)$ by a Butterworth filter.

Synthesis of all-pole transfer functions

5 Realize the following transfer function:

$$H(s) = \frac{1}{s^3 + 6s^2 + 12s + 8}$$

using

(*a*) One third-order building block.
(*b*) Three first-order building blocks.
(*c*) One first-order and one second-order building block.

6 Which of the following transfer functions can be synthesized by the three kinds of building blocks in Sec. 1.2?

$$\text{(i)}\ \frac{1}{s^2 + a} \qquad \text{(iii)}\ \frac{s - 2}{s^3 + s^2 - 4s - 4}$$

$$\text{(ii)}\ \frac{1}{(s+1)(s^2+2)} \qquad \text{(iv)}\ \frac{1}{s^2 + 2s - 1}$$

7 Using only one op amp and passive R, L, and C elements, design a circuit which has the following transfer function:

$$H(s) = \frac{1}{as^2 + 1} \qquad a > 0$$

8 Consider the following transfer function:

$$H(s) = \frac{-1}{s^2 - 1}$$

(*a*) Is it possible to realize this transfer function using an op amp and passive R, L, and C elements? Explain.

(*b*) If R is allowed to be negative, propose a circuit to realize this transfer function.

Renormalization

9 A low-pass filter is described by the following transfer function:

$$H(s) = \frac{1}{s^3 + 3s^2 + 3s + 1}$$

(*a*) Assuming $C_1 = C_2 = C_3 = 1\,\text{F}$, find the values for R_1, R_2, and R_3 to realize this transfer function.

(*b*) Keep R_1, R_2, and R_3 fixed in the circuit obtained in (*a*) and find the new values of C_1, C_2, and C_3 to have a 9-dB attenuation at $\omega = 10\,\text{rad/s}$.

(*c*) Repeat part (*b*) keeping C_1, C_2, and C_3 fixed, and find the new values of R_1, R_2, and R_3.

10 It is desired to use frequency scaling to design a Butterworth filter, $H(j\omega)$, which satisfies

$$A(f) \leq 0.2\,\text{dB} \qquad f \leq 4\,\text{kHz}$$

and

$$A(f) \geq 10\,\text{dB} \qquad f \geq 10\,\text{kHz}$$

where

$$A(f) = 20 \log \left| \frac{H(0)}{H(j2\pi f)} \right|$$

(*a*) Find the minimum order n and the cutoff frequency f_c.
(*b*) Determine $A(4 \times 10^3)$ and $A(10^4)$.
(*c*) If we want $A(10^4) = 10\,\text{dB}$, what is $A(4 \times 10^3)$?

11 Realize the following transfer function using building blocks with all capacitances less than 0.1 μF. Assume $\omega_c = 10^3$ rad/s.

$$H(s) = \frac{1}{s+1} - \frac{s-1}{s^2+1}$$

Sensitivity

12 Prove the following properties of the sensitivity function:

(i) $S_x^{ky} = S_{kx}^y = S_x^y$

(ii) $S_{1/x}^y = S_x^{1/y} = -S_x^y$

(iii) $S_x^{\prod_{i=1}^n y_i} = \sum_{i=1}^n S_x^{y_i}$

(iv) $S_x^{\sum_{i=1}^n y_i} = \dfrac{\sum_{i=1}^n y_i S_x^{y_i}}{\sum_{i=1}^n y_i}$

(v) $S_x^{\ln y} = \dfrac{1}{\ln y} S_x^y$

(vi) $S_x^{u(y_1, y_2, \ldots, y_n)} = \sum_{i=1}^n S_{y_i}^u S_x^{y_i}$

13 Use the properties in Prob. 12 to show:

$$S_x^y = S_x^{|y|} + j\measuredangle y \cdot S_x^{\measuredangle y}$$

where y is a complex quantity and x is real.

14 (*a*) Show that:

$$\frac{d}{dx} S_x^y = \frac{1}{x} S_x^y (1 + S_x^{dy/dx} - S_x^y)$$

(*b*) If y is a linear function of x, what is

$$\frac{d}{dx} S_x^y$$

15 Consider the second-order low-pass filter shown in Fig. P14.15.

(*a*) Find the transfer function $H(s) = V_0(s)/V_i(s)$.

(*b*) Find ω_0 and Q in Eq. (2.8) in terms of R_1, R_2, C_1, and C_2.

(*c*) Compute $S_{C_1}^{\omega_0}$, $S_{C_2}^{\omega_0}$, $S_{R_1}^{\omega_0}$, $S_{R_2}^{\omega_0}$, $S_{C_1}^{Q}$, $S_{C_2}^{Q}$, $S_{R_1}^{Q}$, $S_{R_2}^{Q}$.

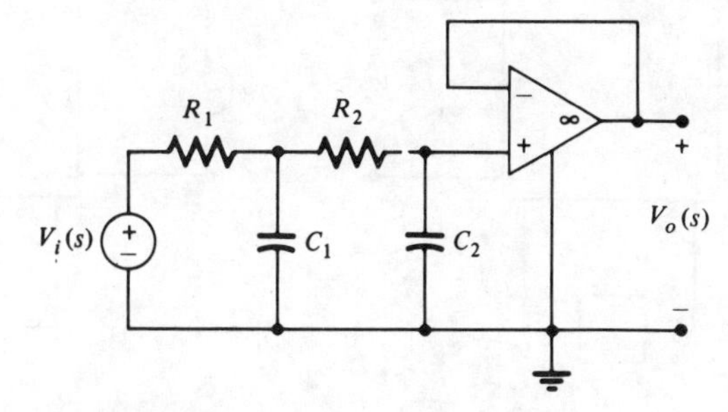

Figure P14.15

16 Given $H(s) = 1/(1 + RCs)$, determine:

(*a*) $S_R^{H(j\omega)}$

(*b*) $S_R^{|H(j\omega)|}$

(*c*) $S_R^{\measuredangle H(j\omega)}$

17 Given $H(s) = 1/(a_3 s^3 + a_2 s^2 + a_1 s + a_0)$, determine the coefficient sensitivities $S_{a_i}^{H(s)}$ for $i = 0, 1, 2, 3$.

Sensitivity and network function

18 Consider the network shown in Fig. P14.18. The following measurements have been made:

$$H_0 = \left.\frac{I_2}{I_1}\right|_{\substack{V_2=0\\Z=0}} \qquad H_\infty = \left.\frac{I_2}{I_1}\right|_{\substack{V_2=0\\Z=\infty}}$$

and

$$Z_3 = \left.\frac{V_3}{I_3}\right|_{\substack{I_1=0\\V_2=0}}$$

(*a*) Find $H = \left.\frac{I_2}{I_1}\right|_{V_2=0}$ in terms of H_0, H_∞, Z_3, and Z.
(*b*) Determine S_Y^H, $Y = 1/Z$.

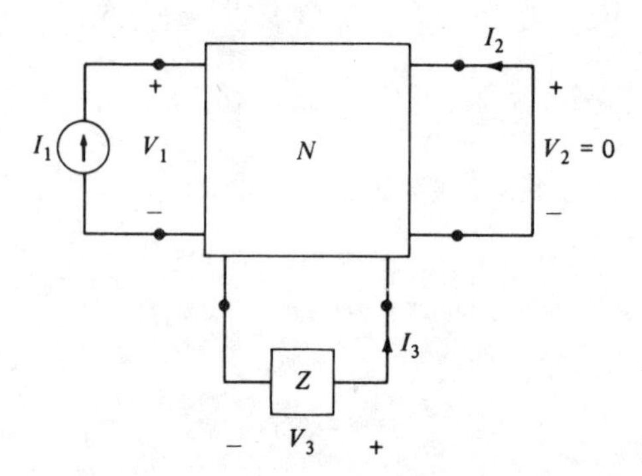

Figure P14.18

19 Consider the networks shown in Fig. P14.19.

(*a*) Determine $H_1 = \left.\frac{V_2}{I_1}\right|_{I_2=0}$ and $H_2 = \left.\frac{I_2}{V_1}\right|_{V_2=0}$.

(*b*) Determine $S_Z^{H_1}$ and $S_Z^{H_2}$.

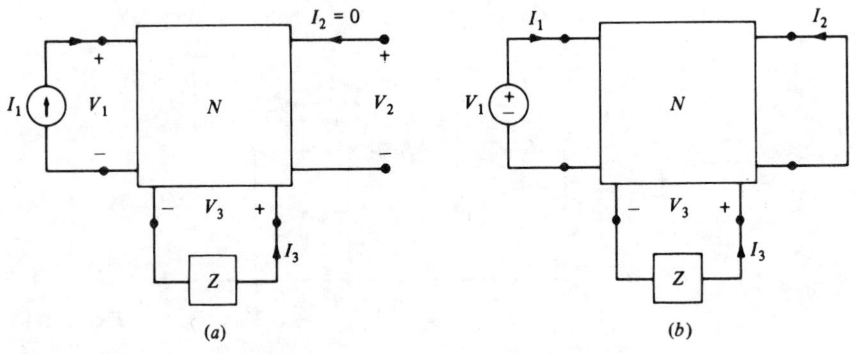

Figure P14.19

Sensitivity and node admittance matrix

20 For the circuit shown in Fig. P14.20 the element values are $G_1 = 2$ S, $G_2 = 1$ S, $G_3 = 3$ S, and $g_m = 2$ S,
(*a*) Calculate $S_{G_3}^{E_1}$ and $S_{G_3}^{E_2}$.
(*b*) Calculate $S_{g_m}^{V_1}$ and S_{g_m}.

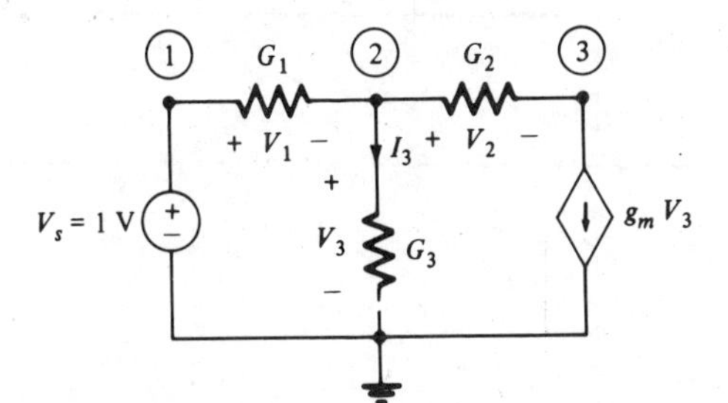

Figure P14.20

Adjoint circuit and adjoint equation

21 Consider the circuit shown in Fig. P14.21. Let $G_1 = 2$ S, $G_2 = 1$ S, $G_3 = 3$ S, $C = 4$ F, and $g_m = 2$ S.

(*a*) Draw the adjoint circuit.

(*b*) For $\mathbf{I}_s^a = [0 \quad 0 \quad -1]^T$, determine $\mathbf{E}^a$.

(*c*) Let $H(j\omega) = V_3/I_s$, calculate $S_{G_3}^H$, and S_C^H.

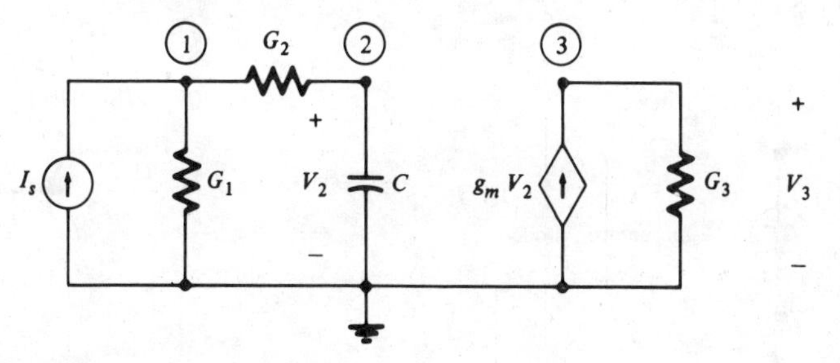

Figure P14.21

22 Consider the circuit shown in Fig. P14.22, where $R_1 = 2\,\Omega$, $R_2 = 10\,\Omega$, $R_3 = 1\,\Omega$, $g_m = 10$ S, $L = 1$ H, $C = 1$ F, and $I_s = 2\cos\omega_0 t$.

(*a*) Draw the adjoint circuit.

(*b*) Let $H(j\omega) = E_4/I_s$, determine S_L^H, S_C^H, $S_{R_1}^H$, $S_{R_2}^H$, $S_{R_3}^H$, and $S_{g_m}^H$.

(*c*) Determine $S_L^{V_2}$ and $S_C^{V_2}$ via the adjoint equation method.

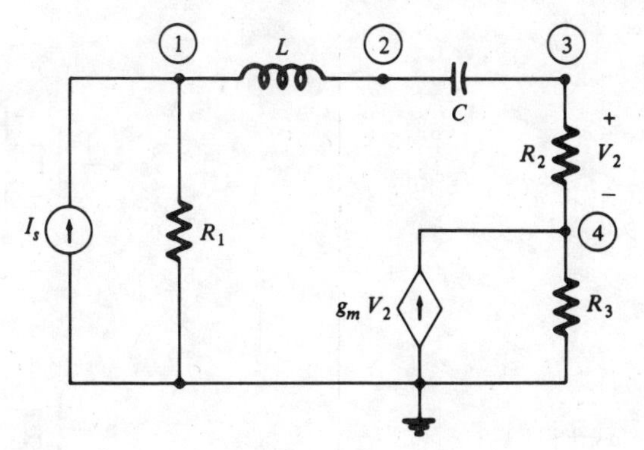

Figure P14.22

LU decomposition

23 (*a*) Perform an **LU** decomposition of **Y** from the circuit in Prob. 22.

(*b*) Let $\mathbf{I}_s^a = [0\,0\,-1]^T$; calculate $\mathbf{E}^a$.

GLOSSARY

A

SYMBOLS

I DEVICE SYMBOLS

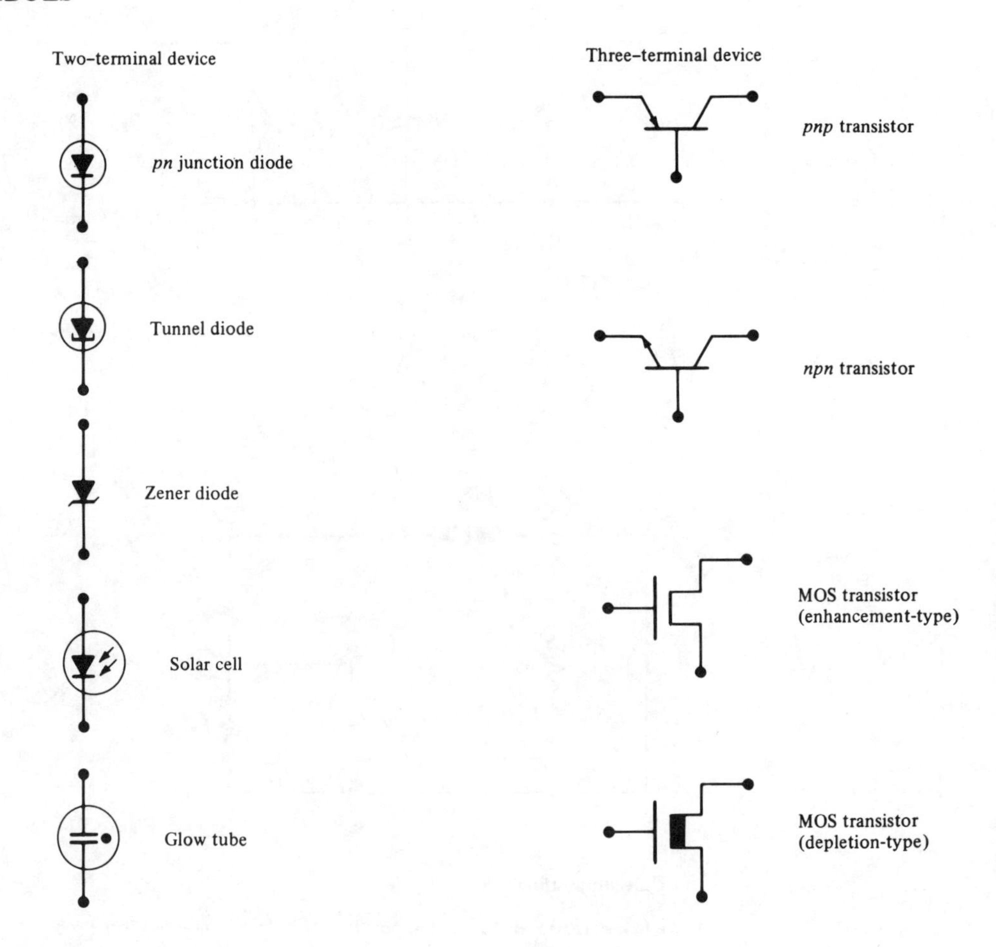

II CIRCUIT ELEMENT SYMBOLS

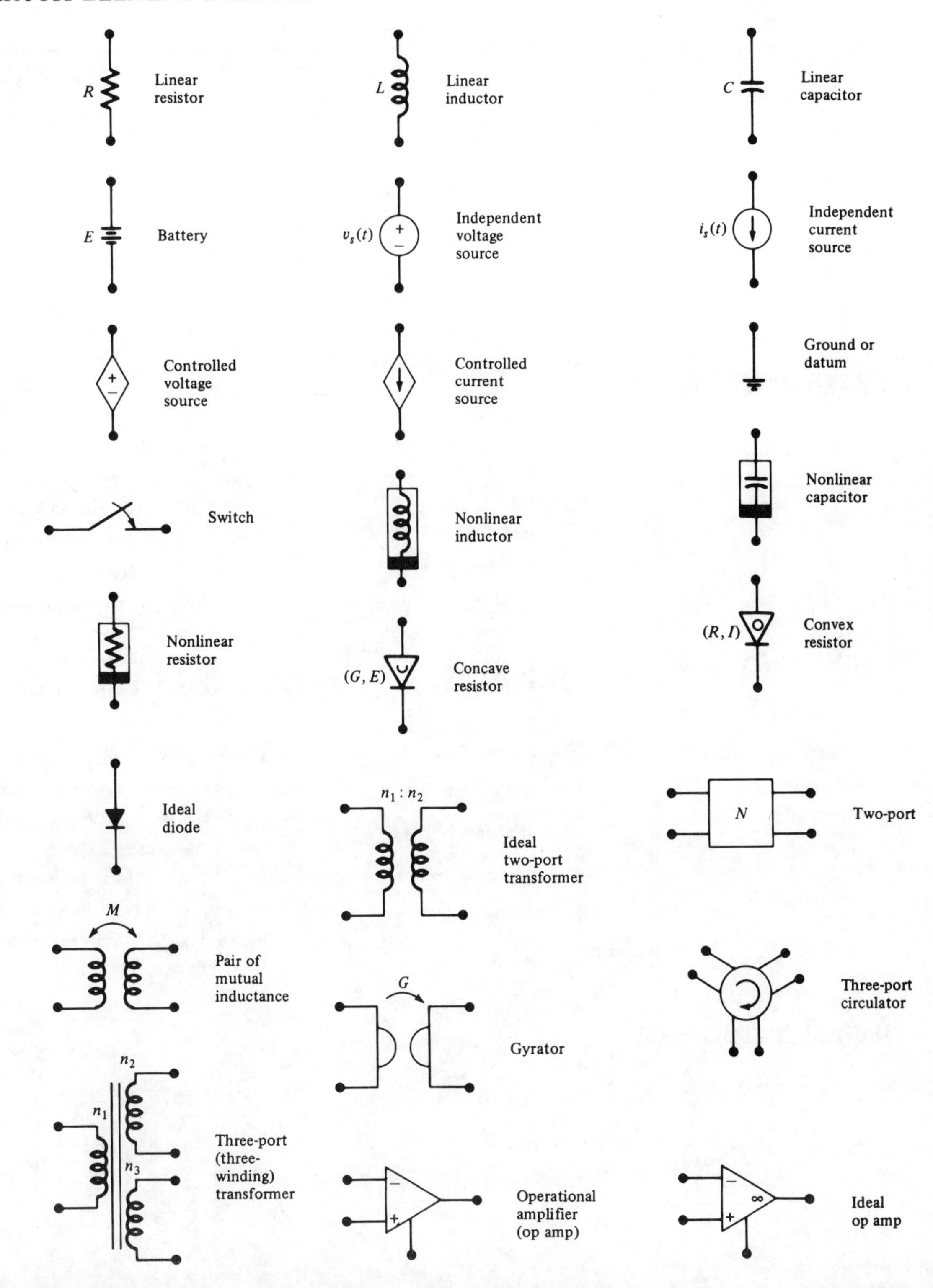

GLOSSARY

B

NOTATIONS

I MATHEMATICAL

$\triangleq$	Definition
$\Longrightarrow$	Implies
$\Longleftarrow$	Is implied by
$\Longleftrightarrow$	Equivalent to
iff	If and only if
ϵ	Is a member of
x-y plane	x = abscissa; y = ordinate
x-y characteristic	curve plotted in the x-y plane
$\perp$	Orthogonal
$\measuredangle$	Phase angle
$\star$	Convolution operation
⬆	Impulse (delta function)
Bold capital letter	Matrix
Bold lower-case letter	Vector
$\mathbf{M}^T$	Transpose of matrix $\mathbf{M}$
$\mathbf{x}^T$	Transpose of vector $\mathbf{x}$
$\mathbb{R}$	Real numbers
$\mathbb{R}^n$	n-dimensional space
T	Trace
Δ	Determinant
$z = x + jy$	Complex number

$j = \sqrt{-1}$

Re z = Real part of $z \triangleq x + jy = x$

Im z = Imaginary part of $z \triangleq x + jy = y$

$\bar{z}$ = Complex conjugate of $z \triangleq x + jy = x - jy$

s = Complex frequency

s_1, s_2	Natural frequencies
λ_1, λ_2	Eigen values
$\boldsymbol{\eta} = \boldsymbol{\eta}_r + j\boldsymbol{\eta}_i$	Complex eigen vector
$\boldsymbol{\eta}_r = \text{Re}\,\boldsymbol{\eta}$	$\boldsymbol{\eta}_i = \text{Im}\,\boldsymbol{\eta}$

II CIRCUIT–THEORETIC

v	Voltage
V	Volts
Ω	Ohm
i	Current
A	Ampere
KΩ	kiloohm

q	Charge
F	Farad
S	Siemens
ϕ	Flux linkage
H	Henry
mA	Milliampere
p	Power
$p(t)$	Instantaneous power
Q	Quality factor (figure of merit)
$\mathbf{Z}$	Open circuit impedance matrix
$\mathbf{Y}$	Short-circuit admittance matrix
$\mathbf{T}$	Tableau matrix
$\textcircled{k}$	Node k
e_k	Voltage from node $\textcircled{k}$ to datum
$\mathbf{e}$	Node-to-datum voltage vector
$1(t)$	Unit step function
$s(t)$	Step response
$\delta(t)$	Unit impulse
$h(t)$	Impulse response
τ	Time constant
$x(t_\infty)$	Equilibrium state
$x(t_0)$	Initial state
Q	Operating point
$\mathbf{x}_Q$	Coordinates of operating point Q
$\tilde{\mathbf{x}} = \mathbf{x} - \mathbf{x}_Q$	Small-signal component of $\mathbf{x}(t)$ about the operating point $\mathbf{x}_Q$
S_x^H	Relative sensitivity of H with respect to x
$\dot{x}$	Derivative of x with respect to *time t*
$\hat{f}(\cdot)$	A function
$\hat{f}'(x)$	Derivative of $\hat{f}(x)$ with respect to x

III GRAPH–THEORETIC

$\mathcal{G}$	Graph
$\mathscr{C}$	Cut set
$\mathcal{S}$	Gaussian surface
n	Number of nodes
b	Number of branches
$\mathbf{v}_t$	Twig voltage vector
$\mathbf{i}_l$	Link current vector
$\mathbf{A}_a$	Incidence matrix
$\mathbf{A}$	Reduced incidence matrix
$\mathbf{B}$	Fundamental loop matrix
$\mathbf{Q}$	Fundamental cut set matrix

INDEX